## Functions

| | |
|---|---|
| Linear | $f(x) = mx + b$ |
| Constant | $f(x) = b$ |
| Identity | $f(x) = x$ |
| Quadratic | $f(x) = ax^2 + bx + c$ |
| Polynomial | $f(x) = a_n x^n + a_{n-1}x^{n-1} + \cdots + a_1 x + a_0$ |
| Rational | $f(x) = \dfrac{P(x)}{Q(x)}$, $P(x)$ and $Q(x)$ polynomial functions |

## Logarithms

$x = \log_a y$ means $a^x = y$.

1. Product Rule $\quad \log_a xy = \log_a x + \log_a y$
2. Quotient Rule $\quad \log_a \dfrac{x}{y} = \log_a x - \log_a y$
3. Power Rule $\quad \log_a x^c = c \log_a x$
4. Change of Base $\quad \log_b x = \dfrac{\log_a x}{\log_a b}$

## Arithmetic Sequences

$$a_n = a_1 + (n-1)d$$

$$S_n = \frac{n}{2}[2a_1 + (n-1)d] = \frac{n}{2}[a_1 + a_n]$$

## Geometric Sequences

$$a_n = a_1 r^{n-1}$$

$$S_n = \frac{a_1 - a_1 r^n}{1-r} = \frac{a_1 - r a_n}{1-r}$$

$$S = \frac{a_1}{1-r}, \text{ if } |r| < 1$$

## Permutations

$${}_nP_r = \frac{n!}{(n-r)!} \qquad {}_nP_n = n!$$

## Combinations

$${}_nC_r = \frac{n!}{r!(n-r)!} \qquad {}_nC_n = 1 \qquad {}_nC_0 = 1$$

## Binomial Theorem

$$(a+b)^n = \binom{n}{0}a^n + \binom{n}{1}a^{n-1}b^1 + \binom{n}{2}a^{n-2}b^2 + \cdots + \binom{n}{r}a^{n-r}b^r + \cdots + \binom{n}{n}b^n \text{ where } \binom{n}{r} = \frac{n!}{r!(n-r)!}$$

## Angular Measure

$\pi$ radians $= 180$ degrees $\qquad$ 1 degree $= \dfrac{\pi}{180}$ degrees $\qquad$ 1 radian $= \dfrac{180}{\pi}$ degrees

# COLLEGE ALGEBRA AND TRIGONOMETRY WITH APPLICATIONS

L. MURPHY JOHNSON
ARNOLD R. STEFFENSEN
*Northern Arizona University*

**Scott, Foresman and Company**
*Glenview, Illinois*
*Boston*
*London*

**For the Student**

*To help you study and understand the course material, a* Solutions and Study Guide, *by Joseph Mutter, is available from your college bookstore. This book provides complete, step-by-step solutions to more than half of the odd-numbered exercises in the text, detailed chapter summaries, and practice chapter tests with complete solutions.*

*To Barbara, Barbara, Becky, Cindy, and Pam*

Unless otherwise acknowledged, all photos are the property of Scott, Foresman and Company.

Page 1 Dean Abramson; 2 NASA; 41(t) Dean Abramson; 41(b) Mike Mazzaschi/Stock Boston; 111(t) Dean Abramson; 111(b) Jean-Claude Lejeune; 166(t) Art Pahlke; 166(b) Erich Hartmann/Magnum; 208(t) Michael Weisbrot/Stock Boston; 208(b) © Andrew Brilliant; 250(b) U.S. Forest Service; 250(t) Peter Menzel/Stock Boston; 318(t) Dean Abramson; 318(b) David R. Frazier/Frazier Photolibrary; 372 Focus on Sports; 427(t) NASA; 427(b) Cameramann International, Ltd.; 481 Art Pahlke; 531 Thomas England; 532 Frank Siteman/Stock Boston; 588(t) Laimute Druskis/Taurus Photos; 588(b) Alan Carey/ The Image Works



**Library of Congress Cataloging-in-Publication Data**

Johnson, L. Murphy (Lee Murphy)
College algebra and trigonometry with applications.

Includes index.
1. Algebra. 2. Trigonometry. I. Steffensen, Arnold R. II. Title.
QA154.2.J56 1988 512.9 87-13917
ISBN 0-673-18355-6


Printed in the United States of America.
1 2 3 4 5 6—RRC—92 91 90 89 88 87

# Preface

*College Algebra and Trigonometry with Applications* is designed to provide comprehensive coverage of the usual topics in algebra and trigonometry needed by students for later courses in mathematics, engineering, business, statistics, or the natural sciences. Students with two years of high school algebra or its equivalent should have the necessary prerequisite skills. The text is organized for maximum instructional flexibility. More than enough material is included for a one-semester or two-quarter course.

Chapter 1 provides a review of basic algebra, which some classes may cover quickly or skip altogether. Chapters 2 through 5 present the major topics in college algebra including equations and inequalities, functions and their graphs, the theory of polynomials, and polynomial, rational, exponential, and logarithmic functions. Chapters 6 through 8 provide a thorough treatment of trigonometry; the trigonometric functions are introduced via right triangles, followed by the unit circle approach. For added flexibility, topics in analytic geometry are presented separately in Chapter 9. The text concludes with a variety of other algebraic topics including systems of equations and inequalities, matrices, determinants, sequences, series, and probability. Applications provide practical motivation throughout the book.

## FEATURES

The text is written informally; explanations are carefully worded to ensure student comprehension. Second color is used pedagogically to highlight important steps and emphasize methods and terminology. The many figures and graphs are labeled for easy reference and employ color to clarify the concepts presented. Cautions warn students of common mistakes and special problems, while Notes provide additional explanations or other pertinent information.

### Examples

The text contains over 650 carefully selected examples with detailed step-by-step solutions and helpful side annotations.

### Exercises

There are over 4600 exercises in the text. The exercise sets are carefully graded and begin with paired routine problems that are followed by a variety of challenging extension problems and numerous applications. A set of For Review exercises is included at the end of most exercise sets to help students review previously covered material or prepare for the next section. A collection of review exercises concludes each chapter. Answers to odd-numbered section exercises and to all For Review and Chapter Review exercises are included at the back of the book.

## Applications

To demonstrate the usefulness and practicality of mathematics, applications have been given special attention in this text. Over 700 relevant applied problems from such diverse areas as business, engineering, geology, physics, chemistry, medicine, forestry, and agriculture are included in the chapter introductions, examples, and exercises.

## Calculators

Calculators are discussed at appropriate places throughout the text, and illustrations are included for both Algebraic Logic and Reverse Polish Notation. Calculator exercises are not specifically marked, however, since students should learn when to use and when not to use calculators. Appendices on the use of logarithmic and trigonometric function tables are provided for instructors who prefer that their students learn these techniques.

# SUPPLEMENTS

## For the Instructor

The **Instructor's Guide** contains a Placement Test, four different but equivalent tests for each chapter, two final examinations, an extensive bank of additional problems, and answers to all test items and even-numbered text exercises. As an alternative to the tests in the Instructor's Guide, the **Computer-Assisted Testing System (CATS)** can be used with Apple and IBM computers to construct and print tests. More than 50 **overhead transparencies** featuring key figures from the text are also provided for classroom lectures and presentations.

## For the Student

The **Solutions and Study Guide** contains complete, step-by-step solutions to more than half of the odd-numbered text exercises, detailed chapter summaries, and practice chapter tests with complete solutions.


# ACKNOWLEDGMENTS

We extend our sincere gratitude to those who helped develop this book by reviewing all or part of the manuscript. We have implemented many of your suggestions to the great benefit of the text.

Richard D. Armstrong, St. Louis Community College
Edgar M. Chandler, Phoenix College
Ben Cornelius, Oregon Institute of Technology
Mary Coughlin, University of Toledo
Michael B. Curry, Pima Community College
John A. Dersch, Jr., Grand Rapids Junior College
Karen R. Fawcett, University of Southern Mississippi
Linda Holden, Indiana University
Charles M. Lindsay, Coe College
Carolyn F. Neptune, Johnson County Community College
Charles V. Peele, Marshall University
William Radulovich, Florida Community College
James V. Stewart, Oregon Institute of Technology
Eric Tellenbach, Citrus Community College

We extend special appreciation to Joseph Mutter for reviewing the entire manuscript and offering numerous suggestions for improvements, for writing the supplements, and for checking the problems. Thanks go to Diana Denlinger Vanlandingham and Gail Dickerson for typing the manuscript and supplements.

To everyone at Scott, Foresman and Company we are greatly indebted. Special thanks go to Jack Pritchard, Steve Quigley, Terry McGinnis, Sarah Joseph, and Ellen Pettengell.

Finally, we are deeply indebted to our families and in particular to our wives, Barbara and Barbara, who have given us unceasing support, time, and encouragement over the years.

L. Murphy Johnson

Arnold R. Steffensen

# Contents

# To the Student

During the past several years we have taught college algebra and trigonometry to more than 1500 students having a variety of career choices. Some were taking mathematics to satisfy graduation requirements, while others were preparing for more advanced courses in mathematics, science, business, or engineering. Regardless of your educational goals, this text has been written with you, the student, in mind. The material is introduced gradually, building from basic to more advanced skills. We have tried to demonstrate the relevance and usefulness of mathematics throughout the text by including practical everyday applications. As you begin this course, keep in mind these guidelines that are both necessary and helpful.

## GENERAL GUIDELINES

**1.** Mastering algebra and trigonometry requires motivation and dedication. Just as an athlete does not improve without commitment to his or her goal, an algebra student must be prepared to work hard and spend time studying.

**2.** Algebra and trigonometry are not learned simply by watching, listening, or reading; *they are learned by doing*. Use your pencil and practice. When your thoughts are organized and written in a neat and orderly way, you have taken a giant step toward success. Be complete and write out all details. The following are samples of two students' work on an applied problem. Can you tell which one was more successful in the course?

**Student A**

*Let* $n$ = *number of units produced during the month*

$C$ = *total cost per month*

$$C = 10n^2 - 100n - 2000$$

$$10{,}000 = 10n^2 - 100n - 2000$$

$$0 = 10n^2 - 100n - 12{,}000$$

$$0 = n^2 - 10n - 1200$$

$$0 = (n-40)(n+30)$$

$$n - 40 = 0 \quad \text{or} \quad n + 30 = 0$$

$$n = 40 \quad \text{or} \quad n = -30$$

*Thus, 40 units were produced.*

**Student F**

$$10(10{,}000)^2 - 100(10{,}000) - 2000$$

$$=$$

$n$ = *units*

$$n = 10n^2 - 100n - 2000$$

$$0 = 10n^2 - 101n - 2000$$

3. A calculator is useful in any course in algebra or trigonometry. Become familiar with your calculator by consulting your owner's manual. Use the calculator as a time-saving device for work with decimals or complicated functions, but do not become so dependent that you use it for simple calculations that can be done mentally. Learn when to use and when not to use your calculator. See "A Word About Calculators" for more information about how calculators can be used with this text.

## SPECIFIC GUIDELINES

1. As you begin to study each section, look through the material for a preview of what is coming.
2. Return to the beginning of the section and study the text and examples carefully. The side comments in color will help you if something is not clear.
3. Periodically you will encounter a CAUTION or a NOTE. The CAUTIONs warn you of common mistakes and special problems to avoid. The NOTEs provide pertinent information or additional explanations.
4. After you have completed the material in the section, check your mastery of the skills and apply what you have learned by working the exercises assigned by your instructor. Answers to the odd-numbered problems are at the back of the text. Complete worked-out solutions to selected odd-numbered problems are also available in the *Solutions and Study Guide*.
5. Exercises marked For Review, located at the end of most exercise sets, help keep previously covered materials fresh in your mind and often prepare you for the next section.
6. After you have completed a chapter, review each section and work the CHAPTER REVIEW EXERCISES. Answers to all these exercises are at the back of the text. To help you prepare for tests, additional chapter review material is also included in the *Solutions and Study Guide*.

If you follow these suggestions and work closely with your instructor, you will greatly improve your chances for success in the course.

# A Word About Calculators

It is assumed that most students will have a hand calculator in this course. Although it is not absolutely essential, your work will be easier if you use a calculator. The major difference between the types of calculators is in the way they perform various operations. Perhaps more desirable at this level, since the order of operations is the same as in algebra, is the type that uses Algebraic Logic (ALG). The alternative system, Reverse Polish Notation (RPN), is preferred by many mathematicians and professionals, however. Each system, with its advantages and disadvantages, will perform the calculations necessary in this course. Throughout the text we illustrate both systems using ALG for Algebraic Logic and RPN for Reverse Polish Notation. As an example, we show the sequence of steps used in each system to compute

$$\frac{(2)(4.5) - (1.3)^2}{5\sqrt{3}}.$$

| | Keystrokes | Display |
|---|---|---|
| ALG: | 2 [×] 4.5 [−] 1.3 [$x^2$] [=] [÷] 5 [÷] 3 [√] [=] → | 0.8440861 |
| RPN: | 2 [ENTER] 4.5 [×] 1.3 [$x^2$] [−] 5 [÷] 3 [√] [÷] → | 0.8440861 |

Notice that RPN calculators use an [ENTER] key instead of the [=] key found on ALG calculators. This is an essential difference between the two operating systems. Other variations in the types of keys are strictly notational. For example, to change the sign of a number (for entering negative numbers), some calculators have a [+/−] key, while others have a [CHS] key. Also, one calculator uses the [STO] key to place a number in memory, while another has an [M] key. We will try to point out some of the differences that arise as we consider various computations. However, since it is impossible to mention all of these differences, the best advice is to read your owner's manual.

With calculators, slight variations in accuracy due to rounding differences are bound to occur. Most of these will appear in the seventh or eighth decimal place and should not be of much concern. Throughout the text we have not rounded results until the final step, holding calculated values in memory. Even with this agreement, small variations due to individual calculator differences may arise. Don't panic if your calculator gives an answer that disagrees slightly with what we have shown.

Finally, keep in mind that a calculator is a tool for doing complicated computations; it does not think and only reacts to your input. Do not become so dependent on your calculator that you reach for it to make simple computations that can be made mentally. You must learn when a calculator should and should not be used and when your results are reasonable and appropriate.

CHAPTER

# 1 REVIEW OF FUNDAMENTAL CONCEPTS

A knowledge of algebra not only gives us a foundation for the study of more advanced mathematics but also provides the tools for solving many applied problems in business, science, and engineering. Consider the following applications.

**BUSINESS ▶**

A payroll office administrator needs a formula for calculating the new salary of employees who have received an 8% raise.

Let $x$ = employee's previous salary,

$0.08x$ = employee's raise.

The new salary is found by adding the raise to the old salary.

$$\text{new salary} = x + 0.08x = 1.08x$$

For example, if an employee's old salary $x$ was \$32,000 per year, the new salary $1.08x$ would be (1.08)(\$32,000) = \$34,560.

◀ ENGINEERING

The height in feet of a rocket $t$ seconds after firing is given by the expression $-16t^2 + 180t$. Find the height 4 seconds into its flight.

$$\begin{aligned} \text{height} &= -16t^2 + 180t \\ \text{height at 4 sec} &= -16(4)^2 + 180(4) \\ &= -256 + 720 \\ &= 464 \text{ ft} \end{aligned}$$

In this chapter we review number systems and their basic properties, which serve as a foundation for our work.

## 1.1 The Real Number System

Because sets of numbers are fundamental to our study of algebra, we begin by reviewing them briefly. Remember that a **set** is a collection of objects called **elements.** The elements of a set are listed within braces, { }. The most basic sets of numbers are given here.

$N = \{1, 2, 3, \ldots\}$ **Natural (counting) numbers**

$W = \{0, 1, 2, 3, \ldots\}$ **Whole numbers**

$I = \{\ldots, -3, -2, -1, 0, 1, 2, 3, \ldots\}$ **Integers**

$Q = \left\{\frac{a}{b} \text{ such that } a \text{ and } b \text{ are integers with } b \neq 0\right\}$ **Rational numbers**

$P = \{x \text{ such that } x \text{ is not rational}\}$ **Irrational numbers**

$R = \{x \text{ such that } x \text{ is rational or irrational}\}$ **Real numbers**

In addition to these sets of numbers, we often have occasion to refer to the set of *negative integers,* $\{\ldots, -3, -2, -1\}$, and the set of *positive integers,* $\{1, 2, 3, \ldots\}$. Notice that the three dots indicate that the pattern continues.

The set of **rational numbers** includes the set of integers together with all quotients of integers. Every rational number can be written as a fraction or a decimal. For example, $\frac{3}{8}$ can be written as 0.375 (dividing 3 by 8) and $\frac{3}{11}$ as 0.2727... (dividing 3 by 11). The decimal 0.375 is called a **terminating decimal** because the sequence of digits comes to an end, while 0.2727... is a **repeating decimal** because the block of digits 27 repeats indefinitely. Repeating decimals are often written with a bar over the block of digits that repeats. For example,

$$\frac{3}{11} = 0.\overline{27} \quad \text{and} \quad \frac{1}{3} = 0.\overline{3}.$$

Every rational number has a decimal form that either terminates or repeats. This property is sometimes used to define the set of rational numbers.

Numbers that are not rational, that is, that *cannot* be written as a quotient of integers, are called **irrational numbers.** An irrational number cannot be written as a terminating or repeating decimal. One of the best known irrational numbers is $\pi$, the ratio of the circumference of any circle to its diameter. Numbers such as $\sqrt{2}$ and $\sqrt{26}$, square roots of positive integers that are not perfect squares, are also irrational.

## The Set of Real Numbers

The set of **real numbers** consists of the rational numbers together with the irrational numbers. The relationships among the sets of numbers we have discussed is displayed in Figure 1.

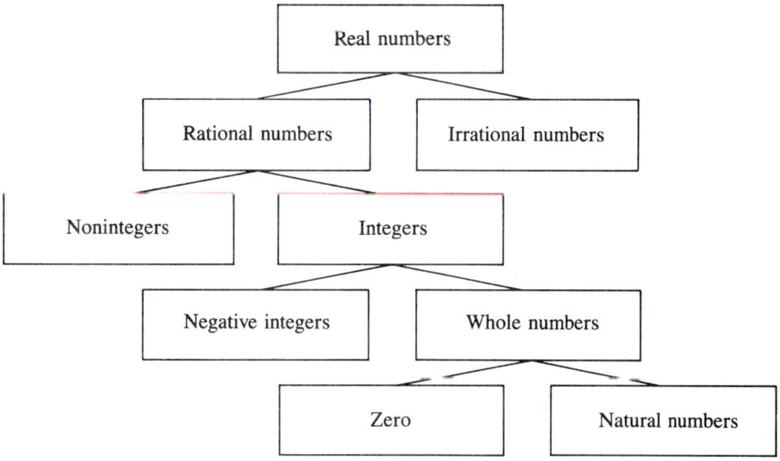


**Figure 1 The Real Number System**

An excellent means of displaying numbers and showing some of their important properties is a **number line,** as shown in Figure 2. The **origin** is labeled zero and unit lengths in both directions are marked off. Points to the right of zero are identified with positive numbers, while points to the left of zero correspond to negative numbers.

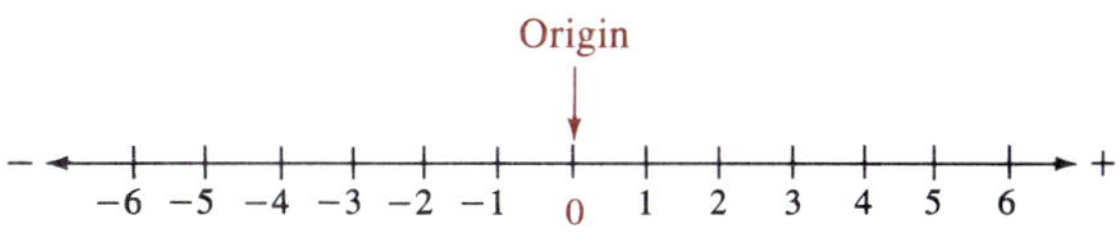


**Figure 2 Number Line**

Every real number can be identified with exactly one point on a number line, and every point on a number line corresponds to exactly one real number. Figure 3 shows a number line with points corresponding to several real numbers plotted on it.

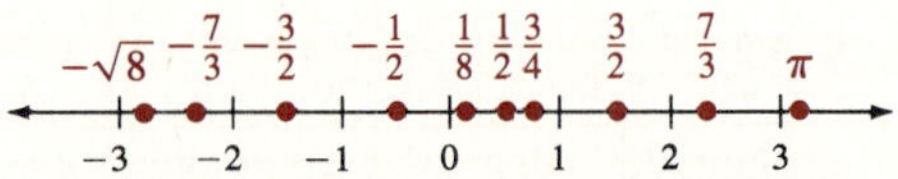


**Figure 3 Points on a Number Line**

Numbers that are **equal** ($a = b$) correspond to the same point on the number line. If $a$ is to the left of $b$, we say $a$ is **less than** $b$ and write $a < b$. We also say that $b$ is **greater than** $a$ and write $b > a$. More formally, $a < b$ or $b > a$ if $b - a$ is a positive number ($b - a > 0$).

If $a$ and $b$ are any real numbers the **trichotomy** property states that exactly one of the following holds.

$a > b$, $a = b$, or $a < b$ **Trichotomy property**

The **transitive property** for inequalities states that

if $a < b$ and $b < c$, then $a < c$. **Transitive property**

Both the trichotomy and transitive properties are **axioms,** properties that are accepted without verification.

The axioms of equality that are used in solving equations and simplifying algebraic expressions are summarized next.

**Axioms of Equality**

Let $a$, $b$, and $c$ be real numbers.

| | |
|---|---|
| **Reflexive property** | $a = a$ |
| **Symmetric property** | If $a = b$, then $b = a$. |
| **Transitive property** | If $a = b$ and $b = c$, then $a = c$. |
| **Substitution property** | If $a = b$, then either may replace the other in any statement without affecting the truth of the statement. |

## EXAMPLE 1

State the property illustrated.

**(a)** If $x = 5$, then $5 = x$. Symmetric property

**(b)** $-2 < -\frac{1}{2}$ and $-\frac{1}{2} < 3$ implies $-2 < 3$. Transitive property for $<$

**(c)** If $x = 7$ and $y + x = 6$, then $y + 7 = 6$. Substitution property

**(d)** If $y = 6$ and $6 = z$, then $y = z$. Transitive property for = ■

## Operations on Real Numbers

The basic operations on the real numbers are **addition** and **multiplication.** When these operations are performed on two real numbers we always obtain a real number. This is the first of the axioms involving operations on real numbers.

### Axioms of the Real Numbers

Let $a$, $b$, and $c$ be real numbers.

| | |
|---|---|
| **Closure properties** | $a + b$ is a real number |
| | $ab$ is a real number |
| **Commutative properties** | $a + b = b + a$ |
| | $ab = ba$ |
| **Associative properties** | $a + (b + c) = (a + b) + c$ |
| | $a(bc) = (ab)c$ |
| **0 is the additive identity** | $a + 0 = a = 0 + a$ |
| **1 is the multiplicative identity** | $a \cdot 1 = a = 1 \cdot a$ |
| **Negatives (additive inverses)** | $a + (-a) = 0 = (-a) + a$ |
| **Reciprocals (multiplicative inverses)** | $a\left(\frac{1}{a}\right) = 1 = \left(\frac{1}{a}\right)a \qquad (a \neq 0)$ |
| **Distributive property** | $a(b + c) = ab + ac$ |

## EXAMPLE 2

State the property illustrated.

**(a)** $3(x + y) = 3x + 3y$ Distributive property

**(b)** $7 + (-7) = 0$ Existence of negatives

**(c)** $5(xy) = (5x)y$ Associative property of multiplication ■

The operations of *subtraction* and *division* are defined in terms of addition and multiplication respectively. **Subtraction** is defined as adding a negative, while **division** is defined as multiplying by a reciprocal.

| | |
|---|---|
| $a - b = a + (-b)$ | **Subtraction** |
| $a \div b = \frac{a}{b} = a\left(\frac{1}{b}\right) \qquad (b \neq 0)$ | **Division** |

Properties that can be proved using axioms and definitions are called **theorems.** We present a proof of the first of the following theorems leaving some of the remaining proofs as exercises.

### Theorems Involving Zero

If $a$ is any real number,

**1.** $a \cdot 0 = 0 \cdot a = 0.$

**2.** $a - 0 = a$ and $0 - a = -a.$

**3.** $0 \div a = \dfrac{0}{a} = 0 \quad (a \neq 0).$ $\left(\dfrac{a}{0} \text{ is undefined.}\right)$

PROOF OF 1

| | |
|---|---|
| $0 = a + (-a)$ | Additive inverse |
| $= a \cdot 1 + (-a)$ | Multiplicative identity |
| $= a \cdot (0 + 1) + (-a)$ | Additive identity |
| $= (a \cdot 0 + a \cdot 1) + (-a)$ | Distributive property |
| $= (a \cdot 0 + a) + (-a)$ | Multiplicative identity |
| $= a \cdot 0 + (a + (-a))$ | Associative property of addition |
| $= a \cdot 0 + 0$ | Additive inverse |
| $= a \cdot 0$ | Additive identity |

Therefore, by the transitive property of equality, $0 = a \cdot 0$. By the symmetric property of equality, $a \cdot 0 = 0$. Since $0 \cdot a = a \cdot 0$ by the commutative property of multiplication, by the substitution property of equality, $0 \cdot a = 0$. ■

The following theorems, easily proved using the axioms of equality, are important for solving equations in Chapter 2.

### Theorems of Equality

Let $a$, $b$, and $c$ be real numbers.

| | |
|---|---|
| **1.** If $a = b$ then $a + c = b + c$. | **Addition property** |
| **2.** If $a = b$ then $ac = bc$. | **Multiplication property** |

The proofs of several of the following properties of negatives follow from the axioms and definitions and are left as exercises.

**Theorems Involving Negatives**

Let $a$ and $b$ be real numbers.

**1.** $-(-a) = a$

**2.** $(-a)b = -(ab) = a(-b)$

**3.** $(-a)(-b) = ab$

**4.** $(-1)a = -a$

**5.** $-(a + b) = -a - b$

**6.** $-(a - b) = -a + b$

**7.** $\frac{-a}{b} = -\frac{a}{b} = \frac{a}{-b}$ $(b \neq 0)$

**8.** $\frac{-a}{-b} = -\frac{-a}{b} = -\frac{a}{-b} = \frac{a}{b}$ $(b \neq 0)$

## EXAMPLE 3

Use the theorems involving zero and negatives.

**(a)** $3 - (-x + y) = 3 + x - y$

**(b)** $0 - 8x = -8x$

**(c)** $\frac{x - y}{y - x} = \frac{x - y}{-(x - y)} = -1$

**(d)** $(4x + 3y) \cdot 0 = 0$ ■

## Absolute Value

The number of units that a number $a$ is from zero on the number line is called the *absolute value* of $a$ and is denoted $|a|$. See Figure 4.

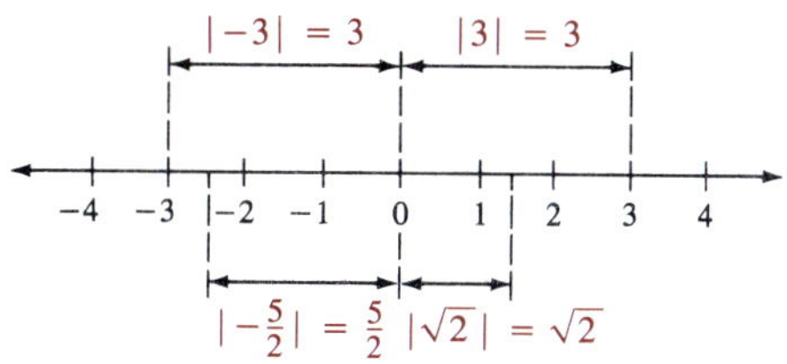


**Figure 4 Finding Absolute Values**

Note that for any real number $a$, $|-a| = |a|$ and that $|a| \geq 0$ ($\geq$ is read "greater than or equal to"). These properties can be proved using the following definition.

Let $a$ be any real number. The **absolute value** of $a$ is denoted $|a|$ and

$$|a| = \begin{cases} a, & \text{when } a \geq 0; \\ -a, & \text{when } a < 0. \end{cases}$$

## EXAMPLE 4

Find the absolute value.

**(a)** $|-9| = -(-9) = 9$ $\quad -9 < 0$

**(b)** $|0| = 0$ $\quad 0 \geq 0$

**(c)** $|\sqrt{5}| = \sqrt{5}$ $\quad \sqrt{5} \geq 0$ ■

We can use the notion of absolute value to define the distance between two numbers.

Let $a$ and $b$ be any real numbers. The **distance between $a$ and $b$** is denoted $d(a, b)$ and given by

$$d(a, b) = |a - b|.$$

**EXAMPLE 5** Find the distance between the numbers.

**(a)** 6 and 13 $|6 - 13| = |-7| = -(-7) = 7$

**(b)** 7 and $-8$ $|7 - (-8)| = |7 + 8| = |15| = 15$ ■

## 1.1 EXERCISES

Answer true or false in Exercises 1–24. If the answer is false, explain why.

**1.** 3 is a rational number.

**2.** $\frac{3}{2}$ is a rational number.

**3.** $-5$ is a natural number.

**4.** 0 is a natural number.

**5.** $\sqrt{5}$ is a rational number.

**6.** $0.\overline{281}$ is a rational number.

**7.** $0.\overline{65}$ is a real number.

**8.** $2\sqrt{3} - 1$ is a real number.

**9.** Every integer is a rational number.

**10.** Every rational number is an integer.

**11.** Some irrational numbers are integers.

**12.** Some rational numbers are integers.

**13.** The set of whole numbers is closed with respect to addition.

**14.** The set of whole numbers is closed with respect to multiplication.

**15.** The set of whole numbers is closed with respect to division.

**16.** The set of whole numbers is closed with respect to subtraction.

**17.** For every real number $a$, $|a| \geq 0$.

**18.** For every real number $a$, $|a| > 0$.

**19.** For all real numbers $a$ and $b$, $|a - b| = |b - a|$.

**20.** For all real numbers $a$ and $b$, $|a| + |b| = |a + b|$.

**21.** For all real numbers $x$ and $y$, $x - y = -(y - x)$.

**22.** For all real numbers $x$ and $y$, $-x + y = -(x - y)$.

**23.** For every real number $x$, $\frac{1}{x}$ is a real number.

**24.** For every real number $x$, $-x$ is a negative number.

In Exercises 25–30 state the illustrated property.

**25.** $x + 6 = 6 + x$

**26.** $3(xy) = (3x)y$

**27.** If $x = 7$, then $7 = x$.

**28.** $y < 3$ or $y = 3$ or $y > 3$

**29.** $(5x + y) + 0 = 5x + y$

**30.** $(x^2 + 1)\left(\frac{1}{x^2 + 1}\right) = 1$

In Exercises 31–36 complete each statement using the specified law.

**31.** Associative property of addition: $x + (3 + 7) =$ ______.

**32.** Distributive property: $3a + 3w =$ ______.

**33.** Transitive property for $<$: If $a < b$ and $b < 9$, then ______.

**34.** Transitive property for $=$: If $x = 4$ and $4 = w$, then ______.

**35.** Existence of negatives: $15 +$ ______ $= 0$.

**36.** Existence of multiplicative identity: $1 \cdot 28 =$ ______.

In Exercises 37–44 use the properties of negatives to simplify each expression.

**37.** $-(-(-3))$

**38.** $-[-(1 - x)]$

**39.** $6 - (x - 4)$

**40.** $-[x - (x + y)]$

**41.** $\left(-\frac{1}{9}\right) + \left(-\frac{1}{3}\right)$

**42.** $\frac{2}{3} \quad \left( \quad \frac{3}{4}\right)$

**43.** $\left( \quad \frac{2}{3}\right)\left(\frac{3}{4}\right)$

**44.** $\left(-\frac{2}{3}\right) \div \left(-\frac{3}{4}\right)$

In Exercises 45–56 rewrite each expression without using absolute value notation.

**45.** $\left|-\frac{3}{4}\right|$

**46.** $-|-2|$

**47.** $|x - y|$ if $x < y$

**48.** $|x + y|$ if $x < 0$ and $y < 0$

**49.** $|-7 + 4|$

**50.** $|-7| + |4|$

**51.** $|-5| - |-8|$

**52.** $|-5 - (-8)|$

**53.** $\frac{|-12|}{|3|}$

**54.** $\left|\frac{-12}{3}\right|$

**55.** $|(-3)(-2) - (-4)|$

**56.** $|(-2)(4) - (-5)|$

In Exercises 57–62 find the distance between the given numbers.

**57.** $-3$ and $2$

**58.** $-\frac{1}{2}$ and $-\frac{1}{3}$

**59.** $7$ and $y$ if $y > 7$

**60.** $-6$ and $x$ if $x < -6$

**61.** $x$ and $y$ if $x < 0$ and $y > 0$

**62.** $x$ and $y$ if $x > y$

**63.** Find real numbers $a$, $b$, and $c$ such that $(a - b) - c \neq a - (b - c)$. This shows that subtraction is not associative.

**64.** Find real numbers $a$, $b$, and $c$ such that $a + (b \cdot c) \neq (a + b) \cdot (a + c)$. This shows that addition does not distribute over multiplication.

Use the properties of the real numbers to prove the following where $a$, $b$, and $c$ are real.

**65.** $0 - a = -a$

**66.** $0 \div a = 0$

**67.** If $a = b$ then $a + c = b + c$.

**68.** If $a = b$ then $ac = bc$.

**69.** $-(-a) = a$

**70.** $(-1)a = -a$

**71.** $-(a + b) = -a - b$

**72.** $-(a - b) = -a + b$

## 1.2 Integer Exponents

For multiplying a number or expression by itself several times, **exponential notation** can be used to avoid long strings of factors. For example,

$$\underbrace{3 \cdot 3 \cdot 3 \cdot 3}_{\textbf{4 factors}} = 3^4. \begin{matrix}\leftarrow \textbf{exponent} \\ \uparrow \textbf{base}\end{matrix}$$

### Exponential Notation

If $a$ is any real number and $n$ is a positive integer,

$$a^n = \underbrace{a \cdot a \cdot a \ . \ . \ . \ a}_{n \text{ factors}},$$

where $a$ is the **base,** $n$ is the **exponent,** and $a^n$ is the **exponential expression.**

CAUTION An exponent applies only to the factor next to it unless parentheses are used. For example,

$$2y^3 = 2 \cdot y \cdot y \cdot y \quad \text{and} \quad (2y)^3 = (2y)(2y)(2y) = 8y^3.$$

Also, $-x^2 \neq (-x)^2$. Thus, $-4^2 = -16$ but $(-4)^2 = 16$. ■

### Properties of Exponents

When we perform operations involving exponential expressions, our work can be simplified by using the basic properties of exponents. For example,

$$a^3 \cdot a^4 = \underbrace{(a \cdot a \cdot a)}_{\text{3 factors}}\underbrace{(a \cdot a \cdot a \cdot a)}_{\text{4 factors}} = \underbrace{a \cdot a \cdot a \cdot a \cdot a \cdot a \cdot a}_{\text{7 factors}} = a^7.$$

Thus, in general, if $m$ and $n$ are positive integers,

$$a^m \cdot a^n = \underbrace{a \cdot a \ . \ . \ . \ a}_{m \text{ factors}} \cdot \underbrace{a \cdot a \ . \ . \ . \ a}_{n \text{ factors}} = \underbrace{a \cdot a \ . \ . \ . \ a}_{m + n \text{ factors}} = a^{m+n}.$$

We have just proved the theorem

$$a^m \cdot a^n = a^{m+n}. \qquad \text{Product rule}$$

Similarly, we can show that if $m$ and $n$ are positive integers and $m - n$ is positive,

$$\frac{a^m}{a^n} = a^{m-n}. \qquad \text{Quotient rule}$$

When raising a power to a power, we multiply exponents. For example,

$$(a^2)^3 = \underbrace{(a^2)(a^2)(a^2)}_{\text{3 factors}} = \underbrace{a \cdot a \cdot a \cdot a \cdot a \cdot a}_{\text{6 factors}} = a^6.$$

Thus,

$$(a^m)^n = a^{mn}. \qquad \text{Power rule}$$

**EXAMPLE 1** Simplify using the rules of exponents.

**(a)** $2x^6 \cdot x^3 = 2x^{6+3} = 2x^9 \qquad a^m \cdot a^n = a^{m+n}$

**(b)** $\dfrac{b^5}{b^2} = b^{5-2} = b^3 \qquad \dfrac{a^m}{a^n} = a^{m-n}$

**(c)** $(3^2)^3 = 3^{2 \cdot 3} = 3^6 \qquad (a^m)^n = a^{mn}$ ■

Rules also exist for raising products or quotients to a power. For example,

$$(4x)^3 = (4x)(4x)(4x) = 4 \cdot 4 \cdot 4 \cdot x \cdot x \cdot x = 4^3x^3$$

$$\text{and} \quad \left(\frac{4}{x}\right)^3 = \left(\frac{4}{x}\right)\left(\frac{4}{x}\right)\left(\frac{4}{x}\right) = \frac{4 \cdot 4 \cdot 4}{x \cdot x \cdot x} = \frac{4^3}{x^3}.$$

Thus, for $n$ a positive integer,

$$(ab)^n = a^nb^n \quad \text{and} \quad \left(\frac{a}{b}\right)^n = \frac{a^n}{b^n} \quad (b \neq 0).$$

**EXAMPLE 2** Simplify using the rules of exponents.

**(a)** $(3x^4y^3)^4 = 3^4 \cdot (x^4)^4(y^3)^4 = 3^4x^{16}y^{12}$

**(b)**
$$\left(\frac{2w^2}{z^4}\right)^3 = \frac{(2w^2)^3}{(z^4)^3} \qquad \left(\frac{a}{b}\right)^n = \frac{a^n}{b^n}$$
$$= \frac{2^3(w^2)^3}{(z^4)^3} \qquad (ab)^n = a^nb^n$$
$$= \frac{8w^6}{z^{12}} \qquad (a^m)^n = a^{mn}$$ ■

In order to extend the quotient rule to the case where $m = n$, consider

$$\frac{a^m}{a^m} = a^{m-m} = a^0.$$

Since $a^m/a^m = 1$, this suggests the following definition. If $a$ is any real number except zero,

$$a^0 = 1.$$

To come to a definition for negative exponents, consider extending the quotient rule to $n > m$. For example, let $m = 2$ and $n = 5$.

$$\frac{a^m}{a^n} = \frac{a^2}{a^5} = a^{2-5} = a^{-3}$$

Also $$\frac{a^m}{a^n} = \frac{a^2}{a^5} = \frac{\cancel{a} \cdot \cancel{a}}{\cancel{a} \cdot \cancel{a} \cdot a \cdot a \cdot a} = \frac{1}{a \cdot a \cdot a} = \frac{1}{a^3}.$$

This suggests that $a^{-3} = 1/a^3$. Thus, if $a \neq 0$ and $n$ is a positive integer, then

$$\boldsymbol{a^{-n} = \frac{1}{a^n}}.$$

**EXAMPLE 3** Simplify using the rules of exponents.

**(a)** $8^0 = 1$

**(b)** $(2x^2y^2)^0 = 1 \quad (x \neq 0 \text{ and } y \neq 0)$

**(c)** $(-2)^{-3} = \dfrac{1}{(-2)^3} = -\dfrac{1}{8}$

**(d)** $\dfrac{1}{3^{-2}} = \dfrac{1}{\frac{1}{3^2}} = 3^2 = 9$ ■

**CAUTION** Do not write $5^{-2}$ as $-5^2$ or $(-2)(5)$. These are all different numbers.

$$5^{-2} = \frac{1}{5^2} = \frac{1}{25}, \qquad -5^2 = -25, \qquad (-2)(5) = -10$$ ■

We can extend the rules developed for positive integer exponents to negative integers. These rules are now summarized for all integer exponents.

### Rules of Exponents

Let $a$ and $b$ be any real numbers, $m$ and $n$ any integers.

1. $a^m a^n = a^{m+n}$
2. $\dfrac{a^m}{a^n} = a^{m-n}$
3. $(a^m)^n = a^{mn}$
4. $(ab)^n = a^n b^n$
5. $\left(\dfrac{a}{b}\right)^n = \dfrac{a^n}{b^n} \quad (b \neq 0)$
6. $a^0 = 1 \quad (a \neq 0)$
7. $a^{-n} = \dfrac{1}{a^n} \quad (a \neq 0)$
8. $\dfrac{1}{a^{-n}} = a^n \quad (a \neq 0)$

**EXAMPLE 4** Simplify and write without negative exponents.

**(a)** $(3x)^{-1} = \dfrac{1}{(3x)^1} = \dfrac{1}{3x}$ $\qquad a^{-n} = \dfrac{1}{a^n}$

**(b)** $3x^{-1} = 3\left(\dfrac{1}{x}\right) = \dfrac{3}{x}$ $\qquad$ Compare with **(a)**

**(c)** $\left(\dfrac{x^2y^3}{xy^4}\right)^{-2} = (x^{2-1}y^{3-4})^{-2}$ $\qquad \dfrac{a^m}{a^n} = a^{m-n}$

$= (xy^{-1})^{-2}$

$= x^{-2}y^2$ $\qquad (ab)^n = a^nb^n$ and $(a^m)^n = a^{mn}$

$= \dfrac{y^2}{x^2}$ $\qquad a^{-n} = \dfrac{1}{a^n}$

**(d)** $\dfrac{2^0}{a^{-2} + b^{-2}} = \dfrac{1}{\dfrac{1}{a^2} + \dfrac{1}{b^2}}$ $\qquad a^0 = 1$ and $a^{-n} = \dfrac{1}{a^n}$

$= \dfrac{1}{\dfrac{b^2}{a^2b^2} + \dfrac{a^2}{a^2b^2}}$ $\qquad$ Obtain a common denominator

$= \dfrac{1}{\dfrac{b^2 + a^2}{a^2b^2}} = \dfrac{a^2b^2}{b^2 + a^2}$

**(e)** $\left[\left(\dfrac{3x^6y^{-6}}{x^{-5}y^{-3}}\right)^3\right]^{-1} = \left(\dfrac{3x^6y^{-6}}{x^{-5}y^{-3}}\right)^{-3}$ $\qquad (a^m)^n = a^{mn}$

$= (3x^{11}y^{-3})^{-3}$ $\qquad \dfrac{a^m}{a^n} = a^{m-n}$

$= 3^{-3}x^{-33}y^9$ $\qquad (ab)^n = a^nb^n$

$= \dfrac{y^9}{27x^{33}}$ $\qquad a^{-n} = \dfrac{1}{a^n}$ ■

## Scientific Notation

One important application of integer exponents involves *scientific notation*. A number is written in **scientific notation** when it is expressed in the form

$$a \times 10^n \quad \text{where} \quad 1 \le a < 10 \quad \text{and } n \text{ is an integer.}$$

For example, a plant cell might contain 208,000,000,000,000 molecules, which could be written as $2.08 \times 10^{14}$. Also, one estimate for the weight of an oxygen molecule is 0.000 000 000 000 000 000 000 053, which in scientific notation is $5.3 \times 10^{-23}$.

Numbers that are greater than 10 will have positive integer exponents in scientific notation, while numbers less than 1 will show a negative integer power of 10. The exponent is determined by counting decimal places.

## EXAMPLE 5

Write each number in scientific notation.

**(a)** $93{,}000{,}000 = 9.3 \times 10^7$
**7 places**

**(b)** $0.000\,000\,93 = 9.3 \times 10^{-7}$
**7 places**

**(c)** $10 = 1 \times 10^1 = 1 \times 10$
**1 place**

**(d)** $0.1 = 1 \times 10^{-1}$ ■
**1 place**

Calculations in scientific notation can be made using the rules of exponents. For example,

$$\begin{aligned}\frac{(3.2 \times 10^{-8})(8.4 \times 10^{4})}{2.1 \times 10^{16}} &= \frac{(3.2)(8.4)}{(2.1)} \times \frac{10^{-8}10^{4}}{10^{16}}\\ &= 12.8 \times 10^{-8+4-16}\\ &= 12.8 \times 10^{-20}\\ &= 1.28 \times 10^{1} \times 10^{-20}\\ &= 1.28 \times 10^{-19}\end{aligned}$$

Very large or very small numbers must be written in scientific notation in order to be entered into a calculator. The [EE] or [EEX] key allows us to enter numbers in scientific notation. The calculation above would be carried out as follows:

ALG: 3.2 [EE] 8 [+/−] [×] 8.4 [EE] 4 [÷] 2.1 [EE] 16 [=] [1.28 −19]

RPN: 3.2 [EEX] 8 [CHS] [ENTER] 8.4 [EEX] 4 [×] 2.1 [EEX] 16 [÷] [1.28 −19]

Calculator display

The answer in scientific notation is $1.28 \times 10^{-19}$.

**EXAMPLE 6** Convert to scientific notation and use a calculator to perform the indicated operations.

**(a)** $\frac{(0.000\,036)(69{,}000{,}000)}{0.000\,000\,012} = \frac{(3.6 \times 10^{-5})(6.9 \times 10^{7})}{1.2 \times 10^{-8}}$ Convert to scientific notation

ALG: 3.6 [EE] 5 [+/−] [×] 6.9 [EE] 7 [÷] 1.2 [EE] 8 [+/−] [=] [2.07 11]

RPN: 3.6 [EEX] 5 [CHS] [ENTER] 6.9 [EEX] 7 [×] 1.2 [EEX] 8 [CHS] [÷] [2.07 11]

Calculator display

Thus, the result in scientific notation is $2.07 \times 10^{11}$.

**(b)** $\frac{(3.25 \times 10^{9})^2}{(7.86 \times 10^{26})(5.82 \times 10^{-10})}$

ALG: 3.25 [EE] 9 [$x^2$] [÷] 7.86 [EE] 26 [÷] 5.82 [EE] 10 [+/−] [=] [2.309 01]

RPN: 3.25 [EEX] 9 [ENTER] [$x^2$] 7.86 [EEX] 26 [÷] 5.82 [EEX] 10 [CHS] [÷] [2.309 01]

Thus, rounded to two decimal places, the answer is $2.31 \times 10^{1}$. ■

**Significant Digits**

Notice that we rounded the answer in Example 6(b) to the same number of decimal places as the numbers in the original problem. Rounding is necessary when approximate measurement numbers are used in a calculation, since it would be inappropriate to give an answer with a higher degree of accuracy than the values used to compute it. The notion of *significant digits* is often used to describe approximate values. Suppose that a number $x$ has been written in scientific notation as $a \times 10^n$ where $a$ has been rounded to $k$ decimal places. In this case, $x$ is said to be accurate to $k + 1$ **significant digits.** For example, given that $x = 12.3456$, we have the following:

| Approximation of $x$ | 12.346 | 12.35 | 12.3 | 12 | 10 |
|---|---|---|---|---|---|
| Number of significant digits | 5 | 4 | 3 | 2 | 1 |

## 1.2 EXERCISES

In Exercises 1–30 simplify and write each expression without negative exponents. Assume that variables in denominators are nonzero.

**1.** $x^4x^3$ **2.** $a^2a^{-3}a^4$ **3.** $6x^3x^{-2}$ **4.** $\dfrac{x^2x^{-5}}{x^3}$

**5.** $(b^{-3})^{-2}$ **6.** $(b^2)^{-3}$ **7.** $\dfrac{a^3}{b^2}$ **8.** $a^4b^3$

**9.** $(5^0)^3$ **10.** $(7^4)^0$ **11.** $(8x)^{-1}$ **12.** $8x^{-1}$

**13.** $\dfrac{2x^2}{x^{-2}}$ **14.** $\dfrac{9y^{-3}}{3y^3}$ **15.** $\left(\dfrac{6x^2}{y^{-4}}\right)^0$ $(x \neq 0,\ y \neq 0)$ **16.** $\left(\dfrac{8a^{-3}}{9b^{-2}}\right)^0$ $(a, b \neq 0)$

**17.** $\dfrac{1}{x^{-1} + y^{-1}}$ **18.** $\dfrac{1}{(x + y)^{-1}}$ **19.** $\left(\dfrac{2x^2}{y^{-3}}\right)^{-2}$ **20.** $\left(\dfrac{2x^2}{y^{-3}}\right)^2$

**21.** $\dfrac{3x^3y^{-5}}{12x^{-4}y^{-1}}$ **22.** $\dfrac{8a^{-5}b^8}{2a^7b^{-3}}$ **23.** $\left(\dfrac{6^0x^6y}{3x^{-2}y^{-1}}\right)^3$ **24.** $\left(\dfrac{8^0a^{-6}b^{-1}}{2a^4b^{-3}}\right)^4$

**25.** $\left(\dfrac{2x^{-2}y^{-3}}{x^4y^{-1}}\right)^{-2}$ **26.** $\left(\dfrac{3x^4y^{-2}}{x^{-4}y^3}\right)^{-3}$ **27.** $\left[\left(\dfrac{2a^{-2}b^{-3}}{a^{-4}b^2}\right)^{-2}\right]^{-1}$ **28.** $\left[\left(\dfrac{x^{-3}y^4}{2x^{-5}y^{-1}}\right)^{-2}\right]^3$

**29.** $\left(\dfrac{a^6b^2}{a^{-1}b}\right)^2\left(\dfrac{2a^{-1}b^3}{a^{-4}}\right)^{-2}$ **30.** $\left(\dfrac{x^4y^3}{x^{-1}y^2}\right)^{-1}\left(\dfrac{3x^{-4}y}{x^{-1}y^{-1}}\right)^{-2}$

In Exercises 31–34 write each number in scientific notation.

**31.** 456,000,000 **32.** 0.000 000 000 003 21 **33.** 0.01 **34.** 100

In Exercises 35–38 write each number in decimal notation.

**35.** $2.35 \times 10^8$ **36.** $8.62 \times 10^2$ **37.** $4.17 \times 10^{-8}$ **38.** $3.2 \times 10^{-1}$

Use your calculator in Exercises 39–42 to perform the indicated operations. Round to three significant digits and give answers in scientific notation.

**39.** (0.000 273)(428,000) **40.** (0.000 000 018 2)(0.007 26)

**41.** $\dfrac{(86{,}000)^2(0.000\ 000\ 392)}{(0.001\ 57)^2}$ **42.** $\dfrac{(0.000\ 005\ 62)^2(369{,}000)^2}{183{,}000{,}000{,}000}$

Solve using scientific notation.

**43. ECONOMICS** The national debt of a small country is \$6,250,000,000 and the population is 2,120,000. What is the amount of debt per person?

**44. BUSINESS** A company produced 925,000 small appliances in one year and made a profit of \$8,620,000. What was the profit on each appliance?

**45. ASTRONOMY** The earth is approximately 92,900,000 miles from the sun. If 1 mi = $1.61 \times 10^3$ m, what is the distance to the sun in meters?

**46. PHYSICS** If the speed of light is $3.00 \times 10^8$ m/sec, use the answer to Exercise 45 to estimate how long it takes light to get from the sun to the earth.

## For Review

Answer true or false in Exercises 47–50. If the answer is false, explain why.

**47.** Every natural number is a whole number.

**48.** The set of integers is closed with respect to division.

**49.** For all real numbers $x$, $|x| = |-x|$.

**50.** If $x > y$, then $|x - y| = x - y$.

# 1.3 Polynomials and Factoring

Before beginning our work with polynomials, we review several important definitions. Remember that a **variable** is a symbol or letter used to represent an arbitrary element of a set containing more than one element. When a symbol is used to represent a specific element, it is called a **constant.** An **algebraic expression** involves sums, differences, products, and quotients of numbers and variables raised to various powers. A **term** of an algebraic expression involves only the product of a number, called the **numerical coefficient,** and variables raised to various powers. For example,

$$3x^2y - 2xy^2, \quad \sqrt{xy} + 2x + 5y, \quad \text{and} \quad \frac{x+y}{x-y} - \frac{2xy}{3x-y}$$

are three algebraic expressions.

The expression $3x^2y - 2xy^2$ is also a polynomial. A **polynomial** is an algebraic expression with terms that are products of numbers and variables raised to whole-number exponents. In the following table we give examples of polynomials, their terms, and their coefficients.

| Polynomial | Terms | Coefficients |
|---|---|---|
| $3x^4 - 2x^2 + x - 8$ | $3x^4, -2x^2, x, -8$ | $3, -2, 1, -8$ |
| $4x^2y - 3xy^2 - xy$ | $4x^2y, -3xy^2, -xy$ | $4, -3, -1$ |
| $\sqrt{2}x^4y^4 - \frac{3}{4}$ | $\sqrt{2}x^4y^4, -\frac{3}{4}$ | $\sqrt{2}, -\frac{3}{4}$ |
| $6$ | $6$ | $6$ |

Expressions such as $\sqrt{x} + 5$ and $\frac{3}{xy} + y^2$ are not polynomials since variables are not raised to whole-number powers. Also, the expression $3^x$ is not a polynomial since the exponent on 3 is a variable.

A **monomial** is a polynomial with one term, while a **binomial** has two terms and a **trinomial** has three terms. A polynomial with more than three terms is simply called a polynomial. Polynomials are further classified by degree. First we define the **degree of a term** to be the sum of the powers on the variables in the term. The **degree of the polynomial** is the same as that of the term with highest degree. A polynomial which is a constant, such as 5, has degree zero since it can be written $5x^0$. However, the zero polynomial, 0, has no degree. We illustrate the notion of degree in the following table.

| Polynomial | Type | Degree |
|---|---|---|
| $8x^4$ | Monomial | 4 |
| $3x^4 - 6x^7 + 5x$ | Trinomial | 7 |
| $16$ | Monomial | 0 |
| $3x^3y^2 - 7xy^6$ | Binomial | 7 |
| $8a^2b^2c^2 - 6a^5b - 7b^4c^4$ | Trinomial | 8 |

## Addition and Subtraction of Polynomials

To add or subtract polynomials we combine *like terms*. Two terms are like terms if they contain the same variables raised to the same powers. Remember that all signs within parentheses that follow a minus sign are changed when the parentheses are removed.

## EXAMPLE 1

**(a)** Add. $(x^4y - 5x^3y^2 - 3xy^3 + 7y) + (-5x^4y + 8x^2y^3 + 4xy^3 + 3y)$

$= x^4y - 5x^3y^2 - 3xy^3 + 7y - 5x^4y + 8x^2y^3 + 4xy^3 + 3y$ Remove parentheses

$= (1 - 5)x^4y - 5x^3y^2 + 8x^2y^3 + (-3 + 4)xy^3 + (7 + 3)y$ Combine like terms

$= -4x^4y - 5x^3y^2 + 8x^2y^3 + xy^3 + 10y$

**(b)** Subtract $-8y^3 + 6y^2 - 2y - 3$ from $-4y^3 - 6y^2 + 5y - 1$.

$(-4y^3 - 6y^2 + 5y - 1) - (-8y^3 + 6y^2 - 2y - 3)$ Be sure to subtract in the correct order

$= -4y^3 - 6y^2 + 5y - 1 + 8y^3 - 6y^2 + 2y + 3$ Remove parentheses, watch the signs

$= (-4 + 8)y^3 + (-6 - 6)y^2 + (5 + 2)y + (-1 + 3)$ Combine like terms

$= 4y^3 - 12y^2 + 7y + 2$ ■

## Multiplication of Polynomials

Multiplication of polynomials was introduced in Section 1.2 when we multiplied monomials using the product rule for exponents. To multiply a monomial by a polynomial with two or more terms, use the distributive property. To multiply two polynomials, neither of which is a monomial, multiply each term of one by each term of the other and collect like terms. When multiplying two binomials we may use the FOIL method (F stands for First terms, O for Outside, I for Inside, and L for Last).

**EXAMPLE 2** Multiply.

**(a)** $-4a^3b^2(2a^2b^3 - 3a) = (-4a^3b^2)(2a^2b^3) - (-4a^3b^2)(3a)$ Distributive property

$= (-4)(2)(a^3a^2)(b^2b^3) - (-4)(3)(a^3a)b^2$

$= -8a^5b^5 + 12a^4b^2$

**(b)** $(5a + 2b)(3a - 7b) = \overset{\text{F}}{(5a)(3a)} + \overset{\text{O}}{(5a)(-7b)} + \overset{\text{I}}{(2b)(3a)} + \overset{\text{L}}{(2b)(-7b)}$

$= 15a^2 - 35ab + 6ab - 14b^2$

$= 15a^2 - 29ab - 14b^2$ ■

When polynomials with three or more terms are multiplied, we often find it convenient to write one above the other and arrange like terms of the product in vertical columns.

**EXAMPLE 3** Multiply.

$$\begin{array}{rl}
3x^2 - 5xy + 2y^2 & \\
7x - 8y & \\
\hline
21x^3 - 35x^2y + 14xy^2 \phantom{- 16y^3} & 7x \text{ multiplies each term of the top polynomial} \\
- 24x^2y + 40xy^2 - 16y^3 & -8y \text{ multiplies each term of the top polynomial} \\
\hline
21x^3 - 59x^2y + 54xy^2 - 16y^3 & \blacksquare
\end{array}$$

Several special products, useful when multiplying, are important for factoring.

**Product Formulas**

$$(a + b)(a - b) = a^2 - b^2$$
$$(a + b)(a + b) = (a + b)^2 = a^2 + 2ab + b^2$$
$$(a - b)(a - b) = (a - b)^2 = a^2 - 2ab + b^2$$

**EXAMPLE 4** Multiply.

**(a)** $(x + 5)(x - 5) = x^2 - 5^2$ Using $(a + b)(a - b) = a^2 - b^2$

$= x^2 - 25$

**(b)** $(x + 5)(x + 5) = (x + 5)^2$

$= x^2 + 2(x)(5) + 5^2$ Using $(a + b)^2 = a^2 + 2ab + b^2$, *not* $a^2 + b^2$

$= x^2 + 10x + 25$

**(c)** $(3a^2 - 2b)(3a^2 + 2b) = (3a^2)^2 - (2b)^2$

$= 3^2(a^2)^2 - 2^2b^2$ $(ab)^n = a^nb^n$

$= 9a^4 - 4b^2$ $(a^m)^n = a^{mn}$

**(d)** 
$$\begin{aligned}(3a^2 - 2b)(3a^2 - 2b) &= (3a^2 - 2b)^2 \\ &= (3a^2)^2 - 2(3a^2)(2b) + (2b)^2 && \text{Using } (a-b)^2 = a^2 - 2ab + b^2 \\ &= 9a^4 - 12a^2b + 4b^2\end{aligned}$$ ■

## Division of Polynomials

To divide a polynomial by a monomial, we divide each term of the polynomial by the monomial.

## EXAMPLE 5

Divide.

**(a)** $$\frac{5x^3 + 10x^2 - 20x}{5x} = \frac{5x^3}{5x} + \frac{10x^2}{5x} - \frac{20x}{5x} = x^2 + 2x - 4$$

**(b)** $$\begin{aligned}\frac{6a^4b^3 + 9a^2b^5 - 18ab^7 + 6a^2}{6a^2b^3} &= \frac{6a^4b^3}{6a^2b^3} + \frac{9a^2b^5}{6a^2b^3} - \frac{18ab^7}{6a^2b^3} + \frac{6a^2}{6a^2b^3} \\ &= a^2 + \frac{3b^2}{2} - \frac{3b^4}{a} + \frac{1}{b^3}\end{aligned}$$ ■

To divide a polynomial by a binomial, we follow the procedure used in long division of numbers. This is illustrated in the next example.

## EXAMPLE 6

Divide.

**(a)**
$$\begin{array}{rll} & x + 7 & \\ x - 3 \,\big)\!\! & x^2 + 4x - 25 & \\ & \underline{x^2 - 3x} & x - 3 \text{ times } x\text{, the first term of the quotient} \\ & 7x - 25 & \text{Subtract and bring down } -25\text{; note that } 4x - (-3x) = 7x \\ & \underline{7x - 21} & x - 3 \text{ times 7, the new term of the quotient} \\ & -4 & \text{Subtract } 7x - 21 \text{ from } 7x - 25 \text{ to obtain the remainder, } -4 \end{array}$$

The answer is $x + 7$ remainder $-4$ or $x + 7 - \dfrac{4}{x - 3}$.

**(b)**
$$\begin{array}{rl} & y^2 - 4y + 3 \\ y^2 + 3 \,\big)\!\! & y^4 - 4y^3 + 6y^2 \qquad + 8 \\ & \underline{y^4 \qquad + 3y^2} \\ & -4y^3 + 3y^2 \\ & \underline{-4y^3 \qquad - 12y} \\ & 3y^2 + 12y + 8 \\ & \underline{3y^2 \qquad + 9} \\ & 12y - 1 \end{array}$$

The answer is $y^2 - 4y + 3 + \dfrac{12y - 1}{y^2 + 3}$. ■

## Factoring Polynomials

Factoring a polynomial involves writing the polynomial as a product; as such, it is the reverse of multiplying. The first type of factoring problem we consider is a direct application of the distributive property. For example,

$$8x^3y - 12x^2y^2 = 4x^2y(2x - 3y).$$

Here $4x^2y$ is the greatest common monomial factor of the terms of $8x^3y - 12x^2y^2$. When factoring, we rewrite the polynomial as a product of other polynomials. The first step is always to remove the greatest common monomial factor from all terms.

A polynomial is called **prime** when it contains no polynomial factors other than 1 or $-1$. We will restrict factoring to the use of integers. For example, $x^2 - 7$ can be factored as $(x + \sqrt{7})(x - \sqrt{7})$, but since $\sqrt{7}$ is not an integer, $x^2 - 7$ is considered prime.

We can also use the distributive property to remove common binomial factors. For example,

$$3a(2a + b) - 2b(2a + b) = (3a - 2b)(2a + b). \qquad 2a + b \text{ is a common factor}$$

In fact this is the last step in a process called **factoring by grouping.** We group terms, factor each group, and then, if possible, remove any common binomial factors from the result.

## EXAMPLE 7

Factor the polynomial by grouping terms.

**(a)** $3x^2y - 6xy^2 + 4x - 8y = 3xy(x - 2y) + 4(x - 2y)$

$= (3xy + 4)(x - 2y)$ $\quad$ $x - 2y$ is a common factor

**(b)** $7ab^3 - 14a^2b^2 - b + 2a = 7ab^2(b - 2a) - (b - 2a)$ $\quad$ Watch signs

$= (7ab^2 - 1)(b - 2a)$ $\quad$ $b - 2a$ is a common factor ■

We now turn our attention to factoring a trinomial of the form $ax^2 + bxy + cy^2$. First recall the patterns involved in multiplying binomials using FOIL.

$$\begin{aligned}(5x - 3y)(2x + 7y) &= \underset{(F)}{(5x)\cdot(2x)} + \underset{(O)}{(5x)\cdot(7y)} + \underset{(I)}{(-3y)\cdot(2x)} + \underset{(L)}{(-3y)\cdot(7y)}\\ &= (5\cdot 2)x^2 + (5\cdot 7)xy + (-3)(2)xy + (-3)\cdot(7)y^2\\ &= 10x^2 + 35xy - 6xy - 21y^2\\ &= 10x^2 + 29xy - 21y^2\end{aligned}$$

(F) (O) (I) (L)

The first two terms of the binomials multiply (F) to give $10x^2$, the last two terms multiply (L) to give $-21y^2$, and $29xy$ comes from the products (O) and (I), $35xy - 6xy$. Thus, to factor $10x^2 + 29xy - 21y^2$ we must factor $10x^2$, factor $-21y^2$, and determine which products forming the middle terms will add to give $29xy$. To factor $ax^2 + bxy + cy^2$ we must fill the blanks in

Factors of $c$

$$ax^2 + bxy + cy^2 = (\_\_x + \_\_y)(\_\_x + \_\_y).$$

Factors of $a$

We must choose the factors so that the Ⓞ and Ⓘ terms add to give $bxy$.

**EXAMPLE 8** Factor the trinomials.

**(a)** $2x^2 + 11xy + 12y^2$ $(a = 2,\ b = 11,\ c = 12)$

Factors of 12

$$2x^2 + 11xy + 12y^2 = (\_\_x + \_\_y)(\_\_x + \_\_y)$$

Factors of 2

Since all terms of the trinomial are positive we use only positive factors. Factors of $c$ must be tried in both orders with the factors of $a$.

| *Factors of a* | *Factors of c* |
|---|---|
| 1, 2 | 1, 12 or 2, 6 or 3, 4 |

$2x^2 + 11xy + 12y^2$

| | |
|---|---|
| $= (x + \_\_y)(2x + \_\_y)$ | The only factors of $a$ are 1 and 2 |
| $\stackrel{?}{=} (x + 12y)(2x + y)$ | Does not work since $xy + 24xy = 25xy$ |
| $\stackrel{?}{=} (x + y)(2x + 12y)$ | Does not work since $12xy + 2xy = 14xy$ |
| $\stackrel{?}{=} (x + 3y)(2x + 4y)$ | Does not work since $4xy + 6xy = 10xy$ |
| $\stackrel{?}{=} (x + 4y)(2x + 3y)$ | This works since $3xy + 8xy = 11xy$ |

Thus, $2x^2 + 11xy + 12y^2 = (x + 4y)(2x + 3y)$.

**(b)** $-12u^2 - 14uv + 40v^2$

First remove the common factor and make the leading coefficient $a > 0$.

$$-12u^2 - 14uv + 40v^2 = -2(6u^2 + 7uv - 20v^2)$$

Now factor $6u^2 + 7uv - 20v^2$ with $a = 6$, $b = 7$, and $c = -20$. Since $c = -20$, we need to consider both positive and negative factors of $c$.

| | |
|---|---|
| $6u^2 + 7uv - 20v^2 \stackrel{?}{=} (u + 20v)(6u - v)$ | Does not work since $-uv + 120uv = 119uv$ |
| $\stackrel{?}{=} (2u - 5v)(3u + 4v)$ | Does not work since $8uv - 15uv = -7uv$ |
| $\stackrel{?}{=} (2u + 5v)(3u - 4v)$ | This works since $-8uv + 15uv = 7uv$ |

Thus, $-12u^2 - 14uv + 40v^2 = -2(2u + 5v)(3u - 4v)$. ■

Several formulas that may be used in factoring are listed below. The first three were used previously to multiply. The other formulas can be verified by multiplying the right side to obtain the left.

**Factoring Formulas**

| | |
|---|---|
| **1.** $a^2 - b^2 = (a + b)(a - b)$ | Difference of squares |
| **2.** $a^2 + 2ab + b^2 = (a + b)^2$ | Perfect square |
| **3.** $a^2 - 2ab + b^2 = (a - b)^2$ | Perfect square |
| **4.** $a^3 + b^3 = (a + b)(a^2 - ab + b^2)$ | Sum of cubes |
| **5.** $a^3 - b^3 = (a - b)(a^2 + ab + b^2)$ | Difference of cubes |

**EXAMPLE 9**

Factor.

**(a)** $u^2 - 25v^2 = u^2 - 5^2v^2$

$= u^2 - (5v)^2$ Use $a^2 - b^2 = (a + b)(a - b)$

$= (u + 5v)(u - 5v)$

**(b)** $49x^4 - 70x^2y + 25y^2 = 7^2(x^2)^2 - 2 \cdot 7 \cdot 5x^2y + 5^2y^2$

$= (7x^2)^2 - 2(7x^2)(5y) + (5y)^2$ Use $a^2 - 2ab + b^2 = (a - b)^2$

$= (7x^2 - 5y)^2$

**(c)** $u^3 + 125v^3 = u^3 + 5^3v^3$

$= u^3 + (5v)^3$ Use $a^3 + b^3 = (a + b)(a^2 - ab + b^2)$

$= (u + 5v)[u^2 - u(5v) + (5v)^2]$

$= (u + 5v)(u^2 - 5uv + 25v^2)$

**(d)** $16x^3 - 54y^6 = 2(8x^3 - 27y^6)$ Common factor first

$= 2[2^3x^3 - 3^3(y^2)^3]$

$= 2[(2x)^3 - (3y^2)^3]$ Use $a^3 - b^3 = (a - b)(a^2 + ab + b^2)$

$= 2(2x - 3y^2)[(2x)^2 + (2x)(3y^2) + (3y^2)^2]$

$= 2(2x - 3y^2)(4x^2 + 6xy^2 + 9y^4)$

**(e)** $4x^2 - y^2 - 10x + 5y = (4x^2 - y^2) - 5(2x - y)$ Since there are four terms, try the method of grouping

$= (2x + y)(2x - y) - 5(2x - y)$ Use $a^2 - b^2 = (a + b)(a - b)$ in the first term

$= [(2x + y) - 5](2x - y)$

$= (2x + y - 5)(2x - y)$ ■

## 1.3 EXERCISES

In Exercises 1–4 indicate whether each polynomial is a monomial, binomial, or trinomial and give the degree of each.

**1.** $x^3 - 6x$ **2.** $-3a^2b^3$ **3.** $a^3b^3 - 6a^2bc^4 + 8ab$ **4.** $-2$

Add the polynomials in Exercises 5–8.

**5.** $3x^2 + 2x - 5$ and $-2x^2 + 5$

**6.** $-8x^3 + 2x^2 - 1$ and $5x^3 + x - 5$

**7.** $6a^2b^2 - 3ab + 2$ and $4a^2b^2 + 8ab - 5$

**8.** $8a^3b^2 - 3ab + 9$ and $-4a^3b^2 + 4ab - 3$

Subtract the polynomials in Exercises 9–12.

**9.** $-2x^2 + 3x - 4$ from $6x^2 - 4x + 1$

**10.** $8x^3 + 4x - 5$ from $-9x^3 + x^2 + 9$

**11.** $-7a^2b^2 + 6ab - 3$ from $2a^2b^2 - 6ab - 3$

**12.** $4a^4b - 6ab^4 + 2$ from $2a^4b + ab^4 - 7$

Perform the indicated operations in Exercises 13–20.

**13.** $(a^2 + 3a - 4) + (a^3 - a^2 - a + 4)$

**14.** $(x^3 + 4x^4 - 2) + (x^4 - x^3 - x + 1)$

**15.** $(y^5 + 4y^2 - 2) - (2y^2 - y^5 + 6)$

**16.** $(3z^2 + 2z - 8) - (2z^3 - z^2 - z - 2)$

**17.** $(5a^4b^2 - 7a^2b^4 + 3) - (6a^4b^2 + 6a^2b^4 - 7)$

**18.** $(5x^2y^2z^2 - 6xyz) - (-2x^2y^2z^2 + xyz + 1)$

**19.** $(4a^2b^2 - 2ab + 3) + (2a^2b^2 + ab - 7) - (-a^2b^2 - ab - 6)$

**20.** $(-4a^3b^2 - 6a^2b^3 + a^2b^2) - (-8a^2b^3 + 3a^2b^2 - 5) - (7a^3b^2 + a^2b^3 + 8)$

Find each product in Exercises 21–52. Assume all exponents are positive integers.

**21.** $(-6x^2y^3)(-3x^4y)$ **22.** $(5a^2b)(-7a^3b^3)$ **23.** $4ab^3(-3a^2b^2 + 5ab)$

**24.** $-6x^2y(2xy - 8x^3)$ **25.** $(a - b)(a + 2b)$ **26.** $(a - b)(a - 2b)$

**27.** $(5xy + 3)(2xy - 7)$ **28.** $(2ab - c)(3ab + 2c)$ **29.** $(x - 2y)(x + 2y)$

**30.** $(7a + 2b)(7a - 2b)$ **31.** $(x + 2y)^2$ **32.** $(7a + 2b)^2$

**33.** $(x - 2y)^2$ **34.** $(7a - 2b)^2$ **35.** $(3a^2 - b^2)(3a^2 + b^2)$

**36.** $(3a^2 - b^2)(3a^2 - b^2)$ **37.** $(6x - 7y + 2)(2x - 3y)$ **38.** $(-5x + 8y - 4)(4x + 2y)$

**39.** $(3a - 7b)(4a^2 - 2ab + 7b^2)$ **40.** $(4a + 5b)(8a^2 + ab - 3b^2)$ **41.** $(x + 2y - 5)^2$

**42.** $(x - 2y + 5)^2$ **43.** $[(x - 2y) + 5][(x - 2y) - 5]$ **44.** $[x - (2y - 5)][x + (2y - 5)]$

**45.** $x^{2n}y^n(x^n - y^n)$ **46.** $x^{n+2}y^2(x^n + y^n)$ **47.** $(x^n - y^n)(x^n + y^n)$

**48.** $(a^{2n} + b^{2n})(a^{2n} - b^{2n})$ **49.** $(x^{n+1} + y^{n+1})^2$ **50.** $(x^ny^n - z^n)^2$

**51.** $(x^2 - 3xy + 2y^2)(2x^2 + 4xy - 3y^2)$

**52.** $(2a^2 - 5ab + 2b^2)(3a^2 - 6ab - b^2)$

Find each quotient in Exercises 53–62.

**53.** $(5y^5 - 30y^4 + 25y^3 - 5) \div 5y^2$

**54.** $(-22x^4 + 11x^3 - 33x^2) \div (-11x)$

**55.** $(15x^4y^3 - 5x^3y^4 + 10x^5y^5) \div (-5x^2y^3)$

**56.** $(18a^5b^3 - 24a^2b^5 + 6ab) \div (-6a^3b^3)$

**57.** $(a^3 + a - 3) \div (a - 1)$

**58.** $(a^2 - a^3 + 1) \div (a - 2)$

**59.** $(2x^2 + 5xy - 3y^2) \div (2x - y)$

**60.** $(6a^3 + 5a^2b + 4ab^2 + b^3) \div (3a + b)$

**61.** $(x^5 - x^4 - 2x^3 + 4x^2 - 15x + 5) \div (x^2 - 5)$

**62.** $(y^5 - y^4 + y^2 + 3y + 2) \div (y^2 + y + 1)$

Factor each polynomial in Exercises 63–110.

**63.** $6x^2y^2 - 3xy$

**64.** $9a^4b^5c^2 - 18a^4b^3c + 81a^2b^2c^2$

**65.** $5a^2b - 10ab + 3a - 6$

**66.** $3xy^2 - 15xy - 4y + 20$

**67.** $8a^3b^3 - 2a^2b^2 - 12abc^2 + 3c^2$

**68.** $10x^3y^3 + 25x^2y^2z - 12xyz - 30z^2$

**69.** $x^2 + 8x + 7$

**70.** $x^2 - 8x + 7$

**71.** $u^2 + 9u + 20$

**72.** $x^2 - 2x - 35$

**73.** $3x^2 - 5x - 2$

**74.** $4x^2 + 4x - 15$

**75.** $5u^2 + 24uv - 5v^2$

**76.** $3x^2 + 10xy - 8y^2$

**77.** $4x^2 - 9y^2$

**78.** $25u^2 - 16v^2$

**79.** $4x^2 + 12xy + 9y^2$

**80.** $4x^2 - 12xy + 9y^2$

**81.** $27u^3 - v^3$

**82.** $27u^3 + v^3$

**83.** $6x^2 - 6y^2$

**84.** $18u^2 - 8v^2$

**85.** $28x^2 - 58xy - 30y^2$

**86.** $-4u^2 - 34uv - 70v^2$

**87.** $5u^2v^2 - 11uv + 2$

**88.** $6x^2 + 7xy - 10y^2$

**89.** $3u^2 - 60uv + 300v^2$

**90.** $8x^2 - 56xy + 98y^2$

**91.** $32x^3 + 4y^9$

**92.** $250u^6 - 16v^9$

**93.** $x^2 + 2xy + 2$

**94.** $x^2 - xy + y^2$

**95.** $-5u^2 + 70uv - 240v^2$

**96.** $-3x^2 - 12xy + 96y^2$

**97.** $3u^2 - 2uv - 16v^2$

**98.** $15x^2 + 11xy - 56y^2$

**99.** $(u - v)^2 - 16$

**100.** $(u - v)^2 - 4(u - v) + 4$

**101.** $9u^6 - 30u^3v^2 + 25v^4$

**102.** $9u^6 + 30u^3v^3 + 25v^6$

**103.** $15x^2 + 34xy + 15y^2$

**104.** $35x^2 - 74xy + 35y^2$

**105.** $(x + y)^2 - (u + v)^2$

**106.** $(x + 2y)^2 - (x - 2y)^2$

**107.** $x^4 + 2x^2y^2 - 15y^4$

**108.** $u^4 - 10u^2v^2 + 21v^4$

**109.** $8u^6 - 7u^3v^3 - v^6$

**110.** $x^6 + 7x^3y^3 - 8y^6$

In Exercises 111–114 factor assuming $n$ is a positive integer.

**111.** $4x^{2n} - y^{2n}$

**112.** $x^{2n} - 4y^{6n}$

**113.** $27x^{3n} - 8y^{3n}$

**114.** $27x^{3n} + 8y^{3n}$

In Exercises 115–120 factor by grouping using special formulas.

**115.** $x^2 - 4y^2 + 6x - 12y$

**116.** $4x^2 - 9y^2 - 2x - 3y$

**117.** $u^2 + 10uv + 25v^2 - w^2$

**118.** $x^2 - y^2 - 4y - 4$

**119.** $x^3 + y^3 + 2x^2y + 2xy^2$

**120.** $x^3 - y^3 - x^2y + xy^2$

Solve.

**121.** ECONOMICS In a business the manufacturing cost in dollars is given by $M = 4u^2 - u + 5$ and the wholesale cost in dollars is given by $W = 2u^2 + u + 6$, where $u$ is the number of units produced.

**(a)** Determine the polynomial which represents the total cost $C$ of the operation, if $C = M + W$.

**(b)** Determine the total cost when 15 units are produced and sold.

**122.** ECONOMICS The total revenue in dollars from the operation in Exercise 121 is given by $R = 7u^2 - u + 3$.

**(a)** Find the polynomial which represents the profit $P$, if $P = R - C$.

**(b)** Find the profit when 8 units are produced and sold.

**123.** AGRICULTURE A circular garden of radius $R$ has a circular fountain in the center with radius $r$. Find a polynomial which represents the area of the garden which can be planted. **(a)** Factor this polynomial. **(b)** Find the area, correct to two decimal places, when $R = 9.25$ $m$ and $r = 2.05$ $m$. (Use 3.14 for $\pi$.) **(c)** How many plants should be purchased if each plant requires approximately 0.75 $m^2$ to grow properly?

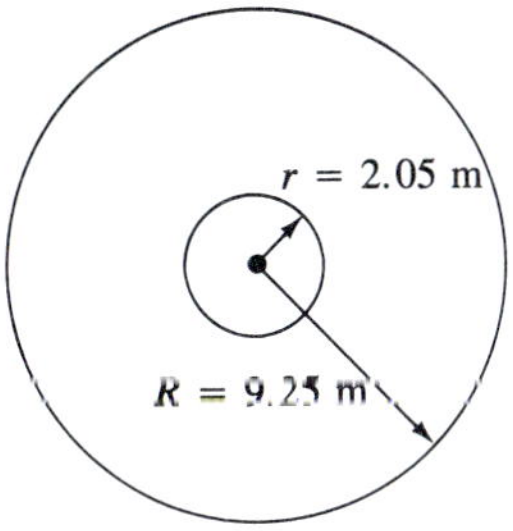


**Exercise 123**

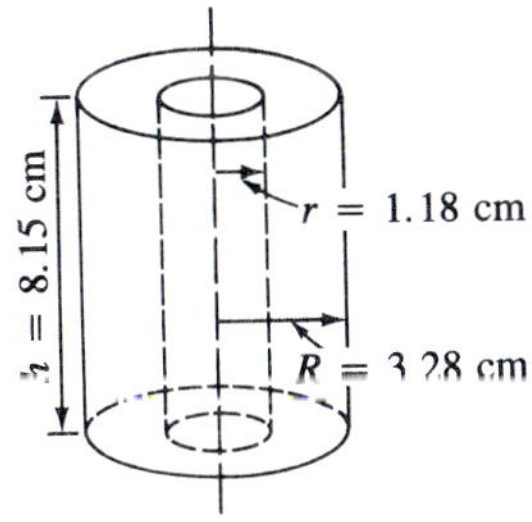


**Exercise 124**

**124.** ENGINEERING A circular cylinder of radius $R$ has its center drilled out to a radius of $r$. If the cylinder is $h$ units high, find a polynomial which represents the volume of the remaining metal. **(a)** Factor this polynomial. **(b)** Find the volume, correct to two decimal places, when $R = 3.28$ cm, $r = 1.18$ cm, and $h = 8.15$ cm. (Use 3.14 for $\pi$.)

## For Review

Simplify and write without negative exponents.

**125.** $\left(\dfrac{3x^3}{y^{-2}}\right)^{-2}$

**126.** $\left(\dfrac{2a^{-4}}{b^6}\right)^{2}$

**127.** $\dfrac{4x^5y^{-2}}{10x^{-2}y^{-4}}$

**128.** $\dfrac{9a^4b^{-6}}{18a^{-2}b^8}$

**129.** $\left(\dfrac{2^0x^4y^{-9}}{3(xy)^{-6}}\right)^{-2}$

**130.** $\dfrac{9(a^2b)^{-3}}{3^0(a^{-4}b^{-2})^{-1}}$

# 1.4 Rational Expressions

A quotient of algebraic expressions is called an **algebraic fraction.** This section considers **rational expressions** which are quotients of two polynomials. For example,

$$\frac{x^2+4}{x-2}, \quad \frac{x+3y}{2x-5y}, \quad \frac{u^2+3v^2}{(x+y)^2}, \quad \frac{3x(x+3)}{(x-5)(x+3)}$$

are all rational expressions.

## Equivalent Expressions

Polynomials are defined for all real numbers, but rational expressions are not defined for values of the variable which make the denominator zero. For example,

$$\frac{3x(x+3)}{(x-5)(x+3)}$$

is not defined for $x = 5$ and $x = -3$. Also, since

$$\frac{3x(x+3)}{(x-5)(x+3)} = \frac{3x}{x-5} \qquad x \neq 5 \text{ and } x \neq -3$$

we say that these two expressions are *equivalent*. In general $a/b$ and $c/d$ are **equivalent** if one can be obtained from the other by multiplying or dividing both numerator and denominator by the same nonzero expression.

The example above is usually written

$$\frac{3x\cancel{(x+3)}}{(x-5)\cancel{(x+3)}} = \frac{3x}{x-5} \qquad \text{Cancel common factors}$$

and we say we **cancel** common factors. If 1 or $-1$ are the only common factors of the numerator and denominator, the rational expression is **reduced to lowest terms.**

**EXAMPLE 1** Reduce each rational expression to lowest terms.

**(a)** $$\frac{8x^4 + 2x}{2x} = \frac{\cancel{2x}(4x^3 + 1)}{\cancel{2x}} \qquad 2x \text{ is a common factor}$$

$$= \frac{4x^3 + 1}{1} = 4x^3 + 1 \qquad \frac{2x}{2x} = 1, \text{ not } 0$$

**(b)** $$\frac{2a^3 + a^2b - ab^2}{a^3 - ab^2} = \frac{a(2a^2 + ab - b^2)}{a(a^2 - b^2)}$$

$$= \frac{\cancel{a(a+b)}(2a-b)}{\cancel{a(a+b)}(a-b)} \qquad \text{Cancel common factors}$$

$$= \frac{2a-b}{a-b} \qquad \text{This is reduced to lowest terms}$$ ■

**CAUTION** Cancel only factors, not terms. For example,

$$\frac{2a-b}{a-b} \text{ is } not \ \frac{2\cancel{a}-\cancel{b}}{\cancel{a}-\cancel{b}}.$$ ■

## Multiplying and Dividing Rational Expressions

If $a/b$ and $c/d$ are rational expressions, with $b \neq 0$ and $d \neq 0$, then

$$\frac{a}{b} \cdot \frac{c}{d} = \frac{ac}{bd}. \qquad \text{Multiplication of rational expressions}$$

Thus, to multiply, we multiply numerators and multiply denominators. However, it is better to reduce the fractions before finding any products.

**EXAMPLE 2** Multiply.

$$\frac{2a^2 + ab - b^2}{6a^2 + 5ab + b^2} \cdot \frac{3a^2 - 2ab - b^2}{2a^2 - ab} = \frac{\cancel{(2a-b)}(a+b)(a-b)\cancel{(3a+b)}}{\cancel{(3a+b)}(2a+b)a\cancel{(2a-b)}} \qquad \text{Indicate product and factor}$$

$$= \frac{(a+b)(a-b)}{a(2a+b)} = \frac{a^2 - b^2}{2a^2 + ab} \qquad \text{Reduce to lowest terms}$$

Either $\dfrac{(a+b)(a-b)}{a(2a+b)}$ or $\dfrac{a^2 - b^2}{2a^2 + ab}$ is an acceptable answer. ■

Division of rational expressions is defined as follows where $b$, $c$, and $d \neq 0$:

$$\frac{a}{b} \div \frac{c}{d} = \frac{a}{b} \cdot \frac{d}{c} \qquad \text{Division of rational expressions}$$

The fraction $d/c$ is called the **reciprocal** of $c/d$. Thus, division is multiplication by the reciprocal of the divisor.

**EXAMPLE 3** Perform the indicated operations.

$$\frac{a^2 - 10ab + 25b^2}{2a^2 - 7ab - 4b^2} \cdot \frac{3a^2 - 17ab + 20b^2}{a^2 - 25b^2} \div \frac{3a^2 - 20ab + 25b^2}{2a^2 - ab - b^2}$$

$$= \frac{a^2 - 10ab + 25b^2}{2a^2 - 7ab - 4b^2} \cdot \frac{3a^2 - 17ab + 20b^2}{a^2 - 25b^2} \cdot \frac{2a^2 - ab - b^2}{3a^2 - 20ab + 25b^2} \qquad \text{Multiply by the reciprocal of the divisor}$$

$$= \frac{\cancel{(a - 5b)}\cancel{(a - 5b)}\cancel{(3a - 5b)}\cancel{(a - 4b)}\cancel{(2a + b)}(a - b)}{\cancel{(a - 4b)}\cancel{(2a + b)}\cancel{(a - 5b)}(a + 5b)\cancel{(3a - 5b)}\cancel{(a - 5b)}} = \frac{a - b}{a + 5b} \qquad \blacksquare$$

## Adding and Subtracting Rational Expressions

Adding or subtracting rational expressions with a common denominator is defined as follows where $c \neq 0$:

$$\frac{a}{c} + \frac{b}{c} = \frac{a + b}{c} \quad \text{and} \quad \frac{a}{c} - \frac{b}{c} = \frac{a - b}{c} \qquad \text{Addition and subtraction}$$

If the denominators are not the same, we must convert the fractions into equivalent fractions that have a common denominator. With rational expressions this is best done by finding the **least common denominator (LCD).**

### To Find the LCD of Two or More Fractions

1. Factor all denominators and reduce each fraction, if possible.
2. Each factor of a denominator appears in the LCD as many times as it appears in the denominator where it is found the greatest number of times.

**EXAMPLE 4** Find the least common denominator (LCD).

**(a)** $\dfrac{ab}{a^2 - 5ab + 6b^2}$ and $\dfrac{a + 2b}{a^3 - 4ab^2}$

$$\frac{ab}{a^2 - 5ab + 6b^2} = \frac{ab}{(a - 2b)(a - 3b)} \qquad \text{Denominator: one } (a - 2b) \text{, one } (a - 3b)$$

$$\frac{a + 2b}{a^3 - 4ab^2} = \frac{\cancel{a + 2b}}{a\cancel{(a + 2b)}(a - 2b)} = \frac{1}{a(a - 2b)} \qquad \text{Denominator: one } a \text{, one } (a - 2b)$$

Thus the factors of the LCD are one $(a - 2b)$, one $(a - 3b)$, and one $a$. The LCD is $a(a - 2b)(a - 3b)$.

**(b)** $\dfrac{3x+y}{x^2-6xy+9y^2}$, $\dfrac{xy}{x^2-9y^2}$, and $\dfrac{x-y}{4x^5-4x^4y+x^3y^2}$

$$\frac{3x+y}{x^2-6xy+9y^2}=\frac{3x+y}{(x-3y)^2}$$ Denominator: two $(x-3y)$

$$\frac{xy}{x^2-9y^2}=\frac{xy}{(x+3y)(x-3y)}$$ Denominator: one $(x+3y)$ one $(x-3y)$

$$\frac{x-y}{4x^5-4x^4y+x^3y^2}=\frac{x-y}{x^3(2x-y)^2}$$ Denominator: three $x$ two $(2x-y)$

Thus, the factors of the LCD include two $(x-3y)$ factors, one $(x+3y)$ factor, three $x$ factors, and two $(2x-y)$ factors. The LCD is

$$x^3(x+3y)(x-3y)^2(2x-y)^2.$$ ■

To add or subtract fractions, each fraction must be changed to an equivalent fraction with the LCD as its denominator. This is done by multiplying numerator and denominator by all factors of the LCD which are missing in the denominator of the particular fraction. Once all fractions have been written as equivalent ones with the LCD as the denominator, then add or subtract the numerators and place the result over the LCD. If possible, reduce the resulting fraction to lowest terms.

**EXAMPLE 5** Perform the indicated operations.

**(a)** $\dfrac{3x+y}{x^2+4xy-21y^2}+\dfrac{x+y}{9y^2-x^2}$

$$=\frac{3x+y}{(x-3y)(x+7y)}+\frac{x+y}{(3y-x)(3y+x)}$$

$$=\frac{3x+y}{(x-3y)(x+7y)}+\frac{x+y}{-(x-3y)(3y+x)}$$ $3y-x=-(x-3y)$

$$=\frac{3x+y}{(x-3y)(x+7y)}-\frac{x+y}{(x-3y)(x+3y)}$$

$$=\frac{(3x+y)\mathbf{(x+3y)}}{(x-3y)(x+7y)\mathbf{(x+3y)}}-\frac{(x+y)\mathbf{(x+7y)}}{(x-3y)(x+3y)\mathbf{(x+7y)}}$$

$$=\frac{(3x+y)(x+3y)-(x+y)(x+7y)}{(x-3y)(x+3y)(x+7y)}$$

$$=\frac{3x^2+10xy+3y^2-(x^2+8xy+7y^2)}{(x-3y)(x+3y)(x+7y)}$$ Use parentheses to avoid sign errors

$$=\frac{3x^2+10xy+3y^2-x^2-8xy-7y^2}{(x-3y)(x+3y)(x+7y)}=\frac{2x^2+2xy-4y^2}{(x-3y)(x+3y)(x+7y)}$$

$$=\frac{2(x+2y)(x-y)}{(x-3y)(x+3y)(x+7y)}$$ No common factors

**(b)** $\dfrac{2xy}{x^2 - y^2} - \dfrac{y}{x + y} + 2$

$$= \frac{2xy}{(x + y)(x - y)} - \frac{y}{x + y} + \frac{2}{1} \qquad 2 = \frac{2}{1}$$

$$= \frac{2xy}{(x + y)(x - y)} - \frac{y(x - y)}{(x + y)(x - y)} + \frac{2(x + y)(x - y)}{(x + y)(x - y)} \qquad \text{LCD} = (x + y)(x - y)$$

$$= \frac{2xy - y(x - y) + 2(x + y)(x - y)}{(x + y)(x - y)}$$

$$= \frac{2xy - yx + y^2 + 2(x^2 - y^2)}{(x + y)(x - y)} = \frac{2xy - xy + y^2 + 2x^2 - 2y^2}{(x + y)(x - y)}$$

$$= \frac{xy - y^2 + 2x^2}{(x + y)(x - y)} = \frac{2x^2 + xy - y^2}{(x + y)(x - y)}$$

$$= \frac{(2x - y)\cancel{(x + y)}}{\cancel{(x + y)}(x - y)} \qquad \text{This can be simplified by canceling common factors in the numerator and denominator}$$

$$= \frac{2x - y}{x - y} \qquad \blacksquare$$

## Complex Fractions

An algebraic fraction which contains at least one fraction in its numerator or denominator is called a **complex fraction.** The following are complex fractions:

$$\frac{\dfrac{3}{13}}{\dfrac{7}{5}}, \qquad \frac{\dfrac{x + y}{y}}{\dfrac{x - y}{x}}, \qquad \frac{\dfrac{1}{a} + \dfrac{1}{b}}{a - b}, \qquad \frac{\dfrac{1}{(x + h)^4} - \dfrac{1}{x^4}}{h}.$$

There are two methods for simplifying complex fractions. In the first method we change the numerator to a single fraction, change the denominator to a single fraction, and then divide. In the second method we multiply the numerator and denominator by the LCD of all the fractions.

**EXAMPLE 6** Simplify each complex fraction.

**(a)** $\dfrac{\dfrac{b}{a + b} - 1}{\dfrac{b}{a - b} + 1} = \dfrac{\dfrac{b}{a + b} - \dfrac{a + b}{a + b}}{\dfrac{b}{a - b} + \dfrac{a - b}{a - b}}$

$$= \frac{\dfrac{b - (a + b)}{a + b}}{\dfrac{b + a - b}{a - b}} = \frac{\dfrac{b - a - b}{a + b}}{\dfrac{b + a - b}{a - b}} = \frac{\dfrac{-a}{a + b}}{\dfrac{a}{a - b}}$$

$$= \frac{-\cancel{a}}{a + b} \cdot \frac{a - b}{\cancel{a}} = \frac{-(a - b)}{a + b} = \frac{b - a}{a + b}$$

**(b)** $$\frac{y-x}{\frac{1}{y^3}-\frac{1}{x^3}} = \frac{(y-x)x^3y^3}{\left(\frac{1}{y^3}-\frac{1}{x^3}\right)x^3y^3}$$ Multiply numerator and denominator by $x^3y^3$

$$= \frac{(y-x)x^3y^3}{\frac{x^3y^3}{y^3}-\frac{x^3y^3}{x^3}}$$ Cancel common factors

$$= \frac{(y-x)x^3y^3}{x^3-y^3} = \frac{(y-x)x^3y^3}{(x-y)(x^2+xy+y^2)}$$

$$= \frac{-(x-y)x^3y^3}{(x-y)(x^2+xy+y^2)} = \frac{-x^3y^3}{x^2+xy+y^2}$$ ■

## 1.4 EXERCISES

In Exercises 1–3 determine the values of the variables for which the rational expression is not defined.

1. $\dfrac{(y-3)(x+2)}{(y^2-5y+6)(x+7)}$
2. $\dfrac{a^2+b^2}{(a^2-10a+25)(b^2-1)}$
3. $\dfrac{6a+b}{a^2+ab}$

In Exercises 4–6 determine whether the given fractions are equivalent where they are defined.

4. $\dfrac{a-b}{a^2-b^2}, \dfrac{1}{a+b}$
5. $\dfrac{x^2(x+2)}{x(x-1)}, \dfrac{x(x+2)}{(x-1)}$
6. $\dfrac{a^2-b^2+c^2}{b-c}, \dfrac{-a^2+b^2-c^2}{c-b}$

Reduce each fraction to lowest terms in Exercises 7–9.

7. $\dfrac{77x^2y^4}{33x^3y^2}$
8. $\dfrac{a-b}{b^2-a^2}$
9. $\dfrac{15x+7x^2-2x^3}{x^2-8x+15}$

In Exercises 10–24 perform the indicated operations and simplify.

10. $\dfrac{200a^2b^3}{36ab}\cdot\dfrac{14a^3b^2}{21}$
11. $\dfrac{3(x+y)}{6x^2y^2}\cdot x^3y^3$
12. $\dfrac{4a+8}{9a+9}\cdot\dfrac{a^2-1}{2a+4}$
13. $\dfrac{3(x^2+xy)}{16(x^2-y^2)}\cdot\dfrac{20y^2}{9(xy+y^2)}$
14. $\dfrac{a^2-b^2}{3a+3b}\div\dfrac{(a-b)^2}{9}$
15. $\dfrac{x^2-x-12}{x^2-9}\div\dfrac{x^2-16}{5x-5}$
16. $\dfrac{a^3-b^3}{a^3+b^3}\div\dfrac{a^2+ab+b^2}{a^2-ab+b^2}$
17. $\dfrac{6x^2-x-35}{2x^2-3x-5}\div\dfrac{3x^2+x-14}{x^2-x-2}$
18. $\dfrac{a^4-b^4}{a^2+ab-2b^2}\div(a^2+b^2)$
19. $\dfrac{y^2-1}{2y+2}\cdot\dfrac{y^2-8y+12}{y^2-4y+4}\cdot\dfrac{y+2}{y-6}$
20. $\dfrac{a^2-4}{a^2-4a+4}\cdot\dfrac{a^2-9a+14}{a^3+2a^2}$
21. $\dfrac{uv-uw+xv-xw}{v-w}\div\dfrac{v^2-2vw+w^2}{xv-xw}$
22. $\dfrac{a^3+64}{2a^2+18a+40}\div\dfrac{a^2-4a+16}{a^2+4a}$
23. $\dfrac{x^2-y^2}{x^2-xy+y^2}\cdot\dfrac{2x^2-3xy-2y^2}{x^2+2xy+y^2}\div\dfrac{2x^2-xy-y^2}{x^3+y^3}$

24. $\dfrac{3u^2 - 5uv - 2v^2}{u^2 + uv - 2v^2} \cdot \dfrac{v^2 + 4vu - 5u^2}{12u^2 + 7uv + v^2} \div \dfrac{5u^2 - 9uv - 2v^2}{8u^2 - 2uv - v^2}$

Find the LCD of the fractions given in Exercises 25–28.

25. $\dfrac{6}{x^2y}$ and $\dfrac{7}{xy^2}$

26. $\dfrac{6x}{x^2 - y^2}$ and $\dfrac{5y}{x^2 + 2xy + y^2}$

27. $\dfrac{x + 7}{x^5 - 27x^2}$ and $\dfrac{x - 3}{5x^3 + 15x^2 + 45x}$

28. $\dfrac{3x + y}{6(x - y)^3}$ and $\dfrac{7}{(y - x)^4}$

Perform the indicated operations in Exercises 29–55.

29. $\dfrac{6y - 2}{y} + \dfrac{3y - 1}{-y}$

30. $\dfrac{3a}{a - 4} + \dfrac{5a}{4 - a}$

31. $\dfrac{a + 2}{a^2 - 4} - \dfrac{3}{a - 2}$

32. $\dfrac{x}{1 - x^2} - \dfrac{1 - 2x}{(x - 1)(x + 1)}$

33. $\dfrac{6}{5x - 10} + \dfrac{3}{x + 2}$

34. $\dfrac{5x}{x^3 - 16x} - \dfrac{4}{x - 4}$

35. $\dfrac{y - 2}{y^2 + y - 6} + \dfrac{2}{y^2 + 4y + 3}$

36. $\dfrac{6a}{a^3 - 27} - \dfrac{4}{2(a^2 + 3a + 9)}$

37. $\dfrac{3x^2 + 2}{x^2 - 7x + 6} - 3$

38. $\dfrac{2ab}{a^2 - b^2} - \dfrac{b}{a - b} + 5$

39. $\dfrac{3}{2x - 4} + \dfrac{2}{3x + 6} - \dfrac{5}{4x + 4}$

40. $\dfrac{2y - 1}{y^2 - 4y + 4} + \dfrac{2y + 3}{4 - y^2}$

41. $\dfrac{1 - \dfrac{x^3}{y^3}}{1 - \dfrac{x}{y}}$

42. $\dfrac{\dfrac{1}{a} + \dfrac{2}{a^2}}{\dfrac{1}{4} - \dfrac{1}{a^2}}$

43. $\dfrac{1 - \dfrac{3}{x} + \dfrac{2}{x^2}}{1 - \dfrac{1}{x}}$

44. $\dfrac{\dfrac{a}{a - b} - 1}{\dfrac{a}{a + b} - 1}$

45. $\dfrac{x + 6 + \dfrac{8}{x}}{x + 4 + \dfrac{4}{x}}$

46. $\dfrac{1}{1 + \dfrac{1}{1 + \dfrac{1}{x}}}$

47. $\dfrac{\dfrac{1}{a - 1} + 1}{\dfrac{1}{a} - \dfrac{1}{a - 1}}$

48. $\dfrac{\dfrac{1}{x + h} - \dfrac{1}{x}}{h}$

49. $\dfrac{\dfrac{1}{(x + h)^2} - \dfrac{1}{x^2}}{h}$

50. $\dfrac{(x^2 + 5)3x^2 - x^3(2x)}{(x^2 + 5)^2}$

51. $\dfrac{(4x + 3)(6x + 2) - (3x^2 + 2x)(4)}{(4x + 3)^2}$

52. $\dfrac{x^{-1} + y^{-1}}{x^{-1}y^{-1}}$

53. $(a^{-1} + b^{-1})^{-1}$

54. $\dfrac{x^{-1} - y^{-1}}{x^{-1} + y^{-1}}$

55. $\dfrac{u^{-2} - v^{-2}}{u^{-1} - v^{-1}}$

In Exercises 56–57 remember the order of operations. Do multiplications and divisions first from left to right. Then do additions and subtractions from left to right.

56. $\dfrac{x}{x^2 - y^2} - \dfrac{y}{x^2 - y^2} \cdot \dfrac{x + y}{x - y} + \dfrac{y(x + y)}{x^2 - 3xy - 4y^2} \div \dfrac{(x + y)^2}{x - 4y}$

57. $\dfrac{ab}{(a - b)^2} + \dfrac{a^2b}{a^2 - 4ab - 21b^2} \cdot \dfrac{(a - 7b)}{a} - \dfrac{b(a + b)}{(a - b)} \div \dfrac{a^2 - b^2}{a}$

**58. ENGINEERING** In an electronic circuit the frequency is given by an expression of the form

$$\frac{\frac{1}{p}\left(1+\frac{p}{q}\right)}{1+(1+p)\frac{q}{p}}.$$

Simplify the complex fraction.

**59. ENGINEERING** In an electronic circuit the current is given by an expression of the form

$$\frac{uv}{u+v-\frac{u}{1+uv}}.$$

Simplify this complex fraction.

## For Review

Factor the polynomials in Exercises 60–62.

**60.** $10x^2 - 1000y^2$

**61.** $6u^2 + uv - 35v^2$

**62.** $27u^3 - 64v^3$

Perform the indicated operations in Exercises 63–65.

**63.** $(7a^3b^2 - 8a^2b^3 + 2) - (3a^3b^2 - 5a^2b^3 - 7)$

**64.** $(3x + 5y)(9x^2 - 15xy + 25y^2)$

**65.** $(8x^4 - 26x^3 + 19x^2 - 20x + 25) \div (2x - 5)$

Answer true or false in Exercises 66–69. If the answer is false, explain why.

**66.** $-18$ is an integer.

**67.** $\sqrt{8}$ is an irrational number.

**68.** Some integers are whole numbers.

**69.** The set of whole numbers is closed with respect to subtraction.

In Exercises 70–72 find **(a)** the square, and **(b)** the cube of each expression.

**70.** $y$

**71.** $2x$

**72.** $3z^2$

# 1.5 Radicals and Rational Exponents

## Roots and Radicals

If $b^2 = a$, then $b$ is a **square root** of $a$, and if $b^3 = a$, then $b$ is a **cube root** of $a$. This concept is generalized in the following definition.

### Principal *k*th Root

Let $k$ be a positive integer greater than 1. Then

$$b = \sqrt[k]{a} \text{ if } b^k = a.$$

If $k$ is even, then $a \geq 0$ and also $b \geq 0$.

The expression $\sqrt[k]{a}$ is called a **radical** ($\sqrt{\phantom{a}}$ is the **radical sign**), $a$ is called the **radicand,** and $k$ the **index.** When $k$ is even, for example $k = 2$, we have $(3)^2 = 9$ and $(-3)^2 = 9$ making both

$$3 = \sqrt{9} \quad \text{and} \quad -3 = -\sqrt{9}$$

square roots of 9. However, only 3 is the principal square root. Thus, $b = \sqrt[k]{a}$ is called the **principal $k$th root of $a$,** and $b$ is nonnegative when $k$ is even. If $k$ is odd, $\sqrt[k]{a}$ is nonnegative when $a \geq 0$ and negative when $a < 0$. We can summarize this as follows:

$$\sqrt[k]{a^k} = |a| \quad \text{if } k \text{ is even;}$$

$$\sqrt[k]{a^k} = a \quad \text{if } k \text{ is odd.}$$

## EXAMPLE 1

**(a)** $\sqrt{(-7)^2} = |-7| = 7$ $\quad$ $k$ is even

**(b)** $\sqrt[3]{(-7)^3} = -7$ $\quad$ $k$ is odd

**(c)** $\sqrt[4]{-16}$ is not a real number

**(d)** $-\sqrt[6]{64} = -\sqrt[6]{2^6} = -|2| = -2$

**(e)** $\sqrt{25x^2} = \sqrt{(5x)^2} = |5x| = 5|x|$ ■

Notice that in Example 1(**e**) we must leave the answer as $5|x|$ since we do not know if $x$ is positive or negative. For the remainder of this section, we will assume that *variables under even index radicals represent positive real numbers,* eliminating the need for absolute values in our answers.

## Simplifying Radicals

Radicals can often be simplified using the following properties.

For $a$ and $b$ real numbers ($a > 0$ and $b > 0$ when the index is even),

1. $\sqrt[k]{a}\sqrt[k]{b} = \sqrt[k]{ab}$, **Product property**
2. $\dfrac{\sqrt[k]{a}}{\sqrt[k]{b}} = \sqrt[k]{\dfrac{a}{b}}$, **Quotient property**
3. $\sqrt[m]{\sqrt[n]{a}} = \sqrt[mn]{a}$. **Index property**

When using these properties, we look for perfect $k$th powers which can be taken from under the radical. Example 2 illustrates this technique.

## EXAMPLE 2

**(a)** 
$$\begin{aligned}\sqrt{32x^5} &= \sqrt{2 \cdot 2^4 \cdot x \cdot x^4} \quad 2^4 \text{ and } x^4 \text{ are perfect squares}\\ &= \sqrt{2^4x^4 \cdot 2x}\\ &= \sqrt{2^4x^4}\sqrt{2x}\\ &= 2^2x^2\sqrt{2x} = 4x^2\sqrt{2x}\end{aligned}$$

**(b)** $\dfrac{\sqrt{27u^3}}{\sqrt{3u}} = \sqrt{\dfrac{27u^3}{3u}}$  $\dfrac{\sqrt{a}}{\sqrt{b}} = \sqrt{\dfrac{a}{b}}$

$= \sqrt{9u^2}$

$= 3u$  Since we assume $u > 0$, $|u| = u$

**(c)** $\sqrt[5]{9a^2b^3}\sqrt[5]{81a^3b^4} = \sqrt[5]{3^6a^5b^7}$  $\sqrt[5]{a}\sqrt[5]{b} = \sqrt[5]{ab}$

$= \sqrt[5]{3^5a^5b^5 \cdot 3b^2}$  Look for perfect fifth powers

$= \sqrt[5]{3^5a^5b^5}\sqrt[5]{3b^2}$

$= 3ab\sqrt[5]{3b^2}$

Notice in the first step that we do not multiply the numbers. Rather, we factor them, looking for perfect fifth powers.

**(d)** $\sqrt[3]{\sqrt{64x^7y^{13}}} = \sqrt[6]{64x^7y^{13}}$  $\sqrt[3]{\sqrt{a}} = \sqrt[6]{a}$

$= \sqrt[6]{2^6x^6y^{12} \cdot xy}$

$= \sqrt[6]{2^6x^6(y^2)^6}\sqrt[6]{xy}$

$= 2xy^2\sqrt[6]{xy}$ ■

## Rational Number Exponents

Some radical expressions can be simplified more easily by using rational number exponents. If we assume that the rules of integer exponents hold for rational exponents, then

$$(a^{1/k})^k = a^{(1/k)k} = a^1 = a.$$

This would indicate that $a^{1/k} = \sqrt[k]{a}$. In general, we have the following definition.

> If $a$ is such that $\sqrt[n]{a}$ is a real number, then
>
> $$a^{m/n} = (\sqrt[n]{a})^m = \sqrt[n]{a^m}.$$

## EXAMPLE 3

Use rational exponents to simplify.

**(a)** $(\sqrt[3]{-8x^9y^6})^2 = (-8x^9y^6)^{2/3}$  $(\sqrt[3]{a})^2 = a^{2/3}$

$= [(-2)^3]^{2/3}(x^9)^{2/3}(y^6)^{2/3}$  $-8 = (-2)^3$

$= (-2)^2x^6y^4 = 4x^6y^4$

**(b)** $\sqrt[4]{\dfrac{16x^9}{81y^8}} = \left(\dfrac{2^4x^9}{3^4y^8}\right)^{1/4}$  $\sqrt[4]{a} = a^{1/4}$

$= \dfrac{(2^4)^{1/4}(x^9)^{1/4}}{(3^4)^{1/4}(y^8)^{1/4}}$

$= \dfrac{2x^{9/4}}{3y^2} = \dfrac{2x^2x^{1/4}}{3y^2}$  $x^{9/4} = x^{8/4+1/4} = x^{8/4}x^{1/4} = x^2x^{1/4}$

$= \dfrac{2x^2\sqrt[4]{x}}{3y^2}$ ■

CAUTION There are no rules for addition and subtraction of radicals similar to the ones for multiplication and division. In general,

$$\sqrt{a+b} \neq \sqrt{a} + \sqrt{b} \text{ and } \sqrt{a-b} \neq \sqrt{a} - \sqrt{b}.$$

We can, however, simplify some expressions by combining like radicals.

## EXAMPLE 4

Add or subtract.

**(a)**
$$\begin{aligned} 6\sqrt{125} + 3\sqrt{20} &= 6\sqrt{25 \cdot 5} + 3\sqrt{4 \cdot 5} \\ &= 6\sqrt{25}\sqrt{5} + 3\sqrt{4}\sqrt{5} \qquad \sqrt{ab} = \sqrt{a}\sqrt{b} \\ &= 6 \cdot 5\sqrt{5} + 3 \cdot 2\sqrt{5} \\ &= 30\sqrt{5} + 6\sqrt{5} = 36\sqrt{5} \end{aligned}$$

**(b)**
$$\begin{aligned} 4\sqrt[3]{8x^4y^5} - 3\sqrt[3]{27x^4y^5} &= 4\sqrt[3]{2^3x^3y^3 \cdot xy^2} - 3\sqrt[3]{3^3x^3y^3 \cdot xy^2} \\ &= 4(2xy)\sqrt[3]{xy^2} - 3(3xy)\sqrt[3]{xy^2} \\ &= 8xy\sqrt[3]{xy^2} - 9xy\sqrt[3]{xy^2} \\ &= -xy\sqrt[3]{xy^2} \end{aligned}$$

## Rationalizing Denominators

For uniformity of answers and simplicity in calculations, it is often desirable to *rationalize denominators*. By **rationalizing the denominator** we mean transforming a fraction into an equivalent expression with no radicals in the denominator. If the fraction has one term in the denominator, the process requires multiplying by a radical which makes the radicand in the denominator a perfect $k$th power. For example, since we want $9x = 3^2x$ to be a perfect cube, we multiply by $3x^2$ to obtain $3^3x^3$ in the following.

$$\sqrt[3]{\frac{5y}{9x}} = \sqrt[3]{\frac{5y}{9x} \cdot \frac{3x^2}{3x^2}} = \sqrt[3]{\frac{15x^2y}{27x^3}} = \frac{\sqrt[3]{15x^2y}}{3x}$$

If the denominator has a binomial factor of the form $c\sqrt{x} + d\sqrt{y}$, then we multiply numerator and denominator by $c\sqrt{x} - d\sqrt{y}$.

## EXAMPLE 5

Rationalize the denominator.

**(a)**
$$\frac{\sqrt{5}}{2\sqrt{7}+\sqrt{5}} = \frac{\sqrt{5}(2\sqrt{7}-\sqrt{5})}{(2\sqrt{7}+\sqrt{5})(2\sqrt{7}-\sqrt{5})}$$

Since the denominator is $2\sqrt{7} + \sqrt{5}$, multiply numerator and denominator by $2\sqrt{7} - \sqrt{5}$

$$= \frac{\sqrt{5}(2\sqrt{7}-\sqrt{5})}{(2\sqrt{7})^2 - (\sqrt{5})^2} \qquad (a+b)(a-b) = a^2 - b^2$$

$$= \frac{2\sqrt{35}-5}{4(7)-5}$$

$$= \frac{2\sqrt{35}-5}{23}$$

**(b)** $$\frac{(x-y)}{\sqrt{x}-\sqrt{y}} = \frac{(x-y)(\sqrt{x}+\sqrt{y})}{(\sqrt{x}-\sqrt{y})(\sqrt{x}+\sqrt{y})}$$ Since the denominator is $\sqrt{x}-\sqrt{y}$, multiply numerator and denominator by $\sqrt{x}+\sqrt{y}$

$$= \frac{(x-y)(\sqrt{x}+\sqrt{y})}{(\sqrt{x})^2-(\sqrt{y})^2}$$

$$= \frac{(x-y)(\sqrt{x}+\sqrt{y})}{x-y} = \sqrt{x}+\sqrt{y} \quad \blacksquare$$

If an approximation is needed for a radical, a calculator can be helpful. For square roots, use the [ $\sqrt{\ }$ ] key. For other roots, use the [ $y^x$ ] key. For example, to find $\sqrt[3]{15}$, consider the following.

ALG: 15 [ $y^x$ ] 3 [ $1/x$ ] [ = ] [ 2.4662121 ]

RPN: 15 [ ENTER ] 1 [ ENTER ] 3 [ ÷ ] [ $y^x$ ] [ 2.4662121 ]

## 1.5 EXERCISES

Assume all variables and algebraic expressions under even index radicals are positive. Simplify each expression in Exercises 1–24.

1. $\sqrt{81}$
2. $\sqrt[3]{-64}$
3. $-\sqrt[4]{625}$
4. $-\sqrt[3]{-27}$
5. $\sqrt[4]{16a^{12}}$
6. $\sqrt[3]{27x^{12}}$
7. $\sqrt[5]{-243a^5b^{10}}$
8. $\sqrt[4]{625x^{16}y^8}$
9. $\sqrt{x^2+10x+25}$
10. $\sqrt{4x^2+8x+4}$
11. $4\sqrt{5}\sqrt{125}$
12. $3\sqrt{7}\sqrt{49}\sqrt{7}$
13. $7\sqrt[3]{128}$
14. $8\sqrt{108}$
15. $\sqrt[3]{2xy^2}\sqrt[3]{4x^4y^2}$
16. $\sqrt[4]{8u^3v^7}\sqrt[4]{8u^2v^8}$
17. $\dfrac{\sqrt{125x^3y}}{\sqrt{5xy^3}}$
18. $\dfrac{\sqrt[3]{81a^4b^5}}{\sqrt[3]{3ab}}$
19. $\dfrac{\sqrt[4]{25x^3y^2}}{\sqrt[4]{5^{-2}x^{-1}y^{-2}}}$
20. $\dfrac{\sqrt[3]{2^{-1}x^{-3}y^{-2}}}{\sqrt[3]{16^{-1}x^{-4}y^{-8}}}$
21. $\sqrt{\sqrt[3]{x^{12}y^{18}}}$
22. $\sqrt[3]{\sqrt[4]{a^{24}b^{48}}}$
23. $\sqrt{\sqrt{16x^6y^9}}$
24. $\sqrt[3]{\sqrt[3]{11u^{25}v^{19}}}$

In Exercises 25–36 write each expression using rational exponents, then simplify. Give answers in radical notation.

25. $\sqrt{125x^2}$
26. $\sqrt[3]{125x^4}$
27. $\sqrt[6]{64a^{18}b^{24}}$
28. $\sqrt[3]{-27x^6y^9}$
29. $\sqrt{75x^5y^7}$
30. $\sqrt[3]{24a^{10}b^{11}}$
31. $\sqrt{\dfrac{25a^3b^4}{4a^{-3}b^2}}$
32. $\sqrt[3]{\dfrac{8x^{-3}y^7}{27x^3y^{-2}}}$
33. $\sqrt[4]{32(x+y)^5}$
34. $\sqrt{5a^2+20a+20}$
35. $\sqrt[4]{(16x^{12}y^{-8})^{-3}}$
36. $\sqrt{(a^3b^2)(a^2b^3)^3}$

Simplify each expression in Exercises 37–57. By simplifying we mean that the radicand contains no factor raised to a power greater than or equal to the index, and there are no fractions in a radicand nor radicals in the denominator.

37. $6\sqrt{147}-3\sqrt{75}$
38. $4\sqrt[3]{81}+5\sqrt[3]{24}$
39. $3\sqrt[4]{48}+5\sqrt[4]{243}$

**40.** $5\sqrt{\dfrac{16}{9}} - 3\dfrac{\sqrt{50}}{\sqrt{32}}$

**41.** $6\sqrt{8x^3y} + 2x\sqrt{8xy}$

**42.** $2\sqrt[3]{27x^4y^6} - 4y\sqrt[3]{125x^4y^3}$

**43.** $\dfrac{8 + \sqrt{48}}{4}$

**44.** $\dfrac{9 - \sqrt{63}}{6}$

**45.** $\sqrt{\dfrac{250}{27}}$

**46.** $\dfrac{\sqrt[3]{6}}{\sqrt[3]{7}}$

**47.** $\dfrac{\sqrt{8x}}{\sqrt{6y}}$

**48.** $\sqrt{\dfrac{9x^3}{2xy}}$

**49.** $\dfrac{\sqrt[3]{3xy}}{\sqrt[3]{2y^2}}$

**50.** $\sqrt{\dfrac{3a^3b^2}{150ab^3}}$

**51.** $\dfrac{\sqrt[4]{32a^5}}{\sqrt[4]{2b^3}}$

**52.** $\sqrt[3]{\dfrac{72x^2}{3xy}}$

**53.** $\dfrac{2}{\sqrt{3} - \sqrt{5}}$

**54.** $\dfrac{3\sqrt{5} - \sqrt{3}}{\sqrt{5} + \sqrt{3}}$

**55.** $\dfrac{\sqrt{x} + 1}{\sqrt{x} - 1}$

**56.** $\dfrac{\sqrt{a} + 2\sqrt{b}}{\sqrt{a} - 2\sqrt{b}}$

**57.** $3\sqrt{20a} - \dfrac{2\sqrt{a}}{\sqrt{5}}$

In Exercises 58–60 rationalize each numerator.

**58.** $\dfrac{1 + \sqrt{3}}{1 - \sqrt{3}}$

**59.** $\dfrac{\sqrt{x} - \sqrt{y}}{\sqrt{xy}}$

**60.** $\dfrac{\sqrt{x + h} - \sqrt{x}}{h}$

ECONOMICS A manufacturing index $I$, related to the number of units in stock $u$, in thousands of units, is calculated by using the expression $I = 0.712(1 + u)^{1/3}$. Use this expression in Exercises 61–62.

**61.** Use your calculator to find $I$, correct to three decimal places, when $u = 0.521$.

**62.** Find the index to three decimal places when $u = -0.102$, that is, when orders exceed the stock available by 102 units.

Use the expression $L = 3.2s^{1/2}t^{2/3}$ in Exercises 63–64.

**63.** What factor is $L$ multiplied by when both $s$ and $t$ are doubled? Give answer to the nearest tenth.

**64.** What factor is $L$ multiplied by when $s$ is multiplied by 0.8 and $t$ by 1.4? Give answer to the nearest tenth.

## For Review

Perform the indicated operations in Exercises 65–70.

**65.** $\dfrac{x^2 - 3xy - 40y^2}{x^2 - 25y^2} \cdot \dfrac{x - 5y}{x^2 - 16xy + 64y^2}$

**66.** $\dfrac{a}{a^3 - 9a} - \dfrac{3}{a^2 - 7a + 12}$

**67.** $\dfrac{\dfrac{1}{x + y} - \dfrac{1}{x}}{\dfrac{1}{x + y} - \dfrac{1}{y}}$

**68.** $\dfrac{u^{-1} + v^{-1}}{1 - (uv)^{-1}}$

**69.** $(3u - 2v + 1)^2$

**70.** $(2x^2y^2 - 5xy^2 + 8xy) + (-3x^2y^2 + 2xy^2 - 4xy)$

# CHAPTER 1 REVIEW EXERCISES

Answer true or false in Exercises 1–4. If the answer is false, explain why.

**1.** $\sqrt{5}$ is a rational number.

**2.** Some integers are natural numbers.

**3.** If $a$ is a rational number, then $\sqrt{a}$ is rational.

**4.** For all positive real numbers $x$ and $y$, $\sqrt{x+y}$ is positive.

State the property illustrated in Exercises 5–6.

**5.** $3x + (y + 2) = (3x + y) + 2$

**6.** If $a < 5$ and $5 < b$, then $a < b$.

Simplify each expression in Exercise 7–9.

**7.** $-[a - (b + a)]$

**8.** $|-6| - |-13|$

**9.** $|-x|$ if $x < 0$

Find the distance between the given numbers in Exercises 10–11.

**10.** $-\frac{1}{5}$ and $\frac{3}{10}$

**11.** $\sqrt{7}$ and $-3\sqrt{7}$

In Exercises 12–15 simplify each expression and write it without negative exponents.

**12.** $5x^3 \cdot x^{-5}$

**13.** $3y^{-1}$

**14.** $\left(\frac{8^0x^{-1}}{y^3}\right)^{-2}$

**15.** $\left(\frac{5x^2y}{x^{-1}y^2}\right)^{-3}$

In Exercises 16–17 use your calculator to perform the indicated operations. Round to three significant digits and give answers in scientific notation.

**16.** $\frac{(0.000\ 571)^2}{992{,}000}$

**17.** $\frac{(0.000\ 021\ 6)(8{,}360{,}000)}{(0.000\ 000\ 011\ 5)}$

**18.** Add $13x^2y^2 - 7xy + 8$ and $-2x^2y^2 + 3xy + 1$.

**19.** Subtract $3x^2 + 2y^2 - 5x + 4y$ from $2x^2 - 5y^2 - 6x + 2y$.

**20.** Perform the indicated operations.
$(8u^3v^2 - 2u^2v^2) + (-4u^2v^3 + 2u^2v^2) - (u^2v^2 - 2)$

Find each product in Exercises 21–26.

**21.** $5x^3y^2(-2xy + 7x^2y)$

**22.** $(7x - 2y)^2$

**23.** $(3a + 8b)(3a - 8b)$

**24.** $(2a^2 + b)(5a^2 - 2b)$

**25.** $(5u + 8v)(2u^2 - 3uv + v^2)$

**26.** $(3x - 2y - 5)^2$

Find each quotient in Exercises 27–28.

**27.** $\frac{14a^3b^4 - 28a^2b^3 + 49ab^5}{-7a^2b^3}$

**28.** $(x^3 - 4x^2y + 2xy^2 - 3y^3) \div (x - 3y)$

Factor each polynomial in Exercises 29–40. Assume $n$ is a positive integer.

**29.** $10x^2y^3 - 15xy^2$

**30.** $2x^2 - 4xy - 3x + 6y$

**31.** $9a^2 - 16b^2$

**32.** $3a^2 + ab - 14b^2$

**33.** $54x^3 + 2y^6$

**34.** $25x^2 + 30xy + 9y^2$

**35.** $8u^2 - 2uv - 15v^2$ **36.** $125x^3 - 64y^3$ **37.** $16x^2 - 56xy + 49y^2$

**38.** $2u^4 - u^2v^2 - v^4$ **39.** $x^2 - y^2 + 10y - 25$ **40.** $9x^{2n} - y^{2n}$

In Exercises 41–42 determine the values of the variable for which the rational expression is not defined.

**41.** $\dfrac{x+7}{x^2 - 3x + 2}$ **42.** $\dfrac{4a^2 + 5b}{(a^2 - 4)(b + 7)}$

Are the fractions given in Exercises 43–44 equivalent where they are defined?

**43.** $\dfrac{x+y}{x^2 - y^2}, \quad \dfrac{1}{x-y}$ **44.** $\dfrac{a^2+1}{a}, \quad \dfrac{a^2+2}{a^2+1}$

In Exercises 45–47 reduce each fraction to lowest terms.

**45.** $\dfrac{36a^2b^2c^2}{75ab^4c^2}$ **46.** $\dfrac{x-5}{5-x}$ **47.** $\dfrac{x^2 + 5x - 24}{x^2 - 10x + 21}$

Perform the indicated operations in Exercises 48–49.

**48.** $\dfrac{a^2 - 5a + 6}{a^2 - 6a + 9} \cdot \dfrac{a^2 + 4a - 21}{a^2 + 2a - 35}$ **49.** $\dfrac{a^3 - b^3}{a^2 + 4a - 5} \div \dfrac{a^2 + ab + b^2}{a^2 + 10a + 25}$

Find the LCD of the fractions given in Exercises 50–51.

**50.** $\dfrac{8ab}{a^3 - b^3}$ and $\dfrac{7}{(a-b)^2}$ **51.** $\dfrac{a+1}{a^2 - 7a + 6}$ and $\dfrac{a+5}{a^2 - 12a + 36}$

Perform the indicated operations in Exercises 52–55.

**52.** $\dfrac{3}{y^2 + y} - \dfrac{2}{y+1}$ **53.** $\dfrac{2xy}{x^2 - y^2} + \dfrac{y}{y-x} + 3$

**54.** $\dfrac{\dfrac{a}{b} - 1}{\dfrac{b}{a} - 1}$ **55.** $\dfrac{a - 7 + \dfrac{10}{a}}{a - 10 + \dfrac{25}{a}}$

Simplify each expression in Exercises 56–70. Rationalize denominators when necessary. Assume all variables represent positive real numbers.

**56.** $\sqrt[3]{-27}$ **57.** $\sqrt[6]{-64}$ **58.** $\sqrt{9a^4}$ **59.** $\sqrt[4]{81a^8b^{20}}$

**60.** $3\sqrt{4x}\sqrt{12xy^2}$ **61.** $\dfrac{\sqrt{20a^3b^7}}{\sqrt{5ab}}$ **62.** $\dfrac{\sqrt[4]{48a^5b^{10}}}{\sqrt[4]{243ab^5}}$ **63.** $4\sqrt{125} + 5\sqrt{80}$

**64.** $4\sqrt[3]{3xy^3} - 6y\sqrt[3]{81x}$ **65.** $\dfrac{\sqrt{8} - \sqrt{3}}{\sqrt{2} + \sqrt{3}}$ **66.** $\dfrac{\sqrt{x} + \sqrt{y}}{2\sqrt{x} - \sqrt{y}}$ **67.** $5\sqrt{40a} + \dfrac{8\sqrt{2a}}{\sqrt{5}}$

**68.** $\dfrac{\sqrt[3]{-81x^3y^{-3}}}{\sqrt[3]{3x^{-3}y^3}}$ **69.** $\left(\dfrac{8a^5b^7}{a^{-4}b^{-3}}\right)^{1/3}$ **70.** $\left(\dfrac{27^{2/3}x^{-5/3}y^2}{x^{-10/3}y^{-2}}\right)^{-3}$

Solve.

71. **PHYSICS** If the speed of light is $3.00 \times 10^5$ km/sec, calculate the value of a light year in kilometers. (A light year is the distance light travels in a year.)

72. **CALCULUS** Simplify the expression that comes from taking a derivative in calculus.

$$\frac{4x^3(4x) - (2x^2 - 1)12x^2}{16x^6}$$

73. **BUSINESS** All employees of Mutter Manufacturing received a 12% raise. Determine a polynomial expression for the new salary if $s$ is the old salary. What was Pat Wood's new salary if she made $24,500 before the raise?

74. **ENGINEERING** The height in feet of a rocket $t$ seconds after firing is $-16t^2 + 240t$. Find the height after 3.25 seconds.

CHAPTER

# 2 EQUATIONS AND INEQUALITIES

Solving equations is an important part of algebra and of mathematics in general. Equations often arise as mathematical models of applied problems. Two examples of the types of problems that we will consider in this chapter follow.

**◀ ECOLOGY**

A park ranger wishes to estimate the number of trout in a lake. He catches 200 fish, tags a fin, and returns them to the lake. After a period of time allowing the tagged fish to mix with the total population, he catches another sample of 200 fish and discovers that 3 of these are tagged. Use a proportion to estimate the number of trout in the lake.

**BUSINESS ▶**

In the manufacture and sale of record albums, the revenue made on the sale of $x$ albums is $\$2.20x$, and the cost of producing $x$ albums is $\$1.30x + \$4500$. In order to make a profit, the revenue received must be greater than the costs of production. For what values of $x$ will a profit be returned?

Solving problems such as these (See Example 8 in Section 2.2 and Example 4 in Section 2.7) involves two steps. First, an equation or inequality describing the problem must be determined; second, the resulting equation or inequality must be solved. We begin by reviewing the basic solution techniques; then we consider a variety of applications involving simple everyday problems as well as more complex problems taken from business and economics, science, music, and other fields.

## 2.1 Linear and Absolute Value Equations

### Equations

A statement that two quantities are equal is called an **equation.** Some equations are true, some are false, and for some, such as

$$x + 3 = 5,$$

the truth value depends on the value of the variable $x$. If the variable can be replaced by a number which makes the equation true, that number is called a **solution** or **root** of the equation. When we **solve** an equation, we determine all solutions or roots to the equation.

Equations such as $x + 3 = 5$ which are true for certain replacements of the variable (when $x = 2$) and false for other replacements are called **conditional equations.** An equation of the type

$$x + 2 = 2 + x$$

which is true for all replacements of the variable (every number is a solution) is called an **identity.** Finally, an equation such as

$$x + 2 = x - 2$$

which has no solutions is called a **contradiction.**

Two equations which have exactly the same solutions are called **equivalent.** Notice that

$$x + 3 = 5 \qquad \text{and} \qquad x = 2$$

are equivalent. To solve an equation, we change it into an equivalent equation which can be solved by direct inspection. This is accomplished by changing the given equation, through a succession of steps, into an equivalent equation with the variable isolated on one side.

### Linear Equations

In this section we concentrate on **linear equations** in one variable where the variable is raised to the first power only. For this reason, linear equations are often called **first-degree equations.** Every linear equation in one variable $x$ can be written as an equivalent equation in the form

$$\mathbf{ax + b = 0,}$$

where $a$ and $b$ are known real numbers, $a \neq 0$.

**EXAMPLE 1** Solve.

**(a)** $3(x - 4) + 5x = 4$

| | |
|---|---|
| $3x - 12 + 5x = 4$ | Clear parentheses first |
| $8x - 12 = 4$ | Collect like terms |
| $8x - 12 + 12 = 4 + 12$ | Add 12 to both sides |
| $8x = 16$ | |
| $\frac{8x}{8} = \frac{16}{8}$ | Divide both sides by 8 |
| $x = 2$ | |

Check: $3(2 - 4) + 5(2) \stackrel{?}{=} 4$

$3(-2) + 10 \stackrel{?}{=} 4$

$-6 + 10 \stackrel{?}{=} 4$

$4 = 4$

The solution is 2.

**(b)** $1.2y - 3.6 = 2.4$

When an equation involves decimals or fractions, solving is easier if they are cleared before proceeding.

| | |
|---|---|
| $10(1.2y - 3.6) = 2.4(10)$ | Multiply both sides by 10 to clear decimals |
| $12y - 36 = 24$ | Clear parentheses |
| $12y - 36 + 36 = 24 + 36$ | Add 36 to both sides |
| $12y = 60$ | |
| $\frac{12y}{12} = \frac{60}{12}$ | Divide both sides by 12 |
| $y = 5$ | |

Check: $1.2(5) - 3.6 \stackrel{?}{=} 2.4$

$6.0 - 3.6 \stackrel{?}{=} 2.4$

$2.4 = 2.4$

The solution is 5.

**(c)** $\frac{5}{2}z + \frac{3}{2} = \frac{3}{2}z - (3 - z)$

| | |
|---|---|
| $2\left[\frac{5}{2}z + \frac{3}{2}\right] = 2\left[\frac{3}{2}z - (3 - z)\right]$ | Multiply both sides by 2, the LCD of the fractions |
| $5z + 3 = 3z - 2(3 - z)$ | Clear brackets |
| $5z + 3 = 3z - 6 + 2z$ | Clear parentheses |
| $5z + 3 = 5z - 6$ | Collect like terms |
| $5z - 5z + 3 = 5z - 5z - 6$ | Subtract $5z$ from both sides |
| $3 = -6$ | |

Whenever a contradiction is obtained, such as $3 = -6$, we know that the original equation is also a contradiction. Thus, there is no solution to the equation. ■

In Example 1(**c**) we saw that whenever a contradiction occurs the equation has no solution. In a similar manner, if at any step an identity occurs, the original equation is also an identity and every real number is a solution.

On first inspection, an equation may not appear to be linear. Such is the case in the next example.

**EXAMPLE 2**

Solve $a^2 - 5 = (a + 2)(a - 3) + 8$.

| | |
|---|---|
| $a^2 - 5 = a^2 - a - 6 + 8$ | Clear parentheses by multiplying the binomials |
| $a^2 - 5 = a^2 - a + 2$ | Collect like terms |
| $-5 = -a + 2$ | Subtract $a^2$ from both sides |
| $-7 = -a$ | Subtract 2 from both sides |
| $7 = a$ | Multiply both sides by $-1$ |

The solution is 7, which does check. ■

## Radical Equations

An equation is called a **radical equation** if the variable appears in a radicand. Some radical equations can be transformed into linear equations. The first step involves using the following theorem.

**Power Theorem**

If $a$ and $b$ are two algebraic expressions, and if $a = b$, then $a^n = b^n$, for any natural number $n$.

When the power theorem is used on an equation involving a variable, the resulting equation may have more solutions than the original. For example, the equation

$$x = 4$$

has only one solution, 4, but

$$x^2 = 16$$

has two solutions, 4 and $-4$. Thus, when the power theorem is used, all possible solutions *must* be checked in the original equation. Solutions which do not check are called **extraneous roots** and must be discarded.

**To Solve an Equation Involving Radicals**

1. Isolate a radical on one side of the equation.
2. Raise both sides to a power equal to the index on that radical.
3. Solve the equation. If a radical remains, isolate it and raise to the power again.
4. Check all possible solutions in the *original* equation.

**EXAMPLE 3** Solve. $\sqrt{x^2+5} - x + 5 = 0$

$$\sqrt{x^2+5} = x - 5 \quad \text{Isolate the radical}$$
$$(\sqrt{x^2+5})^2 = (x-5)^2 \quad \text{Square both sides}$$
$$x^2 + 5 = x^2 - 10x + 25 \quad (x-5)^2 \text{ is } \textit{not } x^2 - 25$$
$$5 = -10x + 25$$
$$-20 = -10x$$
$$2 = x$$

Check:
$$\sqrt{2^2+5} - 2 + 5 \stackrel{?}{=} 0$$
$$\sqrt{9} - 2 + 5 \stackrel{?}{=} 0$$
$$3 - 2 + 5 \neq 0$$

The equation has no solution. ■

CAUTION A common mistake made in solving some radical equations is to square term by term rather than to isolate a radical and square both sides. For example, to solve

$$\sqrt{y+3} - \sqrt{y-2} = 1$$

it is *incorrect* to square term by term and obtain

$$(y+3) - (y-2) = 1.$$

It is easy to see that this technique does not work by considering a numerical example.

$$\sqrt{9} - \sqrt{4} = 3 - 2 = 1$$

However, $(\sqrt{9})^2 - (\sqrt{4})^2 = 9 - 4 = 5$, not $1^2$ or 1.

The correct way to solve this type of equation is illustrated in Example 4. ■

**EXAMPLE 4** Solve. $\sqrt{y+3} - \sqrt{y-2} = 1$

$$\sqrt{y+3} = 1 + \sqrt{y-2} \quad \text{Isolate } \sqrt{y+3}$$
$$(\sqrt{y+3})^2 = (1 + \sqrt{y-2})^2 \quad \text{Use power theorem}$$
$$y + 3 = 1 + 2\sqrt{y-2} + y - 2 \quad \text{Don't forget the middle term, } 2\sqrt{y-2}\text{, when squaring}$$
$$3 = 2\sqrt{y-2} - 1$$
$$4 = 2\sqrt{y-2} \quad \text{Isolate remaining radical}$$
$$2 = \sqrt{y-2} \quad \text{Divide by 2 to simplify}$$
$$4 = y - 2 \quad \text{Square again}$$
$$6 = y$$

Check:
$$\sqrt{6+3} - \sqrt{6-2} \stackrel{?}{=} 1$$
$$\sqrt{9} - \sqrt{4} \stackrel{?}{=} 1$$
$$3 - 2 = 1$$

The solution is 6. ■

## Fractional Equations

An equation that contains one or more algebraic fractions is called a **fractional equation.** We eliminate or clear the fractions by multiplying both sides of the equation by the LCD of all fractions. The resulting equation may be a linear equation that can be solved using previous techniques.

### To Solve a Fractional Equation

1. Find the LCD of all fractions in the equation.
2. Multiply both sides of the equation by the LCD to clear fractions. Make sure that *all* terms are multiplied.
3. Solve this equation.
4. Check your solution in the original equation to be sure that it does not make a denominator zero. If this occurs, the solution must be discarded.

**EXAMPLE 5** Solve. $\dfrac{x+2}{x-1} + \dfrac{3}{x} = 1$

The LCD of the fraction is $x(x-1)$ so we multiply both sides by this expression.

$$x(x-1)\left[\frac{x+2}{x-1} + \frac{3}{x}\right] = x(x-1)[1]$$

$$x\cancel{(x-1)}\frac{x+2}{\cancel{x-1}} + \cancel{x}(x-1)\frac{3}{\cancel{x}} = x(x-1) \quad \text{Clear brackets}$$

$$x(x+2) + 3(x-1) = x(x-1) \quad \text{Divide common factors}$$

$$x^2 + 2x + 3x - 3 = x^2 - x \quad \text{Clear parentheses}$$

$$6x = 3 \quad \text{Collect terms and isolate the variable term}$$

$$x = \frac{1}{2}$$

Check: $$\frac{\frac{1}{2}+2}{\frac{1}{2}-1} + \frac{3}{\frac{1}{2}} \stackrel{?}{=} 1$$

$$\frac{\frac{5}{2}}{-\frac{1}{2}} + \frac{3}{\frac{1}{2}} \stackrel{?}{=} 1$$

$$-5 + 6 \stackrel{?}{=} 1$$

$$1 = 1$$

The solution is $\frac{1}{2}$. ■

CAUTION A common mistake is to confuse addition or subtraction of algebraic fractions with solving a fractional equation. For example, compare the following two problems.

$$\text{Solve } \frac{x+2}{x-1} + \frac{3}{x} = 1. \qquad \text{Add } \frac{x+2}{x-1} + \frac{3}{x} + 1.$$

In each case, the LCD of all fractions must be found; however, it is used in two very different ways. In the first case, since we have an *equation*, we can multiply both sides by the LCD, $x(x-1)$, to find the numerical solution as in Example 5. In the second case, we rewrite each of the three terms with the denominator the LCD and then add numerators.

$$\begin{aligned} \frac{x+2}{x-1} + \frac{3}{x} + 1 &= \frac{x(x+2)}{x(x-1)} + \frac{3(x-1)}{x(x-1)} + \frac{x(x-1)}{x(x-1)} \\ &= \frac{x(x+2) + 3(x-1) + x(x-1)}{x(x-1)} \\ &= \frac{x^2 + 2x + 3x - 3 + x^2 - x}{x(x-1)} \\ &= \frac{2x^2 + 4x - 3}{x(x-1)} \end{aligned}$$

**EXAMPLE 6** Solve. $\dfrac{y^2+9}{y^2-9} - \dfrac{y}{y-3} = \dfrac{3}{y+3}$

Multiply both sides by the LCD of all fractions, $(y-3)(y+3)$.

$$(y-3)(y+3)\left[\frac{y^2+9}{(y-3)(y+3)} - \frac{y}{y-3}\right] = (y-3)(y+3)\left[\frac{3}{y+3}\right]$$

Clear brackets and divide common factors.

$$\begin{aligned} y^2 + 9 - y(y+3) &= 3(y-3) \\ y^2 + 9 - y^2 - 3y &= 3y - 9 \quad \text{Clear parentheses} \\ -6y &= -18 \\ y &= 3 \end{aligned}$$

Check: $\dfrac{3^2+9}{3^2-9} - \dfrac{3}{3-3} \stackrel{?}{=} \dfrac{3}{3+3}$

$$\frac{18}{0} - \frac{3}{0} \stackrel{?}{=} \frac{3}{6}$$

Since division by zero is undefined, we must discard 3. There is no solution. ■

## Absolute Value Equations

Equations that involve absolute values, such as

$$|x| = 4, \quad |3x-1| = 5, \quad \text{and} \quad |1-2x| = 0,$$

are called **absolute value equations.** To solve such an equation, for example $|x| = 4$, we solve two related equations since

$$|x| = \begin{cases} x \text{ if } x \geq 0 \\ -x \text{ if } x < 0. \end{cases}$$

The two equations to solve are

$$x = 4 \quad \text{and} \quad -x = 4.$$

The second equation becomes $x = -4$, giving us the two equations

$$x = 4 \quad \text{and} \quad x = -4,$$

which also gives the two solutions 4 and $-4$.

**EXAMPLE 7** Solve.

**(a)** $|3x - 1| = 5$

We must solve the following two equations.

$$\begin{aligned} 3x - 1 &= 5 \qquad \text{and} \qquad & 3x - 1 &= -5 \\ 3x &= 6 & 3x &= -4 \\ x &= 2 & x &= -\frac{4}{3} \end{aligned}$$

The solutions are 2 and $-\frac{4}{3}$. Check.

**(b)** $|1 + 2x| = |x|$

The only way that the absolute values of two expressions can be equal is if the two expressions are either equal or opposite in sign. Thus, we must solve

$$\begin{aligned} 1 + 2x &= x \qquad \text{or} \qquad & 1 + 2x &= -x. \\ 1 &= -x & 1 &= -3x \\ -1 &= x & -\frac{1}{3} &= x \end{aligned}$$

The solutions are $-1$ and $-\frac{1}{3}$. Check. ■

## Literal Equations

A **literal equation** involves two or more variables. Formulas such as $P = 2l + 2w$ and $d = rt$ are examples of literal equations. Remember that the letters in a literal equation represent numbers. All the rules for solving equations that involve numerals are also applicable for solving literal equations. In the following examples, we solve a similar numerical equation along with each literal equation in order to clarify the solution procedures.

**EXAMPLE 8** (a) Solve $\sqrt{xy + z} = z$ for $x$.

$$\sqrt{xy+z} = z$$
$$(\sqrt{xy+z})^2 = z^2$$
$$xy + z = z^2$$
$$xy = z^2 - z$$
$$\frac{xy}{y} = \frac{z(z-1)}{y}$$
$$x = \frac{z(z-1)}{y}$$

*Similar numerical example*

$$\sqrt{3x+7} = 5$$
$$(\sqrt{3x+7})^2 = 5^2$$
$$3x + 7 = 25$$
$$3x = 18$$
$$\frac{3x}{3} = \frac{18}{3}$$
$$x = 6$$

(b) Solve $\frac{1}{R} = \frac{1}{r_1} + \frac{1}{r_2}$ for $r_1$.

$$\frac{1}{R} = \frac{1}{r_1} + \frac{1}{r_2}$$
$$Rr_1r_2\left(\frac{1}{R}\right) = \left(\frac{1}{r_1} + \frac{1}{r_2}\right)Rr_1r_2$$
$$r_1r_2 = Rr_2 + Rr_1$$
$$r_1r_2 - Rr_1 = Rr_2$$
$$r_1(r_2 - R) = Rr_2$$
$$\frac{r_1(r_2 - R)}{r_2 - R} = \frac{Rr_2}{r_2 - R}$$
$$r_1 = \frac{Rr_2}{r_2 - R}$$

*Similar numerical example*

$$\frac{1}{5} = \frac{1}{r_1} + \frac{1}{3}$$
$$15r_1\left(\frac{1}{5}\right) = \left(\frac{1}{r_1} + \frac{1}{3}\right)15r_1$$
$$3r_1 = 15 + 5r_1$$
$$3r_1 - 5r_1 = 15$$
$$(3 - 5)r_1 = 15$$
$$\frac{-2r_1}{-2} = \frac{15}{-2}$$
$$r_1 = -\frac{15}{2}$$ ■

## 2.1 EXERCISES

Solve each equation in Exercises 1–48.

**1.** $3y + 5 = 7 - y$

**2.** $4z - 3 = 3 - 2z$

**3.** $1.8x - 4.1 = 6.5 + 0.2x$

**4.** $2.5y - 6.1 = 7.4 + 1.3y$

**5.** $\frac{z}{2} + \frac{3}{8} - \frac{z}{4} = \frac{5}{4}$

**6.** $\frac{x}{5} + \frac{1}{10} - \frac{x}{2} = \frac{1}{20}$

**7.** $4(y + 2) = 7(3 - y)$

**8.** $5(z - 1) = 8(z + 2)$

**9.** $5 + 7x = 7x + 4$

**10.** $2(y - 1) = 4 + 2y$

**11.** $5 + 7z = 7z + 5$

**12.** $3(x - 5) = 3(1 + x) - 18$

**13.** $3y - (y - 2) = 8$

**14.** $5z - (1 - z) = 11$

**15.** $2(3x - 2) - 2(4x + 3) = 4$

**16.** $4(2y - 1) - (3y + 2) = 0$

**17.** $z^2 + 2 = (z - 1)(z + 1) + z$

**18.** $(3y + 2)(y - 1) = (y - 4)(3y + 5)$

**19.** $2[z - (1 - 3z)] = 2(3 + 2z)$

**20.** $5[2a - (3 - a)] = 5(a + 3)$

**21.** $3\sqrt{2z - 5} - \sqrt{z + 23} = 0$

**22.** $\sqrt{5a - 1} - 2\sqrt{a + 1} = 0$

**23.** $\sqrt{x^2 + 8} - x - 4 = 0$

**24.** $\sqrt{9y^2 + 5} + 3y - 1 = 0$

**25.** $\sqrt{z-4}+\sqrt{z-8}=-2$

**26.** $\sqrt{a+5}+\sqrt{a-16}=-1$

**27.** $\sqrt{2x+10}-\sqrt{2x-5}=3$

**28.** $\sqrt{9-y}-\sqrt{18-y}=-1$

**29.** $\dfrac{z-2}{z+1}=\dfrac{z}{z-3}$

**30.** $\dfrac{a+1}{a+2}=\dfrac{a+2}{a-1}$

**31.** $\dfrac{2x}{x-3}-\dfrac{3}{x+4}=2$

**32.** $\dfrac{2y}{2y-1}-\dfrac{6}{y+5}=1$

**33.** $\dfrac{3}{z+5}-\dfrac{2}{z-2}=\dfrac{4}{z^2+3z-10}$

**34.** $\dfrac{5}{a-3}-\dfrac{3}{a+3}=\dfrac{7}{a^2-9}$

**35.** $\dfrac{x}{x+4}=\dfrac{4}{x-4}+\dfrac{x^2+16}{x^2-16}$

**36.** $\dfrac{2y}{2y+3}=\dfrac{8y^2+12}{4y^2-9}+\dfrac{2y}{3-2y}$

**37.** $\dfrac{1}{z^2-z-2}-\dfrac{3}{z^2-2z-3}=\dfrac{1}{z^2-5z+6}$

**38.** $\dfrac{2}{a^2+a-6}=\dfrac{2}{a^2-3a+2}-\dfrac{1}{a^2+2a-3}$

**39.** $|x+1|=3$

**40.** $|x-1|=-2$

**41.** $|y+2|=0$

**42.** $|5-2z|=13$

**43.** $|x+3|=|2x-1|$

**44.** $|3y+2|=|y|$

**45.** $|z|=z+1$

**46.** $|1-3x|=2-x$

**47.** $|y-1|=y-1$

**48.** $|z+1|=z+1$

In Exercises 49–54 solve the literal equation for the indicated variable. Assume that all expressions are defined so that division by zero and negative numbers under radicals are avoided.

**49.** $3a+2b=5c$ for $a$

**50.** $3uvw=p$ for $v$

**51.** $\sqrt{\dfrac{x}{y}}=z$ for $y$

**52.** $xy=w-y$ for $y$

**53.** $\sqrt{a+b}+\sqrt{b}=\sqrt{c}$ for $a$

**54.** $1=\dfrac{1}{x}+\dfrac{1}{y}$ for $x$

In Exercises 55–62 solve the formula for the indicated variable.

**55.** $A=\frac{1}{2}(b_1+b_2)h$ for $h$ (formula for the area of a trapezoid)

**56.** $S=2ab+2ac+2bc$ for $a$ (formula for the surface area of a rectangular solid)

**57.** $A=P(1+rt)$ for $r$ (formula for simple interest)

**58.** $C=\frac{5}{9}(F-32)$ for $F$ (formula for temperature conversion)

**59.** $\dfrac{1}{R}=\dfrac{1}{r_1}+\dfrac{1}{r_2}$ for $r_2$ (formula for resistance in an electrical circuit)

**60.** $\dfrac{1}{f}=\dfrac{1}{d_1}+\dfrac{1}{d_2}$ for $f$ (formula for focal length)

**61.** $h=\frac{1}{2}gt^2+vt$ for $g$ (formula for the height of a falling object)

**62.** $A=2\pi rh+2\pi r^2$ for $h$ (formula for the surface area of a cylinder)

In Exercises 63–66 solve the equation for the unknown first, without making any calculations. Then use a calculator to simplify the answer and round it to three decimal places. Store all results in the calculator's memory and do not do any rounding until the final step.

**63.** $9.328(y + 12.465) = 2.003(4.135 - y)$

**64.** $z^2 + 6.557 = (z - 4.055)^2$

**65.** $2\sqrt{x^2 - 4.25} - 2x + 2.8 = 0$

**66.** $\frac{a}{1.035} - \frac{a}{0.541} = 1000$

In Exercises 67–68 find a value of $m$ so that the equations are equivalent.

**67.** $3x - m = x + 8$ and $7(x - 1) = 23 + x$

**68.** $6x + 7 = m + 1$ and $3(x - 5) = 5(2x - 3)$

In Exercises 69–70 find a value of $n$ so that each of the equations has only the solution $-5$.

**69.** $\frac{10}{x} + \frac{15}{x} = n - 3$

**70.** $\sqrt{x + 9} = n + 1$

**71.** Is the equation $a = 9$ equivalent to $\sqrt{a} + 3 = 0$?

**72.** Is the equation $x = 5$ equivalent to $\frac{1}{x - 5} + \frac{1}{x} = \frac{5}{x^2 - 5x}$?

## For Review

In Exercises 73–75 simplify each expression and write it without negative exponents.

**73.** $(b^{-4})^{-3}$

**74.** $(3y)^{-1}$

**75.** $3y^{-1}$

Find each product in Exercises 76–77.

**76.** $(6u - v)(4u + v)$

**77.** $(-4uv + 5)(6u^2v^2 + 7uv - 3)$

Perform the indicated operations in Exercises 78–79.

**78.** $\frac{2x^2 + x - 15}{x^2 - x - 12} \div \frac{2x^2 + 3x - 20}{16x - x^3}$

**79.** $\frac{5 - x}{x^2 - 12x + 35} + \frac{x}{x^2 - 14x + 49}$

**80.** Simplify $\sqrt{\frac{125a^5b^{-2}}{4a^{-2}b^{-6}}}$. Rationalize the denominator and assume that $a$ and $b$ are positive.

# 2.2 Applications of Linear Equations

Most applied problems are expressed in words, not mathematical symbols. Solving such problems using algebra involves two steps. First, the *words* of the problem must be translated into symbols which are then used to write an algebraic equation describing the problem. Second, the resulting equation must be solved.

The unknown quantity in a problem can be represented by a variable. A typical expression and its translation is given below.

| Her salary | increased by | 5% | of | her salary | is | $12,600. |
|---|---|---|---|---|---|---|
| ↓ | ↓ | ↓ | ↓ | ↓ | ↓ | ↓ |
| s | + | (0.05) | · | s | = | 12,600 |

This sentence would probably come from a problem such as:

Mary is presently earning a salary of $12,600. If she received a 5% raise last year, what was her former salary?

Although there is no fixed method that works on all applied problems, solving such problems is easier if these steps are followed.

Step 1. Read the problem to determine the quantity (or quantities) to be found.

Step 2. Represent the unknown quantity (or quantities) symbolically with a letter, and write "Let $x$ = *description of unknown*."

Step 3. Make a sketch or diagram whenever possible.

Step 4. Determine which expressions are equal and write an equation.

Step 5. Solve the equation, and then *state the answer to the problem*.

Step 6. Check the answer or answers to see if the conditions of the original problem are satisfied and if the answer seems reasonable.

To illustrate the problem-solving techniques, we begin with a basic example involving numbers.

**EXAMPLE 1** The sum of three consecutive even integers is 72. Find the integers.

Recall that consecutive even integers are even integers that immediately follow each other in the regular counting order.

Let $x$ = the first even integer,
$x + 2$ = the next consecutive even integer,
$x + 4$ = the third consecutive integer.

$$x + (x + 2) + (x + 4) = 72 \quad \text{Their sum is 72}$$
$$3x + 6 = 72$$
$$3x = 66$$
$$x = 22$$
$$x + 2 = 24$$
$$x + 4 = 26$$

Thus, the integers are 22, 24, and 26. Check by adding. ■

**EXAMPLE 2**
**BUSINESS**

What were the total sales on which Bart received a commission of $692 if the commission rate was 8%?

Recall that the commission received is given by

**commission = (commission rate) · (total sales).**

Let $x$ = Bart's total sales. Then we must solve

$$692 = (0.08)x.$$
$$\frac{692}{0.08} = x$$
$$8650 = x$$

Thus, Bart had total sales amounting to \$8650. ■

**EXAMPLE 3**
**GEOMETRY**

The second angle of a triangle is six times as large as the first angle. If the third angle is 45° more than twice the first, find the measure of each angle.

Make a sketch of the triangle as in Figure 1.

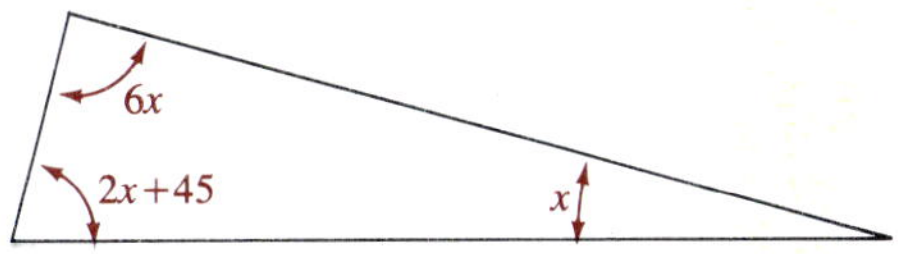


**Figure 1**

Let $x$ = the measure of the first angle,
$6x$ = the measure of the second angle,
$2x + 45$ = the measure of the third angle.

$$x + 6x + (2x + 45) = 180 \quad \text{The sum of their measures is } 180°$$
$$9x + 45 = 180$$
$$9x = 135$$
$$x = 15$$
$$6x = 90,\ 2x + 45 = 75$$

The measures are 15°, 90°, and 75°. ■

Consider the following example. Jim can do a certain job in 3 hours and Dave can do the same job in 7 hours. How long would it take them to do the job if they worked together?

This type of problem is usually called a *work problem,* and three important principles must be kept in mind.

1. The time required to do a job when two individuals work together must be less than the time required for the faster worker to complete the job by himself. The time together is *not* the average of the two times.
2. If a job can be done in $t$ hours, in 1 hour $\frac{1}{t}$ of the job can be completed. For example, since Jim can do the job in 3 hours, in 1 hour he could do $\frac{1}{3}$ of the job.

**3.** The work done by Jim in 1 hour added to the work done by Dave in 1 hour equals the work done by them together in 1 hour. Thus, if $t$ is the time required to complete the job working together,

(amount Jim does in 1 hour) + (amount Dave does in 1 hour)
= (amount done together in 1 hour)

which translates into

$$\frac{1}{3} + \frac{1}{7} = \frac{1}{t}.$$

Solving by multiplying both sides by the LCD, $21t$, we obtain

$$\begin{aligned} \mathbf{21t}\left(\frac{1}{3} + \frac{1}{7}\right) &= \mathbf{21t}\left(\frac{1}{t}\right) \\ 7t + 3t &= 21 \\ 10t &= 21 \\ t &= \frac{21}{10}. \end{aligned}$$

Working together, it would take Jim and Dave 21/10 hours (2 hours, 6 minutes) to complete the job.

## EXAMPLE 4

RATE-MOTION

Two cars leave the same point traveling in opposite directions. The second car travels 15 km/hr faster than the first and after 3 hours they are 465 km apart. How fast is each car traveling?

The distance $d$ that an object travels in a given time $t$ at a constant or uniform rate $r$ is given by the following formula.

$$(\text{distance}) = (\text{rate}) \cdot (\text{time})$$

$$\mathbf{d = rt}$$

Let $d_1$ = the distance the first car travels in km,
$r$ = the rate of the first car in km/hr,
$d_2$ = the distance the second car travels in km,
$r + 15$ = the rate of the second car in km/hr.

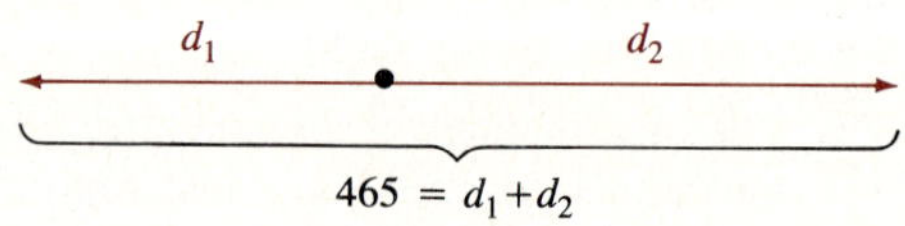


**Figure 2**

$$\begin{aligned} d_1 + d_2 &= 465 \\ 3r + 3(r + 15) &= 465 \qquad d_1 = 3r \text{ and } d_2 = 3(r + 15) \\ 3r + 3r + 45 &= 465 \end{aligned}$$

$$6r = 420$$
$$r = 70$$
$$r + 15 = 85$$

The first car is traveling 70 km/hr and the second 85 km/hr. ■

**EXAMPLE 5**
**RATE-MOTION**

The speed of a stream is 4 mph. A boat travels 48 miles upstream in the same time it takes to travel 72 miles downstream. What is the speed of the boat in still water?

Let $x$ − the speed of the boat in still water,
$x + 4$ = the speed of the boat when traveling downstream,
$x - 4$ = the speed of the boat when traveling upstream,
$\frac{72}{x+4}$ = the time of travel downstream $\left(t = \frac{d}{r}\right)$,
$\frac{48}{x-4}$ = the time of travel upstream.

$$\frac{72}{x+4} = \frac{48}{x-4} \quad \text{Times are equal}$$
$$(x-4)\cancel{(x+4)}\frac{72}{\cancel{x+4}} = \frac{48}{\cancel{x-4}}\cancel{(x-4)}(x+4)$$
$$72(x-4) = 48(x+4)$$
$$72x - 288 = 48x + 192$$
$$24x - 480$$
$$x = 20$$

The speed of the boat in still water is 20 mph. ■

## Proportion Applications

The **ratio** of one number $a$ to another number $b$ is the fraction $\frac{a}{b}$. A **proportion** is an equation that states that two ratios are equal. In the proportion

$$\frac{a}{b} = \frac{c}{d}$$

the first term, $a$, and the fourth term, $d$, are called the **extremes** of the proportion. The second term, $b$, and third term, $c$, are the **means.** If the second and third terms are equal, as in

$$\frac{a}{b} = \frac{b}{c},$$

$b$ is called the **mean proportional** between $a$ and $c$. When one or more terms of a proportion involve a variable, a quick way to solve the equation is to form the **cross-product** equation. For example, the cross-product equation for

$$\frac{a}{b} = \frac{c}{d} \quad \text{is} \quad ad = bc.$$

**EXAMPLE 6** If a man who is 6 ft tall stands in the sunlight and casts a shadow 14 ft long, while a pole casts a shadow 35 ft long, how tall is the pole?

Make a sketch, as in Figure 3.

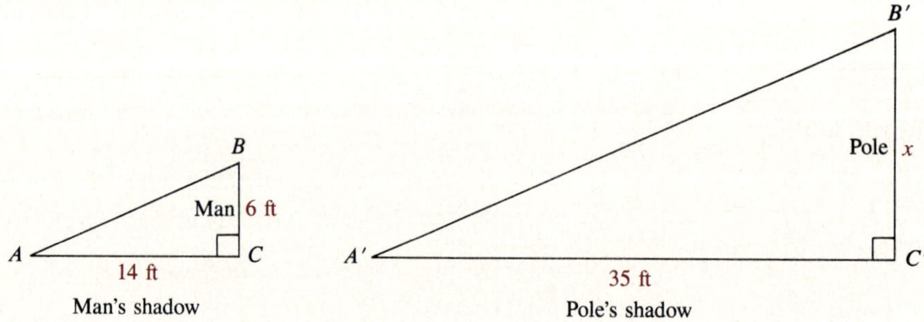


**Figure 3 Similar Triangles**

We assume that $\angle ACB = \angle A'C'B' = 90°$ (a right angle). Since both measurements are taken at the same time, the sun's rays make the same angle with level ground, that is, $\angle BAC = \angle B'A'C'$. Thus, $\triangle ABC$ is similar to $\triangle A'B'C'$ (the corresponding angles are equal) making the corresponding sides proportional. Thus,

$$\frac{AC}{A'C'} = \frac{BC}{B'C'}.$$

$$\frac{14}{35} = \frac{6}{x} \qquad \text{Substituting for } AC,\ BC,\ A'C'$$

$$14x = 35 \cdot 6 \qquad \text{Cross-product equation}$$

$$x = \frac{35 \cdot 6}{14} = 15$$

The pole is 15 ft tall. ■

**CAUTION** When solving applied problems, do not take shortcuts and do not try to do too many parts of the problem in your head. Writing out detailed descriptions of the variables as shown in the examples will save time in the long run and provide organization for your work. ■

We conclude this section by solving the first of the two problems given in the introduction to Chapter 2.

**EXAMPLE 7**
**ECOLOGY**

A park ranger wishes to estimate the number of trout in a lake. He catches 200 fish, tags a fin of each, and returns them to the lake. After a period of time allowing the tagged fish to mix with the total population, he catches another sample of 200 fish and finds that 3 of these are tagged. Use a proportion to estimate the number of trout in the lake.

The key to solving this problem is the proportion

$$\frac{\text{total number trout tagged}}{\text{total number trout in lake}} = \frac{\text{number of tagged trout in sample}}{\text{number of trout in sample}}.$$

Let $x$ = the total number of trout in the lake. Substituting in the above proportion we have:

$$\frac{200}{x} = \frac{3}{200}$$
$$(200)(200) = 3x$$
$$13{,}333.\overline{3} = x$$

Thus, there are approximately 13,333 trout in the lake. ■

## 2.2 EXERCISES

Solve.

**1.** If five times a number is decreased by 7, the result is twice the number, increased by 10. Find the number.

**2.** If twice the sum of a number and 5 is increased by 2, the result is 26 times the number. Find the number.

**3.** The sum of 3 positive, consecutive odd integers is 29 less than four times the first. Find the integers.

**4.** The sum of 4 consecutive integers is 8 more than 3 times the largest. Find the integers.

**5.** EDUCATION If Jorge Ortiz scored 61, 89, and 86 on three tests, what must he score on a fourth test to have an average score of 80?

**6.** SPORTS Three of the four starting linemen on the Arkansas State football team weigh 256 lb, 240 lb, and 242 lb. If the average weight of the starting line is 249.5 lb, how much does the fourth lineman weigh?

**7.** AGE If Sam is twice as old as Harry and the sum of their ages is the same as five times Harry's age, decreased by 28, how old is each?

**8.** CONSTRUCTION A board is 23 feet long. It is to be cut into 3 pieces in such a way that the second piece is twice as long as the first and the third is 3 feet more than the second. Find the length of each piece.

**9.** The numerator of a fraction is 5 less than the denominator. If both numerator and denominator are increased by 3, the result is equal to 2/3. Find the original fraction.

**10.** The denominator of a fraction is 6 less than the numerator. If the denominator is increased by 1, the result is equal to 1/2. Find the original fraction.

**11.** CONSUMER If the tax rate in West, Texas, is 4 percent, what is the tax on a purchase of $52.50?

**12.** SPORTS A quarterback completed 27 passes in 45 attempts. What was his completion percentage?

**13.** BUSINESS If Maria received a 12 percent raise and is now making $25,760 a year, what was her salary before the raise?

14. **RETAILING** A retailer bought a stereo for \$285 and put it on sale at a 70% markup rate. What was the price of the stereo?

15. **ECONOMICS** On Monday an investor bought 100 shares of stock. On Tuesday, the value of the shares went up 6% and on Wednesday, the value fell 5%. How much did the investor pay for the 100 shares on Monday if he sold them on Wednesday for \$1460.15?

16. **DEMOGRAPHY** The population of a town increased 4% during one year, then decreased 7% the next. What was the original population if it was 2418 at the end of these two years?

17. **GEOMETRY** The first angle of a triangle is 5° more than three times the second and the third angle is 10° less than the second. Find the measure of each.

18. **GEOMETRY** If two angles are **supplementary** (their measures total 180°), and the second is 30° less than twice the first, find the measure of each angle.

19. **AGRICULTURE** The perimeter of a rectangular wheat field is 1760 m. The length is 40 m more than the width. Find the dimensions of the field.

20. **GEOMETRY** An **isosceles triangle** has two equal sides called *legs* and a third side called the *base*. If each leg of an isosceles triangle is 2 m less than three times the base and the perimeter is 31 m, find the length of each leg and the length of the base.

21. **CONSTRUCTION** Jeff can frame a shed in 8 days and Barry can do the same job in 9 days. To the nearest tenth of a day, how long would it take them to frame the shed if they worked together?

22. **MANUFACTURING** If a machine built in 1975 can produce 100 units in 25 hours while a machine built this year can produce 100 units in 20 hours, to the nearest tenth of an hour, how long would it take to produce 100 units if the machines worked together?

23. If Bill and Morris can do a job in 18 minutes working together and Bill requires 42 minutes working alone, how long would it take Morris to do the job alone?

24. **SPORTS** If Mike Nesbitt, trainer at NAU, can tape the ankles of a team in 30 minutes, and with the help of a student trainer the job takes 20 minutes, how long would it take the student to do the job by himself?

25. **AGRICULTURE** A 4-inch pipe can fill an irrigation reservoir in 25 days, a 6-inch pipe can fill it in 10 days, and a 9-inch pipe takes 5 days. How long would it take to fill the reservoir if all three pipes were turned on together? (Give answer rounded to the nearest tenth of a day.)

26. If Sue can do a job in 12 hours, Larry can do the same job in 16 hours, and Eve requires 20 hours to do the job, how long would it take if they all worked together? (Give answer rounded to the nearest tenth of an hour.)

27. **RECREATION** A small swimming pool can be filled by an inlet pipe in 2 hours. It can be emptied in 10 hours through a drain pipe. If the two pipes were opened simultaneously, how long would it take to fill an empty pool?

28. **AGRICULTURE** A grain conveyor can fill a silo in 2 days. If it takes 3 days to empty the silo into railroad cars, how long would it take to fill an empty silo if both doors were opened simultaneously?

**29.** AERONAUTICS Two airplanes leave an airport at the same time, one traveling east and the other west. One plane is traveling 50 mph faster than the other. After 3 hours they are 3030 miles apart. How fast is each plane traveling?

**30.** RATE-MOTION Two families agree to meet in Boise for a reunion. They both leave at 8:00 A.M., one travels at a rate 10 mph faster than the other, and they meet in Boise at 12:00 noon. If the total distance they traveled was 480 miles, how fast was each traveling?

**31.** AERONAUTICS A plane flies 480 mi with the wind and 330 mi against the wind in the same time. If the speed of the plane in still air is 135 mph, what is the speed of the wind?

**32.** AERONAUTICS A plane flies 1160 km with the wind and 840 km against the wind in the same length of time. If the speed of the wind is 40 km/hr, what is the speed of the plane in still air?

**33.** COMMUNICATIONS Two printing presses print 31,500 fliers. If their rates of production differ by 30 fliers per minute, what is the rate of each if they both work 210 minutes to complete the job?

**34.** MANUFACTURING A machine made in 1970 has a production rate of 20 units per hour less than a machine made in 1985. If they are turned on at the same time and produce 1120 units in 8 hours, what is the production rate of each?

**35.** CONSTRUCTION If 150 ft of electrical wire weighs 45 lb, what will 270 ft of the same wire weigh?

**36.** CHEMISTRY A chemist wishes to make a mixture of concentrated hydrochloric acid and distilled water so that the ratio of acid to water is 7:3. If he starts with 21 L of acid, how much water should be mixed with the acid?

**37.** A boy 5 ft tall casts a shadow 12 ft long at the same time that a tower casts a shadow 252 ft long. How tall is the tower?

**38.** GEOMETRY Given two similar triangles, the sides of one are 6 inches, 7 inches, and 12 inches. If the shortest side of the other is 18 inches, find the other two sides.

**39.** Find the mean proportional between 25 and 9.

**40.** Find the mean proportional between 36 and 4.

**41.** The fourth term of a proportion is called a **fourth proportional** to the first, second, and third terms. Find the fourth proportional to 20, $-12$, and $-5$.

**42.** What number must be added to each of 20, 8, 32, and 14 to give four numbers which are in proportion in that order?

**43.** EDUCATION In a class of 40 students, the ratio of boys to girls is 5:3. After the first test, a certain number of boys drop the class making the ratio of boys to girls 7:5. How many boys dropped?

**44.** GEOMETRY If a tangent $PA$ and a secant $PB$ are drawn to a circle from an external point, the length of the tangent is the mean proportional between the length of the secant and its external segment $PC$. Find the length of the tangent if the secant is 49 cm and its external segment is 4 cm.

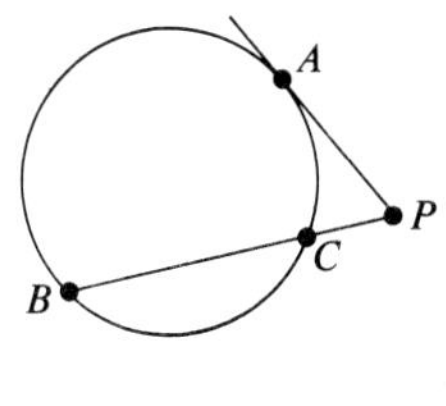

45. **MUSIC** The notes in a major chord in music have frequencies in the ratio of 4 to 5 to 6. If the first note of a chord has a frequency of 264 hertz, what are the frequencies of the other two notes?

46. **SURVEYING** A pole is standing in a lake. The ratios of the length of the pole above the water to the length in the water to the length in the sand at the bottom of the lake are 3 to 4 to 2. If 40 feet of the pole is in the water, how much of the pole is out of the water and how much is in the sand?

47. **INVESTMENT** Jay Arnote invested \$12,000 in an account that pays 12% simple interest. How much additional money must be invested in an account that pays 15% simple interest so that the average return on the two investments amounts to 13%?

48. **RECREATION** In order to balance a teeter-totter, the weight of the first child times his distance from the fulcrum (the central point of the board) must equal the weight of the second child times his distance from the fulcrum. If one child weighing 85 lb is 6 ft from the fulcrum, how far from the fulcrum must the second child be in order to balance the teeter-totter, if he weighs 51 lb?

49. **NAVIGATION** A small boat traveling at 20 knots (a knot is 1 nautical mile per hour) is 15 nautical miles from an island when a Coast Guard cutter initiates pursuit on the same course traveling 30 knots. How long will it take for the cutter to overtake the boat?

50. **GEOMETRY** The length of the altitude drawn to the hypotenuse of a right triangle is the mean proportional between the lengths of the segments of the hypotenuse determined by it. Find the length of the altitude if the segments are 9 inches and 16 inches in length.

## For Review

Solve.

51. $\sqrt{4a^2 + 8} - 2a - 4 = 0$

52. $\dfrac{7}{x - 6} + \dfrac{5}{x - 8} = \dfrac{2}{x^2 - 14x + 48}$

53. $|6 - 5z| = 1$

54. $S = \dfrac{a_n}{1 - r}$ for $r$

Factor and show that each expression in Exercises 55–56 is a perfect square.

55. $x^2 - 12x + 36$

56. $x^2 + \frac{2}{3}x + \frac{1}{9}$

In Exercises 57–58 evaluate $\sqrt{b^2 - 4ac}$ for the given values of $a$, $b$, and $c$.

57. $a = -1$, $b = 5$, and $c = -4$

58. $a = 2$, $b = -3$, and $c = -1$

## 2.3 Quadratic Equations

In Chapter 1 we learned an important property of the real number system stating that if $a = 0$ or $b = 0$, then $ab = 0$. The converse of this property, stated in the next theorem, is also true.

**Zero-Product Rule**

If $a$ and $b$ represent two real numbers, and $ab = 0$, then $a = 0$ or $b = 0$.

PROOF Suppose that $ab = 0$. Then either $a = 0$ or $a \neq 0$. If $a = 0$, then the theorem is proved. Thus, we may assume that $a \neq 0$, in which case $1/a$ is defined. Multiplying both sides of $ab = 0$ by $1/a$ gives $b = 0$. Hence the theorem is true. ■

## Solving by Factoring

The zero-product rule can be used to solve equations that, although not linear, can be reduced to two linear equations. For example, if

$$(x - 2)(x + 3) = 0$$

then each factor in the product can be set equal to zero, and we solve the resulting equations.

$$\begin{aligned} x - 2 &= 0 \qquad \text{or} \qquad & x + 3 &= 0 \\ x &= 2 & x &= -3 \end{aligned}$$

Thus, the solutions are 2 and $-3$. Had we started with the equation

$$x^2 + x - 6 = 0,$$

we could have factored the left side, obtaining

$$(x - 2)(x + 3) = 0,$$

and by following the procedure just outlined, we would have solved a *quadratic equation*. Any equation which can be written in the form

$$ax^2 + bx + c = 0$$

where $x$ is the variable and $a$, $b$, and $c$ are constant real numbers with $a \neq 0$, is called a **quadratic equation** or a **second-degree equation.** If $a = 0$, the $x^2$ term would be missing and we would have the linear equation $bx + c = 0$. We call $ax^2 + bx + c = 0$ the **general form** of a quadratic equation, and usually it is appropriate to write a quadratic equation in this form before attempting to solve it. The following examples illustrate the technique of solving a quadratic equation by the **factoring method.**

## EXAMPLE 1

Solve.

**(a)** $3x^2 - 3x - 18$

$$\begin{aligned} 3x^2 - 3x - 18 &= 0 && \text{Write in general form} \\ 3(x^2 - x - 6) &= 0 && \text{Factor out common factor 3} \\ x^2 - x - 6 &= 0 && \text{Divide both sides by 3} \\ (x - 3)(x + 2) &= 0 && \text{Factor} \end{aligned}$$

$$\begin{aligned} x - 3 &= 0 \qquad \text{or} \qquad & x + 2 &= 0 \qquad \text{Zero-product rule} \\ x &= 3 & x &= -2 \end{aligned}$$

The solutions are 3 and $-2$. Check by substituting in the original equation.

**(b)** $x^2 + 6x = 0$ — No constant term

$x(x + 6) = 0$ — Factor

$x = 0$ or $x + 6 = 0$ — Zero-product rule

$x = -6$

The solutions are 0 and $-6$. Check. ■

**CAUTION** In Example 1**(b),** do not divide both sides of the equation by $x$, or the solution $x = 0$ will be lost. Whenever a quadratic equation has no constant term ($c = 0$), one solution is always 0. ■

**EXAMPLE 2** Solve. $-4x^2 + 2x = \frac{1}{4}$

$-4x^2 + 2x - \frac{1}{4} = 0$ — Write in general form

$16x^2 - 8x + 1 = 0$ — Multiply both sides by $-4$ to clear the fraction and make the coefficient of $x^2$ positive

$(4x - 1)(4x - 1) = 0$

$4x - 1 = 0$ or $4x - 1 = 0$

$4x = 1$ $\qquad$ $4x = 1$

$x = \frac{1}{4}$ $\qquad$ $x = \frac{1}{4}$

There is only one solution, $\frac{1}{4}$. Check. ■

When the two linear equations obtained by using the zero-product rule have the same solution, as in Example 2, the solution is called a **double root** or a **root of multiplicity two.**

## Solving by Taking Roots

If $b = 0$ in a quadratic equation, the equation has no linear term and takes the form

$$ax^2 + c = 0.$$

If we subtract $c$ from both sides, then divide both sides by $a$ (remembering $a \neq 0$), we obtain

$$x^2 = -\frac{c}{a}.$$

Suppose we let $d = -\frac{c}{a}$. Then the given equation becomes

$$x^2 = d.$$

By taking the square root of both sides and recalling that $\sqrt{x^2} = |x|$, we have

$$|x| = \sqrt{d}.$$

Since $|x| = x$ if $x \geq 0$ and $|x| = -x$ if $x < 0$, this equation can be expressed as two equations

$$x = \sqrt{d} \text{ and } x = -\sqrt{d},$$

often written more compactly as

$$x = \pm\sqrt{d}.$$

Thus, the two solutions to the original equation are $\sqrt{d}$ and $-\sqrt{d}$. If $d \geq 0$, then the solutions are real numbers. We shall consider the case when $d < 0$ in Section 2.6. Notice that $d \geq 0$ whenever $c$ and $a$ have opposite signs. The method outlined above and illustrated in the following example is sometimes called the **square root method.**

**EXAMPLE 3** Solve. $4y^2 - 5 = 0$

$$4y^2 = 5$$

$$y^2 = \frac{5}{4}$$

$$y = \pm\sqrt{\frac{5}{4}}$$

$$y = \pm\frac{\sqrt{5}}{2}$$

Thus, the solutions are $\sqrt{5}/2$ and $-\sqrt{5}/2$. Notice that in this case $4y^2 - 5$ cannot be factored in the usual manner using integer coefficients. ■

## Solving by Completing the Square

Factoring and taking roots are the easiest and best methods for solving many quadratic equations. However, since some quadratic equations cannot be solved using these techniques, alternative methods must be found.

Consider the equation

$$(x + 2)^2 = 3.$$

If we extend the method of taking roots and take the square root of both sides,

$$x + 2 = \pm\sqrt{3}$$

$$x = -2 \pm \sqrt{3}.$$

This gives us the two solutions $-2 + \sqrt{3}$ and $-2 - \sqrt{3}$. Suppose, now, that we write out the squared term on the left side of the original equation, then put the result in the general form of a quadratic equation.

$$x^2 + 4x + 4 = 3$$

$$x^2 + 4x + 1 = 0 \quad \text{General form}$$

If we had started with this quadratic equation, by what process could we work backwards to obtain $(x + 2)^2 = 3$, and then the two solutions? The process to use, called **completing the square,** is illustrated below.

| | |
|---|---|
| $x^2 + 4x + 1 \quad = 0$ | |
| $x^2 + 4x \quad = -1$ | Isolate the constant term on the right side, leaving space |
| $x^2 + 4x + (2)^2 = -1 + (2)^2$ | Add (1/2 the coefficient of $x$)$^2$ |
| $x^2 + 4x + 4 = 3$ | Simplify |
| $(x + 2)^2 = 3$ | Left side is a perfect square |

More generally, consider the equation

$$(x + d)^2 = x^2 + 2dx + d^2.$$

In the expression $x^2 + 2dx + d^2$, the coefficient of $x$ is $2d$. If we have only $x^2 + 2dx$, we can complete the square by taking one half the coefficient of $x$, $1/2 \cdot (2d) = d$, squaring it, and adding it to $x^2 + 2dx$. Consider the following examples.

| *To complete square on* | *Add* $\left(\frac{1}{2}\text{ the coefficient of } x\right)^2$ | | *Obtaining the perfect square* |
|---|---|---|---|
| $x^2 + 6x$ | 9 | $\left[9 = \left(\frac{1}{2} \cdot 6\right)^2\right]$ | $x^2 + 6x + 9 = (x + 3)^2$ |
| $x^2 - 10x$ | 25 | $\left[25 = \left(\frac{1}{2} \cdot (-10)\right)^2\right]$ | $x^2 - 10x + 25 = (x - 5)^2$ |
| $x^2 - x$ | $\frac{1}{4}$ | $\left[\frac{1}{4} = \left(\frac{1}{2} \cdot (-1)\right)^2\right]$ | $x^2 - x + \frac{1}{4} = \left(x - \frac{1}{2}\right)^2$ |

**EXAMPLE 4** Solve. $x^2 + 2x - 15 = 0$

We observe that $x^2 + 2x - 15 = (x - 3)(x + 5)$. Using the method of factoring, we have $x - 3 = 0$ or $x + 5 = 0$, giving $x = 3$ and $x = -5$ as solutions. Compare this with the following.

| | |
|---|---|
| $x^2 + 2x \quad = 15$ | Isolate the constant on the right side |
| $x^2 + 2x + \mathbf{1} = 15 + \mathbf{1}$ | Add 1 to both sides ($1 = [1/2 \text{ coefficient of } x]^2$) |
| $(x + 1)^2 = 16$ | |
| $x + 1 = \pm\sqrt{16}$ | Take square root of both sides |
| $x + 1 = \pm 4$ | |
| $x = -1 \pm 4$ | |

$x = -1 + 4$ or $x = -1 - 4$

$x = 3 \qquad x = -5$

The solutions are 3 and $-5$, the same two solutions that were obtained using the factoring method. ■

If the coefficient of $x^2$ in a quadratic equation is not 1, we *cannot* complete the square simply by taking one half the coefficient of $x$ and squaring it. For example, $(2x-3)^2 = 4x^2 - 12x + 9$. However, if we are given $4x^2 - 12x$ and attempt to complete the square by adding $(1/2 \cdot 12)^2$, we obtain 36, not 9. In such instances, we must first divide through by the coefficient of $x^2$, as illustrated in the next example.

**EXAMPLE 5** Solve $2x^2 + 2x - 3 = 0$ by completing the square.

$$2x^2 + 2x = 3 \qquad \text{Isolate the constant}$$

$$x^2 + x = \frac{3}{2} \qquad \text{Divide by 2 to make the coefficient of } x^2 \text{ equal to 1}$$

$$x^2 + x + \mathbf{\frac{1}{4}} = \frac{3}{2} + \mathbf{\frac{1}{4}} \qquad \left(\frac{1}{2}\cdot 1\right)^2 = \frac{1}{4}$$

$$\left(x + \frac{1}{2}\right)^2 = \frac{7}{4}$$

$$x + \frac{1}{2} = \pm\sqrt{\frac{7}{4}}$$

$$x + \frac{1}{2} = \pm\frac{\sqrt{7}}{\sqrt{4}} \qquad \sqrt{\frac{a}{b}} = \frac{\sqrt{a}}{\sqrt{b}}$$

$$x + \frac{1}{2} = \frac{\pm\sqrt{7}}{2}$$

$$x = -\frac{1}{2} \pm \frac{\sqrt{7}}{2} = \frac{-1 \pm \sqrt{7}}{2}$$

The solutions are $\dfrac{-1+\sqrt{7}}{2}$ and $\dfrac{-1-\sqrt{7}}{2}$. ■

## Solving by the Quadratic Formula

Rather than continuing to solve particular quadratic equations by completing the square, we now use this technique to solve a general quadratic equation and in the process derive the quadratic formula. Recall the general form of a quadratic equation.

$$ax^2 + bx + c = 0 \qquad a \neq 0$$

$$ax^2 + bx = \mathbf{-c} \qquad \text{Isolate the constant}$$

$$\frac{ax^2}{a} + \frac{b}{a}x = -\frac{c}{a} \qquad \text{Divide by } a \text{ to make the coefficient of } x^2 \text{ equal to 1}$$

$$x^2 + \frac{b}{a}x + \left(\frac{b}{2a}\right)^2 = -\frac{c}{a} + \left(\frac{b}{2a}\right)^2 \qquad \text{Add } \left(\frac{1}{2}\cdot\frac{b}{a}\right)^2 = \left(\frac{b}{2a}\right)^2$$

$$\left(x + \frac{b}{2a}\right)^2 = -\frac{c}{a} + \frac{b^2}{4a^2}$$

$$\left(x + \frac{b}{2a}\right)^2 = -\frac{4ac}{4a^2} + \frac{b^2}{4a^2} \qquad 4a^2 \text{ is LCD}$$

$$\left(x + \frac{b}{2a}\right)^2 = \frac{b^2 - 4ac}{4a^2} \qquad \text{Subtract}$$

$$x + \frac{b}{2a} = \pm\sqrt{\frac{b^2 - 4ac}{4a^2}}$$

$$x + \frac{b}{2a} = \frac{\pm\sqrt{b^2 - 4ac}}{\sqrt{4a^2}} \qquad \sqrt{\frac{a}{b}} = \frac{\sqrt{a}}{\sqrt{b}}$$

$$x + \frac{b}{2a} = \frac{\pm\sqrt{b^2 - 4ac}}{2a}$$

$$x = -\frac{b}{2a} \pm \frac{\sqrt{b^2 - 4ac}}{2a} \qquad \text{Subtract } \frac{b}{2a}$$

$$\mathbf{x = \frac{-b \pm \sqrt{b^2 - 4ac}}{2a}} \qquad \text{Note that } 2a \text{ is the denominator of the entire expression}$$

This formula is called the **quadratic formula** and it must be memorized. To use it to solve a quadratic equation, identify the constants $a$, $b$, and $c$, substitute them into the quadratic formula, and simplify the numerical expression.

**EXAMPLE 6** Solve using the quadratic formula.

**(a)** $x^2 - 6x + 8 = 0$

The equation is in the simplest general form and therefore $a = 1$, $b = -6$ (not 6), and $c = 8$.

$$\begin{aligned} x &= \frac{-b \pm \sqrt{b^2 - 4ac}}{2a} \\ &= \frac{-(-6) \pm \sqrt{(-6)^2 - 4(1)(8)}}{2(1)} \\ &= \frac{6 \pm \sqrt{36 - 32}}{2} \\ &= \frac{6 \pm \sqrt{4}}{2} = \frac{6 \pm 2}{2} \end{aligned}$$

$$x = \frac{6 + 2}{2} = \frac{8}{2} = 4 \quad \text{or} \quad x = \frac{6 - 2}{2} = \frac{4}{2} = 2.$$

The solutions are 4 and 2.

**(b)** $x^2 - 2x = -\frac{1}{2}$

Write the equation in general form and clear the fraction.

$$x^2 - 2x + \frac{1}{2} = 0$$

$$2x^2 - 4x + 1 = 0$$

Thus, $a = 2$, $b = -4$, and $c = 1$.

$$x = \frac{-b \pm \sqrt{b^2 - 4ac}}{2a}$$

$$= \frac{-(-4) \pm \sqrt{(-4)^2 - 4(2)(1)}}{2(2)}$$

$$= \frac{4 \pm \sqrt{16 - 8}}{4} \quad \text{Multiply 4(2)(1) before subtracting}$$

$$= \frac{4 \pm \sqrt{8}}{4}$$

$$= \frac{4 \pm 2\sqrt{2}}{4} = \frac{\cancel{2}(2 \pm \sqrt{2})}{\cancel{2} \cdot 2} = \frac{2 \pm \sqrt{2}}{2}$$

The solutions are $\frac{2 + \sqrt{2}}{2}$ and $\frac{2 - \sqrt{2}}{2}$. ■

CAUTION Students often make the mistake of canceling the 4's in $\frac{4 \pm 2\sqrt{2}}{4}$ obtaining an incorrect answer, $1 \pm 2\sqrt{2}$. Also, remember that both terms in the numerator must be placed over the denominator. DO NOT write $2 \pm \frac{\sqrt{2}}{2}$ for the answer $\frac{2 \pm \sqrt{2}}{2}$. ■

Factoring and taking roots are the easiest methods for solving quadratic equations. If these techniques are not appropriate, it is usually better to go directly to the quadratic formula rather than to complete the square.

**Approximating Solutions**

To find approximate solutions to a quadratic equation using a calculator, compute $\sqrt{b^2 - 4ac}$ and store this value in memory. Add this value to $-b$ and divide by $2a$ to obtain the first solution. Then subtract this value from $-b$ and divide by $2a$ to obtain the second solution.

**EXAMPLE 7**

Solve. $3.5x^2 - 81.6x + 2.8 = 0$

The solutions are

$$x = \frac{81.6 \pm \sqrt{(81.6)^2 - 4(3.5)(2.8)}}{2(3.5)}.$$

**1.** Sequence of steps to calculate $\sqrt{b^2 - 4ac}$:

ALG: 4 [×] 3.5 [×] 2.8 [=] [+/−] [+] 81.6 [$x^2$] [=] [√] [STO]

RPN: 81.6 [$x^2$] [ENTER] 4 [ENTER] 3.5 [×] 2.8 [×] [−] [√] [STO]

At this point the display will show 81.359449, and this value has also been stored in the memory. Do not clear the display since it is used in the next step.

**2.** Sequence of steps to calculate the first solution:

ALG: [+] 81.6 [=] [÷] 2 [÷] 3.5 [=]

RPN: 81.6 [+] 2 [÷] 3.5 [÷]

The display now shows the first solution, 23.279921.

**3.** Sequence of steps to calculate the second solution:

ALG: 81.6 [−] [RCL] [=] [÷] 2 [÷] 3.5 [=]

RPN: 81.6 [ENTER] [RCL] [−] 2 [÷] 3.5 [÷]

The display now shows the second solution, 0.0343644. ■

## 2.3 EXERCISES

In Exercises 1–10 solve each equation by the factoring method.

**1.** $x^2 - 3x - 10 = 0$ **2.** $y^2 + 3y - 18 = 0$

**3.** $2z^2 - 10z - 12 = 0$ **4.** $3x^2 + 19x - 14 = 0$

**5.** $y^2 + y = -\frac{1}{4}$ **6.** $z^2 = \frac{11}{3}z + \frac{4}{3}$

**7.** $2x^2 + 3x = 0$ **8.** $4y^2 - 7y = 0$

**9.** $5z(z + 2) = 7z$ **10.** $5z(z + 2) = 5(z + 1)^2$

In Exercises 11–16 solve each equation by the square root method.

**11.** $x^2 - 36 = 0$ **12.** $y^2 - 100 = 0$ **13.** $3z^2 - 75 = 0$

**14.** $5x^2 - 80 = 0$ **15.** $(y + 1)^2 = 9$ **16.** $(z - 3)^2 = 25$

What must be added to complete the square in each expression in Exercises 17–20?

**17.** $x^2 - 8x$ **18.** $y^2 + 10y$ **19.** $z^2 - z$ **20.** $z^2 - \frac{1}{4}z$

In Exercises 21–24 solve each equation by completing the square.

**21.** $x^2 - 5x - 24 = 0$ **22.** $x^2 + 3x + 1 = 0$

**23.** $3y^2 - 5y + 1 = 0$ **24.** $2z^2 + 2z - 1 = 0$

In Exercises 25–30 solve each equation by using the quadratic formula.

**25.** $x^2 + 5x - 24 = 0$ **26.** $y^2 + 4y - 21 = 0$

**27.** $3z^2 - 5z + 1 = 0$ **28.** $2x^2 - 3x - 4 = 0$

**29.** $(y + 1)(y - 1) + 2y = 0$ **30.** $(2z - 1)(z - 2) = 5 - 3z$

Solve each equation in Exercises 31–42. (First try factoring or taking roots, and if those fail, use the quadratic formula.)

**31.** $2x^2 + 9x = 5$ **32.** $5y^2 - 14y = 3$ **33.** $\frac{1}{2}z^2 - 8 = 0$

**34.** $\frac{1}{3}x^2 - 12 = 0$ **35.** $y^2 = 3y$ **36.** $2z^2 = -z$

**37.** $3(x^2 + x) = -10x - 4$ **38.** $4(y^2 + 2y) = 3(1 - y)$ **39.** $z^2 = 3(z + 1)$

**40.** $x^2 = 2(x + 3)$ **41.** $y^2 = 4y - 1$ **42.** $3z^2 = z + 1$

Solve for $x$, $y$, or $z$ in Exercises 43–48. Assume that $a$ and $b$ are positive constants.

**43.** $x^2 - 36a^2 = 0$ **44.** $y^2 - 100b^2 = 0$ **45.** $z^2 - (a + b)^2 = 0$

**46.** $x^2 - (2a + b)^2 = 0$ **47.** $y^2 - ay - 2a^2 = 0$ **48.** $z^2 + 2bz - 3b^2 = 0$

In Exercises 49–52 use a calculator to find approximate solutions to the following quadratic equations. Give answers correct to the nearest hundredth.

**49.** $x^2 - 2.1x - 4.3 = 0$ **50.** $y^2 - 5.3y - 1.6 = 0$

**51.** $2.15x^2 + 3.28x - 4.65 = 0$ **52.** $3.25y^2 - 2.17y - 5.07 = 0$

## For Review

Solve.

**53.** EDUCATION If Henry's average score on four quizzes is 76 and he knows that three of his grades are 82, 63, and 92, what grade did he make on the other quiz?

**54.** ECONOMICS What amount of money invested at 4 percent simple interest will increase to $1248 by the end of one year?

**55.** RATE-MOTION Mary Kim drives 12 mph faster than her mother. If Mary travels 310 mi in the same time that her mother travels 250 mi, find the speed of each.

**56.** If 3/4 inch on a map represents 20 miles, how many miles are represented by 12 inches?

**57.** RECREATION If the speed of a stream is 8 km/hr and a boat can travel 105 km downstream in the same time that it travels 49 km upstream, what is the speed of the boat in still water?

**58.** What number when subtracted from 12, 22, 3, and 4 will make the resulting numbers, in that order, proportional?

**59.** $6(x + 5) - 2(4x - 1) = 0$ **60.** $\sqrt{y - 5} - \sqrt{y + 3} = -2$

## 2.4 Equations That Result in Quadratic Equations

### Equations Quadratic in Form

Often, an equation is not quadratic in a particular variable, but rather quadratic in an expression containing the variable. Equations such as this are called **quadratic in form.** For example,

$$x^4 - 5x^2 + 4 = 0$$

is not a quadratic equation. However, if we substitute $u$ for $x^2$, $x^4 - 5x^2 + 4 = 0$ becomes

$$u^2 - 5u + 4 = 0 \qquad u^2 = (x^2)^2 = x^4$$

which is quadratic in the variable $u$. The original equation is quadratic in form, that is, quadratic in the expression $x^2$. Similarly,

$$(z - 4)^2 - 5(z - 4) + 6 = 0$$

is quadratic in $u = z - 4$ since it becomes

$$u^2 - 5u + 6 = 0.$$

**EXAMPLE 1**

Solve. $x^4 - 5x^2 + 4 = 0$

$$u^2 - 5u + 4 = 0 \qquad \text{Let } u = x^2. \text{ Then } u^2 = x^4$$

$$(u - 4)(u - 1) = 0$$

$$u - 4 = 0 \quad \text{or} \quad u - 1 = 0$$

$$u = 4 \qquad\qquad u = 1$$

Since $u = x^2$,

$$x^2 = 4 \qquad\qquad x^2 = 1$$

$$x = \pm\sqrt{4} \qquad\qquad x = \pm\sqrt{1}$$

$$x = \pm 2 \qquad\qquad x = \pm 1$$

Check:

$$x^4 - 5x^2 + 4 = 0 \qquad\qquad x^4 - 5x^2 + 4 = 0$$

$$(2)^4 - 5(2)^2 + 4 \stackrel{?}{=} 0 \qquad\qquad (1)^4 - 5(1)^2 + 4 \stackrel{?}{=} 0$$

$$16 - 20 + 4 \stackrel{?}{=} 0 \qquad\qquad 1 - 5 + 4 \stackrel{?}{=} 0$$

$$0 = 0 \qquad\qquad 0 = 0$$

2 (and clearly, $-2$) checks. 1 (and also $-1$) checks.
The solutions are 2, $-2$, 1, $-1$. ■

CAUTION Do not give 4 and 1 as the solutions to the original equation. Remember that 4 and 1 are solutions to the equation in the variable $u$ and are not solutions relative to the variable $x$. In other words, do not forget to substitute and find the solutions to the original equation when using the $u$-substitution method. ■

**EXAMPLE 2**

Solve. $(z - 4)^2 - 5(z - 4) + 6 = 0$

$$u^2 - 5u + 6 = 0 \qquad \text{Let } u = z - 4. \text{ Then } u^2 = (z - 4)^2$$

$$(u - 2)(u - 3) = 0$$

$$u - 2 = 0 \quad \text{or} \quad u - 3 = 0$$

$$u = 2 \qquad\qquad u = 3$$

Since $u = z - 4$,

$$z - 4 = 2 \qquad\qquad z - 4 = 3$$

$$z = 6 \qquad\qquad z = 7$$

Check: $(z-4)^2 - 5(z-4) + 6 = 0 \qquad (z-4)^2 - 5(z-4) + 6 = 0$

$$(6-4)^2 - 5(6-4) + 6 \stackrel{?}{=} 0 \qquad (7-4)^2 - 5(7-4) + 6 \stackrel{?}{=} 0$$
$$2^2 - 5(2) + 6 \stackrel{?}{=} 0 \qquad 3^2 - 5(3) + 6 \stackrel{?}{=} 0$$
$$4 - 10 + 6 \stackrel{?}{=} 0 \qquad 9 - 15 + 6 \stackrel{?}{=} 0$$
$$0 = 0 \qquad 0 = 0$$

6 checks. 7 checks.

The solutions are 6 and 7. ■

**EXAMPLE 3** Solve. $u^{2/3} + 2u^{1/3} = 15$

$$u^{2/3} + 2u^{1/3} - 15 = 0$$
$$w^2 + 2w - 15 = 0 \qquad \text{Let } w = u^{1/3}\text{. Then } w^2 = u^{2/3}$$
$$(w+5)(w-3) = 0$$
$$w + 5 = 0 \quad \text{or} \quad w - 3 = 0$$
$$w = -5 \qquad w = 3$$

Since $w = u^{1/3}$,

$$u^{1/3} = -5 \qquad u^{1/3} = 3$$
$$(u^{1/3})^3 = (-5)^3 \qquad (u^{1/3})^3 = 3^3$$
$$u = -125 \qquad u = 27$$

The solutions are 27 and $-125$. Check these. ■

## Fractional Equations

In Section 2.1 we solved fractional equations that resulted in linear equations when the fractions were cleared. Recall that the process of clearing the fractions involves multiplying both sides of the equation by the LCD of all fractions. Clearing some fractional equations results in quadratic equations as shown in the next two examples. Remember that any potential solutions to a fractional equation *must always* be checked in the original equation to be sure that they do not make one of the denominators equal to zero.

**EXAMPLE 4** Solve. $\dfrac{y}{y+2} = \dfrac{2}{y-1}$

The LCD $= (y+2)(y-1)$. Multiply both sides by $(y+2)(y-1)$.

$$\cancel{(y+2)}(y-1)\frac{y}{\cancel{(y+2)}} = \frac{2}{\cancel{(y-1)}}(y+2)\cancel{(y-1)}$$
$$(y-1)y = 2(y+2)$$
$$y^2 - y = 2y + 4$$
$$y^2 - 3y - 4 = 0$$
$$(y+1)(y-4) = 0$$
$$y + 1 = 0 \quad \text{or} \quad y - 4 = 0$$
$$y = -1 \qquad y = 4$$

The solutions are $-1$ and 4. Check these to verify that neither makes one of the original denominators equal to zero. ■

**EXAMPLE 5** Solve. $\dfrac{x}{x-2} + 1 = \dfrac{x^2+4}{x^2-4}$

The LCD = $(x-2)(x+2)$.

$$(x-2)(x+2)\left[\frac{x}{x-2} + 1\right] = \left[\frac{x^2+4}{(x-2)(x+2)}\right](x-2)(x+2)$$

$$(x+2)x + (x-2)(x+2) = x^2 + 4$$

$$x^2 + 2x + x^2 - 4 = x^2 + 4$$

$$x^2 + 2x - 8 = 0$$

$$(x+4)(x-2) = 0$$

$$x + 4 = 0 \quad \text{or} \quad x - 2 = 0$$

$$x = -4 \qquad\qquad x = 2$$

Check: $\dfrac{-4}{-4-2} + 1 \stackrel{?}{=} \dfrac{(-4)^2+4}{(-4)^2-4}$ $\qquad$ $\dfrac{2}{2-2} + 1 \stackrel{?}{=} \dfrac{2^2+4}{2^2-4}$

$\dfrac{4}{6} + 1 \stackrel{?}{=} \dfrac{20}{12}$ $\qquad$ $\dfrac{2}{0} + 1 \stackrel{?}{=} \dfrac{8}{0}$

$\dfrac{5}{3} = \dfrac{5}{3}$ $\qquad$ $\dfrac{2}{0}$ and $\dfrac{8}{0}$ are not defined.

The only solution is $-4$. ■

Example 5 illustrates the major reason for checking all possible solutions. Multiplying both sides of the equation by the LCD may introduce extraneous roots that are not solutions to the original problem.

## Radical Equations

Some radical equations become quadratic equations when the radicals are removed by raising both sides to a power. Always check your results since use of the power theorem may introduce extraneous roots.

**EXAMPLE 6** Solve. $\sqrt{2y+11} - y - 4 = 0$

$$\sqrt{2y+11} = y + 4 \qquad \text{Isolate the radical}$$

$$(\sqrt{2y+11})^2 = (y+4)^2 \qquad \text{Square both sides. Don't forget the middle term } 8y$$

$$2y + 11 = y^2 + 8y + 16$$

$$y^2 + 6y + 5 = 0$$

$$(y+1)(y+5) = 0$$

$$y + 1 = 0 \quad \text{or} \quad y + 5 = 0$$

$$y = -1 \qquad\qquad y = -5$$

Check: $\sqrt{2(-1)+11}-(-1)-4 \stackrel{?}{=} 0$ $\quad$ $\sqrt{2(-5)+11}-(-5)-4 \stackrel{?}{=} 0$

$\sqrt{-2+11}+1-4 \stackrel{?}{=} 0$ $\quad$ $\sqrt{-10+11}+5-4 \stackrel{?}{=} 0$

$\sqrt{9}+1-4 \stackrel{?}{=} 0$ $\quad$ $\sqrt{1}+5-4 \stackrel{?}{=} 0$

$3+1-4 \stackrel{?}{=} 0$ $\quad$ $1+5-4 \stackrel{?}{=} 0$

$0=0$ $\quad$ $2 \neq 0$

The only solution is $-1$. ■

**EXAMPLE 7** Solve. $\sqrt{3x+1}-\sqrt{x+9}=2$

$$\sqrt{3x+1} = 2+\sqrt{x+9} \qquad \text{Isolate one radical on the left side}$$

$$(\sqrt{3x+1})^2 = (2+\sqrt{x+9})^2 \qquad \text{Square both sides}$$

$$3x+1 = 4+4\sqrt{x+9}+x+9 \qquad (a+b)^2=a^2+2ab+b^2, \text{ and } 4\sqrt{x+9} \text{ is the } 2ab$$

$$2x-12 = 4\sqrt{x+9} \qquad \text{Isolate the radical}$$

$$x-6 = 2\sqrt{x+9} \qquad \text{Simplify}$$

$$(x-6)^2 = (2\sqrt{x+9})^2 \qquad \text{Square both sides}$$

$$x^2-12x+36 = 4(x+9) \qquad \text{Don't forget to square the 2}$$

$$x^2-12x+36 = 4x+36$$

$$x^2-16x = 0$$

$$x(x-16) = 0$$

$$x=0 \quad \text{or} \quad x-16=0$$

$$x=16$$

Check: $\sqrt{3(0)+1}-\sqrt{0+9} \stackrel{?}{=} 2$ $\quad$ $\sqrt{3(16)+1}-\sqrt{16+9} \stackrel{?}{=} 2$

$\sqrt{1}-\sqrt{9} \stackrel{?}{=} 2$ $\quad$ $\sqrt{49}-\sqrt{25} \stackrel{?}{=} 2$

$1-3 \neq 2$ $\quad$ $7-5=2$

The only solution is 16. ■

## 2.4 EXERCISES

Solve each equation in Exercises 1–42.

**1.** $x^4-10x^2+9=0$

**2.** $y^4-29y^2+100=0$

**3.** $(x+4)^2-5(x+4)+6=0$

**4.** $(y+2)^2-13(y+2)+42=0$

**5.** $x^{2/3}-2x^{1/3}-35=0$

**6.** $y^{2/3}-5y^{1/3}+6=0$

**7.** $(x^2-2x)^2-2(x^2-2x)-3=0$

**8.** $(y^2+4y)^2-(y^2+4y)-20=0$

**9.** $x^{-2}+2x^{-1}-15=0$

**10.** $y^{-2}+y^{-1}-2=0$

**11.** $2\left(\frac{2x-1}{x}\right)^2+7\left(\frac{2x-1}{x}\right)-4=0$

**12.** $3\left(\frac{y+2}{y}\right)^2+\left(\frac{y+2}{y}\right)-2=0$

13. $3x^{1/2} + 12x^{-1/2} = 13$

14. $y^{1/2} + 6y^{-1/2} = 5$

15. $\sqrt[6]{x} + \sqrt[3]{x} - 6 = 0$

16. $4\sqrt[4]{y} + \sqrt{y} - 5 = 0$

17. $1 - \frac{2}{x} = \frac{3}{x^2}$

18. $\frac{5}{y^2 - 14} = \frac{1}{y}$

19. $3y - \frac{2}{y+1} = \frac{4}{y+1}$

20. $y - \frac{3}{y+2} = \frac{-2}{y+2}$

21. $\frac{3x}{2x+1} = \frac{2}{4x^2-1} + \frac{x}{2x-1}$

22. $\frac{y+2}{y-1} = 1 - \frac{y^2-7}{y^2-1}$

23. $\frac{x}{x^2-x-2} - \frac{2}{x^2-5x+6} = \frac{-3}{x^2-2x-3}$

24. $\frac{1}{y^2+2y-3} = \frac{2}{y^2-3y+2} - \frac{y}{y^2+y-6}$

25. $\sqrt{x+4} + 8 = x$

26. $6 + \sqrt{3y+1} = 2y$

27. $\sqrt{x^2+6x} + \sqrt{2x+21} = 0$

28. $2\sqrt{3y-2} + \sqrt{3y^2+2y} = 0$

29. $\sqrt{x^2-7x+15} - \sqrt{4x-13} = 0$

30. $\sqrt{10y-3} - \sqrt{3y^2-7y+7} = 0$

31. $\sqrt{3x+1} - \sqrt{x-4} = 3$

32. $\sqrt{2y+5} + \sqrt{y+2} = 5$

33. $\sqrt[3]{x^2-1} = 2$

34. $\sqrt[4]{y^2-9} = 2$

35. $(2x+11)^{1/2} - x - 4 = 0$

36. $(y+4)^{1/2} - y + 8 = 0$

37. $(3x+3)^{1/2} + (x-1)^{1/2} = 4$

38. $(2y+5)^{1/2} - (y+2)^{1/2} = 1$

39. $\sqrt{x + \sqrt{x-2}} = 2$

40. $\sqrt{y + \sqrt{y+1}} = 1$

41. $\sqrt{x+8} - \sqrt{3x+1} = \sqrt{2x-1}$

42. $\sqrt{x+9} + \sqrt{x+4} = \sqrt{x+25}$

In Exercises 43–46 use a calculator to approximate the solutions to the following equations. Give answers correct to the nearest hundredth.

43. $x^4 - 6x^2 + 1 = 0$

44. $y + \sqrt{y} - 2.45 = 0$

45. $\frac{x}{1.5} + \frac{1}{x} = 3.2$

46. $\sqrt{2y+5.5} + \sqrt{y} = 4.58$

## For Review

Solve.

47. $x^2 + 5x = 21\left(\frac{1}{2} + \frac{x}{2}\right)$

48. $\frac{1}{5}x^2 - 20 = 0$

49. $(y+2)(y-2) + 3y = -4$

50. $y(5y-1) = 1$

## 2.5 Applications of Quadratic Equations

Many applied problems result in quadratic equations. Review the steps for solving applied problems given in Section 2.2. Remember to write out detailed descriptions of the variables, and always check to make sure your answers are reasonable.

**EXAMPLE 1**

One fourth the product of two consecutive positive even integers is 56. Find the integers.

Let $x =$ the first positive even integer,

$x + 2 =$ the second positive even integer,

$\frac{1}{4}x(x + 2) = \frac{x(x + 2)}{4} =$ one fourth their product.

We must solve the following equation.

$$\frac{x(x + 2)}{4} = 56$$

$$x(x + 2) = 224 \quad \text{Clear the fraction}$$

$$x^2 + 2x = 224$$

$$x^2 + 2x - 224 = 0 \quad \text{General form}$$

$$(x - 14)(x + 16) = 0$$

$$x - 14 = 0 \quad \text{or} \quad x + 16 = 0 \quad \text{Zero-product rule}$$

$$x = 14 \qquad\qquad x = -16$$

Since $x$ must be positive, we rule out $-16$. Thus, 14 is the first integer and $14 + 2 = 16$ is the second. Check. ■

To work the next example recall the Pythagorean theorem, which states that the sum of the squares of the legs of a right triangle is equal to the square of the hypotenuse.

**EXAMPLE 2**

GEOMETRY

A ladder 13 ft long leans against a wall. If the base of the ladder is 5 ft from the wall, how far would the lower end have to be pulled out so that the upper end is pulled down the same amount?

Figure 4 shows the ladder in its initial position. Use the Pythagorean theorem to find $y$, the distance from the top of the ladder to ground level.

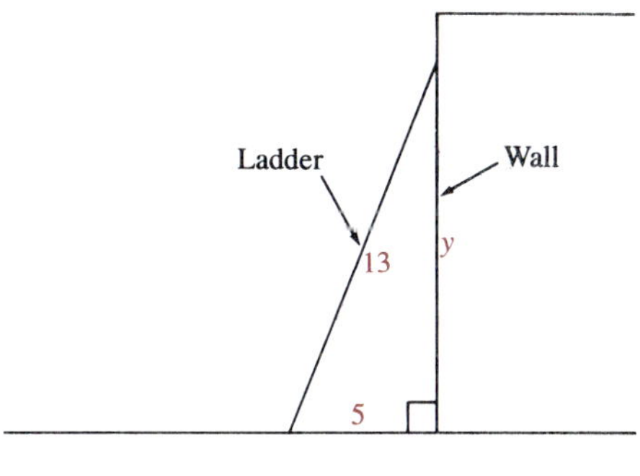


Figure 4

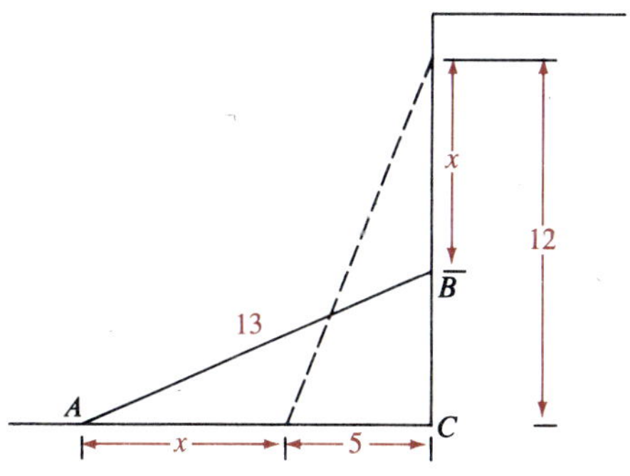


Figure 5

$$\begin{aligned} 5^2 + y^2 &= 13^2 \\ 25 + y^2 &= 169 \\ y^2 &= 144 \\ y &= \pm 12 \end{aligned}$$

Therefore, $y = 12$.

Let $x =$ the distance the ladder is pulled out; then $x$ is also the distance the ladder moves down. The new position of the ladder is given in Figure 5; the dashed line shows the initial position. In triangle $ABC$, $AB = 13$ (the hypotenuse), $AC = x + 5$, and $BC = 12 - x$. Use the Pythagorean theorem again.

$$\begin{aligned} (x + 5)^2 + (12 - x)^2 &= 13^2 \\ x^2 + 10x + 25 + 144 - 24x + x^2 &= 169 \\ 2x^2 - 14x + 169 &= 169 \\ 2x^2 - 14x &= 0 \\ 2x(x - 7) &= 0 \\ x = \cancel{0} \qquad \text{or} \qquad x - 7 &= 0 \\ x &= 7 \end{aligned}$$

If the ladder is pulled out 7 ft, the top moves down the same distance. ■

**EXAMPLE 3**

**INVESTMENT**

The amount of money $A$ that will result if a principal $P$ is invested at $r$ percent interest compounded annually for two years is given by $A = P(1 + r)^2$. Using this formula, what is the rate of interest if \$5000 grows to \$6272 in 2 years?

Let $r =$ the rate of interest. Substituting 6272 for $A$ and 5000 for $P$ in the interest formula gives the equation

$$\begin{aligned} 6272 &= 5000(1 + r)^2. \\ 1.2544 &= (1 + r)^2 \qquad \text{Divide both sides by 5000} \\ \pm\sqrt{1.2544} &= 1 + r \\ \pm 1.12 &= 1 + r \\ -1 \pm 1.12 &= r \\ -2.12,\ 0.12 &= r \end{aligned}$$

But since $r$ cannot be negative, we discard $-2.12$. Thus $r = 0.12 = 12\%$. ■

**EXAMPLE 4**

**CONSUMER**

A group of men are to share equally in the \$210 cost of painting a building. At the last minute, two men drop out. This raises the share of each remaining man \$28. How many men were in the group at the outset?

In a cost problem such as this, the fundamental identity is given by

$$\text{total cost} = (\text{cost per man}) \cdot (\text{number of men}).$$

Let $n =$ the number of men at the outset,
$c =$ the cost per man at the outset.

Then $\quad 210 = nc,$
$n - 2 =$ the number of men after 2 men drop out,
$c + 28 =$ the cost per man after 2 men drop out.

Thus $$210 = (n - 2)(c + 28).$$

Since we want to know the number of men, we need an equation in $n$. We solve each of the above equations for $c$ and set the results equal.

$$c = \frac{210}{n} \qquad c + 28 = \frac{210}{n - 2}$$
$$c = \frac{210}{n - 2} - 28$$

Thus, we must solve

$$\frac{210}{n} = \frac{210}{n - 2} - 28.$$

Multiply both sides by the LCD $= n(n - 2)$.

$$n(n - 2)\frac{210}{n} = \left[\frac{210}{n - 2} - 28\right]n(n - 2)$$
$$210(n - 2) = 210n - 28n(n - 2)$$
$$210n - 420 = 210n - 28n^2 + 56n$$
$$28n^2 - 56n - 420 = 0$$
$$n^2 - 2n - 15 = 0 \quad \text{Divide both sides by 28}$$
$$(n + 3)(n - 5) = 0$$
$$n + 3 = 0 \quad \text{or} \quad n - 5 = 0$$
$$n = -3 \qquad n = 5$$

There were 5 men in the group at the outset. ■

Recall from Section 2.2 that in a work problem, when A and B work together to complete a job, the equation used to solve the problem is:

**(amount done by A in 1 unit of time) +**
**(amount done by B in 1 unit of time) =**
**(amount done together in 1 unit of time)**

**EXAMPLE 5** When each works alone, Sam can do a job in 3 hours less time than Walt. When they work together, it takes them 2 hours to complete the job. How long does it take each to do the job by himself?

Let $\quad t =$ the number of hours required for Walt to do the job,
$t - 3 =$ the number of hours required for Sam to do the job,
$2 =$ the number of hours required to do the job working together,

$$\frac{1}{t} = \text{the amount done by Walt in 1 hour,}$$

$$\frac{1}{t-3} = \text{the amount done by Sam in 1 hour,}$$

$$\frac{1}{2} = \text{the amount done in 1 hour when working together.}$$

We must solve the following equation.

$$\frac{1}{t} + \frac{1}{t-3} = \frac{1}{2}$$

$$2t(t-3)\left[\frac{1}{t} + \frac{1}{t-3}\right] = 2t(t-3)\frac{1}{2} \qquad \text{The LCD} = 2t(t-3)$$

$$2t(t-3)\frac{1}{t} + 2t(t-3)\frac{1}{(t-3)} = 2t(t-3)\frac{1}{2}$$

$$2t - 6 + 2t = t^2 - 3t$$

$$0 = t^2 - 7t + 6$$

$$0 = (t-6)(t-1)$$

$$t - 6 = 0 \quad \text{or} \quad t - 1 = 0$$

$$t = 6 \qquad\qquad t = 1$$

But since $t - 3$ would be negative if $t = 1$, we discard 1 as a possible solution. Thus, Walt can do the job in 6 hours, and Sam can do it in 3 hours. ■

## EXAMPLE 6

**RATE-MOTION**

Two hikers leave the same campground at the same time, one hiking north and the other hiking east. If one is hiking 1 mph faster than the other, and if after 3 hours they are 15 miles apart, at what rate is each hiking?

Let $x$ = rate one is traveling,
$x + 1$ = rate the other is traveling.

Since they both hike for 3 hours,

$3x$ = distance one travels,
$3(x + 1)$ = distance the other travels.

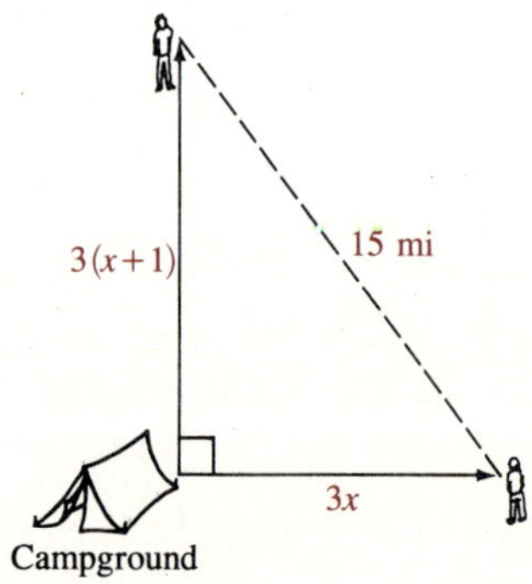


**Figure 6**

Make a sketch describing the problem as in Figure 6. Since the triangle is a right triangle, we use the Pythagorean theorem to write the equation involving $x$.

$$
\begin{aligned}
(3x)^2 + [3(x+1)]^2 &= 15^2 \\
9x^2 + 9x^2 + 18x + 9 &= 225 \\
18x^2 + 18x - 216 &= 0 \\
x^2 + x - 12 &= 0 \\
(x-3)(x+4) &= 0
\end{aligned}
$$

$$x - 3 = 0 \qquad \text{or} \qquad x + 4 = 0$$
$$x = 3 \qquad\qquad\qquad x = -4$$

We can discard $-4$ as a solution since a rate cannot be negative, giving 3 mph and 4 mph as the desired rates. ■

Other motion problems involve finding the height of an object above the ground at a given time after it has been dropped or thrown upward or downward. Let $h$ be the height, measured in feet, of an object above the ground. Let $t$ be the time in seconds following a starting time which is time $t = 0$. Let $v_0$ be the **initial velocity** of the object, measured in feet per second, where $v_0$ is positive if the object is thrown upward and negative if thrown downward at time $t = 0$. And let $h_0$ be the **initial height** of the object above ground level at time $t = 0$. Then the equation

$$\mathbf{h = -16t^2 + v_0 t + h_0}$$

gives the height in terms of the time $t \geq 0$. Notice that if the constants $v_0$ and $h_0$ are known, and if a particular height $h$ is given, we have a quadratic equation in the variable $t$.

**EXAMPLE 7**

**RATE-MOTION**

A rock is dropped from a cliff 420 feet high into a river at the base of the cliff (see Figure 7). How long will it take for the rock to hit the water (to the nearest tenth of a second)?

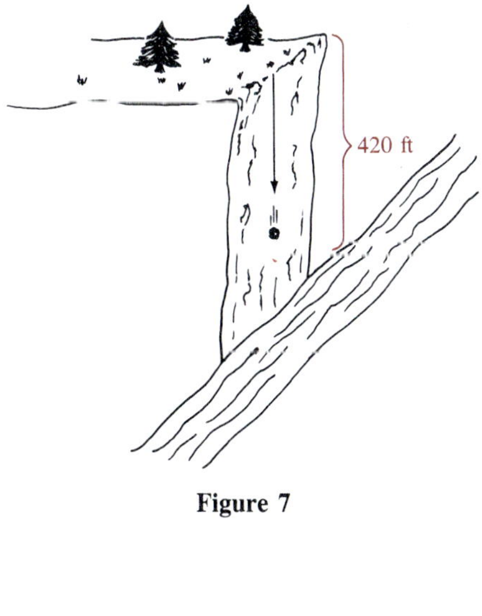


**Figure 7**

Since the initial height is $h_0 = 420$, the initial velocity is $v_0 = 0$ (because the rock is dropped, not thrown, downward), and $h = 0$ when the rock hits the water, we must solve

$$
\begin{aligned}
0 &= -16t^2 + 420. \\
16t^2 &= 420 \\
t^2 &= 26.25 \\
t &= \pm\sqrt{26.25} \approx \pm 5.1
\end{aligned}
$$

Since $t$ must be positive, we discard $-5.1$. Thus, the rock will hit the water in approximately 5.1 seconds. ■

**NOTE** Often the mathematical equation used as a model for a physical problem will have more solutions than are actually applicable. For instance, although $t = -5.1$ is a mathematical solution to the equation in Example 7, it is not a solution to the applied problem since negative time has no meaning in this case. Always check your work and make sure that your answers are meaningful. ■

## 2.5 EXERCISES

Solve.

1. One third the product of two consecutive positive odd integers is 65. Find the integers.
2. The square of 2 more than a number is the same as 4 less than 10 times the number. Find all numbers which satisfy this condition.
3. The principal square root of 4 more than a number is the same as the number less 8. Find the number.
4. A number decreased by 2 is the same as the principal square root of 4 more than the number. Find the number.
5. The sum of a number and its reciprocal is 5/2. Find the number.
6. What two numbers have the property that their sum is 15 and their product is 50?
7. **GEOMETRY** If the length of a rectangle is 11 inches more than the width and the area is 80 $\text{in}^2$, find its dimensions.
8. **GEOMETRY** A window is in the shape of an equilateral triangle with a height of 12 ft. To the nearest tenth of a foot, how long are the sides of the triangle?
9. A ladder 20 ft long leans against a wall. If the base of the ladder is 12 ft from the wall, how far would the lower end have to be pulled out so that the upper end is pulled down the same amount?
10. **GEOMETRY** If the sides of a square are lengthened by 5 yd, the area of the square will be 256 $\text{yd}^2$. Find the length of a side.
11. **MANUFACTURING** A box with no top and a square base is to be made from a square piece of cardboard by cutting out squares with 4-inch sides from each corner and folding up the sides. If the box must hold 400 $\text{in}^3$, what must be the length of the sides of the original piece of cardboard? See the figure.

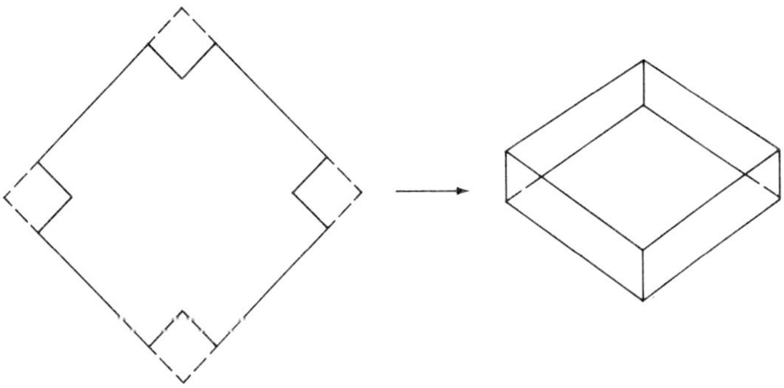

Exercise 11

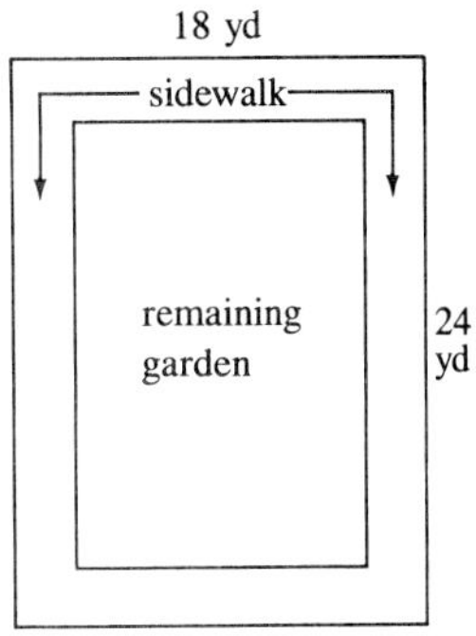


Exercise 12

**12.** AGRICULTURE A rectangular garden with dimensions 18 yd by 24 yd is to have a sidewalk of uniform width placed completely around the inside of its border in such a way that the area of the remaining garden is 216 $\text{yd}^2$. How wide is the sidewalk?

**13.** A piece of wire 100 inches long is cut into two pieces. If each piece is bent into the shape of a square and the combined area of the two squares is 337 $\text{in}^2$, how long is each piece of wire?

**14.** AGRICULTURE A farmer wishes to enclose a rectangular pasture which borders a river. The pasture is to have a length parallel to the river equal to twice the width, no fencing is needed along the side which borders the river, and the enclosed area is to be 39,200 $\text{yd}^2$. How many yards of fence are needed?

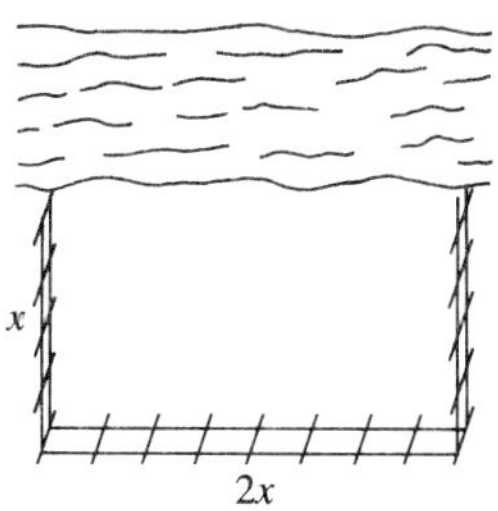


**15.** ECONOMICS Use the formula $A = P(1 + r)^2$ to find the rate of interest necessary for \$10,000 to grow to \$12,321 in two years, if the interest is compounded annually.

**16.** ECONOMICS Using the formula in Exercise 15, what is the rate of interest if \$8000 grows to \$10,305.80 in 2 years?

**17.** CONSUMER A group of women plan to share equally the \$180 cost of having a party. At the last minute, 3 women must be out of town, and this raises the share of each remaining woman by \$10. How many women were in the group at the outset?

**18.** RETAILING A store owner earns \$2000.00 a week on the sale of a particular type of novelty toy. By reducing the price \$2.00 per toy, he can generate more business and sell 50 more toys per week while still earning the same \$2000.00. At what price did he sell each toy originally?

19. **MANUFACTURING** A manufacturer has discovered that the total monthly cost $c$ of operating her plant is given by $c = 10x^2 - 100x - 2000$, where $x$ is the number of items made per month. How many items were produced during a month when the total costs amounted to \$10,000?

20. **MANUFACTURING** The weekly cost $c$ of producing $x$ items is given by the equation $c = 2x^2 + 50x - 30$. What number of items was produced in a week when the cost of production was \$270?

21. **ECONOMICS** The overhead cost for selling a product is \$165 per day. In order to sell $x$ items a day, the price per item must be adjusted to $26 - x$ dollars. In order to break even on a given day, the store owner must sell enough items to pay his overhead costs. What are the break-even points in this situation?

22. **BUSINESS** A wholesaler sells sweaters to a booster club at a rate of \$25 per sweater for all orders of 250 or less. If more than 250 are ordered (up to 300), the price per sweater is reduced at a rate of 5¢ times the total number ordered. If the club is billed \$3080 for an order, how many sweaters were purchased?

23. It takes Jim 9 hours longer to do a job than his father. If they work together, they can do the job in 20 hours. How long would it take each to do the job if he worked alone?

24. **RECREATION** When each valve is turned on separately, a larger pipe can fill a pool in 3 hours less time than a smaller pipe. Whey they are turned on simultaneously, it takes 2 hours to fill it. How long does it take each pipe separately?

25. **COMMUNICATIONS** Two printers can print mailing labels for an alumni group in 10 hours. If the newer printer can do the job in 5 hours less time than the older one, how many hours would it take the older printer to print the labels if the newer one were sent in for repair? (Give answer rounded to the nearest tenth.)

26. **CONSTRUCTION** It takes John 8 days longer to build a redwood deck than his father. If together they can build the deck in fourteen days, how long would it take John's father to build it by himself? (Give answer rounded to the nearest tenth.)

27. **RATE-MOTION** Two cars leave the same city at the same time traveling at right angles to each other. If the first travels 10 mph faster than the second, and if after two hours they are 100 miles apart, find the speed of each car.

28. **NAVIGATION** Two boats leave an island at 12:00 noon, one traveling due south and the other due west. If both are traveling at 30 knots, to the nearest tenth, how many nautical miles apart will they be in 2 hours?

29. **RATE-MOTION** Gary Heintz drove 400 miles at a certain rate in a certain time. If he had gone 10 mph faster, the time of the trip would have been shortened 2 hours. What was his speed?

30. **RATE-MOTION** Marguerite drove from Cleveland to Boston, a distance of 660 miles. She drove 6 mph faster on the return trip, cutting one hour off the trip. At what speed did she drive from Boston to Cleveland?

31. **RATE-MOTION** A child drops a coin from the top of the Empire State Building at a height of 1472 ft. To the nearest tenth of a second, how long will it take for the coin to reach the ground?

32. **RATE-MOTION** A rocket is fired upward from ground level with an initial velocity of 352 ft/sec. How many seconds will pass before it returns to the ground?

**33. RATE-MOTION** A child standing on a bridge 200 ft above the level of a river throws a rock downward with an initial velocity 20 ft/sec. How long will it take for the rock to hit the water? (Give answer rounded to the nearest hundredth of a second.)

**34. RATE-MOTION** To the nearest hundredth of a second, how long would it take for the rock to hit the water if the child on the bridge in Exercise 33 threw the rock upward with initial velocity 20 ft/sec? (Do the two answers to Exercises 33 and 34 seem reasonable?)

**35. RATE-MOTION** A man fires a gun from ground level, with a muzzle velocity of 2400 ft/sec, at a helicopter 3000 ft directly above him. How long (to the nearest hundredth of a second) will it take for the bullet to hit the helicopter?

**36. RATE-MOTION** A man drops an object from a balloon at an elevation of 256 ft. If the balloon is rising vertically at a velocity of 16 ft/sec, how long will it take for the object to hit the ground? (Give answer rounded to the nearest hundredth of a second.)

**37. PHYSICS** The pressure $P$, in pounds per square foot, of wind blowing $V$ miles per hour can be approximated by the equation $P = \frac{3V^2}{1000}$. If the pressure of the wind against the side of a building is 11.25 lb/ft$^2$, what is the approximate velocity of the wind?

**38. AERONAUTICS** The distance $d$ in miles from an object $h$ miles above the surface of the earth to the horizon can be approximated by $d = \sqrt{h(h + 8000)}$. If an airplane is flying such that its distance to the horizon is approximately 220 miles, approximately how high is the plane?

**39. ENGINEERING** When a driver hits the brakes on a car, the distance that the car travels before coming to a stop is called the braking distance. The braking distance $D$, in feet, of a car traveling at a rate of $V$ mph can be approximated by $D = \frac{V(V + 20)}{20}$. If a child runs into the street 100 feet in front of your car and you react instantaneously and hit the brakes, what is the maximum speed you can be traveling and still avoid hitting the child?

**40. GEOMETRY** It can be shown that a polygon with $m$ sides has $\frac{m^2 - 3m}{2}$ diagonals. If a polygon has 27 diagonals, how many sides does it have?

**41.** Prove that the sum of the two solutions to the equation $x^2 + bx + c = 0$ is $-b$.

**42.** Prove that the product of the two solutions to the equation $x^2 + bx + c = 0$ is $c$.

## For Review

Solve.

**43.** $\left(\frac{x - 1}{x + 2}\right)^2 + \frac{x - 1}{x + 2} = 6$

**44.** $x - \frac{20}{2x - 5} = \frac{5}{2x - 5}$

**45.** $\sqrt{x^2 + 1} + \sqrt{x + 1} = 0$

**46.** $\left(\frac{x^2 - 4}{x}\right) + 15\left(\frac{x}{x^2 - 4}\right) - 8 = 0$

$\left[\text{Hint: If } u = \frac{x^2 - 4}{x}, \text{ what is } \frac{1}{u}?\right]$

**47.** $7y^2 - 14y + 2 = 7y(y - 2) + 2$

**48.** $x^2 - 3x = 8$

Rationalize the denominator in Exercises 49–50. This type of process will be used in the next section on complex numbers.

**49.** $\dfrac{1}{5 + 4\sqrt{2}}$

**50.** $\dfrac{2 + 5\sqrt{5}}{1 - 3\sqrt{5}}$

## 2.6 Complex Numbers

### Imaginary Numbers

In the previous sections, we were careful to insure that the quadratic equations considered had real-number solutions. Suppose we attempt to use the quadratic formula to solve $x^2 - 2x + 2 = 0$.

$$x = \frac{-b \pm \sqrt{b^2 - 4ac}}{2a}$$

$$= \frac{-(-2) \pm \sqrt{(-2)^2 - 4(1)(2)}}{2(1)} \qquad a = 1,\ b = -2,\ c = 2$$

$$= \frac{2 \pm \sqrt{-4}}{2}$$

What number squared is $-4$? We know that for any real number $a$, $a^2 \geq 0$ and thus, there is no real number equal to $\sqrt{-4}$. As a result, certain simple equations (such as the one above) have no solution in the system of real numbers. To provide solutions for such equations and square roots for negative numbers, we introduce a new kind of number called an **imaginary number.**

If we agree to follow rules of radicals similar to those for real numbers, we need only one new numeral. We use $i$ as the numeral corresponding to $\sqrt{-1}$. That is,

$$\mathbf{i^2 = -1} \qquad \text{and} \qquad \mathbf{\sqrt{-1} = i.}$$

Every other imaginary number can be expressed as a product of a real number and $i$. For example, $\sqrt{-4} = \sqrt{4(-1)} = \sqrt{4}\sqrt{-1} = 2i$. As a result, the solutions to the equation

$$x^2 - 2x + 2 = 0$$

are

$$x = \frac{2 \pm \sqrt{-4}}{2} = \frac{2 \pm 2i}{2} = \frac{2(1 \pm i)}{2} = 1 \pm i.$$

**Definition of Complex Number**

A number of the form $a + bi$, where $a$ and $b$ are real numbers and $i^2 = -1$, is called a **complex number.** The number $a$ is called the **real part** and $b$ is the **imaginary part.** If $b = 0$, the complex number is simply the real number $a$, while if $a = 0$, it is the **imaginary number** $bi$.

In view of this definition, we see that the complex numbers include both the real numbers and the imaginary numbers. For example, the real number 7 could be represented as $7 + 0i$, and the imaginary number $-2i$ could be written as $0 - 2i$.

**EXAMPLE 1** Simplify.

**(a)** $\sqrt{-25} = \sqrt{25(-1)} = \sqrt{25}\sqrt{-1} = 5i$

**(b)**
$$\begin{aligned}\sqrt{-3}\sqrt{-7} &= \sqrt{3(-1)}\sqrt{7(-1)}\\ &= \sqrt{3}\sqrt{-1}\sqrt{7}\sqrt{-1}\\ &= \sqrt{3}\cdot i\cdot\sqrt{7}\cdot i\\ &= \sqrt{3}\sqrt{7}i^2\\ &= \sqrt{21}(-1) \qquad i^2 = -1\\ &= -\sqrt{21}\end{aligned}$$

**(c)** $\dfrac{\sqrt{-30}}{\sqrt{-5}} = \dfrac{\sqrt{30(-1)}}{\sqrt{5(-1)}} = \dfrac{\sqrt{30}\sqrt{-1}}{\sqrt{5}\sqrt{-1}} = \dfrac{\sqrt{30}\not{i}}{\sqrt{5}\not{i}} = \sqrt{\dfrac{30}{5}} = \sqrt{6}$

**(d)** $\sqrt{-36} + \sqrt{-100} = \sqrt{36}i + \sqrt{100}i = 6i + 10i = 16i$ ■

NOTE An important principle is illustrated in Example 1(**b**). Had we written $\sqrt{-3}\sqrt{-7} = \sqrt{(-3)(-7)} = \sqrt{21}$, we would not obtain the same result. It is important to express the square root of a negative number in the form $i\sqrt{b}$ before making simplifications. Recall that the rule $\sqrt{ab} = \sqrt{a}\sqrt{b}$ was given only for $a \geq 0$ and $b \geq 0$ so that $\sqrt{(-3)(-7)} = \sqrt{-3}\sqrt{-7}$ is incorrect. However, in the special case where $a \geq 0$ and $b = -1$, we can use a similar rule to express square roots of negative numbers in terms of the imaginary number $i$. ■

### Equality of Complex Numbers

Two complex numbers $a + bi$ and $c + di$ are **equal**,

$$a + bi = c + di,$$

if and only if their real parts are equal ($a = c$) and their imaginary parts are equal ($b = d$).

**EXAMPLE 2** **(a)** If $a + bi = 3 + \sqrt{2}i$, then $a = 3$ and $b = \sqrt{2}$.

**(b)** If $c + di = 5$, then $c = 5$ and $d = 0$ since $5 = 5 + 0i$.

**(c)** If $u + vi = -\sqrt{3}i$, then $u = 0$ and $v = -\sqrt{3}$ since $-\sqrt{3}i = 0 - \sqrt{3}i$.

**(d)** If $2x + yi = 3x - 2 + 7i$, then

$$\begin{aligned}2x &= 3x - 2 \quad \text{and} \quad y = 7\\ -x &= -2\\ x &= 2.\end{aligned}$$ ■

CAUTION When a radical appears as a coefficient of $i$ as in $\sqrt{2}i$ and $-\sqrt{3}i$ in Example 2, notice that $i$ is *not* under the radical. Since this is a common mistake, we often write $i\sqrt{2}$ and $-i\sqrt{3}$ to avoid the problem. ■

## Operations on Complex Numbers

### Addition and Subtraction of Complex Numbers

Let $a + bi$ and $c + di$ be two complex numbers. Their **sum** is

$$(a + bi) + (c + di) = (a + c) + (b + d)i,$$

and their **difference** is

$$(a + bi) - (c + di) = (a - c) + (b - d)i.$$

Notice that for addition (or subtraction), we add (or subtract) the real parts and then add (or subtract) the imaginary parts.

**EXAMPLE 3** Perform the indicated operations.

**(a)** $(4 + 6i) + (1 - 5i) = (4 + 1) + (6 - 5)i = 5 + i$

**(b)** $(2 - i\sqrt{5}) - (-7 + 3i\sqrt{5}) = (2 + 7) + (-\sqrt{5} - 3\sqrt{5})i$
$= 9 - 4i\sqrt{5}$ ■

### Multiplication of Complex Numbers

Let $a + bi$ and $c + di$ be two complex numbers. Their **product** is

$$(a + bi)(c + di) = (ac - bd) + (ad + bc)i.$$

Although the multiplication of complex numbers may seem strange at first, it becomes clearer when we multiply as if finding the product of binomials using the FOIL method.

$$\begin{aligned}(a + bi)(c + di) &= \overset{\text{(F)}}{ac} + \overset{\text{(O)}}{adi} + \overset{\text{(I)}}{bci} + \overset{\text{(L)}}{bdi^2}\\ &= ac + bd(-1) + adi + bci \qquad i^2 = -1\\ &= (ac - bd) + (ad + bc)i\end{aligned}$$

In fact, it might be best to use the FOIL method for finding products in specific cases, keeping in mind that $i^2 = -1$.

**EXAMPLE 4** Find the products.

**(a)** $$\begin{aligned}(3 + 6i)(2 - i) &= \overset{\text{(F)}}{6} - \overset{\text{(O)}}{3i} + \overset{\text{(I)}}{12i} - \overset{\text{(L)}}{6i^2}\\ &= 6 - 6i^2 - 3i + 12i\\ &= (6 + 6) + (-3 + 12)i\\ &= 12 + 9i\end{aligned}$$

**(b)** $(4 + i\sqrt{3})(4 - i\sqrt{3}) = 16 - 4i\sqrt{3} + 4i\sqrt{3} - 3i^2$

$$= 16 - 3i^2$$

$$= 16 + 3 = 19 \quad \blacksquare$$

A special product was illustrated in Example 4(**b**). The numbers $4 + i\sqrt{3}$ and $4 - i\sqrt{3}$ are called *conjugates*. In general, the **conjugate** of $a + bi$ is $a - bi$, and the conjugate of $a - bi$ is $a + bi$. When two conjugates are multiplied, the product is always a real number. One application of conjugates is in division.

### Division of Complex Numbers

Let $a + bi$ and $c + di$ be two complex numbers. The **quotient** of $a + bi$ and $c + di$ can be simplified by multiplying the numerator and denominator by the conjugate of the denominator. That is,

$$\frac{a + bi}{c + di} = \frac{(a + bi)(c - di)}{(c + di)(c - di)} = \frac{(ac + bd) + (bc - ad)i}{c^2 + d^2}$$

**EXAMPLE 5** Divide $2 + 5i$ by $1 - 3i$.

$$\frac{2 + 5i}{1 - 3i} = \frac{(2 + 5i)(\mathbf{1 + 3i})}{(1 - 3i)(\mathbf{1 + 3i})} \qquad 1 + 3i \text{ is the conjugate of } 1 - 3i$$

$$= \frac{2 + 6i + 5i + 15i^2}{1 + 3i - 3i - 9i^2}$$

$$= \frac{2 + 11i - 15}{1 + 9} \qquad i^2 = -1$$

$$= \frac{-13 + 11i}{10} = -\frac{13}{10} + \frac{11}{10}i \quad \blacksquare$$

**EXAMPLE 6** Find the reciprocal of $5 + 4i$ and express it in the form $a + bi$.

The reciprocal of $5 + 4i$ is $\frac{1}{5 + 4i}$. Thus, we must divide the complex number $1 = 1 + 0i$ by the complex number $5 + 4i$. In effect, since $i = \sqrt{-1}$, we are rationalizing the denominator.

$$\frac{1}{5 + 4i} = \frac{1(\mathbf{5 - 4i})}{(5 + 4i)(\mathbf{5 - 4i})}$$

$$= \frac{5 - 4i}{25 + 16}$$

$$= \frac{5 - 4i}{41} = \frac{5}{41} - \frac{4}{41}i \quad \blacksquare$$

## Solutions to Quadratic Equations

Now that we have defined the set of complex numbers, every quadratic equation can be solved. To determine the nature of the solutions to a given quadratic equation, suppose we look again at the quadratic formula

$$x = \frac{-b \pm \sqrt{b^2 - 4ac}}{2a}.$$

The number under the radical, $b^2 - 4ac$, called the **discriminant,** can be used to determine the types of solutions to a quadratic equation without actually solving it.

### Types of Solutions to a Quadratic Equation

The quadratic equation $ax^2 + bx + c = 0$ has

**1.** two real-number solutions if the discriminant is positive;

**2.** exactly one real-number solution (sometimes considered as two real but equal solutions) if the discriminant is zero;

**3.** two complex-number solutions if the discriminant is negative.

## EXAMPLE 7

Without solving, determine the nature of the solutions.

**(a)** $2x^2 - 3x - 7 = 0$

$$\begin{aligned} b^2 - 4ac &= (\mathbf{-3})^2 - 4(\mathbf{2})(\mathbf{-7}) \qquad a = 2,\ b = -3,\ c = -7 \\ &= 9 + 56 = 65 > 0 \end{aligned}$$

There are two real solutions.

**(b)** $2x^2 - 3x + 7 = 0$

$$\begin{aligned} b^2 - 4ac &= (\mathbf{-3})^2 - 4(\mathbf{2})(\mathbf{7}) \qquad a = 2,\ b = -3,\ c = 7 \\ &= 9 - 56 = -47 < 0 \end{aligned}$$

There are two complex solutions.

**(c)** $4x^2 - 12x + 9 = 0$

$$\begin{aligned} b^2 - 4ac &= (\mathbf{-12})^2 - 4(\mathbf{4})(\mathbf{9}) \qquad a = 4,\ b = -12,\ c = 9 \\ &= 144 - 144 = 0 \end{aligned}$$

There is one real solution. ■

Some equations, although not quadratic themselves, can be factored and solved using the zero-product rule. A more detailed discussion of these types of equations is given in Chapter 4, but we now consider a special case.

## EXAMPLE 8

Solve. $x^3 - 8 = 0$

Factoring using the difference-of-cubes formula we obtain

$$(x - 2)(x^2 + 2x + 4) = 0.$$

$$x - 2 = 0 \quad \text{or} \quad x^2 + 2x + 4 = 0 \quad \text{Zero-product rule}$$

$$x = 2 \qquad x = \frac{-2 \pm \sqrt{4 - 4(1)(4)}}{2(1)}$$

$$= \frac{-2 \pm \sqrt{-12}}{2}$$

$$= \frac{-2 \pm 2i\sqrt{3}}{2}$$

$$= -1 \pm i\sqrt{3}$$

The three numbers 2, $-1 + i\sqrt{3}$, and $-1 - i\sqrt{3}$ are all cube roots of 8. Prior to this section, we were only aware of the real-number cube root, 2. ■

An interesting property of the solutions to a quadratic equation involves their sum and product. Suppose that

$$x_1 = \frac{-b + \sqrt{b^2 - 4ac}}{2a} \quad \text{and} \quad x_2 = \frac{-b - \sqrt{b^2 - 4ac}}{2a}$$

are the two solutions to the quadratic equation $ax^2 + bx + c = 0$. Then

$$x_1 + x_2 = \frac{-b + \sqrt{b^2 - 4ac}}{2a} + \frac{-b - \sqrt{b^2 - 4ac}}{2a}$$

$$= \frac{-b + \sqrt{b^2 - 4ac} - b - \sqrt{b^2 - 4ac}}{2a}$$

$$= \frac{-2b}{2a} = -\frac{b}{a}.$$

We have just proved the first part of the following theorem. The proof of the second part is left as an exercise.

If $x_1$ and $x_2$ are solutions to $ax^2 + bx + c = 0$, then

$$x_1 + x_2 = -\frac{b}{a} \quad \text{and} \quad x_1x_2 = \frac{c}{a}.$$

**EXAMPLE 9** Find the sum and product of the solutions to $3x^2 - 7x + 8 = 0$.

Since $a = 3$, $b = -7$, and $c = 8$, the sum of the solutions is $-\frac{b}{a} = -\frac{-7}{3} = \frac{7}{3}$, and their product is $\frac{c}{a} = \frac{8}{3}$. ■

The converse of the preceding theorem is also true and can be used to determine quickly whether two numbers are solutions to a given quadratic equation.

**EXAMPLE 10** Determine whether $5 + i$ and $5 - i$ are solutions to the equation $2x^2 - 20x - 52 = 0$.

Since $x_1 + x_2 = (5 + i) + (5 - i) = 10$ and $-\frac{b}{a} = -\frac{-20}{2} = 10$, these numbers pass the first part of the test. However, $x_1x_2 = (5 + i)(5 - i) = 26$, but $\frac{c}{a} = \frac{-52}{2} = -26$. Thus, $x_1x_2 \neq \frac{c}{a}$ so $5 + i$ and $5 - i$ cannot be solutions to the equation. ■

## 2.6 EXERCISES

In Exercises 1–12, simplify and express each result as a real number or in terms of $i$.

**1.** $\sqrt{-8}$ **2.** $\sqrt{-20}$ **3.** $\sqrt{-7}\sqrt{-5}$ **4.** $-\sqrt{-27}\sqrt{-3}$

**5.** $\frac{\sqrt{-35}}{\sqrt{-7}}$ **6.** $\frac{\sqrt{-49}}{\sqrt{-7}}$ **7.** $\frac{-\sqrt{-81}}{9}$ **8.** $\frac{-\sqrt{-121}}{11}$

**9.** $\sqrt{-4 - 25}$ **10.** $\sqrt{-9 - 36}$ **11.** $\sqrt{-4} - \sqrt{-25}$ **12.** $\sqrt{-9} - \sqrt{-36}$

Use the definition of equality of complex numbers in Exercises 13–16 to determine $x$ and $y$.

**13.** $(2x + 1) + 4yi = 3$ **14.** $5x + (3y + 2)i = -i$

**15.** $x + yi = \sqrt{4} + \sqrt{-4}$ **16.** $x + yi = \sqrt{9} - \sqrt{-9}$

Perform the indicated operations in Exercises 17–25.

**17.** $(3 + 2i) + (-1 + 5i)$ **18.** $(-1 - i) - (8 - 3i)$ **19.** $(2 - 5i) + 7i$

**20.** $(1 + 2i)(3 - 4i)$ **21.** $(2 - i)(-3 + 4i)$ **22.** $3i(2 - 4i)$

**23.** $5(2 - 3i)$ **24.** $(-3 - 4i)^2$ **25.** $(-3 - 4i)^3$

Give the conjugate of each complex number in Exercises 26–29.

**26.** $5 - 7i$ **27.** $-8i$ **28.** $7$ **29.** $-3 + i\sqrt{5}$

Find the quotients in Exercises 30–33.

**30.** $\frac{2 + 3i}{1 - i}$ **31.** $\frac{-3 + 5i}{2 + 3i}$ **32.** $\frac{1 + i}{(1 - i)^2}$ **33.** $\frac{2 - 3i}{(1 + i)(3 - 2i)}$

Find the reciprocals of the numbers in Exercises 34–37.

**34.** $2 + i$ **35.** $-3i$ **36.** $-7$ **37.** $\frac{2 + i}{3 - i}$

Solve each equation in Exercises 38–49.

**38.** $x^2 - x + 1 = 0$ **39.** $3y^2 + 2y = -1$ **40.** $2z^2 - z + 7 = 0$

**41.** $x^4 - x^2 - 12 = 0$ **42.** $(y^2 - 1)^2 + 7(y^2 - 1) + 10 = 0$ **43.** $1 + \frac{3}{z^2 - 3z} = \frac{-1}{z^2 - 3z}$

**44.** $x^3 + 8 = 0$ **45.** $y^3 - 1 = 0$ **46.** $z^3 + 1 = 0$

**47.** $x^3 + 4x = 0$ **48.** $y^3 - 4y^2 + 5y = 0$ **49.** $z^4 - 1 = 0$

In Exercises 50–52, use the discriminant to determine the nature of the solutions to the equations (two real, one real, or two complex).

**50.** $3x^2 = 5x - 8$ **51.** $x^2 - 10x + 25 = 0$ **52.** $-5x^2 + 2x + 1 = 0$

Without solving, find the sum and product of the solutions to the equations in Exercises 53–55.

**53.** $x^2 - 10x = -25$ **54.** $-5x^2 + 1 = -2x$ **55.** $\frac{1}{2}x^2 - \frac{1}{3} = x$

Without solving, use the formula for the sum and product of solutions to determine whether $x_1$ and $x_2$ are solutions to the quadratic equations in Exercises 56–59.

**56.** $x^2 - 4x - 12 = 0;\quad x_1 = -2,\ x_2 = 6$ **57.** $x^2 - 3x + 2 = 0;\quad x_1 = 2,\ x_2 = 1$

**58.** $3x^2 + 8x - 35 = 0;\quad x_1 = \frac{4}{3},\ x_2 = -4$ **59.** $4x^2 + 11x - 3 = 0;\quad x_1 = 4,\ x_2 = -\frac{1}{4}$

**60.** Powers of the imaginary number $i$ occurs in cycles of four. For example, $i^1 = i$, $i^2 = -1$, $i^3 = i(-1) = -i$, $i^4 = i^2 \cdot i^2 = (-1)(-1) = 1$, $i^5 = i^4 \cdot i = (1)i = i$, $i^6 = i^4 \cdot i^2 = (1)(-1) = -1$, and so forth. If $n$ is a whole number, prove that $i^{4n} = 1$, $i^{4n+1} = i$, $i^{4n+2} = -1$, and $i^{4n+3} = -i$.

**61.** Use the results of Exercise 60 to find the following powers of $i$.
**(a)** $i^{10}$ **(b)** $i^{15}$ **(c)** $i^{28}$ **(d)** $i^{73}$

**62.** Prove that the product of any complex number and its conjugate is always a positive real number.

**63.** The absolute value of a complex number is defined by $|a + bi| = \sqrt{a^2 + b^2}$. Use this definition to evaluate the following.
**(a)** $|3 + 4i|$ **(b)** $|1 - i|$ **(c)** $|-9i|$

If $x = a + bi$ is a complex number, we often denote the conjugate of $x$ by $\bar{x} = a - bi$. Prove that the following are true for complex numbers $x$ and $y$.

**64.** $\overline{x + y} = \bar{x} + \bar{y}$ **65.** $\overline{x - y} = \bar{x} - \bar{y}$ **66.** $|x|^2 = x \cdot \bar{x}$ **67.** $\overline{x \cdot y} = \bar{x} \cdot \bar{y}$

**68.** Let $a$ be a real number. Show that $|a|$, thinking of $a$ as a real number, is the same as $|a|$ thinking of $a$ as the complex number $a + 0i$.

**69.** Prove that a quadratic equation $ax^2 + bx + c = 0$, where $a$, $b$, and $c$ are integers, has a single rational-number solution if the discriminant is 0.

**70.** Prove that a quadratic equation $ax^2 + bx + c = 0$, where $a$, $b$, and $c$ are integers, has two rational-number solutions if the discriminant is a positive perfect square.

**71.** It can be shown that $ax^2 + bx + c$, with $a$, $b$, and $c$ integers, is factorable (with integer coefficient factors) whenever $ax^2 + bx + c = 0$ has rational-number solutions. Use the results of Exercises 69 and 70 to determine whether the following can be factored.
**(a)** $x^2 + x - 56$ **(b)** $7x^2 - 3x + 9$ **(c)** $36x^2 - 60x + 25$

**72.** Prove that if $x_1$ and $x_2$ are solutions to the quadratic equation $ax^2 + bx + c = 0$, then $x_1x_2 = \frac{c}{a}$.

## For Review

**73.** GEOMETRY The area of a triangle is 22 $\text{cm}^2$. If the altitude is 3 cm more than twice the base, find the base and altitude.

**74.** CONSUMER A man pays \$300 for a number of shares of stock. If each share had cost \$3 less, he could have bought 5 more shares for the same \$300. How many shares did he buy?

**75.** Fred can paint a house in 1 day less than it takes Charlie. If together they can paint it in 3 days, to the nearest tenth of a day, how long would it take Fred to do the job by himself?

**76.** RATE-MOTION A child fires a toy rocket from the ground into the air with an initial velocity of 128 ft/sec.

**(a)** How many seconds later will the rocket return to the ground?

**(b)** After how many seconds will the rocket reach a height of 240 ft? Why are there two answers to this problem?

**(c)** After how many seconds will the rocket reach a height of 256 ft? Why is there only one answer to this problem?

**(d)** After how many seconds will the rocket reach a height of 300 ft? Explain.

Solve.

**77.** $(a^2 - 2)^2 - 6(a^2 - 2) - 7 = 0$

**78.** $y^4 - 2y^2 - 63 = 0$

## 2.7 Linear and Absolute Value Inequalities

### Linear Inequalities

In Chapter 1 we introduced the inequality symbols $<$, $>$, $\leq$, and $\geq$. Recall that if $a$ and $b$ are real numbers, $a < b$ means that $b - a$ is a positive number. Statements such as

$$x + 2 > 7 \qquad 4x \leq 9 \qquad 2x + 1 \geq 3x - 5 \qquad 2(x + 1) < 3(x + 2) + 5,$$

in which $x$ represents a real number, are called **linear inequalities.** We now consider several properties of inequalities which will be used to solve linear inequalities in a manner similar to solving linear equations.

If $x$ is replaced with a real number which makes an inequality true, that number is called a **solution** to the inequality. To solve an inequality is to find all of its solutions. As was the case with solving equations, we transform inequalities into **equivalent inequalities** which have the same solutions, ending with an inequality having the variable isolated on one side. The next theorem summarizes four important properties which are used in the solution process.

**Properties of Inequalities**

Let $a$, $b$, and $c$ be real numbers.

1. If $a < b$, then $a + c < b + c$.
2. If $a < b$, then $a - c < b - c$.
3. If $a < b$ and $c > 0$, then $ac < bc$.
4. If $a < b$ and $c < 0$, then $ac > bc$.

PROOF OF 1: Since $a < b$, by definition $b - a$ is positive. Since $(b + c) - (a + c) = b - a$, $(b + c) - (a + c)$ is positive. Thus, $a + c < b + c$.

PROOF OF 2: Since $a < b$, using part 1, $a + (-c) < b + (-c)$, or, equivalently, $a - c < b - c$.

PROOF OF 3: Since $a < b$, $b - a$ is positive. With $c$ also positive, the product $(b - a)c$ is positive. Then $bc - ac$ is positive so that $ac < bc$.

PROOF OF 4: Since $a < b$, $b - a$ is positive. With $c$ negative, we know that $-c$ is positive. Then the product $(b - a)(-c)$ is positive. Thus $ac - bc$ is positive so that $bc < ac$, or, equivalently, $ac > bc$. ■

NOTE Similar results to those in the preceding theorem are also true for the inequalities $>$, $\leq$, and $\geq$. The only difference between these properties and the corresponding ones for equations occurs when an inequality is multiplied (or divided) on both sides by a negative number. In this case *always remember to reverse the sense of the inequality*. ■

**EXAMPLE 1** Solve. $3x - 1 < 5x - 7$

| | |
|---|---|
| $3x - \mathbf{5x} - 1 < 5x - \mathbf{5x} - 7$ | Subtract $5x$ from both sides |
| $-2x - 1 < -7$ | |
| $-2x - 1 + \mathbf{1} < -7 + \mathbf{1}$ | Add 1 to both sides |
| $-2x < -6$ | |
| $\left(-\frac{1}{2}\right)(-2x) > \left(-\frac{1}{2}\right)(-6)$ | Multiply both sides by $-1/2$ and reverse the inequality |
| $x > 3$ | |

Thus, the solutions are all real numbers greater than 3. ■

NOTE Usually we do not state the solution in a sentence as in Example 1. We simply give the solution as $x > 3$. ■

**EXAMPLE 2** Solve. $y - 3(2 + y) \geq 2(3y - 2) + 2$

| | |
|---|---|
| $y - 6 - 3y \geq 6y - 4 + 2$ | Clear parentheses |
| $-2y - 6 \geq 6y - 2$ | Collect like terms |
| $-8y - 6 \geq -2$ | Subtract $6y$ from both sides |
| $-8y \geq 4$ | Add 6 to both sides |
| $y \leq -\frac{1}{2}$ | Divide both sides by $-8$ and reverse the inequality ■ |

As is the case with equations, some inequalities are **identities** that have every real number as a solution, and others are **contradictions** that have no solution.

**EXAMPLE 3** Solve.

**(a)** $3(x + 2) < 5 + 3x$

$3x + 6 < 5 + 3x$ Clear parentheses

$6 < 5$ Subtract $3x$ from both sides

Since $6 < 5$ is a false inequality equivalent to the original inequality, the original inequality is also false regardless of the choice of replacement for the variable $x$. When this occurs, the inequality has no solution and is a contradiction.

**(b)** $2(x + 5) + x \geq 3(x + 1)$

$2x + 10 + x \geq 3x + 3$

$3x + 10 \geq 3x + 3$

$10 \geq 3$ Subtract $3x$ from both sides

Since $10 \geq 3$ is a true inequality equivalent to the original inequality, the original inequality is also true regardless of the choice of replacement for the variable $x$. When this happens, the inequality has every real number for a solution and is an identity. ■

Some applied problems result in linear inequalities. We now solve the second problem presented in the introduction to this chapter.

**EXAMPLE 4**
BUSINESS

In the manufacture and sale of record albums, the revenue made on the sale of $x$ albums is \$2.20$x$, and the cost of producing $x$ albums is \$1.30$x$ + \$4500. In order to make a profit, the revenue received must be greater than the costs of production. For what values of $x$ will a profit be returned?

Since the revenue received must be greater than the cost of production, we must solve the following inequality.

$$2.20x > 1.30x + 4500$$
$$0.9x > 4500$$
$$x > 5000$$

Thus, a profit will be made when more than 5000 albums are sold. ■

## Graphing Inequalities

To graph an equation, we plot the point or points on a number line which correspond to solutions of the equation. For example, to graph $2x - 1 = 3$, we first solve the equation, obtaining $x = 2$, and then graph the solution as in Figure 8.

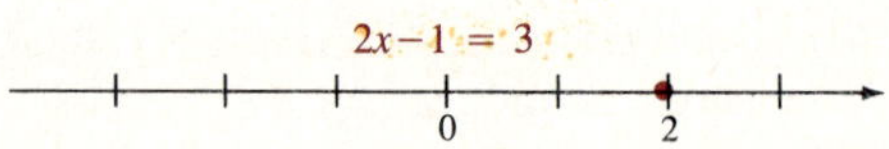


**Figure 8**

Graphing inequalities is somewhat more interesting. For example, to graph $3(x - 2) > 2x - 4$, we solve the inequality, obtaining $x > 2$, and graph the result as shown in Figure 9. The small circle at the left end of the colored arrow indicates that the number 2 is not included while all points to the right of 2 are included. Figure 10 is the graph of $x < 0$ or $0 > x$. The graph of $x \geq -2$ or $-2 \leq x$ is given in Figure 11. The circle at $-2$ is solid to indicate that the number $-2$ is included in the graph. Finally, the graph of $x \leq 0$ is given in Figure 12.

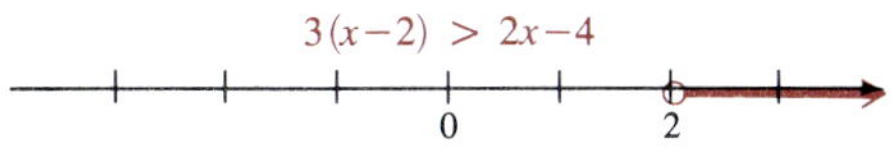


**Figure 9**

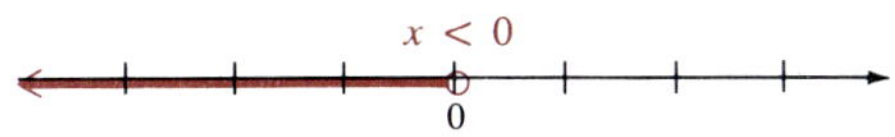


**Figure 10**

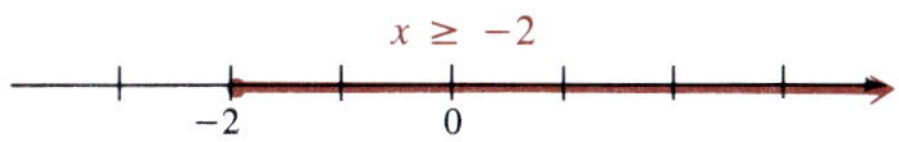


**Figure 11**

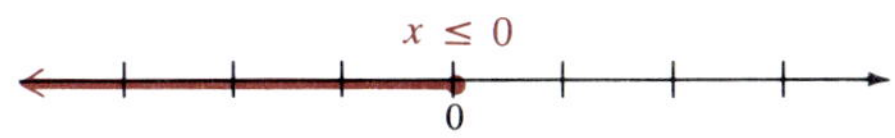


**Figure 12**

## Compound Inequalities

Sometimes two inequalities are combined or joined together to form **compound inequalities** by using the connectives *and* or *or*. For example, the compound inequality

$$x > -3 \text{ and } x < 1$$

could be given by

$$-3 < x \qquad \text{and} \qquad x < 1$$

or by the single chain of inequalities

$$-3 < x < 1.$$

The numbers which satisfy this chain of inequalities are between $-3$ and 1, and the graph is shown in Figure 13.

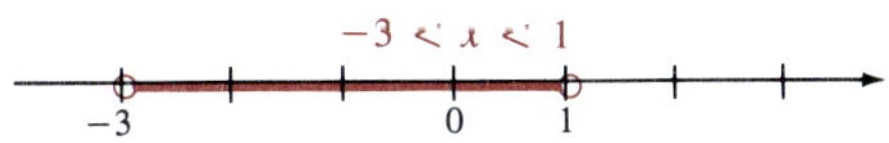


**Figure 13**

## Intervals

Another way to express a compound inequality is to write it in *interval notation.*

If $a$ and $b$ are real numbers with $a < b$, the symbol $(a, b)$, defined by

$$(a, b) = \{x \text{ such that } a < x < b\},$$

is called an **open interval** with **endpoints** $a$ and $b$.

The open interval $(-3, 1)$ is graphed in Figure 13. Notice that the endpoints $-3$ and 1 are not part of the interval. If an endpoint is to be included in an interval, brackets are used instead of parentheses to form a **closed interval,**

$$[a, b] = \{x \text{ such that } a \leq x \leq b\}.$$

**Half-open intervals** are defined similarly.

$$(a, b] = \{x \text{ such that } a < x \le b\} \text{ and}$$
$$[a, b) = \{x \text{ such that } a \le x < b\}.$$

The closed interval $[-2, 3]$ is shown in Figure 14 and the half-open interval $[-1, 2)$ is shown in Figure 15. Notice that brackets and parentheses can be used in the graphs to replace solid circles and open circles, respectively, at the endpoints.

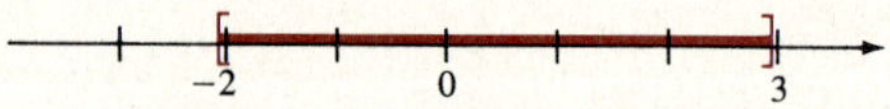


**Figure 14 Closed Interval [−2, 3]**

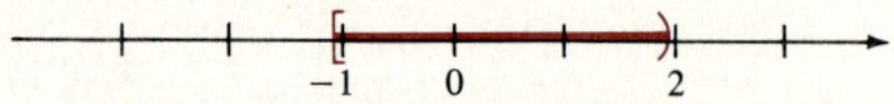


**Figure 15 Half-Open Interval [−1, 2)**

It is sometimes convenient to consider another type of interval, an **infinite interval,** which can be used to describe inequalities such as

$$x < 2, \qquad x > -1, \qquad x \le 3, \qquad \text{and} \qquad x \ge -4.$$

We define

$$(-\infty, a) = \{x \text{ such that } x < a\},$$

for $a$, a real number. The symbol $-\infty$, read "minus infinity," does not represent a real number. It is used simply to denote that all numbers less than $a$ are included in $(-\infty, a)$. In a similar manner, if $a$ is a real number,

$$(-\infty, a] = \{x \text{ such that } x \le a\}$$
$$(a, \infty) = \{x \text{ such that } x > a\}$$
$$[a, \infty) = \{x \text{ such that } x \ge a\}.$$

Using this notation, the four inequalities given above can be represented by the infinite intervals $(-\infty, 2)$, $(-1, \infty)$, $(-\infty, 3]$, and $[-4, \infty)$, respectively. The entire set of real numbers could be written as the interval $(-\infty, \infty)$.

**EXAMPLE 5** Solve and graph. $-1 < 2x + 1 \le 3$

$$-1-1 < 2x + 1-1 \le 3-1 \qquad \text{Subtract 1 throughout each portion}$$
$$-2 < 2x \le 2$$
$$\frac{1}{2}(-2) < \frac{1}{2}(2x) \le \frac{1}{2}(2) \qquad \text{Multiply throughout by 1/2}$$
$$-1 < x \le 1$$

In interval notation, the solution is $(-1, 1]$, and its graph is given in Figure 16. ■

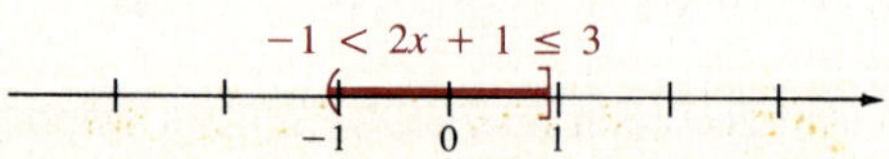


**Figure 16**

The second important type of compound inequality involves the connective *or*, for example,

$$x < -1 \quad or \quad x > 2.$$

Do *not* form a single chain when the word *or* is used. Notice that $2 < x < -1$ is a meaningless expression (2 is not less than $-1$). The graph of an *or* combination is usually two segments of a number line. For example, Figure 17 shows the graph of $x < -1$ or $x > 2$, which in interval notation is $(-\infty, -1)$ or $(2, \infty)$.

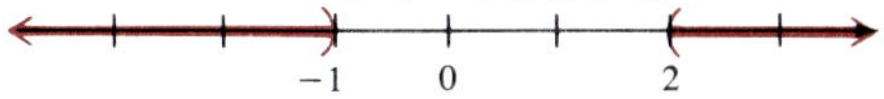


**Figure 17**

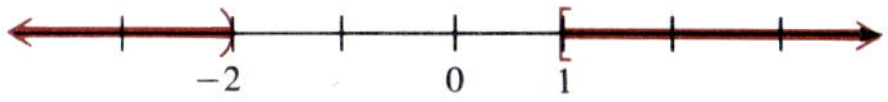


**Figure 18**

Similarly, $x < -2$ or $x \geq 1$, in interval notation $(-\infty, -2)$ or $[1, \infty)$, is graphed in Figure 18.

**EXAMPLE 6** Solve and graph.

$$
\begin{array}{rcrl}
3 - 4x < -1 & \text{or} & 3 - 4x \geq 9 & \\
-4x < -4 & \text{or} & -4x \geq 6 & \text{Subtract 3} \\
x > 1 & \text{or} & x \leq -\dfrac{6}{4} & \text{Reverse} \\
 & & x \leq -\dfrac{3}{2} &
\end{array}
$$

The solution is $x > 1$ or $x \leq -\dfrac{3}{2}$, graphed in Figure 19. Using intervals, the solution is $(1, \infty)$ or $\left(-\infty, -\dfrac{3}{2}\right]$. ■

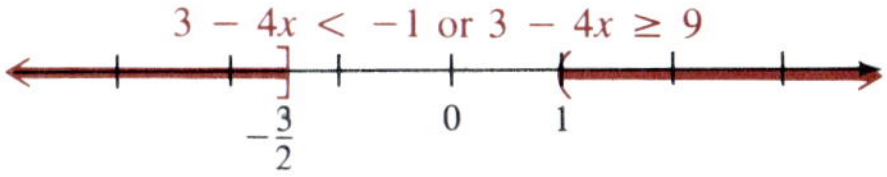


**Figure 19**

NOTE Remember that an *and* combination (such as $-4 < x$ and $x \leq 3$) can often be expressed as a single chain of inequalities ($-4 < x \leq 3$) and generally has as its graph a single segment of a number line. An *or* combination (such as $x < -1$ or $x > 2$) cannot be expressed as a single chain of inequalities (*never write* $2 < x < -1$) and generally has as its graph two segments of a number line. ■

## Absolute Value Inequalities

An inequality which involves absolute value, called an **absolute value inequality,** can be solved only by translating it into an equivalent compound inequality. The method of translating depends on the following theorem.

Let $a$ be a real number, $a > 0$.

**1.** $|x| < a$ is equivalent to $-a < x < a$.

**2.** $|x| > a$ is equivalent to $x < -a$ or $x > a$.

Intuitively, since $|x|$ can be interpreted geometrically as the distance from $x$ to 0 on a number line, if $|x| < a$, then $x$ must be *inside* the interval $(-a, a)$ which means $-a < x < a$. Also, if $|x| > a$, then $x$ must be *outside* the interval $(-a, a)$ which means $x < -a$ or $x > a$. Similar results hold for $|x| \leq a$ and $|x| \geq a$.

**EXAMPLE 7** Solve the following absolute value inequalities.

**(a)** $|x| < 5$
Translated to the equivalent compound inequality, the solution is $-5 < x < 5$, or in interval notation, $(-5, 5)$.

**(b)** $|x| \geq 5$
Translating we obtain the solution $x \leq -5$ or $x \geq 5$; equivalently, $(-\infty, -5]$ or $[5, \infty)$.

**(c)** $|x| < 0$
Since $|x|$ is never negative, this inequality has no solution.

**(d)** $|x| \leq 0$
The only solution is $x = 0$ since $|x|$ cannot be negative.

**(e)** $|x| < -5$
Since $|x| \geq 0$ for every real number $x$, there is no solution. ■

**EXAMPLE 8** Solve.

**(a)** $|2 - 3x| < 11$

| | |
|---|---|
| $-11 < 2 - 3x < 11$ | Equivalent compound inequality |
| $-13 < -3x < 9$ | Subtract 2 throughout |
| $\frac{13}{3} > x > -3$ | Divide through by $-3$ and reverse the inequalities |

Thus, the solution is $-3 < x < \frac{13}{3}$ or $\left(-3, \frac{13}{3}\right)$.

**(b)** $|2x - 1| \geq 5$

| | | | |
|---|---|---|---|
| $2x - 1 \leq -5$ | or | $2x - 1 \geq 5$ | Equivalent compound inequality |
| $2x \leq -4$ | or | $2x \geq 6$ | |
| $x \leq -2$ | or | $x \geq 3$ | |

Thus, the solution is $x \leq -2$ or $x \geq 3$, or $(-\infty, -2]$ or $[3, \infty)$. Do not write $3 \leq x \leq -2$ for the answer! ■

## 2.7 EXERCISES

Solve and graph each inequality in Exercises 1–24. Give answers using both inequalities and intervals.

**1.** $4x - 2 \le 6$

**2.** $3x - 1 \ge 5$

**3.** $2y + 5 < 4y - 9$

**4.** $5(y + 2) - 3 > 3(y - 1)$

**5.** $(z - 1)^2 < z(z + 2)$

**6.** $(x + 2)^2 \ge x(x - 2)$

**7.** $(4y + 1)(y - 3) \ge (2y + 1)(2y - 1)$

**8.** $(z - 1)(z + 2) > (z + 1)^2$

**9.** $\dfrac{3}{2y + 4} > 0$
[Hint: The denominator must be positive.]

**10.** $\dfrac{3}{2y + 4} < 0$

**11.** $(z + 4)^{-1} > 0$

**12.** $(x + 4)^{-1} < 0$

**13.** $\dfrac{y + 1}{3} \le \dfrac{y + 1}{2}$

**14.** $\dfrac{2(z + 2)}{5} < \dfrac{3(z + 2)}{7}$

**15.** $4 \le 2x - 6 < 10$

**16.** $3 \le 5y - 2 \le 18$

**17.** $0 \le \dfrac{z + 1}{2} < 1$

**18.** $0 \le \dfrac{3(x + 1)}{6} < 2$

**19.** $2y - 6 < 4$ or $2y - 6 \ge 10$

**20.** $5z - 2 \le 3$ or $5z - 2 > 18$

**21.** $2z + 1 > z$ or $z + 3 < 0$

**22.** $3 - x > x + 3$ or $1 - x \le x - 3$

**23.** $y < 2y + 1 < 1 - y$
[Hint: Write as two inequalities.]

**24.** $z - 2 < 2z + 1 \le 2 + z$

In Exercises 25–27 determine the values of $x$ that make the given expression positive.

**25.** $3(x + 5)$

**26.** $-3(x + 5)$

**27.** $(x + 5)^2$

In Exercises 28–30 determine the values of $x$ that make the given expression negative.

**28.** $3(x + 5)$

**29.** $-3(x + 5)$

**30.** $-(x + 5)^2$

Solve.

**31.** SCIENCE Fahrenheit ($F$) and Celsius ($C$) temperatures are related by $C = \frac{5}{9}(F - 32)$. If the Celsius temperature is between −15°C and 30°C, inclusive, what is the range of Fahrenheit temperatures?

**32.** EDUCATION In order to receive a grade of C in a course, a student must have an average mark between 70% and 80%, inclusive. If Jonas has 68%, 83%, and 79% on the first three tests this semester, what range of scores on the fourth test would give him a C?

**33.** EDUCATION Burford is in the same class as Jonas (see Exercise 32). Burford has 45%, 32% and 58% on the first three tests. What range of scores on the fourth test would give him a C?

**34.** BUSINESS To make a profit on the sale of video recorders, the total income on sales, $S$, must exceed the total costs involved, $C$. If $x$ represents the number of recorders sold, $S = 200x$, and $C = 150x + 300$, what is the smallest number of recorders that must be sold to make a profit?

**35.** ENGINEERING The power measured in watts $W$ on an electrical circuit is related to the pressure in volts $E$ and the current in amperes $I$ by the equation $W = EI$. If the power demands on a 110-volt circuit in a garage vary between 220 watts and 2310 watts, what is the range of current in the circuit?

**36.** ENGINEERING The force in pounds $F$ required to stretch a spring $x$ inches beyond its normal length is given by $F = 5.5x$. What are the corresponding values for $x$ if $F$ is between 2.75 pounds and 6.60 pounds, inclusive?

Solve each inequality in Exercises 37–51. Give answers using both inequalities and intervals.

**37.** $|x| < 7$ **38.** $|y| > 7$ **39.** $|z| \leq 0$

**40.** $|x| < 0$ **41.** $|y| > 0$ **42.** $|z| \geq 0$

**43.** $|x + 2| > 3$ **44.** $|y - 2| \leq 5$ **45.** $|1 - 5z| < 6$

**46.** $\left|\frac{x + 5}{2}\right| \leq 10$ **47.** $\left|\frac{y + 7}{3}\right| > 3$ **48.** $|2(z + 3) - 1| < 9$

**49.** $|2x - 7| < -1$ **50.** $|2y - 7| \geq 1$ **51.** $|2z - 7| \leq 0$

Use the discriminant to find all values of $m$ in Exercises 52–55 that make the quadratic equation satisfy the given condition.

**52.** $4x^2 - 4x + m = 0$, two real solutions

**53.** $4x^2 - 4x + m = 0$, two complex solutions

**54.** $mx^2 - x + 2 = 0$, two complex solutions

**55.** $mx^2 - x + 2 = 0$, two real solutions

**56.** Prove the transitive law which states that if $a$, $b$, and $c$ are real numbers with $a > b$ and $b > c$, then $a > c$.

**57.** Prove that if $a$ and $b$ are real numbers satisfying $0 < a < b$, then $a^2 < b^2$. Is this still true if the condition $0 < a$ is deleted?

## For Review

In Exercises 58–61 perform the indicated operations and express the result in the form $a + bi$.

**58.** $(-2 - 3i) - (4 - i)$ **59.** $(4 - 2i)(3 + i)$ **60.** $\frac{-2 + 7i}{2 + i}$ **61.** $(3 - 3i)^{-1}$

Solve the equations in Exercises 62–63.

**62.** $3x^2 + 1 = x$ **63.** $2x^3 - 5x^2 + 4x = 0$

**64.** Without solving, find the sum and product of the solutions to the equation $6x^2 - 12x + 1 = 0$.

**65.** Use the discriminant to determine the nature of the solutions (two real, one real, or two complex) to the equation $7x^2 - 3x + 1 = 0$.

## 2.8 Quadratic and Rational Inequalities

### Quadratic Inequalities

Inequalities of the form

$$ax^2 + bx + c < 0, \qquad ax^2 + bx + c > 0,$$
$$ax^2 + bx + c \leq 0, \qquad \text{or} \qquad ax^2 + bx + c \geq 0,$$

where $a$, $b$, and $c$ are real numbers, $a \neq 0$, are called **quadratic inequalities.** Suppose that $x_1$ and $x_2$ are two real-number solutions to the quadratic equation $ax^2 + bx + c = 0$ such that $x_1 < x_2$. Then $x_1$ and $x_2$ separate the number line into three distinct intervals

$$(-\infty, x_1), \qquad (x_1, x_2), \qquad \text{and} \qquad (x_2, \infty).$$

It can be shown that the algebraic sign of $ax^2 + bx + c$ does not change within each of these intervals. This fact is used to solve a quadratic inequality as illustrated in the following example.

### EXAMPLE 1

Solve. $2x^2 + x - 6 < 0$

We first solve the related quadratic equation.

$$2x^2 + x - 6 = 0$$
$$(2x - 3)(x + 2) = 0 \quad \text{Factor}$$
$$2x - 3 = 0 \quad \text{or} \quad x + 2 = 0 \quad \text{Zero-product rule}$$
$$2x = 3 \qquad\qquad x = -2$$
$$x = \frac{3}{2}$$

These solutions separate the number line into three intervals $(-\infty, -2)$, $(-2, 3/2)$, and $(3/2, \infty)$ as shown in Figure 20.

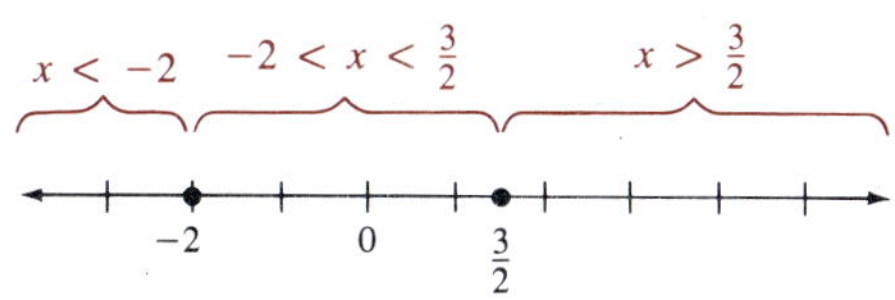


**Figure 20**

We know that the algebraic sign of $2x^2 + x - 6$ will be the same at each point in $(-\infty, -2)$. To determine that sign, we need only select one number in the interval and evaluate $2x^2 + x - 6$ at that value. Suppose we choose $-3$:

$$2x^2 + x - 6 = 2(-3)^2 + (-3) - 6 = 9 > 0$$

Thus, $2x^2 + x - 6 > 0$ for all real numbers $x$ in $(-\infty, -2)$. Similarly, if we choose an arbitrary number in $\left(-2, \frac{3}{2}\right)$, say 0, we have

$$2x^2 + x - 6 = 2(0)^2 + (0) - 6 = -6 < 0.$$

Thus, $2x^2 + x - 6 < 0$ for all real numbers $x$ in $\left(-2, \frac{3}{2}\right)$. Finally, choose 2 in $\left(\frac{3}{2}, \infty\right)$.

$$2x^2 + x - 6 = 2(2)^2 + (2) - 6 = 4 > 0$$

Thus, $2x^2 + x - 6 > 0$ for all real numbers in $\left(\frac{3}{2}, \infty\right)$. These results are summarized in Figure 21.

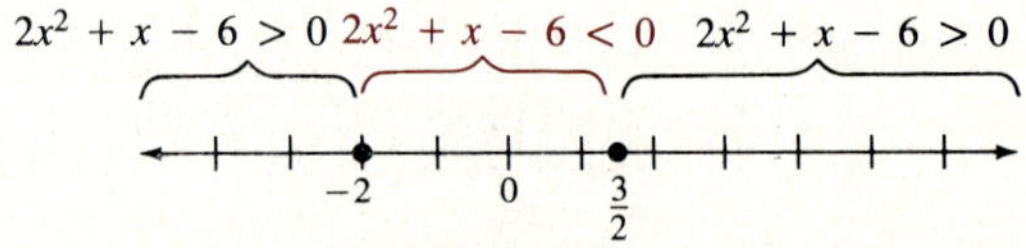


**Figure 21**

In this way, the solutions to $2x^2 + x - 6 < 0$ are found in the interval $\left(-2, \frac{3}{2}\right)$, and we write the solution as $-2 < x < \frac{3}{2}$. ■

To help organize the work of solving a quadratic inequality, we will call the solutions of the related quadratic equation **critical points,** and the arbitrary points selected from each interval, **test points.** A table such as the one below summarizing the work in Example 1, can be of assistance.

| Intervals determined by critical points | Test points | Test value of $2x^2 + x - 6$ | Sign of $2x^2 + x - 6$ |
|---|---|---|---|
| $(-\infty, -2)$ | $-3$ | $2(-3)^2 + (-3) - 6 = 9$ | + |
| $\left(-2, \frac{3}{2}\right)$ | 0 | $2(0)^2 + (0) - 6 = -6$ | − |
| $\left(\frac{3}{2}, \infty\right)$ | 2 | $2(2)^2 + (2) - 6 = 4$ | + |

Notice that the inequality

$$2x^2 + x - 6 > 0$$

would use the exact same table, and we could identify the solution as

$$x < -2 \quad \text{or} \quad x > \frac{3}{2},$$

which is where the sign of $2x^2 + x - 6$ is positive. Similarly, the solution to

$$2x^2 + x - 6 \leq 0$$

is $-2 \leq x \leq \frac{3}{2}$. Since the inequality is $\leq$ and equality holds at the end points of each interval (the critical points), these values are included in the solution this time. Finally, if we solved

$$2x^2 + x - 6 \geq 0,$$

we would obtain $x \leq -2$ or $x \geq \frac{3}{2}$.

**EXAMPLE 2**

Solve. $x^2 + 2x - 15 \geq 0$

Solving the related quadratic equation, $x^2 + 2x - 15 = 0$, we obtain the critical points $x = -5$ and $x = 3$.

| Intervals determined by critical points | Test points | Test values of $x^2 + 2x - 15$ | Sign of $x^2 + 2x - 15$ |
|---|---|---|---|
| $(-\infty, -5)$ | $-6$ | $(-6)^2 + 2(-6) - 15 = 9$ | + |
| $(-5, 3)$ | 0 | $0^2 + 2(0) - 15 = -15$ | − |
| $(3, \infty)$ | 4 | $(4)^2 + 2(4) - 15 = 9$ | + |

Since $x^2 + 2x - 15$ is positive in $(-\infty, -5)$ and in $(3, \infty)$, and is zero at the endpoints of these intervals, the solution is

$$x \leq -5 \quad \text{or} \quad x \geq 3, \quad \text{also expressed} \quad (-\infty, -5] \quad \text{or} \quad [3, \infty).$$ ■

**EXAMPLE 3**

Solve. $(x + 3)(x - 1) \leq -2$

Simplify and write the inequality in general form

$$x^2 + 2x - 1 \leq 0$$

To solve the related quadratic equation $x^2 + 2x - 1 = 0$, we must use the quadratic formula. Solving, we obtain $x = -1 \pm \sqrt{2}$ as the critical points. Using 1.4 as an approximation for $\sqrt{2}$, we obtain the approximations 0.4 and −2.4 for the critical points. We use these values to choose appropriate test points.

| Intervals | Test points | Test values | Sign |
|---|---|---|---|
| $(-\infty, -2.4)$ | $-3$ | $(-3)^2 + 2(-3) - 1 = 2$ | + |
| $(-2.4, 0.4)$ | 0 | $(0)^2 + 2(0) - 1 = -1$ | − |
| $(0.4, \infty)$ | 1 | $(1)^2 + 2(1) - 1 = 2$ | + |

Since the sign is negative in $(-2.4, 0.4)$ or $(-1 - \sqrt{2}, -1 + \sqrt{2})$ and since the endpoints are included in the solution (the inequality is $\leq$), the solution is

$$-1 - \sqrt{2} \leq x \leq -1 + \sqrt{2} \quad \text{or} \quad [-1 - \sqrt{2}, -1 + \sqrt{2}]. \quad \blacksquare$$

In the examples considered thus far, there were two critical points that resulted in three intervals on the number line. Recalling that the critical points are solutions to a quadratic equation and that quadratic equations may also have one or no real solutions, we should be prepared for these special cases.

Suppose we solve $x^2 > 3(2x - 3)$. There is only one critical point, 3, which yields the two intervals $(-\infty, 3)$ and $(3, \infty)$. Test points from these intervals show that the solution is $x < 3$ or $x > 3$. Also, if we solve $x(x + 1) \geq -2$, there are no real solutions to the associated quadratic equation. Whenever there are no real critical points, either every real number is a solution or else there is no real solution at all. In this case, a single test point (0 is the easiest to use) shows that every real number is a solution. In other words, the solution is the interval $(-\infty, \infty)$.

## Rational Inequalities

Inequalities such as

$$\frac{x+2}{x+3} \geq 0, \qquad \frac{2x+1}{x-1} > 3, \qquad \text{and} \qquad \frac{x^2 + 3x - 18}{x+2} < 0$$

are called **rational inequalities.** If we adjust the definition of critical points to include values that make the denominator zero along with the values that make the rational expression zero (the numerator zero), we can solve these inequalities using a technique similar to the one for quadratic inequalities.

**EXAMPLE 4** Solve. $\dfrac{x-2}{x+3} \geq 0$

To find the critical points set $x - 2 = 0$, $x + 3 = 0$, and solve to obtain 2 and $-3$. We can use a table as before.

| Intervals | Test points | Test values | Sign |
|---|---|---|---|
| $(-\infty, -3)$ | $-4$ | $\frac{-4-2}{-4+3} = 6$ | + |
| $(-3, 2)$ | 0 | $\frac{0-2}{0+3} = -\frac{2}{3}$ | − |
| $(2, \infty)$ | 3 | $\frac{3-2}{3+3} = \frac{1}{6}$ | + |

Thus, $\dfrac{x-2}{x+3} > 0$ on the intervals $(-\infty, -3)$ and $(2, \infty)$ where the test values are positive. Since the original inequality symbol is $\geq$, we also include the endpoint 2 where $\dfrac{x-2}{x+3} = 0$. Thus, the solution is

$$x < -3 \text{ or } x \geq 2,$$

or, using intervals,

$$(-\infty, -3) \text{ or } [2, \infty).$$

Notice that the endpoint $-3$ is not part of the solution since the expression is undefined when $x = -3$. ■

**EXAMPLE 5** Solve. $\dfrac{2x+1}{x-1} > 3$

We rewrite the inequality so that 0 is on the right side, and simplify the result.

$$\frac{2x+1}{x-1} - 3 > 0 \quad \text{Subtract 3}$$

$$\frac{2x+1}{x-1} - \frac{3(x-1)}{x-1} > 0 \quad \text{The LCD} = x - 1$$

$$\frac{2x+1-3x+3}{x-1} > 0 \quad \text{Watch the signs when subtracting}$$

$$\frac{-x+4}{x-1} > 0$$

Using the critical points 4 and 1, we can make the table.

| Intervals | Test points | Test values | Sign |
|---|---|---|---|
| $(-\infty, 1)$ | 0 | $\dfrac{-0+4}{0-1} = -4$ | $-$ |
| $(1, 4)$ | 2 | $\dfrac{-2+4}{2-1} = 2$ | $+$ |
| $(4, \infty)$ | 5 | $\dfrac{-5+4}{5-1} = -\dfrac{1}{4}$ | $-$ |

Thus, the solution is

$$1 < x < 4 \quad \text{or} \quad (1, 4).$$

Notice that neither endpoint is included in the solution since the original inequality symbol is $>$. ■

**EXAMPLE 6** Solve. $\frac{x^2 + 3x - 18}{x + 2} < 0$

If we factor $x^2 + 3x - 18$, it is easy to identify the critical points.

$$\frac{(x - 3)(x + 6)}{x + 2} < 0$$

Thus, the critical points are 3, $-6$, and $-2$.

| Intervals | Test points | Test values | Sign |
|---|---|---|---|
| $(-\infty, -6)$ | $-7$ | $\frac{(-7 - 3)(-7 + 6)}{-7 + 2} = -2$ | $-$ |
| $(-6, -2)$ | $-3$ | $\frac{(-3 - 3)(-3 + 6)}{-3 + 2} = 18$ | $+$ |
| $(-2, 3)$ | $0$ | $\frac{(0 - 3)(0 + 6)}{0 + 2} = -9$ | $-$ |
| $(3, \infty)$ | $4$ | $\frac{(4 - 3)(4 + 6)}{4 + 2} = \frac{5}{3}$ | $+$ |

Since $\frac{x^2 + 3x - 18}{x + 2}$ is negative on $(-\infty, -6)$ and on $(-2, 3)$, the solution is

$$x < -6 \quad \text{or} \quad -2 < x < 3. \quad ■$$

Some applied problems result in quadratic inequalities.

**EXAMPLE 7**
**RATE-MOTION**

A rocket is fired upward from a platform, 10 ft off the ground, with an initial velocity of 128 ft/sec. During what time interval will its height exceed 250 ft?

Recall that the formula for the height $h$ of an object propelled upward from an initial height $h_0$ with initial velocity $v_0$ is

$$h = -16t^2 + v_0t + h_0.$$

In our example, $v_0 = 128$ and $h_0 = 10$.
Thus, we are looking for the interval of time for which

$$-16t^2 + 128t + 10 > 250.$$
$$-16t^2 + 128t - 240 > 0$$
$$-16(t^2 - 8t + 15) > 0$$
$$t^2 - 8t + 15 < 0 \quad \text{Divide by } -16 \text{ and reverse the inequality}$$
$$(t - 3)(t - 5) < 0$$

The critical values are 3 and 5.

| Intervals | Test points | Test values | Sign |
|---|---|---|---|
| $(-\infty, 3)$ | 0 | $(0 - 3)(0 - 5) = 15$ | + |
| $(3, 5)$ | 4 | $(4 - 3)(4 - 5) = -1$ | − |
| $(5, \infty)$ | 6 | $(6 - 3)(6 - 5) = 3$ | + |

Thus, the solution to the inequality is

$$3 < t < 5,$$

which means that during the time period from 3 seconds to 5 seconds, the rocket is higher than 250 ft above ground level. ■

## 2.8 EXERCISES

Solve each inequality in Exercises 1–27. Give answers using both inequalities and intervals.

**1.** $x^2 + 2x - 15 > 0$ **2.** $x^2 + 2x - 15 \leq 0$ **3.** $2x^2 + 3x \geq 20$

**4.** $2x^2 + 3x < 20$ **5.** $x^2 - 2x - 2 > 0$ **6.** $x^2 - 2x - 2 \leq 0$

**7.** $3x^2 - x + 2 > 0$ **8.** $3x^2 - x + 2 \leq 0$ **9.** $4x^2 - 20x + 25 \leq 0$

**10.** $4x^2 - 20x + 25 < 0$ **11.** $5x^2 - 2x \leq 1$ **12.** $\frac{x-3}{x+1} < 0$

**13.** $\frac{x-3}{x+1} \geq 0$ **14.** $\frac{2x-1}{x+3} \geq 1$ **15.** $\frac{2x-1}{x+3} < 1$

**16.** $\frac{3x+2}{x-3} < 2$ **17.** $\frac{3x+2}{x-3} \geq 2$ **18.** $(x-2)(x+2)(x-5) < 0$

**19.** $(x-2)(x+2)(x-5) > 0$ **20.** $(x-2)(x+2) < 3x$ **21.** $(x-2)(x+2) \geq 3x$

**22.** $\frac{(x-3)(x+6)}{x-1} \geq 0$ **23.** $\frac{(x-3)(x+6)}{x-1} < 0$ **24.** $(x^2 - 2x)(x^2 + 8x + 15) < 0$

**25.** $(x^2 - 2x)(x^2 + 8x + 15) \geq 0$ **26.** $x^3 < x$ **27.** $x^3 \geq x$

In Exercises 28–31 determine the interval(s) in which the radical expression defines a real number.

**28.** $\sqrt{x^2 - 1}$ **29.** $\sqrt{x(x+5)}$ **30.** $\sqrt{2x^2 - 9x - 5}$ **31.** $\sqrt{\frac{3-x}{x+8}}$

In Exercises 32–33 determine the interval(s) containing $m$ in which the quadratic equation has the desired solutions.

**32.** $x^2 + mx + 4 = 0$; two real solutions **33.** $x^2 + mx + 4 = 0$; two complex solutions

Solve.

**34.** BUSINESS The profit made when $t$ units are sold, $t > 0$, is given by $P = t^2 - 31t + 220$. Determine the number of units to be sold in order for
**(a)** $P = 0$ (the break-even points),
**(b)** $P > 0$ (a profit is made), and
**(c)** $P < 0$ (a loss is taken).

**35.** BUSINESS The cost of producing $t$ units is $C = t^2 + 6t$, and the revenue generated from sales is $R = 2t^2 + t$. Determine the number of units to be sold in order to generate a profit.

**36.** GEOMETRY A rectangular enclosure must have an area of at least 900 yd$^2$. If 200 yd of fencing is to be used, and the width cannot exceed the length, within what limits must the width of the enclosure lie?

**37.** RATE-MOTION A coin is tossed upward from a balcony 200 ft high with an initial velocity of 48 ft/sec. During what interval of time will the coin be at a height of at least 40 ft?

**38.** BUSINESS A retailer has determined that $n$ games can be sold during the month provided the price of each game is $20 - 0.2n$ dollars. If he purchases each game from a wholesaler for \$10, and if he wishes to make a profit of at least \$120 per month on sales of this game, how many games must he sell each month?

**39.** RATE-MOTION If a rocket is propelled upward from ground level, its height in meters after $t$ seconds is given by $h = -9.8t^2 + 147t$. During what interval of time will the rocket be higher than 529.2 m?

**40.** A rectangular sheet of cardboard is 16 inches long and 12 inches wide. A box is to be made from the sheet by cutting out four square pieces, one in each corner, and folding up the sides. If the area of the base of the box must exceed 32 in$^2$, what are the possible heights of the box?

**41.** A signal flare fired from the bottom of a gorge is visible to an observer on the rim of the gorge only when the flare is above the level of the rim. If the flare is fired with an initial velocity of 176 ft/sec, and the gorge is 448 ft deep, during what time interval can the flare be seen?

In Exercises 42–43 use a calculator to solve the inequality. Give answers rounded to the nearest hundredth.

**42.** $2.5x^2 - 3.9x - 4.7 > 0$

**43.** $\dfrac{x - 6.25}{x^2 - 1.5x - 7.3} \le 0$

## For Review

Solve each inequality in Exercises 44–47. Give answers using inequalities and intervals.

**44.** $(y + 5)^2 \le y(y - 15)$

**45.** $\dfrac{2}{3z + 9} < 0$

**46.** $|1 - 4x| > 5$

**47.** $|3(z + 1) + 1| \le 10$

**48.** RETAILING The owner of a pet store wishes to make a profit of at least \$100 on the sale of pet mice during the month. If each mouse sells for \$1.50 and costs the dealer \$0.85, and if the estimated fixed overhead cost related to supplying the mice amounts to \$13.75 a month, what is the minimum number he must sell?

**49.** Working together, Peggy and Beth can paint a garage in 6 days. Working alone, Peggy can do the job in 10 days. How many days would be required for Beth to paint the garage by herself?

**50.** Use a calculator to solve

$$\frac{6.2}{x - 5.8} + \frac{2.3}{x + 7.9} = 47.3.$$

Give the answer rounded to the nearest hundredth.

## CHAPTER 2 REVIEW EXERCISES

Solve each equation or inequality in Exercises 1–24.

**1.** $(3y - 2) - (y + 1) = 0$

**2.** $4(x - 2) + 3(3 - x) = 3x - 5$

**3.** $5\sqrt{x - 1} - 3\sqrt{2x + 5} = 0$

**4.** $\sqrt{z^2 + 7} - z - 1 = 0$

**5.** $\frac{x - 3}{x - 1} - \frac{5}{x} = 1$

**6.** $\frac{z}{z + 2} - \frac{2}{z - 2} = \frac{z^2 + 4}{z^2 - 4}$

**7.** $y(y + 1) = 6(3 - y)$

**8.** $3x^2 - x - 1 = 0$

**9.** $5x^2 + 2x + 1 = 0$

**10.** $x^{2/3} + 2x^{1/3} + 1 = 0$

**11.** $\left(\frac{y - 3}{y + 1}\right)^2 - 6\left(\frac{y - 3}{y + 1}\right) + 8 = 0$

**12.** $2x + \frac{x}{x + 7} = \frac{8}{x + 7}$

**13.** $\frac{3y}{y + 1} = \frac{2}{y^2 - 1} + \frac{y}{y - 1}$

**14.** $\sqrt{x^2 + 2} - \sqrt{3x + 6} = 0$

**15.** $\sqrt{3y + 1} - \sqrt{y + 4} = 1$

**16.** $|2x + 3| = 3$

**17.** $|3 - 4x| = 0$

**18.** $|x + 5| = |5 - x|$

**19.** $\frac{1}{x} - \frac{1}{y} = \frac{1}{z}$ for $y$

**20.** $3(x - 1) + x < 4(x + 1)$

**21.** $(x - 3)(x^2 - 2x - 35) < 0$

**22.** $3x^2 + x \leq 10$

**23.** $x^2 - 3x + 1 \geq 0$

**24.** $3x^2 - 2x + 1 < 0$

Solve and graph each inequality in Exercises 25–28. Give answers using both inequalities and intervals.

**25.** $5(x + 2) - 3x \leq x + 7$

**26.** $\frac{x - 5}{2x - 3} \geq 1$

**27.** $|2 - x| < 1$

**28.** $3x^2 + x - 10 > 0$

Perform the indicated operations in Exercises 29–30.

**29.** $(8 - 6i) - (-3 + i)$

**30.** $\frac{3 - 5i}{-2 + i}$

**31.** Use the discriminant to determine the nature of the solutions to $3x^2 + 3x + 1 = 0$.

**32.** Without solving, find the sum and product of the solutions to the equation $5x^2 - 10x + 25 = 0$.

Solve.

**33.** The sum of three positive, consecutive, even integers is 40 more than twice the largest. Find the integers.

**34.** The average weight of Jane, Bill, Henry, and Sue is 78 kg. If Sue weighs 65 kg, Bill weighs 84 kg, and Henry weighs 88 kg, how much does Jane weigh?

**35.** ECONOMICS If after receiving an 8 percent raise Gloria makes $30,240, what was her former salary?

**36.** GEOMETRY The perimeter of a rectangle is 70 m. If the length is 1 m less than twice the width, find the dimensions of the rectangle.

**37.** GEOMETRY The first angle of a triangle is 5° more than twice the third and the second is 5° less than six times the first. Find the measure of each.

**38.** John can do a job in 24 minutes and Sean can do the same job in 18 minutes. How long will it take if they work together? (Give answer correct to the nearest tenth of a minute.)

**39.** RATE-MOTION Mark and Pam live 159 miles apart. If Mark averages 50 mph driving while Pam averages 56 mph, and they both leave at 8:00 A.M. and drive toward each other, at what time will they meet?

**40.** AERONAUTICS A plane flies 700 km with the wind and 500 km against the wind in the same length of time. If the speed of the wind is 20 km/hr, what is the speed of the plane in still air?

**41.** A wire 64 ft long is cut into two pieces which have the ratio of 11:5. How long is each piece?

**42.** A tree casts a shadow 32 ft long at the same time that a 100-ft tower casts a shadow 64 ft long. How tall is the tree?

**43.** RATE-MOTION Two cars leave the same city at the same time traveling at right angles to each other. If the first travels 15 mph faster than the second, and if after two hours they are 150 miles apart, find the speed of each car.

**44.** RECREATION A group of men plan to share equally the $132 cost of a fishing trip. At the last minute, two men decide not to go, and this raises the share of each remaining man by $11. How many men were planning to go initially?

**45.** Jan can do a job in 12 minutes less time than her mother, and together they can do the job in 8 minutes. How long would it take each to do it alone?

**46.** GEOMETRY The Eiffel Tower in Paris is 984 ft tall. If an observer at the top of the tower on a clear day views the horizon, how far away does she see? Use 4000 mi as the radius of the earth, and consider the figure.

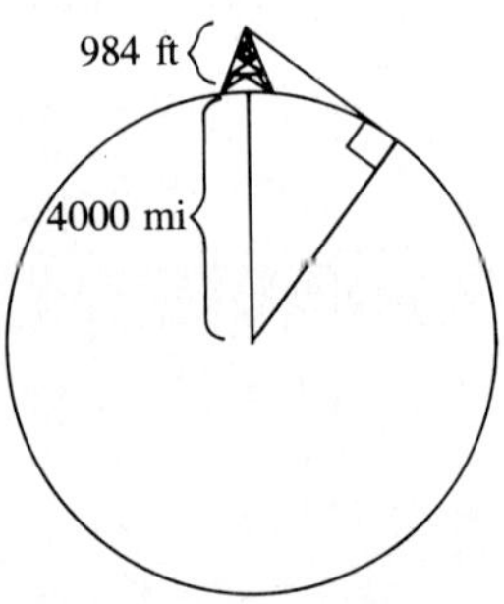


**47.** RATE-MOTION Fernando throws a baseball into the air with an initial velocity of 64 ft/sec. If the ball is released from a height of 6 ft, during what period of time will the ball be at a height greater than 54 ft?

**48.** ECOLOGY A state game officer wishes to estimate the number of antelope in a preserve. He catches 100 antelope, tags an ear, and returns them to the preserve. After allowing his sample to mix thoroughly with the rest of the herd, he catches another sample of 50 antelope and discovers that 5 of these are tagged. Estimate the number of antelope in the game preserve.

**49.** MANUFACTURING In the manufacture and sale of wood-burning stoves, the revenue made on the sale of $x$ stoves is $\$450x$, and the cost of producing $x$ stoves is $\$200x + \$750$ per week. For what values of $x$ will a profit be returned?

**50.** Use a calculator to solve $7.3x^2 - 3.2x - 8.4 = 0$. Give the answer rounded to the nearest hundredth.

CHAPTER

# 3 FUNCTIONS AND GRAPHS

The concept of function is basic to mathematics and the application of mathematics. In business, science, and engineering the idea of one variable depending on another is used extensively. The following are examples of the numerous applications we will consider.

◀ **BUSINESS**

A car rental agency averages 20 customers per day for a model which rents for \$32 a day. A survey shows that the agency could get 2 new customers for each \$1 reduction in the daily rental rate. What daily rate would give the company the maximum revenue?

**ENGINEERING** ▶

The resistance $R$ of a wire at constant temperature varies directly as the length and inversely as the square of the diameter $d$. A section of wire having a diameter of 0.01 inches and a length of 1 foot has a resistance of 8.2 ohms. Find the resistance of a 1-mile long wire with diameter 0.05 inches.

The first of these applications is solved by finding the maximum of a quadratic function. (See Example 4 in Section 3.6.) The second is solved using variation. (See Example 4 in Section 3.7.)

In this chapter we study not only functions and their characteristics but also how to describe them graphically. Linear and quadratic functions are given extensive treatment; the subject of variation concludes the chapter.

## 3.1 The Cartesian Coordinate System

In Chapter 2 we graphed equations and inequalities in one variable on a number line. When a horizontal and vertical number line are superimposed so that the two origins coincide and the lines are perpendicular, the result is a **rectangular coordinate system** or **coordinate plane.** This configuration is also called a **Cartesian coordinate system** after René Descartes, the French mathematician who introduced it.

Figure 1 shows the important features of a rectangular coordinate system. The horizontal number line is called the ***x*-axis** and the vertical number line the ***y*-axis.** The point of intersection of the axes is called the **origin,** and the system is sometimes called an ***x, y*-coordinate system.**

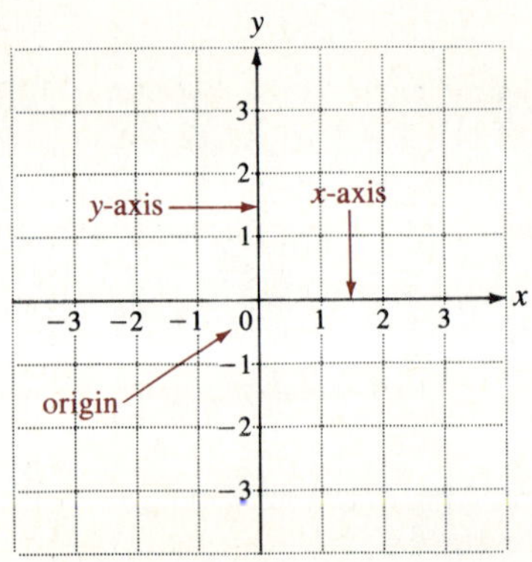


**Figure 1 Rectangular or Cartesian Coordinate System (Coordinate Plane)**

### Plotting Points

Recall that there is one and only one point on a number line associated with each real number. Similarly, there is a unique pair of numbers associated with each point in the coordinate plane. The pair is written $(x, y)$ where $x$ represents units on the $x$-axis and $y$ corresponds to units on the $y$-axis. We call $(x, y)$ an **ordered pair** since the order, $x$ first and $y$ second, must be preserved to properly identify points. The ordered pair $(a, b)$ is plotted in Figure 2 at the intersection of a vertical line through $a$ and a horizontal line through $b$. The number $a$ is called the ***x*-coordinate** or **abscissa** of the point $(a, b)$, and $b$ is called the ***y*-coordinate** or **ordinate** of the point.

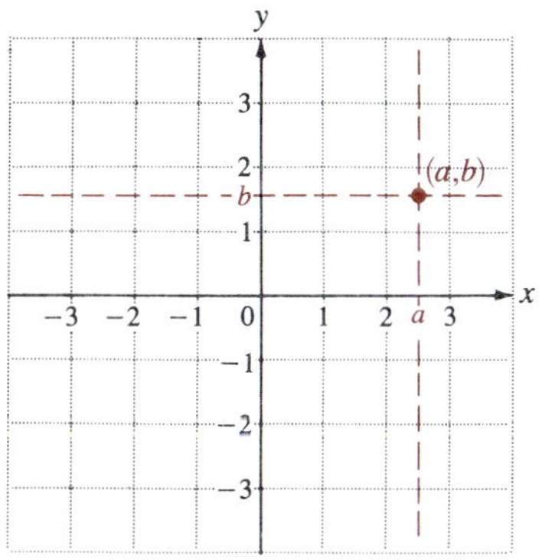


Figure 2 Plotting Points

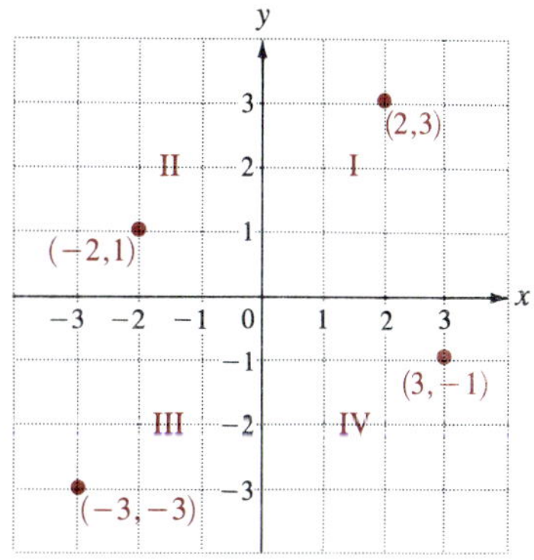


Figure 3 Quadrants and Graphs

In Figure 3 we have plotted the points (2, 3), (−2, 1), (−3, −3), and (3, −1). Notice also that in this figure the Roman numerals I, II, III, and IV are included. These designate the four sections in a rectangular coordinate system called **quadrants.** The point (2, 3) is in the first quadrant, (−2, 1) in the second (−3, −3) in the third, and (3, −1) in the fourth. The signs of the $x$-coordinate and $y$-coordinate in each of the quadrants are as follows.

I: (+, +), II: (−, +), III: (−, −), IV: (+, −)

## Graphs

The points plotted in Figure 3 are called the **graph** of the set {(2, 3), (−2, 1), (−3, −3), (3, −1}. Many graphs need infinitely many points. For example, to graph the set of solutions of $y = -2x + 3$ we must show all the ordered pairs for which the equation is true. In Figure 4 we plotted a few of the solutions, (−1, 5), (0, 3), (1, 1), (2, −1), (3, −3), and (4, −5). We saw that all the points lie in a straight line and constructed the line. The graph in this case consists of the infinitely many points on the line.

| $x$ | $y$ |
|---|---|
| −1 | 5 |
| 0 | 3 |
| 1 | 1 |
| 2 | −1 |
| 3 | −3 |
| 4 | −5 |

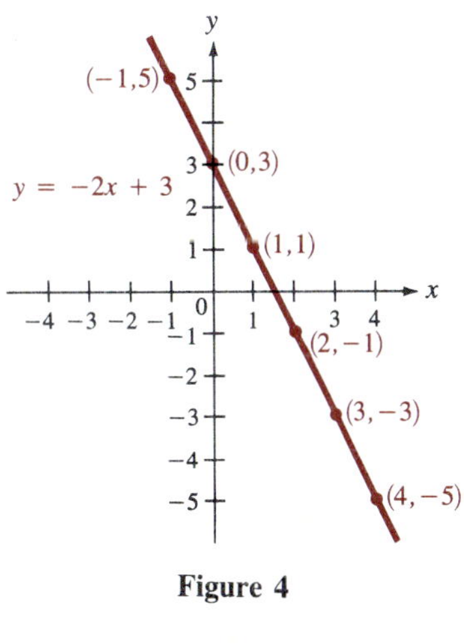


Figure 4

Notice that a table containing the points plotted is given along with the graph in Figure 4. One procedure for graphing an equation is to select values of one of the variables (usually $x$), calculate values of the other variable, and store the results in such a table. We plot enough points to determine the shape of the curve.

## EXAMPLE 1

Graph $y = x^2 - 3$.

Select values of $x$ which give $y$-values that can be plotted in a coordinate system of reasonable size. For example, for

$$x = -3 \qquad y = (-3)^2 - 3 = 6,$$
$$x = -2 \qquad y = (-2)^2 - 3 = 1.$$

Other points are calculated for the table and plotted in Figure 5. Enough points are included to allow us to connect them with the indicated curve. The graph of $y = x^2 - 3$ is all the points on the curve. ■

| $x$ | $y$ |
|---|---|
| −3 | 6 |
| −2 | 1 |
| −1 | −2 |
| 0 | −3 |
| 1 | −2 |
| 2 | 1 |
| 3 | 6 |

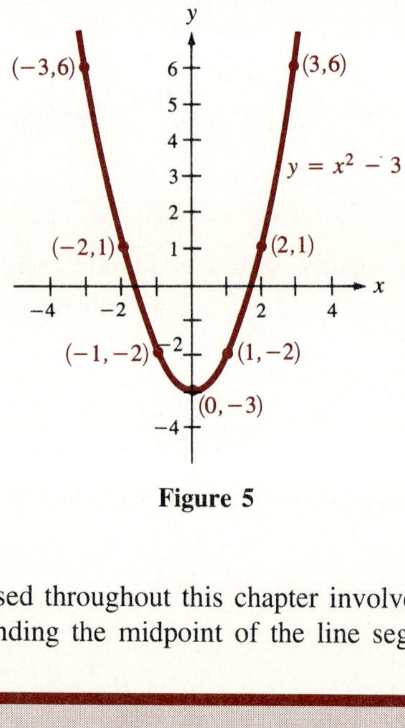


Figure 5

## Distance Formula

Two formulas that are used throughout this chapter involve finding the distance between two points and finding the midpoint of the line segment joining two points.

The distance between two points with coordinates $(x_1, y_1)$ and $(x_2, y_2)$ is given by the **distance formula**

$$d = \sqrt{(x_1 - x_2)^2 + (y_1 - y_2)^2}.$$

To derive this formula, consider Figure 6. Apply the Pythagorean theorem to the right triangle with vertices $P$, $Q$, and $R$ (the sum of the squares of the legs of a right triangle is equal to the square of the hypotenuse).

$$d^2 = (x_1 - x_2)^2 + (y_1 - y_2)^2$$
$$d = \sqrt{(x_1 - x_2)^2 + (y_1 - y_2)^2}$$

Observe that since both $(x_1 - x_2)$ and $(y_1 - y_2)$ are squared, if $P$ is $(x_2, y_2)$ and $Q$ is $(x_1, y_1)$, the formula yields the same result for the distance. Also, if the points are placed in different quadrants, the same formula applies.

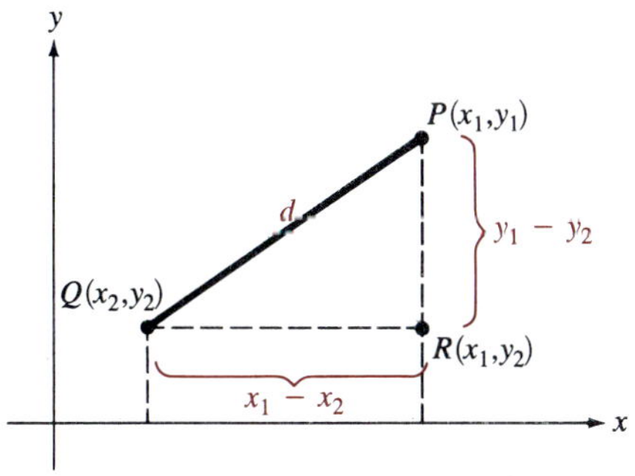


**Figure 6 Distance Formula**

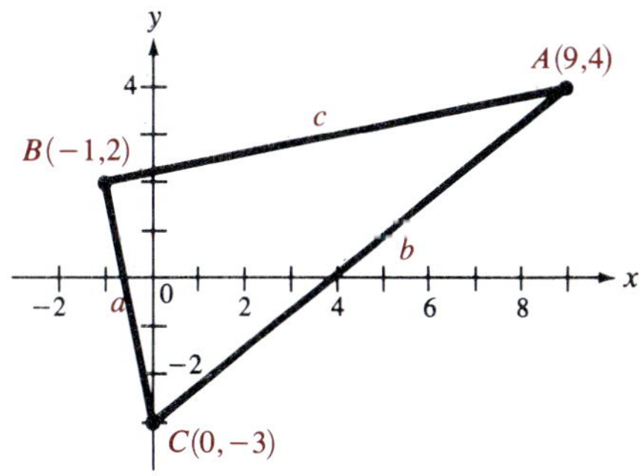


**Figure 7 Right Triangle**

**EXAMPLE 2** Find the lengths of the sides of the triangle $ABC$ with vertices $A(9, 4)$, $B(-1, 2)$, and $C(0, -3)$. Is $ABC$ a right triangle?

First we draw $ABC$ as in Figure 7.

$$a = \sqrt{(0 - (-1))^2 + (-3 - 2)^2} = \sqrt{1^2 + (-5)^2} = \sqrt{26}$$
$$b = \sqrt{(9 - 0)^2 + (4 - (-3))^2} = \sqrt{9^2 + 7^2} = \sqrt{130}$$
$$c = \sqrt{(-1 - 9)^2 + (2 - 4)^2} = \sqrt{(-10)^2 + (-2)^2} = \sqrt{104} = 2\sqrt{26}.$$

If $ABC$ is a right triangle, $\sqrt{130}$ must be the length of the hypotenuse (it is the longest side), which together with the legs, must satisfy the conditions of the Pythagorean theorem.

$$a^2 + c^2 = (\sqrt{26})^2 + (\sqrt{104})^2 = 26 + 104 = 130 = (\sqrt{130})^2 = b^2$$

Thus, the triangle is a right triangle. ■

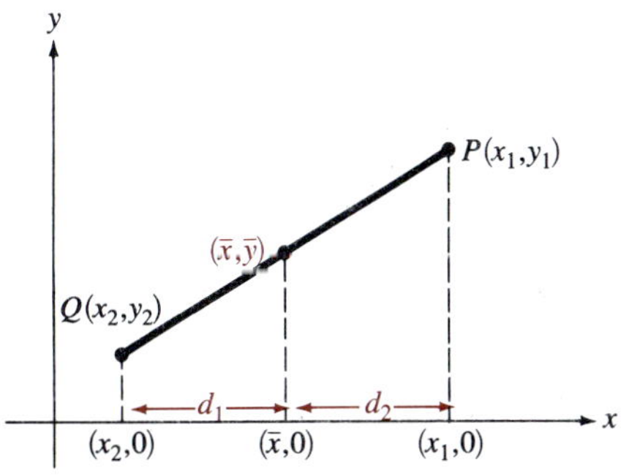


**Figure 8 Midpoint Formula**

## Midpoint Formula

The midpoint formula can be derived by considering Figure 8. Let $(\bar{x}, \bar{y})$ (read $x$-bar, $y$-bar) be the midpoint of the line segment joining $P(x_1, y_1)$ and $Q(x_2, y_2)$. From geometry we know that the point $(\bar{x}, 0)$ is the midpoint between $(x_2, 0)$ and $(x_1, 0)$. Thus, the distance $d_1$ on the $x$-axis between $(x_2, 0)$ and $(\bar{x}, 0)$ is the same as the distance $d_2$ between $(\bar{x}, 0)$ and $(x_1, 0)$.

$$\begin{aligned} d_1 &= d_2 \\ |\bar{x} - x_2| &= |x_1 - \bar{x}| \\ \bar{x} - x_2 &= x_1 - \bar{x} \qquad \text{Since } \bar{x} > x_2 \text{ and } x_1 > \bar{x} \\ 2\bar{x} &= x_1 + x_2 \\ \bar{x} &= \frac{x_1 + x_2}{2} \end{aligned}$$

Similarly, by considering distance on the $y$-axis, we can show $\bar{y} = \frac{1}{2}(y_1 + y_2)$. This gives the following theorem.

The coordinates $(\bar{x}, \bar{y})$ of the midpoint of the line segment joining $(x_1, y_1)$ and $(x_2, y_2)$ are given by the **midpoint formula**

$$(\bar{x}, \bar{y}) = \left(\frac{x_1 + x_2}{2}, \frac{y_1 + y_2}{2}\right).$$

Notice that the midpoint is determined by finding the average of the $x$-coordinates and the average of the $y$-coordinates.

**EXAMPLE 3** Find the midpoint between the points $(-5, 2)$ and $(7, 3)$.

$$\begin{aligned} (\bar{x}, \bar{y}) &= \left(\frac{x_1 + x_2}{2}, \frac{y_1 + y_2}{2}\right) \\ &= \left(\frac{-5 + 7}{2}, \frac{2 + 3}{2}\right) = \left(\frac{2}{2}, \frac{5}{2}\right) = \left(1, \frac{5}{2}\right) \end{aligned}$$ ■

## 3.1 EXERCISES

Graph the set of ordered pairs in Exercises 1–2.

**1.** $\{(2, 5), (-1, 3)\}$

**2.** $\{(-3, 4), (-3, 2), (-3, 0), (-3, -2), (-3, -3)\}$

In Exercises 3–6 give the quadrant in which the point is located.

**3.** $(-2, 1)$

**4.** $(\sqrt{2}, -\sqrt{2})$

**5.** $\left(-\frac{1}{2}, -\frac{1}{4}\right)$

**6.** $(88, 90)$

Graph the solutions to each equation (graph the equation) in Exercises 7–14.

**7.** $y - x = 0$

**8.** $y = -\frac{1}{2}x + 4$

**9.** $2x = -5$

**10.** $2y = 7$

**11.** $y + x^2 = 0$

**12.** $y^2 = x + 2$

**13.** $y = \sqrt{x + 2}$

**14.** $y = x^3$

In Exercises 15–20 find the distance between the points and the midpoint of the line segment joining them.

**15.** (4, 3) and (1, 7) **16.** (5, 10) and (−3, −2) **17.** (7, −1) and (5,8)

**18.** (−3, 6) and (2,7) **19.** (−6, 2) and (−6, −5) **20.** (4, −8) and (6, −8)

**21.** If (3, 2) is the midpoint of the line segment joining $(a, 8)$ and $(7, b)$, determine $a$ and $b$.

**22.** If $\left(\frac{7}{2}, -7\right)$ is the midpoint of the line segment joining $(a, b)$ and (3, −5), determine $a$ and $b$.

**23.** Find all numbers $a$ such that the distance from $(a, -2)$ to (8, 1) is equal to 5.

**24.** Find all points $(1, b)$ which are at a distance of 10 units from (7, 6).

**25.** Use the distance formula and the Pythagorean theorem to determine if triangle $ABC$ with vertices $A(5, 6)$, $B(-2, -1)$, and $C(2, -3)$ is a right triangle.

**26.** Determine if (6, −1), (1, −6), (−7, 2), and (−2, 7) are the vertices of a rectangle. [Hint: What can be said about the lengths of the diagonals of a rectangle?]

**27.** **AGRICULTURE** A large ranch is marked with electronic devices at the intersection of each mile line to aid in the location of herds of cattle. If one herd is at the point (4, 7) and another at (9, 3), find the distance between them, to the nearest tenth of a mile.

**28.** **AGRICULTURE** The herds of cattle in Exercise 27 are to be combined at a point midway between them. To what point should the ranch foreman fly in his helicopter to assist in combining the herds?

**29.** **ECOLOGY** If rainfall was 2.5 inches in January, 3.7 inches in February, and 6.3 inches in March, use J, F, and M for the months to list ordered pairs that represent this data.

**30.** **ECOLOGY** In 1970 the number of deer in a forest management area was 2800. By 1975 the number had dropped to 2100, but was back up to 2600 in 1980 and 3100 in 1985. Present this data as a set of ordered pairs.

## For Review

Solve each inequality in Exercises 31–34. Give answers using both inequalities and intervals.

**31.** $x^2 + 5x + 6 > 0$ **32.** $4x^2 - 20x + 25 \geq 0$ **33.** $\frac{x - 5}{2x + 3} \leq 0$ **34.** $|4x - 3| \geq 1$

**35.** Consider the equation $x + 2y - 6 = 0$. **(a)** Assume that $x = 0$ and solve for $y$. **(b)** Assume that $y = 0$ and solve for $x$. **(c)** Solve the equation for $y$ in terms of $x$.

# 3.2 Linear Equations

## General Form

A **linear equation** is an equation in two variables $x$ and $y$ that has as its graph a straight line. Such equations can be written in the **general form**

$$ax + by + c = 0$$

where $a$, $b$, and $c$ are real-number constants, $a$ and $b$ not both zero. Since the variables occur raised to the first power only, linear equations are also called **first-degree equations.** As examples,

$$3x + 2y = 5, \quad -4x + 7y = 0, \quad 5y = 6, \quad x - 2 = 0$$

are linear equations while

$$x^2 + y^2 = 6, \quad 3xy = 5, \quad \frac{1}{x} + y = 5, \quad y = x^3$$

are *not* linear equations. Their graphs are curves that are not straight lines.

## Intercepts

In Section 3.1 we graphed a linear equation by plotting several points. Now that we know that the graph of a linear equation is a straight line, we need only plot two points. For most equations the **intercepts,** the points where the graph crosses the axes, are used. The intersection with the $x$-axis is called the **$x$-intercept** and with the $y$-axis is called the **$y$-intercept.** Any $x$-intercept will have the form $(a, 0)$ and can be found for an equation by letting $y = 0$ and solving for $x$. Likewise, the $y$-intercept, $(0, b)$, is found by letting $x = 0$ and solving for $y$.

For example, to graph $x - 2y - 4 = 0$, let $x = 0$ to find the $y$-intercept $(0, -2)$, and let $y = 0$ to find the $x$-intercept $(4, 0)$. The graph is shown in Figure 9.

| $x$ | $y$ |
|---|---|
| 0 | −2 |
| 4 | 0 |

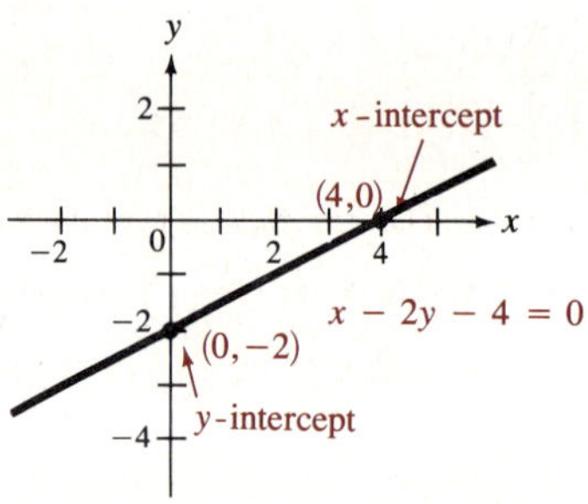


**Figure 9 Using Intercepts**

| $x$ | $y$ |
|---|---|
| 0 | 0 |
| 3 | −2 |

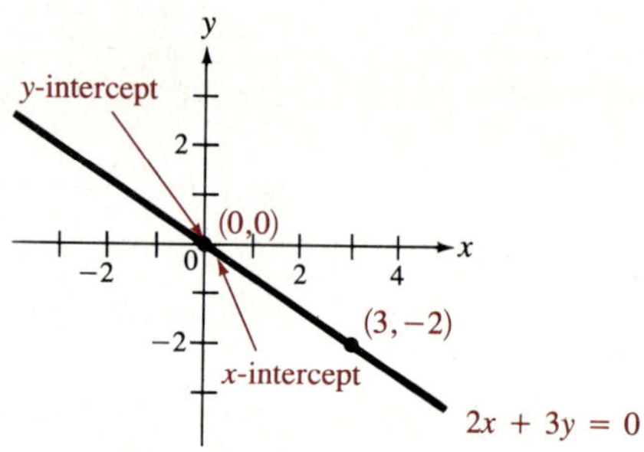


**Figure 10 Line through Origin**

For the equation $2x + 3y = 0$, one point $(0, 0)$, is both $x$-intercept and $y$-intercept. The line goes through the origin and we need one other point to graph the equation. If $x = 3$, then $y = -2$. The graph is shown in Figure 10. Thus, an equation of the form $ax + by = 0$ has as its graph a line through the origin.

**$ax + by = 0$** Line through the origin

Two other special cases should be considered. It can be shown that an equation of the form $x = a$ has a vertical line as its graph.

**$x = a$** Vertical line

Also, $y = b$ has a horizontal line as its graph.

$$y = b \qquad \text{Horizontal line}$$

For example, $2x - 4 = 0$ can be written as $x = 2$ and has the graph shown in Figure 11. The graph of $y = -\frac{5}{2}$ is shown in Figure 12.

| $x$ | $y$ |
|---|---|
| 2 | 0 |
| 2 | −3 |
| 2 | 4 |

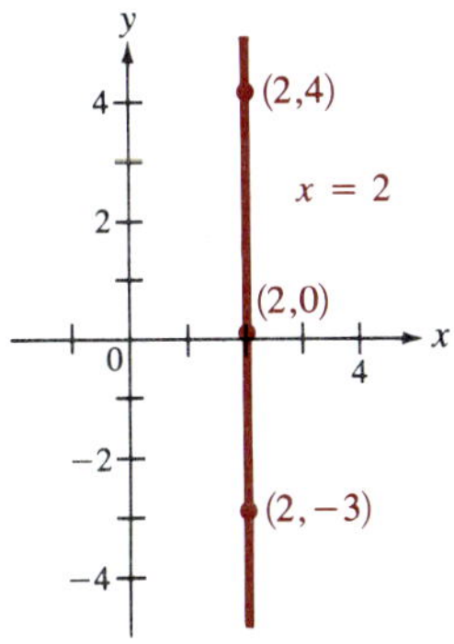


Figure 11 Vertical Line

| $x$ | $y$ |
|---|---|
| 0 | −5/2 |
| −3 | −5/2 |
| 4 | −5/2 |

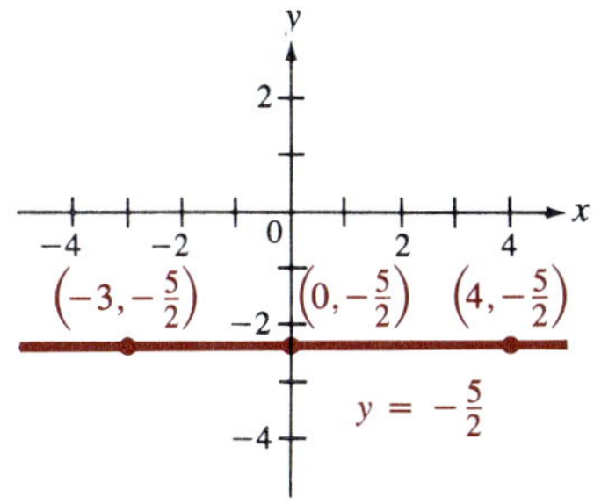


Figure 12 Horizontal Line

## Slope

Notice that the graph in Figure 9 slopes upward as $x$ increases, while in Figure 10 the graph slopes downward for increasing $x$. Figure 11 shows a vertical line, and Figure 12 a horizontal line. We can describe these differences better by defining *slope* of a line. Consider Figure 13 to help understand the definition.

Let $(x_1, y_1)$ and $(x_2, y_2)$ be two distinct points on a nonvertical line. The **slope** $m$ of the line is given by

$$m = \frac{y_2 - y_1}{x_2 - x_1}.$$

The slope of a vertical line is undefined.

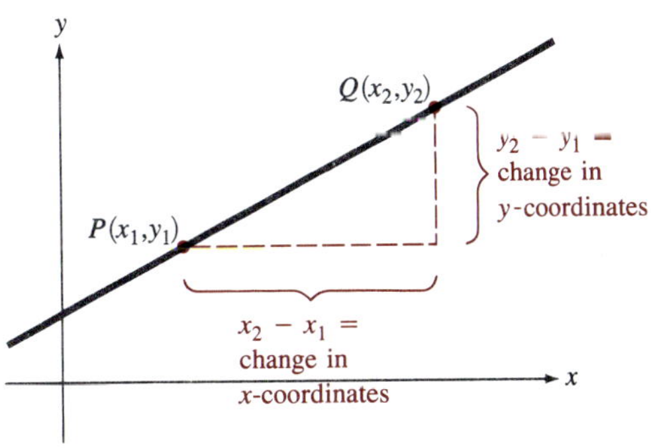


Figure 13 Slope of a Line

NOTE Since $y_2 - y_1$ is the change in $y$ and $x_2 - x_1$ is the change in $x$, slope is a measure of the steepness of a line. Also, since

$$\frac{y_2 - y_1}{x_2 - x_1} = \frac{y_1 - y_2}{x_1 - x_2},$$

the slope does not depend on which point is called $P(x_1, y_1)$ and which one is called $Q(x_2, y_2)$. ■

**EXAMPLE 1** Find the slope of the line passing through the given points.

**(a)** (6, 3) and (−2, 1)
Let $P(x_1, y_1)$ be identified with (6, 3) and $Q(x_2, y_2)$ with (−2, 1).

$$\begin{aligned} m &= \frac{y_2 - y_1}{x_2 - x_1} \\ &= \frac{1 - 3}{-2 - 6} = \frac{-2}{-8} = \frac{1}{4} \end{aligned}$$

**(b)** (−2, 3) and (1, −4)
Let $P(x_1, y_1)$ be (−2, 3) and $Q(x_2, y_2)$ be (1, −4).

$$\begin{aligned} m &= \frac{y_2 - y_1}{x_2 - x_1} \\ &= \frac{-4 - 3}{1 - (-2)} = \frac{-7}{1 + 2} = -\frac{7}{3} \end{aligned}$$

**(c)** (4, 3) and (−2, 3)
Let $(x_1, y_1) = (4, 3)$ and $(x_2, y_2) = (-2, 3)$.

$$\begin{aligned} m &= \frac{y_2 - y_1}{x_2 - x_1} \\ &= \frac{3 - 3}{-2 - 4} = \frac{0}{-6} = 0 \end{aligned}$$

**(d)** (−2, 0) and (−2, −3)
Let $(x_1, y_1) = (-2, 0)$ and $(x_2, y_2) = (-2, -3)$.

$$\begin{aligned} m &= \frac{y_2 - y_1}{x_2 - x_1} \\ &= \frac{-3 - 0}{-2 - (-2)} = \frac{-3}{0} \end{aligned}$$

Since $\frac{-3}{0}$ is not defined, the slope is undefined and the line is a vertical line. ■

The lines through the points in Example 1 are graphed in Figure 14. These summarize the four types of possible outcomes in determining slope.

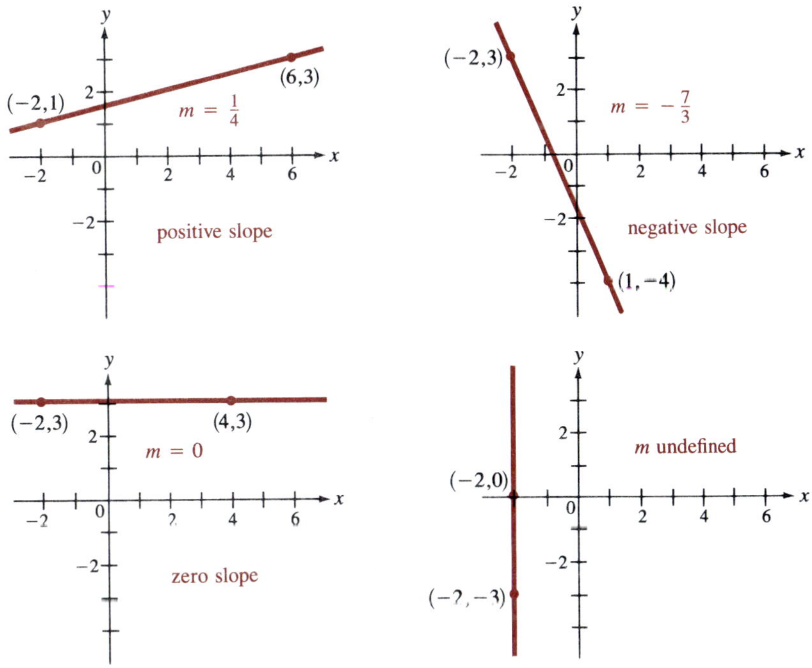


**Figure 14 Types of Slopes**

### Forms of the Equation of a Line

By using the notion of slope, other forms of the equation of a line can be developed. Let $P(x_1, y_1)$ be a fixed point on the line with slope $m$. If $(x, y)$ is any other point on the line with $x \neq x_1$, then

$$m = \frac{y - y_1}{x - x_1}$$

or

$$y - y_1 = m(x - x_1).$$

**Point-Slope Form**

An equation for the line with slope $m$ passing through the point $(x_1, y_1)$ is

$$y - y_1 = m(x - x_1).$$

## EXAMPLE 2

Find the general form of the equation of a line passing through the points (3, −4) and (6, 1).

First determine the slope of the line.

$$m = \frac{1 - (-4)}{6 - 3} = \frac{5}{3} \qquad (3, -4) = (x_1, y_1) \text{ and } (6, 1) = (x_2, y_2)$$

Now use the point-slope form and change to the general form.

$$y - y_1 = m(x - x_1) \qquad \text{Point-slope form}$$

$$y - (-4) = \frac{5}{3}(x - 3) \qquad m = \frac{5}{3} \text{ and } (x_1, y_1) = (3, -4)$$

$$y + 4 = \frac{5}{3}(x - 3)$$

$$3(y + 4) = 5(x - 3) \qquad \text{Multiply by 3}$$

$$3y + 12 = 5x - 15$$

$$-5x + 3y + 27 = 0$$

$$5x - 3y - 27 = 0 \qquad \text{General form}$$ ■

By substituting the $y$-intercept $(0, b)$ for $(x_1, y_1)$ in the point-slope form, we obtain

$$y - b = m(x - 0)$$

$$y = mx + b.$$

**Slope-Intercept Form**

An equation for the line with slope $m$ and $y$-intercept $(0, b)$ is

$$y = mx + b.$$

Thus, if $m$ and $(0, b)$ are given, we can find an equation for the line. For example, if $m = \frac{2}{3}$ and the $y$-intercept is $(0, -5)$, then

$$y = mx + b$$

$$y = \frac{2}{3}x - 5.$$

To put the equation in general form,

$$3y = 2x - 15 \qquad \text{Multiply by 3}$$

$$-2x + 3y + 15 = 0$$

$$2x - 3y - 15 = 0. \qquad \text{General form}$$

This shows that we can obtain the general form from the slope-intercept form.

Conversely, if we are given the general form, $ax + by + c = 0$, we can determine the slope and $y$-intercept by writing the equation in slope-intercept form.

**EXAMPLE 3** Find the slope and $y$-intercept of the line $3x + 7y - 14 = 0$.

Since the equation is in general form, we first convert it to the slope-intercept form, and then read $m$ and $b$ directly.

$$3x + 7y - 14 = 0$$
$$7y = -3x + 14$$
$$y = -\frac{3}{7}x + \frac{14}{7}$$
$$y = -\frac{3}{7}x + 2$$

Thus, $m = -\frac{3}{7}$ (not $-\frac{3}{7}x$) and $(0, b) = (0, 2)$. ■

### Parallel and Perpendicular Lines

By writing each of two different lines in slope-intercept form, we can determine if they are *parallel* (they never intersect) or *perpendicular* (they intersect at right angles). The following theorem is needed.

Given two distinct lines with slopes $m_1$ and $m_2$:

1. The lines are **parallel** if and only if $m_1 = m_2$ (the slopes are equal).
2. The lines are **perpendicular** if and only if $m_1 = -\frac{1}{m_2}$ or $m_1m_2 = -1$ (the slopes are negative reciprocals).

### EXAMPLE 4

**(a)** Determine if the lines $3x - 2y + 5 = 0$ and $10x + 15y = 7$ are parallel or perpendicular.

$$3x - 2y + 5 = 0 \qquad\qquad 10x + 15y = 7$$
$$-2y = -3x - 5 \qquad\qquad 15y = -10x + 7$$
$$y = \frac{3}{2}x + \frac{5}{2} \qquad\qquad y = -\frac{10}{15}x + \frac{7}{15}$$
$$\text{slope} = \frac{3}{2} \qquad\qquad y = -\frac{2}{3}x + \frac{7}{15}$$
$$\text{slope} = -\frac{2}{3}$$

Since the slopes are $\frac{3}{2}$ and $-\frac{2}{3}$ (negative reciprocals) the lines are perpendicular.

**(b)** Determine if the lines $6x + 3y + 2 = 0$ and $6y = -12x - 4$ are parallel or perpendicular.

$$6x + 3y + 2 = 0 \qquad\qquad 6y = -12x - 4$$
$$3y = -6x - 2 \qquad\qquad y = -\frac{12}{6}x - \frac{4}{6}$$
$$y = -\frac{6}{3}x - \frac{2}{3} \qquad\qquad y = -2x - \frac{2}{3}$$
$$y = -2 - \frac{2}{3}$$

Since the slopes are the same (both $-2$), we are tempted to conclude that the lines are parallel. However, we observe that the $y$-intercepts are also equal (both are $(0, -\frac{2}{3})$). Thus, the equations determine the same line. ■

**EXAMPLE 5**

Find the general form of the equation of a line with $y$-intercept $(0, -2)$ perpendicular to a line with slope $-\frac{4}{5}$.

$$y = mx + b$$
$$y = \frac{5}{4}x - 2 \qquad m = -\frac{1}{-\frac{4}{5}} = \frac{5}{4} \text{ and } b = -2$$
$$4y = 5x - 8 \qquad \text{Multiply by 4}$$
$$-5x + 4y + 8 = 0$$
$$5x - 4y - 8 = 0 \qquad ■$$

**EXAMPLE 6**

**ENGINEERING**

In the design of a machine part an engineer needs to find the equation of the perpendicular bisector of the line segment joining the points $(-3, 2)$ and $(3, -6)$.

The perpendicular bisector is the line perpendicular to the segment and passing through its midpoint. First calculate the slope of the line segment.

$$m = \frac{-6 - 2}{3 - (-3)} = \frac{-8}{6} = -\frac{4}{3}$$

The slope of the desired line is the negative reciprocal of $-\frac{4}{3}$.

$$-\frac{1}{m} = -\frac{1}{-\frac{4}{3}} = \frac{3}{4}$$

Now determine the midpoint.

$$(\bar{x}, \bar{y}) = \left(\frac{x_1 + x_2}{2}, \frac{y_1 + y_2}{2}\right) = \left(\frac{-3 + 3}{2}, \frac{2 + (-6)}{2}\right) = (0, -2)$$

Use the point-slope form of the equation of a line with $(0, -2)$ as the point and $\frac{3}{4}$ as the slope to find the equation needed in the design.

$$y - (-2) = \frac{3}{4}(x - 0)$$
$$y + 2 = \frac{3}{4}x$$
$$4(y + 2) = 3x \qquad \text{Multiply by 4}$$
$$4y + 8 = 3x$$
$$-3x + 4y + 8 = 0$$
$$3x - 4y - 8 = 0 \qquad \text{General form} \qquad ■$$

## 3.2 EXERCISES

In Exercises 1–10 find the intercepts and graph the equation.

**1.** $3x - 2y - 6 = 0$ **2.** $2x + 3y = 6$ **3.** $4x + 3y = 12$

**4.** $2x - 5y = 10$ **5.** $2x - 3 = 0$ **6.** $3x + 9 = 0$

**7.** $y = 3$

**8.** $3y = -12$

**9.** $3x - 2y = 0$

**10.** $5x = -10y$

Find the slope of the line through the given points in Exercises 11–16.

**11.** $(6, -2)$ and $(4, 1)$

**12.** $(5, 3)$ and $(-2, -6)$

**13.** $(-3, 2)$ and $(-3, -4)$

**14.** $(7, -2)$ and $(-3, -2)$

**15.** $(-4, 5)$ and $(1, 5)$

**16.** $(4, -5)$ and $(4, 8)$

In Exercises 17–30 find the general form of the equation of the line satisfying the given conditions.

**17.** With slope $-2$ passing through $(-3, 5)$.

**18.** With slope $\frac{1}{3}$ passing through $(2, -1)$.

**19.** With slope $\frac{4}{3}$ and $x$-intercept $(3, 0)$.

**20.** With slope $\frac{1}{7}$ and $y$-intercept $(0, 5)$.

**21.** Horizontal line through $(6, -5)$.

**22.** Vertical line through $(6, -5)$.

**23.** Passing through $(-3, 4)$ and $(2, 6)$.

**24.** Passing through $(3, 0)$ and $(0, -5)$.

**25.** Passing through $(-3, 7)$ and perpendicular to $5x - 10y + 3 = 0$.

**26.** Passing through $(2, 1)$ and parallel to $4x + 8y - 5 = 0$.

**27.** Passing through $(2, 3)$ and parallel to $2y - 3 = 0$.

**28.** Passing through $(2, 3)$ and perpendicular to $2y - 3 = 0$.

**29.** With $x$-intercept $(-5, 0)$ and parallel to $7x + 4 = 0$.

**30.** With $y$-intercept $(0, -5)$ and perpendicular to $7x + 4 = 0$.

Find the slope and $y$-intercept of each line in Exercises 31–36.

**31.** $2x - 7y + 5 = 0$

**32.** $5x - 2y + 8 = 0$

**33.** $8x = -3y + 7$

**34.** $-3x - 2y = 6$

**35.** $-3y + 6 = 0$

**36.** $4x + 11 = 0$

Determine if the pair of lines in Exercises 37–42 are parallel or perpendicular.

**37.** $6x - 2y + 7 = 0$ and $x + 3y - 8 = 0$

**38.** $2x - 5 = 0$ and $3x + 1 = 0$

**39.** $4x - 2y = -6$ and $10x = 5y + 8$

**40.** $3x - 8y + 7 = 0$ and $4x + 6y - 15 = 0$

**41.** $3y - 8 = 0$ and $4x + 2 = 0$

**42.** $7x - 21y + 8 = 0$ and $12x + 4y - 7 = 0$

**43.** Use slope to determine if the points $P(4, 3)$, $Q(2, 0)$, and $R(-2, -6)$ are **collinear** (on the same line).

**44.** Use slope to determine if the triangle with vertices $A(-2, 7)$, $B(-3, 3)$, and $C(-6, 8)$ is a right triangle.

**45.** Beginning with the point-slope formula, derive the **two-point form** of the equation of a line,

$$y - y_1 = \frac{y_2 - y_1}{x_2 - x_1}(x - x_1).$$

Use the two-point form to find the general equation of a line through $(-3, 6)$ and $(2, -4)$.

**46.** Show that the **two-intercept form** of the equation of a line,

$$\frac{x}{a} + \frac{y}{b} = 1,$$

where $(a, 0)$ is the $x$-intercept and $(0, b)$ is the $y$-intercept, can be written in general form. Use the two-intercept form to find the general form of the equation of the line through $(-2, 0)$ and $(0, 5)$.

**47.** Find the equation of the perpendicular bisector of the line segment joining $(1, 7)$ and $(-3, -5)$.

**48.** Find the general form of the equation of the perpendicular bisector of the portion of the line $x + y = 6$ which lies in the first quadrant.

**49.** CONSUMER A new house was purchased for \$85,000. After 5 years the value of the house was \$105,000. Assume that the appreciation in value is given by a linear equation. Find the equation, and find the value of the house 7 years after it was originally purchased.

**50.** ECONOMICS A new car was purchased for \$14,500, and 3 years later it was worth \$9100. Assume that the depreciation in value is given by a linear equation. Find the equation, and find the value of the car at the end of 6 years.

**51.** If the house in Exercise 49 is now worth \$125,000, how many years ago was it purchased?

**52.** How many years will it take for the car in Exercise 50 to have a value of only \$100?

**53.** ECONOMICS If \$1200 is borrowed at 9% simple interest, the sum to be paid at the end of $n$ years is given by the linear equation

$$S = 1200(1 + 0.09n).$$

**(a)** Find $S$ when $n = 3$. **(b)** Find $n$ when $S = \$1740$.

**54.** ECONOMICS If \$1200 is borrowed at $x$ percent simple interest (in decimal form), the sum to be paid at the end of 2 years is

$$S = 1200(1 + 2x).$$

**(a)** Find $S$ when $x = 0.12$ (12% interest).
**(b)** Find the interest rate when $S = \$1428$.

**55.** CHEMISTRY In a laboratory experiment 3 grams of sulfur were produced in 16 minutes and 8 grams in 36 minutes.

**(a)** Use the number of grams ($g$) for $y$ and time ($t$) for $x$ to find a linear equation which fits this data.
**(b)** Use the equation to find the number of grams produced in 20 minutes.
**(c)** Why can you conclude that this equation is not accurate when $t = 0$? This illustrates the fact that empirical equations are often valid only for restricted values of the variable.

**56.** ECOLOGY An environmental scientist found 30 deer on 800 acres in one area and 50 deer on 1200 acres in another area.

**(a)** Use the number of deer ($d$) for $y$ and the number of acres ($a$) for $x$ to find a linear equation relating the number of deer to the number of acres.

**(b)** Find the number of deer to be expected on 960 acres.

**(c)** Use 20 acres to explain why this equation is not accurate for a small number of acres.

## For Review

In Exercises 57–58 find the distance between the two points and the midpoint of the line segment joining them.

**57.** $(-1, 2)$ and $(4, 6)$

**58.** $(4, -5)$ and $(-4, 2)$

In Exercises 59–60 determine the values of $x$ for which the given expression *is not* a real number.

**59.** $\dfrac{2}{(x+1)(x-5)}$

**60.** $\dfrac{1}{\sqrt{x+3}}$

# 3.3 Functions

Relationships or correspondences are all around us. For example,

1. To each person there corresponds a height.
2. To each company there corresponds a number of employees.
3. For a given amount of money borrowed there corresponds an amount to be paid back after one year.
4. To each number in $\{1, 2, 3, 4\}$ there corresponds one or more numbers in the set which are greater than or equal to the number.

### Definition of Relation

A correspondence between a first set of objects, the **domain,** and a second set of objects, the **range,** is called a **relation** if to each element of the domain there corresponds *one or more* elements in the range.

The four examples given above are all relations. Notice that in the first three relations only one element in the range is related to each element in the domain. These are examples of *functions*.

### Definition of Function

A **function** is a relation with the property that to each element in the domain there corresponds *one and only one* (exactly one) element in the range.

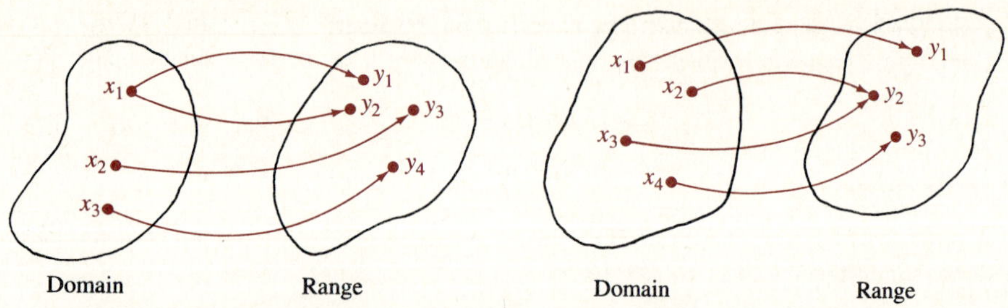


Relation that is *not* a function

Relation that is a function

**Figure 15 Relations and Functions**

Figure 15 illustrates the difference between a relation that is not a function and a relation that is a function.

To each person there corresponds one and only one height, and thus that relation is a function. Notice however, that the relation $\leq$ (less than or equal to) on the set $\{1, 2, 3, 4\}$ is not a function since, for example,

$$2 \leq 2, \quad 2 \leq 3, \quad 2 \leq 4.$$

That is, 2 is related to three numbers, 2, 3, and 4.

All the linear equations in Section 3.2, except those representing vertical lines, are examples of functions. If the equation can be written in the slope intercept form, $y = mx + b$, then for each $x$ there corresponds one and only one $y$. Letters such as $f$ and $g$ are frequently used for the names of functions. For example, for $y = 2x - 3$, we write

$$f(x) = 2x - 3$$

where $f(x)$ is read "$f$ of $x$" or "the value of $f$ at $x$." The instruction to find $y$ or $f(x)$ when $x = -4$ is written "find $f(-4)$." Thus,

$$f(-4) = 2(-4) - 3 = -8 - 3 = -11.$$

This means that corresponding to the number $x = -4$ is the number $y = -11$ or $f(-4) = -11$. Of course, this may also be symbolized using the ordered pair, $(-4, -11)$. Since $f(x)$ or $y$ depends on $x$, we call $x$ the **independent variable** and $y$ the **dependent variable.**

## Types of Functions

The general class of functions that have as their graph a straight line and can be written

$$f(x) = mx + b \qquad \text{Linear function}$$

are called **linear functions.** Included in these functions are the **constant functions**

$$f(x) = b \qquad \text{Constant function}$$

and the **identity function**

$$f(x) = x. \qquad \text{Identity function}$$

Another function, studied in detail in Section 3.6, is the **quadratic function.**

$$f(x) = ax^2 + bx + c \qquad \text{Quadratic function}$$

**EXAMPLE 1** Find $f(0)$, $f(-3)$, $f(a)$, and $f(x+h)$ for each of the following functions.

**(a)** $f(x) = -5x + 3$

$f(0) = -5(0) + 3 = 3$ $\qquad$ $f(a) = -5(a) + 3 = -5a + 3$

$f(-3) = -5(-3) + 3 = 18$ $\qquad$ $f(x+h) = -5(x+h) + 3 = -5x - 5h + 3$

**(b)** $f(x) = 6$

$f(0) = 6$ $\qquad$ $f(a) = 6$

$f(-3) = 6$ $\qquad$ $f(x+h) = 6$

**(c)** $f(x) = x$

$f(0) = 0$ $\qquad$ $f(a) = a$

$f(-3) = -3$ $\qquad$ $f(x+h) = x + h$

**(d)** $f(x) = -2x^2 + 5x - 4$

$f(0) = -2(0)^2 + 5(0) - 4 = -4$

$f(-3) = -2(-3)^2 + 5(-3) - 4 = -18 - 15 - 4 = -37$

$f(a) = -2(a)^2 + 5(a) - 4 = -2a^2 + 5a - 4$

$$\begin{aligned} f(x+h) &= -2(x+h)^2 + 5(x+h) - 4 = -2(x^2 + 2xh + h^2) + 5x + 5h - 4 \\ &= -2x^2 - 4xh - 2h^2 + 5x + 5h - 4 \\ &= -2x^2 + (5 - 4h)x - 2h^2 + 5h - 4 \quad ■ \end{aligned}$$

**Domain**

To define completely a function we need not only to give the correspondence, but also to specify the domain. When the domain is not stated, we shall assume that it consists of all real numbers for which the function is defined. The linear and quadratic functions are defined for all real numbers. However, if

$$g(x) = \sqrt{x+2}, \qquad \text{Defined for } x \geq -2$$

then the domain of $g$ is all real numbers $x$ such that $x + 2 \geq 0$ or $x \geq -2$. Notice that the range of $g$ is the set of real numbers $g(x) \geq 0$.

**EXAMPLE 2** Find the set of all real numbers for which the function is defined.

**(a)** $h(x) = \dfrac{1}{x+2}$

This function is defined for all real numbers except when $x + 2 = 0$. Thus, the domain is the set of all real numbers except $x = -2$.

**(b)** $k(x) = \dfrac{x-3}{\sqrt{x^2+1}}$

This function is defined for all real numbers since $x^2 + 1 > 0$ for all $x$. Thus, the domain is the set of all real numbers.

**(c)** $f(x) = \dfrac{\sqrt{3-x}}{(x-1)(x+2)}$

There are three conditions: $3 - x \geq 0$ or $3 \geq x$

$x - 1 \neq 0$ or $x \neq 1$

$x + 2 \neq 0$ or $x \neq -2$

Thus, the domain is the set of all real numbers such that $x \leq 3$ and $x \neq 1$ and $x \neq -2$. ■

CAUTION As Example 2 illustrates, we must exclude from a domain numbers that make a denominator zero or give a negative value under an even-indexed radical. ■

## EXAMPLE 3

RETAILING

A retailer received a shipment of shirts which cost him \$32.00 each. In order to find the total price that his customers will have to pay for a given markup, he wants a relation between total price $T$ and the percent markup $x$. **(a)** Find the total price function $T$ if the markup is $x$ percent and there is a 5% sales tax on the marked price. **(b)** Find the total price when the markup is 25%.

**(a)** Let $x$ = percent markup in decimal notation.
The selling price of each shirt is the cost plus markup.

$$32 + 32x = 32(1 + x) \qquad \text{Selling price}$$

The total price is $32(1 + x)$ plus 5% of $32(1 + x)$.

$$\begin{aligned} T(x) &= 32(1 + x) + 0.05(32)(1 + x) && \text{Selling price plus tax} \\ &= 32(1 + x)[1 + 0.05] \\ &= 32(1.05)(1 + x) \\ &= 33.6(1 + x) && \text{Total price as a function of the markup rate} \end{aligned}$$

**(b)** $T(x) = 33.6(1 + x)$

$$\begin{aligned} T(0.25) &= 33.6(1 + 0.25) && 25\% = 0.25 \\ &= \$42.00 && \text{Price with tax} \end{aligned}$$

Thus, if each shirt is marked up 25% the total price is \$42.00. ■

## EXAMPLE 4

ENGINEERING

Cylindrical storage tanks that are 10 m long and have a radius of $r$ meters are to be painted (including both ends). **(a)** If the paint job costs \$0.22 per square meter, express the cost $C$ of painting a tank as a function of the radius $r$. **(b)** Find the cost of painting a tank with radius 2.6 m.

**(a)** Let $r$ = radius of the cylindrical tank.
Then $2\pi r^2$ = area of both ends of the tank,
$2\pi r(10)$ = area of sides of the tank,
$2\pi r^2 + 20\pi r$ = total surface area of the tank.
The total cost $C$ is the cost per square meter times the surface area.

$$\begin{aligned} C(r) &= 0.22[2\pi r^2 + 20\pi r] \\ &= 0.22(2\pi)r(r + 10) \\ &\approx 1.38r(r + 10) && \text{Required function} \end{aligned}$$

**(b)** Now find $C(2.6)$.

$$C(2.6) = 1.38(2.6)(2.6 + 10)$$
$$\approx \$45.21$$

The cost of painting a cylinder with radius 2.6 m is approximately \$45.21. ■

## 3.3 EXERCISES

Determine which of the relations given in Exercises 1–6 is a function if $x$ is the independent variable and $y$ the dependent variable.

**1.** $x = 2y + 5$ **2.** $y = 5x^2$ **3.** $y = \sqrt{x - 1}$

**4.** $x = 5y^2$ **5.** $y = \dfrac{3x^2 - 2}{x + 1}$ **6.** $x^2 = \dfrac{y - 1}{y + 1}$

In Exercises 7–12 the function is described completely by giving the relation and the domain. Determine the range.

**7.** $f(x) = 5x - 3$; $\{1, 2, 3\}$ **8.** $f(x) = x^2 - 1$; $\{-1, 0, 1, 2\}$

**9.** $g(x) = \sqrt{x + 2}$; $x \geq -2$ **10.** $g(x) = \dfrac{1}{x}$; $x > 0$

**11.** $h(x) = \dfrac{1}{x - 3}$; all real numbers except 3 **12.** $h(x) = \dfrac{-1}{x^2}$; all real numbers except 0

Determine all the real numbers for which the function in Exercises 13–18 is defined.

**13.** $f(x) = 7x - 8$ **14.** $f(x) = 3x^2 + 2x + 1$ **15.** $g(x) = \sqrt{2 - x}$

**16.** $h(x) = \dfrac{(x + 5)(x - 5)}{x^2 - 25}$ **17.** $k(x) = \dfrac{\sqrt{x + 8}}{(x + 2)(x - 3)}$ **18.** $k(x) = \dfrac{\sqrt{8 - x}}{x(x + 5)}$

Use the most appropiate word, *identity, constant, linear,* or *quadratic,* to describe the function in Exercises 19–24.

**19.** $f(x) = x$ **20.** $g(x) = x^2 + 2$ **21.** $h(x) = 1 - x$

**22.** $h(x) = -x$ **23.** $k(x) = 14$ **24.** $k(x) = (x + 1)(x - 1)$

In Exercises 25–30 determine $f(0)$, $f(1)$, $f(-3)$, $f(a)$, and $f(a - 1)$.

**25.** $f(x) = 5x - 3$ **26.** $f(x) = x^2 - 10$ **27.** $f(x) = 7$

**28.** $f(x) = -8$ **29.** $f(x) = |x - 5|$ **30.** $f(x) = |x| - 5$

In Exercises 31–34 determine $h(a^2)$, $[h(a)]^2$, $h\left(\dfrac{1}{a}\right)$, and $\dfrac{1}{h(a)}$.

**31.** $h(x) = 3x - 7$ **32.** $h(x) = 2x^2 - 5$ **33.** $h(x) = \dfrac{1}{x} + 6$ **34.** $h(x) = \dfrac{1}{x + 6}$

In Exercises 35–38 determine $f(1 + h)$, $\dfrac{f(1 + h) - f(1)}{h}$, $f(x + h)$, and $\dfrac{f(x + h) - f(x)}{h}$.

**35.** $f(x) = x + 5$ **36.** $f(x) = -3x - 4$ **37.** $f(x) = x^2 + 2$ **38.** $f(x) = 3x^2 - 5x - 7$

**39.** **RETAILING** A retailer wishes to mark up dresses by 28.5%. **(a)** If $c$ is the cost of each dress, determine the function $S$ that represents the selling price. **(b)** What is the selling price of a dress that costs \$56?

**40.** **RETAILING** Lori's Shop has discounted sweaters by 12.5%. **(a)** If $s$ is the selling price of a sweater, find a function $P$ that represents the discounted price. **(b)** What would be the sale price of a sweater that was originally priced at \$44.00?

**41.** **RETAILING** The Pant Shop usually has a markup of 25% on pants. Those that do not sell are discounted a percent based on the selling price. **(a)** If $x$ represents this percent as a decimal, determine a function $S$ for the discount price of pants which cost \$40 per pair. **(b)** What would be a function $T$ representing the total price if a 5% sales tax were added? **(c)** Find the percent discount that makes the discount price (excluding sales tax) the same as the original cost.

**42.** **RETAILING** The Appliance Dealer uses a markup of $x$ percent ($x$ in decimal notation) on all products. A sale with markdowns of 50% starts next month. **(a)** Find a function $S$ that represents the sale price of a toaster that originally cost \$30. **(b)** What would be a function $T$ representing the total price if a 6% sales tax were added? **(c)** What would have to be the markup if the sale price (excluding sales tax) were to be \$3.00 below the original cost?

**43.** **ENGINEERING** Spherical tanks of radius $r$ are to be insulated at a cost of \$1.25 per square foot. **(a)** Find the total cost $T$ for the insulation job as a function of $r$. **(b)** Find the total cost when $r = 9.6$ ft.

**44.** **AGRICULTURE** It costs 0.03¢ (0.0003 dollars) per liter to fill a spherical tank of radius $r$ meters with liquid fertilizer. **(a)** Find the cost $C$ (in dollars) as a function of $r$. **(b)** Find $C$ for a tank with radius 5.2 m. [Hint: $1 \text{ m}^3 = 1000$ liters]

**45.** **MINING** Empirical data has shown that the cost in dollars of extracting $x$ percent of a metal from a ton of ore is given by the function $C(x) = 265.9x^2 - 36.8$, where $0.75 \le x \le 0.95$. **(a)** Find the range of this function over the domain $0.75 \le x \le 0.95$. **(b)** Use the number $x = 0.30$ outside the domain to show why the function is restricted to its domain.

**46.** **ECOLOGY** The cost of cleaning up a small oil spill has been found to be given by the function $C(x) = \dfrac{582.6}{1 - x} - 726.4$, $0.5 \le x \le 0.9$, where $x$ is the percent of the spill removed. **(a)** Determine the range of the function for the domain $0.5 \le x \le 0.9$. **(b)** Use a cleanup of 15% to show why the function is restricted to its domain.

## For Review

**47.** Find the equation of the line passing through $(-1, 3)$ and $(2, 4)$.

**48.** Find the equation of the line with $y$-intercept $(0, -3)$ and perpendicular to $4x + 2y - 7 = 0$.

**49.** Find the equation of the line passing through $(8, 5)$ and perpendicular to the $x$-axis.

**50.** Find the equation of the perpendicular bisector of the line segment joining the intercepts of $5x - 2y = 10$.

**51.** Find the intercepts and graph the equation $2x - 5 = 0$.

**52.** Find the slope of the line through the two points $(-2, 4)$ and $(-8, -3)$.

## 3.4 Properties of Functions and Graphs

### Vertical Line Test

We now investigate the nature of the graph of a function. By definition a function has the property that for each $x$ in the domain there is one and only one $y$ in the range. Thus, if we draw a vertical line through any point on the $x$-axis it will cross the graph of a function in at most *one* point. Using this **vertical line test** on the graphs in Figure 16, we see that **(a)** and **(b)** are the graphs of functions but **(c)** and **(d)** are not.

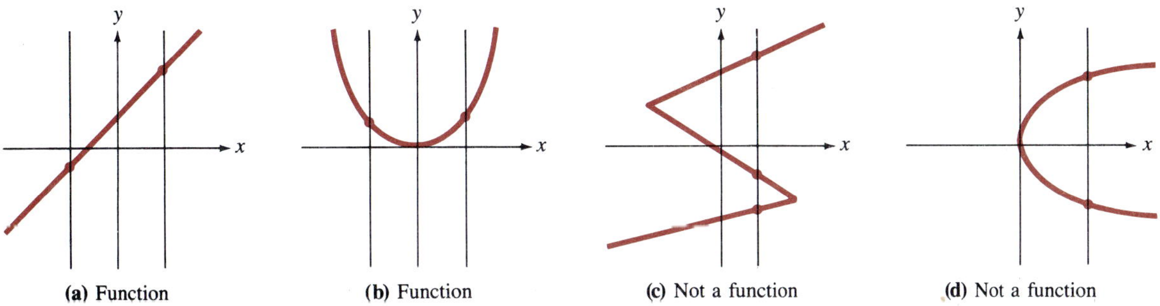


**Figure 16 Vertical Line Test**

### Increasing, Decreasing, and Constant Functions

The graph of a function can be classified using the words *increasing, decreasing,* and *constant*. The graph of a linear function will be increasing if the slope is positive, decreasing if the slope is negative, and constant if the slope is zero. The graphs of $f(x) = \frac{1}{2}x - 3$, $g(x) = -2x + 4$, and $h(x) = 3$ are used to illustrate these three cases in Figure 17.

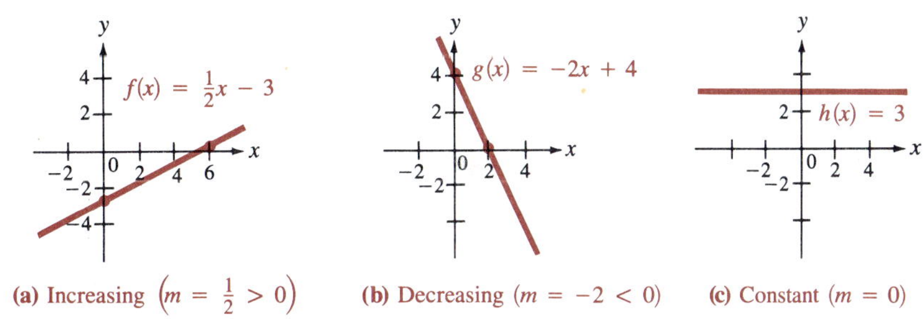


**Figure 17 Linear Functions**

If $a$ and $b$ are any numbers in an interval $I$ in the domain of $f$, then

**1.** $f$ is **increasing** on $I$ if whenever $a < b$, then $f(a) < f(b)$.

**2.** $f$ is **decreasing** on $I$ if whenever $a < b$, then $f(b) < f(a)$.

**3.** $f$ is **constant** on $I$ if $f(a) = f(b)$ for all $a$ and $b$ in $I$.

Thus, for an increasing function, $y$ increases as $x$ increases, and for a decreasing function, $y$ decreases as $x$ increases.

The linear functions in Figure 17 had one of these properties for all values of $x$. Some functions will increase on one interval and decrease on another. Consider the **absolute value function,**

$$f(x) = |x| = \begin{cases} x, & x \geq 0 \\ -x, & x < 0. \end{cases}$$

To graph this function, graph $f(x) = x$ for all $x \geq 0$ and $f(x) = -x$ for all $x < 0$. The graph is shown in Figure 18.

| $x$ | $f(x)$ |
|---|---|
| −5 | 5 |
| −1 | 1 |
| 0 | 0 |
| 1 | 1 |
| 5 | 5 |

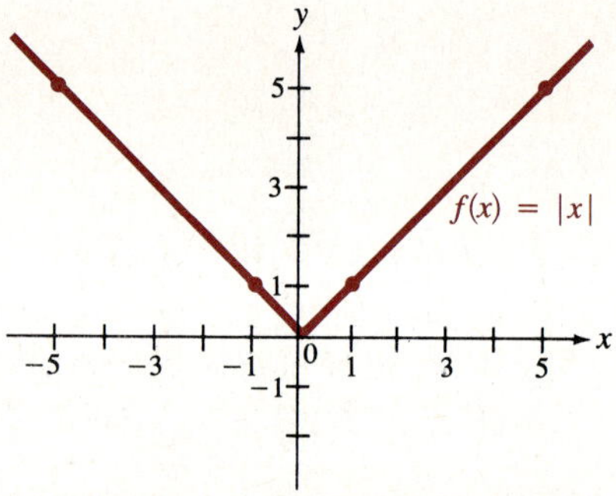


**Figure 18 Absolute Value Function**

Notice that for $x \leq 0$ the function is decreasing while for $x \geq 0$ it is increasing.

## EXAMPLE 1

Graph $g(x) = -x^2 + 3$ and determine the values of $x$ for which $g$ is increasing and for which $g$ is decreasing.

| $x$ | $g(x)$ |
|---|---|
| 0 | 3 |
| 1 | 2 |
| −1 | 2 |
| 2 | −1 |
| −2 | −1 |
| 3 | −6 |
| −3 | −6 |

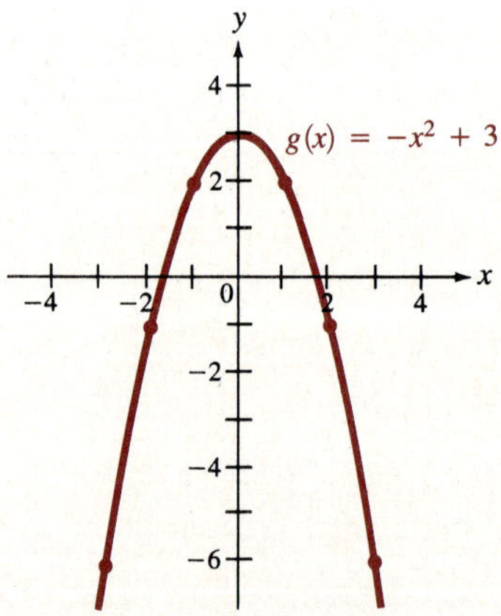


**Figure 19 Quadratic Function**

As Figure 19 indicates, $g$ is increasing for $x \leq 0$ and decreasing for $x \geq 0$. ■

## Shifts in Graphs

We now use the absolute value function shown in Figure 18 to discuss shifts in graphs. Suppose we graph the functions $g(x) = |x| + 2$ and $h(x) = |x| - 3$ in Figure 20. The graph of $f(x) = |x|$ is shown with dashed lines.

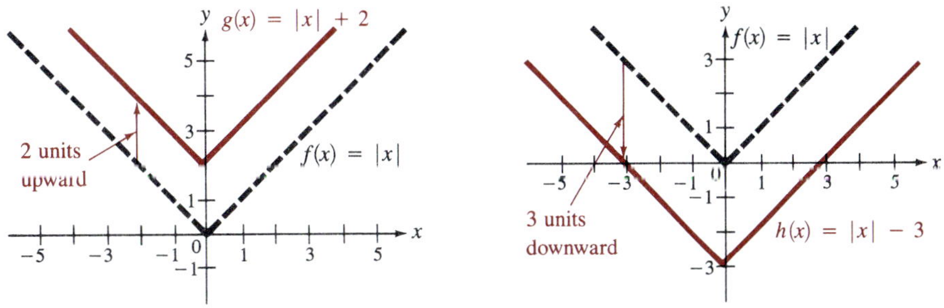


(a) Upward shift (b) Downward shift

**Figure 20**

Notice that the graph of $g(x) = |x| + 2$ in Figure 20(**a**) is the graph of $f(x) = |x|$ shifted upward 2 units, and $h(x) = |x| - 3$ in Figure 20(**b**) is shifted downward 3 units. Thus, $g$ can be obtained from $f$ by a vertical shift of 2 units upward. The function $h$ requires a vertical shift of 3 units downward.

### Vertical Shifts in a Function

Let $c$ be a positive number.

1. To obtain the graph of $f(x) + c$ shift the graph of $f(x)$ upward $c$ units.
2. To obtain the graph $f(x) - c$ shift the graph of $f(x)$ downward $c$ units.

Vertical shifts are a type of change in a graph called a **transformation.** Another transformation is a horizontal shift illustrated by the graphs of the functions $g(x) = |x + 2|$ and $h(x) = |x - 3|$ in Figure 21. The graph of $f(x) = |x|$ is shown with dashed lines.

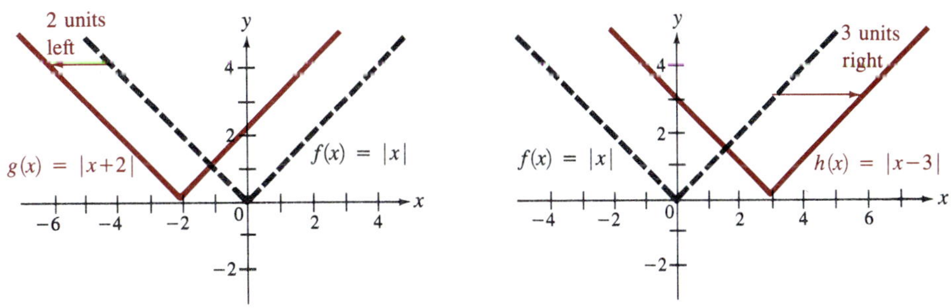


(a) Shift left (b) Shift right

**Figure 21**

Figure 21(**a**) shows that the graph of $g(x) = |x + 2|$ is shifted 2 units to the left of $f(x) = |x|$. In Figure 21(**b**), $h(x) = |x - 3|$ is shifted 3 units to the right of $f(x) = |x|$.

### Horizontal Shifts in a Function

Let $c$ be a positive number.

**1.** To obtain the graph of $f(x + c)$ shift the graph of $f(x)$ $c$ units to the left.

**2.** To obtain the graph of $f(x - c)$ shift the graph of $f(x)$ $c$ units to the right.

## EXAMPLE 2

Graph each function by using vertical and horizontal shifts of $f(x) = |x|$.

**(a)** $g(x) = |x + 1| - 2$
This is a function of the form $f(x + c_1) - c_2$ for $c_1 = 1$ and $c_2 = 2$. Thus, the graph of $f(x)$ is shifted one unit to the left and two units downward as in Figure 22(**a**).

**(b)** $h(x) = |x - 2| + 1$
This is a function of the form $f(x - c_1) + c_2$ where $c_1 = 2$ and $c_2 = 1$. This means that the graph of $f(x)$ is shifted two units to the right and one unit upward as shown in Figure 22(**b**). ■

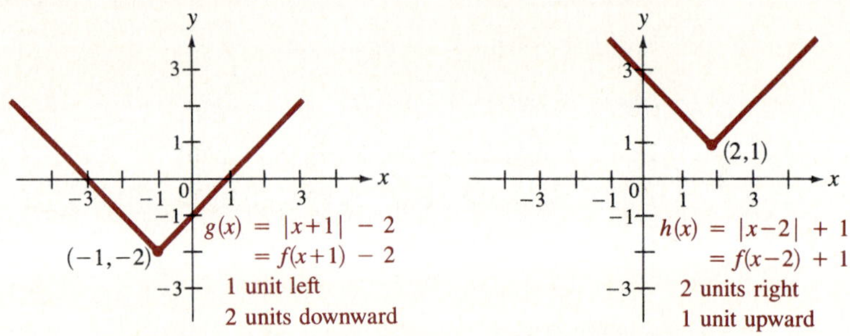


**(a)** Shift left and down **(b)** Shift right and up

**Figure 22**

## EXAMPLE 3

Using the graph of $f(x) = x^2$ in Figure 23(**a**), find the graph of the given function.

**(a)** $g(x) = -x^2$
Notice that each ordinate of $g(x) = -x^2$ is the negative of the corresponding ordinate of $f(x) = x^2$. Thus, the graph is the reflection of $f(x)$ across the $x$-axis. See Figure 23(**b**).

**(b)** $h(x) = 3x^2$
Each ordinate of $f(x) = x^2$ is multiplied by a factor of 3. Thus, for any $x$ the ordinate of $h(x)$ is 3 times the ordinate of $f(x)$. See Figure 23(**c**).

**(c)** $k(x) = \frac{1}{2}x^2$

Each ordinate of $f(x) = x^2$ is multiplied by 1/2 or divided by 2. Thus, for any $x$ the graph of $h(x)$ is 1/2 the height of $f(x)$. See Figure 23(**d**). ■

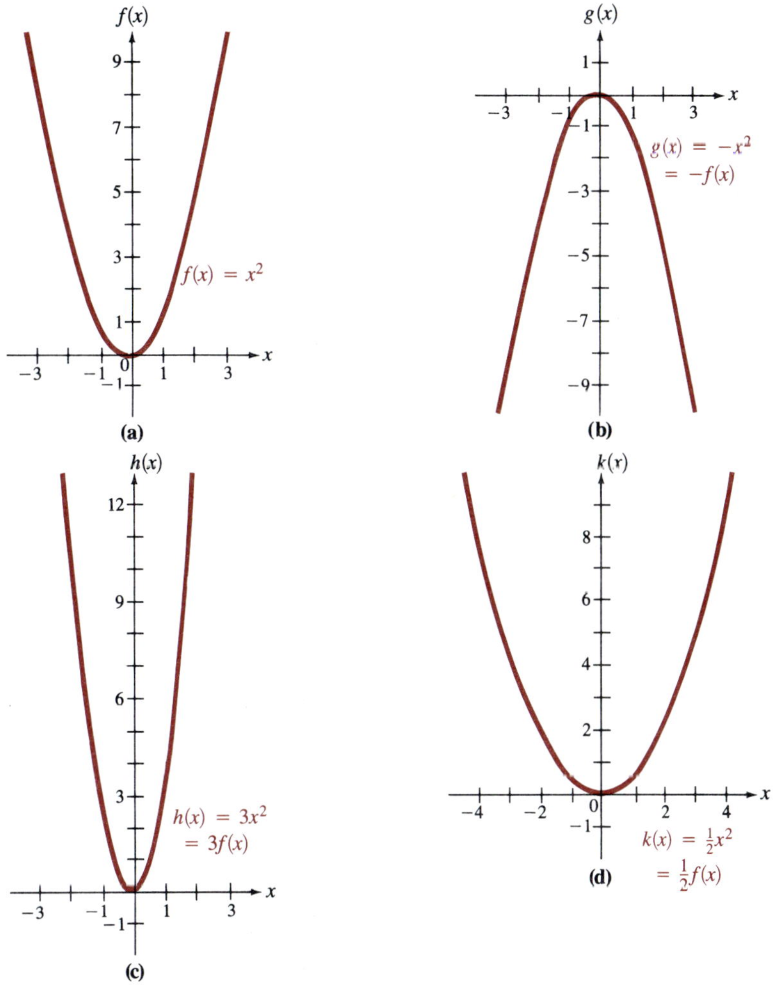


**Figure 23**

## Even and Odd Functions

Each of the graphs in Figure 23 is symmetric with respect to the $y$-axis. Notice that $f(x) = f(-x)$, that is, the graph has the same ordinate for $x$ and $-x$.

If for all $x$ in the domain of $f$,

1. $f(x) = f(-x)$ then $f$ is an **even function;**
2. $f(x) = -f(-x)$ then $f$ is an **odd function.**

**EXAMPLE 4** Determine if the given function is even or odd.

**(a)** $f(x) = -x^2 + 3$
We must determine if $f(x) = f(-x)$ or $f(x) = -f(-x)$.

$$f(-x) = -(-x)^2 + 3 = -x^2 + 3 = f(x)$$

Thus $f$ is even. The graph shown in Figure 24(**a**) is symmetric with respect to the $y$-axis. Thus, for $(x, f(x))$ on the graph $(-x, f(x))$ is also on the graph.

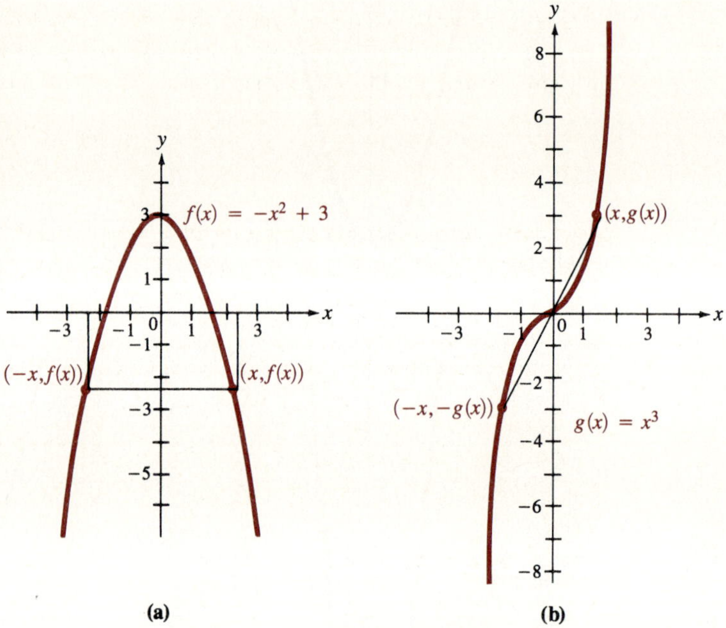


**Figure 24 Even and Odd Functions**

**(b)** $g(x) = x^3$

$$g(-x) = (-x)^3 = -x^3 = -g(x)$$

Since $g(-x) = -g(x)$ is equivalent to $g(x) = -g(-x)$, the function is odd. Notice that the graph of $g$ in Figure 24(**b**) is symmetric with respect to the origin. That is, for $(x, g(x))$ on the graph $(-x, -g(x))$ is also on the graph.

**(c)** $h(x) = x^3 + x^2 - 1$

$$h(-x) = (-x)^3 + (-x)^2 - 1 = -x^3 + x^2 - 1$$

Since $h(-x) \neq h(x)$ and $h(-x) \neq -h(x)$, the function is neither even nor odd. ■

## Tests for Symmetry

For relations in general we test symmetry with respect to the $y$-axis by replacing $x$ with $-x$. If the equation is unchanged the graph is symmetric with respect to the $y$-axis. For example, consider the equation of a circle with center at the origin and radius 3, $x^2 + y^2 = 9$. Replace $x$ with $-x$.

$$(-x)^2 + y^2 = 9$$
$$x^2 + y^2 = 9$$

Since $(-x)^2 = x^2$ the graph is symmetric with respect to the $y$-axis. Replace $y$ with $-y$.

$$x^2 + (-y)^2 = 9$$
$$x^2 + y^2 = 9$$

This means that the graph is symmetric with respect to the $x$-axis. Observe that a graph that is symmetric with respect to the $x$-axis cannot be the graph of a function.

The graph of $x^2 + y^2 = 9$ is also symmetric with respect to the origin. Replace $x$ with $-x$ and $y$ with $-y$ to check this.

$$(-x)^2 + (-y)^2 = 9$$
$$x^2 + y^2 = 9$$

**Tests for Symmetry**

If the equation remains unchanged when:

1. $x$ is replaced by $-x$, there is symmetry with respect to the $y$-axis.
2. $y$ is replaced by $-y$, there is symmetry with respect to the $x$-axis.
3. $x$ is replaced by $-x$ and $y$ is replaced by $-y$, there is symmetry with respect to the origin.

**EXAMPLE 5** Test the symmetry of each relation.

**(a)** $y = |x| + 6$

Replace $x$ by $-x$.

$y = 2|-x| + 6 = 2|x| + 6.$ Symmetric with respect to the $y$-axis

Replace $y$ by $-y$.

$-y = 2|x| + 6$ Not symmetric with respect to $x$-axis

Replace $x$ by $-x$ and $y$ by $-y$.

$-y = 2|-x| + 6 = 2|x| + 6$ Not symmetric with respect to the origin

**(b)** $y^2 = x - 5$

Replace $x$ by $-x$.

$y^2 = (-x) - 5 = -x - 5$ Not symmetric with respect to $y$-axis

Replace $y$ by $-y$.

$(-y)^2 = x - 5$

$y^2 = x - 5$ Symmetric with respect to $x$-axis

Considering the first two tests together, we can see that the relation is not symmetric with respect to the origin.

**(c)** $y^4 + x^2 = 5$

Replace $x$ by $-x$ and $y$ by $-y$.

$(-y)^4 + (-x)^2 = 5$

$y^4 + x^2 = 5$ Symmetric with respect to the origin

Notice that the graph is symmetric with respect to the $y$-axis and $x$-axis also.

**(d)** $y^4 + x^2 = xy$

Replace $x$ by $-x$ and $y$ by $-y$.

$(-y)^4 + (-x)^2 = (-x)(-y)$

$y^4 + x^2 = xy$ Symmetric with respect to the origin

Replace $x$ by $-x$.

$y^4 + (-x)^2 = (-x)y$

$y^4 + x^2 = -xy$ Not symmetric with respect to $y$-axis

Similarly, the graph is not symmetric with respect to the $x$-axis. ■

## Graphs That Are Not Continuous Curves

We conclude this section by considering functions whose graphs are not continuous curves. For example, consider the following functions.

$$f(x) = \begin{cases} -2, & x \leq -3 \\ x - 1, & -3 < x < 0 \\ x^2, & x \geq 0 \end{cases}$$

This function is defined by different relations on different intervals of its domain. In Figure 25 the solid dot at breaks in the graph shows the value of the function. There is an open dot on the other part of the curve at this value of $x$.

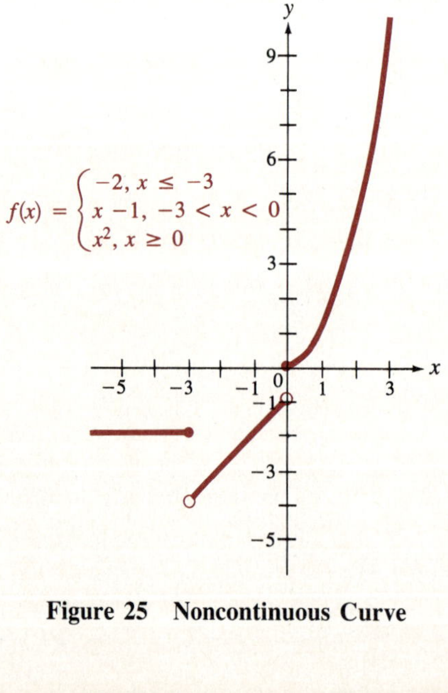


**Figure 25** **Noncontinuous Curve**

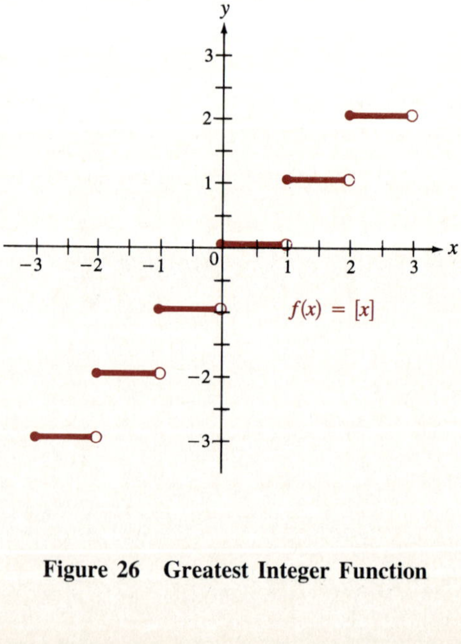


**Figure 26** **Greatest Integer Function**

The graph in Figure 26 shows the **greatest integer function.** It is denoted by

$$f(x) = [x] \qquad \text{Greatest integer function}$$

and is defined to be the largest integer $n$ such that $n \le x$. Thus $f(x) = [x] = -2$ for $-2 \le x < -1$ and $f(x) = [x] = 1$ for $1 \le x < 2$. Note that $[\sqrt{2}] = 1$, $[-1.4] = -2$, and $[\pi] = 3$.

## 3.4 EXERCISES

Use the vertical line test in Exercises 1–6 to determine if each graph is the graph of a function.

**1.**

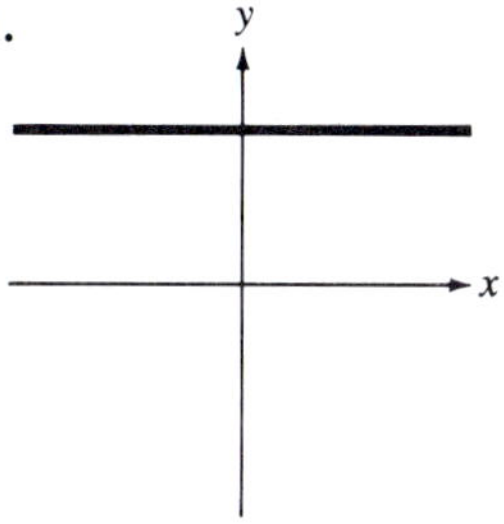


**2.**

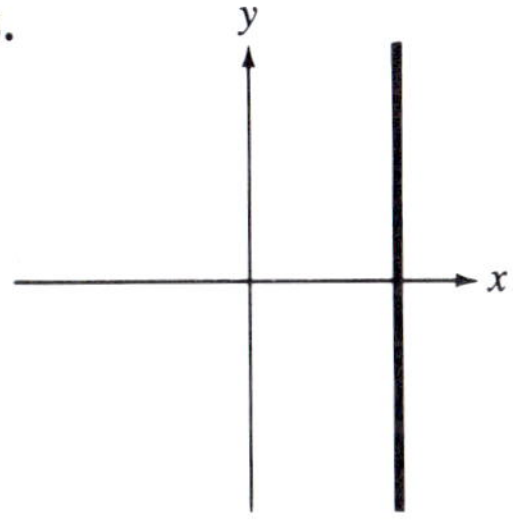


**3.**

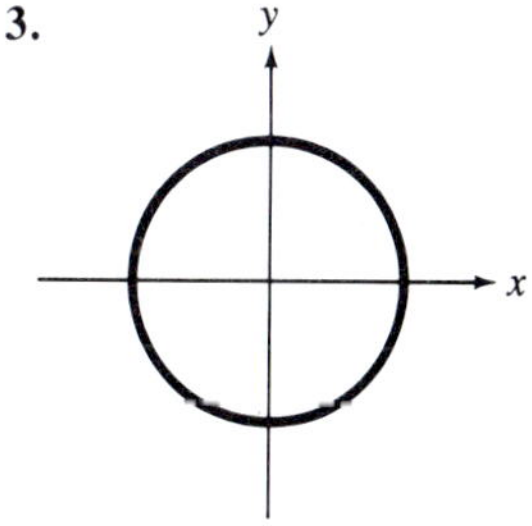


**4.**

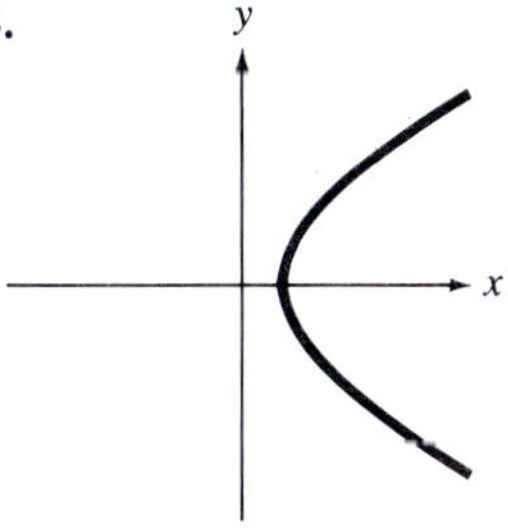


**5.**

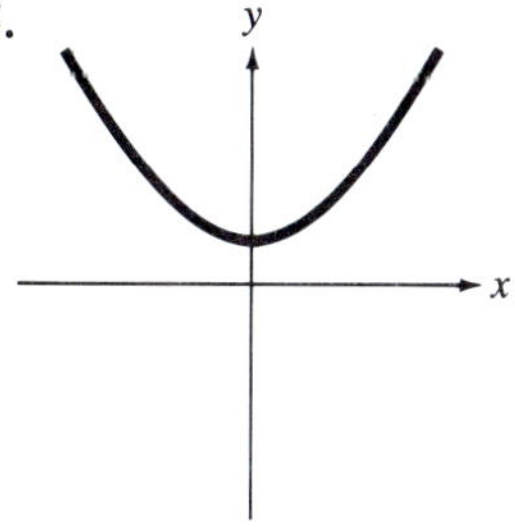


**6.**

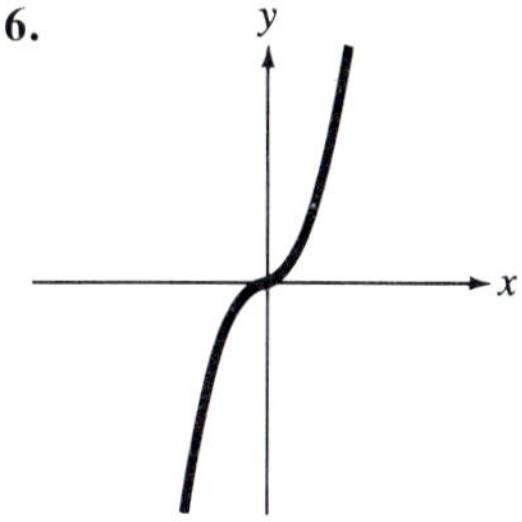


In Exercises 7–14 graph the function and determine the values of $x$ for which the function increases and the values for which it decreases.

**7.** $f(x) = 5x - 3$ **8.** $g(x) = -2x$ **9.** $h(x) = -4$ **10.** $h(x) = 4$

**11.** $k(x) = |x| - 2$ **12.** $f(x) = |x + 1|$ **13.** $g(x) = -(x - 3)^2$ **14.** $g(x) = x(x - 2)$

The graph of $f$ is given in the figure. Use this to sketch the graph of each function in Exercises 15–22.

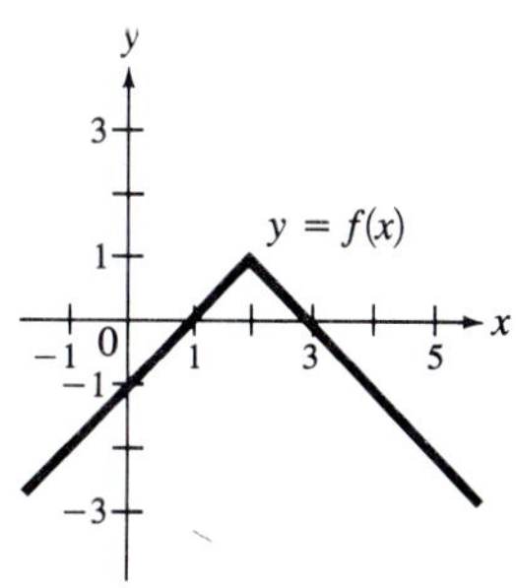


**15.** $y = f(x) + 3$ **16.** $y = f(x) - 2$ **17.** $y = f(x + 3)$

**18.** $y = f(x - 2)$ **19.** $y = -f(x)$ **20.** $y = 2f(x)$

**21.** $y = f(x + 3) - 2$ **22.** $y = f(x - 3) + 2$

The graph of $g$ is given in the figure. Use this to sketch the graph of each function in Exercises 23–30.

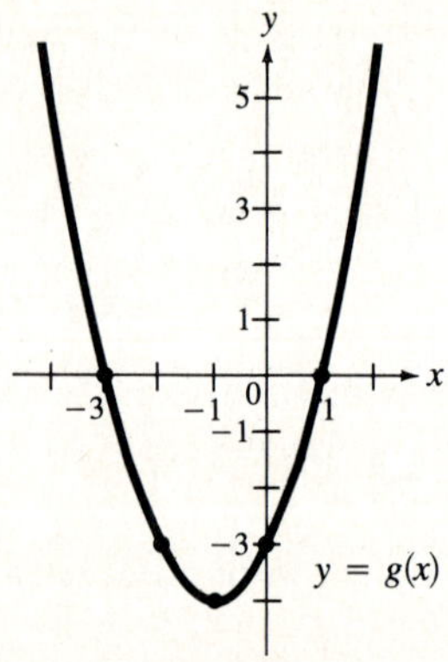


**23.** $y = g(x) - 1$ **24.** $y = g(x) + 4$ **25.** $y = g(x - 3)$

**26.** $y = g(x + 1)$ **27.** $y = \frac{1}{2}g(x)$ **28.** $y = -\frac{3}{4}g(x)$

**29.** $y = g(x - 2) + 3$ **30.** $y = g(x + 1) - 1$

Use the graphs of $f$, $g$, and $h$, given in the figure, for Exercises 31–34.

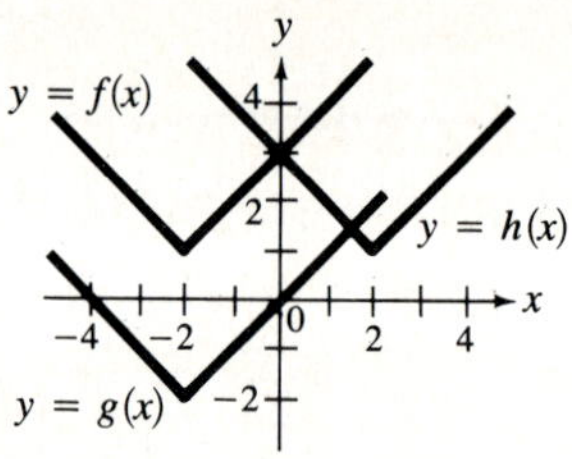


**31.** Determine $g(x)$ in terms of $f(x)$. **32.** Determine $h(x)$ in terms of $f(x)$.

**33.** Determine $h(x)$ in terms of $g(x)$. **34.** Determine $g(x)$ in terms of $h(x)$.

In Exercises 35–40 determine if each function is even or odd.

**35.** $f(x) = 2x^3 - 3x$ **36.** $f(x) = -4x^6 + x^4$ **37.** $h(x) = 0$

**38.** $h(x) = 3$ **39.** $k(x) = 3|x| - 5$ **40.** $k(x) = 3x^2 - 2x + 1$

Without graphing, determine the symmetry of the graphs in Exercises 41–48.

**41.** $y^2 + x^2 = 8$ **42.** $xy = 5$ **43.** $y = 3x^2 + x^4$ **44.** $x^2 + xy^2 = 5$

**45.** $y = -3x^2 + 2x$ **46.** $y = \sqrt{x}$ **47.** $x^2y + xy^2 = 0$ **48.** $(x + y)(x - y) = 8$

Graph the functions in Exercises 49–54. The greatest integer function is written $[x]$.

**49.** $f(x) = \begin{cases} 3, & x \leq 1 \\ -2, & x > 1 \end{cases}$ **50.** $g(x) = \begin{cases} -2, & x < -2 \\ x^2, & -2 \leq x \leq 2 \\ 4, & x > 2 \end{cases}$ **51.** $h(x) = \begin{cases} \dfrac{x^2 - 4}{x + 2}, & x \neq -2 \\ -1, & x = -2 \end{cases}$

**52.** $h(x) = \begin{cases} \dfrac{x^2 - 1}{x - 1}, & x \neq 1 \\ 3, & x = 1 \end{cases}$ **53.** $k(x) = [x] + 2$ **54.** $k(x) = [x - 2]$

## For Review

In Exercises 55–56 determine $f(-2)$, $f(x - 1)$, and $f(x^2)$.

**55.** $f(x) = x^2 - 3$ **56.** $f(x) = (x - 3)^2$

**57.** **MANAGEMENT** All employees of a chemical company were given an 8% increase in salary plus a \$1200 increase. **(a)** Find a function $S$ that represents the new salary in terms of the previous salary $x$. **(b)** If a laboratory technician was making \$18,500, what is her new salary?

**58.** MANUFACTURING The bottom and sides of shipping boxes are made by cutting squares of length $x$ inches from the corners of a piece of cardboard that is 12 inches by 16 inches. **(a)** Find a function $V(x)$ that represents the volume of the box. **(b)** What is the volume when $x = 2$ in?

**59.** Find the distance between the two points $(-2, 6)$ and $(5, 3)$.

**60.** Find the midpoint of the line segment joining the two points $(4, -8)$ and $(2, -2)$.

## 3.5 Composite and Inverse Functions

### Composite Functions

Consider $f(x) = 3x - 7$ and $g(x) = 5x^2 + 4$. Let $f$ be evaluated at $x = 3$, $f(3) = 3(3) - 7 = 2$. Now evaluate $g$ at 2, $g(2) = 5(2)^2 + 4 = 24$. We have determined $g(f(3))$.

If the range of $f$ is contained in the domain of $g$, then the **composite function** $g \circ f$, read "$g$ composed with $f$," is defined by

$$(g \circ f)(x) = g(f(x)).$$

We generally use the notation $g(f(x))$, read "g of $f$ of $x$." If $h(x) = (g \circ f)(x) = g(f(x))$, then $h$ is a function that takes $x$ in the domain of $f$ directly to the range of $g$. This process is illustrated in Figure 27.

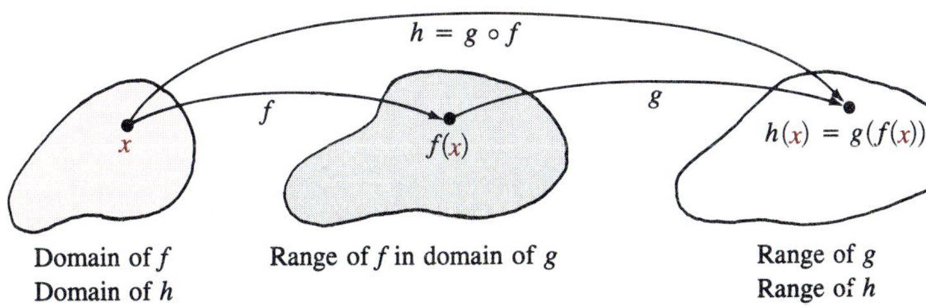


**Figure 27 Composite Function**

**EXAMPLE 1** Let $f(x) = 2x + 1$ and $g(x) = x^2 - 4$. Determine $g(f(x))$ and $f(g(x))$. Are the two functions equal?

$$\begin{aligned}(g \circ f)(x) = g(f(x)) &= g(2x + 1) && f(x) = 2x + 1\\ &= (2x + 1)^2 - 4 && g(x) = x^2 - 4\\ &= 4x^2 + 4x - 3\end{aligned}$$

$$\begin{aligned}(f \circ g)(x) = f(g(x)) &= f(x^2 - 4) && g(x) = x^2 - 4\\ &= 2(x^2 - 4) + 1 && f(x) = 2x + 1\\ &= 2x^2 - 7\end{aligned}$$

Thus $g(f(x)) \neq f(g(x))$ or $g \circ f \neq f \circ g$. Hence, composition of functions is not commutative. ■

## EXAMPLE 2

**ENVIRONMENTAL SCIENCE**

The radius of an oil spill is increasing with time and given by the function $r(t) = \sqrt{t}$ for $3 \le t \le 48$, where $t$ is in hours and $r$ is in miles. The cost $C$ in dollars of cleaning up the spill is given by $C(r) = 2630r^2 + 580r + 9320$.

**(a)** Find the cost $C$ as a function of time.
We must find the composite function $C(r(t))$.

$$\begin{aligned}C(r(t)) &= 2630(\sqrt{t})^2 + 5820(\sqrt{t}) + 9320\\ &= 2630t + 5820\sqrt{t} + 9320\end{aligned}$$

**(b)** Find the approximate cost if the cleaning process is started when $t = 40$ hr.

$$C(40) = 2630(40) + 5820\sqrt{40} + 9320 \approx \$151{,}000$$ ■

## Inverse Functions

Consider the functions $h(x) = x^2$ for $x \ge 0$ and $k(x) = \sqrt{x}$. Suppose we find $h(k(x))$ and $k(h(x))$.

$$\begin{aligned}h(k(x)) &= h(\sqrt{x}) = (\sqrt{x})^2 = x\\ k(h(x)) &= k(x^2) = \sqrt{x^2} = x && \text{Since } x \ge 0\end{aligned}$$

Not only is $h(k(x)) = k(h(x))$ but they are equal to the identity function. The functions $h$ and $k$ are *inverses* of each other.

A function $g$ is the **inverse** of $f$ if

$$f(g(x)) = x \quad \text{for all } x \text{ in the domain of } g$$

and

$$g(f(x)) = x \quad \text{for all } x \text{ in the domain of } f.$$

The inverse of $f$ is usually denoted by $f^{-1}$.

NOTE Using the $f^{-1}$ notation, we have

$$f(f^{-1}(x)) = x \quad \textbf{and} \quad f^{-1}(f(x)) = x.$$ ■

## One-to-One Functions

Before finding the inverse of a function $f$, we must determine which functions have inverses. To show that $h(x) = x^2$ and $k(x) = \sqrt{x}$ were inverses, we had to restrict the domain of $h$ to $x \ge 0$. This restriction makes $h$ a *one-to-one function*.

A function $f$ is a **one-to-one function** if for $x_1$ and $x_2$ in the domain of $f$, $x_1 \neq x_2$ implies $f(x_1) \neq f(x_2)$.

Another way of describing the one-to-one property is to say that for each $y$ in the range of $f$ there is one and only one $x$ in the domain of $f$. Graphically, if any horizontal line crosses the graph of $f$ in more than one point, $f$ is not one-to-one. This **horizontal line test** in Figure 28 shows that $h(x) = x^2$ with domain $x \geq 0$ is one-to-one, while $g(x) = x^2$ with domain all real $x$ is not one-to-one.

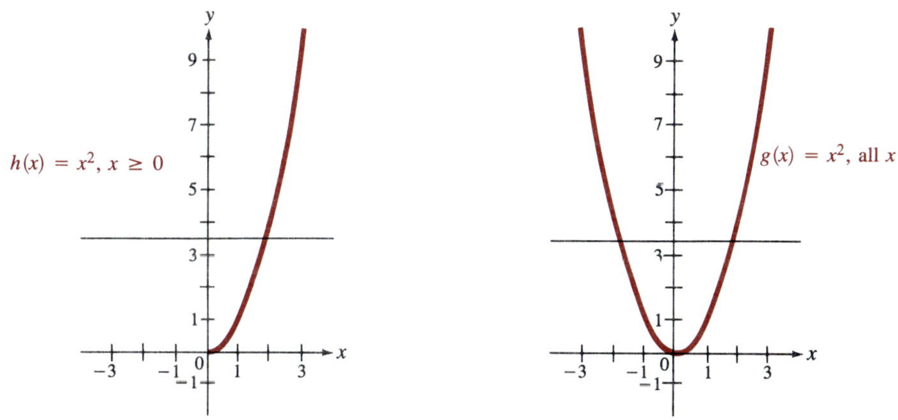


**(a)** One-to-one function **(b)** Not a one-to-one function

**Figure 28 Horizontal Line Test**

The next theorem helps us determine which functions are one-to-one.

If $f$ is an increasing function for all $x$ in its domain or if $f$ is a decreasing function for all $x$ in its domain, then $f$ is one to one.

To see why this is true for increasing functions, consider two different points $x_1$ and $x_2$ in the domain of $f$. If $x_1 < x_2$, then $f(x_1) \neq f(x_2)$ since the increasing property requires $f(x_1) < f(x_2)$. Similarly, the theorem is true for $x_2 < x_1$ and for decreasing functions. This theorem implies that any linear function $f(x) = mx + b$, $m \neq 0$, is a one-to-one function. Only the constant linear functions are not one-to-one.

**EXAMPLE 3** Determine if the given function is one-to-one.

**(a)** $f(x) = x^3$, with domain all $x$

Since the function is increasing for all $x$ (see Figure 24**(b)** on page 138), $f$ is one-to-one.

**(b)** $f(x) = |x|$, with domain all $x$

Since $|x| = -x$ if $x < 0$ and $|x| = x$ if $x \geq 0$, $f$ is decreasing for $x \leq 0$ and increasing for $x \geq 0$ (see Figure 18 on page 134). Thus $f$ is not one-to-one. ■

This leads to the following theorem.

If $f$ is a one-to-one function, then there is a function $f^{-1}$ such that

$$f(f^{-1}(x)) = x = f^{-1}(f(x)).$$

Now that we know which functions have inverses, a technique is needed for finding the inverse of a given function. Consider again the function $h(x) = x^2$ $(x \geq 0)$ and $k(x) = \sqrt{x}$. Figure 29 shows that $h$ and $k$ are symmetric with respect to the line $y = x$. For example, (2, 4) is on the graph of $h$ and the point (4, 2) is on the graph of $k$. In general, if $(x, y)$ is on the graph of $f$, then $(y, x)$ is on the graph of $f^{-1}$.

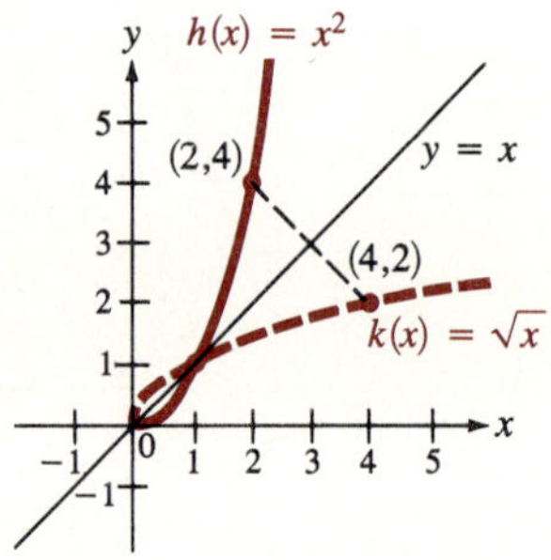


**Figure 29 Inverse Function**

This discussion suggests the following procedure for finding the inverse of a function $f$.

### Procedure for Finding Inverses

1. Write the function using $y$ for $f(x)$.
2. Interchange $x$ and $y$ in the equation.
3. Obtain the expression for $f^{-1}$ by solving the equation for $y$.

We can always prove our results by showing $f(f^{-1}(x)) = x = f^{-1}(f(x))$.

**EXAMPLE 4** Find the inverse of the given function if it exists.

**(a)** $f(x) = 5x - 8$

$f^{-1}$ exists since $f$ is a linear function that is increasing.

$$\begin{aligned} y &= 5x - 8 && \text{Write } y \text{ for } f(x) \\ x &= 5y - 8 && \text{Interchange } x \text{ and } y \\ x + 8 &= 5y \\ y &= \frac{x+8}{5} && \text{Solve for } y \\ f^{-1}(x) &= \frac{x+8}{5} && y \text{ is } f^{-1}(x) \end{aligned}$$

As a check, determine if $f(f^{-1}(x)) = x$ and $f^{-1}(f(x)) = x$.

$$f(f^{-1}(x)) = f\left(\frac{x+8}{5}\right) = 5\left(\frac{x+8}{5}\right) - 8 = x + 8 - 8 = x$$

$$f^{-1}(f(x)) = f^{-1}(5x - 8) = \frac{(5x-8)+8}{5} = \frac{5x - 8 + 8}{5} = x$$

**(b)** $g(x) = \sqrt{x - 3}, \quad x \geq 3$
Since $g$ is an increasing function where it is defined, it has an inverse.

$$\begin{aligned} y &= \sqrt{x-3} && \text{Write } y \text{ for } g(x) \\ x &= \sqrt{y-3} && \text{Interchange } x \text{ and } y \\ x^2 &= y - 3 && \text{Square both sides} \\ y &= x^2 + 3 && \text{Solve for } y \\ g^{-1}(x) &= x^2 + 3 && y \text{ is } g^{-1}(x) \end{aligned}$$

The check will indicate something about the domain of $g^{-1}$.

$$g(g^{-1}(x)) = g(x^2 + 3) = \sqrt{x^2 + 3 - 3} = \sqrt{x^2} = x \quad \text{If } x \geq 0$$
$$g^{-1}(g(x)) = g^{-1}(\sqrt{x-3}) = (\sqrt{x-3})^2 + 3 = x$$

Notice that for $g(g^{-1}(x))$ to be $x$, we required $x \geq 0$. Thus, the domain of $g^{-1}$ is $x \geq 0$. In general, for any function $f$, the domain of $f^{-1}$ is the range of $f$ and the domain of $f$ is the range of $f^{-1}$.

**(c)** $h(x) = x^4 + 2, \quad$ all real $x$
Since $h(x)$ is decreasing for $x \leq 0$ and increasing for $x \geq 0$, it does not have an inverse.

**(d)** $k(x) = 4 - x^3$
Since $k$ is decreasing for all $x$, it has an inverse.

$$\begin{aligned} y &= 4 - x^3 && \text{Write } y \text{ for } k(x) \\ x &= 4 - y^3 && \text{Interchange } x \text{ and } y \\ y^3 &= 4 - x \\ y &= \sqrt[3]{4-x} && \text{Solve for } y \\ k^{-1}(x) &= \sqrt[3]{4-x} && y \text{ is } k^{-1}(x) \end{aligned}$$

Check: $k(k^{-1}(x)) = k(\sqrt[3]{4-x}) = 4 - (\sqrt[3]{4-x})^3 = 4 - (4 - x) = x$
$k^{-1}(k(x)) = k^{-1}(4 - x^3) = \sqrt[3]{4 - (4 - x^3)} = \sqrt[3]{x^3} = x$ ■

## 3.5 EXERCISES

Determine $f(g(2))$, $g(f(-1))$, $f(g(x))$, and $g(f(x))$ for each pair of functions in Exercises 1–8.

**1.** $f(x) = 5x$ and $g(x) = 3x - 2$

**2.** $f(x) = 1 - x$ and $g(x) = 5x + 1$

**3.** $f(x) = 2x^2$ and $g(x) = -x$

**4.** $f(x) = \sqrt{x^2 + 5}$ and $g(x) = x^4$

**5.** $f(x) = -2x + 1$ and $g(x) = x^2 + 2$

**6.** $f(x) = 2x + 3$ and $g(x) = (x - 3)^2$

**7.** $f(x) = x^3 + 1$ and $g(x) = \sqrt[3]{x - 1}$

**8.** $f(x) = \dfrac{x}{1 - x}$ and $g(x) = \dfrac{x}{x + 1}$

In Exercises 9–14 determine which functions are one-to-one.

**9.** $f(x) = -2x + 6$

**10.** $g(x) = 2 - 3x^2$

**11.** $h(x) = |x + 1|$

**12.** $h(x) = \dfrac{1}{x}$

**13.** $k(x) = x^3 + 1$

**14.** $k(x) = \sqrt{x - 2}$

In Exercises 15–18 determine if the functions are inverses of each other and graph them in the same coordinate system.

**15.** $f(x) = \sqrt{x - 2}$ and $g(x) = x^2 + 2, \quad x \geq 0$

**16.** $f(x) = -2x + 5$ and $g(x) = \dfrac{5 - x}{2}$

**17.** $f(x) = \sqrt{x + 4}$ and $g(x) = x^2 - 4, \quad x \geq 0$

**18.** $f(x) = \dfrac{1}{2}x^3$ and $g(x) = \sqrt[3]{2x}$

Find the inverse of each function in Exercises 19–26.

**19.** $f(x) = 2x + 3$

**20.** $g(x) = 1 - x^2, \quad x \geq 0$

**21.** $h(x) = \sqrt{x - 3}$

**22.** $h(x) = \sqrt{x + 2}$

**23.** $k(x) = 2x^3$

**24.** $k(x) = \sqrt[3]{x - 3}$

**25.** $g(x) = \sqrt{1 - x^2}, \quad 0 \leq x \leq 1$

**26.** $g(x) = \sqrt{x^2 - 1}, \quad x \geq 1$

**27.** Find the inverse of $f(x) = mx + b$. Are there any restrictions on $m$ or $b$?

**28.** Find the inverse of $g(x) = ax^2 + c$, $x \geq 0$. Are there any restrictions on $a$ or $c$?

**29.** ECOLOGY The radius of a circular grass fire is increasing with time according to the equation $r(t) = 2t^2 - 1$. **(a)** Find the circumference $C$ of the fire as a function of time. **(b)** Find the area burned $A$ as a function of time.

**30.** METEOROLOGY The radius of a weather balloon is increasing with time according to the equation $r(t) = \sqrt{t + 2}$. **(a)** Find the volume $V$ of the balloon as a function of time. **(b)** Find the surface area $S$ of the balloon as a function of time.

**31.** ECOLOGY An oil spill is moving down a river which is 22 m wide. The distance in kilometers that it has traveled at time $t$ in hours is given by $d(t) = 3t^2 + 2t + 1$. **(a)** Find the area $A$, in square meters, covered by the spill as a function of time. **(b)** Find the area after 2 hours.

**32.** ENGINEERING The cost of cleaning the spill in Exercise 31 is given by the equation $C(d) = 2220d + 9530$, where $C$ is in dollars and $d = 3t^2 + 2t + 1$ is in kilometers. **(a)** Find $C$ as a function of time. **(b)** Find $C$ if the spill was stopped after 2 hr.

### For Review

In Exercises 33–34 determine what effect the transformation has on the graph of $f$.

**33.** $y = f(x - 4) + 2$

**34.** $y = f(x + 4) - 2$

Determine whether each function in Exercises 35–36 is even or odd.

**35.** $f(x) = 4x^3 + 2x$ **36.** $g(x) = |x^3|$

**37.** Find the equation of the line passing through $(-6, 2)$ and parallel to the line $3x - 2y + 5 = 0$.

**38.** Find the equation of the perpendicular bisector of the line segment joining $(4, 6)$ and $(-8, -2)$.

**39.** What must be added to complete the square in each expression? **(a)** $x^2 - 8x$ **(b)** $x^2 + 3x$

**40.** Solve. $2x^2 + 9x - 5 = 0$

## 3.6 Quadratic Functions

In Section 3.4 graphs of functions of the form $f(x) = ax^2 + c$ were used to study various properties of functions. Now we consider this type of function in more detail.

A **quadratic function** is a function of the form

$$f(x) = ax^2 + bx + c$$

where $a$, $b$, and $c$ are constant and $a \neq 0$.

In Figure 30 the graphs of $f(x) = x^2 + 2x - 3$ and $g(x) = -x^2 + 2x + 3$ are determined by plotting points.

| $x$ | $f(x)$ |
|---|---|
| 0 | −3 |
| 1 | 0 |
| −1 | −4 |
| 2 | 5 |
| −2 | −3 |
| −3 | 0 |
| −4 | 5 |

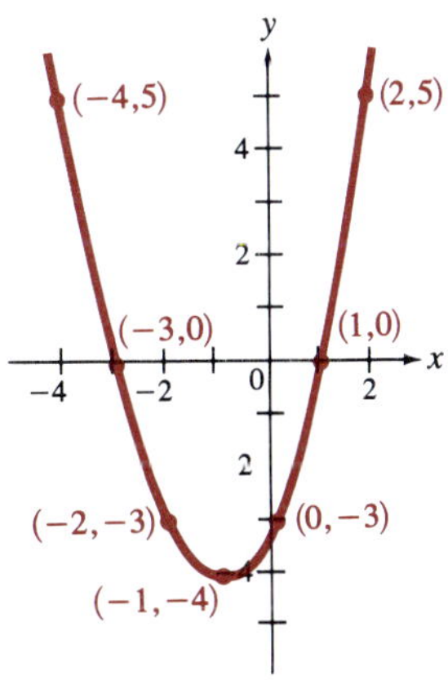


(a) $f(x) = x^2 + 2x - 3$

| $x$ | $g(x)$ |
|---|---|
| 0 | 3 |
| 1 | 4 |
| −1 | 0 |
| 2 | 3 |
| −2 | −5 |
| 3 | 0 |
| 4 | −5 |

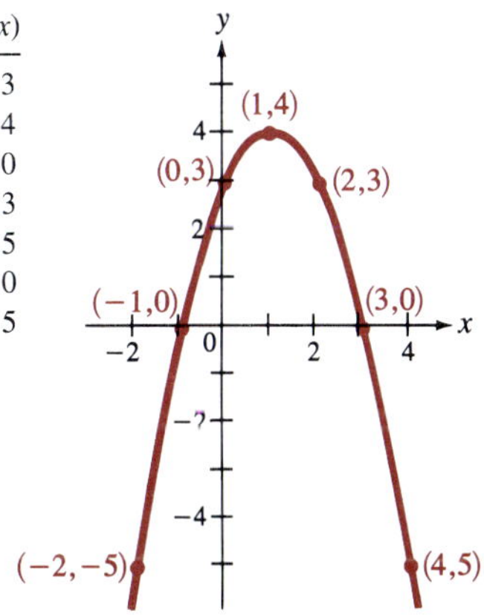


(b) $g(x) = -x^2 + 2x + 3$

**Figure 30 Parabolas**

These graphs are typical of quadratic functions. All are *parabolas* that open up if $a > 0$ as in Figure 30(**a**) and open down if $a < 0$ as in Figure 30(**b**). Notice also that each has a minimum or maximum point called the **vertex.** For example, the vertex in Figure 30(**a**) is $(-1, -4)$.

## Standard Form

Graphing quadratic functions is made easier by finding the vertex before plotting points. Suppose we complete the square on the following function.

$$\begin{aligned} f(x) &= 2x^2 + 12x + 13 \\ &= 2(x^2 + 6x \qquad) + 13 \\ &= 2(x^2 + 6x + 9) + 13 - 18 && \left(\tfrac{1}{2} \cdot 6\right) = 9 \text{ and } 9 \cdot 2 = 18 \\ &= 2(x + 3)^2 - 5 && x^2 + 6x + 9 = (x + 3)^2 \\ &= 2[x - (-3)]^2 - 5 \end{aligned}$$

Thus, the minimum value of $f$ will occur when $[x - (-3)]^2$ is 0, that is, when $x = -3$. Therefore, the vertex is $(h, k) = (-3, -5)$. If we write any quadratic function in **standard form,**

$$f(x) = a(x - h)^2 + k,$$

the vertex can easily be determined. Rather than completing the square on each quadratic function, we can obtain a formula for the vertex $(h, k)$ by considering a general quadratic function.

$$\begin{aligned} f(x) &= ax^2 + bx + c \\ &= a\left(x^2 + \frac{b}{a}x \qquad\right) + c \\ &= a\left(x^2 + \frac{b}{a}x + \frac{b^2}{4a^2}\right) + c - \frac{b^2}{4a} && \left(\frac{1}{2} \cdot \frac{b}{2}\right)^2 = \frac{b^2}{4a^2} \text{ and } a\left(\frac{b^2}{4a^2}\right) = \frac{b^2}{4a} \\ &= a\left(x + \frac{b}{2a}\right)^2 + c - \frac{b^2}{4a} && x^2 + \frac{b}{a}x + \frac{b^2}{4a^2} = \left(x + \frac{b}{2a}\right)^2 \\ &= a\left[x - \left(-\frac{b}{2a}\right)\right]^2 + \frac{4ac - b^2}{4a} \end{aligned}$$

This proves the following theorem.

### Vertex Formula

If $f(x) = ax^2 + bx + c$ is a quadratic function, then the vertex of the graph is

$$(h, k) = \left(-\frac{b}{2a}, f\left(-\frac{b}{2a}\right)\right).$$

## EXAMPLE 1

Find the vertex of $f(x) = 3x^2 - 2x - 2$.

First find $h$.

$$h = -\frac{b}{2a} = -\frac{(-2)}{2(3)} = \frac{1}{3}$$

Now find $f(h)$.

$$f\left(\frac{1}{3}\right) = 3\left(\frac{1}{3}\right)^2 - 2\left(\frac{1}{3}\right) - 2$$
$$= \frac{1}{3} - \frac{2}{3} - \frac{6}{3} = -\frac{7}{3}$$

The vertex is $\left(\frac{1}{3}, -\frac{7}{3}\right)$. ■

**Intercepts**

Before using the vertex when graphing, we find the ***x*-intercepts.** These are the points where $f(x) = 0$, and can be found by solving the quadratic equation

$$ax^2 + bx + c = 0.$$

From Section 2.3,

$$x = \frac{-b \pm \sqrt{b^2 - 4ac}}{2a}.$$

Three cases occur as shown in Figure 31.

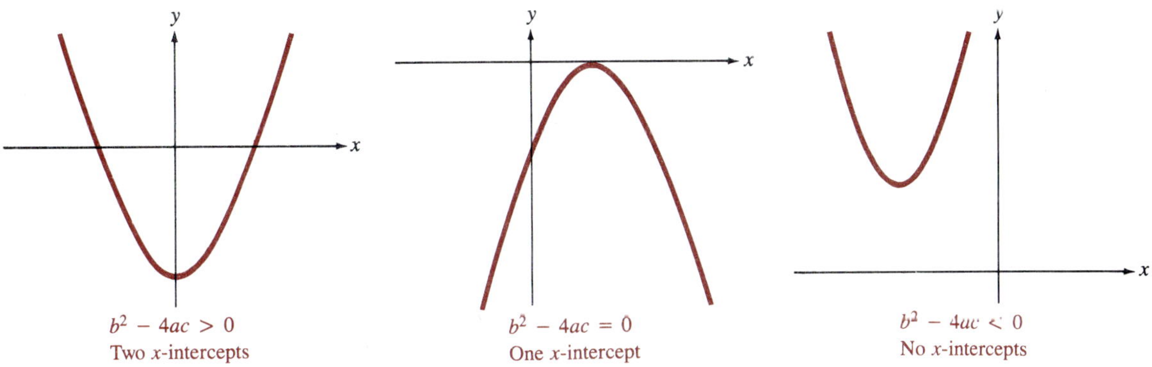


**Figure 31 Intercepts**

All this information is used to graph a quadratic function. First find the vertex and any $x$-intercepts. Then determine as many additional points as necessary to sketch the parabola.

**EXAMPLE 2**

Find the vertex, find any $x$-intercepts, and graph the quadratic function.

**(a)** $f(x) = x^2 - 6x + 5$

Note that $a = 1 > 0$, so the graph opens upward. Now we find the vertex $(h, k)$.

$$h = -\frac{b}{2a} = -\frac{-6}{2(1)} = 3$$
$$k = f\left(-\frac{b}{2a}\right) = f(3) = 3^2 - 6(3) + 5 = -4$$

The vertex is $(3, -4)$. We solve the equation $x^2 - 6x + 5 = 0$ to find the $x$-intercepts.

$$\begin{aligned} x^2 - 6x + 5 &= 0 \\ (x-1)(x-5) &= 0 \\ x - 1 = 0 \quad &\text{or} \quad x - 5 = 0 \\ x = 1 \quad & \quad\quad\; x = 5 \end{aligned}$$

**Line of Symmetry**

The $x$-intercepts are $(1, 0)$ and $(5, 0)$. Now choosing an $x < 1$ and an $x > 5$, we plot two additional points $(0, 5)$ and $(6, 5)$. Notice that the graph shown in Figure 32 is symmetric with respect to the **line of symmetry** $x = 3$.

| $x$ | $f(x)$ |
|---|---|
| 3 | −4 |
| 1 | 0 |
| 5 | 0 |
| 0 | 5 |
| 6 | 5 |

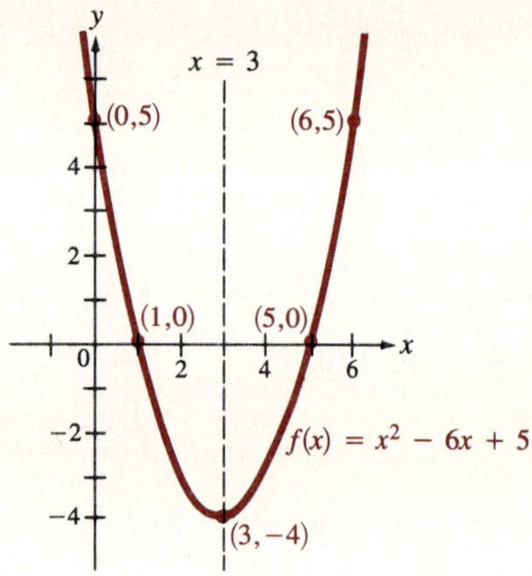


**Figure 32**

**(b)** $g(x) = -3x^2 + 9x - 4$

The graph opens downward since $a = -3 < 0$. Now we find $(h, k)$.

$$h = -\frac{b}{2a} = -\frac{9}{2(-3)} = \frac{3}{2}$$

$$k = f\left(-\frac{b}{2a}\right) = f\left(\frac{3}{2}\right) = (-3)\left(\frac{3}{2}\right)^2 + 9\left(\frac{3}{2}\right) - 4 = \frac{11}{4}$$

The vertex is $\left(\frac{3}{2}, \frac{11}{4}\right)$. Now we use the quadratic formula to solve $-3x^2 + 9x - 4 = 0$ and find the $x$-intercepts.

$$\begin{aligned} x &= \frac{-b \pm \sqrt{b^2 - 4ac}}{2a} \\ &= \frac{-9 \pm \sqrt{(9)^2 - 4(-3)(-4)}}{2(-3)} \\ &= \frac{-9 \pm \sqrt{81 - 48}}{-6} = \frac{-9 \pm \sqrt{33}}{-6} = \frac{9 \pm \sqrt{33}}{6} \end{aligned}$$

We use a calculator to find $x \approx 0.54$ and $x \approx 2.46$. Thus, the $x$-intercepts are $(0.54, 0)$ and $(2.46, 0)$. We plot the additional points and graph the function in

Figure 33. Notice that the line of symmetry is $x = \frac{3}{2}$. Symmetry should be used when determining the additional points to plot. ■

| $x$ | $g(x)$ |
|---|---|
| $\frac{3}{2}$ | $\frac{11}{4}$ |
| 1 | 2 |
| 2 | 2 |
| 0.54 | 0 |
| 2.64 | 0 |
| 0 | −4 |
| 3 | −4 |

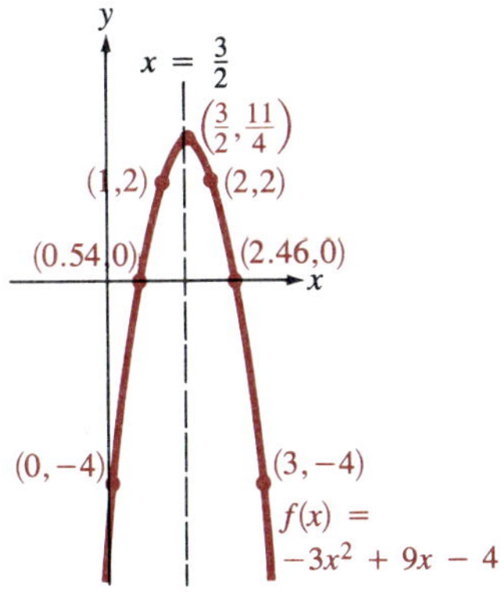


**Figure 33**

The function $f(x) = \frac{1}{2}x^2 + 5x + 13$ illustrates the case when there are no $x$-intercepts. Notice that

$$b^2 \quad 4ac = (5)^2 - 4\left(\frac{1}{2}\right)(13) = 25 - 26 = -1 < 0.$$

Since the discriminant is negative there are no $x$-intercepts. The graph opens upward, and the vertex is $(-5, 0.5)$. Use symmetry with respect to $x = -5$ and plot enough points to determine the graph in Figure 34.

| $x$ | $f(x)$ |
|---|---|
| −5 | 0.5 |
| −4 | 1 |
| −6 | 1 |
| −3 | 2.5 |
| −7 | 2.5 |
| −2 | 5 |
| −8 | 5 |
| −1 | 8.5 |
| −9 | 8.5 |

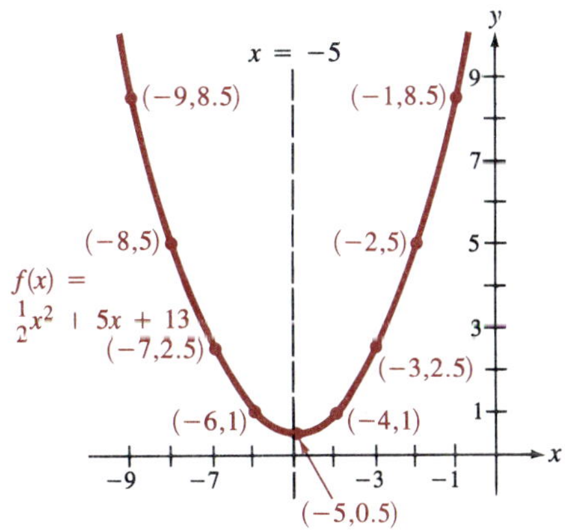


**Figure 34**

Many applications involving quadratic functions are solved by finding the maximum value (if $a < 0$) or the minimum value (if $a > 0$) using the vertex.

## EXAMPLE 3

**AGRICULTURE**

A rancher needs to build a holding pen for cattle along an existing fence. She has available 600 m of fencing to build the 3 sides of the rectangular pen shown in Figure 35. Find the dimensions of the pen that give the maximum area and also find the maximum area.

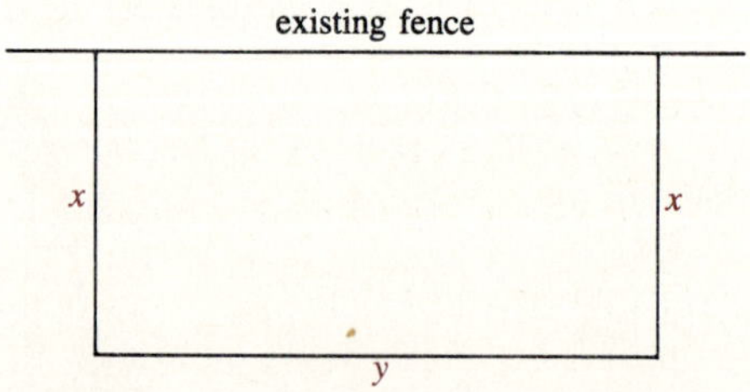


**Figure 35**

Let $x$ = length of sides perpendicular to the existing fence,
$y$ = length of side parallel to the existing fence.
Since 600 m of fencing is available,

$$2x + y = 600 \qquad \text{Length of the sides is 600 m}$$
$$y = 600 - 2x. \qquad \text{Write } y \text{ in terms of } x$$

The quantity to be maximized is the area $A$ of the pen.

$$\begin{aligned} A &= xy \\ &= x(600 - 2x) && y = 600 - 2x \\ &= 600x - 2x^2 \\ &= -2x^2 + 600x && A \text{ is a quadratic function} \end{aligned}$$

Notice that in the function, $a = -2 < 0$, and thus the parabola opens downward and has a maximum. We can find the value of $x$ that gives the maximum by finding $h$ in the vertex $(h, k)$.

$$h = -\frac{b}{2a} = -\frac{600}{2(-2)} = 150$$

Hence, each side perpendicular to the existing fence should be 150 m to give the maximum area. Now find $y$.

$$y = 600 - 2x = 600 - 2(150) = 300$$

The side parallel to the existing fence is 300 m. The maximum area can be found by finding $k$ in $(h, k)$ or by finding the product of $x$ and $y$.

$$A = xy = (150)(300) = 45{,}000 \text{ m}^2$$

The maximum area is 45,000 $\text{m}^2$. ■

The following application was first presented in the introduction to this chapter.

## EXAMPLE 4

**BUSINESS**

A car rental agency averages 20 customers per day for a model which rents for \$32 per day. A survey shows that the agency could get 2 new customers for each \$1 reduction in the daily rental rate. What daily rate would give the company the maximum revenue?

Let $n =$ the number of dollars reduction in the daily rental fee. Then, since there are 2 new customers for each dollar reduction, $2n$ is the number of new customers. Thus,

$$20 + 2n = \text{number of customers after decreasing the daily fee by } n \text{ dollars},$$
$$32 - n = \text{new daily rental fee.}$$

The product $(20 + 2n)(32 - n)$ is the daily revenue that is to be maximized.

$$\begin{aligned}(20 + 2n)(32 - n) &= 640 + 44n - 2n^2 \\ &= -2n^2 + 44n + 640\end{aligned}$$

Since $a = -2 < 0$, the quadratic function has a maximum. Find $h$ in $(h, k)$.

$$h = -\frac{b}{2a} = -\frac{44}{2(-2)} = 11$$

Thus the daily rental fee should be reduced by \$11 to give the maximum revenue. This would make the new fee \$21. ■

## 3.6 EXERCISES

Without graphing, find the vertex and any $x$-intercepts of the quadratic functions in Exercises 1–12.

**1.** $f(x) = x^2 - 5x + 6$

**2.** $g(x) = -x^2 - 6x - 8$

**3.** $h(x) = x^2 + 2x + 2$

**4.** $h(x) = x^2 - 2x + 1$

**5.** $k(x) = -x^2 - 8x$

**6.** $k(x) = x^2 + 10x$

**7.** $m(x) = 2x^2 - 4$

**8.** $m(x) = -3x^2 + 6$

**9.** $f(x) = 2x^2 + 8x - 5$

**10.** $g(x) = -3x^2 + 4x + 3$

**11.** $h(x) = \frac{1}{3}x^2 - 2x + 5$

**12.** $k(x) = -4x^2 + 12x + 5$

In Exercises 13–24 find the vertex, find any $x$-intercepts, and graph the function.

**13.** $g(x) = -x^2 - 2x + 3$

**14.** $h(x) = x^2 - 7x + 6$

**15.** $k(x) = -x^2 + 5x - 2$

**16.** $k(x) = -x^2 - 4x + 2$

**17.** $f(x) = x^2 + 4x + 5$

**18.** $f(x) = x^2 - 3x - 5$

**19.** $g(x) = \frac{1}{2}x^2 - 7x + 20$

**20.** $h(x) = -\frac{2}{3}x^2 - \frac{4}{3}x + 4$

**21.** $k(x) = -4x^2 + 8x - 4$

**22.** $k(x) = \frac{1}{2}x^2 + 4x + 8$

**23.** $m(x) = 2(x + 1)^2 + 2$

**24.** $m(x) = -(x + 2)^2 - 2$

**25.** **AGRICULTURE** A rancher needs to build a pen with 1000 yd of fencing. What are the dimensions of the largest rectangular pen that he can enclose? What is the maximum area?

**26. AGRICULTURE** A rectangular cattle pen is to be built along a straight river bank. If no fence is needed along the river and there are 240 ft of fencing available, what are the dimensions of the largest rectangular pen possible? What is the maximum area?

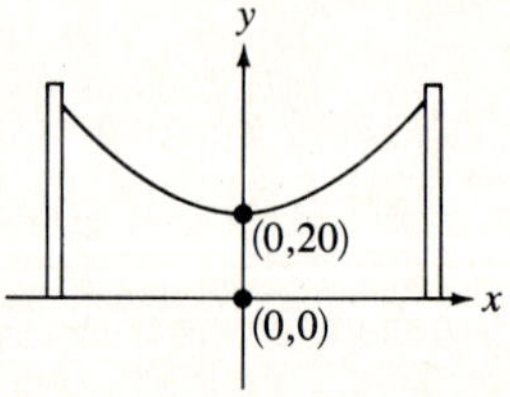


**27. ENGINEERING** The towers for a suspension bridge are 100 ft apart. If a coordinate system is set up as indicated in the figure, the cable between the towers follows the curve $f(x) = \frac{1}{100}x^2 + 20$. At what height on each tower is the cable connected?

**28. ENGINEERING** If the cable in the figure must be connected 30 ft up each tower and $f(x) = \frac{1}{100}x^2 + 20$, how far apart must the towers be placed?

**29. PHYSICS** The height in feet of a small rocket $t$ seconds after it is fired is given by the function $h(t) = -16t^2 + 256t$. **(a)** At what time will the rocket reach its maximum height? **(b)** What is the maximum height? **(c)** At what time will it reach the ground again?

**30. MINING** Coal is dropped from a conveyor belt to the bed on a truck. The height in feet of coal above the bed of the truck is described by $h(t) = -16t^2 + 16$ where $t$ is the number of seconds after the coal leaves the belt. If the velocity of the coal is given by $v(t) = -32t$ in ft/sec, what is the velocity with which the first coal hits the bed of the truck?

**31. MANUFACTURING** The daily profit in a manufacturing plant is given by $P(x) = -x^2 + 120x - 1000$, where $x$ is the number of machine parts produced. What is the number of parts that will give the maximum profit? What is the maximum profit?

**32. MANUFACTURING** The daily manufacturing cost of a small plant is given by $C(x) = 2x^2 - 180x + 5200$ where $x$ is the number of pairs of shoes produced. What is the number of pairs of shoes which will minimize the cost? What is this minimum cost?

**33. MANUFACTURING** The cost in a manufacturing operation is given by a quadratic function. The minimum cost occurs at the point (30, 200). If the curve also goes through the point (0, 2000), find the quadratic function.

**34. RETAILING** The profit in a retail operation is given by a quadratic function. The maximum profit occurs at the point (100, 400). If the curve also goes through the point (80, 200), find the quadratic function.

**35. BUSINESS** Electronic components are sold by the wholesaler for \$40 each on orders of 200 or less. For larger orders the price per component is reduced at a rate of \$0.20 times the number ordered over 200. What size order would give the wholesaler the largest revenue?

## For Review

In Exercises 36–37 determine $f(g(0))$, $g(f(-2))$, $f(g(x))$, and $g(f(x))$.

**36.** $f(x) = -3x + 4$ and $g(x) = 2x - 5$

**37.** $f(x) = |x + 5|$ and $g(x) = x^2$

In Exercises 38–40 determine the inverse of each function.

**38.** $f(x) = 7x + 8$

**39.** $g(x) = \sqrt{x + 4}$

**40.** $h(x) = \sqrt[3]{x + 1}$

In Exercises 41–42 graph the function and determine the values of $x$ for which the function increases and the values for which it decreases.

**41.** $g(x) = -5x - 3$

**42.** $f(x) = |x| + 3$

## 3.7 Variation

### Direct Variation

Expressions such as "the cost of production *varies* with the number of units produced" and "the volume of a gas *varies* with the temperature" are frequently used. *Variation* can be defined precisely with one variable represented as a function of one or more other variables. For example, if a car is moving at a constant rate of 55 miles per hour, the distance traveled is a function of the time.

$$d = 55t$$

Expressed in terms of variation we say "the distance *varies directly* as the time." This is an example of the first type of variation called *direct variation*.

If two variables $x$ and $y$ are so related that

$$y = cx$$

where $c$ is a constant, we say that $y$ **varies directly as** $x$, $y$ **is proportional to** $x$, or simply $y$ **varies as** $x$. We call $c$ the **constant of variation** or **constant of proportionality.**

Notice from the definition that the subject of variation is closely related to proportion studied in Section 2.2. In fact, variation is another way of considering proportion problems.

**EXAMPLE 1** Determine the equation of variation.

**(a)** $y$ varies directly as $x$, and when $x = \frac{3}{2}$ then $y = 10$.

First we give the equation decribed by *y varies directly as x*.

$$y = cx$$

Determine the value of the constant by using the information $y = 10$ when $x = \frac{3}{2}$.

$$10 = c \cdot \frac{3}{2}$$

$$\frac{20}{3} = c$$

Thus, the equation of variation is $y = \frac{20}{3}x$.

**(b)** The volume $V$ of a gas under constant pressure varies directly as the temperature $T$. The volume is 220 cm$^3$ when the temperature is 320°K (degrees Kelvin). The equation of variation is

$$V = cT.$$

Using $V = 220 \text{ cm}^3$ when $T = 320°\text{K}$, we have

$$220 = c(320).$$

$$c = \frac{220}{320} = \frac{11}{16}$$

Thus $$V = \frac{11}{16}T.$$ ■

Notice in Example 1(**b**) that after it has been determined that $V = \frac{11}{16}T$, we can find the volume for any temperature. For example, if $T = 480°\text{K}$, then

$$V = \frac{11}{16}(480) = 330 \text{ cm}^3.$$

Also, if $V = 77 \text{ cm}^3$, we can find $T$.

$$77 = \frac{11}{16}T$$

$$T = \frac{16}{11}(77) = 112°\text{K}$$

In the direct variation relation, $y = cx$, the variable $y$ is a linear function of $x$. We can also consider $y$ varying as powers of $x$. The following table gives several possibilities.

| Variation Statement | Equation |
|---|---|
| $y$ varies as the square of $x$ | $y = cx^2$ |
| $u$ varies as the square root of $v$ | $u = c\sqrt{v}$ |
| the volume $V$ of a sphere varies as the cube of the radius $r$ | $V = cr^3 \quad \left(c = \frac{4}{3}\pi\right)$ |

**EXAMPLE 2**

PHYSICS

The period $P$ of a simple pendulum varies directly as the square root of the length $l$. If the period is 3.2 sec when the length is 12.5 cm, find the period when $l = 20.2$ cm.

The equation of variation is

$$P = c\sqrt{l}.$$

To find $c$, use $P = 3.2$ sec when $l = 12.5$ cm.

$$3.2 = c\sqrt{12.5}$$

$$c = \frac{3.2}{\sqrt{12.5}} \approx 0.91$$

Thus, $$P = 0.91\sqrt{l}.$$

When $l = 20.2$,

$$P = (0.91)\sqrt{20.2} \approx 4.1 \text{ sec.}$$ ■

## Inverse Variation

Consider now a second kind of variation called *inverse variation*.

If two variables $x$ and $y$ are so related that

$$y = \frac{c}{x}$$

where $c$ is a constant, then we say that $y$ **varies inversely as** $x$ or $y$ is **inversely proportional to** $x$. Again, $c$ is called the **constant of variation.**

In direct variation, $y$ increases as $x$ increases when $c > 0$. In contrast, for inverse variation $y$ decreases as $x$ increases ($c > 0$). For example, over a fixed distance of 100 miles the time $t$ varies inversely as the rate $r$ of travel.

$$t = \frac{100}{r}$$

Thus, as the rate $r$ increases, the time $t$ to travel the 100 mi decreases.

## EXAMPLE 3

The intensity of illumination $I$ of a light source varies inversely as the square of the distance $d$ from the source. When $d = 20$ ft, then $I = 5000$ candlepower. Find $I$ when $d$ is 60 ft.

First write the equation of variation and then solve for $c$.

$$I = \frac{c}{d^2}$$

$$5000 = \frac{c}{(20)^2} \qquad I = 5000 \text{ candlepower when } d = 20 \text{ ft}$$

$$c = 2{,}000{,}000$$

$$I = \frac{2{,}000{,}000}{d^2} \qquad \text{Equation of variation}$$

$$I = \frac{2{,}000{,}000}{(60)^2} \qquad \text{Let } d = 60 \text{ ft}$$

$$\approx 556 \text{ candlepower}$$ ■

## Joint Variation

A variable $z$ can vary with two or more variables. This is called *joint variation*.

If three variables $x$, $y$, and $z$ are so related that

$$z = cxy$$

where $c$ is a constant, we say that $z$ varies directly as $x$ and directly as $y$ or that $z$ **varies jointly** as $x$ and $y$.

Other variation equations can be found by combining the various types of variation. The following table gives several examples.

| Variation Statement | Equation |
|---|---|
| $y$ varies jointly with $x$ and $z$ and inversely as $w$ | $y = \dfrac{cxz}{w}$ |
| the volume $V$ of a gas varies directly as the temperature $T$ and inversely as the pressure $P$ | $V = \dfrac{cT}{P}$ |
| the resistance $R$ of a wire varies directly as the length $l$ and inversely as the square of the diameter $d$ | $R = \dfrac{cl}{d^2}$ |

Example 4 considers one of the applications given in the introduction to this chapter.

**EXAMPLE 4**

**ENGINEERING**

The resistance $R$ of a wire at a constant temperature varies directly as the length $l$ and inversely as the square of the diameter $d$. A section of wire having a diameter of 0.01 inches and a length of 1 foot has a resistance of 8.2 ohms. Find the resistance of a 1-mile-long wire with diameter 0.05 inches.

The equation is $R = cl/d^2$. Substitute the first set of values.

$$\mathbf{8.2} = \frac{c \cdot \mathbf{1}}{(\mathbf{0.01})^2}$$

$$(0.01)^2(8.2) = c$$

$$0.00082 = c$$

Thus, the variation equation is $R = 0.00082l/d^2$. Substitute $d = 0.05$ and $l = 5280$ ft (1 mi = 5280 ft).

$$R = \frac{(0.00082)(\mathbf{5280})}{(\mathbf{0.05})^2} \approx 1732$$

The resistance is approximately 1732 ohms. ■

## 3.7 EXERCISES

Find the equation of variation in Exercises 1–10.

1. $y$ varies directly as $x$, and when $x = 16$ then $y = 8$.
2. $p$ varies directly as $q$, and when $q = 40$ then $p = 400$.
3. $z$ varies inversely as $w$, and $z = 12$ when $w = 5$.
4. $u$ varies inversely as $v$, and $u = 0.025$ when $v = 0.4$.
5. $z$ varies jointly as $x$ and $y$. When $x = 2$ and $y = 7$, then $z = 42$.
6. $m$ varies jointly as $p$ and $q$. When $p = 0.25$ and $q = 9.5$ then $m = 8.55$.
7. $a$ is directly proportional to the square of $b$ and inversely proportional to the square root of $d$. When $b = 3$ and $d = 25$, then $a = 405$.

8. $P$ is directly proportional to the cube root of $R$ and inversely proportional to the fourth power of $T$. When $R = 0.027$ and $T = 0.1$, then $P = 3300$.

9. $w$ varies jointly as the square of $x$ and the cube root of $y$ and inversely as the square root of $z$. If $x = 40$, $y = 125$, and $z = 400$, then $w = 330$.

10. $L$ varies directly as the square root of $M$ and inversely as the square of $N$ and the cube of $K$. If $M = 0.09$, $N = 0.2$, and $K = 2$, then $L = 1.5$.

11. CONSUMER The simple interest $I$ paid on a loan for one year varies directly as the amount borrowed $A$. If \$250 interest was paid on a loan of \$2000, what interest must be paid on a loan of \$9250?

12. CONSUMER The yearly interest $I$ received on savings varies directly as the interest rate $r$. If \$468 in interest was received when the rate was 9%, what interest would have been received if the rate were 7%?

13. MANUFACTURING The daily cost $C$ of production varies directly as the number $n$ of units produced, when $3000 \leq n \leq 7000$. If the cost is \$25,200 when 4100 are produced, what will be the cost if production is increased to 6400?

14. BUSINESS The daily profit $P$ varies directly as the number of units sold $n$, if $200 \leq n \leq 1000$. If the profit is \$6520 when 580 units are sold, what is the profit when only 340 units are sold?

15. MANUFACTURING In a manufacturing operation the productivity $P$ varies directly as the square of the efficiency $e$ and inversely as the square root of the downtime $d$. If $P = 7.6$ when $e = 7$ and $d = 16$, find $P$ for $e = 8$ and $d = 9$.

16. MANAGEMENT In a wholesale operation the movement factor $M$ is directly proportional to the square root of the number $n$ of workers on the job and inversely proportional to the square of the break time $t$. If $M = 5.8$ when $n = 9$ and $t = 2.2$, find $M$ when $n = 16$ and $t = 3.1$.

17. PHYSICS The volume $V$ of a gas varies directly as the temperature $T$ and inversely as the pressure $P$. If $V = 16\text{ in}^3$ when $T = 480°\text{K}$ and $P = 45\text{ lb/in}^2$, find $V$ when $T = 1200°\text{K}$ and $P = 15\text{ lb/in}^2$.

18. PHYSICS Use the variation equation determined in Exercise 17 to find the pressure $P$ when $V = 6\text{ in}^3$ and $T = 360°\text{K}$.

19. SCIENCE The weight $w$ of a body above the surface of the earth is inversely proportional to the square of its distance $d$ from the center of the earth. What is the effect on the weight when the distance is doubled?

20. NAVIGATION The horsepower $P$ required to move a ship varies directly as the cube of the speed $s$. What is the effect on $P$ when the speed is cut in half?

21. PHYSICS The period of vibration $P$ of a pendulum varies directly as the square root of the length $l$. If the period of vibration is 1.56 sec when the length is 9.25 cm, what is the period when $l = 16.8$ cm?

22. SCIENCE The gravitational attraction $A$ between two masses varies inversely as the square of the distance between them. The force of attraction is 17.6 lb when the masses are 12.4 ft apart. What is the attraction when the masses are 34.6 ft apart?

23. PHYSICS The intensity $I$ of light varies inversely as the square of the distance $d$ from the source. If the intensity must be doubled, what must be done to the distance?

**24.** **PHYSICS** The force $F$ of attraction between two charged particles varies jointly as the charges $q_1$ and $q_2$ and inversely as the square of the distance $d$ between them. What happens to the force when $q_1$ is doubled, $q_2$ is divided by 3, and $d$ is doubled?

**25.** **MANUFACTURING** The time $T$ necessary to make an enlargement of a photo negative varies directly as the area $A$ of the enlargement. If 25 seconds are required to make a 5-by-7 enlargement, find the time required for an 11-by-14 enlargement.

**26.** **PHYSICS** The weight $w$ of a liquid varies directly as its volume $v$. If the weight of the liquid in a cylinder of radius 2 cm and height 12 cm is 142 gm, find the weight in a sphere of radius 5 cm.

**27.** **ENGINEERING** The speed $s$ of a car when the brakes are applied can be estimated by measuring the length $l$ of the skid marks. This is possible since $s$ is directly proportional to the square root of $l$. If a car traveling 45 mph makes a skid mark of 65 feet, estimate the speed of a car that leaves a mark 120 ft long.

**28.** **ENGINEERING** Use the equation of variation in Exercise 27 to estimate the length of a skid mark of a car traveling 90 mph when the brakes are applied.

**29.** **ENGINEERING** The maximum safe load $m$ of a horizontal beam varies jointly with the width $w$ and the square of the thickness $t$ and inversely as the length $l$. If a second beam has width $W = 5w$ thickness $T = \frac{1}{2}t$, and length $L = 3l$, what is the relationship between the safe load $M$ on the second beam and the load on the first?

**30.** **ENGINEERING** The total pressure $P$ of the wind on a flat surface varies jointly as the area $A$ of the surface and the square of the velocity of the wind $v$. If the area of the surface is doubled, what must happen to the speed of the wind to keep $P$ the same?

## For Review

In Exercises 31–36 find the vertex and any $x$-intercepts of each quadratic function.

**31.** $f(x) = x^2 - 7x + 6$

**32.** $f(x) = -x^2 - 12x - 11$

**33.** $g(x) = -2x^2 + 3x + 20$

**34.** $g(x) = 3x^2 + 4x - 4$

**35.** $h(x) = 4x^2 - 4x + 1$

**36.** $h(x) = -4x^2 + 2x - 2$

In Exercises 37–40 find the general form of the equation of the line satisfying the given conditions.

**37.** With slope $-8$ and $x$-intercept $(-4, 0)$

**38.** Passing through $(4, -2)$ and parallel to a line with slope $\frac{2}{3}$

**39.** With $x$-intercept $(3, 0)$ and perpendicular to $2x - 5y = 4$

**40.** With $y$-intercept $(0, -8)$ and parallel to $3x - 9y = 5$

# CHAPTER 3 REVIEW EXERCISES

**1.** Graph the solutions to the equation $3x - 2y = 6$.

**2.** Find the distance between the points $(-2, 6)$ and $(7, 3)$.

**3.** If $(7, 2)$ is the midpoint of the line segment joining $(a, b)$ and $(8, 5)$, find $(a, b)$.

**4.** Find all points $(a, 2)$ which are 5 units from $(2, -1)$.

**5.** Find the slope of the line through the points $(-2, 8)$ and $(-3, -1)$.

Find the general form of the equation of a line in Exercises 6–7.

**6.** Passing through $(-3, -7)$ and $(-1, 2)$

**7.** With $x$-intercept $(-2, 0)$ and parallel to $4x - 3y + 6 = 0$

**8.** Find the slope and $y$-intercept of $7x - 2y = 4$.

Determine whether the lines are parallel or perpendicular in Exercises 9–10.

**9.** $8x + 4y - 7 = 0$ and $-12x - 6y + 5 = 0$

**10.** $x + 3y - 2 = 0$ and $6x - 2y + 7 = 0$

**11.** Is the triangle with vertices $A(5, 6)$, $B(-2, 1)$, and $C(3, -6)$ a right triangle?

**12.** Determine the general form of the perpendicular bisector of the line segment joining $(-2, -4)$ and $(4, 8)$.

In Exercises 13–14 determine whether the relation is a function if $x$ is the independent variable and $y$ the dependent variable.

**13.** $y = 2x^2 + 3$

**14.** $x = 2y^2 + 3$

Determine the real numbers for which each function is defined in Exercises 15–16.

**15.** $f(x) = \dfrac{1}{\sqrt{x + 1}}$

**16.** $g(x) = \dfrac{x + 5}{x^2 - 9}$

Use the most appropriate word, *identity, constant, linear,* or *quadratic* for each function in Exercises 17–18.

**17.** $g(x) = 1 - x$

**18.** $h(x) = 16$

**19.** If $f(x) = 2x^2 + 3x$, determine $f(0)$, $f(-2)$, $f\left(\frac{1}{2}\right)$, $f(a)$, and $f(a + 1)$.

**20.** If $g(x) = 3x - 5$, determine $g(a^3)$, $g\left(\frac{1}{a}\right)$, and $g(\sqrt{a})$.

Use the vertical line test in Exercises 21–22 to determine if the graph is the graph of a function.

**21.**

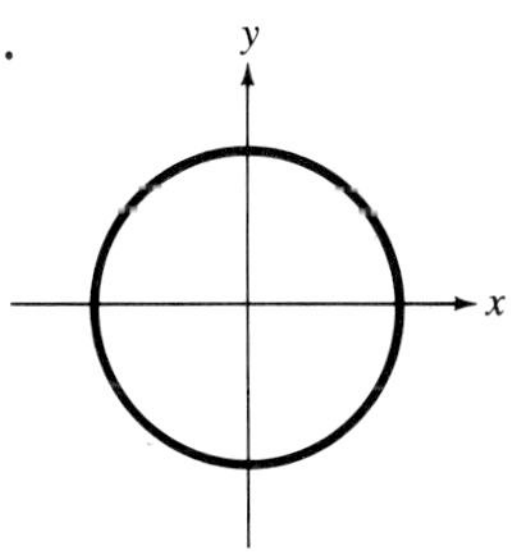


**22.**

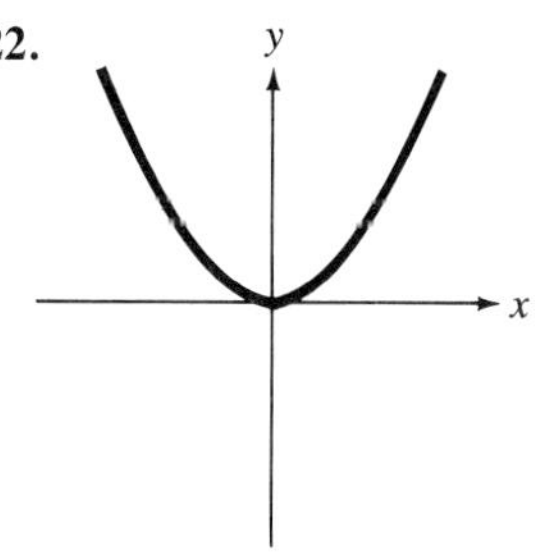


Determine the values of $x$ for which the function increases and the values for which it decreases in Exercises 23–24.

**23.** $g(x) = -\frac{1}{3}x - 2$

**24.** $k(x) = \dfrac{4}{x}$

The graph of $f$ is given in the figure. Use this to sketch the graphs of each function in Exercises 25–28.

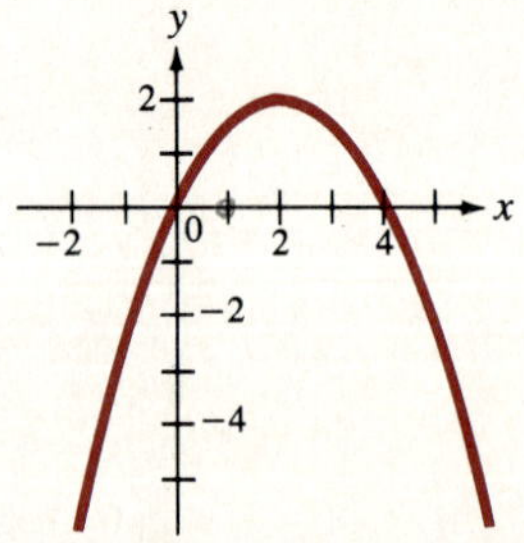


**25.** $y = -f(x)$ **26.** $y = 2f(x)$ **27.** $y = f(x + 1) - 2$ **28.** $y = f(x - 1) + 2$

In Exercises 29–31 determine if the function is even or odd.

**29.** $f(x) = 2x^2 - 3$ **30.** $g(x) = 4x^3 + 2x$ **31.** $h(x) = |x| + x$

In Exercises 32–34 determine the symmetry of the graphs without graphing the relation.

**32.** $3x^2 + 2y^2 = 5$ **33.** $8xy = 1$ **34.** $y^2 + 2x = 5$

Graph each function in Exercises 35–36.

**35.** $f(x) = \begin{cases} -x, & x \leq -2 \\ -2, & -2 < x \leq 1 \\ 2x - 1, & x > 1 \end{cases}$ **36.** $g(x) = \frac{1}{2}[x]$

Determine $f(g(3))$, $g(f(3))$, $f(g(x))$, and $g(f(x))$ for each pair of functions in Exercises 37–38.

**37.** $f(x) = 3x^2 + 2$ and $g(x) = x - 5$ **38.** $f(x) = |x - 2|$ and $g(x) = 1 - x^2$

Determine if the functions in Exercises 39–40 are one-to-one.

**39.** $f(x) = \sqrt{x - 4}$ **40.** $f(x) = |x - 4|$

Find the inverse of each function in Exercises 41–44.

**41.** $f(x) = \frac{1}{2}x + 3$ **42.** $g(x) = \sqrt{x + 5}$

**43.** $h(x) = \frac{1}{2}x^2 - 2, \quad x \geq 0$ **44.** $k(x) = \frac{1}{2}x^3$

In Exercises 45–46 find the vertex and any $x$-intercepts of the quadratic function.

**45.** $f(x) = 2x^2 + 9x - 5$ **46.** $g(x) = -3x^2 - 2x + 8$

In Exercises 47–48 find the vertex and any $x$-intercepts, and graph the function.

**47.** $f(x) = -2x^2 + 8x - 5$ **48.** $g(x) = \frac{1}{2}x^2 + 3x - 2$

**49. CONSUMER** If $x$ dollars are borrowed for 3 years at 11% simple interest and $S$ is the amount to be paid back, find a linear equation that relates $S$ to $x$. What amount must be paid back when \$2200 is borrowed?

**50.** RETAILING A retailer has a markup of 30% and then a markdown of $x$ percent. If sales tax is 5%, find a function $T$ that represents the total price of a coat that originally cost \$120. What is the total price if the markdown is 40%?

**51.** BUSINESS Two outdoor storage areas are to be built along an existing fence as indicated in the figure. If 120 ft of fencing is to be used, what are the dimensions $x$ and $y$ that will enclose the maximum area?

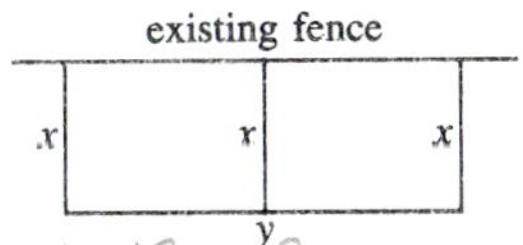


**52.** BUSINESS The daily cost of a warehouse operation is given by $C(x) = 2x^2 - 80x + 2200$ where $x$ is the number of trucks loaded. What number of trucks gives the minimum cost? What is the minimum cost?

**53.** MANUFACTURING The profit in a small manufacturing operation is given by a quadratic function. The maximum profit occurs at the point (150, 850). If the curve also goes through the point (130, 50), find the quadratic function.

**54.** RETAILING The revenue $R$ received from sales varies directly as the number $n$ of units sold. If $R =$ \$2500 on sales of 320 units, what would be the revenue on 920 units?

**55.** ECOLOGY The radius $r$ of an oil spill is increasing with time $t$ in hours according to the function $r(t) = 2t + 1$. If the cost in dollars of cleaning the spill is $C(r) = 30r^2 + 8000$, find $C$ as a function of time. Determine the cost when $t = 3$ hr.

**56.** MINING Empirical data gives the cost $C$ in dollars of extracting $x$ percent of a metal from a ton of ore as $C(x) = 923.6x^2 - 1756x + 46.8$. Find the percent that minimizes the cost.

**57.** BUSINESS Spherical storage tanks of radius $r$ are to be painted at a cost of \$3.20 per square meter. Find the cost $C$ as a function of $r$. Find the cost when $r = 3.2$ m.

**58.** AERONAUTICS The height in feet of a small rocket, $t$ seconds after firing, is given by the function $h(t) = -16t^2 + 96t$. What is the maximum height the rocket will reach and at what time will it hit the ground?

**59.** ENGINEERING The time $T$ required for an elevator to lift a weight $w$ varies jointly as the weight and the distance $d$ to be lifted and inversely as the power $p$ of the motor. If it takes 24 seconds for a 3-horsepower motor to lift 500 lb a distance of 60 ft, what power would be needed to lift 700 lb a distance of 100 ft in 35 sec?

**60.** PHYSICS The period $P$ of a pendulum varies directly as the square root of the length $l$. How would the length have to be changed to make the period three times as long?

CHAPTER

# 4 POLYNOMIAL AND RATIONAL FUNCTIONS

In Chapter 1 we reviewed the various operations on polynomial expressions. A polynomial in one variable defines a function with domain the set of all real numbers. Such functions have played a major role in mathematics. Functions formed by taking quotients of polynomials, called rational functions, also have numerous applications. Consider the following applications of polynomial and rational functions.

**ENGINEERING ▶**

A petroleum delivery truck has a tank in the shape of a cylinder with a hemisphere attached at each end. If the total length of the tank is 20 ft, and the total storage capacity is $162\pi$ ft$^3$, what is the radius of the cylinder?

**◀ MANUFACTURING**

During an eight-hour shift, the number of items produced in a factory is given by $n(t) = t^2 + 16t$ where $0 \le t \le 8$. The total cost in dollars of producing $n(t)$ items is given by $c(t) = 120t + 960$. The average cost of production is a function of $t$, given by $a(t) = \dfrac{c(t)}{n(t)}$. Graph $a(t)$, concentrate on the portion for $0 < t \le 8$, and interpret the results.

Once we have developed the theory of polynomials and polynomial equations and considered the classical polynomial theorems, we will have the tools for solving the first of these problems (see Example 5 in Section 4.3) as well as for graphing polynomial functions. Finally, a study of rational functions and their graphs will give us the theory for solving the second problem (see Example 7 in Section 4.5).

## 4.1 Polynomials and Synthetic Division

### Polynomials

In Chapter 2 we reviewed the technique for solving equations that can be transformed into a linear or first-degree equation of the form

$$ax + b = 0 \quad (a \neq 0).$$

We also learned how to solve quadratic or second-degree equations of the form

$$ax^2 + bx + c = 0 \quad (a \neq 0).$$

One of the primary objectives of the present chapter is to learn how to solve equations of degree higher than the second. Formulas exist for solving **third-degree** or **cubic equations** such as

$$ax^3 + bx^2 + cx + d = 0 \quad (a \neq 0)$$

and **fourth-degree** or **quartic equations** such as

$$ax^4 + bx^3 + cx^2 + dx + e = 0 \quad (a \neq 0).$$

However, they are seldom used since they are extremely complicated. No formulas for solving equations of degree higher than the fourth exist. Consequently, when solving general equations of degree higher than the second, we search for solutions based on various theorems. Often solutions are only approximated using numerical techniques

#### Polynomials with Real Coefficients

An expression of the form

$$a_nx^n + a_{n-1}x^{n-1} + a_{n-2}x^{n-2} + \cdots + a_2x^2 + a_1x + a_0$$

where $a_n, a_{n-1}, a_{n-2}, \ldots, a_2, a_1, a_0$ are constant real numbers and $n$ is a nonnegative integer, is called a **polynomial with real coefficients in the variable $x$.** Each of the expressions $a_nx^n, a_{n-1}x^{n-1}, \ldots, a_1x$, and $a_0$ is called a **term** of the polynomial. The numbers

$$a_n, \quad a_{n-1}, \ldots, \quad a_1, \quad a_0$$

are called **coefficients,** and $a_n \neq 0$ is called the **leading coefficient.** If all coefficients are zero, we call the polynomial the **zero polynomial.** The polynomial $a_nx^n + a_{n-1}x^{n-1} + \cdots + a_1x + a_0$ $(a_n \neq 0)$ has **degree $n$** or is **of $n$th degree,** and the zero polynomial does not have a degree.

For example, the polynomial $2x^4 - 3x^3 - x + 1$ is a fourth-degree polynomial with leading coefficient 2. The coefficient of the missing $x^2$-term is 0. The zero polynomial, denoted by 0, has no leading coefficient and no degree. A polynomial of degree zero is simply the constant $a_0$. For example, 15 is a polynomial of degree zero since 15 can be thought of as $15x^0$.

## Polynomial Equations

A **solution** to a **polynomial equation** of the form

$$a_nx^n + a_{n-1}x^{n-1} + \cdots + a_2x^2 + a_1x + a_0 = 0 \quad (a_n \neq 0)$$

is a number (either real or complex) which, when substituted for the variable, makes the equation true. The use of functional notation with polynomials can shorten our work. For example, if we let

$$P(x) = a_nx^n + a_{n-1}x^{n-1} + \cdots + a_2x^2 + a_1x + a_0,$$

the phrase "evaluate the polynomial $P(x)$ when $x$ is equal to 2" can be shortened to "find $P(2)$." The corresponding polynomial equation is abbreviated $P(x) = 0$.

## EXAMPLE 1

Let $P(x) = x^3 - 2x^2 + 3x - 1$.

**(a)** Find the value of $P(x)$ when $x$ is 2, that is, find $P(2)$.

$$P(\mathbf{2}) = (\mathbf{2})^3 - 2(\mathbf{2})^2 + 3(\mathbf{2}) - 1$$
$$= 8 - 8 + 6 - 1 = 5$$

**(b)** Find $P(-1)$.

$$P(\mathbf{-1}) = (\mathbf{-1})^3 - 2(\mathbf{-1})^2 + 3(\mathbf{-1}) - 1$$
$$= -1 - 2 - 3 - 1 = -7$$ ■

### Zeros of a Polynomial

If $P(x)$ is a polynomial and $r$ is a number (real or complex) such that $P(r) = 0$, then $r$ is called a **zero** of the polynomial $P(x)$ and a **solution** or **root** of the polynomial equation $P(x) = 0$.

The letter $P$ used to represent a polynomial has no special importance; other letters can also be used, just as letters other than $x$ can be used for the variable. Thus, we might see polynomials such as $Q(y)$, $F(t)$, or $f(x)$.

## EXAMPLE 2

Evaluate the polynomial $Q(t) = t^3 + 3t^2 + t + 3$ when $t = -3$, $i$, and $-i$.

$$Q(-3) = (-3)^3 + 3(-3)^2 + (-3) + 3 = -27 + 27 - 3 + 3 = 0$$
$$Q(i) = (i)^3 + 3(i)^2 + (i) + 3 = -i - 3 + i + 3 = 0$$
$$Q(-i) = (-i)^3 + 3(-i)^2 + (-i) + 3 = i - 3 - i + 3 = 0$$

Thus, $-3$, $i$, and $-i$ are zeros of $Q(t)$, and solutions to the polynomial equation $Q(t) = 0$. ■

## Synthetic Division

Solutions to polynomial equations may be real or complex numbers as shown in Example 2. At this point we can see whether a given number is a solution to a polynomial equation; however, we do not yet have a method for finding possible solutions. Although not apparent now, the ability to divide quickly a polynomial $P(x)$ by a binomial of the form $x - r$ is an extremely helpful tool for solving equations. The method of long division, reviewed in Chapter 1, can be shortened with a process called **synthetic division.** Suppose we divide $P(x) = 2x^3 - 3x^2 + 3x - 4$ by $x - 2$.

$$
\begin{array}{rrl}
 & 2x^2 + x + 5 \phantom{xxxxxxxx} & \leftarrow \textbf{The quotient } Q(x) \\
\textbf{The divisor} \longrightarrow x - 2 & \overline{)\,2x^3 - 3x^2 + 3x - 4} & \leftarrow \textbf{The dividend } P(x) \\
 & \underline{2x^3 - 4x^2} \phantom{+ 3x - 4} & \\
 & x^2 + 3x \phantom{- 4} & \\
 & \underline{x^2 - 2x} \phantom{- 4} & \\
 & 5x - 4 & \\
 & \underline{5x - 10} & \\
 & 6 & \leftarrow \textbf{The remainder } R
\end{array}
$$

We can shorten our work by writing only the coefficients.

$$
\begin{array}{rrl}
 & 2 + 1 + 5 \phantom{- 4} & 2 + 1 + 5 \text{ corresponds to } 2x^2 + x + 5 \\
1 - 2 & \overline{)\,2 - 3 + 3 - 4} & \\
 & \underline{\mathbf{2} - 4} \phantom{+ 3 - 4} & \\
 & 1 + 3 \phantom{- 4} & \\
 & \underline{\mathbf{1} - 2} \phantom{- 4} & \\
 & 5 - 4 & \\
 & \underline{\mathbf{5} - 10} & \\
 & 6 &
\end{array}
$$

If $P(x)$ is of degree $n$, then $Q(x)$ is of degree $n - 1$. In this case, since $2x^3 - 3x^2 + 3x - 4$ is of degree 3, $Q(x)$ is of degree 2.

The first term in each line (in color above) is unnecessary since it is subtracted in the next step. Hence, we can abbreviate by eliminating these numbers.

$$
\begin{array}{rr}
 & 2 + 1 + 5 \phantom{- 4} \\
1 - 2 & \overline{)\,2 - 3 + 3 - 4} \\
 & \underline{\mathbf{- 4}} \phantom{+ 3 - 4} \\
 & 1 + 3 \phantom{- 4} \\
 & \underline{\mathbf{- 2}} \phantom{- 4} \\
 & 5 - 4 \\
 & \underline{\mathbf{- 10}} \\
 & 6
\end{array}
$$

The remainder $R$ is always a constant, for if $R$ involves the variable $x$ to some power, the process of division is continued until $x$ does not appear.

The 1 in the divisor $1 - 2$ is unnecessary since we are no longer writing the first number in other lines. Also, "bringing down" the next term merely wastes space since all subtractions can be performed in one line, as shown in color below.

$$
\begin{array}{rr}
 & 2 + 1 + 5 \phantom{- 4} \\
-2 & \overline{)\,2 - 3 + 3 - 4} \\
 & \underline{\mathbf{- 4 - 2 - 10}} \\
 & 1 + 5 + 6
\end{array}
$$

The initial 2 in the quotient $2 + 1 + 5$ is always the same as the initial term in the dividend $2 - 3 + 3 - 4$, and the remaining terms in the quotient $(1 + 5)$ are duplicated in front of the remainder (which is 6) in the bottom line $1 + 5 + 6$. Thus, we could abbreviate further by writing the quotient only in the bottom line by first "bringing down" the 2.

$$\begin{array}{r|l} -2 & 2 - 3 + 3 - \ \ 4 \\ & \underline{\ \ \ - 4 - 2 - 10} \\ & 2 + 1 + 5 + \ \ 6 \end{array}$$

Finally, since the division lines are unnecessary, we omit them. Also, since most of us can add more quickly than we subtract, by changing the sign of the divisor (from $-2$ to $+2$) we can obtain the coefficients in the quotient by addition instead of subtraction.

$$\begin{array}{r|l} +2 & 2 - 3 + 3 - \ \ 4 \\ & \underline{\ \ \ + 4 + 2 + 10} \\ & 2 + 1 + 5 + \ \ 6 \end{array}$$

The first 2 is brought down below the line and multiplied by the divisor $+2$, and the product, 4, is placed above the line. Add $-3$ to 4, obtaining 1, multiply 1 by the divisor $+2$, and place the product, 2, above the line. Add 3 to 2, obtaining 5, multiply 5 by the divisor $+2$, and place the product, 10, above the line. The final addition of $-4$ and 10 gives 6, the last entry below the line, which is the remainder while $2 + 1 + 5$ corresponds to the quotient $2x^2 + x + 5$.

**CAUTION** When dividing a polynomial $P(x)$ by a binomial $x - r$ using synthetic division, always arrange $P(x)$ in descending powers of $x$, and insert a zero for all terms with zero coefficients. ■

## EXAMPLE 3

Divide $x^4 - 2x^2 + x + 3$ by $x - 1$ using synthetic division.

The divisor $x - r$ is $x - 1$ so that $r = 1$. Be sure to supply the zero in the position of the missing $x^3$-term. The polynomial is $x^4 + 0x^3 - 2x^2 + x + 3$.

$$\begin{array}{r|l} 1 & 1 + 0 - 2 + 1 + 3 \\ & \underline{\ \ \ \ \ 1 + 1 - 1 + 0} \\ & 1 + 1 - 1 + 0 + 3 \end{array}$$

The quotient is $x^3 + x^2 - x$ and the remainder is 3. ■

## EXAMPLE 4

Divide $3x^5 - 2x^3 + 7 - x + x^2$ by $x + 2$ using synthetic division.

The divisor $x - r$ is $x + 2$ so we have $x + 2 = x - (-2)$ making $r = -2$. Be sure to supply the coefficient zero and rearrange the terms in descending order. The polynomial is $3x^5 + 0x^4 - 2x^3 + x^2 - x + 7$.

$$\begin{array}{r|l} -2 & 3 + 0 - \ \ 2 + \ \ 1 - \ \ 1 + \ \ 7 \\ & \underline{\ \ \ - 6 + 12 - 20 + 38 - 74} \\ & 3 - 6 + 10 - 19 + 37 - 67 \end{array}$$

The quotient is $3x^4 - 6x^3 + 10x^2 - 19x + 37$, and the remainder is $-67$. ■

With practice, the process of synthetic division becomes mechanical, and can be performed in a short time. With the aid of a calculator, we can increase our speed even more, especially when the coefficients of the polynomial are large in absolute value or are decimals. This is illustrated in the next example.

**EXAMPLE 5** Divide $6x^4 - 48x^3 + 210x^2 - 457x - 5043$ by $x + 3$ using synthetic division and a calculator.

| $-3$ | 6 | $-48$ | 210 | $-457$ | $-5043$ |
|---|---|---|---|---|---|
| | | $-18$ | 198 | 1224 | 5043 |
| | 6 | $-66$ | 408 | $-1681$ | 0 |

FIRST:
- ALG: 3 [+/−] [STO] [×] 6 [+] 48 [+/−] [=]
- RPN: 3 [CHS] [STO] [ENTER] 6 [×] 48 [CHS] [+]

SECOND:
- ALG: [DISPLAY] [×] [RCL] [+] 210 [=]
- RPN: [DISPLAY] [RCL] [×] 210 [+]

THIRD:
- ALG: [DISPLAY] [×] [RCL] [+] 457 [+/−] [=]
- RPN. [DISPLAY] [RCL] [×] 457 [CHS] [+]

FOURTH:
- ALG: [DISPLAY] [×] [RCL] [+] 5043 [+/−] [=]
- RPN: [DISPLAY] [RCL] [×] 5043 [CHS] [+]

The numbers $-18$, 198, $-1224$, 5043 shown in color are not written when the calculator steps are performed. They are provided here simply for verifying the steps used. Also, the symbol [DISPLAY] does not represent a calculator entry. Rather it is the number shown in the display at the end of each step. Notice how the sequence

ALG: [×] [RCL] [+] Number [+/−] [=]

RPN: [RCL] [×] Number [CHS] [+]

occurs repeatedly where [+/−] or [CHS] is used only in the case of a negative coefficient. ■

**NOTE** Synthetic division of a polynomial $P(x)$ by a binomial was developed only for binomials of the form $x - r$. If we wish to divide by a binomial of the form $ax - r$, $a \neq 1$, we must first change the form of $P(x)$ and $ax - r$. For example, since

$$\frac{9x^3 + 18x^2 - 3x + 6}{3x - 1} = \frac{\left(\frac{1}{3}\right)(9x^3 + 18x^2 - 3x + 6)}{\left(\frac{1}{3}\right)(3x - 1)} = \frac{3x^3 + 6x^2 - x + 2}{x - \frac{1}{3}},$$

we can transform the given problem into the correct form for synthetic division. ■

## 4.1 EXERCISES

For each polynomial in Exercises 1–9 determine **(a)** its degree, **(b)** the leading coefficient, **(c)** $P(2)$, and **(d)** $P(-1)$.

**1.** $P(x) = x^3 - 3x^2 + x - 5$

**2.** $P(z) = -z^4 - 3z^2 + 5$

**3.** $P(x) = -2x + \sqrt{2}$

**4.** $P(u) = \sqrt{5} - u^3$

**5.** $P(y) = 2y^2 - y^4 - 1$

**6.** $P(x) = 3x^3 - x^5 + 2$

**7.** $P(t) = 8$

**8.** $P(t) = 0$

**9.** $P(x) = 3x^{12} - x^4 - 5x^2 + 3$

**10.** Determine whether 1, 2, and 3 are zeros of the polynomial $P(x) = x^3 - 2x^2 - 2x - 3$.

**11.** Determine whether 1, $-1$, and 2 are zeros of the polynomial $f(x) = 3x^4 - 2x^2 - 1$.

**12.** Determine whether 3, $1 + i$, and $1 - i$ are zeros of the polynomial $Q(x) = x^3 - 5x^2 + 8x - 6$.

**13.** Determine whether 0, 1, and $-1$ are solutions to the polynomial equation $2x^4 - 5x^3 + 3x = 0$.

**14.** Determine whether 1, $2i$, and $-2i$ are solutions to the polynomial equation $x^3 - x^2 + 4x - 4 = 0$.

**15.** Determine whether $-1$, $\dfrac{3 + i\sqrt{3}}{2}$, and $\dfrac{3 - i\sqrt{3}}{2}$ are solutions to the polynomial equation $x^3 - 3x^2 + 3x = 0$.

**16.** BUSINESS The revenue in dollars brought in on the sale of $x$ items is given by $R(x) = 200x$. The cost in dollars of producing $x$ of these items is given by $C(x) = 1000 + 100x$. The zeros of the polynomial $P(x) = R(x) - C(x)$ are called break-even points. Find $P(x)$ and use it to determine whether 10 is a break-even point.

**17.** RATE-MOTION The height in feet of an object which is propelled upward with initial velocity 240 ft/sec from an initial height of 1600 ft can be expressed as a polynomial in the variable $t$ (time) by $h(t) = -16t^2 + 240t + 1600$. Show that 20 is a zero of $h(t)$ and interpret this result.

Find each quotient and remainder in Exercises 18–29 using synthetic division.

**18.** $(x^3 - x^2 + 2x - 3) \div (x - 1)$

**19.** $(3y^3 - 2y^2 + y - 5) \div (y - 2)$

**20.** $(2x^3 - 3x + 5) \div (x - 2)$

**21.** $(3z^4 - 2z^2 + z - 3) \div (z - 3)$

**22.** $(t^4 - 2t^3 + t^2 - t + 5) \div (t + 2)$

**23.** $(2x^5 - 3x^4 - 6x^3 + x + 8) \div (x - 3)$

**24.** $(z^2 - z + 2 + z^4) \div (z - 2)$

**25.** $(y^3 - y + y^4 - 3) \div (y + 1)$

**26.** $(6t^3 - t^2 + 2) \div \left(t + \frac{1}{3}\right)$

**27.** $(6x^3 - x + 4) \div \left(x - \frac{1}{2}\right)$

**28.** $(2y^3 + 3y^2 + 7y + 3) \div (2y + 1)$

**29.** $(t^3 + 3t^2 + t + 3) \div (t + i)$

Use a calculator and synthetic division to find each quotient and remainder in Exercises 30–33.

**30.** $(5x^4 - 30x^3 + 180x^2 - 640x + 800) \div (x - 3)$

**31.** $(7x^5 - 25x^4 + 80x^3 - 170x^2 + 305x + 2600) \div (x + 2)$

**32.** $(2.1x^3 - 3.45x^2 + 6.865x + 7.5312) \div (x - 1.2)$

**33.** $(6.5x^3 - 2.77x^2 + 11.413x - 456.329) \div (x + 3.6)$

**34.** Is it possible for a polynomial to have no zeros?

**35.** Show that the polynomial $P(x) = -x^7 + x^6 - x^5 + x^4 - x^3 + x^2 - x + 1$ cannot have a zero which is a negative real number.

**36.** Show that the polynomial $P(x) = x^7 + x^6 + x^5 + x^4 + x^3 + x^2 + x + 1$ cannot have a zero which is a positive real number.

**37.** Find a value of $m$ so that $x^3 + x^2 - x + m$ has a remainder of 0 when divided by $x - 1$.

**38.** Find a value of $m$ so that $x^3 - 2x^2 + mx + 1$ has a remainder of 5 when divided by $x + 2$.

**39.** Let $P(x) = x^3 - 2x^2 + 3x + 1$. **(a)** Find $P(1)$. **(b)** Find the remainder when $P(x)$ is divided by $x - 1$. **(c)** Compare the results of parts **(a)** and **(b)**.

**40.** Let $P(x) = 2x^4 - 4x^2 + x - 2$. **(a)** Find $P(-2)$. **(b)** Find the remainder when $P(x)$ is divided by $x + 2$. **(c)** Compare the results of parts **(a)** and **(b)**.

## For Review

**41.** Find the midpoint of the line segment joining $(6, -1)$ and $(3, 5)$.

**42.** Find the slope of the line passing through the points $(4, -3)$ and $(7\ -3)$.

**43.** Find the general form of the equation of the line passing through $(6, 1)$ and perpendicular to $3x + 6y - 2 = 0$.

**44.** Find the quotient and remainder when the polynomial $3x^3 - 4x^2 + x - 7$ is divided by $x - 2$.

**45.** What is the conjugate of each complex number? **(a)** $3 - 2i$ **(b)** $5 + i\sqrt{7}$

# 4.2 Polynomial Theorems

## The Division Algorithm

A familiar property from arithmetic, called the **division algorithm,** states that if $p$, $d$, $q$, and $r$ are whole numbers with $d \neq 0$, such that $p \div d = q$ with a remainder of $r$, then

$$p = qd + r.$$

For example, if we divide 13 by 4, the quotient is 3 with a remainder of 1, and

$$13 = 3 \cdot 4 + 1.$$

A similar result holds for polynomials.

**The Division Algorithm**

Let $P(x)$ and $D(x)$ be polynomials with $D(x) \neq 0$ (the zero polynomial). Then there exist unique polynomials $Q(x)$ and $R(x)$ such that

$$P(x) = Q(x)D(x) + R(x)$$

is true for all real numbers $x$. Also $R(x)$ is either the zero polynomial or else the degree of $R(x)$ is less than the degree of $D(x)$.

**EXAMPLE 1** Let $P(x) = 3x^3 - 4x^2 + x - 7$ and $D(x) = x - 2$.

Suppose we divide $P(x)$ by $D(x)$ using synthetic division.

$$\begin{array}{r|rrrr} 2 & 3 & -4 & +1 & -7 \\ & & 6 & +4 & +10 \\ \hline & 3 & +2 & +5 & +3 \end{array}$$

Thus, the quotient is $Q(x) = 3x^2 + 2x + 5$, and the remainder is $R(x) = 3$. Then

$$\begin{aligned} P(x) &= Q(x)D(x) + R(x) \\ &= (3x^2 + 2x + 5)(x - 2) + 3. \end{aligned}$$ ■

## The Remainder Theorem

When dividing a polynomial $P(x)$ by a binomial $x - r$, the remainder $R(x)$ must be the zero polynomial 0, or must have degree 0 since its degree must be less than 1, the degree of $x - r$. As a result, we write $R$ for $R(x)$ in many such cases emphasizing that $R$ is a constant ($R(x)$ is a constant polynomial). Suppose that $P(x)$ is a polynomial of degree $n$, and we divide $P(x)$ by $x - r$. Then $Q(x)$ is of degree $n - 1$, $R$ is a constant, and

$$P(x) = Q(x)(x - r) + R.$$

With $P(x)$ expressed in this form, suppose we find $P(r)$. Substituting,

$$\begin{aligned} P(r) &= Q(r)(r - r) + R \\ &= Q(r) \cdot 0 + R \\ &= R. \end{aligned}$$

Thus, the remainder when $P(x)$ is divided by $x - r$ is $P(r)$. We have just proved the following theorem.

**The Remainder Theorem**

If $P(x)$ is a polynomial, $r$ is a complex number, and $P(x)$ is divided by $x - r$, then the remainder obtained is equal to $P(r)$.

**EXAMPLE 2** Use the remainder theorem to find the value of the polynomial $P(x) = x^4 + 3x^3 + x^2 + x - 6$ for the given number.

**(a)** $r = 2$

We divide $P(x)$ by $(x - 2)$ using synthetic division.

$$\begin{array}{r|rrrrr} 2 & 1 & +3 & +1 & +1 & -6 \\ & & 2 & +10 & +22 & +46 \\ \hline & 1 & +5 & +11 & +23 & +40 \end{array} \leftarrow \text{Remainder}$$

Thus, by the remainder theorem, $P(2) = 40$. We can check this result by direct substitution.

$$\begin{aligned} P(2) &= 2^4 + 3(2)^3 + (2)^2 + (2) - 6 \\ &= 16 + 24 + 4 + 2 - 6 = 40 \end{aligned}$$

**(b)** $r = -3$

Divide $P(x)$ by $x + 3$ using synthetic division.

$$\begin{array}{r|l} -3 & 1 + 3 + 1 + 1 - 6 \\ & \quad - 3 + 0 - 3 + 6 \\ \hline & 1 + 0 + 1 - 2 + \mathbf{0} \leftarrow \textbf{Remainder} \end{array}$$

Thus, by the remainder theorem, $P(-3) = 0$. This means that $-3$ is a zero of $P(x)$, and since

$$P(x) = Q(x)(x - r) + R,$$

we have

$$\begin{aligned} P(x) &= (x^3 + x - 2)(x + 3) + 0 \\ &= (x^3 + x - 2)(x + 3). \quad \blacksquare \end{aligned}$$

**The Factor Theorem**

This example leads us to the following theorem.

### The Factor Theorem

Let $P(x)$ be a polynomial, and let $r$ be a complex number. Then $r$ is a zero of $P(x)$ if and only if $x - r$ is a factor of $P(x)$.

**PROOF** Since the factor theorem is an "if and only if" theorem, there are two things to prove. *First,* we assume that $r$ is a zero of $P(x)$ and then show that $x - r$ is a factor. If $P(x)$ is divided by $x - r$, we obtain $P(x) = Q(x)(x - r) + R$. But by the remainder theorem, $R = P(r)$, and since $P(r) = 0$, $P(x) = Q(x)(x - r)$ so that $x - r$ is a factor of $P(x)$. *Second,* suppose that $x - r$ is a factor of $P(x)$. Then $P(x) = Q(x)(x - r)$, and $P(r) = Q(r)(r - r) = Q(r) \cdot 0 = 0$. Thus, $r$ is a zero of $P(x)$. $\blacksquare$

**EXAMPLE 3** Determine whether $x - 5$ is a factor of

$$P(x) = x^3 - 7x^2 + 13x - 15.$$

Using the factor theorem, we must determine whether $P(5) = 0$. The best way to do this is to use synthetic division.

$$\begin{array}{r|l} 5 & 1 - 7 + 13 - 15 \\ & \quad + 5 - 10 + 15 \\ \hline & 1 - 2 + \ 3 + \ \mathbf{0} \leftarrow \textbf{Remainder} = \boldsymbol{P(5)} \end{array}$$

Since $P(5) = 0$, $x - 5$ is a factor of $P(x)$. By using synthetic division, not only have we discovered that $P(5) = 0$, but we also know that the quotient when $P(x)$ is divided by $x - 5$ is $Q(x) = x^2 - 2x + 3$. Thus we can represent $P(x)$ in the factored form

$$P(x) = (x^2 - 2x + 3)(x - 5). \quad \blacksquare$$

As soon as we have found one zero of a polynomial, we can use the factor theorem, and look for other zeros that will be zeros of a polynomial of degree 1 less than the original. For example, to solve the polynomial equation

$$x^3 - 7x^2 + 13x - 15 = 0,$$

if we have discovered that 5 is a solution (see Example 3), we can write the equation in the factored form

$$(x^2 - 2x + 3)(x - 5) = 0,$$

and use the zero-product rule to obtain

$$x^2 - 2x + 3 = 0 \qquad \text{or} \qquad x - 5 = 0.$$

Using the quadratic formula to solve $x^2 - 2x + 3 = 0$, we obtain

$$x = \frac{2 \pm \sqrt{4 - 4(3)}}{2} = \frac{2 \pm \sqrt{-8}}{2} = \frac{2 \pm 2i\sqrt{2}}{2} = 1 \pm i\sqrt{2}.$$

Thus, the solutions to the original equation are 5, $1 + i\sqrt{2}$, and $1 - i\sqrt{2}$.

## Fundamental Theorem of Algebra

We know that every linear equation of the form

$$a_1x + a_0 = 0 \quad (a_1 \neq 0)$$

has one solution. Likewise, every quadratic equation of the form

$$a_2x^2 + a_1x + a_0 = 0 \quad (a_2 \neq 0)$$

has at least one solution and at most two solutions that are complex numbers. The next theorem is a generalization of these observations and is a result of the **fundamental theorem of algebra** which guarantees that every polynomial of degree $n \geq 1$ has at least one zero in the system of complex numbers. (Remember that real numbers are also complex numbers.)

### Fundamental Theorem of Algebra

Every polynomial equation

$$a_nx^n + a_{n-1}x^{n-1} + \cdots + a_2x^2 + a_1x + a_0 = 0 \quad (a_n \neq 0)$$

where $n \geq 1$, has at least one and at most $n$ distinct solutions which are complex numbers.

Notice that this theorem guarantees that solutions exist; it does not tell us how to find them.

## EXAMPLE 4

Determine the possible number of distinct solutions to each polynomial equation.

**(a)** $3x^3 - x^2 + 5 = 0$ has at least one and at most three solutions ($n = 3$).

**(b)** $12x^7 - 5x^4 + x^3 - 2x^2 + 7x - 1 = 0$ has at least one and at most seven solutions ($n = 7$). ■

The results of the preceding theorem can be made more precise by defining **multiple roots.** For example, the polynomial equation

$$P(x) = (x - 1)(x - 1)(x + 2) = x^3 - 3x + 2 = 0$$

has at least one and at most three solutions. In fact, we can find the solutions by noting that when $x = 1$ or $x = -2$, $P(x) = 0$. Thus, 1 and $-2$ are solutions. Since $(x - 1)$ is a double factor of $P(x)$, 1 is a **double root** of $P(x)$, or a **root of multiplicity two.** Thus, counting multiplicities, $P(x)$ has *exactly* three zeros: 1, 1, and $-2$.

**EXAMPLE 5**

The polynomial $f(x) = (x + 1)^2(x - 5)^3(x + 3)$ is a polynomial of degree 6 ($n = 6$), and therefore has at least one and at most six zeros (the polynomial equation $f(x) = 0$ has at least one and at most six solutions). Counting multiplicities, we can say that $f(x)$ has *exactly* six zeros: $-1$, $-1$, 5, 5, 5, and $-3$. We call $-1$ a zero of multiplicity two, 5 a zero of multiplicity three, and $-3$ a zero of multiplicity one. ■

Example 5 serves to motivate the next theorem.

Every polynomial $P(x)$ of degree $n$ has exactly $n$ zeros (counting multiplicities) which are real or complex numbers. That is, the equation

$$P(x) = a_nx^n + a_{n-1}x^{n-1} + \cdots + a_1x + a_0 = 0 \quad (a_n \neq 0)$$

has exactly $n$ solutions or roots (counting multiplicities).

It is easy to construct a polynomial equation that has a specified collection of roots by using the factor theorem.

**EXAMPLE 6**

Determine a polynomial equation that has $-2$ as a double root (a root of multiplicity 2), 4 as a triple root (a root of multiplicity 3), and 1 as a single root (a root of multiplicity 1).

Counting multiplicities, we must obtain a polynomial of degree six ($2 + 3 + 1 = 6$), which has factors $(x - (-2))^2 = (x + 2)^2$, $(x - 4)^3$, and $(x - 1)$. Since

$$(x + 2)^2(x - 4)^3(x - 1) = 0,$$

we can multiply the factors to get

$$x^6 - 9x^5 + 12x^4 + 76x^3 - 144x^2 - 192x + 256 = 0,$$

a polynomial equation having the desired roots. ■

## Complex and Irrational Roots

The polynomial (quadratic) equation $x^2 - 2x + 5 = 0$ has roots $1 + 2i$ and $1 - 2i$ (found using the quadratic formula). Notice that these solutions are conjugates. This is a result of the following general theorem, which we do not prove.

If the complex number $a + bi$ is a root of the polynomial equation with *real* coefficients,

$$P(x) = a_nx^n + a_{n-1}x^{n-1} + \cdots + a_1x + a_0 = 0 \quad (a_n \neq 0,\ n \geq 1),$$

then its conjugate $a - bi$ is also a root. That is, complex roots always occur in conjugate pairs.

**EXAMPLE 7**

Determine a polynomial, $P(x)$, of degree three with real coefficients such that the equation $P(x) = 0$ has 3 and $1 + i$ as roots.

Since $1 + i$ is a root, by the above theorem, $1 - i$ is also a root. Thus, $(x - 3)$, $(x - (1 + i))$, and $(x - (1 - i))$ are factors of $P(x)$ by the factor theorem.

$$\begin{aligned} P(x) &= (x - 3)\mathbf{(x - 1 - i)(x - 1 + i)} \\ &= (x - 3)(x^2 - 2x + 2) \\ &= x^3 - 5x^2 + 8x - 6 \end{aligned}$$

$$\begin{array}{lllll} x & -\ 1 & -\ i & & \\ x & -\ 1 & +\ i & & \\ \hline x^2 & -\ x & -\ xi & & \\ & -\ x & & +\ 1 + i & \\ & & +\ xi & & -\ i - i^2 \\ \hline x^2 & -\ 2x & & +\ 1 & +\ 1 \end{array}$$

■

**EXAMPLE 8**

Determine the polynomial, $P(x)$, of degree four with real coefficients such that the equation $P(x) = 0$ has $2 + i$ as a root of multiplicity two and 7 as a root of multiplicity one.

Since $2 + i$ is a double root, $2 - i$ must also be a double root. If 7 is also a root, the polynomial has five roots, and cannot be of degree four. Thus, no polynomial exists that satisfies the given conditions. ■

The preceding theorem is true only when the polynomial in question has real coefficients. If an additional restriction of *rational* coefficients is imposed, there is a similar theorem for certain real roots.

> If the real number $a + b\sqrt{c}$, where $a$ and $b$ are rational but $\sqrt{c}$ is irrational, is a root of the polynomial equation with *rational* coefficients,
>
> $$P(x) = a_n x^n + a_{n-1}x^{n-1} + \cdots + a_1 x + a_0 = 0 \quad (a_n \neq 0,\ n \geq 1),$$
>
> then $a - b\sqrt{c}$ is also a root.

**EXAMPLE 9**

Determine a polynomial, $P(x)$, of degree four with rational coefficients such that $P(x) = 0$ has $1 + \sqrt{2}$ and $i\sqrt{3}$ as roots.

Since $1 + \sqrt{2}$ is a root, $1 - \sqrt{2}$ is a second root. Also, since $i\sqrt{3}$ is a root, $-i\sqrt{3}$ is the fourth root. Thus,

$$\begin{aligned} P(x) &= (x - (1 + \sqrt{2}))(x - (1 - \sqrt{2}))(x - i\sqrt{3})(x + i\sqrt{3}) \\ &= (x^2 - 2x - 1)(x^2 + 3) \\ &= x^4 - 2x^3 + 2x^2 - 6x - 3. \end{aligned}$$

■

## 4.2 EXERCISES

In Exercises 1–4 use synthetic division and the remainder theorem to find each value of $P(x)$ where $P(x) = x^3 - 3x^2 + x + 1$.

**1.** $P(1)$ **2.** $P(-1)$ **3.** $P(5)$ **4.** $P(-2)$

In Exercises 5–8 use synthetic division and the remainder theorem to find each value of $P(x)$ where $P(x) = x^5 + 3x^3 - x + 1$.

**5.** $P(-1)$ **6.** $P(2)$ **7.** $P(-3)$ **8.** $P(1)$

In Exercises 9–12 use synthetic division to find the remainder when $P(x) = x^3 - 27$ is divided by each binomial.

**9.** $x - 3$ **10.** $x + 2$ **11.** $x - 4$ **12.** $x + 3$

In Exercises 13–16 use synthetic division to find the remainder when $P(x) = x^4 + x^2 - 3$ is divided by each binomial.

**13.** $x + 2$ **14.** $x - 1$ **15.** $x - 3$ **16.** $x + 4$

In Exercises 17–20 use synthetic division and the division algorithm to express $P(x) = x^3 - 6x^2 - 9x + 14$ in the form $P(x) = Q(x)D(x) + R(x)$ for each of the given binomial divisors, $D(x)$.

**17.** $x - 7$ **18.** $x + 2$ **19.** $x + 1$ **20.** $x - 3$

In Exercises 21–24 use synthetic division and the factor theorem to determine whether $P(x) = 2x^3 - x^2 - 15x + 18$ has the given binomial as a factor.

**21.** $x + 3$ **22.** $x - 2$ **23.** $x - 1$ **24.** $x - \frac{3}{2}$

**25.** If a polynomial $P(x)$ is divided by $x - r$, what is the value of the remainder?

**26.** If a polynomial $P(x)$ is divided by $x + r$, what is the value of the remainder?

**27.** If $r$ is a zero of polynomial $P(x)$, give one factor of $P(x)$.

**28.** If $-r$ is a zero of polynomial $P(x)$, give one factor of $P(x)$.

**29.** A polynomial of degree $n \geq 1$ has at least how many distinct zeros?

**30.** A polynomial of degree $n \geq 1$ has at most how many distinct zeros?

**31.** If 5 is a zero of multiplicity 2 of polynomial $P(x)$, give a quadratic factor of $P(x)$.

**32.** If $-2$ is a zero of multiplicity 3 of a polynomial $P(x)$, give a cubic factor of $P(x)$.

**33.** Counting multiplicities, a polynomial of degree 5 has exactly how many real or complex zeros?

**34.** Counting multiplicities, a polynomial of degree 0 has exactly how many real or complex zeros?

**35.** If $3 + 2i$ is a zero of polynomial $P(x)$, with real coefficients, give another zero of $P(x)$.

**36.** If $5 - 6i$ is a zero of polynomial $P(x)$, with real coefficients, give another zero of $P(x)$.

**37.** If $1 - 3\sqrt{2}$ is a zero of polynomial $P(x)$, with rational coefficients, give another zero of $P(x)$.

**38.** If $2 + 3\sqrt{5}$ is a zero of polynomial $P(x)$, with rational coefficients, give another zero of $P(x)$.

**39.** Show that 3 is a solution to $x^3 - 3x^2 - x + 3 = 0$, and find the remaining solutions.

**40.** Show that $-2$ is a solution to $x^3 + 2x^2 + 4x + 8 = 0$, and find the remaining solutions.

**41.** What is the greatest number of distinct solutions the equation $x^5 - 3x^3 + x^2 - 5x + 7 = 0$ can have? The fewest? Exactly how many solutions does it have, counting multiplicities?

**42.** What is the greatest number of distinct solutions the equation $x^7 + x^4 - 3x^3 + 7x^2 + 1 = 0$ can have? The fewest? Exactly how many solutions does it have, counting multiplicities?

In Exercises 43–46 determine a polynomial $P(x)$ with rational coefficients of least degree such that the given numbers are roots of $P(x) = 0$.

**43.** 3 is a double root and $1 + 2i$ is a single root

**44.** 2 and $3 + \sqrt{2}$ are single roots

**45.** 0 is a double root and $1 - \sqrt{3}$ is a single root

**46.** $-2$, $-\sqrt{2}$, and $-i\sqrt{2}$ are single roots

In Exercises 47–48 use a calculator, synthetic division, and the remainder theorem to find each polynomial value.

**47.** $P(x) = 6.5x^5 - 4.1x^4 + 3.25x^3 - 7.51x - 521.4674;\quad P(-2.6)$

**48.** $P(x) = 7.3x^4 - 8.5x^3 + 2.153x^2 - 4.16x - 3219.0574;\quad P(5.3)$

**49.** The solutions to $x^3 - 1 = 0$ are called cube roots of 1. How many cube roots of 1 are there? Find them.

**50.** If $P(x) = a_nx^n + a_{n-1}x^{n-1} + \cdots + a_1x + a_0$ is a polynomial such that $P(x) = 0$ for every real number $x$, prove that $a_n, a_{n-1}, \ldots, a_1, a_0$ are all zero, that is, prove that $P(x)$ is the zero polynomial.

**51.** If $P(x) = a_nx^n + a_{n-1}x^{n-1} + \cdots + a_1x + a_0$, $Q(x) = b_nx^n + b_{n-1}x^{n-1} + \cdots + b_1x + b_0$, and $P(x) = Q(x)$ for every real number $x$, prove that $a_i = b_i$, for $i = 0, 1, 2, \ldots, n$.

**52.** Show that $1 + i$ is a zero of $P(x) = x^2 + (1 - i)x - 2 - 2i$, but $1 - i$ is not. Does this contradict a theorem given in this section?

**53.** Show that $1 + \sqrt{2}$ is a zero of $P(x) = x^2 - \sqrt{2}x - 1 - \sqrt{2}$, but that $1 - \sqrt{2}$ is not. Does this contradict a theorem given in this section?

## For Review

**54.** Consider the polynomial $P(x) = x^8 - 3x^4 + x^2 - x^9 + 7$. **(a)** What is the degree of $P(x)$? **(b)** What is the leading coefficient of $P(x)$? **(c)** Find $P(-1)$.

**55.** **BUSINESS** The revenue in dollars brought in by selling $x$ items is given by $R(x) = 125x$. The cost in dollars of producing $x$ items is given by $C(x) = 600 + 85x$. If $P(x) = R(x) - C(x)$, find the break-even points by finding the zeros of $P(x)$.

**56.** Write the polynomial $2x^2 - 7x^5 + 5 - 3x^3 + x^7 + 8x^6$ in descending order.

**57.** Consider the polynomial $P(x) = 15x^4 - 3x^3 + x^2 + 7x - 6$. **(a)** Find $P(-x)$. **(b)** List all the factors of the leading coefficient 15. **(c)** List all the factors of the constant term $-6$.

# 4.3 Polynomial Equations

The theorems in the preceding section gave us information about the nature of the solutions to a polynomial equation. Although a specific rule for finding all solutions to a general polynomial equation is not available, in certain situations we may be able to find some or even all solutions. In this section we consider three theorems that will help with this problem. The first theorem gives us information about the number of real solutions; the second tells us how to locate an interval in which real solutions, if they exist, must be found; and the third will determine all possible rational-number solutions.

## Descartes' Rule of Signs

The first theorem we consider was developed in 1636 by the French mathematician René Descartes, the father of analytic geometry. It allows us to determine the number of positive and negative real-number solutions to a polynomial equation, and depends on the "variation of sign" from term to term in a polynomial written in descending order. For example, consider the polynomial equation

$$P(x) = 2x^5 - x^4 + 3x - 6 = 0.$$

There are three variations in the signs of the terms of $P(x)$, as shown below:

$$P(x) = \overset{1}{2x^5 - x^4} \overset{2}{\;+\;} \overset{3}{3x - 6} = 0$$

Similarly, there are 4 variations in the signs of $Q(x)$ in the equation

$$Q(x) = \overset{1}{5x^5 - 3x^4} - \overset{2}{2x^3 + x^2} \overset{3}{-} \overset{4}{7x + 10}.$$

Thus, a "variation of signs" occurs whenever successive terms have opposite signs. The theorem also depends on the variation of signs of the equation formed by replacing $x$ with $-x$ throughout. For example,

$$P(-x) = -2x^5 - x^4 - 3x - 6$$

has no variation in signs, while

$$Q(-x) = -5x^5 - \overset{1}{3x^4 + 2x^3} + x^2 + 7x + 10$$

has one variation in signs. Descartes' theorem is presented without proof since a proof is beyond the scope of this course.

### Descartes' Rule of Signs

Let $P(x)$ be a polynomial with real coefficients having a non-zero constant term.

1. Positive real solutions: The number of positive real solutions of the equation $P(x) = 0$ is either equal to the number of variations of signs in $P(x)$ or less than the number of variations by an even number.
2. Negative real solutions: The number of negative real solutions of the equation $P(x) = 0$ is either equal to the number of variations of signs in $P(-x)$ or less than the number of variations by an even number.

**EXAMPLE 1** What can be said about the number of positive and negative real solutions to the given polynomial equation?

**(a)** $P(x) = 5x^4 - 6x^3 + x - 9 = 0$

$$P(x) = 5x^4 - 6x + x - 9 \qquad \text{Three variations}$$

$$P(-x) = 5x^4 + 6x^3 - x - 9 \qquad \text{One variation}$$

Thus, the equation has 1 or 3 positive real solutions and 1 negative solution. We can use a table to summarize the different possibilities. Remember that since the degree of $P(x)$ is 4, there must be 4 solutions to the equation.

| Total number of solutions | Number of negative solutions | Number of positive solutions | Number of nonreal solutions |
|---|---|---|---|
| 4 | 1 | 1 | 2 |
| 4 | 1 | 3 | 0 |

**(b)** $Q(x) = 4x^6 + 3x^4 - 7x + 8 = 0$

$$Q(x) = 4x^6 + 3x^4 - 7x + 8 \qquad \text{Two variations}$$

$$Q(-x) = 4x^6 + 3x^4 + 7x + 8 \qquad \text{No variations}$$

Thus, the equation has 2 or 0 positive real solutions and 0 negative solutions.

| Total number of solutions | Number of negative solutions | Number of positive solutions | Number of nonreal solutions |
|---|---|---|---|
| 6 | 0 | 0 | 6 |
| 6 | 0 | 2 | 4 |

**(c)** $R(x) = 7x^6 - 2x^4 - 5x^3 + 2x^2 - x = 0$

Since the constant term is zero, Descartes' rule does not apply to $R(x)$. However, notice that we can factor out an $x$ and apply the theorem to the remaining factor. As a result, clearly 0 is one solution to the equation

$$R(x) = x(7x^5 - 2x^3 - 5x^2 + 2x - 1) = 0.$$

Let

$$S(x) = 7x^5 - 2x^3 - 5x^2 + 2x - 1. \qquad \text{Three variations}$$

$$S(-x) = -7x^5 + 2x^3 - 5x^2 - 2x - 1 \qquad \text{Two variations}$$

Thus, $S(x) = 0$ has 3 or 1 positive real solutions and 2 or 0 negative real solutions. The possible solutions to $S(x) = 0$ are summarized in the table.

| Total number of solutions | Number of negative solutions | Number of positive solutions | Number of nonreal solutions |
|---|---|---|---|
| 5 | 0 | 1 | 4 |
| 5 | 0 | 3 | 2 |
| 5 | 2 | 1 | 2 |
| 5 | 2 | 3 | 0 |

**(d)** $W(x) = x^6 + 5x^4 + 2x^2 + 8 = 0$
Since there are no variations of sign in either $W(x)$ or $W(-x)$, there are no positive nor negative real solutions. Thus, the equation must have 6 complex solutions which occur in three conjugate pairs. ■

## Bounding Solutions

The next theorem helps us to determine upper and lower bounds for the real solutions of a polynomial equation. If $P(x) = 0$ has no real root greater than the real number $a$, $a$ is called an **upper bound** for the real roots. Similarly, if $P(x) = 0$ has no real roots less than the real number $b$, $b$ is called a **lower bound** for the real roots. That is, $a$ and $b$ are upper and lower bounds for the real roots of $P(x) = 0$ if

$$b \leq \text{every real root} \leq a.$$

The roots of the polynomial equation $P(x) = (x - 1)(2x - 1)(2x + 1) = 0$ are $-1/2$, $1/2$, and 1. Thus, any number $b$ such that $b \leq -1/2$ is a lower bound for the roots, and any number $a$ such that $a \geq 1$ is an upper bound. Suppose we had started with the factors of $P(x)$ multiplied out,

$$P(x) = 4x^3 - 4x^2 - x + 1 = 0,$$

and did not know the values of the three roots. Let us use synthetic division to divide $P(x)$ by $x + 2$, $x + 1$, $x - 0$, $x - 1$, $x - 2$, and $x - 3$.

$$\begin{array}{r|rrrr} -2 & 4 & -4 & -1 & +1 \\ & & -8 & +24 & -46 \\ \hline & 4 & -12 & +23 & -45 \end{array} \qquad \begin{array}{r|rrrr} -1 & 4 & -4 & -1 & +1 \\ & & -4 & +8 & -7 \\ \hline & 4 & -8 & +7 & -6 \end{array} \qquad \begin{array}{r|rrrr} 0 & 4 & -4 & -1 & +1 \\ & & 0 & +0 & +0 \\ \hline & 4 & -4 & -1 & +1 \end{array}$$

$$\begin{array}{r|rrrr} 1 & 4 & -4 & -1 & +1 \\ & & +4 & +0 & -1 \\ \hline & 4 & +0 & -1 & +0 \end{array} \qquad \begin{array}{r|rrrr} 2 & 4 & -4 & -1 & +1 \\ & & +8 & +8 & +14 \\ \hline & 4 & +4 & +7 & +15 \end{array} \qquad \begin{array}{r|rrrr} 3 & 4 & -4 & -1 & +1 \\ & & +12 & +24 & +69 \\ \hline & 4 & +8 & +23 & +70 \end{array}$$

Notice that when we divided by $-2$ and $-1$, the signs in the row below the line alternated from + to − and back again. This always happens when we divide by a negative number that is a lower bound for the real roots. Also, when we divided by 2 and 3, the signs in the row below the line were all positive. This always happens when we divide by a positive number that is an upper bound for the real roots. We summarize these results in the following theorem, which is given without proof.

Suppose that the polynomial $P(x)$ is divided by $x - r$ (using synthetic division).

1. If $r$ is negative and the terms in the third row (below the line) alternative from positive to negative (0 can be considered either positive or negative), then $P(x) = 0$ has no root less than $r$.
2. If $r$ is positive and the terms in the third row are all positive or zero, then $P(x) = 0$ has no root greater than $r$.

**EXAMPLE 2** Verify that $-2$ is a lower bound and 3 is an upper bound for the real roots of $3x^4 - 5x^3 + 19x^2 - 35x - 14 = 0$.

We use synthetic division and divide by $-2$ and 3.

$$\begin{array}{r|rrrrr} -2 & 3 & -5 & +19 & -35 & -14 \\ & & -6 & +22 & -82 & +234 \\ \hline & 3 & -11 & +41 & -117 & +220 \\ & + & - & + & - & + \end{array} \qquad \begin{array}{r|rrrrr} 3 & 3 & -5 & +19 & -35 & -14 \\ & & +9 & +12 & +93 & +174 \\ \hline & 3 & +4 & +31 & +58 & +160 \\ & + & + & + & + & + \end{array}$$

Since the signs alternate in the first case and are all plus in the second, we know that $-2$ is a lower bound and 3 is an upper bound for the roots of the equation. ■

## Rational Root Theorem

The process of determining all rational solutions to a polynomial equation having integer coefficients is greatly simplified by the preceding theorem and the next theorem.

**Rational Root Theorem**

Suppose that

$$P(x) = a_n x^n + a_{n-1}x^{n-1} + \cdots + a_2x^2 + a_1x + a_0 \quad (a_n \neq 0)$$

is a polynomial for which all coefficients, $a_n, a_{n-1}, \cdots, a_1, a_0$, are integers. If $p/q$ is a rational number reduced to lowest terms, such that $p/q$ is a root of $P(x) = 0$ (that is $P(p/q) = 0$), then $p$ is a factor of $a_0$ and $q$ is a factor of $a_n$.

This theorem does not directly give us the zeros of a polynomial $P(x)$ (the solutions to the equation $P(x) = 0$). However, it does give us a limited list of possible rational-number solutions. The rational root theorem is used together with the process of synthetic division and the following result.

Suppose that $p/q$ is a root; then $x - p/q$ is a factor, and

$$P(x) = \left(x - \frac{p}{q}\right) \cdot Q(x),$$

where $Q(x)$ is a polynomial of degree one less than $P(x)$. Additional solutions to $P(x) = 0$ must then be solutions to

$$Q(x) = 0.$$

We then apply the rational root theorem to solve $Q(x) = 0$. Also, it is wise to consider the number of possible positive and negative real solutions first.

**EXAMPLE 3** List the possible rational-number solutions to the equation $P(x) = 2x^4 - 3x^3 + 2x^2 - 6x - 4 = 0$, and determine all solutions.

Since $P(x)$ has three variations of sign, the equation has 3 or 1 positive real solutions. Also, since $P(-x)$ has one variation of sign, there must be 1 negative real solution. Suppose we summarize these results for easy reference.

| Total number of solutions | Number of negative solutions | Number of positive solutions | Number of nonreal solutions |
|---|---|---|---|
| 4 | 1 | 3 | 0 |
| 4 | 1 | 1 | 2 |

If $p/q$ is a rational-number solution to this equation, then $p$ is a factor of $-4$ $(a_0 = -4)$ and $q$ is a factor of 2 $(a_n = 2)$. Thus, the possibilities for $p$ and $q$ are

$$p\colon\ 1,\ -1,\ 2,\ -2,\ 4,\ -4; \qquad q\colon\ 1,\ -1,\ 2,\ -2$$

so that the possible rational-number solutions are

$$\frac{p}{q}\colon\ 1,\ -1,\ 2,\ -2,\ 4,\ -4,\ \frac{1}{2},\ -\frac{1}{2}.$$

If we use synthetic division to find one solution, we will also discover the polynomial of degree 3 that is the other factor of $P(x)$. We can then concentrate on finding its zeros.

Now we begin to try the eight possible solutions.

$$\frac{p}{q} = 1: \qquad \begin{array}{r|rrrrr} 1 & 2 & -3 & +2 & -6 & -4 \\ & & 2 & -1 & +1 & -5 \\ \hline & 2 & -1 & +1 & -5 & -9 \end{array}$$

Thus, $P(p/q) = P(1) = -9 \neq 0$ so that 1 is not a solution.

$$\frac{p}{q} = -1: \qquad \begin{array}{r|rrrrr} -1 & 2 & -3 & +2 & -6 & -4 \\ & & -2 & +5 & -7 & +13 \\ \hline & 2 & -5 & +7 & -13 & +9 \end{array}$$

Thus, $P(p/q) = P(-1) = 9$ so that $-1$ is not a solution. Moreover, since the signs in the third row alternate, there is no need to try $-2$ or $-4$ since $-1$ is a lower bound for the roots.

$$\frac{p}{q} = 2: \qquad \begin{array}{r|l} 2 & 2 - 3 + 2 - 6 - 4 \\ & \quad\ \ 4 + 2 + 8 + 4 \\ \hline & 2 + 1 + 4 + 2 + 0 \end{array}$$

Thus, $P(p/q) = P(2) = 0$ so that 2 is a solution, and by the factor theorem, $x - 2$ is a factor of $P(x)$, with $Q(x) = 2x^3 + x^2 + 4x + 2$ as the other factor. Thus,

$$P(x) = (x - 2)(2x^3 + x^2 + 4x + 2) = 0,$$

which means that we must solve

$$2x^3 + x^2 + 4x + 2 = 0.$$

Any rational solution $p/q$ of this equation must still be found in the previous list, but now we know that $p$ must be a factor of 2 ($a_0$ is now 2), and $q$ must be a factor of 2 ($a_n$ is now 2). The only possibilities are

$$\frac{p}{q}: \ 2, \ \frac{1}{2}, \ -\frac{1}{2}$$

since 1, $-1$, and $-2$ have already been discarded.

We now try $p/q = 2$. Although 2 was found to be a solution, it must be tried again since it could be a solution to the new polynomial equation $2x^3 + x^2 + 4x + 2 = 0$, that is, it could be a double root of the original equation.

$$\frac{p}{q} = 2: \qquad \begin{array}{r|l} 2 & 2 + 1 + \ \ 4 + \ \ 2 \\ & \quad\ \ 4 + 10 + 28 \\ \hline & 2 + 5 + 14 + 30 \end{array}$$

Therefore, 2 is not a double root. Since the signs are all positive, 2 is an upper bound for the real solutions to the equation. The only possibilities remaining are $-1/2$ and $1/2$.

$$\frac{p}{q} = -\frac{1}{2}: \qquad \begin{array}{r|l} -\frac{1}{2} & 2 + 1 + 4 + 2 \\ & \quad -1 + 0 - 2 \\ \hline & 2 + 0 + 4 + 0 \end{array}$$

Thus, $P(p/q) = P(-1/2) = 0$, so that $-1/2$ is a solution, and by the factor theorem, $(x + 1/2)$ is a factor of $2x^3 + x^2 + 4x + 2$ with $2x^2 + 4$ as the other factor. Thus, we have resolved the original equation into

$$(x - 2)\left(x + \frac{1}{2}\right)(2x^2 + 4) = 0$$

and we know that 2 and $-1/2$ are two rational solutions. We now solve the quadratic equation $2x^2 + 4 = 0$.

$$\begin{aligned} 2x^2 + 4 &= 0 \\ x^2 + 2 &= 0 \\ x^2 &= -2 \\ x &= \pm\sqrt{-2} = \pm\sqrt{2}i \end{aligned}$$

The four solutions to the original equation are 2, $-1/2$, $\sqrt{2}i$, and $-\sqrt{2}i$. Notice that we obtained one negative real solution, one positive real solution, and two complex solutions. ■

**EXAMPLE 4**

Find all rational solutions to $P(x) = x^3 - 5x^2 + x + 2 = 0$.

Since $P(x)$ has two variations of sign, the equation has 2 or 0 positive real solutions. Also, since $P(-x)$ has one variation of sign, the equation has 1 negative real solution. Thus, the equation has 1 negative real solution and 2 positive real solutions or 1 negative real solution and 2 nonreal solutions.

The possibilities for $p$ are the factors of 2 ($a_0 = 2$), and the possibilities for $q$ are the factors of 1 ($a_n = 1$).

$$p\colon\ 1, -1, 2, -2; \qquad q\colon\ 1, -1; \qquad \frac{p}{q}\colon\ 1, -1, 2, -2$$

Whenever the leading coefficient is 1 (as in this case), the only possible values for $q$ are 1 and $-1$. In these cases, the possibilities for $p/q$ will always be integers (the factors of $a_0$).

$$\begin{array}{r|l} 2 & 1 - 5 + 1 + \ \ 2 \\ & \quad\ \ 2 - 6 - 10 \\ \hline & 1 - 3 - 5 - \ \ 8 \end{array} \qquad \begin{array}{r|l} 1 & 1 - 5 + 1 + 2 \\ & \quad\ \ 1 - 4 - 3 \\ \hline & 1 - 4 - 3 - 1 \end{array} \qquad \begin{array}{r|l} -1 & 1 - 5 + 1 + 2 \\ & \quad - 1 + 6 - 7 \\ \hline & 1 - 6 + 7 - 5 \end{array}$$

Since the signs in the third row alternate when we try $-1$, there is no need to try $-2$. The equation has no rational solutions. ■

The theory of equations and the study of methods of solving polynomial equations is an extensive area in mathematics that we have only briefly considered. However, this introduction will be helpful to the student who pursues advanced topics in mathematics.

We conclude this section by solving the first of the applied problems presented in the introduction to this chapter.

**EXAMPLE 5**

**ENGINEERING**

A petroleum delivery truck has a tank in the shape of a cylinder with a hemisphere attached at each end. If the total length of the tank is 20 ft, what is the radius of the cylinder if the total storage capacity is $162\pi$ ft$^3$?

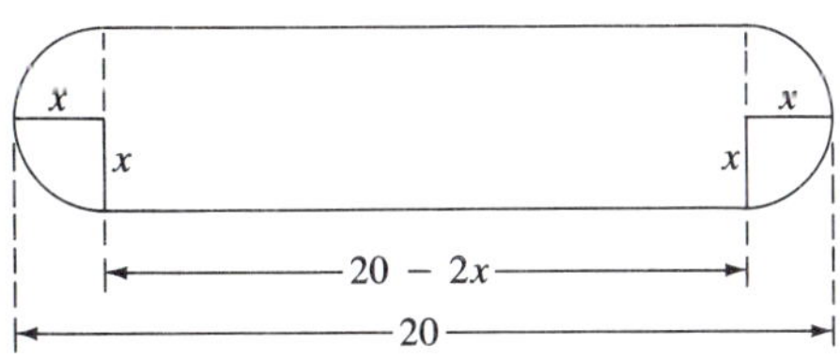


**Figure 1**

Let $x$ = the radius of the cylinder. Consider the sketch of the tank in Figure 1. Notice that the radius of the cylinder is the same as the radius of each hemisphere. The volume of the cylinder is $\pi r^2 h = \pi x^2(20 - 2x)$ and the total volume of the two hemispheres is $\frac{4}{3}\pi r^3 = \frac{4}{3}\pi x^3$. We must solve

$$\pi x^2(20 - 2x) + \frac{4}{3}\pi x^3 = 162\pi.$$

Dividing through by $\pi$ and simplifying we have

$$x^3 - 30x^2 + 243 = 0.$$

Since the leading coefficient is 1, the only possible rational solutions must be integers which are divisors of 243. Since $243 = 3^5$, the only possibilities are 3, 9, 27, 81, and 243 (the negative solutions do not need to be considered).

$$\begin{array}{r|rrrr} 3 & 1 & -30 & +0 & +243 \\ & & +3 & -81 & -243 \\ \hline & 1 & -27 & -81 & 0 \end{array}$$

Therefore, 3 is a solution. The remaining solutions must solve $x^2 - 27x - 81 = 0$, but it is easy to show that one of them is negative, and the other is far too big to satisfy the conditions of the problem. Thus, the radius of the cylinder is 3 ft. ■

## 4.3 EXERCISES

In Exercises 1–6 use Descartes' rule of signs to find the possible number of negative, positive, and nonreal solutions to each equation. Do not try to find the solutions.

**1.** $x^3 - x^2 + 2x + 1 = 0$
**2.** $x^3 + 3x^2 - x - 9 = 0$
**3.** $-x^4 + x^3 - 2x^2 + x - 5 = 0$
**4.** $3x^5 - x^4 + x^2 + x + 6 = 0$
**5.** $2x^6 - 4x^4 + x^2 - 3 = 0$
**6.** $8x^8 + 6x^6 + 4x^4 + 2x^2 + 1 = 0$

In Exercises 7–12 use the upper and lower bound test to find the smallest positive integer upper bound and the largest negative integer lower bound for the real solutions to each equation.

**7.** $4x^4 + 7x^2 - 2 = 0$
**8.** $2x^3 - 3x^2 + 6x - 9 = 0$
**9.** $x^4 + x^3 - 7x^2 - 5x + 10 = 0$
**10.** $9x^4 - 6x^3 + 10x^2 - 30x - 165 = 0$
**11.** $6x^4 - 7x^3 - 26x^2 + 7x + 20 = 0$
**12.** $2x^4 - 15x^3 + 41x^2 - 48x + 20 = 0$

In Exercises 13–20 find all rational solutions (if they exist). If possible, find the other solutions.

**13.** $x^3 + 2x^2 - x - 2 = 0$
**14.** $x^3 - x^2 - 14x + 24 = 0$
**15.** $x^3 - 5x^2 + x + 12 = 0$
**16.** $x^3 - 6x^2 + 4x - 24 = 0$
**17.** $3x^3 + x^2 + 12x + 4 = 0$
**18.** $x^4 - 4x^3 + x^2 + 8x - 6 = 0$
**19.** $9x^5 + 12x^4 + 10x^3 + x^2 - 2x = 0$
**20.** $2x^5 - x^4 - x^3 + 4x^2 - 2x = 0$

**21.** Explain why 1/2 could not be a solution to the equation $x^3 + 3x^2 - x + 7 = 0$.

**22.** Explain why a polynomial with real coefficients of degree 3 must have at least one real zero.

**23.** If $a$ is a positive real number, explain why $x^4 + a = 0$ cannot have a real root.

**24.** If $a$ is a positive real number, explain why $x^4 - a = 0$ must have exactly two real roots.

25. **BUSINESS** A shipping carton is to be made out of cardboard in such a way that the length is three times the width and the height is 2 inches less than the width. Find the dimensions of the carton if its volume is 2400 in$^3$.

26. **CONSTRUCTION** The framework for a rectangular shipping crate is to be made from 36 ft of steel rod in such a way that the ends of the crate form a square. Find the dimensions of the crate if its volume is 20 ft$^3$.

27. **AGRICULTURE** A corn storage silo is in the shape of a right circular cylinder topped by a hemisphere. If the total height of the silo is 20 ft and it will hold $1377\pi$ ft$^3$ of corn, what is the radius of the cylinder?

28. **AGRICULTURE** A cattle range is in the shape of a right triangle with the hypotenuse 4 miles longer than one of the legs. Find the dimensions of the range if its area is 24 mi$^2$.

29. **MANUFACTURING** Suppose that the cost of producing $x$ items is given by $C(x) = 1000 - x^3$ and the revenue made on the sale of $x$ items is $R(x) = 200x - 125$. Find the number of items that serves as a break-even point.

## For Review

30. Use synthetic division and the remainder theorem to find $P(-3)$ where $P(x) = 4x^4 - x^2 + 3x + 8$.

31. Use synthetic division and the factor theorem to determine whether $x - 5$ is a factor of $P(x) = x^5 - 5x^4 + x^2 - 6x + 5$.

32. What is the greatest number of distinct solutions the equation $x^6 + 3x^4 - x^2 + 5x - 7 = 0$ can have? The fewest? Exactly how many solutions does it have, counting multiplicities?

33. If $5 - 2i$ is a zero of a polynomial $P(x)$ with real coefficients, give another zero of $P(x)$.

34. If $1 + \sqrt{7}$ is a zero of a polynomial $P(x)$ with rational coefficients, give another zero of $P(x)$

35. Determine a polynomial $P(x)$ with rational coefficients of least degree such that 2 is a double root, $1 + i$ is a single root, and $1 + \sqrt{2}$ is a single root of $P(x) = 0$.

# 4.4 Graphing Polynomial Functions

### Nature of Graphs

A function of the form

$$y = P(x) = a_nx^n + a_{n-1}x^{n-1} + \cdots + a_1x + a_0 \quad (a_n \neq 0),$$

with real numbers as coefficients, is called a **polynomial function (with real coefficients).** Linear functions of the form $f(x) = a_1x + a_0$ and quadratic functions of the form $g(x) = a_2x^2 + a_1x + a_0$ are special cases of polynomial functions whose graphs are straight lines and parabolas, respectively. In more advanced courses in mathematics, it is shown that the graph of a polynomial function $y = P(x)$ is a "smooth" (no sharp corners) and "continuous" (no breaks or holes) curve.

The graphs in Figure 2 are typical graphs of polynomials. The graphs in Figure 3 cannot be graphs of polynomials.

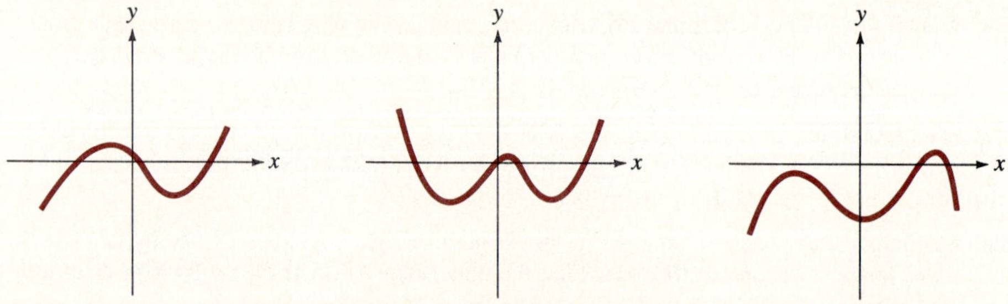


**Figure 2 Polynomial Graphs**

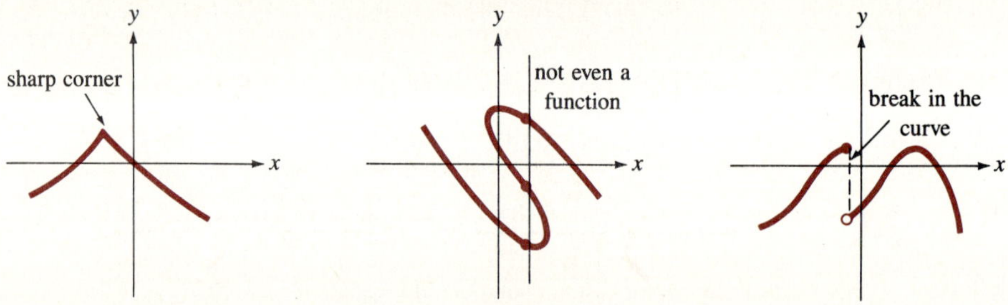


**Figure 3 Nonpolynomial Graphs**

When graphing polynomials, it is helpful to know the behavior of the graph for large values of $|x|$. Consider the polynomial

$$P(x) = a_n x^n + a_{n-1}x^{n-1} + \cdots + a_1 x + a_0 \quad (a_n \neq 0).$$

When $|x|$ is very large, the first term of the polynomial, $a_n x^n$, is larger in absolute value than the sum of the remaining terms. As a result, the sign of the first term determines the nature of the graph when $|x|$ is extremely large. The graph of a polynomial will eventually "take off" and increase or decrease, depending on the sign of the leading

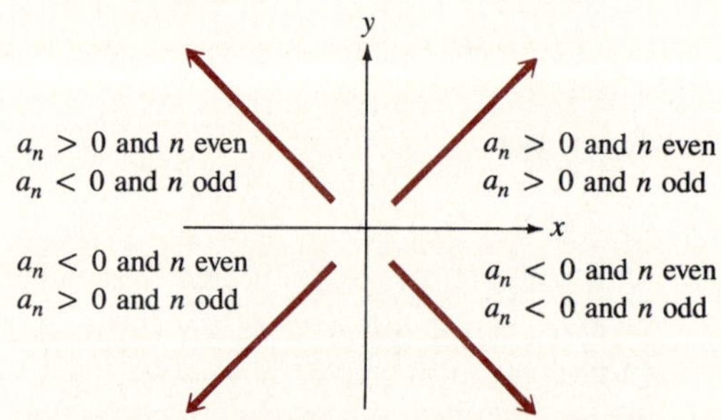


**Figure 4 Nature of $P(x)$ for Large $|x|$**

coefficient, $a_n$, and whether $n$ is even or odd. The possibilities are summarized in Figure 4 in which the arrow indicates the eventual nature of the graph.

**EXAMPLE 1** Discuss the eventual nature of the graph for large $|x|$.

**(a)** $P(x) = x^3 - 3x^2 + x - 1$
Since $n = 3$ ($n$ is odd) and $a_3 = 1 > 0$, the graph of $P(x)$ will eventually go up to the right and down to the left, as shown in Figure 5(**a**).

**(b)** $P(x) = -2x^3 + x^2 - 5$.
Since $n = 3$ ($n$ is odd) and $a_3 = -2 < 0$, the graph of $P(x)$ will eventually go down to the right and up to the left, as shown in Figure 5(**b**).

**(c)** $P(x) = 2x^4 - 3x^2 + 7$.
Since $n = 4$ ($n$ is even) and $a_4 = 2 > 0$, the graph of $P(x)$ will eventually go up to the right and also up to the left, as in Figure 5(**c**). ■

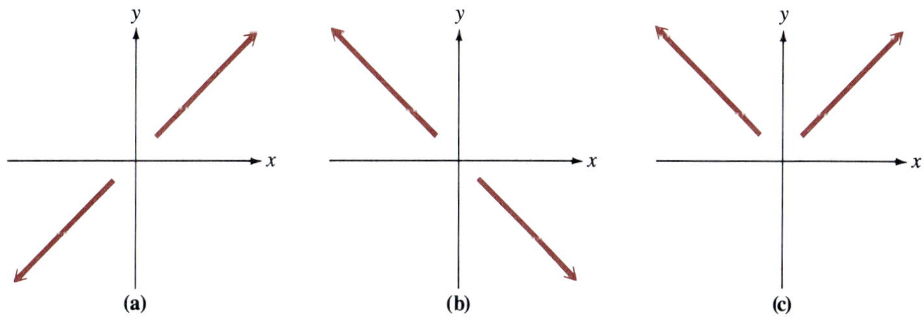


**Figure 5**

## Turning Points

Now that we know how to determine the eventual nature of the graph of a polynomial, both left and right, we turn our attention to the ''middle'' portion of the graph. Remember that the graph must be a smooth curve. The points on the graph at which the curve ''turns'' smoothly from upward to downward or downward to upward are called **turning points** of the graph. The next theorem, the proof of which is given in calculus, states an important fact about turning points.

### Number of Turning Points

Let $P(x)$ be a polynomial of degree $n$. The graph of $y = P(x)$ can have at most $n - 1$ turning points.

Thus, for example, a third-degree polynomial can have at most two turning points, although it may have none at all. Two typical third-degree polynomials are sketched in Figure 6(**a**) and (**b**) on the following page.

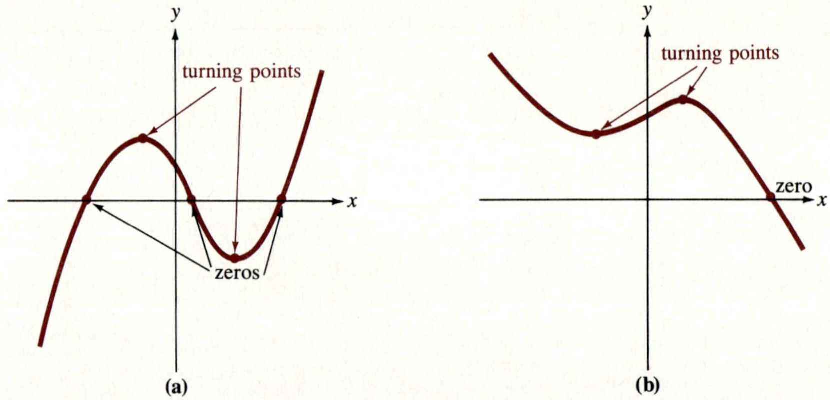


**Figure 6 Third-Degree Polynomials**

The zeros of a polynomial correspond to the $x$-intercepts of the graph of the polynomial. Thus, a polynomial of degree $n$ can cross the $x$-axis at most at $n$ points. The third-degree polynomial in Figure 6(**a**) has three distinct real zeros (three $x$-intercepts) while the one in Figure 6(**b**) has only one real zero.

For graphing polynomials, it is helpful to use information such as turning points, zeros, eventual appearance for large values of $|x|$, and the fact that all polynomial graphs are "smooth" and "continuous." Other helpful bits of information include the $y$-intercept and possible symmetries with respect to the $y$-axis or the origin.

**EXAMPLE 2**

Graph $y = P(x) = x^3 + 2x^2 - x - 2$.

Since the degree of $P(x)$ is 3, the graph has at most two turning points. Also, since 3 is odd and $a_3 = 1 > 0$, the curve must eventually go up to the right and down to the left. One easy point to determine is the $y$-intercept, where $x = 0$. Since $P(0) = -2$, $(0, -2)$ is one point on the graph. Other point(s) of interest are the zeros of $P(x)$. Since $P(x)$ has one variation of signs, there will be 1 positive real zero. Also, since $P(-x)$ has two variations of sign, there will either be no negative zeros or 2 negative zeros. We begin by searching for any rational zeros. The possibilities are listed below.

$$p\text{: } \pm 1, \pm 2 \qquad q\text{: } \pm 1 \qquad \frac{p}{q}\text{: } \pm 1, \pm 2$$

$$\begin{array}{r|l} 1 & 1 + 2 - 1 - 2 \\ & \quad\;\; 1 + 3 + 2 \\ \hline & 1 + 3 + 2 + 0 \end{array}$$

Thus, $P(x) = (x - 1)(x^2 + 3x + 2) = (x - 1)(x + 1)(x + 2)$ and its zeros are $1, -1$, and $-2$. All of this information is helpful in sketching the graph in Figure 7. ■

**EXAMPLE 3**

Graph $y = P(x) = x^4 - 4x^3 - 3x^2 + 14x - 8$.

Since the degree of $P(x)$ is 4, the graph has at most three turning points. Also, since 4 is even and $a_4 = 1 > 0$, the curve must eventually go up to the right and also up to the left. The $y$-intercept is $(0, -8)$. Since there are three variations of sign, $P(x)$ must have 1 or 3 positive zeros, and since there is one variation in sign of $P(-x)$, there must

| $x$ | $P(x)$ |
|---|---|
| $-1$ | 0 |
| 1 | 0 |
| $-2$ | 0 |
| 2 | 12 |
| 0 | $-2$ |
| $-3$ | $-8$ |
| $-3/2$ | 5/8 |

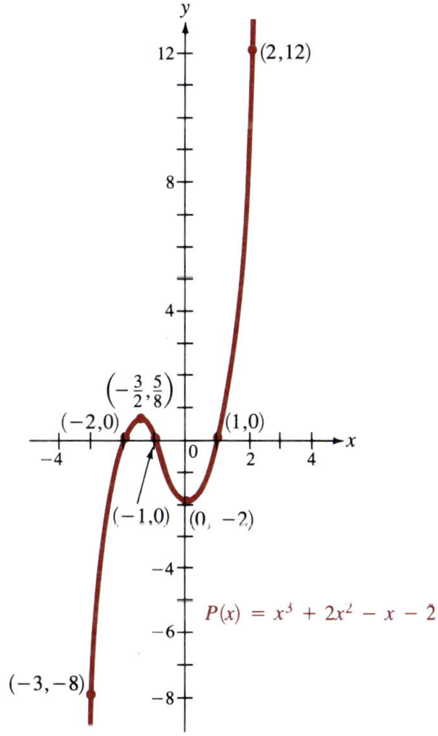


**Figure 7**

be 1 negative zero. Using the rational root theorem, it is not difficult to show that 1 is a zero of multiplicity two, and that $-2$ and 4 are the other two zeros. Thus, $P(x) = (x-1)^2(x+2)(x-4)$. Using this information and plotting several additional points we can sketch the graph as given in Figure 8. ■

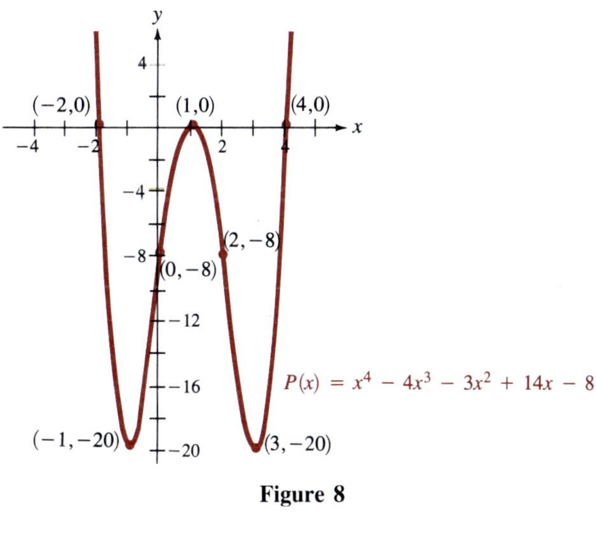


**Figure 8**

**NOTE** The graph of the polynomial in Example 3 crossed the $x$-axis at the two zeros of multiplicity 1 [at $(-2, 0)$ and $(4, 0)$] and was tangent to the $x$-axis at the zero of multiplicity 2 [at $(1, 0)$]. This is true in general. That is, the graph of a polynomial will cross the $x$-axis at a zero with an odd multiplicity, and it will be tangent to the $x$-axis at a zero with an even multiplicity. This information can be helpful when sketching a graph. Note that, although we know the approximate position of a turning point, we have no way of determining it precisely with methods of this course. ■

**EXAMPLE 4**

Graph $y = P(x) = x^4 - 4$.

The graph will have at most three turning points and four $x$-intercepts since the degree of $P(x)$ is 4. Also, since 4 is even and $a_4 = 1 > 0$, the curve must go up to the right and up to the left. Since both $P(x)$ and $P(-x)$ have one variation of signs, there will be 1 positive zero and 1 negative zero. However, a search for rational zeros yields none so the best we can do is construct a table of values. Notice that since $P(-x) = P(x)$, the graph, given in Figure 9, is symmetric with respect to the $y$-axis. ■

| $x$ | $P(x)$ |
|---|---|
| −2 | 12 |
| −1 | −3 |
| 0 | −4 |
| 1 | −3 |
| 2 | 12 |

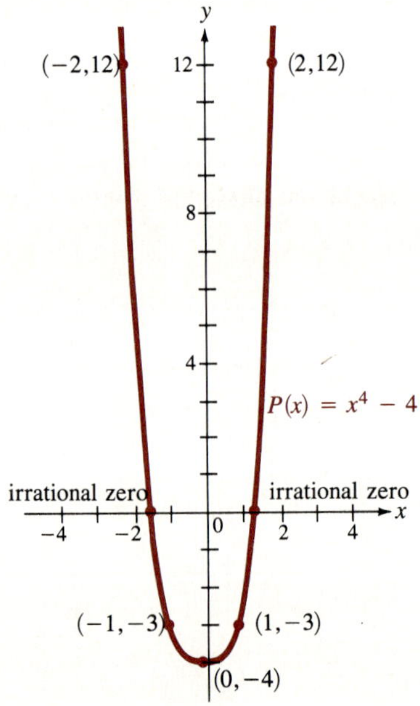


Figure 9

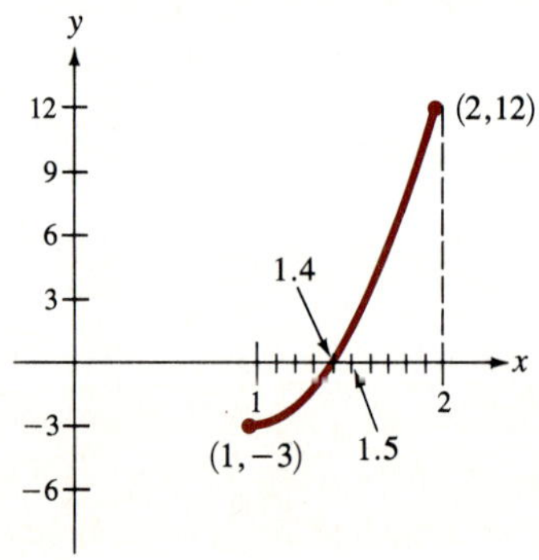


Figure 10

## The Intermediate Value Theorem

We knew that the polynomial $P(x) = x^4 - 4$, considered in Example 4, had one positive zero and one negative zero. Since the rational root theorem failed to identify any rational zeros, we know that the two zeros are irrational. In fact, from the graph in Figure 9, we know even more: the positive zero must be a number between 1 and 2 while the negative zero is between $-2$ and $-1$. This follows since at 2, $P(2) = 12 > 0$ and at 1, $P(1) = -3 < 0$ so at some value $r$ between 1 and 2, $P(r) = 0$. This is a special case of a theorem which is proved in calculus.

**The Intermediate Value Theorem**

If $P(x)$ is a polynomial, $a$ and $b$ are real numbers with $a < b$, and $P(a) \neq P(b)$, then $P(x)$ takes on every value between $P(a)$ and $P(b)$ in the interval $[a, b]$.

In particular, for $P(x) = x^4 - 4$, since $P(1) = -3$ and $P(2) = 12$, there is at least one real number $r$ in $[1, 2]$ such that $P(r) = 0$. Although the intermediate value theorem guarantees the existence of a zero for $P(x)$ between 1 and 2, it does not indicate how to find it. Since the zero is irrational, often the best we can do is approximate it by rational numbers using a numerical approximation technique. One of the simplest methods, which can be easily programmed on a computer, involves determining successive polynomial values between a known positive value and a known negative value. For example, Figure 10 shows the portion of the preceding graph between 1 and 2, enlarged. We divide the segment from 1 to 2 into ten equal parts, with division points at

$$1, \quad 1.1, \quad 1.2, \quad 1.3, \quad 1.4, \quad 1.5, \quad 1.6, \quad 1.7, \quad 1.8, \quad 1.9, \quad 2.$$

The irrational zero that we wish to approximate must be located between two of these values. Using synthetic division or substitution, we calculate the value of the polynomial at the numbers in the above list until the value is less than zero at one value and greater than zero at the next. For example, we find that

$$P(1.4) \approx -0.1584 \qquad \text{and} \qquad P(1.5) \approx 1.0625.$$

Thus, the desired zero is between 1.4 and 1.5. For greater accuracy, the process can be repeated. Dividing the segment from 1.4 to 1.5 into ten equal parts gives the values

$$1.40, \quad 1.41, \quad 1.42, \quad 1.43, \quad 1.44, \quad 1.45, \quad 1.46, \quad 1.47, \quad 1.48, \quad 1.49, \quad 1.50.$$

Again, by synthetic division or direct substitution, we would discover that

$$P(1.41) \approx -0.0475 \qquad \text{and} \qquad P(1.42) \approx 0.0659.$$

Thus, the desired zero is between 1.41 and 1.42. This procedure can be repeated until we attain any desired degree of accuracy for our approximation.

In this particular example, we could have found the zeros more directly by factoring, since $P(x) = x^4 - 4 = (x^2 - 2)(x^2 + 2)$. The zeros of $P(x)$ are found by setting each factor equal to zero.

$$\begin{aligned} x^2 - 2 &= 0 & \quad\text{or}\quad && x^2 + 2 &= 0 \\ x^2 &= 2 &&& x^2 &= -2 \\ x &= \pm\sqrt{2} &&& x &= \pm\sqrt{2}i \end{aligned}$$

We have two imaginary zeros and two real zeros. The zero which we were approximating is $\sqrt{2} \approx 1.4142135$.

## 4.4 EXERCISES

In Exercises 1–4 match each polynomial with one of the graphs **(a)**, **(b)**, **(c)**, or **(d)**.

**1.** $P(x) = x^2 - 5x + 4$

**2.** $P(x) = x^3 - x^2 - 4x + 4$

**3.** $P(x) = x^3 - 4x^2 - x + 4$

**4.** $P(x) = x^4 - 5x^2 + 4$

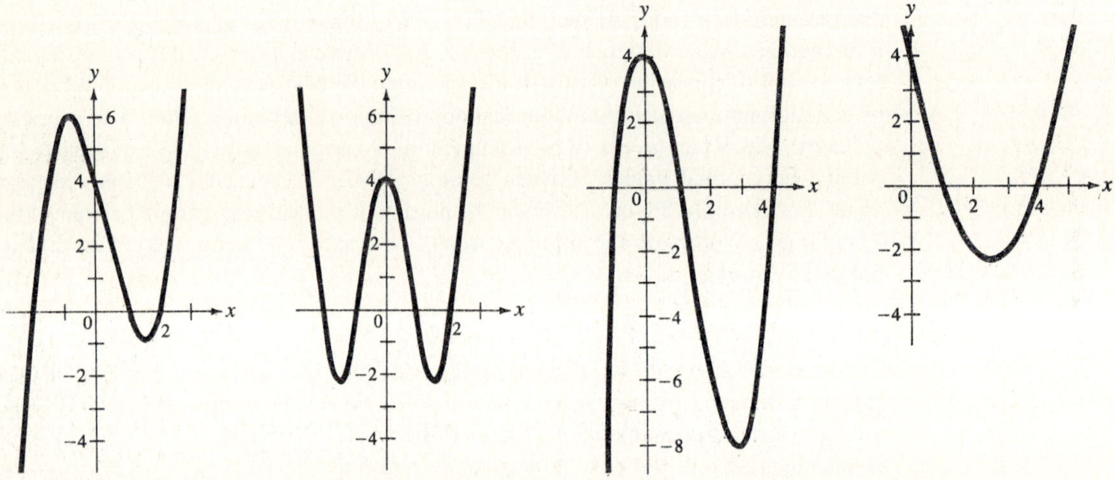

**(a)** **(b)** **(c)** **(d)**

For each of the polynomial functions in Exercises 5–12 determine **(a)** the maximum number of real zeros (the maximum number of $x$-intercepts), **(b)** the maximum number of turning points, **(c)** the $y$-intercept, **(d)** the direction of the graph for large positive values of $x$, and **(e)** the direction of the graph for small negative values of $x$.

**5.** $P(x) = x^3 + 3x^2 - x + 7$

**6.** $P(x) = 2x^5 + x^3 - 3x^2 + x - 2$

**7.** $P(x) = -2x^3 + x^2 - 5x + 12$

**8.** $P(x) = -5x^5 + 4x^4 - 3x^2 + 2x + 3$

**9.** $P(x) = 5x^4 + x^3 - 3x - 1$

**10.** $P(x) = -4x^6 + 2x^5 - x^3 + 3x - 6$

**11.** $P(x) = -3x^6 + 4x^4 + 2x^2 - x - 5$

**12.** $P(x) = 7x^8 + 2x^6 - 3x^4 + 4x^2 + 1$

In Exercises 13–20 sketch the graph of each polynomial function using the various techniques of this section.

**13.** $y = P(x) = -x^3 + 4x$

**14.** $y = P(x) = x^3 - 9x$

**15.** $y = P(x) = x^3 - 2x$

**16.** $y = P(x) = x^3 - 2x^2 - x + 2$

**17.** $y = P(x) = x^3 + x^2 + x + 1$

**18.** $y = P(x) = -x^3 + 2x^2 - 2x + 4$

**19.** $y = P(x) = x^4 - 4x^2$

**20.** $y = P(x) = -x^4 + 9x^2$

In Exercises 21–24 show that $P(x)$ has a zero between the given values $a$ and $b$.

**21.** $P(x) = x^3 + 5x^2 - 2x - 10$; $a = 1, b = 2$

**22.** $P(x) = x^3 - 4x^2 - 2x + 8$; $a = -2, b = -1$

**23.** $P(x) = x^4 - 9x^2 + 8$; $a = -3, b = -2$

**24.** $P(x) = x^5 - 8x^3 + x^2 - 8$; $a = 2, b = 3$

In Exercises 25–28 the given polynomial has an irrational zero between $a$ and $b$. Approximate this zero to the nearest hundredth.

**25.** $P(x) = x^3 - 2x^2 - 3x + 6$; $a = 1$, $b = 2$

**26.** $P(x) = x^3 + 4x^2 - 5x - 20$; $a = -3$, $b = -2$

**27.** $P(x) = x^3 + 2x^2 - 5x - 10$; $a = 2$, $b = 3$

**28.** $P(x) = x^3 + 4x^2 - 10x - 40$; $a = 3$, $b = 4$

**29.** Sketch the graph of $P(x) = x^3$. Use this graph to sketch each of the following graphs and interpret the results.
**(a)** $P(x) = x^3 + 2$
**(b)** $P(x) = x^3 - 3$
**(c)** $P(x) = (x + 1)^3$
**(d)** $P(x) = (x - 3)^3$

**30.** Sketch the graph of $P(x) = x^4$. Use this graph to sketch each of the following graphs and interpret the results.
**(a)** $P(x) = x^4 + 2$
**(b)** $P(x) = x^4 - 3$
**(c)** $P(x) = (x + 1)^4$
**(d)** $P(x) = (x - 3)^4$

**31.** Find the value of $m$ so that the graph of $y = x^4 - x^2 + x + m$ passes through the point $(2, -1)$.

**32.** Find the value of $m$ so that the graph of $y = x^5 + 4x^4 - x^2 + x + m$ has $y$-intercept $(0, 5)$.

## For Review

Find all solutions of each polynomial equation in Exercises 33–34.

**33.** $x^3 - 5x^2 + x - 5 = 0$

**34.** $2x^4 - 5x^3 - 13x^2 + 25x + 15 = 0$

**35.** Use synthetic division to find the quotient and remainder when $y^5 - 3y^4 + y^3 - 7y^2 + y - 2$ is divided by $y + 1$.

**36.** Use a calculator, synthetic division, and the remainder theorem to find $P(4.1)$ if $P(x) = 1.2x^3 - 3.7x^2 - 5.35x + 1.4268$.

**37.** The solutions to $x^3 + 8 = 0$ are called cube roots of $-8$. How many cube roots of $-8$ are there? Find them.

What is the domain of each function in Exercises 38–39? This type of function is considered in the next section.

**38.** $f(x) = \dfrac{x^2 - 5x - 6}{x + 3}$

**39.** $g(x) = \dfrac{3x + 1}{x^2 + 5}$

## 4.5 Graphing Rational Functions

A function of the form

$$f(x) = \frac{P(x)}{Q(x)}$$

where $P(x)$ and $Q(x)$ are polynomials and $Q(x)$ is not the zero polynomial, is called a **rational function.** For example,

$$f(x) = \frac{x^2 + 2x - 3}{x - 2}, \qquad a(x) = \frac{x - 7}{x^2 + x + 3}, \qquad b(x) = \frac{2x^3 - 5x^2 + x - 7}{3}$$

are rational functions. Every polynomial function is also a rational function since

$$P(x) = \frac{P(x)}{1} = \frac{P(x)}{Q(x)} \quad \text{where } Q(x) = 1.$$

In this section we are only concerned with rational functions whose denominators are not constants (otherwise, they could be reduced to polynomial functions). Unlike polynomial functions which are defined for every real number $x$, rational functions are not defined at any value of $x$ which makes the denominator zero. For example,

$$f(x) = \frac{x^2 + 2x - 3}{x - 2}$$

is not defined when $x = 2$, since

$$f(2) = \frac{(2)^2 + 2(2) - 3}{(2) - 2} = \frac{5}{0},$$

which is undefined. However, $f(x)$ is a number if $x$ is any number other than 2, and thus the domain of $f(x)$ is all real numbers except 2.

## Asymptotes

If a curve gets closer and closer to a line when the values of $x$ get either closer and closer to a given number or larger and larger positively or negatively, the line is called an **asymptote** of the curve. Many rational functions have asymptotes. Examples are shown in Figure 11.

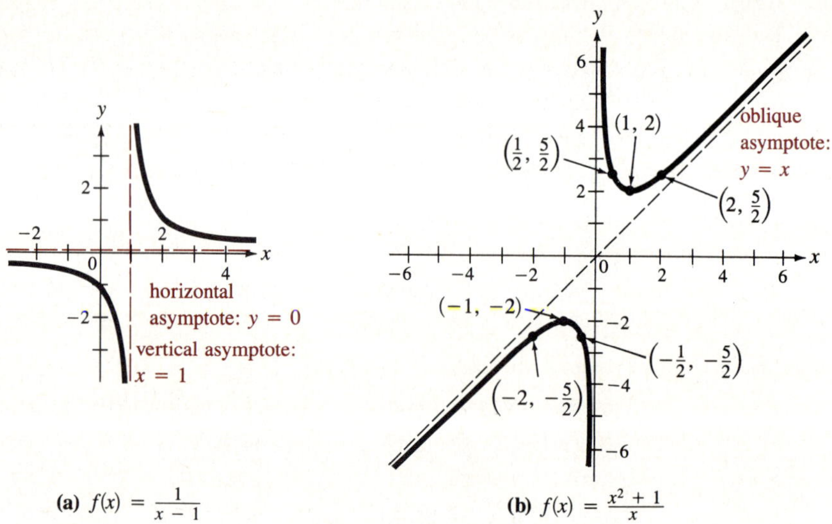


**Figure 11** Asymptotes

There are three types of asymptotes: **vertical asymptotes** which are equal to or parallel to the $y$-axis (like $x = 1$ in Figure 11(**a**)), **horizontal asymptotes** which are equal to or parallel to the $x$-axis (like $y = 0$ in Figure 11(**a**)), and **oblique asymptotes,** which are lines not parallel to either coordinate axis (like $y = x$ in Figure 11(**b**)).

**To Find the Vertical Asymptotes of a Rational Function**

$$f(x) = \frac{P(x)}{Q(x)},$$

where $P(x)$ and $Q(x)$ have no common factor, set $Q(x) = 0$ and solve. If $r$ is a real zero of $Q(x)$, then the line $x = r$ is a vertical asymptote.

**EXAMPLE 1** Determine the vertical asymptotes of each function.

**(a)** $f(x) = \dfrac{x^2 + 2x - 3}{x - 2}$

Setting the denominator $x - 2$ equal to zero, we obtain $x = 2$ as a vertical asymptote.

**(b)** $g(x) = \dfrac{x + 1}{x^2 - 4}$

To solve $x^2 - 4 = 0$, we factor and obtain $x = 2$ and $x = -2$ as the vertical asymptotes.

**(c)** $h(x) = \dfrac{2x^2}{x^2 + 1}$

Since $x^2 + 1 = 0$ has no real solutions, there are no vertical asymptotes. ■

Horizontal and oblique asymptotes, which describe what happens to the graph when $|x|$ becomes large, can be discovered by considering the degrees of the polynomials in the numerator and denominator. Suppose that the degree of the polynomial in the numerator of a rational function is less than the degree of the polynomial in the denominator, as in the following function, $k(x)$.

$$k(x) = \frac{3x + 5}{x^2 - 2x + 7}$$

Multiply both numerator and denominator by $1/x^2$, or equivalently, divide each term by $x^2$.

$$k(x) = \frac{\dfrac{3x}{x^2} + \dfrac{5}{x^2}}{\dfrac{x^2}{x^2} - \dfrac{2x}{x^2} + \dfrac{7}{x^2}} = \frac{\dfrac{3}{x} + \dfrac{5}{x^2}}{1 - \dfrac{2}{x} + \dfrac{7}{x^2}}$$

As $x$ becomes larger, $3/x$ becomes smaller (for example, if $x = 10$, $3/x = 3/10 = 0.3$; if $x = 100$, $3/x = 3/100 = 0.03$; if $x = 1000$, $3/x = 3/1000 = 0.003$). Likewise, when $x$ gets large, $5/x^2$, $2/x$, and $7/x^2$ all approach zero. Thus, the fraction approaches

$$\frac{0 + 0}{1 - 0 + 0} = \frac{0}{1} = 0.$$

The same is true when $x$ gets larger and larger in the negative direction. Thus, as $|x|$ increases, $k(x)$ approaches 0 so that the $x$-axis, $y = 0$, is a horizontal asymptote.

If the degree of the numerator is equal to the degree of the denominator, the rational function also has a horizontal asymptote, but this time it is a line parallel to the $x$-axis. For example, consider

$$s(x) = \frac{2x^2 - x + 3}{5x^2 + 1}.$$

Again, we multiply numerator and denominator by $1/x^2$.

$$s(x) = \frac{2 - \frac{1}{x} + \frac{3}{x^2}}{5 + \frac{1}{x^2}}$$

As $|x|$ increases, $1/x$, $3/x^2$, and $1/x^2$ all approach zero and the fraction approaches

$$\frac{2 - 0 + 0}{5 + 0} = \frac{2}{5}.$$

Thus, as $|x|$ increases, $k(x)$ approaches 2/5 which means that $y = 2/5$ is a horizontal asymptote.

**Horizontal Asymptote of a Rational Function**

Let $f(x) = \dfrac{P(x)}{Q(x)} = \dfrac{a_nx^n + a_{n-1}x^{n-1} + \cdots + a_1x + a_0}{b_mx^m + b_{m-1}x^{m-1} + \cdots + b_1x + b_0}$, $(a_n \neq 0,\ b_m \neq 0)$ be a rational function in which $P(x)$ and $Q(x)$ have no common factor.

1. If $n < m$, then the $x$-axis, $y = 0$, is a horizontal asymptote of the graph of $f(x)$.
2. If $n = m$, then the line $y = a_n/b_m$ is a horizontal asymptote.
3. If $n > m$, there is no horizontal asymptote.

**EXAMPLE 2** Determine the horizontal asymptotes of each function.

**(a)** $f(x) = \dfrac{x^2 + 2x - 3}{x - 2}$

Since $P(x) = x^2 + 2x - 3$ has degree $n = 2$ and $Q(x) = x - 2$ has degree $m = 1$, $n > m$ so there is no horizontal asymptote.

**(b)** $g(x) = \dfrac{x + 1}{x^2 - 4}$

Since $P(x) = x + 1$ has degree $n = 1$ and $Q(x) = x^2 - 4$ has degree $m = 2$, $n < m$ so that the $x$-axis, $y = 0$, is a horizontal asymptote.

**(c)** $h(x) = \dfrac{2x^2}{x^2 + 1}$

Since $P(x)$ and $Q(x)$ are both of degree 2 ($n = m$), with $a_n = 2$ and $b_m = 1$, the line $y = \dfrac{a_n}{b_m} = \dfrac{2}{1} = 2$ is a horizontal asymptote. ■

Finally, when the degree of the numerator of a rational function exceeds the degree of the denominator by one, the graph has an oblique asymptote.

**EXAMPLE 3** In the function

$$f(x) = \frac{x^2 + 2x - 3}{x - 2},$$

the degree of the numerator is 2, which is one more than the degree of the denominator. In such cases, we divide $x^2 + 2x - 3$ by $x - 2$ and get the quotient $x + 4$ and remainder 5. Thus,

$$f(x) = \frac{x^2 + 2x - 3}{x - 2} = x + 4 + \frac{5}{x - 2}.$$

As $|x|$ gets larger and larger, $5/(x - 2)$ approaches zero so that $f(x)$ approaches $x + 4$. Hence, the line $y = x + 4$ is an oblique asymptote of the graph. ■

### Oblique Asymptote of a Rational Function

Let
$$f(x) = \frac{P(x)}{Q(x)}$$

be a rational function such that the degree of $P(x)$ is $n$ and the degree of $Q(x)$ is $m$.

1. If $n = m + 1 (m > 0)$, and if the quotient of $P(x)$ and $Q(x)$ is $g(x)$ and the remainder is not zero, then the equation
$$y = g(x)$$
is an oblique asymptote of the graph.
2. If $n > m + 1$, there are no oblique asymptotes.

## Zeros

A **zero of a rational function** is a value of $x$ that makes the numerator zero while the denominator is nonzero. Of course, the zeros are the $x$-intercepts of the graph. To graph a rational function $f(x) = \frac{P(x)}{Q(x)}$, determine the asymptotes, the $x$-intercepts, the $y$-intercept (found by setting $x = 0$), any symmetries, and consider whether $f(x)$ is positive or negative in the intervals with endpoints determined by the zeros of $P(x)$ (the zeros of the rational function) and the zeros of $Q(x)$ (the values that determine vertical asymptotes). In addition, it may be necessary to plot several points to obtain a reasonably accurate sketch. We illustrate this technique in the following examples.

**EXAMPLE 4** Graph $y = f(x) = \dfrac{x^2 + 2x - 3}{x - 2}$.

| | | |
|---|---|---|
| Vertical asymptotes: | $x = 2$ | See Example 1 |
| Horizontal asymptotes: | none | See Example 2 |
| Oblique asymptotes: | $y = x + 4$ | See Example 3 |

We can factor the numerator.

$$\frac{x^2 + 2x - 3}{x - 2} = \frac{(x + 3)(x - 1)}{x - 2} = 0$$

Therefore, $-3$ and $1$ are zeros of $f(x)$.

$x$-intercepts: $(-3, 0)$ and $(1, 0)$

Substitute 0 for $x$ to find the $y$-intercept.

$$f(0) = \frac{(0)^2 + 2(0) - 3}{0 - 2} = \frac{-3}{-2} = \frac{3}{2}$$

$y$-intercept: $(0, 3/2)$
Symmetries: none (The tests for symmetry with respect to the $y$-axis and origin both fail.)

The intervals determined by the zeros of $f(x)$, $-3$ and $1$, and the vertical asymptote, $x = 2$, are $(-\infty, -3)$, $(-3, 1)$, $(1, 2)$, and $(2, \infty)$. A table can be used to summarize the results.

| Interval | $(-\infty, -3)$ | $(-3, 1)$ | $(1, 2)$ | $(2, \infty)$ |
|---|---|---|---|---|
| Test point $a$ | $-4$ | $0$ | $\frac{3}{2}$ | $3$ |
| Value of $f(a)$ | $f(-4) = -\frac{5}{6}$ | $f(0) = \frac{3}{2}$ | $f\left(\frac{3}{2}\right) = -\frac{9}{2}$ | $f(3) = 12$ |
| Sign of $f(x)$ | $-$ | $+$ | $-$ | $+$ |
| Location of the graph of $f(x)$ | below the $x$-axis | above the $x$-axis | below the $x$-axis | above the $x$-axis |

The graph is shown in Figure 12. ■

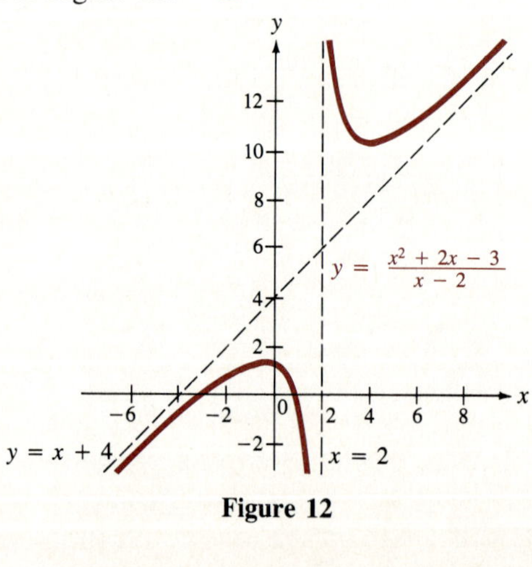


**Figure 12**

**EXAMPLE 5** Graph $y = h(x) = \dfrac{2x^2}{x^2 + 1}$.

Vertical asymptotes: none
Horizontal asymptotes: $y = 2$
Oblique asymptotes: none
$x$-intercepts: $(0, 0)$
$y$-intercept: $(0, 0)$
Symmetries: $y$-axis

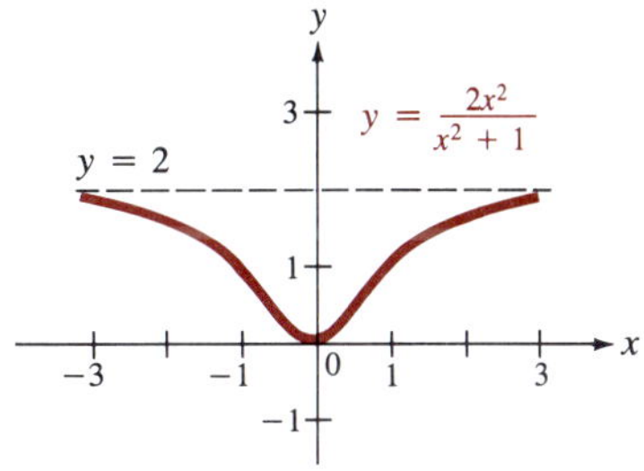


**Figure 13**

This time there are only two intervals to consider, $(-\infty, 0)$ and $(0, \infty)$. Since the graph is symmetric with respect to the $y$-axis, the sign of $h(x)$ will be the same on both intervals. Using the test point 1 in $(0, \infty)$, we have $h(1) = 1 > 0$, so the sign of $h(x)$ is $+$ on $(0, \infty)$ and also on $(-\infty, 0)$. Thus, the graph of $h(x)$, given in Figure 13, is above the $x$-axis on both intervals. ■

**EXAMPLE 6** Graph $y = F(x) = \dfrac{x^2 + 1}{x^2 - 1}$.

Vertical asymptotes: $x = 1$ and $x = -1$
Horizontal asymptotes: $y = 1$
Oblique asymptotes: none
$x$-intercepts: none
$y$-intercept: $(0, -1)$
Symmetries: $y$-axis

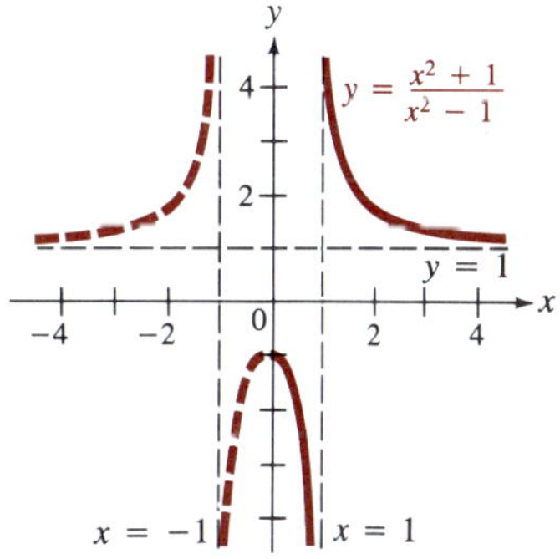


**Figure 14**

With no $x$-intercepts, the intervals, determined by the vertical asymptotes $x = 1$ and $x = -1$, are

$$(-\infty, -1), \quad (-1, 1), \quad \text{and} \quad (1, \infty).$$

Using the symmetry with respect to the $y$-axis, the sign of $F(x)$ on $(-\infty, -1)$ will be the same as the sign on $(1, \infty)$.

| Interval | $(-1, 1)$ | $(1, \infty)$ |
|---|---|---|
| Test point $a$ | 0 | 2 |
| Value of $F(a)$ | $F(0) = -1$ | $F(2) = \frac{5}{3}$ |
| Sign of $F(x)$ | $-$ | $+$ |
| Location of the graph of $F(x)$ | below the $x$-axis | above the $x$-axis |

Thus, the graph of $F(x)$ is above the $x$-axis on $(-\infty, -1)$ and $(1, \infty)$, and below the $x$-axis on $(-1, 1)$. In Figure 14, the solid portion of the curve was plotted first with the dashed portion resulting from using symmetry with respect to the $y$-axis. ■

CAUTION Graphs of rational functions, like polynomial functions, consist of curves that are smooth with no sharp "points" or turns. This fact is illustrated by the graphs in the preceding examples. ■

We conclude this section by solving the second of the two applied problems given in the introduction to this chapter.

## EXAMPLE 7

**MANUFACTURING**

During an eight-hour shift, the number of items produced in a factory is given by $n(t) = t^2 + 16t \quad (0 \le t \le 8)$. The total cost in dollars of producing $n(t)$ items is given by $c(t) = 120t + 960$. The average cost of production is a function of $t$, given by $a(t) = \frac{c(t)}{n(t)}$. Graph $a(t)$ concentrating on the portion for $0 < t \le 8$ and interpret the results.

In order to determine the asymptotes we write $a(t)$ in factored form.

$$y = a(t) = \frac{120t + 960}{t(t + 16)}$$

Vertical asymptotes: $t = 0$ and $t = -16$
Horizontal asymptotes: $y = 0$
Oblique asymptotes: none
$x$-intercepts: $(-8, 0)$
$y$-intercepts: none
Symmetries: none

The graph, given in Figure 15, shows the portion of interest $(0 < t \le 8)$ with a solid line.

| $t$ | $a(t)$ |
|---|---|
| −8 | 0 |
| −14 | 25.7 |
| −2 | −25.7 |
| −18 | −33.3 |
| −20 | −18 |
| −24 | −10 |
| 2 | 33.3 |
| 4 | 18 |
| 8 | 10 |
| 12 | 7.1 |

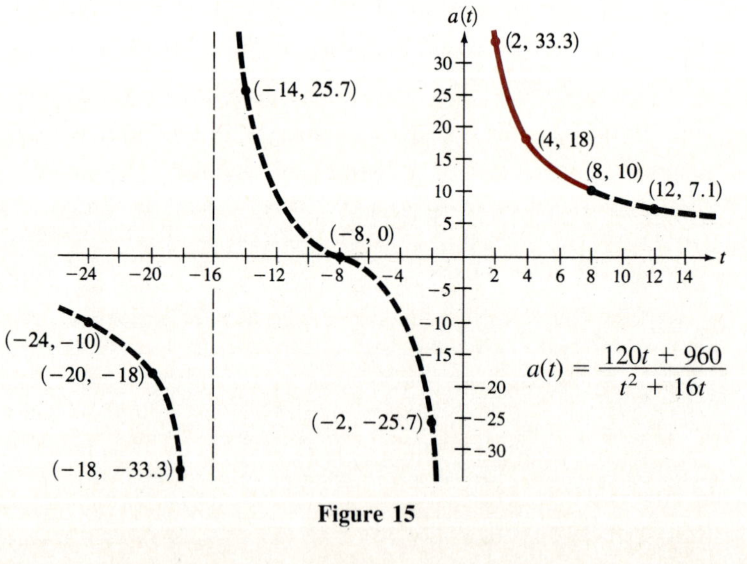


Figure 15

The average cost of production decreases to a minimum after 8 hours ($t = 8$), and amounts to \$10 per item at that time. ■

## 4.5 EXERCISES

In Exercises 1–10 give the **(a)** vertical asymptotes, **(b)** horizontal asymptotes, **(c)** oblique asymptotes, **(d)** $x$-intercepts, **(e)** $y$-intercept, and **(f)** symmetries of the graph of each rational function.

**1.** $f(x) = \dfrac{5}{x+2}$ **2.** $g(x) = \dfrac{-3}{x-4}$ **3.** $F(x) = \dfrac{4x}{x-5}$ **4.** $G(x) = \dfrac{-2x}{2x+1}$

**5.** $f(x) = \dfrac{x+1}{x^2+x-2}$ **6.** $g(x) = \dfrac{2(3x-1)}{2x^2+9x-5}$ **7.** $F(x) = \dfrac{x^2+x-6}{x-5}$

**8.** $G(x) = \dfrac{x^2-16}{x+1}$ **9.** $f(x) = \dfrac{5x^4}{x^4+1}$ **10.** $F(x) = \dfrac{2x^3}{x^3-1}$

In Exercises 11–22 graph each rational function using information about asymptotes, intercepts, symmetries, and the intervals where the function is above or below the $x$-axis.

**11.** $f(x) = \dfrac{2}{x-2}$ **12.** $F(x) = \dfrac{4}{4-x}$ **13.** $g(x) = \dfrac{x}{x+1}$ **14.** $G(x) = \dfrac{x}{1-x}$

**15.** $F(x) = \dfrac{1}{x^2-4}$ **16.** $f(x) = \dfrac{-1}{x^2-9}$ **17.** $G(x) = \dfrac{x}{x^2-4}$ **18.** $g(x) = \dfrac{x}{1-x^2}$

**19.** $f(x) = \dfrac{2x^2+1}{x^2-1}$ **20.** $F(x) = \dfrac{x^2+4}{4-x^2}$ **21.** $g(x) = \dfrac{5}{x^2+5}$ **22.** $f(x) = \dfrac{x^2+x-6}{x+2}$

**23.** MANUFACTURING During a 12-hour shift, the number of items produced in a factory is given by $n(t) = t^2 + 24t$ $(0 \le t \le 12)$. The total cost in dollars of producing $n(t)$ items is given by $c(t) = 144t + 1728$. Graph the rational function which expresses the average cost of production for $0 < t \le 12$ and interpret the results.

**24.** SCIENCE The power $p$ in watts consumed by an electrical circuit is given by $p = \dfrac{10r}{(1+r)^2}$ where $r \ge 0$ is the resistance in ohms. Sketch the graph of this function and estimate the resistance which yields a maximum power.

**25.** Give a rational function whose graph has one vertical asymptote at $x = 1$, one horizontal asymptote at $y = 0$, and a $y$-intercept $(0, -3)$.

**26.** Give a rational function whose graph has one vertical asymptote at $x = 2$, one horizontal asymptote at $y = -1$, and $x$-intercept $(3, 0)$.

**27.** Can the graph of a rational function ever cross a horizontal asymptote?

**28.** Can the graph of a rational function ever cross a vertical asymptote?

### For Review

**29.** Graph the polynomial function $y = P(x) = 2x^2 - x^4$.

**30.** Graph the polynomial function $y = P(x) = x^3 - x^2 - 16x - 20$.

**31.** Use Descartes' rule of signs to find the possible number of negative, positive, and nonreal solutions to the equation $x^4 + 5x^3 + 2x^2 + x + 2 = 0$.

**32.** Use the upper and lower bound test to find the smallest positive integer upper bound and the largest negative integer lower bound for the real solutions to the equation $8x^3 + 2x^2 - 15x = 0$.

## CHAPTER 4 REVIEW EXERCISES

**1.** Answer the following for the polynomial $P(x) = 3x^4 - 2x^2 + 7x - 8$.
**(a)** What is the degree of $P(x)$?
**(b)** What is the leading coefficient?
**(c)** What is $P(1)$?
**(d)** What is $P(-1)$?

**2.** Determine whether 2, $3i$, and $-3i$ are zeros of $Q(x) = x^3 - 2x^2 + 9x - 18$.

**3.** Use synthetic division to find the quotient and remainder when $x^4 - 3x^3 + x - 2$ is divided by $x + 3$.

**4.** If a polynomial $P(x)$ is divided by $x - 5$, what will be the value of the remainder?

**5.** What is the name of the theorem used to answer Exercise 4?

**6.** If $P(x)$ is a polynomial and $P(-3) = 0$, give one factor of $P(x)$.

**7.** What is the name of the theorem used to answer Exercise 6?

**8.** BUSINESS The revenue in dollars brought in on the sale of $x$ books is given by $R(x) = 10.95x$. The cost in dollars of producing $x$ books is given by $C(x) = 500 + 5.95x$. The zeros of $P(x) = R(x) - C(x)$ are called break-even points. Show that 100 is a break-even point.

**9.** Find a value of $m$ so that $x^4 - 2x^3 + mx - 60$ has a remainder of 0 when divided by $x + 3$. What is $P(-3)$?

**10.** If $P(x) = 2x^4 - 3x^2 + x - 8$, use synthetic division and the remainder theorem to find the following.
**(a)** $P(1)$ **(b)** $P(-1)$ **(c)** $P(2)$

**11.** If $P(x) = x^4 - 4x^3 - 3x^2 + 10x + 8$, use synthetic division and the factor theorem to determine if each of the following is a factor of $P(x)$.
**(a)** $x - 4$ **(b)** $x + 1$ **(c)** $x + 2$

**12.** If $-4$ is a zero of multiplicity 3 of a polynomial $P(x)$, give a cubic factor of $P(x)$.

**13.** If $2 - 7i$ is a zero of polynomial $P(x)$, with real coefficients, give another zero of $P(x)$.

**14.** If $3 + \sqrt{11}$ is a zero of a polynomial $P(x)$, with rational coefficients, give another zero of $P(x)$.

**15.** What is the greatest number of distinct solutions the equation $5x^7 + 3x^5 - 2x^4 + x^2 - 9 = 0$ can have? The fewest? Exactly how many solutions does it have, counting multiplicities?

**16.** Determine a polynomial $P(x)$ with rational coefficients of least degree such that $-2$ is a double root, $1 + \sqrt{2}$ is a single root, and $1 - i\sqrt{2}$ is a single root.

In Exercises 17–18 use Descartes' rule of signs to find the possible number of negative, positive, and nonreal solutions to each equation. Do not try to find the solutions.

**17.** $2x^6 - 4x^5 - 2x^3 + 3x^2 + x - 7 = 0$

**18.** $3x^5 + 7x^3 + 10x + 1 = 0$

In Exercises 19–20 use the upper and lower bound test to find the smallest positive integer upper bound and the largest negative integer lower bound for the real solutions to each equation.

**19.** $8x^4 - 16x^3 - 2x^2 - 16x - 10 = 0$

**20.** $3x^6 + 13x^5 + 14x^4 + 6x^2 + 26x + 28 = 0$

**21.** Suppose $\frac{a}{b}$ is a rational solution to $4x^4 - 3x^3 + 2x^2 - x + 5 = 0$.

**(a)** $a$ is a factor of which coefficient?

**(b)** $b$ is a factor of which coefficient?

**22.** Explain why any possible rational solutions of $x^5 - 3x^3 + x^2 + x - 9 = 0$ must be integers.

In Exercises 23–24 find all rational solutions (if they exist). If possible, find all solutions.

**23.** $6x^3 - 11x^2 + 9x - 2 = 0$

**24.** $x^3 - 3x^2 - 9x - 5 = 0$

**25.** **GEOMETRY** Find the radius of a sphere whose surface area and volume are equal.

For each of the polynomial functions in Exercises 26–27 determine **(a)** the maximum number of real zeros (the maximum number of $x$-intercepts), **(b)** the maximum number of turning points, **(c)** the $y$-intercept, **(d)** the direction of the graph for large positive values of $x$, and **(e)** the direction of the graph for small negative values of $x$.

**26.** $P(x) = x^3 - 7x^2 + x - 5$

**27.** $P(x) = -x^4 + 3x^2 - x + 8$

In Exercises 28–29 sketch the graph of each polynomial function using information about the $x$-intercepts, the $y$-intercept, the turning points, and the eventual direction of the graph.

**28.** $y = P(x) = -x^3 - x^2 + 2x$

**29.** $y = P(x) = 5x^4 - 5$

**30.** Given that the polynomial $P(x) = x^3 + 5x^2 - 11x - 55$ has an irrational zero between 3 and 4, approximate this zero to the nearest hundredth.

In Exercises 31–32 give the **(a)** vertical asymptotes, **(b)** horizontal asymptotes, **(c)** oblique asymptotes, **(d)** $x$-intercepts, **(e)** $y$-intercept, **(f)** symmetries of the graph of each rational function.

**31.** $f(x) = \dfrac{10}{x - 9}$

**32.** $g(x) = \dfrac{6x^2}{9x^2 - 1}$

In Exercises 33–34 graph each rational function using information about asymptotes, intercepts, and symmetries.

**33.** $y = f(x) = \dfrac{1}{x^2 - 2x + 1}$

**34.** $y = f(x) = \dfrac{x^3 + 1}{x^2}$

**35.** **SCIENCE** The nose section of a missile is in the shape of a right circular cylinder with a hemisphere attached to one end. If the total length of the nose section is 40 inches, and the total volume of the section is 5019 in$^3$, what is the radius of the cylinder (to the nearest tenth of an inch)? [Hint: The radius is between 6 inches and 7 inches in length.]

CHAPTER

# 5 EXPONENTIAL AND LOGARITHMIC FUNCTIONS

Historically, logarithms were developed to help carry out complicated numerical calculations. With the advent of computers and hand calculators, logarithmic calculations are no longer of much interest. However, logarithmic and exponential equations and functions remain important, with numerous applications in mathematics today. The following examples illustrate two types of problems we can solve using exponentials and logarithms.

**BUSINESS ▶**

In order to purchase a new home a young couple borrows $60,000 at an annual interest rate of 12%. It is to be paid back over a 30-year period in equal monthly payments. What is their monthly payment, and what is the total interest paid over the period of the loan?

**◀ MEDICINE**

Audiologists generally agree that continuous exposure to a sound level in excess of 90 decibels for more than 5 hours daily may cause long-term hearing problems. If a student listens to a walkman that produces an intensity of $2 \times 10^{-3}$ watt/m$^2$ for long periods of time daily, is there a potential danger to his hearing?

The first of these problems can be solved using an exponential function (see Example 2 in Section 5.6), and the second requires a logarithmic function (see Example 8 in Section 5.6). Our study of these functions begins with a brief introduction to logarithms. Following this, we develop the basic properties of logarithms, consider common and natural logarithms and their relationship to equations, and conclude with an investigation of a variety of applied problems. For working with logarithms, since calculations and answers often involve decimals, a calculator is an invaluable tool. Generally, all computations will be made and stored on a calculator with no rounding until the final answer. Even with this agreement there can be slight variations resulting from rounding differences among various calculators. If you use a table instead of a calculator, your answers may differ somewhat but should be close enough to verify the accuracy of your work.

## 5.1 Logarithms

In previous chapters we learned various methods for solving equations that involved sums, differences, products, quotients, powers, and roots of algebraic expressions. In none of these equations was the variable present in an exponent. When a variable does appear in an exponent, as in the equation $2^x = 8$, the equation-solving rules we have already learned are of little or no help. At present, we must discover the solution by inspection. In this case, the problem is not too difficult, and we would probably recognize that when 2 is raised to the *third* power, the result is 8. Thus, the solution to the equation is

$$x = 3. \qquad 2^3 = 8$$

Alternatively, we might give the solution in words.

$x$ is the exponent on 2 that gives the number 8

The word *logarithm* can be used instead of *exponent,* and the solution is written

$x$ is the logarithm on 2 that gives the number 8.

Since 2 is called the *base* in the expression $2^x$, the solution can also be written in the form

$x$ is the logarithm of 8 using 2 as the base

or $x$ is the logarithm to the base 2 of 8.

This final statement is usually symbolized by

$$x = \log_2 8.$$

Basically, we have shown that

$$2^x = 8 \qquad \text{and} \qquad x = \log_2 8$$

are two forms of the same equation, with $2^x = 8$ called the **exponential form** and $x = \log_2 8$ the **logarithmic form.** That is, they are equivalent equations, with the logarithmic form having the desirable quality of being "solved" for the variable $x$. In this case, one additional equivalent equation is

$$x = 3. \qquad 3 = \log_2 8$$

For now, however, we cannot always make this additional simplification.

### Definition of Logarithm

Let $a$, $x$, and $y$ be real numbers, $a > 0$, $a \neq 1$, that satisfy the exponential equation

$$a^x = y.$$

Then $x$ is the **logarithm to the base *a*** of $y$, and the equivalent logarithmic equation is

$$x = \log_a y.$$

The restriction $a > 0$ is necessary since if $a < 0$, then $a^x$ would not be a real number for values of $x$ such as 1/2. Also, if $a = 1$, $a^x = 1$ for all values of $x$.

**EXAMPLE 1** Write each equation in an equivalent form.

**(a)** $9^x = d$ The logarithmic form of this equation is $x = \log_9 d$.

**(b)** $w = \log_3 7$ The exponential form of this equation is $3^w = 7$. ■

When converting an equation from exponential to logarithmic form, or from logarithmic to exponential form, remember that in both cases the base is written below the level of the logarithm (exponent), as indicated in the following diagram.

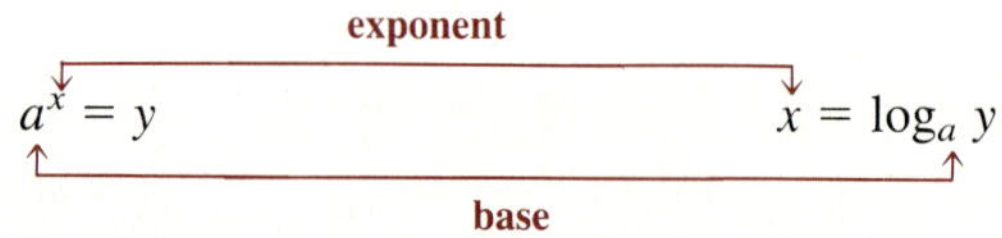


Solutions to simple exponential equations can often be determined by direct inspection; solutions to some logarithmic equations are best found by converting to exponential form.

**EXAMPLE 2** **(a)** Determine the numerical value of $x$ in $x = \log_4 64$.
We first convert $x = \log_4 64$ to exponential form.

$$4^x = 64$$

At this point, it might be clear that $x$ must be 3 since $4^3 = 64$. By writing 64 as a power of 4, $64 = 4^3$, we have

$$4^x = 4^3.$$

In this form, it is now obvious that $x$ is 3 since the bases on both sides of the equation are 4, forcing the exponents on both sides to be equal.

**(b)** Determine the numerical value of $x$ in $\log_x \frac{1}{8} = -3$.

Convert to exponential form.

$$x^{-3} = \frac{1}{8}$$

$$x^{-3} = 2^{-3} \qquad \frac{1}{8} = \frac{1}{2^3} = 2^{-3}$$

It is now clear that $x$ is 2 since the exponents on both sides of the equation are $-3$, forcing the bases on both sides to be equal. ■

Two properties of exponentials, used in Examples 2(**a**) and 2(**b**), are presented below. These properties are intuitively clear and will become even more apparent as we consider exponential functions in the next section.

### Properties of Exponentials

Let $x$ and $y$ be real numbers, $a$ and $b$ positive real numbers with $a \neq 1$ and $b \neq 1$.

1. $a^x = a^y$ if and only if $x = y$.
2. $a^x = b^x$ if and only if $a = b$.

**EXAMPLE 3** Solve each equation.

**(a)** $3^{x-1} = 27$

Since $27 = 3^3$, $3^{x-1} = 3^3$. Thus,

$$x - 1 = 3$$

$$x = 4.$$

**(b)** $(x - 1)^4 = 16$

Since $16 = 2^4$, $(x - 1)^4 = 2^4$. Thus,

$$x - 1 = 2$$

$$x = 3.$$

**(c)** $\log_x 36 = 2$

Convert to exponential form.

$$x^2 = 36 = 6^2$$

$$x = 6$$

Notice that $x \neq -6$ since the base for a logarithm must be positive.

**(d)** $\log_3 x = 1$

$$x = 3^1 = 3 \qquad \text{Exponential form}$$

**(e)** $\log_3 1 = x$

$$3^x = 1 \quad \text{Exponential form}$$
$$3^x = 3^0 \quad 3^0 = 1$$
$$x = 0 \quad ■$$

Parts **(d)** and **(e)** of Example 3 illustrate the following theorem.

**Properties of Logarithms**

For any base $a$ ($a > 0$ and $a \neq 1$)

**1.** $\log_a a = 1$ **2.** $\log_a 1 = 0$.

The proof of this theorem is shown by changing to exponential form since $\log_a a = 1$ is equivalent to $a^1 = a$ and $\log_a 1 = 0$ is equivalent to $a^0 = 1$.

This brief introduction to exponentials and logarithms will help to clarify the formal development of exponential and logarithmic functions given in the next section.

## 5.1 EXERCISES

In Exercises 1–8, convert each of the following to logarithmic form.

**1.** $2^3 = 8$ **2.** $u^5 = 25$ **3.** $5^v = 9$ **4.** $a^b = 7$

**5.** $a^{-3} = c$ **6.** $u^v = w$ **7.** $u^{-v} = w$ **8.** $w^{x+1} = z$

In Exercises 9–16, convert each of the following to exponential form.

**9.** $\log_3 9 = 2$ **10.** $\log_5 \frac{1}{25} = -2$ **11.** $\log_a b = 7$ **12.** $\log_a 64 = 3$

**13.** $\log_3 \frac{1}{27} = b$ **14.** $\log_b 8 = \frac{1}{3}$ **15.** $\log_a 6 = c$ **16.** $\log_a u = v$

Solve each equation in Exercises 17–36.

**17.** $3^y = 243$ **18.** $2^z = \frac{1}{8}$ **19.** $b^{-3} = \frac{1}{125}$ **20.** $c^3 = 216$

**21.** $25^x = 5$ **22.** $27^x = 3$ **23.** $a^{1/2} = 7$ **24.** $b^{1/3} = 6$

**25.** $c^{-1/2} = 6$ **26.** $x^{-1/3} = 2$ **27.** $\log_2 8 = y$ **28.** $\log_a 9 = 2$

**29.** $\log_3 x = 4$ **30.** $\log_7 x = -3$ **31.** $\log_a \frac{1}{27} = -3$ **32.** $\log_x 2 = \frac{1}{3}$

**33.** $\log_8 0.125 = w$ **34.** $\log_4 0.25 = x$ **35.** $2^{x^2} = 16$ **36.** $3^{x^2} = 81$

**37.** **ENGINEERING** An electrical engineer uses the formula $D = 10 \log_{10} \frac{S}{S_0}$ to calculate the gain on an amplifier.

**(a)** What is the value of $D$ when $S$ is 100 and $S_0$ is 10?

**(b)** What is the value of $D$ when $S$ is 600 and $S_0$ is 6?

**38.** BUSINESS A financial advisor uses the formula $A = 100(1 + 0.05)^n$ to calculate interest compounded semiannually.
**(a)** What is the value of $A$ when $n$ is 2? **(b)** What is the value of $A$ when $n$ is 2.5?

## For Review

**39.** Given the polynomial function $P(x) = -3x^5 + x^3 + x^2 - 4x - 9$, determine **(a)** the maximum number of real zeros (the maximum number of $x$-intercepts), **(b)** the maximum number of turning points, **(c)** the $y$-intercept, **(d)** the direction of the graph for large positive values of $x$, and **(e)** the direction of the graph for small negative values of $x$.

**40.** ECOLOGY The annual cost of removing a given percentage $x$ of the pollutants from the smoke of a power plant increases tremendously as the percentage approaches 100%. The rational function $C(x) = \dfrac{1000x}{100 - x}$ approximates the cost of this operation for a particular plant in Arizona. Sketch the graph of $C(x)$ for $0 \le x < 100$, and find the cost of removing 95% of the pollutants.

# 5.2 Exponential and Logarithmic Functions

## Exponential Functions

Polynomial functions have constants used as exponents and a variable for a base. In this section we consider *exponential functions* that have variable exponents and a constant for a base.

**Exponential Function**

Let $a$ be a real number, $a > 0$ and $a \neq 1$. The function

$$f(x) = a^x$$

is called an **exponential function** with **base $a$.**

Evaluating an exponential function when $x$ is an integer $n$ or a rational number $\frac{p}{q}$ poses no problem since $a^n$ and $a^{p/q} = \sqrt[q]{a^p}$ have already been defined. Although we have not defined $a^x$ for irrational $x$ (such as $\sqrt{2}$ or $\pi$), we assume that such values could be approximated by using rational number approximations for $x$. For example, since $\sqrt{2} \approx 1.414$, $a^{\sqrt{2}} \approx a^{1.414} = a^{1414/1000} = \sqrt[1000]{a^{1414}}$. By using more and more accurate approximations of $\sqrt{2}$, we could obtain closer and closer approximations of $a^{\sqrt{2}}$. The notion of defining irrational exponents is treated more thoroughly in calculus, but we will assume that irrational exponents do make sense and that all of the properties of rational exponents, such as the product, quotient, and power rules, also apply to irrational exponents. When we graph an exponential function such as $f(x) = 2^x$, our assumptions about irrational exponents will appear reasonable.

**EXAMPLE 1** Graph the function $y = 2^x$.

Selecting values of $x$ and calculating the corresponding values of $y$, we obtain a table of values. We can use 1.41 as an approximation for $\sqrt{2}$. From the graph in Figure 1, we see that 2.7 is an approximation for $2^{\sqrt{2}}$. Similarly, using 3.14 as an approximation for $\pi$, we can estimate $2^{\pi}$ to be about 8.8. ■

| $x$ | $y$ |
|---|---|
| 0 | 1 |
| −1 | 1/2 |
| 1 | 2 |
| −2 | 1/4 |
| 2 | 4 |
| −3 | 1/8 |
| 3 | 8 |
| −4 | 1/16 |

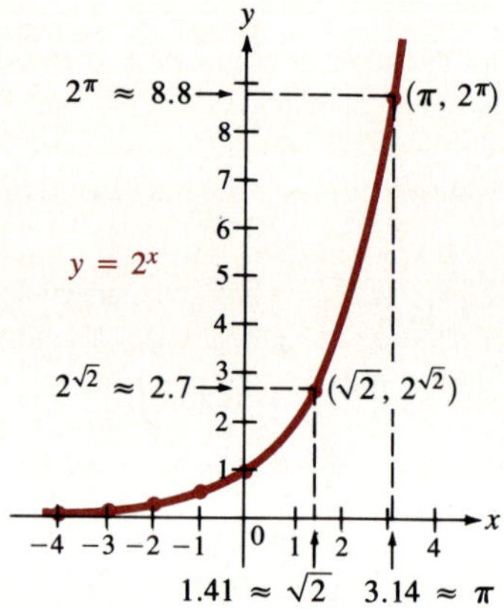


**Figure 1**

To compare exponential functions, we shall graph several such functions in the same coordinate system. This will help us to generalize about the behavior of these functions.

**EXAMPLE 2** Graph $y = 2^x$, $y = \left(\frac{1}{2}\right)^x$, $y = 3^x$, $y = \left(\frac{1}{3}\right)^x$, $y = 10^x$, and $y = \left(\frac{1}{10}\right)^x$.

First we construct a table of values.

| $x$ | 0 | −1 | 1 | −2 | 2 | −3 | 3 | −4 | 4 |
|---|---|---|---|---|---|---|---|---|---|
| $y = 2^x$ | 1 | $\frac{1}{2}$ | 2 | $\frac{1}{4}$ | 4 | $\frac{1}{8}$ | 8 | $\frac{1}{16}$ | 16 |
| $y = \left(\frac{1}{2}\right)^x$ | 1 | 2 | $\frac{1}{2}$ | 4 | $\frac{1}{4}$ | 8 | $\frac{1}{8}$ | 16 | $\frac{1}{16}$ |
| $y = 3^x$ | 1 | $\frac{1}{3}$ | 3 | $\frac{1}{9}$ | 9 | $\frac{1}{27}$ | 27 | $\frac{1}{81}$ | 81 |
| $y = \left(\frac{1}{3}\right)^x$ | 1 | 3 | $\frac{1}{3}$ | 9 | $\frac{1}{9}$ | 27 | $\frac{1}{27}$ | 81 | $\frac{1}{81}$ |
| $y = 10^x$ | 1 | $\frac{1}{10}$ | 10 | $\frac{1}{100}$ | 100 | $\frac{1}{1000}$ | 1000 | $\frac{1}{10000}$ | 10000 |
| $y = \left(\frac{1}{10}\right)^x$ | 1 | 10 | $\frac{1}{10}$ | 100 | $\frac{1}{100}$ | 1000 | $\frac{1}{1000}$ | 10000 | $\frac{1}{10000}$ |

The scale for our graphs in Figure 2 does not allow us to plot all of the points listed in the table, but the values are included for purposes of comparison. ■

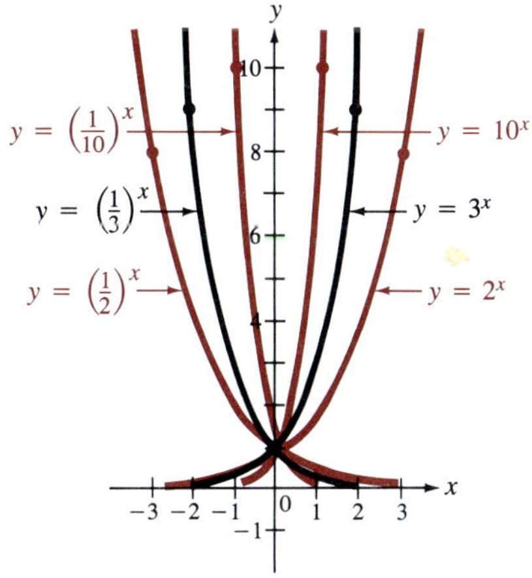


**Figure 2 Exponential Functions**

## Properties of Exponential Functions

Several observations can be made in view of Example 2.

1. If $a > 1$, $y = a^x$ increases as $x$ increases, and if $a < 1$, $y = a^x$ decreases as $x$ increases. That is, $f(x) = a^x$ is an increasing function when $a > 1$, and a decreasing function when $a < 1$.
2. The graphs of $y = a^x$ and $y = \left(\frac{1}{a}\right)^x$ are reflections of each other about the $y$-axis.
3. The graphs of $y = a^x$ and $y = \left(\frac{1}{a}\right)^x$ pass through $(0, 1)$ since $a^0 = 1$ for any $a > 0$.
4. The domain (the possible values of $x$) of every exponential function is the set of all real numbers.
5. The range (the possible values of $y$) of every exponential function is the set of all positive real numbers.

## Approximating Exponentials

A calculator with a $\boxed{y^x}$ key can be extremely helpful for work with exponential functions. For example, suppose we wish to approximate $2^{\sqrt{5}}$. The sequence of keys to use on a calculator is:

ALG: 2 $\boxed{y^x}$ 5 $\boxed{\sqrt{\phantom{x}}}$ $\boxed{=}$ RPN: 2 $\boxed{\text{ENTER}}$ 5 $\boxed{\sqrt{\phantom{x}}}$ $\boxed{y^x}$.

If this sequence is followed, the display will show 4.7111, correct to four decimal places. Since $\sqrt{5} \approx 2.2$, locate this value for $x$ on the graph of $y = 2^x$ given in Figure 2, and notice that the value of $y$ is about 4.7.

**EXAMPLE 3** Use a calculator to approximate each number correct to four decimal places.

**(a)** $(1.01)^{50}$

ALG: 1.01 [$y^x$] 50 [=] RPN: 1.01 [ENTER] 50 [$y^x$]

Thus, $(1.01)^{50} \approx 1.6446$.

**(b)** $\pi^{\sqrt{11}}$

ALG: [$\pi$] [$y^x$] 11 [$\sqrt{\ }$] [=] RPN: [$\pi$] [ENTER] 11 [$\sqrt{\ }$] [$y^x$]

Thus, $\pi^{\sqrt{11}} \approx 44.5512$. ■

More complex exponential functions of the form $f(x) = a^{p(x)}$, where $p(x)$ is some function of $x$, frequently occur in applied problems and have interesting graphs as illustrated in the next example.

**EXAMPLE 4** Sketch the graph of $f(x) = 2^{x^2}$.

A table of values is helpful for graphing functions of this type. Since $f(x) = f(-x)$, the graph is symmetric with respect to the $y$-axis and is given in Figure 3. ■

| $x$ | $f(x)$ |
|---|---|
| 0 | 1 |
| 1 | 2 |
| 2 | 16 |
| 3 | 512 |

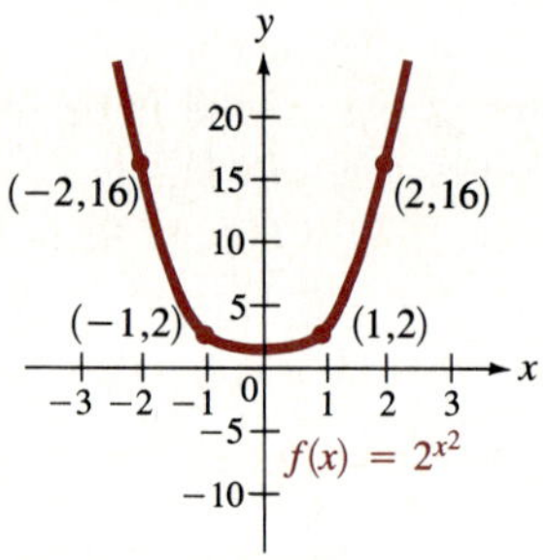


**Figure 3**

## Logarithmic Functions

Since each exponential function $y = a^x$ ($a > 0$ and $a \neq 1$) is either increasing or decreasing, it must have an inverse. To find the equation of the inverse, interchange $x$ and $y$ and solve the result for $y$. Starting with

$$y = a^x$$

we interchange $x$ and $y$ and obtain

$$x = a^y.$$

To solve this equation for $y$, we convert it to logarithmic form, $y = \log_a x$. Thus, the inverse of the exponential function $y = a^x$ is the **logarithmic function** defined by

$$y = \log_a x.$$

To graph $y = \log_a x$, we need only graph $x = a^y$. Alternatively, since the graphs of inverse functions are symmetric with respect to the line $y = x$, we could graph $y = a^x$ and use it to find the graph of $y = \log_a x$.

**EXAMPLE 5** Graph $y = 2^x$ and $y = \log_2 x$ $(x = 2^y)$ in the same coordinate system.

The tables of values for these functions are below and their graphs are given in Figure 4. ■

| $x$ | $y = 2^x$ |
|---|---|
| 0 | 1 |
| −1 | 1/2 |
| 1 | 2 |
| −2 | 1/4 |
| 2 | 4 |
| −3 | 1/8 |
| 3 | 8 |

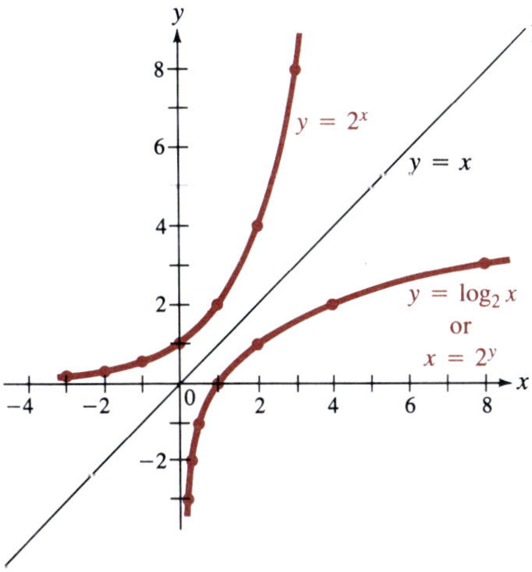


| $x = 2^y$ | $y$ |
|---|---|
| 1 | 0 |
| 1/2 | −1 |
| 2 | 1 |
| 1/4 | −2 |
| 4 | 2 |
| 1/8 | −3 |
| 8 | 3 |

**Figure 4**

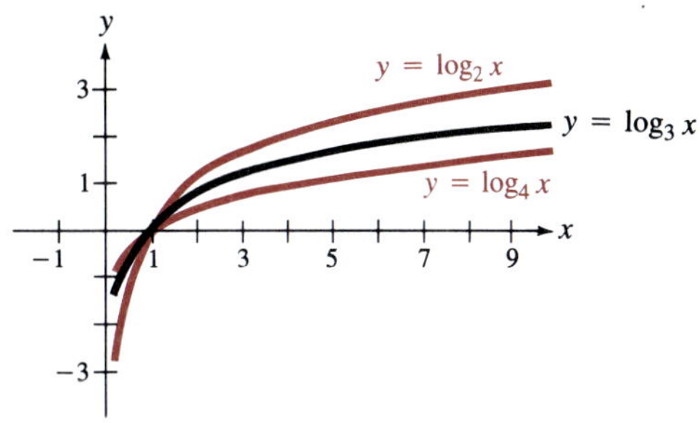


**Figure 5 Logarithmic Functions**

The graphs of the three logarithmic functions $y = \log_4 x$, $y = \log_3 x$, and $y = \log_2 x$ are given in Figure 5. If we agree to restrict our discussion of logarithmic functions to bases larger than 1, these graphs lead to the observations which follow.

**Properties of Logarithmic Functions**

1. If $a > 1$, $y = \log_a x$ increases as $x$ increases. That is, $f(x) = \log_a x$ is an increasing function when $a > 1$.
2. The graph of $y = \log_a x$ passes through $(1, 0)$ for every $a > 1$. That is, $\log_a 1 = 0$ for every $a > 1$.
3. The domain (the possible $x$ values) of every logarithmic function is the set of all positive real numbers.
4. The range (the set of all possible values of $y$) of every logarithmic function is the set of all real numbers.

**Inverse Function Properties**

The fact that logarithmic and exponential functions are inverses gives two special equations. Remember that the inverse of a function $f$ is denoted by $f^{-1}$ and that $f(f^{-1}(x)) = x$ and $f^{-1}(f(x)) = x$. These properties lead to the next theorem.

For any base $a$, $a > 0$ and $a \neq 1$,

1. $\log_a a^x = x$, for any real number $x$.
2. $a^{\log_a x} = x$, for any real number $x > 0$.

We use the fact that $f(x) = a^x$ and $f^{-1}(x) = \log_a x$ are inverses of each other to prove the theorem.

**1.** $x = f^{-1}(f(x))$
$= f^{-1}(a^x)$
$= \log_a a^x$

**2.** $x = f(f^{-1}(x))$
$= f(\log_a x)$
$= a^{\log_a x}$

**EXAMPLE 6**

Simplify.

**(a)** $\log_3 3^2 = 2$ $\quad$ $\text{Log}_a\, a^x = x$ for any $x$

**(b)** $\log_{100} 10^4 = \log_{100} (10^2)^2 = \log_{100} 100^2 = 2$

**(c)** $\log_{10} 0.01 = \log_{10} 10^{-2} = -2$

**(d)** $3^{\log_3 5} = 5$ $\quad$ $a^{\log_a x} = x$ for any positive $x$

**(e)** $4^{\log_4 (5x+2)} = 5x + 2$ ■

As with exponential functions, we are often interested in more complex logarithmic functions of the form $f(x) = \log_a p(x)$. Keep in mind that $p(x)$ must be positive for such a function to exist.

**EXAMPLE 7**

Sketch the graph of $y = f(x) = \log_2 x^2$ for $x \neq 0$.

Since $f(-x) = f(x)$, the graph is symmetric with respect to the $y$-axis. A table of values for $2^y = x^2$ is helpful, and the graph is given in Figure 6. ■

| $y$ | $x$ |
|---|---|
| 0 | $\pm 1$ |
| 1 | $\pm\sqrt{2} \approx \pm 1.4$ |
| 2 | $\pm 2$ |
| 4 | $\pm 4$ |
| $-2$ | $\pm 1/2$ |

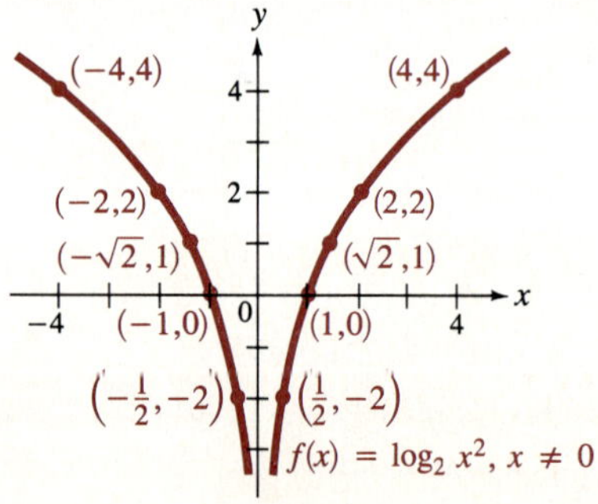


**Figure 6**

## 5.2 EXERCISES

In Exercises 1–18 sketch the graph of each function.

**1.** $f(x) = 5^x$

**2.** $f(x) = 8^x$

**3.** $f(x) = \left(\frac{1}{5}\right)^x$

**4.** $f(x) = \left(\frac{1}{8}\right)^x$

**5.** $f(x) = -5^x$

**6.** $f(x) = -8^x$

**7.** $f(x) = 2 + 5^x$

**8.** $f(x) = 3^{x+1}$

**9.** $f(x) = 2^{x^2+1}$

**10.** $f(x) = 3^{x^2+1}$

**11.** $f(x) = 2^x - 2^{-x}$

**12.** $f(x) = \log_4 x$

**13.** $f(x) = \log_8 x$

**14.** $f(x) = 1 + \log_4 x$

**15.** $f(x) = \log_8 (-x)$ for $x < 0$

**16.** $f(x) = \log_2 (x + 1)$

**17.** $f(x) = \log_2 |x|$ for $x \neq 0$

**18.** $f(x) = |\log_2 x|$

In Exercises 19–24 simplify each expression.

**19.** $\log_5 5^3$

**20.** $\log_{10} 1000$

**21.** $\log_{10} 0.001$

**22.** $7^{\log_7 1.5x}$

**23.** $e^{\log_e (x+7)}$

**24.** $\pi^{\log_\pi \sqrt{x}}$

In Exercises 25–32 use a calculator to approximate each number correct to four decimal places.

**25.** $5^{\sqrt{3}}$

**26.** $-7^{\sqrt{6}}$

**27.** $-\pi^{\sqrt{3}}$

**28.** $(\sqrt{3})^{\pi}$

**29.** $6^{-\sqrt{2}}$

**30.** $9^{-\pi}$

**31.** $(1.02)^{60}$

**32.** $(1.02)^{-60}$

**33.** The graph of every exponential function passes through which point?

**34.** The graph of every logarithmic function passes through which point?

**35.** What is the domain of every exponential function?

**36.** What is the range of every exponential function?

**37.** What is the domain of every logarithmic function?

**38.** What is the range of every logarithmic function?

**39.** What is the base of the logarithmic function $y = \log_a x$ if its graph passes through the point (3, 1)?

**40.** What is the base of the exponential function $y = a^x$ if its graph passes through the point $\left(\frac{1}{2}, 8\right)$?

**41.** Why is 1 omitted as a base for a logarithmic function?

**42.** Why is every negative number omitted as a base for a logarithmic function?

In Exercises 43–46 find the inverse of each function.

**43.** $f(x) = 2^{x+1} + 1$

**44.** $f(x) = 3^{x-1} + 2$

**45.** $f(x) = 2 \log_3 (x - 4)$

**46.** $f(x) = 5 \log_2 (x + 3)$

**47.** BIOLOGY The number of bacteria $N$ in a culture is given in terms of time $t$, in hours, by the formula $N = (5000)2^t$.
**(a)** How many bacteria are present at the start of the experiment (when $t = 0$)?
**(b)** How many bacteria are present in 5 hours?
**(c)** How many bacteria are present in 3.25 hours?

**48.** PHYSICS The atmospheric pressure $P$ on an object, in pounds per square inch, can be approximated using the formula $P = 14.7(2.7)^{-0.2x}$ where $x$ is the height of the object above sea level in miles. What is the pressure on an object **(a)** 1 mile high? **(b)** 5 miles high? **(c)** 8.5 miles high?

**49.** CHEMISTRY A radioactive isotope has a half-life of 4 years. This means that in 4 years, one half of any given amount will change to another substance due to radioactive decay. If 50 grams of this isotope are allowed to decay in a reactor, the amount $A$ that will remain at the end of $t$ years is given by $A = 50\left(\frac{1}{2}\right)^{t/4}$. How many grams remain at the end of **(a)** 2 years? **(b)** 100 years? **(c)** 5.5 years?

**50.** **BUSINESS** The amount $A$ of money in a savings account at the end of $t$ years if \$1000 is invested originally at an annual interest rate of 10% compounded quarterly, is given by $A = 1000(1.025)^{4t}$. What amount of money will be in the account at the end of **(a)** 1 year? **(b)** 10 years? **(c)** 3.5 years?

In Exercises 51–52 use a calculator and graph each function.

**51.** $f(x) = 1.4 + (2.5)^x$

**52.** $f(x) = 3.8 + \log_{2.5} x$

## For Review

In Exercises 53–58 solve each equation.

**53.** $\log_a 32 = 5$

**54.** $\log_5 125 = y$

**55.** $\log_{10} x = -2$

**56.** $\log_7 7^{2x-3} = 5$

**57.** $10^{\log_{10}(2x+5)} = 7$

**58.** $\log_3 x^2 = 4$

To prepare for the next section, we review the properties of exponents. Answer true or false in Exercises 59–62.

**59.** $a^m a^n = a^{m+n}$

**60.** $(a^m)^n = a^{m+n}$

**61.** $\dfrac{a^m}{a^n} = a^{m-n} \quad (a \neq 0)$

**62.** $\left(\dfrac{a^m a^p}{a^q}\right)^k = a^{mk+pk-qk} \quad (a \neq 0)$

## 5.3 Properties of Logarithms

In this section we prove a number of important theorems related to logarithmic functions that provide valuable tools for solving logarithmic equations. Recall from Section 5.1 that if $a > 0$ and $a \neq 1$, then

$$a^x = a^y \quad \text{if and only if} \quad x = y.$$

The next theorem is the logarithmic equivalent of this.

Let $a$ be a base for logarithms, $a > 0$ and $a \neq 1$, with $x$ and $y$ positive real numbers. Then

$$\log_a x = \log_a y \quad \text{if and only if} \quad x = y.$$

We have two things to prove. First, if $x = y$, then clearly $\log_a x = \log_a y$ since $f(x) = \log_a x$ is a function. Conversely, if $\log_a x = \log_a y$, then $a^{\log_a x} = a^{\log_a y}$ by the exponential theorem cited above. Thus, since $x = a^{\log_a x}$ and $y = a^{\log_a y}$, we have $x = y$.

The following example illustrates the usefulness of this theorem for solving certain equations involving logarithms.

**EXAMPLE 1** Solve. $\log_3 (4x + 5) = \log_3 x^2$

Using the theorem above with $a = 3$, we have $4x + 5 = x^2$, which is a quadratic equation in $x$.

$$x^2 - 4x - 5 = 0$$
$$(x + 1)(x - 5) = 0$$
$$x + 1 = 0 \quad \text{or} \quad x - 5 = 0$$
$$x = -1 \qquad\qquad x = 5$$

Check: $x = -1$: $\log_3 (4(-1) + 5) \stackrel{?}{=} \log_3 (-1)^2$
$$\log_3 1 = \log_3 1$$
$x = 5$: $\log_3 (4(5) + 5) \stackrel{?}{=} \log_3 5^2$
$$\log_3 25 = \log_3 25$$

Thus, $-1$ and 5 are both solutions. ■

Recall that the word *logarithm* means *exponent*. The next theorem presents three laws of logarithms that are based on three familiar laws of exponents. The proof of the theorem will help to illustrate the equivalence of logarithms and exponents.

**Properties of Logarithms**

For any base $a$, $a > 0$ and $a \neq 1$, any real number $c$, and positive real numbers $x$ and $y$,

1. $\log_a xy = \log_a x + \log_a y$ — The log of a product is the sum of the logs.
2. $\log_a \frac{x}{y} = \log_a x - \log_a y$ — The log of a quotient is the difference of the logs.
3. $\log_a x^c = c \log_a x$. — The log of a number to a power is the power times the log of the number.

PROOF OF 1: Let $u = \log_a x$ and $v = \log_a y$. Then $a^u = x$ and $a^v = y$.

$$xy = a^u \cdot a^v$$
$$xy = a^{u+v}$$

From the definition of a logarithm, the last equation $xy = a^{u+v}$ becomes $\log_a xy = u + v$. Since $u = \log_a x$ and $v = \log_a y$, we have

$$\log_a xy = \log_a x + \log_a y.$$

This is sometimes called the **product rule** for logarithms. ■

CAUTION $\log_a (x + y) \neq \log_a x + \log_a y$.
If these were equal, we would have $\log_a (x + y) = \log_a xy$, so that $x + y = x \cdot y$, which is clearly not true. ■

**PROOF OF 2:** Proceed as before and let $u = \log_a x$ and $v = \log_a y$. Then $a^u = x$ and $a^v = y$.

$$\frac{x}{y} = \frac{a^u}{a^v} \qquad y \neq 0 \text{ since the domain of the log function is the set of positive numbers}$$

$$\frac{x}{y} = a^{u-v}$$

$$\log_a \frac{x}{y} = u - v = \log_a x - \log_a y.$$

This is called the **quotient rule** for logarithms. ■

**CAUTION** $\log_a (x - y) \neq \log_a x - \log_a y$.
If they were equal, $x - y$ would equal $x/y$. ■

**PROOF OF 3:** Begin by letting $u = \log_a x$. Then $a^u = x$, and

$$x^c = (a^u)^c$$

$$x^c = a^{uc}$$

$$\log_a x^c = uc = cu = c \log_a x.$$

This is called the **power rule** for logarithms. ■

**EXAMPLE 2** Use the properties of logarithms to expand the following.

**(a)** $\log_a \frac{x}{yz} = \log_a x - \log_a yz$ — Quotient rule

$= \log_a x - (\log_a y + \log_a z)$ — Product rule, use parentheses to avoid a sign error

$= \log_a x - \log_a y - \log_a z$

**(b)** $\log_a \sqrt{\frac{uv}{w^2}} = \log_a \left(\frac{uv}{w^2}\right)^{1/2}$ — $\sqrt{a} = a^{1/2}$

$= \frac{1}{2} \log_a \frac{uv}{w^2}$ — Power rule

$= \frac{1}{2}(\log_a uv - \log_a w^2)$ — Quotient rule

$= \frac{1}{2}(\log_a u + \log_a v - 2 \log_a w)$ — Product rule, power rule

$= \frac{1}{2} \log_a u + \frac{1}{2} \log_a v - \log_a w$ — Distribute ■

**EXAMPLE 3** Express as a single logarithm.

**(a)** $\log_a x - \mathbf{2} \log_a y = \log_a x - \log_a y^2$ — Make each coefficient an exponent using power rule

$= \log_a \frac{x}{y^2}$ — The difference of logs is the log of a quotient

**(b)** $\frac{1}{2}\log_a u + 3\log_a v = \log_a u^{1/2} + \log_a v^3$ Make coefficients exponents

$= \log_a u^{1/2}v^3$ Sums of logs become log of product ■

**EXAMPLE 4** Suppose that $\log_a 2 = 0.6309$ and $\log_a 5 = 1.4650$. Approximate the numerical value of each of the following.

**(a)** $\log_a \sqrt{20}$

$$\log_a \sqrt{20} = \log_a (2^2 \cdot 5)^{1/2} = \frac{1}{2}\log_a (2^2 \cdot 5) \quad \text{Power rule}$$

$$= \frac{1}{2}[\log_a 2^2 + \log_a 5] \quad \text{Product rule}$$

$$= \frac{1}{2}[2\log_a 2 + \log_a 5] \quad \text{Power rule}$$

$$= \frac{1}{2}[2(0.6309) + 1.4650] \quad \text{Substitute}$$

$$= 1.3634$$

**(b)** $\log_a \frac{5a}{\sqrt[3]{2}}$

$$\log_a \frac{5a}{\sqrt[3]{2}} = \log_a 5a - \log_a \sqrt[3]{2}$$

$$= \log_a 5 + \log_a a - \frac{1}{3}\log_a 2$$

$$= 1.4650 + 1 - \frac{1}{3}(0.6309)$$

$$= 2.2547$$ ■

The final theorem of this section deals with the relationship between logarithms with different bases. Suppose we consider $\log_a x$ and $\log_b x$, and let

$$u = \log_b x$$
$$b^u = x.$$

Take the logarithm to the base $a$ of both sides.

$$\log_a b^u = \log_a x \quad \text{If } y = z \text{ then } \log_a y = \log_a z$$

$$u\log_a b = \log_a x \quad \text{Power rule}$$

$$(\log_b x)(\log_a b) = \log_a x \quad u = \log_b x$$

$$\log_b x = \frac{\log_a x}{\log_a b} \quad \textit{Not } \log_a \frac{x}{b} = \log_a x - \log_a b$$

A special case of this formula is developed by letting $x = a$.

$$\log_b x = \frac{\log_a x}{\log_a b}$$

$$\log_b a = \frac{\log_a a}{\log_a b} \qquad \text{Letting } x = a$$

$$\log_b a = \frac{1}{\log_a b} \qquad \text{Log}_a\, a = 1$$

We have just proved the following.

**Base Conversion Formula**

If $a > 0,\ b > 0,\ a \neq 1,\ b \neq 1,$ and $x > 0$, then

$$\log_b x = \frac{\log_a x}{\log_a b} \qquad \text{and} \qquad \log_b a = \frac{1}{\log_a b}.$$

Notice that the base conversion formula states that logarithms to the base $b$ are simply multiples of logarithms to the base $a$ where the multiplier is the constant $\frac{1}{\log_a b}$.

**EXAMPLE 5**

**(a)** Find $\log_{16} 8$ by changing to logarithms base 2.

$$\log_{16} 8 = \frac{\log_2 8}{\log_2 16} = \frac{\log_2 2^3}{\log_2 2^4} = \frac{3}{4}$$

Notice that by changing to exponential form, $\log_{16} 8 = \frac{3}{4}$ becomes $16^{3/4} = 8$, and $(\sqrt[4]{16})^3$ is indeed 8.

**(b)** Find $\log_9 3$ by changing to logarithms base 3.

$$\log_9 3 = \frac{1}{\log_3 9} = \frac{1}{\log_3 3^2} = \frac{1}{2}$$

Notice that $9^{1/2} = \sqrt{9}$ does indeed equal 3. ■

## 5.3 EXERCISES

Solve each equation in Exercises 1–6.

**1.** $\log_2 (3x + 1) = \log_2 (5x - 7)$

**2.** $\log_5 (1 + x) = \log_5 (1 - x)$

**3.** $\log_5 x^2 = \log_5 (3x + 4)$

**4.** $\log_3 (x^2 + 4) = \log_3 4x$

**5.** $\frac{1}{3} \log_2 x^2 = \log_8 2x$

**6.** $\frac{1}{2} \log_3 2x = \log_9 (5x)$

In Exercises 7–24 evaluate each logarithm given that $\log_{10} 2 = 0.3010$, $\log_{10} 3 = 0.4771$, and $\log_{10} 5 = 0.6990$.

**7.** $\log_{10} 500$ **8.** $\log_{10} 25$ **9.** $\log_{10} 27$ **10.** $\log_{10} \frac{5}{2}$

**11.** $\log_{10} \frac{2}{5}$ **12.** $\log_{10} \frac{6}{5}$ **13.** $\log_{10} \frac{1}{9}$ **14.** $\log_{10} \sqrt{5}$

**15.** $\log_{10} \sqrt{6}$ **16.** $\log_{10} 3^7$ **17.** $\log_{10} \sqrt[7]{3}$ **18.** $\log_{10} \sqrt{150}$

**19.** $\log_5 3$ **20.** $\log_3 2$ **21.** $\dfrac{\log_{10} 2^7}{\log_{10} 5}$ **22.** $\dfrac{\log_{10} \sqrt[7]{5}}{\log_{10} \sqrt[3]{2}}$

**23.** $\log_{10} 1.8$ **24.** $\log_{10} 7.5$

In Exercises 25–36 use the properties of logarithms to write each logarithm as a sum, difference, or multiple of logarithms.

**25.** $\log_a xyz$ **26.** $\log_a x^2y^3z^4$ **27.** $\log_a \dfrac{xz^2}{y}$ **28.** $\log_a \left(\dfrac{xz}{y}\right)^2$

**29.** $\log_a \dfrac{y\sqrt{x}}{z^3}$ **30.** $\log_a \dfrac{(x+y)^2}{z}$ **31.** $\log_a z(x+1)^3$ **32.** $\log_a ax^2$

**33.** $\log_a x^2\sqrt{\dfrac{y}{z}}$ **34.** $\log_a \dfrac{1}{ax^2}$ **35.** $\log_a \sqrt{x\sqrt{y}}$ **36.** $\log_a \sqrt{\dfrac{\sqrt{xy}}{z^2}}$

In Exercises 37–48 use the properties of logarithms to write each expression as a single logarithm.

**37.** $\log_a x + \frac{1}{2} \log_a y$ **38.** $\log_a x - \frac{1}{2} \log_a y$ **39.** $3 \log_a x - 2 \log_a xy$

**40.** $4 \log_a z + 5 \log_a yx$ **41.** $\frac{1}{2} \log_a x - 3 \log_a yz + 6 \log_a z$ **42.** $\frac{3}{2} \log_a z + \frac{1}{2} \log_a y$

**43.** $x \log_a y - 3 \log_a z$ **44.** $y \log_a z - z \log_a y$ **45.** $-\log_a (x-1) + \log_a (x^2-1)$

**46.** $2[\log_a x - \log_a (y+1) + \log_a z]$ **47.** $\frac{1}{2}\left[\log_a x - \log_a y - \frac{1}{2} \log_a z\right]$ **48.** $2 \log_a x - y \log_a z + \frac{1}{3} \log_a xy$

In Exercises 49–58 determine whether the given equation is true or false.

**49.** $\log_a uv = \log_a u + \log_a v$ **50.** $\log_a uv = (\log_a u)(\log_a v)$

**51.** $\dfrac{\log_a u}{\log_a v} = \log_a \dfrac{u}{v}$ **52.** $\log_a (u+v) = \log_a uv$

**53.** $\log_a u - \log_a v = \log_a \frac{u}{v}$ **54.** $\log_a (u-v) = \log_a \frac{u}{v}$ **55.** $\log_a u^v = v \log_a u$

**56.** $\log_a au = 1 + \log_a u$ **57.** $\log_a \sqrt{u} = \sqrt{\log_a u}$ **58.** $\frac{1}{2} \log_a u = \sqrt{\log_a u}$

In Exercises 59–64 use the base conversion formulas to simplify each logarithm relative to the given base.

**59.** $\log_{32} 8$; use base 2 **60.** $\log_{32} 16$; use base 2 **61.** $\log_{27} 3$; use base 3

**62.** $\log_{27} 9$; use base 3 **63.** $\log_9 27$; use base 3 **64.** $\log_{25} 5$; use base 5

**65.** If $P = \log_2 (m + n)$, calculate the value of $P$ when $m = 8$ and $n = 8$, and show that $\log_2 (m + n) \neq \log_2 m + \log_2 n$.

**66.** If $M = \log_3 \frac{m}{n}$, calculate the value of $M$ when $m = 27$ and $n = 3$, and show that $\log_3 \frac{m}{n} \neq \frac{\log_3 m}{\log_3 n}$.

**67.** If $R = \log_3 mn$, calculate the value of $R$ when $m = 1$ and $n = 9$, and show that $\log_3 mn \neq (\log_3 m)(\log_3 n)$.

**68.** If $T = \log_2 (m - n)$, calculate the value of $T$ when $m = 8$ and $n = 4$, and show that $\log_2 (m - n) \neq \log_2 m - \log_2 n$.

**69.** Express $\log_a x - \log_a y - z = 0$ in exponential form without using logarithms.

**70.** Express $\log_a x + \log_a y - z = 0$ in exponential form without using logarithms.

## For Review

Graph each function.

**71.** $f(x) = 2^{x^2} + 2$

**72.** $f(x) = \log_3 |x|$

Simplify each expression in Exercises 73–74.

**73.** $\log_\pi \pi^{3x+2}$

**74.** $10^{\log_{10} 5x^2}$

**75.** Evaluate $10^m$ if $m = \log_{10} 35$.

# 5.4 Common and Natural Logarithms

### Common Logarithms

Recall that every positive number, with the exception of the number 1, can be used as a base for logarithms. However, two logarithm bases occur more frequently in applications of science and mathematics than others, base 10 and base $e$ ($e$ is an irrational number, approximately 2.71828, which will be discussed later in this section). Historically, since our number system is based on 10, logarithms to the base 10 were commonly used for computation. As a result, base 10 logarithms are called **common logarithms.** For convenience, we omit the base number 10 in common logarithmic expressions and write, for example,

$$\log 427 \quad \text{instead of} \quad \log_{10} 427.$$

Thus, when a logarithm is written without the base, the base is understood to be 10.

Before common logarithms can be used to solve problems, we must be able to find the approximate value of the logarithm of a given number. Prior to the ready availability of calculators, tables similar to Table 1 at the back of this book were used to approximate logarithms. Appendix A explains how to use the table of common logarithms for those who might not wish to use a calculator. The presentation given in this section and throughout the rest of the chapter will utilize calculators. Keep in mind, that finding the value of log 427, for example, means looking for the exponent on 10 which results in 427. Thus, if $a = \log 427$, $a$ is an exponent and in fact, $10^a = 427$. With $10^2 = 100$, $10^3 = 1000$, and $100 < 427 < 1000$, we would suspect that since

$$10^2 = 100 < 427 = 10^a < 1000 = 10^3,$$

$a$ must be a real number satisfying $2 < a < 3$. This number is found in Example 1**(a)**.

## EXAMPLE 1

Use a calculator to find the following common logarithms.

**(a)** log 427
The key labeled [LOG] will give the desired result.

ALG & RPN: 427 [LOG] → [2.6304279]   The display shows log 427

**(b)** log 0.000427

ALG & RPN: 0.000427 [LOG] → [−3.3695721]

**(c)** log (−427)

ALG: 427 [+/−] [LOG] → [Error]   RPN: 427 [CHS] [LOG] → [Error 0]

In this case the display will show *error* since logarithms of negative numbers are not defined.

**(d)** log (0.000 000 004 27)
Since most calculator displays only allow eight digits, very small or very large numbers must be entered in scientific notation.

ALG: 4.27 [EE] 9 [+/−] [LOG] → [−8.3696 00]
RPN: 4.27 [EEX] 9 [CHS] [LOG] → [−8.3696 00]

The display shows the logarithm −8.3696, correct to four decimal places. ■

## Logarithms to Bases Other Than 10

In Section 5.3, we introduced the base conversion formula

$$\log_b x = \frac{\log_a x}{\log_a b},$$

which can be used in conjunction with a calculator to find logarithms to any base $b$ by letting $a = 10$.

## EXAMPLE 2

Use a calculator and the base conversion formula to find the following logarithms.

**(a)** $\log_2 59.4 = \dfrac{\log 59.4}{\log 2}$   Remember that $a = 10$ and $\log_{10} 59.4 = \log 59.4$

Notice that the right side of this equation is a quotient of logs and must be calculated by dividing, not by subtracting.

ALG: 59.4 [LOG] [÷] 2 [LOG] [=] → [5.8923910]

RPN: 59.4 [LOG] [ENTER] 2 [LOG] [÷] → [5.8923910]

The final digit in the display may vary slightly from calculator to calculator. However,

$$\log_2 59.4 = 5.8924,$$

correct to four decimal places.

**(b)** $\log_3 0.00845$

$$\log_3 0.00845 = \frac{\log 0.00845}{\log 3}$$

ALG: 0.00845 [LOG] [÷] 3 [LOG] [=] → [−4.3451078]

RPN: 0.00845 [LOG] [ENTER] 3 [LOG] [÷] → [−4.3451078]

Thus, correct to four decimal places,

$$\log_3 0.00845 = -4.3451. \quad ■$$

Always remember that a logarithm is an exponent. In Example 2**(a)**, we found that $\log_2 59.4 \approx 5.8924$. In exponential form, this means that $2^{5.8924} \approx 59.4$. We can use the [$y^x$] key on a calculator to verify this result.

ALG: 2 [$y^x$] 5.8924 [=] → [59.400369]

RPN: 2 [ENTER] 5.8924 [$y^x$] → [59.4003695]

Similarly, in Example 1**(a)**, we found that $\log 427 \approx 2.6304$. Thus $10^{2.6304} \approx 427$.

ALG: 10 [$y^x$] 2.6304 [=] → [426.97259]

RPN: 10 [ENTER] 2.6304 [$y^x$] → [426.9725940]

Notice that due to round-off errors, we do not obtain 427 in the display, rather an approximation of it. Another way to obtain this result is by using a different key or set of keys. Some calculators have a [$10^x$] key while others use two keys, [INV] followed by [LOG]. Both sequences appear below.

ALG: 2.6304 [INV] [LOG] → [426.97259]

RPN: 2.6304 [$10^x$] → [426.9725940]

**Antilogarithm**

We can use one of these sequences to find a number from its logarithm (the reverse of what we did previously). A number that has a given logarithm is called its **antilogarithm (antilog).** In the equation

$$\log n = x,$$

$x$ is a logarithm of $n$ and $n$ is an antilogarithm of $x$. In exponential form, $n = 10^x$.

**EXAMPLE 3** Use a calculator to find the following antilogs.

**(a)** $\log x = 0.5623$
We must find $x$ where $x = 10^{0.5623}$.

ALG: 0.5623 [INV] [LOG] → [3.6500600]

RPN: 0.5623 [$10^x$] → [3.6500600]

Thus, $x = 3.65$, correct to three significant digits.

**(b)** $\log x = -3.4215$

ALG: 3.4215 [+/−] [INV] [LOG] → [0.0003789]

RPN: 3.4215 [CHS] [$10^x$] → [0.0003789]

Thus, $x = 0.000379$, correct to three significant digits. ■

## Natural Logarithms

The number $e$ as a base for logarithms arises naturally in many applications, such as population growth, continuously compounding interest, and radioactive decay. As a result, logarithms to the base $e$ are called **natural logarithms.** The full significance of this number cannot be realized without more advanced study. One way to define $e$ is to find the value of

$$\left(1 + \frac{1}{n}\right)^n$$

as $n$ increases through larger and larger values. To evaluate this expression we use the following sequence on a calculator.

ALG: $n$ [1/x] [+] 1 [=] [$y^x$] $n$ [=] → [Display]

RPN: $n$ [1/x] [ENTER] 1 [+] [ENTER] $n$ [$y^x$] → [Display]

The table below shows this computation for increasing values of $n$.

| $n$ | $\left(1 + \frac{1}{n}\right)^n$ |
|---:|---|
| 1 | 2.00000 |
| 10 | 2.59374 |
| 100 | 2.70481 |
| 1000 | 2.71692 |
| 10,000 | 2.71815 |
| 100,000 | 2.71827 |
| 1,000,000 | 2.71828 |
| ↓ $\infty$ | ↓ $e$ |

The number $e$ was named for the eighteenth-century mathematician Leonhard Euler. So important is this number that we often call $f(x) = e^x$ *the* exponential function, elevating it above all other exponential functions with other bases $a$. Since logarithms to the base $e$, **natural logarithms,** occur so frequently in mathematics, a special notation, $\ln x$, instead of $\log_e x$, is used. Most calculators have a [ln x] or [LN] key along with the common logarithm key [LOG].

**EXAMPLE 4** Use a calculator to find the following natural logarithms.

**(a)** ln 53.6

ALG & RPN: 53.6 [ln x] → [3.9815491]

Thus, ln 53.6 = 3.9815, correct to four decimal places.

**(b)** ln 0.000 000 062 5

ALG: 6.25 [EE] 8 [+/−] [ln x] → [−1.6588 01]

RPN: 6.25 [EEX] 8 [CHS] [LN] → [−16.5880993]

Thus, ln 0.000 000 062 5 = −16.588, correct to three decimal places. ■

Finding antilogarithms involving natural logarithms uses either the [$e^x$] key or the [INV] and [ln x] keys in a manner similar to that for common logarithms.

**EXAMPLE 5** Use a calculator to find the following antilogs.

**(a)** $\ln x = 5.9002$

We must find $x$ where $\ln x = 5.9002$.

ALG: 5.9002 [INV] [ln x] → [365.11048]

RPN: 5.9002 [$e^x$] → [365.1104827]

Thus, $x = 365.1$, correct to four significant digits.

**(b)** $\ln x = -2.1258$

ALG: 2.1258 [+/−] [INV] [ln x] → [0.1193375]

RPN: 2.1258 [CHS] [$e^x$] → [0.1193375]

Thus, $x = 0.1193$, correct to four decimal places. ■

**EXAMPLE 6** Graph the inverse functions $y = \ln x$ and $y = e^x$ in the same coordinate system.

Representative values of each function are given in the tables with the graphs shown in Figure 7. ■

| $x$ | $\ln x$ |
|---|---|
| 0.5 | −0.7 |
| 1 | 0 |
| 2 | 0.7 |
| 3 | 1.1 |
| 4 | 1.4 |
| 5 | 1.6 |

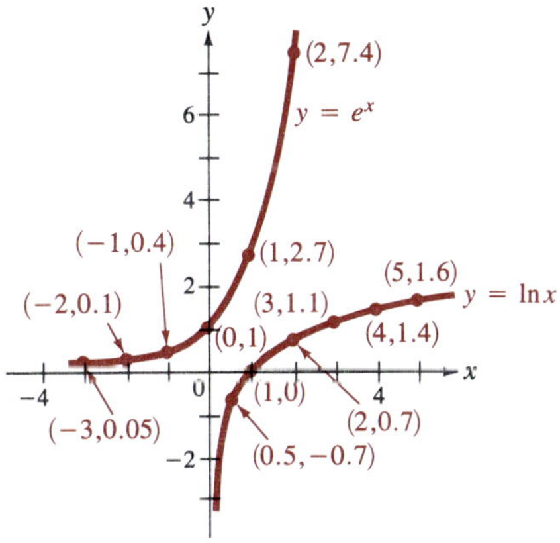


| $x$ | $e^x$ |
|---|---|
| −3 | 0.05 |
| −2 | 0.1 |
| −1 | 0.4 |
| 0 | 1 |
| 1 | 2.7 |
| 2 | 7.4 |

**Figure 7**

## 5.4 EXERCISES

In Exercises 1–10 find each common logarithm, correct to four decimal places.

**1.** log 625 **2.** log 0.00311 **3.** log 2.00046 **4.** log 694,000

**5.** log 0.000 066 9 **6.** log 71.00045 **7.** log $(5.12 \times 10^{11})$

**8.** log $(3.39 \times 10^{-9})$ **9.** log 0.000 000 007 41 **10.** log 229,000,000

Use a calculator and the base conversion formula to find each logarithm, correct to four decimal places, in Exercises 11–16.

**11.** $\log_4 15.6$ **12.** $\log_{11} 0.566$ **13.** $\log_{3.5} 468$

**14.** $\log_{4.6} 2.63$ **15.** $\log_{\sqrt{2}} 348.2$ **16.** $\log_\pi e$

In Exercises 17–26 find each common antilogarithm, correct to three significant digits.

**17.** $\log x = 1.3565$ **18.** $\log x = 4.9907$ **19.** $\log x = -3.2117$ **20.** $\log x = -2.0075$

**21.** $\log x = 14.6531$ **22.** $\log x = 18.8309$ **23.** $\log x = -0.0035$ **24.** $\log x = -0.0007$

**25.** $\log x = -35.2264$ **26.** $\log x = 42.5761$

In Exercises 27–36 find each natural logarithm, correct to four decimal places.

**27.** ln 846 **28.** ln 1.001 **29.** ln 0.000 525 **30.** ln 605.34

**31.** ln 0.000 038 7 **32.** ln 215,000,000 **33.** ln $(2.66 \times 10^{15})$ **34.** ln $(5.07 \times 10^{-12})$

**35.** $\ln \pi$ **36.** $\ln \sqrt{2}$

In Exercises 37–46 find each natural antilogarithm, correct to three significant digits.

**37.** $\ln x = 2.4308$ **38.** $\ln x = 7.4992$ **39.** $\ln x = -4.1553$ **40.** $\ln x = -3.0055$

**41.** $\ln x = 21.7439$ **42.** $\ln x = -11.6545$ **43.** $\ln x = -0.0070$ **44.** $\ln x = 0.0300$

**45.** $\ln x = -41.4239$ **46.** $\ln x = 38.8572$

In Exercises 47–52 sketch the graph of each function.

**47.** $y = e^{2x}$ **48.** $y = e^{x/2}$ **49.** $y = e^{-x}$

**50.** $y = \ln (x + 1)$ **41.** $y = \ln x^2$ **52.** $y = \ln |x|$

**53. BUSINESS** The demand equation for $x$ units of a product relative to the price per unit, $P$, is given by $P = 100 - 0.2e^{0.005x}$.

**(a)** Find the price if the demand is 100 units. **(b)** Find the price if the demand is 1000 units.
**(c)** Find the demand if the price is \$80.00. **(d)** Find the demand if the price is \$60.00.

**54. CHEMISTRY** In chemistry, the **pH (hydrogen potential)** of a substance is defined by

$$\text{pH} = -\log [\text{H}^+],$$

where $[\text{H}^+]$ is the concentration of hydrogen ions in the solution, measured in moles per liter. The scale of the pH of a solution varies from 0 to 14, with the pH of distilled water equal to 7. If the pH of a substance is less than 7, the substance is called an *acid,* and if the pH of a substance is greater than 7, it is called a *base*.

**(a)** Find the pH of a certain brand of coffee whose $[\text{H}^+]$ is $3.89 \times 10^{-7}$ moles per liter.
**(b)** Find the pH of a sample of seawater whose $[\text{H}^+]$ is $3.15 \times 10^{-9}$ moles per liter.
**(c)** Find the hydrogen ion concentration of distilled water.
**(d)** Find the hydrogen ion concentration of rain water which has a pH of 5.6.

**55. PHYSICS** The atmospheric pressure $P$, measured in pounds per square inch, at an altitude $x$, measured in miles, above sea level can be approximated by the formula

$$P = 14.7e^{-0.2x}.$$

**(a)** What is the atmospheric pressure at sea level?
**(b)** What is the atmospheric pressure on the outside of a jet airplane cruising at an altitude of 6.25 miles?
**(c)** At what altitude will the atmospheric pressure be one-half the pressure at sea level?
**(d)** At what altitude will the atmospheric pressure be one-fourth the pressure at sea level?

**56. ASTRONOMY** Stars are categorized by brightness on a scale measured in magnitudes, with the faintest star still visible to the naked eye assigned a magnitude of 6. Telescopes are designed to view stars beyond a magnitude of 6. Every telescope has a limiting magnitude denoted by $L$ which depends on the diameter in inches, $d$, of its lens. The formula which relates these two quantities is

$$L = 8.8 + 2.2 \ln d.$$

**(a)** What is the limiting magnitude of a telescope with a 2-inch lens?
**(b)** What is the limiting magnitude of a telescope with a 30-inch lens?
**(c)** What is the diameter of the lens in a telescope with a limiting magnitude of 10?
**(d)** What is the approximate diameter of the lens in a human eye?

**57. MEDICINE** When a drug is introduced into the human circulatory system, the concentration of the drug decreases as it is eliminated by the liver and kidneys. If an initial dose in the amount of 15 mg is taken orally, the amount $A$ remaining in the system after $x$ hours can be approximated by the formula

$$A = 15e^{-0.3x}.$$

**(a)** How much of the drug remains in the system after 3 hours?
**(b)** How much of the drug remains in the system after 1 day?
**(c)** For the drug to be effective, at least 5 mg must remain in the system. How long will the drug be effective?

**(d)** What is the half-life of the drug (the time at which one-half of the initial dose remains in the system)?

**58.** PSYCHOLOGY In order to determine retention of concepts learned, a group of students was given an exam and retested each month thereafter using an equivalent exam. The average decrease in score $D$ was found to satisfy the formula

$$D = 80 - 12 \ln (x + 1),$$

where $x$ is the time in months.

**(a)** What was the initial average score on the exam?
**(b)** What was the average score at the end of 1 year?
**(c)** How many months did it take for the average score to fall below 60?
**(d)** How many months did it take for the average score to fall below 40?

## For Review

Use the properties of logarithms in Exercises 59–61 to write each logarithm as a sum, difference, or multiple of logarithms.

**59.** $\ln \dfrac{e\sqrt{z}}{\sqrt[3]{x}}$ **60.** $\ln \sqrt{\dfrac{x\sqrt{z}}{y^4}}$ **61.** $\ln \sqrt{e\sqrt{y}}$

Use the properties of logarithms in Exercises 62–64 to write each expression as a single logarithm.

**62.** $3 \ln xy + \frac{1}{2} \ln z$ **63.** $-\frac{1}{2} \ln (x + y) + \frac{1}{2} \ln (x^2 - y^2)$ **64.** $3 \ln x - \frac{1}{2} \ln y - 5 \ln z$

**65.** Express $2 \ln x + \ln y - 5 = 0$ in exponential form without using logarithms.

**66.** Solve $\ln (3x - 2) = \ln x^2$.

In Exercises 67–68 sketch the graph of each function.

**67.** $f(x) = 2^x + 2^{-x}$ **68.** $f(x) = \log_4 (-x)$ for $x < 0$

Solve each equation in Exercises 69–70.

**69.** $7^x = 7^{2x+3}$ **70.** $\log_3 x = \log_3 (x^2 - 2)$

# 5.5 Exponential and Logarithmic Equations

### Solving Exponential Equations

In previous sections we solved several simple equations involving exponentials or logarithms either by inspection or by using a direct application of one of the two theorems

$$a^x = a^y \text{ if and only if } x = y$$

or

$$\log_a x = \log_a y \text{ if and only if } x = y.$$

These methods are now expanded to include more complex equations. We begin by considering **exponential equations** in which the variable appears in one or more exponents. In some cases it may be possible to express both sides of an exponential equation as powers using the same base and equate the exponents.

**EXAMPLE 1** Solve. $3^{5x} = 9^{x-6}$

$$3^{5x} = 9^{x-6}$$

$$3^{5x} = (3^2)^{x-6} \qquad 9 = 3^2$$

$$3^{5x} = 3^{2(x-6)} \qquad (a^m)^n = a^{m \cdot n}$$

$$5x = 2(x - 6) \qquad \text{Since both sides are powers of 3, we equate exponents}$$

$$5x = 2x - 12$$

$$3x = -12$$

$$x = -4$$

Check: $3^{5(-4)} \stackrel{?}{=} 9^{-4-6}$

$3^{-20} = 9^{-10} = 3^{-20}$ The solution is $-4$ ∎

In some equations it may be difficult to express both sides as exponentials with the same base. When this occurs, we take the logarithm of both sides, using a suitable base, and apply the power rule to remove variables from exponents.

**EXAMPLE 2** Solve. $3^{2x} = 7^{x+5}$

$$3^{2x} = 7^{x+5}$$

$$\log 3^{2x} = \log 7^{x+5} \qquad \text{Take log of both sides}$$

$$2x \log 3 = (x + 5) \log 7 \qquad \text{Power rule}$$

$$2x \log 3 = x \log 7 + 5 \log 7 \qquad \text{Distributive law}$$

$$2x \log 3 - x \log 7 = 5 \log 7 \qquad \text{Write both } x\text{-terms on left}$$

$$(2 \log 3 - \log 7)x = 5 \log 7 \qquad \text{Factor out } x$$

$$x = \frac{5 \log 7}{2 \log 3 - \log 7} \qquad \text{Divide by } (2 \log 3 - \log 7)$$

ALG: 2 [×] 3 [LOG] [=] [−] 7 [LOG] [=] [STO] 5 [×] 7 [LOG] [=] [÷] [RCL] [=] → [38.714652]

RPN: 3 [LOG] [ENTER] 2 [×] 7 [LOG] [−] [STO] 7 [LOG] [ENTER] 5 [×] [RCL] [÷] → [38.714652]

Thus, $x \approx 38.7$, correct to three significant digits. Check. ∎

**To Solve an Exponential Equation**

1. Try to express each side as a power using the same base, and equate the resulting exponents.
2. If this fails, take the logarithm of each side and use the power rule to eliminate the variable exponents.
3. Solve the resulting equation and check in the original.

## Solving Logarithmic Equations

Equations that contain logarithms of expressions involving a variable are called **logarithmic equations.** Solving such equations requires a thorough knowledge of the basic properties of logarithms.

CAUTION Always check possible solutions in the original equation and remember that any value that results in the logarithm of a negative number must be omitted.

**EXAMPLE 3** Solve. $\log_2 (x + 1) - \log_2 (x - 6) = 3$

$$\log_2 (x + 1) - \log_2 (x - 6) = 3$$

$$\log_2 \left(\frac{x+1}{x-6}\right) = 3 \quad \text{The difference of logs is the log of a quotient}$$

$$\left(\frac{x+1}{x-6}\right) = 2^3 \quad \text{Convert to exponential form}$$

$$\left(\frac{x+1}{x-6}\right) = 8$$

$$(x + 1) = 8(x - 6) \quad \text{Multiply by the LCD, } (x - 6)$$

$$x + 1 = 8x - 48$$

$$-7x = -49$$

$$x = 7$$

Check: $\log_2 (7 + 1) - \log_2 (7 - 6) \stackrel{?}{=} 3$

$$\log_2 8 - \log_2 1 \stackrel{?}{=} 3$$

$$3 - 0 = 3$$

The solution is 7. ■

**EXAMPLE 4** Solve. $\ln (3x - 1) + \ln (x + 1) = \ln (4x + 7)$

$$\ln (3x - 1) + \ln (x + 1) = \ln (4x + 7)$$

$$\ln (3x - 1)(x + 1) = \ln (4x + 7) \quad \ln u + \ln v = \ln uv$$

$$(3x - 1)(x + 1) = 4x + 7 \quad \ln u = \ln v \text{ means } u = v$$

$$3x^2 + 2x - 1 = 4x + 7$$

$$3x^2 - 2x - 8 = 0$$

$$(3x + 4)(x - 2) = 0$$

$$3x + 4 = 0 \quad \text{or} \quad x - 2 = 0$$

$$x = -\frac{4}{3} \qquad x = 2$$

The only solution is 2 since $-\frac{4}{3}$ does not check. (Why?) ■

**EXAMPLE 5** Solve. $\log_2 x + \log_3 x = 1$

$\log_2 x + \log_3 x = 1$ — These expressions cannot be combined since the bases are different

$\dfrac{\log x}{\log 2} + \dfrac{\log x}{\log 3} = 1$ — Convert to common logarithms

$\log x\left[\dfrac{1}{\log 2} + \dfrac{1}{\log 3}\right] = 1$ — Factor out $\log x$

$\log x = \dfrac{1}{\dfrac{1}{\log 2} + \dfrac{1}{\log 3}}$ — Solve for $\log x$

$\log x = 0.1845757$ — Evaluate using a calculator

$x \approx 1.53$ — Find the antilog of 0.1845757, correct to three significant digits

Check this result by substituting and then converting to common logarithms. ■

**To Solve a Logarithmic Equation**

1. Obtain a single logarithmic expression using the same base on one side of the equation, or write each side as a logarithm using the same base.
2. Convert the result to an exponential equation, or use the fact that $\log_a u = \log_a v$ implies $u = v$, and solve.
3. Check all possible solutions in the original equation. (Negative numbers do not have logarithms.)

## 5.5 EXERCISES

Solve each equation in Exercises 1–48. When appropriate, give answer correct to three significant digits.

**1.** $4^x = 8$ **2.** $9^{2x} = 27$ **3.** $2^{5x} = 8^{x+5}$ **4.** $27^{x-4} = 9^{x+2}$

**5.** $2^{x^2} = 16^x$ **6.** $125^{2x-1} = 25^{x+8}$ **7.** $2^{3x} = 7$ **8.** $3^{x+2} = 4$

**9.** $5^{x+1} = 3^{2x}$ **10.** $7^{x+2} = 4^{3x}$ **11.** $(1.08)^x = 100$ **12.** $(1.12)^{x/2} = 1000$

**13.** $8^{\log_2 x} = 125$ **14.** $9^{\log_3 x} = 4$ **15.** $2^{x^2}3^{x^2} = 6^{5x-6}$

**16.** $3^{x^2}5^{x^2} = 15^{4x+5}$ **17.** $7^{2x}3^x = 10$ **18.** $2^{4x}3^x = 100$

**19.** $\log_2(x + 2) = 3$ **20.** $\log_5(2x + 1) = 2$ **21.** $\log_3 x + \log_3 9 = 5$

**22.** $\log_2(x + 3) - \log_2(x - 6) = \log_2(x - 5)$ **23.** $\ln(x + 1) + \ln(x - 1) = \ln(4x + 4)$

**24.** $\log_5(x + 2) + \log_5(x - 2) = \log_5(2x - 1)$ **25.** $\log_5(2x - 1) - \log_5(x - 5) = 1$

**26.** $\log_3(2x + 5) - \log_3(x - 1) = 2$ **27.** $\log_5 x + \log_7 x = 2$

**28.** $\log_3 x + \log_9 x = 5$ **29.** $\log_4 4x = \log_2(x + 1)$ **30.** $\log 3^x + \log 4^{2x} = 6$

**31.** $(\log_5 x)^2 = 2 \log_5 x$

**32.** $(\ln x)^2 = 5 \ln x$

**33.** $\ln x^3 = (\ln x)^3$

**34.** $(\log x)^2 = \log x^2$

**35.** $x^{\ln x} = e^2 x$

**36.** $2^{\log x} = 4x$

**37.** $\log (\log x) = 1$

**38.** $\ln (\ln x) = 2$

**39.** $\dfrac{e^x + e^{-x}}{2} = 1$

**40.** $\dfrac{e^x - 3e^{-x}}{2} = 1$

**41.** $\log \sqrt{x} = \sqrt{\log x}$

**42.** $\ln \sqrt[3]{x} = \sqrt[3]{\ln x}$

**43.** $\ln |x| = 1$

**44.** $\log |x| = 2$

**45.** $|\log x| = 2$

**46.** $|\ln x| = 3$

**47.** $\dfrac{e^x + e^{-x}}{e^x - e^{-x}} = 2$

**48.** $\dfrac{e^x - e^{-x}}{e^x + e^{-x}} = 1$

PHYSICS Newton's law of cooling relates the time $t$, in minutes, that it takes for an object heated to a temperature $T_0$ to cool to a temperature $T$ when placed in an area of constant temperature $T_c$. The formula is

$$T = T_c + (T_0 - T_c)e^{-kt},$$

where the constant $k$ depends on the nature of the object considered. Use this information in Exercises 49–54.

**49.** An iron rod, with constant $k = 0.025$, is heated to 250°F and placed in a room with a constant temperature of 70°F. What will be the temperature of the rod after 10 minutes?

**50.** An iron ball, with constant $k = 0.03$, is heated to 320°F and placed in a room with a constant temperature of 75°F. What will be the temperature of the ball after 5 minutes?

**51.** An object in a room at a constant temperature of 80°F cools from 400°F to 200°F in 15 minutes. Find the value of the constant $k$, and use it to determine the temperature of the object after 30 minutes.

**52.** An object in a room at a constant temperature of 70°F cools from 120°F to 100°F in 30 minutes. Find the value of the constant $k$, and use it to determine the temperature of the object after 1 hour.

**53.** Solve Newton's law of cooling for $t$ in terms of $T$.

**54.** CRIMINOLOGY When a coroner arrived at the scene of a murder at 12:00 noon, he discovered that the body temperature of the victim was 80°F. The room in which the victim was found had a temperature of 70°F. Assuming that the constant $k$ is 0.0044, use the formula in Exercise 53 to determine the approximate time of death. [Hint: The normal temperature of a living person is 98.6°F.]

OCEANOGRAPHY When a beam of light passes through water, the intensity of the light $I$ is diminished by the number of feet $x$ from the surface of the water. The relationship between these two variables is given by

$$I = I_0 e^{-kx},$$

where $I_0$ is the intensity of the light at the surface of the water, and $k$ is the coefficient of extinction depending on the particular water under consideration. Use this formula in Exercises 55–56.

**55.** If the coefficient of extinction of Lake Louise is 0.05, at what depth (to the nearest foot) will the intensity of a light be reduced to 50% of that at the surface?

**56.** If the coefficient of extinction of Lake Powell is 0.009, at what depth (to the nearest foot) will the intensity of a light be reduced to 10% of that at the surface?

## For Review

In Exercises 57–58 sketch the graph of each function.

**57.** $y = \ln(x^2 + 1)$

**58.** $y = e^{-0.5x}$

**59.** CHEMISTRY Use the pH equation, $\text{pH} = -\log [\text{H}^+]$, to find the pH of a soft drink with a hydrogen ion concentration of $3.25 \times 10^{-7}$.

**60.** BUSINESS Use the demand equation $P = 250 - 0.3e^{0.007x}$ to find the unit price of an item when the demand is 200 units.

# 5.6 Applications

We have already considered a variety of applied problems that use exponentials or logarithms. In this section we concentrate on special applications with broad interest and appeal.

## Compound Interest

Suppose that \$5000 is deposited in a savings account that pays 12% interest per year. At the end of one year, the amount in the account will be

$$\$5000 + 0.12(\$5000) = (1.12)(\$5000) = \$5600.$$

In order to see a pattern we write \$5600 as (1.12)(5000) and note that at the end of the second year, the amount will grow to

$$\begin{aligned}\$5600 + (0.12)(\$5600) &= (1.12)(\$5000) + (0.12)[(1.12)(\$5000)]\\ &= [1 + 0.12](1.12)(\$5000)\\ &= (1.12)^2(\$5000) = \$6272.\end{aligned}$$

Continuing in this manner, after $t$ years, the value of the account will be

$$\$5000(1.12)^t.$$

In general, if an initial amount $P$, called the **principal,** is invested for $t$ years at an annual rate of interest $r$, compounded annually, the amount $A$ will grow to

$$A = P(1 + r)^t.$$

For example, after 30 years, the original \$5000 would grow to

$$A = \$5000(1 + 0.12)^{30} = \$5000(1.12)^{30} = \$149{,}799.61.$$

Calculator steps are given below.

ALG: 1.12 [$y^x$] 30 [=] [×] 5000 [=] → [149799.61]

RPN: 1.12 [ENTER] 30 [$y^x$] 5000 [×] → [149799.61]

Remember that calculator values may vary slightly.

Often deposits earn interest by compounding semiannually (twice a year) or quarterly (four times a year). The rate of interest per period is found by dividing the annual interest by the number of compounding periods per year. If \$5000 is deposited in an account that pays 12% interest compounded semiannually, the amount in the account at the end of one year (2 compounding periods) is

$$A = \$5000(1 + 0.06)^2 = \$5000(1.06)^2 = \$5618.$$

Notice that by compounding semiannually instead of annually, we increased the value of the account by \$18. If compounding were quarterly, the amount at the end of one year would be

$$A = \$5000(1 + 0.03)^4 = \$5000(1.03)^4 = \$5627.55.$$

In general, if a principal $P$ is deposited in an account at an annual rate of interest $r$, compounded $k$ times a year for $t$ years, the amount in the account is given by the following formula.

**Compound Interest Formula**

$$A = P\left(1 + \frac{r}{k}\right)^{kt} \quad (k \text{ compounding periods a year})$$

Suppose now that we let the number $k$ of compounding periods become larger and larger without bound. The result is called **continuous compounding.** Write the compound interest formula in terms of $\frac{k}{r}$ and substitute $n$ for $\frac{k}{r}$.

$$A = P\left(1 + \frac{1}{\frac{k}{r}}\right)^{\left(\frac{k}{r}\right)rt} = P\left(1 + \frac{1}{n}\right)^{nrt} = P\left[\left(1 + \frac{1}{n}\right)^n\right]^{rt}.$$

Since $n = \frac{k}{r}$ and $r$ is a fixed constant, as $k$ gets larger so also does $n$. But from our development of the number $e$ in Section 5.4, we know that

$$\left(1 + \frac{1}{n}\right)^n \quad \text{approaches} \quad e.$$

Thus, for continuous compounding, we have the following formula.

**Compound Interest Formula**

$$A = Pe^{rt} \quad \text{(continuous compounding)}$$

**EXAMPLE 1**

**BUSINESS**

If \$10,000 is deposited in an account that pays an annual interest rate of 12%, find the amount in the account at the end of 5 years if compounding is **(a)** quarterly and **(b)** continuous.

**(a)** Since compounding is quarterly, $k = 4$. Substitute 10,000 for $P$, 0.12 for $r$, and 5 for $t$ in the compound interest formula.

$$A = P\left(1 + \frac{r}{k}\right)^{kt} = 10{,}000\left(1 + \frac{0.12}{4}\right)^{(4)(5)}$$
$$= 10{,}000(1.03)^{20}$$
$$= \$18{,}061.10$$

**(b)** For continuous compounding, substitute 10,000 for $P$, 0.12 for $r$, and 5 for $t$.

$$A = Pe^{rt} = 10{,}000e^{(0.12)(5)}$$
$$= 10{,}000e^{0.6}$$
$$= \$18{,}221.19$$

Notice that with continuous compounding, the account would earn \$160.09 more over the 5-year period than with quarterly compounding. ■

## Amortizing a Loan

When an amount $P$ of money is borrowed from a bank and is to be paid back over $t$ years in equal regular payments of $p$ dollars $k$ times a year, this is called **amortizing** the loan. That is, a loan is **amortized** when a portion of each payment is used to pay the interest and the rest is used to reduce the outstanding principal. Since each payment reduces the principal, that portion of each payment designated for interest will decrease as the payments are made. The formulas for the amount $p$ of each payment and the total interest $I$ charged are given below.

### Amortization Formulas

$$p = \frac{Pr}{k\left[1 - \left(1 + \frac{r}{k}\right)^{-kt}\right]} \quad \text{and} \quad I = kp\left[t - \frac{1 - \left(1 + \frac{r}{k}\right)^{-kt}}{r}\right]$$

We now solve the first applied problem given in the introduction to this chapter.

## EXAMPLE 2

CONSUMER

In order to purchase a new home a young couple borrows \$60,000 at an annual interest rate of 12%. It is to be paid back over a 30-year period in equal monthly payments. What is their monthly payment, and what is the total interest paid over the period of the loan?

To find the monthly payment, substitute 60,000 for $P$, 12 for $k$, 0.12 for $r$, and 30 for $t$ in the first amortization formula.

$$p = \frac{Pr}{k\left[1 - \left(1 + \frac{r}{k}\right)^{-kt}\right]} = \frac{(60{,}000)(0.12)}{12\left[1 - \left(1 + \frac{0.12}{12}\right)^{-12(30)}\right]}$$
$$= 617.17 \quad \text{To nearest cent}$$

The sequence used to calculate $p$ is given below.

ALG: 12 [×] 30 [=] [+/−] [STO] 0.12 [÷] 12 [+] 1 [=] [$y^x$] [RCL] [=] [+/−] [+] 1 [=] [×] 12 [=] [1/x] [×] 60,000 [×] 0.12 [=] → [617.16756]

RPN: 12 [ENTER] 30 [×] [CHS] [STO] 0.12 [ENTER] 12 [÷] 1 [+] [RCL] [$y^x$] [CHS] 1 [+] 12 [×] [1/x] 60,000 [×] 0.12 [×] → [617.16756]

Thus, the monthly payment is \$617.17. To find the total interest paid, substitute the same values for $k$, $r$, and $t$, along with 617.17 for $p$ in the second amortization formula.

$$I = kp\left[t - \frac{1 - \left(1 + \frac{r}{k}\right)^{-kt}}{r}\right]$$

$$= (12)(617.17)\left[30 - \frac{1 - \left(1 + \frac{0.12}{12}\right)^{-(12)(30)}}{0.12}\right]$$

$$= 162{,}180.96$$

Over 30 years, the couple will pay \$162,180.96 in interest. ■

## Population Growth

The growth rate of a population of organisms which is not influenced by a lack of food, living space, or predators can be approximated exponentially. The model used to describe the number $N$ of organisms in terms of the initial population $I$ and time $t$ is given by the formula

$$N = Ie^{kt},$$

where $k$ is called the growth constant. In effect, population growth is another application of continuous compounding, and is called the **Malthusian model** after Thomas Malthus who, in the 18th century, made important observations and predictions about growing populations. The value of $k$ is the percent of growth per unit of time $t$.

## EXAMPLE 3

**DEMOGRAPHY**

How long will it take the present population of the United States to double if it continues to grow at an annual rate of 0.9% per year?

If $I$ is the initial population of the United States, we must find $t$ when $N = 2I$ and $k = 0.009$ in the formula $N = Ie^{kt}$.

$$2I = Ie^{0.009t}$$

$$2 = e^{0.009t} \quad \text{Divide both sides by } I$$

$$\ln 2 = \ln e^{0.009t} \quad \text{Take the natural log of both sides}$$

$$\ln 2 = 0.009t$$

Thus,

$$t = \frac{\ln 2}{0.009} \approx 77 \text{ years.}$$ ■

**EXAMPLE 4**
**DEMOGRAPHY**

Suppose the population of West, Texas was 385 people in 1960 and 510 people in 1970. Assuming that the population growth of West can be approximated by the Malthusian model, find the value of $k$ (correct to three decimal places), and use it to determine the number of people who will live there in the year 2000.

When $t = 0$, the initial population $I$ is 385. When $t = 10$, the population is 510. With this information we can find the value of the growth constant $k$.

$$510 = 385e^{k(10)}$$

$$\left[\frac{510}{385}\right]^{1/10} = e^k$$

Thus,
$$k = \ln\left[\frac{510}{385}\right]^{1/10} \approx 0.028.$$

Using this value of $k$, we can calculate the population when $t = 40$.

$$N = 385e^{(0.028)(40)} \approx 1180$$

In the year 2000, West will have a population of approximately 1180 people. ■

## Radioactive Decay

Radioactive substances give off radiation (alpha particles, beta particles, and gamma rays) at a rate that depends on the amount of radioactive material remaining in the substance. As the substance gives off radiation, the atomic structure of the substance is changed so that uranium, for example, changes into thorium, radium, and finally lead. Since the rate of change is slowed by the decreasing amount of material present, the time for the material to decay completely cannot be determined theoretically. However, it is possible to find the **half-life** of a substance, the time needed for one-half the substance to decay. The amount $A$ of radioactive material remaining from an initial amount $I$, at a given time $t$, can be approximated by

$$\mathbf{A = Ie^{kt}},$$

where $k$ is a negative constant determined by the particular nature of the material.

**EXAMPLE 5**
**CHEMISTRY**

Nuclear reactors use radioactive strontium-90 for fuel. If the radioactive decay of strontium-90 can be approximated by

$$A = Ie^{-0.025t},$$

determine its half-life.

We must solve the equation for $t$ when $A = 0.5I$.

$$0.5I = Ie^{-0.025t}$$

$$0.5 = e^{-0.025t}$$

$$\ln 0.5 = \ln e^{-0.025t}$$

$$\ln 0.5 = -0.025t$$

Thus,
$$t = \frac{\ln 0.5}{-0.025} \approx 27.7 \text{ years.}$$ ■

Radioactive carbon-14 ($C^{14}$), an isotope present in the atmosphere, is absorbed by all living organisms through the intake of carbon dioxide. The level of carbon-14 in the organism remains relatively constant while the organism is alive. However, once the organism dies, carbon-14 is not absorbed, and the slow process of decay begins. The decrease in carbon-14 after the death of an organism is used by archaeologists to date very old artifacts by a process called **carbon dating.** The amount $A$ of carbon-14 left after $t$ years is approximated by the formula

$$A = Ie^{-0.000125t},$$

where $I$ is the original amount present.

## EXAMPLE 6

ARCHAEOLOGY

A tool made of bone was found at an archaeological site near Page, Arizona in 1980. It was determined that approximately 85% of the original carbon-14 still remained. Approximately when was the tool made?

Since 85% of the carbon-14 remained, we must solve the following equation for $t$.

$$\begin{aligned} 0.85I &= Ie^{-0.000125t} \\ 0.85 &= e^{-0.000125t} \\ \ln 0.85 &= \ln e^{-0.000125t} \\ \ln 0.85 &= -0.000125t \\ t &= \frac{\ln 0.85}{-0.000125} \approx 1300 \text{ years} \end{aligned}$$

The tool was made about 1300 years previous to 1980, or in about 680 A.D. ■

## Earthquake Magnitudes

For years geologists tried to agree on a scale for measuring the magnitude of an earthquake. Today the scale used most often is called the **Richter scale,** after American seismologist Charles Richter. To establish the scale, a reference earthquake at level zero was defined as any earthquake with a greatest seismic wave of 0.001 millimeter registered on a seismograph placed 100 kilometers from the epicenter of the quake. The **magnitude** of an earthquake is given by

$$M = \log \frac{A}{A_0},$$

where $A$ is the measurement of the seismic wave of the earthquake, and $A_0$ is the measurement of seismic wave of a level zero earthquake with the same epicenter. The ratio $\frac{A}{A_0}$ is called the **intensity** of the earthquake.

## EXAMPLE 7

GEOLOGY

The most powerful earthquake ever recorded occurred in 1906 in South America and was $10^{8.9}$ times more powerful than a level zero earthquake. What was the magnitude of this earthquake on the Richter scale?

Since $A = 10^{8.9}A_0$,

$$\begin{aligned} M = \log \frac{A}{A_0} &= \log \frac{10^{8.9}A_0}{A_0} \\ &= \log 10^{8.9} = 8.9. \end{aligned}$$

The magnitude of the earthquake was 8.9 on the Richter scale. ■

### Intensity of Sound

Shortly after Alexander Graham Bell invented the telephone, it became apparent that some means of measuring the intensity of the signal was necessary. The unit of measurement was called a *bel* in his honor. A tenth of a bel, or **decibel,** is approximately the smallest intensity of sound that the human ear can detect. It has been determined that the threshold of human hearing is crossed when a sound wave produces approximately $S_0 = 10^{-12}$ watt/m$^2$. This, then, is the zero level for decibels. If $S$ is the number of watt/m$^2$ produced by a particular sound wave, the loudness of the sound in decibels is given by

$$D = 10 \log \frac{S}{S_0}.$$

As a frame of reference, a whisper measures about 20 decibels, a gunshot measures about 100 decibels, and sounds measuring over 125 decibels can cause pain.

We can now solve the second applied problem given at the beginning of this chapter.

### EXAMPLE 8

**AUDIOLOGY**

Audiologists generally agree that continuous exposure to a sound level in excess of 90 decibels for more than 5 hours daily may cause long-term hearing problems. If a student listens to a walkman that produces an intensity of $2 \times 10^{-3}$ watt/m$^2$ for long periods of time daily, is there a potential danger to his hearing?

We must find $D$ when $S = 2 \times 10^{-3}$ watt/m$^2$ and $S_0 = 10^{-12}$ watt/m$^2$.

$$\begin{aligned} D = 10 \log \frac{S}{S_0} &= 10 \log \frac{2 \times 10^{-3}}{10^{-12}} \\ &= 10 \log (2 \times 10^9) \\ &\approx 93 \text{ decibels} \end{aligned}$$

Thus, there is a potential danger of hearing loss over a long period of time. ■

**NOTE** In several examples we converted an absolute scale of measurement to a more manageable logarithmic scale. Numbers on an absolute scale from $1 = 10^0$ to $10{,}000{,}000{,}000 = 10^{10}$ translate with common logarithms onto a scale from 0 to 10. ■

## 5.6 EXERCISES

**BUSINESS** In Exercises 1–10 use the compound interest formula for $k$ compounding periods, the compound interest formula for continuous compounding, or one of the two amortization formulas to solve each problem.

1. Suppose \$1000 is deposited in an account which pays an annual interest rate of 9%. What is the value of the account at the end of 4 years if compounding is **(a)** annual, **(b)** semiannual, **(c)** quarterly, **(d)** monthly, **(e)** weekly (use 52 weeks per year), **(f)** daily (use 365 days per year), and **(g)** continuously?
2. Repeat Exercise 1 if the rate of interest is 10% and the number of years is 5.
3. How long will it take an amount of money to double if it is invested at an interest rate of 12% compounded annually?

4. How long will it take an amount of money to triple if it is invested at an interest rate of 9% compounded annually?

5. How long will it take an amount of money to double if it is invested at an interest rate of 12% compounded semiannually?

6. How long will it take an amount of money to triple if it is invested at an interest rate of 9% compounded semiannually?

7. How long will it take an amount of money to double if it is invested at an interest rate of 12% compounded continuously?

8. How long will it take an amount of money to triple if it is invested at an interest rate of 9% compounded continuously?

9. **CONSUMER** Jim Kirk purchased a new truck and camper. He borrowed \$18,000 at an annual interest rate of 12%, to be paid back over a 5-year period in equal monthly payments. What is his monthly payment and what is the total interest paid over the period of the loan?

10. The Hagoods borrowed \$45,000 to purchase a vacation home. If the term of the loan is 20 years and the annual interest rate is 10%, what is their monthly payment, and how much interest will they pay over the 20-year period?

In Exercises 11–18 use the Malthusian model $N = Ie^{kt}$ to solve each problem.

11. **DEMOGRAPHY** The population of a city is growing at an annual rate of 2.1%. How long will it take for the population to double?

12. **DEMOGRAPHY** How long will it take for the population of a small Caribbean nation to triple if its annual growth rate is 1.8%?

13. **DEMOGRAPHY** The population of the United States was 237 million in 1985. If the annual growth rate is approximately 0.9%, what will be the population in the year 2000?

14. **BIOLOGY** The rodent population in a border town is estimated to be 75,000 and growing at an annual rate of 12%. When will the population reach 1 million?

15. **ECOLOGY** An endangered species of African antelope is estimated to number 8000 and is decreasing at a rate of 6.2% per year. In how many years will the population decline to 2000?

16. **ECONOMICS** When the interstate freeway bypassed a small town in Arizona, the population began to decline from 12,000 residents at an annual rate of 4.1%. How many years will it take for the population to reach 8000?

17. **DEMOGRAPHY** The population of Alpine, Colorado was 800 in 1960 and 1150 in 1980. Assuming the same growth rate find the value of $k$ (correct to three decimal places) and use it to predict the population of Alpine in the year 2000.

18. **DEMOGRAPHY** The population of the southwestern United States is expected to double in 30 years. What will be the approximate annual growth rate of this region?

**CHEMISTRY** Use the formula for radioactive decay, $A = Ie^{kt}$, to solve Exercises 19–26.

19. A certain radioactive isotope decays according to the equation $A = Ie^{-0.03t}$ where $t$ is measured in years. Determine the half-life of this isotope.

20. Determine the half-life of carbon-14 if its radioactive decay is given by the formula $A = Ie^{-0.000125t}$.

**21.** The half-life of radioactive radium is approximately 1650 years. How many years would be required for a given amount of radium to decay to three-fourths that amount?

**22.** An isotope of thorium has a half-life of approximately 8.5 days. How long will it take 75% of a given amount to decay?

**23.** ARCHAEOLOGY A wooden death mask is discovered in an ancient burial site. If it contains 55% of the carbon-14 ($k = -0.000125$) it originally contained, what is the approximate age of the mask?

**24.** ARCHAEOLOGY The mummified remains of a woman were discovered in 1960 in a cave in Texas. If approximately 73% of the original carbon-14 ($k = -0.000125$) was still present in the mummy, in approximately what year did she die?

**25.** MEDICINE A radioactive isotope with a half-life of 3.2 hours is used as a tracer in a medical test. If 30 milligrams are injected into the system at 9:00 A.M., how many milligrams remain in the system at 6:00 P.M. the same day?

**26.** METEOROLOGY A radioactive isotope is used to power a weather station in a remote location. If the rate of decay reduces the power by 0.025% daily, how many days will the station be functional if it becomes inoperable when the power reaches a level at 40% of the original supply?

GEOLOGY In Exercises 27–30 use the formula for determining the magnitude of an earthquake, $M = \log \frac{A}{A_0}$.

**27.** A California earthquake in 1983 was 2,511,890 times more powerful than a reference level zero quake. What was the magnitude of the California earthquake?

**28.** The San Francisco earthquake of 1906 was $1.995 \times 10^8$ times more powerful than a reference level zero quake. What was the magnitude of the San Francisco earthquake?

**29.** If an earthquake measured 7.2 on the Richter scale, what was the intensity of the earthquake?

**30.** The magnitude of an earthquake in Mexico measured 7.8 on the Richter scale. Fourteen years later another quake measured 4.8 on the Richter scale. The first earthquake was how many times more intense than the second?

AUDIOLOGY In Exercises 31–36 use the sound equation $D = 10 \log \frac{S}{S_0}$, where $S_0 = 10^{-12}$ watt/m$^2$ is the threshold of human hearing.

**31.** If the conversation level in a restaurant produces $4.15 \times 10^{-6}$ watt/m$^2$ of power, what is this decibel level?

**32.** When a jet airplane takes off the noise produces 0.5 watt/m$^2$ of power. What is the decibel level of this noise?

**33.** What is the intensity in watt/m$^2$ of a noise measured at 60 decibels?

**34.** The intensity of a sound is $2.5 \times 10^{-3}$ watt/m$^2$. If the intensity is doubled, will the decibel rate also double?

**35.** The intensity of a sound is $4.6 \times 10^{-7}$ watt/m$^2$. If the intensity is multiplied by 100, will the decibel rate also be multiplied by 100?

**36.** A sound measuring 125 decibels can cause pain to the human ear. What is the approximate intensity of such a sound in watt/m$^2$?

## For Review

Solve each equation in Exercises 37–40.

**37.** $5^{2x} = 7^{x+1}$

**38.** $\log_2 (x + 3) + \log_2 (x - 4) = 3$

**39.** $(\log x)^2 = \log x^5$

**40.** $|\ln x| = 5$

In Exercises 41–42 find $x$ correct to four decimal places.

**41.** $x = \log 4{,}820{,}000$

**42.** $x = \ln (-0.0045)$

In Exercises 43–44 find $x$ correct to three significant digits.

**43.** $\log x = -9.4994$

**44.** $\ln x = 13.2317$

**45.** Use the properties of logarithms to write $\ln \dfrac{x\sqrt{y}}{z^4}$ as a sum, difference, or multiple of logarithms.

# CHAPTER 5 REVIEW EXERCISES

**1.** If $a$ is any base for logarithms, what is the value of **(a)** $\log_a a$, and **(b)** $\log_a 1$?

**2.** Convert $2^x = 5$ to logarithmic form.

**3.** Convert $\log_3 a = w$ to exponential form.

Solve each equation in Exercises 4–5.

**4.** $\log_a 64 = 3$

**5.** $125^x = 5$

**6.** For any base $a$ and any positive real number $x$, simplify $a^{\log_a x}$.

**7.** For any base $a$ and any real number $x$, simplify $\log_a a^x$.

In Exercises 8–10 sketch the graph of each function.

**8.** $f(x) = -3^x$

**9.** $f(x) = e^{-x^2}$

**10.** $f(x) = \log_5 (-x)$ for $x < 0$

In Exercises 11–12 use a calculator to approximate each number correct to four decimal places.

**11.** $(\sqrt{5})^\pi$

**12.** $(1.02)^{-50}$

In Exercises 13–16 evaluate each logarithm given that $\log_a 3 = 0.5283$ and $\log_a 7 = 0.9358$.

**13.** $\log_a 21$

**14.** $\log_a \dfrac{14}{6}$

**15.** $\log_a \left(\dfrac{7}{3}\right)^{3/2}$

**16.** $\log_3 7$

In Exercises 17–18 use the properties of logarithms to write each logarithm as a sum, difference, or multiple of logarithms.

**17.** $\log_a \left(\dfrac{xy}{z}\right)^{2/3}$

**18.** $\log_a y^5 \sqrt[3]{\dfrac{x^2}{z^2}}$

In Exercises 19–20 use the properties of logarithms to write each expression as a single logarithm.

**19.** $3 \log_a xy + \frac{1}{2} \log_a z$

**20.** $2 \log_a (x^2 - y^2) - \log_a (x + y)$

In Exercises 21–22 determine whether the given equation is true or false.

**21.** $\log_a (u + v) = \log_a u + \log_a v$

**22.** $\log_a \frac{u}{a} = \log_a u - 1$

Find each logarithm in Exercises 23–24.

**23.** $\ln (2.58 \times 10^{-7})$

**24.** $\log_2 0.0425$

Find each antilogarithm in Exercises 25–26.

**25.** $\log x = -8.6219$

**26.** $\ln x = 41.6219$

Solve each equation in Exercises 27–34.

**27.** $81^{x-2} = 9^{x+5}$

**28.** $\log_2 (3x + 1) + \log_2 (x - 1) = \log_2 (10x + 14)$

**29.** $\log_3 (8x + 1) - \log_3 (x - 7) = 3$

**30.** $7^{x-5} = 3^{x+7}$

**31.** $(1.05)^x(1.02)^{2x} = 1000$

**32.** $\log_2 16^{2x+1} = 8$

**33.** $(\log_5 x)(\log_2 x) = \log x$

**34.** $\ln (\ln x) = 0$

**35.** BIOLOGY The number of bacteria $N$ in a culture is given in terms of time $t$, in hours, by $N = (3000)2^t$. **(a)** How many bacteria are present in 6 hours? **(b)** In how many hours will the bacteria count reach 3,072,000?

**36.** PHYSICS The atmospheric pressure $P$ in pounds per square inch is approximated by $P = 14.7e^{-0.2x}$, where $x$ is the altitude of the object above sea level in miles. **(a)** What is the pressure on an object 6 miles high? **(b)** A pressure reading of 4.89 lb/in$^2$ is taken on the surface of an airplane. How high is the plane?

**37.** BUSINESS The demand equation for $x$ units of a product relative to the price per unit $P$ is given by $P = 1000 - 0.5e^{0.003x}$. **(a)** Find the price if the demand is 500 units. **(b)** Find the demand if the price is \$850.

**38.** CHEMISTRY The pH of a substance is given by pH $= -\log [H^+]$, where $[H^+]$ is the concentration of hydrogen ions in the solution. **(a)** Find the pH of a mixed drink with hydrogen ion concentration $2.75 \times 10^{-7}$ moles per liter. **(b)** Find the hydrogen ion concentration of the water in a lake which has a pH of 5.1.

**39.** ASTRONOMY The limiting magnitude $L$ of a telescope is related to the diameter $d$ of the lens by $L = 8.8 + 2.2 \ln d$. **(a)** What is the limiting magnitude of a telescope with a 3.5-inch lens? **(b)** What is the diameter of the lens of a telescope with a limiting magnitude of 13?

**40.** MEDICINE The amount of a drug remaining in the circulatory system $x$ hours after a 15 mg dose is taken is approximated by $A = 15e^{-0.25x}$. **(a)** How much of the drug remains in the system after 4 hours? **(b)** For the drug to be effective, at least 10 mg must remain in the system. How long will the dose be effective?

**41.** PSYCHOLOGY The formula $D = 75 - 15 \ln (x + 1)$ gives the average decrease in a test score when the test was repeated on monthly intervals. **(a)** What was the average score 6 months after the test was first given? **(b)** How many months did it take for the average score to fall below 30?

**42.** PHYSICS An object placed in a room with a constant temperature of 72°F cools from 130°F to 105°F in 10 minutes. Find the value of the constant $k$ (to three decimal places) in Newton's law of cooling, $T = T_c + (T_0 - T_c)e^{-kt}$, and use it to determine the temperature of the object after 30 minutes.

**43.** OCEANOGRAPHY The intensity $I$ of a light is diminished as it is directed through water according to the formula $I = I_0e^{-kx}$, where $I_0$ is the intensity of the light at the surface of the water and $x$ is the depth in feet below the surface. If the intensity of a light is reduced by 50% at 20 ft below the surface, find the coefficient of extinction $k$ of the light in water.

**44.** INVESTMENT If \$2000 is deposited in an account that pays an annual interest rate of 14%, what is the value of the account at the end of 2 years if compounding is **(a)** semiannually, **(b)** monthly, **(c)** continuously?

**45.** INVESTMENT How long will it take an amount of money to quadruple if it is invested at an interest rate of 10% compounded continuously?

**46.** BUSINESS If \$30,000 is borrowed at an annual interest rate of 12%, and is to be paid back over a 15-year period in equal monthly payments, what is the amount of each payment? What is the total interest paid over the 15-year period?

**47.** DEMOGRAPHY Use the Malthusian model $N = Ie^{kt}$ with $k$ computed to three decimal places, to determine the population of Winter, Minnesota in the year 2000 if in 1970 the population was 25,000 and in 1980 the population was 28,000.

**48.** CHEMISTRY Use the formula for radioactive decay, $A = Ie^{kt}$, to determine the half-life of a radioactive isotope when $t$ is measured in years and $k = -0.0046$.

**49.** GEOLOGY An earthquake was measured to have an intensity $2.65 \times 10^5$ times more powerful than a reference level zero earthquake. Use the formula $M = \log \frac{A}{A_0}$ to determine the magnitude of the earthquake using the Richter scale.

**50.** AUDIOLOGY A rock band's performance produces an intensity of $5.5 \times 10^{-2}$ watt/m$^2$ during a four-hour concert. If the band performs daily, is there a potential danger of hearing loss to the members of the band?

CHAPTER

# 6 TRIGONOMETRIC FUNCTIONS

The word *trigonometry* is derived from Greek words which mean *three-angle measurement* or *triangle measurement*. Historically, trigonometry was developed as a tool for finding the measurements of parts (sides or angles) of a triangle when other parts were known. As a result, it became indispensable in areas such as navigation and surveying. More recently the study of trigonometry has taken a different direction. Although applications involving triangles remain important, viewing trigonometry in terms of functions of real numbers has expanded its usefulness. A broader spectrum of physical applications have been found, such as harmonic motion and orbits of atomic particles. Two types of problems that can be solved with trigonometry follow.

### GEOLOGY ▶

A geologist, standing on one rim of the Grand Canyon in Arizona, looks across the canyon at the Coconino Sandstone layer, a formation approximately 350 ft in depth. If the sandstone layer, top to bottom, subtends an angle of $00°29'$ to the geologist's eye, what is the approximate distance from one rim of the canyon to the other?

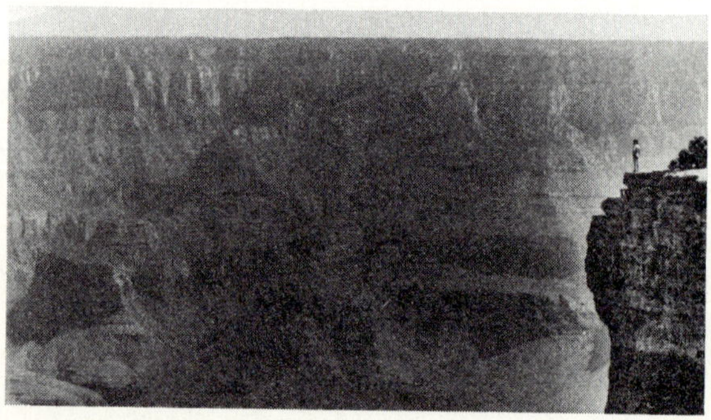

### ◀ FORESTRY

A small forest fire is sighted due south of a fire lookout tower on Woody Mountain. From a second tower, located 11.0 miles due east of the first, the bearing of the fire is S51°10′W. How far is the fire from the tower on Woody Mountain?

The first of these problems can be solved by using an arc length formula (see Example 2 in Section 6.2), and the second by using properties of right triangles (see Example 4 in Section 6.4).

The approach we have taken to the study of trigonometry includes both the traditional triangle approach and the real-number functional approach. We begin with a discussion of angles and angular measure and proceed to the trigonometric functions of acute angles. We conclude by considering the trigonometric functions of real numbers, their graphs, and several fundamental identities.

## 6.1 Angles and Angular Measure

### Angles

An **angle** is generated by rotating a **half-line** or **ray** about its end point, called the **vertex** of the angle. See Figure 1. The initial position of the ray, denoted by $OA$, is called the **initial side** of the angle, and the terminal position of the ray, denoted by $OB$, called the **terminal side.** The symbol $\angle$ represents the word *angle,* and the angle is denoted by $\angle AOB$.

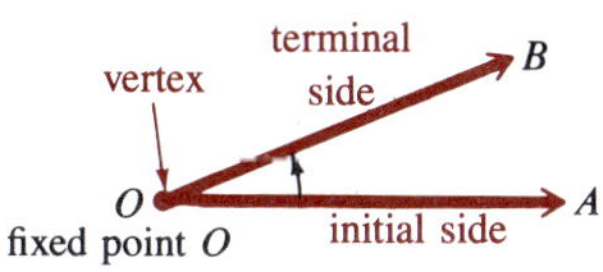


**Figure 1 Positive Angle**

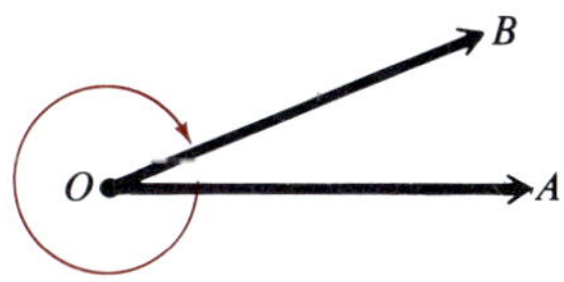


**Figure 2 Negative Angle**

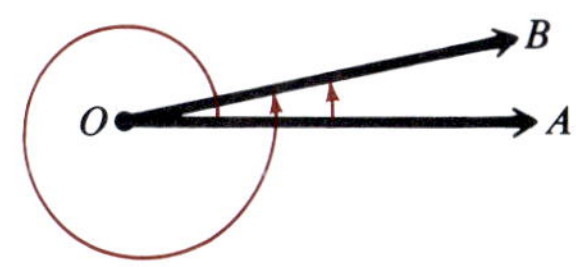


**Figure 3 Coterminal Angles**

If the rotation of the terminal side of an angle is counterclockwise (as in Figure 1), the angle is **positive.** If the rotation of the terminal side is clockwise (as in Figure 2), the angle is **negative.** In both cases an arrow indicates the direction of rotation. It is also possible for the terminal side to make one complete rotation and continue, as shown in color in Figure 3. Thus, the rays $OA$ and $OB$ may be the initial and terminal sides of many different angles, called **coterminal angles.**

### Degree Measure

If the initial side of an angle is placed along the positive $x$-axis in a Cartesian coordinate system, with its vertex at the origin $(0, 0)$, the angle is said to be in **standard position.** Then, if the initial side of an angle in standard position is allowed to make one complete revolution so that the initial and terminal sides coincide, the resulting angle has measure 360 **degrees.** The symbol ° is used to represent degrees. An angle of measure 180° is called a **straight angle,** and an angle of measure 90° is called a **right angle.** Figure 4, on the following page, shows these angles in standard position along with other positive and negative angles.

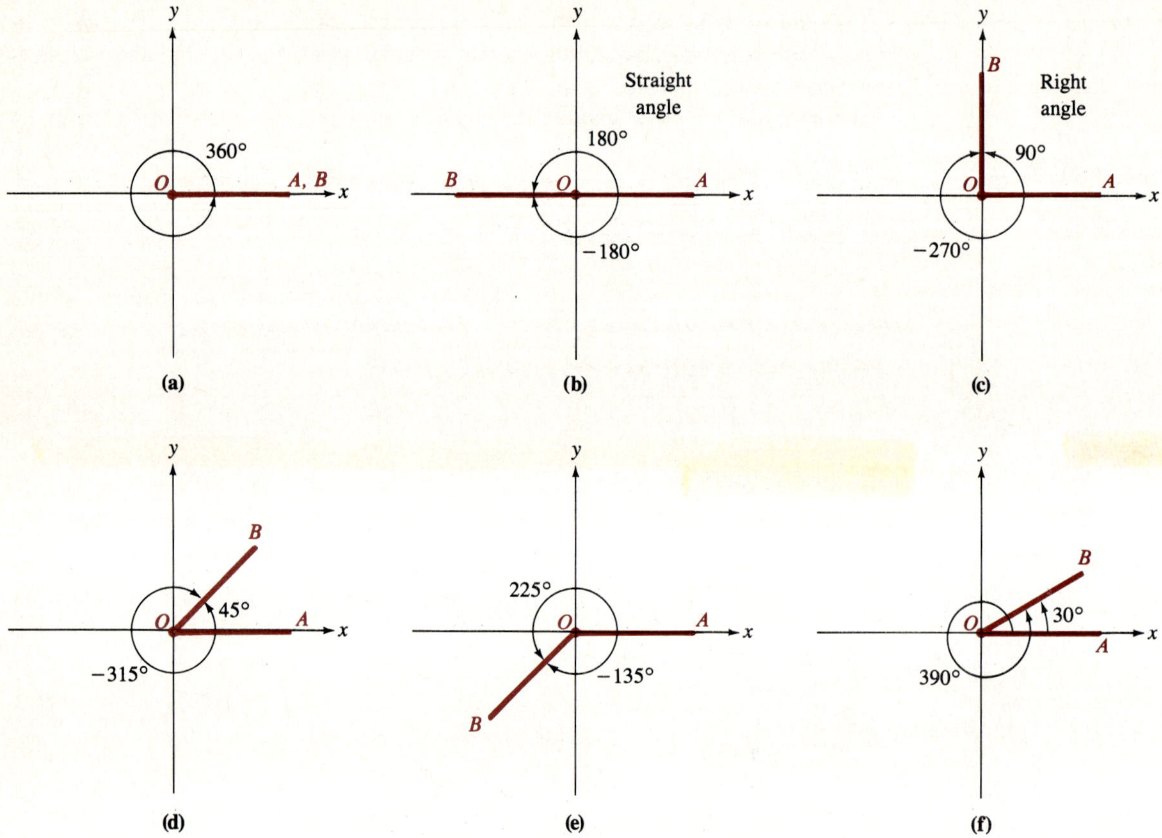


**Figure 4 Angles in Standard Position**

If an angle in standard position has measure between 0° and 90°, its terminal side is in quadrant I. If the angle is between 90° and 180°, its terminal side is in quadrant II, and so forth. If the terminal side of an angle lies on one of the coordinate axes, such as angles of measure 0°, 90°, 180°, and −90°, the angle is called a **quadrantal angle.** An angle measuring between 0° and 90° is **acute,** and one measuring between 90° and 180° is **obtuse.** Two acute angles are called **complementary** if the sum of their measures is 90°, and two positive angles are **supplementary** if their measures add to 180°. The angles 35° and 55° are complementary, and 35° and 145° are supplementary.

**EXAMPLE 1** Give two angles that are coterminal with a 30° angle, one positive and one negative.

Figure 4(**f**) shows that the positive angle 390° is coterminal with 30°. The angle −330° is also coterminal with 30°. In fact, to find any positive angle coterminal with 30°, we can add any positive multiple of 360° to 30°. Similarly, any negative multiple of 360° added to 30° will result in a negative coterminal angle. ■

When a greater degree of precision is required, 1 degree can be divided into 60 equal parts, called **minutes** (denoted by ′), and one minute can be divided into 60 equal parts, called **seconds** (denoted by $''$). Thus, the angle $45°15'35''$ has measure 45 degrees, 15 minutes, and 35 seconds. Another way to obtain more precise measurement is to express parts of a degree as decimals. As we shall see, with the use of calculators, this method is often more convenient.

**EXAMPLE 2**

**(a)** Express $45°15'35''$ as degrees and the decimal part of a degree.
Since $15' = \frac{15}{60°}$, dividing gives 0.25. Also, since $35'' = \frac{35'}{60} = \frac{35°}{3600}$, dividing gives 0.0097222. Adding these two decimals to 45 yields the required value. The value can be obtained with a calculator, starting with $35''$, by following these steps.

ALG: 35 [÷] 3600 [+] 15 [÷] 60 [+] 45 [=] → [45.259722]

RPN: 35 [ENTER] 3600 [÷] 15 [ENTER] 60 [÷] [+] 45 [+] → [45.259722]

Thus, $45°15'35'' = 45.26°$, correct to two decimal places.

**(b)** Express $123.67°$ using degrees, minutes, and seconds.
To convert $0.67°$ into minutes we multiply by 60.

$$(60)(0.67°) = 40.2'$$

To convert $0.2'$ into seconds we multiply by 60.

$$(60)(0.2') = 12''$$

Thus, $123.67° = 123°40'12''$. ■

## Radian Measure

Angular measure using degrees is appropriate for solving problems involving angles. However, most modern applications of trigonometry use an alternative form of measure called **radian measure.** Consider a circle with radius $r$ centered at the origin of a Cartesian coordinate system, as in Figure 5. Starting at the point $A$ with coordinates $(r, 0)$, mark off length $r$ in a counterclockwise direction at point $B$. Join $B$ to the origin forming angle $\theta$ (theta—Greek letters such as $\theta$ are often used to abbreviate angles such as $\angle AOB$). The **central angle** $\theta$ is opposite to or **subtended by** the arc $\widehat{AB}$ of length $r$, and $\theta$ has measure 1 **radian.** This is restated in the definition on the following page.

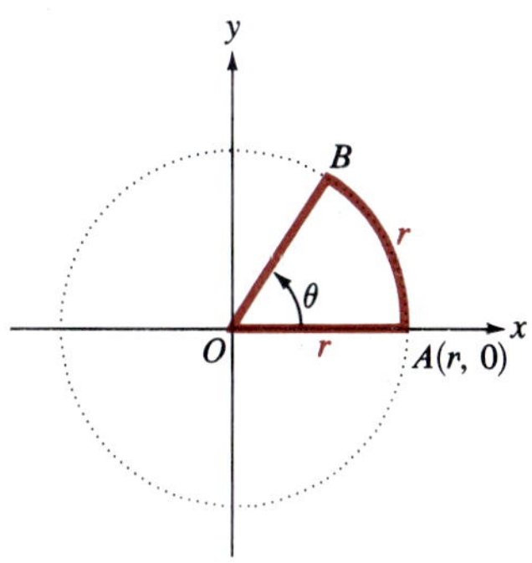


**Figure 5 Angle with Measure 1 Radian**

> An angle with vertex at the center of a circle subtended by an arc on the circle equal in length to the radius of the circle has measure 1 **radian.**

If two lengths of the radius of a circle are marked off in a counterclockwise direction, the angle subtended by the resulting arc has measure 2 radians. If, however, two lengths of the radius are marked off in a clockwise direction, the resulting angle has measure $-2$ radians. There is no symbol for a radian, and often the word radian is omitted. Any angle measure given without the degree symbol is understood to be in radians. Thus, when we discuss angles of measure 2 or $-2$, we will mean 2 radians or $-2$ radians. In Figure 6, we see that $\alpha = 2$ and $\beta = -2$.

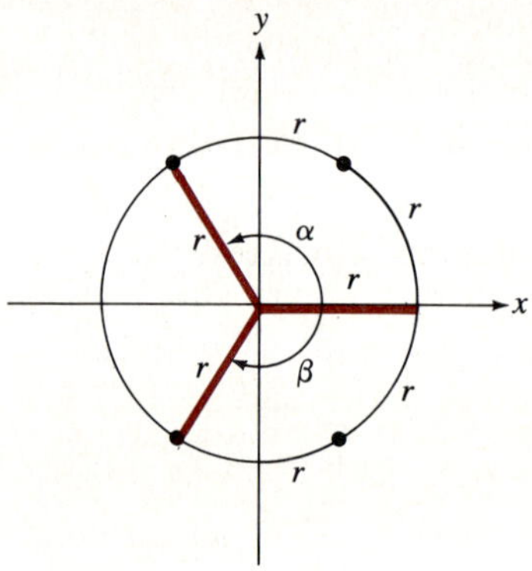


**Figure 6 Angles Measuring 2 and −2 Radians**

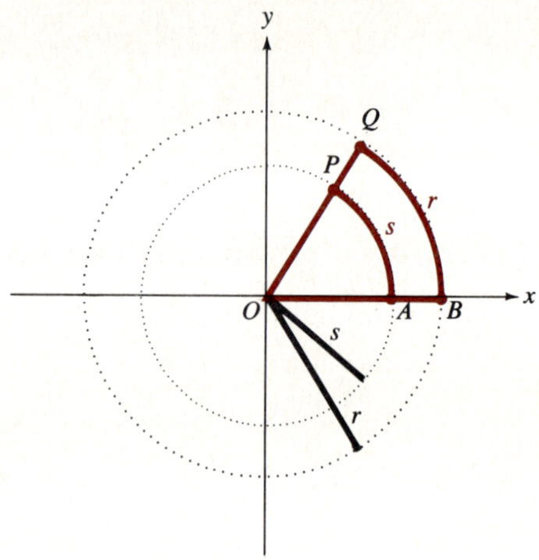


**Figure 7 Radian Measure: Different Circles**

A 1-radian angle is independent of the size of the circle used to measure it. Suppose that two concentric circles with radii $r$ and $s$ are placed with their center at the origin of a coordinate system as in Figure 7.

## The Unit Circle

The point $P$, $s$ units from $A$, and the point $Q$, $r$ units from $B$, can be shown to lie on the same straight line through the point $O$. Thus $\angle POA$ and $\angle QOB$ are the same angle with measure 1 radian. As a result, it is often convenient to focus attention on a **unit circle,** a circle with radius 1. On a unit circle, the length of an arc is numerically equal to the number of radians in the angle subtended by the arc. Since a unit circle has circumference $2\pi$ ($c = 2\pi r$ and $r = 1$), an angle of measure 360° subtends an arc of length $2\pi$ making

$$\mathbf{360° = 2\pi \text{ radians.}}$$

## Conversion Between Degrees and Radians

Also, a straight angle of measure 180° determines an arc of length $\pi$, equal to half the circumference of the circle, making

$$\mathbf{180° = \pi \text{ radians.}}$$

Solving this equation for 1° we obtain

$$\mathbf{1° = \frac{\pi}{180}\ radians,}$$

and solving for 1 radian we have

$$\mathbf{\frac{180°}{\pi} = 1\ radian.}$$

These results are summarized below.

### Conversion Rule

**1.** To convert degrees to radians, multiply by $\frac{\pi}{180}$.

**2.** To convert radians to degrees, multiply by $\frac{180}{\pi}$.

## EXAMPLE 3

Convert the following from radians to degrees.

**(a)** $\frac{\pi}{4}$ radians $\qquad \frac{\pi}{4}\text{ radians} = \frac{\cancel{\pi}}{4} \cdot \frac{180°}{\cancel{\pi}} = 45°$

**(b)** $-2\pi$ radians $\qquad 2\pi\text{ radians} = (-2\cancel{\pi})\left(\frac{180°}{\cancel{\pi}}\right) = \mathbf{-360°}$

**(c)** 3 radians $\qquad 3\text{ radians} = 3 \cdot \frac{180°}{\pi} = \frac{540°}{\pi} \approx 171.89°$ ■

## EXAMPLE 4

Convert the following from degrees to radians.

**(a)** 30° $\qquad 30° = 30 \cdot \frac{\pi}{180}\ \mathbf{radians} = \frac{\pi}{6}\text{ radians}$

**(b)** 47° $\qquad 47° = 47 \cdot \frac{\pi}{180}\text{ radians} \approx 0.820\text{ radians}$

**(c)** 23°46′55″
First convert 23°46′55″ to a decimal.

ALG: 55 [:] 3600 [+] 46 [÷] 60 [+] 23 [−] → [23.781944]
RPN: 55 [ENTER] 3600 [÷] 46 [ENTER] 60 [÷] [+] 23 [+] → [23.781944]

Then multiply by $\frac{\pi}{180}$.

ALG: 23.781944 [×] [π] [÷] 180 [=] → [0.4150732]
RPN: 23.781944 [π] [×] 180 [÷] → [0.4150732]

Thus, 23°46′55″ ≈ 0.4151 radians. ■

## 6.1 EXERCISES

Are the angles given in Exercises 1–8 coterminal?

**1.** 45°, −45°  **2.** 90°, −270°  **3.** $\frac{\pi}{6}, \frac{-11\pi}{6}$  **4.** $\frac{\pi}{4}, \frac{9\pi}{4}$

**5.** 45°, $\frac{\pi}{4}$  **6.** 30°, $\frac{\pi}{6}$  **7.** 390°, $\frac{13\pi}{6}$  **8.** 420°, $\frac{-5\pi}{3}$

In Exercises 9–12 express each angle as degrees and the decimal part of a degree, correct to the nearest hundredth.

**9.** 38°41′13″  **10.** 135°50′25″  **11.** −52°35′21″  **12.** −247°27′52″

In Exercises 13–16 express each angle using degrees, minutes, and seconds.

**13.** 85.42°  **14.** 135.65°  **15.** −48.18°  **16.** −390.76°

In Exercises 17–22 convert each angle from radians to degrees, correct to the nearest hundredth of a degree.

**17.** $3\pi$  **18.** $\frac{-\pi}{2}$  **19.** 7  **20.** −3.5

**21.** −1.4236  **22.** 7.2113

In Exercises 23–26 convert each angle from radians to degrees, minutes, and seconds.

**23.** 5  **24.** −3  **25.** −0.8525  **26.** 6.1135

In Exercises 27–32 convert each angle from degrees to radians (correct to four decimal places when appropriate).

**27.** 60°  **28.** 270°  **29.** −120°  **30.** −135°

**31.** 38.5°  **32.** −105.6°

In Exercises 33–36 convert each angle from degrees, minutes, and seconds to radians, correct to four decimal places.

**33.** 55°20′00″  **34.** −41°10′00″  **35.** −110°50′28″  **36.** 250°49′15″

## 6.2 Arc Length and Angular Velocity

In this section we develop formulas for finding the length of an arc on a circle and for determining the change in an angle over time. These formulas are helpful in a variety of applications.

### Arc Length

Suppose $s$ and $t$ are two arcs on a circle of radius $r$ which subtend angles $\theta$ and $\phi$ (Greek letter phi) respectively (see Figure 8). A theorem in plane geometry states that

$$\frac{s}{t} = \frac{\theta}{\phi}$$

where $\theta$ and $\phi$ are measured in radians.

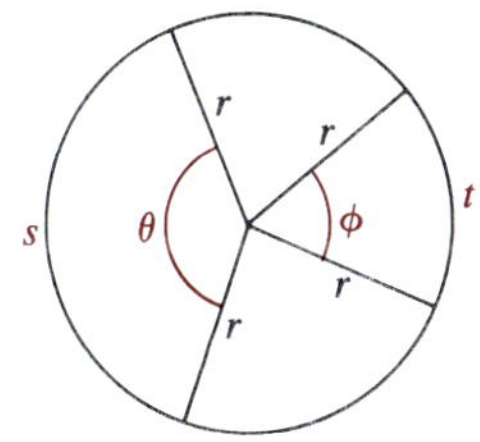


**Figure 8 Comparing Arcs**

In the special case where $\phi$ has measure 1 radian, then $t = r$. Substituting, we obtain

$$\frac{s}{r} = \frac{\theta}{1},$$

which simplifies to $s = r\theta$. We have just established the following formula.

### Arc Length Formula

In a circle of radius $r$, suppose that an arc of length $s$ subtends a central angle $\theta$ measured in radians. Then

$$s = r\theta.$$

CAUTION When working with the arc length formula, remember that $\theta$ must be measured in radians. If $\theta$ is given in degrees, we must convert to radians before using the formula. ■

**EXAMPLE 1** Find the length of an arc on a circle with radius 2.3 ft which subtends a central angle of 72°15′.

We first put 15′ in decimal form. Since $\frac{15}{60} = 0.25$, 72°15′ = 72.25°. Now convert to radians using the following steps:

ALG: 72.25 [×] [π] [÷] 180 → [1.2610004]

RPN: 72.25 [ENTER] [π] [×] 180 [÷] → [1.2610004]

Since $s = r\theta$, we multiply the number in the display, $\theta$ by 2.3, $r$.

ALG: [1.2610004] [×] 2.3 [=] → [2.9003009]

RPN: [1.2610004] 2.3 [×] → [2.9003009]

The length of the arc is approximately 2.9 ft. ■

The arc length formula provides the tool for solving the first problem in the introduction to this chapter.

## EXAMPLE 2

GEOLOGY

A geologist, standing on one rim of the Grand Canyon in Arizona, looks across the canyon at the Coconino Sandstone layer, a formation approximately 350 ft in depth. If the sandstone layer, top to bottom, subtends an angle of 00°29′ to the geologist's eye, what is the approximate distance from one rim of the canyon to the other?

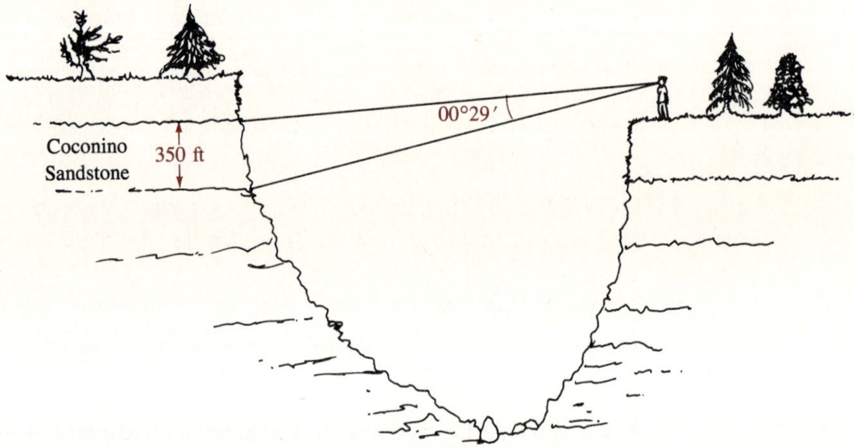


**Figure 9**

See Figure 9 (drawing not to scale). When a central angle of a circle is quite small, the length of an arc and the length of the chord formed by the arc, in this case the depth of the Coconino Sandstone formation, are approximately equal. Thus, we can use the arc length formula $s = r\theta$ to approximate the solution. We must find $r$ when $s = 350$ ft and $\theta = 0.0084$ ($00°29' \approx 0.0084$ radians).

$$r = \frac{s}{\theta} = \frac{350}{0.0084} \approx 41{,}667 \text{ ft}$$

Dividing by 5280 to convert to miles, we have $r \approx 7.9$. The canyon is approximately 7.9 miles across at this point. ■

## Linear and Angular Velocity

The arc length formula can be used to obtain a relationship between two kinds of velocity, linear and angular. If a point moves with a constant **linear velocity** $v$ over a period of time $t$, the distance traveled $s$ is given by $s = vt$, and the linear velocity by $v = s/t$. If the point is moving along a circle, the radius from the point to the center of the circle is sweeping out an angle $\theta$, and the change in $\theta$ over time is called the **angular velocity** of the point. Since the length of an arc is given by

$$s = r\theta,$$

dividing both sides by $t$ we obtain

$$\frac{s}{t} = \frac{r\theta}{t} = r \cdot \frac{\theta}{t}.$$

Letting $v = s/t$ (the linear velocity of the point), and $a = \theta/t$ (the angular velocity of the point), we obtain the following formula.

**Velocity Formula**

The linear velocity of a moving point on a circle of radius $r$ is given by

$$v = ra$$

where $a$ is the angular velocity given in radians per unit of time.

**CAUTION** Although the angular velocity formula appears simple, care must be taken to see that all units of measure agree. Since $v$ is linear distance per unit of time, the linear unit for $v$ must be the same as the unit for $r$, and the time unit for $v$ must be the same as the time unit for $a$. Also, angular velocity given as revolutions per unit of time must be converted to radians per unit of time before the angular velocity formula can be used. Since 1 revolution equals $2\pi$ radians, multiply by $2\pi$. ■

**EXAMPLE 3** A wheel with a 16-cm diameter is rotating at 12 revolutions per minute (rpm). What is the linear velocity in cm/sec of a point on the rim?

Remember to divide the diameter by 2 to obtain the radius,

$$r = 8 \text{ cm},$$

and to convert 12 revolutions per minute to radians per second.

$$\begin{aligned} a &= 12(2\pi) \text{ radians per minute} \\ &= 24\pi \text{ radians per minute} \\ &= \frac{24\pi}{60} \text{ radians per second} \end{aligned}$$

Substituting, we have

$$\begin{aligned} v = ra &= 8 \cdot \frac{24\pi}{60} \text{ cm/sec} \\ &= 3.2\pi \text{ cm/sec} \\ &\approx 10.1 \text{ cm/sec.} \end{aligned}$$ ■

## 6.2 EXERCISES

Solve.

1. Find the length of an arc on a circle of radius 4.0 cm which subtends a central angle of $\frac{\pi}{3}$.
2. Find the length of an arc on a circle of radius 5.3 cm which subtends a central angle of 5 radians.

**3.** Find the length of an arc on a circle of diameter 10.00 ft which subtends a central angle of 135°25′.

**4.** Find the radius of a circle on which a central angle of 48°10′ subtends an arc of 90.00 inches.

**5.** ENGINEERING The wheel on a locomotive has diameter 5.00 ft. Through what angle (to the nearest minute) does it turn when the locomotive rolls 4.45 ft?

**6.** PHYSICS If a pendulum of length 40.0 cm swings 2°25′ on each side of its vertical position, what is the length of the arc through which the end of the pendulum swings?

**7.** GEOGRAPHY Assuming that the earth is a perfect sphere of radius 4000 miles, a cross-section of the earth is a perfect circle of radius 4000 miles. Latitude lines are circles on the surface of the earth formed by intersecting the earth with a plane parallel to the equator. Each latitude line is identified with an angle between 0° and 90° according to the number of degrees it lies north or south of the equator. For example, the latitude of Sun Key, Florida is 28°N (see the figure below). How far is Sun Key from the North Pole?

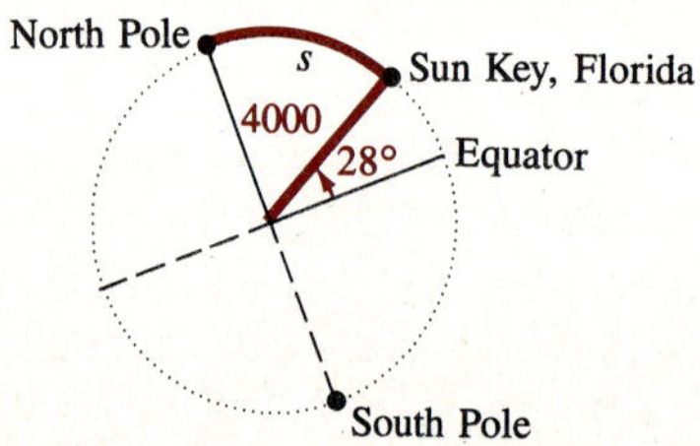


**8.** GEOGRAPHY Great Forks, Michigan, has latitude 44°10′N. How far is Great Forks from the equator? From the North Pole? From the South Pole?

**9.** GEOGRAPHY Snow Peak, Montana, is 3450 miles from the equator. Find the latitude of Snow Peak (to the nearest minute).

**10.** GEOGRAPHY Montreal, Canada (latitude 45°40′N) is due north of Santiago, Chile (latitude 35°20′S). How far is Santiago from Montreal?

**11.** RECREATION A bicycle with a 30-inch wheel (diameter) travels a distance of one mile. How many revolutions does the wheel make (to the nearest revolution)?

**12.** RECREATION A child's car has 14-inch (in diameter) wheels. To the nearest revolution, how many revolutions does the wheel make when the car travels half a mile?

**13.** FORESTRY At a distance of 1000 ft, a tree subtends an angle of 2.5° to the eye of a forest ranger. What is the approximate height of the tree?

**14.** ASTRONOMY An astronomer measures the angle subtended by the width of the sun and obtains 0.5°. What is the approximate diameter of the sun if the distance between the sun and the earth is $9.3 \times 10^7$ miles?

**15.** ASTRONOMY Assume that the earth is stationary and that the moon travels in a circular orbit with radius 240,000 mi around the earth. What distance does the moon travel when a line joining the earth and moon sweeps out an angle of 30°?

**16.** ENGINEERING A pulley with a 3-inch radius is connected to a pulley with a 5-inch radius by a belt. Assuming no slippage, how many revolutions will the smaller pulley make if the larger revolves 1 time?

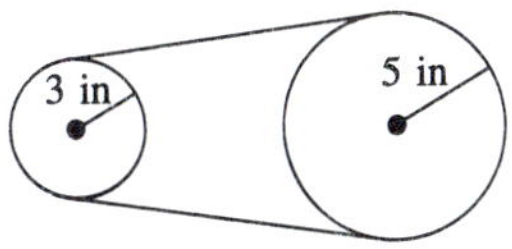


**Exercise 16**

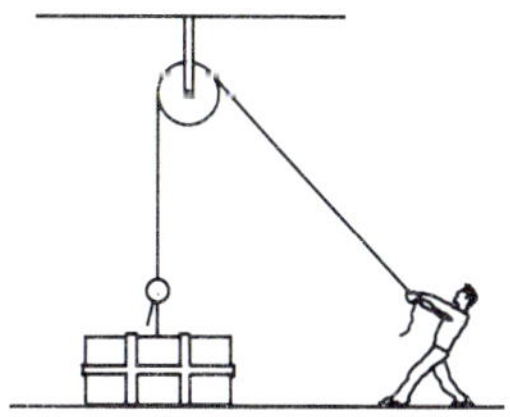

**Exercise 17**

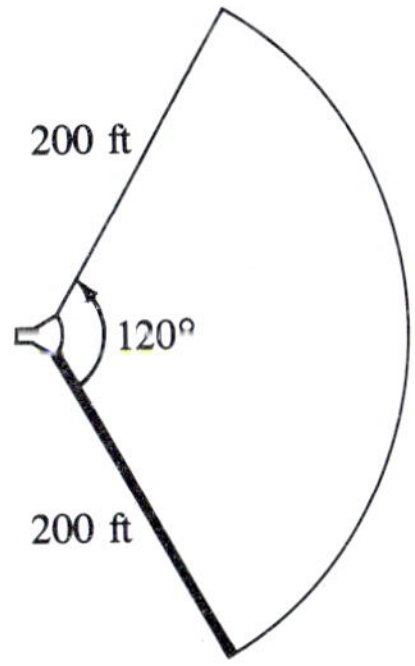


**Exercise 20**

**17.** ENGINEERING A man pulls 15 ft of rope through a pulley 16 inches in diameter in order to lift a metal crate. Approximately how many revolutions will the pulley make if there is no slippage?

**18.** NAVIGATION Most forms of navigation use *nautical miles* instead of *statute miles* (5280 ft). A nautical mile is the length of an arc on the surface of the earth subtended by a central angle of measure $1'$. Assuming that the radius of the earth is 4000 statute miles, what is the approximate length of a nautical mile in feet?

**19.** GEOMETRY Let $A$ and $B$ be two points on a circle with center $O$ and radius $r$. The **sector** $AOB$ is the region bounded by the arc $\overset{\frown}{AB}$ and the two radii $OA$ and $OB$. If the central angle, $\angle AOB$, has measure $\theta$ radians, show that the area of the sector is given by $A = \frac{1}{2}r^2\theta$.

**20.** PHYSICS A floodlight has a range of 200 ft and casts its beam through an angle of $\frac{2\pi}{3}$ radians. Approximately how many square feet are illuminated by the light?

**21.** ENGINEERING A wheel with a 30.0-inch diameter is rotating at $5\pi$ radians per second. What is the linear velocity of a point on the rim of the wheel in feet per second?

**22.** RECREATION A child on the outside horse is 15.0 ft from the center of a merry-go-round. If the merry-go-round is turning at a rate of 12 rpm, at what velocity, in feet per second, is the child traveling?

**23.** ENGINEERING A flywheel is 6.0 ft in diameter and is turning at a rate of 30 rpm. Find the speed of the belt that drives the wheel, in feet per second, if there is no slippage.

**24.** ENGINEERING A pulley 12 inches in diameter is driven by a belt moving at 2000 ft/min. How many revolutions per minute are made by the pulley?

**25.** GEOGRAPHY Find the linear speed due to the rotation of the earth of a point near the Amazon River located at the equator. Use 4000 miles as the radius of the equator and express the result in miles per hour correct to the nearest mile.

**26.** ENGINEERING One gear of radius $r$ drives a second gear of radius $R$. Show that the angular velocity of the second gear is $\frac{r}{R}$ times the angular velocity of the driving gear. See the figure.

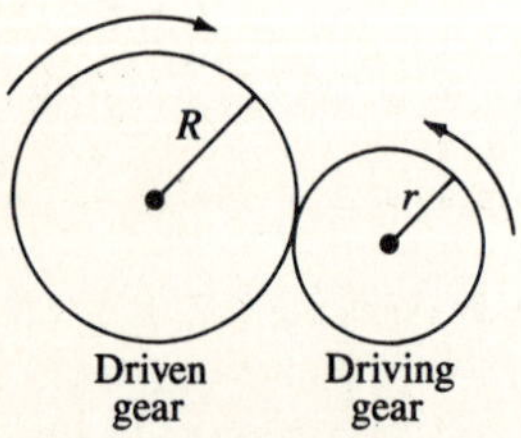


**Exercise 26**

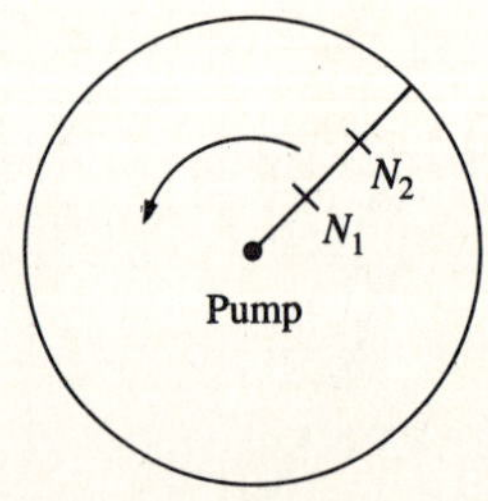


**Exercise 27**

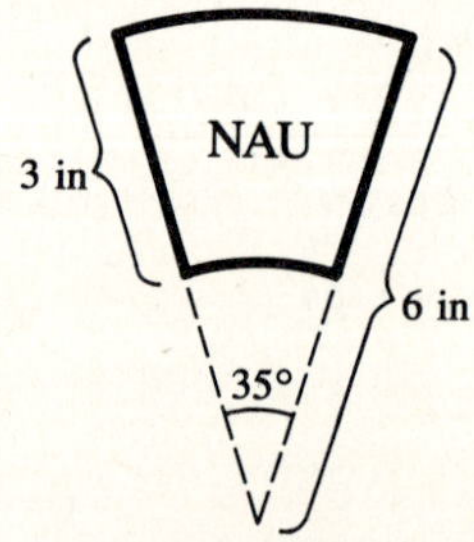


**Exercise 28**

**27.** AGRICULTURE The arm on a circular sprinkler irrigation system (see the sketch) on a Kansas farm revolves around a pump, making 1 revolution every 5 hours. If distances from the pump to two nozzles $N_1$ and $N_2$ are 50 yds and 75 yds, respectively, what is the linear velocity of each nozzle in yards per minute?

**28.** MANUFACTURING A company that manufactures team logo patches receives an order for 500 patches from the NAU booster club. If each patch is to be made according to the sketch, and is to have gold ribbon stitched around its border, how many feet of ribbon are needed to fill the order?

## For Review

**29.** Convert 4.5 radians to degrees, minutes, and seconds.

**30.** Convert $-85°42'18''$ to radians, correct to four decimal places.

# 6.3 Trigonometric Functions of an Acute Angle

## Right Triangles

In section 6.1 we developed two different ways of measuring angles, degrees and radians. Now we are ready to define six trigonometric functions of an acute angle in a right triangle. A **right triangle** is a triangle containing a right angle (90°); the side opposite the right angle is called the **hypotenuse** of the triangle. The remaining two angles are complementary acute angles, and the sides opposite these angles are called **legs** of the triangle.

We often label the angles of a triangle with capital letters. The sides of the triangle are indicated by lower-case letters corresponding to the capital letter of the angle opposite the side. For example, the right triangle with angles $A$, $B$, and $C$ and sides $a$, $b$, and $c$ is shown in Figure 10. We shall agree (unless specified otherwise) to identify the right angle in a triangle with the letter $C$ (⊿ is used to denote a right angle). The side **opposite** $A$ is $a$, and the side **adjacent** to $A$ is $b$.

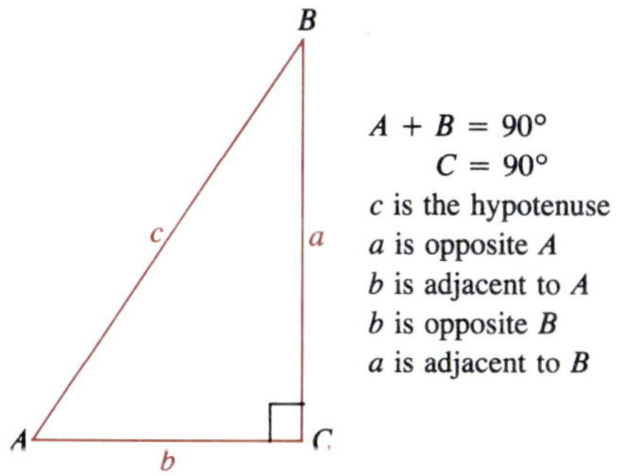


**Figure 10** **Right Triangle**

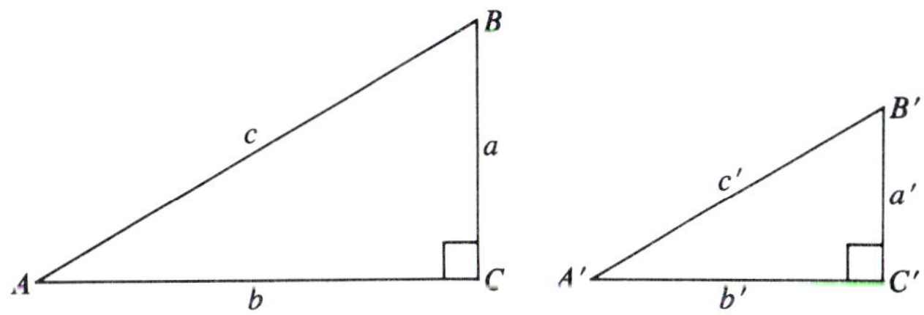


**Figure 11** **Similar Right Triangles**

## Similar Right Triangles

Two triangles are **similar** if the angles of one triangle are equal to the corresponding angles of the other. As a result, two right triangles are similar if an acute angle of one is equal to an acute angle of the other. A theorem in plane geometry states that the corresponding sides of similar triangles are proportional. Suppose that two right triangles $ABC$ and $A'B'C'$ are similar as shown in Figure 11. Then $a/c = a'/c'$, $b/c = b'/c'$, $a/b = a'/b'$, $b/a = b'/a'$, $c/b = c'/b'$, and $c/a = c'/a'$.

## The Six Trigonometric Functions

The six possible ratios of the lengths of the sides of a right triangle are given particular names relative to the acute angle $A$. Since the six ratios are independent of the size of the right triangle containing $A$, these ratios can be used to define six **trigonometric functions** of an acute angle, the **sine, cosine, tangent, cotangent, secant,** and **cosecant** functions.

### The Six Trigonometric Functions

Let $A$ be an acute angle in right triangle $ABC$. (See Figure 12.)

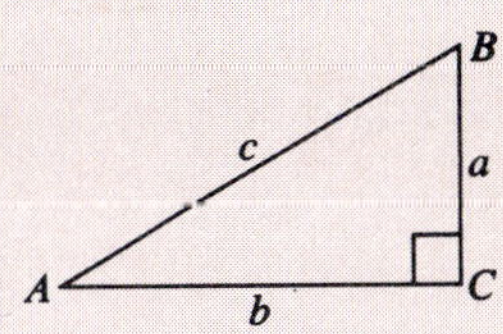


**Figure 12**

| Function name | Abbreviation | Value of function at $A$ |
|---|---|---|
| **sine** | **sin** | $\sin A = \dfrac{\text{opposite side}}{\text{hypotenuse}} = \dfrac{a}{c}$ |
| **cosine** | **cos** | $\cos A = \dfrac{\text{adjacent side}}{\text{hypotenuse}} = \dfrac{b}{c}$ |
| **tangent** | **tan** | $\tan A = \dfrac{\text{opposite side}}{\text{adjacent side}} = \dfrac{a}{b}$ |
| **cotangent** | **cot** | $\cot A = \dfrac{\text{adjacent side}}{\text{opposite side}} = \dfrac{b}{a}$ |
| **secant** | **sec** | $\sec A = \dfrac{\text{hypotenuse}}{\text{adjacent side}} = \dfrac{c}{b}$ |
| **cosecant** | **csc** | $\csc A = \dfrac{\text{hypotenuse}}{\text{opposite side}} = \dfrac{c}{a}$ |

Since $\sin A = \frac{a}{c}$ and $\csc A = \frac{c}{a}$

$\sin A$ and $\csc A$ are reciprocals. That is,

$$\sin A = \frac{a}{c} = \frac{1}{\frac{c}{a}} = \frac{1}{\csc A}$$

and

$$\csc A = \frac{c}{a} = \frac{1}{\frac{a}{c}} = \frac{1}{\sin A}.$$

Similarly, $\cos A = 1/\sec A$ and $\sec A = 1/\cos A$, and also $\tan A = 1/\cot A$ and $\cot A = 1/\tan A$. Thus,

sine and cosecant are reciprocals,
cosine and secant are reciprocals,
tangent and cotangent are reciprocals.

Since the sides of a triangle may be labeled with different letters, it is best to memorize the ratios using the names for the sides rather than the letters associated with the sides. Thus, relative to acute angle $B$ in the triangle in Figure 12,

$$\sin B = \frac{\text{opposite side}}{\text{hypotenuse}} = \frac{b}{c} \qquad \csc B = \frac{\text{hypotenuse}}{\text{opposite side}} = \frac{c}{b}$$

$$\cos B = \frac{\text{adjacent side}}{\text{hypotenuse}} = \frac{a}{c} \qquad \sec B = \frac{\text{hypotenuse}}{\text{adjacent side}} = \frac{c}{a}$$

$$\tan B = \frac{\text{opposite side}}{\text{adjacent side}} = \frac{b}{a} \qquad \cot B = \frac{\text{adjacent side}}{\text{opposite side}} = \frac{a}{b}.$$

Since $\frac{b}{c}$ is both $\sin B$ and $\cos A$, it is clear that ratios should not be memorized using letters only.

**EXAMPLE 1** In right triangle $ABC$, $a = 1$ and $b = 3$. Determine the six trigonometric ratios for $A$ and $B$.

First, make a sketch like the one in Figure 13 describing the problem. We can calculate the hypotenuse $c$ using the Pythagorean theorem. Substitute $a = 1$ and $b = 3$ into

$$c^2 = a^2 + b^2.$$

$$c^2 = \mathbf{1}^2 + \mathbf{3}^2$$

$$= 1 + 9 = 10$$

$$c = \sqrt{10} \quad c \text{ cannot be } -\sqrt{10}$$

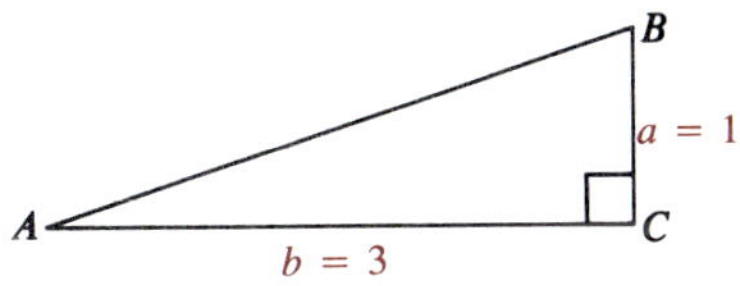


**Figure 13**

$$\sin A = \frac{\text{opp}}{\text{hyp}} = \frac{1}{\sqrt{10}} \qquad \sin B = \frac{\text{opp}}{\text{hyp}} = \frac{3}{\sqrt{10}}$$

$$\cos A = \frac{\text{adj}}{\text{hyp}} = \frac{3}{\sqrt{10}} \qquad \cos B = \frac{\text{adj}}{\text{hyp}} = \frac{1}{\sqrt{10}}$$

$$\tan A = \frac{\text{opp}}{\text{adj}} = \frac{1}{3} \qquad \tan B = \frac{\text{opp}}{\text{adj}} = \frac{3}{1}$$

$$\csc A = \frac{\text{hyp}}{\text{opp}} = \frac{\sqrt{10}}{1} \qquad \csc B = \frac{\text{hyp}}{\text{opp}} = \frac{\sqrt{10}}{3}$$

$$\sec A = \frac{\text{hyp}}{\text{adj}} = \frac{\sqrt{10}}{3} \qquad \sec B = \frac{\text{hyp}}{\text{adj}} = \frac{\sqrt{10}}{1}$$

$$\cot A = \frac{\text{adj}}{\text{opp}} = \frac{3}{1} \qquad \cot B = \frac{\text{adj}}{\text{opp}} = \frac{1}{3}$$ ■

**EXAMPLE 2** Suppose $\theta$ is an acute angle and $\tan \theta = \frac{3}{4}$. Find the remaining trigonometric function values of $\theta$.

Make a sketch of a right triangle with the property that the side opposite $\theta$ is 3 and the side adjacent to $\theta$ is 4 as in Figure 14. By the Pythagorean theorem,

$$(\text{hyp})^2 = 3^2 + 4^2 = 25$$
$$\text{hyp} = 5.$$

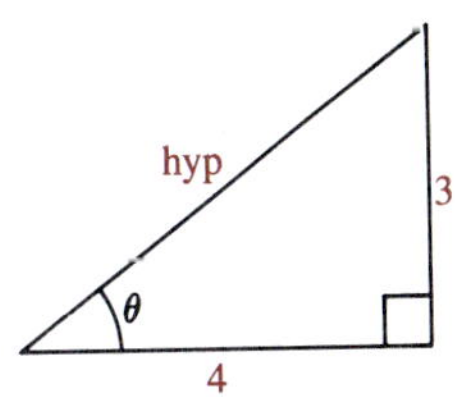


**Figure 14**

The definitions of the trigonometric functions give

$$\sin \theta = \frac{3}{5} \qquad \csc \theta = \frac{5}{3}$$

$$\cos \theta = \frac{4}{5} \qquad \sec \theta = \frac{5}{4}$$

$$\tan \theta = \frac{3}{4} \qquad \cot \theta = \frac{4}{3}.$$ ■

Three special acute angles, 30° $\left(\frac{\pi}{6} \text{ radians}\right)$, 45° $\left(\frac{\pi}{4} \text{ radians}\right)$, and 60° $\left(\frac{\pi}{3} \text{ radians}\right)$ have trigonometric function values that are easy to find. These values are given in the following two examples.

**EXAMPLE 3** Determine the six trigonometric function values of 45°.

Suppose we sketch a right triangle with a 45° acute angle. Since the other acute angle is the complement of 45°, it too is 45°, making the triangle **isosceles** (having two sides equal). Since the size of the triangle is unimportant, suppose the two equal legs have measure 1, as shown in Figure 15. Using the Pythagorean theorem,

$$1^2 + 1^2 = c^2$$
$$2 = c^2$$
$$\sqrt{2} = c.$$

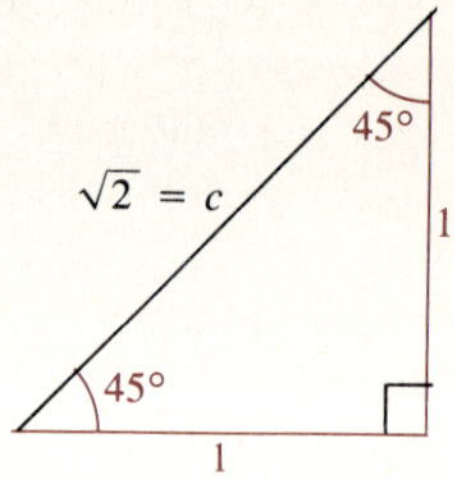


Figure 15

Thus,

$$\sin 45° = \frac{1}{\sqrt{2}} = \frac{\sqrt{2}}{2} \qquad \csc 45° = \frac{\sqrt{2}}{1} = \sqrt{2}$$

$$\cos 45° = \frac{1}{\sqrt{2}} = \frac{\sqrt{2}}{2} \qquad \sec 45° = \frac{\sqrt{2}}{1} = \sqrt{2}$$

$$\tan 45° = \frac{1}{1} = 1 \qquad \cot 45° = \frac{1}{1} = 1$$ ■

**EXAMPLE 4** Determine the six trigonometric function values of 30° and 60°.

To obtain a triangle with acute angles 30° and 60°, start with an equilateral triangle (having all sides equal) with sides 2 units in length. The median from one vertex to the opposite side separates the triangle into two right triangles with the desired acute angles as shown in Figure 16.

The length of the median, $\sqrt{3}$, can be found using the Pythagorean theorem.

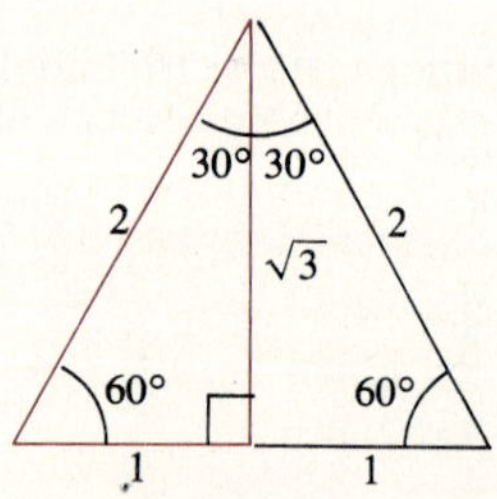


Figure 16

Concentrating on the 30°-60° right triangle shown in color in Figure 16, we can use the definitions of the six trigonometric functions to obtain

$$\sin 30° = \frac{1}{2} = \cos 60° \qquad \csc 30° = 2 = \sec 60°$$

$$\cos 30° = \frac{\sqrt{3}}{2} = \sin 60° \qquad \sec 30° = \frac{2}{\sqrt{3}} = \frac{2\sqrt{3}}{3} = \csc 60°$$

$$\tan 30° = \frac{1}{\sqrt{3}} = \frac{\sqrt{3}}{3} = \cot 60° \qquad \cot 30° = \sqrt{3} = \tan 60°$$ ■

**NOTE** The reason that the sine and **co**sine, secant and **co**secant, and tangent and **co**tangent are defined as **co**functions becomes clear in view of Example 4. The value of a trigonometric function of an acute angle (such as 30°) is the same as the value of the cofunction of its complement (which is 60°). ■

From the definitions of the functions and the fact that the hypotenuse of a right triangle is longer than either leg, the following theorem is obvious.

Let $\theta$ be an acute angle. Then

| | | | |
|---|---|---|---|
| **1.** $\sin \theta < 1$ | **2.** $\cos \theta < 1$ | **3.** $\csc \theta > 1$ | **4.** $\sec \theta > 1$ |

One way to describe a function is to construct a table of values. Table 2 (at the back of this book) gives the values of the six trigonometric functions of angles in 10′ increments from 0° to 90°. Prior to the availability of calculators such tables provided the most convenient way to find trigonometric function values of angles other than the special angles of 30°, 45°, and 60°. Since calculators have made tables almost obsolete, our presentation is geared toward calculator use. For the student who wishes to use Table 2, Appendix A gives a brief introduction to that method.

To find trigonometric function values with a calculator, place it in the correct mode for degrees or radians. On many calculators, this will involve pressing the [DRG] key repeatedly until the desired mode "DEG" for degrees or "RAD" for radians appears near the bottom of the display (ignore "GRAD" and press [DRG] again to return to "DEG"). Some calculators have both [DEG] and [RAD] keys. Refer to your owner's manual if your calculator does not have a [DRG] key. In addition, many calculators compute degrees only in decimal form and cannot accept minutes and seconds. The [sin], [cos], and [tan] keys are used to find the sine, cosine, and tangent of an angle (in either degree or radian mode). To find the cosecant, secant or cotangent of an angle, use the [1/x] key, since

$$\csc \theta = \frac{1}{\sin \theta}, \quad \sec \theta = \frac{1}{\cos \theta}, \quad \text{and } \cot \theta = \frac{1}{\tan \theta}.$$

The next example will make the procedure clear.

**EXAMPLE 5** Use a calculator to find each trigonometric function value.

**(a)** sin 60°
First we make sure the calculator has been set in degree mode.

ALG & RPN: 60 [sin] → [0.8660254]

We obtain a seven-place decimal approximation of the exact value, $\frac{\sqrt{3}}{2}$, obtained in Example 4.

**(b)** cos 29°42′
We must first convert 29°42′ to decimal form.

ALG: 42 [÷] 60 [+] 29 [=] [cos] → [0.8686315]

RPN: 42 [ENTER] 60 [÷] 29 [+] [cos] → [0.8686315]

**(c)** sec 71°
First find cos 71° and then take its reciprocal.

ALG & RPN: 71 [cos] [1/x] → [3.0715535]

Thus, sec 71° ≈ 3.0715535. Another calculator gives 3.0715534. This is due to internal rounding differences in the calculators and should not be of concern.

**(d)** tan 0.8563
First place the calculator in radian mode.

ALG & RPN: 0.8563 [tan] → [1.1529009]

Again, a different calculator might give a slightly different value. ■

We now know how to find a trigonometric function value of a given acute angle. Suppose we reverse the process and find the acute angle with a given trigonometric function value. The process is similar to the method in Section 5.4 for finding antilogarithms. While most calculators use the [INV] key followed by the appropriate [sin], [cos], or [tan] key, some calculators use a key labeled [$\sin^{-1}$] or [arcsin] for the two keys [INV] and [sin]. Similar remarks apply to the cosine and tangent functions.

**EXAMPLE 6** Use a calculator to find each acute angle with the given trigonometric function value.

**(a)** $\sin \theta = 0.5948$
First decide whether $\theta$ is to be in degrees or radians and set the appropriate mode. Suppose we find $\theta$ in degrees first. The following sequence yields the desired result.

ALG & RPN: 0.5948 [INV] [sin] → [36.498376]

Thus, $\theta \approx 36°30'$. Setting the calculator in radian mode, the same sequence gives $\theta = 0.6370$, correct to four decimal places.

**(b)** $\csc\theta = 1.2011$
Keeping the calculator in degree mode, the first step is to use the [1/x] key to find $\sin\theta$.

ALG & RPN: 1.2011 [1/x] [INV] [sin] → [56.363666]

Thus, $\theta = 56.363666° = 56°21'49''$

**(c)** $\cot\theta = 3.5339$
This time we find $\theta$ in radians.

ALG & RPN: 3.5339 [1/x] [INV] [tan] → [0.2757639]

Thus, $\theta = 0.2758$, correct to four decimal places. ■

## 6.3 EXERCISES

In Exercises 1–6 two sides of a right triangle $ABC$ ($C$ is the right angle) are given. Determine the six trigonometric function values of angles $A$ and $B$.

**1.** $a = 5$ and $b = 12$

**2.** $a = 3$ and $b = 10$

**3.** $a = 7$ and $c = 15$

**4.** $b = 7$ and $c = 20$

**5.** $a = 0.6$ and $b = 0.8$

**6.** $a = \sqrt{5}$ and $b = \sqrt{11}$

In Exercises 7–12 $\theta$ is an acute angle with the given trigonometric function value. Sketch an appropriate right triangle and find the six trigonometric function values of $\theta$ and $90°$-$\theta$.

**7.** $\sin\theta = \frac{5}{7}$

**8.** $\csc\theta = \frac{5}{2}$

**9.** $\sec\theta = 7$

**10.** $\cot\theta = 6$

**11.** $\cos\theta = \frac{\sqrt{2}}{3}$

**12.** $\sin\theta = \frac{\sqrt{105}}{21}$

**13.** Without using a calculator, find the trigonometric function values of $30°\left(\frac{\pi}{6}\right)$, $45°\left(\frac{\pi}{4}\right)$, and $60°\left(\frac{\pi}{3}\right)$, and complete the following table.

| Function | $30°\left(\frac{\pi}{6}\right)$ | $45°\left(\frac{\pi}{4}\right)$ | $60°\left(\frac{\pi}{3}\right)$ |
|---|---|---|---|
| sine | | | |
| cosine | | | |
| tangent | | | |
| cosecant | | | |
| secant | | | |
| cotangent | | | |

Without using a calculator, evaluate each expression in Exercises 14–19.

**14.** $\sin 30° + \cos 30°$ **15.** $\sin 60° + \cos 60°$ **16.** $(\sin 60°)(\csc 60°)$

**17.** $(\cos 45°)(\sec 45°)$ **18.** $(\tan 45°)(\cot 45°)$ **19.** $(\sin 45°)(\cos 45°)$

In Exercises 20–23 $\theta$ is an acute angle. Determine whether the given statement is true or false.

**20.** $0 < \cos \theta < 1$ **21.** $\csc \theta < 1$ **22.** $\sec \theta > 1$

**23.** If $0 < \theta < \frac{\pi}{4}$, then $\tan \theta < 1$.

In Exercises 24–32 sketch a right triangle $ABC$ with corresponding legs $a$ and $b$ and hypotenuse $c$. Use this triangle to determine whether the following are true or false.

**24.** $\tan A = \cot B$ **25.** $c = a \sin A$ **26.** $a = b \tan A$

**27.** $\sin A \csc A = 1$ **28.** $\cos B \sec B = 1$ **29.** $\tan A \cot B = 1$

**30.** $\sin^2 A + \cos^2 A = 1$
[Note: $\sin^2 A$ represents $(\sin A)^2$] **31.** $1 + \tan^2 B = \sec^2 B$ **32.** $1 + \cot^2 A = \csc^2 A$

In Exercises 33–44 use a calculator to find each trigonometric function value correct to four decimal places.

**33.** $\sin 65°$ **34.** $\cos 5°$ **35.** $\cot 72°$ **36.** $\csc 18°$

**37.** $\cos 42°38'$ **38.** $\tan 71°14'$ **39.** $\sec 0.7159$ **40.** $\sin 0.8439$

**41.** $\tan 1.4327$ **42.** $\cot 1.2305$ **43.** $\csc \frac{\pi}{5}$ **44.** $\sec \frac{\pi}{7}$

In Exercises 45–52 use a calculator to find each acute angle with the given trigonometric function value. Give the angle to the nearest hundredth of a degree.

**45.** $\sin \theta = 0.8319$ **46.** $\cos \theta = 0.9758$ **47.** $\tan \theta = 4.0015$ **48.** $\csc \theta = 2.3541$

**49.** $\sec \theta = 3.2711$ **50.** $\cot \theta = 7.1559$ **51.** $\cos \theta = 0.5511$ **52.** $\sin \theta = 0.8995$

In Exercises 53–60 use a calculator to find each acute angle with the given trigonometric function value. Give the answer in radians, correct to four decimal places.

**53.** $\sin \theta = 0.5718$ **54.** $\cos \theta = 0.1295$ **55.** $\tan \theta = 3.1560$ **56.** $\csc \theta = 2.5619$

**57.** $\sec \theta = 4.4554$ **58.** $\cot \theta = 6.1397$ **59.** $\cos \theta = 0.8415$ **60.** $\sin \theta = 0.2219$

In Exercises 61–64 use a calculator to find the six trigonometric function values of the given acute angle, correct to four decimal places.

**61.** $\theta = 47°13'54''$ **62.** $\theta = 13°05'27''$ **63.** $\theta = 82°31'46''$ **64.** $\theta = 2°52'27''$

In Exercises 65–66 use a calculator to find **(a)** $\sin^2 \theta$, **(b)** $\cos^2 \theta$, **(c)** $\tan^2 \theta$, **(d)** $\csc^2 \theta$, **(e)** $\sec^2 \theta$, and **(f)** $\cot^2 \theta$, correct to four decimal places.

**65.** $\theta = 0.8563$ **66.** $\theta = 35°18'58''$

**67.** **PHYSICS** A ramp of length $L$ ft is inclined at an angle $\theta$, measured in degrees, with the horizontal. The work $W$, in foot-pounds, required to move an object which weighs $P$ lb up the ramp is given by $W = PL \sin \theta$. How much work is involved in pushing a 100-pound storage chest up a 30-foot ramp inclined at an angle of $15°30'$? See the figure.

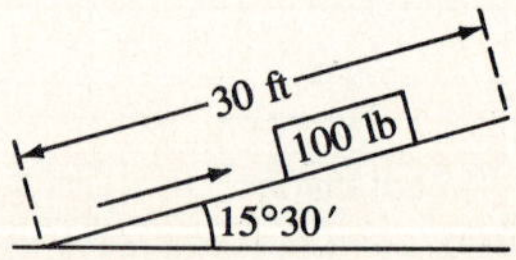

**68.** ASTRONOMY The rate $R$, in miles per second, at which the earth is traveling around the sun can be approximated by $R = L \cot 89.99°$, where $L$ is the speed of light, approximately 186,000 mi/sec. Find the speed of the earth around the sun.

## For Review

**69.** Find the length of an arc on a circle with radius 2.3 ft which subtends a central angle of $72°50'$.

**70.** Find the number of degrees (to the nearest hundredth) in the central angle which subtends an arc of 6 inches on a circle of diameter 8 ft.

**71.** AERONAUTICS The propeller of an airplane is 7 ft in diameter. What is the linear velocity in ft/sec of the tip of a propeller blade if the angular velocity of the propeller is 45 radians per second?

# 6.4 Right Triangles and Applications

## Solving Right Triangles

The six parts of any triangle (three angles and three sides) can often be determined when the measures of certain of these parts are given. The process is called **solving the triangle.** A right triangle can be *solved* when two of its sides are known or when one side and one acute angle are known. Simply knowing the angles of a triangle is not enough for solving the triangle since any similar triangle will have the same angles but, more than likely, different sides.

## EXAMPLE 1

Solve the right triangle in which $a = 3$ and $b = 7$.

First, draw a reasonably accurate sketch of the triangle and label the parts as in Figure 17. Since $C$ is always the 90° angle, we must find $A$, $B$, and $c$.

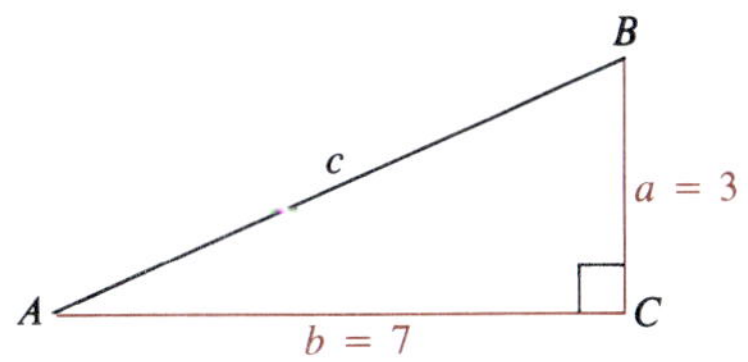


**Figure 17**

*To find A:* $\tan A = \dfrac{\text{opp}}{\text{adj}} = \dfrac{3}{7}$

ALG: 3 [÷] 7 [=] [INV] [tan] → [23.198591]

RPN: 3 [ENTER] 7 [÷] [INV] [tan] → [23.198591]

To the nearest degree, $A = 23°$.

*To find B:* $B = 90° - A = 90° - 23° = 67°$

Angle $B$ could also be found by using, for example, $\tan B = 7/3$, $B = 67°$ (to the nearest degree). Obviously it is easier to find $B$ using the fact that $A$ and $B$ are complementary. However, if an error is made in finding $A$, we will compound the error if $B$ is found using the value of $A$.

*To find c:* We can find $c$ using the Pythagorean theorem.

$$c^2 = a^2 + b^2$$
$$= 3^2 + 7^2 = 58$$
$$c = \sqrt{58} \approx 7.6 \quad ■$$

## Accuracy of Answers

Example 1 illustrates the procedure for solving a right triangle. It also points out a situation that requires some agreement before we proceed. Giving $c$ as 7.6 (rounded to nearest tenth) seems a bit presumptuous since the values of $a$ and $b$ were given only to the nearest integer. If these numbers were approximate measurements, for example, we would be claiming that our calculated value is more accurate than the values used to calculate it!

A similar problem exists for the values of the acute angles. What level of precision should be required (nearest degree, nearest 10′, nearest minute) for an angle, considering the sides from which it is calculated? In the example, angles $A$ and $B$ were given to the nearest degree. We could have given the angles to a greater degree of accuracy, but would this have made sense in view of the measures given for the sides? To be consistent, we will agree to the following rule.

In the solution to a right triangle, sides and angles should be given according to the following:

1. All sides are rounded to the same number of decimal places as the least precise measurement.
2. Angles will be given in degrees and minutes according to the table below.

| Decimal places on a side | Examples | Angles to the nearest | Examples |
|---|---|---|---|
| 0 | 6 cm | degree | 23° |
| 1 | 4.3 ft | 10′ or 0.1° | 41°20′ or 41.3° |
| 2 | 7.41 m | 1′ or 0.01° | 48°25′ or 48.42° |

## EXAMPLE 2

Solve the right triangle in which $c = 8.12$ and $B = 72°35′$.

We must find $a$, $b$, and $A$. The triangle is sketched in Figure 18.

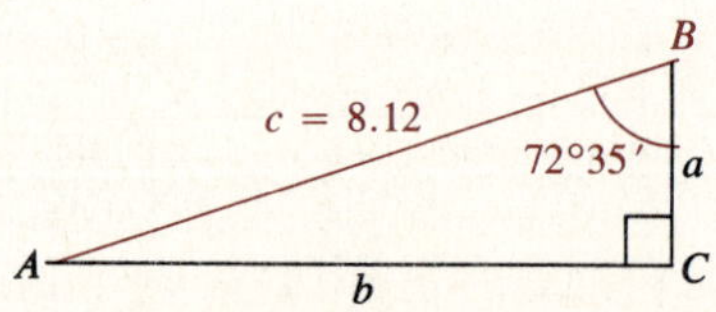


**Figure 18**

*To find A:* $A = 90°00' - B$
$= 90°00' - 72°35'$
$= 17°25'$

*To find a:* $\cos B = \dfrac{a}{c}$

$\cos 72°35' = \dfrac{a}{8.12}$

Thus, $a = (8.12)\cos 72°35' = 2.43$

ALG: 35 [÷] 60 [+] 72 [=] [STO] [cos] [×] 8.12 [=] → [2.4304651]
RPN: 35 [ENTER] 60 [÷] 72 [+] [STO] [cos] 8.12 [×] → [2.4304651]

*To find b:* $\sin 72°35' = \dfrac{b}{8.12}$
Thus, $b = (8.12)\sin 72°35' = 7.75$.

ALG: [RCL] [sin] [×] 8.12 [=] → [7.7477248]
RPN: [RCL] [sin] 8.12 [×] → [7.7477248] ■

NOTE When we found $a$ in Example 2, the calculator step [STO] placed the angle 72°35′ in memory. This value was recalled, [RCL], and used for finding $b$ in the next step. ■

## Applications of Right Triangles

Next we investigate a variety of applied problems, all of which depend in some way on finding unknown parts of a right triangle. In most situations we shall be interested in a process which might be called *indirect measurement:* finding a particular length, distance, or angle without directly applying some measuring instrument such as a tape measure or protractor.

If an observer must look up to view an object, as in Figure 19, the angle formed by the line of sight and the horizontal is called the **angle of elevation** of the object. If an observer must look down to view an object, as in Figure 20, the angle formed by the line of sight and the horizontal is called the **angle of depression** of the object.

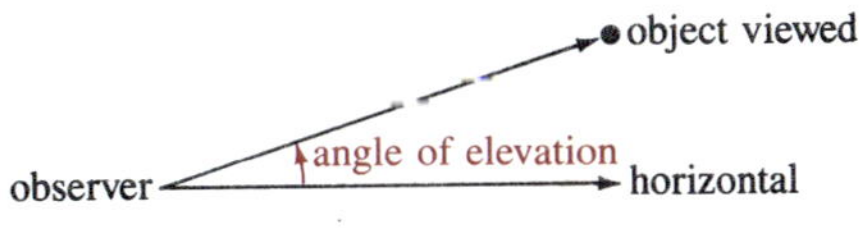


**Figure 19 Angle of Elevation**

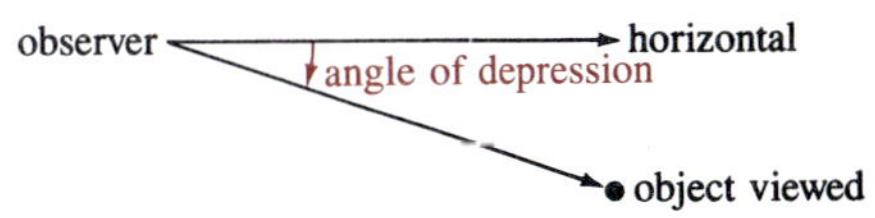


**Figure 20 Angle of Depression**

## EXAMPLE 3

A building casts a shadow 200 ft long when the angle of elevation of the top of the building, as seen from the end of the shadow, is 23°10′. Find the height of the building. What is the angle of elevation of the sun?

In Figure 21, $A$ is the angle of elevation of the top of the building, and also the angle of elevation of the sun. We must find side $a$ (the height of the building).

$$\tan A = \frac{a}{b} \quad \text{We know } A \text{ and } b \text{ so } a \text{ can be found}$$

$$\tan 23°10' = \frac{a}{200}$$

$$a = (200)\tan 23°10' \approx 85.58$$

The building is approximately 86 ft high. ■

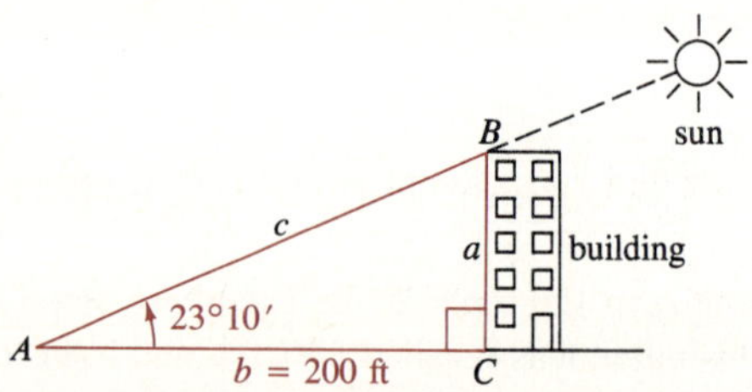


**Figure 21**

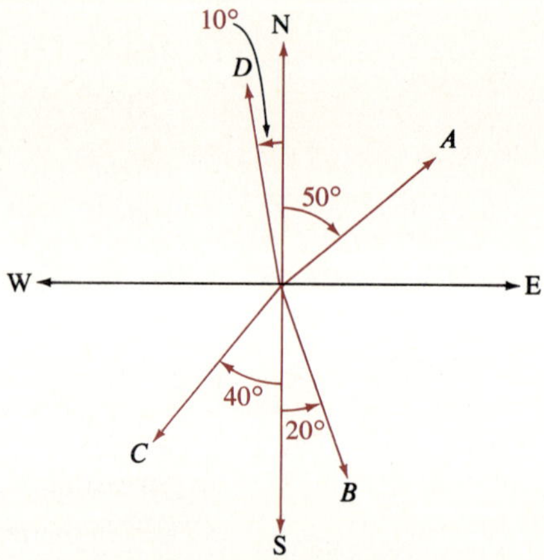


**Figure 22 Bearing**

One method of describing a particular direction from a given point is to use the notion of *bearing*. The **bearing** of an object from a point $O$ is the acute angle made by the north-south line and the line joining the object to $O$. To give the bearing, we first write the letter N or S, then the angle of deviation from north or south, and then the letter E or W.

In Figure 22 the bearing of point $A$ from $O$ is N50°E, the bearing of point $B$ from $O$ is S20°E, the bearing of point $C$ from $O$ is S40°W, and the bearing of point $D$ from $O$ is N10°W. Remember to start in the direction of the first letter (N or S) and sweep out the angle in the direction of the second letter (E or W).

We now solve the second problem in the introduction to this chapter.

## EXAMPLE 4

**FORESTRY**

A small forest fire is sighted due south of a fire lookout tower on Woody Mountain. From a second tower, located 11.0 mi due east of the first, the bearing of the fire is S51°10′W. How far is the fire from the tower on Woody Mountain?

Figure 23 exhibits the information given. Notice that S51°10′W designates $\angle CBF$ to be 51°10′, and $\angle FAB$ is a right angle. In triangle $ABF$, $\angle ABF = 38°50'$, since it is complementary to $\angle CBF$. The length of $AF$ gives the distance from the fire to the lookout on Woody Mountain.

$$\tan 38°50' = \frac{AF}{11.0}$$

$$AF = (11.0)\tan 38°50' \approx 8.9$$

Thus, the fire is approximately 8.9 miles from the lookout on Woody Mountain. ■

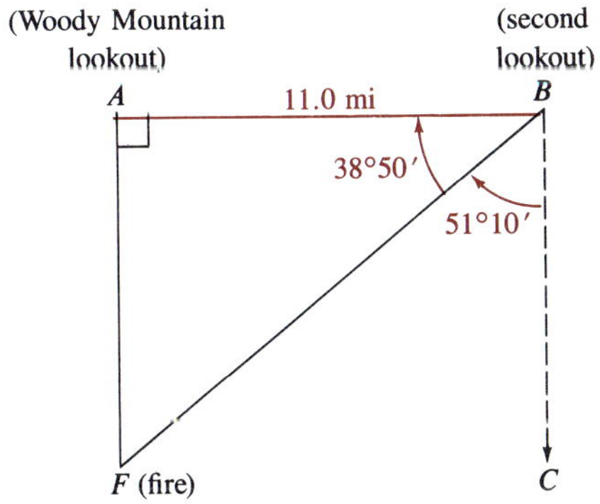


**Figure 23**

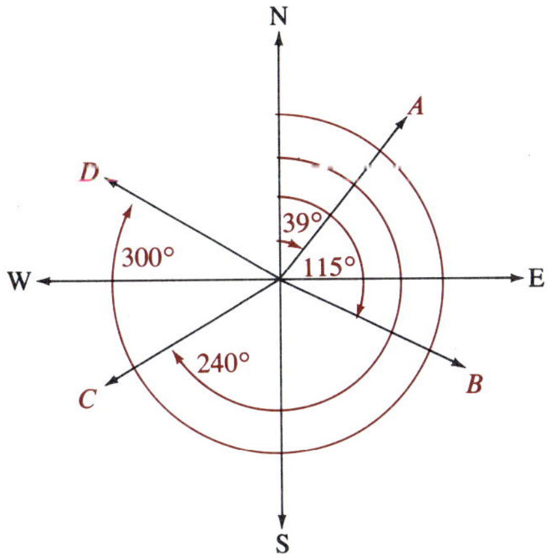


**Figure 24** **Aerial or Naval Navigation**

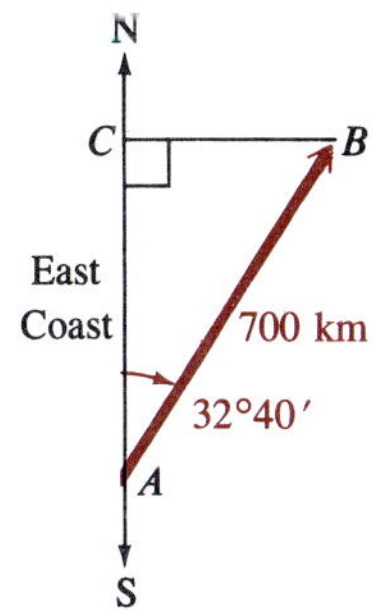


**Figure 25**

In **aerial** or **naval navigation,** directions are given using a single angle measured clockwise from the north. Thus a direction of 90° would be due east, 180° would be due south, 270° would be due west, and 0° would be due north.

In Figure 24, $A$ is in the direction 39°, $B$ is in the direction 115°, $C$ is in the direction 240°, and $D$ is in the direction 300°. The relationship between the navigational direction and the bearing of each of the points is shown in the following table.

| *Navigational Direction* | *Bearing* |
|---|---|
| 39° | N39°E |
| 115° | S65°E (180° − 115° = 65°) |
| 240° | S60°W (240° − 180° = 60°) |
| 300° | N60°W (360° − 300° = 60°) |

## EXAMPLE 5
### NAVIGATION

An airplane leaves the East Coast of the United States and flies for two hours at 350 km/hr in a direction of 32°40′. Assuming that the East Coast is a straight north-south line, how far is the airplane from the coast line?

The sketch in Figure 25 shows the essentials of the problem. Length $AB$ is 700 km ($d = rt = (350 \text{ km/hr})(2 \text{ hr}) = 700$ km), and we are asked to find the length of $BC$.

$$\sin 32°40' = \frac{BC}{700}$$

$$BC = 700 \cdot \sin 32°40' \approx 378$$

The airplane is approximately 378 km from the Coast. ■

When a boat moves through water at a constant rate of speed, waves are generated which extend back from the bow (front) of the boat at a fixed angle, as shown in Figure 26. If we are given any two of the three quantities—the velocity of the boat, the velocity of the **bow wave,** or the angle between the two waves—we can determine the third.

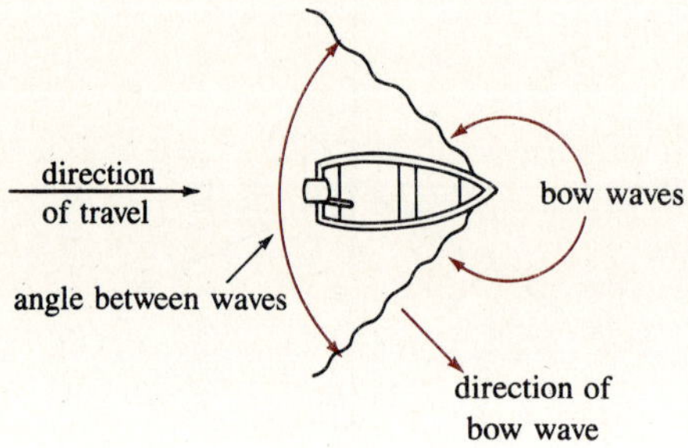


**Figure 26**

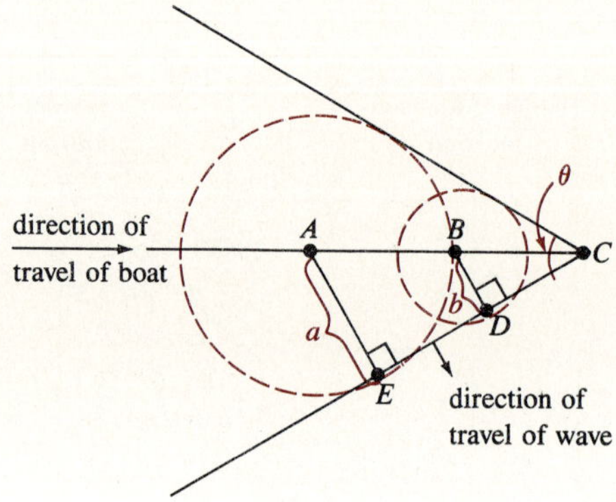


**Figure 27**

Assume that the wave radiates outward in a circular pattern and that the boat is moving in a straight line from $A$ to $B$ to $C$, as in Figure 27. By the time the boat reaches position $C$, the wave originating at $B$ will have traveled $b$ units and the wave originating at $A$ will have traveled $a$ units. Since the velocity of the boat, $v_b$, is constant, the velocity of the wave, $v_w$, is also constant, which means that points $C$, $D$, and $E$ are on the same straight line, so that the waves form straight lines which intersect in an angle of $\theta$.

If $t$ is the time of travel from $A$ to $C$, using the distance formula, $d = rt$, we have

$$AE = v_w \cdot t \quad \text{and} \quad AC = v_b \cdot t.$$

Since triangle $AEC$ is a right triangle and $\angle ACE = \frac{\theta}{2}$,

$$\sin \frac{\theta}{2} = \frac{AE}{AC} = \frac{v_w \cdot t}{v_b \cdot t} = \frac{v_w}{v_b}.$$

**EXAMPLE 6**

**NAVIGATION**

A boat is moving at a velocity of 30.0 mph in such a way that the bow waves are traveling at 14.0 mph. Find the angle formed by the waves.

We have $v_b = 30.0$ mph and $v_w = 14.0$ mph so that

$$\sin \frac{\theta}{2} = \frac{v_w}{v_b} = \frac{14}{30} = 0.4666667$$

$$\frac{\theta}{2} \approx 27°50'$$

Thus, the angle formed by the waves is $55°40'$. ■

**EXAMPLE 7**
NAVIGATION

A boat moving at a velocity of 40.0 km/hr generates bow waves which make an angle of 65°00′. Find the velocity of the waves.

Then $v_b = 40$ km/hr and $\theta = 65°00'$.

$$\sin \frac{65°00'}{2} = \frac{v_w}{40}$$

$$\sin 32°30' = \frac{v_w}{40}$$

$$40 \sin 32°30' = v_w$$

$$21.5 \approx v_w$$

Thus, the wave velocity is approximately 21.5 km/hr. ■

In exactly the same way, an airplane flying faster than the speed of sound generates sound waves, which form a cone behind the plane. When the cone intersects the surface of the Earth, the line of intersection approximates the shape of a hyperbola, as shown in Figure 28, along which a sonic boom is heard. Just as with the boat problem, if $\theta$ is the angle of the cone, $v_a$ the velocity of the airplane, and $v_s$ the velocity of the sound wave, then

$$\sin \frac{\theta}{2} = \frac{v_s}{v_a}.$$

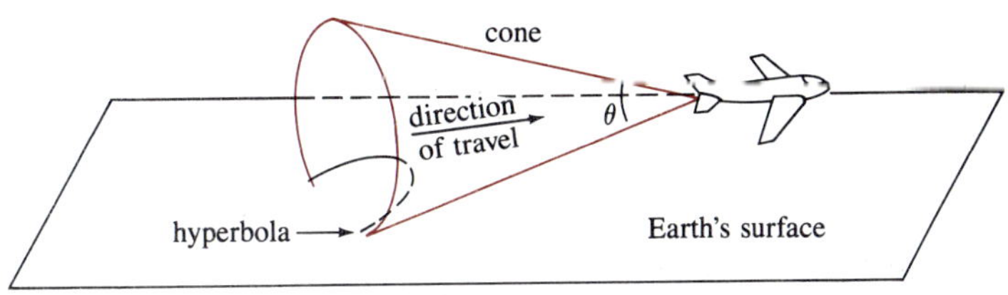


**Figure 28**

We all have experienced the sensation of inserting a straight object, such as a pole or pencil, into a body of water and observing that the object seems to bend at the point of intersection with the surface of the water. This phenomenon, due to refraction, or bending, of light waves, can be described by a simple equation involving the sine of two angles. Suppose that $a$ is the speed of light in air, $w$ is the speed of light in water, angles $A$ and $W$ are given as in Figure 29 on the next page, and light waves are traveling from point $P$ to point $Q$. René Descartes (1596–1650) showed that

$$\frac{a}{w} = \frac{\sin A}{\sin W}.$$

By using the concept of the **index of refraction,** the ratio of the speed of light in a vacuum to the speed of light in a given substance (such as air or water), it was later shown by Willebrod Snell (1591–1626) that if $I_a$ is the index of refraction for air and $I_w$ is the index of refraction for water, then

$$\frac{I_w}{I_a} = \frac{\sin A}{\sin W}. \quad \text{Note the change of order}$$

**EXAMPLE 8** Suppose that the index of refraction for water is 1.3 and for air is 1.0003. If a flashlight is shining on a lake in such a way that angle $A$ is 21°40′ (Figure 29), find angle $W$.

We have that $I_w = 1.3$, $I_a = 1.0003$, and $A = 21°40'$. Substitute these values and solve for $W$.

$$\frac{1.3}{1.0003} = \frac{\sin 21°40'}{\sin W}$$

$$\sin W = \frac{1.0003 \sin 21°40'}{1.3}$$

$$\sin W = 0.2840899$$

$$W \approx 16°30' \quad ■$$

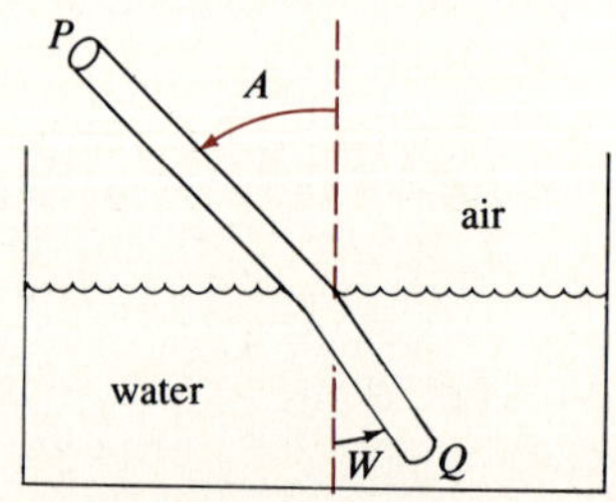

**Figure 29**

## 6.4 EXERCISES

Solve each right triangle in Exercises 1–12 by using given parts rather than calculated parts. Remember that $C = 90°$.

1. $a = 19$ and $b = 48$
2. $a = 4.3$ and $b = 6.1$
3. $a = 9.2$ and $c = 10.1$
4. $b = 7$ and $c = 15$
5. $A = 26°40'$ and $c = 12.0$
6. $A = 62°10'$ and $b = 7.25$
7. $b = 3.2$ and $B = 10°50'$
8. $c = 16.55$ and $B = 12°45'$
9. $A = 22°30'$ and $B = 67°30'$
10. $B = 41°20'$ and $A = 48°40'$
11. $c = 21.85$ and $B = 81°42'$
12. $a = 19.25$ and $A = 26°34'$

Solve.

13. When the angle of elevation of the sun is 48°50′, a pole casts a shadow 40.0 ft in length. What is the height of the pole?
14. GEOLOGY From a point in the Grand Canyon 3 miles from the base of the Redwall Limestone, but in the same horizontal plane, a geologist sights the top of the Redwall with angle of elevation 2°10′. How many feet thick is the Redwall Limestone formation at this point?
15. SURVEYING Two buildings are on either side of a street 70.0 ft apart. From the top of one, which is 80.0 ft high, the angle of elevation of the top of the second is 37°30′. How tall is the second building?
16. SURVEYING From a point 1200.0 ft from the base of Devil's Tower, the angle of elevation to the top of the columnar rock is 35°50′. How high is the tower?
17. NAVIGATION From the top of a tower 30 m above the level of a river and on the bank 15 m from the edge of the river, the angle of depression of a point directly across the river is 15°40′. Find the width of the river.
18. RECREATION From the top of Tower Butte, two boats are sighted (in the same straight line as the butte) on Lake Powell. The angles of depression of the two boats are 12°30′ and 26°10′. How far apart are the boats if Tower Butte rises 1580 ft above the water level?

19. **FORESTRY** From the top of a 100-ft tower, the angle of depression of a small forest fire is 5°40′. How far from the base of the tower is the fire?

20. **SURVEYING** The angle of depression of a house due north of a point on the ground directly below a balloon is 24°30′, while the angle of depression of another house due south of the point below the balloon is 42°10′. If the balloon is 1375 ft high, how far apart are the houses?

21. **FORESTRY** From Fire Lookout B, located 8.0 miles due south of Lookout A, a fire is observed. If the fire is due east of A, and has a bearing of N31°40′E from B, how far is the fire from B?

22. **GEOGRAPHY** Phoenix is 140.0 miles due south of Flagstaff, and Winslow is 60.0 miles due east of Flagstaff. What is the bearing of Winslow from Phoenix? Of Phoenix from Winslow?

23. **AERONAUTICS** An airplane leaves an airport and flies 200 km in a direction 140°. How far east of the airport is the airplane at this time? How far south?

24. **NAVIGATION** A ship sails 30.0 miles in a direction 322°20′. How far north has it sailed? How far west?

Two other angles related to aerial navigation are the **angle of climb** (the angle between an upward path of flight and a horizontal course) and the **angle of glide** (the angle between a downward flight and a horizontal course). The figure illustrates these concepts.

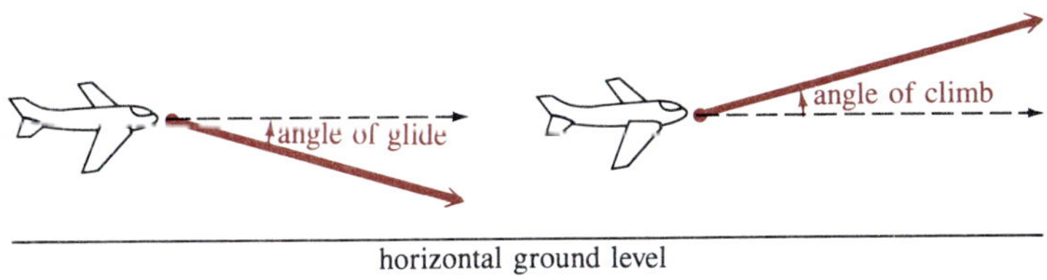


25. **AERONAUTICS** At an altitude of 2500 ft, the engine on a small plane suddenly fails. What angle of glide is needed to reach an airport runway 3 miles away?

26. **AERONAUTICS** The air speed of an airplane is 600 mph and its angle of climb is 12°20′. What is its ground speed? (Its speed relative to the ground is its horizontal speed.)

27. **AERONAUTICS** At an altitude of 2.0 miles, an airplane sets an angle of glide of 5°30′. How far will the plane travel before reaching the ground?

The acute angle made by a straight line and a horizontal line, as in the figure, is called the **angle of inclination** of the line.

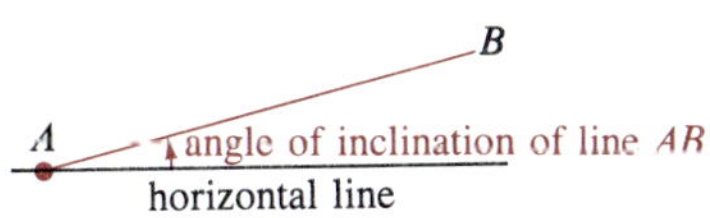


28. **ENGINEERING** A train tunnel through a mountain is 2000 m in length. If the angle of inclination of the track is 2°10′, what is the change in elevation from one end of the tunnel to the other?

**29. CONSUMER** If the distance between the two levels in the MetroCenter Shopping Center is 16.8 ft and the angle of inclination of the escalator is 32°10′, how far would a person travel while riding from the first to the second level?

**30. RECREATION** The chairlift at a ski resort has an average angle of inclination of 17°30′ with a vertical rise of 800 m. What is the approximate length of the ride to the top of the lift?

**31. NAVIGATION** A ship traveling at a constant velocity produces bow waves traveling at a rate of 8.25 km/hr. If the angle between the waves is 42°40′, find the velocity of the ship.

**32. RECREATION** A speedboat traveling in a straight line at a velocity of 25.0 mph generates bow waves traveling at 10.0 mph. What is the angle between the waves?

**33. AERONAUTICS** A sound wave produced by an airplane traveling at a supersonic speed travels at 820 mph. If the angle of the sonic cone is 78°40′, how fast is the airplane flying?

**34. AERONAUTICS** If an aircraft is flying at Mach two (twice the speed of sound), what will be the angle of the sonic cone? [Hint: $v_a = 2v_s$]

For Exercises 35–38 use 1.3 for the index of refraction of water, 1.0003 for the index of refraction of air, and Figure 29 as a reference.

**35. PHYSICS** If a light hits the surface of water in a pond such that $A = 35°50'$, find $W$.

**36. OCEANOGRAPHY** A diver using an underwater spotlight shines the light toward the surface in such a way that $W = 50°00'$. Find $A$.

**37.** Repeat Exercise 36 with $W = 50°10'$.

**38.** Repeat Exercise 36 with $W = 50°20'$. Explain what happens in this case and in cases when $W$ is even larger.

**39. SURVEYING** Two points $A$ and $B$ are on opposite sides of a lake as shown in the figure. A surveyor standing at point $P$, 1500 feet from $B$, measures $\angle APB$ and obtains 42°40′. How far apart are $A$ and $B$?

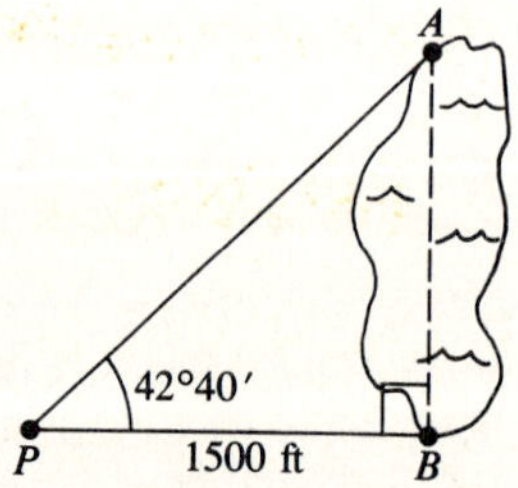


**40. METEOROLOGY** A beam of light is directed towards a cloud directly above a beacon on the ground. At a point 200 m from the beacon, the angle of elevation to the point where the light hits the cloud is 80°20′. How high is the cloud?

**41. NAVIGATION** A small boat leaves an island and sails due north for 2 hours at an average rate of 28.0 mph. It then changes course, sailing due east for 1 hour at the same rate before it runs aground on a reef. If a rescue ship sails from the island, what course should it set in order to reach the incapacitated boat? How long will it take to reach the boat if it travels 30.0 mph (to the nearest minute)?

**42.** A trucker is traveling due west on level ground at an elevation of 5000 ft. As he approaches the Rocky Mountains at a speed of 60 mph he observes that in a 15-minute interval the angle of elevation of a peak changes from 2°00′ to 3°00′. What is the approximate elevation of the peak?

## For Review

In Exercises 43–44 use a calculator to find **(a)** $\sin\theta$, **(b)** $\cos\theta$, **(c)** $\tan\theta$, **(d)** $\csc\theta$, **(e)** $\sec\theta$, **(f)** $\cot\theta$, **(g)** $\sin^2\theta$, and **(h)** $\sec^2\theta$, correct to four decimal places.

**43.** $\theta = 1.2535$ **44.** $\theta = 62°41'25''$

In Exercises 45–46 use a calculator to find each angle with the given trigonometric function value **(a)** to the nearest hundredth of a degree, **(b)** in radians correct to four decimal places, and **(c)** in degrees, minutes, and seconds.

**45.** $\sin\theta = 0.2569$ **46.** $\cot\theta = 4.1558$

Solve.

**47.** RECREATION The diameter of a travel trailer tire is 16 inches. When the trailer is being towed at 60 mph, how many revolutions does the wheel make in 1 second?

**48.** To review for the next section, give the definition of a function with domain the set of real numbers.

# 6.5 Trigonometric Functions of Real Numbers

## The Unit Circle

Thus far we have considered the trigonometric functions of acute angles measured in degrees or in radians. Now we extend the development of trigonometric functions to functions of a real number. To accomplish this we use the **unit circle,** a circle centered at the origin (0, 0) of a Cartesian coordinate system which has radius 1 unit in length. The equation of the unit circle is

$$\mathbf{x^2 + y^2 = 1.}$$

Notice that neither $x$ nor $y$ can be greater than 1 or less than $-1$ for $(x, y)$ to satisfy $x^2 + y^2 = 1$.

When $x$ is replaced with $-x$, when $y$ is replaced with $-y$, and when $x$ and $y$ are replaced simultaneously with $-x$ and $-y$, the equation $x^2 + y^2 = 1$ remains unchanged. As a result, the graph of $x^2 + y^2 = 1$ is symmetric with respect to the $y$-axis, the $x$-axis, and the origin. Suppose that a point on the unit circle has coordinates $(a, b)$. Since the graph is symmetric with respect to the $y$-axis, the point $(-a, b)$ must also be on the graph. Since the graph is symmetric with respect to the $x$-axis, the point $(a, -b)$ must also be on the graph. Since the graph is symmetric with respect to the origin, the point $(-a, -b)$ must also be on the graph. See Figure 30 on the next page.

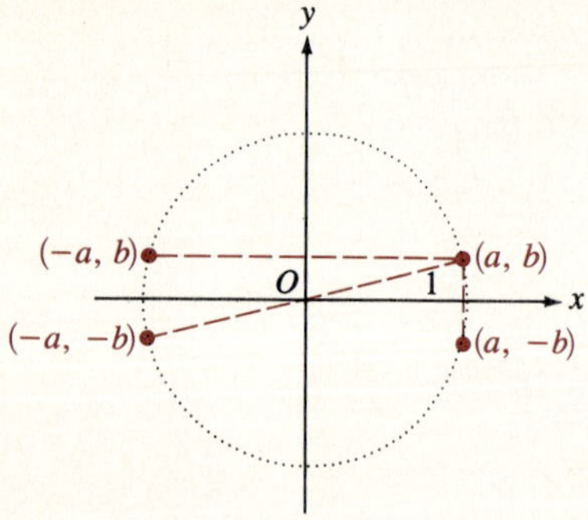


Figure 30 Unit Circle

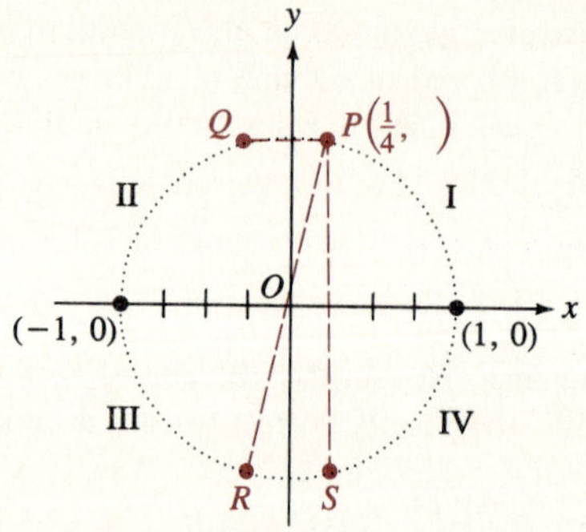


Figure 31

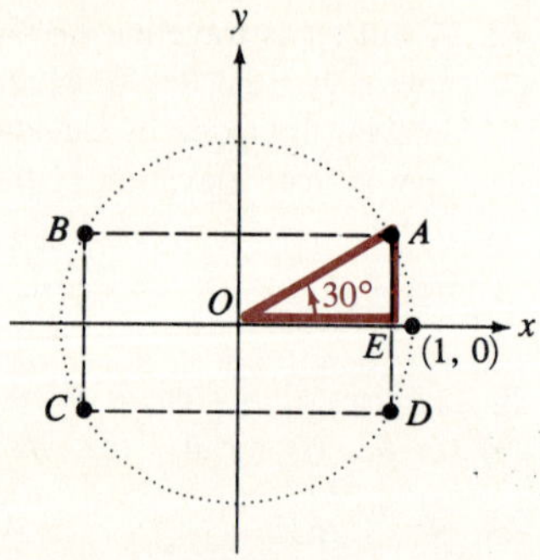


Figure 32

## EXAMPLE 1

Determine the coordinates of three points on the unit circle which are symmetric to the point $P$, with $x$-coordinate 1/4, on the circle in quadrant I.

We first determine the $y$-coordinate of $P$ on the unit circle in Figure 31 by substituting $x = 1/4$ into $x^2 + y^2 = 1$.

$$\left(\frac{1}{4}\right)^2 + y^2 = 1$$

$$y^2 = 1 - \frac{1}{16} = \frac{15}{16}$$

$$y = \pm\frac{\sqrt{15}}{4}$$

$P$ has coordinates $(1/4, \sqrt{15}/4)$, and the three points that are symmetric to $P$ are $Q(-1/4, \sqrt{15}/4)$, $R(-1/4, -\sqrt{15}/4)$, and $S(1/4, -\sqrt{15}/4)$ in quadrants II, III, and IV, respectively. That is, $P$ is symmetric to $Q$ with respect to the $y$-axis, to $R$ with respect to the origin, and to $S$ with respect to the $x$-axis. ■

## EXAMPLE 2

Determine the coordinates of points $B$, $C$, and $D$ on the unit circle in Figure 32.

To determine the coordinates of $B$, $C$, and $D$, we first find the coordinates of $A$, which are $OE$ (the $x$-coordinate) and $EA$ (the $y$-coordinate). In right triangle $AOE$, $OA = 1$ ($OA$ is a radius of the unit circle), and since $EA$ is the side opposite the 30° angle in triangle $AOE$, $EA$ must be 1/2 (one half the hypotenuse). Using the Pythagorean theorem,

$$(OE)^2 + (EA)^2 = 1^2$$

$$(OE)^2 + \left(\frac{1}{2}\right)^2 = 1$$

$$(OE)^2 = 1 - \frac{1}{4} = \frac{3}{4}$$

$$OE = \frac{\sqrt{3}}{2}.$$ The negative root does not apply since $A$ is in quadrant I

Thus the coordinates of $A$ are $(\sqrt{3}/2, 1/2)$. By the various symmetries, the coordinates of $B$ are $(-\sqrt{3}/2, 1/2)$, the coordinates of $C$ are $(-\sqrt{3}/2, -1/2)$, and the coordinates of $D$ are $(\sqrt{3}/2, -1/2)$. ■

## Distances on the Unit Circle

Recall that the formula for the circumference, $c$, of a circle is

$$c = \pi d$$

where $d$ is the diameter. For the unit circle, the radius $r$ is 1 so that $d = 2r = 2$. Thus, the circumference of the unit circle is $2\pi$ units.

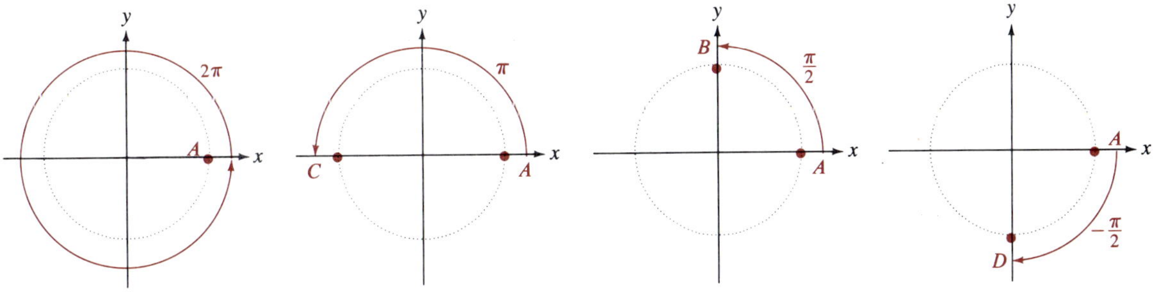


**Figure 33 Distances on the Unit Circle**

If a point starts at position $A$ (Figure 33) and moves counterclockwise all the way around the circle and stops again at $A$, it has traveled a distance of $2\pi$ units. If the point travels in a counterclockwise direction from position $A$ to position $C$, it has traveled $\pi$ units. A point moving counterclockwise from position $A$ to position $B$ travels $\pi/2$ units. If a point moves clockwise from position $A$ to position $D$, we say that the distance traveled is $-\pi/2$ units. The sign indicates a direction or **directed distance.** A positive distance corresponds to motion in a counterclockwise direction, and a negative distance corresponds to a clockwise direction. Thus, if we say that a point moves $\pi$ units (from $A$) on the unit circle, it travels counterclockwise from $A$ to $C$. If we say that a point moves $-\pi$ units (from $A$) on the unit circle, it travels clockwise from $A$ to $C$. In all of our work we will assume that the starting point is $A$ (1, 0).

## EXAMPLE 3

**(a)** When a point travels from $A$ (1, 0) one eighth of the way around the circle in a counterclockwise direction it moves $(1/8)(2\pi) = \pi/4$ units and stops at position $B$. When the point travels one eighth of the way around the circle in a clockwise direction, it moves $-\pi/4$ units and stops at position $C$. See Figure 34.

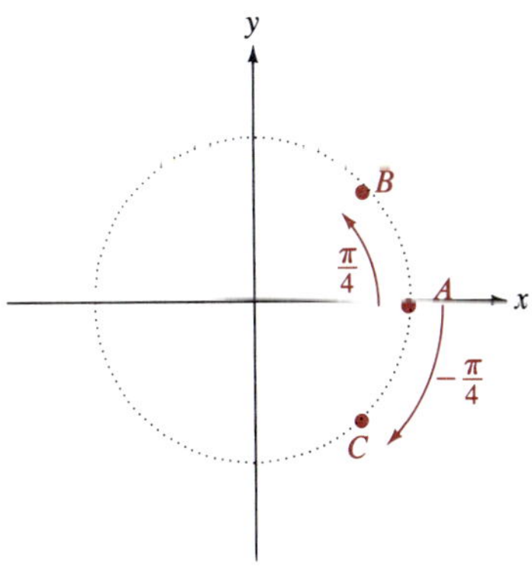


**Figure 34**

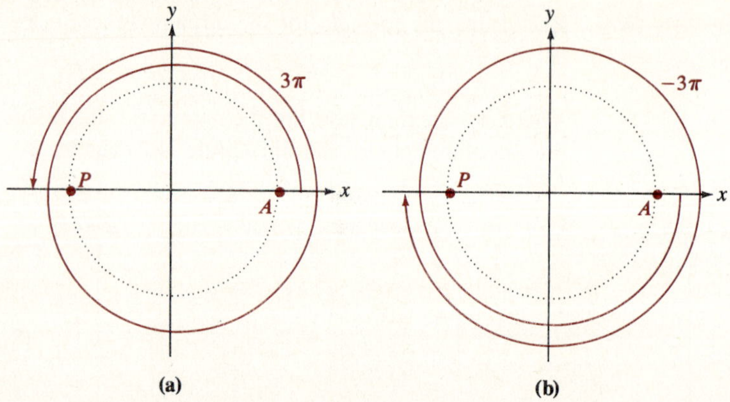


Figure 35

**(b)** If a point travels one complete circuit and then one half the way around again, it travels $2\pi + (1/2)(2\pi) = 2\pi + \pi = 3\pi$ units (counterclockwise direction), shown in Figure 35**(a)**, or $(-2\pi) + (-\pi) = -3\pi$ units (clockwise direction), shown in 35**(b)**. ■

## Determining Points on the Unit Circle

We have considered special cases of **distances on the unit circle.** In the examples we specified a particular fractional part of the way around the circle (counterclockwise or clockwise) and located the point $P$ at which the motion stopped. Actually, any real number can determine a point on the unit circle.

### To Determine a Point on a Unit Circle

Let $t$ be a real number. Beginning at (1, 0), move $|t|$ units around the circle in a counterclockwise direction if $t$ is positive and a clockwise direction if $t$ is negative. Once the distance has been "marked off," the terminal point is identified with the number $t$.

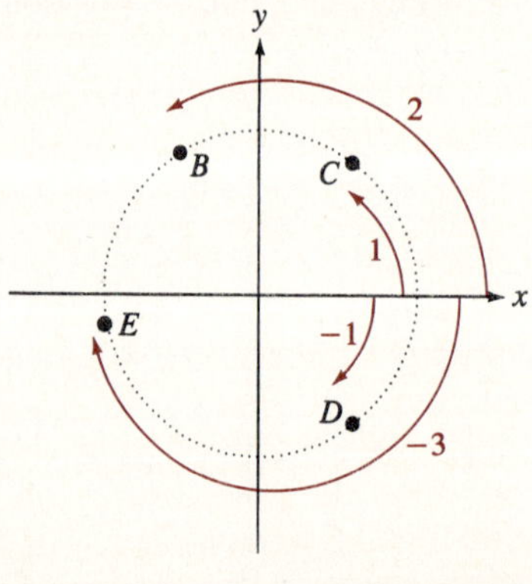


**Figure 36 Points on the Unit Circle**

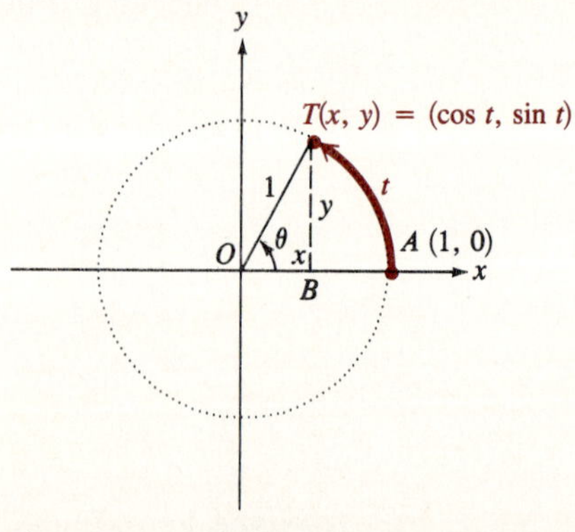


**Figure 37 Trigonometric Functions of Real Number $t$**

When distances on the unit circle are determined by real numbers, the "nice" real numbers turn out to be fractional parts of $\pi$ since the circumference of the circle is $2\pi$. If integers are used to determine points on the unit circle the points are not as "nice." For example, in Figure 36, $-3$ determines point $E$, which is just short of $(-1, 0)$, the point that corresponds to $-\pi$.

## Trigonometric Functions of a Real Number

Since every real number $t$ corresponds to a unique point on the unit circle, we can define the trigonometric functions of a real number.

**Trigonometric Functions of the Real Number $t$**

Let $t$ be any real number and $T(x, y)$ the point on the unit circle identified with $t$.

1. The $y$-coordinate of $T(x, y)$ is the **sine** of $t$, that is,
$$\sin t = y.$$
2. The $x$-coordinate of $T(x, y)$ is the **cosine** of $t$, that is,
$$\cos t = x.$$
3. The **tangent** of $t$, given by $\tan t = \frac{y}{x}$, is defined for all real numbers $t$ for which $x \neq 0$.
4. The **cotangent** of $t$, given by $\cot t = \frac{x}{y}$, is defined for all real numbers $t$ for which $y \neq 0$.
5. The **secant** of $t$, given by $\sec t = \frac{1}{x}$, is defined for all real numbers $t$ for which $x \neq 0$.
6. The **cosecant** of $t$, given by $\csc t = \frac{1}{y}$, is defined for all real numbers $t$ for which $y \neq 0$.

The above definitions generalize the definitions of trigonometric functions of an acute angle found in Section 6.3. In right triangle $OTB$ in Figure 37, notice that if angle $\theta$ is measured in radians, then $\theta = t$, and

$$\sin \theta = \frac{y}{1} = y = \sin t \qquad \cos \theta = \frac{x}{1} = x = \cos t$$

$$\tan \theta = \frac{y}{x} = \tan t \ (x \neq 0) \qquad \cot \theta = \frac{x}{y} = \cot t \ (y \neq 0)$$

$$\sec \theta = \frac{1}{x} = \sec t \ (x \neq 0) \qquad \csc \theta = \frac{1}{y} = \csc t \ (y \neq 0).$$

Thus, the trigonometric functions of the real number $t$ agree with the trigonometric functions of the acute angle $\theta$ with radian measure $t$.

**EXAMPLE 4** Using the unit circles in Figure 38, calculate the six trigonometric function values for the values of $t$ in **(a)** through **(f)**.

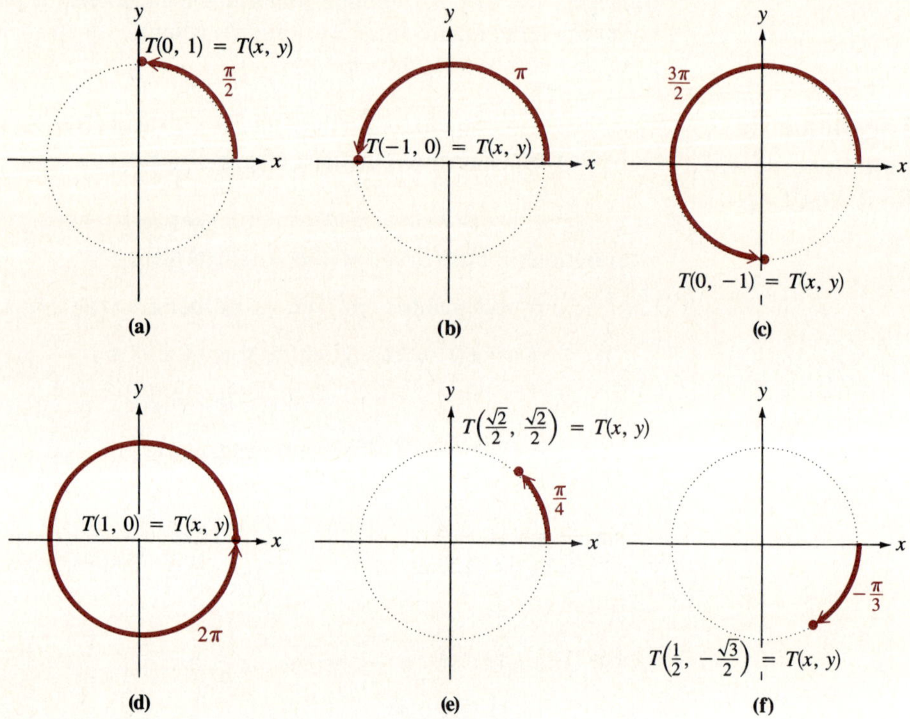


**Figure 38**

**(a)** $t = \frac{\pi}{2}\left(\text{also } \frac{5\pi}{2}, \frac{9\pi}{2}, \cdots \text{ and } -\frac{3\pi}{2}, -\frac{7\pi}{2}, \cdots\right)$

$\sin\frac{\pi}{2} = y = 1$ $\qquad$ $\csc\frac{\pi}{2} = \frac{1}{y} = \frac{1}{1} = 1$

$\cos\frac{\pi}{2} = x = 0$ $\qquad$ $\sec\frac{\pi}{2} = \frac{1}{x} = \frac{1}{0}$ is undefined

$\tan\frac{\pi}{2} = \frac{y}{x} = \frac{1}{0}$ is undefined $\qquad$ $\cot\frac{\pi}{2} = \frac{x}{y} = \frac{0}{1} = 0$

**(b)** $t = \pi$ (also $3\pi, 5\pi, \cdots$ and $-\pi, -3\pi, -5\pi, \cdots$)

$\sin\pi = y = 0$ $\qquad$ $\csc\pi = \frac{1}{y} = \frac{1}{0}$ is undefined

$\cos\pi = x = -1$ $\qquad$ $\sec\pi = \frac{1}{x} = \frac{1}{-1} = -1$

$\tan\pi = \frac{y}{x} = \frac{0}{-1} = 0$ $\qquad$ $\cot\pi = \frac{x}{y} = \frac{-1}{0}$ is undefined

**(c)** $t = \frac{3\pi}{2}\left(\text{also } \frac{7\pi}{2}, \frac{11\pi}{2}, \cdots \text{ and } -\frac{\pi}{2}, -\frac{5\pi}{2}, -\frac{9\pi}{2}, \cdots\right)$

$$\sin \frac{3\pi}{2} = y = -1 \qquad \csc \frac{3\pi}{2} = \frac{1}{y} = \frac{1}{-1} = -1$$

$$\cos \frac{3\pi}{2} = x = 0 \qquad \sec \frac{3\pi}{2} = \frac{1}{x} = \frac{1}{0} \text{ is undefined}$$

$$\tan \frac{3\pi}{2} = \frac{y}{x} = \frac{-1}{0} \text{ is undefined} \qquad \cot \frac{3\pi}{2} = \frac{x}{y} = \frac{0}{-1} = 0$$

**(d)** $t = 2\pi$ (also $0, 4\pi, 6\pi, \cdots$ and $-2\pi, -4\pi, -6\pi, \cdots$)

$$\sin 2\pi = y = 0 \qquad \csc 2\pi = \frac{1}{y} = \frac{1}{0} \text{ is undefined}$$

$$\cos 2\pi = x = 1 \qquad \sec 2\pi = \frac{1}{x} = \frac{1}{1} = 1$$

$$\tan 2\pi = \frac{y}{x} = \frac{0}{1} = 0 \qquad \cot 2\pi = \frac{x}{y} = \frac{1}{0} \text{ is undefined}$$

**(e)** $t = \frac{\pi}{4}$ $\left(\text{also } \frac{9\pi}{4}, \frac{17\pi}{4}, \cdots \text{ and } -\frac{7\pi}{4}, -\frac{15\pi}{4}, -\frac{23\pi}{4}, \cdots\right)$

$$\sin \frac{\pi}{4} = y = \frac{\sqrt{2}}{2} \qquad \csc \frac{\pi}{4} = \frac{1}{y} = \frac{1}{\frac{\sqrt{2}}{2}} = \sqrt{2}$$

$$\cos \frac{\pi}{4} = x = \frac{\sqrt{2}}{2} \qquad \sec \frac{\pi}{4} = \frac{1}{x} = \frac{1}{\frac{\sqrt{2}}{2}} = \sqrt{2}$$

$$\tan \frac{\pi}{4} = \frac{y}{x} = \frac{\frac{\sqrt{2}}{2}}{\frac{\sqrt{2}}{2}} = 1 \qquad \cot \frac{\pi}{4} = \frac{x}{y} = \frac{\frac{\sqrt{2}}{2}}{\frac{\sqrt{2}}{2}} = 1$$

**(f)** $t = -\frac{\pi}{3}$ $\left(\text{also } -\frac{7\pi}{3}, -\frac{13\pi}{3}, \cdots \text{ and } \frac{5\pi}{3}, \frac{11\pi}{3}, \frac{17\pi}{3}, \cdots\right)$

$$\sin\left(-\frac{\pi}{3}\right) = y = -\frac{\sqrt{3}}{2} \qquad \csc\left(-\frac{\pi}{3}\right) = \frac{1}{y} = \frac{1}{-\frac{\sqrt{3}}{2}} = -\frac{2\sqrt{3}}{3}$$

$$\cos\left(-\frac{\pi}{3}\right) = x = \frac{1}{2} \qquad \sec\left(-\frac{\pi}{3}\right) = \frac{1}{x} = \frac{1}{\frac{1}{2}} = 2$$

$$\tan\left(-\frac{\pi}{3}\right) = \frac{y}{x} = \frac{-\frac{\sqrt{3}}{2}}{\frac{1}{2}} = -\sqrt{3} \qquad \cot\left(-\frac{\pi}{3}\right) = \frac{x}{y} = \frac{\frac{1}{2}}{-\frac{\sqrt{3}}{2}} = -\frac{\sqrt{3}}{3}$$

■

## Signs of the Trigonometric Functions

Keep in mind that the coordinates of points on the unit circle have signs, by quadrants, as follows:

$$\text{I: } (+, +), \quad \text{II: } (-, +), \quad \text{III: } (-, -), \quad \text{IV: } (+, -).$$

With this information we know the signs of the trigonometric functions of real number $t$ from the quadrant containing the point $T(x, y)$ determined by $t$.

**Rule of Signs for Trigonometric Functions**

| Quadrant | sin $t$ | cos $t$ | tan $t$ | cot $t$ | sec $t$ | csc $t$ |
|---|---|---|---|---|---|---|
| I | + | + | + | + | + | + |
| II | + | − | − | − | − | + |
| III | − | − | + | + | − | − |
| IV | − | + | − | − | + | − |

## Evaluating Trigonometric Functions

It is easy to find the trigonometric function value of any real number $t$ with a calculator set in radian mode. Simply enter the real number and press the appropriate function key, or, using the reciprocal relationships, the appropriate function key followed by [1/x]. This is illustrated in the next example.

## EXAMPLE 5

Find each trigonometric function value.

**(a)** sin 5
The sine of the real number 5 is found as follows:

ALG & RPN: 5 [sin] → [−0.9589243]

Thus, sin 5 = −0.9589, correct to four decimal places. Remember that the calculator must be set for radians, *not* degrees.

**(b)** tan (−27)

ALG & RPN: 27 [+/−] [tan] → [3.2737038]

Thus, tan (−27) = 3.2737, correct to four decimal places. Remember that different calculators may give different values and that some calculators use [CHS] instead of [+/−]. For example, another calculator gives 3.2737134 for tan (−27). Both answers round to 3.2737.

**(c)** cot 0

ALG & RPN: 0 [tan] [1/x] → [Error]

Notice that the error message is displayed because cot 0 is undefined. When $t = 0$, $T(x, y) = (1, 0)$, and since $y = 0$, $\cot t = \dfrac{x}{y}$ is undefined. ■

## 6.5 EXERCISES

In Exercises 1–2 determine the coordinates of the four points $P$, $Q$, $R$, and $S$ in the figure.

**1.**

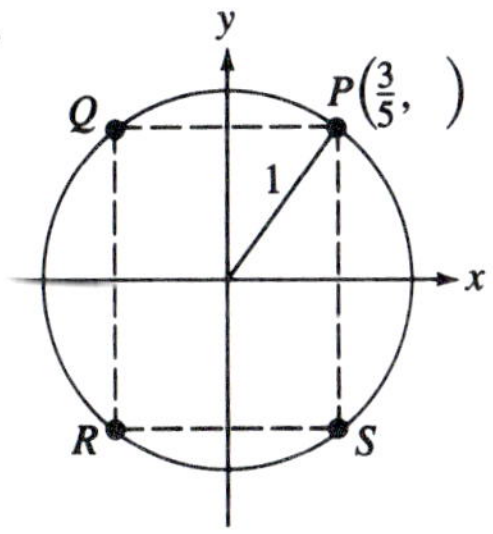


**2.**

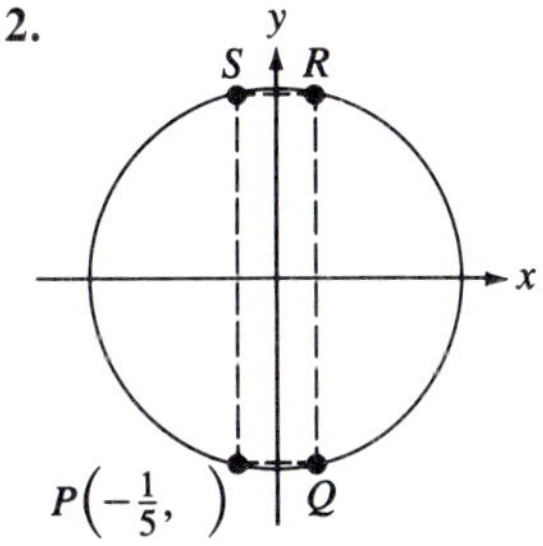


In Exercises 3–4 give the coordinates of the points $A$, $B$, $C$, $D$, and $E$ in the figure.

**3.**

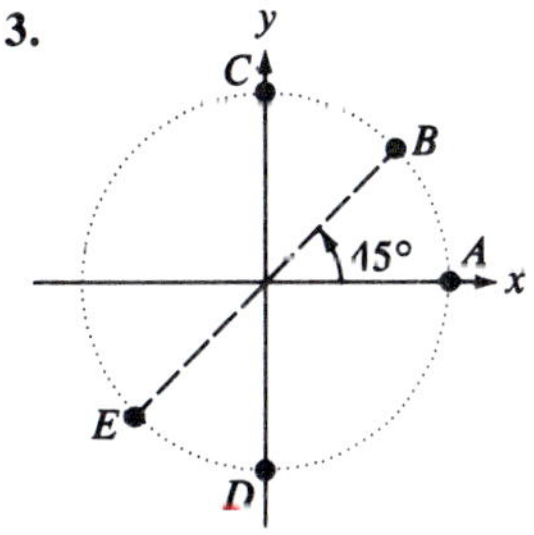


**4.**

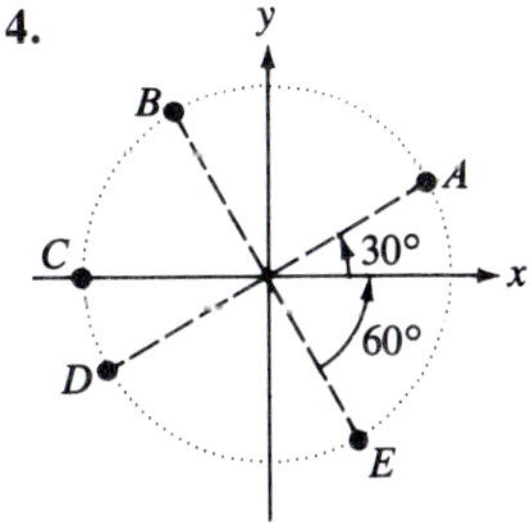


**5.** What is the circumference of the unit circle?

**6.** What are the coordinates of the starting point used to determine directed distances on the unit circle?

**7.** A positive distance is marked off in what direction on the unit circle?

**8.** A negative distance is marked off in what direction on the unit circle?

**9.** If a real number $s$ determines a point $T(a, b)$ on the unit circle, what are the coordinates of the point determined by **(a)** $-s$, **(b)** $\pi - s$, **(c)** $\pi + s$, **(d)** $s - \pi$, **(e)** $s + \frac{\pi}{2}$, and **(f)** $s - \frac{\pi}{2}$?

**10.** If a real number $s$ determines a point $T(-a, -b)$ on the unit circle, what are the coordinates of the point determined by **(a)** $-s$, **(b)** $\pi - s$, **(c)** $\pi + s$, **(d)** $s - \pi$, **(e)** $s + \frac{\pi}{2}$, and **(f)** $s - \frac{\pi}{2}$?

In Exercises 11–30 find the coordinates of the point on the unit circle determined by the given real number.

**11.** $\frac{\pi}{2}$ **12.** $\pi$ **13.** $-\frac{\pi}{2}$ **14.** $-\pi$

**15.** $-\frac{3\pi}{2}$ **16.** $\frac{3\pi}{2}$ **17.** $2\pi$ **18.** $-2\pi$

**19.** $\frac{\pi}{4}$ **20.** $\frac{3\pi}{4}$ **21.** $\frac{5\pi}{4}$ **22.** $\frac{7\pi}{4}$

23. $\frac{\pi}{3}$ 24. $\frac{2\pi}{3}$ 25. $\frac{4\pi}{3}$ 26. $\frac{5\pi}{3}$

27. $\frac{\pi}{6}$ 28. $\frac{5\pi}{6}$ 29. $\frac{7\pi}{6}$ 30. $\frac{11\pi}{6}$

In Exercises 31–32 determine the shortest distance that a point must move on the unit circle, counterclockwise, if it goes from point $A$ to point **(a)** $A$, **(b)** $B$, **(c)** $C$, **(d)** $D$, **(e)** $E$, **(f)** $F$, **(g)** $G$, and **(h)** $H$.

**31.**

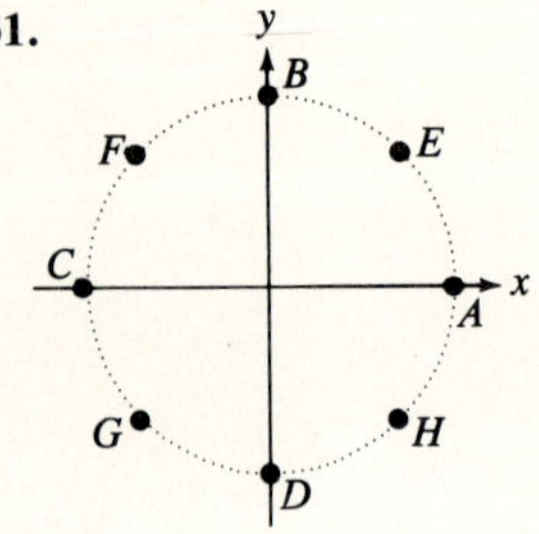


**32.**

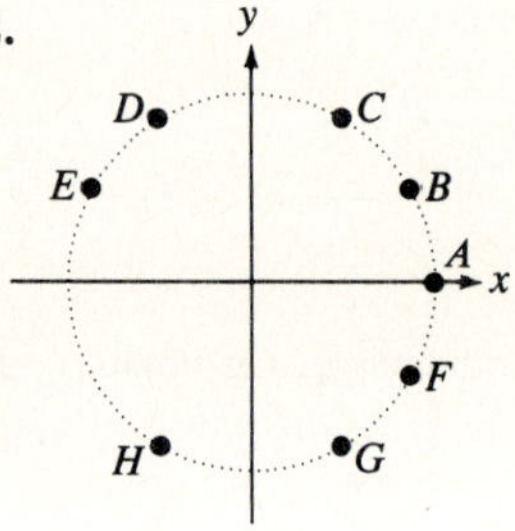


**33.** Repeat Exercise 31 with the point moving in a clockwise direction.

**34.** Repeat Exercise 32 with the point moving in a clockwise direction.

In Exercises 35–36 give two real numbers between $-2\pi$ and $2\pi$, inclusive, which determine the points **(a)** $A$, **(b)** $B$, **(c)** $C$, **(d)** $D$, **(e)** $E$, **(f)** $F$, **(g)** $G$, **(h)** $H$, **(i)** $I$.

**35.**

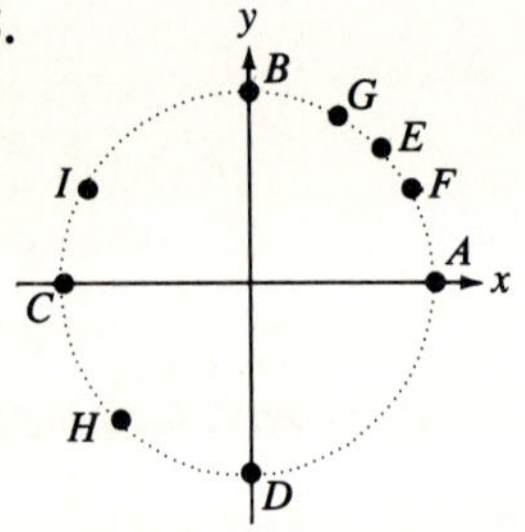


**36.**

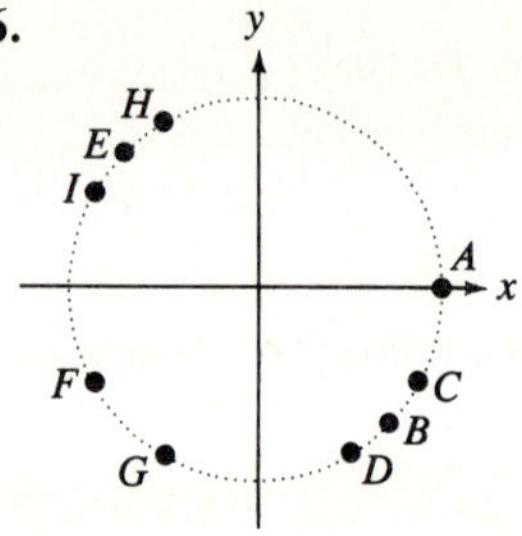


In Exercises 37–52 use the unit circle (no calculator values) and find **(a)** $\sin t$, **(b)** $\cos t$, **(c)** $\tan t$, **(d)** $\csc t$, **(e)** $\sec t$, and **(f)** $\cot t$, for the given real number $t$.

37. $-\frac{\pi}{2}$ 38. $-\pi$ 39. $-\frac{3\pi}{2}$ 40. 0

41. $\frac{9\pi}{4}$ 42. $-\frac{3\pi}{4}$ 43. $\frac{5\pi}{4}$ 44. $-\frac{\pi}{4}$

45. $\frac{\pi}{3}$ 46. $-\frac{2\pi}{3}$ 47. $-\frac{4\pi}{3}$ 48. $\frac{5\pi}{3}$

49. $\frac{\pi}{6}$ 50. $\frac{5\pi}{6}$ 51. $-\frac{\pi}{6}$ 52. $-\frac{5\pi}{6}$

In Exercises 53–56 assume that $t$ determines a point in the given quadrant. Which of the six trigonometric function values are positive in that quadrant?

53. I 54. II 55. III 56. IV

In Exercises 57–64 use a calculator (set in radian mode) to find **(a)** sin $t$, **(b)** cos $t$, **(c)** tan $t$, **(d)** csc $t$, **(e)** sec $t$, **(f)** cot $t$, correct to four decimal places, for the given real number $t$.

**57.** $t = 7$ **58.** $t = -3$ **59.** $t = -12.5$ **60.** $t = 38.2$

**61.** $t = \pi$ **62.** $t = \frac{\pi}{2}$ **63.** $t = -\frac{\pi}{4}$ **64.** $t = \frac{\pi}{3}$

## For Review

**65.** Solve the right triangle in which $B = 54°27'$ and $a = 9.5$ by using the given parts rather than calculated parts. Remember that $C = 90°$.

Solve.

**66.** AERONAUTICS An airplane flying at an altitude of 5 miles passes directly over an observer. One minute later, the angle of elevation of the plane is $29°30'$. Approximately how fast is the plane flying?

**67.** NAVIGATION A ship sails for 2 hours at rate of 28.0 mph in a direction of $148°40'$. How far south has it sailed?

**68.** NAVIGATION A boat traveling at a constant velocity produces a bow wave traveling at a rate of 7.75 mph. If the angle between the waves is $52°40'$, use the bow wave equation $\sin \frac{\theta}{2} = \frac{v_w}{v_b}$ to find the velocity of the boat.

In Exercises 69–70, to review for the next section, determine the values of $x$ for which the function is increasing and those for which it is decreasing.

**69.** $f(x) = (x - 1)^2 - 3$ **70.** $f(x) = -|x + 2| + 5$

# 6.6 Graphing Trigonometric Functions

## Periodic Functions

In previous chapters our study of various kinds of functions included finding their graphs. In this section we consider the graphs of the six trigonometric functions, beginning with the sine and cosine functions. Recall that a given real number $t$ determines a point $T$ on the unit circle which has coordinates (cos $t$, sin $t$). Since the unit circle has circumference $2\pi$, the point determined by $t + 2\pi$ is the same as the point determined by $t$. That is,

$$\cos(t + 2\pi) = \cos t \text{ and } \sin(t + 2\pi) = \sin t,$$

for every real number $t$. Thus, the values of the sine and cosine functions repeat in a cyclic manner on intervals of length $2\pi$ making these functions *periodic* with *period* $2\pi$ according to the definition on the following page.

A function is a **periodic function** if there exists a positive real number $p$ such that

$$f(t + p) = f(t)$$

for every real number $t$ in the domain of $f$. The smallest number $p$ for which a function $f$ is periodic is called the **period** of $f$.

To graph a periodic function with period $p$, we can concentrate on one interval of length $p$, plot the graph on that interval, and use periodicity to obtain the remainder of the graph. For sine and cosine functions we can concentrate on the interval $[0, 2\pi]$.

## Graphing the Sine and Cosine Functions

Suppose we make a combined table of values for the sine and cosine functions,

$$y = \sin x \qquad \text{and} \qquad y = \cos x.$$

A calculator set in radian mode makes it easy to find the values. Notice that $x$ and $y$ are now being used as the variables defining these functions and *not* as the coordinates of a point on the unit circle determined by a real number $t$.

| $x$ | $0$ | $\frac{\pi}{4}$ | $\frac{\pi}{2}$ | $\frac{3\pi}{4}$ | $\pi$ | $\frac{5\pi}{4}$ | $\frac{3\pi}{2}$ | $\frac{7\pi}{4}$ | $2\pi$ |
|---|---|---|---|---|---|---|---|---|---|
| $\sin x$ | 0 | 0.7 | 1 | 0.7 | 0 | −0.7 | −1 | −0.7 | 0 |
| $\cos x$ | 1 | 0.7 | 0 | −0.7 | −1 | −0.7 | 0 | 0.7 | 1 |

It is clear that values of the sine and cosine functions follow a pattern.

| *When x varies from* | *sin x* | *cos x* |
|---|---|---|
| $0$ to $\frac{\pi}{2}$ | increases from 0 to 1 | decreases from 1 to 0 |
| $\frac{\pi}{2}$ to $\pi$ | decreases from 1 to 0 | decreases from 0 to −1 |
| $\pi$ to $\frac{3\pi}{2}$ | decreases from 0 to −1 | increases from −1 to 0 |
| $\frac{3\pi}{2}$ to $2\pi$ | increases from −1 to 0 | increases from 0 to 1 |

The table of values for the sine and cosine functions does not include integer values (except for 0) of $x$. Instead, the table contains the "nice" values relative to these trigonometric functions, the fractional parts of $\pi$.

The graph of $y = \sin x$ in Figure 39 shows the nine points in the table in color. The extension of the graph is in black. Because the sine function is periodic with period $2\pi$, we know the graph continues in both directions as shown. The interval $[0, 2\pi]$ is one period or cycle of the sine function.

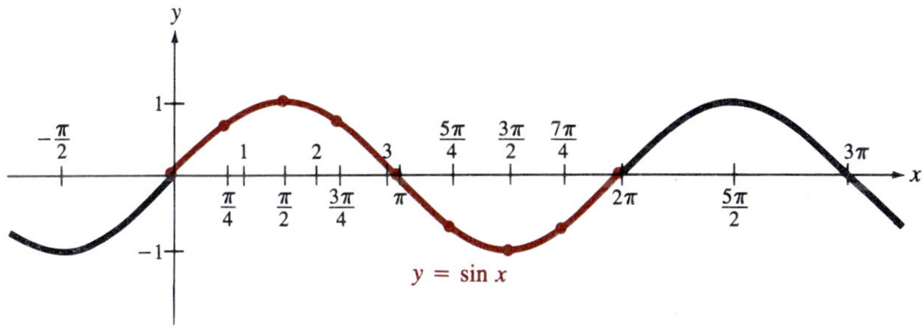


**Figure 39**

Similar remarks apply to the graph of the cosine function in Figure 40. Knowing that the sine and cosine functions are periodic, we can sketch their graphs quickly, plotting only the points determined by the real numbers 0, $\frac{\pi}{2}$, $\pi$, $\frac{3\pi}{2}$, and $2\pi$.

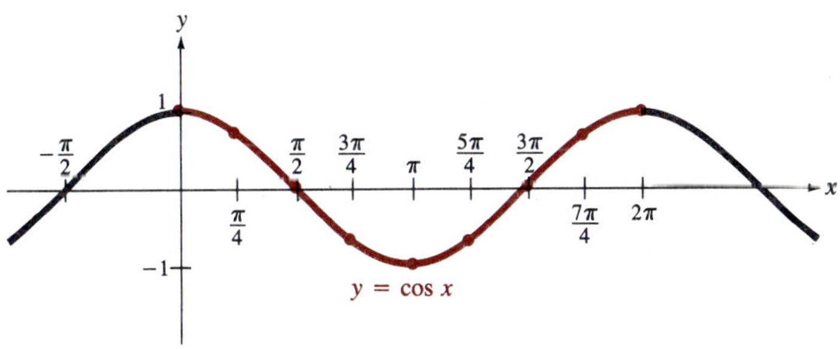


**Figure 40**

## Graphs of the Remaining Trigonometric Functions

Unlike the sine and cosine functions which are defined for all real numbers, the four remaining functions are not defined for certain real numbers. For example, if the real number $t$ determines point $T(x, y)$ on the unit circle,

$$\tan t = \frac{y}{x} \quad \text{and} \quad \sec t = \frac{1}{x}.$$

These two functions are not defined when $x = 0$, that is, when $T(x, y)$ is on the $y$-axis. Thus, the values

$$\pm\frac{\pi}{2}, \quad \pm\frac{3\pi}{2}, \quad \pm\frac{5\pi}{2}, \quad \pm\frac{9\pi}{2}, \cdots$$

are not included in the domains of the tangent and secant functions. Likewise, since

$$\cot t = \frac{x}{y} \quad \text{and} \quad \csc t = \frac{1}{y},$$

these two functions are not defined when $y = 0$, that is, when $T(x, y)$ is on the $x$-axis. Thus, the values

$$0, \quad \pm\pi, \quad \pm 2\pi, \quad \pm 3\pi, \quad \pm 4\pi, \cdots$$

are not included in the domains of the cotangent and cosecant functions.

As $t$ varies from 0 to $\pi/2$, tan $t$ starts at 0 and continues to get larger and larger as $t$ gets close to $\pi/2$ ($x$ gets close to zero and $y$ gets close to 1). We say that tan $t$ **increases without bound,** or approaches $+\infty$ as $t$ approaches $\pi/2$. Likewise, as $t$ approaches $\pi/2$ from values above $\pi/2$ ($\tan t = y/x$ is negative since $y > 0$ and $x < 0$), tan $t$ **decreases without bound,** or approaches $-\infty$. Thus, as $t$ varies from 0 to $\pi/2$, tan $t$ varies from 0 to $+\infty$, and as $t$ varies from $\pi/2$ to $\pi$, tan $t$ varies from $-\infty$ to 0. The variation of tan $t$ as $t$ varies from 0 to $2\pi$ is summarized in the following table.

| *When t varies from* | *tan t* |
|---|---|
| 0 to $\frac{\pi}{2}$ | increases from 0 to $+\infty$ |
| $\frac{\pi}{2}$ to $\pi$ | increases from $-\infty$ to 0 |
| $\pi$ to $\frac{3\pi}{2}$ | increases from 0 to $+\infty$ |
| $\frac{3\pi}{2}$ to $2\pi$ | increases from $-\infty$ to 0 |

Thus, as $t$ varies from $\pi/2$ to $3\pi/2$, tan $t$ varies over the entire set of real numbers. Then on the next interval from $3\pi/2$ to $5\pi/2$, tan $t$ again varies over the entire set of real numbers. In fact, on any interval of length $\pi$ units, the tangent function behaves exactly the same way. This means that the tangent function is periodic with period $\pi$. The character of this function is better understood when its graph is analyzed.

A table of values for the tangent function is presented below.

| $x$ | 0 | $\frac{\pi}{6}$ | $\frac{\pi}{4}$ | $\frac{\pi}{3}$ | $\frac{\pi}{2}$ | $\frac{2\pi}{3}$ | $\frac{3\pi}{4}$ | $\frac{5\pi}{6}$ | $\pi$ | $\frac{5\pi}{4}$ | $\frac{3\pi}{2}$ | $\frac{7\pi}{4}$ | $2\pi$ |
|---|---|---|---|---|---|---|---|---|---|---|---|---|---|
| tan $x$ | 0 | 0.6 | 1 | 1.7 | undef. | $-1.7$ | $-1$ | $-0.6$ | 0 | 1 | undef. | $-1$ | 0 |

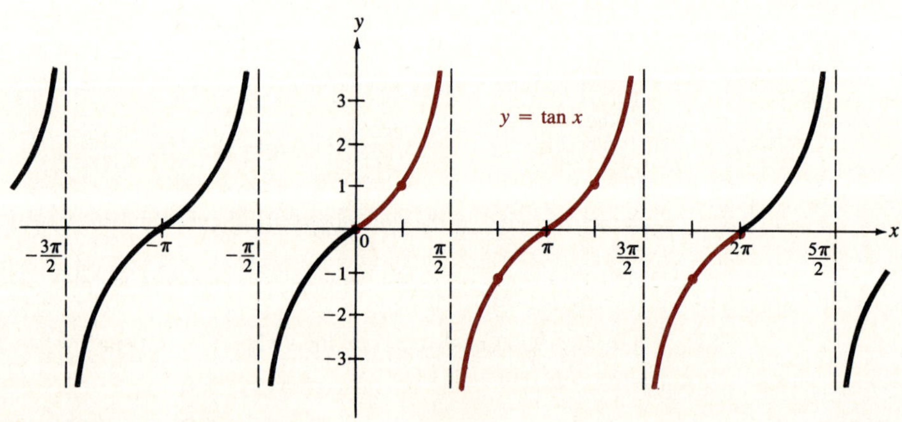


**Figure 41**

In the graph of the tangent function in Figure 41, the vertical dotted lines, called **vertical asymptotes,** are lines that the graph approaches but never crosses. These lines occur at values of $x$ for which $\tan x$ is undefined ($\pm\pi/2$, $\pm 3\pi/2$, $\pm 5\pi/2, \cdots$). The portion of the graph corresponding to the values in the table is plotted in color. The extension of the graph is in black. Since the function is periodic, the graph continues to the right and to the left with the same appearance as it has between $\pi/2$ and $3\pi/2$, repeated again and again.

The graph of the cotangent function can be determined by using the identity $\cot x = 1/\tan x$. For secant and cosecant we use $\sec x = 1/\cos x$ and $\csc x = 1/\sin x$. A table of values for the cotangent, secant, and cosecant functions is constructed using these identities and the graphs are given in Figures 42, 43, and 44.

| $x$ | $0$ | $\frac{\pi}{3}$ | $\frac{\pi}{2}$ | $\frac{2\pi}{3}$ | $\pi$ | $\frac{4\pi}{3}$ | $\frac{3\pi}{2}$ | $\frac{5\pi}{3}$ | $2\pi$ |
|---|---|---|---|---|---|---|---|---|---|
| $\cot x$ | undef. | 0.6 | 0 | −0.6 | undef. | 0.6 | 0 | −0.6 | undef. |
| $\sec x$ | 1 | 2 | undef. | 2 | −1 | −2 | undef. | 2 | 1 |
| $\csc x$ | undef. | 1.2 | 1 | 1.2 | undef. | −1.2 | −1 | −1.2 | undef. |

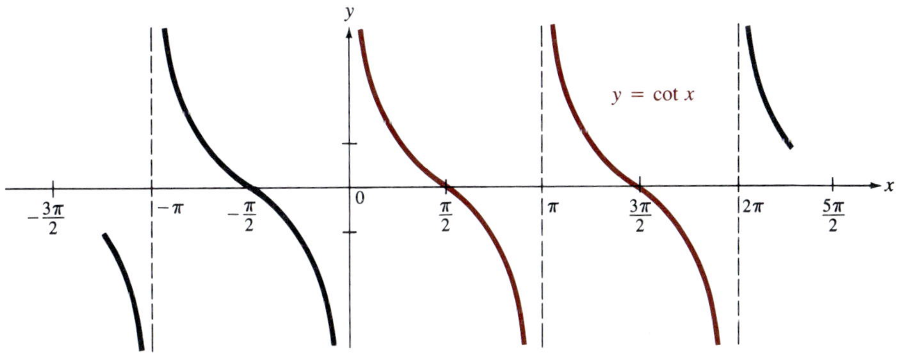


**Figure 42**

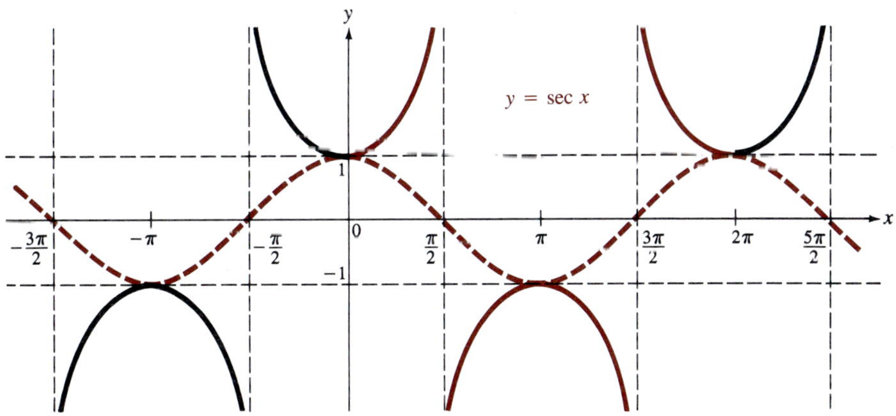


**Figure 43**

Recall that the secant function is the reciprocal of the cosine function. By graphing $y = \cos x$ in the same system as $y = \sec x$ (as in Figure 43) this reciprocal relationship can be observed. Notice that at values where the cosine is zero (such as $-\pi/2$ and $3\pi/2$), the secant is undefined, at values where the cosine is 1 or $-1$ (such as 0 or $\pi$), the secant has the same values, and as the cosine decreases from 1 to 0, its reciprocal increases from 1 to $\infty$.

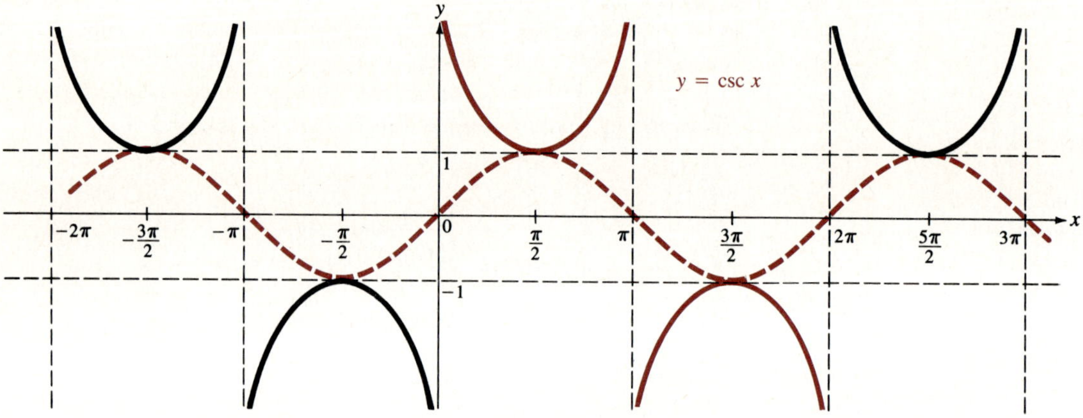


**Figure 44**

Since the cosecant function is the reciprocal of the sine function, we emphasize this relationship by graphing $y = \sin x$ in the same system as $y = \csc x$ in Figure 44.

## Graphing Variations of the Trigonometric Functions

Variations of the six basic trigonometric functions can be graphed with information we already have. Recall that if the graph of a function $y = f(x)$ is known, the graph of $y = f(x) + D$ can be sketched quickly by moving the graph of $y = f(x)$ up $|D|$ units if $D > 0$ or down $|D|$ units if $D < 0$.

### EXAMPLE 1

Graph the functions $y = \sin x - 3$ and $y = \sin x + 1$ starting with the graph of $y = \sin x$.

Notice that $\sin x - 3 \neq \sin (x - 3)$. To graph $y = \sin x - 3$, shift the graph of $y = \sin x$ down 3 units, and to graph $y = \sin x + 1$, shift the graph of $y = \sin x$ up 1 unit. See Figure 45. ■

If the graph of a function $y = f(x)$ is known, the graph of $y = Af(x)$ can be sketched by multiplying each $y$-value in $y = f(x)$ by the constant $A$. If $A > 1$, the graph of $y = Af(x)$ will be a stretched-out version of $y = f(x)$. If $0 < A < 1$, the graph of $y = Af(x)$ will be a compressed version of $y = f(x)$. If $A < 0$, then $-A > 0$ and the graph of $y = Af(x)$ will be the reflection of $y = -Af(x)$ in the $x$-axis.

### EXAMPLE 2

Graph the functions $y = 2 \sin x$ and $y = -3 \sin x$ by starting with the graph of $y = \sin x$.

It may be helpful to construct a table of values for these two functions. See Figure 46. ■

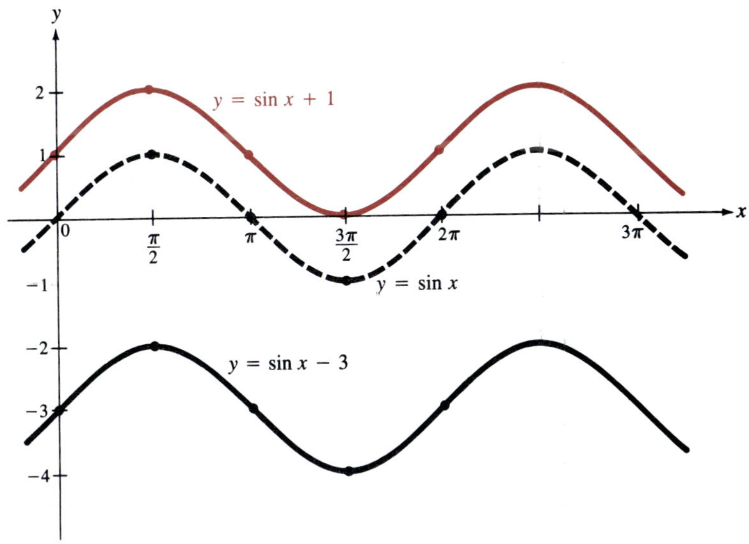


Figure 45

| $x$ | $\sin x$ | $2\sin x$ | $-3\sin x$ |
|---|---|---|---|
| 0 | 0 | 0 | 0 |
| $\frac{\pi}{4}$ | 0.707 | 1.414 | $-2.121$ |
| $\frac{\pi}{2}$ | 1 | 2 | $-3$ |
| $\frac{3\pi}{4}$ | 0.707 | 1.414 | $-2.121$ |
| $\pi$ | 0 | 0 | 0 |
| $\frac{3\pi}{2}$ | $-1$ | $-2$ | 3 |
| $2\pi$ | 0 | 0 | 0 |

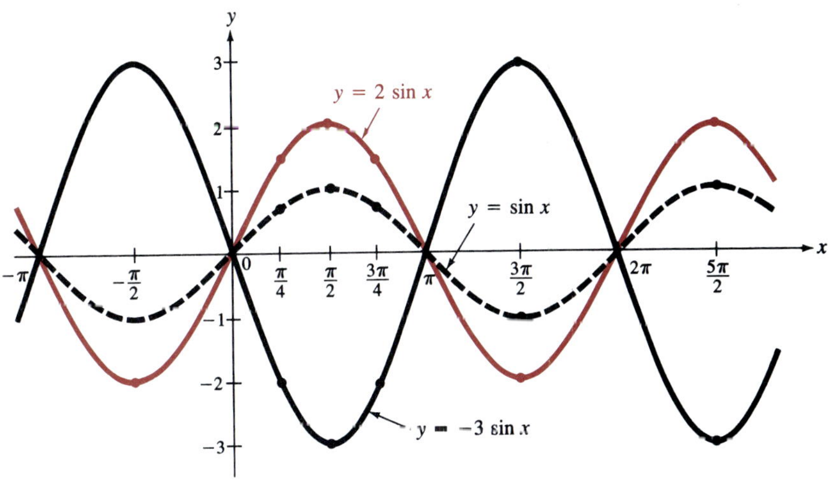


Figure 46

### Amplitude of Sine and Cosine

The number $|A|$ in the functions

$$y = A\sin x \text{ and } y = A\cos x$$

is called the **amplitude** of the function.

Since $y = \sin x = 1 \cdot \sin x$ and $y = \cos x = 1 \cdot \cos x$, the amplitude of the sine and cosine functions is 1. Intuitively we see the amplitude of a function indicates how high and how low the functional values rise and fall. The amplitude of $y = 2 \sin x$ is 2, and the amplitude of $y = -3 \sin x$ is 3.

Next we investigate graphs of trigonometric functions in which the variable $x$ is multiplied by a constant $B$.

## EXAMPLE 3

Graph the functions $y = \sin 2x$ and $y = \sin \frac{1}{2}x$ and compare them with the graph of $y = \sin x$.

Consider the table of values given at the left. The graphs of the two functions are shown in Figure 47. ■

| $x$ | $\sin 2x$ | $\sin \frac{1}{2}x$ |
|---|---|---|
| 0 | 0 | 0 |
| $\frac{\pi}{4}$ | 1 | 0.707 |
| $\frac{\pi}{2}$ | 0 | 1 |
| $\frac{3\pi}{4}$ | −1 | 0.707 |
| $\pi$ | 0 | 0 |
| $\frac{5\pi}{4}$ | 1 | −0.707 |
| $\frac{3\pi}{2}$ | 0 | −1 |
| $\frac{7\pi}{4}$ | −1 | −0.707 |
| $2\pi$ | 0 | 0 |

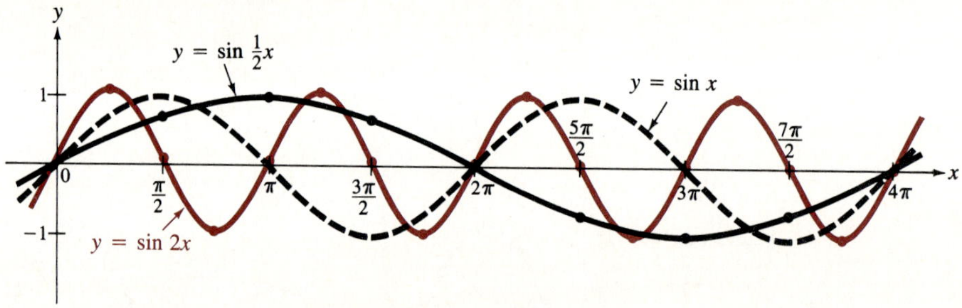


**Figure 47**

From Example 3 we can see that with functions of the form

$$y = \sin Bx, \quad B > 0,$$

the amplitude is the same as that of $y = \sin x$, but the period is decreased if $B > 1$ and increased if $B < 1$. For example, when $B = 2$, the period of $y = \sin 2x$ is $\pi$, whereas when $B = \frac{1}{2}$, the period of $y = \sin \frac{1}{2}x$ is $4\pi$. These are special cases of the following theorem.

### Period of Sine and Cosine

The **period** of the functions

$$y = \sin Bx \quad \text{and} \quad y = \cos Bx, \quad B > 0,$$

is $\frac{2\pi}{B}$.

The graphs of functions of the form

$$y = A \sin Bx \quad \text{and} \quad y = A \cos Bx$$

can be obtained by considering our previous discussions of amplitude and periodicity.

**EXAMPLE 4** Graph the function $y = 4 \cos \frac{1}{3}x$ and compare it with the graph of $y = \cos x$.

The amplitude of $y = 4 \cos \frac{1}{3}x$ is $|A| = |4| = 4$, and the period is $\frac{2\pi}{B} = \frac{2\pi}{1/3} = 6\pi$. Thus, to graph the function we must consider values of $x$ in the interval $[0, 6\pi]$. See Figure 48. Notice that the cosine function begins at the high point of the period, hits its low point one-half the way through the period, and crosses 0 one-fourth and three-fourths of the way through the period. This will be true whatever the length of the period. ■

| $x$ | $4 \cos \frac{1}{3} x$ |
|---|---|
| 0 | 4 |
| $\frac{\pi}{2}$ | 3.46 |
| $\pi$ | 2 |
| $\frac{3\pi}{2}$ | 0 |
| $2\pi$ | −2 |
| $3\pi$ | −4 |
| $4\pi$ | −2 |
| $\frac{9\pi}{2}$ | 0 |
| $6\pi$ | 4 |

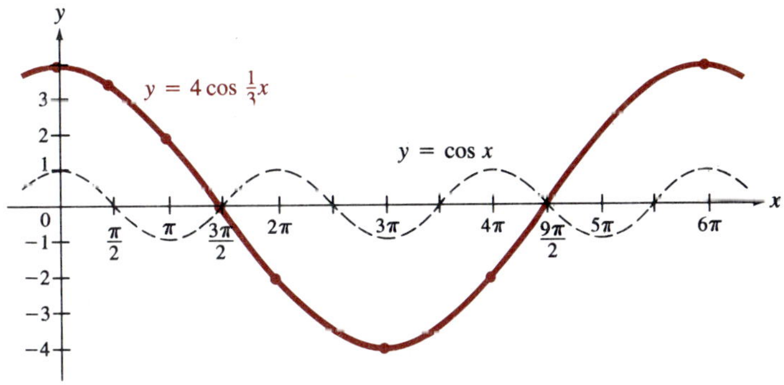


**Figure 48**

## 6.6 EXERCISES

Sketch the graph of each function in Exercises 1–6.

**1.** $y = \sin x - 2$ **2.** $y = \cos x - 2$ **3.** $y = \cos x + 3$

**4.** $y = \sin x + 3$ **5.** $y = \tan x + 1$ **6.** $y = \tan x - 2$

In Exercises 7–30 **(a)** give the amplitude, **(b)** give the period, and **(c)** sketch the graph of each function.

**7.** $y = 3 \sin x$ **8.** $y = 3 \cos x$ **9.** $y = -2 \cos x$ **10.** $y = -2 \csc x$

**11.** $y = \frac{1}{2} \sin x$ **12.** $y = \frac{1}{2} \cos x$ **13.** $y = -\frac{1}{2} \cos x$ **14.** $y = -\frac{1}{2} \sec x$

**15.** $y = \sin 4x$ **16.** $y = \cos 4x$ **17.** $y = \cos \frac{1}{3}x$ **18.** $y = \sin \frac{1}{3}x$

**19.** $y = 3 \tan x$ **20.** $y = \frac{1}{4} \cot x$ **21.** $y = -\frac{1}{4} \tan x$ **22.** $y = -3 \tan x$

**23.** $y = \tan \frac{1}{2}x$ **24.** $y = \tan 2x$ **25.** $y = 3 \sin \frac{1}{2}x$ **26.** $y = 3 \cos \frac{1}{2}x$

**27.** $y = 4 \cos 3x$ **28.** $y = 4 \sin 3x$ **29.** $y = -2 \sin \frac{1}{3}x$ **30.** $y = -2 \cos \frac{1}{3}x$

For each function in Exercises 31–36 give **(a)** the domain of the function, and **(b)** the range of the function.

**31.** $y = \sin x$ **32.** $y = \cos x$ **33.** $y = \tan x$

**34.** $y = \cot x$ **35.** $y = \sec x$ **36.** $y = \csc x$

Recall that a function $f$ is even if $f(-x) = f(x)$, and odd if $f(-x) = -f(x)$. Use this information in Exercises 37–42.

**37.** Prove that the cosine function is even.

**38.** Prove that the sine function is odd.

**39.** Prove that the secant function is even.

**40.** Prove that the cosecant function is odd.

**41.** Prove that the tangent function is odd.

**42.** Prove that the cotangent function is odd.

**43.** Prove that the sine function has period $2\pi$.

**44.** Prove that the cosine function has period $2\pi$.

In Exercises 45–50 consider the given function on the interval $[0, 2\pi]$. On what subintervals is the function **(a)** increasing, and **(b)** decreasing?

**45.** $y = \sin x$ **46.** $y = \cos x$ **47.** $y = \tan x;\ x \neq \frac{\pi}{2}, \frac{3\pi}{2}$

**48.** $y = \cot x;\ x \neq 0, \pi, 2\pi$ **49.** $y = \sec x;\ x \neq \frac{\pi}{2}, \frac{3\pi}{2}$ **50.** $y = \csc x;\ x \neq 0, \pi, 2\pi$

## For Review

In Exercises 51–53 find **(a)** the coordinates of the point on the unit circle determined by the given real number, **(b)** $\sin t$, **(c)** $\cos t$, **(d)** $\tan t$, **(e)** $\csc t$, **(f)** $\sec t$, **(g)** $\cot t$.

**51.** $\frac{3\pi}{4}$ **52.** $\frac{2\pi}{3}$ **53.** $\frac{7\pi}{6}$

In Exercises 54–55 use a calculator (set in radian mode) to find **(a)** $\sin t$, **(b)** $\cos t$, **(c)** $\tan t$, **(d)** $\csc t$, **(e)** $\sec t$, **(f)** $\cot t$.

**54.** $t = -5.9$ **55.** $t = -\frac{11\pi}{3}$

**56.** **PHYSICS** A light hits the surface of the water in a lake such that $A = 40°30'$. Use 1.3 for the index of refraction for water, 1.0003 for the index of refraction for air, and the light wave equation $I_w/I_a = \sin A/\sin W$ to find $W$.

## 6.7 More on Graphing Trigonometric Functions

### Sinusoidal Functions

In Section 6.6 we discussed the graphs of the sine and cosine functions along with variations of these functions when amplitudes and periods are changed. In this section we will develop the tools for graphing two important types of functions, that are also variations of the sine and cosine functions, called **sinusoidal functions.** They take the form

$$y = A \sin B(x - C) \quad \text{and} \quad y = A \cos B(x - C).$$

Sinusoidal functions have numerous applications to physical problems. We have seen that the graphs of

$$y = \sin x + D \quad \text{and} \quad y = \cos x + D$$

can be obtained from the graphs of $y = \sin x$ and $y = \cos x$ by moving these graphs up or down $|D|$ units, depending on whether $D$ is positive or negative, respectively. In a similar manner, the graphs of

$$y = \sin (x - C) \quad \text{and} \quad y = \cos (x - C)$$

can be obtained from the graphs of $y = \sin x$ and $y = \cos x$ by shifting these graphs right or left $|C|$ units depending on whether $C$ is positive or negative, respectively.

**EXAMPLE 1** Graph the function $y = \sin\left(x - \frac{\pi}{2}\right)$ and compare it with the graph of $y = \sin x$.

A table of values is given below and the graph is shown in Figure 49. Notice that the graph of $y = \sin\left(x - \frac{\pi}{2}\right)$ can be obtained from the graph of $y = \sin x$ by moving it right $\frac{\pi}{2}$ units so that one cycle begins at $\left(\frac{\pi}{2}, 0\right)$ and ends at $\left(2\pi + \frac{\pi}{2}, 0\right) = \left(\frac{5\pi}{2}, 0\right)$. ■

| $x$ | $\sin\left(x - \frac{\pi}{2}\right)$ |
|---|---|
| $0$ | $-1$ |
| $\frac{\pi}{2}$ | $0$ |
| $\pi$ | $1$ |
| $\frac{3\pi}{2}$ | $0$ |
| $2\pi$ | $-1$ |
| $\frac{5\pi}{2}$ | $0$ |

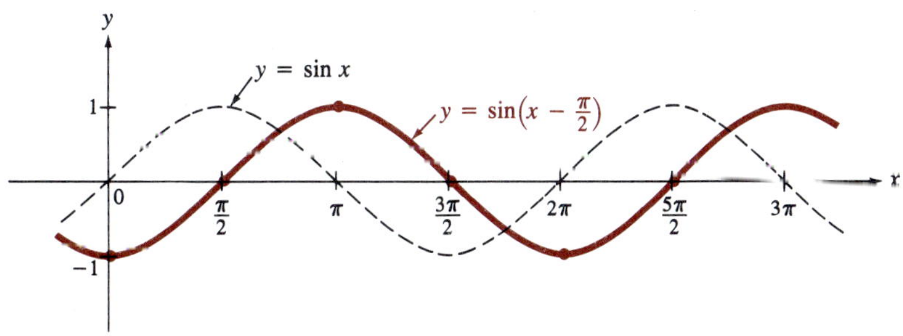


Figure 49

**EXAMPLE 2** Graph the function $y = \cos\left(x + \frac{\pi}{4}\right)$ and compare it with the graph of $y = \cos x$.

The graph is given in Figure 50. Notice that the graph of $y = \cos x$ could be moved $\frac{\pi}{4}$ units to the left to obtain the graph of $y = \cos\left(x + \frac{\pi}{4}\right)$ so that one cycle begins at $\left(-\frac{\pi}{4}, 1\right)$ and ends at $\left(-\frac{\pi}{4} + 2\pi, 1\right) = \left(\frac{7\pi}{4}, 1\right)$. ■

| $x$ | $\cos\left(x + \frac{\pi}{4}\right)$ |
|---|---|
| $-\frac{\pi}{4}$ | 1 |
| 0 | 0.7 |
| $\frac{\pi}{4}$ | 0 |
| $\frac{\pi}{2}$ | −0.7 |
| $\frac{3\pi}{4}$ | −1 |
| $\frac{5\pi}{4}$ | 0 |
| $\frac{7\pi}{4}$ | 1 |

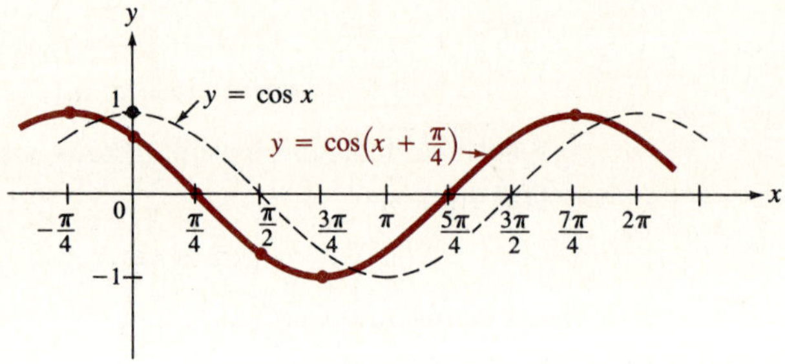


**Figure 50**

---

**Phase Shift**

Consider the functions

$$y = \sin(x - C) \text{ and } y = \cos(x - C)$$

where $C$ is a real number. The number $|C|$ is called the **phase shift** of each function.

---

We can now sketch the graph of a function such as

$$y = 3 \sin(2x - \pi).$$

Begin by factoring 2 from $(2x - \pi)$ and writing the function as the sinusoidal function

$$y = 3 \sin 2\left(x - \frac{\pi}{2}\right).$$

In this form we can identify the amplitude as $|3| = 3$, the period as $\frac{2\pi}{2} = \pi$, and the phase shift as $\left|\frac{\pi}{2}\right| = \frac{\pi}{2}$. One cycle of the graph will begin at $\left(\frac{\pi}{2}, 0\right)$ and end $\pi$ units later at $\left(\frac{3\pi}{2}, 0\right)$. First graph $y = \sin 2x$ (see Figure 47 in Section 6.6), use it to find the graph of $y = 3 \sin 2x$, and then shift this graph to the *right* $\frac{\pi}{2}$ units (since $C =$

$\frac{\pi}{2} > 0$, the shift is to the right$\Big)$ to obtain the graph of $y = 3 \sin 2\left(x - \frac{\pi}{2}\right)$. The graph is given in Figure 51. The adjacent table of values serves as a check.

| $x$ | $3 \sin 2\left(x - \frac{\pi}{2}\right)$ |
|---|---|
| 0 | 0 |
| $\frac{\pi}{4}$ | $-3$ |
| $\frac{\pi}{2}$ | 0 |
| $\frac{3\pi}{4}$ | 3 |
| $\pi$ | 0 |

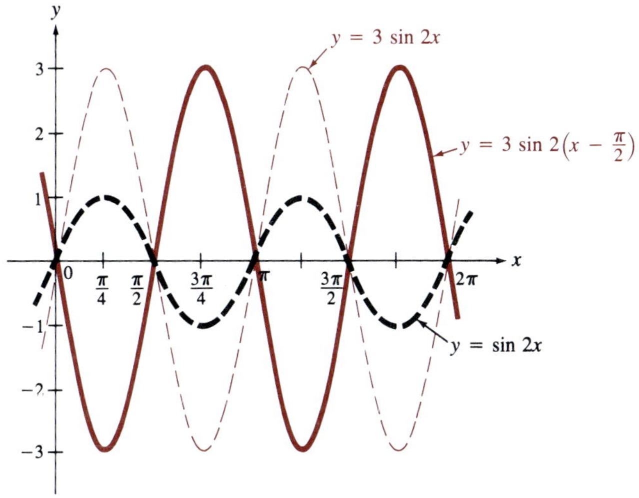


**Figure 51**

**EXAMPLE 3** Graph the function $y = 2 \cos\left(\frac{1}{2}x + \frac{\pi}{2}\right)$.

Write
$$y = 2 \cos \frac{1}{2}(x + \pi) = 2 \cos \frac{1}{2}[x - (-\pi)].$$

Then see that the amplitude is 2, the period is $2\pi/(1/2) = 4\pi$, and the phase shift is $\pi$ units to the left. Thus, one cycle of the graph will begin at $(-\pi, 2)$ and end $4\pi$ units later at $(3\pi, 2)$.

Plot $y = \cos \frac{1}{2}x$ and use this to plot $y = 2 \cos \frac{1}{2}x$. When the graph of $y = 2 \cos \frac{1}{2}x$ is shifted $\pi$ units to the left, we obtain the graph of $y = 2 \cos \frac{1}{2}(x + \pi)$ shown in Figure 52. Once again the few points found in the table give a check. ■

| $x$ | $2 \cos \frac{1}{2}(x + \pi)$ |
|---|---|
| 0 | 0 |
| $\pi$ | $-2$ |
| $2\pi$ | 0 |
| $3\pi$ | 2 |
| $4\pi$ | 0 |

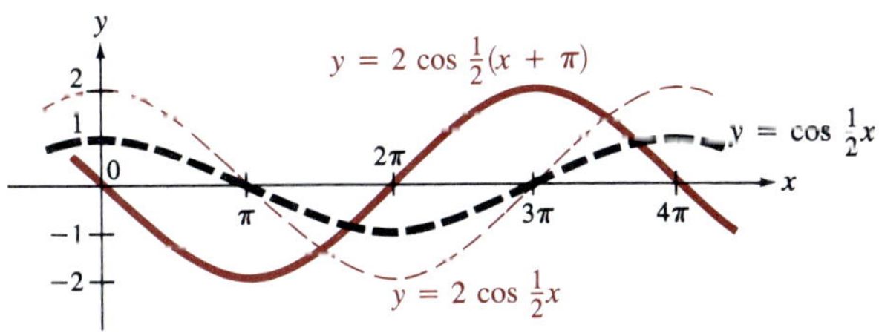


**Figure 52**

Let us summarize the results illustrated in the examples.

**Sinusoidal Function**

If a function can be written as a sinusoidal function in the form

$$y = A \sin B(x - C) \qquad \text{or} \qquad y = A \cos B(x - C) \quad (B > 0)$$

then

$$\text{amplitude} = |A|$$

$$\text{period} = \frac{2\pi}{B}$$

$$\text{phase shift} = |C|.$$

The graph is shifted to the right if $C > 0$ and to the left if $C < 0$.

## Graphing by Adding Ordinates

To graph sums or differences of sine and cosine functions plot each of the two functions separately and add ordinates ($y$ values) graphically. This procedure, illustrated in Examples 4 and 5, is sometimes called the **addition of ordinates** method.

**EXAMPLE 4** Graph the function $y = \sin 2x + \cos 2x$.

| $x$ | $\sin 2x$ | $\cos 2x$ | $\sin 2x + \cos 2x$ |
|---|---|---|---|
| 0 | 0 | 1 | 1 |
| $\frac{\pi}{8}$ | 0.7 | 0.7 | 1.4 |
| $\frac{\pi}{4}$ | 1 | 0 | 1 |
| $\frac{3\pi}{8}$ | 0.7 | −0.7 | 0 |
| $\frac{\pi}{2}$ | 0 | −1 | −1 |
| $\frac{5\pi}{8}$ | −0.7 | −0.7 | −1.4 |
| $\frac{3\pi}{4}$ | −1 | 0 | −1 |
| $\frac{7\pi}{8}$ | −0.7 | 0.7 | 0 |
| $\pi$ | 0 | 1 | 1 |

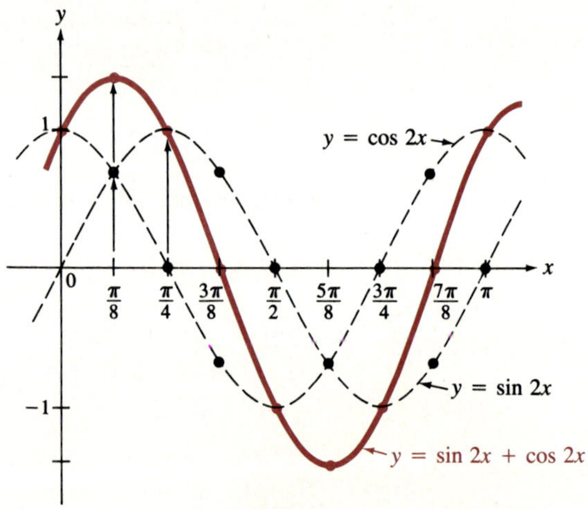


Figure 53

We can see from the graph in Figure 53 that for $x = \pi/8$, $\sin 2x$ and $\cos 2x$ have the same ordinate, so their sum is twice that value. Also, at $x = \pi/4$, $\cos 2x$ is 0 and thus the sum curve crosses the $\sin 2x$ curve. When $x = 3\pi/8$, $\sin 2x = -\cos 2x$ making the sum 0 and thus the graph of their sum crosses the $x$-axis. ■

In Example 4 the sum curve looks like a sine curve (and indeed it is). In the next example the curve is more difficult to determine and many points are plotted. Once this graphing procedure is perfected we can reduce the number of points we need to plot.

**EXAMPLE 5** Graph the function $y = 3 \sin 2x - 2 \cos x$.

| $x$ | $3 \sin 2x$ | $-2 \cos x$ | $3 \sin 2x + (-2 \cos x)$ |
|---|---|---|---|
| 0 | 0 | −2 | −2 |
| $\frac{\pi}{8}$ | 2.12 | −1.85 | 0.27 |
| $\frac{\pi}{4}$ | 3 | −1.41 | 1.59 |
| $\frac{3\pi}{8}$ | 2.12 | −0.77 | 1.35 |
| $\frac{\pi}{2}$ | 0 | 0 | 0 |
| $\frac{5\pi}{8}$ | −2.12 | 0.77 | −1.35 |
| $\frac{3\pi}{4}$ | −3 | 1.41 | −1.59 |
| $\frac{7\pi}{8}$ | −2.12 | 1.85 | −0.27 |
| $\pi$ | 0 | 2 | 2 |
| $\frac{9\pi}{8}$ | 2.12 | 1.85 | 3.97 |
| $\frac{5\pi}{4}$ | 3 | 1.41 | 4.41 |
| $\frac{11\pi}{8}$ | 2.12 | 0.77 | 2.89 |
| $\frac{3\pi}{2}$ | 0 | 0 | 0 |
| $\frac{13\pi}{8}$ | −2.12 | −0.77 | −2.89 |
| $\frac{7\pi}{4}$ | −3 | −1.41 | −4.41 |
| $\frac{15\pi}{8}$ | −2.12 | −1.85 | −3.97 |
| $2\pi$ | 0 | −2 | −2 |

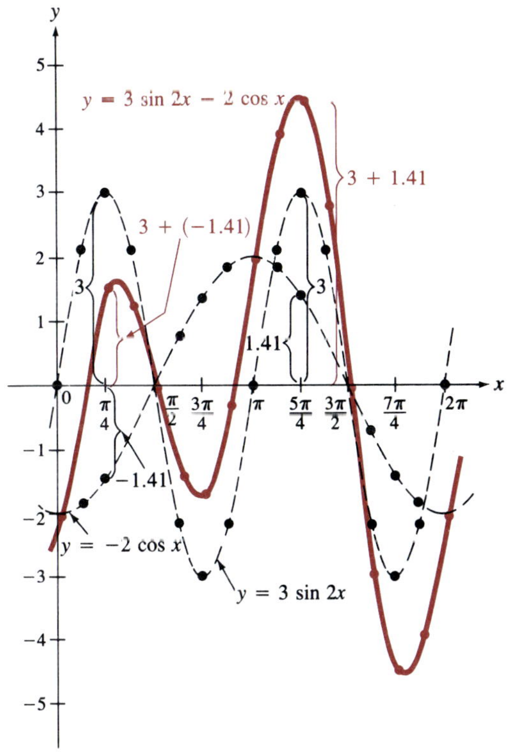


Figure 54

Note in Figure 54 that at $x = \pi/4$ thc point on the curve is determined by adding the positive ordinate of the ($3 \sin 2x$)-curve to the negative ordinate of the ($-2 \cos x$)-curve. The point on the curve when $x = 5\pi/4$ comes from adding the two positive values of the curves at this point. ■

## Graphing Damped Functions

Other more complex variations of trigonometric functions, called **damped functions,** involve a product of functions as illustrated in Example 6.

### EXAMPLE 6

Graph the function $f(x) = \frac{x}{2} \sin x$.

The function $f$ is the product of the two functions $g(x) = \frac{x}{2}$ and $h(x) = \sin x$. The absolute value of $f(x)$ will give a clue as to finding the graph. Notice that $|f(x)| = \left|\frac{x}{2}\right| \left|\sin x\right| = \frac{x}{2} |\sin x|$, for $x > 0$, and since $|\sin x| \le 1$, it follows that $|f(x)| \le \frac{x}{2}$, $x > 0$. Hence,

$$-\frac{x}{2} \le f(x) \le \frac{x}{2}$$

which means that its graph lies between the graphs of $y = \frac{x}{2}$ and $y = -\frac{x}{2}$. Points on the graph of $f(x)$ will coincide with points on these lines whenever $|\sin x| = 1$, that is, when $x = \pm\frac{\pi}{2}, \pm\frac{3\pi}{2}, \pm\frac{5\pi}{2}, \ldots$ . Also, the $x$-intercepts of $f(x)$ occur when $\frac{x}{2} = 0$ and when $\sin x = 0$, that is, at $x = 0, \pm\pi, \pm 2\pi, \pm 3\pi, \ldots$ . Using this information we can sketch the graph in Figure 55. ■

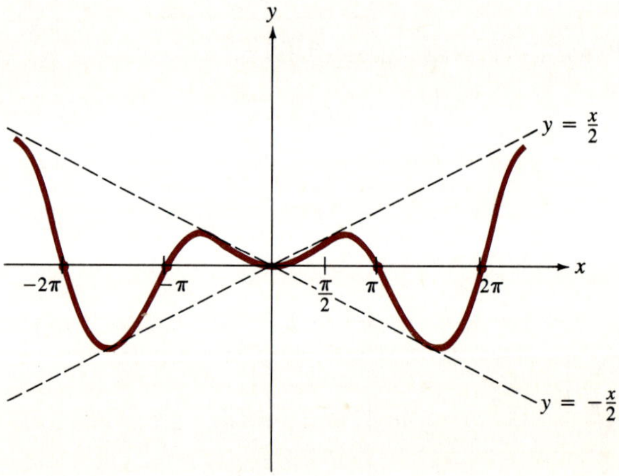


**Figure 55 Damped Function**

The function $f(x) = \frac{x}{2} \sin x$ is an example of a **damped sine wave** with $\frac{x}{2}$ called the **damping factor.** In general, sine (or cosine) waves can be compressed or expanded by using assorted damping factors. The graphs of such functions are important in engineering and physics.

## 6.7 EXERCISES

In Exercises 1–14 **(a)** give the amplitude, **(b)** give the period, **(c)** give the phase shift, and **(d)** sketch the graph of each function over an interval of length one period.

**1.** $y = \sin\left(x + \frac{\pi}{2}\right)$

**2.** $y = \cos\left(x - \frac{\pi}{2}\right)$

**3.** $y = \cos\left(x - \frac{\pi}{4}\right)$

**4.** $y = \sin\left(x + \frac{\pi}{4}\right)$

**5.** $y = 2\sin\left(x + \frac{\pi}{2}\right)$

**6.** $y = 3\cos\left(x - \frac{\pi}{2}\right)$

**7.** $y = -3\cos\left(x - \frac{\pi}{2}\right)$

**8.** $y = -2\sin\left(x + \frac{\pi}{4}\right)$

**9.** $y = 3\cos 2\left(x - \frac{\pi}{2}\right)$

**10.** $y = 2\cos 2\left(x + \frac{\pi}{2}\right)$

**11.** $y = 4\sin(2x + \pi)$

**12.** $y = 4\cos(2x - \pi)$

**13.** $y = 3\cos\left(\frac{1}{2}x - \frac{\pi}{2}\right)$

**14.** $y = 3\cos\frac{\pi}{2}x$

In Exercises 15–28 use the method of adding ordinates to sketch the graph of each function.

**15.** $y = \sin x + \cos x$

**16.** $y = \sin x - \cos x$

**17.** $y = \sin x - 2\cos x$

**18.** $y = 2\sin x + \cos x$

**19.** $y = \sin 2x + \cos x$

**20.** $y = 2\sin 2x - \cos 2x$

**21.** $y = x + \sin x$

**22.** $y = x - \cos x$

**23.** $y = \sin 2x + \sin x$

**24.** $y = \sin 2x - \sin x$

**25.** $y = \sin \pi x - \cos \pi x$

**26.** $y = \sin \pi x + \cos \pi x$

**27.** $y = \sin\frac{\pi x}{2} + \cos \pi x$

**28.** $y = |x| - \cos x$

In Exercises 29–38 sketch the graph of each function.

**29.** $y = x\sin x$

**30.** $y = 2^{-x}\sin x$

**31.** $y = |x|\sin x$

**32.** $y = \frac{1}{4}x^2\cos x$

**33.** $y = 2^{-x/2}\cos x$

**34.** $y = e^x\sin x$

**35.** $y = |\sin x|$

**36.** $y = |\sin x| + 1$

**37.** $y = \sin|x|$

**38.** $y = 2\tan\frac{1}{2}x$

Solve.

**39.** MANAGEMENT Over a period of years the owner of a small business in a springtime tourist area has discovered that his sales can be approximated by

$$S = 8500 + 3700\sin\frac{\pi x}{6}$$

where $x$ is the time in months, with $x = 1$ corresponding to January. Sketch the graph of this function over the period of one year and use it to determine the month with the least sales.

**40.** METEOROLOGY Using data collected over a number of years, the weather bureau has determined that the average monthly low temperature reading in degrees Fahrenheit for a desert town in southern California can be approximated by

$$t = 24.8\sin\left(\frac{\pi x}{6} - \frac{\pi}{2}\right) + 69.8,$$

where $x$ is the time in months, with $x = 0$ corresponding to January. Graph this function and use it to determine the average low temperature during the hottest month of the year.

## For Review

**41. AERONAUTICS** An airplane takes off with an angle of climb of 8°45′. If its average air speed is 320 mph, how long will it take to reach a cruising altitude of 32,000 ft?

**42. AERONAUTICS** If an airplane is flying at Mach three (three times the speed of sound), use the sound wave equation $\sin \frac{\theta}{2} = \frac{v_s}{v_a}$ to determine the angle of the sonic cone.

# 6.8 Fundamental Identities

Recall from Chapter 2 that an equation is called an identity if it is true for every replacement of the variable. For example,

$$2x + 2 = 2(x + 1)$$

is an identity, since no matter what number is used to replace $x$, the resulting equation is true. The equal sign is often replaced with $\equiv$ which is read *is identical to* or *identically equals*. Thus, we would write

$$2x + 2 \equiv 2(x + 1).$$

There are many identities involving the trigonometric functions, some of which are considered now. Some identities should be memorized, while others, if the method is understood, can easily be derived when needed. In the study of trigonometry and its application to higher mathematics, recognition of identities and their algebraic variations is essential.

We have grouped the fundamental identities into several categories, the first of which is already familiar from previous work. For now we will concentrate on identities involving trigonometric functions of a real number, because their verification is easy with the unit circle. However, since the trigonometric function of a real number can be thought of as the trigonometric function of an angle in radian measure, the

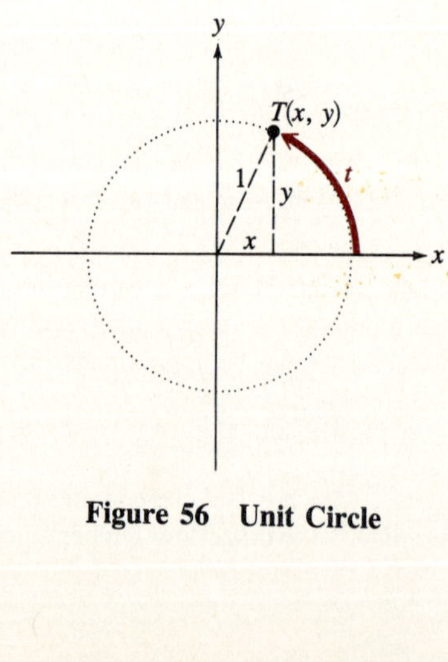


**Figure 56 Unit Circle**

identities apply to angles as well as real numbers. In the next chapter we will consider in detail trigonometric functions of arbitrary angles.

One fact to remember is that an identity is true only for those real numbers for which the functions are defined. For example, an identity involving the tangent function excludes numbers such as $\frac{\pi}{2}$ where the tangent is undefined. A review of the definitions of the trigonometric functions of a real number in Section 6.5 would be helpful before proceeding.

## Reciprocal Identities

The first group of identities is sometimes called the **reciprocal identities.**

| Reciprocal Identities | Algebraic Variations |
|---|---|
| $\csc t \equiv \frac{1}{\sin t}$ | $\sin t \equiv \frac{1}{\csc t}$, $\sin t \csc t \equiv 1$ |
| $\sec t \equiv \frac{1}{\cos t}$ | $\cos t \equiv \frac{1}{\sec t}$, $\cos t \sec t \equiv 1$ |
| $\cot t \equiv \frac{1}{\tan t}$ | $\tan t \equiv \frac{1}{\cot t}$, $\tan t \cot t \equiv 1$ |

We prove the first identity and leave verification of the other two as exercises. To show that $\csc t = \frac{1}{\sin t}$, consider the unit circle with real number $t$ associated with the point $T(x, y)$ in Figure 56. We defined $\sin t = y$ and $\csc t = \frac{1}{y}$. Thus, $\csc t = \frac{1}{y} = \frac{1}{\sin t}$, and the identity has been proved.

## Ratio Identities

The next category of identities are often called the **ratio identities.**

| Ratio Identities | Algebraic Variations |
|---|---|
| $\tan t \equiv \frac{\sin t}{\cos t}$ | $\sin t \equiv \tan t \cos t$, $\cos t \equiv \frac{\sin t}{\tan t}$ |
| $\cot t \equiv \frac{\cos t}{\sin t}$ | $\cos t \equiv \cot t \sin t$, $\sin t \equiv \frac{\cos t}{\cot t}$ |

These identities can be proved easily using the unit circle, and the proofs are left as exercises.

## Pythagorean Identities

Another group of identities is called the **Pythagorean Identities,** a name which will be justified after the first one is proved.

| **Pythagorean Identities** | **Algebraic Variations** |
|---|---|
| $\sin^2 t + \cos^2 t \equiv 1$ | $1 - \cos^2 t \equiv \sin^2 t, \quad 1 - \sin^2 t \equiv \cos^2 t$ |
| $1 + \tan^2 t \equiv \sec^2 t$ | $\sec^2 t - 1 \equiv \tan^2 t, \quad \sec^2 t - \tan^2 t \equiv 1$ |
| $1 + \cot^2 t \equiv \csc^2 t$ | $\csc^2 t - 1 \equiv \cot^2 t, \quad \csc^2 t - \cot^2 t \equiv 1$ |

Recall that the equation of the unit circle is $x^2 + y^2 = 1$. Since $\cos t = x$ and $\sin t = y$,

$$x^2 + y^2 = \cos^2 t + \sin^2 t = 1,$$

so that $\sin^2 t + \cos^2 t = 1$, by the commutative law of addition. Remember that we write $\sin^2 x$ and $\cos^2 x$ for $(\sin x)^2$ and $(\cos x)^2$, respectively. By considering the right triangle with legs $x$ and $y$ and hypotenuse 1 in Figure 56, it is also clear by the Pythagorean theorem that $x^2 + y^2 = 1$. This is why these identities have been called *Pythagorean identities*. Verification of the remaining identities is left for the exercises.

## Cofunction Identities

As their names suggest, the sine and cosine, secant and cosecant, and tangent and cotangent functions are often called **cofunctions** of each other. Several identities relate the cofunctions. We concentrate on the sine and cosine since the other identities can be obtained using the reciprocal and ratio identities.

**Cofunction Identities for Sine and Cosine Functions**

| | |
|---|---|
| $\sin\left(t + \frac{\pi}{2}\right) \equiv \cos t$ | $\cos\left(t + \frac{\pi}{2}\right) \equiv -\sin t$ |
| $\sin\left(t - \frac{\pi}{2}\right) \equiv -\cos t$ | $\cos\left(t - \frac{\pi}{2}\right) \equiv \sin t$ |
| $\sin\left(\frac{\pi}{2} - t\right) \equiv \cos t$ | $\cos\left(\frac{\pi}{2} - t\right) \equiv \sin t$ |

The cofunction identities can be verified by sketching a unit circle and a general value for $t$. We determine the point associated with each variant of $t$ and apply the definitions of the sine and cosine functions. Two of the cofunction identities are verified in the following examples with the remaining four left as exercises.

### EXAMPLE 1

Verify that $\cos\left(t + \frac{\pi}{2}\right) \equiv -\sin t$.

Sketch a unit circle, identify $t$ and $t + \frac{\pi}{2}$ with associated points $T$ and $S$, respectively, and assume that $T$ has coordinates $(x, y) = (\cos t, \sin t)$ as in Figure 57. Since triangle $OSQ$ is congruent to triangle $OTP$, the coordinates of $S$ are $(-y, x)$. Then $\cos\left(t + \frac{\pi}{2}\right) = -y = -\sin t$. Thus, $\cos\left(t + \frac{\pi}{2}\right) \equiv -\sin t$. ■

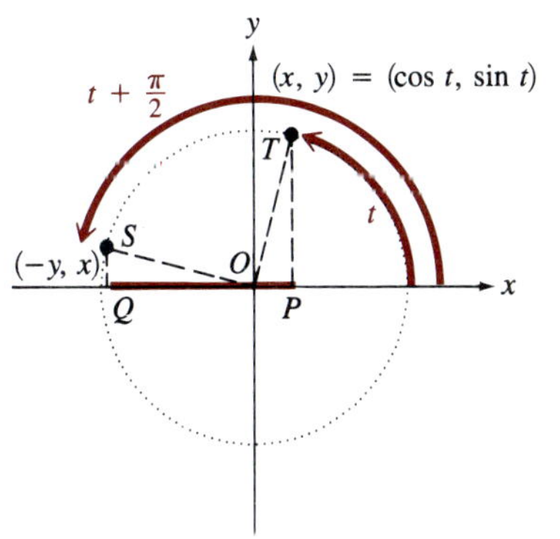


Figure 57

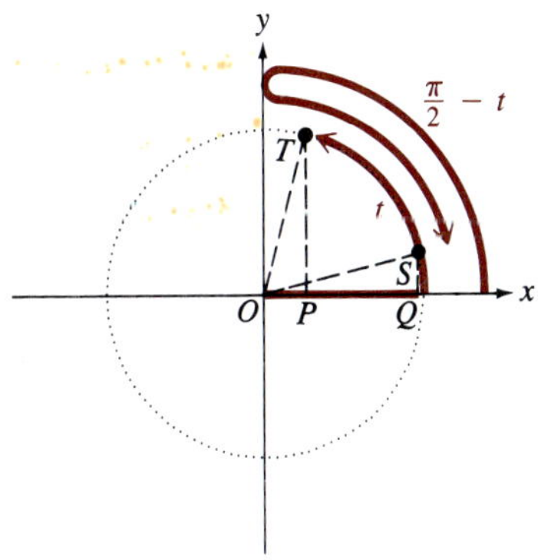


Figure 58

**EXAMPLE 2** Verify that $\sin\left(\frac{\pi}{2} - t\right) \equiv \cos t$.

Sketch a unit circle, identify $t$ and $\frac{\pi}{2} - t$ with associated points $T(x, y) = (\cos t, \sin t)$ and $S$, respectively, as shown in Figure 58. With triangle $OTP$ congruent to triangle $OSQ$, the coordinates of $S$ are $(y, x)$. Then $\sin\left(\frac{\pi}{2} - t\right) = x = \cos t$. Thus, $\sin\left(\frac{\pi}{2} - t\right) \equiv \cos t$. ■

As we mentioned previously, the remaining cofunction identities need not be memorized since they are easily obtainable using previous identities and the cofunction identities for the sine and cosine functions. The next example illustrates this.

**EXAMPLE 3** Verify that $\tan\left(t - \frac{\pi}{2}\right) \equiv -\cot t$.

$$\tan\left(t - \frac{\pi}{2}\right) = \frac{\sin\left(t - \frac{\pi}{2}\right)}{\cos\left(t - \frac{\pi}{2}\right)} = \frac{-\cos t}{\sin t} = -\frac{\cos t}{\sin t} = -\cot t \quad ■$$

## Miscellaneous Identities

The next group of identities, which can be proved using the technique in Examples 1 and 2, completes the list of fundamental identities. Some of them were already considered when we discussed even and odd functions in Exercises 6.6, and periodicity in Section 6.6. Proofs of several of these are part of the exercises.

**Miscellaneous Identities**

Let $t$ be any real number.

| | |
|---|---|
| $\sin(-t) \equiv -\sin t$ | $\cos(-t) \equiv \cos t$ |
| $\sin(t + 2\pi) \equiv \sin t$ | $\cos(t + 2\pi) \equiv \cos t$ |
| $\sin(t - 2\pi) \equiv \sin t$ | $\cos(t - 2\pi) \equiv \cos t$ |
| $\sin(t + \pi) \equiv -\sin t$ | $\cos(t + \pi) \equiv -\cos t$ |
| $\sin(t - \pi) \equiv -\sin t$ | $\cos(t - \pi) \equiv -\cos t$ |
| $\sin(\pi - t) \equiv \sin t$ | $\cos(\pi - t) \equiv -\cos t$ |

Many times, a complex expression involving the trigonometric functions can be simplified by algebraic manipulation together with an application of an identity. Two examples will help to make this process clear.

**EXAMPLE 4** Multiply and simplify each expression.

**(a)** $(\sin t + \cos t)^2$

$$\begin{aligned}(\sin t + \cos t)^2 &= \sin^2 t + 2\sin t\cos t + \cos^2 t && \text{Use } (a+b)^2 = a^2 + 2ab + b^2\\ &= (\sin^2 t + \cos^2 t) + 2\sin t\cos t\\ &= 1 + 2\sin t\cos t && \sin^2 t + \cos^2 t = 1\end{aligned}$$

**(b)** $\sin\left(\frac{\pi}{2} - t\right)[\sec t - \cos t]$

$$\begin{aligned}\sin\left(\frac{\pi}{2} - t\right)[\sec t - \cos t] &= \cos t[\sec t - \cos t] && \sin\left(\frac{\pi}{2} - t\right) = \cos t\\ &= \cos t\sec t - \cos^2 t\\ &= 1 - \cos^2 t && \text{Reciprocal identity}\\ &= \sin^2 t && \text{Pythagorean identity}\end{aligned}$$ ■

**EXAMPLE 5** Factor each expression and simplify.

**(a)** $\sin^2 t\cos^2 t + \sin^4 t$

$$\begin{aligned}\sin^2 t\cos^2 t + \sin^4 t &= \sin^2 t(\cos^2 t + \sin^2 t) && \text{Factor out the common factor}\\ &= \sin^2 t(1) && \text{Pythagorean identity}\\ &= \sin^2 t\end{aligned}$$

**(b)** $\cos^4 t - \sin^4 t$

$$\begin{aligned}\cos^4 t - \sin^4 t &= (\cos^2 t - \sin^2 t)(\cos^2 t + \sin^2 t) && a^4 - b^4 = (a^2 - b^2)(a^2 + b^2)\\ &= (\cos^2 t - \sin^2 t)(1)\\ &= (\cos t - \sin t)(\cos t + \sin t)\end{aligned}$$ ■

Sometimes it is helpful to transform a given trigonometric function into another trigonometric function. The procedure to use is illustrated in the next example.

**EXAMPLE 6** Express $\cot t$ in terms of $\sin t$, where $t$ is a real number that determines a point $T$ on the unit circle in quadrant I.

Recall that $\cot t = \dfrac{\cos t}{\sin t}$. Also, $\cos^2 t = 1 - \sin^2 t$, and by taking the square root of both sides $\cos t = \pm\sqrt{1 - \sin^2 t}$. But since $t$ determines a point in quadrant I, $\cos t$ is positive so we have $\cos t = \sqrt{1 - \sin^2 t}$. Substituting,

$$\cot t = \frac{\sqrt{1 - \sin^2 t}}{\sin t}. \quad \blacksquare$$

**EXAMPLE 7** Suppose that $t$ determines a point in quadrant III, and that $\csc t = -\frac{5}{4}$. Use the fundamental identities to find $\cot t$.

Since $1 + \cot^2 t = \csc^2 t$, we have that

$$\cot^2 t = \csc^2 t - 1.$$

Therefore,

$$\cot t = \pm\sqrt{\csc^2 t - 1}.$$

But since $t$ determines a point in quadrant III, and $\cot t$ is positive in that quadrant, use the positive value of the root. Substituting, we obtain

$$\begin{aligned} \cot t &= \sqrt{\csc^2 t - 1} \\ &= \sqrt{\left(-\frac{5}{4}\right)^2 - 1} \\ &= \sqrt{\frac{9}{16}} = \frac{3}{4}. \quad \blacksquare \end{aligned}$$

NOTE Although we have used the variable $t$ throughout this section, as with any functional notation, any letter can be used for the variable. In the exercises that follow, different variables will be used. ■

## 6.8 EXERCISES

Prove the identities in Exercises 1–20. In Exercises 1–10, assume that $T(x, y)$ is the point on the unit circle determined by the real number $t$.

**1.** $\sec t \equiv \dfrac{1}{\cos t}$

**2.** $\cot t \equiv \dfrac{1}{\tan t}$

**3.** $1 + \tan^2 t \equiv \sec^2 t$

**4.** $1 + \cot^2 t \equiv \csc^2 t$

**5.** $\tan t \equiv \dfrac{\sin t}{\cos t}$

**6.** $\cot t \equiv \dfrac{\cos t}{\sin t}$

**7.** $\sin\left(t + \dfrac{\pi}{2}\right) \equiv \cos t$

**8.** $\sin\left(t - \dfrac{\pi}{2}\right) \equiv -\cos t$

**9.** $\cos\left(t - \dfrac{\pi}{2}\right) \equiv \sin t$

**10.** $\cos\left(\dfrac{\pi}{2} - t\right) \equiv \sin t$

**11.** $\cot\left(t - \dfrac{\pi}{2}\right) \equiv -\tan t$

**12.** $\sec\left(t + \dfrac{\pi}{2}\right) \equiv -\csc t$

**13.** $\sec\left(\frac{\pi}{2} - t\right) \equiv \csc t$

**14.** $\sin(t + \pi) \equiv -\sin t$

**15.** $\cos(t + \pi) \equiv -\cos t$

**16.** $\sin(t - \pi) \equiv -\sin t$

**17.** $\cos(\pi - t) \equiv -\cos t$

**18.** $\sin(t - 2\pi) \equiv \sin t$

**19.** $\sin(-t) \equiv -\sin t$

**20.** $\cos(-t) \equiv \cos t$

In Exercises 21–39 complete each sentence so that the result is an identity.

**21.** $\sin^2 t +$ ______ $\equiv 1$

**22.** $1 + \tan^2 t \equiv$ ______

**23.** ______ $+ \cot^2 t \equiv \csc^2 t$

**24.** $1 -$ ______ $\equiv \cos^2 x$

**25.** $\sec^2 x -$ ______ $\equiv \tan^2 x$

**26.** ______ $- 1 \equiv \cot^2 x$

**27.** ______ $- \cos^2 y \equiv \sin^2 y$

**28.** $-\sin^2 y - \cos^2 y \equiv$ ______

**29.** $\csc^2 y - 1 \equiv$ ______

**30.** $\tan z \cdot \cot z \equiv$ ______

**31.** ______ $\equiv \dfrac{1}{\sin t}$

**32.** ______ $\equiv \dfrac{1}{\cos t}$

**33.** ______ $\equiv \dfrac{1}{\tan t}$

**34.** $\dfrac{\sin u}{\cos u} \equiv$ ______

**35.** $\dfrac{\cos u}{\sin u} \equiv$ ______

**36.** $\sin\left(t + \frac{\pi}{2}\right) \equiv$ ______

**37.** $\cos\left(\frac{\pi}{2} - t\right) \equiv$ ______

**38.** $\sin\left(t - \frac{\pi}{2}\right) \equiv$ ______

**39.** $\cos\left(t - \frac{\pi}{2}\right) \equiv$ ______

In Exercises 40–47 multiply and simplify each expression.

**40.** $(\cos t - \sin t)^2$

**41.** $(\sec x + \cos x)^2$

**42.** $(\sin y + \csc y)(\sin^2 y - 1 + \csc^2 y)$

**43.** $(\cot t + 1)^2$

**44.** $(\csc x + \sec x)(\cos x + \sin x)$

**45.** $(\sec u - 1)(\sec u + 1)$

**46.** $(\cos y - \sec y)(\cos^2 y + 1 + \sec^2 y)$

**47.** $\cos\left(x - \frac{\pi}{2}\right)\cos\left(\frac{\pi}{2} - x\right) + \sin\left(x + \frac{\pi}{2}\right)\sin\left(\frac{\pi}{2} - x\right)$

In Exercises 48–55 factor and simplify each expression.

**48.** $\sin^2 t + \sin^2 t \cot^2 t$

**49.** $\tan^4 y - \sec^4 y$

**50.** $1 - 2\sin^2 x + \sin^4 x$

**51.** $1 - \sin^3 t$

**52.** $\cos^3 y + \sec^3 y$

**53.** $\tan^2 x \csc^2 x - \tan^2 x$

**54.** $\cos\left(y - \frac{\pi}{2}\right)\tan y + \sin y \cot y$

**55.** $\sin(\pi - z) - \cos\left(\frac{\pi}{2} - z\right)\tan z$

**56.** Express $\tan t$ in terms of $\cos t$, where $t$ determines a point in quadrant I.

**57.** Express $\sec t$ in terms of $\sin t$, where $t$ determines a point in quadrant II.

**58.** Express $\csc t$ in terms of $\cos t$, where $t$ determines a point in quadrant III.

**59.** Express $\cot t$ in terms of $\cos t$, where $t$ determines a point in quadrant IV.

**60.** Suppose that $t$ determines a point in quadrant II and that $\sec t = -\frac{5}{3}$. Use the fundamental identities to find **(a)** $\tan t$, **(b)** $\sin t$, and **(c)** $\csc t$.

**61.** Suppose that $t$ determines a point in quadrant III and that $\tan t = \frac{3}{4}$. Use the fundamental identities to find **(a)** $\sin t$, **(b)** $\cos t$, and **(c)** $\sec t$.

In Exercises 62–65 prove that each equation is an identity.

**62.** $\log \sin x = -\log \csc x$

**63.** $\log \tan x = \log \sin x - \log \cos x$

**64.** $\ln \cot x + \ln \sin x = \ln \cos x$

**65.** $\ln \sec x = -\ln \cos x$

## For Review

For each function in Exercises 66–67 determine the **(a)** amplitude, **(b)** period, and **(c)** phase shift.

**66.** $y = 3 \sin (4x - 12\pi)$

**67.** $y = -\frac{1}{2} \cos \left(\frac{x}{2} + 3\pi\right)$

Graph each function in Exercises 68–70.

**68.** $y = 2 \sin \left(x - \frac{\pi}{2}\right)$

**69.** $y = 3 + 2 \sin \frac{1}{3}x$

**70.** $y = 3^x \cos x$

# CHAPTER 6 REVIEW EXERCISES

**1.** Are the angles $\frac{\pi}{3}$ and 30° coterminal?

**2.** Express 148°39′42″ as degrees and the decimal part of a degree, correct to the nearest hundredth.

**3.** Convert 29.6° to radians, correct to four decimal places.

**4.** Convert 7.1359 radians to degrees, minutes, and seconds.

**5.** Find the length of an arc on a circle of diameter 12.00 m if the arc subtends a central angle of measure 80°49′.

**6.** GEOGRAPHY The island of Wake in the Pacific Ocean has latitude 19°20′N. How far is Wake from the North Pole? [Use 4000 mi for the radius of the earth.]

**7.** MANUFACTURING A flywheel is turning at the rate of 50 rpm. What is the radius of the wheel if a point on the rim is moving at $200\pi$ ft/min?

**8.** ENGINEERING A pulley with a 16-inch diameter is driven by a belt which is moving at 1500 ft/min. To the nearest revolution, how many revolutions per minute are made by the pulley?

**9.** Suppose $\theta$ is an acute angle in a right triangle such that $\sin \theta = \frac{\sqrt{5}}{3}$. Sketch an appropriate triangle and find the five remaining trigonometric function values of $\theta$ and $90° - \theta$.

In Exercises 10–12 sketch an appropriate right triangle and give the six trigonometric function values of the given angle. (Do not use a calculator.)

**10.** 30°

**11.** 45°

**12.** 60°

In Exercises 13–15 use a calculator to find each trigonometric function value correct to four decimal places.

**13.** cos 59°26′

**14.** tan 1.3267

**15.** $\csc \frac{\pi}{10}$

In Exercises 16–18 use a calculator to find each acute angle with the given trigonometric function value **(a)** to the nearest hundredth of a degree, and **(b)** in radians correct to four decimal places.

**16.** $\sin \theta = 0.7915$

**17.** $\cot \theta = 3.2541$

**18.** $\sec \theta = 2.1547$

**19.** Solve the right triangle with $a = 5.0$ and $b = 15.0$ by using given parts rather than calculated parts. Remember that $C = 90°$.

**20.** When the angle of elevation of the sun is 64°20′, a pole casts a shadow 28.0 m in length. What is the height of the pole?

**21.** FORESTRY A fire is sighted due west of Lookout A. The bearing of the fire from Lookout B, 5.2 miles due south of A, is N51°40′W. How far is the fire from B? From A?

**22.** NAVIGATION A boat sails for 2 hours at 25 mph in a direction 200°. How far west has it sailed? How far south?

**23.** AERONAUTICS The air speed of an airplane is 550 km/hr and its angle of climb is 9°20′. What is its ground speed?

**24.** RECREATION A pleasure boat produces bow waves which travel at 16 mph. If the angle between the waves is 62°40′, find the speed of the boat. $\left[\text{Hint: Use the equation } \sin\frac{\theta}{2} = \frac{v_w}{v_b}.\right]$

**25.** PHYSICS If a light hits the surface of a lake such that $A = 44°30'$, find $W$ assuming that the index of refraction for water is 1.3 and for air is 1.0003. $\left[\text{Hint: Use the equation } \frac{I_w}{I_a} = \frac{\sin A}{\sin W}.\right]$

**26.** Determine the shortest distance that a point must move on the unit circle, counterclockwise, if it goes from point $A$ to point **(a)** $A$, **(b)** $B$, **(c)** $C$, **(d)** $D$, **(e)** $E$, **(f)** $F$, **(g)** $G$, **(h)** $H$, **(i)** $I$, and **(j)** $J$.

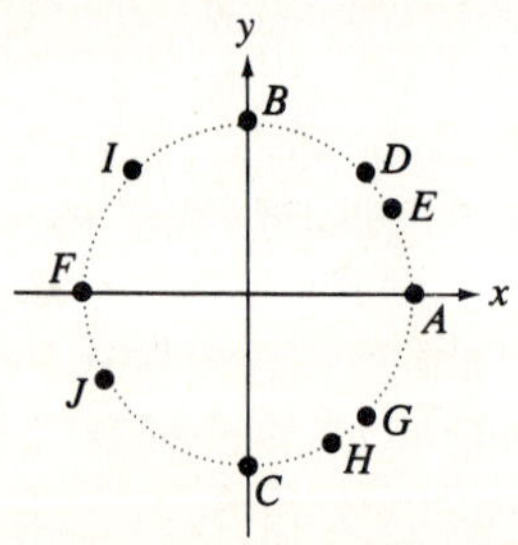


**27.** Repeat Exercise 26 with the point moving in a clockwise direction.

In Exercises 28–30 use the unit circle (no calculator values) and find **(a)** $\sin t$ **(b)** $\cos t$ **(c)** $\tan t$ **(d)** $\csc t$ **(e)** $\sec t$, and **(f)** $\cot t$, for the given real number $t$.

**28.** $\frac{3\pi}{2}$ **29.** $-\frac{\pi}{4}$ **30.** $\frac{7\pi}{6}$

In Exercises 31–32 give the quadrant which contains the point $T$ determined by real number $t$ with the specified function value signs.

**31.** $\tan t > 0$ and $\cos t < 0$ **32.** $\csc t < 0$ and $\cot t < 0$

Sketch the graph of each function in Exercises 33–38.

**33.** $y = \cos x + 1$

**34.** $y = -4 \sin x$

**35.** $y = \tan 4x$

**36.** $y = \frac{1}{2}x + \sin x$

**37.** $y = |\cos x|$

**38.** $y = 2^x \sin x$

In Exercises 39–40 **(a)** give the amplitude, **(b)** give the period, **(c)** give the phase shift, and **(d)** sketch the graph of each function.

**39.** $y = -3 \sin (2x - \pi)$

**40.** $y = 3 \cos \left(\frac{1}{2}t + \frac{\pi}{2}\right)$

In Exercises 41–52 determine whether the given statement is true or false.

**41.** $1 + \sec^2 t \equiv \tan^2 t$

**42.** $1 - \sin^2 t \equiv \cos^2 t$

**43.** $\sin t \equiv \dfrac{\cos t}{\cot t}$

**44.** $\dfrac{1}{\sin t} \equiv \csc t$

**45.** $\cos (-t) \equiv -\cos t$

**46.** $\csc^2 t - \cot^2 t \equiv 1$

**47.** $\sin (t - \pi) \equiv \sin (t + \pi)$

**48.** $\sin (\pi - t) \equiv -\sin (t + \pi)$

**49.** $\dfrac{1}{\cot t} \equiv \csc t$

**50.** $\sin \left(t + \frac{\pi}{2}\right) \equiv \sin \left(\frac{\pi}{2} - t\right)$

**51.** $\cos^2 \left(\frac{\pi}{2} - t\right) + \sin^2 \left(t + \frac{\pi}{2}\right) \equiv 1$

**52.** $\dfrac{1}{\sin \left(t - \frac{\pi}{2}\right)} \equiv \sec t$

In Exercises 53–56 multiply and simplify each expression.

**53.** $\cos \left(\frac{\pi}{2} - x\right)[\csc x - \sin x]$

**54.** $\sin y \tan y(\cot y - \csc y)$

**55.** $(\tan y + \cot y)(\cot^2 y + \tan^2 y - 1)$

**56.** $(\sin x + \cos x)^2$

In Exercises 57–60 factor and simplify each expression.

**57.** $\sin (\pi - x) + \tan x \cos \left(x + \frac{\pi}{2}\right)$

**58.** $\sec^4 y + \sec^2 y \tan^2 y - 2 \tan^4 y$

**59.** $\dfrac{\sin^2 y - 1}{\sin y + 1}$

**60.** $\dfrac{\sin^2 x + 2 \cos \left(x - \frac{\pi}{2}\right) + 1}{\cos \left(\frac{\pi}{2} - x\right) + 1}$

CHAPTER

# 7 APPLICATIONS OF TRIGONOMETRY

Numerous applied problems can be solved by finding the angle or the length of a side of a triangle. In Section 6.4 we considered applications involving right triangles. In this chapter we extend the theory to include triangles that do not contain a right angle. The following two examples illustrate the types of problems that can arise.

**CONSUMER ▶**

A mountain cabin has a wall in the shape of a triangle with two sides of length 26 ft and a third measuring 18 ft. The wall is 6 inches thick and contains a window measuring 4 ft by 5 ft. If insulation costs \$0.90 a cubic foot, what would be the approximate cost of insulating the wall?

**◀ NAVIGATION**

Two docks are located on opposite sides of a river which flows from east to west with a current of 3.5 mph. The bearing of the dock on the north bank from the one on the south bank is N10°20′E. Allowing for drift due to the current, in what direction should the captain of a ferry boat sail at 7 mph, in order to cross directly from the southern dock to the northern dock?

The first of these problems can be solved using a formula for the area of a triangle (see Example 4 in Section 7.4), and the second can be interpreted using the notion of vectors (see Example 7 in Section 7.6). We begin this chapter by considering the trigonometric functions of arbitrary angles, that is, any angle of any measure in degrees or radians. We then solve arbitrary triangles using the law of sines and the law of cosines and conclude by discussing area of triangles and vector applications.

## 7.1 Trigonometric Functions of Arbitrary Angles

The trigonometric functions of acute angles were considered in Section 6.3. In a sense, by thinking of an angle as measured in radians, we have already considered the trigonometric functions of arbitrary angles when the trigonometric functions of a real number were defined in Section 6.5. However, since numerous applied problems involve angles that are not acute, it is appropriate to consider the trigonometric functions of angles in general in more detail. To do this suppose that $A$ is an angle in standard position in a rectangular coordinate system, and point $P(a, b)$ is an arbitrary point on the terminal side of $A$ as shown in Figure 1.

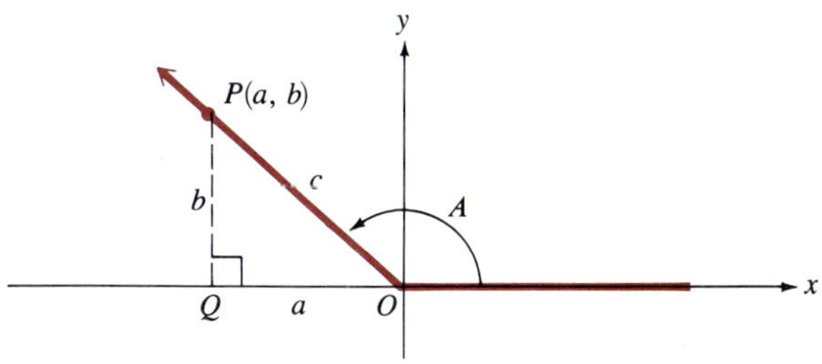


**Figure 1 Angle in Standard Position**

Using the distance formula with the points $(a, b)$ and $(0, 0)$, we can calculate $c$, the hypotenuse of right triangle $PQO$, and obtain $c = \sqrt{a^2 + b^2}$.

### Trigonometric Functions of an Arbitrary Angle

Let $A$ be an arbitrary angle in standard position in a rectangular coordinate system with $P(a, b)$ an arbitrary point other than the origin on the terminal side of $A$. If $c = \sqrt{a^2 + b^2}$, then

$$\sin A = \frac{b}{c} \qquad \csc A = \frac{c}{b} \quad (b \neq 0)$$

$$\cos A = \frac{a}{c} \qquad \sec A = \frac{c}{a} \quad (a \neq 0)$$

$$\tan A = \frac{b}{a} \quad (a \neq 0) \qquad \cot A = \frac{a}{b} \quad (b \neq 0).$$

If $R(a', b')$ is another point on the terminal side of $A$, then by similar triangles

$$\frac{b}{c} = \frac{b'}{c'}, \quad \frac{c}{b} = \frac{c'}{b'}, \quad \frac{a}{c} = \frac{a'}{c'}, \quad \frac{c}{a} = \frac{c'}{a'}, \quad \frac{b}{a} = \frac{b'}{a'}, \text{ and } \frac{a}{b} = \frac{a'}{b'}.$$

Thus the trigonometric function values do not depend on the choice of the point $P(a, b)$.

Recall from Section 6.3 that for the acute angle $A$ in triangle $POQ$ (see Figure 2),

$$\sin A = \frac{\text{opposite side}}{\text{hypotenuse}} = \frac{b}{c}.$$

This agrees with sin $A$ of an arbitrary angle as defined in the screened box above. If we let $c = 1$ as in Figure 3, then $P(a, b)$ is a point on the unit circle. Remember from Section 6.5 that the measure of angle $A$ in radians is $t$ radians. Then, using the sine function again as an example,

$$\sin A = \frac{b}{c} = \frac{b}{1} = b = \sin t.$$

Thus, the definitions of trigonometric functions for arbitrary angles agree with the definitions of trigonometric functions for real numbers as well as those for acute angles.

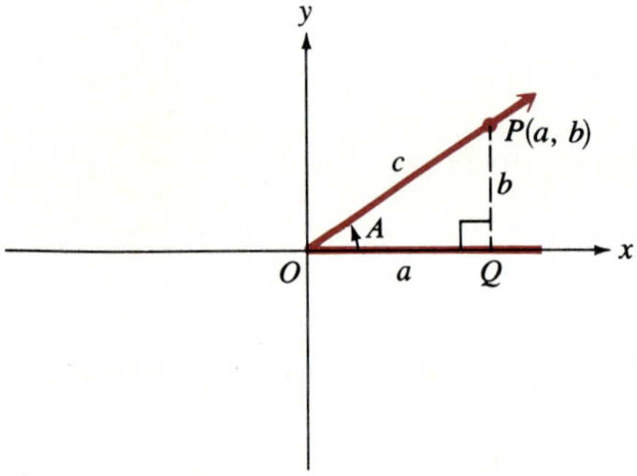


**Figure 2 Acute Angle**

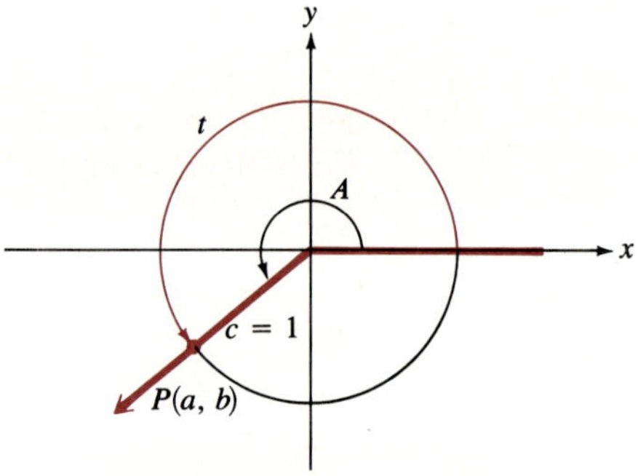


**Figure 3 Arbitrary Angle**

Since the definitions agree, all the fundamental identities that we have used are valid for any angles. Also, the signs of the trigonometric function values agree with the rule of signs first presented in Section 6.5 and summarized again in the following table.

**Table of Signs of Trigonometric Functions of Angle *A***

| Quadrant containing the terminal side of $A$ | Positive trigonometric functions of $A$ | Negative trigonometric functions of $A$ |
|---|---|---|
| **I** | All | None |
| **II** | sin, csc | cos, sec, tan, cot |
| **III** | tan, cot | sin, csc, cos, sec |
| **IV** | cos, sec | sin, csc, tan, cot |

**EXAMPLE 1** Suppose $A$ is an angle in standard position in a rectangular coordinate system, and $P(-5, 2)$ is a point on the terminal side of $A$. Find the six trigonometric function values of $A$.

First make a sketch of angle $A$ with point $P$ as in Figure 4.

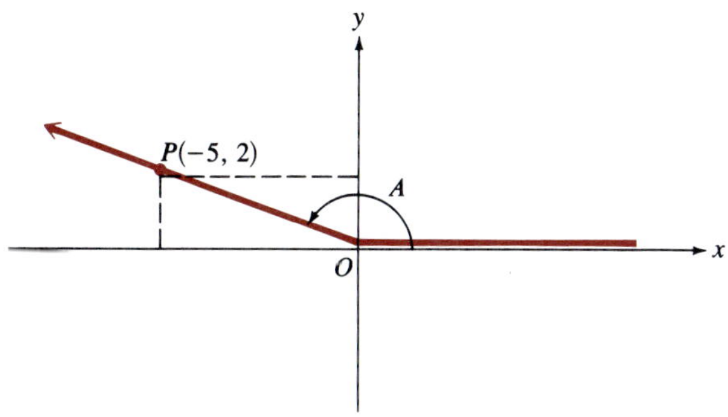


**Figure 4**

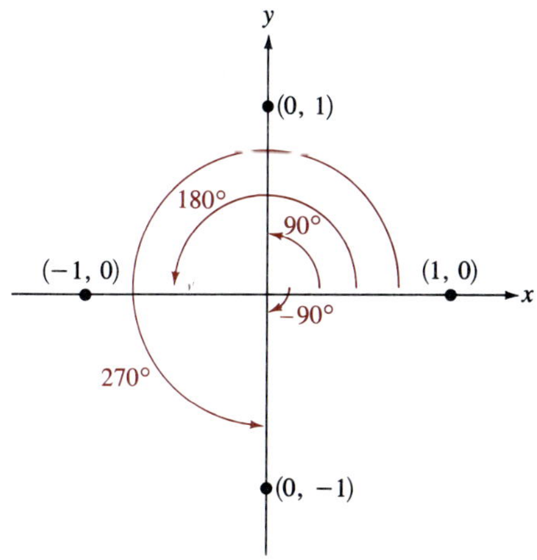


**Figure 5 Quadrantal Angles**

Use the distance formula to obtain $c$.

$$c = \sqrt{(-5)^2 + 2^2} = \sqrt{25 + 4} = \sqrt{29}$$

Then

$$\sin A = \frac{b}{c} = \frac{2}{\sqrt{29}} = \frac{2\sqrt{29}}{29} \qquad \csc A = \frac{c}{b} = \frac{\sqrt{29}}{2}$$

$$\cos A = \frac{a}{c} = -\frac{5}{\sqrt{29}} = -\frac{5\sqrt{29}}{29} \qquad \sec A = \frac{c}{a} = -\frac{\sqrt{29}}{5}$$

$$\tan A = \frac{b}{a} = -\frac{2}{5} \qquad \cot A = \frac{a}{b} = -\frac{5}{2}. \quad \blacksquare$$

## Quadrantal Angles

Angles that have their terminal side on one of the axes are called **quadrantal angles.** Every quadrantal angle passes through one of the points (1, 0), (0, 1), (−1, 0), or (0, −1), like the angles shown in Figure 5. Since we already know how to compute the trigonometric function values of numbers such as $\pi/2$, $\pi$, $3\pi/2$, and $-\pi/2$, we simply identify each trigonometric function of a quadrantal angle given in degrees with the trigonometric function of the number corresponding to the radian measure of the angle. The table on the next page summarizes the trigonometric function values of the quadrantal angles from 0° to 360°, inclusive.

| $A$ (radians) | $A$ (degrees) | $\sin A$ | $\cos A$ | $\tan A$ | $\csc A$ | $\sec A$ | $\cot A$ |
|---|---|---|---|---|---|---|---|
| 0 | 0° | 0 | 1 | 0 | undefined | 1 | undefined |
| $\frac{\pi}{2}$ | 90° | 1 | 0 | undefined | 1 | undefined | 0 |
| $\pi$ | 180° | 0 | $-1$ | 0 | undefined | $-1$ | undefined |
| $\frac{3\pi}{2}$ | 270° | $-1$ | 0 | undefined | $-1$ | undefined | 0 |
| $2\pi$ | 360° | 0 | 1 | 0 | undefined | 1 | undefined |

## Reference Angles

A useful relationship exists between a nonquadrantal angle and a related acute angle, called its *reference angle*.

### Reference Angle

Suppose $A$ is a nonquadrantal angle in standard position in a rectangular coordinate system. The **reference angle** for $A$ is the acute angle $R$ formed by the terminal side of $A$ and the $x$-axis.

Figure 6 shows several angles $A$ along with $R$, the reference angle for $A$. Consider the right triangle $PQO$ containing the reference angle $R$. For angle $A$ in each of the four quadrants as shown in Figure 6, if $c = \sqrt{a^2 + b^2}$, then $c$ is the hypotenuse of triangle $PQO$. The legs of the triangle have length $a$ and $b$ in quadrant I, $|a|$ and $b$ in quadrant II, $|a|$ and $|b|$ in quadrant III, and $a$ and $|b|$ in quadrant IV. In all cases, the trigonometric functions of $A$ are the same as the trigonometric functions of the acute angle $R$ in right triangle $PQO$ *except* possibly for the sign. For example, with $A$ in quadrant II, $a < 0$ so that $|a| = -a$. Thus,

$$\sin R = \frac{b}{c} = \sin A \qquad \csc R = \frac{c}{b} = \csc A$$

$$\cos R = \frac{|a|}{c} = \frac{-a}{c} = -\cos A \qquad \sec R = \frac{c}{|a|} = \frac{c}{-a} = -\sec A$$

$$\tan R = \frac{b}{|a|} = \frac{b}{-a} = -\tan A \qquad \cot R = \frac{|a|}{b} = \frac{-a}{b} = -\cot A.$$

Similar results hold for $A$ in the other quadrants. This leads to the following theorem.

### Reference Angle Theorem

Suppose $A$ is a nonquadrantal angle in standard position in a rectangular coordinate system with reference angle $R$. To find a trigonometric function value of $A$, find the corresponding function value of $R$ and attach the appropriate sign (+ or −). Use the table of signs of the function values of $A$ given in this section.

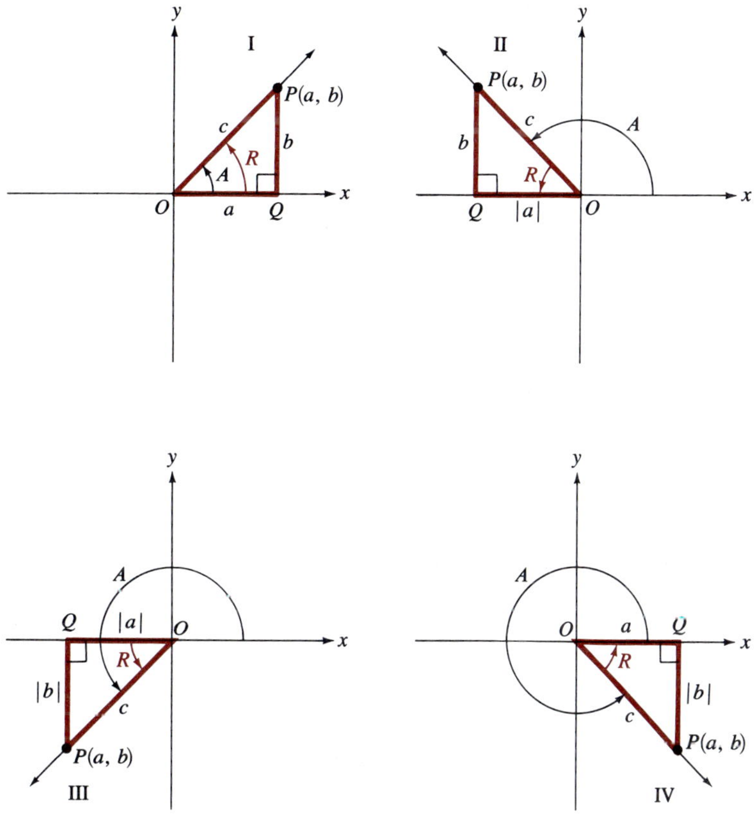


**Figure 6 Reference Angle**

**EXAMPLE 2** Find the six trigonometric function values of 330° by using the reference angle for 330°.

The reference angle for 330° is 30° as shown in Figure 7. Recall that in a 30°-60° right triangle, the side opposite the 30° angle is one-half the hypotenuse. Suppose we sketch such a triangle with hypotenuse of length 2. Then the side opposite the 30° angle is 1, and by the Pythagorean theorem, the side opposite the 60° angle is $\sqrt{3}$, as shown in Figure 8.

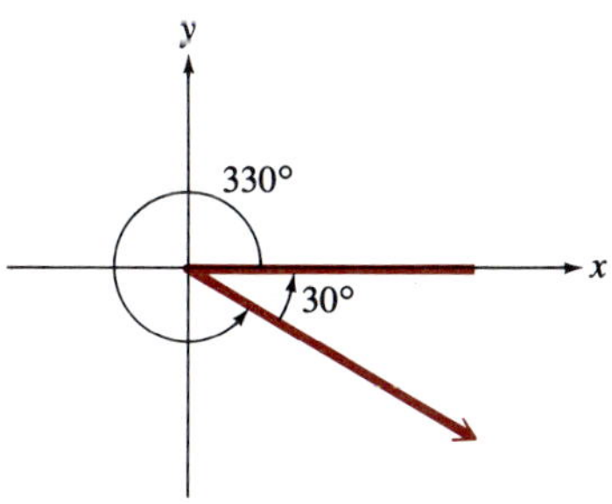


**Figure 7**

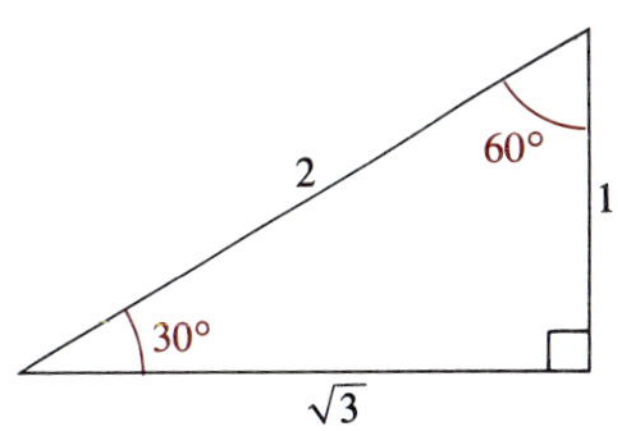


**Figure 8 30°-60° Right Triangle**

Since 330° is in quadrant IV, the only trigonometric function values that are positive are the cosine and secant. With this information we can find the desired values.

$$\sin 330° = -\sin 30° = -\frac{\text{opp}}{\text{hyp}} = -\frac{1}{2}$$

$$\cos 330° = \cos 30° = \frac{\text{adj}}{\text{hyp}} = \frac{\sqrt{3}}{2}$$

$$\tan 330° = -\tan 30° = -\frac{\text{opp}}{\text{adj}} = -\frac{1}{\sqrt{3}} = -\frac{\sqrt{3}}{3}$$

$$\csc 330° = -\csc 30° = -\frac{\text{hyp}}{\text{opp}} = -\frac{2}{1} = -2$$

$$\sec 330° = \sec 30° = \frac{\text{hyp}}{\text{adj}} = \frac{2}{\sqrt{3}} = \frac{2\sqrt{3}}{3}$$

$$\cot 330° = -\cot 30° = -\frac{\text{adj}}{\text{opp}} = -\frac{\sqrt{3}}{1} = -\sqrt{3}$$ ■

It is not necessary to use reference angles to find trigonometric functions of arbitrary angles when a calculator is used. For example, to find cos 330° with a calculator, we use the following sequence (making sure the calculator is in degree mode).

ALG & RPN: 330 [cos] → [0.8660254]

If we find cos 30° with a calculator,

ALG & RPN: 30 [cos] → [0.8660254]

so that cos 330° = cos 30° = 0.8660254, which is an approximation of the exact value, $\frac{\sqrt{3}}{2}$. However, to find an angle from a given functional value, knowledge of reference angles becomes extremely important. For example, suppose we were given that $\cos A = 0.8660254$, and were asked to find $A$. If we follow the steps below,

ALG & RPN: 0.8660254 [INV] [cos] → [30]

we obtain $A = 30°$. However, this is only one possible value of $A$ since, as we saw above, $A$ could also be 330°. To find other values of $A$, we must recognize that there are many angles whose cosine is 0.8660254. They all have reference angle 30°, and, since this cosine is positive, they are all in quadrant I or quadrant IV. One such angle is 330°.

It is helpful to be aware of another constraint with calculators. Most calculators, when the [INV] [cos] keys are used, give the results in the range 0° to 180°, or 0 to $\pi$ radians. However, [INV] [sin] or [INV] [tan] calculations give results in the range −90° to 90°, or $-\frac{\pi}{2}$ to $\frac{\pi}{2}$ radians. The following table summarizes the values of an angle $A$ that are obtained with the typical calculator. These constraints will become more apparent when inverse trigonometric functions are considered in detail in Section 8.5.

| Calculator entry | Gives $A$ in degrees | in radians |
|---|---|---|
| INV sin | $-90° \leq A \leq 90°$ | $-\frac{\pi}{2} \leq A \leq \frac{\pi}{2}$ |
| INV cos | $0° \leq A \leq 180°$ | $0 \leq A \leq \pi$ |
| INV tan | $-90° < A < 90°$ | $-\frac{\pi}{2} < A < \frac{\pi}{2}$ |

For example, we saw above that

$$\sin 330° = -\frac{1}{2} = -0.5.$$

Suppose we work in reverse following the steps below.

ALG: 0.5 [+/−] [INV] [sin] → [−30]
RPN: 0.5 [CHS] [INV] [sin] → [−30]

Instead of 330°, the calculator gives the coterminal angle −30°. As a result, if the desired angle was known to be between 0° and 360°, we would need to subtract 30°, the reference angle, from 360° to obtain 330°.

**EXAMPLE 3** Find angle $A$, to the nearest hundredth of a degree, if $0 \leq A < 360°$.

**(a)** $\sin A = -0.8155$
Since $\sin A$ is negative, there are two values for $A$, one in quadrant III and the other in quadrant IV.

ALG: 0.8155 [+/−] [INV] [sin] → [−54.63683]
RPN: 0.8155 [CHS] [INV] [sin] → [−54.63683]

This value of $A$, although in quadrant IV, is not between 0° and 360°. The coterminal angle in this range has reference angle 54.63683°, so that

$$A = 360° - 54.63683° = 305.36317° \approx 305.36°.$$

Also, the angle in quadrant III with reference angle 54.63683° is

$$A = 180° + 54.63683° = 234.63683° \approx 234.64°.$$

**(b)** $\cot A = -3.9665$

Since $\cot A$ is negative, the two values for $A$ are in quadrant II and quadrant IV.

ALG: 3.9665 [+/−] [1/x] [INV] [tan] → [−14.150047]

RPN: 3.9665 [CHS] [1/x] [INV] [tan] → [−14.150047]

The angle in quadrant IV with reference angle 14.150047° and between 0° and 360° is

$$A = 360° - 14.150047° = 345.84995° \approx 345.85°.$$

The angle in quadrant II with the same reference angle is

$$A = 180° - 14.150047° = 165.84995° \approx 165.85°. \quad ■$$

## 7.1 EXERCISES

In Exercises 1–14 $A$ is an angle in standard position and satisfies the given conditions. Find the six trigonometric function values of $A$. Do not use a calculator.

**1.** $P(3, 4)$ is on the terminal side of $A$.

**2.** $P(-3, 4)$ is on the terminal side of $A$.

**3.** $P(-3, -4)$ is on the terminal side of $A$.

**4.** $P(3, -4)$ is on the terminal side of $A$.

**5.** The terminal side of $A$ is in quadrant I and lies on the line $y = 5x$.

**6.** The terminal side of $A$ is in quadrant III and lies on the line $y = 5x$.

**7.** The terminal side of $A$ is in quadrant II and is parallel to the line $2x + y - 3 = 0$.

**8.** The terminal side of $A$ is in quadrant IV and is parallel to the line $2x + y - 3 = 0$.

**9.** The terminal side of $A$ is in quadrant I and is perpendicular to the line $2x + y - 3 = 0$.

**10.** The terminal side of $A$ is in quadrant III and is perpendicular to the line $2x + y - 3 = 0$.

**11.** The terminal side of $A$ is in quadrant II and parallel to the line passing through the points $(-4, 1)$ and $(2, -2)$.

**12.** The terminal side of $A$ is in quadrant III and perpendicular to the line passing through the points $(-4, 1)$ and $(2, -2)$.

**13.** The terminal side of $A$ bisects the first quadrant.

**14.** The terminal side of $A$ bisects the fourth quadrant.

In Exercises 15–22 find the six trigonometric function values of the given angle $A$.

**15.** $A = 0° = 0$

**16.** $A = 90° = \frac{\pi}{2}$

**17.** $A = 180° = \pi$

**18.** $A = 270° = \frac{3\pi}{2}$

**19.** $A = -360° = -2\pi$

**20.** $A = -270° = -\frac{3\pi}{2}$

**21.** $A = -180° = -\pi$

**22.** $A = -90° = -\frac{\pi}{2}$

In Exercises 23–28 determine the quadrant containing the terminal side of $A$ under the given conditions.

**23.** $\tan A > 0$ and $\cos A > 0$ **24.** $\tan A > 0$ and $\sec A < 0$ **25.** $\sin A > 0$ and $\cos A < 0$

**26.** $\sin A < 0$ and $\sec A > 0$ **27.** $\csc A > 0$ and $\cot A < 0$ **28.** $\cos A > 0$ and $\csc A < 0$

In Exercises 29–44 determine the reference angle for angle $A$.

**29.** $A = 137°$ **30.** $A = 23°$ **31.** $A = -40.2°$ **32.** $A = -512.8°$

**33.** $A = -325°15'$ **34.** $A = 749°25'$ **35.** $A = -178°45'$ **36.** $A = 862°55'$

**37.** $A = \frac{3\pi}{4}$ **38.** $A = -\frac{7\pi}{4}$ **39.** $A = \frac{5\pi}{6}$ **40.** $A = -\frac{11\pi}{6}$

**41.** $A = -\frac{\pi}{3}$ **42.** $A = -\frac{\pi}{6}$ **43.** $A = -\frac{8\pi}{3}$ **44.** $A = -\frac{11\pi}{3}$

In Exercises 45–48 find the six trigonometric function values of $A$ (correct to four decimal places) by using the reference angle for $A$. Check your work by finding the functional values directly using a calculator.

**45.** $A = -50°$ **46.** $A = 235°$ **47.** $A = 920°$ **48.** $A = -948°$

In Exercises 49–54 find angle $A$ to the nearest minute, if $0 \le A < 360°$.

**49.** $\sin A = -0.4215$ **50.** $\cos A = -0.7259$ **51.** $\tan A = 2.2175$

**52.** $\sec A = 2.4135$ **53.** $\csc A = -4.3277$ **54.** $\cot A = -3.2553$

In Exercises 55–60 find the remaining five trigonometric function values of angle $A$ with the given value and condition.

**55.** $\cos A = \frac{4}{5}$ and $A$ is in quadrant IV. **56.** $\tan A = -\frac{5}{12}$ and $A$ is in quadrant II.

**57.** $\sin A = -\frac{2}{3}$ and $A$ is in quadrant III. **58.** $\cot A = \frac{12}{5}$ and $A$ is in quadrant III

**59.** $\sin A = \frac{3}{2}$ and $A$ is in quadrant I. **60.** $\sec A = -\frac{1}{3}$ and $A$ is in quadrant II.

## For Review

In the next section we solve triangles that are not right triangles. Exercise 61 gives a review of solutions to right triangles.

**61.** **SURVEYING** Two cabins in a valley are viewed from the top of a cliff. Assume that the cabins and the base of the cliff are on a line and the observer is 2000 ft above the valley floor. The angles of depression to the cabins are 31°20′ and 48°10′. How far apart are the cabins?

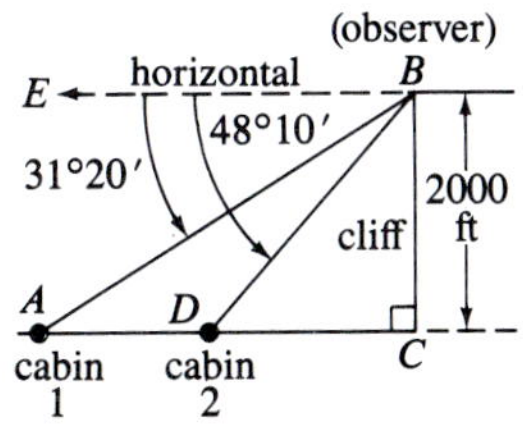

## 7.2 The Law of Sines

### Solving Oblique Triangles

In Chapter 6 we solved right triangles using the trigonometric functions. Triangles that do not contain a right angle, called **oblique triangles,** can also be solved trigonometrically. If three parts of a triangle are known, the remaining parts can be found. At least one part must be a side, for there is an unlimited number of similar triangles having three given angles.

We shall study two methods for solving oblique triangles, the *law of sines* and the *law of cosines*. The method to be used depends upon which parts of the triangle are known. Four possible cases exist.

Case I. One side and two angles are given *(AAS)*. In this case we know all three angles since the third angle can be found by subtracting the sum of the other two from 180°. This case is solvable by the law of sines.

Case II. Two sides and an angle opposite one of these sides are given *(SSA)*. This case can lead to an ambiguous situation where there may be no solution, exactly one solution, or two solutions to the triangle. The law of sines is applicable when there is a solution (or solutions).

Case III. Two sides and the angle between them are given *(SAS)*. This case requires the law of cosines, usually followed by the law of sines.

Case IV. Three sides are given *(SSS)*. This case requires the law of cosines, usually followed by the law of sines.

### Proof of the Law of Sines

We now derive the law of sines using an **acute triangle,** in which all angles are acute. The derivation using an **obtuse triangle,** in which one angle is obtuse, is similar.

In acute triangle $ABC$, draw an altitude from vertex $C$ as in Figure 9. Then, relative to triangles $ACP$ and $BCP$, we have

$$\frac{h}{b} = \sin A \qquad \text{and} \qquad \frac{h}{a} = \sin B.$$

Then $h = b \sin A$ and $h = a \sin B$ so that

$$b \sin A = a \sin B.$$

Divide both sides by $\sin A \sin B$ (which is never zero since $A$ and $B$ are neither 0° nor 180°).

$$\frac{b}{\sin B} = \frac{a}{\sin A}$$

Similarly, if we draw an altitude from vertex $A$, we obtain

$$\frac{c}{\sin C} = \frac{b}{\sin B}.$$

When these results are combined, we obtain the law of sines.

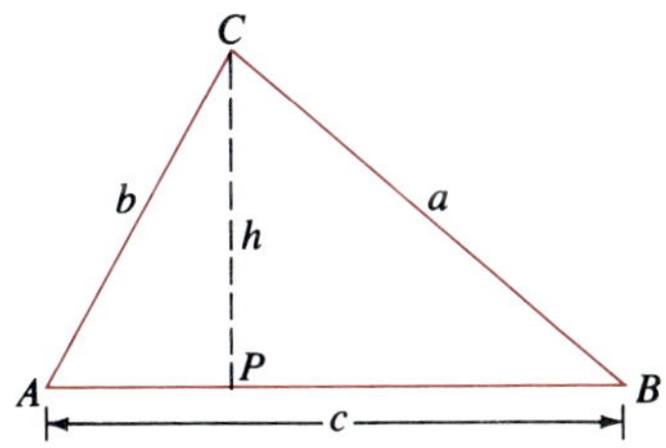


Figure 9 Law of Sines

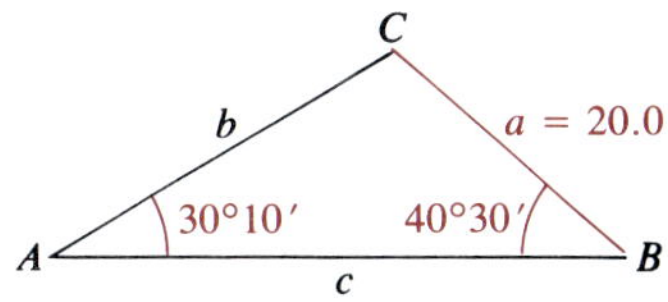


Figure 10

**The Law of Sines**

In any triangle $ABC$,

$$\frac{a}{\sin A} = \frac{b}{\sin B} = \frac{c}{\sin C}.$$

That is, the sides are proportional to the sines of the angles opposite them.

## Solving Triangles

We now solve a Case I triangle, with one side and two angles given *(AAS)*. As with solving right triangles, we follow the precision rule in Section 6.4. We give all sides to the same decimal place as the least precise side, and all angles to the nearest degree, nearest ten minutes, or nearest minute depending on the precision of the sides.

**EXAMPLE 1** Solve triangle $ABC$ in which $a = 20.0$, $A = 30°10'$, and $B = 40°30'$.

Make a fairly accurate sketch, like Figure 10, and identify the parts to be found.

*To find C:* Since $A + B + C = 180°$,

$$30°10' + 40°30' + C = 180°$$
$$C = 180° - 70°40' = 109°20'.$$

*To find b:* Use the equation in the law of sines having three of the known parts and $b$.

$$\frac{a}{\sin A} = \frac{b}{\sin B}$$
$$\frac{\mathbf{20.0}}{\sin \mathbf{30°10'}} = \frac{b}{\sin \mathbf{40°30'}}$$
$$b = \frac{(20.0)\sin 40°30'}{\sin 30°10'}$$
$$\approx 25.8$$

The sequence of calculator steps needed to obtain $b$ is given below.

ALG: 10 [÷] 60 [+] 30 [=] [sin] [STO] 30 [÷] 60 [+] 40 [=] [sin] [×] 20 [=] [÷] [RCL] [=]

RPN: 10 [ENTER] 60 [÷] 30 [+] [sin] [STO] 30 [ENTER] 60 [÷] 40 [+] [sin] 20 [×] [RCL] [÷]

The display will then show [25.847802]. Do not clear the memory because the value of sin 30°10′ can be used in the next step.

*To find c:* $\dfrac{a}{\sin A} = \dfrac{c}{\sin C}$

$$\frac{20.0}{\sin 30°10'} = \frac{c}{\sin 109°20'}$$

$$c = \frac{(20.0)(\sin 109°20')}{\sin 30°10'}$$

$$\approx 37.6$$

ALG: 20 [÷] 60 [+] 109 [=] [sin] [×] 20 [÷] [RCL] [=] → [37.555283]

RPN: 20 [ENTER] 60 [÷] 109 [+] [sin] 20 [×] [RCL] [÷] → [37.555283] ■

## The Ambiguous Case

For Case II, the ambiguous case involving two sides and an angle opposite one of the sides *(SSA)*, there are four possibilities. Each of these is considered in the next four examples.

## EXAMPLE 2

Solve triangle $ABC$ in which $a = 10.0$, $b = 20.0$, and $A = 50°10'$.

When we attempt to draw the triangle in Figure 11, side $a$ does not appear long enough to reach the third side. However, with the information given, we attempt to find $B$.

$$\frac{a}{\sin A} = \frac{b}{\sin B}$$

$$\frac{\mathbf{10.0}}{\sin \mathbf{50°10'}} = \frac{\mathbf{20.0}}{\sin B}$$

$$\sin B = \frac{20.0(\sin 50°10')}{10.0}$$

$$\approx 1.5358$$

Since the sine of an angle can be no larger than 1, there can be no angle $B$ with sine 1.5358. Using the following calculator steps we obtain "error" in the display for the value of $B$.

ALG: 10 [÷] 60 [+] 50 [=] [sin] [×] 20 [÷] 10 [=] [INV] [sin] → [Error]

RPN: 10 [ENTER] 60 [÷] 50 [+] [sin] 20 [×] 10 [÷] [INV] [sin] → [Error]

Thus, there is no triangle that satisfies the given conditions. ■

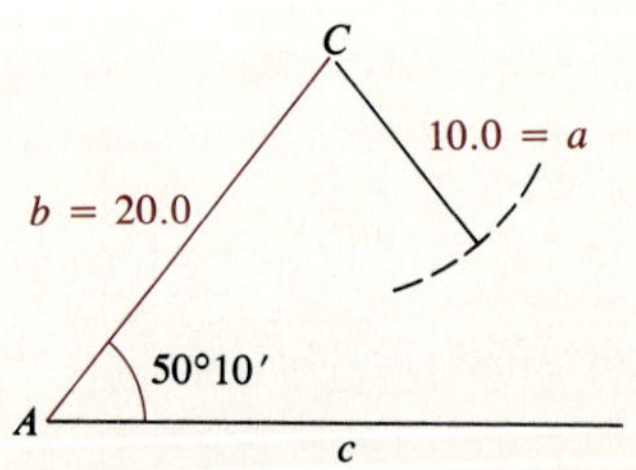


**Figure 11**

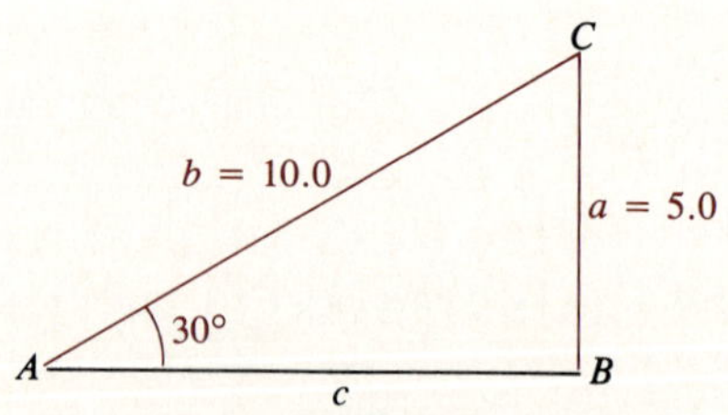


**Figure 12**

**EXAMPLE 3** Solve triangle $ABC$ in which $a = 5.0$, $b = 10.0$, and $A = 30°00'$, as shown in Figure 12.

Substitute into $\dfrac{a}{\sin A} = \dfrac{b}{\sin B}$.

$$\frac{\mathbf{5.0}}{\sin \mathbf{30°}} = \frac{\mathbf{10.0}}{\sin B}$$

$$\sin B = \frac{(10.0)(\sin 30°)}{5.0} = 1$$

$$B = 90°$$

In this case we have exactly one triangle, a right triangle, which satisfies the conditions. We may use the Pythagorean theorem to find $c = \sqrt{75} \approx 8.7$. Also, $C = 90° - A = 90° - 30° = 60°$. ■

**EXAMPLE 4** Solve triangle $ABC$ in which $a = 16.0$, $b = 21.0$, and $A = 30°10'$.

In this case, if we make an accurate sketch, it appears that two different triangles may fit the given information (see Figure 13). Substitute into

$$\frac{a}{\sin A} = \frac{b}{\sin B}.$$

$$\frac{\mathbf{16.0}}{\sin \mathbf{30°10'}} = \frac{\mathbf{21.0}}{\sin B}$$

$$\sin B = \frac{(21.0)(\sin 30°10')}{16.0} \approx 0.6596$$

$$B = 41°20' \quad \text{or} \quad B = 180° - 41°20' = 138°40'$$

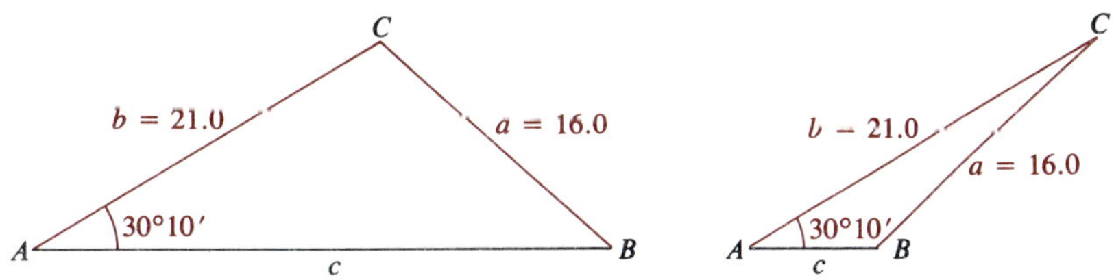


**Figure 13**

Remember that there are *two* angles between 0° and 180° with sine equal to 0.6596, one in the first quadrant (41°20′) and the other in the second quadrant (138°40′). Thus, there are two completely different solutions, corresponding to the figures above.

*First solution:* When $A = 30°10'$, $B = 41°20'$, $a = 16.0$, and $b = 21.0$,

$$C = 180° - (A + B) = 180° - 71°30' = 108°30'.$$

Also,

$$\frac{a}{\sin A} = \frac{c}{\sin C}$$

$$\frac{\mathbf{16.0}}{\sin \mathbf{30°10'}} = \frac{c}{\sin \mathbf{108°30'}}$$

$$c = \frac{(16.0)(\sin 108°30')}{\sin 30°10'} \approx 30.2.$$

*Second solution:* When $A = 30°10'$, $B = 138°40'$, $a = 16.0$, and $b = 21.0$,

$$C = 180° - (A + B) = 180° - 168°50' = 11°10'.$$

Also,

$$\frac{a}{\sin A} = \frac{c}{\sin C}$$

$$\frac{\mathbf{16.0}}{\sin \mathbf{30°10'}} = \frac{c}{\sin \mathbf{11°10'}}$$

$$c = \frac{(16.0)(\sin 11°10')}{\sin 30°10'} \approx 6.2. \quad ■$$

**EXAMPLE 5** Solve triangle $ABC$ in which $a = 24.0$, $b = 9.0$, and $A = 41.5°$.

From the sketch in Figure 14, it appears that only one triangle fits the information.

$$\frac{a}{\sin A} = \frac{b}{\sin B}$$

$$\frac{\mathbf{24.0}}{\sin \mathbf{41.5°}} = \frac{\mathbf{9.0}}{\sin B}$$

$$\sin B = \frac{(9.0)(\sin 41.5°)}{24.0} \approx 0.2485$$

$$B \approx 14.4° \quad \text{or} \quad B = 180° - 14.4° = 165.6°$$

Again we have two angles between 0° and 180° with sine equal to 0.2485. This time, however, $B$ cannot be the larger angle, 165.6°, since then

$$A + B = 41.5° + 165.6° = 207.1°,$$

which is more than the 180° possible for the sum of the angles in the triangle. Thus we have only one solution. When $A = 41.5°$, $B = 14.4°$, $a = 24.0$, and $b = 9.0$, then

$$C = 180° - (A + B) = 180° - 55.9° = 124.1°.$$

Also,

$$\frac{a}{\sin A} = \frac{c}{\sin C}.$$

$$\frac{24.0}{\sin 41.5°} = \frac{c}{\sin 124.1°}$$

$$c = \frac{(24.0)(\sin 124.1°)}{\sin 41.5°} \approx 30.0 \quad ■$$

The key to removing the *ambiguity* from the ambiguous case involving two sides $a$ and $b$ and an angle, $A$, opposite one of them is to calculate $\sin B$.

1. If $\sin B > 1$, there is no solution since no angle has sine $> 1$.
2. If sine $B = 1$, there is exactly one solution, a right triangle, since a 90° angle has sine $= 1$.
3. If $\sin B < 1$, there are one or two solutions, since there are two angles between 0° and 180° with sine $< 1$. If $B$ is the first-quadrant angle and $B'$ is the second-quadrant angle, one solution exists with $B$, and a second solution exists with $B'$ if $B' + A < 180°$.

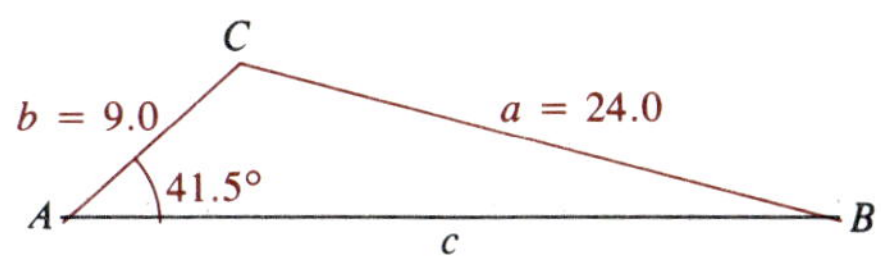


Figure 14

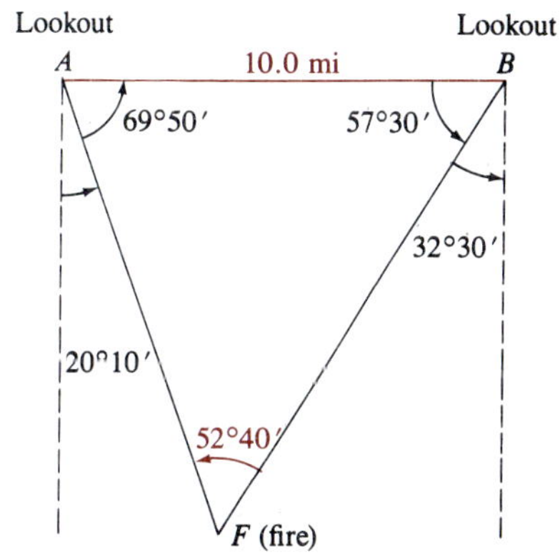


Figure 15 Bearing

NOTE Many applied problems result in oblique triangles. The same steps used to solve problems applying right triangles are also used for other triangle problems.

1. Draw a sketch and label the figure.
2. Indicate the unknown part(s) with variable(s).
3. Find an equation that shows a relationship between known quantities and an unknown part of the triangle. ■

## EXAMPLE 6
FORESTRY

Lookout station $B$ is located 10.0 miles due east of station $A$. The bearing of a fire from $A$ is S20°10′E and the bearing from $B$ is S32°30′W. How far is the fire from $A$?

The sketch in Figure 15 gives the pertinent information.

$$\angle BAF = 90° - 20°10' = 69°50'$$

$$\angle ABF = 90° - 32°30' = 57°30'$$

Thus $$\angle AFB = 180° - (69°50' + 57°30') = 52°40'.$$

Since we have $AAS$, we use the law of sines.

$$\frac{AF}{\sin ABF} = \frac{AB}{\sin AFB}$$

$$\frac{AF}{\sin 57°30'} = \frac{10.0}{\sin 52°40'}$$

$$AF = \frac{(10.0)(\sin 57°30')}{\sin 52°40'} \approx 10.6$$

The fire is about 10.6 miles from lookout station $A$. ■

## EXAMPLE 7
AERONAUTICS

An airplane leaves airport $A$ and flies in a direction 60°20′. At a given time, its direction from city $B$, located 200 miles due west of $A$, is 78°10′. At that time, how far is the airplane from $B$? [Remember navigational direction is measured clockwise from north.]

Referring to the sketch in Figure 16,

$$\angle PBA = 90° - 78°10' = 11°50'$$

$$\angle PAB = 90° + 60°20' = 150°20'.$$

Thus

$$\angle BPA = 180° - (11°50' + 150°20')$$

$$= 180° - 162°10' = 17°50'.$$

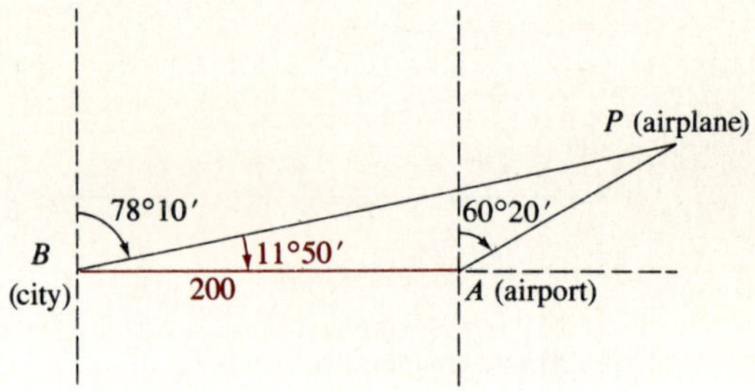


**Figure 16 Navigational Direction**

We must find $PB$, the distance from the plane to city $B$. Using the law of sines,

$$\frac{PB}{\sin PAB} = \frac{BA}{\sin BPA}$$

$$\frac{PB}{\sin 150°20'} = \frac{200}{\sin 17°50'}.$$

$$PB = \frac{(200)(\sin 150°20')}{\sin 17°50'} \approx 323$$

The airplane is approximately 323 miles from city $B$. ■

## 7.2 EXERCISES

In Exercises 1–14 solve each triangle with the given parts.

**1.** $A = 42°30'$, $B = 50°20'$, $a = 10.5$

**2.** $A = 37°10'$, $C = 100°20'$, $b = 5.0$

**3.** $B = 52.75°$, $C = 61.25°$, $a = 12.45$

**4.** $A = 91.92°$, $B = 22.58°$, $c = 6.18$

**5.** $A = 30.0°$, $b = 12.0$, $a = 6.0$

**6.** $B = 45.0°$, $b = 5.0$, $c = 5.0$

**7.** $A = 47°20'$, $b = 18.5$, $a = 3.5$

**8.** $C = 60°30'$, $b = 12.2$, $c = 8.7$

**9.** $B = 28°50'$, $b = 8.0$, $c = 13.5$

**10.** $A = 32°40'$, $b = 17.3$, $a = 10.7$

**11.** $C = 51°10'$, $c = 30.2$, $b = 12.3$

**12.** $B = 48°40'$, $b = 24.3$, $a = 5.6$

**13.** $C = 20°42'$, $b = 11.35$, $c = 7.06$

**14.** $B = 12°18'$, $b = 18.47$, $a = 7.95$

Solve.

**15.** **FORESTRY** Lookout station B is 15.0 km due east of lookout station A. A fire is sighted from B with bearing N23°10′W, while the bearing of the fire from A is N51°20′E. How far is the fire from lookout station A?

**16. FORESTRY** Lookout tower A is located 20.0 mi due north of tower B. The bearing of a fire from A is N41.0°E and the bearing of the fire from B is N22.0°E. How far is the fire from A? From B?

**17. NAVIGATION** A boat sails a course of 20.0° from an island which is located 8.0 mi due east of a lighthouse. In 3 hours its position from the lighthouse is 31.0°. At that time, how far is it from the lighthouse? From the island?

**18. AERONAUTICS** Statton Air Station is located due west of Jackson Field. An airplane leaves Statton and flies in a direction 125°40′. At the instant it is 350 mi from Statton, the direction to Jackson is 64°20′. What is the distance from Statton Air Station to Jackson Field?

**19. SURVEYING** Points $A$ and $B$ are on opposite sides of a canyon. Point $C$ is 1.00 mi from $A$. The measure of $\angle BAC$ is 95°20′ and the measure of $\angle ACB$ is 34°50′. What is the distance between $A$ and $B$?

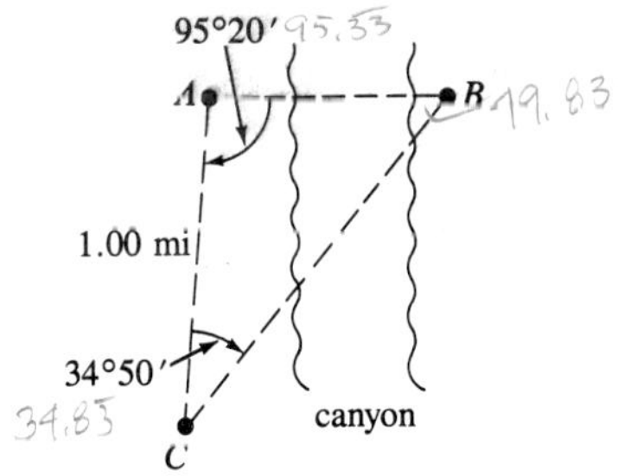


**Exercise 19**

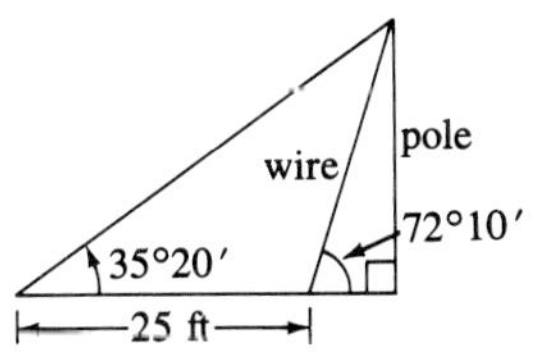


**Exercise 20**

**20. ENGINEERING** A guy wire to the top of a pole makes an angle of 72°10′ with level ground. At a point 25 ft farther from the base of the pole than the guy wire, but in line with the base of the wire, the angle of elevation to the top of the pole is 35°20′. What is the length of the guy wire?

**21. SURVEYING** From a point on one island, 1500 ft from the shore, the angle of depression of the shore of a second island is 10°30′. If the angle of depression of the shore of the first island from the same point is 32°40′, how far apart are the two islands?

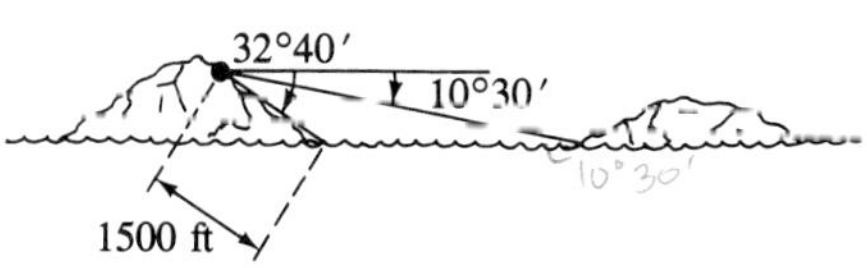


**Exercise 21**

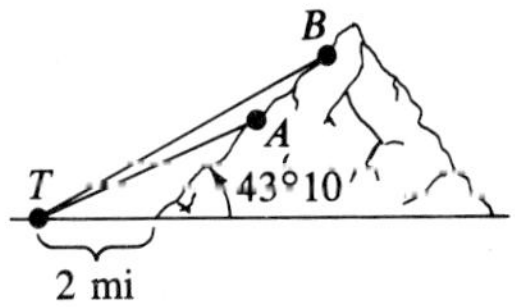


**Exercise 22**

**22. ENGINEERING** The side of a mountain makes an angle of 43°10′ with the horizontal. The angles of elevation of two points $A$ and $B$ on the side of the mountain from a town at point $T$, 2 miles from the base of the mountain, are 10°30′ and 11°10′. What is the approximate distance between $A$ and $B$?

**23. NAVIGATION** Two ships, the USS Lincoln and the USS Wilson, are 120.0 nautical miles apart when they receive a distress signal from the pilot of a downed aircraft. The direction of the signal is 35°15′ from the Lincoln, the direction of the Wilson is 82°35′ from the Lincoln, and the direction of the signal from the Wilson is 319°55′. Which ship should reach the pilot first if they both can travel at a rate of 38 knots? How long will it take the first rescue ship to arrive?

**24.** **ASTRONOMY** Assume that the orbits of Mercury and Earth are circular in the same plane, Earth is $9.3 \times 10^7$ miles from the sun, and Mercury is $3.6 \times 10^7$ miles from the sun. Suppose that Mercury is sighted from Earth, and the angle formed by the sun, Earth, and Mercury, with Earth at the vertex, is 8.35°. **(a)** How far is Mercury from Earth at this time? **(b)** What would be the maximum angle possible between the sun, Earth, and Mercury?

**25.** **ENGINEERING** A radio transmitter tower was blown over during Hurricane Bruce and ended up in the position shown in the figure below. What was the original height of the tower?

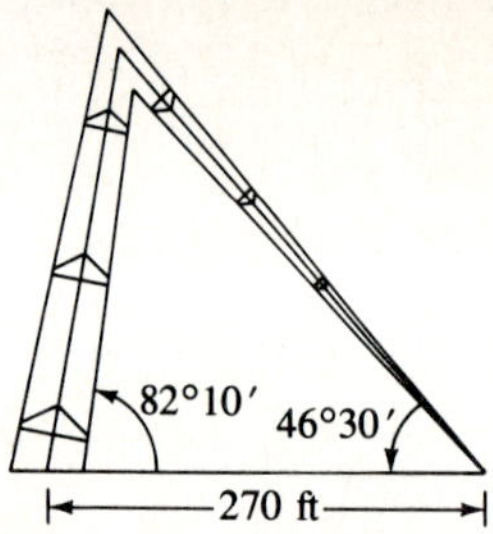


**26.** The formula

$$(a - b) \cos \frac{C}{2} = c \sin \left(\frac{A - B}{2}\right),$$

called **Mollweide's formula,** has the property that all six parts of any triangle are used. For this reason, the formula can be used to check the solution of the triangle. That is, if the six values are substituted and the left side does not equal the right, then an error must have been made in solving the triangle. Use Mollweide's formula to check the solution to the triangle in Exercise 1.

## For Review

**27.** What is the reference angle for $A = -251°40'$?

**28.** Find the six trigonometric function values of angle $A$ in standard position if the terminal side of $A$ in quadrant II is parallel to the line $4x + y - 8 = 0$.

**29.** Find angle $A$, to the nearest minute, if $0 \le A < 360°$ and $\tan A = -4.1395$.

**30.** Suppose that $A$ is in quadrant IV and $\cot A = -\frac{4}{5}$. Find the remaining function values of $A$.

## 7.3 The Law of Cosines

In Section 7.2 we solved oblique triangles when one side and two angles were given *(AAS)* and when two sides and an angle opposite of one of them were given *(SSA)*. When two sides and the included angle *(SAS)* or when three sides *(SSS)* are known, the law of sines is not applicable since there will always be two unknowns in any form of the equation. In these two cases, the *law of cosines* can be used.

## Proof of the Law of Cosines

Draw an altitude from vertex $C$ to the opposite side of triangle $ABC$ in Figure 17. Applying the Pythagorean theorem to right triangles $ACD$ and $BCD$ we have

$$b^2 = x^2 + h^2 \quad \text{and} \quad a^2 = h^2 + (c - x)^2.$$

Solve both equations for $h^2$.

$$h^2 = b^2 - x^2 \quad \text{and} \quad h^2 = a^2 - (c - x)^2$$

Set the results equal and solve for $a^2$.

$$b^2 - x^2 = a^2 - (c - x)^2$$
$$b^2 - x^2 = a^2 - c^2 + 2cx - x^2$$
$$b^2 = a^2 - c^2 + 2cx$$
$$a^2 = b^2 + c^2 - 2cx$$

In triangle $ACD$, $\cos A = x/b$ so that $x = b \cos A$. Substituting, we have

$$a^2 = b^2 + c^2 - 2bc \cos A.$$

Similarly, $$b^2 = a^2 + c^2 - 2ac \cos B$$

and $$c^2 = a^2 + b^2 - 2ab \cos C$$

as can be seen by drawing an altitude from the other vertices.

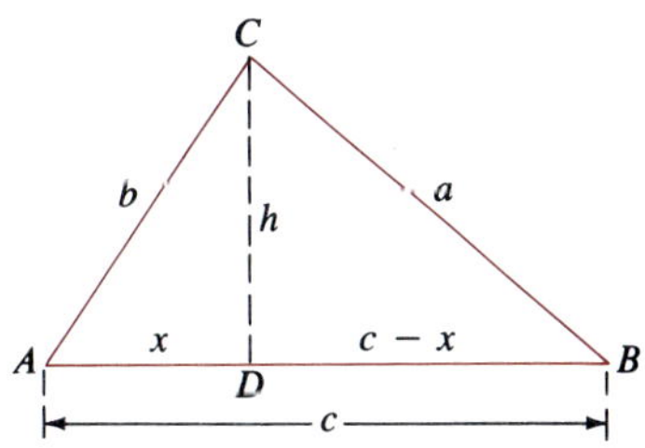


**Figure 17 Law of Cosines**

**Law of Cosines**

In a triangle $ABC$,

$$a^2 = b^2 + c^2 - 2bc \cos A$$
$$b^2 = a^2 + c^2 - 2ac \cos B$$
$$c^2 = a^2 + b^2 - 2ab \cos C.$$

It is best to remember the law of cosines in one sentence that covers all three equations.

**Law of Cosines (Alternate Form)**

The square of the length of any side of a triangle is equal to the sum of the squares of the lengths of the other two sides minus twice the product of these lengths and the cosine of the angle between them.

The law of cosines is often called a generalization of the Pythagorean theorem since if the angle in the formula is a right angle, its cosine is 0 and the term $-2bc \cos A$ (or $-2ac \cos B$ or $-2ab \cos C$) drops out, leaving the Pythagorean theorem.

## Solving Triangles

We first consider an example of Case III (see Section 7.2) with two sides and the included angle *(SAS)*.

**EXAMPLE 1**

Solve triangle $ABC$ in which $A = 58°30'$, $b = 10.0$, and $c = 15.0$.

Make a sketch of the triangle as in Figure 18 and observe that two sides and the included angle *(SAS)* are given. Notice that if we attempt to use the law of sines,

$$\frac{a}{\sin 58°30'} = \frac{10}{\sin B} \quad \text{or} \quad \frac{a}{\sin 58°30'} = \frac{15}{\sin C},$$

we obtain an equation with two unknowns. This shows why the law of cosines is necessary.

*To find a:* Substitute into the formula

$$a^2 = b^2 + c^2 - 2bc \cos A.$$
$$a^2 = (\mathbf{10})^2 + (\mathbf{15})^2 - 2(\mathbf{10})(\mathbf{15}) \cos \mathbf{58°30'}$$
$$\approx 168.25$$
$$a \approx \sqrt{168.25} \approx 13.0.$$

The calculator steps used to find $a$ follow.

ALG: 30 [÷] 60 [+] 58 [=] [cos] [×] 15 [×] 10 [×] 2 [+/−] [+] 15 [$x^2$] [+] 10 [$x^2$] [=] [$\sqrt{\ }$]

RPN: 30 [ENTER] 60 [÷] 58 [+] [cos] 15 [×] 10 [×] 2 [×] [CHS] 15 [$x^2$] [+] 10 [$x^2$] [+] [$\sqrt{\ }$]

The display shows [12.971138]. Rounding to the nearest tenth, we obtain $a \approx 13.0$. Once $a$ is known, it is probably easier to continue by using the law of sines.

*To find B:*

$$\frac{\mathbf{10.0}}{\sin B} = \frac{\mathbf{13.0}}{\sin \mathbf{58°30'}} \qquad \text{Use the law of sines}$$
$$\sin B = \frac{(10.0)(\sin 58°30')}{13.0}$$
$$\approx 0.6559$$
$$B = 41°00' \quad \text{or} \quad 139°00'$$

But since $139°00' + 58°30' = 197°30' > 180°$, the larger angle $139°00'$ must be discarded. So the only solution is

$$B = 41°00'.$$

*To find C:* $C = 180° - (A + B) = 180° - 99°30' = 80°30'$. ■

Next we consider Case IV in which the three sides are known *(SSS)*.

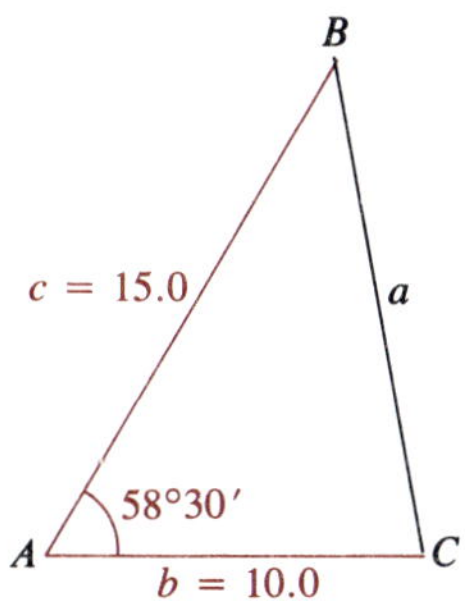


**Figure 18**

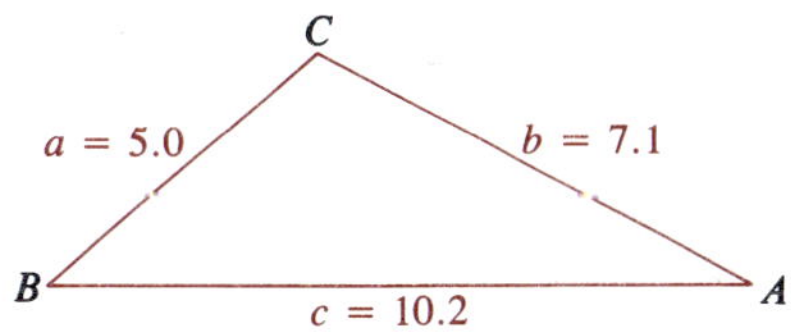


**Figure 19**

**EXAMPLE 2** Solve triangle $ABC$ in which $a = 5.0$, $b = 7.1$, and $c = 10.2$.

First make a sketch of the appropriate triangle, as in Figure 19. Clearly, with no angle given, the law of sines is not applicable.

*To find A:* Substitute into

$$a^2 = b^2 + c^2 - 2bc \cos A.$$

It is best to solve for $\cos A$ first.

$$\cos A = \frac{a^2 - b^2 - c^2}{-2bc} = \frac{b^2 + c^2 - a^2}{2bc}$$

Now substitute the values of $a$, $b$, and $c$.

$$\cos A = \frac{(7.1)^2 + (10.2)^2 - (5.0)^2}{2(7.1)(10.2)} \approx 0.8937$$

Thus, $A \approx 26.7°$.

The steps used to calculate $A$ are given below.

ALG: 7.1 [$x^2$] [+] 10.2 [$x^2$] [−] 5 [$x^2$] [=] [÷] 2 [÷] 7.1 [÷] 10.2 [=] [INV] [cos]

RPN: 7.1 [$x^2$] 10.2 [$x^2$] [+] 5 [$x^2$] [−] 2 [÷] 7.1 [÷] 10.2 [÷] [INV] [cos]

The display will show [26.652341], which gives $A = 26.7°$, correct to the nearest tenth of a degree.

*To find C:* $C$ could be found with the law of sines, but to show what happens when the formula yields a negative cosine, suppose we use

$$c^2 = a^2 + b^2 - 2ab \cos C.$$

First solve for $\cos C$ and then substitute the given values.

$$\begin{aligned}\cos C &= \frac{a^2 + b^2 - c^2}{2ab} \\ &= \frac{(5.0)^2 + (7.1)^2 - (10.2)^2}{2(5.0)(7.1)} \\ &\approx -0.4032\end{aligned}$$

Thus, $C \approx 113.8°$ To the nearest tenth of a degree

*To find B:* $B = 180° - (A + C) = 180° - 140.5° = 39.5°$. ■

Many applied problems require the law of cosines for their solution.

**EXAMPLE 3**
**AERONAUTICS**

Town $A$ is 200.0 miles due south of town $B$. A pilot, flying from $A$ to $B$, flies the first 120.0 miles off course in a direction 352°20′. How far is he from $B$ at this time?

From the sketch in Figure 20 we see that $\angle BAP$ is $360° - 352°20' = 7°40'$ and the desired distance is $BP$. In triangle $PBA$, two sides are given, $AP = 120$ and $AB = 200$, and the included angle, $\angle BAP = 7°40'$. Substitute into the law of cosines.

$$\begin{aligned}(BP)^2 &= (AP)^2 + (AB)^2 - 2(AP)(AB) \cos BAP \\ &= (120)^2 + (200)^2 - 2(120)(200) \cos 7°40' \\ &\approx 6829.1\end{aligned}$$

Thus, $BP \approx \sqrt{6829.1} \approx 82.6.$

The pilot is approximately 82.6 miles from town $B$. ■

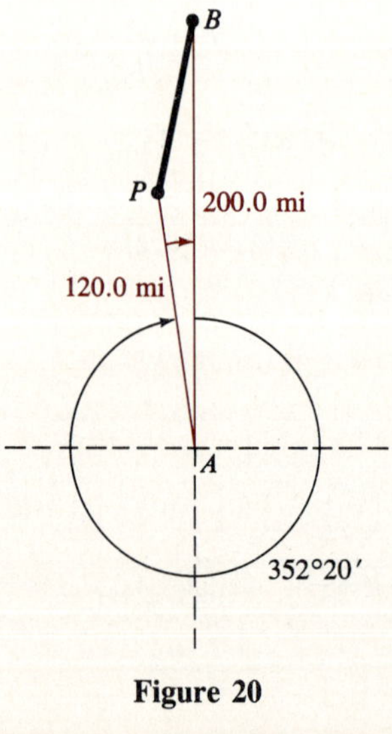


Figure 20

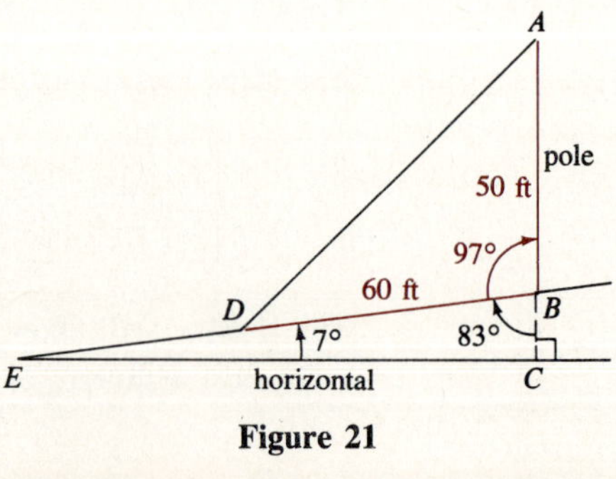


Figure 21

**EXAMPLE 4**
ENGINEERING

A hill is inclined 7° to the horizontal. How long will a guy wire need to be to reach the top of a 50-ft vertical pole if it is attached 60 ft downhill from the base of the pole?

See the sketch in Figure 21. We must determine $AD$. In right triangle $ECB$, $\angle EBC = 90° - 7° = 83°$. Thus, $\angle ABD$ in triangle $ABD$ is equal to $180° - 83° = 97°$. Hence we have two sides ($AB = 50$ and $DB = 60$) and the included angle ($\angle ABD = 97°$). Substitute into the law of cosines.

$$\begin{aligned}(AD)^2 &= (AB)^2 + (DB)^2 - 2(AB)(DB)\cos ABD \\ &= (50)^2 + (60)^2 - 2(50)(60)\cos 97° \\ &\approx 6831.2\end{aligned}$$

Thus, $\qquad AD \approx 83.$

The guy wire must be approximately 83 ft long. ■

**CAUTION** The preceding examples suggest that the law of cosines should be memorized in the alternate form, since in an applied problem the letters $a$, $b$, and $c$ will not be specified. ■

## 7.3 EXERCISES

In Exercises 1–12 solve each triangle with the given parts.

1. $A = 42.5°$, $b = 7.8$, $c = 9.3$
2. $B = 121°20'$, $a = 8.4$, $c = 12.3$
3. $a = 9.0$, $b = 12.0$, $c = 20.0$
4. $a = 15.35$, $b = 8.24$, $c = 22.76$
5. $a = 15.0$, $b = 15.0$, $c = 13.0$
6. $a = 3.25$, $b = 7.85$, $c = 7.85$
7. $a = 2.2$, $b = 3.7$, $c = 12.4$
8. $a = 35.1$, $b = 7.3$, $c = 10.4$
9. $A = 12°40'$, $b = 10.0$, $C = 130°10'$
10. $B = 20°10'$, $b = 18.4$, $c = 9.2$
11. $a = 11.3$, $c = 8.2$, $B = 120.4°$
12. $a = 5.45$, $b = 6.89$, $c = 7.01$

In Exercises 13–14 use the fact that adjacent angles in a parallelogram are supplementary to find the lengths of the diagonals of a parallelogram with sides $a$ and $b$ and one angle $C$.

13. $a = 5.5$, $b = 8.2$, $C = 42°30'$
14. $a = 15.35$, $b = 21.62$, $C = 105°38'$

A **rhombus** is a parallelogram with four sides of equal length. In Exercises 15–16 find the lengths of the diagonals of a rhombus with sides of length $a$ and one angle $A$.

15. $a = 7.28$, $A = 121°55'$
16. $a = 4.3$, $A = 28°40'$
17. AERONAUTICS Jefferson City is 300 miles due south of Willville. A pilot flying from Jefferson City to Willville flies the first 200 miles off course in a direction of 9°00′. How far from Willville is he at this time?
18. AERONAUTICS An airplane leaves an airport and flies due west for 200 miles and then 150 miles in a direction 230°. How far is it then from the airport?
19. NAVIGATION Two boats leave an island at the same time. The first sails 20 mph in a direction 330.0°, and the second sails 25 mph in a direction 200.0°. After 2 hours, how far apart are the boats?

**20.** NAVIGATION A boat sails 24.2 miles from port in a direction 31.0°, and then 16.4 miles in a direction of 142.0°. At this time, how far is the boat from port?

**21.** SURVEYING Two points $A$ and $B$ are on opposite sides of a building. A surveyor chooses a third point $C$ 45 m from $A$ and 65 m from $B$ with $\angle ACB = 52°30'$. How far apart are $A$ and $B$?

**22.** SURVEYING A triangular plot of ground has one corner angle measuring 69°40′. If the sides of the plot which form this angle measure 180 m and 125 m, what is the approximate length of the third side?

**23.** ENGINEERING A vertical tower 500 ft high stands on top of a hill whose side makes an angle of 12° with the horizontal. A guy wire is fastened to the top of the tower and is attached to the ground 600 ft down the hill from the base of the tower. How long is the wire?

**24.** ENGINEERING Two trusses have lengths 14.0 ft and 12.0 ft, and meet at an angle of 62.0°. What is the length of the span supported by the trusses?

**25.** SPORTS The distance from home plate to dead center field in Sun Devil Stadium is 408 ft. If a baseball diamond is a square and the distance from home plate to first base is 90 ft, how far is it from dead center field to first base?

**26.** SPORTS A golfer tees off on a hole and slices his drive 21°00′ to the right of the line from the tee to the hole. If the distance from the tee to the hole is 360 yd and his drive went 215 yd, how long will his second shot need to be to reach the hole?

**27.** AEROSPACE A tracking station sights a satellite above the surface of the earth. If the tracking dish is aimed 41.0° above the horizon, and the satellite is 123.0 miles from the station, how high is the satellite from the surface of the earth? Use 4000 miles for the radius of the earth.

**28.** GEOMETRY Three circles with radii 2.8 inches, 4.1 inches, and 7.3 inches are externally tangent to each other as shown in the figure. What are the angles of the triangle formed by joining their centers?

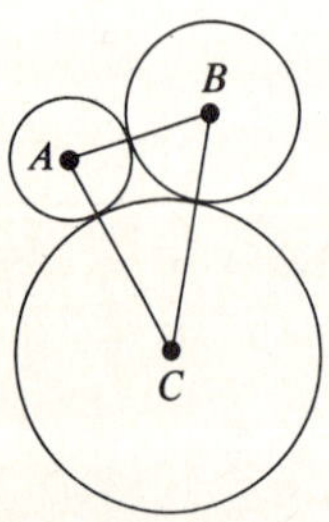


**Exercise 28**

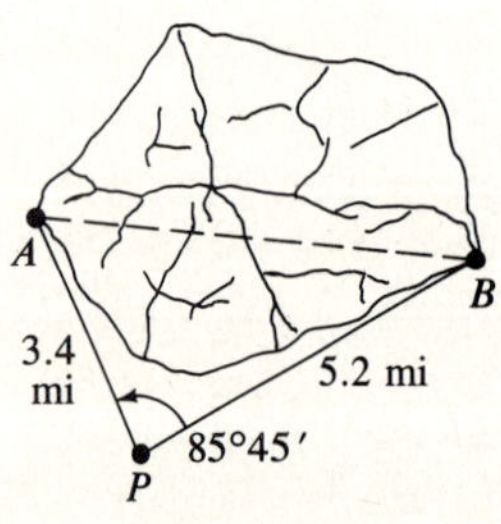


**Exercise 30**

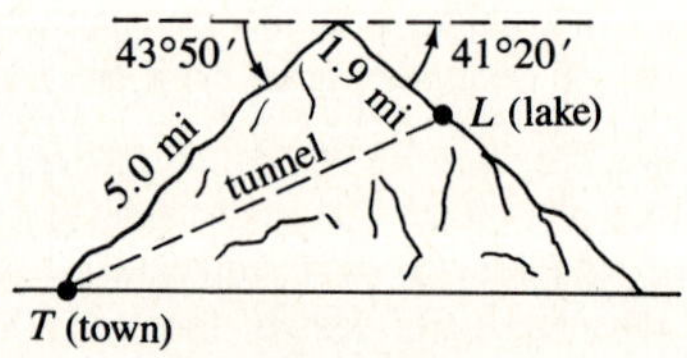


**Exercise 31**

**29.** GEOMETRY Three points $A(1, 9)$, $B(3, -2)$, and $C(-4, 1)$ are the vertices of a triangle. Point $D$ is the midpoint of side $AB$. Find $\angle ADC$ correct to the nearest ten minutes.

**30.** ENGINEERING A tunnel is to be constructed through a mountain from point $A$ to point $B$ as shown in the figure above. What is the length of the tunnel?

**31.** ENGINEERING A water diversion tunnel is to be constructed from a lake on the western slope of a mountain to a town at the base of the mountain on the eastern slope. An engineer takes the measurements shown in the figure above. **(a)** What will be the length of the tunnel? **(b)** If the construction crew starts at the town and begins tunneling through the mountain, what angle with the horizontal should they use to reach the lake?

**32.** NAVIGATION A ship smuggling contraband sails at a rate of 20 knots in a direction of 40°00′ from Devil's Island which is located 10 nautical miles due north of a port on the mainland. A coastguard cutter initiates pursuit from the port at the same instant traveling at 35 knots. In what direction should the cutter head to intercept the ship, and how long will it take to reach it? [Hint: Let $x$ be the number of hours required for interception and consider the triangle with sides of length 10, $20x$, and $35x$.]

## For Review

In Exercises 33–34 solve each triangle with the given parts.

**33.** $A = 25°20'$, $B = 72°40'$, $b = 3.6$

**34.** $A = 48°50'$, $b = 8.3$, $a = 14.7$

**35.** AEROSPACE A UFO is sighted above level ground directly between two men who are 1500 m apart. If the angles of elevation from the two men to the UFO are 32°20′ and 44°40′, how high is the object above the ground?

**36.** SURVEYING The angle of elevation to the top of Mount Humphrey from a point with elevation 8000 ft is 35°30′. From a second point 500 ft closer to the mountain, but still at an elevation 8000 ft, the angle of elevation is 37°40′. Approximately how high is Mount Humphrey?

# 7.4 Area of a Triangle

The most common formula for the area of a triangle involves one side, called the **base,** and the altitude from the opposite vertex to that base, called the **height.** The area of the triangle in Figure 22 is given by

$$\textbf{Area} = \frac{1}{2}bh.$$

To verify this formula, consider Figure 23. Clearly, the areas of triangles I and II are equal, as are the area of triangles III and IV. Thus, we have

$$\begin{aligned}\text{Area of rectangle } ACDE &= \text{I} + \text{II} + \text{III} + \text{IV}\\ &= \text{II} + \text{II} + \text{III} + \text{III} \qquad \text{I} = \text{II and III} = \text{IV}\\ &= 2(\text{II}) + 2(\text{III})\\ &= 2(\text{II} + \text{III})\\ &= 2(\text{Area of triangle } ABC).\end{aligned}$$

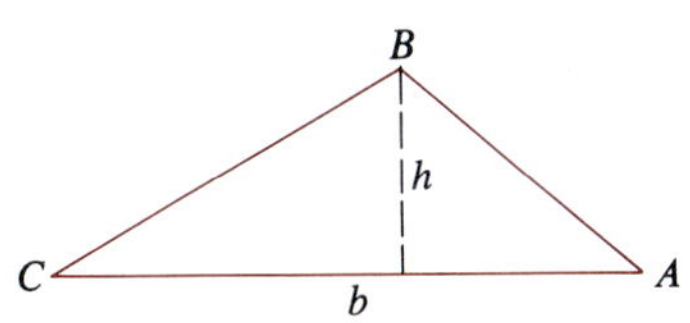


**Figure 22**

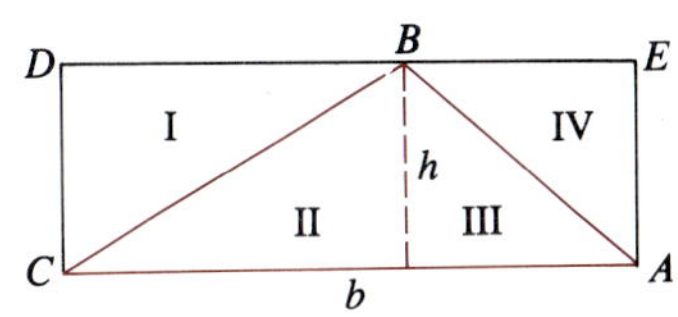


**Figure 23**

Dividing both sides by 2 we obtain

$$\text{Area of triangle } ABC = \frac{1}{2} \text{ Area of rectangle } ACDE$$
$$= \frac{1}{2}bh.$$

In this section we develop three additional formulas for finding the area of a triangle when facts other than a base and the corresponding height are given. The three situations are as follows.

1. Two sides and the included angle *(SAS)* are given.
2. Two angles (hence really all three) and a side (*ASA* or *AAS*) are given.
3. Three sides *(SSS)* are given.

## Area of Triangle for *SAS*

Consider triangle $ABC$ in Figure 24, and suppose that $a$, $b$, and $C$ are given. Let $h$ be the altitude drawn to base $AC$ of the triangle, intersecting $AC$ at point $E$. Then in right triangle $BEC$,

$$\sin C = \frac{h}{a}$$
$$a \sin C = h.$$

Since the area of triangle $ABC$ is

$$\text{Area} = \frac{1}{2}bh,$$

substituting for $h$ gives

$$\text{Area} = \frac{1}{2}ba \sin C.$$

This formula involves only the known parts of triangle $ABC$. The derivation uses a figure in which $C$ is acute. If $C$ is a right angle, then $\sin C = 1$ and $a = h$ (see Figure 25), so that Area $= \frac{1}{2}bh$ still yields the same result. If $C$ is obtuse, as in Figure 26, we extend $AC$ and again draw the altitude, $h$, from $B$. Letting $\theta$ be $\angle BCE$, then

$$\sin \theta = \frac{h}{a}$$
$$a \sin \theta = h.$$

But $\theta = 180° - C$, and since

$$\sin \theta = \sin (180° - C) = \sin C,$$

again we obtain

$$\text{Area} = \frac{1}{2}bh$$
$$= \frac{1}{2}ba \sin \theta$$
$$= \frac{1}{2}ba \sin C.$$

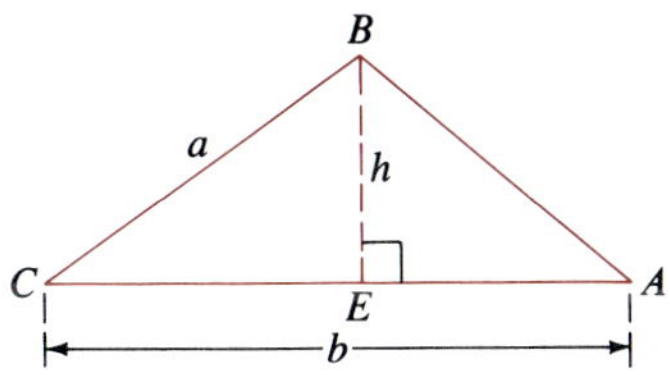


Figure 24

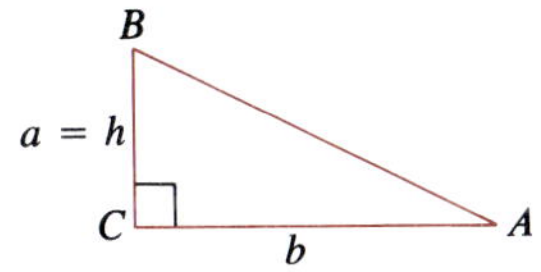


Figure 25

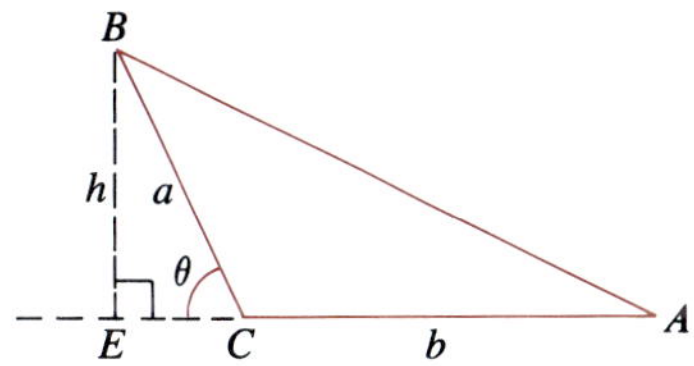


Figure 26

### Area of a Triangle

For any triangle $ABC$ with sides $a$, $b$, and $c$, when two sides and the included angle are known, the area is given by:

$$\text{Area} = \frac{1}{2}ab \sin C$$

$$\text{Area} = \frac{1}{2}bc \sin A$$

$$\text{Area} = \frac{1}{2}ac \sin B.$$

That is, the area of any triangle is one-half the product of the lengths of any two sides and the sine of the included angle.

**EXAMPLE 1** Find the area of the triangle given in Figure 27.

$$\begin{aligned}\text{Area} &= \frac{1}{2}ab \sin C\\ &= \frac{1}{2}(7.0)(5.0) \sin 50°20'\\ &\approx 13.5 \text{ in}^2\end{aligned}$$

■

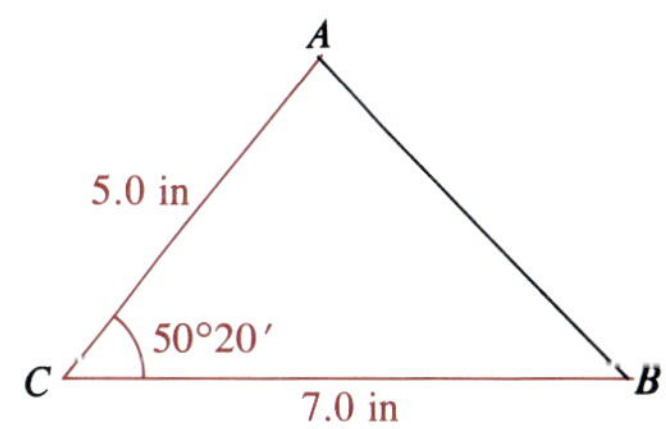


Figure 27

**Area of Triangle for ASA or AAS**

By using the law of sines, the formula Area $= \frac{1}{2}ab \sin C$ can be modified to cover the situation in which two angles are known. Since

$$\frac{a}{\sin A} = \frac{b}{\sin B},$$

we have

$$b = \frac{a \sin B}{\sin A}.$$

Substitute this value for $b$ in Area $= \frac{1}{2}ab \sin C$.

$$\begin{aligned}\text{Area} &= \frac{1}{2}a\left(\frac{a \sin B}{\sin A}\right) \sin C\\ &= \frac{1}{2}a^2 \frac{\sin B \sin C}{\sin A}\end{aligned}$$

Again, variations on the labeling of the triangle yield the three formulas in the next theorem.

**Area of a Triangle**

For any triangle $ABC$ with sides $a$, $b$, $c$, when two angles are known, the area is given by:

$$\text{Area} = \frac{1}{2}a^2 \frac{\sin B \sin C}{\sin A}$$

$$\text{Area} = \frac{1}{2}b^2 \frac{\sin A \sin C}{\sin B}$$

$$\text{Area} = \frac{1}{2}c^2 \frac{\sin A \sin B}{\sin C}.$$

That is, the area of any triangle is one-half the product of the square of the length of any side and the sines of the including angles divided by the sine of the opposite angle.

**EXAMPLE 2** Find the area of the triangle given in Figure 28.

First find $C$.

$$C = 180° - 23°50' - 27°20'$$
$$= 128°50'$$

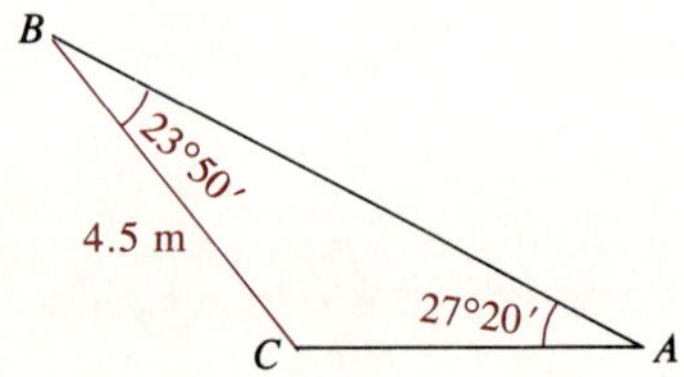

**Figure 28**

Then with $a = 4.5$ m, $A = 27°20'$, $B = 23°50'$, and $C = 128°50'$, substitute to find the area.

$$\text{Area} = \frac{1}{2}a^2 \frac{\sin B \sin C}{\sin A}$$

$$= \frac{1}{2}(\mathbf{4.5})^2 \frac{\sin \mathbf{23°50'} \sin \mathbf{128°50'}}{\sin \mathbf{27°20'}}$$

$$\approx 6.9 \text{ m}^2$$ ■

## Area of Triangle for *SSS*

Once again, we can begin with

$$\text{Area} = \frac{1}{2}ab \sin C$$

and derive a formula for computing the area of a triangle when the three sides are given. First square both sides.

$$(\text{Area})^2 = \frac{1}{4}a^2b^2 \sin^2 C$$

$$= \frac{1}{4}a^2b^2(1 - \cos^2 C) \qquad \text{Since } \sin^2 C + \cos^2 C = 1$$

By the law of cosines,

$$c^2 = a^2 + b^2 - 2ab \cos C.$$

Solve for cos $C$.

$$\cos C = \frac{a^2 + b^2 - c^2}{2ab}$$

Substitute for cos $C$ in the above expression for $(\text{Area})^2$.

$$\begin{aligned}
(\text{Area})^2 &= \frac{1}{4}a^2b^2\left[1 - \left(\frac{a^2 + b^2 - c^2}{2ab}\right)^2\right] \\
&= \frac{1}{4}a^2b^2\left[1 - \frac{(a^2 + b^2 - c^2)^2}{4a^2b^2}\right] \\
&= \frac{1}{4}a^2b^2\left[\frac{4a^2b^2 - (a^2 + b^2 - c^2)^2}{4a^2b^2}\right] \\
&= \frac{1}{16}[2ab - (a^2 + b^2 - c^2)][2ab + (a^2 + b^2 - c^2)] \quad \text{Factor the difference of squares} \\
&= \frac{1}{16}[c^2 - (a - b)^2][(a + b)^2 - c^2] \quad \text{Clear parentheses, use commutative and associative laws, and factor} \\
&= \frac{1}{16}[c - (a - b)][c + (a - b)][(a + b) - c][(a + b) + c] \\
&= \frac{1}{16}[c - a + b][c + a - b][a + b - c][a + b + c]
\end{aligned}$$

Take the square root of both sides and apply the commutative law.

$$\text{Area} = \frac{1}{4}\sqrt{(a + b + c)(a + b - c)(a + c - b)(b + c - a)}$$

To remember the factors under the radical, recall that the first one is the sum of the lengths of the three sides and that the other three are formed by adding the lengths of two sides and then subtracting the length of the third. By defining the **semiperimeter** of a triangle with sides $a$, $b$, and $c$ as

$$s = \frac{a + b + c}{2},$$

a more familiar form of this area formula can be obtained.

$$\begin{aligned}
\text{Area} &= \frac{1}{4}\sqrt{(a + b + c)(a + b - c)(a + c - b)(b + c - a)} \\
&= \sqrt{\frac{1}{16}(a + b + c)(a + b - c)(a + c - b)(b + c - a)} \\
&= \sqrt{\frac{(a + b + c)}{2} \cdot \frac{(a + b - c)}{2} \cdot \frac{(a + c - b)}{2} \cdot \frac{(b + c - a)}{2}} \\
&= \sqrt{s(s - c)(s - b)(s - a)}
\end{aligned}$$

This well-known formula for the area of a triangle was first stated by the Greek mathematician Heron (approximately 75 A.D.) and is known as **Heron's formula** (sometimes called Hero's formula).

**Heron's Formula**

For any triangle $ABC$ with sides $a$, $b$, and $c$ given, the semiperimeter is

$$s = \frac{a+b+c}{2},$$

and the area is

$$\text{Area} = \sqrt{s(s-a)(s-b)(s-c)}.$$

**EXAMPLE 3**

Find the area of a triangle with sides 3 ft, 4 ft, and 5 ft.

The semiperimeter is $s = \frac{3+4+5}{2} = \frac{12}{2} = 6$ ft. Thus,

$$\begin{aligned}\text{Area} &= \sqrt{6(6-3)(6-4)(6-5)}\\ &= \sqrt{6(3)(2)(1)}\\ &= \sqrt{36}\\ &= 6 \text{ ft}^2.\end{aligned}$$

You might recall that a triangle with sides of 3 ft, 4 ft, and 5 ft is actually a right triangle with hypotenuse 5 ft. Thus, the legs can be thought of as the base and height so that the area is $\frac{1}{2}bh = \frac{1}{2} \cdot 3 \cdot 4 = 6$ ft$^2$. Thus we are provided with a check of Heron's formula in this particular case (only because it is a right triangle). ■

Some applied problems involve finding the area of a triangle. We conclude this section with the first application in the introduction to this chapter.

**EXAMPLE 4**
CONSUMER

A mountain cabin has a wall in the shape of a triangle with two sides of length 26 ft and a third measuring 18 ft. The wall is 6 inches thick and contains a window measuring 4 ft by 5 ft. If insulation costs \$0.90 a cubic foot, what would be the approximate cost of insulating the wall?

We must find the volume of the triangular wall and diminish it by the volume taken out by the window. Consider the sketch in Figure 29.

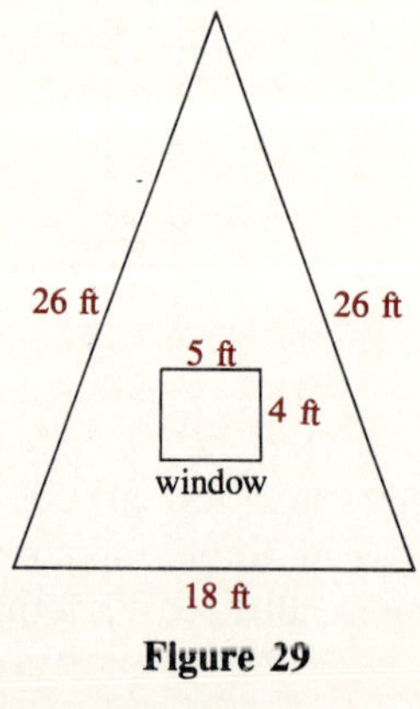


**Figure 29**

The volume taken out by the window is

$$V_1 = (5)(4)(0.5) = 10 \text{ ft}^3.$$

The volume of the triangular wall is

$$\begin{aligned} V_2 &= 0.5\sqrt{s(s-a)(s-b)(s-c)} \\ &= 0.5\sqrt{35(35-26)(35-26)(35-18)} \qquad s = \frac{26+26+18}{2} = 35 \\ &= 0.5\sqrt{48{,}195} \\ &\approx 109.77 \text{ ft}^3. \end{aligned}$$

Thus, the volume to be insulated is

$$V = V_2 - V_1 \approx 99.77 \text{ ft}^3,$$

and the cost is $(99.77)(0.90) \approx \$89.79$. ■

## 7.4 EXERCISES

In Exercises 1–16 find the area of triangle $ABC$ with the given parts.

1. $a = 20.0$ in. $b = 8.0$ in, $C = 35°20'$
2. $b = 7.35$ cm, $c = 3.25$ cm, $A = 41°40'$
3. $a = 11.75$ yd, $c = 8.90$ yd, $B = 110°45'$
4. $a = 32.3$ m, $b = 11.5$ m, $A = 147°30'$
5. $A = 40°50'$, $B = 32°40'$, $c = 16.2$ ft
6. $A = 61°35'$, $C = 42°55'$, $b = 19.10$ cm
7. $B = 32°48'$, $C = 54°24'$, $b = 15.45$ in
8. $B = 41°50'$, $C = 127°30'$, $a = 13.9$ in
9. $a = 2.35$ yd, $b = 3.18$ yd, $c = 4.09$ yd
10. $a = 16.0$ cm, $b = 16.0$ cm, $c = 16.0$ cm
11. $a = 4.1$ m, $b = 4.1$ m, $c = 4.1$ m
12. $a = 8.72$ ft, $b = 10.84$ ft, $c = 15.62$ ft
13. $B = 27°20'$, $a = 6.7$ ft, $b = 15.0$ ft [Hint: Use the law of sines first.]
14. $A = 102°40'$, $a = 15.3$ cm, $b = 5.2$ cm
15. $A = 30.3°$, $B = 27.7°$, $C = 122.0°$
16. $A = 60.0°$, $B = 60.0°$, $C = 60.0°$

Solve.

17. **CONSUMER** A triangular-shaped lot contains a circular pond with diameter 18 yd. If the sides of the lot measure 75 yd, 55 yd, and 40 yd and sod costs \$2.10 per square yard, how much will it cost to put sod on the lot?
18. **CONSUMER** A room in the shape of a triangle has sides of length 15.2 yd, 12.4 yd, and 11.8 yd. If carpeting costs \$19.95 a square yard, padding costs \$2.25 a square yard, and installation is included at no charge, how much, to the nearest dollar, will it cost to carpet the room over the padding, assuming there is no waste?
19. **CONSUMER** A sign painter wishes to make a sign in the shape of an isosceles triangle with equal sides of 12.0 yd and base angles of 65°. What is the area of the sign?

**20.** AGRICULTURE A garden is in the shape of a **quadrilateral** (a four-sided figure) as shown in the figure. What is the area of the garden?

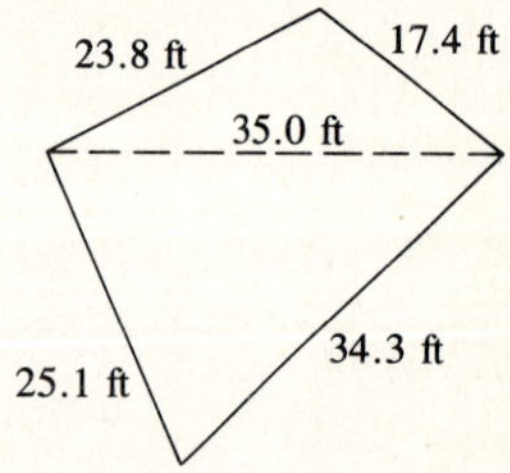


**Exercise 20**

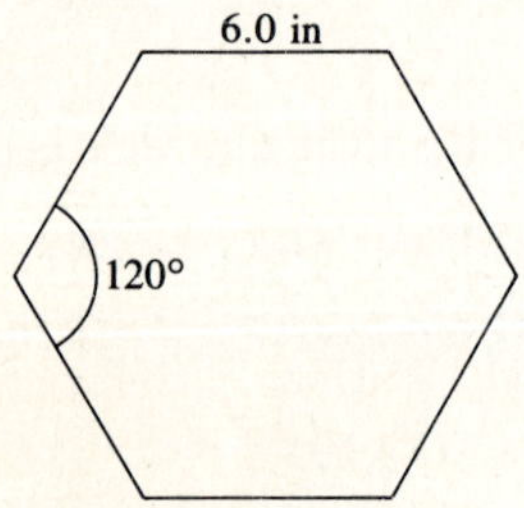


**Exercise 21**

**21.** GEOMETRY Find the area of the **regular hexagon** in the figure.

**22.** GEOMETRY What is the area of the triangle formed by the centers of three circles that are externally tangent and have radii 4.0 cm, 6.0 cm, and 10.0 cm?

**23.** ENGINEERING An aluminum plate, 0.5 in thick, is triangular in shape and has sides of 11.0 inches, 15.5 inches, and 18.7 inches. What is the volume of the aluminum in the plate?

**24.** ENGINEERING If a hole 2.5 inches in diameter is drilled in the plate in Exercise 23, what is the volume of the remaining aluminum?

**25.** ENGINEERING A solar reflector is made from 32 triangular mirrors, each with sides 2.2 m, 2.2 m, and 0.4 m. What is the total surface area of the reflector?

**26.** ARCHAEOLOGY A pyramid in Mexico has four congruent triangular faces. If each face is an isosceles triangle with a base measuring 180.0 m and equal sides measuring 165.0 m, what is the total surface area of the pyramid (including the base)?

**27.** GEOMETRY Verify that the area of the parallelogram shown below is given by

$$\text{Area} = bd \sin A.$$

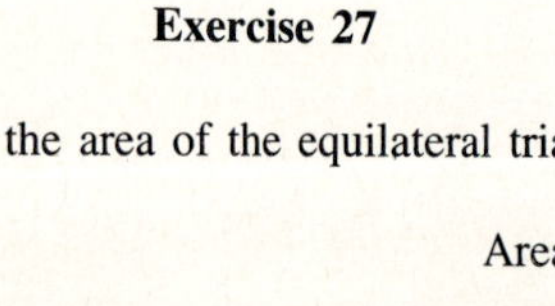


**Exercise 27**

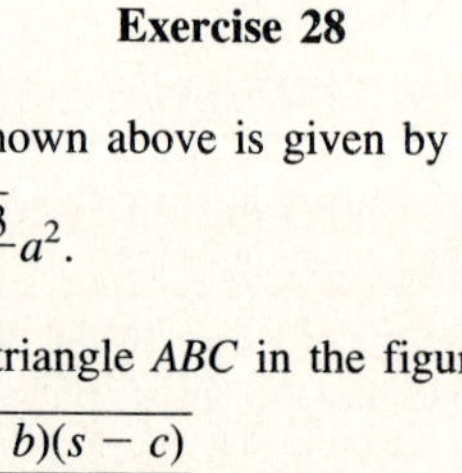


**Exercise 28**

**28.** Verify that the area of the equilateral triangle shown above is given by

$$\text{Area} = \frac{\sqrt{3}}{4}a^2.$$

**29.** Verify that the radius of the circle inscribed in triangle $ABC$ in the figure is given by

$$r = \sqrt{\frac{(s-a)(s-b)(s-c)}{s}},$$

where $s$ is the semiperimeter of the triangle.

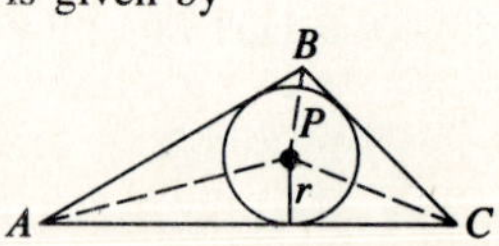

**30.** RECREATION A recreation director wishes to build a circular jogging track in a triangular park with sides 350.0 yd, 420.0 yd, and 570.0 yd. What is the length of the largest track possible?

## For Review

**31.** GEOLOGY Two points $A$ and $B$ are located on the extreme opposite sides of a crater in such a way that a direct measurement between the two is impossible. A third point $C$ is chosen in such a way that $AC$ is 80.0 m, $BC$ is 75.0 m, and $\angle ACB$ is 41.5°. How wide is the crater?

**32.** NAVIGATION A boat leaves Key West, Florida and sails for 4 hours at 20 knots in a direction 310°30′. It then changes course, and heads in a direction 215°30′ traveling for 5 hours and at 22 knots. How far is the boat from Key West (to the nearest nautical mile)?

Do Exercises 33–34 to prepare for the study of simple harmonic motion in the next section. Find the **(a)** amplitude, **(b)** period, and **(c)** phase shift of each function.

**33.** $y = -3 \sin (2x - \pi)$

**34.** $y = 5 \cos \left(4x + \frac{\pi}{2}\right)$

# 7.5 Simple Harmonic Motion

Numerous applied problems involve an object which is moving up and down (or back and forth) in a rhythmic motion. Such motions occur when the strings on a musical instrument are strummed, when a piston in an engine moves, when a pendulum swings back and forth, when waves in the ocean roll toward shore, and when a person's vocal chords make sounds. Vibratory motion of these types is called **harmonic motion.** When all frictional forces are neglected, **simple harmonic motion** results, and can be described nicely using trigonometric functions.

A straightforward example of simple harmonic motion is the wave motion of a length of rope fastened at one end to a wall and made to oscillate by moving the other end up and down in a rhythmic pattern. Waves appear to originate at the moving end and pass along the rope toward the fixed end. (See Figure 30.) A knot tied in the rope moves up and down in a rhythmic manner, and in fact is in a state of simple harmonic motion.

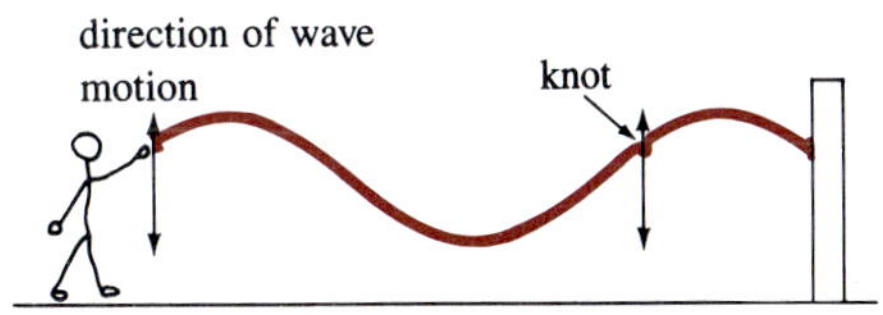


**Figure 30 Simple Harmonic Motion**

## Derivation of Equations

To establish the relationship between the sine and cosine functions and simple harmonic motion, consider a point $P(x, y)$ moving along the circumference of a circle at a constant rate. In Figure 31 the angle $\theta$ is opening $B$ radians/second. Assume that the point starts at the position with coordinates $(A, 0)$, where $A$ is the radius of the circle with center $(0, 0)$. $H(x, 0)$ is called the **horizontal projection** of $P$ onto the $x$-axis, and $V(0, y)$ is the **vertical projection** of $P$ onto the $y$-axis. As $P$ moves continuously around the circle starting at $(A, 0)$, $H$ varies from $(A, 0)$ to $(-A, 0)$ and back. The distance from $H$ to the origin $(0, 0)$ is between 0 and $A$, inclusive. If we interpret negative units to be to the left of the origin, then the motion of $P$ determines the point $H(x, 0)$ where $-A \le x \le A$. Similarly the motion determines $V(0, y)$ where $-A \le y \le A$. We can describe the measure of angle $\theta$ in terms of the variable $t$, representing time, by

$$\theta = Bt,$$

since the angular velocity is constant. Then

$$\cos \theta = \frac{x}{A} \quad \text{and} \quad \sin \theta = \frac{y}{A}.$$

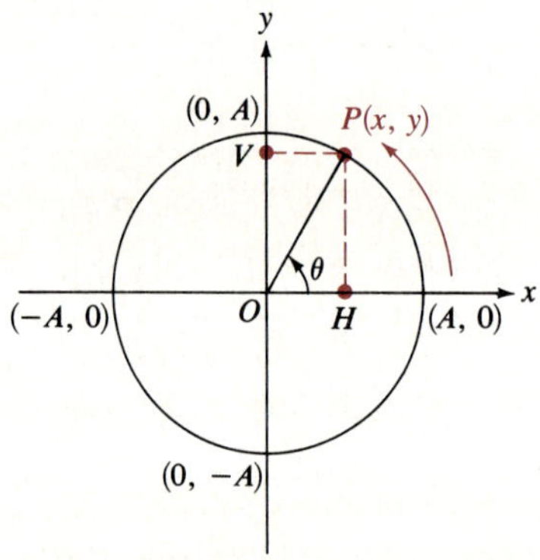


**Figure 31 Motion Around a Circle**

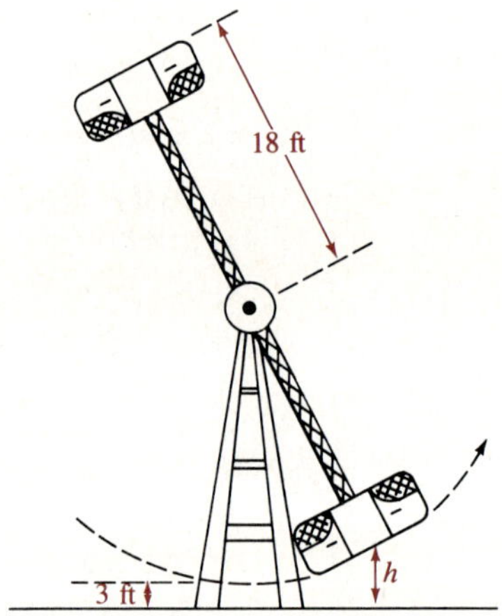


**Figure 32 Loop-O-Plane**

Substitute $Bt$ for $\theta$ and solve for $x$ and for $y$.

$$x = A \cos Bt \quad \text{and} \quad y = A \sin Bt$$

In our discussion, we have assumed that the point $P$ started at $(A, 0)$. One final variation gives us a more general result. If at time $t = 0$ (the point begins its movement at time zero) the point $P$ is at an angle $\theta = D$ radians, then the angle $\theta$ can be described in terms of time $t$ by

$$\theta = Bt + D. \qquad \text{When } t = 0, \text{ then } \theta = D$$

Thus, the horizontal motion and vertical motion of the points $H$ and $V$ are described, respectively, by

$$x = A \cos (Bt + D) \quad \text{and} \quad y = A \sin (Bt + D).$$

Recall from Section 6.7 that to graph these relationships, we factor out $B$, obtaining

$$x = A \cos B\left(t + \frac{D}{B}\right) \quad \text{and} \quad y = A \sin B\left(t + \frac{D}{B}\right),$$
$$= A \cos B(t - C) \qquad\qquad = A \sin B(t - C). \quad \text{Where } C = -\frac{D}{B}$$

Thus the amplitudes of each are $|A|$, the periods are $2\pi/B$, and the phase shifts are $C = -(D/B)$. In this setting, the period $2\pi/B$ is the length of time required to complete one cycle of motion. Another useful quantity is the **frequency,** $B/2\pi$, of the motion, which is the number of cycles that are completed in one second. (For example, if the period is 1/3, then one cycle is completed in 1/3 second and the frequency is 3 $(3 = 3/1)$, since 3 cycles are completed in 1 second.)

## EXAMPLE 1

**RECREATION**

The Loop-O-Plane at an amusement park shown in Figure 32 has an 18-ft radius and turns at 15 rpm. Suppose that the passenger compartment is 3 ft above ground at the bottom of the arc. Find an equation that gives the height of a passenger above the ground as a function of time $t$ in seconds, and use the equation to estimate the height of a passenger 2.5 seconds into the ride.

Since the ride makes 15 rpm, this is 1/4 revolution per second, or 1 revolution every 4 seconds. Also, the maximum height is 39 ft and the minimum height is 3 ft. The minimum height occurs when $t = 0, 4, 8, 12, \ldots$, and the maximum height occurs when $t = 2, 6, 10, 14, \ldots$. We can describe the relationship using an equation of the form

$$h = A \cos B\left(t + \frac{D}{B}\right) + E,$$

where $E$ is the equilibrium point. If at $t = 0$ the passenger is at the bottom of the ride, then $D = 0$ (the initial angle is 0 radians). Since $h$ varies from 3 to 39, the amplitude is $|A| = \frac{1}{2}(39 - 3) = \frac{1}{2}(36) = 18$, the radius of the ride. The period, $2\pi/B$, is 4 so that $B = \pi/2$. The value of $E$ is 21 since the center of the rotation is at 21 ft, and we must set $A = -18$ to obtain the correct position at time $t = 0$. Thus, the equation is

$$h = -18 \cos \frac{\pi}{2} t + 21.$$

The table below shows $h$ at several critical times during one revolution and verifies that the equation does indeed describe the motion.

| $t$ (seconds) | $h$ (feet) |
|---|---|
| 0 | 3 |
| 1 | 21 |
| 2 | 39 |
| 3 | 21 |
| 4 | 3 |

Thus, 2.5 seconds into the ride the height of the passenger is

$$h = -18 \cos \frac{\pi}{2}(2.5) + 21 \approx -18(-0.7071) + 21 \approx 34 \text{ ft.}$$ ■

## Spring-Mass Systems

One of the best examples of harmonic motion involves **spring-mass systems,** in which a weight is hanging on a spring attached to a ceiling. The system is in a **state of equilibrium** when the suspended weight is completely motionless. If the weight is pulled down (or pushed up) a certain distance and released, it will continue to oscillate up and down indefinitely, if air resistance and friction are ignored. If a coordinate system is superimposed so that the equilibrium position of the weight coincides with the origin, an equation of the type

$$y = A \sin (Bt + D)$$

gives the distance between the weight and the equilibrium position at any time $t$. (See Figure 33.) The constants $A$, $B$, and $D$ will all depend, of course, on the size of the weight, the tension of the spring, and the distance the weight is drawn down initially.

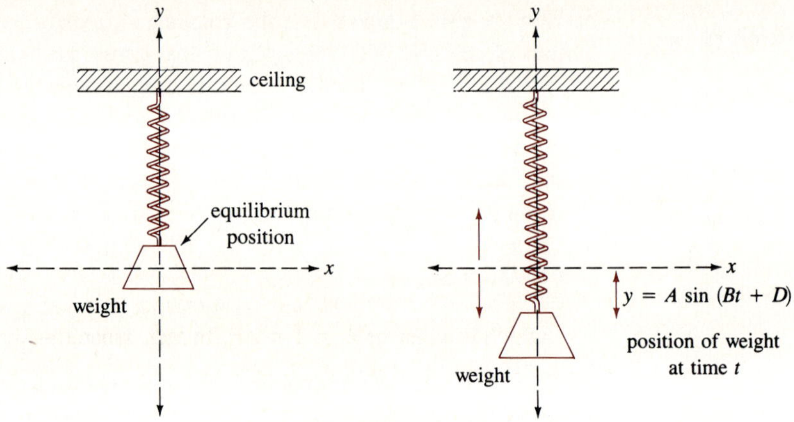


**Figure 33 Spring-Mass System**

## EXAMPLE 2

**SPRING-MASS**

Suppose that the motion of a spring-mass system is described by

$$y = 4 \sin \left(\pi t + \frac{\pi}{2}\right),$$

where $y$ is the distance in feet from the weight to the equilibrium position (negative is below and positive is above the equilibrium position) and $t$ is the time measured in seconds. Further, assume that the weight is 15 ft from the ceiling in a state of equilibrium.

**(a)** At time $t = 0$, the weight, which has been pushed upward to the starting position, is released. How far is it from the ceiling at this time?

Substituting $t = 0$, we have

$$y = 4 \sin \left(\pi \cdot \mathbf{0} + \frac{\pi}{2}\right)$$
$$= 4 \sin \frac{\pi}{2} = 4.$$

Since the weight is 4 ft above the equilibrium point, which is 15 ft from the ceiling, the weight is released from a position 11 ft from the ceiling.

**(b)** When is the first time the weight reaches the equilibrium position?

This will occur when $y = 0$. Thus we solve

$$0 = 4 \sin\left(\pi t + \frac{\pi}{2}\right)$$

$$0 = \sin\left(\pi t + \frac{\pi}{2}\right). \quad \text{Divide through by 4}$$

Since the sine function is zero at $0, \pi, 2\pi, 3\pi, \ldots$, we must let $\pi t + \pi/2$ be one of these values. If $\pi t + \pi/2 = 0$, then $t = -\frac{1}{2}$, which is meaningless. Thus we try $\pi$.

$$\pi t + \frac{\pi}{2} = \pi$$

$$\pi t = \frac{\pi}{2}$$

$$t = \frac{1}{2}$$

When $t = \frac{1}{2}$,

$$y = 4 \sin\left(\pi \cdot \frac{1}{2} + \frac{\pi}{2}\right) = 4 \sin \pi = 4 \cdot 0 = 0.$$

Therefore, $\frac{1}{2}$ second after the weight is released, it will pass through the equilibrium position. The next time it is in this location will be at time $t = \frac{3}{2}$ seconds, which is when $\pi t + \pi/2 = 2\pi$. In fact, when $t = \frac{1}{2}, \frac{3}{2}, \frac{5}{2}, \frac{7}{2}, \frac{9}{2}, \ldots$, the weight will pass the equilibrium position.

**(c)** What is the farthest the weight will ever be from the ceiling?

This will occur when $|y|$ is as large as possible, but $y$ is negative. We have seen that the amplitude of a function such as this is $|A|$. Since $A = 4$, the smallest value of $y$ will be $-4$. With the equilibrium point at 15 ft, the farthest the weight ever gets from the ceiling is 19 ft. In order for $y$ to be $-4$,

$$-4 = 4 \sin\left(\pi t + \frac{\pi}{2}\right)$$

$$-1 = \sin\left(\pi t + \frac{\pi}{2}\right).$$

Thus, $\pi t + \frac{\pi}{2}$ must be $\frac{3\pi}{2}, \frac{7\pi}{2}, \frac{11\pi}{2}, \ldots$. Using the value $\frac{3\pi}{2}$,

$$\pi t + \frac{\pi}{2} = \frac{3\pi}{2}$$

$$\pi t = \pi$$

$$t = 1.$$

Therefore, 1 second after the weight is released, it will reach the low point of 19 ft from the ceiling. It will return to this position at time 3 seconds, 5 seconds, 7 seconds, and so forth. ■

### Damped Motion and Resonance

In Example 2 we assumed a situation which cannot really happen. Actually, the motion of the weight would be such that the maximum distance from the equilibrium point (amplitude) would continue to decrease until the weight came to rest in a state of equilibrium again. In such a case, we say that the harmonic motion is **damped.** Damped harmonic motion is common in physical applications. The opposite effect, called **resonance,** occurs when the maximum distance from the equilibrium point continues to increase as time increases. Resonance can create a dangerous situation, and the classic example of this is the collapse of a suspension bridge in 1940 in the state of Washington. Any suspension bridge, similar to the rope we considered earlier, will wave up and down because of wind and traffic conditions. If this motion is resonant, the bridge will literally wave itself apart and collapse. Such was the case with Tacoma Narrows Bridge over Puget Sound. As a result, engineers recognized that it is essential that the inherent harmonic motion of such things as bridges, buildings, automobiles, and aircraft be damped. Trigonometry is important in analyzing problems of this type.

## 7.5 EXERCISES

Solve.

**1.** RECREATION The ferris wheel at the Beach Front Amusement Park has a 60 ft diameter, turns at a rate of 6 rpm, and is 5 ft off the ground at the low point.
**(a)** Find an equation which gives the height of a passenger in terms of time $t$.
**(b)** What is the height of a passenger 35 seconds into the ride?

**2.** Repeat Exercise 1 for the ferris wheel in a traveling carnival which has a 30 ft diameter, turns at a rate of 10 rpm, and is 4 ft off the ground at the low point.

In Exercises 3–8 the motion of a spring-mass system is described by the given equation where $y$ is the distance in feet from the equilibrium position and $t$ is time in seconds. If the weight is $K$ ft from the ceiling in a state of equilibrium, determine the following. **(a)** The starting position of the weight (at $t = 0$), **(b)** the first time at which the weight passes the equilibrium position, **(c)** the closest the weight will ever be to the ceiling, **(d)** the farthest the weight will ever be from the ceiling, **(e)** the distance from the ceiling at time $t = 2.5$, **(f)** the amplitude of the motion, **(g)** the period of the motion, **(h)** the frequency of the motion, and **(i)** the phase shift of the motion.

**3.** $y = 5 \sin\left(\pi t - \frac{\pi}{2}\right)$, $K = 10$

**4.** $y = 5 \sin\left(\pi t + \frac{\pi}{2}\right)$, $K = 4$

**5.** $y = 5 \sin\left(\pi t - \frac{\pi}{2}\right)$, $K = 4$

**6.** $y = 7 \sin\left(\pi t - \frac{\pi}{2}\right)$, $K = 15$

**7.** $y = 8 \sin\left(\frac{1}{4}\pi t + \frac{\pi}{2}\right)$, $K = 20$

**8.** $y = 12 \sin\left(\frac{1}{4}\pi t - \frac{\pi}{2}\right)$, $K = 30$

PHYSICS An object in a spring-mass system described by $y = A \sin(Bt + D)$ is moving with a velocity given by $v = AB \cos(Bt + D)$. Use this information in Exercises 9–10.

**9.** What is the velocity of the object described in Exercise 3 when $t$ is **(a)** 0, **(b)** 1, **(c)** 0.5, and **(d)** 1.3?

**10.** What is the velocity of the object described in Exercise 8 when $t$ is **(a)** 0, **(b)** 1, **(c)** 0.5, and **(d)** 1.3?

11. OCEANOGRAPHY A buoy in the ocean oscillates up and down as waves move past. Suppose the buoy moves a total of 4 ft from low point to high point and returns to the low point every 12 seconds. Assuming that at $t = 0$ the buoy is at its low point, find an equation describing this harmonic motion.

12. Repeat Exercise 11 this time assuming that the buoy is at its high point at $t = 0$.

13. ELECTRONICS The drop in voltage across two terminals in an electrical circuit using alternating current is given by the equation $E = 180 \sin(120\pi t)$, where $t$ is measured in seconds.
    (a) What is the maximum voltage drop?
    (b) What is the period for the circuit?
    (c) What is the frequency for the circuit?

14. MEDICINE The wave generated by an X-ray machine has the form $y = A \sin(2\pi \cdot 10^{18} t)$, where $t$ is measured in seconds. What is the frequency of the wave?

15. ENGINEERING A connecting rod attached to the rim of a wheel by means of prongs drives a piston up and down within a cylinder as shown in the figure. The radius of the wheel is 1.5 ft and the connecting rod is 4 ft in length. Find an equation that describes the distance $d$ between the bottom of the piston and the $x$-axis in terms of time $t$, in seconds. Assume that at $t = 0$, the point $A$ is at (1.5, 0) and that the wheel is turning at a rate of 2 revolutions/second. [Hint: A radius of the wheel is sweeping out an angle which is opening $4\pi$ radians/second.]

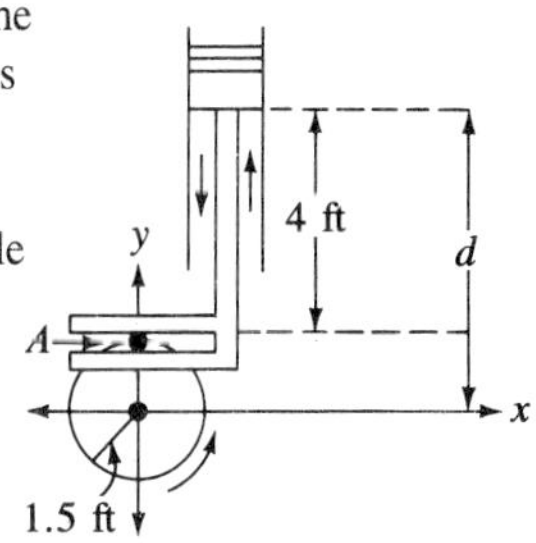


16. When is the first time the piston in Exercise 15 will reach its high point? When is the first time it will reach its low point?

17. The **stroke** of a piston moving within a cylinder is the difference between the high point and the low point. What is the stroke of the piston described in Exercise 15?

18. ECOLOGY Ecologists are interested in studying predator-prey systems, in which the number of predators and the number of prey tend to behave in a cyclic or periodic manner. For example, suppose a given region is populated by coyotes and rabbits. As the rabbit population increases, the coyote population, having a more plentiful food supply, will also increase. When the region can no longer support the increasing number of rabbits, that is, when the rabbit food supply is overextended, the rabbit population will begin to decline. This decline will, in a short time, cause a decline in the coyote population. Once the number of rabbits has reached a low point, enabling their food supply to reestablish itself, the rabbit population will again increase (the number of coyotes is now low), which will be followed by an increasing population of coyotes. This cycle will continue indefinitely if no other outside forces are in effect. It has been determined that the rabbit population in the Lone Pine River Valley has varied (approximately) according to the formula

$$r = 500 + 120 \sin\left(\frac{\pi}{2}t\right).$$

In this equation, $r$ is the number of rabbits at time $t$, $t$ is measured in years, and time $t = 0$ was July 1, 1975.
    (a) What was the rabbit population on July 1, 1975?
    (b) What was the maximum rabbit population?
    (c) On what date was the maximum rabbit population first reached?
    (d) What was the minimum rabbit population?
    (e) On what date was the minimum rabbit population first reached?
    (f) What was the rabbit population on January 1, 1985?

**19. ECOLOGY** The number of coyotes in the Lone Pine River Valley has varied (approximately) according to the formula

$$c = 100 + 40 \sin\left(\frac{\pi}{2}t - \frac{\pi}{4}\right),$$

where $t$ is measured in years and $t = 0$ was July 1, 1975.

**(a)** What was the coyote population on July 1, 1975?
**(b)** What was the maximum coyote population?
**(c)** On what date was the maximum coyote population first reached?
**(d)** What was the minimum coyote population?
**(e)** On what date was the minimum coyote population first reached?
**(f)** What was the coyote population on January 1, 1985?
**(g)** Compare the equations for $c$ and $r$ (in Exercise 18), and interpret the phase shift of $\frac{1}{2}$ for $c$.

## For Review

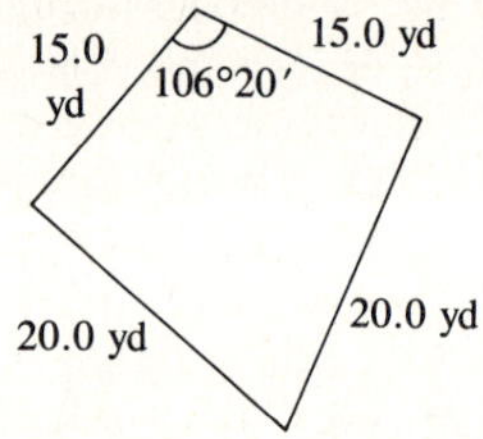


**20. CONSUMER** A recreation room in the shape of a quadrilateral, as shown in the figure, is to be covered with indoor-outdoor carpeting which costs \$18.50 per yd$^2$. Assuming no waste and that the installation fee is included in the price of the carpet, approximately how much will the project cost?

**21. GEOMETRY** In the next section we will need to determine the length of a diagonal of a parallelogram in order to find the magnitude of a vector. If one angle of a parallelogram is 38.2° and the two sides are 16.7 cm and 29.3 cm, use the law of cosines to determine the length of the longer of the two diagonals.

# 7.6 Vectors and Applications

Many quantities in applied problems such as cost, length, time, and area can be described in magnitude by a single number scaled to some unit of measure. These quantities are called **scalar quantities** and the real numbers used to describe them are called **scalars.** Other quantities, such as force, velocity, and movement, require both magnitude and direction for their description. These quantities, called **vector quantities,** are described using **vectors.** Geometrically, a vector is a directed line segment where the length of the segment indicates the magnitude of the vector and an arrowhead indicates the direction of the vector, as illustrated in Figure 34. For example, if a ship sailing in a direction 80° with velocity 25 knots, the vector in Figure 35 could represent this motion.

## Magnitude and Direction

If a vector begins at point $A$ (the **initial point**) and ends at point $B$ (the **terminal point**), we use $\overrightarrow{AB}$ to denote the vector. The length of the segment from $A$ to $B$ is the **magnitude** of $\overrightarrow{AB}$, denoted by $|\overrightarrow{AB}|$, and the **direction** of the vector is the direction of the directed line segment. If a second vector $\overrightarrow{XY}$ has the same magnitude and direction as

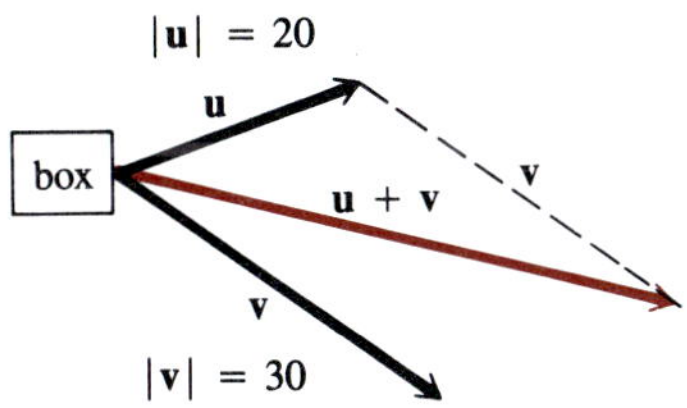


**Figure 40 Resultant Vector**

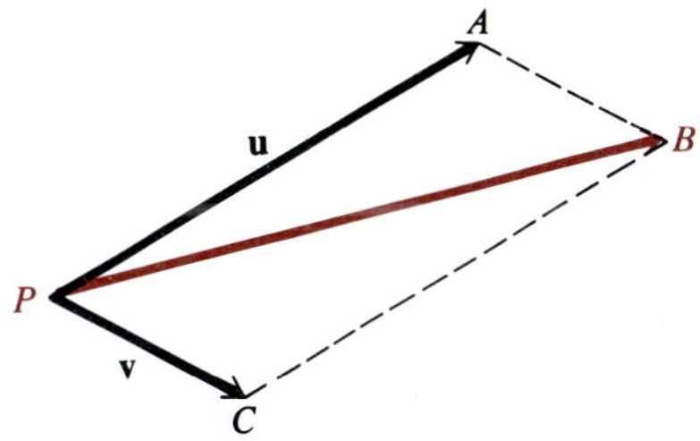


**Figure 41 Parallelogram Law**

## Vector Notation

This brief geometrical introduction to vectors provides the necessary tools for solving a variety of applied problems. Before doing so, however, it is appropriate to consider vectors from an algebraic standpoint. Since a vector can be moved to various positions without altering its direction and magnitude, a given vector can be placed with its starting point at the origin of a rectangular coordinate system. Using this scheme to provide a model for considering vectors, the terminal point of an arbitrary vector can be identified with a point $P$ in the plane as shown in Figure 42. Thus there is a one-to-one correspondence between vectors and ordered pairs of real numbers $(a, b)$. To avoid confusion with notation, we agree to use $\langle a, b\rangle$ as the symbol for the vector with terminal point $(a, b)$. The real numbers $a$ and $b$ are **components** of the vector $\langle a, b\rangle$.

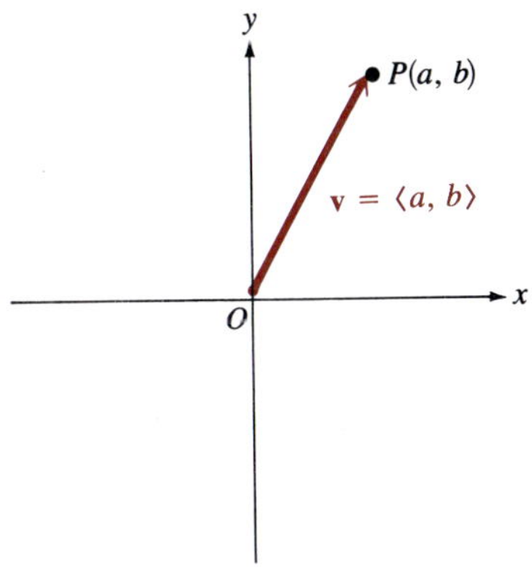


**Figure 42**

### Magnitude and Direction of a Vector

Let $\mathbf{v} = \langle a, b\rangle$ be a vector. The magnitude of $\mathbf{v}$ is

$$|\mathbf{v}| = \sqrt{a^2 + b^2},$$

and the direction is the smallest positive angle $\theta$ from the positive $x$-axis to the vector.

The above definition seems reasonable in view of the geometrical interpretation of a vector given earlier since $\sqrt{a^2 + b^2}$ is the distance from $(a, b)$ to $(0, 0)$, the length of the line segment $OP$. Our knowledge of trigonometry provides the necessary tools for finding the direction of a given vector $\langle a, b\rangle$ as illustrated in Example 1.

**EXAMPLE 1** Find the magnitude and direction of each vector.

**(a)** $\mathbf{v} = \langle -1, 2\rangle$

Consider Figure 43.

$$|\mathbf{v}| = \sqrt{(-1)^2 + (2)^2} = \sqrt{5}$$

$$\tan\theta = \frac{2}{-1} = -2$$

$$\theta = 116°34'$$

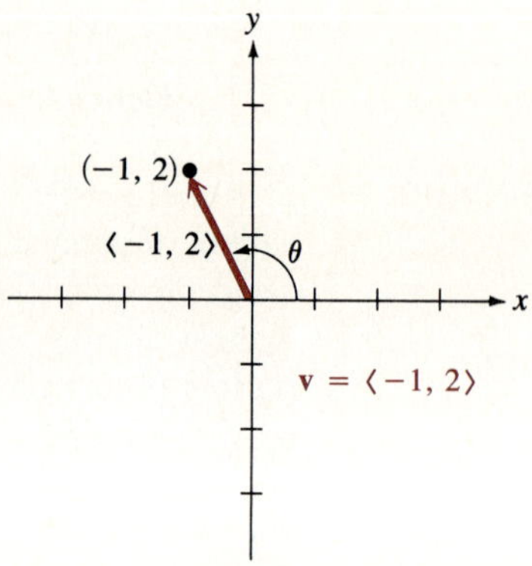


**Figure 43**

**(b)** $\mathbf{u} = \langle 0, -3\rangle$

Consider Figure 44.

$$|\mathbf{u}| = \sqrt{0^2 + (-3)^2} = \sqrt{9} = 3$$

$\tan\theta$ is undefined.

$$\theta = 270°00' \quad ■$$

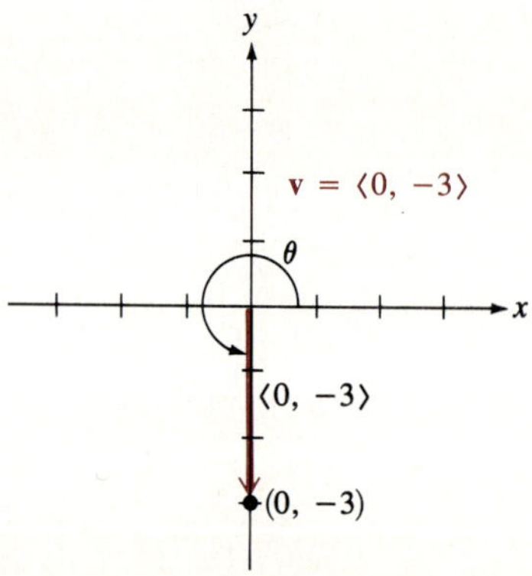


**Figure 44**

The operations of scalar multiplication and vector addition can be defined using the ordered pair notation.

### Operations on Ordered Pairs

Let $\mathbf{u} = \langle a, b\rangle$ and $\mathbf{v} = \langle c, d\rangle$ be two vectors with $k$ a scalar.

$\mathbf{u} + \mathbf{v} = \langle a, b\rangle + \langle c, d\rangle = \langle a + c, b + d\rangle$ Add components

$k\mathbf{u} = k\langle a, b\rangle = \langle ka, kb\rangle$ Multiply each component by $k$

Vector addition using the ordered pair notation agrees with the geometric definition of addition given earlier. If $\mathbf{u} = \langle a, b \rangle$ and $\mathbf{v} = \langle c, d \rangle$ are two vectors as shown in Figure 45, then by the parallelogram law, $\mathbf{u} + \mathbf{v} = \langle a + c, b + d \rangle$.

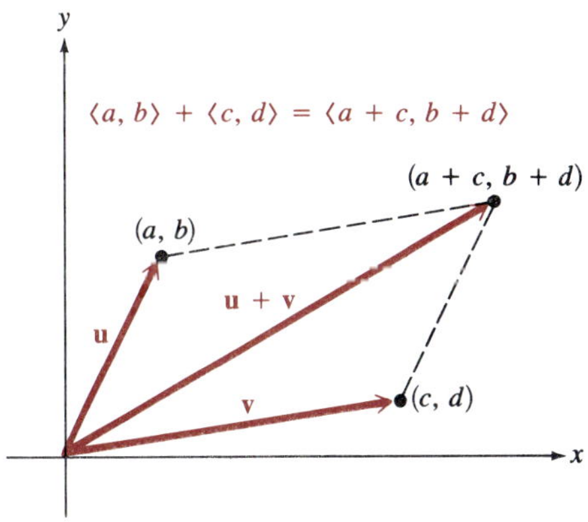


**Figure 45**

## EXAMPLE 2

Let $\mathbf{u} = \langle 2, -3 \rangle$ and $\mathbf{v} = \langle 1, 4 \rangle$. Find **(a)** $\mathbf{u} + \mathbf{v}$, **(b)** $4\mathbf{u}$, and **(c)** $3\mathbf{u} + 5\mathbf{v}$.

**(a)** $\mathbf{u} + \mathbf{v} = \langle 2, -3 \rangle + \langle 1, 4 \rangle = \langle 3, 1 \rangle$

**(b)** $4\mathbf{u} = 4\langle 2, -3 \rangle = \langle 8, -12 \rangle$

**(c)** $$\begin{aligned} 3\mathbf{u} + 5\mathbf{v} &= 3\langle 2, -3 \rangle + 5\langle 1, 4 \rangle \\ &= \langle 6, -9 \rangle + \langle 5, 20 \rangle \\ &= \langle 11, 11 \rangle \end{aligned}$$ ■

### Zero and Negative Vectors

The vector $\mathbf{0} = \langle 0, 0 \rangle$ is called the **zero vector.** If $\mathbf{v} = \langle a, b \rangle$, the vector $-\mathbf{v} = \langle -a, -b \rangle$ is called the **additive inverse** or **negative** of $\mathbf{v}$.

## Properties of Operations

The following properties of vector addition can easily be proved using the definition of vector addition and properties of real numbers.

### Properties of Vector Addition

Let $\mathbf{u}$, $\mathbf{v}$, and $\mathbf{w}$ be three vectors.

1. $\mathbf{u} + \mathbf{v} = \mathbf{v} + \mathbf{u}$ — Commutative law
2. $\mathbf{u} + (\mathbf{v} + \mathbf{w}) = (\mathbf{u} + \mathbf{v}) + \mathbf{w}$ — Associative law
3. $\mathbf{u} + \mathbf{0} = \mathbf{u}$ — Existence of the zero vector
4. $\mathbf{u} + (-\mathbf{u}) = \mathbf{0}$ — Existence of negatives

As with subtraction of real numbers, **subtraction of vectors** is defined by adding the negative of a vector.

### Subtraction of Vectors

Let $\mathbf{u} = \langle a, b\rangle$ and $\mathbf{v} = \langle c, d\rangle$ be two vectors. Then

$$\mathbf{u} - \mathbf{v} = \mathbf{u} + (-\mathbf{v}),$$

and

$$\mathbf{u} - \mathbf{v} = \langle a - c, b - d\rangle.$$

**EXAMPLE 3** Let $\mathbf{u} = \langle 2, -3\rangle$ and $\mathbf{v} = \langle 1, 4\rangle$. Find **(a)** $\mathbf{u} - \mathbf{v}$, **(b)** $\mathbf{v} - \mathbf{u}$, and **(c)** $5\mathbf{u} - 3\mathbf{v}$.

**(a)** $\mathbf{u} - \mathbf{v} = \langle 2, -3\rangle - \langle 1, 4\rangle = \langle 1, -7\rangle$

**(b)** $\mathbf{v} - \mathbf{u} = \langle 1, 4\rangle - \langle 2, -3\rangle = \langle -1, 7\rangle$

**(c)** $5\mathbf{u} - 3\mathbf{v} = 5\langle 2, -3\rangle - 3\langle 1, 4\rangle$
$= \langle 10, -15\rangle - \langle 3, 12\rangle = \langle 7, -27\rangle$ ■

The following properties are easily verified using the definition of scalar multiples and properties of real numbers.

### Properties of Scalar Multiples

Let $\mathbf{u}$ and $\mathbf{v}$ be two vectors, with $k$ and $n$ two real numbers.

1. $0\mathbf{u} = \mathbf{0}$ and $k\mathbf{0} = \mathbf{0}$
2. $1\mathbf{u} = \mathbf{u}$
3. $(kn)\mathbf{u} = k(n\mathbf{u})$
4. $(k + n)\mathbf{u} = k\mathbf{u} + n\mathbf{u}$
5. $k(\mathbf{u} + \mathbf{v}) = k\mathbf{u} + k\mathbf{v}$

An alternative way to represent vectors uses two special vectors,

$$\mathbf{i} = \langle 1, 0\rangle \quad \text{and} \quad \mathbf{j} = \langle 0, 1\rangle,$$

called **unit basis vectors.** Note that $|\mathbf{i}| = 1$ and $|\mathbf{j}| = 1$ (both are 1 unit in length), and $\mathbf{i}$ is directed along the positive $x$-axis while $\mathbf{j}$ is directed along the positive $y$-axis. If $\mathbf{v} = \langle a, b\rangle$, then

$$\mathbf{v} = \langle a, b\rangle = \langle a, 0\rangle + \langle 0, b\rangle = a\langle 1, 0\rangle + b\langle 0, 1\rangle = a\mathbf{i} + b\mathbf{j}.$$

We call $a\mathbf{i} + b\mathbf{j}$ a **linear combination** of the vectors $\mathbf{i}$ and $\mathbf{j}$. Thus, every vector in the plane can be represented as a linear combination of the basis vectors $\mathbf{i}$ and $\mathbf{j}$. We call $a$ the **horizontal component** of $\mathbf{v} = a\mathbf{i} + b\mathbf{j} = \langle a, b\rangle$, and $b$ its **vertical component.** It will be shown in the exercises that if $\theta$ is the direction of $\mathbf{v}$, then $a = |\mathbf{v}| \cos \theta$ and

$b = |\mathbf{v}| \sin \theta$. All of the rules for addition, subtraction, and scalar multiplication can be stated in this unit basis notation.

**EXAMPLE 4**

Write $2\mathbf{u} - 7\mathbf{v}$ as a linear combination of unit basis vectors $\mathbf{i}$ and $\mathbf{j}$ when $\mathbf{u} = -\mathbf{i} + 5\mathbf{j}$ and $\mathbf{v} = 2\mathbf{i} + 3\mathbf{j}$.

$$\begin{aligned} 2\mathbf{u} - 7\mathbf{v} &= 2(-\mathbf{i} + 5\mathbf{j}) - 7(2\mathbf{j} + 3\mathbf{j}) \\ &= (-2\mathbf{i} + 10\mathbf{j}) - (14\mathbf{i} + 21\mathbf{j}) \\ &= -16\mathbf{i} - 11\mathbf{j} \quad ■ \end{aligned}$$

We conclude this section by considering a variety of applied problems that can be solved with the geometrical notion of a vector. For convenience in notation, if vector $\mathbf{v} = \overrightarrow{AB}$, we will simply use $v$ or $AB$ to represent the magnitude of $\mathbf{v}$, $|\mathbf{v}|$.

**EXAMPLE 5**

PHYSICS

Two forces, one 30 lb and the other 40 lb, act on an object at right angles to each other. Find the magnitude of the resultant force and the angle that the resultant makes with the larger force.

Draw a sketch like the one in Figure 46, with two vectors corresponding to the forces of 30 lb and 40 lb. Using the parallelogram law, indicate the resultant vector $\mathbf{r} = \overrightarrow{PB}$. The length of $\mathbf{r}$ is the length of segment $PB$, which can be found with the Pythagorean theorem.

$$\begin{aligned} PB^2 &= BC^2 + PC^2 \\ &= 30^2 + 40^2 \\ &= 900 + 1600 = 2500 \\ PB &= 50 \text{ lb} \end{aligned}$$

The direction of $\mathbf{r}$ is given by $\angle BPC$ (the angle between $\mathbf{r}$ and the longest vector). To find $\angle BPC$ we use the tangent function in the right triangle $BPC$.

$$\tan BPC = \frac{BC}{PC} = \frac{30}{40} = \frac{3}{4} = 0.75$$

Then $$\angle BPC = 37°,$$

which is given to the nearest degree since the forces were given to the nearest unit. ■

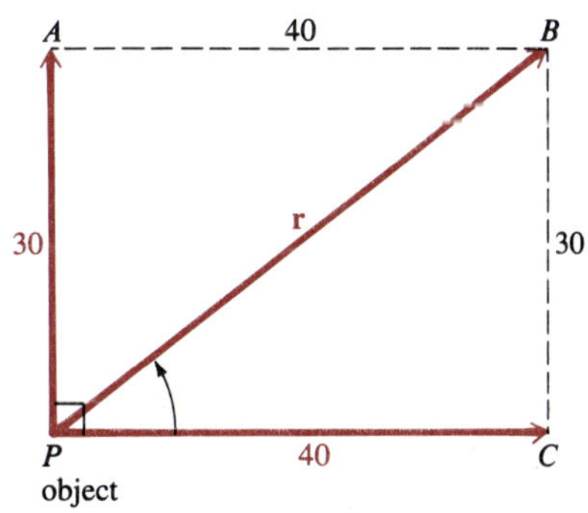


**Figure 46**

## EXAMPLE 6

**PHYSICS**

What is the minimum force required to prevent a 16.0-lb steel ball from rolling down a ramp which is inclined 12.5° to the horizontal? Also, what is the force of the ball against the ramp?

The weight or force of the ball acting vertically downward, $\overrightarrow{PB}$ can be resolved into two components, one parallel to the ramp ($\overrightarrow{PA}$) and the other perpendicular to the ramp ($\overrightarrow{PC}$), as shown in Figure 47. The force required to prevent the ball from rolling down the ramp (the force of the ball down the ramp) corresponds to $PA$, while the force of the ball against the ramp corresponds to $PC$.

Since $\angle APB$ and 12.5° are complementary, $\angle ABP = 12.5°$.

$$\sin ABP = \sin 12.5° = \frac{PA}{PB} = \frac{PA}{16.0}$$

$$PA = 16.0 \sin 12.5° \approx 3.5$$

$$\cos ABP = \cos 12.5° = \frac{AB}{PB} = \frac{PC}{PB} = \frac{PC}{16.0}$$

$$PC = 16.0 \cos 12.5° \approx 15.6$$

Thus, it takes a force of 3.5 lb to prevent the ball from rolling, and the ball exerts a force of 15.6 lb against the ramp. (Does this seem reasonable? How would $PA$ and $PC$ change as the angle of inclination of the ramp increases?) ■

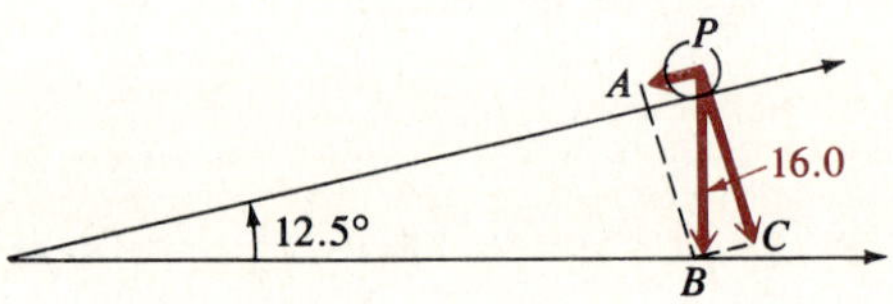


Figure 47

The final example in this section is the second applied problem in the introduction to this chapter.

## EXAMPLE 7

**NAVIGATION**

Two docks are located on opposite sides of a river which flows from east to west with a current of 3.5 mph. The bearing of the dock on the north bank from the one on the south bank is N10°20′E. Allowing for drift due to the current, in what direction should

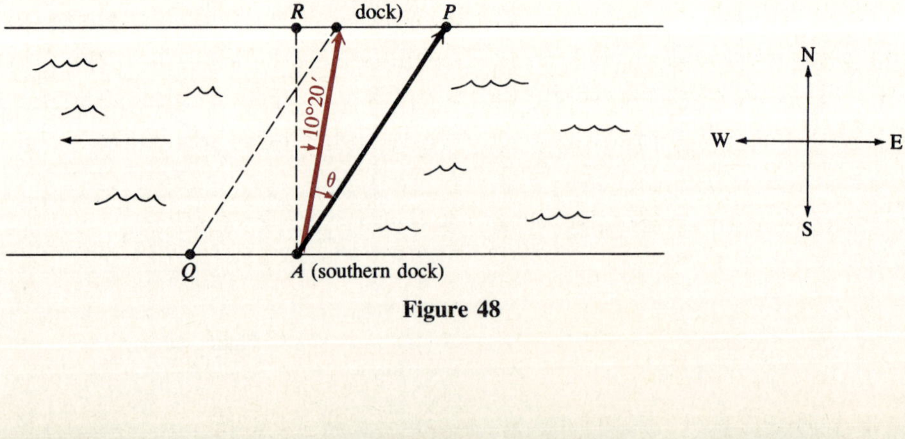


Figure 48

the captain of a ferry boat sail at 7 mph, in order to cross directly from the southern dock to the northern dock?

Make a sketch as in Figure 48. Since the bearing from $A$ to $B$ is N10°20′E, $\angle BAR = 10°20'$. Vector $\overrightarrow{AP}$ corresponds to the velocity vector of the ferry, and $\overrightarrow{PB}$ corresponds to the velocity of the current. We are looking for the direction of $\overrightarrow{AP}$, which can be found once angle $\theta$ is known. Since $\angle PBA$ and $\angle BAQ$ are alternate interior angles, from geometry we know they are equal. Thus $\angle PBA = \angle BAQ = 10°20' + 90°00' = 100°20'$. With $AP = 7$ and $PB = 3.5$, we can use the law of sines in triangle $BAP$ to find $\theta$.

$$\frac{AP}{\sin 100°20'} = \frac{PB}{\sin \theta}$$

$$\frac{7}{\sin 100°20'} = \frac{3.5}{\sin \theta}$$

$$\sin \theta = \frac{3.5 \sin 100°20'}{7} \approx 0.4919$$

$$\theta \approx 29°30'$$

Then $\angle RAP = 10°20' + 29°30' = 39°50'$. Thus, the captain must head in a direction 39°50′ in order to sail directly to the northern dock. ■

## 7.6 EXERCISES

In Exercises 1–8 find **(a)** $|\mathbf{v}|$, **(b)** the direction of $\mathbf{v}$ (to the nearest ten minutes), **(c)** $\mathbf{u} + \mathbf{v}$, **(d)** $\mathbf{u} - \mathbf{v}$, **(e)** $3\mathbf{v}$, **(f)** $2\mathbf{u} + 3\mathbf{v}$, **(g)** $5\mathbf{u} - 2\mathbf{v}$, **(h)** $\mathbf{u} + \mathbf{0}$, and **(i)** $-\mathbf{v}$.

**1.** $\mathbf{u} = \langle 1, 3\rangle$ and $\mathbf{v} = \langle 3, 4\rangle$

**2.** $\mathbf{u} = \langle -1, 2\rangle$ and $\mathbf{v} = \langle 4, 3\rangle$

**3.** $\mathbf{u} = \langle 0, 2\rangle$ and $\mathbf{v} = \langle -4, 0\rangle$

**4.** $\mathbf{u} = \langle -1, -1\rangle$ and $\mathbf{v} = \langle 1, 1\rangle$

**5.** $\mathbf{u} = 2\mathbf{i} + 3\mathbf{j}$ and $\mathbf{v} = -\mathbf{i} + 5\mathbf{j}$

**6.** $\mathbf{u} = \mathbf{i} + \mathbf{j}$ and $\mathbf{v} = -\mathbf{i} - \mathbf{j}$

**7.** $\mathbf{u} = -2\mathbf{j}$ and $\mathbf{v} = 5\mathbf{i}$

**8.** $\mathbf{u} = 2\mathbf{i} + 2\mathbf{j}$ and $\mathbf{v} = 2\mathbf{i} - 2\mathbf{j}$

In Exercises 9–20 let $\mathbf{u} = \langle a, b\rangle$, $\mathbf{v} = \langle c, d\rangle$, and $\mathbf{w} = \langle e, f\rangle$ be vectors with $k$ and $n$ scalars. Prove each property.

**9.** $\mathbf{u} + \mathbf{v} = \mathbf{v} + \mathbf{u}$

**10.** $\mathbf{u} + (\mathbf{v} + \mathbf{w}) = (\mathbf{u} + \mathbf{v}) + \mathbf{w}$

**11.** $\mathbf{u} + \mathbf{0} = \mathbf{u}$

**12.** $\mathbf{u} + (-\mathbf{u}) = \mathbf{0}$

**13.** $0\mathbf{u} = \mathbf{0}$

**14.** $1\mathbf{u} = \mathbf{u}$

**15.** $(kn)\mathbf{u} = k(n\mathbf{u})$

**16.** $(k + n)\mathbf{u} = k\mathbf{u} + n\mathbf{u}$

**17.** $k(\mathbf{u} + \mathbf{v}) = k\mathbf{u} + k\mathbf{v}$

**18.** If $k\mathbf{u} = \mathbf{0}$ and $k \neq 0$, then $\mathbf{u} = \mathbf{0}$.

**19.** $|k\mathbf{v}| = |k||\mathbf{v}|$

**20.** If $\mathbf{u} + \mathbf{v} = \mathbf{u}$ then $\mathbf{v} = \mathbf{0}$.

**21.** Let $\mathbf{v} = \langle a, b\rangle$ be a vector and $\theta$ the direction of $\mathbf{v}$. Show that $a = |\mathbf{v}| \cos \theta$ and $b = |\mathbf{v}| \sin \theta$.

**22.** Let $\mathbf{v} = \langle a, b\rangle$ be a vector. Show that the direction $\theta$ of $\mathbf{v}$ satisfies $\tan \theta = \dfrac{b}{a}$.

**23.** Let $\mathbf{u} = \langle a, b\rangle$ and $\mathbf{v} = \langle c, d\rangle$ be two vectors. The **dot product** of $\mathbf{u}$ and $\mathbf{v}$ is the scalar (a number *not* a vector)

$$\mathbf{u} \cdot \mathbf{v} = ac + bd.$$

Use this definition to find $\mathbf{u} \cdot \mathbf{v}$ for the following vectors.

**(a)** $\mathbf{u} = \langle 1, 3\rangle$ and $\mathbf{v} = \langle -2, 4\rangle$
**(b)** $\mathbf{u} = \langle 2, 5\rangle$ and $\mathbf{v} = \langle -5, 2\rangle$
**(c)** $\mathbf{u} = -3\mathbf{i} + 7\mathbf{j}$ and $\mathbf{v} = 4\mathbf{i} - 2\mathbf{j}$
**(d)** $\mathbf{u} = -2\mathbf{i}$ and $\mathbf{v} = \mathbf{i} - 4\mathbf{j}$

**24.** Let $\mathbf{u} = \langle a, b\rangle$ and $\mathbf{v} = \langle c, d\rangle$ be two vectors with $\theta$ the angle between $\mathbf{u}$ and $\mathbf{v}$. Use the law of cosines and the definition given in Exercise 23 to prove that

$$\mathbf{u} \cdot \mathbf{v} = |\mathbf{u}||\mathbf{v}| \cos \theta.$$

**25.** Use the results of Exercise 24 to prove that two nonzero vectors $\mathbf{u}$ and $\mathbf{v}$ are perpendicular if and only if $\mathbf{u} \cdot \mathbf{v} = 0$.

**26.** Let $\mathbf{u} = \langle a, b\rangle$, $\mathbf{v} = \langle c, d\rangle$, and $\mathbf{w} = \langle e, f\rangle$ be three vectors with $k$ a scalar. Use the definition of dot product given in Exercise 23 to prove the following properties.

**(a)** $\mathbf{u} \cdot \mathbf{v} = \mathbf{v} \cdot \mathbf{u}$
**(b)** $\mathbf{u} \cdot (\mathbf{v} + \mathbf{w}) = \mathbf{u} \cdot \mathbf{v} + \mathbf{u} \cdot \mathbf{w}$
**(c)** $k(\mathbf{u} \cdot \mathbf{v}) = (k\mathbf{u}) \cdot \mathbf{v}$
**(d)** $\mathbf{u} \cdot \mathbf{u} = |\mathbf{u}|^2$

In Exercises 27–30 $\mathbf{v} = \langle a, b\rangle$ is a force vector with the given magnitude and direction. Find $a$ and $b$, correct to the nearest tenth.

**27.** 30.0 lb; N25°30′E
**28.** 45.0 kg; N19°40′W
**29.** 12.2 kg; S15°40′W
**30.** 51.8 lb; S42°10′E

In Exercises 31–34 the magnitudes and directions of two forces acting on an object are given. Find the magnitude of the resultant (correct to the nearest tenth), and the direction of the resultant (correct to the nearest ten minutes).

**31.** 25.0 lb; N40°00′E and 42.0 lb; S50°00′E
**32.** 16.1 kg; N35°00′W and 12.5 kg; S55°00′W
**33.** 15.3 kg; N10°20′E and 5.9 kg; N71°30′E
**34.** 22.0 lb; S15°20′E and 14.2 lb; S25°40′W

In Exercises 35–36 find the magnitudes of the horizontal and vertical components of the given resultant vector, correct to the nearest tenth.

**35.** **SPORTS** A basketball player shoots the ball with a velocity of 20.0 ft/sec at an angle of 40° with the horizontal.

**36.** A child pulls a wagon by exerting a force of 15.0 lb at an angle of 25° with the horizontal.

Solve.

**37.** **PHYSICS** Two forces, one of 70.0 lb and the other of 25.0 lb, act on an object at right angles to each other. Find the magnitude of the resultant of the two forces and the angle that the resultant makes with the larger force.

**38.** **ENGINEERING** What is the minimum force required to prevent a 16.0-lb steel ball from rolling down a ramp which is inclined 18°30′ to the horizontal? Also, what is the force of the ball against the ramp?

**39.** **ENGINEERING** A man wishes to roll a barrel weighing 400 lb up a plank and onto a platform 3 ft above the ground. If the plank is 12 ft long, and if the man can exert a pushing force of 120 lb, will he be able to roll the barrel onto the platform? [Hint: First find the angle of inclination of the plank.]

**40.** **RECREATION** When a child pulls a wagon with a rope, the rope makes an angle of 32°00′ with the horizontal. If a pull of 10.0 lb on the rope is needed to move the wagon, what is the horizontal component force (the force required to move the wagon if pulled horizontally)?

**41.** **AERONAUTICS** An airplane with **air speed** (speed in still air) of 275.0 mph is headed due east. If a northerly wind is blowing at 30.0 mph, find the direction of travel and the **ground speed** (speed relative to the ground) of the airplane.

**42.** **AERONAUTICS** An airplane with air speed of 300 mph is headed due west. If the wind is from the south and blowing at 25 mph, find the direction of travel and the ground speed of the airplane.

**43.** **AERONAUTICS** An airplane is headed in a direction of 140° and flying at a speed of 240 mph. If the wind is blowing from a direction of 40° at a rate of 30 mph, what is the direction of travel and the ground speed of the airplane?

**44.** **NAVIGATION** A ship sails N65°00′W for 180 miles, and then S55°00′W for 220 miles. How far is the ship from its starting point and in what direction?

**45.** **RECREATION** The speed of a river flowing due east is 4 mph. A rowboat is headed directly north across the river and is rowed in such a way as to have a speed of 7 mph in still water. Find the magnitude of the velocity of the boat and the direction of travel (to the nearest ten minutes).

**46.** **NAVIGATION** A boat, on the south bank of a river that flows due west, can travel at a speed of 20 km/hr. The driver wishes to cross the river whose current is 5 km/hr. In order to make the crossing at right angles, in what direction should the boat be headed (to the nearest ten minutes)?

**47.** **AERONAUTICS** A jet plane has velocity 600 mph traveling in a direction 310°. How far north has it traveled in one hour?

**48.** **METEOROLOGY** A weather balloon is rising at 15 ft/sec while the wind is blowing 20 ft/sec. What is the speed of the balloon and the angle of ascent made with the horizontal (to the nearest ten minutes)?

**49.** **RECREATION** A backpacker hangs his food sack from the center of a 30-ft line between two trees. If the weight of the sack is 6.0 lb, the center of the line is pulled down 2 ft. What is the force in pounds on the line? [Hint: Use the figure below. The force in pounds on the line is the magnitude of $t$.]

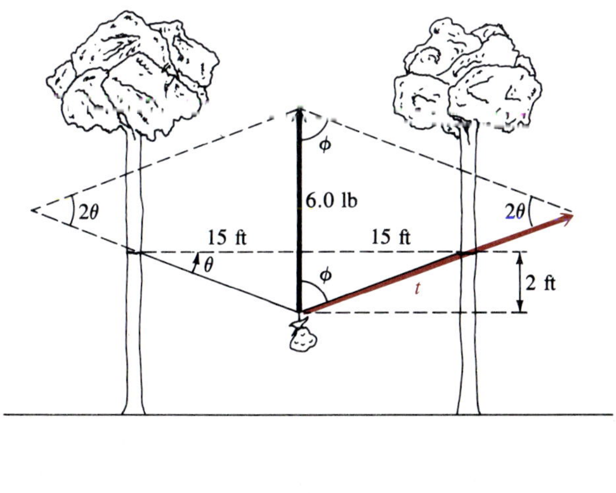

## For Review

**50.** The motion in a spring-mass system is described by $y = 2.5 \sin\left(\pi t - \frac{\pi}{2}\right)$ where $y$ is the distance in feet from the equilibrium position and $t$ is time in seconds. Suppose the weight is 5 ft from the ceiling in a state of equilibrium. Determine the following.

**(a)** The starting position of the weight,
**(b)** The first time the weight passes the equilibrium position,
**(c)** The closest the weight will ever be to the ceiling,
**(d)** The farthest the weight will ever be from the ceiling,
**(e)** The distance from the ceiling at time $t = 1.5$,
**(f)** The amplitude of the motion,
**(g)** The period of the motion,
**(h)** The frequency of the motion,
**(i)** The phase shift of the motion.

# CHAPTER 7 REVIEW EXERCISES

In Exercises 1–2, $A$ is an angle in standard position and satisfies the given conditions. Find the six trigonometric function values of $A$. Do not use a calculator.

**1.** $P(-2, 6)$ is on the terminal side of $A$.

**2.** The terminal side of $A$ is in quadrant III and is perpendicular to the line $3x + 6y - 7 = 0$.

**3.** Which quadrant contains the terminal side of angle $A$ if $\sin A < 0$ and $\tan A > 0$?

In Exercises 4–7 determine the reference angle for angle $A$.

**4.** $A = 205°$ **5.** $A = -304°20'$ **6.** $A = -\frac{2\pi}{3}$ **7.** $A = \frac{17\pi}{4}$

In Exercises 8–9 find angle $A$, to the nearest minute, if $0 \le A \le 360°$.

**8.** $\tan A = 3.1579$ **9.** $\sec A = 4.2765$

**10.** If $A$ is in quadrant III and $\tan A = \frac{4}{3}$, find the remaining trigonometric function values of $A$.

In Exercises 11–14 solve each triangle with the given parts.

**11.** $A = 38°20'$, $B = 61°20'$, $a = 13.4$ **12.** $B = 51°20'$, $a = 20.2$, $b = 7.4$

**13.** $A = 71°45'$, $b = 4.20$, $c = 5.15$ **14.** $a = 16.5$, $b = 12.2$, $c = 23.7$

Solve.

**15.** **SURVEYING** Points $A$ and $B$ are on opposite sides of a lake. A point $C$ is 100 yards from $A$. The measure of $\angle BAC$ is $62°20'$ and the measure of $\angle ACB$ is $36°40'$. How far apart are $A$ and $B$?

**16.** **FORESTRY** Lookout station B is 20.0 km due west of lookout station A. The bearing of a fire from A is S24°20′W and the bearing from B is S51°20′E. How far is the fire from B?

**17.** NAVIGATION Phoenix is 220 miles due south of the Grand Canyon. A pilot flying from Phoenix to the Grand Canyon flies 150 miles off course in a direction of 352°20′ before discovering his error. At this time, how far is he from the Grand Canyon?

**18.** NAVIGATION Two boats leave an island at the same time. The first sails 25 mph in a direction of 315°00′ and the second sails 30 mph in a direction of 200°00′. How far apart are the boats after two hours?

In Exercises 19–20 find the area of triangle $ABC$ with the given parts.

**19.** $a = 3.1$ yd, $b = 5.8$ yd, $C = 38°20'$

**20.** $B = 120°20'$, $C = 25°40'$, $a = 12.20$ in

**21.** CONSUMER The Hugheses purchased a lot in the shape of a quadrilateral with measurements as shown. What is the area of the lot in acres, if 1 acre = 43,560 ft$^2$?

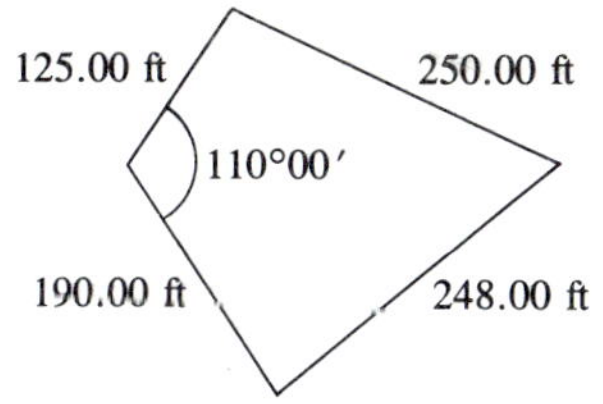


**22.** AGRICULTURE A triangular garden with sides 31.5 yd, 27.1 yd, and 25.8 yd is to be fertilized. If one bag of fertilizer covers 2000 ft$^2$, how many bags will be needed for the project?

**23.** ELECTRONICS The drop in voltage across two terminals in an electrical circuit is given by the equation $E = 120 \sin (100\pi t)$, where $t$ is measured in seconds. **(a)** What is the maximum voltage drop? **(b)** What is the period of the circuit? **(c)** What is the frequency of the circuit?

**24.** PHYSICS A point $P$ on a vibrating surface is in simple harmonic motion, with frequency of 1/4 oscillation per second and amplitude 10 cm. Express the motion of $P$ with an equation of the form $y = A \sin Bt$.

In Exercises 25–26 find **(a)** $|\mathbf{u}|$, **(b)** the direction of $\mathbf{u}$, **(c)** $\mathbf{u} + \mathbf{v}$, **(d)** $\mathbf{u} - \mathbf{v}$, **(e)** $-2\mathbf{u}$, **(f)** $4\mathbf{u} - 3\mathbf{v}$, **(g)** $\mathbf{v} + \mathbf{0}$, and **(h)** $-\mathbf{u}$.

**25.** $\mathbf{u} = \langle -5, 12 \rangle$ and $\mathbf{v} = \langle 1, -2 \rangle$

**26.** $\mathbf{u} = 7\mathbf{i}$ and $\mathbf{v} = -\mathbf{i} + \mathbf{j}$

**27.** SPORTS A quarterback throws a pass with an initial velocity of 60.0 ft/sec at an angle of 33° with the horizontal. Find the magnitudes of the horizontal and vertical components (correct to the nearest tenth).

**28.** AERONAUTICS An airplane with air speed of 400 mph is headed due south. If the wind is blowing from 45°00 at 25 mph, find the direction of travel and the ground speed of the airplane.

**29.** METEOROLOGY A weather balloon is rising 18 ft/sec while the wind is blowing 30 ft/sec. What are the speed of the balloon and the angle of ascent made with the horizontal?

**30.** PHYSICS What is the minimum force required to prevent a 200-lb barrel from rolling down a ramp inclined 20°30′ with the horizontal? Also, what is the force of the barrel against the ramp?

CHAPTER

# 8 ANALYTIC TRIGONOMETRY

We have used trigonometry to solve a variety of applied problems involving distances, heights, velocities, bearings, and so on. Another, more abstract use is the application of the theory of analytic trigonometry to transform complex trigonometric expressions into simpler ones. Although analytic trigonometry is applied primarily in advanced mathematics, some practical problems apply the theory, as illustrated by the following.

◀ SPORTS

A football player kicks a field goal at an angle of inclination with the playing field equal to $\theta$. It can be shown that the distance the ball will travel in the air if it is not touched is given in feet by

$$d = \frac{v^2 \sin 2\theta}{32},$$

where $v$ is the initial velocity of the ball in ft/sec. **(a)** Find the distance that the ball will travel if $\theta = 40°$ and $v = 50$ ft/sec. **(b)** If $v$ is constant, discuss why the maximum distance will be attained when $\theta = 45°$.

HYDROLOGY ▶

A Colorado River water diversion project in Arizona uses a canal in the shape of an isosceles trapezoid which is wider at the top than at the base. If the bottom and the two sides of the canal measure 8 yd, and $\theta$ is the acute angle formed by a side and level ground, show that the cross-sectional area of the canal, which determines the water-carrying capacity, is given by

$$A = 64 \sin \theta(1 + \cos \theta).$$

Use this formula to find $\theta$ when $A = 48\sqrt{3}$ ft$^2$.

The first of these problems involves a function of a double-angle (see Example 9 in Section 8.3), and the second requires the ability to solve trigonometric equations (see Example 10 in Section 8.5).

We begin this chapter with a discussion of trigonometric identities and their proofs before proceeding to numerous identities involving functions of sums, differences, and multiples of angles (or real numbers). We then investigate product and factor identities which are useful in calculus, develop techniques for solving trigonometric equations, and conclude the chapter with the study of inverse trigonometric functions.

## 8.1 Trigonometric Identities

Recall that there are three types of equations: contradictions (that are never true for any value of the variable), conditional equations (that are true for some and false for other values of the variable), and identities (that are true for all values of the variable). In this section we are concerned with **trigonometric identities,** identities involving the trigonometric functions. A number of fundamental trigonometric identities have already been considered in Section 6.8 and are repeated on the next page for easy reference. These identities, along with the basic rules of algebra for manipulating and simplifying algebraic expressions, can be employed to prove that a variety of other equations are also trigonometric identities.

### Suggestions for Proving Identities

Most of the identities throughout this section are not very important by themselves. What is important are the techniques used to prove them and the mental maturity that comes from constructing the proofs. To prove an identity, transform one side into an expression identical to the other side, or into an expression obtained from the other side

### The Fundamental Trigonometric Identities

**Reciprocal Identities**

$$\csc A \equiv \frac{1}{\sin A} \qquad \sec A \equiv \frac{1}{\cos A} \qquad \cot A \equiv \frac{1}{\tan A}$$

**Ratio Identities**

$$\tan A \equiv \frac{\sin A}{\cos A} \qquad \cot A \equiv \frac{\cos A}{\sin A}$$

**Pythagorean Identities**

$$\sin^2 A + \cos^2 A \equiv 1 \qquad 1 + \tan^2 A \equiv \sec^2 A \qquad 1 + \cot^2 A \equiv \csc^2 A$$

**Cofunction Identities**

$$\sin\left(A + \frac{\pi}{2}\right) \equiv \cos A \qquad \sin\left(A - \frac{\pi}{2}\right) \equiv -\cos A \qquad \sin\left(\frac{\pi}{2} - A\right) \equiv \cos A$$

$$\cos\left(A + \frac{\pi}{2}\right) \equiv -\sin A \qquad \cos\left(A - \frac{\pi}{2}\right) \equiv \sin A \qquad \cos\left(\frac{\pi}{2} - A\right) \equiv \sin A$$

$$\tan\left(A + \frac{\pi}{2}\right) \equiv -\cot A \qquad \tan\left(A - \frac{\pi}{2}\right) \equiv -\cot A \qquad \tan\left(\frac{\pi}{2} - A\right) \equiv \cot A$$

**Miscellaneous Identities**

$$\sin(-A) \equiv -\sin A \qquad \cos(-A) \equiv \cos A$$

$$\sin(A \pm 2\pi) \equiv \sin A \qquad \cos(A \pm 2\pi) \equiv \cos A$$

$$\sin(A \pm \pi) \equiv -\sin A \qquad \cos(A \pm \pi) \equiv -\cos A$$

$$\sin(\pi - A) \equiv \sin A \qquad \cos(\pi - A) \equiv -\cos A$$

by similar procedures. The ability to prove identities comes only with extensive practice and sometimes with trial and error. Several suggestions are helpful.

1. Memorize the fundamental identities and be able to recognize variations of them.
2. Work with only one side of an identity at a time, starting with the side that appears more complex.
3. Remember that various algebraic operations, such as factoring and collecting like terms, also apply to terms involving trigonometric functions.
4. When working with one side always keep the other side in mind, to remember where you are headed.
5. If one side has a single trigonometric function, try to eliminate all functions, except that one, from the other side.
6. It sometimes helps to transform all functions into sines and cosines.
7. If a numerator or denominator contains an expression such as $1 + \sin A$ (or $1 - \sin A$), try multiplying both numerator and denominator by $1 - \sin A$ (or $1 + \sin A$).

8. Perhaps most important of all is to try something. Even if what is attempted fails, something has been learned. Learning what does not work helps to eliminate possibilities until you discover what does work.

### Format for Proving Identities

The format we use in the proof of an identity has a vertical line separating the two sides. This is a reminder not to work with both sides simultaneously, at least initially. Remember that any variable can be used in an identity.

### EXAMPLE 1

Prove the identity. $\sec x \cot x \equiv \csc x$

| | |
|---|---|
| $\sec x \cot x$ | $\csc x$ |
| Work on the left side. Change to sines and cosines. | |
| $\frac{1}{\cancel{\cos x}} \cdot \frac{\cancel{\cos x}}{\sin x}$ | |
| $\frac{1}{\sin x}$ | |
| $\csc x$ | |

Since the left side is identical to the right side, the identity has been proved. ■

### EXAMPLE 2

Prove the identity. $\cos^2 \theta \cot^2 \theta \equiv \cot^2 \theta - \cos^2 \theta$

| | |
|---|---|
| $\cos^2 \theta \cot^2 \theta$ | $\cot^2 \theta - \cos^2 \theta$ |
| | The right side seems more complex. Use $\cot \theta = \frac{\cos \theta}{\sin \theta}$. |
| | $\frac{\cos^2 \theta}{\sin^2 \theta} - \cos^2 \theta$ |
| | $\cos^2 \theta\left(\frac{1}{\sin^2 \theta} - 1\right)$ |
| | $\cos^2 \theta\left(\frac{1 - \sin^2 \theta}{\sin^2 \theta}\right)$ |
| | Since $\sin^2 \theta + \cos^2 \theta = 1$, then $1 - \sin^2 \theta = \cos^2 \theta$. |
| | $\cos^2 \theta \cdot \frac{\cos^2 \theta}{\sin^2 \theta}$ |
| | $\cos^2 \theta \cot^2 \theta$ |

Since the right side is the same as the left, we have verified the identity. ■

**EXAMPLE 3** Prove the identity. $\dfrac{\tan A + \cot A}{\csc A} \equiv \sec A$

| | |
|---|---|
| $\dfrac{\tan A + \cot A}{\csc A}$ | $\sec A$ |
| Since three functions are involved, change everything to sines and cosines. | |
| $\dfrac{\dfrac{\sin A}{\cos A} + \dfrac{\cos A}{\sin A}}{\dfrac{1}{\sin A}}$ | |
| Multiply numerator and denominator by the LCD of all fractions in the complex fraction. | |
| $\dfrac{\dfrac{\sin A}{\cos A} + \dfrac{\cos A}{\sin A}}{\dfrac{1}{\sin A}} \cdot \dfrac{\sin A \cos A}{\sin A \cos A}$ | |
| $\dfrac{\sin^2 A + \cos^2 A}{\cos A}$ | |
| $\sin^2 A + \cos^2 A = 1$ | |
| $\dfrac{1}{\cos A}$ | |
| $\sec A$ | |

Since the left side has been transformed into the right side, the identity is proved. ■

To prove some identities we may need to do some work on each side. This is illustrated in the next example.

**EXAMPLE 4** Prove the identity. $\dfrac{1}{\sin x \cos x} - \dfrac{\cos x}{\sin x} \equiv \dfrac{\sin x \cos x}{1 - \sin^2 x}$

| | |
|---|---|
| $\dfrac{1}{\sin x \cos x} - \dfrac{\cos x}{\sin x}$ | $\dfrac{\sin x \cos x}{1 - \sin^2 x}$ |
| With two terms on the left and one on the right, we begin on the left by subtracting the two fractions. | Remember that $1 - \sin^2 x = \cos^2 x$ |
| | $\dfrac{\sin x \cos x}{\cos^2 x}$ |
| | Divide out $\cos x$. |
| $\dfrac{1}{\sin x \cos x} - \dfrac{\cos x}{\sin x} \cdot \dfrac{\cos x}{\cos x}$ | $\dfrac{\sin x}{\cos x}$ |
| $\dfrac{1 - \cos^2 x}{\sin x \cos x}$ | $\tan x$ |

Recall that $1 - \cos^2 x = \sin^2 x$.

$$\frac{\sin^2 x}{\sin x \cos x}$$

Divide out $\sin x$.

$$\frac{\sin x}{\cos x}$$

$$\tan x$$

With this side simplified, now work on the right side.

Since both sides are identical to $\tan x$, we have proven the given identity. ■

**EXAMPLE 5** Prove the identity. $\dfrac{1 - \cos B}{\sin B} \equiv \dfrac{\sin B}{1 + \cos B}$

| | |
|---|---|
| $\dfrac{1 - \cos B}{\sin B}$ | $\dfrac{\sin B}{1 + \cos B}$ |

Since the expression contains $1 - \cos B$, multiply numerator and denominator by $1 + \cos B$.

$$\frac{(1 - \cos B)(1 + \cos B)}{\sin B(1 + \cos B)}$$

$$\frac{1 - \cos^2 B}{\sin B(1 + \cos B)}$$

$$\frac{\sin^2 B}{\sin B(1 + \cos B)}$$

Divide out $\sin B$.

$$\frac{\sin B}{1 + \cos B}$$

This time the left side has been transformed directly into the right side, proving the identity. We could have started on the right and multiplied numerator and denominator by $1 - \cos B$. This would have given the left side following steps similar to those above. ■

The next example illustrates that we must be careful when taking square roots of trigonometric expressions.

**EXAMPLE 6** Since $\cos^2 x = 1 - \sin^2 x$, for what real numbers $x$, $0 \le x < 2\pi$, will $\cos x = \sqrt{1 - \sin^2 x}$?

When taking the square root of both sides of an equation, remember that there are actually two roots, one positive and one negative. Since

$$\cos^2 x = 1 - \sin^2 x,$$

taking the square root of both sides,

$$\sqrt{\cos^2 x} = \sqrt{1 - \sin^2 x}$$
$$|\cos x| = \sqrt{1 - \sin^2 x}$$

is true. However, in order for

$$\cos x = \sqrt{1 - \sin^2 x}$$

to be true, $\cos x$ must be nonnegative (since $\sqrt{1 - \sin^2 x}$ is nonnegative). This will happen when $0 \le x \le \frac{\pi}{2}$ or $\frac{3\pi}{2} \le x < 2\pi$. ■

Some algebraic expressions containing a radical can be transformed to an equivalent expression without the radical by making a **trigonometric substitution.** Such substitutions play an important role in calculus.

**EXAMPLE 7**

**CALCULUS**

Change $\sqrt{a^2 - x^2}$ to an equivalent trigonometric expression free of radicals by substituting $a \cos \theta$ for $x$. Assume that $a > 0$ and that $0 \le \theta \le \pi$.

Let $x = a \cos \theta$.

$$\begin{aligned} \sqrt{a^2 - x^2} &= \sqrt{a^2 - (a \cos \theta)^2} \\ &= \sqrt{a^2 - a^2 \cos^2 \theta} \\ &= \sqrt{a^2(1 - \cos^2 \theta)} \\ &= \sqrt{a^2 \sin^2 \theta} \\ &= a \sin \theta \end{aligned}$$

Notice that this is true when $a > 0$ (then $\sqrt{a^2} = a$) and when $0 \le \theta \le \pi$ (then $\sin \theta > 0$ and $\sqrt{\sin^2 \theta} = \sin \theta$). ■

CAUTION Always remember to write the variable associated with a trigonometric function when proving identities. Do not get lazy and write "sin" for "sin $x$" since "sin" is meaningless unless the variable is specified. ■

## 8.1 EXERCISES

Prove the identities in Exercises 1–54.

1. $\dfrac{\tan x}{\sec x} \equiv \sin x$

2. $\dfrac{\csc A}{\sec A} \equiv \cot A$

3. $\dfrac{1 - \sin^2 x}{\cos x} \equiv \cos x$

4. $\dfrac{1 + \tan^2 A}{\sec A} \equiv \sec A$

5. $\dfrac{1}{1 - \sin^2 x} \equiv 1 + \tan^2 x$

6. $\dfrac{\sec^2 A - 1}{\sec^2 A} = \sin^2 A$

7. $1 - 2 \sin^2 y \equiv 2 \cos^2 y - 1$

8. $\sin^2 \theta - \tan^2 \theta \equiv -\sin^2 \theta \tan^2 \theta$

9. $\dfrac{\csc^2 x}{1 + \tan^2 x} \equiv \cot^2 x$

10. $\dfrac{1 + \cot A}{\csc A} \equiv \sin A + \cos A$

11. $\dfrac{1 + \cos^2 x}{\sin^2 x} \equiv 2 \csc^2 x - 1$

12. $\sin^2 A \sec A - \sec A \equiv -\cos A$

**13.** $\cos y(\tan y + \cot y) \equiv \csc y$

**14.** $\sec \theta - \cos \theta \equiv \tan \theta \sin \theta$

**15.** $\dfrac{\cos y}{1 - \sin y} \equiv \dfrac{1 + \sin y}{\cos y}$

**16.** $\dfrac{\sec \theta + 1}{\tan \theta} \equiv \dfrac{\tan \theta}{\sec \theta - 1}$

**17.** $\dfrac{1 + \csc x}{\sec x} - \cot x \equiv \cos x$

**18.** $\dfrac{\cot A - 1}{\cot A + 1} \equiv \dfrac{1 - \tan A}{1 + \tan A}$

**19.** $\sin^4 y + 2 \sin^2 y \cos^2 y + \cos^4 y \equiv 1$

**20.** $\dfrac{\sin A}{\tan A} + \dfrac{\cos A}{\cot A} \equiv \sin A + \cos A$

**21.** $\dfrac{1 + \sec y}{\sin y + \tan y} \equiv \csc y$

**22.** $\dfrac{1 + \sin \theta}{1 - \sin \theta} \equiv (\sec \theta + \tan \theta)^2$

**23.** $\dfrac{\cos^3 y - \sin^3 y}{\cos y - \sin y} \equiv 1 + \sin y \cos y$

**24.** $(\sin \theta + \cos \theta)^2 + (\sin \theta - \cos \theta)^2 \equiv 2$

**25.** $\cos^4 x - \sin^4 x \equiv 1 - 2 \sin^2 x$

**26.** $\dfrac{\sin \theta - 1}{\cos \theta} \equiv \dfrac{\cos \theta - \cot \theta}{\cos \theta \cot \theta}$

**27.** $\sec^4 x - \sec^2 x \equiv \tan^4 x + \tan^2 x$

**28.** $\cos^4 A + \sin^2 A \equiv \sin^4 A + \cos^2 A$

**29.** $\csc y + \cot y \equiv \dfrac{\sin y}{1 - \cos y}$

**30.** $\cos^2 \theta + \sin \theta \cos \theta \equiv \dfrac{\cos \theta(1 + \cot \theta)}{\csc \theta}$

**31.** $\dfrac{\sin y}{1 + \cos y} + \dfrac{1 + \cos y}{\sin y} \equiv 2 \csc y$

**32.** $\dfrac{\sin \theta - \cos \theta}{\cos^2 \theta} \equiv \dfrac{\tan^2 \theta - 1}{\sin \theta + \cos \theta}$

**33.** $\dfrac{\sin\left(\dfrac{\pi}{2} - x\right)}{1 - \sin x} - \sec x \equiv \tan x$

**34.** $\dfrac{\sin\left(\dfrac{\pi}{2} - A\right)}{1 - \tan A} + \dfrac{\sin (\pi - A)}{1 - \cot A} \equiv \cos A + \sin A$

**35.** $(\sin y + \cos y)^2 - (\sin (-y) + \cos (-y))^2 \equiv 4 \sin y \cos y$

**36.** $\dfrac{1}{\tan x(\csc x + 1)} \equiv \tan x(\csc x - 1)$

**37.** $\sin x \tan x + \cos x \cot x \equiv \sec x + \csc x - \cos x - \sin x$

**38.** $\dfrac{\sin A + \cos A + 1}{\sin A + \cos A - 1} \equiv \csc A \sec A + \csc A + \sec A + 1$

**39.** $\dfrac{1 - 2 \cos^2 y - \sin y}{2 \sin y + 1} \equiv \sin y - 1$

**40.** $\dfrac{8 - 2 \tan \theta - 5 \sec^2 \theta}{3 - 5 \tan \theta} \equiv 1 + \tan \theta$

**41.** $\dfrac{1 + \sin x}{|\cos x|} \equiv \sqrt{\dfrac{1 + \sin x}{1 - \sin x}}$

**42.** $\dfrac{|\sin A|}{1 - \cos A} \equiv \sqrt{\dfrac{1 + \cos A}{1 - \cos A}}$

**43.** $\sec y - \tan y \equiv \sqrt{\dfrac{1 - \sin y}{1 + \sin y}}$ for $0 < x < \dfrac{\pi}{2}$

**44.** $\dfrac{\sec \theta - 1}{\tan \theta} \equiv \sqrt{\dfrac{\sec \theta - 1}{\sec \theta + 1}}$ for $0 < x < \dfrac{\pi}{2}$

**45.** $\log 10^{\sin x} \equiv \sin x$

**46.** $10^{\log |\cos A|} \equiv |\cos A|$

**47.** $\ln |\sec x| \equiv -\ln |\cos x|$

**48.** $\ln |\tan x| \equiv \ln |\sin x| - \ln |\cos x|$

**49.** $\ln |\sec x + \tan x| \equiv -\ln |\sec x - \tan x|$

**50.** $\ln |\csc A + \cot A| \equiv -\ln |\csc A - \cot A|$

**51.** $e^{\ln |\sin y|} \equiv |\sin y|$

**52.** $e^{\ln |\tan \theta|} \equiv |\tan \theta|$

**53.** $\ln e^{\sin^2 x + \cos^2 x} \equiv 1$

**54.** $\ln e^{\tan^2 x - \sec^2 x} \equiv -1$

In Exercises 55–60 determine all real numbers $x$, for $0 \le x < 2\pi$, that make the given equation true.

**55.** $\sin x = \sqrt{1 - \cos^2 x}$

**56.** $\sqrt{1 + \tan^2 x} = \sec x$

**57.** $\csc x = \sqrt{1 + \cot^2 x}$

**58.** $\sqrt{\sec^2 x - \tan^2 x} = 1$

**59.** $\log (\tan x) = \log (\sin x) - \log (\cos x)$

**60.** $\ln (\sin x) = -\ln (\csc x)$

**CALCULUS** In Exercises 61–63 change the given expression to an equivalent trigonometric expression free of radicals by substituting $a \cos \theta$ for $x$. Assume that $a > 0$ and that $0 < \theta < \frac{\pi}{2}$.

**61.** $(a^2 - x^2)^{5/2}$

**62.** $\dfrac{x}{\sqrt{a^2 - x^2}}$

**63.** $\dfrac{\sqrt{a^2 - x^2}}{x^2}$

In Exercises 64–66 change the given expression to an equivalent trigonometric expression free of radicals by substituting $a \tan \theta$ for $x$. Assume that $a > 0$ and that $-\frac{\pi}{2} < \theta < \frac{\pi}{2}$.

**64.** $\sqrt{a^2 + x^2}$

**65.** $\dfrac{1}{\sqrt{a^2 + x^2}}$

**66.** $(a^2 + x^2)^{5/2}$

In Exercises 67–70 write each expression in terms of the given function only. Assume that $0 < \theta < \frac{\pi}{2}$.

**67.** $\cot \theta \sec^2 \theta$; $\cos \theta$

**68.** $\dfrac{1 + \tan^2 \theta}{\tan^2 \theta}$; $\sin \theta$

**69.** $\sin (\pi - \theta) \tan \theta$; $\cos \theta$

**70.** $\cos \left(\frac{\pi}{2} - \theta\right) \tan \theta$; $\sin \theta$

In Exercises 71–74 show that the given equation is *not* an identity by finding values of $A$ and $B$ that make the equation false.

**71.** $\cos (A - B) = \cos A - \cos B$

**72.** $\cos (A + B) = \cos A + \cos B$

**73.** $\sin (A + B) = \sin A + \sin B$

**74.** $\sin (A - B) = \sin A - \sin B$

## For Review

In Exercises 75–76 let $\mathbf{u} = \langle 2, -5 \rangle$, $\mathbf{v} = \langle -1, 4 \rangle$, and write each vector in terms of the unit basis vectors **i** and **j**.

**75.** $3\mathbf{u} - 2\mathbf{v}$

**76.** $4\mathbf{u} + \mathbf{v}$

## 8.2 Sum and Difference Identities

The fundamental identities reviewed in the previous section involved trigonometric functions of a single real-number variable or a single angle measured in degrees or radians. In this section we consider identities involving sums or differences of two variables or angles, such as $\cos (A - B) = \cos A \cos B + \sin A \sin B$. We can show by example that $\cos (A - B)$ is not $\cos A - \cos B$. Consider $A = 60°$ and $B = 30°$.

$$\cos (60° - 30°) \neq \cos 60° - \cos 30°$$

since $$\cos (60° - 30°) = \cos 30° = \frac{\sqrt{3}}{2} \approx 0.8660,$$

while $$\cos 60° - \cos 30° = \frac{1}{2} - \frac{\sqrt{3}}{2} = \frac{1 - \sqrt{3}}{2} \approx -0.3660.$$

The formulas that we derive hold true for trigonometric functions of real numbers, angles in radians, and angles in degrees.

### Cosine of a Difference

$$\cos (A - B) \equiv \cos A \cos B + \sin A \sin B$$

**Derivation of Identity for $\cos (A - B)$**

To verify this identity, consider the unit circle in Figure 1 showing two angles $A$ and $B$ in standard position. Let $P$ be the point of intersection of the terminal side of $A$ and the unit circle, with $Q$ the similar point relative to $B$. Then the coordinates of $P$ are $(\cos A, \sin A)$, and the coordinates of $Q$ are $(\cos B, \sin B)$. Using the distance formula, the square of the distance between $P$ and $Q$ is

$$\begin{aligned} d^2 &= (\cos A - \cos B)^2 + (\sin A - \sin B)^2 \\ &= \cos^2 A - 2 \cos A \cos B + \cos^2 B + \sin^2 A - 2 \sin A \sin B + \sin^2 B \\ &= \underbrace{(\sin^2 A + \cos^2 A)} + \underbrace{(\sin^2 B + \cos^2 B)} - 2 \cos A \cos B - 2 \sin A \sin B \\ &= \qquad 1 \qquad + \qquad 1 \qquad - 2 \cos A \cos B - 2 \sin A \sin B \\ &- 2 \quad 2 \cos A \cos B - 2 \sin A \sin B. \end{aligned}$$

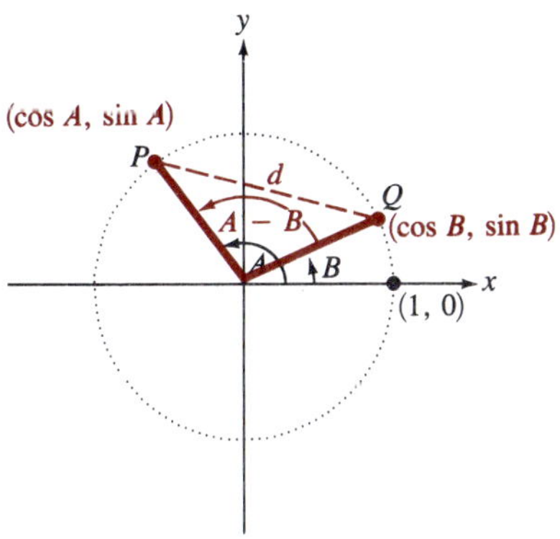


**Figure 1 Angles $A$ and $B$**

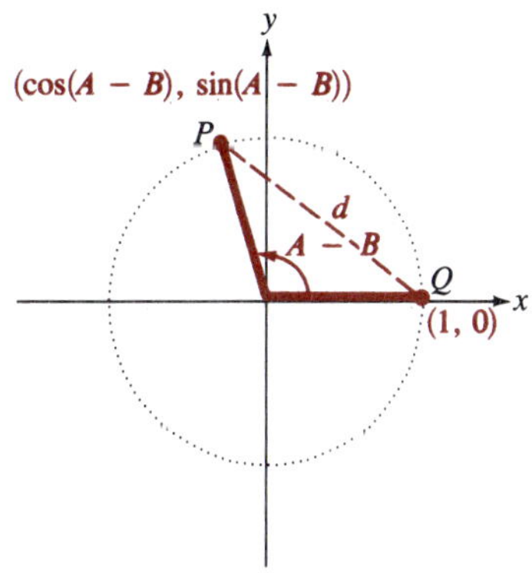


**Figure 2 Angle $A - B$**

Suppose that we rotate the unit circle in a clockwise direction until $Q$ coincides with $(1, 0)$, as in Figure 2. That is, we place angle $A - B$ in standard position. The point $P$ now has coordinates $(\cos (A - B), \sin (A - B))$.

Find the square of the distance between $P$ and $Q$ using the new coordinates.

$$\begin{aligned} d^2 &= [1 - \cos (A - B)]^2 + [0 - \sin (A - B)]^2 \\ &= 1 - 2 \cos (A - B) + \cos^2 (A - B) + \sin^2 (A - B) \\ &= 1 - 2 \cos (A - B) + \underbrace{\cos^2 (A - B) + \sin^2 (A - B)} \\ &= 1 - 2 \cos (A - B) + \qquad 1 \\ &= 2 - 2 \cos (A - B) \end{aligned}$$

Since the distance between $P$ and $Q$ remains the same regardless of the rotation of the circle, we have that

$$2 - 2\cos(A - B) = 2 - 2\cos A\cos B - 2\sin A\sin B.$$

Subtract 2 and multiply through by $-1/2$.

$$\mathbf{\cos(A - B) = \cos A\cos B + \sin A\sin B}$$

NOTE It may be easier to remember this identity in words: the cosine of the difference of two angles (numbers) is the sum of the product of the cosines of the angles (numbers) and the product of the sines of the two angles (numbers). ■

## EXAMPLE 1

Find the exact value of $\cos 15°$ using the fact that $15° = 60° - 45°$.

$$\begin{aligned}\cos 15° = \cos(60° - 45°) &= \cos 60°\cos 45° + \sin 60°\sin 45°\\ &= \left(\frac{1}{2}\right)\left(\frac{\sqrt{2}}{2}\right) + \left(\frac{\sqrt{3}}{2}\right)\left(\frac{\sqrt{2}}{2}\right)\\ &= \frac{\sqrt{2} + \sqrt{3}\sqrt{2}}{4}\\ &= \frac{\sqrt{2}(1 + \sqrt{3})}{4}\end{aligned}$$ ■

We should also be able to recognize the difference formula in the reverse order.

## EXAMPLE 2

Simplify $\cos 73°\cos 13° + \sin 73°\sin 13°$.

$$\begin{aligned}\cos 73°\cos 13° + \sin 73°\sin 13° &= \cos(73° - 13°)\\ &= \cos 60° = \frac{1}{2}\end{aligned}$$ ■

## Identity for $\cos(A + B)$

The identity for the sum of two angles (real numbers) is obtained from the difference identity recognizing that $A + B = A - (-B)$.

$$\begin{aligned}\cos(A + B) &= \cos(A - (-B))\\ &= \cos A\cos(-B) + \sin A\sin(-B)\\ &= \cos A\cos B + \sin A[-\sin B] &&\cos(-B) \equiv \cos B \text{ and } \sin(-B) \equiv -\sin B\\ &= \cos A\cos B - \sin A\sin B\end{aligned}$$

**Cosine of a Sum**

$$\cos(A + B) \equiv \cos A\cos B - \sin A\sin B$$

NOTE As with the difference identity, it may be better to remember the sum identity in words: The cosine of the sum of two angles (numbers) is the product of the cosines of the angles (numbers) minus the product of the sines of the angles (numbers). ■

**EXAMPLE 3** Find the exact value of $\cos 75°$ using the fact that $75° = 45° + 30°$.

$$\cos 75° = \cos (45° + 30°) = \cos 45° \cos 30° - \sin 45° \sin 30°$$
$$= \left(\frac{\sqrt{2}}{2}\right)\left(\frac{\sqrt{3}}{2}\right) - \left(\frac{\sqrt{2}}{2}\right)\left(\frac{1}{2}\right) = \frac{\sqrt{2}\sqrt{3} - \sqrt{2}}{4} = \frac{\sqrt{2}(\sqrt{3} - 1)}{4} \quad ■$$

Many of the identities that were verified earlier using the unit circle can be proved now using the sum and difference identities.

**EXAMPLE 4** Verify the identity. $\cos\left(A + \frac{\pi}{2}\right) \equiv -\sin A$

$$\cos\left(A + \frac{\pi}{2}\right) \equiv \cos A \cos\frac{\pi}{2} - \sin A \sin\frac{\pi}{2}$$
$$\equiv \cos A \cdot (0) \quad - \sin A \cdot (1) \qquad \cos\frac{\pi}{2} = 0 \text{ and } \sin\frac{\pi}{2} = 1$$
$$\equiv \quad 0 \quad - \sin A$$
$$\equiv -\sin A \quad ■$$

**EXAMPLE 5** Given that $\sin A = 3/5$, with $A$ in quadrant II, and that $\cos B = -12/13$, with $B$ in quadrant III, find $\cos (A - B)$ and $\cos (A + B)$.

To solve a problem like this, sketch the two angles as in Figure 3 and determine $\cos A$ and $\sin B$. Since $\cos B$ is $-12/13$, $x = -12$, and $r = 13$, and

$$y^2 + (-12)^2 = 13^2,$$
$$y = -5.$$

Thus $$\sin B = -\frac{5}{13}.$$

Similarly, $$\cos A = -\frac{4}{5}.$$

Substitute these values into the identities.

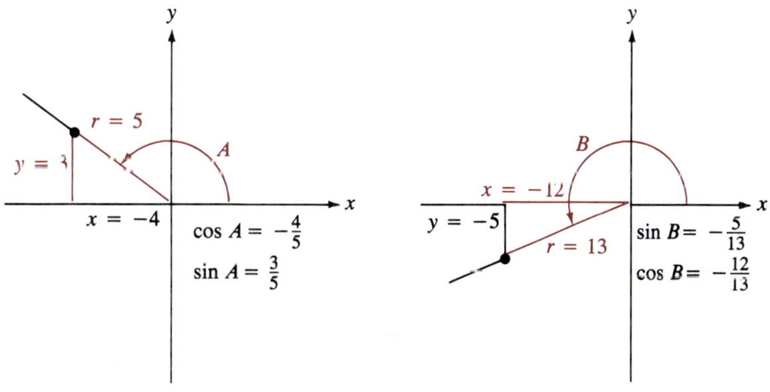


**Figure 3**

$$\cos(A - B) = \cos A \cos B + \sin A \sin B$$
$$= \left(-\frac{4}{5}\right)\left(-\frac{12}{13}\right) + \left(\frac{3}{5}\right)\left(-\frac{5}{13}\right) = \frac{48}{65} - \frac{15}{65} = \frac{33}{65}$$
$$\cos(A + B) = \cos A \cos B - \sin A \sin B$$
$$= \left(-\frac{4}{5}\right)\left(-\frac{12}{13}\right) - \left(\frac{3}{5}\right)\left(-\frac{5}{13}\right) = \frac{48}{65} + \frac{15}{65} = \frac{63}{65}$$

Notice that it was not necessary to find the values of angles $A$ and $B$ to determine $\cos(A - B)$ and $\cos(A + B)$. ■

### Derivation of Identity for sin (A + B)

To derive the identities for the sine of a sum or difference of two angles (numbers), we use the identity for the cosine of a difference together with the identities

$$\sin A \equiv \cos\left(\frac{\pi}{2} - A\right) \quad \text{and} \quad \sin\left(\frac{\pi}{2} - A\right) \equiv \cos A.$$

Substitute $A + B$ for $A$ in the first.

$$\sin(\mathbf{A} + \mathbf{B}) \equiv \cos\left[\frac{\pi}{2} - (\mathbf{A} + \mathbf{B})\right] \qquad \sin x = \cos\left(\frac{\pi}{2} - x\right)$$
$$\equiv \cos\left[\left(\frac{\pi}{2} - A\right) - B\right]$$
$$\equiv \cos\left(\frac{\pi}{2} - A\right)\cos B + \sin\left(\frac{\pi}{2} - A\right)\sin B \qquad \cos(x - y) = \cos x \cos y + \sin x \sin y$$
$$\equiv \sin A \cos B + \cos A \sin B$$

### Sine of a Sum

$$\sin(A + B) \equiv \sin A \cos B + \cos A \sin B$$

**NOTE** Remember this identity in words: the sine of the sum of two angles (numbers) is the sine of the first times the cosine of the second plus the cosine of the first times the sine of the second. ■

**EXAMPLE 6** Find the exact value of $\sin \frac{19\pi}{12}$ using the fact that $\frac{19\pi}{12} = \frac{3\pi}{4} + \frac{5\pi}{6}$.

$$\sin\frac{19\pi}{12} = \sin\left(\frac{3\pi}{4} + \frac{5\pi}{6}\right) = \sin\frac{3\pi}{4}\cos\frac{5\pi}{6} + \cos\frac{3\pi}{4}\sin\frac{5\pi}{6}$$
$$= \left(\frac{\sqrt{2}}{2}\right)\left(-\frac{\sqrt{3}}{2}\right) + \left(-\frac{\sqrt{2}}{2}\right)\left(\frac{1}{2}\right)$$
$$= \frac{-\sqrt{2}\sqrt{3} - \sqrt{2}}{4} = \frac{-\sqrt{2}(\sqrt{3} + 1)}{4}$$ ■

**EXAMPLE 7** Simplify $\sin 68° \cos 22° + \cos 68° \sin 22°$.

$$\sin 68° \cos 22° + \cos 68° \sin 22° = \sin(68° + 22°) = \sin 90° = 1$$ ■

**Identity for $\sin(A - B)$**

If we use the identity $\sin(A + B) = \sin A \cos B + \cos A \sin B$ and replace $B$ with $-B$, we can obtain the identity for the sine of a difference.

$$\begin{aligned}\sin(A - B) &\equiv \sin(A + (-B))\\ &\equiv \sin A \cos(-B) + \cos A \sin(-B)\\ &\equiv \sin A \cos B + \cos A(-\sin B) \quad \cos(-B) \equiv \cos B \text{ and } \sin(-B) \equiv -\sin B\\ &\equiv \sin A \cos B - \cos A \sin B\end{aligned}$$

**Sine of a Difference**

$$\sin(A - B) \equiv \sin A \cos B - \cos A \sin B$$

NOTE Remember this identity in words: the sine of a difference of two angles (numbers) is the sine of the first times the cosine of the second minus the cosine of the first times the sine of the second. ■

**EXAMPLE 8** Find the exact value of $\sin 15°$ using the fact that $15° = 60° - 45°$.

$$\begin{aligned}\sin 15° = \sin(60° - 45°) &= \sin 60° \cos 45° - \cos 60° \sin 45°\\ &= \left(\frac{\sqrt{3}}{2}\right)\left(\frac{\sqrt{2}}{2}\right) - \left(\frac{1}{2}\right)\left(\frac{\sqrt{2}}{2}\right)\\ &= \frac{\sqrt{3}\sqrt{2} - \sqrt{2}}{4}\\ &= \frac{\sqrt{2}(\sqrt{3} - 1)}{4}\end{aligned}$$ ■

**EXAMPLE 9** Given that $\sin A = 3/5$, with $A$ in quadrant II, and that $\cos B = -12/13$, with $B$ in quadrant III, find $\sin(A + B)$ and $\sin(A - B)$.

Sketch the two angles as in Figure 4 and show that $\cos A = -\frac{4}{5}$ and $\sin B = -\frac{5}{13}$.

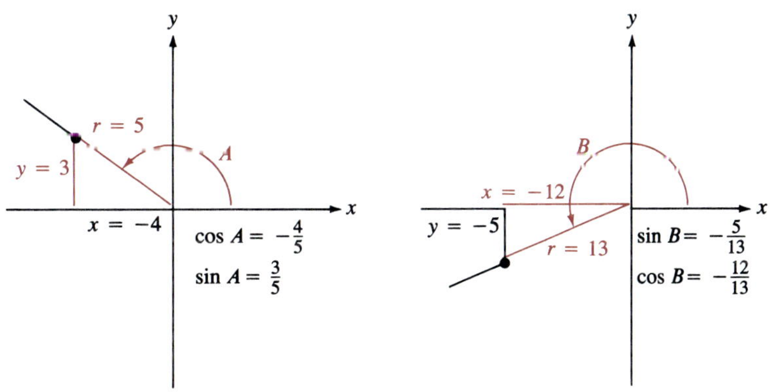


Figure 4

Substitute into the identities.

$$\begin{aligned}\sin(A+B) &= \sin A\cos B + \cos A\sin B\\ &= \left(\frac{3}{5}\right)\left(-\frac{12}{13}\right) + \left(-\frac{4}{5}\right)\left(-\frac{5}{13}\right)\\ &= -\frac{36}{65} + \frac{20}{65} = -\frac{16}{65}\end{aligned}$$

$$\begin{aligned}\sin(A-B) &= \sin A\cos B - \cos A\sin B\\ &= \left(\frac{3}{5}\right)\left(-\frac{12}{13}\right) - \left(-\frac{4}{5}\right)\left(-\frac{5}{13}\right)\\ &= -\frac{36}{65} - \frac{20}{65} = -\frac{56}{65}\end{aligned}$$

■

**Derivation of Identity for $\tan(A + B)$**

The identity for the tangent of the sum of two angles or numbers can be derived using the sum identities for sine and cosine functions with the identity $\tan x \equiv \dfrac{\sin x}{\cos x}$.

$$\begin{aligned}\tan(A+B) &\equiv \frac{\sin(A+B)}{\cos(A+B)}\\ &\equiv \frac{\sin A\cos B + \cos A\sin B}{\cos A\cos B - \sin A\sin B}\\ &\equiv \frac{\dfrac{\sin A\cos B}{\cos A\cos B} + \dfrac{\cos A\sin B}{\cos A\cos B}}{\dfrac{\cos A\cos B}{\cos A\cos B} - \dfrac{\sin A\sin B}{\cos A\cos B}} && \text{Multiply each term in the numerator and denominator by } \frac{1}{\cos A\cos B}\\ &\equiv \frac{\dfrac{\sin A}{\cos A} + \dfrac{\sin B}{\cos B}}{1 - \dfrac{\sin A}{\cos A}\cdot\dfrac{\sin B}{\cos B}}\\ &\equiv \frac{\tan A + \tan B}{1 - \tan A\tan B}\end{aligned}$$

**Tangent of a Sum**

$$\tan(A+B) \equiv \frac{\tan A + \tan B}{1 - \tan A\tan B}$$

**Identity for $\tan(A - B)$**

If $B$ is replaced by $-B$ in the tangent of a sum identity, we obtain the identity for the tangent of a difference.

$$\begin{aligned}\tan(A-B) \equiv \tan(A+(-B)) &\equiv \frac{\tan A + \tan(-B)}{1 - \tan A\tan(-B)}\\ &\equiv \frac{\tan A - \tan B}{1 + \tan A\tan B} && \tan(-B) = -\tan B\end{aligned}$$

**Tangent of a Difference**

$$\tan(A - B) \equiv \frac{\tan A - \tan B}{1 + \tan A \tan B}$$

**EXAMPLE 10** Simplify $\dfrac{\tan \frac{\pi}{4} + \tan \frac{7\pi}{4}}{1 - \tan \frac{\pi}{4} \tan \frac{7\pi}{4}}$.

$$\frac{\tan \frac{\pi}{4} + \tan \frac{7\pi}{4}}{1 - \tan \frac{\pi}{4} \tan \frac{7\pi}{4}} \equiv \tan\left(\frac{\pi}{4} + \frac{7\pi}{4}\right) = \tan 2\pi = 0$$ ■

## 8.2 EXERCISES

In Exercises 1–6 find the exact value of the **(a)** sine, **(b)** cosine, and **(c)** tangent of the given angle using the sum and difference identities.

**1.** $105° = 60° + 45°$

**2.** $165° = 135° + 30°$

**3.** $255° = 300° - 45°$

**4.** $195° = 225° - 30°$

**5.** $\frac{\pi}{12} = \frac{\pi}{3} - \frac{\pi}{4}$

**6.** $\frac{7\pi}{12} = \frac{3\pi}{4} - \frac{\pi}{6}$

In Exercises 7–22 simplify each expression by writing it as a function of a single angle.

**7.** $\cos S \cos T + \sin S \sin T$

**8.** $\cos M \cos N - \sin M \sin N$

**9.** $\sin T \cos S + \cos T \sin S$

**10.** $\sin M \cos N - \cos M \sin N$

**11.** $\dfrac{\tan S - \tan T}{1 + \tan S \tan T}$

**12.** $\dfrac{\tan M + \tan N}{1 - \tan M \tan N}$

**13.** $\cos 190° \cos 20° + \sin 190° \sin 20°$

**14.** $\cos \frac{7\pi}{3} \cos \frac{2\pi}{3} - \sin \frac{7\pi}{3} \sin \frac{2\pi}{3}$

**15.** $\sin 190° \cos 20° + \cos 190° \sin 20°$

**16.** $\sin \frac{7\pi}{3} \cos \frac{2\pi}{3} - \cos \frac{7\pi}{3} \sin \frac{2\pi}{3}$

**17.** $\cos 2x \cos 5x + \sin 2x \sin 5x$

**18.** $\sin 5x \cos 2x - \cos 5x \sin 2x$

**19.** $\sin(-3) \cos 7 + \cos(-3) \sin 7$

**20.** $\cos(-3) \cos 7 - \sin(-3) \sin 7$

**21.** $\dfrac{\tan 5 + \tan(-2)}{1 - \tan 5 \tan(-2)}$

**22.** $\dfrac{\tan 35° - \tan 15°}{1 + \tan 35° \tan 15°}$

**23.** Given that $\sin A = \frac{4}{5}$, with $A$ in quadrant I, and that $\cos B = -\frac{3}{5}$, with $B$ in quadrant III. Find **(a)** $\cos(A + B)$, **(b)** $\cos(A - B)$, **(c)** $\sin(A + B)$, **(d)** $\sin(A - B)$, **(e)** $\tan(A + B)$, and **(f)** $\tan(A - B)$.

**24.** Given that $\cos A = \frac{5}{13}$, with $A$ in quadrant IV, and that $\sin B = \frac{-3}{5}$, with $B$ in quadrant III. Find **(a)** $\cos(A + B)$, **(b)** $\cos(A - B)$, **(c)** $\sin(A + B)$, **(d)** $\sin(A - B)$, **(e)** $\tan(A + B)$, and **(f)** $\tan(A - B)$.

Verify each identity in Exercises 25–54.

**25.** $\cos\left(\frac{\pi}{2} - A\right) \equiv \sin A$

**26.** $\sin\left(\frac{\pi}{2} - A\right) \equiv \cos A$

**27.** $\sin(A + 180^\circ) \equiv -\sin A$

**28.** $\cos(A + 180^\circ) \equiv -\cos A$

**29.** $\cos(A - 2\pi) \equiv \cos A$

**30.** $\sin(A - 2\pi) \equiv \sin A$

**31.** $\sin(A + 360^\circ) \equiv \sin A$

**32.** $\cos(A + 360^\circ) \equiv \cos A$

**33.** $\cos\left(A - \frac{\pi}{2}\right) \equiv \sin A$

**34.** $\sin\left(A - \frac{\pi}{2}\right) \equiv -\cos A$

**35.** $\sin\left(A + \frac{\pi}{4}\right) \equiv \left(\frac{\sqrt{2}}{2}\right)(\sin A + \cos A)$

**36.** $\cos\left(A + \frac{\pi}{4}\right) \equiv \left(\frac{\sqrt{2}}{2}\right)(\cos A - \sin A)$

**37.** $\tan(A + 45^\circ) \equiv \dfrac{1 + \tan A}{1 - \tan A}$

**38.** $\tan(A - 45^\circ) \equiv \dfrac{\tan A - 1}{1 + \tan A}$

**39.** $\sin(A + B)\sin(A - B) \equiv \sin^2 A - \sin^2 B$

**40.** $\cos(A + B)\cos(A - B) \equiv \cos^2 A - \sin^2 B$

**41.** $\sin(A + B) + \sin(A - B) \equiv 2\sin A\cos B$

**42.** $\cos(A + B) + \cos(A - B) \equiv 2\cos A\cos B$

**43.** $\cos(A + B + C) \equiv \cos A\cos B\cos C - \cos A\sin B\sin C - \sin A\cos B\sin C - \sin A\sin B\cos C$

**44.** $\sin(A + B + C) \equiv \sin A\cos B\cos C + \cos A\sin B\cos C + \cos A\cos B\sin C - \sin A\sin B\sin C$

**45.** $\sin 2A \equiv 2\sin A\cos A$

**46.** $\cos 2A \equiv \cos^2 A - \sin^2 A$

**47.** $\tan A - \tan B \equiv \dfrac{\sin(A - B)}{\cos A\cos B}$

**48.** $\cot A - \tan B \equiv \dfrac{\cos(A + B)}{\sin A\cos B}$

**49.** $\dfrac{\sin(A + B)}{\cos(A - B)} \equiv \dfrac{\tan A + \tan B}{1 + \tan A\tan B}$

**50.** $\dfrac{\sin(A + B)}{\sin(A - B)} \equiv \dfrac{\tan A + \tan B}{\tan A - \tan B}$

**51.** $\dfrac{\cos(A + B)}{\sin(A - B)} \equiv \dfrac{\cot A - \tan B}{1 - \cot A\tan B}$

**52.** $\dfrac{\cos(A + B)}{\cos(A - B)} \equiv \dfrac{1 - \tan A\tan B}{1 + \tan A\tan B}$

**53.** $\cot(A + B) \equiv \dfrac{\cot A\cot B - 1}{\cot A + \cot B}$

**54.** $\cot(A - B) \equiv \dfrac{\cos A\cot B + 1}{\cot B - \cot A}$

**CALCULUS** The expression $\dfrac{f(x + h) - f(x)}{h}$, where $f(x)$ is a given function and $h$ is a nonzero real number, plays an important role in calculus. Use this expression in Exercises 55–56.

**55.** If $f(x) = \sin x$, show that

$$\frac{f(x + h) - f(x)}{h} = \sin x\left(\frac{\cos h - 1}{h}\right) + \cos x\left(\frac{\sin h}{h}\right).$$

**56.** If $f(x) = \cos x$, show that

$$\frac{f(x + h) - f(x)}{h} = \cos x\left(\frac{\cos h - 1}{h}\right) - \sin x\left(\frac{\sin h}{h}\right).$$

**57. ANALYTIC GEOMETRY** Suppose that $l_1$ and $l_2$ are two lines with slopes $m_1$ and $m_2$, respectively, as shown in the figure. It can be shown that $\tan\theta_1 = m_1$ and $\tan\theta_2 = m_2$. Use this information to show that $\tan(\theta_2 - \theta_1) = \dfrac{m_2 - m_1}{1 + m_1 m_2}$, and that $l_1$ and $l_2$ are perpendicular if and only if $m_1 = -\dfrac{1}{m_2}$.

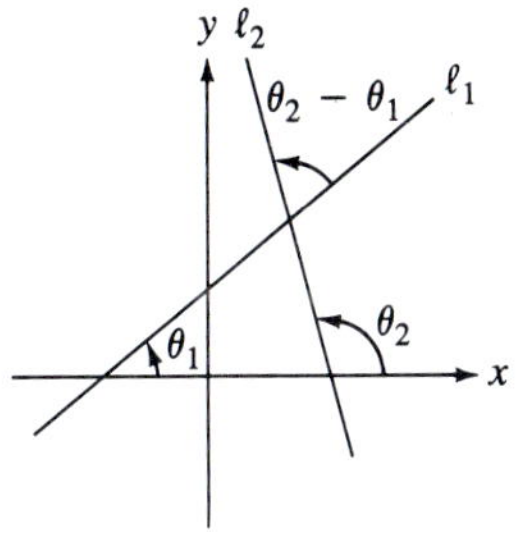


Exercise 57

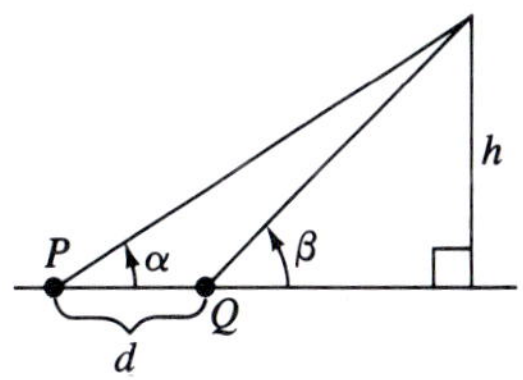


Exercise 58

**58.** SURVEYING The height $h$ of an object when viewed from two points $P$ and $Q$ in a straight line with the object can be calculated using the angles of elevation $\alpha$, from $P$, and $\beta$, from $Q$ (see the figure). Show that

$$h = \frac{d \sin \alpha \sin \beta}{\sin (\beta - \alpha)}$$

where $d$ is the distance between $P$ and $Q$. Find the height of a mountain, to the nearest foot, if $\alpha = 25°00'$, $\beta = 29°30'$, and $d = 2500$ ft.

## For Review

Prove the identities in Exercises 59–60.

**59.** $\dfrac{\tan^2 x}{\sec x + 1} \equiv \dfrac{1 - \cos x}{\cos x}$

**60.** $\cos^4 y - \sin^4 y + 1 \equiv 2 \cos^2 y$

# 8.3 Multiple-Angle Identities

## Derivation of Double-Angle Identities

By letting $B = A$ in the identities for $\sin (A + B)$, $\cos (A + B)$ and $\tan (A + B)$, we obtain identities for $\sin 2A$, $\cos 2A$, and $\tan 2A$.

$$\begin{aligned} \sin (2A) = \sin (A + A) &= \sin A \cos A + \cos A \sin A \\ &= 2 \sin A \cos A \\ \cos (2A) = \cos (A + A) &= \cos A \cos A - \sin A \sin A \\ &= \cos^2 A - \sin^2 A \end{aligned}$$

Two additional forms for $\cos 2A$ can be found, one by substituting $1 - \sin^2 A$ for $\cos^2 A$ and the other by substituting $1 - \cos^2 A$ for $\sin^2 A$ in $\cos 2A = \cos^2 A - \sin^2 A$.

$$\cos 2A = \mathbf{\cos^2 A} - \sin^2 A = \mathbf{(1 - \sin^2 A)} - \sin^2 A = 1 - 2 \sin^2 A$$

$$\cos 2A = \cos^2 A - \mathbf{\sin^2 A} = \cos^2 A - \mathbf{(1 - \cos^2 A)} = 2 \cos^2 A - 1$$

$$\tan (2A) = \tan (A + A) = \frac{\tan A + \tan A}{1 - \tan A \tan A} = \frac{2 \tan A}{1 - \tan^2 A}$$

**Double-Angle Identities**

$$\sin 2A \equiv 2 \sin A \cos A$$

$$\cos 2A \equiv \cos^2 A - \sin^2 A \equiv 1 - 2 \sin^2 A \equiv 2 \cos^2 A - 1$$

$$\tan 2A \equiv \frac{2 \tan A}{1 - \tan^2 A}$$

As with sum and difference identities, it is better to memorize the double-angle identities with words, not with the particular letter $A$. Also, remember the first form for $\cos 2A$ and develop the other two using $\sin^2 A + \cos^2 A \equiv 1$ and substitution.

**EXAMPLE 1**

Given that $\sin A = \frac{3}{5}$ with $A$ in quadrant II, find $\sin 2A$, $\cos 2A$, and $\tan 2A$, and determine the quadrant in which $2A$ is located.

From the sketch in Figure 5 with $A$ in quadrant II and $\sin A = \frac{3}{5}$, we find that $\cos A = \frac{-4}{5}$. Substitute

$$\sin 2A = 2 \sin A \cos A = 2\left(\frac{3}{5}\right)\left(-\frac{4}{5}\right) = 2\left(-\frac{12}{25}\right) = -\frac{24}{25}$$

$$\cos 2A = \cos^2 A - \sin^2 A = \left(-\frac{4}{5}\right)^2 - \left(\frac{3}{5}\right)^2 = \frac{16}{25} - \frac{9}{25} = \frac{7}{25}$$

$$\tan 2A = \frac{2 \tan A}{1 - \tan^2 A} = \frac{2\left(\frac{-3}{4}\right)}{1 - \left(-\frac{3}{4}\right)^2} = -\frac{24}{7}$$

Since $\sin 2A$ is negative and $\cos 2A$ is positive, $2A$ is in quadrant IV. ■

**EXAMPLE 2**

Given that $\csc Y = -4$ with $Y$ in quadrant III, find $\sin 2Y$, $\cos 2Y$, and $\tan 2Y$, and determine the quadrant in which $2Y$ is located.

With $\csc Y = -4$, we know that $\sin Y = \frac{1}{\csc Y} = -\frac{1}{4}$. From the sketch in Figure 6, we find that $\cos Y = -\frac{\sqrt{15}}{4}$. Substituting,

$$\sin 2Y = 2 \sin Y \cos Y = 2\left(-\frac{1}{4}\right)\left(-\frac{\sqrt{15}}{4}\right) = \frac{2\sqrt{15}}{16} = \frac{\sqrt{15}}{8},$$

$$\cos 2Y = \cos^2 Y - \sin^2 Y = \left(-\frac{\sqrt{15}}{4}\right)^2 - \left(-\frac{1}{4}\right)^2 = \frac{15}{16} - \frac{1}{16} = \frac{14}{16} = \frac{7}{8},$$

$$\tan 2Y = \frac{2 \tan Y}{1 - \tan^2 Y} = \frac{2\left(\frac{\sqrt{15}}{15}\right)}{1 - \left(\frac{\sqrt{15}}{15}\right)^2} = \frac{\sqrt{15}}{7}.$$

Since $\sin 2Y$ and $\cos 2Y$ are both positive, $2Y$ is in quadrant I. ■

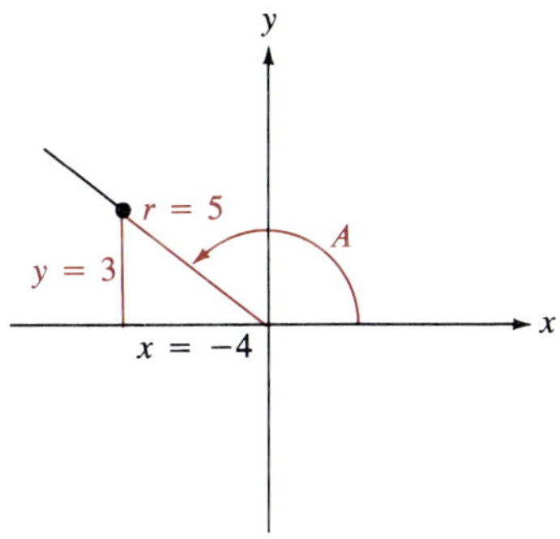


**Figure 5** Angle $A$

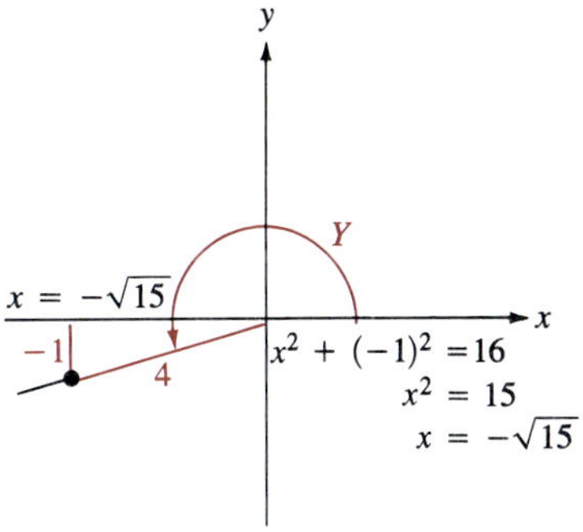


**Figure 6** Angle $Y$

## EXAMPLE 3

Find an expression equal to sin 4$B$ which involves only functions of $B$.

$$\begin{aligned}\sin 4B &= \sin 2(\mathbf{2B}) \\ &= 2 \sin \mathbf{2B} \cos \mathbf{2B} \quad \text{Using } 2B \text{ for } A \text{ in the double-angle formula} \\ &= 2(2 \sin B \cos B)(\cos^2 B - \sin^2 B) \quad \text{Using both double-angle formulas} \\ &= 4 \sin B \cos^3 B - 4 \sin^3 B \cos B\end{aligned}$$

Two equivalent forms could have been derived by substituting $1 - 2 \sin^2 B$ or $2 \cos^2 B - 1$ instead of $\cos^2 B - \sin^2 B$ for cos 2$B$. ■

## Derivation of Power-Reduction Identities

The double-angle identities for the cosine function can be used to obtain three power-reduction identities useful in calculus. The first two are obtained from $\cos 2A = 1 - 2 \sin^2 A$ and $\cos 2A \equiv 2 \cos^2 A - 1$ by solving for $\sin^2 A$ and $\cos^2 A$, respectively. The third follows directly from the first two and the fact that $\tan^2 A = \dfrac{\sin^2 A}{\cos^2 A}$.

### Power-Reduction Identities

$$\sin^2 A \equiv \frac{1 - \cos 2A}{2} \qquad \cos^2 A \equiv \frac{1 + \cos 2A}{2} \qquad \tan^2 A \equiv \frac{1 - \cos 2A}{1 + \cos 2A}$$

## EXAMPLE 4

Express $\cos^4 Y$ in terms of the first power of the cosine function.

$$\begin{aligned}\cos^4 Y = (\cos^2 Y)^2 &= \left(\frac{1 + \cos 2Y}{2}\right)^2 \\ &= \frac{1}{4}(1 + 2 \cos 2Y + \mathbf{\cos^2 2Y}) \\ &= \frac{1}{4}\left(1 + 2 \cos 2Y + \mathbf{\frac{1 + \cos 4Y}{2}}\right) \\ &= \frac{1}{4} + \frac{1}{2} \cos 2Y + \frac{1}{8} + \frac{1}{8} \cos 4Y \\ &= \frac{3}{8} + \frac{1}{2} \cos 2Y + \frac{1}{8} \cos 4Y\end{aligned}$$ ■

## Half-Angle Identities

The half-angle identities are obtained from the power-reduction identities by replacing $A$ with $\frac{A}{2}$ and taking the square root of both sides. The correct sign is chosen according to the quadrant in which $\frac{A}{2}$ is located.

**Half-Angle Identities**

$$\sin \frac{A}{2} \equiv \pm\sqrt{\frac{1-\cos A}{2}} \qquad \cos \frac{A}{2} \equiv \pm\sqrt{\frac{1+\cos A}{2}} \qquad \tan \frac{A}{2} \equiv \pm\sqrt{\frac{1-\cos A}{1+\cos A}}$$

**EXAMPLE 5** Given that $\sin A = -\frac{3}{5}$ with $A$ in quadrant III, find $\sin \frac{A}{2}$ and $\cos \frac{A}{2}$.

From the sketch in Figure 7, with $A$ in quadrant III, $\cos A = -\frac{4}{5}$. Substituting,

$$\sin \frac{A}{2} = \pm\sqrt{\frac{1-\cos A}{2}} = \pm\sqrt{\frac{1-\left(-\frac{4}{5}\right)}{2}}$$

$$= \pm\sqrt{\frac{1+\frac{4}{5}}{2}} = \pm\sqrt{\frac{\frac{9}{5}}{2}} = \pm\sqrt{\frac{9}{10}}$$

$$= \pm\frac{3}{\sqrt{10}} = \pm\frac{3\sqrt{10}}{10}.$$

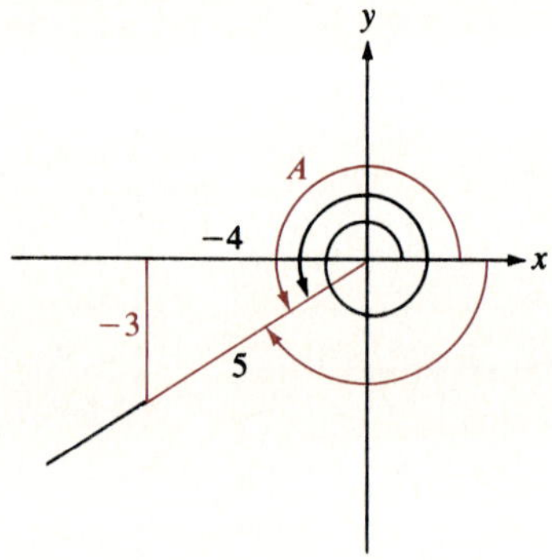


Figure 7

Substituting again,

$$\cos \frac{A}{2} = \pm\sqrt{\frac{1+\cos A}{2}} = \pm\sqrt{\frac{1+\left(-\frac{4}{5}\right)}{2}}$$

$$= \pm\sqrt{\frac{1-\frac{4}{5}}{2}} = \pm\sqrt{\frac{\frac{1}{5}}{2}} = \pm\sqrt{\frac{1}{10}}$$

$$= \pm\frac{1}{\sqrt{10}} = \pm\frac{\sqrt{10}}{10}.$$

Without additional information on the size of angle $A$ ($A$ could be negative or greater than 360°), hence the location of $\frac{A}{2}$, neither the sine nor the cosine of $\frac{A}{2}$ can be specified as positive or negative. ■

Two alternative forms for the $\tan \frac{A}{2}$ can be derived. Suppose we start with the identity for $\tan^2 \frac{A}{2}$ and multiply the numerator and denominator by $1 - \cos A$.

$$\begin{aligned}\tan^2 \frac{A}{2} &= \frac{(1 - \cos A)(1 - \cos A)}{(1 + \cos A)(1 - \cos A)} \\ &= \frac{(1 - \cos A)^2}{1 - \cos^2 A} \\ &= \frac{(1 - \cos A)^2}{\sin^2 A} \qquad 1 - \cos^2 A = \sin^2 A\end{aligned}$$

Take the square root of both sides.

$$\tan \frac{A}{2} = \pm \frac{1 - \cos A}{\sin A}$$

Since $-1 \leq \cos A \leq 1$ for every $A$, $1 - \cos A$ is never negative, and thus the sign of $\frac{1 - \cos A}{\sin A}$ is the same as the sign of $\sin A$. Also, $\tan \frac{A}{2}$ and $\sin A$ will always have the same sign as shown in the following table.

| $A$ | $\frac{A}{2}$ | $\sin A$ | $\tan \frac{A}{2}$ |
|---|---|---|---|
| $0 < A < \pi$ | $0 < \frac{A}{2} < \frac{\pi}{2}$ | + | + |
| $\pi < A < 2\pi$ | $\frac{\pi}{2} < \frac{A}{2} < \pi$ | − | − |
| $2\pi < A < 3\pi$ | $\pi < \frac{A}{2} < \frac{3\pi}{2}$ | + | + |
| $3\pi < A < 4\pi$ | $\frac{3\pi}{2} < \frac{A}{2} < 2\pi$ | − | − |

As a result, we can ignore the $\pm$ sign and use the positive sign only. Thus,

$$\tan \frac{A}{2} = \frac{1 - \cos A}{\sin A}.$$

By multiplying the numerator and denominator by $1 + \cos A$ and using a similar argument, we can obtain

$$\tan \frac{A}{2} \equiv \frac{\sin A}{1 + \cos A}.$$

**Half-Angle Identities for Tangent (Alternative Form)**

$$\tan \frac{A}{2} \equiv \frac{1 - \cos A}{\sin A} \qquad \tan \frac{A}{2} \equiv \frac{\sin A}{1 + \cos A}$$

**EXAMPLE 6** Given that $\sin A = -\frac{3}{5}$ with $A$ in quadrant III, find $\tan \frac{A}{2}$.

From the sketch of angle $A$ in Figure 7, we again have that $\cos A = -\frac{4}{5}$. Substituting,

$$\tan \frac{A}{2} = \frac{1 - \cos A}{\sin A} = \frac{1 - \left(-\frac{4}{5}\right)}{-\frac{3}{5}} = \frac{\frac{9}{5}}{-\frac{3}{5}} = -3. \quad ■$$

**EXAMPLE 7** Find the exact value of $\sin 22.5°$, $\cos 22.5°$, and $\tan 22.5°$ using the fact that $22.5° = \frac{45°}{2}$.

Since $22.5°$ is in the first quadrant, all three functional values are positive.

$$\sin 22.5° = \sin \frac{45°}{2} = \sqrt{\frac{1 - \cos 45°}{2}}$$

$$= \sqrt{\frac{1 - \frac{\sqrt{2}}{2}}{2}} = \sqrt{\frac{\frac{2 - \sqrt{2}}{2}}{2}}$$

$$= \sqrt{\frac{2 - \sqrt{2}}{4}} = \frac{\sqrt{2 - \sqrt{2}}}{2}$$

$$\cos 22.5° = \cos \frac{45°}{2} = \sqrt{\frac{1 + \cos 45°}{2}}$$

$$= \sqrt{\frac{1 + \frac{\sqrt{2}}{2}}{2}} = \frac{\sqrt{2 + \sqrt{2}}}{2}$$

$$\tan 22.5° = \tan \frac{45°}{2} = \frac{1 - \cos 45°}{\sin 45°}$$

$$= \frac{1 - \frac{\sqrt{2}}{2}}{\frac{\sqrt{2}}{2}}$$

$$= \frac{2 - \sqrt{2}}{\sqrt{2}} = \frac{2\sqrt{2} - 2}{2} = \sqrt{2} - 1 \quad ■$$

**EXAMPLE 8** Prove the identity. $\cos^2 \frac{x}{2} - \sin^2 \frac{x}{2} \equiv \cos x$

We might recognize that the left side takes the form of the double-angle identity with $x/2$ playing the role of $A$. As a result,

$$\cos^2 \frac{x}{2} - \sin^2 \frac{x}{2} \equiv \cos 2\left(\frac{x}{2}\right) \equiv \cos x.$$

If we did not recognize this relationship, we could start with the left side and substitute directly for $\cos \frac{x}{2}$ and $\sin \frac{x}{2}$.

$$\cos^2 \frac{x}{2} - \sin^2 \frac{x}{2} \equiv \left(\pm\sqrt{\frac{1 + \cos x}{2}}\right)^2 - \left(\pm\sqrt{\frac{1 - \cos x}{2}}\right)^2$$
$$\equiv \frac{1 + \cos x}{2} - \frac{1 - \cos x}{2}$$
$$\equiv \frac{2 \cos x}{2} \equiv \cos x \quad ■$$

We conclude this section with the first applied problem from the chapter introduction.

**EXAMPLE 9**

**SPORTS**

A football player kicks a field goal at an angle of inclination with the playing field equal to $\theta$. It can be shown that the distance the ball will travel in the air if it is not touched is given in feet by

$$d = \frac{v^2 \sin 2\theta}{32},$$

where $v$ is the initial velocity of the ball in ft/sec. **(a)** Find the distance that the ball will travel if $\theta = 40°$ and $v = 50$ ft/sec. **(b)** If $v$ is constant, discuss why the maximum distance will be attained when $\theta = 45°$.

**(a)** Substitute 40° for $\theta$ and 50 for $v$ in the formula

$$d = \frac{(50)^2 \sin 2(40°)}{32} = \frac{2500 \sin 80°}{32}$$
$$\approx 76.9 \text{ ft.}$$

**(b)** Since $d = \frac{v^2 \sin 2\theta}{32}$, and $v^2$ and 32 are positive constants, the maximum value of $d$ will occur when $\sin 2\theta$ is as large as possible. The maximum value of the sine function is 1, which occurs at 90°. Thus, the maximum value of $d$ will be achieved when $2\theta = 90°$ or when $\theta = 45°$. ■

## 8.3 EXERCISES

In Exercises 1–8 use the given information to find **(a)** $\sin 2A$, **(b)** $\cos 2A$, **(c)** $\tan 2A$, and **(d)** the quadrant in which $2A$ is located.

**1.** $\sin A = \frac{3}{5}$ with $A$ in quadrant I.

**2.** $\sin A = \frac{12}{13}$ with $A$ in quadrant I

**3.** $\cos A = -\frac{12}{13}$ with $A$ in quadrant II

**4.** $\cos A = -\frac{3}{5}$ with $A$ in quadrant II

**5.** $\tan A = \frac{3}{4}$ with $A$ in quadrant III

**6.** $\tan A = \frac{5}{12}$ with $A$ in quadrant III

**7.** $\csc A = -\frac{13}{12}$ with $A$ in quadrant IV

**8.** $\sec A = \frac{5}{4}$ with $A$ in quadrant IV

In Exercises 9–14 find the exact value of **(a)** $\sin \frac{A}{2}$, **(b)** $\cos \frac{A}{2}$, and **(c)** $\tan \frac{A}{2}$ for the given value of $A$.

**9.** $30°$ **10.** $330°$ **11.** $\frac{\pi}{4}$

**12.** $\frac{\pi}{6}$ **13.** $210°$ **14.** $135°$

In Exercises 15–16 use the given information to find $\tan \frac{A}{2}$.

**15.** $\sin A = \frac{4}{5}$ with $A$ in quadrant II **16.** $\cos A = -\frac{12}{13}$ with $A$ in quadrant III

Using the half-angle identities to simplify each expression in Exercises 17–20.

**17.** $\sqrt{\frac{1 + \cos 4x}{2}}$ **18.** $\sqrt{\frac{1 - \cos 6y}{2}}$ **19.** $-\sqrt{\frac{1 - \cos 2x}{1 + \cos 2x}}$ **20.** $-\sqrt{\frac{1 + \cos (x^2 + 1)}{2}}$

In Exercises 21–24 find an expression equal to the given expression which involves only functions of $A$.

**21.** $\cos 4A$ **22.** $\sin 3A$ **23.** $\cos 3A$ **24.** $\sin 6A$

In Exercises 25–28 use the power-reduction identities to write each expression in terms of the first power of the cosine function.

**25.** $\sin^4 A$ **26.** $\cos^2 A \sin^2 A$ **27.** $\sin^2 A \cos^4 A$ **28.** $\sin^4 A \cos^2 A$

Verify each identity in Exercises 29–48.

**29.** $\sin 8A \equiv 2 \sin 4A \cos 4A$ **30.** $\cos 8A \equiv \cos^2 4A - \sin^2 4A$ **31.** $\cos^2 5x - \sin^2 5x \equiv \cos 10x$

**32.** $2 \sin \frac{x}{4} \cos \frac{x}{4} \equiv \sin \frac{x}{2}$ **33.** $\csc 2A \equiv \frac{\csc A}{2 \cos A}$ **34.** $\sec 2A \equiv \frac{\sec^2 A}{2 - \sec^2 A}$

**35.** $\cos^4 x - \sin^4 x \equiv \cos 2x$ **36.** $(\sin x + \cos x)^2 \equiv 1 + \sin 2x$ **37.** $\sec \frac{A}{2} \equiv \pm\sqrt{\frac{2 \tan A}{\tan A + \sin A}}$

**38.** $\csc \frac{A}{2} \equiv \pm\sqrt{\frac{2 \tan A}{\tan A - \sin A}}$ **39.** $\frac{1 - \cos 2x}{\tan x} \equiv \sin 2x$ **40.** $\frac{1 - \cos 2x}{\sin x} \equiv 2 \sin x$

**41.** $\cos 2A + \sin 2A \tan A \equiv 1$ **42.** $\frac{\sin 2A}{1 - \sin^2 A} - \tan A \equiv \tan A$ **43.** $\frac{\tan A - \sin A}{2 \tan A} \equiv \sin^2 \frac{A}{2}$

**44.** $\frac{\sin^3 A + \cos^3 A}{\sin A + \cos A} \equiv \frac{2 - \sin 2A}{2}$ **45.** $2 \cos x - 2 \cos^3 x \equiv \sin x \sin 2x$ **46.** $\frac{\sin 2x}{\sin x} - \frac{\cos 2x}{\cos x} \equiv \sec x$

**47.** $\cos 2A \equiv \frac{1 - \tan^2 A}{1 + \tan^2 A}$ **48.** $\tan \frac{A}{2} \equiv \csc A - \cot A$

**49. SPORTS** A golfer hits a ball from the tee at an angle of inclination with the horizontal equal to 35°. If the initial velocity of the ball is 125 ft/sec, what distance, to the nearest foot, will the ball travel in the air? [Hint: Refer to Example 9.]

**50. ENGINEERING** The cross-sectional view of a domed stadium with height $d$ and diameter $D$ is shown in the figure. Show that the angle $\phi$, subtended by the portion of the roof that is a circular arc is related to $d$ and $D$ by the equation $\sin \frac{\phi}{2} = \frac{2d}{D}$. What is the value of $\phi$, to the nearest tenth of a degree, when the height of the dome is 175 ft and the diameter is 500 ft?

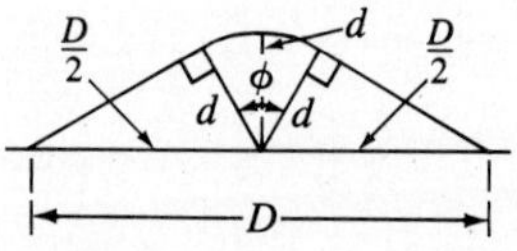

## For Review

**51.** Assume that $\sin A = 3/5$, with $A$ in quadrant II, and that $\cos B = 12/13$, with $B$ in quadrant IV. Find **(a)** $\cos(A + B)$, **(b)** $\cos(A - B)$, **(c)** $\sin(A + B)$, **(d)** $\sin(A - B)$, **(e)** $\tan(A + B)$, and **(f)** $\tan(A - B)$.

Verify each identity in Exercises 52–53.

**52.** $\dfrac{\sin(A + B)}{\cos A \cos B} \equiv \tan A + \tan B$

**53.** $\tan\left(\dfrac{\pi}{4} - \theta\right) \equiv \dfrac{1 - \tan\theta}{1 + \tan\theta}$

Exercises 54–55 lead to material in the next section. Expand each function using formulas from the preceding section.

**54.** $y = A\cos B(x - c)$

**55.** $y = A\sin B(x + c)$

# 8.4 Product and Addition Identities

In this section we develop several identities for writing a product of sine and cosine functions as a sum or difference of such functions, and conversely. These identities are helpful for applications in calculus and for graphing certain more complicated functions. They all result directly from the sum and difference identities for the sine and cosine functions.

**Product Identities**

$$\sin A \cos B \equiv \frac{1}{2}[\sin(A + B) + \sin(A - B)]$$

$$\cos A \sin B \equiv \frac{1}{2}[\sin(A + B) - \sin(A - B)]$$

$$\cos A \cos B \equiv \frac{1}{2}[\cos(A + B) + \cos(A - B)]$$

$$\sin A \sin B \equiv -\frac{1}{2}[\cos(A + B) - \cos(A - B)]$$

## Verification of Product Identities

The first product identity is verified by adding the left and right sides of the identities for $\sin(A + B)$ and $\sin(A - B)$.

$$\begin{aligned} \sin(A + B) &\equiv \sin A \cos B + \cos A \sin B \\ \sin(A - B) &\equiv \sin A \cos B - \cos A \sin B \\ \hline \sin(A + B) + \sin(A - B) &\equiv 2 \sin A \cos B \end{aligned}$$

Dividing both sides by 2, we obtain

$$\sin A \cos B = \frac{1}{2}[\sin(A + B) + \sin(A - B)].$$

The remaining three identities are proved in a similar manner.

**EXAMPLE 1** Express each of the following as a sum.

**(a)** $\sin 3x \cos x$

Use the first product identity with $A = 3x$ and $B = x$.

$$\sin \mathbf{3x} \cos \mathbf{x} \equiv \frac{1}{2}[\sin (\mathbf{3x + x}) + \sin (\mathbf{3x - x})]$$

$$\equiv \frac{1}{2}[\sin 4x + \sin 2x]$$

We also could have used the second identity with $A = x$ and $B = 3x$.

$$\cos \mathbf{x} \sin \mathbf{3x} \equiv \frac{1}{2}[\sin (\mathbf{x + 3x}) - \sin (\mathbf{x - 3x})]$$

$$\equiv \frac{1}{2}[\sin 4x - \sin (-2x)]$$

Since $\sin (-2x) = -\sin 2x$, the representations are identical.

**(b)** $\cos 2y \cos 5y$

Use the third product identity with $A = 2y$ and $B = 5y$.

$$\cos \mathbf{2y} \cos \mathbf{5y} \equiv \frac{1}{2}[\cos (\mathbf{2y + 5y}) + \cos (\mathbf{2y - 5y})]$$

$$\equiv \frac{1}{2}[\cos 7y + \cos (-3y)]$$

This could also be written as $\frac{1}{2}[\cos 7y + \cos 3y]$ since $\cos (-3y) \equiv \cos 3y$. ■

The product identities can also be used to evaluate trigonometric functions as illustrated in the next example.

**EXAMPLE 2** Evaluate $\sin 45° \sin 15°$ using an appropriate product identity.

Use the fourth product identity with $A = 45°$ and $B = 15°$.

$$\sin \mathbf{45°} \sin \mathbf{15°} = -\frac{1}{2}[\cos (\mathbf{45° + 15°}) - \cos (\mathbf{45° - 15°})]$$

$$= -\frac{1}{2}[\cos 60° - \cos 30°]$$

Since $\cos 30° = \sqrt{3}/2$ and $\cos 60° = 1/2$, this can be simplified further.

$$\sin 45° \sin 15° = -\frac{1}{2}\left[\frac{1}{2} - \frac{\sqrt{3}}{2}\right]$$

$$= -\frac{1}{2}\left[\frac{1 - \sqrt{3}}{2}\right] = \frac{\sqrt{3} - 1}{4}$$ ■

## Graphing

Recall from Section 6.7 that to graph sums or differences of sine and cosine functions, we graph each function separately and add ordinates graphically. Thus, to graph a function that is a product of sine or cosine functions, first use one of the product identities to express the function as a sum or difference, and then use the addition of ordinates method.

**EXAMPLE 3**

Graph the function $y = 4 \sin x \cos^2 x$.

First write the function as $y = 2\ (2 \sin x \cos x) \cos x$, and then use the double-angle formula $2 \sin x \cos x = \sin 2x$ to write $y$ in the form

$$y = 2 \sin 2x \cos x.$$

Then by the formula for $\sin A \cos B$, with $A = 2x$ and $B = x$, we have

$$\begin{aligned} y &= 2 \cdot \frac{1}{2}[\sin (2x + x) + \sin (2x - x)] \\ &= \sin 3x + \sin x. \end{aligned}$$

We graph $y = \sin 3x$ and $y = \sin x$ in the same coordinate system and add ordinates to obtain the graph of $y = \sin 3x + \sin x = 4 \sin x \cos^2 x$, as shown in Figure 8. ■

| $x$ | $\sin x$ | $\sin 3x$ | $\sin 3x + \sin x$ |
|---|---|---|---|
| 0 | 0 | 0 | 0 |
| $\frac{\pi}{6}$ | 0.5 | 1 | 1.5 |
| $\frac{\pi}{3}$ | 0.866 | 0 | 0.866 |
| $\frac{\pi}{2}$ | 1 | −1 | 0 |
| $\frac{2\pi}{3}$ | 0.866 | 0 | 0.866 |
| $\frac{5\pi}{6}$ | 0.5 | 1 | 1.5 |
| $\pi$ | 0 | 0 | 0 |

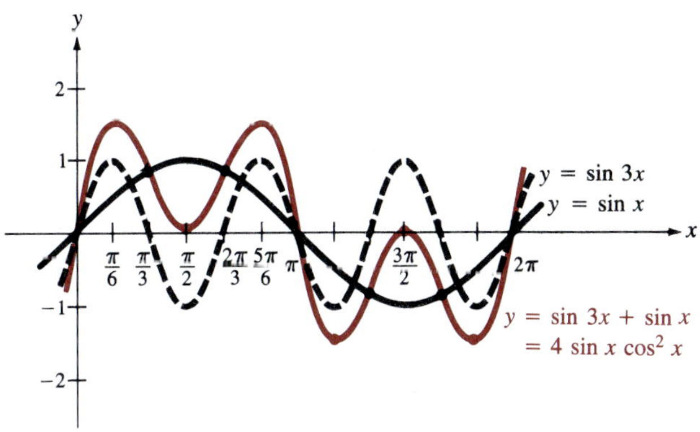


**Figure 8**

## Derivation of Addition Identities

The product identities can be used to obtain four addition identities. Suppose we let $A + B = C$ and $A - B = D$. Solving this system of two equations in the two variables $A$ and $B$, we obtain

$$A = \frac{C + D}{2} \quad \text{and} \quad B = \frac{C - D}{2}.$$

We substitute these values of $A$ and $B$ in the identity for $\sin A \cos B$.

$$\begin{aligned} \sin \left(\frac{C + D}{2}\right) \cos \left(\frac{C - D}{2}\right) \\ &\equiv \frac{1}{2}\left[\sin \left(\frac{C + D}{2} + \frac{C - D}{2}\right) + \sin \left(\frac{C + D}{2} - \frac{C - D}{2}\right)\right] \\ &\equiv \frac{1}{2}[\sin C + \sin D] \end{aligned}$$

Thus, $\quad \sin C + \sin D \equiv 2 \sin \left(\frac{C + D}{2}\right) \cos \left(\frac{C - D}{2}\right).$

Three additional identities can be obtained in a similar manner, and all four are given below.

**Addition Identities**

$$\sin C + \sin D \equiv 2 \sin \left(\frac{C + D}{2}\right) \cos \left(\frac{C - D}{2}\right)$$

$$\sin C - \sin D \equiv 2 \cos \left(\frac{C + D}{2}\right) \sin \left(\frac{C - D}{2}\right)$$

$$\cos C + \cos D \equiv 2 \cos \left(\frac{C + D}{2}\right) \cos \left(\frac{C - D}{2}\right)$$

$$\cos C - \cos D \equiv -2 \sin \left(\frac{C + D}{2}\right) \sin \left(\frac{C - D}{2}\right)$$

**EXAMPLE 4** Express each of the following as a product.

**(a)** $\sin 2A - \sin 6A$

Use the second addition identity with $C = 2A$ and $D = 6A$.

$$\begin{aligned}\sin 2A - \sin 6A &= 2 \cos \left(\frac{2A + 6A}{2}\right) \sin \left(\frac{2A - 6A}{2}\right) \\ &= 2 \cos 4A \sin (-2A) \\ &= -2 \cos 4A \sin 2A\end{aligned}$$

**(b)** $\cos 2x - \cos x$

Use the fourth addition identity with $C = 2x$ and $D = x$.

$$\begin{aligned}\cos 2x - \cos x &= -2 \sin \left(\frac{2x + x}{2}\right) \sin \left(\frac{2x - x}{2}\right) \\ &= -2 \sin \frac{3x}{2} \sin \frac{x}{2} \quad \blacksquare\end{aligned}$$

**EXAMPLE 5** Evaluate $\sin 105° + \sin 195°$ using an appropriate addition identity.

Use the first addition identity with $C = 105°$ and $D = 195°$.

$$\begin{aligned}\sin 105° + \sin 195° &= 2 \sin \left(\frac{105° + 195°}{2}\right) \cos \left(\frac{105° - 195°}{2}\right) \\ &= 2 \sin 150° \cos (-45°)\end{aligned}$$

Since $\sin 150° = \frac{1}{2}$ and $\cos (-45°) = \frac{\sqrt{2}}{2}$, this can be simplified further.

$$\sin 105° + \sin 195° = 2\left(\frac{1}{2}\right)\left(\frac{\sqrt{2}}{2}\right) = \frac{\sqrt{2}}{2}. \quad \blacksquare$$

**EXAMPLE 6** Verify the identity. $\dfrac{\cos 4x - \cos 6x}{\sin 4x + \sin 6x} \equiv \tan x$

Use the fourth and first addition identities on the numerator and denominator of the left side.

$$\frac{\cos 4x - \cos 6x}{\sin 4x + \sin 6x} \equiv \frac{-2\sin\left(\dfrac{4x+6x}{2}\right)\sin\left(\dfrac{4x-6x}{2}\right)}{2\sin\left(\dfrac{4x+6x}{2}\right)\cos\left(\dfrac{4x-6x}{2}\right)}$$

$$\equiv \frac{-2\sin 5x \sin(-x)}{2\sin 5x \cos(-x)}$$

$$\equiv -\frac{\sin(-x)}{\cos(-x)}$$

$$\equiv -\frac{-\sin(x)}{\cos x}$$

$$\equiv \frac{\sin x}{\cos x} \equiv \tan x \quad ■$$

## Derivation of Reduction Formula

Many applied problems in physics (involving electricity, heat, and spring-mass systems, for example) can be described using a function such as

$$y = M\cos Bx + N\sin Bx.$$

This representation of $y$ can be transformed into the sinusoidal form

$$y = A\cos B(x - C).$$

In this form, it is easier to determine amplitude, $|A|$, period, $2\pi/B$, phase shift, $|C|$, and the graph. In order to establish this relationship, suppose we begin by expanding $\cos B(x - C) = \cos(Bx - BC)$, using the cosine of a difference formula.

$$\begin{aligned} y = A\cos(Bx - BC) &= A[\cos Bx \cos BC + \sin Bx \sin BC] \\ &= A\cos BC \cos Bx + A\sin BC \sin Bx \end{aligned}$$

For $y$ in this form to be equal to $M\cos Bx + N\sin Bx$,

$$A\cos BC = M \qquad \text{and} \qquad A\sin BC = N.$$

Squaring both sides gives

$$A^2\cos^2 BC = M^2 \qquad \text{and} \qquad A^2\sin^2 BC = N^2.$$

Adding,

$$\begin{aligned} A^2\cos^2 BC + A^2\sin^2 BC &= M^2 + N^2 \\ A^2(\cos^2 BC + \sin^2 BC) &= M^2 + N^2 \\ A^2 &= M^2 + N^2 \qquad \cos^2 BC + \sin^2 BC \equiv 1 \\ A &= \sqrt{M^2 + N^2}. \end{aligned}$$

In addition, taking ratios we have

$$\frac{N}{M} = \frac{A\sin BC}{A\cos BC} = \frac{\sin BC}{\cos BC} = \tan BC.$$

With

$$\cos BC = \frac{M}{A} = \frac{M}{\sqrt{M^2 + N^2}} \quad \text{and} \quad \sin BC = \frac{N}{A} = \frac{N}{\sqrt{M^2 + N^2}},$$

angle $BC$ must be chosen such that

$$\tan BC = \frac{N}{M}$$

and the point $(M, N)$ is on its terminal side. (See Figure 9.)

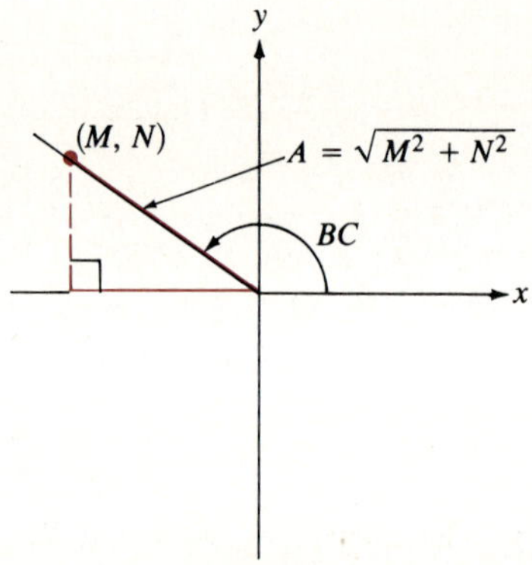


**Figure 9**

These remarks are summarized in the next theorem.

### Reduction Formula

If $y = M \cos Bx + N \sin Bx$, then $y$ can be represented by

$$y = A \cos B(x - C) = A \cos (Bx - BC)$$

where $A = \sqrt{M^2 + N^2}$, $\tan BC = N/M$ with the point $(M, N)$ on its terminal side, and $BC$ is chosen so that $|BC|$ is as small as possible.

**EXAMPLE 7** Express $y = \cos 2x + \sqrt{3} \sin 2x$ in the form $y = A \cos B(x - C)$ and give its amplitude, period, and phase shift.

Since $M = 1$ and $N = \sqrt{3}$, we locate the point $(M, N) = (1, \sqrt{3})$ and determine $BC$. (See Figure 10.) Since $\tan BC = \sqrt{3}/1$, we recognize $BC$ to be $\pi/3$. Also, since $A = \sqrt{M^2 + N^2}$, $A = \sqrt{1 + 3} = \sqrt{4} = 2$. Then

$$y = 2 \cos \left(2x - \frac{\pi}{3}\right),$$

or

$$y = 2 \cos 2\left(x - \frac{\pi}{6}\right).$$

Thus, the amplitude is 2, the period is $\pi$, and the phase shift is $\pi/6$ to the right. ■

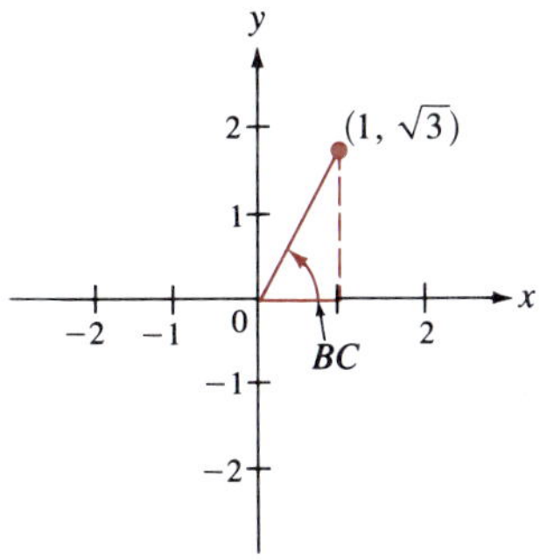


Figure 10

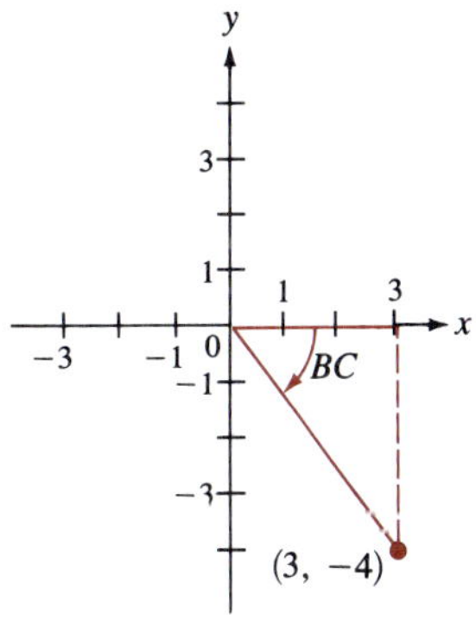


Figure 11

**EXAMPLE 8**

Express $y = 3 \cos 2\pi x - 4 \sin 2\pi x$ in the form $y = A \cos B(x - C)$ and give its amplitude, period, and phase shift.

Since $M = 3$ and $N = -4$, we find the point $(M, N) = (3, -4)$ and determine $BC$. (See Figure 11.) Then $A = \sqrt{M^2 + N^2} = \sqrt{9 + 16} = \sqrt{25} = 5$. Since

$$\tan BC = \frac{N}{M} = -\frac{4}{3} \approx -1.333,$$

using a calculator we find that $BC = -0.927$ radians. Then

$$y = 5 \cos (2\pi x + 0.927),$$

or

$$y = 5 \cos 2\pi\left(x + \frac{0.927}{2\pi}\right).$$

The amplitude is 5, the period is 1, and the phase shift is $0.927/2\pi \approx 0.148$ units to the left. ■

**EXAMPLE 9**

Graph the function $y = \sin 2x + \cos 2x$ with $M = N = 1$, $A = \sqrt{M^2 + N^2} = \sqrt{2}$.

Since $\tan BC = 1$ and $(M, N) = (1, 1)$ is in the first quadrant, $BC = \pi/4$. Then

$$y = \sin 2x + \cos 2x = \sqrt{2} \cos\left(2x - \frac{\pi}{4}\right) = \sqrt{2} \cos 2\left(x - \frac{\pi}{8}\right).$$

The graph is given in Section 6.7, Example 4 (Figure 53). At that time we found the graph by adding ordinates. However, once the function has been reduced to its sinusoidal form, it would be easier to find its graph using the amplitude $\sqrt{2}$, the period $\pi$, and the phase shift $\pi/8$ units to the right. ■

## 8.4 EXERCISES

In Exercises 1–6 express each product as a sum.

**1.** $\sin 2x \cos x$

**2.** $\cos 3y \sin 2y$

**3.** $2 \cos 4u \cos u$

**4.** $4 \sin v \sin 3v$

**5.** $4 \cos 3x \sin 4x$

**6.** $2 \sin 4y \cos 5y$

In Exercises 7–12 evaluate each expression using an appropriate product identity.

**7.** $\sin 75° \cos 15°$ **8.** $\cos 75° \cos 15°$ **9.** $\sin 15° \sin 105°$

**10.** $\cos 15° \sin 105°$ **11.** $2 \cos 105° \cos 165°$ **12.** $2 \sin 195° \sin 75°$

In Exercises 13–16 graph each function by changing it to a sum and adding ordinates.

**13.** $y = 2 \cos 2x \sin x$ **14.** $2 \cos x \sin 3x$

**15.** $y = 4 \sin^2 x \cos x$ **16.** $y = 2 \cos^2 x \sin x - 2 \sin^3 x$

In Exercises 17–22 write each expression as a product.

**17.** $\sin 5x - \sin x$ **18.** $\sin 7y - \sin 3y$ **19.** $\cos 2u + \cos u$

**20.** $\sin 5v + \sin 2v$ **21.** $\cos 4x - \cos 3x$ **22.** $\cos y - \cos 3y$

In Exercises 23–28 evaluate each expression using an appropriate addition identity.

**23.** $\sin 75° + \sin 15°$ **24.** $\cos 75° - \cos 15°$ **25.** $\cos 15° + \cos 105°$

**26.** $\sin 15° - \sin 105°$ **27.** $\frac{1}{2}[\cos 105° - \cos 165°]$ **28.** $\frac{1}{2}[\sin 195° + \sin 75°]$

In Exercises 29–34 express the given function in the sinusoidal form $y = A \cos B(x - C)$ and give its amplitude, period, and phase shift.

**29.** $y = \cos x - \sin x$ **30.** $y = \cos x + \sin x$ **31.** $y = \cos 2x - \sqrt{3} \sin 2x$

**32.** $y = \sqrt{3} \cos \pi x + \sin \pi x$ **33.** $y = 4 \sin 2\pi x - 3 \cos 2\pi x$ **34.** $y = 12 \sin x + 5 \cos x$

In Exercises 35–38 graph each function by first reducing it to sinusoidal form.

**35.** $y = \sqrt{3} \sin x + \cos x$ **36.** $y = \sin x + \sqrt{3} \cos x$

**37.** $y = \sin x + \cos x$ **38.** $y = \cos x - \sin x$

Verify each identity in Exercises 39–46.

**39.** $\dfrac{\sin x + \sin 3x}{\cos x - \cos 3x} \equiv \cot x$ **40.** $\dfrac{\cos 3x - \cos 5x}{\sin 5x + \sin 3x} \equiv \tan x$

**41.** $\dfrac{\sin 8x + \sin 3x}{\cos 8x - \cos 3x} \equiv -\cot\left(\dfrac{5}{2}x\right)$ **42.** $\dfrac{\sin 8x - \sin 3x}{\cos 8x + \cos 3x} \equiv \tan\left(\dfrac{5}{2}x\right)$

**43.** $\dfrac{\cos x - \cos 3x}{\cos x + \cos 3x} \equiv \tan x \tan 2x$ **44.** $\dfrac{\sin x + \sin 3x}{\sin x - \sin 3x} \equiv -\cot x \tan 2x$

**45.** $\dfrac{\sin x + \sin y}{\sin x - \sin y} \equiv \tan\left(\dfrac{x + y}{2}\right) \cot\left(\dfrac{x - y}{2}\right)$ **46.** $\dfrac{\cos x - \cos y}{\cos x + \cos y} \equiv -\tan\left(\dfrac{x + y}{2}\right) \tan\left(\dfrac{x - y}{2}\right)$

**47.** Verify the product identity $\cos A \cos B \equiv \frac{1}{2}[\cos (A + B) + \cos (A - B)]$.

**48.** Verify the product identity $\sin A \sin B \equiv -\frac{1}{2}[\cos (A + B) - \cos (A - B)]$.

**49.** Verify the addition identity $\cos C + \cos D \equiv 2 \cos\left(\dfrac{C + D}{2}\right) \cos\left(\dfrac{C - D}{2}\right)$.

**50.** Verify the addition identity $\cos C - \cos D \equiv -2 \sin\left(\dfrac{C + D}{2}\right) \sin\left(\dfrac{C - D}{2}\right)$.

In Exercises 51–52 use a calculator and verify the identity $\sin A \cos B \equiv \frac{1}{2}[\sin (A + B) + \sin (A - B)]$ for the given values of $A$ and $B$.

**51.** $A = 27°42'$ and $B = 171°31'$

**52.** $A = 0.413$ and $B = 2.437$

In Exercises 53–54 use a calculator and verify the identity $\sin C + \sin D \equiv 2 \sin \left(\frac{C + D}{2}\right) \cos \left(\frac{C - D}{2}\right)$ for the given values of $C$ and $D$.

**53.** $C = 1.305$ and $D = 0.567$

**54.** $C = 123°52'$ and $D = 12°15'$

## For Review

**55.** Suppose $\cos A = -3/5$ with $A$ in quadrant III. Find **(a)** $\sin 2A$, **(b)** $\cos 2A$, **(c)** $\tan 2A$, and **(d)** the quadrant in which $2A$ is located.

**56.** Suppose $\sin A = 12/13$ with $A$ in quadrant II. Find **(a)** $\sin \frac{A}{2}$, **(b)** $\cos \frac{A}{2}$, **(c)** $\tan \frac{A}{2}$, and **(d)** the quadrant in which $\frac{A}{2}$ is located.

# 8.5 Trigonometric Equations

### Nature of Solutions

Much of our work in trigonometry has been concerned with identities, that is, equations that are true for all replacements for the variable. In this section, we consider conditional equations, equations that are true for specific replacements of the variable. A conditional equation that involves one or more trigonometric functions is called a **trigonometric equation.** A **solution** to a trigonometric equation is a value for the variable (angle) that makes the equation true. For example, the equation

$$\sin x = 1$$

has $x = 90°$ as a solution. In addition, any other angle of the form

$$90° \pm k \cdot 360°$$

is also a solution. When solving trigonometric equations, we agree to specify only those solutions that are greater than or equal to 0° and less than 360° since all solutions could then be obtained by adding an integer multiple of 360°. Solutions can also be given using radian measure. For example, $x = \frac{\pi}{2} \pm 2k\pi$ are solutions to the above equation.

### Method of Solution

Unlike some types of equations, such as linear, quadratic, fractional, and radical equations, there is no solution method that works for all trigonometric equations. However, the next rule gives several guidelines that might be helpful.

### To Solve a Trigonometric Equation

1. If a single function of a single angle is involved, isolate the function and solve.
2. If all terms can be written on one side of the equation and the resulting expression factored, set each factor equal to zero (using the zero-product rule) and solve.
3. If more than one function of a single variable is present, use identities to rewrite the expression in terms of a single function of a single variable, then apply Step 1 or Step 2 (if possible).
4. If several angles are involved, use identities to rewrite the expression in terms of functions of a single angle, then apply Steps 1, 2, or 3 (if possible).

The next four examples illustrate, in order, these four suggestions.

**EXAMPLE 1** Solve $2 \cos A - 1 = 0$.

Since we have a single function of a single angle, we solve for $\cos A$.

$$2 \cos A = 1$$

$$\cos A = \frac{1}{2} = 0.5$$

Since $\cos A > 0$, $A$ is an angle in quadrant I or in quadrant IV. We should recognize that $A = 60°$ is the solution in quadrant I, and that $A = 300°$ is the solution in quadrant IV since the reference angle of 300° is 60°. Thus, the solutions are 60° and 300°, or, in radians, $\pi/3$ and $5\pi/3$. ■

**EXAMPLE 2** Solve $2 \sin^2 x = \sqrt{3} \sin x$.

Write all terms on the left side of the equation and factor.

$$2 \sin^2 x - \sqrt{3} \sin x = 0$$

$$\sin x(2 \sin x - \sqrt{3}) = 0 \qquad \text{Do not divide by } \sin x$$

Apply the zero-product rule.

$$\sin x = 0 \quad \text{or} \quad 2 \sin x - \sqrt{3} = 0$$

$$2 \sin x = \sqrt{3}$$

$$\sin x = \frac{\sqrt{3}}{2}$$

Once again we should recognize immediately that 0° and 180° are solutions to the first equation. Since $\sin x = \sqrt{3}/2 > 0$, solutions to the second equation must be in quadrants I and II. We should recall that $\sin 60° = \sqrt{3}/2$, so that 60° and 120° (with reference angle 60°) are the other two solutions. Thus, the solutions are 0°, 60°, 120°, and 180°, or, in radians, 0, $\pi/3$, $2\pi/3$, and $\pi$. ■

**CAUTION** Had we incorrectly divided out $\sin x$ in Example 2, two of the solutions, 0° and 180°, would have been lost. *Never* divide both sides of an equation by an expression containing the variable. ■

**EXAMPLE 3**

Solve $\cos^2 B - \sin^2 B - \sin B = 1$.

In this case there is more than one function, but a single variable. By substituting $1 - \sin^2 B$ for $\cos^2 B$, we obtain

$$1 - \sin^2 B - \sin^2 B - \sin B = 1$$
$$-2 \sin^2 B - \sin B = 0.$$

Factoring,
$$\sin B(-2 \sin B - 1) = 0.$$

Set each factor equal to zero.

$$\sin B = 0 \quad \text{or} \quad -2 \sin B - 1 = 0$$
$$-2 \sin B = 1$$
$$\sin B = -\frac{1}{2} = -0.5$$

The solutions to the first equation are 0° and 180°. The second equation has solutions in quadrants III and IV where the sine function is negative. Since the reference angle is 30°, the angle in quadrant III is 210° (180° + 30° = 210°) and in quadrant IV is 330° (360° − 30° = 330°). Thus, the solutions are 0°, 180°, 210°, and 330°, or 0, $\pi$, $7\pi/6$, and $11\pi/6$. ■

**EXAMPLE 4**

Solve $\sin 2y + 2 \cos y = 0$.

This equation involves two different angles, $2y$ and $y$. We must first rewrite the equation using functions of a single angle. Substituting $2 \sin y \cos y$ for $\sin 2y$,

$$2 \sin y \cos y + 2 \cos y = 0.$$

Factoring,
$$2 \cos y(\sin y + 1) = 0.$$

Apply the zero-product rule.

$$2 \cos y = 0 \quad \text{or} \quad \sin y + 1 = 0$$
$$\cos y = 0 \qquad \sin y = -1$$
$$y = \frac{\pi}{2} \text{ or } \frac{3\pi}{2} \qquad y = \frac{3\pi}{2}$$

The solutions are $\pi/2$ and $3\pi/2$ or 90° and 270°. ■

**EXAMPLE 5**

Solve $2 \sin^2 A = 1$.

First solve for $\sin A$.

$$2 \sin^2 A = 1$$
$$\sin^2 A = \frac{1}{2}$$
$$\sin A = \pm\sqrt{\frac{1}{2}} = \pm\frac{1}{\sqrt{2}} = \pm\frac{\sqrt{2}}{2}$$

With $\sin A$ both positive and negative, there are solutions in all four quadrants. The four angles whose sine is either $\sqrt{2}/2$ or $-\sqrt{2}/2$ are 45°, 135°, 225°, and 315° or, in radians $\pi/4$, $3\pi/4$, $5\pi/4$, and $7\pi/4$. ■

Sometimes a trigonometric equation is quadratic in form in one of the trigonometric functions. The method to use in such cases is illustrated in the next example.

**EXAMPLE 6**

Solve $3 \cos^2 x + 2 \cos x - 1 = 0$.

The left side of this equation is the quadratic form

$$3u^2 + 2u - 1 = 0,$$

where $u = \cos x$. Factoring, we obtain

$$(3u - 1)(u + 1) = 0$$

which becomes

$$(3 \cos x - 1)(\cos x + 1) = 0$$

when $\cos x$ is substituted for $u$. Set each factor equal to zero and solve.

$$\begin{aligned} 3 \cos x - 1 &= 0 \quad \text{or} \quad & \cos x + 1 &= 0 \\ 3 \cos x &= 1 & \cos x &= -1 \\ \cos x &= \frac{1}{3} \end{aligned}$$

Solutions to the first equation are in quadrants I and IV (where the cosine function is positive). Here we need a calculator.

ALG: 1 [÷] 3 [=] [INV] [cos] → [70.528779]

RPN: 1 [ENTER] 3 [÷] [INV] [cos] → [70.528779]

To the nearest minute, the solutions are 70°32′ and 289°28′ or 1.2310 and 5.0522 radians. The only solution to the second equation is 180° or $\pi$ radians. ■

Some trigonometric equations can be solved using different methods as shown in the following example.

**EXAMPLE 7**

Solve $\sin x = 1 - \cos x$.

*First Method:* By squaring both sides we can write the equation in terms of $\cos x$ only.

$$\begin{aligned} \sin^2 x &= (1 - \cos x)^2 \\ \sin^2 x &= 1 - 2 \cos x + \cos^2 x \\ 1 - \cos^2 x &= 1 - 2 \cos x + \cos^2 x && \sin^2 x = 1 - \cos^2 x \\ 0 &= 2 \cos^2 x - 2 \cos x && \text{Collect terms} \\ 0 &= 2 \cos x(\cos x - 1) && \text{Factor} \end{aligned}$$

$$\begin{aligned} 2 \cos x &= 0 \quad \text{or} \quad & \cos x - 1 &= 0 \qquad \text{Zero-product rule} \\ \cos x &= 0 & \cos x &= 1 \\ x &= 90°, 270° & x &= 0° \end{aligned}$$

Recall from solving radical equations that when both sides of an equation are squared, extraneous roots may have been introduced. Thus, we must check all possible solutions. It is easy to verify that 90° and 0° check but 270° does not. Thus 0° and 90° are the only solutions.

*Second Method:* First we write the equation in the form

$$\sin x + \cos x = 1,$$

and then use the reduction formula given in Section 8.4 to write the left side as $\sqrt{2}\cos\left(x - \frac{\pi}{4}\right)$ or $\sqrt{2}\cos(x - 45°)$ (see Exercise 37). Thus, we must solve

$$\sqrt{2}\cos(x - 45°) = 1$$

$$\cos(x - 45°) = \frac{1}{\sqrt{2}} = \frac{\sqrt{2}}{2}.$$

The solutions for $x - 45°$ are 45° and 315° so that $x = 90°$ and $x = 360°$. But since we want the solutions that are greater than or equal to 0° and less than 360°, we use 0° instead of 360°. With this method the problem of possible extraneous roots is avoided. ■

Trigonometric equations involving a single factor of a multiple angle require special consideration.

**EXAMPLE 8** Solve $2 \sin 3x = 1$.

First solve for $\sin 3x$.

$$2 \sin 3x = 1$$

$$\sin 3x = \frac{1}{2}$$

Rather than trying to write $\sin 3x$ in terms of functions of $x$, we first solve for $3x$. To obtain all solutions for $x$ satisfying $0° \le x < 360°$, we must find all solutions for $3x$ satisfying $0° \le 3x < 3(360)° = 1080°$. Since $\sin 3x = 1/2$,

$$3x = 30°, 390°, \text{ and } 750°, \qquad 390° = 30° + 1 \cdot 360° \text{ and } 750° = 30° + 2 \cdot 360°$$

and

$$3x = 150°, 510°, \text{ and } 870°. \qquad 510° = 150° + 1 \cdot 360° \text{ and } 870° = 150° + 2 \cdot 360°$$

Dividing each of these solutions by 3, we obtain

$$x = 10°, 50°, 130°, 170°, 250°, \text{ and } 290°.$$ ■

Some trigonometric equations can be simplified using one of the factoring identities presented in Section 8.4.

**EXAMPLE 9** Solve $\sin x - \sin 3x + \cos 2x = 0$.

First use the addition identity on $\sin x - \sin 3x$.

$$\begin{aligned}\sin x - \sin 3x &\equiv 2 \cos\left(\frac{x + 3x}{2}\right) \sin\left(\frac{x - 3x}{2}\right)\\ &\equiv 2 \cos 2x \sin(-x)\\ &\equiv -2 \cos 2x \sin x\end{aligned}$$

Substitute in the original equation.

$$\begin{aligned}-2 \cos 2x \sin x + \cos 2x &= 0\\ \cos 2x(-2 \sin x + 1) &= 0 \qquad \text{Factor}\\ \cos 2x = 0 \text{ or } -2 \sin x + 1 &= 0\\ \sin x &= \frac{1}{2}\end{aligned}$$

The first equation has solutions for $2x$

$$90°,\ 270°,\ 450°, \text{ and } 630°.$$

The values for $x$ are found by dividing each of these by 2. The solutions of the second equation are 30° and 150°. Thus, all the solutions are 30°, 45°, 135°, 150°, 225°, and 315°. ■

We conclude this section with the second applied problem presented in the introduction to this chapter.

## EXAMPLE 10

**HYDROLOGY**

A Colorado River water diversion project in Arizona uses a canal in the shape of an isosceles trapezoid which is wider at the top than at the base. If the bottom and the two sides of the canal measure 8 yds, and $\theta$ is the acute angle formed by a side and level ground, show that the cross-sectional area of the canal, which determines the water-carrying capacity, is given by

$$A = 64 \sin \theta(1 + \cos \theta).$$

Use this formula to find $\theta$ when $A = 48\sqrt{3}$ ft$^2$. The sketch in Figure 12 shows the cross-sectional view of the canal.

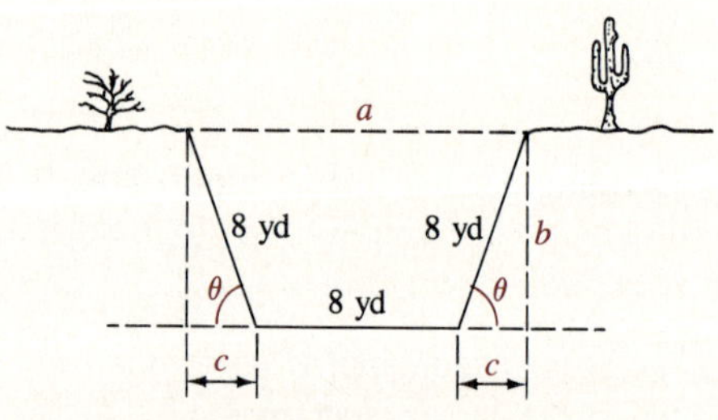


**Figure 12**

If we use the labels in the figure, the area of the trapezoidal cross-section is

$$A = ab - 2\left(\frac{1}{2}bc\right) = ab - bc = b(a - c),$$

where $\frac{1}{2}bc$ is the area of each of the triangles. Also,

$$a = 8 + 2c,$$

$$b = 8 \sin \theta, \qquad \frac{b}{8} = \sin \theta$$

and

$$c = 8 \cos \theta. \qquad \frac{c}{8} = \cos \theta$$

Thus,

$$a = 8 + 2c = 8 + 16 \cos \theta.$$

Substituting in the equation for $A$,

$$\begin{aligned} A &= b(a - c) \\ &= 8 \sin \theta(8 + 16 \cos \theta - 8 \cos \theta) \\ &= 8 \sin \theta(8 + 8 \cos \theta) \\ &= 64 \sin \theta(1 + \cos \theta). \end{aligned}$$

Let $A = 48\sqrt{3}$ and solve for $\theta$.

$$48\sqrt{3} = 64 \sin \theta(1 + \cos \theta)$$

$$\frac{3\sqrt{3}}{4} = \sin \theta(1 + \cos \theta)$$

Square both sides.

$$\frac{27}{16} = \sin^2 \theta(1 + \cos \theta)^2$$

$$\frac{27}{16} = \sin^2 \theta(1 + 2 \cos \theta + \cos^2 \theta)$$

$$\frac{27}{16} = (1 - \cos^2 \theta)(1 + 2 \cos \theta + \cos^2 \theta) \qquad \sin^2 \theta = 1 - \cos^2 \theta$$

After multiplying and collecting like terms this will simplify to

$$16 \cos^4 x + 32 \cos^3 x - 32 \cos x + 11 = 0,$$

which is the polynomial equation in $u = \cos x$ given by

$$16u^4 + 32u^3 - 32u + 11 = 0.$$

The possible rational-number solutions to this equation are of the form $p/q$ where $q$ divides 16 and $p$ divides 11. Since $u$ is $\cos x$ and we are only interested in solutions for $x$ between 0° and 90° ($x$ is an acute angle), only those values of $p/q$ that are positive and less than 1 need be considered. These possibilities are: 11/16, 1/16, 1/8, 1/4, and 1/2. Using synthetic division it is easy to show that 1/2 is the only rational solution. In fact, 1/2 is a double root and the only real solution. Thus, $u = \cos \theta = 1/2$ which means that $\theta = 60°$. ■

## 8.5 EXERCISES

In Exercises 1–50 solve each equation. Give all solutions greater than or equal to 0 and less than $2\pi$ in Exercises 1–20. Give all solutions greater than or equal to 0° and less than 360° in Exercises 21–50. Convert answers to the nearest minute, where appropriate.

**1.** $2 \sin A - 1 = 0$

**2.** $2 \cos A + \sqrt{3} = 0$

**3.** $\tan x - \sqrt{3} = 0$

**4.** $\csc B + \sqrt{2} = 0$

**5.** $2 \cos^2 y = \sqrt{3} \cos y$

**6.** $\tan^2 y - \tan y = 0$

**7.** $\sin^2 x - \cos^2 x - \cos x = 1$

**8.** $\tan^2 x + 2 \sec^2 x - 3 = 0$

**9.** $\sin 2B + 2 \sin B = 0$

**10.** $\cos 2B - \cos B = 0$

**11.** $\tan^2 A - 1 = 0$

**12.** $3 \csc^2 A - 4 = 0$

**13.** $2 \sin^2 B + \sin B - 1 = 0$

**14.** $2 \cos^2 B - 3 \cos B + 1 = 0$

**15.** $4 \cos^2 y + 3 \cos y - 1 = 0$

**16.** $5 \sin^2 y + 4 \sin y - 1 = 0$

**17.** $\cos 2A + \sin^2 A = 1$

**18.** $\cos 2A - 2 = 3 \sin A$

**19.** $\cos^4 x = \sin^2 x + 1$

**20.** $\sin^4 x = \cos^2 x - 1$

**21.** $2 \cos (B - 90°) = \sin^2 B + \cos^2 B$

**22.** $2 \sin (B + 90°) = \sec^2 B - \tan^2 B$

**23.** $2 \cos A \sin A - \cos^2 A \sin A = 0$

**24.** $\sin^2 A \cos A + \sin A \cos A = 0$

**25.** $\csc x = 1 - \cot x$

**26.** $\sin x = 1 + \cos x$

**27.** $\sin 2B = \frac{1}{2}$

**28.** $2 \cos 2B = \sqrt{3}$

**29.** $\cot 3y = -1$

**30.** $2 \sin 3y = -\sqrt{3}$

**31.** $\cos^2 3A - 3 \cos 3A + 2 = 0$

**32.** $\sin^2 4A + 3 \sin 4A - 4 = 0$

**33.** $2 + \sin x = 2.5640$

**34.** $4 - \cos x = 4.5664$

**35.** $4 \sin B \cos B = 1$

**36.** $2 \cos^2 B = 2 \sin^2 B + 1$

**37.** $\sin 3y + \sin 5y = 0$

**38.** $\cos 3y - \cos y = 0$

**39.** $2 \sin \left(\frac{3A}{2}\right) \cos \left(\frac{A}{2}\right) = \sin A$

**40.** $2 \cos \left(\frac{3A}{2}\right) \cos \left(\frac{A}{2}\right) = \cos 2A$

**41.** $\sin 4x = -\sin 2x$

**42.** $\cos 9x = -\cos 3x$

**43.** $2 \cos^3 B + \cos^2 B - 2 \cos B - 1 = 0$

**44.** $2 \sin^3 B - \sin^2 B - 2 \sin B + 1 = 0$

**45.** $2 \sin^2 y = 5 - 2 \cos^2 y$

**46.** $3 \csc^2 y + 1 = 3 \cot^2 y$

**47.** $12 \cos^2 A + 13 \cos A + 3 = 0$

**48.** $12 \sin^2 A - 5 \sin A - 2 = 0$

**49.** $2 \tan^2 x - 9 \tan x - 5 = 0$

**50.** $3 \sec^2 x + 5 \sec x - 12 = 0$

**51.** CALCULUS Using a technique developed in calculus, an engineer could obtain the equation $\cos^2 \theta - \sin^2 \theta + \cos \theta = 0$ from the formula $A = 64 \sin \theta(1 + \cos \theta)$, which gave the cross-sectional area of the canal described in Example 10. The solution to this equation gives the value of $\theta$ that determines the maximum carrying capacity of the canal. Find this value of $\theta$.

**52.** **PHYSICS** The motion in a spring-mass system is described by $y = 4 \sin\left(\pi t + \frac{\pi}{2}\right)$, where $t$ is time in seconds. The times when the weight reaches the equilibrium point occur when $y = 0$. Find all such times $t$, for $0 \le t \le 2$.

**53.** **ELECTRONICS** A generator produces an alternating current according to the equation $I = 40 \sin 120\pi t$, where $t$ is time in seconds and $I$ is the current in amperes. What is the smallest time $t$ such that $I = 20$?

**54.** **BUSINESS** The weekly sales in thousands of items of a product which has a seasonal sales record is approximated by $n = 65.25 + 30.0 \sin \frac{\pi t}{26}$, where $t$ is time in weeks with $t = 1$ corresponding to the first week in the year. During which two weeks will the sales equal 72,500 items?

## For Review

**55.** Express $2 \cos 6x \sin 2x$ as a sum.

**56.** Express $\sin 7x + \sin x$ as a product.

**57.** Express $y = -\cos 2x + \sqrt{3} \sin 2x$ in the sinusoidal form $y = A \cos B(x - C)$ and give its amplitude, period, and phase shift.

**58.** Verify the identity $\dfrac{\cos 7x - \cos x}{\sin 7x + \sin x} \equiv -\tan 3x$.

Inverse trigonometric functions will be discussed in the next section. As a review of inverse functions, find the inverse of each function in Exercises 59–60.

**59.** $f(x) = 5x - 3$

**60.** $f(x) = \sqrt{x + 3}$

## 8.6 Inverse Trigonometric Functions

In Section 3.5 the notion of an inverse function was introduced. Recall that in order for a function to have an inverse, the function must be one-to-one on its domain. A function $f$ is one-to-one if every element in the range of $f$ has one and only one corresponding element in its domain. Since the trigonometric functions are periodic, they are not one-to-one and cannot have inverse functions as such. For example, since

$$\sin 0 = 0 \quad \text{and} \quad \sin \pi = 0,$$

two numbers in the domain of the sine function (0 and $\pi$) correspond to the same number 0 in the range. However, it is possible to place a restriction on the domain of each trigonometric function so that on this interval the restricted function is one-to-one and does have an inverse. These inverse functions, called the inverse trigonometric functions, are discussed separately, beginning with the *inverse sine function*.

## Inverse Sine

Recall that the graph of the inverse of a function can be obtained by reflecting the graph of the given function in the line $y = x$. Consider the graph of the sine function $y = \sin x$ in Figure 13. By reflecting it in the line $y = x$, that is, plotting the point $(b, a)$ for each point $(a, b)$, we obtain the graph in Figure 14.

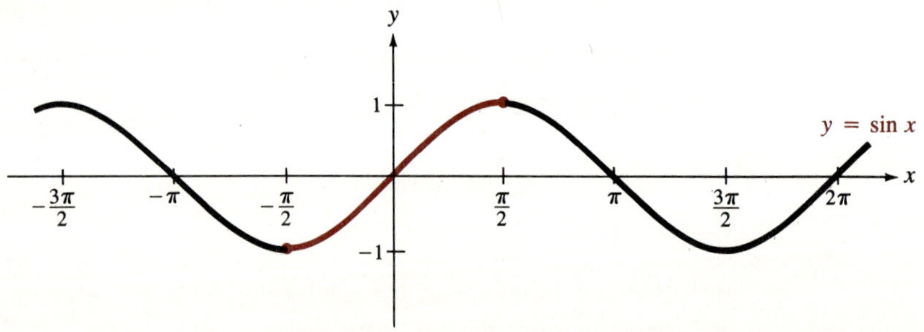


Figure 13 Sine Function

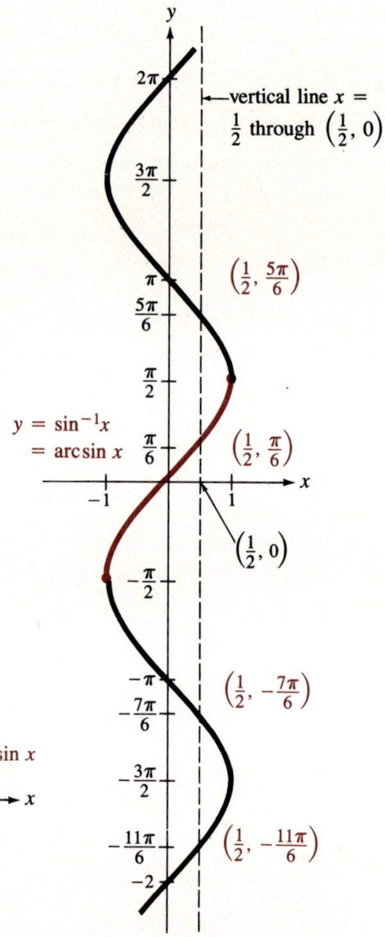


Figure 14 Inverse Sine

The resulting graph is *not* the graph of a function since a given value of $x$ between $-1$ and 1 corresponds to many values of $y$. For example, a vertical line through $\left(\frac{1}{2}, 0\right)$ passes through the graph at infinitely many points, four of which are

$$\left(\frac{1}{2}, \frac{\pi}{6}\right), \quad \left(\frac{1}{2}, \frac{5\pi}{6}\right), \quad \left(\frac{1}{2}, -\frac{7\pi}{6}\right), \quad \text{and} \quad \left(\frac{1}{2}, -\frac{11\pi}{6}\right).$$

However, we will restrict the domain of the sine function to the interval $\left[-\frac{\pi}{2}, \frac{\pi}{2}\right]$. The graph of this restricted sine function is the colored curve in Figure 13, and the corresponding graph, shown in color in Figure 14, is now the graph of a function with domain $[-1, 1]$.

### Inverse Sine Function

The **inverse sine (arcsine) function,** denoted by $\sin^{-1}$ or by arcsin, is given by

$$y = \sin^{-1} x \text{ if and only if } x = \sin y \quad \text{and } y \text{ satisfies } -\frac{\pi}{2} \leq y \leq \frac{\pi}{2}.$$

The domain of $\sin^{-1}$ is the closed interval $[-1, 1]$, and its range is the closed interval $\left[-\frac{\pi}{2}, \frac{\pi}{2}\right]$.

The notation $\sin^{-1} x$ should not be confused with exponential notation; it is the inverse function notation (recall that we used $f^{-1}$ for the inverse of $f$). The arcsin notation originates from the unit circle since arcsin $x$ is the "arc" on the circle whose sine is $x$. Although the domain and range of the function $\sin^{-1}$ are sets of real numbers, we often think of $\sin^{-1} x$ as "the angle whose sine is $x$." Of course, there is no problem of thinking this way since the range of $\sin^{-1}$ can be thought of as angles in radian measure. In fact, when it is appropriate, we may even think of the range as angles measured in degrees.

We have already introduced the use of the keys $\boxed{\sin^{-1}}$ or $\boxed{\text{INV}}$ $\boxed{\sin}$ or $\boxed{\text{ARC}}$ $\boxed{\sin}$ on a calculator, and now the reason for this notation should be apparent.

**EXAMPLE 1** Evaluate the following expressions.

**(a)** $\sin^{-1} \frac{1}{2}$

Remember that $\sin^{-1} \frac{1}{2}$ represents "the number between $-\frac{\pi}{2}$ and $\frac{\pi}{2}$ whose sine is $\frac{1}{2}$." That is, if $x = \sin^{-1} \frac{1}{2}$, then $\sin x = \frac{1}{2}$ and $x = \frac{\pi}{6}$. Keep in mind that there are infinitely many numbers (angles) whose sine is $\frac{1}{2}$ $\left(\frac{\pi}{6}, \frac{5\pi}{6}, \frac{-7\pi}{6}, \text{ and so forth}\right)$, but only $\frac{\pi}{6}$ is in the range of the inverse sine function $\left(\text{in the interval } \left[-\frac{\pi}{2}, \frac{\pi}{2}\right]\right)$.

**(b)** $\arcsin \frac{\sqrt{2}}{2}$

The number in $\left[-\frac{\pi}{2}, \frac{\pi}{2}\right]$ whose sine is $\frac{\sqrt{2}}{2}$ is $\frac{\pi}{4}$. Hence,

$$\arcsin \frac{\sqrt{2}}{2} = \frac{\pi}{4}. \qquad \sin \frac{\pi}{4} = \frac{\sqrt{2}}{2}$$

Sin $\frac{3\pi}{4}$ is also $\frac{\sqrt{2}}{2}$, but $\frac{3\pi}{4}$ is not in the range of the inverse sine function.

**(c)** $\sin^{-1}\left(-\frac{\pi}{4}\right)$

Use care with expressions such as this since it is easy to misread it and find $\sin\left(-\frac{\pi}{4}\right)$ instead. We use a calculator set in radian mode to evaluate $\sin^{-1}\left(-\frac{\pi}{4}\right)$ as follows:

ALG: [π] [÷] 4 [+/−] [=] [INV] [sin] → [−0.9033391]

RPN: [π] 4 [÷] [CHS] [INV] [sin] → [−0.9033391].

Thus, the number whose sine is $-\frac{\pi}{4}$ is approximately −0.9033, correct to four decimal places. That is,

$$\sin^{-1}\left(-\frac{\pi}{4}\right) \approx -0.9033.$$

**(d)** arcsin (0.2153)

ALG & RPN: 0.2153 [INV] [sin] → [0.216999.]

Thus, $$\arcsin(0.2153) \approx 0.2170. \quad ■$$

Recall that for inverse functions $f$ and $f^{-1}$,

$$f(f^{-1}(x)) = x \quad \text{and} \quad f^{-1}(f(x)) = x.$$

Thus, for the sine and inverse sine functions,

$$\mathbf{\sin(\sin^{-1} x) \equiv x \text{ if } -1 \leq x \leq 1}$$

and $$\mathbf{\sin^{-1}(\sin x) \equiv x \text{ if } -\frac{\pi}{2} \leq x \leq \frac{\pi}{2}}.$$

Using the alternative notation these identities become

$$\mathbf{\sin(\arcsin x) \equiv x} \quad \text{and} \quad \mathbf{\arcsin(\sin x) \equiv x}.$$

**EXAMPLE 2** Evaluate the following expressions.

**(a)** $\sin(\sin^{-1} 0.5)$

Since 0.5 satisfies $-1 \leq 0.5 \leq 1$, using the identity given above,

$$\sin(\sin^{-1} 0.5) = 0.5.$$

**(b)** $\arcsin\left(\sin \frac{5\pi}{4}\right)$

Since $\frac{5\pi}{4}$ is not between $-\frac{\pi}{2}$ and $\frac{\pi}{2}$, we cannot use the identity above. However, since $\sin \frac{5\pi}{4} = \sin\left(-\frac{\pi}{4}\right)$, and $-\frac{\pi}{2} \leq -\frac{\pi}{4} \leq \frac{\pi}{2}$,

$$\arcsin\left(\sin \frac{5\pi}{4}\right) = \arcsin\left(\sin\left(-\frac{\pi}{4}\right)\right) = -\frac{\pi}{4}.$$

We can also do this by first writing $\sin \frac{5\pi}{4} = -\frac{\sqrt{2}}{2}$.

$$\arcsin\left(\sin \frac{5\pi}{4}\right) = \arcsin\left(-\frac{\sqrt{2}}{2}\right) = -\frac{\pi}{4}$$

A calculator can be used to do such problems.

ALG: 5 [×] [π] [÷] 4 [=] [sin] [INV] [sin] → [−0.7853982]

RPN: 5 [ENTER] [π] [×] 4 [÷] [sin] [INV] [sin] → [−0.7853982]

The display shows an approximation for $-\frac{\pi}{4}$.

**(c)** $\sin^{-1}\left(\cot \frac{3\pi}{4}\right)$

Since $\cot \frac{3\pi}{4} = -1$, we have

$$\sin^{-1}\left(\cot \frac{3\pi}{4}\right) = \sin^{-1}(-1) = -\frac{\pi}{2}.$$

**(d)** $\tan\left(\arcsin\left(-\frac{1}{2}\right)\right)$

Since $\sin\left(-\frac{\pi}{6}\right) = -\frac{1}{2}$, $\arcsin\left(-\frac{1}{2}\right) = -\frac{\pi}{6}$. Thus,

$$\tan\left(\arcsin\left(-\frac{1}{2}\right)\right) = \tan\left(-\frac{\pi}{6}\right) = -\frac{1}{\sqrt{3}} = -\frac{\sqrt{3}}{3}.$$

With a calculator,

ALG: 1 [÷] 2 [+/−] [=] [INV] [sin] [tan] → [−0.5773503]

RPN: 1 [ENTER] 2 [÷] [CHS] [INV] [sin] [tan] → [−0.5773503]

The display shows an approximation for $-\frac{\sqrt{3}}{3}$. ■

## Inverse Cosine

To define the inverse cosine function, we must restrict the domain to an interval where $y = \cos x$ is one-to-one. The interval chosen is $[0, \pi]$. The graph of $y = \cos x$ is given in Figure 15 with the restricted portion in color. Reflecting this graph in the line $y = x$ gives the graph of the inverse relation in Figure 16 with the portion in color showing the graph of the inverse cosine function.

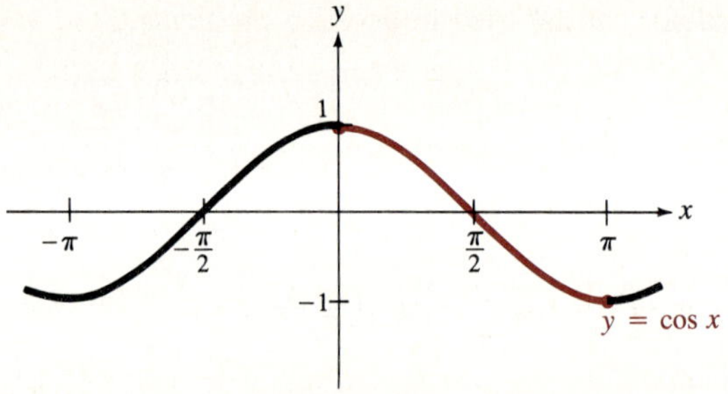


**Figure 15**

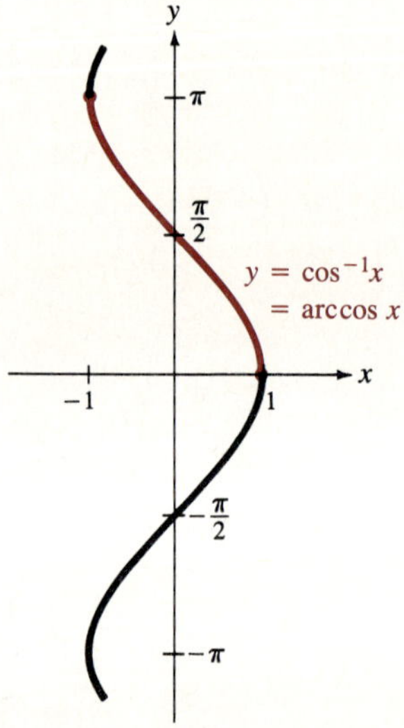


**Figure 16**

### Inverse Cosine Function

The **inverse cosine (arccosine) function,** denoted by $\cos^{-1}$ or arccos, is given by

$$\mathbf{y = \cos^{-1} x \text{ if and only if } x = \cos y} \quad \text{and } y \text{ satisfies } 0 \le y \le \pi.$$

The domain of $\cos^{-1}$ is the closed interval $[-1, 1]$, and its range is the closed interval $[0, \pi]$.

As with the sine and inverse sine functions, the following are identities:

$$\cos(\cos^{-1} x) \equiv x \text{ if } -1 \le x \le 1$$

and

$$\cos^{-1}(\cos x) \equiv x \text{ if } 0 \le x \le \pi.$$

**EXAMPLE 3** Evaluate the following expressions.

**(a)** $\cos^{-1} \dfrac{1}{2}$

If $x = \cos^{-1} \dfrac{1}{2}$, this is equivalent to $\cos x = \dfrac{1}{2}$ where, by definition, $0 \le x \le \pi$. Thus, $x = \dfrac{\pi}{3}$ and $\cos^{-1} \dfrac{1}{2} = \dfrac{\pi}{3}$.

**(b)** $\cos^{-1} \frac{\pi}{6}$

Do not misread this as $\cos \frac{\pi}{6}$. Use a calculator set in radian mode.

ALG: [$\pi$] [$\div$] 6 [$=$] [INV] [cos] $\rightarrow$ [1.0197267]

RPN: [$\pi$] 6 [$\div$] [INV] [cos] $\rightarrow$ [1.0197267]

Thus, the number whose cosine is $\frac{\pi}{6}$ is approximately 1.0197, correct to four decimal places. That is,

$$\cos^{-1} \frac{\pi}{6} \approx 1.0197.$$

**(c)** $\cos(\cos^{-1} 0.2515)$

Since 0.2515 satisfies $-1 \leq 0.2515 \leq 1$, using the identity above,

$$\cos(\cos^{-1} 0.2515) = 0.2515.$$

**(d)** $\arccos\left(\cos \frac{3\pi}{2}\right)$

Since $\frac{3\pi}{2}$ is not between 0 and $\pi$, the above identity does not apply. However, since $\cos \frac{3\pi}{2} = \cos \frac{\pi}{2}$, and $0 \leq \frac{\pi}{2} \leq \pi$,

$$\arccos\left(\cos \frac{3\pi}{2}\right) = \arccos\left(\cos \frac{\pi}{2}\right) = \frac{\pi}{2}.$$

With a calculator,

ALG: 3 [$\times$] [$\pi$] [$\div$] 2 [$=$] [cos] [INV] [cos] $\rightarrow$ [1.5707963]

RPN: 3 [ENTER] [$\pi$] [$\times$] 2 [$\div$] [cos] [INV] [cos] $\rightarrow$ [1.5707963]

The display shows an approximation of $\frac{\pi}{2}$.

**(e)** $\cos^{-1}\left(\tan\left(-\frac{5\pi}{6}\right)\right)$

With a calculator,

ALG: 5 [$\times$] [$\pi$] [$\div$] 6 [$+/-$] [$=$] [tan] [INV] [cos] $\rightarrow$ [0.9553166]

RPN: 5 [ENTER] [$\pi$] [$\times$] 6 [$\div$] [CHS] [tan] [INV] [cos] $\rightarrow$ [0.9553166]

the display gives an approximation of $\cos^{-1}\left(\tan\left(-\frac{5\pi}{6}\right)\right)$ as 0.9553, correct to four decimal places. ■

## Inverse Tangent

By restricting the domain of the tangent function to the interval $\left(-\frac{\pi}{2}, \frac{\pi}{2}\right)$, we obtain a one-to-one function which has an inverse. The graph of $y = \tan x$ is given in Figure

17 with restricted portion in color. Reflecting this graph in the line $y = x$ gives the graph of the inverse relation in Figure 18 with the portion in color showing the graph of the inverse tangent function.

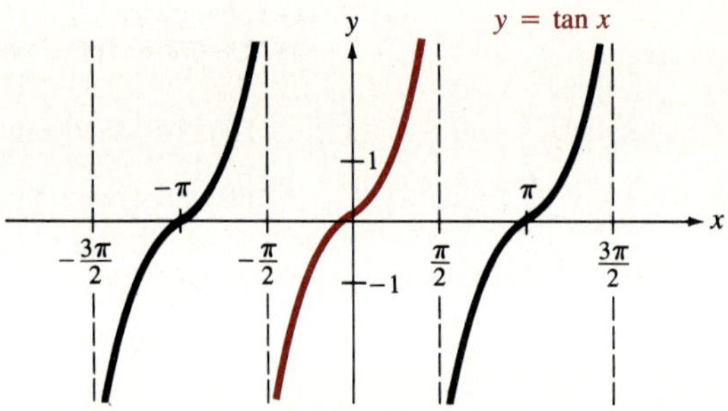


Figure 17

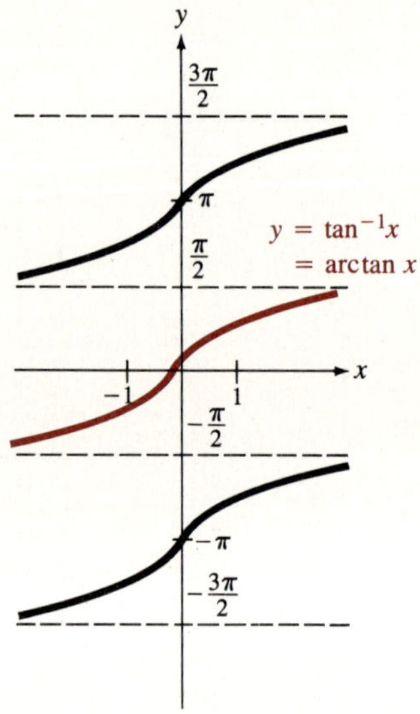


Figure 18

### Inverse Tangent Function

The **inverse tangent (arctangent) function,** denoted by $\tan^{-1}$ or arctan, is given by

$$\mathbf{y = \tan^{-1} x \text{ if and only if } x = \tan y} \quad \text{and } y \text{ satisfies } -\frac{\pi}{2} < y < \frac{\pi}{2}.$$

The domain of $\tan^{-1}$ is the entire set of real numbers, and its range is the open interval $\left(-\frac{\pi}{2}, \frac{\pi}{2}\right)$.

The inverse tangent identities are:

$$\tan(\tan^{-1} x) \equiv x \text{ for every real number } x$$

and

$$\tan^{-1}(\tan x) \equiv x \text{ if } -\frac{\pi}{2} < x < \frac{\pi}{2}.$$

**EXAMPLE 4** Evaluate the following expressions.

**(a)** $\tan^{-1}(-1)$

If $x = \tan^{-1}(-1)$, this is equivalent to $\tan x = -1$, where $-\frac{\pi}{2} < x < \frac{\pi}{2}$ by definition. Thus, $x = -\frac{\pi}{4}$ and $\tan^{-1}(-1) = -\frac{\pi}{4}$.

**(b)** $\sin\left(\arctan \frac{\sqrt{3}}{3}\right)$

If $x = \arctan \frac{\sqrt{3}}{3}$, then $\tan x = \frac{\sqrt{3}}{3}$ where $-\frac{\pi}{2} < x < \frac{\pi}{2}$, and $x = \frac{\pi}{6}$. Since $\sin \frac{\pi}{6} = \frac{1}{2}$, $\sin\left(\arctan \frac{\sqrt{3}}{3}\right) = \frac{1}{2}$. We can verify this using a calculator.

ALG: 3 [√] [÷] 3 [=] [INV] [tan] [sin] → [0.5]

RPN: 3 [√] 3 [÷] [INV] [tan] [sin] → [0.5]

**(c)** $\tan^{-1}\left(\sec \frac{\pi}{6}\right)$

Using a calculator,

ALG: [π] [÷] 6 [=] [cos] [1/x] [INV] [tan] → [0.8570719]

RPN: [π] 6 [÷] [cos] [1/x] [INV] [tan] → [0.8570719]

The display shows an approximation of $\tan^{-1}\left(\sec \frac{\pi}{6}\right)$ as 0.8571, correct to four decimal places. ■

The remaining three trigonometric functions can have inverses defined in a similar manner. However, they are not widely used and there is no general agreement about the domains of the inverse secant and inverse cosecant functions. One possibility for each is given in Exercises 64–66.

Certain expressions can be evaluated by a technique that does not involve a calculator. This is illustrated in the following two examples.

**EXAMPLE 5** Without using a calculator, evaluate $\sin\left(\arccos \frac{4}{5}\right)$.

First sketch a right triangle as in Figure 19 and label it in such a way that one of the acute angles, say $A$, has cosine 4/5. That is,

$$\arccos \frac{4}{5} = A. \qquad \cos A = \frac{4}{5}$$

Using the Pythagorean theorem, we find that side $a = 3$. Thus,

$$\sin\left(\arccos \frac{4}{5}\right) = \sin A = \frac{3}{5}. \quad ■$$

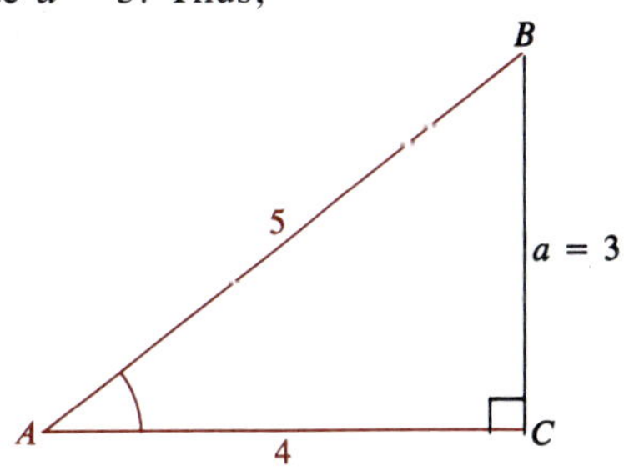


**Figure 19**

**EXAMPLE 6** Simplify $\cos(\tan^{-1} x)$.

Let $A = \tan^{-1} x$ and sketch a right triangle like the one in Figure 20 in which $\tan A = x = \frac{x}{1}$. By the Pythagorean theorem,

$$c^2 = 1^2 + x^2$$
$$c = \sqrt{1 + x^2}.$$

Thus,

$$\cos(\tan^{-1} x) = \cos A = \frac{1}{\sqrt{1 + x^2}}.$$ ■

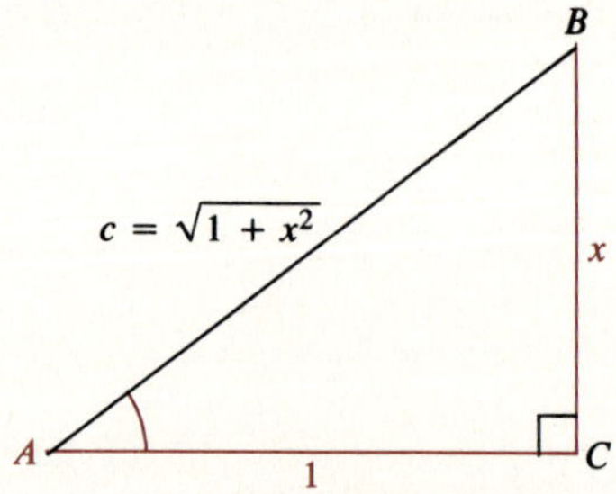

**Figure 20**

In some problems, sum, difference, half-angle, or double-angle formulas must be applied.

**EXAMPLE 7** Evaluate sin [arctan 1/2 + arcsin 4/5].

Let $A = \arctan 1/2$ and $B = \arcsin 4/5$. Then

$$\sin\left[\arctan\frac{1}{2} + \arcsin\frac{4}{5}\right] = \sin[A + B]$$
$$= \sin A \cos B + \cos A \sin B$$

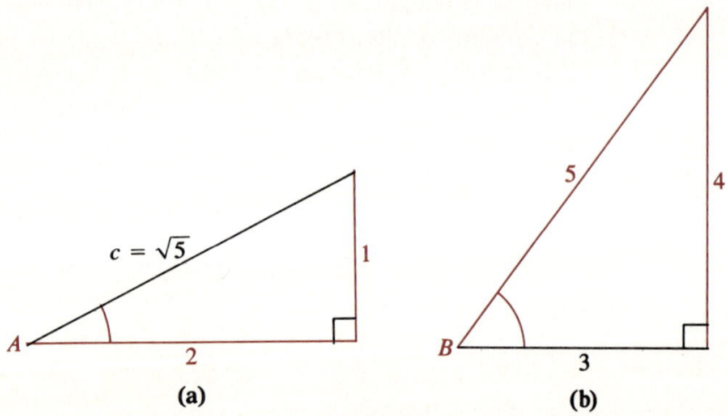

**Figure 21**

From Figure 21(**a**) and 21(**b**) $\sin A = \frac{1}{\sqrt{5}}$, $\cos B = \frac{3}{5}$, $\cos A = \frac{2}{\sqrt{5}}$, and $\sin B = \frac{4}{5}$. Substituting,

$$\sin\left[\arctan\frac{1}{2} + \arcsin\frac{4}{5}\right] = \left(\frac{1}{\sqrt{5}}\right)\left(\frac{3}{5}\right) + \left(\frac{2}{\sqrt{5}}\right)\left(\frac{4}{5}\right)$$
$$= \frac{3}{5\sqrt{5}} + \frac{8}{5\sqrt{5}} = \frac{11}{5\sqrt{5}} = \frac{11\sqrt{5}}{25}.$$ ■

## 8.6 EXERCISES

Evaluate each expression in Exercises 1–30.

**1.** $\arcsin \frac{1}{2}$
**2.** $\arccos \frac{1}{2}$
**3.** $\cos^{-1}\left(-\frac{\sqrt{2}}{2}\right)$
**4.** $\tan^{-1} 1$

**5.** $\sin^{-1}\left(-\frac{\sqrt{3}}{2}\right)$
**6.** $\cos^{-1}(\sqrt{3})$
**7.** $\arcsin \sqrt{3}$
**8.** $\arccos\left(-\frac{\sqrt{3}}{2}\right)$

**9.** $\tan^{-1}\sqrt{3}$
**10.** $\arctan(-\sqrt{3})$
**11.** $\sin^{-1}\frac{\pi}{6}$
**12.** $\cos^{-1}\frac{\pi}{6}$

**13.** $\arccos(0.7254)$
**14.** $\arctan(0.7254)$
**15.** $\sin\left(\sin^{-1}\frac{\sqrt{3}}{2}\right)$
**16.** $\cos\left(\cos^{-1}\frac{\sqrt{3}}{2}\right)$

**17.** $\tan(\tan^{-1}\sqrt{3})$
**18.** $\sin^{-1}\left(\sin\frac{7\pi}{6}\right)$
**19.** $\cos^{-1}\left(\cos\frac{7\pi}{6}\right)$
**20.** $\tan^{-1}\left(\tan\frac{7\pi}{6}\right)$

**21.** $\sin(\cos^{-1} 0.4251)$
**22.** $\cos(\tan^{-1} 5)$
**23.** $\sin^{-1}\left(\sin\frac{\pi}{3}\right)$
**24.** $\tan^{-1}\left(\cos\left(-\frac{\pi}{6}\right)\right)$

**25.** $\sin^{-1}\left(\csc\frac{\pi}{4}\right)$
**26.** $\cos^{-1}\left(\sec\left(-\frac{5\pi}{4}\right)\right)$
**27.** $\cot(\sin^{-1} 0.4556)$
**28.** $\sec(\tan^{-1} 2.4560)$

**29.** $\sin^{-1}(\arcsin 0.1135)$
**30.** $\cos^{-1}(\arccos 0.7885)$

Without using a calculator, evaluate each expression in Exercises 31–49.

**31.** $\cos\left(\arcsin\frac{4}{5}\right)$
**32.** $\sin\left(\arccos\frac{3}{5}\right)$
**33.** $\tan\left(\arcsin\frac{3}{5}\right)$
**34.** $\tan\left(\arccos\frac{4}{5}\right)$

**35.** $\cos\left(\arcsin\left(-\frac{3}{5}\right)\right)$
**36.** $\sin\left(\arccos\left(-\frac{4}{5}\right)\right)$
**37.** $\sin\left(\arccos\frac{1}{3}\right)$
**38.** $\cos(\arctan 2)$

**39.** $\tan\left(\arcsin\frac{1}{2}\right)$
**40.** $\sin\left[2\arccos\frac{3}{5}\right]$
**41.** $\cos\left[2\arcsin\frac{3}{5}\right]$
**42.** $\tan\left[2\arccos\frac{3}{5}\right]$

**43.** $\cos\left[\frac{1}{2}\arcsin\frac{1}{2}\right]$
**44.** $\sin\left[\frac{1}{2}\arcsin\frac{1}{2}\right]$
**45.** $\tan\left[\frac{1}{2}\arccos\frac{1}{2}\right]$

**46.** $\sin\left[\arccos\frac{1}{2}+\arcsin\frac{\sqrt{3}}{2}\right]$
**47.** $\sin\left[\arccos\frac{1}{2}-\arcsin\frac{\sqrt{3}}{2}\right]$

**48.** $\sin\left[\arctan\frac{4}{3}+\arccos\frac{4}{5}\right]$
**49.** $\cos\left[\arctan\frac{4}{3}-\arcsin\frac{3}{5}\right]$

Simplify each expression in Exercises 50–55.

**50.** $\sin(\cos^{-1} x)$
**51.** $\sin(\arctan x)$
**52.** $\tan(\sin^{-1} x)$

**53.** $\cos(\sin^{-1} x)$
**54.** $\cos(\tan^{-1} x)$
**55.** $\tan(\cos^{-1} x)$

Verify the identities in Exercises 56–59.

**56.** $\arcsin x + \arccos x \equiv \frac{\pi}{2}$
**57.** $\arctan x + \arccos\frac{x}{\sqrt{1+x^2}} \equiv \frac{\pi}{2}$

**58.** $\sin^{-1}(-x) \equiv -\sin^{-1} x$
**59.** $\tan^{-1}(-x) \equiv -\tan^{-1} x$

Give an example to show that each equation in Exercises 60–63 is not an identity.

**60.** $(\sin^{-1} x)^2 + (\cos^{-1} x)^2 = 1$
**61.** $\sin^{-1} x = \frac{1}{\sin x}$

**62.** $\cos^{-1} x = \dfrac{1}{\cos x}$

**63.** $\dfrac{\sin^{-1} x}{\cos^{-1} x} = \tan^{-1} x$

**64.** Define the inverse cotangent function by restricting the domain to the open interval $(0, \pi)$.

**65.** Define the inverse secant function by restricting the domain to the intervals $\left[-\pi, \dfrac{-\pi}{2}\right)$ or $\left[0, \dfrac{\pi}{2}\right)$.

**66.** Define the inverse cosecant function by restricting the domain to the intervals $\left(-\pi, -\dfrac{\pi}{2}\right]$ or $\left(0, \dfrac{\pi}{2}\right]$.

**67.** Use a calculator to evaluate arcsin (sin 2). Explain why the answer is not 2.

**68.** Use a calculator to find arccos (cos (−2)). Explain why the answer is not −2.

**69.** Use a calculator to find $\sin^{-1}$ (sin (−1)). Explain why the answer must be −1. Compare this result with Exercise 67.

**70.** Use a calculator to find $\cos^{-1}$ (cos (1)). Explain why the answer must be 1. Compare this result with Exercise 68.

## For Review

In Exercises 71–74 solve each equation. Give all solutions greater than or equal to 0° and less than 360°. Convert answers to the nearest minute, where appropriate.

**71.** $4 \cos^2 A = 3$

**72.** $\cos 2B = \cos^2 B$

**73.** $\tan y - 1 = 1.539$

**74.** $20 \sin^2 y - \sin y - 1 = 0$

**75.** **MILITARY** A gun on a destroyer has a muzzle velocity of 1500 ft/sec. The distance that the shell will travel is given by $d = \dfrac{v^2 \sin 2\theta}{32}$, where $v$ is the initial velocity of the shell and $\theta$ is the angle of elevation of the gun. At what angle should the gun be aimed in order to hit an enemy carrier located 1 mile from the destroyer?

# CHAPTER 8 REVIEW EXERCISES

Prove the identities in Exercises 1–14.

**1.** $\dfrac{1}{1 + \sin x} + \dfrac{1}{1 - \sin x} \equiv \dfrac{2}{\cos^2 x}$

**2.** $\tan x + \cos x \equiv 2 \csc 2x$

**3.** $\dfrac{\sec y}{\tan y + \cot y} \equiv \sin y$

**4.** $\dfrac{1}{\cot B - \tan B} \equiv \dfrac{\sin 2B}{4 \cos^2 B - 2}$

**5.** $\cot \dfrac{x}{2} \equiv \dfrac{\sin x}{1 - \cos x}$

**6.** $\dfrac{1 - \cos 2A}{2 \tan^2 A} \equiv \dfrac{\sin^2 A + \cos^2 A}{1 + \tan^2 A}$

**7.** $\dfrac{\sin 2y}{1 - \cos^2 y} \equiv \dfrac{2}{\tan y}$

**8.** $\sin 2B - \tan B \cos 2B \equiv \dfrac{\sin 2B}{2 \cos^2 B}$

**9.** $\dfrac{\sin (x - y)}{\sin x \sin y} \equiv \cot y - \cot x$

**10.** $\sin 3x + \sin x \equiv 2 \sin 2x \cos x$

**11.** $4 \sin A \cos A \cos 2A \equiv \sin 4A$

**12.** $\dfrac{1 - \tan B}{1 + \tan B} \equiv \dfrac{\cot B - 1}{\cot B + 1}$

**13.** $\dfrac{\tan y + \sin y}{\csc y + \cot y} \equiv \dfrac{\sin y}{\cot y}$

**14.** $\dfrac{2 \tan x + \sec^2 x}{\sin^2 x} \equiv (\csc x + \sec x)^2$

**15.** Express $\tan \theta \csc^2 \theta$ in terms of $\sin \theta$ only. Assume that $\theta$ satisfies $0 < \theta < \pi/2$.

In Exercises 16–27 assume that $\sin A = 12/13$, with $A$ in quadrant II, and that $\cos B = 4/5$, with $B$ in quadrant IV. Find the requested trigonometric function value.

**16.** $\sin (A + B)$ **17.** $\cos (A + B)$ **18.** $\tan (A + B)$ **19.** $\sin (A - B)$

**20.** $\cos (A - B)$ **21.** $\tan (A - B)$ **22.** $\sin 2A$ **23.** $\cos 2A$

**24.** $\tan 2A$ **25.** $\sin \dfrac{A}{2}$ **26.** $\cos \dfrac{A}{2}$ **27.** $\tan \dfrac{A}{2}$

In Exercises 28–29 express each product as a sum, and simplify if possible.

**28.** $4 \sin 5x \cos 3x$ **29.** $\cos 15° \cos 45°$

In Exercises 30–31 write each expression as a product and simplify if possible.

**30.** $\cos 5x - \cos 3x$ **31.** $\sin \dfrac{17\pi}{12} - \sin \dfrac{11\pi}{12}$

**32.** Express $y = \sqrt{3} \cos x - \sin x$ in the sinusoidal form $y = A \cos B(x - C)$ and give its amplitude, period, and phase shift.

Solve each equation in Exercises 33–40.

**33.** $\sqrt{3} \tan A - 1 = 0$ **34.** $\sqrt{2} \cos x \sin x = \cos x$ **35.** $\cos 2B = \sin B$

**36.** $4 \sin^2 2\theta = 1$ **37.** $\tan y + 4 = 6.194$ **38.** $\sec^2 x - \sec x - 2 = 0$

**39.** $\cos 3x - \cos 5x = 0$ **40.** $\csc x + \cot x = 1$

Evaluate each expression in Exercises 41–52.

**41.** $\arctan 1$ **42.** $\arcsin \left(-\dfrac{\sqrt{2}}{2}\right)$ **43.** $\arccos \left(-\dfrac{1}{2}\right)$

**44.** $\sin^{-1} \left(\dfrac{\sqrt{3}}{2}\right)$ **45.** $\tan^{-1} (-\sqrt{3})$ **46.** $\cos^{-1} \left(-\dfrac{\sqrt{2}}{2}\right)$

**47.** $\sin \left(\arccos \dfrac{1}{2}\right)$ **48.** $\arccos (\cos 0)$ **49.** $\cos (\tan^{-1} 3)$

**50.** $\sin \left[2 \arcsin \dfrac{1}{3}\right]$ **51.** $\cos \left[\dfrac{1}{2} \arcsin \dfrac{\sqrt{3}}{2}\right]$ **52.** $\sin \left[\arctan \dfrac{3}{4} - \arccos \dfrac{3}{5}\right]$

Simplify each expression in Exercises 53–54.

**53.** $\sin (\tan^{-1} 2x)$ **54.** $\cos (\sin^{-1} 2x)$

**55.** SPORTS A baseball player hits a fly ball at an angle of inclination with the horizontal equal to $\theta$. The distance the ball travels in the air is given by $d = \dfrac{v^2 \sin 2\theta}{32}$, where $v$ is the initial velocity of the ball in ft/sec. Find $\theta$ if the ball lands at the base of the centerfield fence 420 ft from home plate when hit with initial velocity of 125 ft/sec.

CHAPTER

# 9 TOPICS IN ANALYTIC GEOMETRY

The study of **analytic geometry** involves the description of geometric figures in terms of algebraic equations. Our work will require us to solve two types of problems.

1. Given a figure or curve, find an equation that describes the curve.
2. Given an equation, graph the corresponding curve in a coordinate system.

Our primary interest is in curves called **conic sections.** The name comes from the fact that these curves can be obtained by intersecting a right circular cone with a plane. In this chapter we will describe the circle, ellipse, hyperbola, and parabola geometrically, and derive an equation for each that fits the description.

Another approach to the study of the conic sections is to describe them as particular cases of a general equation. This equation is the **second-degree equation** or **quadratic equation** in two variables

$$Ax^2 + Bxy + Cy^2 + Dx + Ey + F = 0,$$

where at least one of $A$, $B$, or $C$ is not zero. In Chapter 3 we studied the parabola as the graph of the quadratic function $y = ax^2 + bx + c$, which is a particular case of the general quadratic equation.

The study of conics dates back to Menaechmus, a geometer in Plato's academy. The Greeks probably used these curves primarily to solve construction problems, but applications of the conics continue to grow in modern times. The reflecting properties of a parabolic surface allow it to be used in telescopes, radar units, and navigational systems. Planets and satellites follow elliptical orbits, and atomic particles travel in hyperbolic paths.

Two applications follow that we will study in detail in this chapter.

## AEROSPACE ▶

A communications company has determined that the most economical orbit for one of its space labs is an ellipse which has a maximum distance above the surface of the earth of 200 miles and a minimum distance of 100 miles. Use 4000 miles as the radius of the earth and determine the equation of the ellipse. Estimate the distance traveled by the space lab in one revolution.

## ◀ ARCHITECTURE

A domed ceiling is a parabolic surface. For the best lighting on the floor, a light source is to be placed at the focus of the surface. If 10.0 m down from the top of the dome the ceiling is 15.0 m wide, find the best location for the light source.

The first of these problems applies the ellipse (see Example 3 in Section 9.2), while the second shows one of the many uses of the parabola (see Example 4 in Section 9.4).

In this chapter each conic section is studied and an equation derived. Next, general equations are transformed into standard form to determine which types of conic they represent. This process includes rotation of axes. Polar coordinates are introduced to expand our discussion of graphing. Finally, we use polar coordinates to describe additional properties of complex numbers.

# 9.1 The Circle

## Conics

Figure 1 shows how the circle, ellipse, hyperbola, and parabola can be obtained by intersecting a double-napped right circular cone with a plane.

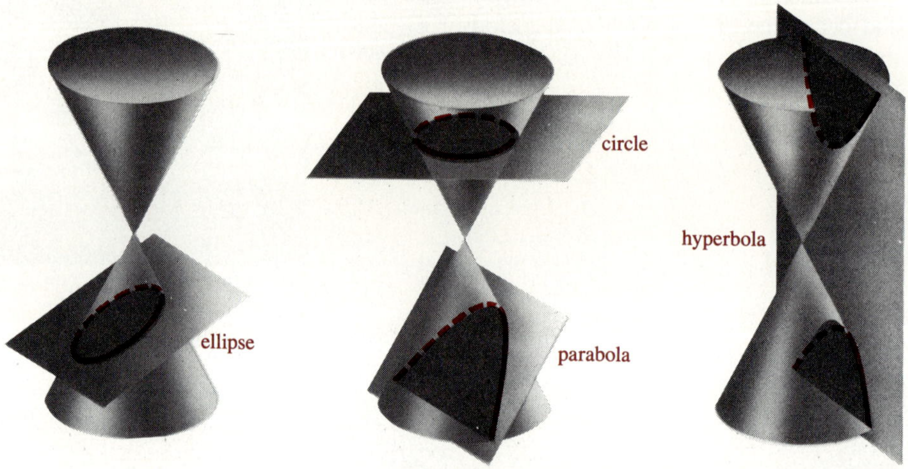


**Figure 1 Conics**

## Degenerate Conics

The circle is a special case of the ellipse in which the plane is perpendicular to the axis of the cone. Notice that none of the planes in Figure 1 pass through the vertex of the cone. When the intersecting plane does pass through the vertex, we obtain the **degenerate conic sections** shown in Figure 2. Our discussion is centered on the conics of Figure 1.

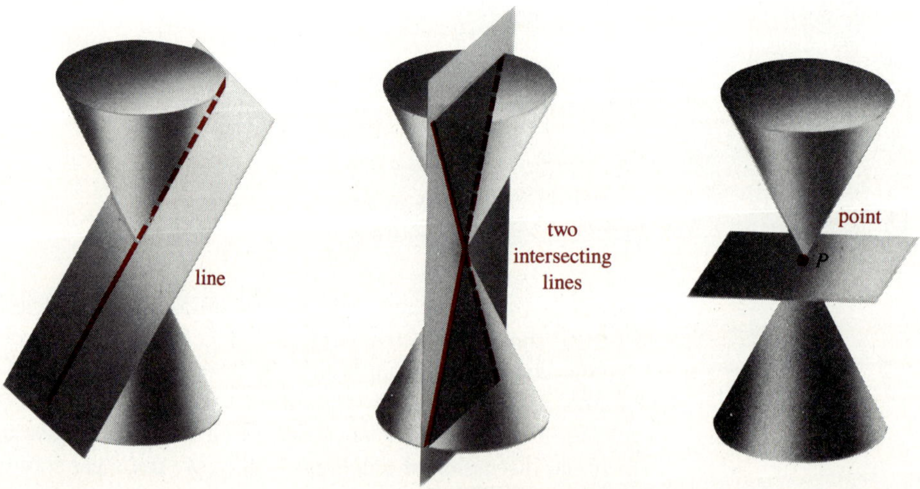


**Figure 2 Degenerate Conics**

The first conic section to be discussed is the *circle*. We define a circle geometrically and then derive an equation.

### The Circle

Let $(h, k)$ be any point in the plane and $r$ be any positive real number. The collection of all points in the plane that are $r$ units from $(h, k)$ is a **circle** with **center** $(h, k)$ and **radius** $r$.

## Equation of a Circle

The distance formula

$$d = \sqrt{(x_1 - x_2)^2 + (y_1 - y_2)^2},$$

which gives the distance between two points with coordinates $(x_1, y_1)$ and $(x_2, y_2)$, is used to derive the equation of a circle. Let $(x, y)$ be an arbitrary point on the circle with radius $r$ and center at $(h, k)$. See Figure 3.

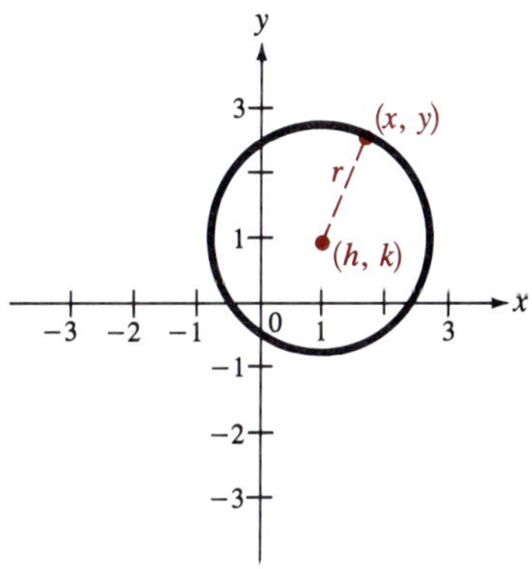


**Figure 3 Circle**

Since every point $(x, y)$ is $r$ units from $(h, k)$,

$$r = \sqrt{(x - h)^2 + (y - k)^2}$$

$$r^2 = (x - h)^2 + (y - k)^2. \quad \text{Square both sides}$$

The equation

$$(x - h)^2 + (y - k)^2 = r^2$$

is called the **standard form** of the equation of the circle with center $(h, k)$ and radius $r$.

When $(h, k) = (0, 0)$, then the equation of a circle with center at the origin is $x^2 + y^2 = r^2$. If the standard form is expanded, the result is the *general form* of the equation of a circle.

$$(x - h)^2 + (y - k)^2 = r^2$$

$$x^2 - 2hx + h^2 + y^2 - 2ky + k^2 = r^2$$

$$x^2 + y^2 - 2hx - 2ky + (h^2 + k^2 - r^2) = 0$$

This is a second-degree equation with **general form**

$$x^2 + y^2 + Dx + Ey + F = 0.$$

**EXAMPLE 1**

Find the standard and general forms of the equation of a circle with center $(2, -1)$ and radius 3.

Substitute $h = 2$, $k = -1$, and $r = 3$ in the standard form.

$$(x - h)^2 + (y - k)^2 = r^2$$
$$(x - \mathbf{2})^2 + (y - (\mathbf{-1}))^2 = \mathbf{3^2} \quad \text{Watch the sign on } k$$
$$(x - 2)^2 + (y + 1)^2 = 9 \quad \text{Standard form}$$

Expanding,

$$x^2 - 4x + 4 + y^2 + 2y + 1 = 9$$
$$x^2 + y^2 - 4x + 2y - 4 = 0. \quad \text{General form}$$ ■

As Example 1 illustrates, obtaining the general form from the standard form is simply a matter of clearing the parentheses. However, to find the standard form from the general form of the equation, we must complete the squares in both $x$ and $y$.

**EXAMPLE 2**

Determine the standard form of the equation $x^2 + y^2 - 6x + 2y - 6 = 0$, and then graph the circle.

First write the $x$ terms together and the $y$ terms together, and place the constant term on the right, leaving space to write in numbers to complete both squares.

$$x^2 - 6x \qquad + y^2 + 2y \qquad = 6$$

To complete the square on $x$, add the square of one-half the coefficient of $x$, namely $(-3)^2$ or 9. Similarly, add $(1)^2$ or 1 to complete the square on $y$. We have added 9 and 1 to the left side of the equation and $9 + 1$ or 10 to the right side.

$$x^2 - 6x \mathbf{+ 9} + y^2 + 2y \mathbf{+ 1} = 6 \mathbf{+ 10}$$
$$(x - 3)^2 + (y + 1)^2 = 16$$
$$(x - 3)^2 + (y + 1)^2 = 4^2$$

Since $x - h = x - 3$, $h = 3$. Also $y - k = y + 1 = y - (-1)$ implies that $k = -1$. Thus, the center of the circle is $(3, -1)$ and the radius is 4. See Figure 4. ■

Generally the quadratic equation in two variables of the form

$$Ax^2 + Cy^2 + Dx + Ey + F = 0,$$

where $A = C \neq 0$, can be written in the standard form of the equation of a circle by dividing by $A$ and completing the squares. For example,

$$3x^2 + 3y^2 + 12x - 24y - 15 = 0$$
$$x^2 + y^2 + 4x - 8y - 5 = 0 \quad \text{Divide by 3}$$
$$x^2 + 4x + \mathbf{4} + y^2 - 8y + \mathbf{16} = 5 + \mathbf{20} \quad \text{Complete the squares}$$
$$(x + 2)^2 + (y - 4)^2 = 5^2. \quad \text{Standard form}$$

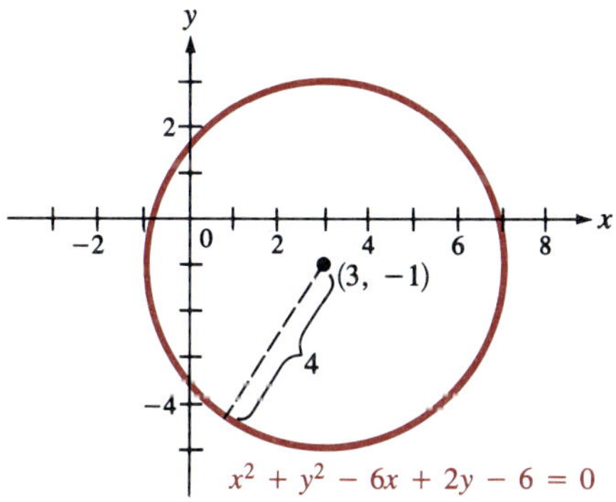


**Figure 4**

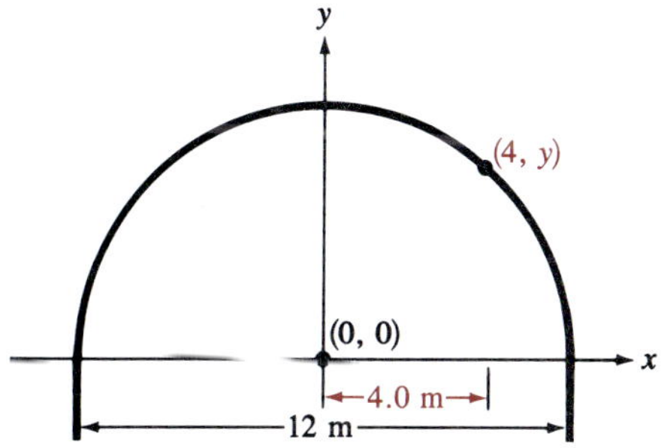


**Figure 5 Highway Tunnel**

CAUTION There are equations that appear to be equations of a circle that are actually degenerate conics. For example, when the squares are completed in $x^2 + y^2 - 4x + 6y + 13 = 0$, we obtain

$$(x - 2)^2 + (y + 3)^2 = 0.$$

Since $(2, -3)$ is the only solution to this equation, the graph is the single point $(2, -3)$. This can be thought of as a point circle. There are no points that satisfy the equation $2x^2 + 2y^2 + 4x - 2y + 3 = 0$ since by completing the squares we obtain

$$(x + 1)^2 + \left(y - \frac{1}{2}\right)^2 = -\frac{1}{4}.$$

Clearly the left side of this equation is nonnegative and thus the equation has no real solutions. ■

We conclude this section by considering an application of the circle.

**EXAMPLE 3**
**ENGINEERING**

A highway tunnel is to be designed in the shape of a semicircle with diameter 12.0 m. Find an equation for the semicircle and determine the vertical clearance at a point 4.0 m from the centerline.

Set a coordinate system with the centerline of the roadway at (0, 0) as shown in Figure 5. The equation of the circle of which the semicircle is the top half becomes

$$(x - 0)^2 + (y - 0)^2 = 6^2$$
$$x^2 + y^2 = 36.$$

Solving for $y$,

$$y^2 = 36 - x^2$$
$$y = \pm\sqrt{36 - x^2}.$$

Using only the positive value of $y$ for the top half of the circle,

$$y = \sqrt{36 - x^2}. \quad \text{Equation of semicircle}$$

The vertical clearance 4.0 m from the centerline can be found by letting $x = 4.0$ m.

$$y = \sqrt{36 - (4.0)^2} = \sqrt{36 - 16} = \sqrt{20} \approx 4.5$$

Thus, the vertical clearance is approximately 4.5 m. ■

## 9.1 EXERCISES

Find the standard and general forms of the equation of each circle described in Exercises 1–6.

**1.** Center $(-3, 2)$, radius 1

**2.** Center $(5, -3)$, radius 2

**3.** Center $\left(\frac{1}{4}, -1\right)$, radius 3

**4.** Center $\left(-\frac{1}{2}, 0\right)$, radius 1

**5.** Center $(0, 0)$, radius 1

**6.** Center $\left(0, \frac{1}{5}\right)$, radius 4

Find the standard form of the equation of each circle described in Exercises 7–18.

**7.** Center at $(0, 0)$, passing through $(0, 5)$

**8.** Center at $(0, 0)$, passing through $(6, 8)$

**9.** Center at $(1, -5)$, passing through $(7, 3)$

**10.** Center at $(-1, -2)$, passing through $(4, -3)$

**11.** Center at $(4, -1)$, radius $\sqrt{3}$

**12.** Center at $(-3, 5)$, radius $\sqrt{6}$

**13.** Center at $(6, 8)$, tangent to the $x$-axis

**14.** Center at $(-5, 3)$, tangent to the $y$-axis

**15.** Endpoints of a diameter at $(1, -4)$ and $(5, 2)$

**16.** Endpoints of a diameter at $(0, 6)$ and $(6, 0)$

**17.** Endpoints of a diameter at $(-7, 8)$ and $(4, 7)$

**18.** Endpoints of a diameter at $(3, -2)$ and $(10, -9)$

In Exercises 19–24 graph each circle.

**19.** $x^2 + y^2 = 16$

**20.** $(x - 2)^2 + (y + 3)^2 = 16$

**21.** $x^2 + y^2 + 4y = 5$

**22.** $x^2 + y^2 - 4x - 5 = 0$

**23.** $x^2 + y^2 - 6x - 8y + 21 = 0$

**24.** $x^2 + y^2 + 10x + 2y + 10 = 0$

In Exercises 25–30 determine the standard form of each equation.

**25.** $x^2 + y^2 - 4x - 2y + 4 = 0$

**26.** $x^2 + y^2 + 2x - 10y + 17 = 0$

**27.** $4x^2 + 4y^2 - 4x + 24y - 63 = 0$

**28.** $16x^2 + 16y^2 - 8x - 8y - 62 = 0$

**29.** $9x^2 + 9y^2 + 6x - 6y - 142 = 0$

**30.** $3x^2 + 3y^2 - 18x + 6y + 30 = 0$

Solve.

**31.** **ENGINEERING** A canal with cross section a semicircle is 10 ft deep at the center. Find an equation for the semicircle and use it to find the depth 4 ft from the edge.

**32.** **ENGINEERING** A highway tunnel in the shape of a semicircle is to have a vertical clearance of 6.2 m at a distance of 4.8 m from the center line. To the nearest tenth of a meter, what must be the width of the tunnel measured across the roadway? Give an equation for the semicircle.

# 9.2 The Ellipse

Consider a string attached to two fixed points on a piece of paper. Now draw the string taut with a pencil, and move the pencil around the points keeping the string taut to construct an *ellipse*. See Figure 6.

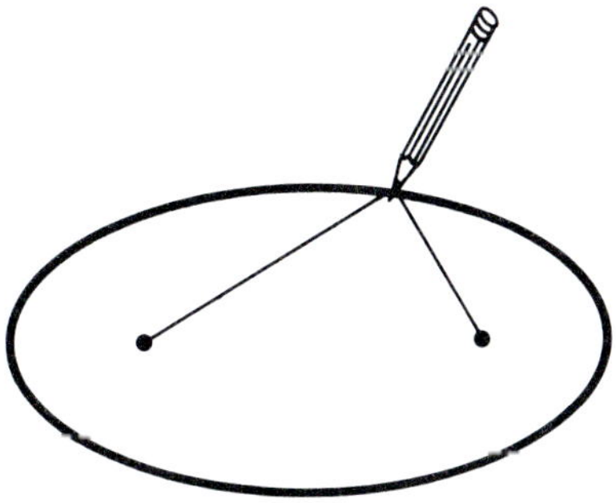

**Figure 6 Construction of Ellipse**

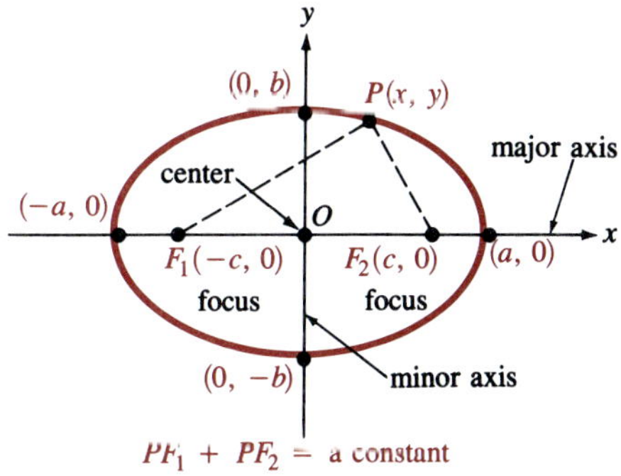


**Figure 7 Ellipse**

Notice that since the string does not change length in this process, the sum of the distances from the fixed points to each point on the curve remains constant. This leads to the definition of an ellipse.

### The Ellipse

Let $F_1$ and $F_2$ be two fixed points in a plane. An **ellipse** with **foci** (plural of **focus**) $F_1$ and $F_2$ is the collection of all points in the plane the sum of whose distances from $F_1$ and $F_2$ is a constant. The midpoint of the line segment joining $F_1$ and $F_2$ is called the **center** of the ellipse.

## Equation of an Ellipse

The equation of an ellipse is developed by considering the ellipse with center at the origin shown in Figure 7. The points $(-a, 0)$ and $(a, 0)$ are called the **vertices** of the ellipse. The line segment joining the vertices which passes through the foci and has length $2a$ is called the **major axis.** The line segment through the center of the ellipse with endpoints $(0, -b)$ and $(0, b)$ is called the **minor axis.** Notice that the minor axis has length $2b$ and the distance between the foci, $F_1(-c, 0)$ and $F_2(c, 0)$ is $2c$.

If $P(x, y)$ is any point on the ellipse, then by definition

$$PF_1 + PF_2 = \text{a constant.}$$

If point $P$ is at $(a, 0)$, then $PF_1 = a + c$ and $PF_2 = a - c$ so that

$$PF_1 + PF_2 = (a + c) + (a - c) = 2a.$$

Thus, the sum of the distances must be $2a$. Using the point $P(x, y)$, by the distance formula

$$PF_1 = \sqrt{(x + c)^2 + (y - 0)^2} = \sqrt{(x + c)^2 + y^2}$$

and
$$PF_2 = \sqrt{(x - c)^2 + (y - 0)^2} = \sqrt{(x - c)^2 + y^2}.$$

Thus,
$$PF_1 + PF_2 = \sqrt{(x + c)^2 + y^2} + \sqrt{(x - c)^2 + y^2} = 2a$$

Isolate one radical and square both sides.

$$\begin{aligned}
\sqrt{(x + c)^2 + y^2} &= 2a - \sqrt{(x - c)^2 + y^2} \\
(x + c)^2 + y^2 &= 4a^2 - 4a\sqrt{(x - c)^2 + y^2} + (x - c)^2 + y^2 \\
x^2 + 2xc + c^2 + y^2 &= 4a^2 - 4a\sqrt{(x - c)^2 + y^2} + x^2 - 2xc + c^2 + y^2 \\
4xc - 4a^2 &= -4a\sqrt{(x - c)^2 + y^2} \\
xc - a^2 &= -a\sqrt{(x - c)^2 + y^2}
\end{aligned}$$

Square both sides again.

$$\begin{aligned}
x^2c^2 - 2xca^2 + a^4 &= a^2[(x - c)^2 + y^2] \\
&= a^2[x^2 - 2xc + c^2 + y^2] \\
&= a^2x^2 - 2a^2xc + a^2c^2 + a^2y^2 \\
x^2(c^2 - a^2) - a^2y^2 &= -a^2(a^2 - c^2)
\end{aligned}$$

Multiply both sides by $-1$.

$$x^2(a^2 - c^2) + a^2y^2 = a^2(a^2 - c^2)$$

If point $P$ is at $(0, b)$, as in Figure 8, $PF_1 = PF_2 = a$ (since $PF_1 + PF_2 = 2a$ and $PF_1 = PF_2$). Since $PO = b$ and $F_2O = c$, by the Pythagorean theorem,

$$a^2 = b^2 + c^2 \qquad \text{so that} \qquad a^2 - c^2 = b^2.$$

Substitute $b^2$ for $a^2 - c^2$ in the equation above, then divide through by $a^2b^2$.

$$\begin{aligned}
x^2\mathbf{b^2} + a^2y^2 &= a^2\mathbf{b^2} \\
\frac{x^2b^2}{a^2b^2} + \frac{a^2y^2}{a^2b^2} &= \frac{a^2b^2}{a^2b^2} \\
\frac{x^2}{a^2} + \frac{y^2}{b^2} &= 1
\end{aligned}$$

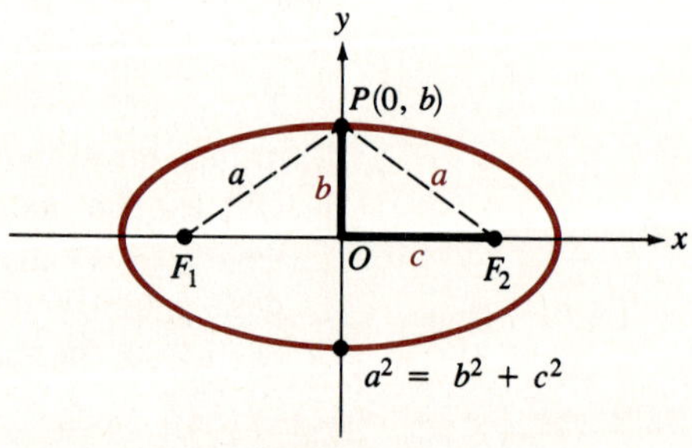


Figure 8

This equation is the **standard form** of the equation of an ellipse centered at the origin with $x$-intercepts $(a, 0)$ and $(-a, 0)$ and $y$-intercepts $(0, b)$ and $(0, -b)$. Remember that the foci are on the $x$-axis at $(-c, 0)$ and $(c, 0)$ and that $a > b$.

**Ellipse with Horizontal Major Axis**

The equation

$$\frac{x^2}{a^2} + \frac{y^2}{b^2} = 1,$$

where $a > b$, is an ellipse with graph shown in Figure 9. $a$, $b$, and $c$ are related by $b^2 = a^2 - c^2$.

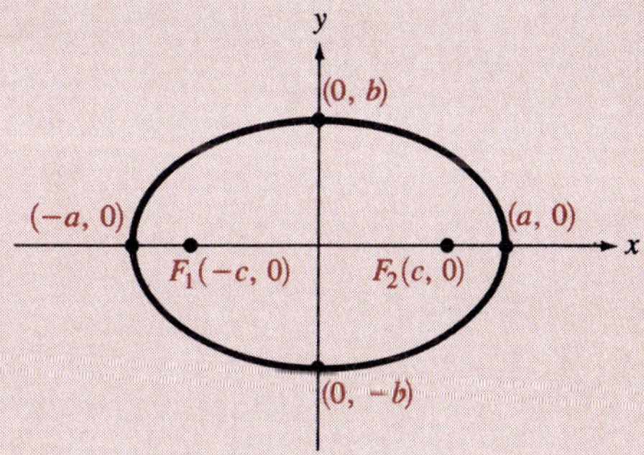


Figure 9

**EXAMPLE 1** Find the equation of the ellipse with foci at $(-\sqrt{5}, 0)$ and $(\sqrt{5}, 0)$ and $x$-intercepts $(-3, 0)$ and $(3, 0)$.

Since the foci are on the $x$ axis, we know $a = 3$. Thus,

$$b^2 = a^2 - c^2 = (3)^2 - (\sqrt{5})^2 = 4$$
$$b = 2.$$

Substitute $a = 3$ and $b = 2$ in the standard form.

$$\frac{x^2}{3^2} + \frac{y^2}{2^2} = 1$$
$$\frac{x^2}{9} + \frac{y^2}{4} = 1 \quad \blacksquare$$

An ellipse may have its major axis on the $y$-axis with endpoints $(0, -a)$ and $(0, a)$. The foci are at $(0, -c)$ and $(0, c)$; endpoints of the minor axis $(-b, 0)$ and $(b, 0)$.

**Ellipse with Vertical Major Axis**

The equation

$$\frac{x^2}{b^2} + \frac{y^2}{a^2} = 1,$$

where $a > b$, is an ellipse with graph shown in Figure 10. $a$, $b$, and $c$ are related by $b^2 = a^2 - c^2$.

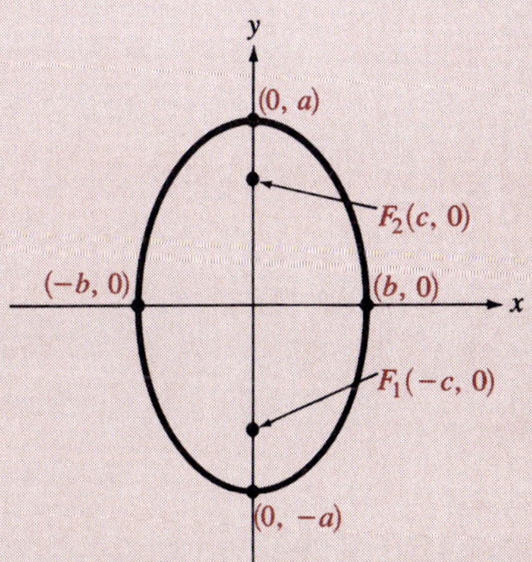


Figure 10

**EXAMPLE 2** Graph the equation

$$25x^2 + 16y^2 = 400.$$

Divide by 400.

$$\frac{25x^2}{400} + \frac{16y^2}{400} = 1$$

$$\frac{x^2}{16} + \frac{y^2}{25} = 1 \quad \text{Standard form}$$

Observe that since $25 > 16$ this is the standard form of an ellipse with vertical major axis. Also, $a^2 = 25$, or $a = 5$, and $b^2 = 16$ which means $b = 4$. The $x$-intercepts are $(-4, 0)$ and $(4, 0)$ while the $y$-intercepts are $(0, -5)$ and $(0, 5)$.

Another way to find the $x$-intercepts is to set $y = 0$ in the original equation and solve for $x$.

$$\frac{x^2}{16} + \frac{0^2}{25} = 1$$

$$\frac{x^2}{16} = 1$$

$$x^2 = 16$$

$$x = \pm 4$$

Thus, the $x$-intercepts are $(4, 0)$ and $(-4, 0)$. Similarly, by setting $x = 0$, we obtain the $y$-intercepts $(0, -5)$ and $(0, 5)$. The graph of this ellipse is given in Figure 11. ■

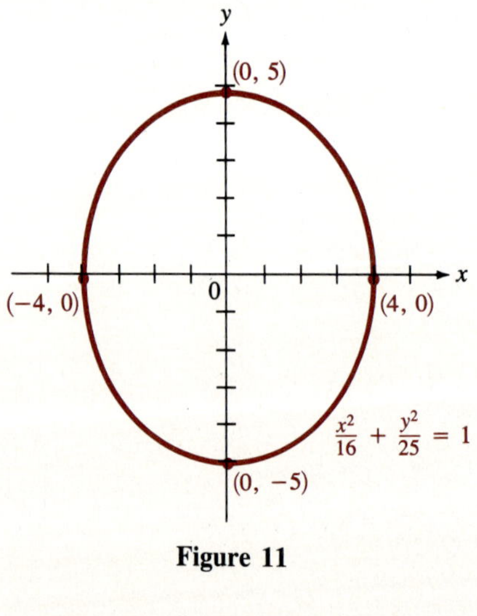


**Figure 11**

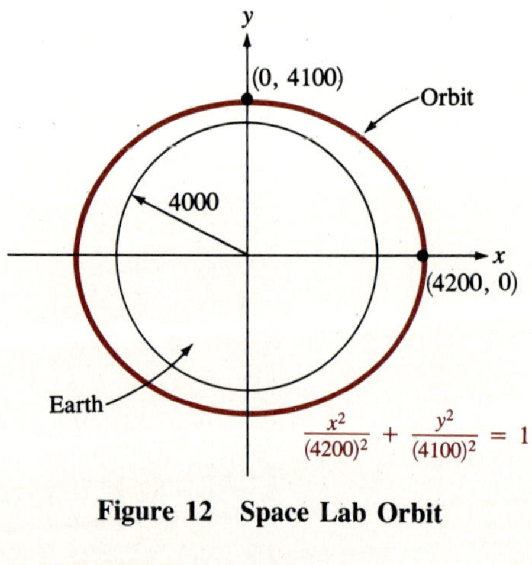


**Figure 12 Space Lab Orbit**

We now solve the first problem in the introduction to this chapter.

**EXAMPLE 3**
**AEROSPACE**

A communications company has determined that the most economical orbit for one of its space labs is an ellipse which has a maximum distance above the surface of the earth of 200 miles and a minimum distance of 100 miles. Use 4000 miles as the radius of the earth and determine the equation of the ellipse. Estimate the distance traveled by the space lab in one revolution.

The coordinate system in Figure 12 has been set to have the maximum of the orbit occur on the $x$-axis. The minimum is then on the $y$-axis. Thus, $a = 4200$ and $b = 4100$. The equation is

$$\frac{x^2}{(4200)^2} + \frac{y^2}{(4100)^2} = 1.$$

Notice that for this ellipse $a$ and $b$ are relatively the same size. If $a = b$, the equation would be a circle. With this information we have an easy way to estimate the distance traveled by the space lab in one revolution. Calculate the circumference of a circle with radius $r = \dfrac{a + b}{2}$.

$$C = 2\pi r = 2\pi\left(\frac{4200 + 4100}{2}\right)$$
$$\approx 26{,}100 \text{ mi}$$

Thus, the space lab travels approximately 26,100 mi in each revolution. ■

## Translation of Axes

To discuss an ellipse with center at any point $(h, k)$, we use *translation of axes*. Consider the ellipse in Figure 13 with center at $(h, k)$ in an $xy$-coordinate system.

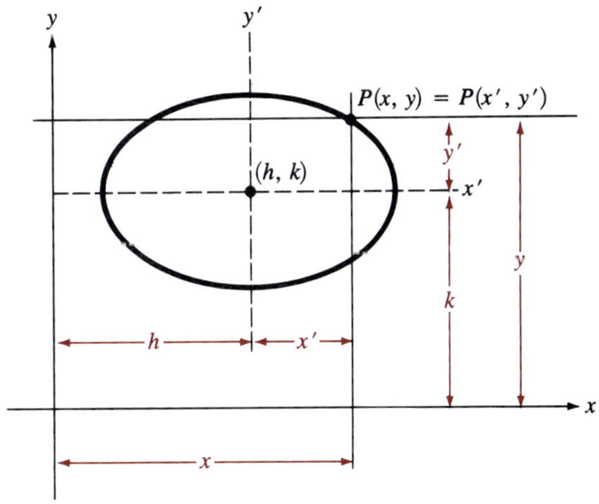


**Figure 13 Translation of Axes**

Now establish a new $x'y'$-coordinate system with origin at $(h, k)$ in the $xy$-coordinate system. Notice in Figure 13 that the coordinates of a point on the ellipse, $P(x, y) = P(x', y')$, satisfy

$$x = x' + h$$
$$y = y' + k$$

or

$$x' = x - h$$
$$y' = y - k.$$

These are transformation equations which hold when a **translation of axes** takes place. This means a new coordinate system is established in such a way that the $x'$-axis is parallel to the $x$-axis and the $y'$-axis is parallel to the $y$-axis.

The equation for the ellipse in Figure 13 is of the form

$$\frac{x'^2}{a^2} + \frac{y'^2}{b^2} = 1$$

in the $x'y'$-coordinate system. Using $x' = x - h$ and $y' = y - k$, we have in the $xy$-coordinate system,

$$\frac{(x-h)^2}{a^2} + \frac{(y-k)^2}{b^2} = 1 \qquad a > b$$

as the standard form of the equation of the ellipse with center at $(h, k)$ and horizontal major axis. If the major axis is vertical, the equation is

$$\frac{(x-h)^2}{b^2} + \frac{(y-k)^2}{a^2} = 1. \qquad a > b$$

**EXAMPLE 4**

Find the standard form and graph the ellipse that has $b = 4$ and foci at $(-1, 3)$ and $(5, 3)$.

The center of the ellipse is the midpoint of the line segment joining the foci.

$$h = \frac{-1+5}{2} = 2 \quad \text{and} \quad k = \frac{3+3}{2} = 3$$

Thus $(h, k) = (2, 3)$. Also $2c = 5 - (-1) = 6$, so $c = 3$.

$$a^2 = b^2 + c^2 = (4)^2 + (3)^2 = 25$$

Therefore, $a = 5$ and the standard form of the ellipse is

$$\frac{(x-2)^2}{25} + \frac{(y-3)^2}{16} = 1.$$

The graph intersects the line $y = 3$ at points $a = 5$ units on each side of the center $(2, 3)$, and the line $x = 2$ at points $b = 4$ units above and below the center $(2, 3)$. See Figure 14. ■

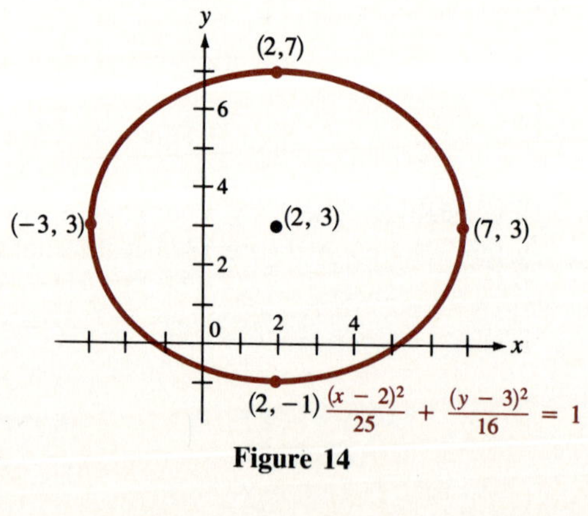


**Figure 14**

If the parentheses are cleared in the standard form, we obtain the **general form of the equation of an ellipse.**

$$Ax^2 + Cy^2 + Dx + Ey + F = 0 \quad A \neq C,\ A \text{ and } C \text{ same signs}$$

For the equation to represent an ellipse, $A$ and $C$ must have the same signs and must not be equal. In order to convert from general form to standard form, proceed as with the circle and complete the squares.

**EXAMPLE 5** Write $36x^2 + 64y^2 + 108x - 128y - 431 = 0$ in standard form.

$$36x^2 + 108x \quad + 64y^2 - 128y \quad = 431 \qquad \text{Rewrite}$$

$$36(x^2 + 3x \quad) + 64(y^2 - 2y \quad) = 431 \qquad \text{Factor out coefficients of squared terms}$$

$$36\left(x^2 + 3x + \frac{9}{4}\right) + 64(y^2 - 2y + 1) = 431 + 36\left(\frac{9}{4}\right) + 64(1) \qquad \text{Complete the squares}$$

Do not just add 9/4 and 1 to the right side since they are multiplied on the left side by 36 and 64, respectively.

$$36\left(x + \frac{3}{2}\right)^2 + 64(y - 1)^2 = 576$$

$$\frac{36\left(x + \frac{3}{2}\right)^2}{576} + \frac{64(y - 1)^2}{576} = \frac{576}{576} \qquad \text{Divide by 576}$$

$$\frac{\left(x + \frac{3}{2}\right)^2}{16} + \frac{(y - 1)^2}{9} = 1$$

The equation is an ellipse with center at $\left(-\frac{3}{2}, 1\right)$, $a = 4$, and $b = 3$. ■

## 9.2 EXERCISES

In Exercises 1–10 find the equation of each ellipse with center at the origin.

1. Intercepts $(\pm 7, 0)$ and $(0, \pm 6)$
2. Intercepts $(\pm 6, 0)$ and $(0, \pm 7)$
3. Foci $(\pm 3, 0)$ and $x$-intercepts $(\pm 5, 0)$
4. Foci $(\pm\sqrt{7}, 0)$ and $y$-intercepts $(0, \pm 4)$
5. Foci $(0, \pm 3)$ and $x$-intercepts $(\pm 4, 0)$
6. Foci $(0, \pm\sqrt{7})$ and $y$-intercepts $(0, \pm\sqrt{11})$
7. Length of major axis 12 and $y$-intercepts $(0, \pm 5)$
8. Length of minor axis 6 and foci $(0, \pm 6)$
9. Major axis intercepts $(\pm 10, 0)$ and passing through $(5, \sqrt{3})$ [Hint: Use the equation in standard form to find $b$.]
10. Minor axis intercepts $(\pm 2, 0)$ and passing through $(\sqrt{2}, 2\sqrt{2})$

In Exercises 11–16 find the equation of each ellipse.

**11.** Center at $(5, -2)$, $a = 4$, and foci $(3, -2)$ and $(7, -2)$

**12.** Center at $(-2, 5)$, $b = 4$, and foci $(-2, 3)$ and $(-2, 7)$

**13.** Foci $(\pm 4, 3)$ and $y$-intercepts $(0, 0)$ and $(0, 6)$

**14.** Foci $(5, \pm 3)$ and $x$-intercepts $(4, 0)$ and $(6, 0)$

**15.** Foci $(7, -1)$ and $(7, -7)$ and $a = 8$

**16.** Foci $(-2, -3)$ and $(4, -3)$ and $b = 7$

Graph each ellipse in Exercises 17–24.

**17.** $\frac{x^2}{49} + \frac{y^2}{25} = 1$

**18.** $\frac{x^2}{4} + \frac{y^2}{25} = 1$

**19.** $\frac{(x-2)^2}{16} + \frac{(y+2)^2}{4} = 1$

**20.** $\frac{(x+3)^2}{49} + \frac{(y-1)^2}{16} = 1$

**21.** $4x^2 + 25y^2 - 24x + 50y - 39 = 0$

**22.** $4x^2 + 9y^2 - 16x + 36y + 16 = 0$

**23.** $16x^2 + 4y^2 + 32x - 20y + 5 = 0$

**24.** $36x^2 + 25y^2 - 36x - 50y - 191 = 0$

In Exercises 25–30 write each equation in the form $\frac{(x-h)^2}{m} + \frac{(y-k)^2}{n} = 1$.

**25.** $8x^2 + 7y^2 - 56 = 0$

**26.** $10x^2 + 12y^2 - 60 = 0$

**27.** $9x^2 + 50y^2 + 54x - 300y + 306 = 0$

**28.** $16x^2 + 21y^2 - 128x + 210y + 753 = 0$

**29.** $4x^2 + 9y^2 + 16x + 88 = 0$

**30.** $100x^2 + 4y^2 - 300x + 16y + 241 = 0$

Solve.

**31.** ENGINEERING An elliptical riding path is to be built on a rectangular piece of property that measures 3 mi by 6 mi. Find an equation for the ellipse if the path is to touch the center of the property line on all four sides.

**32.** AEROSPACE A satellite is to be put into an elliptical orbit around one of the moons of a planet. The moon is a sphere with radius of 900 km. Determine an equation for the ellipse if the distance of the satellite from the surface of the moon varies from 200 km to 300 km. Estimate the distance traveled by the satellite in one revolution.

**33.** ENGINEERING A railroad tunnel is in the shape of a semiellipse with horizontal major axis. The height of the tunnel at the center is 30 ft and the vertical clearance must be 25 ft at a point 20 ft from the center. Find an equation for the ellipse.

**34.** ENGINEERING A bridge over a canal is a semiellipse. The equation of the ellipse is $\frac{x^2}{16} + \frac{y^2}{9} = 1$ for a coordinate system centered at the middle of the canal and horizontal $x$-axis across the canal. Solve the equation for $y$ and determine the appropriate vertical clearance for boats 3.0 m from the edge of the canal.

**35.** CARPENTRY A rectangular board is $m$ by $n$ units where $m > n$. To construct the largest elliptical tabletop the foci of the ellipse must be determined. How far from the short side of the board will the foci be located?

**36.** A string is connected to the foci in Exercise 35 and pulled taut by a pencil in order to draw the ellipse. What length string must be used?

## For Review

Find the equation of each circle described in Exercises 37–38.

**37.** Center at $(2, -5)$, passing through $(4, -1)$

**38.** Endpoints of a diameter at $(-3, 2)$ and $(6, 1)$

In Exercises 39–40 write the equation in standard form.

**39.** $x^2 + y^2 + 4x - 16y + 59 = 0$

**40.** $36x^2 + 36y^2 - 36x + 24y - 131 = 0$

# 9.3 The Hyperbola

The definition of the hyperbola is similar to the definition of the ellipse given in the previous section. The one difference is that the hyperbola is a *difference* of distances instead of a *sum*.

### The Hyperbola

Let $F_1$ and $F_2$ be two fixed points in a plane. A **hyperbola** with foci $F_1$ and $F_2$ is the collection of all points in the plane the difference of whose distances from $F_1$ and $F_2$ is a constant. The midpoint of the line segment joining $F_1$ and $F_2$ is called the **center** of the hyperbola.

## Equation of a Hyperbola

The equation of a hyperbola is developed by considering the graph in Figure 15.

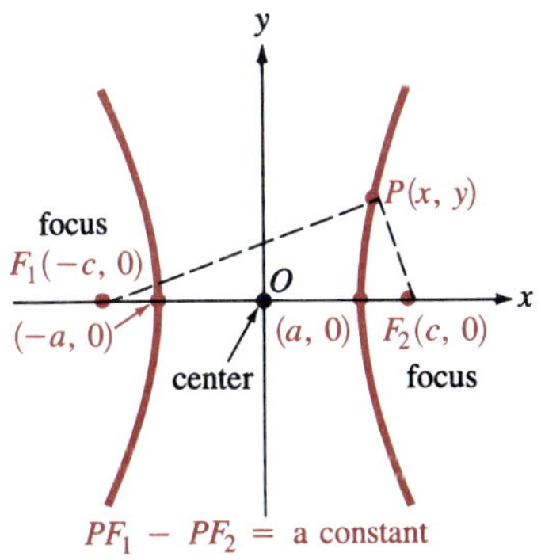


**Figure 15 Hyperbola**

It can be shown that the difference in the distances is the constant $2a$, and thus

$$PF_1 - PF_2 = 2a.$$

If a point on the other side or **branch** of the hyperbola is chosen, we write

$$PF_2 - PF_1 = 2a.$$

In either case it is true that

$$|PF_1 - PF_2| = 2a.$$

Using this relation and techniques similar to those employed for the ellipse, we can derive the standard form of the equation of a hyperbola. Note that $a < c$ for the hyperbola, and we define $b$ using

$$b^2 = c^2 - a^2.$$

The points $(a, 0)$ and $(-a, 0)$ are called **vertices** (plural of **vertex**) and the line segment joining the vertices is called the **transverse axis** of the hyperbola. The line through $(0, -b)$ and $(0, b)$ is the **conjugate axis.**

### Hyperbola with Horizontal Transverse Axis

The equation

$$\frac{x^2}{a^2} - \frac{y^2}{b^2} = 1$$

has as its graph a hyperbola with center at $(0, 0)$, foci at $(\pm c, 0)$, and vertices at $(\pm a, 0)$. $a$, $b$, and $c$ are related by $b^2 = c^2 - a^2$.

**EXAMPLE 1** Find the equation of a hyperbola with foci $(\pm 5, 0)$ and $b = 2$.

Since $c = 5$ and $b = 5$ we can determine $a$.

$$a^2 = c^2 - b^2$$
$$a^2 = 5^2 - 2^2 = 21$$

Thus, the vertices are $(\sqrt{21}, 0)$ and $(-\sqrt{21}, 0)$ and the equation is

$$\frac{x^2}{21} - \frac{y^2}{4} = 1. \quad ■$$

Graphing a hyperbola requires a different approach from that for the ellipse. Notice that the hyperbola has only two intercepts. Setting $y = 0$, we show that the vertices are the $x$-intercepts.

$$\frac{x^2}{a^2} - \frac{0}{b^2} = 1$$
$$x = \pm a$$

However, if we set $x = 0$, the equation $-\dfrac{y^2}{b^2} = 1$ has no solution, and there are no $y$-intercepts.

**Asymptotes** To gain more information for graphing, solve

$$\frac{x^2}{a^2} - \frac{y^2}{b^2} = 1$$

for $y$.

$$\frac{y^2}{b^2} = \frac{x^2}{a^2} - 1 = \frac{x^2 - a^2}{a^2} \qquad \text{Common denominator is } a^2$$

$$y^2 = \frac{b^2}{a^2}(x^2 - a^2) \qquad \text{Multiply by } b^2$$

$$y = \pm\frac{b}{a}\sqrt{x^2 - a^2} \qquad \text{Take square root of both sides}$$

For $|x|$ very large $x^2 - a^2 \approx x^2$ and the hyperbola will be close to the lines

$$y = \pm\frac{b}{a}x.$$

Since the hyperbola approaches the lines $y = \frac{b}{a}x$ and $y = -\frac{b}{a}x$ as $|x|$ becomes large, we call these lines **asymptotes.**

To graph a hyperbola, draw the asymptotes as shown in Figure 16. The best procedure is to construct a rectangle with sides parallel to the axes and passing through the points $(a, 0)$, $(-a, 0)$, $(0, b)$, and $(0, -b)$. The asymptotes will then pass through the corners of this rectangle. Now plot the vertices $(a, 0)$ and $(-a, 0)$. The hyperbola passes through the vertices and approaches the asymptotes.

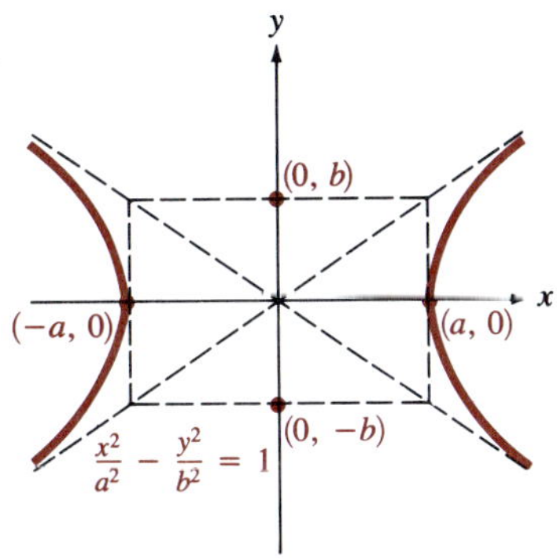


**Figure 16 Asymptotes**

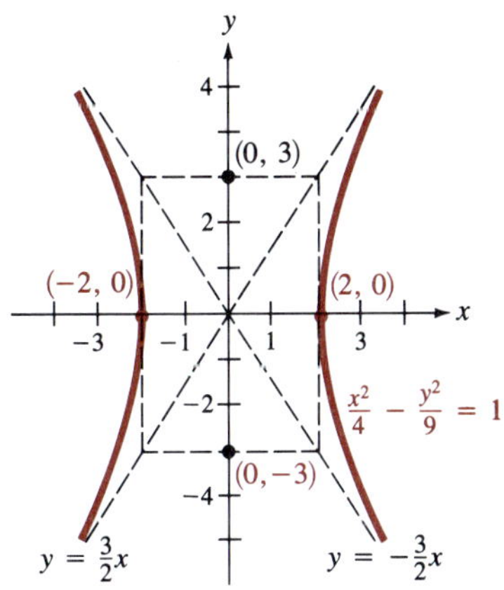


**Figure 17**

**EXAMPLE 2** Graph the hyperbola with equation $\frac{x^2}{4} - \frac{y^2}{9} = 1$.

Since the equation can be written $\frac{x^2}{2^2} - \frac{y^2}{3^2} = 1$, $a = 2$ and $b = 3$. (Note that $a$ need not be greater than $b$ as was the case with the ellipse.) Now construct the rectangle with sides through the points $(2, 0)$, $(-2, 0)$, $(0, 3)$, and $(0, -3)$. Draw the asymptotes through the corners of the rectangle. The hyperbola passes through the vertices $(2, 0)$ and $(-2, 0)$ and approaches the asymptotes as in Figure 17. ■

Only hyperbolas with transverse axis along the $x$-axis and conjugate axis along the $y$-axis have been discussed in the preceding paragraphs. If the $y^2$-term is positive and the $x^2$-term negative in the standard form, the hyperbola will have transverse axis along the $y$-axis. The standard form of the equation of these hyperbolas is given in the next theorem.

**Hyperbola with Vertical Transverse Axis**

The equation

$$\frac{y^2}{a^2} - \frac{x^2}{b^2} = 1$$

has as its graph a hyperbola with center at (0, 0), foci at $(0, \pm c)$, and vertices at $(0, \pm a)$. $a$, $b$, and $c$ are related by $b^2 = c^2 - a^2$.

The asymptotes for these hyperbolas are $y = \pm \frac{a}{b}x$, but the procedure for constructing them is the same. See Figure 18.

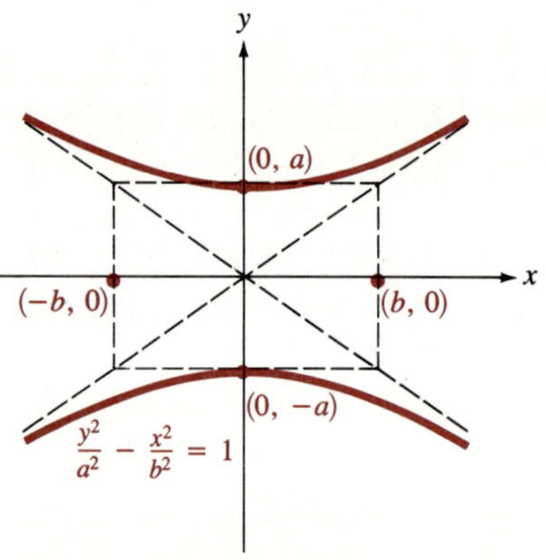


**Figure 18**

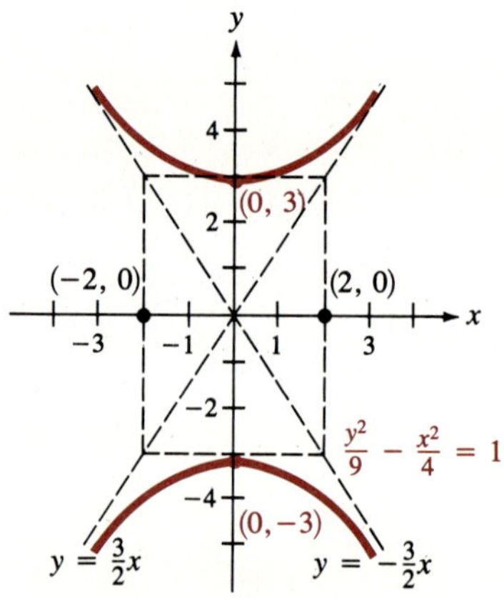


**Figure 19**

**EXAMPLE 3** Graph the hyperbola with foci at $(0, \pm\sqrt{13})$ and vertices $(0, \pm 3)$.

Knowing $c = \sqrt{13}$ and $a = 3$, find $b$.

$$b^2 = c^2 - a^2$$
$$= (\sqrt{13})^2 - (3)^2 = 13 - 9 = 4$$
$$b = 2$$

The equation of the hyperbola is

$$\frac{y^2}{9} - \frac{x^2}{4} = 1.$$

Construct a rectangle through (2, 0), (−2, 0), (0, 3), and (0, −3) and draw the asymptotes through the corners, (2, 3), (−2, 3), (−2, −3), and (2, −3). The hyperbola, shown in Figure 19, passes through vertices (0, 3) and (0, −3) and approaches the asymptotes. ■

If the axes are translated, we can discuss hyperbolas centered at points other than the origin. The standard form of a hyperbola with center $(h, k)$ and horizontal transverse axis is

$$\frac{(x-h)^2}{a^2} - \frac{(y-k)^2}{b^2} = 1.$$

If the hyperbola has a vertical transverse axis, the standard form is

$$\frac{(y-k)^2}{a^2} - \frac{(x-h)^2}{b^2} = 1.$$

**EXAMPLE 4** Graph the hyperbola

$$\frac{(x+1)^2}{9} - \frac{(y-2)^2}{16} = 1.$$

Since $$\frac{(x-(-1))^2}{3^2} - \frac{(y-2)^2}{4^2} = 1,$$

$(h, k) = (-1, 2)$, $a = 3$, and $b = 4$. First locate the center, $(-1, 2)$, and then sketch the lines $x = -1$ and $y = 2$. Think of these lines as new coordinate axes, with the given hyperbola centered at the origin of this system. We use $a = 3$ and $b = 4$, relative to these new axes, to sketch the rectangle and the asymptotes. Since the hyperbola has a horizontal transverse axis, we know it passes through the points 3 units right and left of $(-1, 2)$, that is, through (2, 2) and (−4, 2). Using this information we can sketch the graph in Figure 20. ■

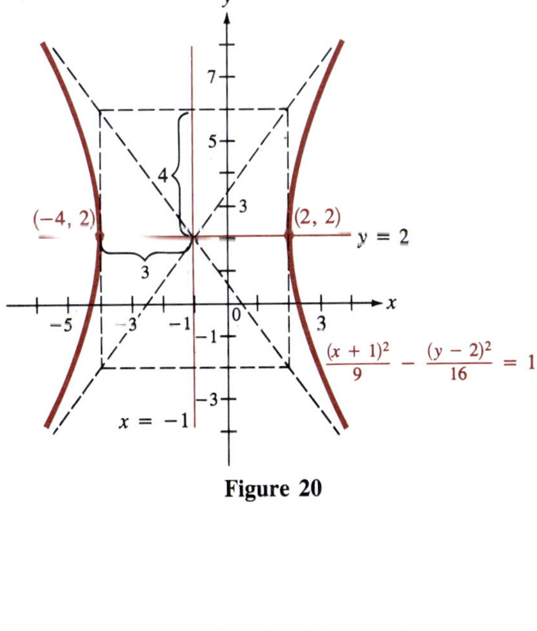


**Figure 20**

**NOTE** If the equation in Example 4 is expanded, the result is the general form of the hyperbola.

$$16x^2 - 9y^2 + 32x + 36y - 164 = 0$$

All of the hyperbolas discussed in this section can be written in the general form

$$Ax^2 + Cy^2 + Dx + Ey + F = 0,$$

where $A$ and $C$ have opposite signs. If the general form is given, complete the squares to find the standard form. ■

**EXAMPLE 5** Write the equation $25x^2 - 9y^2 - 100x - 54y - 206 = 0$ in standard form.

$$\begin{aligned} 25x^2 - 100x \quad - 9y^2 - 54y \quad &= 206 \\ 25(x^2 - 4x \quad ) - 9(y^2 + 6y \quad ) &= 206 \\ 25(x^2 - 4x + 4) - 9(y^2 + 6y + 9) &= 206 + 100 - 81 \\ 25(x - 2)^2 - 9(y + 3)^2 &= 225 \\ \frac{(x - 2)^2}{9} - \frac{(y + 3)^2}{25} &= 1 \\ \frac{(x - 2)^2}{3^2} - \frac{(y + 3)^2}{5^2} &= 1 \end{aligned}$$

■

Considering the most general quadratic equation

$$Ax^2 + Bxy + Cy^2 + Dx + Ey + F = 0$$

leads to another type of equation with a hyperbola as its graph. Example 6 discusses the case when all the constants are zero except $B$ and $F$.

**EXAMPLE 6** Graph $2xy - 6 = 0$.

Simplifying we have $xy = 3$, which is graphed by plotting points in Figure 21. The graph is a hyperbola with the coordinate axes as asymptotes. Observe that the graph appears to be "rotated" with respect to the graphs we have studied previously. ■

| $x$ | $y$ |
|---|---|
| 1 | 3 |
| 2 | 3/2 |
| 3 | 1 |
| −1 | −3 |
| −2 | −3/2 |
| −3 | −1 |

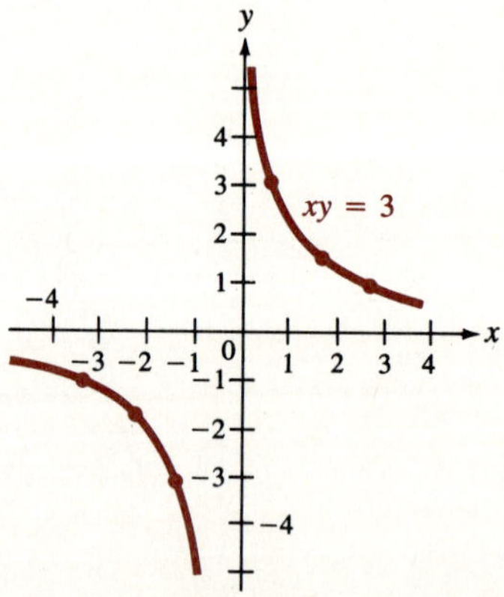


**Figure 21**

## 9.3 EXERCISES

Find the standard form of the equation of each hyperbola in Exercises 1–12.

1. Foci at $(\pm 5, 0)$ and vertices at $(\pm 3, 0)$
2. Foci at $(0, \pm 5)$ and vertices at $(0, \pm 3)$
3. $a = b = 3$, transverse axis along the $y$-axis, and center at origin
4. $a = b = 5$, transverse axis along the $x$-axis, and center at origin
5. Foci $(\pm 7, 0)$ and length of the transverse axis is 10
6. Foci $(0, \pm 8)$ and length of the conjugate axis is 4
7. Vertices $(0, \pm 6)$ and asymptotes $y = \pm \frac{3}{4}x$
8. Vertices $(\pm 5, 0)$ and asymptotes $y = \pm x$
9. Center at $(-3, 2)$, foci $(-3, 8)$ and $(-3, -4)$, and vertices $(-3, 6)$ and $(-3, -2)$
10. Center at $(5, 3)$, foci $(0, 3)$ and $(10, 3)$, and length of the transverse axis is 6
11. Vertices $(\pm 4, 0)$ and passing through $(8, 6)$
12. Vertices $(0, \pm 2)$ and passing through $(6, 4)$

Sketch the graph of each hyperbola in Exercises 13–20.

13. $\dfrac{x^2}{25} - \dfrac{y^2}{4} = 1$
14. $\dfrac{y^2}{4} - \dfrac{x^2}{25} = 1$
15. $\dfrac{(x + 3)^2}{9} - \dfrac{(y + 1)^2}{16} = 1$
16. $\dfrac{(y + 1)^2}{16} - \dfrac{(x + 3)^2}{9} = 1$
17. $x^2 - y^2 - 4x + 2y + 12 = 0$
18. $4x^2 - y^2 + 32x + 2y + 59 = 0$
19. $xy = -2$
20. $5xy - 10 = 0$

In Exercises 21–26, write each equation in the form $\dfrac{(x - h)^2}{m} - \dfrac{(y - k)^2}{n} = 1$ or $\dfrac{(y - k)^2}{m} - \dfrac{(x - h)^2}{n} = 1$.

21. $x^2 - y^2 + 25 = 0$
22. $2y^2 - 3x^2 + 6 = 0$
23. $16y^2 - 9x^2 - 192y - 54x + 351 = 0$
24. $18x^2 - 7y^2 + 180x + 28y + 359 = 0$
25. $4x^2 - y^2 + 32x + 2y + 63 = 0$
26. $25x^2 - 9y^2 - 100x - 54y + 19 = 0$

Solve.

27. **ARCHITECTURE** The roof of a building is in the shape of the hyperbola $y^2 - x^2 = 25$, where $x$ and $y$ are in meters. Refer to the figure and determine the height $h$ of the outside walls.

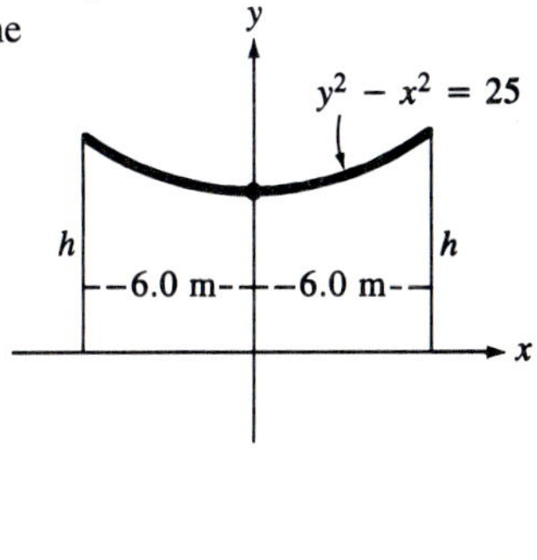

**28.** ASTRONOMY A comet follows the hyperbolic path described by the equation $\frac{x^2}{9} - \frac{y^2}{16} = 1$, where $x$ and $y$ are in millions of miles. If, as shown in the figure, the sun is the focus of this path, how close to the sun is the vertex of the comet's path?

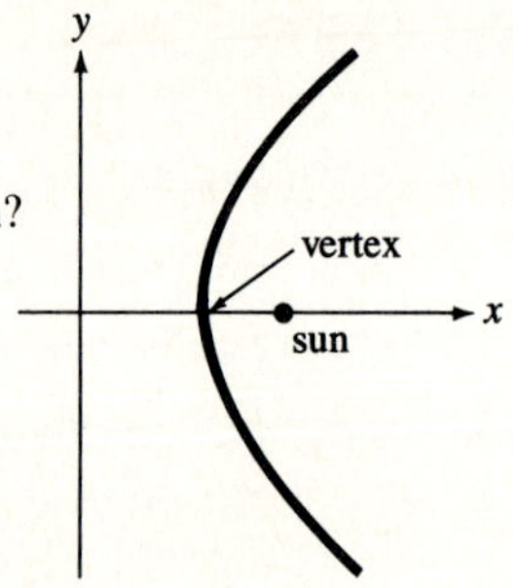


**29.** Determine the coefficients of the general quadratic equation $Ax^2 + Bxy + Cy^2 + Dx + Ey + F = 0$ when the standard form of the equation is $\frac{(x-h)^2}{a^2} - \frac{(y-k)^2}{b^2} = 1$.

**30.** Determine the equations of the intersecting lines that result from the degenerate hyperbola $\frac{(y-k)^2}{a^2} - \frac{(x-h)^2}{b^2} = 0$.

## For Review

Determine the standard form of the equation of each ellipse in Exercises 31–32.

**31.** Foci $(0, \pm\sqrt{11})$ and length of minor axis 4

**32.** Foci $(\pm 6, 2)$ and $y$-intercepts $(0, -1)$ and $(0, 5)$

In Exercises 33–34 write each equation in the form $\frac{(x-h)^2}{m} + \frac{(y-k)^2}{n} = 1$.

**33.** $8x^2 + 4y^2 - 24y + 4 = 0$

**34.** $15x^2 + 7y^2 - 60x + 56y + 67 = 0$

# 9.4 The Parabola

In Section 3.6 parabolas were discussed as graphs of quadratic functions. We now give a geometric definition of the parabola.

### The Parabola

Let $F$ be a fixed point and $L$ a fixed line in the plane. A **parabola** with **focus** $F$ and **directrix** $L$ is the set of all points in the plane equidistant from $F$ and $L$.

## Equation of a Parabola

The line through the focus of a parabola perpendicular to the directrix is called the **axis of symmetry,** and the point of intersection of the parabola and the axis of symmetry is called the **vertex.** Figure 22 will be used as an aid in developing an equation for the parabola. Observe that this parabola opens to the right and has vertex at the origin (0, 0), focus $(p, 0)$ on the positive $x$-axis, and directrix the line $x = -p$.

To derive the equation for the above parabola, first note that

$$PF = \sqrt{(x-p)^2 + (y-0)^2} = \sqrt{(x-p)^2 + y^2} \quad \text{and} \quad PQ = x + p.$$

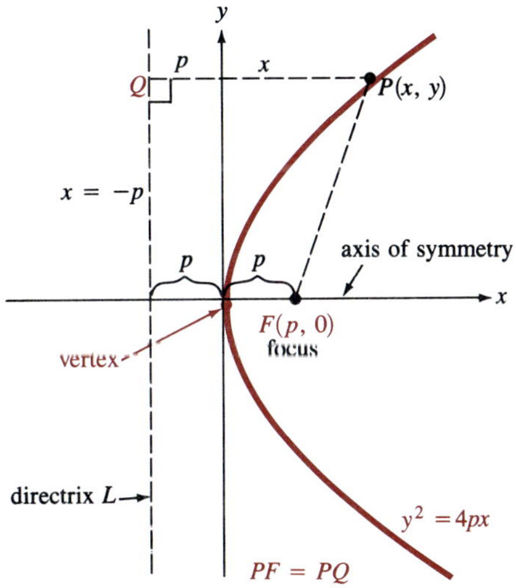


**Figure 22 Parabola**

By definition of a parabola, $PF = PQ$.

$$\sqrt{(x-p)^2 + y^2} = x + p$$

$$(x-p)^2 + y^2 = (x+p)^2 \qquad \text{Squaring both sides}$$

$$x^2 - 2px + p^2 + y^2 = x^2 + 2px + p^2$$

$$y^2 = 4px$$

In a similar manner, it can be shown that the three parabolas in Figure 23 have the given standard-form equations.

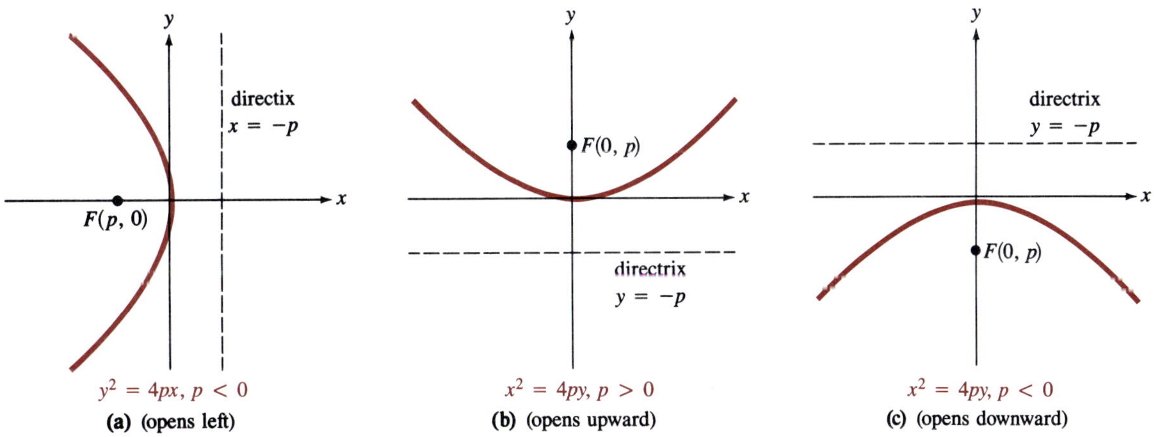


**Figure 23 Parabolas**

### Standard Form of Parabola with Vertex at Origin

A parabola with vertex at the origin and axis of symmetry one of the coordinate axes has one of the following standard-form equations.

1. $y^2 = 4px$ Focus is $(p, 0)$ and directrix is $x = -p$. Opens right if $p > 0$ and left if $p < 0$.
2. $x^2 = 4py$ Focus is $(0, p)$ and directrix is $y = -p$. Opens up if $p > 0$ and down if $p < 0$.

**EXAMPLE 1**

Give the standard form of the equation of a parabola with vertex at the origin which opens upward and has focus at (0, 2).

By the theorem the equation to use is $x^2 = 4py$. This is the case shown in Figure 23(**b**). Since the focus is (0, 2), $p = 2$. Thus,

$$x^2 = 4(2)y = 8y.$$

The desired equation is $x^2 = 8y$. ■

The vertex, focus, and directrix can also be determined when the equation of a parabola is given.

**EXAMPLE 2**

Determine the vertex, focus, and directrix, and graph the parabola with equation $y^2 = -6x$.

First write the equation in the form $y^2 = 4px$.

$$4p = -6$$

$$p = -\frac{3}{2}$$

Thus the equation is $y^2 = 4\left(-\frac{3}{2}\right)x$. From the theorem the vertex is (0, 0), the focus is $\left(-\frac{3}{2}, 0\right)$, and the directrix is $x = -\left(-\frac{3}{2}\right) = \frac{3}{2}$. To graph the parabola in Figure 24 we have plotted two additional points, $\left(-\frac{3}{2}, 3\right)$ and $\left(-\frac{3}{2}, -3\right)$. ■

**Vertex at (*h*, *k*)**

To study parabolas with vertex at a point $(h, k)$, translate the axes. The resulting equations are given in the next theorem.

### Standard Form of Parabola with Vertex at (*h*, *k*)

A parabola with vertex at $(h, k)$ and axis of symmetry parallel to one of the coordinate axes has one of the following standard-form equations.

1. $(y - k)^2 = 4p(x - h)$ Focus is $(h + p, k)$ and directrix is $x = h - p$. Opens right if $p > 0$ and left if $p < 0$.
2. $(x - h)^2 = 4p(y - k)$ Focus is $(h, k + p)$ and directrix is $y = k - p$. Opens up if $p > 0$ and down if $p < 0$.

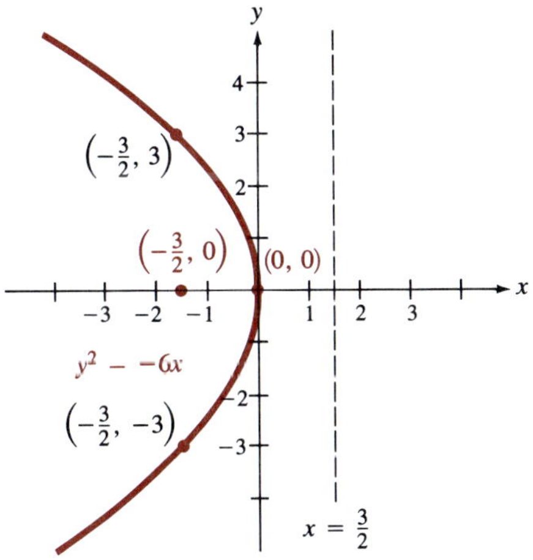


**Figure 24**

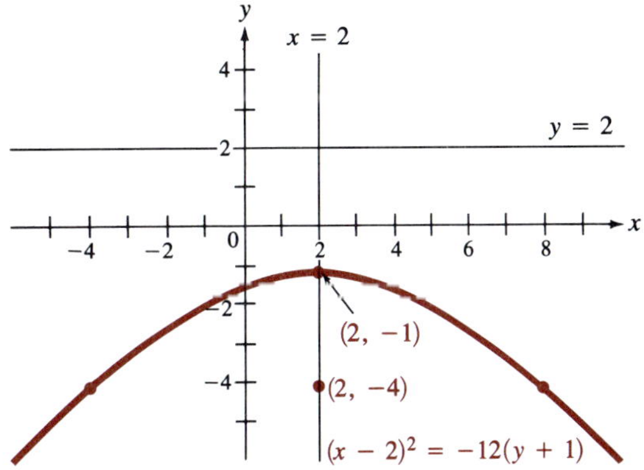


**Figure 25**

**EXAMPLE 3** Determine the vertex, focus, and directrix, and graph the parabola with equation $(x - 2)^2 = -12(y + 1)$.

Write the equation as $(x - 2)^2 = -12[y - (-1)]$ to show that $h = 2$ and $k = -1$. Thus the vertex is $(2, -1)$. Since

$$4p = -12$$
$$p = -3,$$

the focus is $(h, k + p) = (2, -1 - 3) = (2, -4)$. That is, the focus is 3 units below the vertex. The directrix is the horizontal line $y = k - p = -1 - (-3) = 2$. To sketch the graph plot the additional points $(-4, -4)$ and $(8, -4)$. See Figure 25. ■

**EXAMPLE 4** Determine the equation of a parabola with vertex $(2, -3)$, axis of symmetry $y = -3$, and passing through the point $(4, -1)$.

Since the axis of symmetry $y = -3$ is parallel to the $x$-axis, the equation of the parabola has the form $(y - k)^2 = 4p(x - h)$. Since the vertex $(h, k)$ is $(2, -3)$, the equation becomes

$$[y - (-3)]^2 = 4p(x - 2) \qquad h = 2 \text{ and } k = -3$$
$$(y + 3)^2 = 4p(x - 2).$$

Now the only unknown is $p$. Use the fact that the parabola passes through $(4, -1)$ to find $p$.

$$(y + 3)^2 = 4p(x - 2)$$
$$(-1 + 3)^2 = 4p(4 - 2) \qquad x = 4 \text{ and } y = -1$$
$$4 = 8p$$
$$\frac{1}{2} = p$$

Thus, the equation is completely determined.

$$(y + 3)^2 = 4\left(\frac{1}{2}\right)(x - 2) \qquad p = \frac{1}{2}$$

$$(y + 3)^2 = 2(x - 2) \quad \blacksquare$$

We now solve the second applied problem from the beginning of this chapter.

**EXAMPLE 5**

**ARCHITECTURE**

A domed ceiling is a parabolic surface. For the best lighting on the floor, a light source is to be placed at the focus of the surface. If 10.0 m down from the top of the dome the ceiling is 15.0 m wide, find the best location for the light source.

Think of a plane cutting through the dome to form the parabola in Figure 26. The vertex of the parabola is at (0, 0) and the parabola opens downward. Thus, the equation is $x^2 = 4py$. Since when $y$ is $-10.0$ m the width is 15.0 m, the points $(-7.5, -10)$ and $(7.5, -10)$ are on the parabola. Use $(7.5, -10)$ in the equation to find $p$.

$$x^2 = 4py$$
$$(7.5)^2 = 4p(-10)$$
$$p \approx -1.4$$

Thus, the light should be placed 1.4 m down from the top of the dome. ■

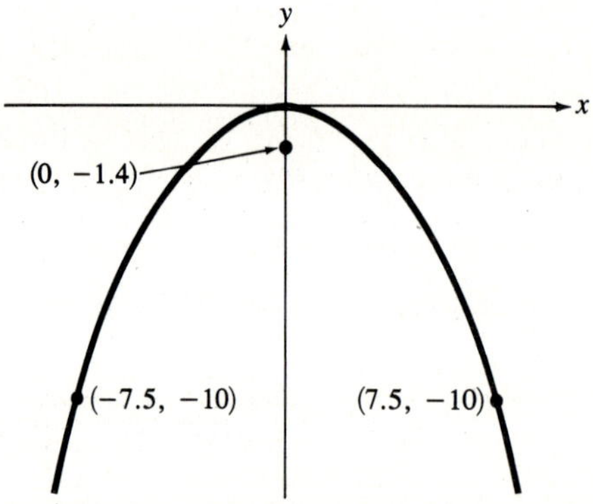


Figure 26

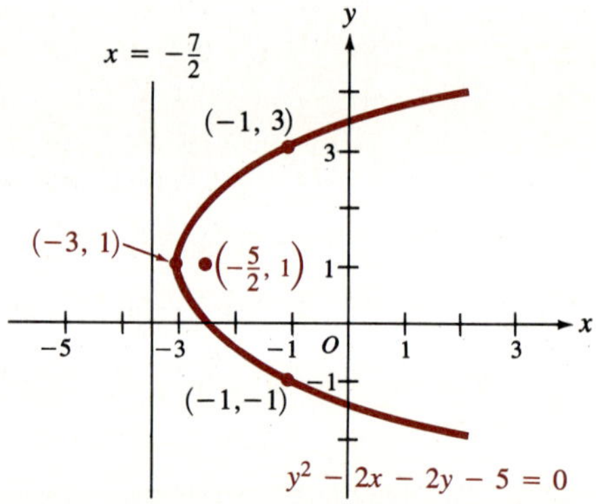


Figure 27

If the equation of a parabola is expanded, the general form of the equation results. For example,

$$(y + 1)^2 = 5(x - 3)$$
$$y^2 + 2y + 1 = 5x - 15$$
$$y^2 - 5x + 2y + 16 = 0.$$

In general form a parabola can be written as

$$Ax^2 + Dx + Ey + F = 0 \quad (A, E \neq 0) \quad \text{or} \quad Cy^2 + Dx + Ey + F = 0 \quad (C, D \neq 0).$$

To convert from general form to standard form complete the square on the squared variable.

**EXAMPLE 6** Write the equation $y^2 - 2x - 2y - 5 = 0$ in standard form and give the vertex, focus, and directrix. Graph the equation.

Complete the square on $y$.

$$y^2 - 2y = 2x + 5 \quad \text{Isolate the } y\text{-terms}$$
$$y^2 - 2y + 1 = 2x + 5 + 1 \quad \text{Add 1 to both sides}$$
$$(y - 1)^2 = 2x + 6$$
$$(y - 1)^2 = 2(x + 3)$$

The parabola opens to the right and the vertex is $(-3, 1)$. Since $4p = 2$, $p = \frac{1}{2}$. Thus, the focus is $\frac{1}{2}$ unit to the right of $(-3, 1)$. The focus is

$$(h + p, k) = \left(-3 + \frac{1}{2}, 1\right) = \left(-\frac{5}{2}, 1\right).$$

The directrix is the vertical line $x = h - p = -3 - \frac{1}{2} = -\frac{7}{2}$. Plot the additional points $(-1, -1)$ and $(-1, 3)$ to graph the parabola. See Figure 27. ■

## 9.4 EXERCISES

In Exercises 1–12 determine the equation of each parabola.

**1.** Vertex at the origin and focus $\left(0, \frac{3}{2}\right)$

**2.** Vertex at $(0, 0)$ and focus $(-3, 0)$

**3.** Vertex at $(0, 0)$ and focus $\left(\frac{7}{2}, 0\right)$

**4.** Vertex at the origin and focus $(0, -1)$

**5.** Directrix $x = -3$ and focus $(3, 0)$

**6.** Directrix $y = 5$ and focus $(0, -5)$

**7.** Directrix $y = -1$ and focus $(4, 5)$

**8.** Directrix $x = 8$ and focus $(-2, 6)$

**9.** Vertex $(4, -6)$, axis of symmetry $x = 4$, and passing through $(0, -8)$

**10.** Vertex $(-5, -1)$, axis of symmetry $y = -1$, and passing through $(-2, -4)$

**11.** Vertex $(8, -7)$, axis of symmetry parallel to the $x$-axis, and passing through $(6, -8)$

**12.** Vertex $\left(\frac{1}{2}, 1\right)$, axis of symmetry parallel to the $y$-axis, and passing through $\left(-\frac{9}{2}, 6\right)$

Sketch the graph of each parabola in Exercises 13–20.

**13.** $y^2 = 2x$

**14.** $x^2 = -4y$

**15.** $(x - 2)^2 = -(y + 1)$

**16.** $(y - 1)^2 = 8(x + 3)$

**17.** $x^2 - 4y + 8 = 0$

**18.** $x^2 - 4x - 2y = 0$

**19.** $y^2 + 4x + y - \frac{47}{4} = 0$

**20.** $y^2 + 4x - 6y + 15 = 0$

In Exercises 21–26 write each parabola in the form $(y - k)^2 = 4p(x - h)$ or $(x - h)^2 = 4p(y - k)$ and determine the vertex, focus, and directrix.

**21.** $y^2 - 12x + 24 = 0$

**22.** $x^2 - 10x - 6y + 25 = 0$

**23.** $x^2 + 8x + 9y + 25 = 0$

**24.** $y^2 + 5x - 16y + 74 = 0$

**25.** $4x^2 - 12x - 2y - 3 = 0$

**26.** $25y^2 - 75x - 20y - 11 = 0$

In Exercises 27–34 complete the squares to determine if the equation represents a circle, ellipse, hyperbola, parabola, or degenerate conic.

**27.** $5x^2 - 2y^2 + x - 7 = 0$

**28.** $-3x^2 + 2x - y + 8 = 0$

**29.** $-x^2 - y^2 + x + y + 12 = 0$

**30.** $4x^2 + 4y^2 - x + 3y - 15 = 0$

**31.** $4x^2 - 49y^2 + 8x + 4 = 0$

**32.** $64x^2 + 25y^2 - 200y + 400 = 0$

**33.** $20x^2 + 10y^2 - 5x + 15y - 75 = 0$

**34.** $36x^2 - 100y^2 - 36x - 400y - 1291 = 0$

Solve.

**35. ARCHITECTURE** A modern building has an entry in the shape of a parabolic arch which is 32.5 ft high and 24.2 ft wide at the base. Find an equation for the parabola (see the figure) if the vertex is put at the origin of the coordinate system.

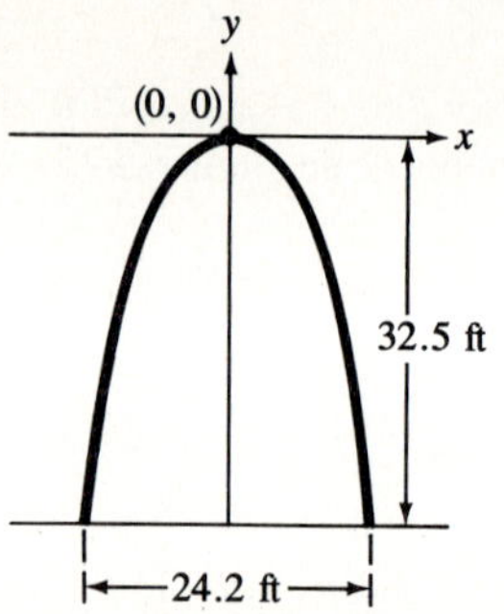


Exercise 35

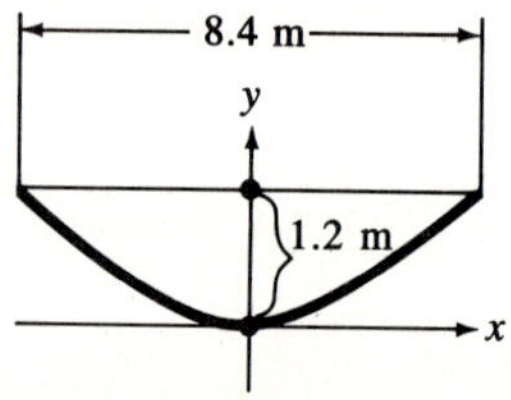


Exercise 36

**36. ASTRONOMY** A radiotelescope has a parabolic surface. If it is 1.2 m deep and 8.4 m wide, how far is the focus from the vertex? See the figure.

**37. ENGINEERING** A tunnel is in the shape of a parabola. The maximum height is 12.8 m and it is 10.2 m wide at the base. What is the vertical clearance 1.5 m from the edge of the tunnel?

**38. AGRICULTURE** A cross-section of an irrigation canal is a parabola. If the surface of the water is 34.6 ft wide and the canal is 22.0 ft deep at the center, how deep is it 5.0 ft from the edge?

## For Review

Determine the equation of each hyperbola in Exercises 39–40.

**39.** Center at (4, −2), foci (4, −7) and (4, 3), and length of the transverse axis is 6

**40.** Vertices (±5, 0) and passing through (10, 6)

In Exercises 41–42 write each equation in the form $\dfrac{(x-h)^2}{m} - \dfrac{(y-k)^2}{n} = 1$ or $\dfrac{(y-k)^2}{m} - \dfrac{(x-h)^2}{n} = 1$.

**41.** $6x^2 - 5y^2 - 24x - 30y - 51 = 0$

**42.** $10y^2 - 8x^2 - 144x + 100y - 478 = 0$

## 9.5 Rotation of Axes

Translation of axes has been used in the preceding sections to derive equations of conics that did not have center or vertex at the origin. If a conic has axes that are not parallel to the coordinate axes, a **rotation of axes** can help identify and graph the equation.

Consider starting with an $x,y$-coordinate system and rotating the axes about the origin through an angle $\theta$. Label the new axes $x'$ and $y'$ and indicate an arbitrary point $P(x, y) = P(x', y')$. Thus, the point will have names in both coordinate systems as indicated in Figure 28.

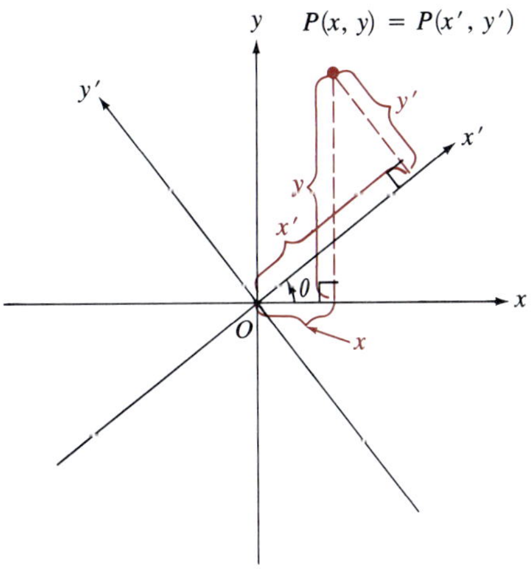


**Figure 28** Rotation of Axes

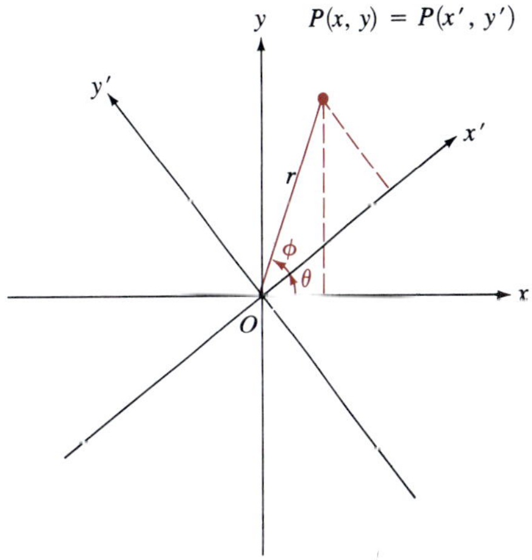


**Figure 29** Angles Involved in Rotation

### Transformation of $(x', y')$ to $(x, y)$

Draw the line from the origin, $O$, to point $P$ and use $r$ for its length. See Figure 29. Let $\phi$ be the angle between this line and the positive $x'$-axis. Then

$$\cos \phi = \frac{x'}{r} \text{ and } \sin \phi = \frac{y'}{r}.$$

Solving for $x'$ and $y'$,

$$x' = r \cos \phi \qquad \text{and} \qquad y' = r \sin \phi.$$

In a similar manner, since the angle between $OP$ and the positive $x$-axis is $\theta + \phi$,

$$\cos (\theta + \phi) = \frac{x}{r} \qquad \text{and} \qquad \sin (\theta + \phi) = \frac{y}{r}.$$

Solving for $x$ and $y$,

$$x = r \cos (\theta + \phi) \qquad \text{and} \qquad y = r \sin (\theta + \phi).$$

Expanding this by using the formula for the sum of two angles,

$$x = r(\cos \theta \cos \phi - \sin \theta \sin \phi) \qquad \text{and} \qquad y = r(\sin \theta \cos \phi + \cos \theta \sin \phi),$$

which can be written in the form

$$x = r\cos\theta\cos\phi - r\sin\theta\sin\phi \quad \text{and} \quad y = r\sin\theta\cos\phi + r\cos\theta\sin\phi.$$

But since $\qquad r\cos\phi = x' \quad$ and $\quad r\sin\phi = y'$,

by substituting we obtain

$$x = x'\cos\theta - y'\sin\theta \quad \text{and} \quad y = x'\sin\theta + y'\cos\theta.$$

### Transformation Equations

If the coordinates of a point $P$ are $(x', y')$ in a rotated coordinate system with angle of rotation $\theta$, the coordinates of $P$ in the original system are $(x, y)$, given by

$$x = x'\cos\theta - y'\sin\theta$$
$$y = x'\sin\theta + y'\cos\theta.$$

## EXAMPLE 1

A coordinate system is rotated through an angle of 30°. If the coordinates of a point are (2, 1) in the new, rotated system, find the coordinates of the point in the original system.

Figure 30 shows (2, 1) as the coordinates of $P$ in the $x',y'$-coordinate system. To find the $x,y$-coordinates of $P$ we substitute $x' = 2$, $y' = 1$, and $\theta = 30°$ into the equations of the theorem.

$$x = x'\cos\theta - y'\sin\theta = \mathbf{2}\cdot\cos\mathbf{30°} - \mathbf{1}\cdot\sin\mathbf{30°}$$
$$= 2\cdot\frac{\sqrt{3}}{2} - 1\cdot\frac{1}{2} = \sqrt{3} - \frac{1}{2} = \frac{2\sqrt{3}-1}{2}$$
$$y = x'\sin\theta + y'\cos\theta = \mathbf{2}\cdot\sin\mathbf{30°} + \mathbf{1}\cdot\cos\mathbf{30°}$$
$$= 2\cdot\frac{1}{2} + 1\cdot\frac{\sqrt{3}}{2} = 1 + \frac{\sqrt{3}}{2} = \frac{2+\sqrt{3}}{2}$$

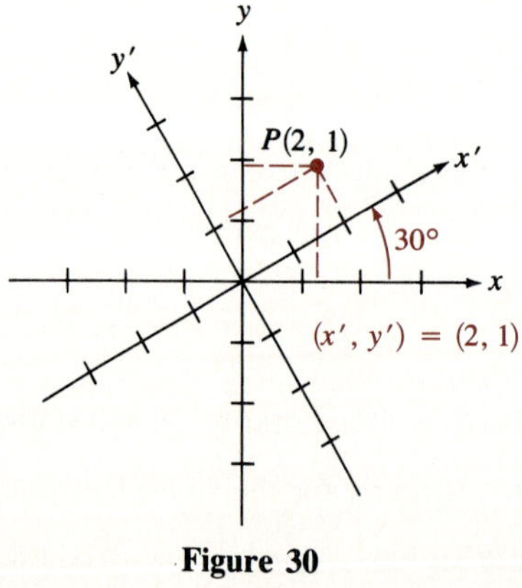


Figure 30

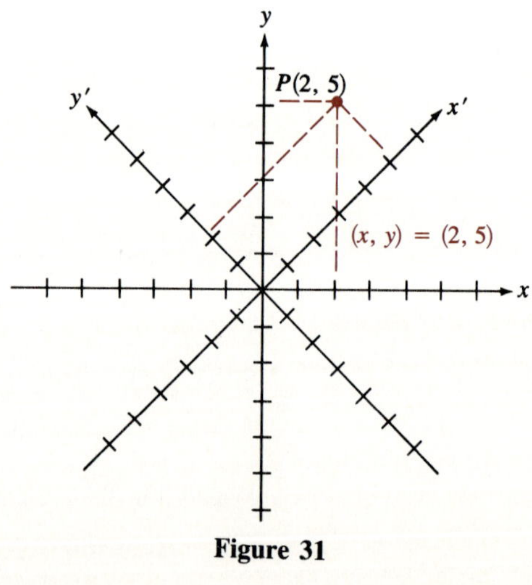


Figure 31

Approximating $\sqrt{3}$ with 1.7,

$$x \approx \frac{2(1.7) - 1}{2} = 1.2 \qquad \text{and} \qquad y \approx \frac{2 + 1.7}{2} = 1.85.$$

Referring to Figure 30, the coordinates (1.2, 1.85) of $P$ in the $x,y$-coordinate system appear to be correct. ■

**Transformation of $(x, y)$ to $(x', y')$**

The equations relating coordinates in the original and rotated systems may also be solved for $x'$ and $y'$ in terms of $x$ and $y$.

### Transformation Equations

If the coordinates of a point $P$ are $(x, y)$ in a given coordinate system, the coordinates of $P$ in an $x',y'$-coordinate system rotated through an angle $\theta$ are $(x', y')$ given by

$$x' = x \cos \theta + y \sin \theta$$
$$y' = \quad x \sin \theta + y \cos \theta.$$

## EXAMPLE 2

A coordinate system is rotated through an angle of 45°. If the coordinates of a point are (2, 5) in the original system, find the coordinates of the point in the rotated system.

Figure 31 shows the two pairs of axes and the point $P(2, 5)$. To find the $x',y'$-coordinates of $P$ we substitute $x = 2$, $y = 5$, and $\theta = 45°$ into the equations of the theorem.

$$x' = x \cos \theta + y \sin \theta = 2 \cdot \cos 45° + 5 \cdot \sin 45°$$
$$= 2 \cdot \frac{1}{\sqrt{2}} + 5 \cdot \frac{1}{\sqrt{2}} = \frac{2}{\sqrt{2}} + \frac{5}{\sqrt{2}} = \frac{7}{\sqrt{2}}$$
$$y' = -x \sin \theta + y \cos \theta = -2 \cdot \sin 45° + 5 \cdot \cos 45°$$
$$= -2 \cdot \frac{1}{\sqrt{2}} + 5 \cdot \frac{1}{\sqrt{2}} = \frac{-2}{\sqrt{2}} + \frac{5}{\sqrt{2}} = \frac{3}{\sqrt{2}}$$

Approximating $\sqrt{2}$ with 1.4,

$$x' \approx \frac{7}{1.4} = 5 \qquad \text{and} \qquad y' \approx \frac{3}{1.4} \approx 2.14.$$

The coordinates (5, 2.14) of $P$ in the $x'y'$-coordinate system appear to be correct as shown in Figure 31. ■

**Simplifying Equations**

The real value of rotating axes is realized when it is used to change a particular equation into an equivalent but simpler equation with coordinates in a rotated system. An equation whose graph might be unknown on inspection may be easily recognized in its new form. For example, consider the second-degree equation

$$5x^2 - 6xy + 5y^2 = 8.$$

Suppose we rotate the axes through an angle of 45° and determine the $(x', y')$ coordinates of a point $(x, y)$ on the graph of this equation.

$$\begin{aligned} x &= x' \cos \theta - y' \sin \theta = x' \cos 45^\circ - y' \sin 45^\circ \\ &= x' \cdot \frac{1}{\sqrt{2}} - y' \cdot \frac{1}{\sqrt{2}} = \frac{x' - y'}{\sqrt{2}} \\ y &= x' \sin \theta + y' \cos \theta = x' \sin 45^\circ + y' \cos 45^\circ \\ &= x' \cdot \frac{1}{\sqrt{2}} + y' \cdot \frac{1}{\sqrt{2}} = \frac{x' + y'}{\sqrt{2}} \end{aligned}$$

Substitute these values of $x$ and $y$ in the equation $5x^2 - 6xy + 5y^2 = 8$.

$$5\left(\frac{x' - y'}{\sqrt{2}}\right)^2 - 6\left(\frac{x' - y'}{\sqrt{2}}\right)\left(\frac{x' + y'}{\sqrt{2}}\right) + 5\left(\frac{x' + y'}{\sqrt{2}}\right)^2 = 8$$

$$5\left(\frac{x'^2 - 2x'y' + y'^2}{2}\right) - 6\left(\frac{x'^2 - y'^2}{2}\right) + 5\left(\frac{x'^2 + 2x'y' + y'^2}{2}\right) = 8$$

$$5x'^2 - 10x'y' + 5y'^2 - 6x'^2 + 6y'^2 + 5x'^2 + 10x'y' + 5y'^2 = 16$$

$$4x'^2 + 16y'^2 = 16$$

$$\frac{x'^2}{4} + \frac{y'^2}{1} = 1$$

In the $x',y'$-coordinate system, we recognize that the graph of this equation is an ellipse with $a = 2$ and $b = 1$. See Figure 32.

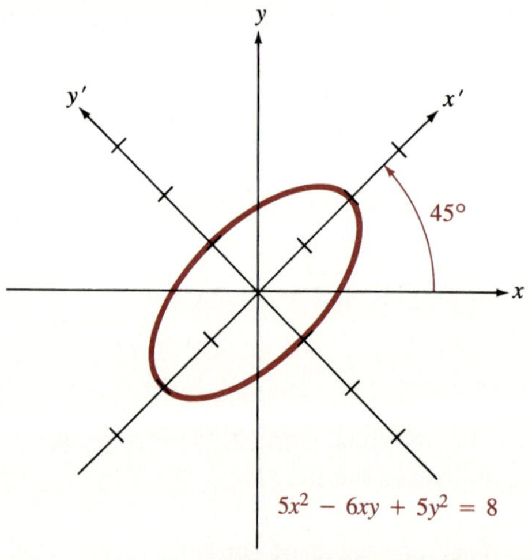


**Figure 32**

## Rotation to Eliminate the xy-Term

Thus, to graph $5x^2 - 6xy + 5y^2 = 8$, we eliminated the $xy$-term by rotating the axes and then obtained an equation in $x'$ and $y'$ which was already familiar to us. This is a specific case of a general procedure. Any second-degree equation with an $xy$-term can be transformed into an equation with no $x'y'$-term. To accomplish this, we must determine the angle of rotation. Suppose we concentrate on a second-degree equation of the form

$$Ax^2 + Bxy + Cy^2 + F = 0.$$

Substitute $x' \cos \theta - y' \sin \theta$ for $x$ and $x' \sin \theta + y' \cos \theta$ for $y$.

$$\begin{aligned} &A(x' \cos \theta - y' \sin \theta)^2 + B(x' \cos \theta - y' \sin \theta)(x' \sin \theta + y' \cos \theta) \\ &\quad + C(x' \sin \theta + y' \cos \theta)^2 + F = 0 \end{aligned}$$

$$\begin{aligned} &A(x'^2 \cos^2 \theta - 2x'y' \cos \theta \sin \theta + y'^2 \sin^2 \theta) \\ &\quad + B(x'^2 \cos \theta \sin \theta - x'y' \sin^2 \theta + x'y' \cos^2 \theta - y'^2 \sin \theta \cos \theta) \\ &\quad + C(x'^2 \sin^2 \theta + 2x'y' \sin \theta \cos \theta + y'^2 \cos^2 \theta) + F = 0 \end{aligned}$$

$$\begin{aligned} &(A \cos^2 \theta + B \cos \theta \sin \theta + C \sin^2 \theta)x'^2 \\ &\quad + (-2A \cos \theta \sin \theta - B \sin^2 \theta + B \cos^2 \theta + 2C \sin \theta \cos \theta)x'y' \\ &\quad + (A \sin^2 \theta - B \sin \theta \cos \theta + C \cos^2 \theta)y'^2 + F = 0 \end{aligned}$$

Thus, we obtain

$$A'x'^2 + B'x'y' + C'y'^2 + F = 0,$$

where

$$\begin{aligned} A' &= A \cos^2 \theta + B \cos \theta \sin \theta + C \sin^2 \theta \\ B' &= -2A \cos \theta \sin \theta - B \sin^2 \theta + B \cos^2 \theta + 2C \sin \theta \cos \theta \\ &= (C - A)2 \sin \theta \cos \theta + B(\cos^2 \theta - \sin^2 \theta) \\ C' &= A \sin^2 \theta - B \sin \theta \cos \theta + C \cos^2 \theta. \end{aligned}$$

Our goal was to obtain an equation with the $x'y'$-term missing. This will happen if $B' = 0$. This means that

$$(C - A)2 \sin \theta \cos \theta + B(\cos^2 \theta - \sin^2 \theta) = 0.$$

But by the double-angle formulas, we have

$$(C - A) \sin 2\theta + B \cos 2\theta = 0,$$

or

$$B \cos 2\theta = (A - C) \sin 2\theta.$$

Dividing both sides by $B \sin 2\theta$, we have

$$\frac{\cos 2\theta}{\sin 2\theta} = \frac{A - C}{B},$$

or

$$\cot 2\theta = \frac{A - C}{B}.$$

We know that $B \neq 0$ because otherwise we would not have been interested in rotating the axes in the first place. The value of $\theta$ which solves this equation is the desired angle of rotation.

### Eliminating the xy-Term

To eliminate the $xy$-term in

$$Ax^2 + Bxy + Cy^2 + F = 0,$$

substitute

$$x = x' \cos \theta - y' \sin \theta$$
$$y = x' \sin \theta + y' \cos \theta,$$

where $\theta$ is a solution to the equation

$$\cot 2\theta = \frac{A - C}{B}.$$

**EXAMPLE 3**

Eliminate the $xy$-term in $8x^2 + 4xy + 5y^2 = 36$, and graph.

With $A = 8$, $B = 4$, and $C = 5$, we have

$$\cot 2\theta = \frac{A - C}{B} = \frac{8 - 5}{4} = \frac{3}{4} = 0.75.$$

Using a calculator, we can determine $2\theta$ and hence $\theta$ to be about 26°30′. On the other hand, since what we really need is $\sin \theta$ and $\cos \theta$, we can find these values without finding $\theta$. Sketch a triangle for which $\cot 2\theta = \frac{3}{4}$, as in Figure 33. The length of the hypotenuse, 5, can be found using the Pythagorean theorem. Then $\cos 2\theta = \frac{3}{5}$. By the half-angle formulas,

$$\sin \theta = \pm\sqrt{\frac{1 - \cos 2\theta}{2}} = \pm\sqrt{\frac{1 - \frac{3}{5}}{2}} = \pm\sqrt{\frac{1}{5}} = \pm\frac{1}{\sqrt{5}}$$

$$\cos \theta = \pm\sqrt{\frac{1 + \cos 2\theta}{2}} = \pm\sqrt{\frac{1 + \frac{3}{5}}{2}} = \pm\sqrt{\frac{4}{5}} = \pm\frac{2}{\sqrt{5}}.$$

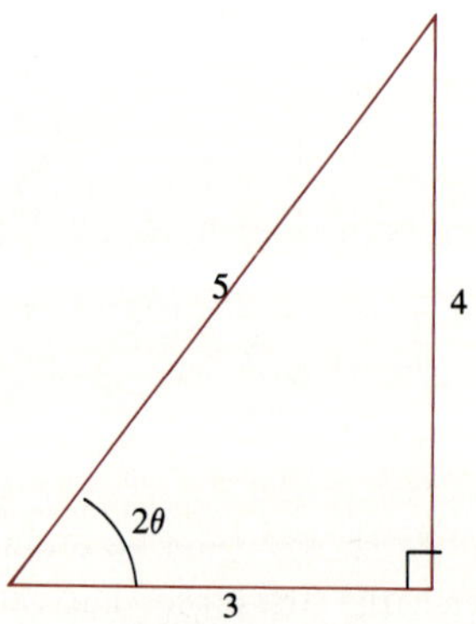


**Figure 33**

When rotating axes, it is simpler for us to choose a value of $2\theta$ between 0 and $\pi$ so that $\theta$ is an angle between 0 and $\pi/2$. As a result, the values of $\sin\theta$ and $\cos\theta$ can be chosen to be positive. Thus,

$$x = x'\cos\theta - y'\sin\theta = x'\cdot\frac{2}{\sqrt{5}} - y'\cdot\frac{1}{\sqrt{5}} = \frac{2x' - y'}{\sqrt{5}},$$

and
$$y = x'\sin\theta + y'\cos\theta = x'\cdot\frac{1}{\sqrt{5}} + y'\cdot\frac{2}{\sqrt{5}} = \frac{x' + 2y'}{\sqrt{5}}.$$

Substitute these values into $8x^2 + 4xy + 5y^2 = 36$.

$$8\left(\frac{2x' - y'}{\sqrt{5}}\right)^2 + 4\left(\frac{2x' - y'}{\sqrt{5}}\right)\left(\frac{x' + 2y'}{\sqrt{5}}\right) + 5\left(\frac{x' + 2y'}{\sqrt{5}}\right)^2 = 36$$

$$8\left(\frac{4x'^2 - 4x'y' + y'^2}{5}\right) + 4\left(\frac{2x'^2 + 3x'y' - 2y'^2}{5}\right) + 5\left(\frac{x'^2 + 4x'y' + 4y'^2}{5}\right) = 36$$

$$32x'^2 - 32x'y' + 8y'^2 + 8x'^2 + 12x'y' - 8y'^2 + 5x'^2 + 20x'y' + 20y'^2 = 180$$

$$45x'^2 + 20y'^2 = 180$$

Dividing through by 180, we obtain

$$\frac{x'^2}{4} + \frac{y'^2}{9} = 1,$$

and recognize the graph as an ellipse with $b = 2$ and $a = 3$. See Figure 34. ■

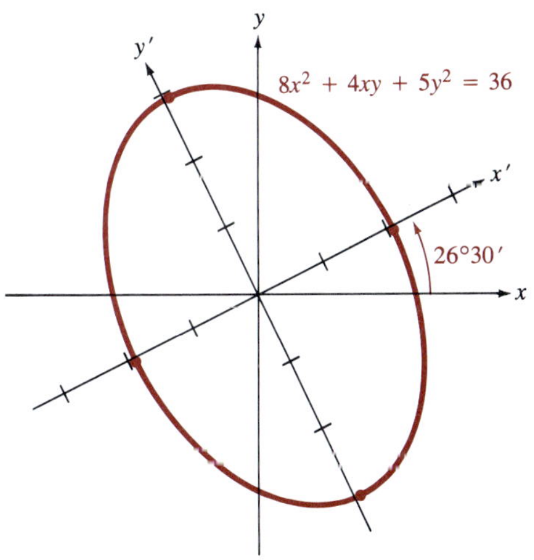


**Figure 34**

**EXAMPLE 4** Eliminate the $xy$-term in $24xy - 7x^2 = 144$, and graph.

Writing $A = -7$, $B = 24$, and $C = 0$, we have

$$\cot 2\theta = \frac{A - C}{B} = \frac{-7 - 0}{24} = \frac{-7}{24}.$$

Since $\cot 2\theta$ is negative, $2\theta$ is an angle in the second quadrant and $\cos 2\theta = -\frac{7}{25}$. See Figure 35. Then $\theta$ is in the first quadrant (approximately 53°10′), and

$$\sin \theta = \sqrt{\frac{1 - \left(-\frac{7}{25}\right)}{2}} = \sqrt{\frac{32}{50}} = \sqrt{\frac{16}{25}} = \frac{4}{5}$$

$$\cos \theta = \sqrt{\frac{1 + \left(-\frac{7}{25}\right)}{2}} = \sqrt{\frac{18}{50}} = \sqrt{\frac{9}{25}} = \frac{3}{5}.$$

Thus, $$x = x'\cos \theta - y' \sin \theta = x' \cdot \frac{3}{5} - y' \cdot \frac{4}{5} = \frac{3x' - 4y'}{5},$$

and $$y = x' \sin \theta + y' \cos \theta = x' \cdot \frac{4}{5} + y' \cdot \frac{3}{5} = \frac{4x' + 3y'}{5}.$$

Substituting, we obtain

$$24\left(\frac{3x' - 4y'}{5}\right)\left(\frac{4x' + 3y'}{5}\right) - 7\left(\frac{3x' - 4y'}{5}\right)^2 = 144,$$

which simplifies to

$$\frac{x'^2}{16} - \frac{y'^2}{9} = 1.$$

We now recognize the graph as a hyperbola with $a = 4$ and $b = 3$. The graph is given in Figure 36. ■

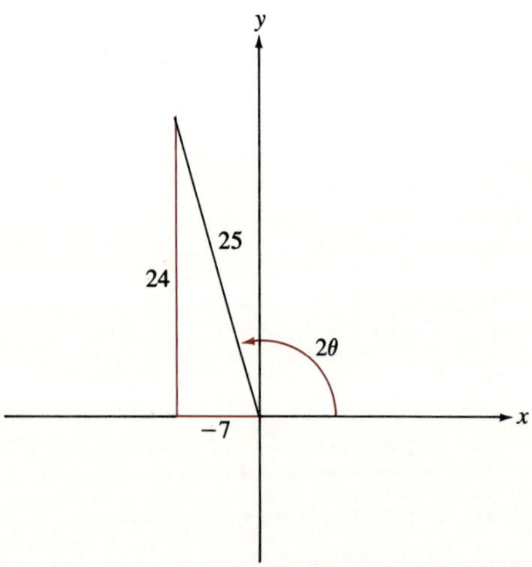

Figure 35

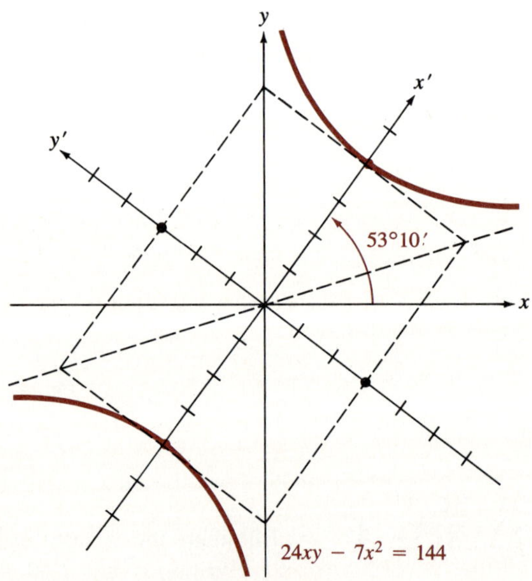

Figure 36

NOTE We have only considered equations of the form $Ax^2 + Bxy + Cy^2 + F = 0$. The equation $Ax^2 + Bxy + Cy^2 + Dx + Ey + F = 0$ may require a translation of axes after rotation in order to obtain an equation in the simplest form.

## 9.5 EXERCISES

In Exercises 1–4 the angle of rotation and the coordinates in the new system are given. Determine the coordinates in the original system.

**1.** 30°; (4, 3) **2.** 45°; (1, 5) **3.** 45°; (3, −2) **4.** 60°; (3, 4)

In Exercises 5–8 the angle of rotation and the coordinates in the original system are given. Determine the coordinates in the rotated system.

**5.** 45°; (2, 4) **6.** 30°; (1, 3) **7.** 60°; (−3, 1) **8.** 45°; $\left(\frac{5\sqrt{2}}{2}, \frac{\sqrt{2}}{2}\right)$

In Exercises 9–12 determine the new equation by rotating through the given angle.

**9.** $x^2 - y^2 = 1$; 90°

**10.** $\frac{x^2}{25} + \frac{y^2}{16} = 1$; 90°

**11.** $x^2 + 4xy + y^2 - 16 = 0$; 45°

**12.** $4xy - 3x^2 = 10$; $\theta = \tan^{-1} 2$

In Exercises 13–22 eliminate the $xy$-term and graph each equation.

**13.** $xy = -1$

**14.** $xy = 12$

**15.** $2x^2 - 4xy + 5y^2 = 36$

**16.** $x^2 - \sqrt{3}xy + 2y^2 = 10$

**17.** $9x^2 + 24xy + 16y^2 = 100$

**18.** $3x^2 + 2xy + 3y^2 + 2 = 0$

**19.** $x^2 - \sqrt{3}xy + 12 = 0$

**20.** $x^2 + 2\sqrt{3}xy - y^2 = 4$

**21.** $x^2 - 2xy + y^2 - 4\sqrt{2}x - 4\sqrt{2}y + 8\sqrt{2} = 0$

**22.** $x^2 + 2xy + y^2 + 4x - 4y = -8$

**23.** Show that for any rotation of coordinates the equation $x^2 + y^2 = r^2$ becomes $x'^2 + y'^2 = r^2$.

**24.** It can be shown that for any rotation, $B'^2 - 4A'C' = B^2 - 4AC$. Thus for a rotation that eliminates the $xy$-term, we have $-4A'C' = B^2 - 4AC$. Use this to prove that if $Ax^2 + Bxy + Cy^2 + Dx + Ey + F = 0$ is not a degenerate conic, then it is a parabola if $B^2 - 4AC = 0$, an ellipse if $B^2 - 4AC < 0$, and a hyperbola if $B^2 - 4AC > 0$.

Use the result of Exercise 24 to determine the nature of the graphs of the conics in Exercises 25–28

**25.** $x^2 - \sqrt{3}xy + 2y^2 = 10$

**26.** $2x^2 - 4xy + 5y^2 = 36$

**27.** $x^2 + 2\sqrt{3}xy - y^2 = 4$

**28.** $x^2 + 2xy + y^2 + 4x - 4y = -8$

### For Review

Determine the equation of each parabola in Exercises 29–30.

**29.** Directrix $y = 4$ and focus $(3, -6)$

**30.** Vertex $(4, -2)$, axis of symmetry $y = -2$, and passing through $(6, 2)$

In Exercises 31–32 write each equation in the form $(y - k)^2 = 4p(x - h)$ or $(x - h^2) = 4p(y - k)$.

**31.** $x^2 - 10x - 20y + 65 = 0$

**32.** $y^2 + 2x + 16y + 70 = 0$

In Exercises 33–36 determine the nature of the graph of each equation.

**33.** $2x^2 - 2y^2 + x - y - 5 = 0$

**34.** $2x^2 + 2y^2 + x - y - 5 = 0$

**35.** $x^2 + 2y^2 - 3x + y + 1 = 0$

**36.** $2y^2 - 2x + y + 4 = 0$

## 9.6 Polar Coordinates and Graphs

Consider the point (4, 3) in a rectangular coordinate system that is $\sqrt{4^2 + 3^2} = 5$ units from the origin. Draw a line to the point from the origin making an angle $\theta$ with the positive $x$-axis. The angle $\theta$ can be determined by finding $\tan^{-1} \frac{y}{x}$. See Figure 37.

$$\theta = \tan^{-1} \frac{y}{x} = \tan^{-1} \frac{3}{4} \approx 36°50'$$

The point (4, 3) is 5 units from the origin on a line that makes an angle of 36°50′ with the positive $x$-axis. In *polar coordinates* the point is called (5, 36°50′).

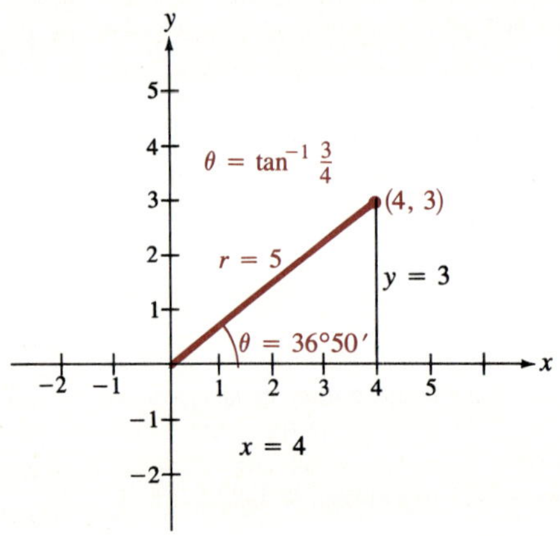


Figure 37

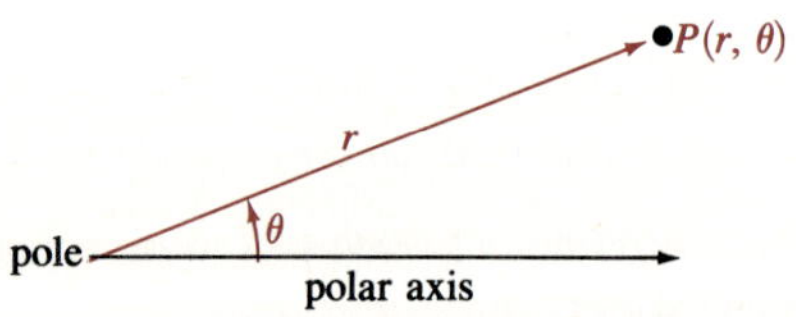


**Figure 38 Polar Coordinates**

### Construction of Polar Coordinate System

To construct a **polar coordinate system,** select a point called the **pole** or **origin** and a ray extending to the right called the **polar axis.** See Figure 38. The point $P$ is named $(r, \theta)$ as with the example using the point (4, 3). The angle $\theta$ is positive when measured in a counterclockwise direction and negative when measured clockwise. The value of $r$ is positive when measured along the ray determined by $\theta$ and negative when measured in the opposite direction. The point (4, 30°) in Figure 39 is named using several possible polar coordinates.

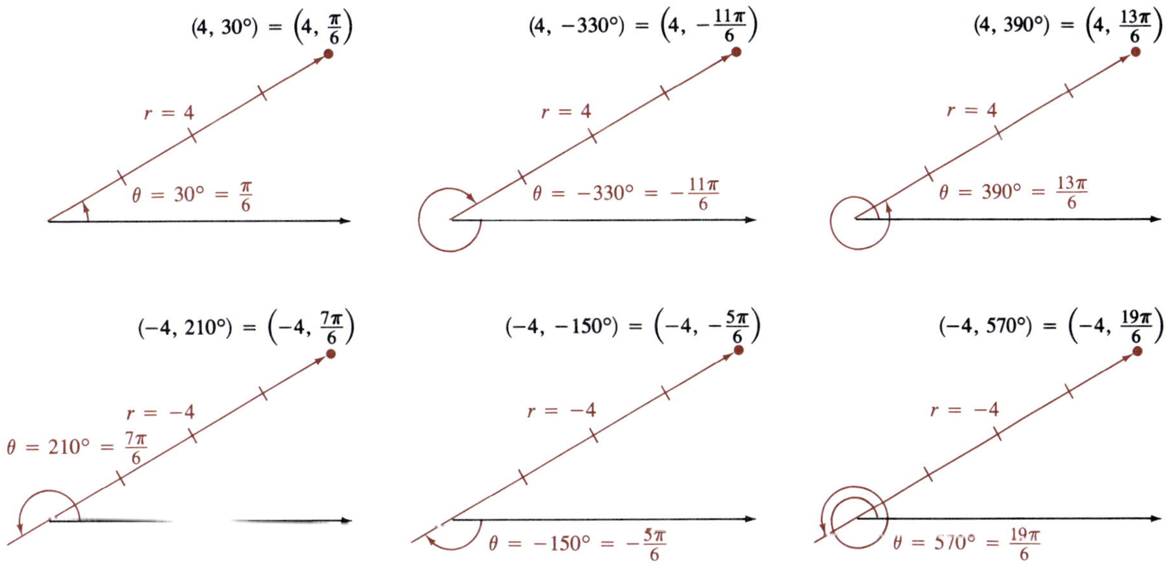


**Figure 39**

It is obvious that the name of a point in polar coordinates is not unique. In fact if $(r, \theta)$ is one name then $(r, \theta \pm k \cdot 360°)$, $k = 0, 1, 2, 3, \ldots$ is also a name. Furthermore $(-r, \theta + 180°)$ is a name along with $(-r, \theta + 180° \pm k \cdot 360°)$ for $k = 0, 1, 2, \ldots$. Thus, there are infinitely many names for a point in polar coordinates.

**EXAMPLE 1** Graph the points $(5, 60°)$, $(3, \pi/2)$, $(-3, -\pi/2)$, $(3, -\pi/2)$, $(-3, \pi/2)$, $(5, 420°)$, $(5, -300°)$, $(-5, 240°)$, $(3, 5\pi/4)$, $(1, \pi)$, $(-4, 180°)$, $(1, -\pi)$, $(4, 150°)$, and $(4, 720°)$ in a polar coordinate system.

The points are plotted in Figure 40. ■

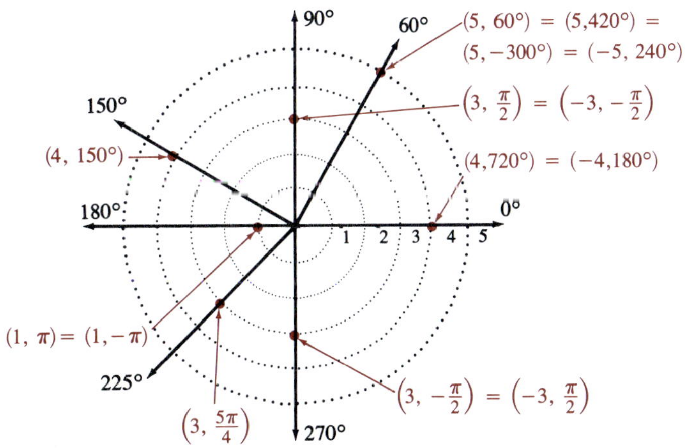


**Figure 40 Graphing Points**

## Conversion Between Systems

We now investigate the problem of conversion from one coordinate system to the other. Consider the triangle in Figure 41. Since

$$\cos\theta = \frac{x}{r} \quad \text{and} \quad \sin\theta = \frac{y}{r},$$

solving for $x$ and $y$,

$$x = r\cos\theta \quad \text{and} \quad y = r\sin\theta.$$

Also, since

$$r^2 = x^2 + y^2 \quad \text{and} \quad \tan\theta = \frac{y}{x},$$

$$r = \sqrt{x^2 + y^2} \quad \text{and} \quad \theta = \arctan\frac{y}{x}.$$

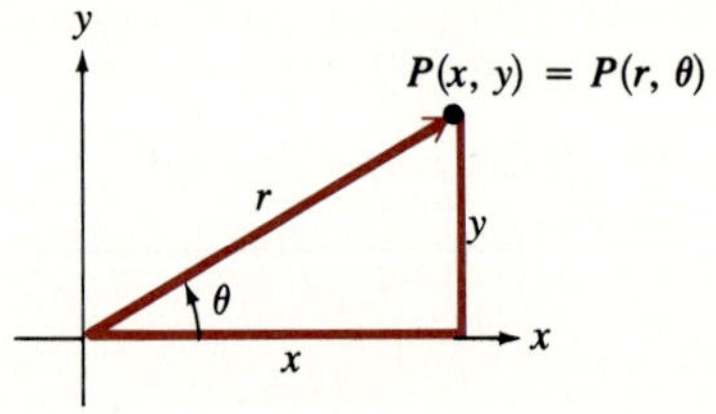


**Figure 41**

### Coordinate Conversion Equations

The following equations are used for conversion between coordinate systems.

| Polar to rectangular | Rectangular to polar |
|---|---|
| $x = r\cos\theta$ | $r^2 = x^2 + y^2$ |
| $y = r\sin\theta$ | $\tan\theta = \dfrac{y}{x}$ |

**EXAMPLE 2** Convert $(3, 90°)$, $\left(5, \frac{\pi}{6}\right)$, $\left(2, -\frac{\pi}{4}\right)$, and $(7, 900°)$ to rectangular coordinates.

| *Polar coordinates* | *Conversion* | *Rectangular coordinates* |
|---|---|---|
| $(3, 90°)$ | $x = r\cos\theta = 3\cos 90° = 0$ <br> $y = r\sin\theta = 3\sin 90° = 3$ | $(0, 3)$ |
| $\left(5, \frac{\pi}{6}\right)$ | $x = r\cos\theta = 5\cos\frac{\pi}{6} = \frac{5\sqrt{3}}{2}$ <br> $y = r\sin\theta = 5\sin\frac{\pi}{6} = \frac{5}{2}$ | $\left(\frac{5\sqrt{3}}{2}, \frac{5}{2}\right)$ |
| $\left(2, -\frac{\pi}{4}\right)$ | $x = r\cos\theta = 2\cos\left(-\frac{\pi}{4}\right) = \frac{2}{\sqrt{2}} = \sqrt{2}$ <br> $y = r\sin\theta = 2\sin\left(-\frac{\pi}{4}\right) = -\frac{2}{\sqrt{2}} = -\sqrt{2}$ | $(\sqrt{2}, -\sqrt{2})$ |
| $(7, 900°)$ | $x = 7\cos 900° = 7\cos 180° = -7$ <br> $y = 7\sin 900° = 7\sin 180° = 0$ ■ | $(-7, 0)$ |

**EXAMPLE 3** Convert (1, 1), (5, 0), (−2, 3), and ($\sqrt{3}$, −1) to polar coordinates.

| *Rectangular coordinates* | *Conversion* | *Polar coordinates* |
|---|---|---|
| (1, 1) | $r = \sqrt{x^2 + y^2} = \sqrt{(1)^2 + (1)^2} = \sqrt{2}$ <br> $\tan \theta = \frac{y}{x} = \frac{1}{1} = 1;\ \theta = \frac{\pi}{4} \pm 2k\pi$ | $\left(\sqrt{2}, \frac{\pi}{4} \pm 2k\pi\right)$ |
| (5, 0) | $r = \sqrt{x^2 + y^2} = \sqrt{(5)^2 + (0)^2} = 5$ <br> $\tan \theta = \frac{y}{x} = \frac{0}{5} = 0;\ \theta = 0° \pm k \cdot 360°$ | $(5, 0° \pm k \cdot 360°)$ |
| (−2, 3) | $r = \sqrt{x^2 + y^2} = \sqrt{(-2)^2 + (3)^2} = \sqrt{13}$ <br> $\tan \theta = \frac{y}{x} = -\frac{3}{2};\ \theta = \tan^{-1}\left(-\frac{3}{2}\right) \pm 2k\pi$ | $\left(\sqrt{13}, \tan^{-1}\left(-\frac{3}{2}\right) \pm 2k\pi\right)$ |
| ($\sqrt{3}$, −1) | $r = \sqrt{x^2 + y^2} = \sqrt{(\sqrt{3})^2 + (-1)^2} = 2$ <br> $\tan \theta = \frac{y}{x} = -\frac{1}{\sqrt{3}};\ \theta = -\frac{\pi}{6} \pm 2k\pi$ ■ | $\left(2, -\frac{\pi}{6} \pm 2k\pi\right)$ |

## Converting Equations

If an equation is given in one coordinate system, it can be changed to the other system by using the appropriate conversion relations.

**EXAMPLE 4** Convert to rectangular coordinates.

**(a)** $r = 6 \cos \theta$

$$r = 6 \cos \theta$$

$$r^2 = 6r \cos \theta \qquad \text{Multiply both sides by } r$$

$$x^2 + y^2 = 6x$$

$$x^2 - 6x + y^2 = 0$$

$$x^2 - 6x + 9 + y^2 = 9 \qquad \text{Complete the square}$$

$$(x - 3)^2 + y^2 = 3^2 \qquad \text{Equation of a circle}$$

**(b)** $r \cos \theta = 6$

$$r \cos \theta = 6$$

$$x = 6 \qquad \text{Equation of a vertical line}$$ ■

**EXAMPLE 5** Convert to polar coordinates.

**(a)** $3x + 2y = 6$

$$3x + 2y = 6 \qquad \text{Equation of a line}$$

$$3r \cos \theta + 2r \sin \theta = 6$$

$$r(3 \cos \theta + 2 \sin \theta) = 6 \qquad \text{Equation of the line in polar coordinates}$$

**(b)** $(x^2 + y^2)^2 = 3x^2 + 3y^2 + 5x + 6$

$$\begin{aligned}(x^2 + y^2)^2 &= 3x^2 + 3y^2 + 5x + 6\\ (r^2)^2 &= 3r^2 \cos^2 \theta + 3r^2 \sin^2 \theta + 5r \cos \theta + 6\\ r^4 &= 3r^2(\cos^2 \theta + \sin^2 \theta) + 5r \cos \theta + 6\\ r^4 &= 3r^2(1) + 5r \cos \theta + 6\\ r^4 &= 3r^2 + 5r \cos \theta + 6 \quad ■\end{aligned}$$

To graph equations in polar coordinates, in some situations it may be helpful to convert to rectangular coordinates since the curve may then be recognized from previous work.

## Graphing

## EXAMPLE 6

Graph the equation $r = 6$.

We note that no matter what $\theta$ is, $r$ is 6. Thus, the graph is a circle with center at the origin and radius 6. Also, since

$$\begin{aligned}r &= 6,\\ r^2 &= 36\\ x^2 + y^2 &= 36.\end{aligned}$$

Thus, the equation in rectangular coordinates confirms that the graph is the circle shown in Figure 42. Several points have also been plotted. ■

| $r$ | $\theta$ |
|---|---|
| 6 | 0° |
| 6 | 60° |
| 6 | 270° |

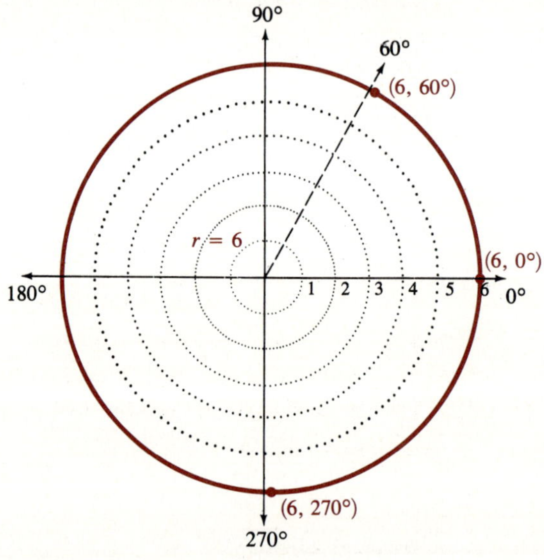


**Figure 42**

## EXAMPLE 7

Graph the equation $r = 6 \cos \theta$.

From Example 4**(a)**, we know that this equation in rectangular coordinates is $(x - 3)^2 + y^2 = 9$ which represents a circle with center at (3, 0) and radius 3. However, we shall plot it in a polar coordinate system. We make a table of values, plot the points, and draw the curve in Figure 43.

| $\theta$ | $0$ | $\frac{\pi}{6}$ | $-\frac{\pi}{6}$ | $\frac{\pi}{4}$ | $-\frac{\pi}{4}$ | $\frac{\pi}{3}$ | $-\frac{\pi}{3}$ | $\frac{\pi}{2}$ | $-\frac{\pi}{2}$ |
|---|---|---|---|---|---|---|---|---|---|
| $r$ | 6 | $3\sqrt{3}$ | $3\sqrt{3}$ | $3\sqrt{2}$ | $3\sqrt{2}$ | 3 | 3 | 0 | 0 |

Due to the periodicity of the cosine function, if we let $\theta$ increase through all real numbers, we obtain the same points over and over again. ■

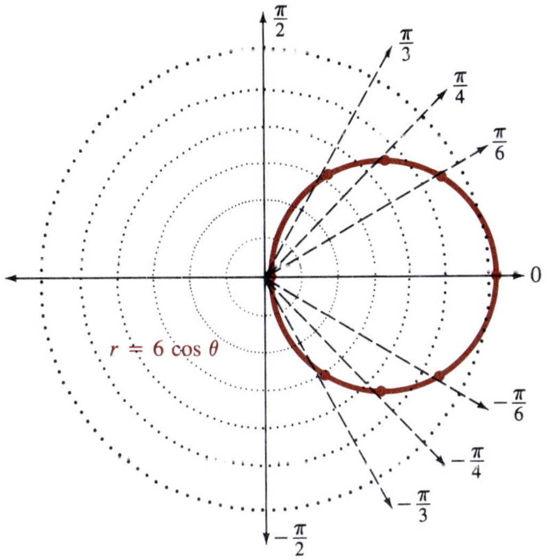


Figure 43

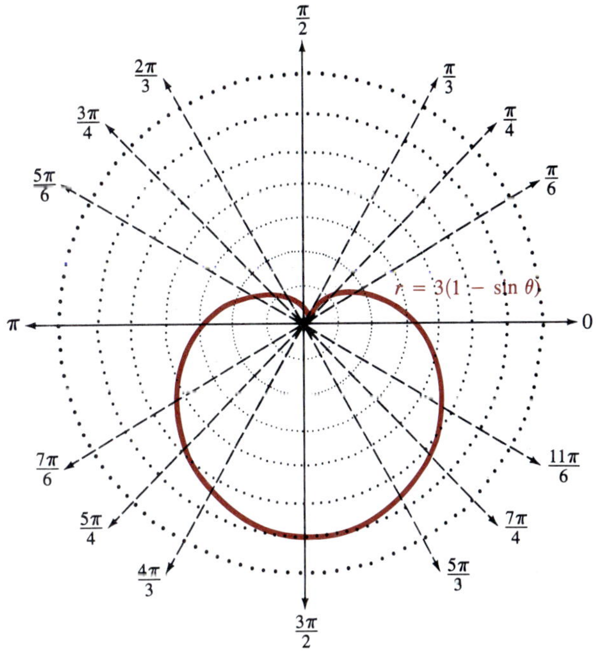


Figure 44

**EXAMPLE 8** Graph the equation $r = 3(1 - \sin\theta)$.

Conversion to rectangular coordinates does not help in graphing this equation. Thus we concentrate on plotting points. To make the graphing easier give decimal approximations for $r$.

| $\theta$ | $0$ | $\frac{\pi}{6}$ | $\frac{\pi}{4}$ | $\frac{\pi}{3}$ | $\frac{\pi}{2}$ | $\frac{2\pi}{3}$ | $\frac{3\pi}{4}$ | $\frac{5\pi}{6}$ | $\pi$ | $\frac{7\pi}{6}$ | $\frac{5\pi}{4}$ | $\frac{4\pi}{3}$ | $\frac{3\pi}{2}$ | $\frac{5\pi}{3}$ | $\frac{7\pi}{4}$ | $\frac{11\pi}{6}$ |
|---|---|---|---|---|---|---|---|---|---|---|---|---|---|---|---|---|
| $r$ | 3 | 1.5 | 0.9 | 0.4 | 0 | 0.4 | 0.9 | 1.5 | 3 | 4.5 | 5.1 | 5.6 | 6 | 5.6 | 5.1 | 4.5 |

The graph shown in Figure 44 is called a **cardioid** since it has the shape of a heart. ■

## 9.6 EXERCISES

Graph the points given in polar coordinates in Exercises 1–12.

**1.** $(3, 30°)$ **2.** $(4, -30°)$ **3.** $(-3, 30°)$ **4.** $(-4, -30°)$

**5.** $\left(2, \frac{\pi}{2}\right)$ **6.** $\left(4, -\frac{\pi}{2}\right)$ **7.** $\left(-2, \frac{\pi}{4}\right)$ **8.** $\left(-3, \frac{3\pi}{4}\right)$

**9.** $(4, 540°)$ **10.** $(3, -840°)$ **11.** $(-1, 7\pi)$ **12.** $\left(0, -\frac{7\pi}{8}\right)$

Change from polar coordinates to rectangular coordinates in Exercises 13–20.

**13.** $(6, 30°)$ **14.** $(-6, 30°)$ **15.** $\left(8, \frac{\pi}{4}\right)$ **16.** $\left(8, -\frac{\pi}{4}\right)$

**17.** $\left(2, \frac{4\pi}{3}\right)$ **18.** $\left(3, \frac{23\pi}{6}\right)$ **19.** $(5, 450°)$ **20.** $(-5, -750°)$

Change from rectangular coordinates to polar coordinates in Exercises 21–28.

**21.** $(-3, 4)$ **22.** $(8, -6)$ **23.** $(0, -6)$ **24.** $(-4, 0)$

**25.** $(-7, -7)$ **26.** $(-2, 2)$ **27.** $(12, -5)$ **28.** $(-4, 2)$

In Exercises 29–34 change each equation to polar coordinates.

**29.** $x^2 + y^2 = 25$ **30.** $3x - 4y = 6$ **31.** $y^2 = 3x$

**32.** $x^2 + xy + y^2 = 3$ **33.** $x^2 + y^2 + 8y = 0$ **34.** $x^2 = y^2 + 9$

In Exercises 35–40 change each equation to rectangular coordinates.

**35.** $r \sin \theta = 3$ **36.** $r = 6$ **37.** $r = 4 \cos \theta$

**38.** $r(5 \cos \theta - 3 \sin \theta) = 15$ **39.** $\theta = \frac{\pi}{3}$ **40.** $r \sin^2 \theta = 4 \cos \theta$

In Exercises 41–52 graph each equation using polar coordinates.

**41.** $r \sin \theta = 3$ **42.** $r \cos \theta = -2$ **43.** $r = 4 \cos \theta$

**44.** $r = -4 \cos \theta$ **45.** $r(5 \cos \theta - 3 \sin \theta) = 15$ **46.** $\theta = -\frac{\pi}{4}$

**47.** $r = 6 \sin \theta$ **48.** $r^2 = 9 \cos 2\theta$ (lemniscate) **49.** $r = \theta$ $(\theta \geq 0)$ (spiral)

**50.** $r = -\theta$ $(\theta \geq 0)$ (spiral) **51.** $r = 3(1 - \cos \theta)$ (cardioid) **52.** $r = 4 \cos 2\theta$ (rose)

### For Review

In Exercises 53–54 eliminate the *xy*-term.

**53.** $x^2 - 6xy + y^2 = 5$ **54.** $x^2 + 2xy + y^2 - \sqrt{2}x + \sqrt{2}y = 0$

## 9.7 Polar Form of Complex Numbers and DeMoivre's Theorem

If each complex number $z = x + yi$ is associated with the corresponding ordered pair $(x, y)$ in a rectangular coordinate system, we have the **complex plane.** See Figure 45. The $x$-axis is called the **real axis,** and the $y$-axis is called the **imaginary axis.**

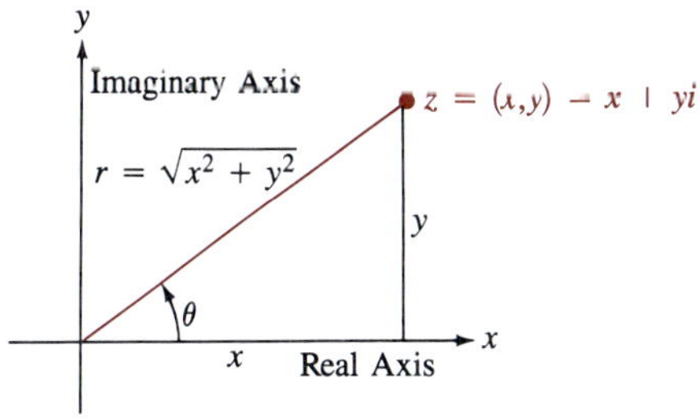


**Figure 45 Polar Form**

The **absolute value** of a complex number $z = x + yi$ is denoted by $|z|$ and is defined to be

$$|z| = \sqrt{x^2 + y^2} = r.$$

From Section 9.6, $x$ and $y$ are related to $r$ and $\theta$ by the equations

$$x = r\cos\theta \qquad \text{and} \qquad y = r\sin\theta.$$

Thus, $z$ can be written in the following **polar** or **trigonometric form,**

$$\begin{aligned} z &= x + yi \\ &= r\cos\theta + ir\sin\theta \\ &= r(\cos\theta + i\sin\theta), \end{aligned}$$

where $r$ is the absolute value of $z$, and $\theta$ is an **argument** of $z$. We say *an* argument since $\theta + 2k\pi$ is also an argument of $z$, for any integer $k$.

**Polar Form of Complex Number**

The polar form of a complex number, $z = x + yi$, is given by

$$z = r(\cos\theta + i\sin\theta).$$

**EXAMPLE 1** Determine $r$ and $\theta$ and write the following complex numbers in polar form.

**(a)** $z = 1 - i\sqrt{3}$

$$r = \sqrt{(1)^2 + (-\sqrt{3})^2} = \sqrt{1 + 3} = \sqrt{4} = 2$$

$$z = 1 - i\sqrt{3} = 2\left(\frac{1}{2} - \frac{\sqrt{3}}{2}i\right) \qquad \text{Factor out } r$$

To have polar form, $\cos \theta = 1/2$ and $\sin \theta = -\sqrt{3}/2$, and therefore,

$$\theta = \frac{5\pi}{3} \pm 2k\pi, \quad k = 0, 1, 2, \ldots .$$

Thus, $|z| = 2$, an argument of $z$ is $5\pi/3$, and the polar form is

$$\begin{aligned} z &= r(\cos \theta + i \sin \theta) \\ &= 2\left(\cos \frac{5\pi}{3} + i \sin \frac{5\pi}{3}\right). \end{aligned}$$

Figure 46 is the graph of $z$.

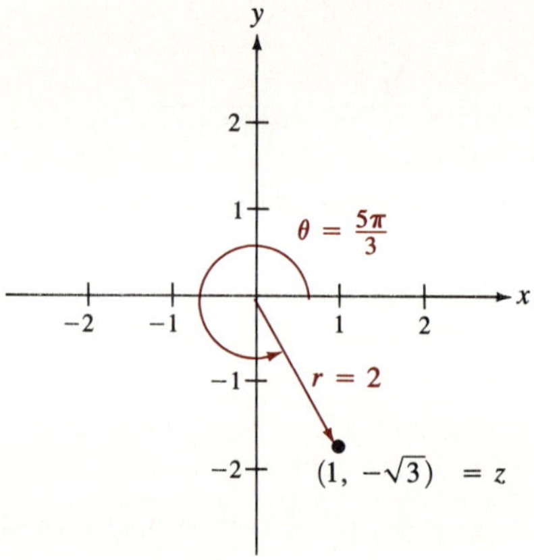


**Figure 46** $z = 1 - i\sqrt{3}$

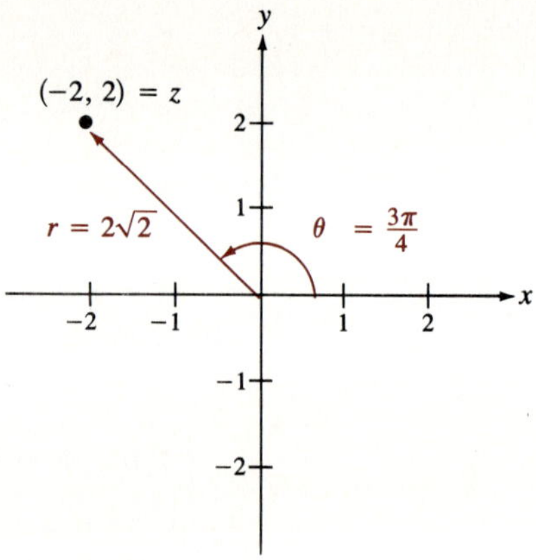


**Figure 47** $z = -2 + 2i$

**(b)** $z = -2 + 2i$

$$\begin{aligned} r &= \sqrt{(-2)^2 + (2)^2} = \sqrt{4 + 4} = \sqrt{8} = 2\sqrt{2} \\ z &= -2 + 2i = \frac{-2\sqrt{2}}{\sqrt{2}} + \frac{2\sqrt{2}}{\sqrt{2}} i \\ &= 2\sqrt{2}\left(-\frac{1}{\sqrt{2}} + \frac{1}{\sqrt{2}} i\right) \quad \text{Factor out } r \end{aligned}$$

From this, $\cos \theta = -\dfrac{1}{\sqrt{2}}$ and $\sin \theta = \dfrac{1}{\sqrt{2}}$.

Thus, $$\theta = \frac{3\pi}{4} \pm 2k\pi, \quad k = 0, 1, 2, \ldots,$$

and $$z = 2\sqrt{2}\left(\cos \frac{3\pi}{4} + i \sin \frac{3\pi}{4}\right).$$

Figure 47 is the graph of $z$. ■

**EXAMPLE 2** Write $z = 4(\cos 30° + i \sin 30°)$ in rectangular form.

$$\begin{aligned} z &= 4(\cos 30° + i \sin 30°) \\ &= 4\left(\frac{\sqrt{3}}{2} + i\frac{1}{2}\right) \\ &= 2\sqrt{3} + 2i \quad ■ \end{aligned}$$

## Product and Quotient in Polar Form

The advantage of the polar form is that it allows us to multiply, divide, and take powers and roots of complex numbers more easily. Suppose we have two complex numbers:

$$z_1 = r_1(\cos \theta_1 + i \sin \theta_1) \quad \text{and} \quad z_2 = r_2(\cos \theta_2 + i \sin \theta_2).$$

Multiply the two numbers.

$$\begin{aligned} z_1 z_2 &= r_1(\cos \theta_1 + i \sin \theta_1) \cdot r_2(\cos \theta_2 + i \sin \theta_2) \\ &= r_1 r_2(\cos \theta_1 + i \sin \theta_1)(\cos \theta_2 + i \sin \theta_2) \\ &= r_1 r_2(\cos \theta_1 \cos \theta_2 + i \cos \theta_1 \sin \theta_2 + i \sin \theta_1 \cos \theta_2 + i^2 \sin \theta_1 \sin \theta_2) \\ &= r_1 r_2[(\cos \theta_1 \cos \theta_2 - \sin \theta_1 \sin \theta_2) + i(\sin \theta_1 \cos \theta_2 + \cos \theta_1 \sin \theta_2)] \\ &= r_1 r_2[\cos (\theta_1 + \theta_2) + i \sin (\theta_1 + \theta_2)] \end{aligned}$$

In a similar way we can show that if $r_2 \neq 0$,

$$\frac{z_1}{z_2} = \frac{r_1}{r_2}[\cos (\theta_1 - \theta_2) + i \sin (\theta_1 - \theta_2)].$$

### Product and Quotient of Complex Numbers

Given the complex numbers $z_1 = r_1(\cos \theta_1 + i \sin \theta_1)$ and $z_2 = r_2(\cos \theta_2 + i \sin \theta_2)$,

$$z_1 z_2 = r_1 r_2[\cos (\theta_1 + \theta_2) + i \sin (\theta_1 + \theta_2)].$$

Thus, $r_1 r_2$ is the absolute value and $\theta_1 + \theta_2$ is an argument of the product.

$$\frac{z_1}{z_2} = \frac{r_1}{r_2}[\cos (\theta_1 - \theta_2) + i \sin (\theta_1 - \theta_2)].$$

Thus, $r_1/r_2$ is the absolute value and $\theta_1 - \theta_2$ is an argument of the quotient.

**EXAMPLE 3** Find the product and quotient of

$$z_1 = 2\left(\cos \frac{\pi}{6} + i \sin \frac{\pi}{6}\right) \quad \text{and} \quad z_2 = 6\left(\cos \frac{\pi}{3} + i \sin \frac{\pi}{3}\right).$$

$$\begin{aligned} z_1 z_2 &= r_1 r_2[\cos (\theta_1 + \theta_2) + i \sin (\theta_1 + \theta_2)] \\ &= (2)(6)\left[\cos \left(\frac{\pi}{6} + \frac{\pi}{3}\right) + i \sin \left(\frac{\pi}{6} + \frac{\pi}{3}\right)\right] \\ &= 12\left[\cos \frac{\pi}{2} + i \sin \frac{\pi}{2}\right] \\ &= 12[0 + i(1)] = 12i \end{aligned}$$

As a check, we could multiply in rectangular form.

$$z_1 = 2\left(\cos\frac{\pi}{6} + i\sin\frac{\pi}{6}\right) \quad \text{and} \quad z_2 = 6\left(\cos\frac{\pi}{3} + i\sin\frac{\pi}{3}\right)$$
$$= 2\left(\frac{\sqrt{3}}{2} + i\cdot\frac{1}{2}\right) \qquad\qquad = 6\left(\frac{1}{2} + i\cdot\frac{\sqrt{3}}{2}\right)$$
$$= \sqrt{3} + i \qquad\qquad = 3 + 3i\sqrt{3}$$

$$\begin{aligned}
z_1z_2 &= (\sqrt{3} + i)(3 + 3i\sqrt{3})\\
&= 3\sqrt{3} + \sqrt{3}(3\sqrt{3})i + 3i + 3\sqrt{3}i^2\\
&= 3\sqrt{3} + 9i + 3i + 3\sqrt{3}(-1)\\
&= 12i
\end{aligned}$$

$$\begin{aligned}
\frac{z_1}{z_2} &= \frac{r_1}{r_2}[\cos(\theta_1 - \theta_2) + i\sin(\theta_1 - \theta_2)]\\
&= \frac{2}{6}\left[\cos\left(\frac{\pi}{6} - \frac{\pi}{3}\right) + i\sin\left(\frac{\pi}{6} - \frac{\pi}{3}\right)\right]\\
&= \frac{1}{3}\left[\cos\left(-\frac{\pi}{6}\right) + i\sin\left(-\frac{\pi}{6}\right)\right]\\
&= \frac{1}{3}\left[\frac{\sqrt{3}}{2} + i\left(-\frac{1}{2}\right)\right]\\
&= \frac{\sqrt{3}}{6} - \frac{i}{6}
\end{aligned}$$

To check this we divide in rectangular form.

$$\begin{aligned}
\frac{z_1}{z_2} &= \frac{\sqrt{3} + i}{3 + 3i\sqrt{3}}\\
&= \frac{(\sqrt{3} + i)}{(3 + 3i\sqrt{3})}\cdot\frac{(3 - 3i\sqrt{3})}{(3 - 3i\sqrt{3})} \qquad \text{Multiply by the conjugate}\\
&= \frac{3\sqrt{3} - \sqrt{3}(3\sqrt{3})i + 3i - 3\sqrt{3}i^2}{(3)^2 + (3\sqrt{3})^2}\\
&= \frac{6\sqrt{3} - 6i}{36} = \frac{\sqrt{3}}{6} - \frac{i}{6} \quad ■
\end{aligned}$$

## Powers in Polar Form

If we multiply $z = r(\cos\theta + i\sin\theta)$ by itself to find $z^2$, we obtain

$$\begin{aligned}
z^2 &= r\cdot r[\cos(\theta + \theta) + i\sin(\theta + \theta)]\\
&= r^2[\cos 2\theta + i\sin 2\theta].
\end{aligned}$$

Also,

$$\begin{aligned}
z^3 &= z^2\cdot z\\
&= r^2\cdot r[\cos(2\theta + \theta) + i\sin(2\theta + \theta)]\\
&= r^3[\cos 3\theta + i\sin 3\theta].
\end{aligned}$$

This process can be continued to obtain a formula for $z^n$, which is the subject of DeMoivre's theorem [named in honor of French mathematician Abraham DeMoivre (1667–1754)].

**DeMoivre's Theorem**

For any positive integer $n$,

$$z^n = r^n[\cos n\theta + i \sin n\theta].$$

**EXAMPLE 4** If $z = 2 + 2i$, determine $z^6$ and $z^8$.

Since $r = \sqrt{(2)^2 + (2)^2} = \sqrt{4 + 4} = \sqrt{8} = 2\sqrt{2}$,

we can write

$$\begin{aligned} z &= 2 + 2i \\ &= \frac{2\sqrt{2}}{\sqrt{2}} + \frac{2\sqrt{2}}{\sqrt{2}}i \\ &= 2\sqrt{2}\left(\frac{1}{\sqrt{2}} + \frac{1}{\sqrt{2}}i\right) \\ &= 2\sqrt{2}\left(\cos\frac{\pi}{4} + i \sin\frac{\pi}{4}\right). \end{aligned}$$

Thus,

$$\begin{aligned} z^6 &= r^6(\cos 6\theta + i \sin 6\theta) \\ &= (2\sqrt{2})^6\left(\cos 6\,\frac{\pi}{4} + i \sin 6\,\frac{\pi}{4}\right) \\ &= (64)(8)\left(\cos\frac{3\pi}{2} + i \sin\frac{3\pi}{2}\right) \\ &= 512(0 + i(-1)) \\ &= -512i. \end{aligned}$$

Also,

$$\begin{aligned} z^8 &= (2\sqrt{2})^8\left(\cos 8\,\frac{\pi}{4} + i \sin 8\,\frac{\pi}{4}\right) \\ &= (256)(16)(\cos 2\pi + i \sin 2\pi) \\ &= 4096(1 + i(0)) \\ &= 4096. \end{aligned}$$ ■

**EXAMPLE 5** Determine $z^{20}$ if $z = \frac{1}{2} - \frac{\sqrt{3}}{2}i$.

Convert to polar form.

$$r = \sqrt{\left(\frac{1}{2}\right)^2 + \left(-\frac{\sqrt{3}}{2}\right)^2} = \sqrt{\frac{1}{4} + \frac{3}{4}} = \sqrt{1} = 1$$

Thus,

$$z = (1)\left[\frac{1}{2} - \frac{\sqrt{3}}{2}i\right]$$

gives

$$\cos\theta = \frac{1}{2} \quad \text{and} \quad \sin\theta = -\frac{\sqrt{3}}{2},$$

which means that $\theta = \frac{5\pi}{3} \pm 2k\pi, \quad k = 0, 1, 2, \ldots.$

$$\begin{aligned}
z^{20} &= (1)^{20}\left(\cos 20 \cdot \frac{5\pi}{3} + i \sin 20 \cdot \frac{5\pi}{3}\right) \\
&= (1)\left(\cos \frac{100\pi}{3} + i \sin \frac{100\pi}{3}\right) \\
&= \cos \frac{100\pi}{3} + i \sin \frac{100\pi}{3} \\
&= \cos\left(33\pi + \frac{1}{3}\pi\right) + i \sin\left(33\pi + \frac{1}{3}\pi\right) && \frac{100}{3} = 33\frac{1}{3} \\
&= \cos\left(32\pi + \pi + \frac{1}{3}\pi\right) + i \sin\left(32\pi + \pi + \frac{1}{3}\pi\right) \\
&= \cos\left[2(16)\pi + \frac{4\pi}{3}\right] + i \sin\left[2(16)\pi + \frac{4\pi}{3}\right] \\
&= \cos \frac{4\pi}{3} + i \sin \frac{4\pi}{3} && \text{Since } \cos(2n\pi + \theta) = \cos\theta \text{ and } \sin(2n\pi + \theta) = \sin\theta \\
&= -\frac{1}{2} - \frac{\sqrt{3}}{2}i \quad \blacksquare
\end{aligned}$$

## Roots in Polar Form

An ***n*th root of a complex number** $z$ is a complex number $w$ with the property that

$$w^n = z.$$

DeMoivre's theorem can be used to prove that there are $n$ distinct $n$th roots of a given complex number $z$.

### Roots of Complex Numbers

The $n$ distinct $n$th roots of $z = r(\cos\theta + i \sin\theta)$ are given by

$$r^{1/n}\left[\cos\left(\frac{\theta}{n} + \frac{2k\pi}{n}\right) + i \sin\left(\frac{\theta}{n} + \frac{2k\pi}{n}\right)\right]$$

where $k = 0, 1, 2, \ldots, n - 1$.

## EXAMPLE 6

Find the two distinct square roots of $z = 2\sqrt{3} + 2i$. Let $w_1$ be the root for $k = 0$, and $w_2$ the root for $k = 1$.

$$\begin{aligned}
r &= \sqrt{(2\sqrt{3})^2 + (2)^2} = \sqrt{4 \cdot 3 + 4} = \sqrt{16} = 4 \\
z &= 4\left(\frac{\sqrt{3}}{2} + \frac{i}{2}\right) \\
&= 4\left(\cos\frac{\pi}{6} + i \sin\frac{\pi}{6}\right) \\
w_1 &= (4)^{1/2}\left[\cos\left(\frac{1}{2} \cdot \frac{\pi}{6} + 0\right) + i \sin\left(\frac{1}{2} \cdot \frac{\pi}{6} + 0\right)\right] \\
&= 2\left(\cos\frac{\pi}{12} + i \sin\frac{\pi}{12}\right)
\end{aligned}$$

$$
\begin{aligned}
w_2 &= (4)^{1/2}\left[\cos\left(\frac{1}{2}\cdot\frac{\pi}{6}+\frac{2\pi}{2}\right)+i\sin\left(\frac{1}{2}\cdot\frac{\pi}{6}+\frac{2\pi}{2}\right)\right] \\
&= 2\left[\cos\left(\frac{\pi}{12}+\pi\right)+i\sin\left(\frac{\pi}{12}+\pi\right)\right] \\
&= 2\left(\cos\frac{13\pi}{12}+i\sin\frac{13\pi}{12}\right) \quad ■
\end{aligned}
$$

When $z = 1$, the $n$ distinct $n$th roots of $z$, called the ***n*th roots of unity,** are of special interest in mathematics. In the next example we find the three cube (third) roots of 1.

**EXAMPLE 7** Find the three cube roots of $z = 1$: $w_1$, $w_2$, and $w_3$. Note that $z = 1(\cos 0 + i\sin 0)$.

$$
\begin{aligned}
w_1 &= (1)^{1/3}\left[\cos\left(\frac{1}{3}\cdot 0+0\right)+i\sin\left(\frac{1}{3}\cdot 0+0\right)\right] \qquad k=0 \\
&= 1(\cos 0 + i\sin 0) \\
&= 1 \\
w_2 &= (1)^{1/3}\left[\cos\left(\frac{1}{3}\cdot 0+\frac{2(1)\pi}{3}\right)+i\sin\left(\frac{1}{3}\cdot 0+\frac{2(1)\pi}{3}\right)\right] \qquad k=1 \\
&= \cos\frac{2\pi}{3}+i\sin\frac{2\pi}{3} \\
&= -\frac{1}{2}+\frac{\sqrt{3}}{2}i \\
w_3 &= (1)^{1/3}\left[\cos\left(\frac{1}{3}\cdot 0+\frac{2(2)\pi}{3}\right)+i\sin\left(\frac{1}{3}\cdot 0+\frac{2(2)\pi}{3}\right)\right] \qquad k=2 \\
&= \cos\frac{4\pi}{3}+i\sin\frac{4\pi}{3} \\
&= -\frac{1}{2}-\frac{\sqrt{3}}{2}i \quad ■
\end{aligned}
$$

If the three cube roots of unity are graphed in a complex plane as shown in Figure 48, we observe that they are all equally spaced on the unit circle. This will always be the case for the $n$th roots of unity.

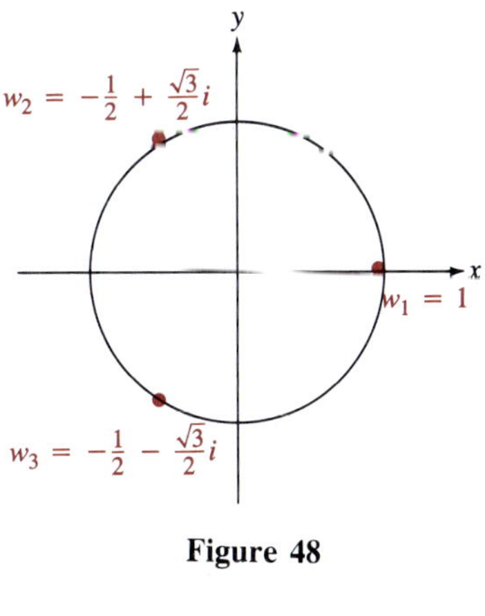


**Figure 48**

## 9.7 EXERCISES

In Exercises 1–12 determine how each complex number is represented geometrically in the complex plane. Use rectangular coordinates.

**1.** $z = 5 + 2i$ **2.** $z = -5 + 2i$ **3.** $z = -(4 - 6i)$ **4.** $z = 3(-2 + 3i)$

**5.** $z = i(1 + 2i)$ **6.** $z = -i(-3 + 4i)$ **7.** $z = (2i)^2$ **8.** $z = (2i)(-2i)$

**9.** $z = i^5$ **10.** $z = -i^7$ **11.** $z = (2 + i)^2$ **12.** $z = (4 + i)(4 - i)$

In Exercises 13–24 determine $r$ and $\theta$ and write $z$ in polar form.

**13.** $z = 1$ **14.** $z = -6$ **15.** $z = i$ **16.** $z = -4i$

**17.** $z = 1 + i$ **18.** $z = -2 - 2i$ **19.** $z = 2\sqrt{2} - 2\sqrt{2}i$ **20.** $z = -4\sqrt{2} + 4\sqrt{2}i$

**21.** $z = 2\sqrt{3} + 2i$ **22.** $z = -2 + 2\sqrt{3}i$ **23.** $z = -\frac{1}{4} + \frac{\sqrt{3}}{4}i$ **24.** $z = 5\sqrt{3} - 5i$

Change $z$ to rectangular form in Exercises 25–30.

**25.** $z = 3\left(\cos\frac{\pi}{3} + i\sin\frac{\pi}{3}\right)$ **26.** $z = \sqrt{2}\left(\cos\frac{3\pi}{4} + i\sin\frac{3\pi}{4}\right)$

**27.** $z = 4(\cos 210° + i\sin 210°)$ **28.** $z = 5(\cos 720° + i\sin 720°)$

**29.** $z = \sqrt{3}\left[\cos\left(-\frac{\pi}{4}\right) + i\sin\left(-\frac{\pi}{4}\right)\right]$ **30.** $z = \frac{1}{2}\left(\cos\frac{17\pi}{2} + i\sin\frac{17\pi}{2}\right)$

In Exercises 31–38 determine $z_1z_2$ and $z_1/z_2$ using polar form. Give answers in rectangular form.

**31.** $z_1 = 2(\cos 30° + i\sin 30°)$ and $z_2 = 3(\cos 60° + i\sin 60°)$

**32.** $z_1 = \frac{1}{2}\left(\cos\frac{\pi}{4} + i\sin\frac{\pi}{4}\right)$ and $z_2 = 6\left(\cos\frac{\pi}{2} + i\sin\frac{\pi}{2}\right)$

**33.** $z_1 = \sqrt{3}\left(\cos\frac{13\pi}{6} + i\sin\frac{13\pi}{6}\right)$ and $z_2 = \sqrt{3}\left(\cos\frac{5\pi}{6} + i\sin\frac{5\pi}{6}\right)$

**34.** $z_1 = 8\left(\cos\frac{3\pi}{4} + i\sin\frac{3\pi}{4}\right)$ and $z_2 = 6\left[\cos\left(-\frac{5\pi}{4}\right) + i\sin\left(-\frac{5\pi}{4}\right)\right]$

**35.** $z_1 = 3 - 3i$ and $z_2 = 2i$ **36.** $z_1 = -\sqrt{3} + i$ and $z_2 = \sqrt{3} + i$

**37.** $z_1 = 2 - i$ and $z_2 = 2 + i$ **38.** $z_1 = 4 - 3i$ and $z_2 = 4 - 3i$

In Exercises 39–46 use DeMoivre's theorem to determine the given power of the complex number. Give answers in rectangular form.

**39.** $i^{10}$ **40.** $(-i)^{15}$ **41.** $(1 + i)^6$ **42.** $(1 - i)^{12}$

**43.** $(\sqrt{3} - i)^8$ **44.** $\left(-\frac{\sqrt{3}}{2} - \frac{1}{2}i\right)^{20}$ **45.** $(4 - 4i)^4$ **46.** $(-\sqrt{2} + \sqrt{2}i)^{15}$

Determine the $n$ distinct roots in Exercises 47–54.

**47.** Fourth roots of 1 **48.** Fourth roots of $-1$ **49.** Cube roots of $i$

**50.** Cube roots of $-i$ **51.** Fourth roots of $8\sqrt{2} + 8\sqrt{2}i$ **52.** Fourth roots of $32\sqrt{2} - 32\sqrt{2}i$

**53.** Square roots of $-1 - \sqrt{3}i$ **54.** Square roots of $\sqrt{3} + i$

Find all solutions to the equations in Exercises 55–58.

**55.** $z^2 + 1 = 0$ **56.** $z^3 + 1 = 0$ **57.** $z^2 + i = 0$ **58.** $z^3 - 8i = 0$

## For Review

In Exercises 59–61 change the equation to polar coordinates.

**59.** $4x^2 + 4y^2 = 1$ **60.** $x = 2y$ **61.** $x^2 + y^2 - 2x - 2y = 0$

In Exercises 62–64 change the equation to rectangular coordinates.

**62.** $r \cos \theta + 5r \sin \theta = 2$ **63.** $r = 6 \sin \theta$ **64.** $3r^2 \cos 2\theta = 1$

# CHAPTER 9 REVIEW EXERCISES

Find the equation of each conic in Exercises 1–8.

**1.** Circle with center $\left(-3, \frac{2}{3}\right)$ and radius 4

**2.** Ellipse centered at the origin and intercepts $(\pm 3, 0)$ and $(0, \pm 7)$

**3.** Hyperbola with foci $(\pm 4, 0)$ and length of the transverse axis 6

**4.** Parabola with directrix $x = -3$ and focus $(5, 2)$

**5.** Circle with endpoints of a diameter at $(-2, -3)$ and $(6, 7)$

**6.** Ellipse with foci $(2, -5)$ and $(2, -1)$, and $b = 1$

**7.** Hyperbola with vertices $(0, \pm 5)$ and passing through $(3, 10)$

**8.** Parabola with vertex $(1, -2)$, axis of symmetry $x = 1$, and passing through $(5, 2)$

Sketch the graph of the conic in Exercises 9–16.

**9.** $x^2 + y^2 = 9$ **10.** $\frac{x^2}{25} + \frac{y^2}{4} = 1$ **11.** $y^2 - x^2 = 4$ **12.** $x^2 + 2y = 0$

**13.** $x^2 + y^2 - 4x + 8y + 11 = 0$ **14.** $4x^2 + y^2 + 4x - 6y - 6 = 0$

**15.** $4x^2 - y^2 + 16x - 2y - 21 = 0$ **16.** $y^2 + 6x - 4y + 22 = 0$

In Exercises 17–22 complete the squares to determine if the equation represents a circle, ellipse, hyperbola, parabola, or degenerate conic.

**17.** $3x^2 + 3y^2 - 6x + 12y - 40 = 0$ **18.** $3x^2 - 3y^2 - 6x + 12y - 80 = 0$

**19.** $3x^2 + 3y^2 - 6x - 6y + 6 = 0$ **20.** $x^2 - 3x = 0$

**21.** $3x^2 - 6x + 12y - 48 = 0$ **22.** $3x^2 + 2y^2 - 6x + 12y - 60 = 0$

**23.** A coordinate system is rotated through an angle of 30°. If the coordinates of a point in the new system are $(-1, 5)$, find the coordinates of the point in the original system.

**24.** A coordinate system is rotated through an angle of 45°. If the coordinates of a point in the original system are $(4, -5)$, find the coordinates of the point in the rotated system.

In Exercises 25–26 eliminate the $xy$-term and graph each equation.

**25.** $13x^2 + 10xy + 13y^2 = 72$

**26.** $61x^2 - 96xy + 89y^2 = 0$

Change from polar coordinates to rectangular coordinates in Exercises 27–28.

**27.** $(4, 30°)$

**28.** $\left(2, \frac{5\pi}{3}\right)$

Change from rectangular coordinates to polar coordinates in Exercises 29–30.

**29.** $(-4, 3)$

**30.** $(-5, -5)$

**31.** Change $2x - 5y + 1 = 0$ to polar coordinates.

**32.** Change $r^2 \sin 2\theta = 7$ to rectangular coordinates.

In Exercises 33–34 graph the equation in polar coordinates.

**33.** $r \cos \theta = 2$

**34.** $r = 8 \sin \theta$

In Exercises 35–36 write $z$ in polar form.

**35.** $z = 1 - i$

**36.** $z = -2\sqrt{3} + 2i$

**37.** Write $z = 6(\cos 210° + i \sin 210°)$ in rectangular form.

**38.** Determine $z_1z_2$ and $\frac{z_1}{z_2}$ if $z_1 = 2(\cos 60° + i \sin 60°)$ and $z_2 = 5(\cos 30° + i \sin 30°)$. Give answers in rectangular form.

**39.** Use DeMoivre's theorem to find $z^8$ if $z = -1 + i$.

**40.** Determine the cube roots of $z = -8i$.

**41.** ENGINEERING A highway tunnel is in the shape of a semiellipse with maximum height 12.8 m and width 16.2 m. Find the vertical clearance 4.4 m from the edge of the tunnel.

**42.** ARCHITECTURE A domed ceiling is to be constructed with a parabolic vertical cross-section. Plans call for a 16-ft beam across the dome at a point 6 ft below the top. If the highest point of the dome is 20 ft from the floor and the origin of a coordinate system is put on the floor below the high point, find the equation of the parabola.

CHAPTER

# 10 SYSTEMS OF EQUATIONS AND INEQUALITIES

In Chapter 2 we reviewed the techniques for solving equations and inequalities in one variable. Many applied problems, however, can be solved more easily if we use several equations or inequalities in two or more variables, called a *system of equations or inequalities*. Two examples of this type of problem follow.

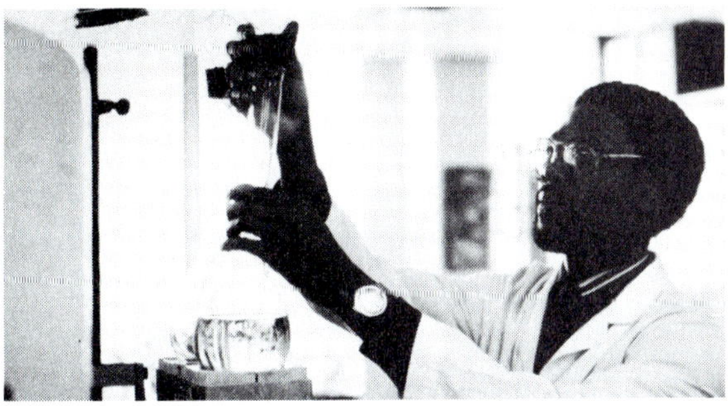

◀ CHEMISTRY

A chemist has two solutions, each containing a certain percentage of acid. If one solution is 5% acid and the other is 15% acid, how much of each should be mixed to obtain 20 L of a solution that is 12% acid?

BUSINESS ▶

Ponderosa Productions manufactures two different novelty items, Sam the Skunk and Willie the Weasel. For each item, three different phases of production are required, A, B, and C. To make one skunk, phase A takes 1 minute, phase B takes 4 minutes, and phase C takes 6 minutes. To complete one weasel, A takes 3 minutes, B takes 3 minutes, and C takes 1 minute. Due to maintenance conditions, phase A is available only 33 minutes each hour, while phases B and C are each in operation only 42 minutes every hour. The profit on each skunk is \$2.00, and the profit on each weasel is \$3.00. Find the number of units of each that should be produced hourly to maximize profit.

The first of these problems can be solved with two equations in two variables (see Example 7 in Section 10.1), and the second requires linear programming (see Example 2 in Section 10.4).

We begin our study of the topics in this chapter by considering linear equations in two and three variables and a variety of applied problems solvable using systems. Next we develop systems of inequalities and use these in linear programming. Nonlinear systems of equations and inequalities, followed by partial fractions, which have applications in calculus, conclude our work in this chapter.

## 10.1 Linear Systems in Two Variables

In Chapter 3, we defined a linear equation in two variables $x$ and $y$ to be an equation of the form

$$ax + by = c,$$

where $a$, $b$, and $c$ are constant real numbers with $a$ and $b$ not both equal to zero. Such equations have infinitely many solutions, ordered pairs of numbers $(x, y)$ that make the equation true.

A **system of two linear equations in two variables,** usually called simply a **system of equations,** is a pair of linear equations

$$\begin{aligned} a_1x + b_1y &= c_1 \\ a_2x + b_2y &= c_2, \end{aligned}$$

where $a_1$, $b_1$, $c_1$, $a_2$, $b_2$, and $c_2$ are constant real numbers. A **solution** to a system of equations is an ordered pair of numbers that is a solution to *both* equations. For example, (2, 1) is a solution to both $x - 3y = -1$ and $2x + y = 5$. Thus, (2, 1) is a solution to the system

$$\begin{aligned} x - 3y &= -1 \\ 2x + y &= 5. \end{aligned}$$

Suppose we graph each equation in this system in the same rectangular coordinate system, as in Figure 1. The solution to the system, (2, 1), corresponds to the point of intersection of the two lines. Thus, a system can be solved by graphing the lines to find the point of intersection.

### Types of Systems

Before considering two algebraic techniques for solving a system, we examine the three possible situations that can arise. If two linear equations are graphed in the same rectangular coordinate system, one of the following occurs:

1. The lines coincide, and the system of equations has infinitely many solutions. The system is said to be **dependent.**
2. The lines are parallel, and the system of equations has no solution. The system is said to be **inconsistent.**
3. The lines intersect in exactly one point, and the system of equations has exactly one solution. The system is said to be **consistent** and **independent.**

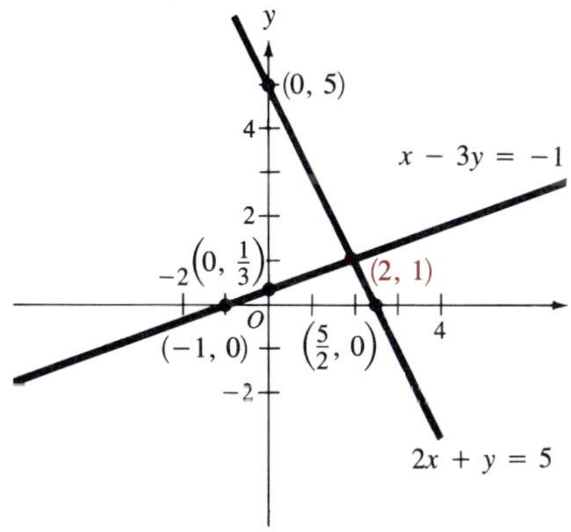


**Figure 1 Graph of a Linear System**

The following systems of equations and their corresponding graphs in Figure 2 illustrate these three cases.

**(A)** $3x - y = -3$ **(C)** $3x - y = -3$ **(E)** $3x - y = -3$

**(B)** $6x - 2y = -6$ **(D)** $3x - y = 1$ **(F)** $3x + 2y = 6$

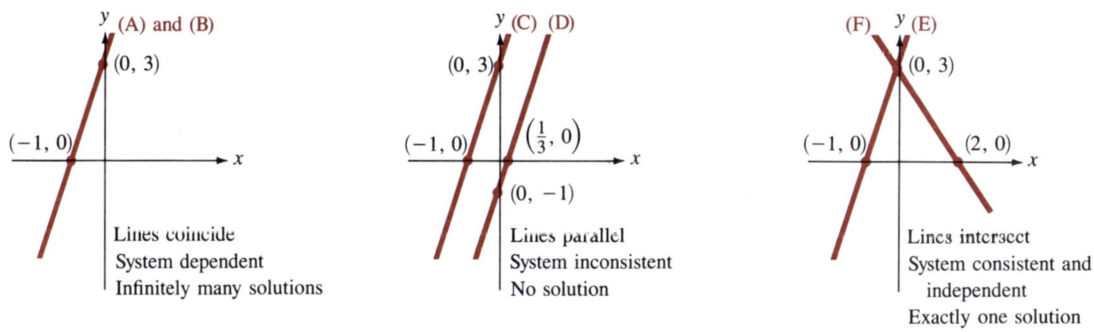


**Figure 2 Types of Linear Systems**

If the equations in a system are written in slope-intercept form, it is easy to determine which of the three cases exists. Recall that parallel lines have equal slopes and that lines with equal slopes and the same $y$-intercept must coincide. If the slopes are unequal, the lines must intersect.

**EXAMPLE 1** Without solving the system, determine the number of solutions.

**(a)** $2x - y = 1$

$-4x + 2y = -1$

Write each equation in slope-intercept form, as shown on the next page.

$$\begin{aligned} 2x - y &= 1 \\ -y &= -2x + 1 \\ y &= 2x - 1 \end{aligned} \qquad \begin{aligned} -4x + 2y &= -1 \\ 2y &= 4x - 1 \\ y &= 2x - \frac{1}{2} \end{aligned}$$

Since both slopes are 2 but the $y$-intercepts are unequal, the lines are parallel. As a result, the system is inconsistent and has no solution.

**(b)** $4x - 3y + 2 = 0$

$-8x + 6y - 4 = 0$

Write each equation in slope-intercept form.

$$\begin{aligned} 4x - 3y + 2 &= 0 \\ -3y &= -4x - 2 \\ y &= \frac{4}{3}x + \frac{2}{3} \end{aligned} \qquad \begin{aligned} -8x + 6y - 4 &= 0 \\ 6y &= 8x + 4 \\ y &= \frac{4}{3}x + \frac{2}{3} \end{aligned}$$

Since both slopes are $\frac{4}{3}$ and both $y$-intercepts are $\left(0, \frac{2}{3}\right)$, the lines coincide. As a result, any solution to either equation is a solution to the system. Since each equation has infinitely many solutions, the system is dependent and has infinitely many solutions.

**(c)** $2x - 4 = 0$

$2x + y = 3$

Since the first equation can be written in the form $x = 2$, its graph is a line parallel to the $y$-axis. The second equation has as its graph a line with slope $-2$. Thus, the two lines intersect in one point so that the system has exactly one solution and is consistent and independent. ■

When one or both of the equations in a system are of the form $ax = c$ (as in Example 1(**c**)) or $by = c$, determining the number of solutions is accomplished by inspection. Recall that the graphs of these equations are lines parallel to one of the axes so that parallel or coinciding lines will result only if both equations in the system are of the same type.

## The Substitution Method

Although we can now determine the number of solutions to a system of equations, we still need a good method for finding a particular solution. Solving a system by graphing takes too much time and hinges on the accuracy of our graphing techniques. We now turn our attention to an algebraic method for solving a system, the **substitution method.** This method is based on the substitution axiom of equality, which guarantees that any quantity can be substituted for its equal.

## EXAMPLE 2

Solve by the substitution method.

$$\begin{aligned} 2x - y &= -6 \\ 5x + 2y &= 3 \end{aligned}$$

We can solve either equation for either variable; however, we can avoid fractions by solving the first equation for $y$.

$$y = 2x + 6$$

Substitute $2x + 6$ for $y$ in the second equation.

$$\begin{aligned} 5x + 2(2x + 6) &= 3 \\ 5x + 4x + 12 &= 3 \\ 9x + 12 &= 3 \\ 9x &= -9 \\ x &= -1 \end{aligned}$$

Substitute $-1$ for $x$ in the first equation.

$$\begin{aligned} 2(-1) - y &= -6 \\ -2 - y &= -6 \\ -y &= -4 \\ y &= 4 \end{aligned}$$

The solution is $(-1, 4)$. Check by substituting $x = -1$ and $y = 4$ in each of the original equations. ■

## The Elimination Method

Suppose we are given the following system.

$$\begin{aligned} 3x + 5y &= -2 \\ 5x + 3y &= 2 \end{aligned}$$

If we use the substitution method in this case, it is impossible to avoid fractions. An alternative method of solving systems such as this, the **elimination method,** is based on *equivalent* systems. Two systems are **equivalent** if they have exactly the same solutions. Recall that if both sides of an equation are multiplied by the same nonzero number, or if the same expression is added to or subtracted from both sides, the resulting equation is equivalent to the original. When these operations are applied to one or both equations in a system, the resulting system is equivalent to the original. Transforming equations in this way is the substance of the elimination method, as illustrated in the next example.

**EXAMPLE 3** Solve by the elimination method.

$$\begin{aligned} 3x + 5y &= -2 \\ 5x + 3y &= 2 \end{aligned}$$

Multiply the first equation by $-5$ and the second by 3.

$$\begin{aligned} -15x - 25y &= 10 && -5 \text{ times the first equation} \\ 15x + 9y &= 6 && 3 \text{ times the second equation} \\ \hline -16y &= 16 && \text{Add to eliminate } x \\ y &= -1 \end{aligned}$$

Substitute $-1$ for $y$ in the first equation.

$$\begin{aligned} 3x + 5(-1) &= -2 \\ 3x - 5 &= -2 \\ 3x &= 3 \\ x &= 1 \end{aligned}$$

The solution is $(1, -1)$. Check by substituting $x = 1$ and $y = -1$ in each of the original equations. ■

**EXAMPLE 4** Solve by either method.

$$\begin{aligned} 3x - 2y &= 5 \\ -3x + 2y &= -5 \end{aligned}$$

We might observe that adding directly will eliminate $x$.

$$\begin{array}{r} 3x - 2y = \phantom{-}5 \\ \underline{-3x + 2y = -5} \\ 0 = 0 \end{array}$$

Unfortunately, we eliminated both variables and obtained the identity $0 = 0$. Suppose we apply the substitution technique, and solve the first equation for $x$.

$$\begin{aligned} 3x &= 2y + 5 \\ x &= \frac{2}{3}y + \frac{5}{3} \end{aligned}$$

Substitute into the second equation.

$$\begin{aligned} -3\left(\frac{2}{3}y + \frac{5}{3}\right) + 2y &= -5 \\ -2y - 5 + 2y &= -5 \\ 0 &= 0 \end{aligned}$$

Again we obtain the identity $0 = 0$. Upon closer examination of the original system, we see that the graphs of the equations are coinciding lines $\left[\text{both have slope } \frac{3}{2} \text{ and } y\text{-intercept } \left(0, -\frac{5}{2}\right)\right]$. As a result, the system is dependent and there are infinitely many solutions. Any pair of numbers that satisfies either equation is a solution to the system. In order to express the solution in such cases, solve one of the equations for $y$.

$$y = \frac{3}{2}x - \frac{5}{2}$$

Then the solutions to the system are of the form $\left(x, \frac{3}{2}x - \frac{5}{2}\right)$ for $x$ any real number. In particular, $\left(0, -\frac{5}{2}\right)$, $(-1, -4)$, and $\left(2, \frac{1}{2}\right)$ are three solutions found by letting $x$ equal 0, $-1$, and 2, respectively. ■

Suppose we attempt to solve the system

$$\begin{aligned} 2x - y &= 5 \\ -4x + 2y &= -3. \end{aligned}$$

Solve the first equation for $y$ and substitute into the second.

$$-4x + 2(2x - 5) = -3 \qquad y = 2x - 5$$
$$-4x + 4x - 10 = -3$$
$$-10 = -3$$

We obtain a contradiction. Similarly, if we multiply the first equation by 2 and add the result to the second, we obtain a contradiction.

$$\begin{array}{r} 4x - 2y = 10 \\ -4x + 2y = -3 \\ \hline 0 = 7 \end{array}$$

Whenever a contradiction results, the graphs of the equations of the system are parallel lines and the system has no solution.

**When Solving a System by Substitution or Elimination**

1. If an identity results, the system is dependent and has infinitely many solutions. The solutions are of the form $(x, y)$ where $x$ is any real number and $y$ is an expression in terms of $x$ obtained by solving either equation for $y$.
2. If a contradiction results, the system is inconsistent and has no solution.

Many applied problems can be translated into a system of linear equations. We now consider a variety of these applications.

**EXAMPLE 5**

**RECREATION**

By going 20 mph for one period of time and then 30 mph for another, a boater traveled from Glen Canyon Dam to the end of Lake Powell, a distance of 180 miles. If he had gone 21 mph throughout the same period of time, he would only have reached Hite Marina, a distance of 147 miles. How many hours did he travel at each speed?

Let $x$ = the number of hours traveled at 20 mph,
$y$ = the number of hours traveled at 30 mph.

Since
$$(\text{distance}) = (\text{rate}) \cdot (\text{time})$$

and the total distance traveled is the sum of the two distances traveled at the two rates, the first equation is

$$20x + 30y = 180.$$

At a rate of 21 mph for the total time, $x + y$, he traveled 147 miles. Thus, the second equation is

$$21(x + y) = 147.$$

Simplify the equations before attempting to solve, dividing the first equation by 10 and the second by 7.

$$2x + 3y = 18$$
$$3x + 3y = 21$$

Subtracting the first from the second, we have

$$x = 3.$$

Then $2(3) + 3y = 18$ so that $3y = 12$, or $y = 4$. He traveled 3 hours at 20 mph and 4 hours at 30 mph. ■

Some applications that result in systems of equations involve combining quantities, with different equations for different combinations. In many such combination problems the total value must be determined. Recall that

**total value = (value per unit) · (number of units).**

In a special type of combination problem, the mixture problem, two quantities are mixed together. The equations in the system might be called the *quantity equation* and the *value equation*.

**EXAMPLE 6**

**BUSINESS**

The owner of a candy store wishes to make a party mix by combining candy that sells for \$1.50 per pound with nuts selling for \$1.00 per pound. If the mixture is to sell for \$1.20 per pound, how many pounds of each must be used to obtain 50 pounds of the party mix?

Let $x$ = the number of pounds of candy,
$y$ = the number of pounds of nuts.

The first equation we obtain is the *quantity equation:*

(number of lb of candy) + (number of lb of nuts) = (number of lb of mixture),

which translates to $$x + y = 50.$$

Next we obtain the *value equation:*

(total value of candy) + (total value of nuts) = (total value of mixture).

That is, the value of a mixture is equal to the sum of the values of its parts. To avoid decimals, convert all monetary units to cents.

$150 \cdot x$ = total value of candy = (150¢ per lb) · (number of lb)
$100 \cdot y$ = total value of nuts = (100¢ per lb) · (number of lb)
$120 \cdot 50$ = total value of mixture = (120¢ per lb) · (number of lb)

Thus, the value equation is

$$150x + 100y = 120 \cdot 50$$

or $$3x + 2y = 120.$$ Dividing through by 50

We must solve the following system.

$$\begin{aligned} x + y &= 50 \\ 3x + 2y &= 120 \end{aligned}$$

Solve the first equation for $x$ and substitute.

$$\begin{aligned} 3(\mathbf{50 - y}) + 2y &= 120 \qquad x = 50 - y \\ 150 - 3y + 2y &= 120 \\ -y &= -30 \\ y &= 30 \\ x = 50 - y = 50 - \mathbf{30} &= 20 \end{aligned}$$

The man must use 20 lb of candy and 30 lb of nuts to obtain the desired mixture of 50 lb. ■

We conclude this section by solving the first applied problem presented in the introduction to this chapter.

**EXAMPLE 7**

**CHEMISTRY**

A chemist has two solutions, each containing a certain percentage of acid. If one solution is 5% acid and the other is 15% acid, how much of each should be mixed to obtain 20 L of a solution that is 12% acid?

Let $x$ = number of liters of the 5% acid solution,
$y$ = number of liters of the 15% acid solution.

The quantity equation is

$$x + y = 20.$$

The value equation equates the amount of acid in the solutions.

$$\begin{aligned} 0.05x &= \text{the amount of acid in } x \text{ L of the 5\% solution} \\ 0.15y &= \text{the amount of acid in } y \text{ L of the 15\% solution} \\ (0.12)(20) &= \text{the amount of acid in 20 L of the 12\% mixture} \end{aligned}$$

Thus, we have

$$\begin{aligned} 0.05x + 0.15y &= (0.12)(20) \\ 5x + 15y &= 240 \qquad \text{Clear all decimals} \\ x + 3y &= 48. \qquad \text{Divide both sides by 5} \end{aligned}$$

Solve the first equation for $x$ and substitute in the second.

$$\begin{aligned} \mathbf{20 - y} + 3y &= 48 \qquad x = 20 - y \\ 2y &= 28 \\ y &= 14 \end{aligned}$$

Then since $x = 20 - y$, $x = 20 - 14 = 6$. The chemist should mix 6 L of the 5% solution with 14 L of the 15% solution. ■

## 10.1 EXERCISES

**1.** Determine whether $(2, -3)$ is a solution to the given system.

**(a)** $x + y = -1$
$3x - y = 9$

**(b)** $2x - 4 = 0$
$y + 3 = 0$

**2.** Determine whether $(-1, 5)$ is a solution to the given system.

**(a)** $3x - 2y = -13$; $x + 3y = 16$

**(b)** $4x - y = -9$; $4x + 4 = 0$

Without solving, in Exercises 3–6, **(a)** determine the number of solutions to the given system, **(b)** tell whether the lines are parallel, coincide, or intersect, and **(c)** state whether the system is inconsistent, dependent, or consistent and independent.

**3.** $3x - y = 2$; $x - 3y = 2$

**4.** $3x + 2y = -5$; $6x + 4y = 5$

**5.** $5x - 3y + 2 = 0$; $-5x + 3y - 2 = 0$

**6.** $x + 5 = 0$; $5 + y = 0$

In Exercises 7–10 solve each system by the substitution method.

**7.** $x + 2y = 7$; $2x + y = 2$

**8.** $4x - y = -3$; $3x + 5y = 15$

**9.** $5x - 2y = 3$; $-10x + 4y = -6$

**10.** $3x - y = 2$; $2y - 8 = 0$

In Exercises 11–14 solve each system by the elimination method.

**11.** $2x - 5y = 6$; $4x + 3y = 12$

**12.** $2s - 11t = 3$; $-4s + 22t = -1$

**13.** $2a - 7b = 3$; $3a + 2b = -8$

**14.** $0.3x + 1.2y = 0.3$; $0.2x - 4.8y = 5.8$

In Exercises 15–24 solve each system by either the substitution method or the elimination method, whichever seems more appropriate.

**15.** $5x - 7y = -2$; $7x - 5y = 2$

**16.** $3y - 12 = 0$; $3x + y = 1$

**17.** $5u - 3v = 2$; $-5u + 3v = 1$

**18.** $7u - 7v = 7$; $-3u + 3v = -3$

**19.** $\frac{3}{2}x - \frac{1}{3}y = \frac{9}{20}$; $\frac{3}{4}x + \frac{2}{9}y = \frac{11}{10}$

**20.** $0.02x + 1.05y = -1.07$; $0.1x - 0.6y = 0.5$

**21.** $2(x + y) - 5(x - y) = 24$; $3(x + y) + (x - y) = 2$

**22.** $3(x - 2y) - (2x + y) = 14$; $5(x - 2y) - 2(2x + y) = 24$

**23.** $\frac{1}{x} + \frac{1}{y} = -1$; $\frac{3}{x} - \frac{6}{y} = 24$ $\left[\text{Hint: Let } u = \frac{1}{x} \text{ and } v = \frac{1}{y}.\right]$

**24.** $\frac{4}{x-1} + \frac{1}{y+2} = \frac{17}{4}$; $\frac{3}{x-1} - \frac{2}{y+2} = \frac{5}{2}$

In Exercises 25–26 solve each system for $x$ and $y$ assuming that $a$ and $b$ are nonzero constants.

**25.** $ax + by = 1$; $3ax - by = -5$

**26.** $ax - y = 2b$; $ax + 2y = -b$

In Exercises 27–28 find $a$ and $b$ such that $(-1, 3)$ is a solution to each system.

**27.** $ax - by = 11$; $-2ax - by = 14$

**28.** $3ax + by = -15$; $ax - 4by = 60$

In Exercises 29–43 solve using a system of equations.

**29.** The sum of two numbers is 44 and their difference is 20. Find the numbers.

**30.** Twice the sum of Sam's and Joe's ages is 40. In 2 years, Sam will be the same age as Joe is now. How old is each?

**31.** By traveling 40 km/hr for one period of time and then 50 km/hr for another, Mario traveled 370 km. Had he gone 10 km/hr faster throughout, he would have traveled 450 km. How many hours did he travel at each rate?

**32.** RECREATION A cruise boat travels 48 miles downstream in 2 hours and returns the 48 miles upstream in 3 hours. Find the speed of the boat and the speed of the stream.

**33.** GEOMETRY Two angles are supplementary, and one is 4° more than seven times the other. Find the angles.

**34.** GEOMETRY In a right triangle, one acute angle is 6° more than twice the other. Find each acute angle.

**35.** CONSUMER Clyde bought 5 shirts of the same value and 4 pairs of socks of the same value for \$87. He returned to the same store later and purchased (at the same prices) 2 shirts and 6 pairs of socks for \$48. What was the price of each?

**36.** RECREATION If there were 600 people at a play, the total receipts were \$980, and the admission price was \$2.00 for adults and \$1.00 for children, how many adults and how many children were in attendance?

**37.** RETAILING A candy mix sells for \$1.10 per pound. If the mix is composed of two kinds of candy, one worth 90¢ per pound and the other worth \$1.50 per pound, how many pounds of each would be in a 60-pound mixture?

**38.** RETAILING The owner of the Coffee Mill wishes to mix two blends of coffee, one selling for \$1.80 per pound and the other for \$2.40 per pound, to obtain a 40-pound mixture selling for \$2.10 per pound. How many pounds of each must he use?

**39.** A collection of dimes and quarters is worth \$4.20. If the total number of coins in the collection is 30, how many of each are there?

**40.** Cindy has 40 coins consisting of nickels and dimes. If the total value of the collection is \$3.35, find the number of nickels and the number of dimes in the collection.

**41.** CHEMISTRY A chemist has one solution that is 25% acid and another that is 50% acid. How much of each should be used to make 25 L of a 40% acid solution?

**42.** CHEMISTRY A lab technician obtains 10 gallons of a 30% saline solution by mixing some 20% solution with some 50% solution. How much of each must he use?

**43.** NUTRITION A lab technician wishes to place the animals in an experiment on a specific diet of 15 grams of protein and 5 grams of fat. She is able to purchase two food mixes, Special Diet, which is 12% protein and 2% fat, and Control K, which is 20% protein and 8% fat. How many grams of each mix (to the nearest tenth of a gram) should she combine to obtain the right diet mixture for her animals?

**44.** STATISTICS A statistician uses the **least squares regression line**, $y = ax + b$, as the line best fitted to represent a collection of points $\{(x_1, y_1), (x_2, y_2), (x_3, y_3)\}$. The values for $a$ and $b$ are determined by the system of equations

$$\begin{aligned} (x_1 + x_2 + x_3)a + \qquad\qquad\quad 3b &= (y_1 + y_2 + y_3) \\ (x_1^2 + x_2^2 + x_3^3)a + (x_1 + x_2 + x_3)b &= (x_1y_1 + x_2y_2 + x_3y_3). \end{aligned}$$

Find the least squares regression line for the points $\{(-1, 0), (1, 1), (2, 2)\}$.

**45.** SPORTS To stay in good physical condition, a professor wants to jog or play tennis 15 times each month. If he estimates that a total of 26 hr can be spent on these activities each month, each time he jogs he spends 1.2 hours, and each time he plays tennis he spends 2.8 hours, how many times should he pursue each activity during the month in order to use the entire 26 hours?

**46.** Find the value(s) for $m$ so that the system

$$\begin{aligned} x - 4y &= m \\ -2x + 8y &= 4 \end{aligned}$$

has no solution.

**47.** Show that the system

$$\begin{aligned} 2x + y &= m \\ x - 4y &= 3 \end{aligned}$$

has exactly one solution for every real number $m$.

**48.** Find the value(s) for $m$ so that the system

$$\begin{aligned} 5x + 2y &= m \\ -15x - 6y &= 9 \end{aligned}$$

has infinitely many solutions.

**49.** Find the values of $a$ and $b$ so that the line with equation $ax + by = 6$ passes through the points with coordinates $(3, 2)$ and $(0, -2)$.

**50.** Show that the graph of the quadratic function $f(x) = x^2 - 4x + 3$ intersects the graph of the linear function $g(x) = -3x + 5$ at the points $(2, -1)$ and $(-1, 8)$.

## For Review

In Exercises 51–54 determine the nature of the conic section by completing the square.

**51.** $5x^2 + 10y^2 - 10x + 20y = 0$

**52.** $3x^2 - 6x + 4y = 4$

**53.** $5x^2 - 10y^2 - 10x + 20y = 0$

**54.** $8x^2 + 8y^2 + 16x + 16y - 64 = 0$

# 10.2 Linear Systems in More Than Two Variables

## Linear Systems in Three Variables

An equation of the form

$$ax + by + cz = d$$

when $a$, $b$, $c$, and $d$ are constant real numbers and $x$, $y$, and $z$ are variables, is called a **linear equation in three variables.** Since the graph of a linear equation in three variables is a plane in space, not a line, the term *linear equation* here is a bit misleading (perhaps *first-degree equation* is better).

A solution to a linear equation such as $2x + y - 3z = 3$ is an **ordered triple** of numbers. For example, $(1, -2, -1)$ is a solution since if $x$, $y$, and $z$ are replaced with 1, $-2$, and $-1$, in that order, a true equation results.

$$2(1) + (-2) - 3(-1) \stackrel{?}{=} 3$$
$$2 - 2 + 3 \stackrel{?}{=} 3$$
$$3 = 3$$

A **system of three linear equations in three variables** (or again simply a system of equations) is a trio of linear equations such as the following.

$$x + y - z = 4$$
$$2x + y + z = 1$$
$$3x - 2y - z = 3$$

A solution to such a system is an ordered triple that is a solution to *all three* equations. It is easy to verify by substitution that $(1, 1, -2)$ is a solution to the above system.

## The Reduction Method

A system of three linear equations in three variables can be solved algebraically by either substitution or elimination. Alternatively, a **reduction method,** using both elimination and substitution, transforms the system into a system of two equations in two variables. This technique is perhaps the easiest to learn and the quickest to apply. We summarize below the steps in the reduction method.

### To Solve a System of Three Equations in Three Variables

1. Select any two of the three equations and eliminate a variable.
2. Use the equation in the original system that was not used in the first step together with either of the other two equations and eliminate the *same* variable.
3. Solve the system of two equations in two variables.
4. Substitute the values of the two variables into any of the original equations to obtain the value of the third variable.
5. If at any step a contradiction is obtained, the system has no solution.

To solve a system of equations, study the system to determine which of the three variables is the easiest to eliminate. This can save time and effort in later steps. We will illustrate the reduction method by solving the system given earlier in this section.

## EXAMPLE 1

Solve the system.

$$\textbf{(A)}\quad x + y - z = 4$$
$$\textbf{(B)}\quad 2x + y + z = 1$$
$$\textbf{(C)}\quad 3x - 2y - z = 3$$

Notice that the multiplication rule is not needed if we eliminate the variable $z$.

Adding equations **(A)** and **(B)** eliminates $z$ and results in an equation in $x$ and $y$. Then we pair equation **(C)** with **(A)** and subtract. This gives a second equation in $x$ and $y$.

$$\begin{array}{rrl} & \textbf{(A)} & x + y - z = 4 \\ & \textbf{(B)} & 2x + y + z = 1 \\ \hline \textbf{(A)} + \textbf{(B)} = & \textbf{(D)} & 3x + 2y \quad = 5 \end{array} \qquad \begin{array}{rrl} & \textbf{(A)} & x + y - z = 4 \\ & \textbf{(C)} & 3x - 2y - z = 3 \\ \hline \textbf{(A)} - \textbf{(C)} = & \textbf{(E)} & -2x + 3y \quad = 1 \end{array}$$

$$\begin{array}{rl} \textbf{(D)} & 3x + 2y = 5 \\ \textbf{(E)} & -2x + 3y = 1 \end{array}$$

We have now reduced the original system to a system of two equations in the two variables $x$ and $y$. To eliminate $x$, we multiply **(D)** by 2 and **(E)** by 3, and add the results.

$$\begin{array}{rr} \textbf{2(D)} & 6x + 4y = 10 \\ \textbf{3(E)} & -6x + 9y = 3 \\ \hline & 13y = 13 \\ & y = 1 \end{array}$$

We can now find the value of $x$ by substituting 1 for $y$ in either **(D)** or **(E).** Suppose we choose **(D).**

$$3x + 2(1) = 5$$
$$3x = 3$$
$$x = 1$$

Finally, to find the value of $z$, we substitute 1 for $x$ and 1 for $y$ in any of the original equations **(A), (B),** or **(C).** Suppose we use **(A).**

$$(1) + (1) - z = 4$$
$$2 - z = 4$$
$$z = -2$$

Thus, the solution to the system is $(1, 1, -2)$. ■

## Systems with No Solution

If a contradiction is obtained in solving a system, then the system has no solution, as shown in the next example.

## EXAMPLE 2

Solve the system.

$$\begin{array}{rl} \textbf{(A)} & x + y - z = 4 \\ \textbf{(B)} & x - y - z = 2 \\ \textbf{(C)} & 3x - y - 3z = -4 \end{array}$$

By adding **(A)** and **(B),** we can eliminate $y$. Then by pairing **(C)** with **(A)** and adding, $y$ is again eliminated.

$$\begin{array}{rrl} & \textbf{(A)} & x + y - z = 4 \\ & \textbf{(B)} & x - y - z = 2 \\ \hline \textbf{(A)} + \textbf{(B)} = & \textbf{(D)} & 2x \quad - 2z = 6 \\ \frac{1}{2}\textbf{(D)} = & \textbf{(E)} & x \quad - z = 3 \end{array} \qquad \begin{array}{rrl} & \textbf{(A)} & x + y - z = 4 \\ & \textbf{(C)} & 3x - y - 3z = -4 \\ \hline \textbf{(A)} + \textbf{(C)} = & \textbf{(F)} & 4x \quad - 4z = 0 \\ \frac{1}{4}\textbf{(F)} = & \textbf{(G)} & x \quad - z = 0 \end{array}$$

$$\begin{aligned} \textbf{(E)}\quad x - z &= 3 \\ \textbf{(G)}\quad x - z &= 0 \end{aligned}$$

Subtracting **(G)** from **(E)** gives the contradiction

$$0 = 3.$$

Thus, the original system has no solution. ■

## Systems with Infinitely Many Solutions

If an identity is obtained in any step of solving a system of three equations in three variables, the system has either infinitely many solutions or no solution. If there are infinitely many solutions, they are expressed as for a system of two equations in two variables. The next example shows how this is done.

## EXAMPLE 3

Solve the system.

$$\begin{aligned} \textbf{(A)}\quad x - 3y + 2z &= 6 \\ \textbf{(B)}\quad 4x - 2y + 3z &= 14 \\ \textbf{(C)}\quad 2x + 4y - z &= 2 \end{aligned}$$

First eliminate $x$ from **(A)** and **(C)**, and then eliminate $x$ from **(B)** and **(C)**.

$$\begin{array}{rrcl} \textbf{2(A)} & 2x - 6y + 4z &=& 12 \\ \textbf{(C)} & 2x + 4y - z &=& 2 \\ \hline \textbf{2(A)} - \textbf{(C)} = \textbf{(D)} & -10y + 5z &=& 10 \\ -\frac{1}{5}\textbf{(D)} = \textbf{(E)} & 2y - z &=& -2 \end{array} \qquad \begin{array}{rrcl} \textbf{(B)} & 4x - 2y + 3z &=& 14 \\ \textbf{2(C)} & 4x + 8y - 2z &=& 4 \\ \hline \textbf{(B)} - \textbf{2(C)} = \textbf{(F)} & -10y + 5z &=& 10 \\ -\frac{1}{5}\textbf{(F)} = \textbf{(G)} & 2y - z &=& -2 \end{array}$$

$$\begin{aligned} \textbf{(E)}\quad 2y - z &= -2 \\ \textbf{(G)}\quad 2y - z &= -2 \end{aligned}$$

When **(E)** and **(G)** are subtracted, the result is the identity

$$0 = 0.$$

To express the solution, select one of the two equivalent equations in $y$ and $z$, say **(E)**, and solve for one of the variables, say $z$.

$$\begin{aligned} 2y - z &= -2 \\ -z &= -2y - 2 \\ z &= 2y + 2 \end{aligned}$$

Substitute this value of $z$ into any of the original equations, say **(A)**, and solve for $x$ in terms of $y$.

$$\begin{aligned} x - 3y + 2(2y + 2) &= 6 \qquad z = 2y + 2 \\ x - 3y + 4y + 4 &= 6 \\ x + y &= 2 \\ x &= 2 - y \end{aligned}$$

Both $x$ and $z$ have now been expressed in terms of $y$. If $y$ is any real number, then the ordered triple $(2 - y, y, 2y + 2)$ is a solution to the system. Particular solutions can be found by selecting a value of $y$ and evaluating the expressions for $x$ ($x = 2 - y$) and for $z$ ($z = 2y + 2$). Three such solutions are $(2, 0, 2)$, $(1, 1, 4)$, and $(3, -1, 0)$, found by letting $y = 0$, 1, and $-1$, respectively. ■

CAUTION In Example 3, if we solved equation **(E)** for $y$ instead of $z$, and substituted into **(B)** to find $x$, then the form of the solution would be $\left(-\frac{z}{2} + 3, \frac{z}{2} - 1, z\right)$, for $z$ any real number. The three particular solutions we found would then be generated by letting $z = 2$, 4, and 0. Also, if we had eliminated one of the other variables instead of $x$ in the very first step, the solution triple might have been expressed using $x$ as any number with $y$ and $z$ in terms of $x$. In other words, there are three ways to represent the solutions when there are infinitely many of them. All three ways yield the same set of solutions. ■

Thus far we have considered systems of two and three equations. We now expand our discussion to include more general systems.

### $n \times n$ Systems

If $x_1, x_2, x_3, \ldots, x_n$ represent $n$ variables, where $n$ is a positive integer, and $c, a_1, a_2, a_3, \ldots, a_n$, represent $n + 1$ constant real numbers such that at least one $a_j \neq 0$, then

$$a_1x_1 + a_2x_2 + a_3x_3 + \cdots + a_nx_n = c$$

is a linear equation in $n$ variables. A system of $n$ linear equations in $n$ variables is called an ***n* × *n* system** or a **square system of order *n*.**

## Nonsquare Systems

In the preceding section we solved $2 \times 2$ systems, and in this section we used the reduction method to solve $3 \times 3$ systems. The reduction method can be used to solve any $n \times n$ system. We reduce it to an $(n - 1) \times (n - 1)$ system, and continue. For example, a $4 \times 4$ system can be reduced to a $3 \times 3$ system, then to a $2 \times 2$ system. Systems of equations that have fewer equations than variables, called **nonsquare systems,** have either infinitely many solutions or no solution, and cannot have a unique solution. We illustrate the method for solving such systems in the next example.

## EXAMPLE 4

Solve the nonsquare systems.

**(a)** **(A)** $x - 2y + z = 3$
**(B)** $x + 4y - z = -7$

By adding **(A)** and **(B),** we can eliminate $z$ and obtain

$$2x + 2y = -4$$
$$x + y = -2$$
$$y = -x - 2. \quad \text{Solve for } y \text{ in terms of } x$$

Substitute $-x - 2$ for $y$ in either of the original equations—we'll choose **(A)**—and solve for $z$ in terms of $x$.

$$x - 2(-x - 2) + z = 3$$
$$x + 2x + 4 + z = 3$$
$$z = -3x - 1$$

Thus, the solutions to the system can be expressed as $(x, -x - 2, -3x - 1)$, for $x$ any real number.

**(b)** **(A)** $-x + 3y + z = 2$
**(B)** $2x - 6y - 2z = -2$

If equation **(A)** is multiplied by 2 and the result added to **(B),** the contradiction $0 = 2$ is obtained. Thus, the system has no solution. ■

## Homogeneous Systems

If all the constant terms in a system of equations are zero, the system is called **homogeneous.** For example,

$$a_1x_1 + a_2x_2 + a_3x_3 = 0$$
$$b_1x_1 + b_2x_2 + b_3x_3 = 0$$
$$c_1x_1 + c_2x_2 + c_3x_3 = 0$$

is a homogeneous $3 \times 3$ system. One obvious solution to a homogeneous system such as this is the **trivial solution,** (0, 0, 0). Homogeneous systems often have nontrivial solutions also, and these can be found by the reduction method.

We conclude this section with an example applying a system of three equations in three variables to a geometry problem.

## EXAMPLE 5

GEOMETRY

In a triangle, the largest angle is 70° more than the smallest angle, and the remaining angle is 10° more than three times the smallest angle. Find the measure of each angle.

Let $x$ = the measure of the smallest angle,
$y$ = the measure of the middle angle,
$z$ = the measure of the largest angle.

Since $z$ is 70° more than $x$,

$$x + 70 = z$$

or **(A)** $x - z = -70.$

Since $y$ is 10° more than 3 times $x$,

$$3x + 10 = y$$

or **(B)** $3x - y = -10.$

Finally, the sum of the measures of the angles of a triangle is 180°.

**(C)** $x + y + z = 180$

Hence, we must solve the following system.

$$\begin{aligned} &\textbf{(A)}\quad x \phantom{{}-y} - z = -70 \\ &\textbf{(B)}\quad 3x - y \phantom{{}+z} = -10 \\ &\textbf{(C)}\quad x + y + z = 180 \end{aligned}$$

Note that $y$ is missing in **(A)**. If we add **(B)** and **(C)** the result is also an equation with $y$ missing.

$$\textbf{(D)}\quad 4x + z = 170 \qquad \text{(B) + (C) = (D)}$$

Thus we must solve the following system.

$$\begin{aligned} &\textbf{(A)}\quad x - z = -70 \\ &\textbf{(D)}\quad 4x + z = 170 \end{aligned}$$

Add **(A)** and **(D)**.

$$\begin{aligned} 5x &= 100 \\ x &= 20 \end{aligned}$$

Substitute 20 for $x$ in **(A)**.

$$\begin{aligned} 20 - z &= -70 \\ -z &= -90 \\ z &= 90 \end{aligned}$$

Substitute 20 for $x$ in **(B)**.

$$\begin{aligned} 3(20) - y &= -10 \\ 60 - y &= -10 \\ -y &= -70 \\ y &= 70 \end{aligned}$$

The angles have measure 20°, 70°, and 90°. ■

## 10.2 EXERCISES

Solve each system of equations in Exercises 1–22.

**1.** $\begin{aligned} x + y + z &= 3 \\ -x + 2y - z &= 0 \\ 3x - y + 2z &= 2 \end{aligned}$

**2.** $\begin{aligned} 3x - y + z &= 10 \\ x + 2y - z &= -2 \\ -2x + y + z &= 0 \end{aligned}$

**3.** $\begin{aligned} x + y + z &= 6 \\ x \phantom{{}+y} - z &= -2 \\ y + 3z &= 11 \end{aligned}$

[Hint: Eliminate $x$ from the first two equations and pair the result with the third.]

**4.** $\begin{aligned} 3x + 2y + z &= 2 \\ x - 2y - z &= 2 \\ 2x - y + z &= 7 \end{aligned}$

[Hint: Adding the first two equations gives the numerical value of $x$. Substitute this value to obtain a $2 \times 2$ system in $y$ and $z$.]

**5.** $\begin{aligned} x + 5y - z &= 2 \\ 4x - y + 3z &= 3 \\ 8x - 2y + 6z &= 7 \end{aligned}$

**6.** $\begin{aligned} x - y + z &= -8 \\ 2x + y + 2z &= -1 \\ x + y + z &= 2 \end{aligned}$

**7.** $\begin{aligned} 3x + y + z &= 0 \\ -5x + 5y + z &= 0 \\ x + 2y + z &= 0 \end{aligned}$

**8.** $\begin{aligned} x + 2y - 3z &= 0 \\ 2y + z &= 0 \\ x + 4y - 2z &= 0 \end{aligned}$

**9.** $\begin{aligned} 2x + y &= 0 \\ x - 3y + z &= 0 \\ 3x + y - z &= 0 \end{aligned}$

**10.** $\begin{aligned} 5x - y + z &= 0 \\ 2x + y - 2z &= 0 \\ x - y - z &= 0 \end{aligned}$

**11.** $\begin{aligned} x - 3y + 2z &= -1 \\ 4x + 3y + 3z &= 6 \end{aligned}$

**12.** $\begin{aligned} 2x - y + 5z &= 3 \\ 2x + y - z &= 1 \end{aligned}$

**13.** $\begin{aligned} x - 3y + 5z &= 2 \\ -2x + 6y - 10z &= 7 \end{aligned}$

**14.** $\begin{aligned} 5x + y - 3z &= 2 \\ -15x - 3y + 9z &= 5 \end{aligned}$

**15.** $\begin{aligned} \frac{1}{4}x - \frac{1}{3}y - \frac{1}{2}z &= -2 \\ \frac{1}{2}x - \frac{1}{2}y + \frac{1}{4}z &= 2 \\ -\frac{1}{4}x + \frac{1}{2}y - \frac{1}{2}z &= -1 \end{aligned}$

**16.** $\begin{aligned} \frac{1}{x} - \frac{2}{y} - \frac{1}{z} &= 2 \\ \frac{2}{x} - \frac{1}{y} + \frac{1}{z} &= 7 \\ \frac{3}{x} + \frac{2}{y} + \frac{1}{z} &= 2 \end{aligned}$

**17.** $\begin{aligned} x - y + z + w &= 2 \\ x + y - z + w &= 4 \\ x + y + z - w &= -2 \\ x - y - z - w &= 0 \end{aligned}$

**18.** $\begin{aligned} 5x - y - 3z + w &= 0 \\ 2x - 3y + z - 4w &= 0 \\ -x + 2y + 5z + w &= 0 \\ 2x + 2y - 3z - w &= 0 \end{aligned}$

**19.** $\begin{aligned} x + y + z - w &= 1 \\ 2x - y + 3z - w &= -2 \\ x - y + 2z - w &= 3 \end{aligned}$

**20.** $\begin{aligned} 2x - y - z + 3w &= -1 \\ x + 5y - 3z + 2w &= 7 \\ -4x + 2y + 2z - 6w &= 5 \end{aligned}$

**21.** $\begin{aligned} x + 2y &= 3 \\ x - 2y &= 5 \\ 3x + y &= 1 \end{aligned}$

**22.** $\begin{aligned} x - y &= 2 \\ x + y &= 0 \\ x - 3y &= 4 \end{aligned}$

In Exercises 23–35 solve using a system of equations.

**23.** The sum of three numbers is 4. The first, plus twice the second, plus the third, is 1. Three times the first, plus the second, minus the third, is −2. Find the three numbers.

**24.** EDUCATION The average of a student's three scores is 74. If the first is 21 points less than the second, and twice the third is 15 more than the sum of the first two, find all three scores.

**25.** The sum of the ages of Milt, Lew, and Jenny is 53. Jenny is 5 years younger than Lew, and in two years, Milt will be the same age as Lew is now. How old is each?

**26.** GEOMETRY The smallest angle of a triangle is one-third the largest angle, and the middle-sized angle is 30° less than the largest angle. Find the measure of each angle.

**27.** SPORTS The total number of seats in a basketball sports arena is 12,000. The arena is divided into three sections, courtside, endzone, and balcony, and there are twice as many balcony seats as courtside seats. For the conference championship game, ticket prices were \$10.00 for courtside, \$8.00 for balcony, and \$7.00 for endzone. If the arena was sold out for the game and the total receipts were \$99,000, how many seats were courtside?

**28.** RETAILING On Wednesday, an appliance dealer sold 3 stoves, 4 refrigerators, and 2 washers for a total of \$4950. On Thursday she sold 2 stoves, 5 refrigerators, and 1 washer for a total of \$4650. On Friday she sold 4 stoves and 2 refrigerators for \$3100. If the stoves, refrigerators, and washers were all of the same value, respectively, what was the price of a stove?

**29.** CONSUMER Becky has a collection of nickels, dimes, and quarters with a total value of \$4.60. Twice the number of nickels is the same as three times the number of quarters, and the total number of coins is 40. How many of each are there?

**30.** A collection of 100 coins consisting of nickels, dimes, and quarters has a value of \$16.00. If the number of nickels plus the number of dimes is equal to the number of quarters, find the number of each.

**31.** **INVESTMENT** Larry has \$5000 split into three separate investments. Part of the money is invested in bonds at 8%, part in certificates at 7%, and the rest is in a mutual fund. If the fund does well and earns 6%, the total earnings from all the investments will amount to \$345. However, if the fund does not do well, he will lose 3% on his investment, and the total earnings from the three will amount to only \$165. How much is invested in each category?

**32.** **INVESTMENT** Gail has \$10,000 split into three separate investments. Part is invested in a mutual fund that earns 8%, part is invested in time certificates that earn 7%, and the rest is invested in a business. If the business does well, it will earn 10% and her total earnings will amount to \$790. If the business loses 2%, her total earnings will amount to only \$550. How much is invested in each category?

**33.** **MANUFACTURING** Aspen Stoves manufactures three models of woodburning stoves, the Sierra, the San Juan, and the Blue Ridge. Each stove must pass through three stages: cutting, welding, and finishing. The total number of production hours weekly is 195 for cutting, 200 for welding, and 190 for finishing. The number of hours required in each stage for each stove is summarized in the table below. How many of each type of stove must be manufactured weekly for the company to operate at full production capacity?

| Stage | Sierra | San Juan | Blue Ridge |
|---|---|---|---|
| Cutting | 5 | 5 | 2 |
| Welding | 4 | 6 | 2 |
| Finishing | 4 | 5 | 3 |

**34.** **RECREATION** When three valves, $A$, $B$, and $C$, are opened simultaneously, a swimming pool can be filled in 1 hr. When $A$ and $B$ are opened (without $C$), the pool can be filled in $\frac{6}{5}$ hr, and when $B$ and $C$ are opened (without $A$), it can be filled in 2 hr. How long would it take each inlet to fill the pool by itself?

**35.** **CURVE-FITTING** The polynomial function $P(x) = x^3 + ax^2 + bx + c$ has the property that $P(1) = 6$, $P(-1) = 6$, and $P(2) = 15$. Find the values of $a$, $b$, and $c$.

**ANALYTIC GEOMETRY** Two distinct points determine a unique straight line. In a similar manner, three noncollinear points, no two of which are on the same vertical line, determine a unique parabola with equation $y = ax^2 + bx + c$. By substituting the coordinates of three given points into this equation, the result is a $3 \times 3$ system in $a$, $b$, and $c$, the solution to which determines the equation. In Exercises 36–37 find the equation of the parabola determined by the given points.

**36.** $(1, 8)$, $(-1, 4)$, $(3, 20)$

**37.** $(1, 0)$, $(-2, -12)$, $(2, 0)$

**38.** STATISTICS A statistician might use the **least squares regression parabola,** $y = ax^2 + bx + c$, as the parabola best fitted to represent a collection of points $\{(x_1, y_1), (x_2, y_2), (x_3, y_3), (x_4, y_4)\}$. The values for $a$, $b$, and $c$ are determined by the system of equations

$$\begin{aligned}
(x_1^2 + x_2^2 + x_3^2 + x_4^2)a + (x_1 + x_2 + x_3 + x_4)b + 4c &= (y_1 + y_2 + y_3 + y_4)\\
(x_1^3 + x_2^3 + x_3^3 + x_4^3)a + (x_1^2 + x_2^2 + x_3^2 + x_4^2)b + (x_1 + x_2 + x_3 + x_4)c &= (x_1y_1 + x_2y_2 + x_3y_3 + x_4y_4)\\
(x_1^4 + x_2^4 + x_3^4 + x_4^4)a + (x_1^3 + x_2^3 + x_3^3 + x_4^3)b + (x_1^2 + x_2^2 + x_3^2 + x_4^2)c &= (x_1^2y_1 + x_2^2y_2 + x_3^2y_3 + x_4^2y_4).
\end{aligned}$$

Find the least squares regression parabola for the points $\{(-1, 0), (0, 2), (0, 1), (-2, 0)\}$.

ANALYTIC GEOMETRY Three noncollinear points determine a unique circle with equation $x^2 + y^2 + ax + by + c = 0$. In Exercises 39–40 find the equation of the circle determined by the given points.

**39.** $(1, 2), (4, -1), (-2, -1)$

**40.** $(-2, 5), (3, 0), (-2, -5)$

## For Review

**41.** Solve the system for $x$ and $y$ assuming that $a$ and $b$ are nonzero constants.

$$\begin{aligned} ax - by &= 2\\ 2ax + by &= 4 \end{aligned}$$

**42.** Find value(s) for $m$ so that the system is dependent.

$$\begin{aligned} 2x - 5y &= 3\\ 6x - 15y &= m \end{aligned}$$

**43.** Find values for $m$ so that the system is inconsistent.

$$\begin{aligned} 3x + 7y &= 2m\\ -6x - 14y &= 5 \end{aligned}$$

**44.** The sum of the digits of a two-digit number is 5. If the digits are reversed, the resulting number is 9 less than the original. Find the number.

**45.** GEOMETRY The perimeter of a rectangle is 50 ft and the length is 1 ft more than twice the width. Find the dimensions of the rectangle.

**46.** CURVE-FITTING Determine the constants $a$ and $b$ for the function $f(x) = ae^x + be^{-x} + 1$ if $f(0) = 2$ and $f(\ln 2) = \dfrac{9}{2}$.

Sketch the graph of each linear equation in Exercises 47–50. These problems will help us prepare for the material in the next section.

**47.** $4x + 2y = -6$

**48.** $3x + 2y = 0$

**49.** $x + 1 = 0$

**50.** $2y - 2 = 0$

# 10.3 Linear Systems of Inequalities

### Linear Inequalities

In Chapter 2 we solved linear inequalities in one variable and graphed the solutions on a number line. We now consider inequalities in two variables in which the variables are raised to the first power. **A linear inequality in two variables** $x$ and $y$ can always be written in one of the forms

$$ax + by < c, \quad ax + by > c, \quad ax + by \leq c, \quad \text{or} \quad ax + by \geq c,$$

where $a$, $b$, and $c$ are constant real numbers with $a$ and $b$ not both zero. As with the linear equations we studied previously, a solution to an inequality is an ordered pair of numbers which, when substituted for $x$ and $y$, makes the inequality true. For example, $(-1, 0)$ is a solution to $2x + y < -1$ since

$$2(-1) + 0 < -1$$
$$-2 < -1 \quad \text{True}$$

is true. On the other hand, $(4, -3)$ is not a solution since

$$2(4) + (-3) < -1$$
$$8 - 3 < -1$$
$$5 < -1 \quad \text{False}$$

is false.

## Graphing Linear Inequalities

One way to identify the solution to a linear inequality is to show its graph, that is, the set of all points in the plane whose coordinates solve the inequality. Graphing a linear inequality is not much more difficult than graphing a linear equation. In fact, to begin we temporarily replace the inequality symbol with an equal sign and graph

$$ax + by = c,$$

which is a straight line. This **boundary line** divides the plane into two regions called **half-planes** as shown in Figure 3. If the inequality is $\leq$ or $\geq$, the points on the boundary line are included with the half-plane, and the region is a **closed half-plane.** If the inequality is $<$ or $>$, the boundary line is not included, and the region is an **open half-plane.** We indicate a closed half-plane with a solid line and an open half-plane with a dashed line.

Since $y_1 > y$, the point $(x, y_1)$ in the upper half-plane in Figure 3 would satisfy the

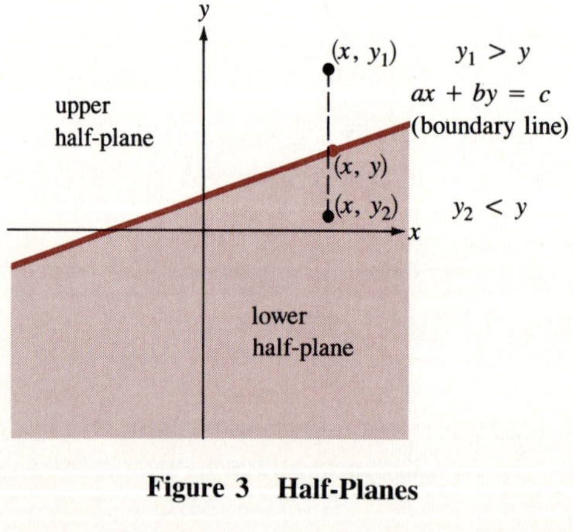


**Figure 3 Half-Planes**

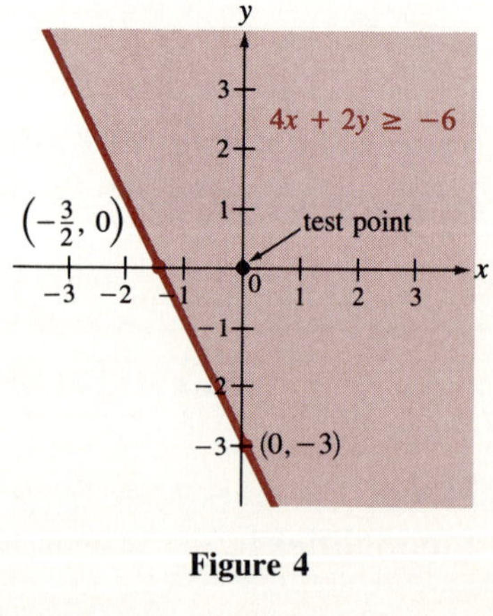


**Figure 4**

inequality $ax + by > c$. Similarly, since $y_2 < y$, the point $(x, y_2)$ in the lower half-plane would satisfy the inequality $ax + by < c$. The solutions of a given inequality correspond to all points in exactly *one* of the half-planes determined by the boundary line.

Thus we have the following method.

### To Graph an Inequality in Two Variables

1. Graph the boundary line using a solid line if the inequality is $\leq$ or $\geq$ or a dashed line if it is $<$ or $>$.
2. Choose any point not on the boundary line (the point (0, 0) is often selected) and use it as a **test point** by substituting its coordinates into the inequality.
3. Shade the half-plane containing the test point if a true inequality is obtained, or shade the half-plane that does not contain the test point if a false inequality results.

**EXAMPLE 1** Graph the inequality $4x + 2y \geq -6$.

First graph the line $4x + 2y = -6$ using a solid line since the inequality is $\geq$. Use the intercepts $(0, -3)$ and $\left(-\frac{3}{2}, 0\right)$ to obtain the graph. Now select a test point; (0, 0) is easy to use.

$$4x + 2y \geq -6$$
$$4(\mathbf{0}) + 2(\mathbf{0}) \geq -6$$
$$0 \geq -6 \qquad \text{This is true}$$

Since we obtained a true inequality, shade the half-plane containing (0, 0) to obtain the graph in Figure 4. ■

In view of our work in Example 1, it is easy to show that the graphs of $4x + 2y \leq -6$, $4x + 2y > -6$, and $4x + 2y < -6$ appear as in Figure 5 **(a)**, **(b)**, and **(c)**, respectively.

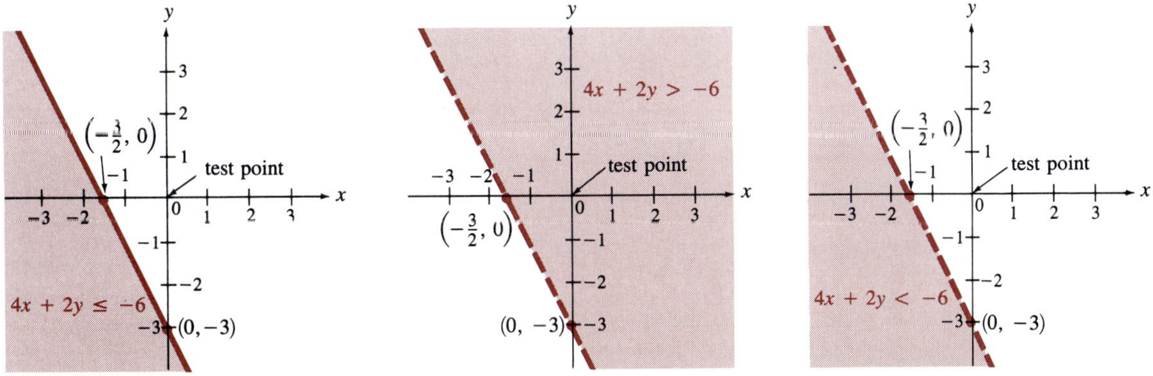


**(a)** $4x + 2y \leq -6$ **(b)** $4x + 2y > -6$ **(c)** $4x + 2y < -6$

**Figure 5**

When $a = 0$ or $b = 0$ in a linear inequality in two variables, the graph of the inequality is an upper or lower half-plane or else a left or right half-plane, respectively.

**EXAMPLE 2** Graph the inequality $2y - 2 < 0$.

First graph the line $2y - 2 = 0$, which simplifies to $y = 1$. Use a dashed line. The test point (0, 0) shows that the graph is the lower, open half-plane in Figure 6. ■

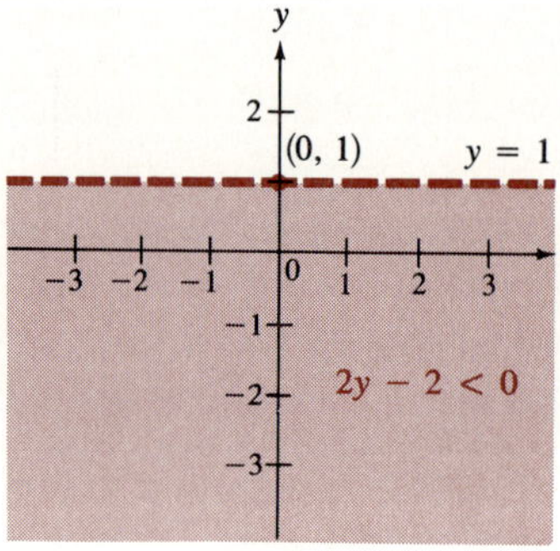


**Figure 6**

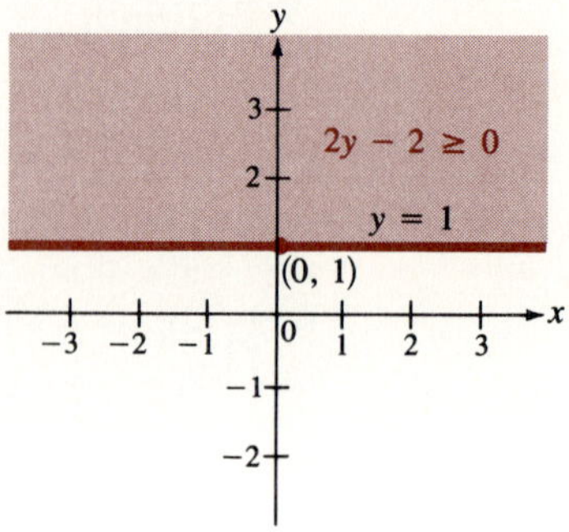


**Figure 7**

In view of Example 2, we know the graph of $2y - 2 \geq 0$ is the upper, closed half-plane in Figure 7.

Similarly, it is easy to verify that the graphs of $3x - 3 \geq 0$ and $3x - 3 < 0$ are those in Figure 8 **(a)** and **(b)**, respectively.

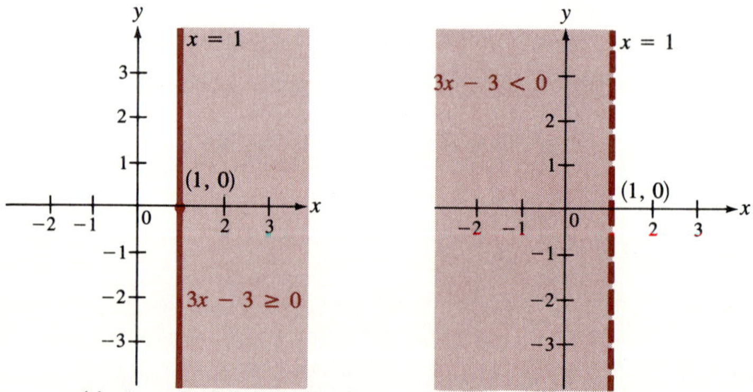


**(a)** $3x - 3 \geq 0$ **(b)** $3x - 3 < 0$

**Figure 8**

## Systems of Linear Inequalities

As you might expect from our previous work, a solution to a system of linear inequalities is an ordered pair of numbers that solves every inequality in the system. The graph of such a system is the set of points in the plane corresponding to these solutions. This graph is found by sketching the graphs of all the inequalities in the system in the same plane and identifying the region where these graphs overlap or intersect.

**EXAMPLE 3** Graph the system.

$$2x + 3y < 6$$
$$-2x + y \geq 1$$

Graph $2x + 3y = 6$ using a dashed line and intercepts (0, 2) and (3, 0), and graph $-2x + y = 1$ using a solid line and intercepts (0, 1) and $\left(-\frac{1}{2}, 0\right)$. The test point (0, 0) can be used for both inequalities.

| | | | |
|---|---|---|---|
| $2x + 3y < 6$ | | $-2x + y \geq 1$ | |
| $2(0) + 3(0) < 6$ | | $-2(0) + 0 \geq 1$ | |
| $0 < 6$ | Is true | $0 \geq 1$ | Is false |

The graph of the system is the region formed by the overlap of the open half-plane below the line $2x + 3y = 6$ and the closed half-plane above the line $-2x + y = 1$. See Figure 9. ■

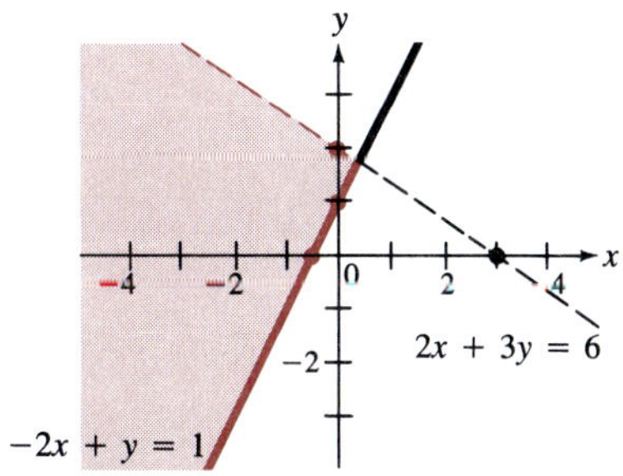


**Figure 9 Graph of a System of Inequalities**

The graph of the system in Example 3 is said to be **unbounded** since it extends infinitely far in some directions. Many systems that have **bounded graphs,** enclosed on all sides by line segments, play an important role in a method of problem-solving discussed in the next section.

**EXAMPLE 4** Graph the system.

$$x \leq 0$$
$$y \geq 0$$
$$2x - 5y \geq -10$$
$$x - y \geq -3$$

Note that the first two inequalities describe the points on the positive $y$-axis, the negative $x$-axis, and in the second quadrant. Turning to the last two inequalities, we graph the lines $2x - 5y = -10$ and $x - y = -3$. The point of intersection of the two lines, $\left(-\frac{5}{3}, \frac{4}{3}\right)$, can be determined by solving the system of equations

$$2x - 5y = -10$$
$$x - y = -3.$$

The region that satisfies the two inequalities $2x - 5y \geq -10$ and $x - y \geq -3$ is cross-shaded in Figure 10 **(a)**. The graph of the original system is the bounded portion of the plane in the second quadrant which also satisfies the two inequalities above, as shown in Figure 10 **(b)**. ■

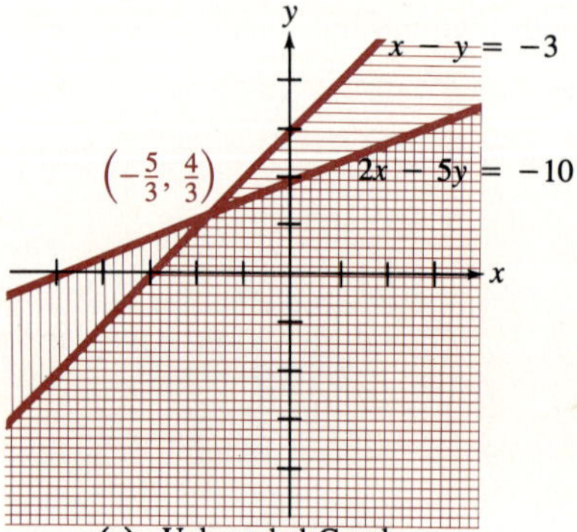


**(a)** Unbounded Graph

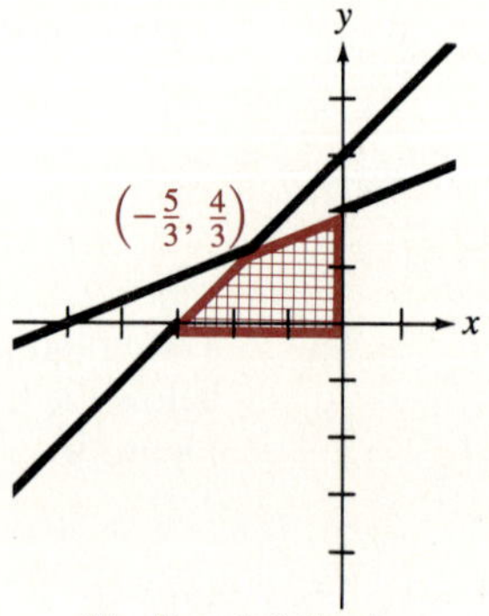


**(b)** Bounded Graph

**Figure 10**

Many applied problems involving two variables can be described by a system of inequalities. The graph of the system shows all possible solutions to the problem. This is illustrated in the following example and will be considered in greater detail in the next section.

**EXAMPLE 5**
**RETAILING**

A retail store owner stocks two models of typewriters, the Executive and the Standard. He has discovered that due to demand it is necessary to have at least twice as many Executive models as Standard models. Also, at all times he must have on hand at least 10 Executive models and 5 Standard models. Finally, due to limitations in space, he has room for no more than 30 typewriters at any time. Graph the system of inequalities that describes this information.

Let $x$ = the number of Executive models in stock,
$y$ = the number of Standard models in stock.

Then the system of inequalities is

$$x \geq 2y$$
$$x \geq 10$$
$$y \geq 5$$
$$x + y \leq 30.$$

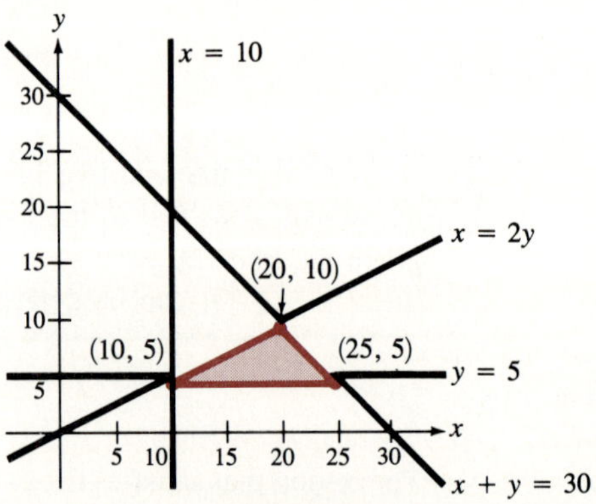


**Figure 11**

The graph of the system is given in Figure 11, where the point (20, 10) was determined by solving the system

$$\begin{aligned} x - 2y &= 0 \\ x + y &= 30. \end{aligned}$$ ■

## 10.3 EXERCISES

In Exercises 1–10, sketch the graph of each inequality.

**1.** $3x - 2y \le 6$ **2.** $3x - 2y > 6$ **3.** $2x + y < -2$ **4.** $2x + y < 3$

**5.** $x - y < -2$ **6.** $x - y \ge -2$ **7.** $3x + 2y < 0$ **8.** $3x + 2y \ge 0$

**9.** $4y - 8 \le 0$ **10.** $3x + 9 > 0$

In Exercises 11–22 sketch the graph of each system of inequalities.

**11.** $\begin{aligned} 2x - 3y &\ge 6 \\ x + y &< -1 \end{aligned}$ **12.** $\begin{aligned} 2x - 3y &< 6 \\ x + y &\le -1 \end{aligned}$ **13.** $\begin{aligned} 2x - 3y &< 6 \\ x + y &\ge -1 \end{aligned}$ **14.** $\begin{aligned} 2x - 3y &\ge 6 \\ x + y &> -1 \end{aligned}$

**15.** $\begin{aligned} 2x - 2y &\ge -4 \\ x - y &< 1 \end{aligned}$ **16.** $\begin{aligned} 2x - 2y &< -4 \\ x - y &\ge 1 \end{aligned}$ **17.** $\begin{aligned} -2x + 3y &\le 6 \\ x + 1 &> 0 \end{aligned}$ **18.** $\begin{aligned} x - 2 &\le 0 \\ y + 1 &> 0 \end{aligned}$

**19.** $\begin{aligned} x &\ge 0 \\ y &\ge 0 \\ 3x + 7y &< 21 \end{aligned}$ **20.** $\begin{aligned} x &\ge 0 \\ y &\le 0 \\ x - y &\le 4 \end{aligned}$ **21.** $\begin{aligned} x &\le 0 \\ y &\ge 0 \\ 2x - y &\ge -6 \\ x + 2y &\le 2 \end{aligned}$ **22.** $\begin{aligned} x &\ge 1 \\ y &> 2 \\ x + 3y &\le 19 \\ 3x + 2y &\le 22 \end{aligned}$

In Exercises 23–24 graph the system described by the given information.

**23.** **BUSINESS** A dealer sells two models of mobile homes in a particular park, the Princess and the Knight. Due to demand, he must have at least three times as many Princess homes available as Knights. At any time he wants at least 6 Princess homes and 2 Knight homes available and ready for occupancy. The Princess model costs the dealer \$30,000, the Knight model costs \$20,000, and the dealer wishes to keep his inventory costs at \$600,000 or less. Form the system of inequalities described by this information and graph the system.

**24.** **MANUFACTURING** A manufacturer of patio picnic tables has two models, the Standard and the Deluxe. The table below shows the information relative to production. The number of each type of table made each week should not exceed the work-hours available in each manufacturing stage. Form the system of inequalities described by this information and graph the system.

| | Standard work-hours per table | Deluxe work-hours per table | Maximum available work-hours per week |
|---|---|---|---|
| Construction stage | 3 | 6 | 96 |
| Finishing stage | 1 | 4 | 36 |

## For Review

Solve.

**25.** CONSUMER If 2 lb of candy and 3 lb of nuts cost a total of \$4.90 and 3 lb of candy and 5 lb of nuts cost a total of \$7.90, find the cost of 1 lb of nuts and of 1 lb of candy.

**26.** CHEMISTRY A chemist has one solution which is 5% salt and a second which is 10% salt. How many liters of each should be mixed to obtain 40 L of an 8% salt solution?

**27.** GEOMETRY The smallest angle of a triangle is 28° less than the largest angle, and the measure of the largest angle less the measure of the middle-sized angle is 2°. Find the measure of each angle.

**28.** INVESTMENT Boris received an inheritance in the amount of \$20,000. He invested the money in three different accounts, one paying 10%, one 11%, and the other 8%, simple interest. If the amount invested at 10% was three times that invested at 8% and his total interest earnings for the year amounted to \$1960, how much was invested at each rate?

**29.** $$\begin{aligned} 3x + y - z &= 0 \\ x - y + 2z &= 0 \\ 7x + y \phantom{+ 2z} &= 0 \end{aligned}$$

**30.** $$\begin{aligned} 5x + 2y - z &= 2 \\ 3x - 3y + z &= -1 \end{aligned}$$

## 10.4 Linear Programming

Systems of linear equations and inequalities provide the tools for a method of problem-solving called **linear programming.** Linear programming, developed by George B. Danzig in the 1940's, was first used by the military to aid in allocating supplies during World War II. Today it is widely used to solve problems of interest to the business community. One such problem involves finding the maximum or minimum value of an expression of the form

$$P = ax + by,$$

called an **objective function,** subject to certain limitations, called **constraints,** on the variables $x$ and $y$. For example, $ax + by$ might denote the profit resulting from "programming" the resources of a production company in such a way that $x$ units of one product and $y$ units of another are produced. The constraints in the problem are given as a system of inequalities which must be graphed.

Suppose we consider the problem of maximizing the objective function $P = 50x + 35y$ subject to the following constraints:

$$\begin{aligned} x &\geq 0 \\ y &\geq 0 \\ x + y &\leq 18 \\ 5x + 3y &\leq 60 \end{aligned}$$

In other words, we want to find the largest value of $50x + 35y$ where $x$ and $y$ satisfy the above system of inequalities. Any pair $(x, y)$ that solves the system of constraints is a **feasible solution** to the problem. The collection of all feasible solutions, shown in

Figure 12, is the **feasible region.** Any feasible solution that determines a maximum value (or minimum value in other problems) is an **optimal solution** to the problem.

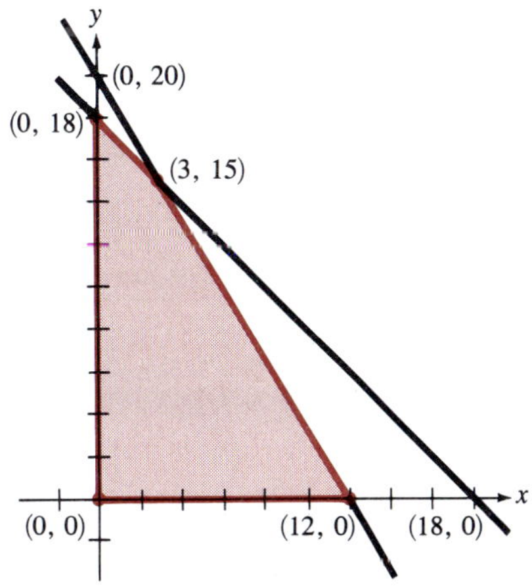


**Figure 12 Feasible Region**

In our work, the feasible region always has a boundary consisting of lines or line segments intersecting in points, each of which is called a **vertex.** The existence of an optimal solution depends on the nature of the feasible region, a consideration beyond the scope of our work. However, under the assumption that an optimal solution will exist, the next theorem tells us where it must be found.

## Fundamental Theorem of Linear Programming

If a linear programming problem has an optimal solution, then at least one vertex of the feasible region will provide that solution.

Suppose we return to the problem of finding the maximum value of the objective function $P = 50x + 35y$ subject to the constraints graphed in Figure 12. The vertices (plural of vertex) are (0, 0), (0, 18), (3, 15), and (12, 0). From the above theorem, an optimal solution must be found among these pairs. The possibilities are summarized in a table.

| Vertex | $50x + 35y$ | |
|---|---|---|
| (0, 0) | 0 | |
| (0, 18) | 630 | |
| (3, 15) | 675 | ← Optimal solution |
| (12, 0) | 600 | |

Thus, considering all possible values of $x$ and $y$ such that $(x, y)$ is in the feasible region, $x = 3$ and $y = 15$ produce the largest value of $50x + 35y$, which is 675.

We can summarize the general procedure in the following algorithm.

### To Find an Optimal Solution to a Linear Programming Problem

1. Graph the system of linear inequalities forming the constraints, and in so doing, identify the feasible region.
2. Determine the vertices of the feasible region by solving, two at a time, the linear equations that describe the boundary of the region.
3. Complete a table of values for the objective function $P = ax + by$ using all of the vertices.
4. If $ax + by$ is to be maximized (minimized), the largest (smallest) value in the table is an optimal solution.

## EXAMPLE 1

BUSINESS

The Mutter Manufacturing Company makes two types of trailer hitches, a standard model and a heavy-duty model. It can produce up to a total of 18 hitches each day using up to 60 total man-hours of work. Experience has shown that it takes 3 man-hours to produce one standard hitch, while 5 man-hours are required to make one heavy-duty hitch. If a heavy-duty model returns a profit of $50 when sold, and a standard model returns a profit of $35, how many of each should be made in a day in order to maximize profit?

First, we must translate the problem into a system of inequalities and identify the objective function.

Let $x$ = the number of heavy-duty hitches produced daily,
$y$ = the number of standard hitches produced daily.

In a problem such as this, $x$ and $y$ cannot be negative, so that

$$x \geq 0$$
$$y \geq 0.$$

Up to a total of 18 hitches can be produced daily.

$$x + y \leq 18$$

Since $5x$ and $3y$ are, respectively, the number of man-hours required daily to produce the desired number of heavy-duty and standard hitches, we have

$$5x + 3y \leq 60.$$

These four inequalities form the constraints for our problem, and we wish to maximize the profit, given by the objective function

$$\text{profit} = 50x + 35y,$$

subject to these constraints.

We now recognize that these conditions are the same as those we analyzed at the beginning of this section so that the maximum profit of $675 will be realized when 3 heavy-duty hitches and 15 standard hitches are produced daily. ■

The next example considers the second applied problem presented at the beginning of this chapter.

## EXAMPLE 2

### PRODUCTION

Ponderosa Productions manufactures two different novelty items, Sam the Skunk and Willie the Weasel. For each item, three different phases of production are required, A, B, and C. To make one skunk, phase A takes 1 minute, phase B takes 4 minutes, and phase C takes 6 minutes. To complete one weasel, A takes 3 minutes, B takes 3 minutes, and C takes 1 minute. Due to maintenance conditions, phase A is available only 33 minutes each hour, while phases B and C are each in operation only 42 minutes every hour. The profit on each skunk is \$2.00, and the profit on each weasel is \$3.00. Find the number of units of each item that should be produced hourly to maximize profit.

Let $x$ = the number of skunks produced each hour,
$y$ = the number of weasels produced each hour.

Each skunk requires 1 minute in phase A. Thus $x$ skunks require $1x = x$ minutes. Each weasel requires 3 minutes in phase A, so that $y$ weasels require $3y$ minutes. Hence, the total number of minutes per hour that phase A is in operation is $x + 3y$. Since A can be used no more than 33 minutes every hour, we have

$$x + 3y \leq 33.$$

Applying the same type of reasoning to phases B and C,

$$4x + 3y \leq 42$$
$$6x + y \leq 42.$$

These three inequalities, together with the obvious ones, $x \geq 0$ and $y \geq 0$, form the constraints to our problem. The feasible region is sketched in Figure 13 where the points (3, 10) and (6, 6) are found, respectively, by solving the systems:

$$\begin{aligned} x + 3y &= 33 \\ 4x + 3y &= 42 \end{aligned} \quad \text{and} \quad \begin{aligned} 4x + 3y &= 42 \\ 6x + y &= 42. \end{aligned}$$

The objective function to be maximized is

$$\text{profit} = 2x + 3y.$$

The possibilities are summarized in the following table.

| Vertex | $2x + 3y$ | |
|---|---|---|
| (0, 11) | 33 | |
| (3, 10) | 36 | ← Optimal solution |
| (6, 6) | 30 | |
| (7, 0) | 14 | |
| (0, 0) | 0 | |

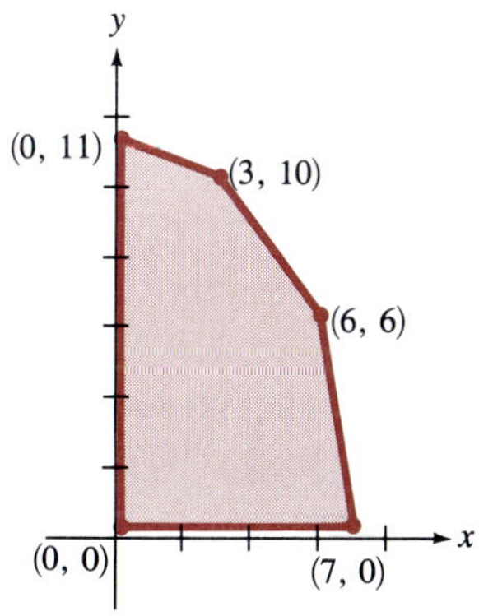


**Figure 13 Feasible Region**

Thus, the maximum profit will be realized when 3 skunks and 10 weasels are produced each hour. ■

**EXAMPLE 3** Find the maximum and minimum values of the objective function $P = 3x + 7y$ subject to the constraints

$$x \geq 2$$
$$y \geq 1$$
$$x + 2y \geq 8$$
$$x + 4y \geq 12.$$

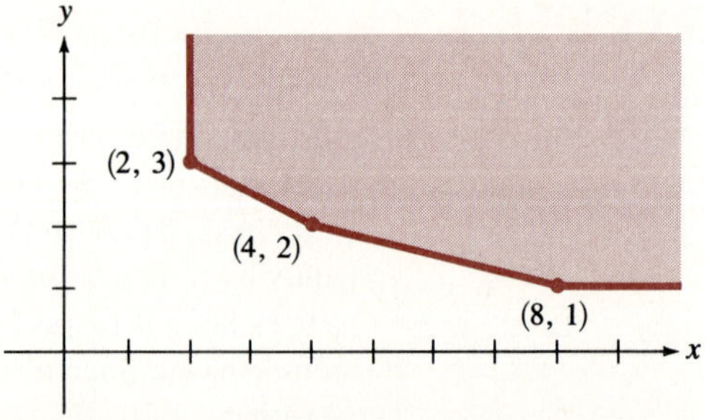


**Figure 14 Unbounded Feasible Region**

The feasible region is shown in Figure 14 where the points (2, 3), (4, 2), and (8, 1) are found, respectively, by solving the following three systems.

$$\begin{array}{ccc} x = 2 & x + 2y = 8 & x + 4y = 12 \\ x + 2y = 8 & x + 4y = 12 & y = 1 \end{array}$$

Notice that in this case, the feasible region is unbounded and that points can be chosen to make $3x + 7y$ as large as we please. In other words, it should be clear that $3x + 7y$ does not assume a maximum value in the region. The minimum value, just as in previous examples, will occur at a vertex.

| Vertex | $3x + 7y$ | |
|---|---|---|
| (2, 3) | 27 | |
| (4, 2) | 26 | ← Optimal solution |
| (8, 1) | 31 | |

Thus, the minimum value of $3x + 7y$ is 26. ■

## 10.4 EXERCISES

In Exercises 1–4 find the maximum and minimum values of the given objective function, subject to the constraints graphed in the feasible region shown.

**1.** $P = x + 5y$

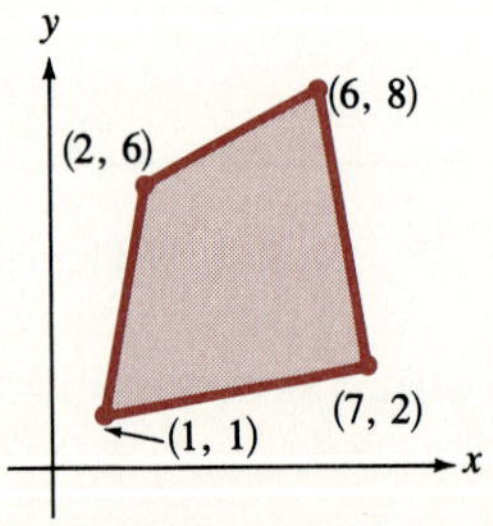


**2.** $P = 5x + 2y$

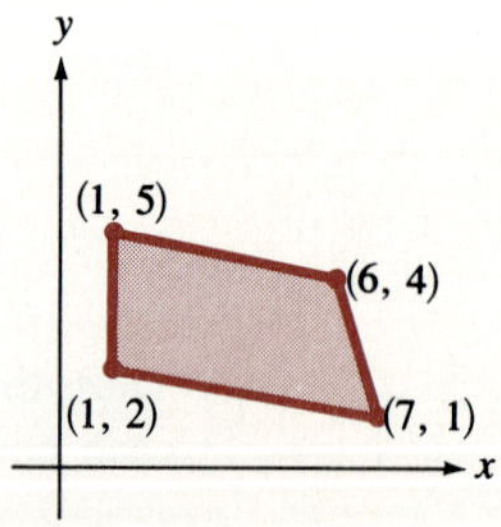

**3.** $P = 75x + 100y$

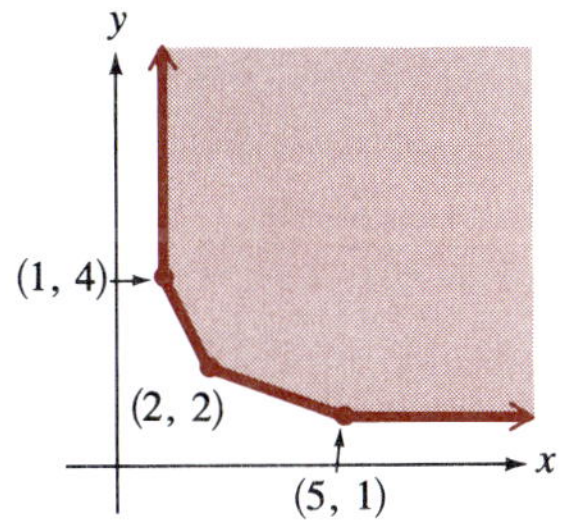


**4.** $P = 15x + 80y$

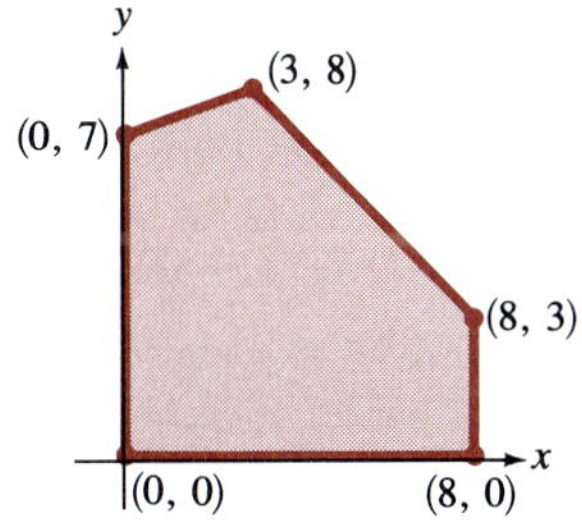


In Exercises 5–8 find the maximum and the minimum values of the objective function $P = 10x + 50y$, subject to the given constraints.

**5.** $x \geq 0$, $y \geq 0$, $x + 2y \leq 6$, $x + y \leq 4$

**6.** $x \geq 1$, $y \geq 1$, $x + 5y \leq 26$, $3x + 2y \leq 26$

**7.** $x \geq 2$, $2y \geq 1$, $5x + 6y \geq 28$

**8.** $7 \geq x \geq 2$, $y \geq 1$, $3y - 2x \leq 11$, $2y + 3x \leq 29$

Solve each linear programming problem in Exercises 9–18.

**9.** **BUSINESS** The Dribble-Well basketball manufacturing firm makes a profit of \$20 on its True-Shot model and \$13 on its True-Bounce model. To meet the demand of wholesalers, the daily production of True-Shots should be between 20 and 100, inclusive, whereas the number of True-Bounce models should be between 10 and 70, inclusive. To maintain quality control, the total number of balls produced each day should not exceed 150. How many of each type should be manufactured daily to maximize profit?

**10.** Repeat Exercise 9 under the assumption that the firm makes the same profit on both models.

**11.** **BUSINESS** An oil refinery can produce up to 5000 barrels of oil per day. Two types are produced, type G, used for gasoline, and type H, used for heating oil. At least 1000 and at most 3500 barrels of type G must be produced each day. If there is a profit of \$7 a barrel for G and \$3 a barrel for H, find the maximum profit per day.

**12.** Repeat Exercise 11 under the assumption that there is a profit of \$4 a barrel for G and \$8 a barrel for H.

**13.** **AGRICULTURE** A dryland farmer has 100 acres on which he can grow corn and wheat. It costs \$5 per acre for seed to plant corn and \$8 per acre for seed to plant wheat. Labor and fuel costs amount to \$20 per acre for corn and \$12 per acre for wheat. If he is fortunate with the weather, he can earn \$220 per acre on the corn and \$250 per acre on the wheat. If he can spend up to \$704 for seed and up to a total of \$1640 for labor and fuel, how many acres of each should he plant to maximize his profit?

**14.** Repeat Exercise 13 under the assumption that the farmer can earn \$270 per acre on corn and \$160 per acre on wheat.

**15.** **MANUFACTURING** A company manufactures two products, A and B. Three different machines, X, Y, and Z, are used to make each product. In order to make one unit of A, machine X must be used for 2 hours, machine Y for 1 hour, and machine Z for 1 hour. The manufacture of one unit of B requires 1 hour of X, 2 hours of Y, and 1 hour of Z. The profit is \$275 on each unit of A and \$180 on each unit of B. Machine X can be used at most 18 hours a day, Y can be used at most 20 hours, and Z at most 11 hours daily. How many of each product should be produced daily to maximize profit?

**16.** Repeat Exercise 15 under the assumption that the profit on each unit of A is \$200 and the profit on each unit of B is \$210.

**17.** **BUSINESS** A publishing company has two distribution centers, one in California (C) and another in Pennsylvania (P). There are 1500 copies of *College Algebra* stored at C and 1200 copies stored at P. Two schools, Central Technology University (CTU) and Mountain College (MC), order 1000 copies and 750 copies, respectively. To ship from C to CTU, the cost is 50¢ per book; from C to MC, the cost is 40¢ per book; from P to CTU, the cost is 45¢ per book; and from P to MC, the cost is 60¢ per book. How should the order be filled to minimize the total shipping cost? [Hint: if $x$ is the number of books shipped to CTU from C, then 1000-$x$ is the number shipped to CTU from P.]

**18.** Repeat Exercise 17 under the assumption that it costs 30¢ per book to ship to CTU from C, 50¢ per book to ship to MC from C, 35¢ per book to ship to CTU from P, and 45¢ per book to ship to MC from P.

## For Review

In Exercises 19–20 sketch the graph of each system of inequalities.

**19.** $2x - 5y \leq 10$
$x + 2y < -2$

**20.** $y \geq x$
$y \geq -x$

**21.** **SPORTS** An auditorium, to be used for a closed-circuit televised championship boxing match, seats 1000 spectators. Promoters plan to charge \$20 for some seats, \$10 for others, and wish to make at least \$16,000 on the event. If at least 300 seats are to be sold for \$10 each, find a system of inequalities that describes this situation and sketch its graph.

In Exercises 22–25 sketch the graph of each equation in a Cartesian coordinate system. These problems will assist us in the material presented in the following section.

**22.** $(x - 2)^2 + y^2 = 4$ **23.** $\frac{x^2}{4} + \frac{y^2}{9} = 1$ **24.** $y = \log_2 x$ **25.** $y = |x - 1| + 1$

# 10.5 Nonlinear Systems

### Interpreting Nonlinear Systems Graphically

When we studied systems of two linear equations, the graphs of the equations gave useful information about the solutions. The same is true for **nonlinear systems.** A solution to a nonlinear system such as

$$\begin{aligned} x^2 + y^2 &= 25 \\ x + y &= 1 \end{aligned}$$

is an ordered pair of numbers that satisfies both equations. The points of intersection of the two graphs correspond to real-number solutions. In the system above, the graph of the first equation is a circle, and the graph of the second is a straight line. In general, the graphs of a circle and a line can be related in one of three different ways, as shown in Figure 15.

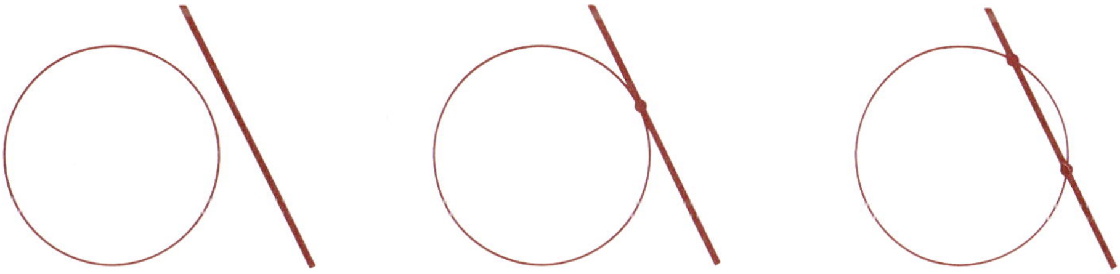

no points of intersection
no real solutions

one point of intersection
one real solution

two points of intersection
two real solutions

**Figure 15 Intersecting a Circle and a Line**

If we graph both of the above equations in the same coordinate system, as in Figure 16, the points of intersection appear to have coordinates $(-3, 4)$ and $(4, -3)$. By substitution, it is easy to show that these pairs of numbers satisfy both equations, hence are solutions to the system.

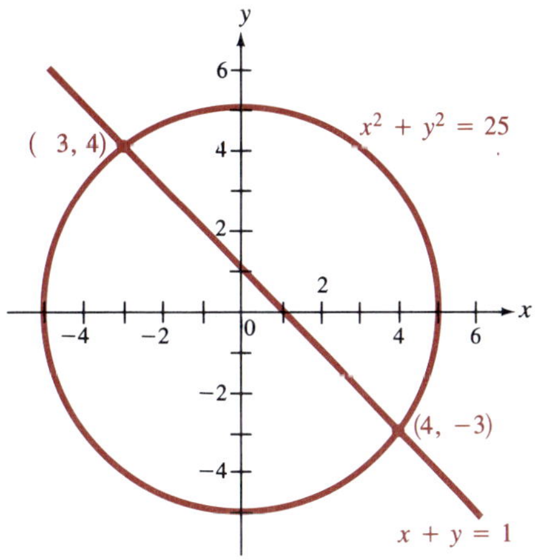


**Figure 16**

## The Substitution Method

It is generally better to solve systems algebraically rather than graphically. When one equation is a first-degree equation, solve it for one of the variables and substitute the result into the other equation. Suppose we use substitution to solve the system given above. Solve $x + y = 1$ for $y$ and substitute in $x^2 + y^2 = 25$.

$$x^2 + (\mathbf{1 - x})^2 = 25 \qquad y = 1 - x$$

$$x^2 + 1 - 2x + x^2 = 25$$

$$2x^2 - 2x - 24 = 0$$

$$x^2 - x - 12 = 0 \qquad \text{Divide out 2}$$

$$(x - 4)(x + 3) = 0$$

Setting each factor equal to zero gives the following.

$$\begin{aligned} x - 4 &= 0 \quad\text{or}\quad & x + 3 &= 0 \\ x &= 4 & x &= -3 \\ y &= 1 - x & y &= 1 - x \\ &= 1 - 4 = -3 & &= 1 - (-3) = 4 \end{aligned}$$

The solutions are $(4, -3)$ and $(-3, 4)$, which were obtained previously by graphing.

**EXAMPLE 1** Solve. $xy = -5$
$x - y = 2$

The first equation is a hyperbola, and the second equation is a line. Three possibilities for the graphs are shown in Figure 17.

**Figure 17 Intersecting a Hyperbola and a Line**

$$\begin{aligned} x &= y + 2 && \text{Solve } x - y = 2 \text{ for } x \\ (y + 2)y &= -5 && \text{Substitute } y + 2 \text{ for } x \text{ in } xy = -5 \\ y^2 + 2y &= -5 \\ y^2 + 2y + 5 &= 0 \\ y &= \frac{-2 \pm \sqrt{4 - 4(1)(5)}}{2} \\ &= \frac{-2 \pm \sqrt{-16}}{2} \\ &= \frac{-2 \pm 4i}{2} = -1 \pm 2i \end{aligned}$$

In this case there are no real solutions; the two graphs do not intersect. However, the complex-number solutions are $(1 + 2i, -1 + 2i)$ and $(1 - 2i, -1 - 2i)$. ■

If the solutions to a system of two second-degree equations are considered graphically, several possibilities exist. Some of these are shown in Figure 18 using a hyperbola and an ellipse.

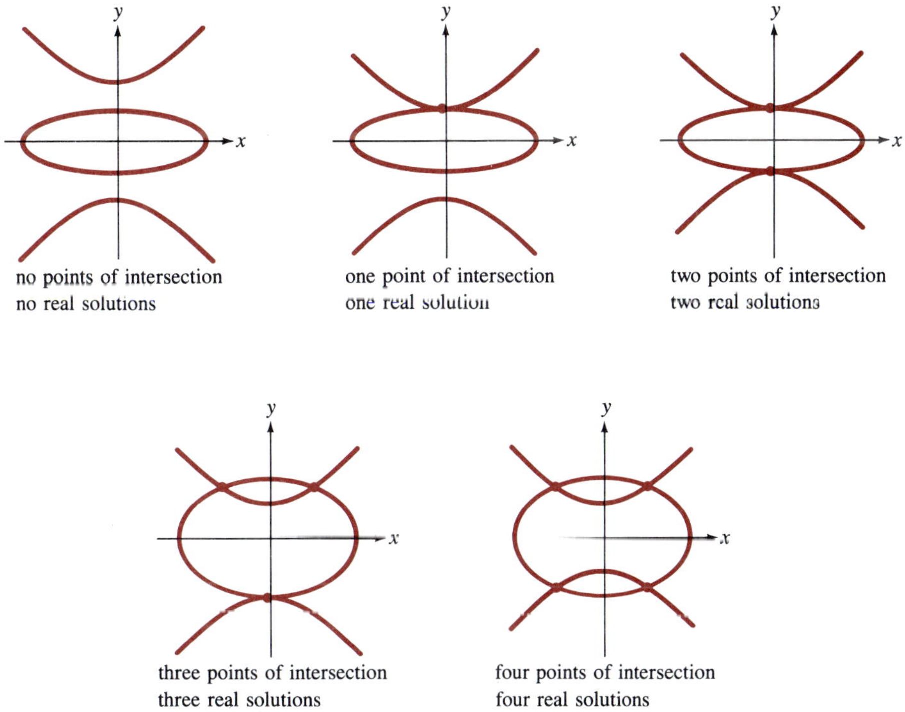


**Figure 18 Intersecting a Hyperbola and an Ellipse**

## The Elimination Method

Many systems of two second-degree equations can be solved by an elimination method similar to the one used for linear systems. We illustrate this technique in the next example.

## EXAMPLE 2

Solve. $x^2 + y^2 = 14$
$x^2 - y^2 = 4$

Add to eliminate $y$.

$$2x^2 = 18$$
$$x^2 = 9$$
$$x = \pm 3$$

Substitute 3 for $x$ in the first equation.

$$(3)^2 + y^2 = 14$$
$$9 + y^2 = 14$$
$$y^2 = 5$$
$$y = \pm\sqrt{5}$$

Two of the solutions are $(3, \sqrt{5})$ and $(3, -\sqrt{5})$. When $-3$ is substituted for $x$ in the first equation we obtain $y = \pm\sqrt{5}$. Thus, there are four solutions: $(3, \sqrt{5})$, $(3, -\sqrt{5})$, $(-3, \sqrt{5})$, and $(-3, -\sqrt{5})$. Check these in both equations. ■

When one of the equations in a system is of the form $xy = a$, it is best to solve this equation for one of the variables and substitute into the other equation. For example, to solve

$$\begin{aligned} x^2 + y^2 &= 25 \\ xy &= 12, \end{aligned}$$

solve the second equation for $x$ and substitute this expression into the first equation.

$$x = \frac{12}{y}$$

$$\frac{144}{y^2} + y^2 = 25$$

Multiply both sides by $y^2$ and solve for $y$.

$$\begin{aligned} 144 + y^4 &= 25y^2 \\ y^4 - 25y^2 + 144 &= 0 \\ (y^2 - 16)(y^2 - 9) &= 0 \\ y^2 = 16 \quad &\text{or} \quad y^2 = 9 \quad \text{Zero-product rule} \\ y = \pm 4 \quad &\text{or} \quad y = \pm 3 \end{aligned}$$

$$y = \mathbf{4}\text{: } x = \frac{12}{\mathbf{4}} = 3 \qquad y = \mathbf{3}\text{: } x = \frac{12}{\mathbf{3}} = 4$$

$$y = \mathbf{-4}\text{: } x = \frac{12}{\mathbf{-4}} = -3 \qquad y = \mathbf{-3}\text{: } x = \frac{12}{\mathbf{-3}} = -4$$

The four solutions are $(4, 3)$, $(-4, -3)$, $(3, 4)$, and $(-3, -4)$, which do check in both equations.

## More Complex Second-Degree Systems

The next example illustrates a technique for obtaining an equation with constant term equal to zero. The resulting quadratic equation can sometimes be solved by factoring.

## EXAMPLE 3

Solve. $\begin{aligned} x^2 + 15xy + 9y^2 &= -5 \\ 7x^2 + 9xy + 27y^2 &= 25 \end{aligned}$

Multiply the first equation by 5 and add the result to the second equation. The resulting equation has constant term zero.

$$\begin{aligned} 12x^2 + 84xy + 72y^2 &= 0 \\ x^2 + 7xy + 6y^2 &= 0 \quad \text{Divide by 12} \\ (x + 6y)(x + y) &= 0 \quad \text{Factor} \\ x = -6y \quad \text{or} \quad x &= -y \end{aligned}$$

$$(-6y)^2 + 15(-6y)y + 9y^2 = -5 \qquad (-y)^2 + 15(-y)y + 9y^2 = -5$$

Substitute for $x$ in the first equation

$$36y^2 - 90y^2 + 9y^2 = -5 \qquad y^2 - 15y^2 + 9y^2 = -5$$

$$-45y^2 = -5 \qquad -5y^2 = -5$$

$$y^2 = \frac{1}{9} \qquad y^2 = 1$$

$$y = \pm\frac{1}{3} \qquad y = \pm 1$$

$$x = -6y = -6\left(\frac{1}{3}\right) = -2 \qquad x = -y = -(+1) = -1$$

$$x = -6y = -6\left(-\frac{1}{3}\right) = 2 \qquad x = -y = -(-1) = 1$$

Substitute for $y$ to find the numerical value of $x$

The solutions are $\left(-2, \frac{1}{3}\right)$, $\left(2, -\frac{1}{3}\right)$, $(-1, 1)$, and $(1, -1)$. ■

## Systems Involving Other Kinds of Functions

A nonlinear system may involve other functions such as logarithmic, exponential, or trigonometric functions. Sometimes these can be solved in a manner similar to that for systems of second-degree equations.

### EXAMPLE 4

Solve.
$$2e^x + y = 4$$
$$e^x + 3y = 7$$

Multiply the first equation by 3 and subtract the second to obtain

$$5e^x = 5$$
$$e^x = 1.$$

Then $x = 0$, and substituting in the first equation,

$$2e^0 + y = 4$$
$$2 + y = 4$$
$$y = 2.$$

The solution is $(0, 2)$. This system could also be solved by substitution. Solve the first equation for $y$ ($y = 4 - 2e^x$) and substitute in the second to obtain $-5e^x = -5$, which also reduces to $e^x = 1$. ■

## Graphing Systems of Nonlinear Inequalities

The technique for graphing nonlinear systems of inequalities parallels that for graphing linear systems of inequalities. We can temporarily replace the inequality symbol with an equal sign and graph the resulting equation with a solid (for $\leq$ or $\geq$) or dashed (for $<$ or $>$) curve. Test points selected from the regions in the plane determined by the curve identify the solution to each inequality. The region common to all solutions is the graph of the system. The four graphs in Figure 19 on the next page illustrate some of the possible nonlinear inequalities that might be found in a system.

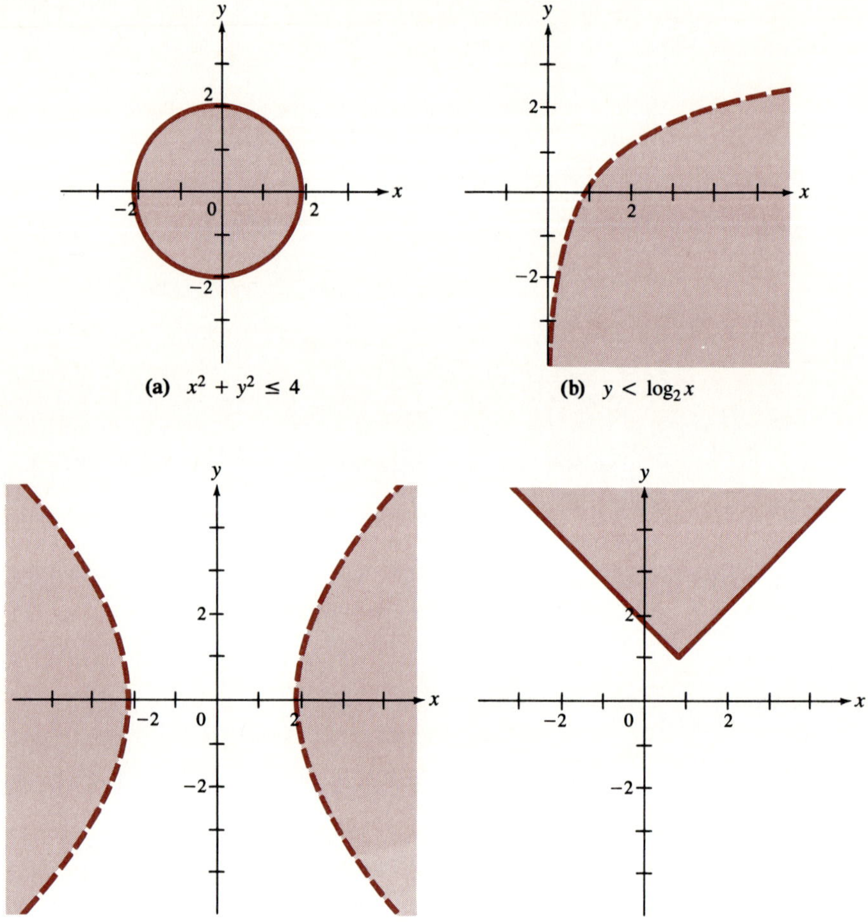


(a) $x^2 + y^2 \leq 4$

(b) $y < \log_2 x$

(c) $\frac{x^2}{4} - \frac{y^2}{9} > 1$

(d) $y - |x - 1| \geq 1$

**Figure 19 Graphs of Inequalities**

**EXAMPLE 5** Graph the system.

$$y \geq x^2 - 1$$
$$y < x + 1$$

Replace the inequality symbols with equal signs. $y = x^2 - 1$ is the equation of a parabola. The graph of $y \geq x^2 - 1$ is all points on or "inside" the parabola. The second inequality has as its graph all points "below" the line $y = x + 1$. Putting this information together, we can obtain the graph of the system in Figure 20. ■

**EXAMPLE 6** Graph the system.

$$y \geq |x - 5|$$
$$y < \log_2 x$$

The graph of $y \geq |x - 5|$ consists of all points on or inside the "vee" of the graph of the absolute value function $y = |x - 5|$. The graph of $y < \log_2 x$ includes all points "below" the graph of the logarithmic function $y = \log_2 x$. See Figure 21. ■

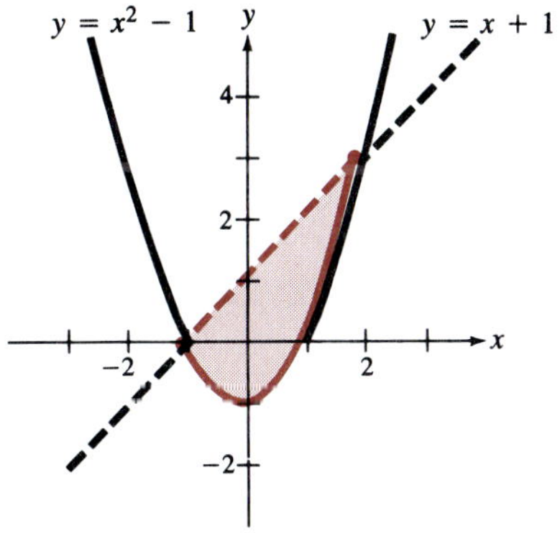


Figure 20

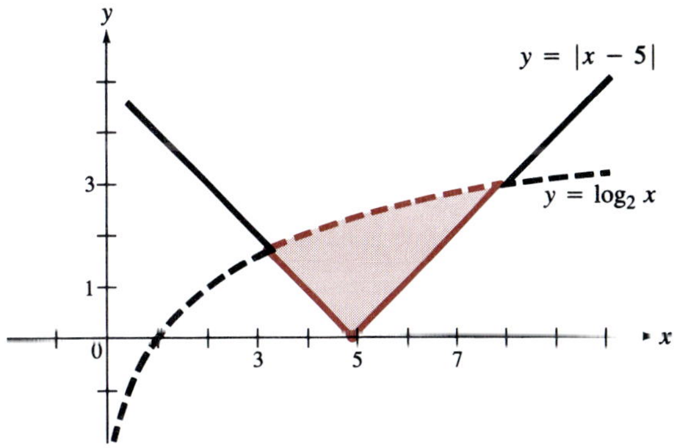


Figure 21

## 10.5 EXERCISES

Solve each system in Exercises 1–20.

**1.** $x^2 + y^2 = 10$
$x - y = 2$

**2.** $2x^2 + y^2 = 9$
$y - 2x = 3$

**3.** $x + y = 3$
$x^2 - y^2 = 3$

**4.** $xy = 12$
$2x - y = -2$

**5.** $5x^2 + xy - y^2 = -1$
$y - 2x = 1$

**6.** $x^2 + 2y^2 = 8$
$x^2 - 2y^2 = 0$

**7.** $x^2 + 3y^2 = 37$
$2x^2 - y^2 = 46$

**8.** $x^2 + 2y^2 = 8$
$2x^2 - y^2 = 1$

**9.** $x^2 + y^2 = 5$
$xy = 2$

**10.** $x^2 + y^2 = 29$
$xy = 10$

**11.** $x^2 + 2xy + y^2 = 9$
$x^2 - 2xy - y^2 = 9$

**12.** $3xy - y^2 = -13$
$2xy + y^2 = -2$

**13.** $x^2 + 2xy + 2y^2 = 10$
$2x^2 + xy + 22y^2 = 50$

**14.** $2x^2 - 5xy + 2y^2 = 20$
$8x^2 - 4xy + y^2 = 20$

**15.** $10^x + y = 11$
$10^x + 2y = 12$

**16.** $2^x - y = 2$
$2^x + y = 6$

**17.** $y + \log_4 (x + 3) = 6$
$y - \log_4 x = 5$

**18.** $\log_3 x - 2y = 2$
$\log_3 x^2 + 2y = 1$

**19.** $|y| + x = 7$
$2|y| - x = 5$

**20.** $|x| + |y| = 5$
$2|x| - |y| = 1$

Graph each system in Exercises 21–32.

**21.** $y \geq x^2$
$y \leq x + 2$

**22.** $y + x^2 \leq 0$
$x + y > -2$

**23.** $y - |x| \geq 0$
$y + |x| < 4$

**24.** $\log_2 x - y > 0$
$2x - y \leq 4$

**25.** $y - 2^x \geq 0$
$y - x \geq 2$

**26.** $x^2 + 4y^2 \leq 16$
$y^2 - x \leq 0$

**27.** $y + |x - 2| < 0$
$x - |y + 1| > 0$

**28.** $x^2 + y^2 \leq 16$
$(x - 2)^2 + y^2 \geq 4$

**29.** $y \geq 0$
$y \leq \sqrt{x - 1}$
$y > x - 3$

**30.** $x \leq 0$
$y \geq 0$
$x + y^2 < 1$

**31.** $x^2 + y^2 > 4$
$x^2 + y^2 < 16$

**32.** $x^2 + y^2 < 9$
$x^2 - y^2 > 1$

Solve.

**33. GEOMETRY** Find the dimensions of a rectangular garden with perimeter 20 yd and area 24 $yd^2$.

**34. ENGINEERING** A rectangular solar heating surface must be 8 inches longer than it is wide. If the area of the surface must be 180 in$^2$, find the dimensions of the surface.

**35. ECONOMICS** The weekly demand equation for a certain product is given by $xy = 100$, where $x$ is the number of people willing to pay $y$ dollars for the product. The supply equation for the product is given by $x^2 - xy = 44$. The **market equilibrium point** occurs at a point where the supply curve intersects the demand curve. Find the market equilibrium point for this product.

**36.** The sum of the squares of the digits of a positive two-digit number is 100, and the tens digit is 2 less than the units digit. Find the number.

## For Review

**37.** Find the maximum and minimum values of $P = 2x - 5y$ subject to the constraints graphed in the feasible region shown.

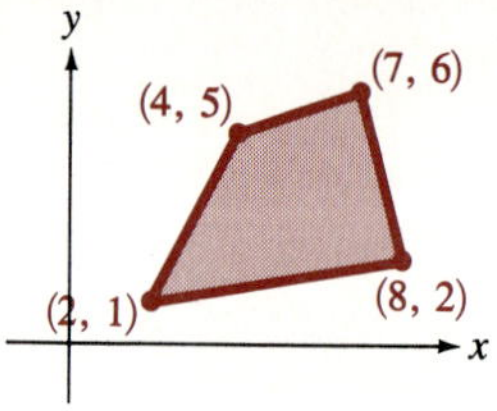


**38.** Find the maximum and minimum values of $P = 10x - 5y$ subject to the given constraints.

$$6 \geq x \geq 0$$
$$y \geq 0$$
$$x + 3y \leq 12$$

**39. BUSINESS** The Backcourt Manufacturing firm makes a profit of \$15 on its Pro-Model tennis racquet and \$10 on its Hacker-Model racquet. To meet the wholesale demand, the daily production of the Pro-Model should be between 10 and 50, inclusive, whereas the Hacker-Model should be between 20 and 40, inclusive. To maintain quality control, the total number of racquets manufactured daily should not exceed 80. How many of each type of racquet should be manufactured daily to maximize profits?

The following exercises help us to prepare for the material presented in the next section.

**40.** What is the degree of the polynomial $P(x) = (x - 2)(x + 3)^2(x^2 - 5)$?

**41.** If $2 + 3i$ is a zero of polynomial $P(x)$, with real coefficients, what is another zero of $P(x)$?

**42.** Suppose $x - 2$ is one factor of $P(x) = x^3 - x^2 - x - 2$, what is another factor?

**43.** Suppose $f(x) = \frac{1}{x-2} + \frac{2}{x+2}$. Find another representation for $f$ which has only one term.

Solve each system of equations in Exercises 44–45.

**44.**
$$\begin{aligned} A + B &= 3 \\ 2A - 2B &= -2 \end{aligned}$$

**45.**
$$\begin{aligned} A \phantom{+B} + C &= 4 \\ B - 2C &= -4 \\ -A + B + C &= 4 \end{aligned}$$

# 10.6 Partial Fractions

In Section 1.5 we reviewed addition of rational expressions such as

$$\frac{1}{x-2}+\frac{2}{x+2}=\frac{3x-2}{x^2-4}.$$

In this section, the reverse procedure is considered. Suppose we are given the rational function

$$f(x)=\frac{3x-2}{x^2-4}$$

and want to express $f(x)$ as the sum of two simpler fractions, called **partial fractions.** The partial fraction decomposition of $f$ is, in fact,

$$f(x)=\frac{1}{x-2}+\frac{2}{x+2}.$$

Decompositions like this are important in calculus and other advanced courses. They involve solving a system of linear equations.

## Conditions for Using Partial Fractions

The partial fraction decomposition of a rational function

$$f(x)=\frac{P(x)}{Q(x)}$$

depends on two basic assumptions:

1. The degree of $P(x)$ is less than the degree of $Q(x)$.
2. The polynomial $Q(x)$ can be factored into a product of powers of linear and quadratic factors in such a way that the quadratic factors have no real zeros.

Neither of these two assumptions should seem unreasonable. In fact, if $f(x)$ does not satisfy the first, we could divide $Q(x)$ into $P(x)$ and obtain a fractional remainder that satisfies this requirement. For example,

$$f(x)=\frac{x^3+2x+1}{x^2-3}$$

can be written as

$$f(x)=x+\frac{5x+1}{x^2-3}. \qquad \text{After dividing by } x^2-3$$

In this form we can concentrate on the fractional part of $f(x)$ which does satisfy the first condition. As for the second assumption, we know from Chapter 4 that a polynomial of degree $n$ has $n$ zeros (real or complex) and, as a result, can be factored into $n$ linear factors. Since the complex zeros occur in conjugate pairs, the linear factors corresponding to these pairs can be multiplied to give one quadratic factor with real coefficients.

## Types of Decomposition

### Partial Fraction Decomposition of a Rational Function

Let $f(x) = \dfrac{P(x)}{Q(x)}$ be a rational function reduced to lowest terms, with the degree of $P(x)$ less than the degree of $Q(x)$.

1. If $(ax + b)^k$ is a factor of $Q(x)$, for $k$ a natural number, then the partial fraction decomposition of $f(x)$ contains terms of the form

$$\frac{A_1}{(ax+b)} + \frac{A_2}{(ax+b)^2} + \cdots + \frac{A_k}{(ax+b)^k},$$

where $A_1, A_2, \ldots, A_k$ are constant real numbers.

2. If $(ax^2 + bx + c)^m$ is a factor of $Q(x)$, for $m$ a natural number, then the partial fraction decomposition of $f(x)$ contains terms of the form

$$\frac{B_1x + C_1}{(ax^2+bx+c)} + \frac{B_2x + C_2}{(ax^2+bx+c)^2} + \cdots + \frac{B_mx + C_m}{(ax^2+bx+c)^m},$$

where $B_1, B_2, \ldots, B_m$ and $C_1, C_2, \ldots, C_m$ are constant real numbers.

Decomposing a fraction into partial fractions, best illustrated by examples, involves finding the linear and quadratic factors of the denominator, expressing the function in terms of general partial fractions, determining a system of linear equations, and solving the system. The system of equations is determined by using the fact that if two polynomials are equal, the coefficients of like terms are equal (see Exercise 56 in Section 4.2). We begin by finding the decomposition of the rational function given earlier.

**EXAMPLE 1** Find the partial fraction decomposition of

$$f(x) = \frac{3x - 2}{x^2 - 4}.$$

Since the degree of the numerator is 1 and the degree of the denominator is 2, we proceed by factoring the denominator.

$$f(x) = \frac{3x - 2}{(x-2)(x+2)}$$

Using the partial fraction decomposition theorem, the decomposition must have two terms $\dfrac{A}{x-2}$ and $\dfrac{B}{x+2}$. Thus,

$$\frac{3x-2}{(x-2)(x+2)} = \frac{A}{x-2} + \frac{B}{x+2},$$

where $A$ and $B$ are constant real numbers, yet to be determined. Multiply both sides of this equation by the LCD, $(x - 2)(x + 2)$.

$$3x - 2 = A(x + 2) + B(x - 2)$$
$$3x - 2 = Ax + 2A + Bx - 2B$$
$$3x - 2 = (A + B)x + (2A - 2B)$$

Since the polynomials on both sides are equal, the coefficients of the like terms must be equal. Equating the coefficients of $x$ and equating the constant terms gives the following system of two linear equations in the two variables $A$ and $B$.

$$A + B = 3$$
$$2A - 2B = -2$$

Solving this system gives $A = 1$ and $B = 2$. Then

$$f(x) = \frac{3x - 2}{x^2 - 4} = \frac{A}{x - 2} + \frac{B}{x + 2}$$

$$= \frac{1}{x - 2} + \frac{2}{x + 2}. \quad \text{Substitute 1 for } A \text{ and 2 for } B$$

Notice that we obtained the sum of fractions given in our introductory remarks. ■

For rational functions such as the one in Example 1 in which the denominator involves only linear factors raised to the first power, there is a quicker way to find $A$ and $B$. By substituting values for $x$ that make the various factors zero, $A$ and $B$ can be determined without solving a system. For example, suppose we begin with

$$3x - 2 = A(x + 2) + B(x - 2).$$

Let $x = -2$: $3(-2) - 2 = A(\mathbf{-2 + 2}) + B(-2 - 2)$ When $x = -2$, $A$ is eliminated

$$-8 = A(\mathbf{0}) + B(-4)$$
$$-8 = -4B$$
$$2 = B$$

Let $x = 2$: $3(2) - 2 = A(2 + 2) + B(\mathbf{2 - 2})$ When $x = 2$, $B$ is eliminated

$$4 = A(4) + B(\mathbf{0})$$
$$4 = 4A$$
$$1 = A$$

**EXAMPLE 2** Find the partial fraction decomposition of

$$f(x) = \frac{4x^2 - 4x + 4}{(x - 1)^2(x + 1)}.$$

Using the partial fraction decomposition theorem,

$$\frac{4x^2 - 4x + 4}{(x - 1)^2(x + 1)} = \frac{A}{x - 1} + \frac{B}{(x - 1)^2} + \frac{C}{x + 1}.$$

Multiply both sides by the LCD, $(x - 1)^2(x + 1)$.

$$4x^2 - 4x + 4 = A(x - 1)(x + 1) + B(x + 1) + C(x - 1)^2.$$

Clear parentheses and collect like terms.

$$4x^2 - 4x + 4 = (A + C)x^2 + (B - 2C)x + (-A + B + C).$$

Equate coefficients of like terms to obtain the following system.

$$\begin{aligned} A \quad\;\; + \;\; C &= \;\; 4 \\ B - 2C &= -4 \\ -A + B + \;\; C &= 4 \end{aligned}$$

Solving this system gives $A = 1$, $B = 2$, and $C = 3$. Thus,

$$f(x) = \frac{4x^2 - 4x + 4}{(x - 1)^2(x - 1)} = \frac{1}{x - 1} + \frac{2}{(x - 1)^2} + \frac{3}{x + 1}. \quad ■$$

The next example illustrates the technique to use when the denominator has a linear factor and a quadratic factor.

**EXAMPLE 3** Find the partial fraction decomposition of

$$f(x) = \frac{-7x - 1}{(x^2 + 2)(x - 3)}.$$

The form of the decomposition is

$$\frac{-7x - 1}{(x^2 + 2)(x - 3)} = \frac{Ax + B}{x^2 + 2} + \frac{C}{x - 3}.$$

$$-7x - 1 = (Ax + B)(x - 3) + C(x^2 + 2)$$ Multiply both sides by the LCD, $(x^2 + 2)(x - 3)$

$$-7x - 1 = (A + C)x^2 + (-3A + B)x + (-3B + 2C)$$ Clear parentheses and collect like terms

Thus, we obtain the system

$$\begin{aligned} A \quad\;\; + \;\; C &= \;\; 0 \\ -3A + \;\; B \qquad &= -7 \\ - 3B + 2C &= -1, \end{aligned}$$

which has solutions $A = 2$, $B = -1$, and $C = -2$. Hence,

$$f(x) = \frac{-7x - 1}{(x^2 + 2)(x - 3)} = \frac{2x - 1}{x^2 + 2} + \frac{-2}{x - 3}. \quad ■$$

**EXAMPLE 4** Find the partial fraction decomposition of

$$f(x) = \frac{x^5 + 5x^3 + x^2 + 5x + 1}{x^4 + 2x^2 + 1}.$$

Since the degree of the numerator exceeds the degree of the denominator, first divide the numerator by the denominator to obtain the following form.

$$f(x) = x + \frac{3x^3 + x^2 + 4x + 1}{x^4 + 2x^2 + 1}$$

We can now concentrate on the fraction which satisfies the hypotheses of the decomposition theorem. First factor the denominator.

$$\frac{3x^3 + x^2 + 4x + 1}{x^4 + 2x^2 + 1} = \frac{3x^3 + x^2 + 4x + 1}{(x^2 + 1)^2}$$

Then

$$\frac{3x^3 + x^2 + 4x + 1}{(x^2 + 1)^2} = \frac{Ax + B}{x^2 + 1} + \frac{Cx + D}{(x^2 + 1)^2}.$$

Multiply both sides by the LCD, $(x^2 + 1)^2$, clear parentheses, and collect like terms.

$$3x^3 + x^2 + 4x + 1 = Ax^3 + Bx^2 + (A + C)x + (B + D)$$

Immediately we recognize that $A = 3$ and $B = 1$. Since $A + C = 4$ and $A = 3$, $C = 1$. Also, with $B + D = 1$ and $B = 1$, $D = 0$. Therefore,

$$f(x) = \frac{x^5 + 5x^3 + x^2 + 5x + 1}{x^4 + 2x^2 + 1} = x + \frac{3x + 1}{x^2 + 1} + \frac{x}{(x^2 + 1)^2}. \quad ■$$

## 10.6 EXERCISES

In Exercises 1–12 use constants $A$, $B$, $C$, and $D$, to express the partial fraction decomposition of each function. Do not solve for these constants.

**1.** $\dfrac{3x + 2}{(x - 1)(x + 5)}$

**2.** $\dfrac{4x^2 + 1}{(x - 2)(x + 2)(x - 5)}$

**3.** $\dfrac{6x^2 - 7}{(x + 3)^2(x + 5)}$

**4.** $\dfrac{x^2 + x + 1}{(x^2 + x + 5)(x + 4)}$

**5.** $\dfrac{x^3 - 5}{(x^2 + 2x + 10)^2}$

**6.** $\dfrac{x^2 - x + 5}{(x - 3)^2(x^2 - 2x + 7)}$

**7.** $\dfrac{x + 1}{x^3 - 2x^2 - 3x}$

**8.** $\dfrac{x^2 + x - 4}{x^3 - 3x^2 - 10x}$

**9.** $\dfrac{x - 3}{x^2(x + 2) - 2x(x + 2) - 3(x + 2)}$

**10.** $\dfrac{x + 2}{x^4 - 16}$

**11.** $\dfrac{x^2 + 5x + 5}{x^4 + 5x^2 + 5}$

**12.** $\dfrac{x^2 - 4x - 5}{x^2(x^2 - 1) - x(x^2 - 1) - 20(x^2 - 1)}$

In Exercises 13–24 determine real numbers $A$, $B$, $C$, $D$, and $E$ so that each rational function $f(x)$ has the given partial fraction decomposition.

**13.** $f(x) = \dfrac{5x - 1}{x^2 - 1} = \dfrac{A}{x - 1} + \dfrac{B}{x + 1}$

**14.** $f(x) = \dfrac{7x + 16}{(x - 2)(x + 4)} = \dfrac{A}{x - 2} + \dfrac{B}{x + 4}$

**15.** $f(x) = \dfrac{1 - x}{(x + 1)^2} = \dfrac{A}{x + 1} + \dfrac{B}{(x + 1)^2}$

**16.** $f(x) = \dfrac{4x - 13}{(x - 3)^2} = \dfrac{A}{x - 3} + \dfrac{B}{(x - 3)^2}$

**17.** $f(x) = \dfrac{6x^2 + 1}{(x^2 + 1)(x - 2)} = \dfrac{Ax + B}{x^2 + 1} + \dfrac{C}{x - 2}$

**18.** $f(x) = \dfrac{7x^2 + 16x + 17}{(x^2 + x + 1)(x + 4)} = \dfrac{Ax + B}{x^2 + x + 1} + \dfrac{C}{x + 4}$

**19.** $f(x) = \dfrac{x^3 + x^2 + 3x + 1}{(x^2 + 2)^2} = \dfrac{Ax + B}{x^2 + 2} + \dfrac{Cx + D}{(x^2 + 2)^2}$

**20.** $f(x) = \dfrac{5x^3 - 15x^2 + 26x - 2}{(x^2 - 3x + 5)^2} = \dfrac{Ax + B}{x^2 - 3x + 5} + \dfrac{Cx + D}{(x^2 - 3x + 5)^2}$

**21.** $f(x) = \dfrac{-2x^2 + 15x + 13}{(x - 2)^2(x + 3)} = \dfrac{A}{x - 2} + \dfrac{B}{(x - 2)^2} + \dfrac{C}{x + 3}$

**22.** $f(x) = \dfrac{3x^3 - 2x^2 + 2x + 19}{(x+1)^2(x-2)^2} = \dfrac{A}{x+1} + \dfrac{B}{(x+1)^2} + \dfrac{C}{x-2} + \dfrac{D}{(x-2)^2}$

**23.** $f(x) = \dfrac{4x}{(x^2+1)^2(x-1)} = \dfrac{Ax+B}{x^2+1} + \dfrac{Cx+D}{(x^2+1)^2} + \dfrac{E}{x-1}$
[Hint: After multiplying by the LCD, clearing parentheses, and combining like terms, substitute 1 for $x$ to find $E$ first.]

**24.** $f(x) = \dfrac{3x^4 + 6x^3 + 3x^2 - 3x}{(x+2)(x^2+x+1)^2} = \dfrac{A}{x+2} + \dfrac{Bx+C}{x^2+x+1} + \dfrac{Dx+E}{(x^2+x+1)^2}$

In Exercises 25–36 find the partial fraction decomposition of each rational function.

**25.** $f(x) = \dfrac{3x-3}{x^2-9}$

**26.** $f(x) = \dfrac{-4x-1}{(x-2)(x+1)}$

**27.** $f(x) = \dfrac{6x^2+20x+19}{(x+2)^2(x+1)}$

**28.** $f(x) = \dfrac{4x^2-8x-8}{x^3-4x}$

**29.** $f(x) = \dfrac{-9x-7}{(x^2+1)(x-5)}$

**30.** $f(x) = \dfrac{3x^2+x+3}{(2x^2+3)(x+1)}$

**31.** $f(x) = \dfrac{7x^2+13x+10}{x^3+2x^2-4x-8}$

**32.** $f(x) = \dfrac{5x^2+4x+1}{x^3-x^2+4x-4}$

**33.** $f(x) = \dfrac{2x^3-x^2+3x-1}{(x^2+1)^2}$

**34.** $f(x) = \dfrac{2x^3+7x+1}{x^4+4x^2+4}$

**35.** $f(x) = \dfrac{x^3-2x^2+7x-2}{x^2-2x+1}$

**36.** $f(x) = \dfrac{2x^3+x^2-22x+17}{x^2+x-12}$

## For Review

Solve each system in Exercises 37–38.

**37.** $\begin{aligned} x^2 + y^2 &= 5 \\ xy &= 2 \end{aligned}$

**38.** $\begin{aligned} 2x^2 - 6xy + 12y^2 &= 8 \\ x^2 - xy + 6y^2 &= 8 \end{aligned}$

Graph each system in Exercises 39–40.

**39.** $\begin{aligned} x &\le 4 - y^2 \\ x &> |y| - 2 \end{aligned}$

**40.** $\begin{aligned} x^2 + y^2 &\ge 4 \\ \frac{x^2}{25} + \frac{y^2}{4} &\le 1 \end{aligned}$

# CHAPTER 10 REVIEW EXERCISES

**1.** Determine whether $(2, -3)$ is a solution to the system.

$$\begin{aligned} 3x + y &= 3 \\ -x + 2y &= -8 \end{aligned}$$

**2.** Solve the following system using substitution.

$$\begin{aligned} 3x + y &= -1 \\ -2x - 3y &= -11 \end{aligned}$$

**3.** Solve the following system using elimination.

$$\begin{aligned} 3x + 5y &= -7 \\ -4x + 2y &= -8 \end{aligned}$$

In Exercises 4–7 solve each system using either substitution or elimination, whichever seems more appropriate.

**4.** $\begin{aligned} 3x - 5y &= 2 \\ -6x + 10y &= -1 \end{aligned}$

**5.** $\begin{aligned} \frac{3}{x} + \frac{1}{y} &= 2 \\ \frac{5}{x} - \frac{3}{y} &= 8 \end{aligned}$

**6.** $\begin{aligned} 2x + 2y &= 6 \\ -3x - 3y &= -9 \end{aligned}$

**7.** $\begin{aligned} 2(x + 3y) + (x + y) &= -28 \\ 3(x + 3y) - (x + y) &= -32 \end{aligned}$

**8.** Find the value(s) of $m$ so that the system

$$\begin{aligned} 2x - 3y &= m \\ -4x + 6y &= 5 \end{aligned}$$

is dependent.

Solve each system in Exercises 9–11.

**9.** $\begin{aligned} 2x + 3y + 4z &= 0 \\ x - 5y + 3z &= -1 \\ 3x + y + z &= 5 \end{aligned}$

**10.** $\begin{aligned} x - 2y + 4z &= 0 \\ x + 2y - 5z &= 0 \\ 2x \qquad - z &= 0 \end{aligned}$

**11.** $\begin{aligned} 3x + 2y \qquad &= 12 \\ -x + y - 5z &= 1 \end{aligned}$

**12.** The parabola with equation $y = ax^2 + bx + c$ passes through the points $(1, 4)$, $(-1, 10)$, and $(3, 14)$. Find the values of $a$, $b$, and $c$.

In Exercises 13–20 solve using a system of equations.

**13.** By going 40 mph for one period of time and then 50 mph for another, a man traveled 370 miles. If he had gone 55 mph throughout the same period of time, he would have traveled 440 miles. How long did he travel at each speed?

**14.** **GEOMETRY** Two angles are complementary, and one is 6° less than seven times the other. Find the measure of each angle.

**15.** **BUSINESS** How many pounds of candy at \$1.60 per lb and how many pounds of candy at \$1.20 per lb must be mixed to obtain 80 pounds of a mixture worth \$1.36 per lb?

**16.** **CHEMISTRY** A chemist has one solution which is 15% acid and another which is 20% acid. How many gallons of each should he mix together to obtain 100 gallons of a solution which is 18% acid?

**17.** A collection of 70 coins consisting of nickels, dimes, and quarters has value \$8.00. If the number of dimes is twice the number of quarters, how many of each type of coin are in the collection?

**18.** **INVESTMENT** A man has \$9000 split into three separate investments. Part is invested in bonds which earn 9%, part is invested in certificates which earn 8%, and part is invested in stocks. If the stocks do well, they will earn 10% and his total earnings will amount to \$780. If the market exhibits a downward trend, his stocks will lose 3%, and his total earnings will amount to \$520. How much is invested in each category?

**19.** **GEOMETRY** The smallest angle of a triangle is one third the middle-sized angle, and the largest angle is 5° more than the middle-sized angle. Find the measure of each angle.

**20.** The polynomial function $P(x) = x^3 + ax + b$ has the property that $P(1) = 4$ and $P(-1) = 12$. Find the values of $a$ and $b$.

In Exercises 21–22 sketch the graph of each system of linear inequalities.

**21.** $2x + y \le 2$
$y - 1 > 0$

**22.** $x \ge 0$
$y \ge 0$
$3x + 2y < 6$

**23.** Find the maximum and minimum values of the objective function $P = 18x + 3y$ subject to the constraints graphed in the feasible region shown.

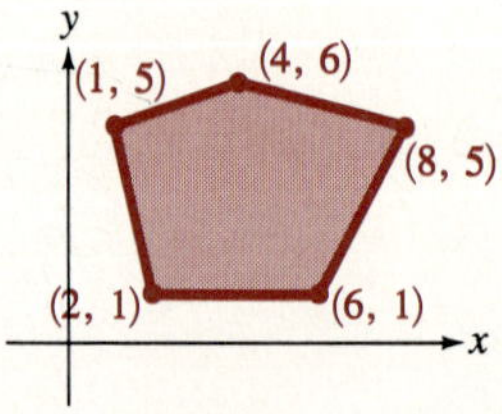


**24.** Find the maximum and the minimum values of the objective function $P = 6x - 10y$ subject to the following constraints.

$$1 \le x \le 5$$
$$y \ge 2$$
$$y - 3x \le 0$$
$$2x + 3y \le 22$$

**25.** **BUSINESS** The owner of a fast-food diner has at most 90 lb of ground beef that can be used to make hamburgers and taco filler daily. Each hamburger contains $\frac{1}{3}$ lb of beef and each taco contains $\frac{1}{6}$ lb of beef. Suppose profit on hamburgers is 42¢, and profit on tacos is 20¢. Labor costs for making a hamburger amount to 18¢, and the labor costs for making a taco are 6¢. If the owner can pay at most \$36 for the labor costs of producing these two items, what number of hamburgers and what number of tacos should be made in order to maximize his profits? What is the maximum profit?

Solve each system in Exercises 26–29.

**26.** $2x^2 + y^2 = 3$
$x - y = 2$

**27.** $x^2 + y^2 = 20$
$xy = 8$

**28.** $3x^2 + xy - 6y^2 = 8$
$x^2 + 2xy - 4y^2 = 4$

**29.** $3^x + y = 10$
$3^x - y = 8$

Graph each system in Exercises 30–31.

**30.** $x^2 + y^2 \le 16$
$y > |x|$

**31.** $y - 3^x \ge 0$
$3x - 4y \ge -9$

In Exercises 32–33 determine real numbers $A$, $B$, $C$, and $D$ so that each rational function $f(x)$ has the given partial fraction decomposition.

**32.** $f(x) = \dfrac{-2x - 5}{(x + 4)^2} = \dfrac{A}{x + 4} + \dfrac{B}{(x + 4)^2}$

**33.** $f(x) = \dfrac{x^3 + 2x^2 + 10x + 7}{(x^2 + x + 7)^2} = \dfrac{Ax + B}{x^2 + x + 7} + \dfrac{Cx + D}{(x^2 + x + 7)^2}$

In Exercises 34–35 find the partial fraction decomposition of each rational function.

**34.** $f(x) = \dfrac{8x^2 - 4x + 16}{(x^2 + 3)(x - 1)}$

**35.** $f(x) = \dfrac{2x^3 - 16x^2 + 28x - 12}{x^2 - 8x + 12}$

CHAPTER

# 11 MATRICES AND DETERMINANTS

Matrices and operations on matrices provide streamlined methods for solving a variety of applied problems. With increased usage of computers, matrices have become more important as tools for organizing and manipulating large sets of data. In this chapter we present a brief introduction to the theory of matrices, an extensive subject to which entire courses are often devoted. Two examples follow of the types of applied problems that we can solve using matrix methods.

POLITICS ▶

A candidate for political office in Arizona hired a public relations firm to promote his campaign in three population centers of the state: Tucson, Phoenix, and Flagstaff. Contacts of potential supporters are to be made by telephone, mail, and direct personal contact. The cost of each of these is estimated to be \$0.50 by telephone, \$0.30 by mail, and \$0.75 for a personal contact. The table below summarizes the desired number of contacts in each category in each area. Use matrices to determine the total cost for the promotion.

| | Telephone | Mail | Personal |
|---|---|---|---|
| Tucson | 5000 | 10,000 | 1000 |
| Phoenix | 8000 | 20,000 | 6000 |
| Flagstaff | 3000 | 5000 | 700 |

◀ **SPORTS**

The Olympic Sports Committee schedules four exhibition basketball games between the U.S. Olympic Team and a team composed of NBA players. Tickets are to be sold for \$5 and \$8 with a given total revenue production specified. Assuming that all seats will be sold in every arena, use the information given in the following table to determine the number of seats at each price that must be sold for each game.

| | Game 1 | Game 2 | Game 3 | Game 4 |
|---|---|---|---|---|
| Total arena seats | 9000 | 12,000 | 8400 | 21,000 |
| Revenue produced | \$63,000 | \$81,000 | \$66,000 | \$114,000 |

These applications are studied in Example 4 in Section 11.2 and Example 3 in Section 11.4. Although both problems can be solved without matrices, more complex problems of a similar nature almost demand the use of matrices.

We begin the study of matrix theory with the basic definitions and operations. Matrices are then used to solve systems of equations using an elimination method and the inverse method. Our presentation concludes with a discussion of determinants, Cramer's rule, and properties of determinants, all of which have numerous applications in mathematics.

## 11.1 Matrices

### Matrix Terminology

Rectangular arrays of numbers, called **matrices** (singular, **matrix**), arise in numerous ways. For example, a table of data, such as the one below showing grade distributions in two different mathematics courses, is in effect a matrix.

| | A | B | C | D | F |
|---|---|---|---|---|---|
| MATH 110 | 8 | 10 | 21 | 4 | 5 |
| MATH 130 | 6 | 12 | 18 | 9 | 3 |

Capital letters denote matrices, and the numbers in the matrix, called **elements** of the matrix, are enclosed in brackets. The matrix $G$, given below, abbreviates the distribution of grades in the table.

$$G = \begin{bmatrix} 8 & 10 & 21 & 4 & 5 \\ 6 & 12 & 18 & 9 & 3 \end{bmatrix}$$

The **order** or **dimension** of a matrix with $m$ **rows** (numbers in horizontal lines) and $n$ **columns** (numbers in vertical lines) is $m \times n$, *read m by n*. The order of matrix $G$ is $2 \times 5$. Consider the following matrices.

$$E = \begin{bmatrix} 1 & 2 & -3 \\ 4 & 0 & 5 \end{bmatrix} \qquad B = \begin{bmatrix} 1 \\ 7 \\ -3 \end{bmatrix} \qquad C = \begin{bmatrix} 2 & 4 \\ -1 & 3 \end{bmatrix} \qquad D = \begin{bmatrix} 2 & 7 & -5 & 1 \end{bmatrix}$$

The order of $E$ is $2 \times 3$, the order of $B$ is $3 \times 1$, the order of $C$ is $2 \times 2$, and the order of $D$ is $1 \times 4$. A matrix of order $n \times n$, such as $C$, is called a **square matrix of order *n*.** A matrix of order $m \times 1$, such as $B$, is called a **column vector of order *m*.** A matrix of order $1 \times n$, such as $D$, is called a **row vector of order *n*.**

A general $m \times n$ matrix is often written in the form

$$A = \begin{bmatrix} a_{11} & a_{12} & a_{13} & \cdots & a_{1n} \\ a_{21} & a_{22} & a_{23} & \cdots & a_{2n} \\ a_{31} & a_{32} & a_{33} & \cdots & a_{3n} \\ \vdots & \vdots & \vdots & & \vdots \\ a_{m1} & a_{m2} & a_{m3} & \cdots & a_{mn} \end{bmatrix}$$

in which the double-subscript notation is used to identify the row and column containing a particular element. The element $a_{ij}$ is found in the $i$th row and the $j$th column. This notation is often abbreviated to

$$A = [a_{ij}]_{m \times n}.$$

## Equality of Matrices

Two matrices $A$ and $B$ are **equal,** written $A = B$, if and only if they have the same order and corresponding elements are equal.

Using the notation for a general matrix, if $A = [a_{ij}]_{m \times n}$ and $B = [b_{ij}]_{m \times n}$, $A = B$ if and only if $a_{ij} = b_{ij}$ for every $i = 1, 2, \ldots m$ and $j = 1, 2, \ldots n$.

**EXAMPLE 1** Consider the following matrices.

$$A = \begin{bmatrix} 2 & 0 & 3^2 \\ (-1)^3 & \sqrt[3]{27} & 1 \end{bmatrix}, \qquad B = \begin{bmatrix} \sqrt{4} & 0 & 9 \\ -1 & 3 & 2^0 \end{bmatrix}, \qquad C = \begin{bmatrix} 1 & 2 \\ 3 & 4 \end{bmatrix},$$

$$D = \begin{bmatrix} u & v \\ w & x \end{bmatrix}, \qquad E = \begin{bmatrix} 1 \\ 2 \\ 3 \\ 4 \end{bmatrix}, \qquad F = [1 \quad 2 \quad 3 \quad 4].$$

**(a)** Since $A$ and $B$ are both of order $2 \times 3$ and corresponding elements are equal, $A = B$.

**(b)** If $C = D$, then $u$ must be 1, $v$ must be 2, $w$ must be 3, and $x$ must be 4.

**(c)** Although $C$, the square matrix of order 2, $E$, the column vector of order 4, and $F$, the row vector of order 4, all contain the same elements, they are not equal since they have different orders. ■

## Operations on Matrices

The sum of two matrices of the same order is found by adding corresponding elements.

### Addition of Matrices

Let $A = [a_{ij}]_{m\times n}$ and $B = [b_{ij}]_{m\times n}$ be two matrices. The sum of $A$ and $B$ is the $m \times n$ matrix given by

$$A + B = [a_{ij} + b_{ij}]_{m\times n}.$$

Addition of matrices with different orders is not defined.

## EXAMPLE 2

Consider the following matrices.

$$A = \begin{bmatrix} 1 & 2 \\ -3 & 4 \end{bmatrix} \quad B = \begin{bmatrix} 0 & -3 \\ 2 & 5 \end{bmatrix} \quad C = \begin{bmatrix} -4 & -2 \\ 3 & 1 \end{bmatrix}$$

$$D = \begin{bmatrix} 0 & -2 & 5 \\ 3 & 7 & -1 \end{bmatrix} \quad E = \begin{bmatrix} -5 & 7 & 2 \\ 4 & -1 & 3 \end{bmatrix}$$

**(a)** $A + B = \begin{bmatrix} 1 & 2 \\ -3 & 4 \end{bmatrix} + \begin{bmatrix} 0 & -3 \\ 2 & 5 \end{bmatrix} = \begin{bmatrix} 1+0 & 2+(-3) \\ -3+2 & 4+5 \end{bmatrix} = \begin{bmatrix} 1 & -1 \\ -1 & 9 \end{bmatrix}$

**(b)** $B + A = \begin{bmatrix} 0 & -3 \\ 2 & 5 \end{bmatrix} + \begin{bmatrix} 1 & 2 \\ -3 & 4 \end{bmatrix} = \begin{bmatrix} 0+1 & -3+2 \\ 2+(-3) & 4+5 \end{bmatrix} = \begin{bmatrix} 1 & -1 \\ -1 & 9 \end{bmatrix}$

**(c)** $A + D$ is not defined since the order of $A$ is $2 \times 2$, which is not the same as the order of $D$, $2 \times 3$.

**(d)** $(A + B) + C = \left(\begin{bmatrix} 1 & 2 \\ -3 & 4 \end{bmatrix} + \begin{bmatrix} 0 & -3 \\ 2 & 5 \end{bmatrix}\right) + \begin{bmatrix} -4 & -2 \\ 3 & 1 \end{bmatrix}$

$$= \begin{bmatrix} 1 & -1 \\ -1 & 9 \end{bmatrix} + \begin{bmatrix} -4 & -2 \\ 3 & 1 \end{bmatrix} = \begin{bmatrix} -3 & -3 \\ 2 & 10 \end{bmatrix}$$

**(e)** $A + (B + C) = \begin{bmatrix} 1 & 2 \\ -3 & 4 \end{bmatrix} + \left(\begin{bmatrix} 0 & -3 \\ 2 & 5 \end{bmatrix} + \begin{bmatrix} -4 & -2 \\ 3 & 1 \end{bmatrix}\right)$

$$= \begin{bmatrix} 1 & 2 \\ -3 & 4 \end{bmatrix} + \begin{bmatrix} -4 & -5 \\ 5 & 6 \end{bmatrix} = \begin{bmatrix} -3 & -3 \\ 2 & 10 \end{bmatrix}$$

**(f)** $D + E = \begin{bmatrix} 0 & -2 & 5 \\ 3 & 7 & -1 \end{bmatrix} + \begin{bmatrix} -5 & 7 & 2 \\ 4 & -1 & 3 \end{bmatrix}$

$$= \begin{bmatrix} 0 + (-5) & -2 + 7 & 5 + 2 \\ 3 + 4 & 7 + (-1) & (-1) + 3 \end{bmatrix} = \begin{bmatrix} -5 & 5 & 7 \\ 7 & 6 & 2 \end{bmatrix} \quad ■$$

Notice in Example 2 **(a)** and **(b)** that $A + B = B + A$, and in Example 2 **(d)** and **(e)** that $(A + B) + C = A + (B + C)$. This is true in general. That is, if $A$, $B$, and $C$ are arbitrary matrices of the same order, then

$$\mathbf{A + B = B + A} \qquad \text{and} \qquad \mathbf{(A + B) + C = A + (B + C).}$$

These properties of matrix addition, the **commutative** and **associative laws of addition,** can easily be proved using the corresponding properties for addition of real numbers.

### The Zero Matrix

If $A = [a_{ij}]_{m \times n}$ is a matrix with the property that $a_{ij} = 0$ for every $i$ and $j$, $A$ is called a **zero matrix** and is denoted by 0. There is a zero matrix of every order.

For example, the $2 \times 3$ zero matrix is

$$0 = \begin{bmatrix} 0 & 0 & 0 \\ 0 & 0 & 0 \end{bmatrix},$$

and if

$$A = \begin{bmatrix} 2 & -1 & 3 \\ 5 & 7 & -4 \end{bmatrix},$$

then

$$A + 0 = \begin{bmatrix} 2 & -1 & 3 \\ 5 & 7 & -4 \end{bmatrix} + \begin{bmatrix} 0 & 0 & 0 \\ 0 & 0 & 0 \end{bmatrix} = \begin{bmatrix} 2 & -1 & 3 \\ 5 & 7 & -4 \end{bmatrix} = A.$$

This is true in general making 0 the **additive identity** for matrix addition. That is, if $A$ is an arbitrary $m \times n$ matrix and 0 is the $m \times n$ zero matrix, then

$$\mathbf{A + 0 = 0 + A = A.}$$

### The Negative of a Matrix

If $A = [a_{ij}]_{m \times n}$ is a matrix, the matrix denoted and given by

$$-A = [-a_{ij}]_{m \times n}$$

is the **negative** or **additive inverse** of $A$.

The negative or additive inverse of the matrix

$$A = \begin{bmatrix} 2 & -1 & 3 \\ 5 & 7 & -4 \end{bmatrix}$$

is

$$-A = \begin{bmatrix} -2 & 1 & -3 \\ -5 & -7 & 4 \end{bmatrix}$$

and

$$A + (-A) = \begin{bmatrix} 2 & -1 & 3 \\ 5 & 7 & -4 \end{bmatrix} + \begin{bmatrix} -2 & 1 & -3 \\ -5 & -7 & 4 \end{bmatrix}$$

$$= \begin{bmatrix} 2 + (-2) & (-1) + 1 & 3 + (-3) \\ 5 + (-5) & 7 + (-7) & (-4) + 4 \end{bmatrix} = \begin{bmatrix} 0 & 0 & 0 \\ 0 & 0 & 0 \end{bmatrix} = 0.$$

In general, if $A$ is an arbitrary $m \times n$ matrix, and 0 is the $m \times n$ zero matrix, then

$$\mathbf{A + (-A) = (-A) + A = 0.}$$

### Subtraction of Matrices

Let $A = [a_{ij}]_{m\times n}$ and $B = [b_{ij}]_{m\times n}$ be two matrices with $-B = [-b_{ij}]_{m\times n}$ the negative of $B$. The **difference** of $A$ and $B$ is the $m \times n$ matrix given by

$$A - B = A + (-B) = [a_{ij} + (-b_{ij})]_{m\times n} = [a_{ij} - b_{ij}]_{m\times n}.$$

Subtraction of two matrices with different orders is not defined.

In effect, if matrices $A$ and $B$ are of the same order, we find $A - B$ by subtracting the elements of $B$ from the corresponding elements of $A$. For example,

$$\begin{bmatrix} 3 & -2 & 7 \\ 4 & 0 & -1 \end{bmatrix} - \begin{bmatrix} 2 & 1 & 3 \\ 0 & 5 & -4 \end{bmatrix} = \begin{bmatrix} 3 - 2 & -2 - 1 & 7 - 3 \\ 4 - 0 & 0 - 5 & -1 - (-4) \end{bmatrix}$$

$$= \begin{bmatrix} 1 & -3 & 4 \\ 4 & -5 & 3 \end{bmatrix}.$$

A matrix can be multiplied by a real number by multiplying every element of the matrix by the number. This operation is called **scalar multiplication.**

### Scalar Multiplication

Let $A = [a_{ij}]_{m\times n}$ be a matrix with $k$ a real number. The **scalar multiple** of $A$ by $k$ is the matrix

$$kA = [ka_{ij}]_{m\times n}.$$

**EXAMPLE 3**

Consider the following matrices:

$$A = \begin{bmatrix} 2 & 5 & -1 \\ 0 & 3 & 7 \end{bmatrix} \qquad B = \begin{bmatrix} -1 & 2 & 4 \\ 3 & -3 & 0 \end{bmatrix}$$

**(a)** $3A = 3\begin{bmatrix} 2 & 5 & -1 \\ 0 & 3 & 7 \end{bmatrix} = \begin{bmatrix} 3(2) & 3(5) & 3(-1) \\ 3(0) & 3(3) & 3(7) \end{bmatrix} = \begin{bmatrix} 6 & 15 & -3 \\ 0 & 9 & 21 \end{bmatrix}$

**(b)** $B - A = \begin{bmatrix} -1 & 2 & 4 \\ 3 & 3 & 0 \end{bmatrix} - \begin{bmatrix} 2 & 5 & -1 \\ 0 & 3 & 7 \end{bmatrix} = \begin{bmatrix} -3 & -3 & 5 \\ 3 & -6 & -7 \end{bmatrix}$

**(c)** $2A - 3B = 2\begin{bmatrix} 2 & 5 & -1 \\ 0 & 3 & 7 \end{bmatrix} - 3\begin{bmatrix} -1 & 2 & 4 \\ 3 & -3 & 0 \end{bmatrix}$

$$= \begin{bmatrix} 4 & 10 & -2 \\ 0 & 6 & 14 \end{bmatrix} - \begin{bmatrix} -3 & 6 & 12 \\ 9 & -9 & 0 \end{bmatrix}$$

$$= \begin{bmatrix} 7 & 4 & -14 \\ -9 & 15 & 14 \end{bmatrix} \quad \blacksquare$$

Matrices can be used in many ways to simplify and manipulate collections of numbers in data bases, often with the help of computers. Keep in mind that our examples are simplified systems involving small data bases but that the same techniques also apply to more realistic larger collections of data.

**EXAMPLE 4**

**BUSINESS**

The State Mutual Insurance Agency has three salespeople, Abbott, Baker, and Chance. Four types of insurance are sold: life, health, automobile, and home. The gross sales in dollars for the months of May and June are given in the following matrices.

$$M = \begin{array}{c} \begin{array}{cccc} \text{Life} & \text{Health} & \text{Automobile} & \text{Home} \end{array} \\ \begin{bmatrix} 20{,}000 & 10{,}000 & 5000 & 40{,}000 \\ 30{,}000 & 5000 & 17{,}000 & 65{,}000 \\ \mathbf{0} & 27{,}000 & 8000 & 32{,}000 \end{bmatrix} \end{array} \begin{array}{l} \text{Abbott} \\ \text{Baker} \\ \text{Chance} \end{array}$$

$$J = \begin{array}{c} \begin{array}{cccc} \text{Life} & \text{Health} & \text{Automobile} & \text{Home} \end{array} \\ \begin{bmatrix} 70{,}000 & 0 & 14{,}000 & 90{,}000 \\ 12{,}000 & \mathbf{20{,}000} & 8000 & 74{,}000 \\ 120{,}000 & 18{,}000 & 11{,}000 & 45{,}000 \end{bmatrix} \end{array} \begin{array}{l} \text{Abbott} \\ \text{Baker} \\ \text{Chance} \end{array}$$

For example, Chance wrote no life insurance policies in May, and Baker sold \$20,000 in health insurance policies in June.

**(a)** The matrix which gives the total sales during the two months for each salesperson in each category is

$$M + J = \begin{bmatrix} 90{,}000 & 10{,}000 & 19{,}000 & \mathbf{130{,}000} \\ 42{,}000 & 25{,}000 & 25{,}000 & 139{,}000 \\ 120{,}000 & 45{,}000 & 19{,}000 & 77{,}000 \end{bmatrix}.$$

For example, Abbott had total sales in home insurance during May and June of \$130,000.

**(b)** The matrix which gives the increase (or decrease) in sales in each category for each salesperson from May to June is

$$J - M = \begin{bmatrix} 50{,}000 & -10{,}000 & 9000 & 50{,}000 \\ -18{,}000 & \mathbf{15{,}000} & -9000 & 9000 \\ 120{,}000 & \mathbf{-9000} & 3000 & 13{,}000 \end{bmatrix}.$$

For example, Baker had an increase in sales of health insurance amounting to \$15,000, while Chance had a decrease in sales of \$9000 in this category.

**(c)** Suppose each salesperson works on an 8% commission rate. The matrix which gives their commission earned in each category during this two-month period is $0.08(M + J)$.

$$0.08(M + J) = 0.08\begin{bmatrix} 90{,}000 & 10{,}000 & 19{,}000 & 130{,}000 \\ 42{,}000 & 25{,}000 & 25{,}000 & 139{,}000 \\ 120{,}000 & 45{,}000 & 19{,}000 & 77{,}000 \end{bmatrix}$$

$$= \begin{bmatrix} (0.08)(90{,}000) & (0.08)(10{,}000) & (0.08)(19{,}000) & (0.08)(130{,}000) \\ (0.08)(42{,}000) & (0.08)(25{,}000) & (0.08)(25{,}000) & (0.08)(139{,}000) \\ (0.08)(120{,}000) & (0.08)(45{,}000) & (0.08)(19{,}000) & (0.08)(77{,}000) \end{bmatrix}$$

$$= \begin{bmatrix} 7200 & 800 & 1520 & \mathbf{10{,}400} \\ 3360 & 2000 & 2000 & 11{,}120 \\ 9600 & 3600 & 1520 & 6160 \end{bmatrix}$$

For example, Abbott earned \$10,400 as a commission for selling home insurance during May and June. ■

**NOTE** If the insurance agency in Example 4 employed 100 salespeople and the number of types of insurance sold was increased considerably, you can begin to recognize and appreciate the power of matrices, especially when a computer is used to perform matrix operations. ■

## 11.1 EXERCISES

Exercises 1–32 refer to the following matrices.

$$A = \begin{bmatrix} 1 & 2 \\ 3 & 4 \end{bmatrix} \quad B = \begin{bmatrix} 5 & -2 \\ 9 & -4 \end{bmatrix} \quad C = \begin{bmatrix} 5^0 & \sqrt{4} \\ |-3| & 2^2 \end{bmatrix} \quad D = \begin{bmatrix} 3 & -1 & 4 \\ 2 & 0 & 5 \end{bmatrix}$$

$$E = \begin{bmatrix} -2 & 3 & 0 \\ 5 & 4 & 7 \end{bmatrix} \quad F = [3 \quad 2 \quad 0 \quad -1] \quad G = \begin{bmatrix} 3 \\ 2 \\ 0 \\ -1 \end{bmatrix}$$

**1.** What is the order of matrix $A$?

**2.** What is the order of matrix $D$?

**3.** What is the order of matrix $F$?

**4.** What is the order of matrix $G$?

**5.** Which of these matrices are square matrices?

**6.** Which of these matrices have order $2 \times 3$?

**7.** Which of these matrices are column vectors?

**8.** Which of these matrices are row vectors?

**9.** Are $A$ and $C$ equal?

**10.** Are $F$ and $G$ equal?

**11.** What element is in the first row and second column of $D$?

**12.** What element is in the second row and second column of $A$?

**13.** Give the zero matrix with the same order as $A$.

**14.** Give the zero matrix with the same order as $E$.

**15.** Identify the element $a_{21}$ in matrix $A$.

**16.** Identify the element $e_{23}$ in matrix $E$.

**17.** Give the matrix $-A$.

**18.** Give the matrix $-D$.

**19.** Find $A + B$.

**20.** Find $D + E$.

**21.** Find $A + D$.

**22.** Find $B + E$.

**23.** Find $D - E$.

**24.** Find $A - B$.

**25.** Find $B - E$.

**26.** Find $A - D$.

**27.** Find $-2A$.

**28.** Find $-4F$.

**29.** Find $a_{12}B$.

**30.** Find $b_{12}A$.

**31.** Find $2A - 3B$.

**32.** Find $3E - 2D$.

In Exercises 33–36 find the values of $a$, $b$, $c$, and $d$.

**33.** $\begin{bmatrix} a & b \\ c & d \end{bmatrix} + \begin{bmatrix} -1 & 3 \\ 0 & 2 \end{bmatrix} = \begin{bmatrix} 6 & -2 \\ 5 & 1 \end{bmatrix}$

**34.** $\begin{bmatrix} a & b \\ c & d \end{bmatrix} - \begin{bmatrix} 2 & 3 \\ 7 & -1 \end{bmatrix} = \begin{bmatrix} 6 & 0 \\ 4 & 2 \end{bmatrix}$

**35.** $\begin{bmatrix} a & 1 \\ 0 & b \end{bmatrix} - \begin{bmatrix} b & -2 \\ 3 & 2a \end{bmatrix} = \begin{bmatrix} -5 & 3 \\ -3 & 8 \end{bmatrix}$

**36.** $\begin{bmatrix} 3 & -b \\ 2 & a \end{bmatrix} + \begin{bmatrix} 1 & 3a \\ -2 & -b \end{bmatrix} = \begin{bmatrix} 4 & 7 \\ 0 & 3 \end{bmatrix}$

**37. SPORTS** Matrices $F$ and $S$ give the points scored, number of rebounds, and minutes played for the starting five in a two-game basketball tournament.

$$F = \begin{array}{c} \text{Points} \quad \text{Rebounds} \quad \text{Minutes} \\ \begin{bmatrix} 21 & 5 & 35 \\ 18 & 9 & 32 \\ 10 & 7 & 28 \\ 8 & 3 & 30 \\ 15 & 12 & 32 \end{bmatrix} \end{array} \begin{array}{l} \text{Hurd} \\ \text{Spencer} \\ \text{Murchison} \\ \text{Payne} \\ \text{Duane} \end{array}$$

$$S = \begin{bmatrix} 19 & 4 & 33 \\ 26 & 8 & 30 \\ 8 & 10 & 26 \\ 8 & 5 & 32 \\ 17 & 12 & 28 \end{bmatrix} \begin{array}{l} \text{Hurd} \\ \text{Spencer} \\ \text{Murchison} \\ \text{Payne} \\ \text{Duane} \end{array}$$

**(a)** Find the matrix that gives the totals for the tournament in each category.
**(b)** Find the matrix $\frac{1}{2}(F + S)$ and discuss what it represents.

**38. BUSINESS** Video Rentals Inc. has two stores in Barstow, California. The total number of VHS cassettes, Beta cassettes, and recorders rented during the months of January and February are given in matrices $J$ and $F$.

$$\begin{array}{c} \\ J = \end{array}\begin{array}{c} \begin{array}{ccc} \text{VHS} & \text{Beta} & \text{Recorders} \end{array} \\ \begin{bmatrix} 2100 & 400 & 55 \\ 3500 & 700 & 60 \end{bmatrix} \end{array}\begin{array}{l} \\ \text{Store 1} \\ \text{Store 2} \end{array}$$

$$F = \begin{bmatrix} 3200 & 750 & 84 \\ 4300 & 900 & 105 \end{bmatrix}\begin{array}{l} \text{Store 1} \\ \text{Store 2} \end{array}$$

**(a)** Find the matrix that gives the two-month totals in each category.
**(b)** Find the matrix $\frac{1}{2}(J + F)$ and discuss what it represents.
**(c)** Find the matrix that gives the increase in rentals from January to February.

## For Review

Solve each system of equations in Exercises 39–44.

**39.** $\begin{aligned} 3x + 5y &= 9 \\ 2x - 7y &= -25 \end{aligned}$

**40.** $\begin{aligned} 2x + 6y &= 12 \\ -x - 3y &= 6 \end{aligned}$

**41.** $\begin{aligned} 4x - y \phantom{- 5z} &= -3 \\ -2y - 5z &= -1 \\ 3x \phantom{- 2y} + 2z &= -2 \end{aligned}$

**42.** $\begin{aligned} x + 5y + 2z &= -37 \\ 3x + 2y - z &= -5 \\ -3x + 3y + z &= 11 \end{aligned}$

**43.** $\begin{aligned} x^2 + y^2 &= 169 \\ x + y &= 17 \end{aligned}$

**44.** $\begin{aligned} x^2 + y^2 &= 25 \\ x^2 - y^2 &= 25 \end{aligned}$

**45.** Graph the system.
$|y| - x \le 0$
$x^2 + y^2 < 16$

# 11.2 Matrix Multiplication

In this section we will define two kinds of matrix multiplication which will probably appear somewhat strange at first. As we shall soon see, however, these operations have numerous applications. Unlike addition and subtraction, the product of two matrices *is not* found by multiplying corresponding elements nor do the two matrices have to be of the same order.

## Scalar Products

We begin by defining the *scalar* or *dot product* of a row vector of order $n$ times a column vector of order $n$. The result is a real number, or scalar, and not a matrix.

**Scalar (Dot) Product of Two Vectors**

Let $A = [a_1\ a_2\ a_3\ .\ .\ .\ a_n]$ and $B = \begin{bmatrix} b_1 \\ b_2 \\ b_3 \\ \vdots \\ b_n \end{bmatrix}$.

The scalar product of $A$ and $B$ is given by

$$\boldsymbol{A \cdot B = a_1b_1 + a_2b_2 + a_3b_3 + \cdots + a_nb_n.}$$

CAUTION The dot written between the two vectors (matrices) in the above definition is very important. If it is omitted, the product is interpreted differently and results in a matrix (not a real number). This is defined shortly. ■

## EXAMPLE 1

Let $A = [3 \quad 2 \quad 1 \quad 5]$ and $B = \begin{bmatrix} 300 \\ 450 \\ 500 \\ 650 \end{bmatrix}$. Find $A \cdot B$.

$$A \cdot B = [3 \quad 2 \quad 1 \quad 5] \cdot \begin{bmatrix} 300 \\ 450 \\ 500 \\ 650 \end{bmatrix} = 3(300) + 2(450) + 1(500) + 5(650)$$

$$= 900 + 900 + 500 + 3250 = 5550 \quad ■$$

This "row times column" method of multiplication may seem a bit artificial at first, but it has numerous applications. For example, suppose that an appliance dealer sells 3 stereos for \$300 each, 2 stereos for \$450 each, 1 stereo for \$500, and 5 stereos for \$650 each. The vector $A$ in Example 1 represents the number of stereos sold at each price, and the vector $B$ represents the price of each type of stereo. The scalar product $A \cdot B$ gives the total revenue produced by selling these stereos, \$5550.

## Multiplying Matrices

Keep in mind that the scalar product of a row vector and a column vector is a real number, not a matrix. We now define multiplication of matrices in such a way that the product will be a matrix. Scalar products are used in this definition.

**Multiplication of Matrices**

Let $A = [a_{ij}]_{m\times n}$ and $B = [b_{ij}]_{n\times p}$ be two matrices with the property that the number of columns of $A$ is the same as the number of rows of $B$. The **product** of $A$ and $B$ is the $m \times p$ matrix

$$AB = [c_{ij}]_{m\times p},$$

where $c_{ij}$ is the scalar product of the $i$th row of $A$ and the $j$th column of $B$.

The diagram below illustrates the restrictions imposed by the definition of multiplication.

$$A_{m\times n} \qquad B_{n\times p} = AB_{m\times p}$$

equal

order of product is $m \times p$

## EXAMPLE 2

Let $A = \begin{bmatrix} 1 & -2 & 3 \\ 4 & 0 & 5 \end{bmatrix}$ and $B = \begin{bmatrix} 2 & -1 \\ -3 & 1 \\ 0 & 4 \end{bmatrix}$. Find $AB$.

Since it is $A_{2\times 3} \qquad B_{3\times 2}$, the product $AB$ is defined and has order $2 \times 2$.

equal

order of $AB$

Thus,

row 1 of $A$ column 1 of $B$ row 1 of $A$ column 2 of $B$

$$AB = \begin{bmatrix} [1 \quad -2 \quad 3] \cdot \begin{bmatrix} 2 \\ -3 \\ 0 \end{bmatrix} & [1 \quad -2 \quad 3] \cdot \begin{bmatrix} -1 \\ 1 \\ 4 \end{bmatrix} \\ [4 \quad 0 \quad 5] \cdot \begin{bmatrix} 2 \\ -3 \\ 0 \end{bmatrix} & [4 \quad 0 \quad 5] \cdot \begin{bmatrix} -1 \\ 1 \\ 4 \end{bmatrix} \end{bmatrix}$$

row 2 of $A$ column 1 of $B$ row 2 of $A$ column 2 of $B$

$$= \begin{bmatrix} (1)(2) + (-2)(-3) + (3)(0) & (1)(-1) + (-2)(1) + (3)(4) \\ (4)(2) + (0)(-3) + (5)(0) & (4)(-1) + (0)(1) + (5)(4) \end{bmatrix}$$

$$= \begin{bmatrix} 8 & 9 \\ 8 & 16 \end{bmatrix}. \quad ■$$

**EXAMPLE 3** Find each product.

**(a)** $\begin{bmatrix} 1 & -2 \\ 0 & -3 \end{bmatrix}_{2\times 2} \begin{bmatrix} 3 & 0 & 4 \\ -1 & 5 & 0 \end{bmatrix}_{2\times 3}$

(inner dimensions: equal; outer dimensions: order of product)

$$= \begin{bmatrix} (1)(3) + (-2)(-1) & (1)(0) + (-2)(5) & (1)(4) + (-2)(0) \\ (0)(3) + (-3)(-1) & (0)(0) + (-3)(5) & (0)(4) + (-3)(0) \end{bmatrix}$$

$$= \begin{bmatrix} 5 & -10 & 4 \\ 3 & -15 & 0 \end{bmatrix}$$

**(b)** $\begin{bmatrix} 3 & 0 & 4 \\ -1 & 5 & 0 \end{bmatrix}_{2\times 3} \begin{bmatrix} 1 & -2 \\ 0 & -3 \end{bmatrix}_{2\times 2}$ This product is not defined.

(inner dimensions: not equal)

**(c)** $[2 \quad 0 \quad -3]_{1\times 3} \begin{bmatrix} 1 & 3 & -2 \\ 4 & -1 & 5 \\ 2 & 3 & -5 \end{bmatrix}_{3\times 3} = [2 + 0 - 6 \quad 6 + 0 - 9 \quad -4 + 0 + 15]$

$(2)(1) + (0)(4) + (-3)(2) = 2 + 0 - 6$

(inner dimensions: equal; outer dimensions: order of product)

$$= [-4 \quad -3 \quad 11]$$

**(d)** $\begin{bmatrix} 2 & -1 \\ 3 & 5 \end{bmatrix}\begin{bmatrix} 4 & 0 \\ 1 & -2 \end{bmatrix} = \begin{bmatrix} 8 - 1 & 0 + 2 \\ 12 + 5 & 0 - 10 \end{bmatrix} = \begin{bmatrix} 7 & 2 \\ 17 & -10 \end{bmatrix}$

**(e)** $\begin{bmatrix} 4 & 0 \\ 1 & -2 \end{bmatrix}\begin{bmatrix} 2 & 1 \\ 3 & 5 \end{bmatrix} = \begin{bmatrix} 8 + 0 & -4 + 0 \\ 2 - 6 & -1 - 10 \end{bmatrix} = \begin{bmatrix} 8 & -4 \\ -4 & -11 \end{bmatrix}$ ■

## Properties of Matrix Multiplication

Matrix multiplication is not a commutative operation. That is, if $A$ and $B$ are matrices, then $AB$ does not necessarily equal $BA$. This is clear if $A$ is of order $2 \times 2$ and $B$ is of order $2 \times 3$, then $AB$ has order $2 \times 2$ but $BA$ is not even defined. (See Example 3 **(a)** and **(b)**.) However, even for square matrices when the products in either order are defined, changing the order of multiplication may result in different products. In Example 3 **(d)** and **(e)**, we saw that

$$\begin{bmatrix} 2 & -1 \\ 3 & 5 \end{bmatrix}\begin{bmatrix} 4 & 0 \\ 1 & -2 \end{bmatrix} \neq \begin{bmatrix} 4 & 0 \\ 1 & -2 \end{bmatrix}\begin{bmatrix} 2 & -1 \\ 3 & 5 \end{bmatrix}.$$

On the other hand, if $A$ is of order $m \times n$, $B$ is of order $n \times p$, and $C$ is of order $p \times q$, it can be shown that

$$(AB)C = A(BC)$$

so that matrix multiplication satisfies the **associative law of multiplication.** Also, for matrices of the appropriate order, the **distributive laws** are true.

$$A(B + C) = AB + AC \qquad \text{and} \qquad (E + F)D = ED + FD$$

Examples of these properties are considered in the exercises at the end of this section.

Consider the following two products.

$$\begin{bmatrix} 2 & -1 \\ 3 & 5 \end{bmatrix}\begin{bmatrix} 1 & 0 \\ 0 & 1 \end{bmatrix} = \begin{bmatrix} 2+0 & 0-1 \\ 3+0 & 0+5 \end{bmatrix} = \begin{bmatrix} 2 & -1 \\ 3 & 5 \end{bmatrix}$$

$$\begin{bmatrix} 1 & 0 \\ 0 & 1 \end{bmatrix}\begin{bmatrix} 2 & -1 \\ 3 & 5 \end{bmatrix} = \begin{bmatrix} 2+0 & -1+0 \\ 0+3 & 0+5 \end{bmatrix} = \begin{bmatrix} 2 & -1 \\ 3 & 5 \end{bmatrix}$$

Notice that in both cases the product is equal to

$$\begin{bmatrix} 2 & -1 \\ 3 & 5 \end{bmatrix}.$$

The matrix

$$\begin{bmatrix} 1 & 0 \\ 0 & 1 \end{bmatrix}$$

is the $2 \times 2$ version of an important group of matrices.

## Identity Matrix

The matrix $I_n$ of order $n \times n$ given by

$$I_n = \begin{bmatrix} 1 & 0 & 0 & \cdots & 0 \\ 0 & 1 & 0 & \cdots & 0 \\ 0 & 0 & 1 & \cdots & 0 \\ \vdots & \vdots & \vdots & & \\ 0 & 0 & 0 & \cdots & 1 \end{bmatrix}$$

with 1's down the **main diagonal** and 0's everywhere else is called the **identity matrix of order *n*.**

Identity matrices play a role in matrix multiplication similar to the role that 1 plays in multiplication of numbers. If $A_{n\times n}$ is an arbitrary matrix, then

$$I_nA_{n\times n} = A_{n\times n}I_n = A_{n\times n}.$$

We have seen that the operations on matrices satisfy many of the properties similar to those for the operations on real numbers. The one exception thus far is the commutative law of multiplication. Another familiar property of real numbers, the zero-product rule, does not carry over to matrix multiplication. Consider the matrices

$$A = \begin{bmatrix} 1 & 4 \\ 1 & 4 \end{bmatrix} \quad \text{and} \quad B = \begin{bmatrix} 4 & 4 \\ -1 & -1 \end{bmatrix}.$$

Then $$AB = \begin{bmatrix} 1 & 4 \\ 1 & 4 \end{bmatrix}\begin{bmatrix} 4 & 4 \\ -1 & -1 \end{bmatrix} = \begin{bmatrix} 0 & 0 \\ 0 & 0 \end{bmatrix} = 0.$$

However, neither $A$ nor $B$ is the zero matrix. We see that matrix multiplication does not satisfy the zero-product rule, since $AB = 0$ while neither $A = 0$ nor $B = 0$.

### Transpose of a Matrix

The **transpose** of an $m \times n$ matrix $A$, denoted by $A^{\mathrm{T}}$, is the $n \times m$ matrix formed by interchanging the rows and columns of $A$. That is, row 1 of $A$ becomes column 1 of $A^{\mathrm{T}}$, row 2 of $A$ becomes column 2 of $A^{\mathrm{T}}$, and so forth.

For example, if $A = \begin{bmatrix} 2 & 3 \\ -1 & 5 \end{bmatrix}$, then $A^{\mathrm{T}} = \begin{bmatrix} 2 & -1 \\ 3 & 5 \end{bmatrix}$, and if $B = \begin{bmatrix} 3 & -1 \\ 0 & 5 \\ 2 & 8 \end{bmatrix}$, then $B^{\mathrm{T}} = \begin{bmatrix} 3 & 0 & 2 \\ -1 & 5 & 8 \end{bmatrix}$. The transpose of a matrix will be used in later sections.

We conclude this section with an application of matrix multiplication given as the first applied problem in the introduction to this chapter.

**EXAMPLE 4**

**POLITICS**

A candidate for political office in Arizona hired a public relations firm to promote his campaign in three population centers of the state: Tucson, Phoenix, and Flagstaff. Contacts of potential supporters are to be made by telephone, mail, and direct personal contact. The cost of each of these is estimated to be \$0.50 by telephone, \$0.30 by mail, and \$0.75 for a personal contact. The table below summarizes the desired number of contacts in each category in each area. Use matrices to determine the total cost for the promotion.

| | Telephone | Mail | Personal |
|---|---|---|---|
| Tucson | 5000 | 10,000 | 1000 |
| Phoenix | 8000 | 20,000 | 6000 |
| Flagstaff | 3000 | 5000 | 700 |

The matrix

$$C = \begin{bmatrix} 0.50 \\ 0.30 \\ 0.75 \end{bmatrix} \begin{matrix} \text{Telephone} \\ \text{Mail} \\ \text{Personal contact} \end{matrix}$$

gives the estimated cost for each type of contact.

The matrix

$$N = \begin{bmatrix} 5000 & 10{,}000 & 1000 \\ 8000 & 20{,}000 & 6000 \\ 3000 & 5000 & 700 \end{bmatrix} \begin{matrix} \text{Tucson} \\ \text{Phoenix} \\ \text{Flagstaff} \end{matrix}$$

(columns: Telephone, Mail, Personal)

gives the number of contacts desired in each area. The product $NC$ gives the column vector with the total costs in each area.

$$NC = \begin{bmatrix} 5000 & 10{,}000 & 1000 \\ 8000 & 20{,}000 & 6000 \\ 3000 & 5000 & 700 \end{bmatrix} \begin{bmatrix} 0.50 \\ 0.30 \\ 0.75 \end{bmatrix} = \begin{bmatrix} 6250 \\ 14{,}500 \\ 3525 \end{bmatrix} \begin{matrix} \text{Total cost in Tucson} \\ \text{Total cost in Phoenix} \\ \text{Total cost in Flagstaff} \end{matrix}$$

The total cost in all three areas can be determined by finding the scalar product of the row vector $T = [1 \quad 1 \quad 1]$ with the column vector $BC$.

$$T \cdot (BC) = [1 \quad 1 \quad 1] \cdot \begin{bmatrix} 6250 \\ 14{,}500 \\ 3525 \end{bmatrix} = 24{,}275$$

Thus, the total cost for the campaign is \$24,275. ■

## 11.2 EXERCISES

Find the scalar products in Exercises 1–4.

**1.** $[1 \quad -2] \cdot \begin{bmatrix} 3 \\ 5 \end{bmatrix}$

**2.** $[-1 \quad 6] \cdot \begin{bmatrix} 0 \\ -7 \end{bmatrix}$

**3.** $[2 \quad 3 \quad 5] \cdot \begin{bmatrix} -1 \\ 0 \\ 2 \end{bmatrix}$

**4.** $[0 \quad 2 \quad 1 \quad -3] \cdot \begin{bmatrix} 4 \\ 1 \\ -2 \\ 5 \end{bmatrix}$

In Exercises 5–8 determine $m$ and $n$ so that each product will be defined.

**5.** $A_{m\times n}B_{3\times 2} = C_{4\times 2}$

**6.** $A_{2\times 3}B_{m\times n} = C_{2\times 5}$

**7.** $A_{4\times m}B_{2\times n} = C_{4\times 3}$

**8.** $A_{m\times 4}B_{n\times 2} = C_{5\times 2}$

Exercises 9–28 refer to the following matrices.

$$A = \begin{bmatrix} 2 & 5 \\ -1 & 3 \end{bmatrix} \quad B = \begin{bmatrix} 0 & -2 \\ 1 & 4 \end{bmatrix} \quad C = \begin{bmatrix} 1 & 1 \\ 1 & 1 \end{bmatrix} \quad D = \begin{bmatrix} 1 & 3 & -2 \\ 4 & 0 & 5 \end{bmatrix}$$

$$E = \begin{bmatrix} 0 & 4 \\ 4 & 0 \end{bmatrix} \quad F = \begin{bmatrix} 3 & 0 \\ -2 & 1 \\ 0 & 5 \end{bmatrix} \quad G = [3 \quad -2] \quad H = \begin{bmatrix} -2 \\ 4 \\ 1 \end{bmatrix}$$

**9.** Find $AB$.
**10.** Find $AC$.
**11.** Find $CD$.
**12.** Find $DE$.
**13.** Find $FD$.
**14.** Find $GE$.
**15.** Find $DH$.
**16.** Find $A^T$.
**17.** Find $B^T$.
**18.** Find $A^2$.
**19.** Find $B^2$.
**20.** Find $D^T$.
**21.** Find $F^T$.
**22.** Find $AA^T$.
**23.** Find $(A^T)^TE$.
**24.** Find $A(BC)$.
**25.** Find $A^T + B^T$.
**26.** Find $B^T + E^T$.
**27.** Find $(A + B)^T$.
**28.** Find $(B + E)^T$.

In Exercises 29–36 use the following matrices to verify each statement.

$$A = \begin{bmatrix} 2 & 1 \\ 0 & 5 \end{bmatrix} \quad B = \begin{bmatrix} 3 & 1 \\ 1 & -2 \end{bmatrix} \quad C = \begin{bmatrix} 0 & 4 \\ -2 & 1 \end{bmatrix}$$

**29.** $AB \neq BA$
**30.** $A(BC) = (AB)C$
**31.** $AI = A$
**32.** $A + (-A) = 0$
**33.** $A + 0 = A$
**34.** $A(B + C) = AB + AC$
**35.** $(B + C)A = BA + CA$
**36.** $A(B + C) \neq (B + C)A$

**37.** **SPORTS** Matrices $F$ and $S$ give the points scored, number of rebounds, and minutes played for the starting five in a two-game basketball tournament.

$$F = \begin{array}{c} \text{Points} \quad \text{Rebounds} \quad \text{Minutes} \\ \begin{bmatrix} 21 & 5 & 35 \\ 18 & 9 & 32 \\ 10 & 7 & 28 \\ 8 & 3 & 30 \\ 15 & 12 & 32 \end{bmatrix} \end{array} \begin{array}{l} \text{Hurd} \\ \text{Spencer} \\ \text{Murchison} \\ \text{Payne} \\ \text{Duane} \end{array}$$

$$S = \begin{bmatrix} 19 & 4 & 33 \\ 26 & 8 & 30 \\ 8 & 10 & 26 \\ 8 & 5 & 32 \\ 17 & 12 & 28 \end{bmatrix} \begin{array}{l} \text{Hurd} \\ \text{Spencer} \\ \text{Murchison} \\ \text{Payne} \\ \text{Duane} \end{array}$$

**(a)** Find $[1 \quad 1 \quad 1 \quad 1 \quad 1]F$ and discuss what it represents.
**(b)** Find $[1 \quad 1 \quad 1 \quad 1 \quad 1](F + S)$ and discuss what it represents.
**(c)** Find $\frac{1}{2}[1 \quad 1 \quad 1 \quad 1 \quad 1](F + S)$ and discuss what it represents.

**38.** **BUSINESS** Video Rentals Inc. has two stores in Barstow, California. The total number of VHS cassettes, Beta cassettes, and recorders rented during the months of January and February are given in matrices $J$ and $F$.

$$J = \begin{array}{c} \text{VHS} \quad \text{Beta} \quad \text{Recorders} \\ \begin{bmatrix} 2100 & 400 & 55 \\ 3500 & 700 & 60 \end{bmatrix} \end{array} \begin{array}{l} \text{Store 1} \\ \text{Store 2} \end{array}$$

$$F = \begin{bmatrix} 3200 & 750 & 84 \\ 4300 & 900 & 105 \end{bmatrix} \begin{array}{l} \text{Store 1} \\ \text{Store 2} \end{array}$$

**(a)** Find $[1 \quad 1]J$ and discuss what it represents.
**(b)** Find $[1 \quad 1](J + F)$ and discuss what it represents.
**(c)** Find $([1 \quad 1](J + F)) \cdot \begin{bmatrix} 1 \\ 1 \\ 1 \end{bmatrix}$ and discuss what it represents.

**39. BUSINESS** Suppose that the rental rates charged by both stores in Video Rentals Inc. of Exercise 38 are \$3 for VHS, \$4 for Beta, and \$10 for a recorder. Use the matrix

$$C = \begin{bmatrix} 3 \\ 4 \\ 10 \end{bmatrix}$$

to answer the following. **(a)** Find $JC$ and interpret the result. **(b)** Find $FC$ and interpret the result. **(c)** Find $(J + F)C$ and interpret the result. **(d)** Find $[1 \quad 1] \cdot ((J + F)C)$ and interpret the result.

**40. VETERINARY SCIENCE** A veterinarian blends two brands of dog food into three different mixes. The amounts of protein and fat, measured in grams per ounce, in each of the two brands are given in matrix $N$. The amounts of each brand, measured in ounces, that he uses in each mix are given in matrix $M$.

$$N = \begin{array}{c} \text{Brand A} \quad \text{Brand B} \\ \begin{bmatrix} 6 & 8 \\ 4 & 3 \end{bmatrix} \end{array} \begin{array}{l} \text{Protein} \\ \text{Fat} \end{array}$$

$$M = \begin{array}{c} \text{Mix 1} \quad \text{Mix 2} \quad \text{Mix 3} \\ \begin{bmatrix} 12 & 18 & 6 \\ 12 & 6 & 18 \end{bmatrix} \end{array} \begin{array}{l} \text{Brand A} \\ \text{Brand B} \end{array}$$

**(a)** Find the matrix $NM = [c_{ij}]_{2\times3}$ and discuss what it represents.

**(b)** What does the element $c_{12}$ represent?

**(c)** What does the element $c_{21}$ represent?

**(d)** Notice that each mix contains 24 ounces of dog food. Find $\frac{1}{24}NM$ and interpret the results.

**(e)** Suppose that Brand A costs \$0.15 per ounce and Brand B costs \$0.20 per ounce. Find the matrix that gives total cost for each mix. Which mix is the most expensive?

## For Review

Use the following matrices in Exercises 41–46.

$$A = \begin{bmatrix} -5 & 0 \\ 2 & 7 \end{bmatrix} \qquad B = \begin{bmatrix} 0 & 3 \\ -2 & 1 \end{bmatrix} \qquad C = \begin{bmatrix} u & v \\ w & z \end{bmatrix}$$

**41.** If $A = C$, find the values of $u$, $v$, $w$, and $z$.

**42.** Identify the element $a_{12}$ in matrix $A$.

**43.** Find $A + 0$. **44.** Find $A + B$. **45.** Find $A - B$. **46.** Find $3A - 7B$.

The systems in Exercises 47–50 will be solved in the next section by matrix methods. Find the solution to each system here by techniques given in the preceding chapter.

**47.** $\begin{aligned} 2x + y &= 1 \\ x - 3y &= 11 \end{aligned}$

**48.** $\begin{aligned} 2x - y + z &= 1 \\ x + 3y - z &= -4 \\ 3x + 2y + z &= 0 \end{aligned}$

**49.** $\begin{aligned} w + x + y + z &= 2 \\ w - x + 3y + z &= 3 \\ x - y - z &= -1 \\ 2w + x + 3y + z &= 0 \end{aligned}$

**50.** $\begin{aligned} x - z &= -3 \\ -2x - y - 2z &= 2 \\ 3x + y + z &= -5 \end{aligned}$

## 11.3 Solving Systems of Equations Using Matrices

### The Gaussian Method

In the previous chapter we solved systems of linear equations using a method of elimination. Solving a system can be simplified by using matrices consisting of the coefficients of the variables along with the constants. The method that we present now, due to famous mathematician Karl Fredrich Gauss (1777–1855), is sometimes referred to as the **Gaussian method.** Consider the system

$$\begin{aligned} 2x + \phantom{3}y &- 1 \\ x - 3y &= 11. \end{aligned}$$

The two matrices associated with this system include the **coefficient matrix**

$$\begin{bmatrix} 2 & 1 \\ 1 & -3 \end{bmatrix}$$

made of the coefficients of the variables, and the **augmented matrix**

$$\begin{bmatrix} 2 & 1 & 1 \\ 1 & -3 & 11 \end{bmatrix}$$

formed by attaching the column of constants to the coefficient matrix. The rows of the augmented matrix correspond to the equations in the system with the variables and the equal sign omitted.

### Elementary Row Operations

Operations on the rows of an augmented matrix which result in a matrix corresponding to equations equivalent to those in the original system (having the same solution as the original) are called **elementary row operations.** The first elementary row operation corresponds to interchanging any two equations in the system, the second corresponds to multiplying both sides of an equation by a nonzero constant, and the third corresponds to adding equal expressions to both sides of an equation. As a result the three elementary row operations all yield a matrix of a system of equations that is equivalent to the original.

**Elementary Row Operations**

Let $k$ be a nonzero real number. The following operations, applied to the augmented matrix of a system of linear equations, yield a matrix equivalent to the original.

1. Interchange any two rows.
2. Multiply all elements in a row by $k$.
3. Add to the elements of one row $k$ times the corresponding elements of another row.

## Reduced Echelon Form

We use elementary row operations to transform the augmented matrix of a system of two linear equations in two variables into an equivalent matrix of the form

$$\begin{bmatrix} 1 & 0 & p \\ 0 & 1 & q \end{bmatrix},$$

called the **reduced echelon form.** Since this matrix corresponds to the system

$$\begin{aligned} x \quad &= p \\ y &= q \end{aligned}$$

it is apparent that the third column gives the solution to this system, $(p, q)$, hence also the solution to the original system.

To transform an augmented matrix for a system of two linear equations into the above form, use the following diagram, which shows the best order for obtaining the 1's and 0's.

**1. Obtain this 1 first** → $\begin{bmatrix} 1 & 0 & p \\ 0 & 1 & q \end{bmatrix}$ **4. Obtain this 0 fourth**

**2. Obtain this 0 second** → **3. Obtain this 1 third**

## EXAMPLE 1

Solve using the Gaussian method.

$$\begin{aligned} 2x + y &= 1 \\ x - 3y &= 11 \end{aligned}$$

The augmented matrix is

$$\begin{matrix} \textbf{(A)} \\ \textbf{(B)} \end{matrix} \begin{bmatrix} 2 & 1 & 1 \\ 1 & -3 & 11 \end{bmatrix}.$$

**1.** Interchange the two rows to obtain a 1 in the upper left corner of the matrix.

$$\begin{matrix} \textbf{(B)} \\ \textbf{(A)} \end{matrix} \begin{bmatrix} 1 & -3 & 11 \\ 2 & 1 & 1 \end{bmatrix} \quad \text{(A) and (B) interchanged}$$

**2.** Keep the first row, multiply it by $-2$, and add the result to the second row.

$$\begin{matrix} \textbf{(B)} \\ \textbf{(C)} \end{matrix} \begin{bmatrix} 1 & -3 & 11 \\ 0 & 7 & -21 \end{bmatrix} \leftarrow \text{(C)} = -2\text{(B)} + \text{(A)}$$

**3.** Multiply the second row by 1/7 to obtain 1 in the second column.

$$\begin{matrix} \textbf{(B)} \\ \textbf{(D)} \end{matrix} \begin{bmatrix} 1 & -3 & 11 \\ 0 & 1 & -3 \end{bmatrix} \leftarrow \text{(D)} = \frac{1}{7}\text{(C)}$$

**4.** To obtain 0 in the second column of the first row, keep the second row, multiply it by 3, and add the result to the first row.

$$\begin{matrix} \textbf{(E)} \\ \textbf{(D)} \end{matrix} \begin{bmatrix} 1 & 0 & 2 \\ 0 & 1 & -3 \end{bmatrix} \begin{matrix} \leftarrow \text{(E)} = 3\text{(D)} + \text{B} \\ \phantom{x} \end{matrix}$$

Since $x = 2$ and $y = -3$, the solution to the system is $(2, -3)$. ■

For a system of three linear equations in three variables, the elementary row operations are used to transform the augmented matrix into the reduced echelon form

$$\begin{bmatrix} 1 & 0 & 0 & p \\ 0 & 1 & 0 & q \\ 0 & 0 & 1 & r \end{bmatrix}$$

which corresponds to the equivalent system

$$\begin{aligned} x \qquad\qquad &= p \\ y \qquad &= q \\ z &= r. \end{aligned}$$

Then the solution to the original system is $(p, q, r)$.

The following diagram shows the best procedure for obtaining 1's and 0's in the augmented matrix of a system of three linear equations. In all cases, once the 1 is obtained in a column, it is used to obtain the two 0's.

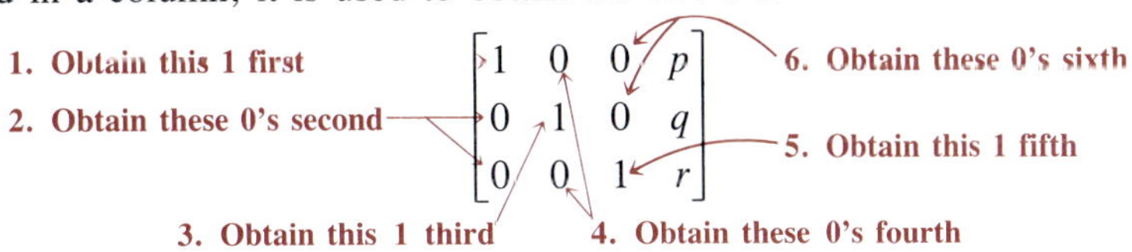


**EXAMPLE 2** Solve using the Gaussian method.

$$\begin{aligned} 2x - y + z &= 1 \\ x + 3y - z &= -4 \\ 3x + 2y + z &= 0 \end{aligned}$$

The augmented matrix is

$$\begin{matrix} \textbf{(A)} \\ \textbf{(B)} \\ \textbf{(C)} \end{matrix} \begin{bmatrix} 2 & -1 & 1 & 1 \\ 1 & 3 & -1 & -4 \\ 3 & 2 & 1 & 0 \end{bmatrix}.$$

1. To obtain 1 in the upper left corner, interchange the first two rows.

$$\begin{matrix} \textbf{(B)} \\ \textbf{(A)} \\ \textbf{(C)} \end{matrix} \begin{bmatrix} 1 & 3 & -1 & -4 \\ 2 & -1 & 1 & 1 \\ 3 & 2 & 1 & 0 \end{bmatrix} \quad \text{(A) and (B) interchanged}$$

2. Keep the first row, multiply it by $-2$, and add the result to the second row. Then multiply the first row by $-3$ and add to the third row. By doing so, we obtain two 0's in the first column.

$$\begin{matrix} \textbf{(B)} \\ \textbf{(D)} \\ \textbf{(E)} \end{matrix} \begin{bmatrix} 1 & 3 & -1 & -4 \\ 0 & -7 & 3 & 9 \\ 0 & -7 & 4 & 12 \end{bmatrix} \begin{matrix} \\ \leftarrow \text{(D)} = -2\text{(B)} + \text{(A)} \\ \leftarrow \text{(E)} = -3\text{(B)} + \text{(C)} \end{matrix}$$

**3.** Obtain 1 in the second column of the second row by multiplying by $-1/7$.

$$\begin{matrix}\textbf{(B)}\\ \textbf{(F)}\\ \textbf{(E)}\end{matrix}\begin{bmatrix}1 & 3 & -1 & -4\\ 0 & 1 & -\frac{3}{7} & -\frac{9}{7}\\ 0 & -7 & 4 & 12\end{bmatrix}\begin{matrix}\\ \leftarrow (F) = -\frac{1}{7}(D)\\ \\ \end{matrix}$$

**4.** Keep the second row, multiply it by $-3$, and add the result to the first row. Then multiply the second row by 7 and add to the third. Again, in one writing we obtain two 0's in the second column.

$$\begin{matrix}\textbf{(G)}\\ \textbf{(F)}\\ \textbf{(H)}\end{matrix}\begin{bmatrix}1 & 0 & \frac{2}{7} & -\frac{1}{7}\\ 0 & 1 & -\frac{3}{7} & -\frac{9}{7}\\ 0 & 0 & 1 & 3\end{bmatrix}\begin{matrix}\leftarrow (G) = -3(F) + (B)\\ \\ \leftarrow (H) = 7(F) + (E)\end{matrix}$$

**5.** In this case, we obtained a 1 in the third row with no additional effort.

**6.** Keep the third row, multiply it by 3/7 and add to the second, and finally, multiply the third row by $-2/7$ and add to the first.

$$\begin{matrix}\textbf{(J)}\\ \textbf{(I)}\\ \textbf{(H)}\end{matrix}\begin{bmatrix}1 & 0 & 0 & -1\\ 0 & 1 & 0 & 0\\ 0 & 0 & 1 & 3\end{bmatrix}\begin{matrix}\leftarrow (J) = -\frac{2}{7}(H) + (G)\\ \leftarrow (I) = \frac{3}{7}(H) + (F)\\ \\ \end{matrix}$$

Since $x = -1$, $y = 0$, and $z = 3$, the solution to the system is $(-1, 0, 3)$. ■

The real effectiveness of the Gaussian method becomes apparent when systems with a large number of variables are involved. As with $2 \times 2$ systems and $3 \times 3$ systems, the augmented matrix of higher order systems is transformed into the reduced echelon form

$$\begin{bmatrix}1 & 0 & 0 & \cdots & 0 & p\\ 0 & 1 & 0 & \cdots & 0 & q\\ 0 & 0 & 1 & \cdots & 0 & r\\ \vdots & \vdots & \vdots & & \vdots & \vdots\\ 0 & 0 & 0 & \cdots & 1 & v\end{bmatrix}$$

by moving from left to right by columns, first obtaining the desired 1 and then the remaining zeros. In fact, it is not difficult to program a computer to carry out these arithmetic operations, obtaining the solution to extremely large systems in a matter of seconds.

**EXAMPLE 3** Solve using the Gaussian method.

$$\begin{aligned} w + x + y + z &= 2\\ w - x + 3y + z &= 3\\ x - y - z &= -1\\ 2w + x + 3y + z &= 0 \end{aligned}$$

The augmented matrix is

$$\begin{matrix}\textbf{(A)}\\ \textbf{(B)}\\ \textbf{(C)}\\ \textbf{(D)}\end{matrix}\begin{bmatrix}1 & 1 & 1 & 1 & 2\\ 1 & -1 & 3 & 1 & 3\\ 0 & 1 & -1 & -1 & -1\\ 2 & 1 & 3 & 1 & 0\end{bmatrix}.$$

**1.** Since there is already a 1 in the upper left corner, use it to obtain 0's in the second and fourth rows.

$$\begin{matrix}\textbf{(A)}\\ \textbf{(E)}\\ \textbf{(C)}\\ \textbf{(F)}\end{matrix}\begin{bmatrix}1 & 1 & 1 & 1 & 2\\ 0 & -2 & 2 & 0 & 1\\ 0 & 1 & -1 & -1 & -1\\ 0 & -1 & 1 & -1 & -4\end{bmatrix}\begin{matrix}\\ \text{(E)} = -\text{(A)} + \text{(B)}\\ \\ \text{(F)} = -2\text{(A)} + \text{(D)}\end{matrix}$$

**2.** Interchange the second and third rows to obtain a 1 in the desired position.

$$\begin{matrix}\textbf{(A)}\\ \textbf{(C)}\\ \textbf{(E)}\\ \textbf{(F)}\end{matrix}\begin{bmatrix}1 & 1 & 1 & 1 & 2\\ 0 & 1 & -1 & -1 & -1\\ 0 & -2 & 2 & 0 & 1\\ 0 & -1 & 1 & -1 & -4\end{bmatrix}\quad \text{(E) and (C) interchanged}$$

**3.** Use the 1 in the second row to obtain 0's in the first, third, and fourth rows.

$$\begin{matrix}\textbf{(G)}\\ \textbf{(C)}\\ \textbf{(H)}\\ \textbf{(I)}\end{matrix}\begin{bmatrix}1 & 0 & 2 & 2 & 3\\ 0 & 1 & -1 & -1 & -1\\ 0 & 0 & 0 & -2 & -1\\ 0 & 0 & 0 & -2 & -5\end{bmatrix}\begin{matrix}\text{(G)} = -\text{(C)} + \text{(A)}\\ \\ \text{(H)} = 2\text{(C)} + \text{(E)}\\ \text{(I)} = \text{(C)} + \text{(F)}\end{matrix}$$

**4.** At this point we might observe that it will be impossible to obtain a 1 in the desired position in the third row. In fact, by replacing the fourth row with $-\text{(H)} + \text{(I)}$, the matrix becomes

$$\begin{matrix}\textbf{(G)}\\ \textbf{(C)}\\ \textbf{(H)}\\ \textbf{(J)}\end{matrix}\begin{bmatrix}1 & 0 & 2 & 2 & 3\\ 0 & 1 & -1 & -1 & -1\\ 0 & 0 & 0 & -2 & -1\\ 0 & 0 & 0 & 0 & -4\end{bmatrix}\begin{matrix}\\ \\ \\ \text{(J)} = -\text{(H)} + \text{(I)}\end{matrix}$$

in which the fourth row corresponds to the contradiction $0w + 0x + 0y + 0z = -4$ or $0 = -4$. As was the case with solution methods in the preceding chapter, this indicates that the original system has no solution. ■

**EXAMPLE 4** Solve by the Gaussian method.

$$\begin{aligned} x \qquad - z &= -3\\ -2x - y - 2z &= 2\\ 3x + y + z &= -5 \end{aligned}$$

The augmented matrix is

$$\begin{matrix}\textbf{(A)}\\ \textbf{(B)}\\ \textbf{(C)}\end{matrix}\begin{bmatrix}1 & 0 & -1 & -3\\ -2 & -1 & -2 & 2\\ 3 & 1 & 1 & -5\end{bmatrix}.$$

**1.** Use the 1 in the first row to obtain 0's in the second and third rows.

$$\begin{matrix}\textbf{(A)}\\ \textbf{(D)}\\ \textbf{(E)}\end{matrix}\begin{bmatrix}1 & 0 & -1 & -3\\ 0 & -1 & -4 & -4\\ 0 & 1 & 4 & 4\end{bmatrix}\quad\begin{matrix}\\ (D) = 2(A) + (B)\\ (E) = -3(A) + (C)\end{matrix}$$

**2.** Multiply the second row by $-1$ to obtain the desired 1, and use it to obtain a 0 in the third row.

$$\begin{matrix}\textbf{(A)}\\ \textbf{(F)}\\ \textbf{(G)}\end{matrix}\begin{bmatrix}1 & 0 & -1 & -3\\ 0 & 1 & 4 & 4\\ 0 & 0 & 0 & 0\end{bmatrix}\quad\begin{matrix}\\ (F) = (-1)(D)\\ (G) = (-1)(F) + E\end{matrix}$$

**3.** At this point we obtain the identity $0 = 0$ in the third row. Once this happens it is probably best to write the equations involving the variables and solve by other methods.

$$\begin{aligned} x \quad - z &= -3\\ y + 4z &= \phantom{-}4\end{aligned}$$

Then $x = z - 3$ and $y = 4 - 4z$ so the system has infinitely many solutions of the form $(z - 3, 4 - 4z, z)$ for $z$ any real number. ■

## 11.3 EXERCISES

In Exercises 1–4 assume that the given matrix is the augmented matrix of a system of equations. Follow the indicated steps and write the matrix in reduced echelon form.

**1.** $\begin{matrix}(A)\\ (B)\end{matrix}\begin{bmatrix}2 & 3 & -4\\ 1 & -2 & 5\end{bmatrix}$

**(a)** Interchange rows (A) and (B).
**(b)** Retain (B) and replace (A) with $(C) = -2(B) + (A)$.
**(c)** Replace (C) with $(D) = \frac{1}{7}(C)$.
**(d)** Retain (D) and replace (B) with $(E) = 2(D) + (B)$.

**2.** $\begin{matrix}(A)\\ (B)\end{matrix}\begin{bmatrix}4 & -5 & 5\\ 1 & 6 & -6\end{bmatrix}$

**(a)** Interchange (A) and (B).
**(b)** Retain (B) and replace (A) with $(C) = -4(B) + (A)$.
**(c)** Replace (C) with $(D) = -\frac{1}{29}(C)$.
**(d)** Retain (D) and replace (B) with $(E) = -6(D) + (B)$.

**3.** $\begin{matrix}(A)\\ (B)\\ (C)\end{matrix}\begin{bmatrix}2 & 1 & -1 & 4\\ 3 & -1 & 0 & 3\\ -4 & 2 & 3 & -10\end{bmatrix}$

**(a)** Replace (A) with $(D) = (B) - (A)$.
**(b)** Retain (D), replace (B) with $(E) = -3(D) + B$, and replace (C) with $(F) = 4(D) + (C)$.

**(c)** Retain (D) and (F) and replace (E) with (G) = $(-1)[(\text{E}) + (\text{F})]$.
**(d)** Retain (G), replace (D) with (H) = $2(\text{G}) + (\text{D})$, and replace (F) with (I) = $6(\text{G}) + (\text{F})$.
**(e)** Retain (H) and (G) and replace (I) with (J) = $-\frac{1}{17}(\text{I})$.
**(f)** Retain (J), replace (G) with (K) = $4(\text{J}) + (\text{G})$, and replace (H) with (L) = $7(\text{J}) + (\text{H})$.

**4.**
$$\begin{matrix}(\text{A})\\(\text{B})\\(\text{C})\end{matrix}\left[\begin{array}{rrr|r} 1 & -2 & 5 & 4 \\ 3 & 1 & 2 & 5 \\ -4 & 3 & -1 & -11 \end{array}\right]$$

**(a)** Retain (A), replace (B) with (D) = $-3(\text{A}) + (\text{B})$, and replace (C) with (E) = $4(\text{A}) + (\text{C})$.
**(b)** Retain (A) and (E) and replace (D) with (F) = $(-1)[2(\text{D}) + 3(\text{E})]$.
**(c)** Retain (F), replace (A) with (G) = $2(\text{F}) + (\text{A})$, and replace (E) with (H) = $5(\text{F}) + (\text{E})$.
**(d)** Retain (G) and (F) and replace (H) with (I) = $-\frac{1}{136}(\text{H})$.
**(e)** Retain (I), replace (F) with (J) = $31(\text{I}) + (\text{F})$, and replace (G) with (K) = $57(\text{I}) + (\text{G})$.

In Exercises 5–26 solve each system by using the Gaussian method.

**5.** $\begin{aligned} x + 3y &= 1 \\ 3x + 7y &= 5 \end{aligned}$

**6.** $\begin{aligned} 4x - y &= 3 \\ x + 3y &= 4 \end{aligned}$

**7.** $\begin{aligned} 4x - 2y &= -3 \\ x + y &= 3 \end{aligned}$

**8.** $\begin{aligned} 3x + 6y &= 4 \\ x - 5y &= -1 \end{aligned}$

**9.** $\begin{aligned} 5x - 3y &= -2 \\ 2x + 7y &= -9 \end{aligned}$

**10.** $\begin{aligned} 2x - 3y &= 1 \\ -4x + 6y &= 2 \end{aligned}$

**11.** $\begin{aligned} 3x + 2y &= 1 \\ -9x - 5y &= -3 \end{aligned}$

**12.** $\begin{aligned} 5x - 4y &= -1 \\ 2x + 8y &= 2 \end{aligned}$

**13.** $\begin{aligned} x + 2y &= 1 \\ -2x - 4y &= -2 \end{aligned}$

**14.** $\begin{aligned} 3x - 5y &= 2 \\ -9x + 15y &= -6 \end{aligned}$

**15.** $\begin{aligned} x + y + z &= 2 \\ -x - y + 3z &= 6 \\ 2x + y - z &= -1 \end{aligned}$

**16.** $\begin{aligned} x - y + z &= 3 \\ 2x + y + 3z &= 3 \\ -x + 2y - 4z &= -4 \end{aligned}$

**17.** $\begin{aligned} x + z &= 1 \\ y + z &= 4 \\ 2x + y &= -3 \end{aligned}$

**18.** $\begin{aligned} x - 2y &= -1 \\ y + z &= 7 \\ x - z &= -2 \end{aligned}$

**19.** $\begin{aligned} 2x - y + 3z &= 0 \\ x + y - z &= 4 \\ 2y + 5z &= -3 \end{aligned}$

**20.** $\begin{aligned} 3x - y + 2z &= -2 \\ 4x + y + z &= 3 \\ 7x - 2y - 2z &= 9 \end{aligned}$

**21.** $\begin{aligned} 3x - y + 2z &= 2 \\ x - z &= 1 \\ 2x - y + 3z &= 1 \end{aligned}$

**22.** $\begin{aligned} 4x - y + 2z &= 4 \\ 3x + y &= 5 \\ x - 2y + 2z &= -3 \end{aligned}$

**23.** $\begin{aligned} w + x - y + z &= 4 \\ w + 2y - z &= -3 \\ w - x + 2z &= 5 \\ 2w + x + 2y - 3z &= -6 \end{aligned}$

**24.** $\begin{aligned} 2w - x - 2y + z &= -1 \\ w + x - y &= -2 \\ w \quad x \quad + 4z &= -1 \\ 2x + y - 3z &= 1 \end{aligned}$

**25.** $\begin{aligned} \sqrt{x} + y^2 - \frac{1}{z} &= 4 \\ 2\sqrt{x} - y^2 + \frac{1}{z} &= 2 \\ -\sqrt{x} - 2y^2 - \frac{1}{z} &= -3 \end{aligned}$

**26.** $\begin{aligned} x^3 - \sqrt{y} + 2^z &= 7 \\ 2x^3 + 3\sqrt{y} - 2^z &= 23 \\ -x^3 + \sqrt{y} + 2^z &= -3 \end{aligned}$

## For Review

Exercises 27–39 refer to the following matrices.

$$A = \begin{bmatrix} 1 & 0 \\ -2 & 5 \end{bmatrix} \quad B = \begin{bmatrix} -2 & 5 \\ 1 & 4 \end{bmatrix} \quad C = \begin{bmatrix} 3 & -1 & 2 \\ 0 & -2 & 4 \end{bmatrix}$$

$$D = [2 \quad 1 \quad -1] \quad E = \begin{bmatrix} 3 \\ 0 \\ 5 \end{bmatrix} \quad F = [4 \quad -3]$$

**27.** What is the order of $C$?

**28.** Identify the element $c_{13}$ in matrix $C$.

**29.** Give the matrix $-B$.

**30.** Give the matrix $C^{\mathrm{T}}$.

**31.** Find $A + B$.

**32.** Find $B - C$.

**33.** Find $2A - 4B$.

**34.** Find $D \cdot E$.

**35.** Find $(AB)^{\mathrm{T}}$.

**36.** Find $FA$.

**37.** Find $AI$.

**38.** Find $C + 0$.

**39.** Find $ED$.

**40.** **BUSINESS** Suppose that $C = [200 \quad 300 \quad 400 \quad 550]$ gives the selling prices of four models of television sets and

$$N = \begin{bmatrix} 5 \\ 3 \\ 4 \\ 2 \end{bmatrix}$$

gives the number of each model sold on Saturday. Find $C \cdot N$ and interpret the result.

# 11.4 The Inverse of a Square Matrix

## Solving Matrix Equations

In our study of the algebra of matrices numerous similarities have been observed between matrices and the algebra of real numbers. Properties of real numbers allow us to solve equations involving real-number variables, and properties of matrices enable us to solve matrix equations. For example, consider the matrix equation

$$A + X = B,$$

where $A = \begin{bmatrix} 1 & -3 \\ 2 & 5 \end{bmatrix}$, $X = \begin{bmatrix} w & x \\ y & z \end{bmatrix}$, and $B = \begin{bmatrix} 0 & 2 \\ -4 & 7 \end{bmatrix}$.

Substitute these matrices into the matrix equation and add the two matrices on the left side.

$$\begin{bmatrix} 1 & -3 \\ 2 & 5 \end{bmatrix} + \begin{bmatrix} w & x \\ y & z \end{bmatrix} = \begin{bmatrix} 0 & 2 \\ -4 & 7 \end{bmatrix}$$

$$\begin{bmatrix} 1 + w & -3 + x \\ 2 + y & 5 + z \end{bmatrix} = \begin{bmatrix} 0 & 2 \\ -4 & 7 \end{bmatrix}$$

Use the definition of equality of matrices to determine $w$, $x$, $y$, and $z$.

$$1 + w = 0 \qquad -3 + x = 2 \qquad 2 + y = -4 \qquad 5 + z = 7$$
$$w = -1 \qquad x = 5 \qquad y = -6 \qquad z = 2$$

Thus, we have solved the matrix equation for the variable matrix $X$ obtaining

$$X = \begin{bmatrix} -1 & 5 \\ -6 & 2 \end{bmatrix}.$$

Using the properties of matrices, the equation would have been solved as follows.

$$\begin{aligned} A + X &= B && \\ -A + (A + X) &= -A + B && \text{Add the negative of } A \\ (-A + A) + X &= -A + B && \text{Associative law} \\ 0 \quad + X &= -A + B && A \text{ and } -A \text{ are additive inverses} \\ X &= -A + B && 0 \text{ is the zero matrix} \end{aligned}$$

We see that obtaining the value for $X$ is merely a matter of performing the matrix operation $-A + B$, or equivalently $B - A$, on the given constant matrices $A$ and $B$.

The next logical step is to consider whether a matrix equation of the form

$$AX = B$$

can be solved in a similar manner. To solve the similar algebraic equation

$$ax = b,$$

we could multiply both sides by the reciprocal or multiplicative inverse of $a$, $\frac{1}{a}$, or $a^{-1}$.

$$\begin{aligned} a^{-1}(ax) &= a^{-1}b \\ (a^{-1}a)x &= a^{-1}b \\ 1 \cdot x &= a^{-1}b \\ x &= a^{-1}b \end{aligned}$$

## The Inverse of a Matrix

In order to solve the similar matrix equation, we must address the problem of finding a multiplicative inverse of the matrix $A$. For reasons that will soon become clear, consideration of inverses is restricted to square matrices. Not every square matrix has an inverse. Those that do are called **nonsingular**, and those that do not are called **singular.** It can be shown that if a matrix is nonsingular, then the matrix $A^{-1}$ is unique.

### The Inverse of a Matrix

Let $A$ be an $n \times n$ square matrix with $I$ the $n \times n$ identity matrix. An $n \times n$ matrix $A^{-1}$ is called the **inverse** of $A$ if and only if

$$AA^{-1} = A^{-1}A = I.$$

**CAUTION** We read the symbol $A^{-1}$ as "$A$ inverse," and do not use the notation $\frac{1}{A}$ since the operation of division has not been defined for matrices. ■

Now that we have defined the inverse of a square matrix, the next consideration is finding that inverse. Suppose we begin with the matrix

$$A = \begin{bmatrix} 3 & 4 \\ 2 & 3 \end{bmatrix}.$$

If $A^{-1}$ exists, then it must be a $2 \times 2$ matrix of the form

$$A^{-1} = \begin{bmatrix} a & b \\ c & d \end{bmatrix},$$

and

$$AA^{-1} = \begin{bmatrix} 3 & 4 \\ 2 & 3 \end{bmatrix}\begin{bmatrix} a & b \\ c & d \end{bmatrix} = \begin{bmatrix} 1 & 0 \\ 0 & 1 \end{bmatrix} = I.$$

Multiply and use the definition of equality to obtain two systems of equations.

$$\begin{bmatrix} 3a + 4c & 3b + 4d \\ 2a + 3c & 2b + 3d \end{bmatrix} = \begin{bmatrix} 1 & 0 \\ 0 & 1 \end{bmatrix}$$

$$3a + 4c = 1 \qquad 3b + 4d = 0$$

$$2a + 3c = 0 \qquad 2b + 3d = 1$$

Solving these systems gives $a = 3$, $b = -4$, $c = -2$, and $d = 3$. Thus,

$$A^{-1} = \begin{bmatrix} 3 & -4 \\ -2 & 3 \end{bmatrix}.$$

It is easy to show that this is the inverse since

$$AA^{-1} = \begin{bmatrix} 3 & 4 \\ 2 & 3 \end{bmatrix}\begin{bmatrix} 3 & -4 \\ -2 & 3 \end{bmatrix} = \begin{bmatrix} 1 & 0 \\ 0 & 1 \end{bmatrix} = \begin{bmatrix} 3 & -4 \\ -2 & 3 \end{bmatrix}\begin{bmatrix} 3 & 4 \\ 2 & 3 \end{bmatrix} = A^{-1}A.$$

It was noted earlier that not all matrices have inverses. It can be shown that

$$B = \begin{bmatrix} 1 & 2 \\ 1 & 2 \end{bmatrix}$$

does not have an inverse. If we follow the same procedure used above for matrix $A$, the two systems of equations in $a$ and $c$ and in $b$ and $d$ are inconsistent and have no solution.

The process of finding the inverse of a matrix, when it exists, can be simplified by reducing a particular augmented matrix to echelon form. This is similar to what was done in the previous section. Consider again the systems obtained when we found $A^{-1}$ above.

$$3a + 4c = 1 \qquad 3b + 4d = 0$$

$$2a + 3c = 0 \qquad 2b + 3d = 1$$

If we solve these systems using the Gaussian method, the augmented matrices are

$$\begin{bmatrix} 3 & 4 & 1 \\ 2 & 3 & 0 \end{bmatrix} \quad \text{and} \quad \begin{bmatrix} 3 & 4 & 0 \\ 2 & 3 & 1 \end{bmatrix}.$$

Since these two matrices are identical in the first two columns, rather than reducing each to echelon form separately, we can speed the process by combining them into a single augmented matrix of the form

$$\begin{array}{c} \textbf{(A)} \\ \textbf{(B)} \end{array} \left[\begin{array}{cc|cc} 3 & 4 & 1 & 0 \\ 2 & 3 & 0 & 1 \end{array}\right].$$

This matrix is of the form $[A|I]$, where $A$ is the original matrix and $I$ the $2 \times 2$ identity matrix. By using the elementary row operations on this augmented matrix, we can transform it to the form $[I|A^{-1}]$, using the following steps.

1. To obtain a 1 in the upper left corner, retain (B) and replace (A) with (C) = $(-1)$(B) + (A).

$$\begin{array}{c} \textbf{(C)} \\ \textbf{(B)} \end{array} \left[\begin{array}{cc|cc} 1 & 1 & 1 & -1 \\ 2 & 3 & 0 & 1 \end{array}\right]$$

2. Retain (C) and replace (B) with (D) = $(-2)$(C) + (B).

$$\begin{array}{c} \textbf{(C)} \\ \textbf{(D)} \end{array} \left[\begin{array}{cc|cc} 1 & 1 & 1 & -1 \\ 0 & 1 & -2 & 3 \end{array}\right]$$

3. Retain (D) and replace (C) with (E) = $(-1)$(D) + (C).

$$\begin{array}{c} \textbf{(E)} \\ \textbf{(D)} \end{array} \left[\begin{array}{cc|cc} 1 & 0 & 3 & -4 \\ 0 & 1 & -2 & 3 \end{array}\right]$$

Notice that the matrix $[I|A^{-1}]$ does give the value of $A^{-1}$ obtained before. Although we have illustrated this method for finding $A^{-1}$ using a $2 \times 2$ matrix $A$, it can be extended to arbitrary $n \times n$ matrices.

### To Find the Inverse of a Square Matrix

Let $A$ be an $n \times n$ matrix.

1. Form the augmented matrix $[A|I]$.
2. Use the elementary row operations to reduce $[A|I]$.
3. If $A^{-1}$ exists, the augmented matrix will be reduced to $[I|A^{-1}]$.
4. If $A^{-1}$ does not exist, a row of zeros will be produced to the left of the vertical line.

## EXAMPLE 1

Find the inverse of $A = \begin{bmatrix} 1 & 0 & -1 \\ -2 & 1 & 0 \\ 2 & -2 & 3 \end{bmatrix}$.

Form the augmented matrix $[A|I]$.

$$\begin{array}{l} \textbf{(A)} \\ \textbf{(B)} \\ \textbf{(C)} \end{array} \left[\begin{array}{rrr|rrr} 1 & 0 & -1 & 1 & 0 & 0 \\ -2 & 1 & 0 & 0 & 1 & 0 \\ 2 & -2 & 3 & 0 & 0 & 1 \end{array}\right]$$

**1.** Retain (A), replace (B) with (D) = 2(A) + (B), and replace (C) with (E) = −2(A) + (C).

$$\begin{array}{l} \textbf{(A)} \\ \textbf{(D)} \\ \textbf{(E)} \end{array} \left[\begin{array}{rrr|rrr} 1 & 0 & -1 & 1 & 0 & 0 \\ 0 & 1 & -2 & 2 & 1 & 0 \\ 0 & -2 & 5 & -2 & 0 & 1 \end{array}\right]$$

**2.** Retain (D) and replace (E) with (F) = 2(D) + (E).

$$\begin{array}{l} \textbf{(A)} \\ \textbf{(D)} \\ \textbf{(F)} \end{array} \left[\begin{array}{rrr|rrr} 1 & 0 & -1 & 1 & 0 & 0 \\ 0 & 1 & -2 & 2 & 1 & 0 \\ 0 & 0 & 1 & 2 & 2 & 1 \end{array}\right]$$

**3.** Retain (F), replace (D) with (G) = 2(F) + (D), and replace (A) with (H) = (F) + (A).

$$\begin{array}{l} \textbf{(H)} \\ \textbf{(G)} \\ \textbf{(F)} \end{array} \left[\begin{array}{rrr|rrr} 1 & 0 & 0 & 3 & 2 & 1 \\ 0 & 1 & 0 & 6 & 5 & 2 \\ 0 & 0 & 1 & 2 & 2 & 1 \end{array}\right]$$

Thus, $A^{-1} = \begin{bmatrix} 3 & 2 & 1 \\ 6 & 5 & 2 \\ 2 & 2 & 1 \end{bmatrix}$. To check show that $AA^{-1} = A^{-1}A = I$. ■

Now that we know how to find inverses, suppose we return to the problem of solving the matrix equation

$$AX = B$$

posed at the beginning of this section. If $A$ is nonsingular, then $A^{-1}$ exists and

$$\begin{aligned} A^{-1}(AX) &= A^{-1}B \\ (A^{-1}A)X &= A^{-1}B \\ IX &= A^{-1}B \\ X &= A^{-1}B. \end{aligned}$$

## Writing Systems as Matrix Equations

Since matrix multiplication is not commutative, $A^{-1}B$ is the correct form for $X$, *not* $BA^{-1}$.

Matrix equations arise in a natural way with systems of linear equations. For example, consider the system

$$\begin{aligned} 2x - \phantom{4}y &= -4 \\ 3x + 4y &= \phantom{-}5. \end{aligned}$$

If we let

$$A = \begin{bmatrix} 2 & -1 \\ 3 & 4 \end{bmatrix}, \quad X = \begin{bmatrix} x \\ y \end{bmatrix}, \quad \text{and} \quad B = \begin{bmatrix} -4 \\ 5 \end{bmatrix},$$

then $A$ is the **coefficient matrix** of the system, $X$ is the **variable matrix,** and $B$ is the **constant matrix.** Consider the matrix equation $AX = B$.

$$AX = B$$

$$\begin{bmatrix} 2 & -1 \\ 3 & 4 \end{bmatrix}\begin{bmatrix} x \\ y \end{bmatrix} = \begin{bmatrix} -4 \\ 5 \end{bmatrix}$$

$$\begin{bmatrix} 2x - y \\ 3x + 4y \end{bmatrix} = \begin{bmatrix} -4 \\ 5 \end{bmatrix}$$

By the definition of equality of matrices,

$$2x - y = -4$$

and

$$3x + 4y = 5.$$

Thus, the matrix equation $AX = B$ is equivalent to the system of two equations in two variables. If $A$ has an inverse $A^{-1}$, the matrix equation can be solved for $X$.

$$X = A^{-1}B$$

Suppose we try to find $A^{-1}$.

$$\begin{matrix} \textbf{(A)} \\ \textbf{(B)} \end{matrix} \left[\begin{array}{cc|cc} 2 & -1 & 1 & 0 \\ 3 & 4 & 0 & 1 \end{array}\right]$$

$$\begin{matrix} \textbf{(C)} \\ \textbf{(B)} \end{matrix} \left[\begin{array}{cc|cc} 1 & 5 & -1 & 1 \\ 3 & 4 & 0 & 1 \end{array}\right] \quad (\text{C}) = (-1)[(-1)(\text{B}) + (\text{A})]$$

$$\begin{matrix} \textbf{(C)} \\ \textbf{(D)} \end{matrix} \left[\begin{array}{cc|cc} 1 & 5 & -1 & 1 \\ 0 & -11 & 3 & -2 \end{array}\right] \quad (\text{D}) = (-3)(\text{C}) + (\text{B})$$

$$\begin{matrix} \textbf{(C)} \\ \textbf{(E)} \end{matrix} \left[\begin{array}{cc|cc} 1 & 5 & -1 & 1 \\ 0 & 1 & -\frac{3}{11} & \frac{2}{11} \end{array}\right] \quad (\text{E}) = -\frac{1}{11}(\text{D})$$

$$\begin{matrix} \textbf{(F)} \\ \textbf{(E)} \end{matrix} \left[\begin{array}{cc|cc} 1 & 0 & \frac{4}{11} & \frac{1}{11} \\ 0 & 1 & -\frac{3}{11} & \frac{2}{11} \end{array}\right] \quad (\text{F}) = (-5)(\text{E}) + (\text{C})$$

Thus, $A^{-1} = \begin{bmatrix} \frac{4}{11} & \frac{1}{11} \\ -\frac{3}{11} & \frac{2}{11} \end{bmatrix}$, which can also be written as $A^{-1} = \frac{1}{11}\begin{bmatrix} 4 & 1 \\ -3 & 2 \end{bmatrix}$, avoiding fractions in the matrix. Then the solution to the matrix equation is

$$X = A^{-1}B = \frac{1}{11}\begin{bmatrix} 4 & 1 \\ -3 & 2 \end{bmatrix}\begin{bmatrix} -4 \\ 5 \end{bmatrix}$$

$$= \frac{1}{11}\begin{bmatrix} -11 \\ 22 \end{bmatrix} = \begin{bmatrix} -1 \\ 2 \end{bmatrix}.$$

Since $X = \begin{bmatrix} x \\ y \end{bmatrix} = \begin{bmatrix} -1 \\ 2 \end{bmatrix}$, we have $x = -1$ and $y = 2$ giving the solution $(-1, 2)$ to the original system.

## The Inverse Method of Solution

Clearly the system of equations above could be solved more quickly using either the substitution or the elimination method discussed in the previous chapter. However, this technique, called the **inverse method,** is important when used in conjunction with a computer or even a programmable calculator.

## EXAMPLE 2

Solve the following system using the inverse method.

$$\begin{aligned} x \qquad\quad - z &= 2 \\ -2x + y \qquad\quad &= -8 \\ 2x - 2y + 3z &= 13 \end{aligned}$$

The system is equivalent to the matrix equation $AX = B$ where

$$A = \begin{bmatrix} 1 & 0 & -1 \\ -2 & 1 & 0 \\ 2 & -2 & 3 \end{bmatrix}, \qquad X = \begin{bmatrix} x \\ y \\ z \end{bmatrix}, \qquad \text{and} \quad B = \begin{bmatrix} 2 \\ -8 \\ 13 \end{bmatrix}.$$

In Example 1, we found the inverse matrix

$$A^{-1} = \begin{bmatrix} 3 & 2 & 1 \\ 6 & 5 & 2 \\ 2 & 2 & 1 \end{bmatrix}.$$

The solution to the matrix equation is

$$\begin{bmatrix} x \\ y \\ z \end{bmatrix} = X = A^{-1}B = \begin{bmatrix} 3 & 2 & 1 \\ 6 & 5 & 2 \\ 2 & 2 & 1 \end{bmatrix}\begin{bmatrix} 2 \\ -8 \\ 13 \end{bmatrix} = \begin{bmatrix} 3 \\ -2 \\ 1 \end{bmatrix},$$

making $x = 3$, $y = -2$, and $z = 1$. Thus, $(3, -2, 1)$ is the solution to the system. ■

We conclude this section with the second problem given in the introduction to this chapter. It shows that the inverse method can be a useful tool for solving several systems of equations, all with the same coefficient matrix.

**EXAMPLE 3**

**SPORTS**

The Olympic Sports Committee schedules four exhibition basketball games between the U.S. Olympic Team and a team composed of NBA players. Tickets are to be sold for \$5 and \$8 with a given total revenue production specified. Assuming that all seats will be sold in every arena, use the information given in the following table to determine the number of seats at each price that must be sold for each game.

| | Game 1 | Game 2 | Game 3 | Game 4 |
|---|---|---|---|---|
| Total arena seats | 9000 | 12,000 | 8400 | 21,000 |
| Revenue produced | \$63,000 | \$81,000 | \$66,000 | \$114,000 |

Let $x =$ the number of tickets sold for \$5,
$y =$ the number of tickets sold for \$8.
In effect, we have four systems of equations to solve, each having the same coefficient matrix $A$. If $n$ is the number of seats in an arena and $r$ is the desired revenue for the game in that arena, then

$$\begin{aligned} x + \phantom{8}y &= n \\ 5x + 8y &= r \end{aligned}$$

represents the general system which must be solved for the four particular values of $n$ and $r$ given in the table. Consider the matrix equation

$$\begin{bmatrix} 1 & 1 \\ 5 & 8 \end{bmatrix}\begin{bmatrix} x \\ y \end{bmatrix} = \begin{bmatrix} n \\ r \end{bmatrix}.$$

The inverse of the coefficient matrix $A$ is

$$A^{-1} = \frac{1}{3}\begin{bmatrix} 8 & -1 \\ -5 & 1 \end{bmatrix}.$$

Then $X = A^{-1}B$ becomes

$$\begin{bmatrix} x \\ y \end{bmatrix} = \frac{1}{3}\begin{bmatrix} 8 & -1 \\ -5 & 1 \end{bmatrix}\begin{bmatrix} n \\ r \end{bmatrix}.$$

To find the number of tickets at each price, substitute the appropriate values for $n$ and $r$ and multiply.

For Game 1: $n = 9000$ and $r = \$63{,}000$.

$$\begin{bmatrix} x \\ y \end{bmatrix} = \frac{1}{3}\begin{bmatrix} 8 & -1 \\ -5 & 1 \end{bmatrix}\begin{bmatrix} 9000 \\ 63{,}000 \end{bmatrix} = \frac{1}{3}\begin{bmatrix} 9000 \\ 18{,}000 \end{bmatrix} = \begin{bmatrix} 3000 \\ 6000 \end{bmatrix}$$

Thus, $x = 3000$ (\$5 seats) and $y = 6000$ (\$8 seats).

For Game 2: $n = 12{,}000$ and $r = \$81{,}000$.

$$\begin{bmatrix} x \\ y \end{bmatrix} = \frac{1}{3}\begin{bmatrix} 8 & -1 \\ -5 & 1 \end{bmatrix}\begin{bmatrix} 12{,}000 \\ 81{,}000 \end{bmatrix} = \frac{1}{3}\begin{bmatrix} 15{,}000 \\ 21{,}000 \end{bmatrix} = \begin{bmatrix} 5000 \\ 7000 \end{bmatrix}$$

Thus, $x = 5000$ ($5 seats) and $y = 7000$ ($8 seats).
For Game 3: $n = 8400$ and $r = \$66{,}000$.

$$\begin{bmatrix} x \\ y \end{bmatrix} = \frac{1}{3}\begin{bmatrix} 8 & -1 \\ -5 & 1 \end{bmatrix}\begin{bmatrix} 8400 \\ 66{,}000 \end{bmatrix} = \frac{1}{3}\begin{bmatrix} 1200 \\ 24{,}000 \end{bmatrix} = \begin{bmatrix} 400 \\ 8000 \end{bmatrix}$$

Thus, $x = 400$ ($5 seats) and $y = 8000$ ($8 seats).
For Game 4: $n = 21{,}000$ and $r = \$114{,}000$.

$$\begin{bmatrix} x \\ y \end{bmatrix} = \frac{1}{3}\begin{bmatrix} 8 & -1 \\ -5 & 1 \end{bmatrix}\begin{bmatrix} 21{,}000 \\ 114{,}000 \end{bmatrix} = \frac{1}{3}\begin{bmatrix} 54{,}000 \\ 9000 \end{bmatrix} = \begin{bmatrix} 18{,}000 \\ 3000 \end{bmatrix}$$

Thus, $x = 18{,}000$ ($5 seats) and $y = 3000$ ($8 seats). ■

## 11.4 EXERCISES

In Exercises 1–4 show that $B = A^{-1}$ by finding $AB$ and $BA$.

**1.** $A = \begin{bmatrix} 2 & 3 \\ -3 & -5 \end{bmatrix}, \quad B = \begin{bmatrix} 5 & 3 \\ -3 & -2 \end{bmatrix}$

**2.** $A = \begin{bmatrix} 1 & 5 \\ 1 & 3 \end{bmatrix}, \quad B = \begin{bmatrix} -\frac{3}{2} & \frac{5}{2} \\ \frac{1}{2} & -\frac{1}{2} \end{bmatrix}$

**3.** $A = \begin{bmatrix} 1 & 0 & 1 \\ 2 & 1 & 1 \\ 3 & 2 & 2 \end{bmatrix}, \quad B = \begin{bmatrix} 0 & 2 & -1 \\ -1 & -1 & 1 \\ 1 & -2 & 1 \end{bmatrix}$

**4.** $A = \begin{bmatrix} 2 & 3 & -3 \\ 0 & -1 & -1 \\ 1 & 4 & 3 \end{bmatrix}, \quad B = \begin{bmatrix} -\frac{1}{4} & \frac{21}{4} & \frac{3}{2} \\ \frac{1}{4} & -\frac{9}{4} & -\frac{1}{2} \\ -\frac{1}{4} & \frac{5}{4} & \frac{1}{2} \end{bmatrix}$

In Exercises 5–16 find the inverse of each matrix, if it exists.

**5.** $\begin{bmatrix} 3 & -4 \\ 4 & -5 \end{bmatrix}$

**6.** $\begin{bmatrix} 1 & -6 \\ -1 & 7 \end{bmatrix}$

**7.** $\begin{bmatrix} 2 & -1 \\ 4 & -3 \end{bmatrix}$

**8.** $\begin{bmatrix} 3 & -3 \\ 2 & -1 \end{bmatrix}$

**9.** $\begin{bmatrix} 2 & 2 \\ 2 & 2 \end{bmatrix}$

**10.** $\begin{bmatrix} 0 & 5 & -1 \\ 2 & 4 & 3 \end{bmatrix}$

**11.** $\begin{bmatrix} 2 & -2 & 3 \\ 1 & 0 & -1 \\ -2 & 1 & 0 \end{bmatrix}$

**12.** $\begin{bmatrix} 3 & 0 & 1 \\ 1 & 2 & 4 \\ 1 & 1 & 2 \end{bmatrix}$

**13.** $\begin{bmatrix} 1 & 2 & 0 \\ 0 & 1 & 1 \\ -1 & 1 & 1 \end{bmatrix}$

**14.** $\begin{bmatrix} 1 & -1 & 1 \\ 2 & 1 & -1 \\ 1 & -2 & 3 \end{bmatrix}$

**15.** $\begin{bmatrix} 1 & 3 & -1 \\ 3 & 1 & 0 \\ -2 & 0 & 1 \end{bmatrix}$

**16.** $\begin{bmatrix} 2 & 1 & -1 \\ 1 & -1 & 0 \\ 0 & 1 & 3 \end{bmatrix}$

In Exercises 17–18 find the values of $x$ and $y$.

**17.** $\begin{bmatrix} x \\ y \end{bmatrix} = \begin{bmatrix} 2 & -1 \\ 3 & 5 \end{bmatrix}\begin{bmatrix} -2 \\ 4 \end{bmatrix}$

**18.** $\begin{bmatrix} x \\ y \end{bmatrix} = \begin{bmatrix} 3 & 0 \\ -4 & 1 \end{bmatrix}\begin{bmatrix} -3 \\ 5 \end{bmatrix}$

In Exercises 19–20 find the values of $x$, $y$, and $z$.

**19.** $\begin{bmatrix} x \\ y \\ z \end{bmatrix} = \begin{bmatrix} 2 & -1 & 3 \\ 4 & 0 & 1 \\ 2 & -2 & 0 \end{bmatrix} \begin{bmatrix} 1 \\ -1 \\ 3 \end{bmatrix}$

**20.** $\begin{bmatrix} x \\ y \\ z \end{bmatrix} = \begin{bmatrix} -3 & 1 & 5 \\ 0 & 2 & 7 \\ 4 & 2 & -2 \end{bmatrix} \begin{bmatrix} 3 \\ 0 \\ 2 \end{bmatrix}$

Use the inverse method to solve each system in Exercises 21–24. The inverse of the coefficient matrix was found in Exercises 5–8.

**21.** $\begin{aligned} 3x - 4y &= 2 \\ 4x - 5y &= 3 \end{aligned}$

**22.** $\begin{aligned} x - 6y &= -23 \\ -x + 7y &= 26 \end{aligned}$

**23.** $\begin{aligned} 2x - y &= 11 \\ 4x - 3y &= 25 \end{aligned}$

**24.** $\begin{aligned} 3x - 3y &= 12 \\ 2x - y &= 6 \end{aligned}$

Use the inverse method to solve each system in Exercises 25–30. The inverse of the coefficient matrix was found in Exercises 11–16.

**25.** $\begin{aligned} 2x - 2y + 3z &= -5 \\ x \qquad\quad - z &= 5 \\ -2x + y \qquad\quad &= -4 \end{aligned}$

**26.** $\begin{aligned} 3x \qquad\quad - z &= 8 \\ x + 2y + 4z &= -1 \\ x + y + 2z &= 1 \end{aligned}$

**27.** $\begin{aligned} x + 2y \qquad &= 4 \\ y + z &= 6 \\ -x + y + z &= 2 \end{aligned}$

**28.** $\begin{aligned} x - y + z &= 5 \\ 2x + y - z &= 11 \\ x - 2y + 3z &= -15 \end{aligned}$

**29.** $\begin{aligned} x + 3y - z &= 13 \\ 3x + y \qquad &= 12 \\ -2x \qquad + z &= -7 \end{aligned}$

**30.** $\begin{aligned} 2x + y - z &= 9 \\ x - y \qquad &= 6 \\ y + 3z &= -1 \end{aligned}$

In Exercises 31–32 use the inverse method to solve each system for the given values of $a$ and $b$.

**31.** $\begin{aligned} 2x + y &= a \\ 5x + 3y &= b \end{aligned}$
**(a)** $a = 1$, $b = 1$ **(b)** $a = 6$, $b = 18$
**(c)** $a = -1$, $b = 0$

**32.** $\begin{aligned} 7x - 2y &= a \\ 5x - 2y &= b \end{aligned}$
**(a)** $a = 22$, $b = 14$ **(b)** $a = -7$, $b = -5$
**(c)** $a = 18$, $b = 14$

In Exercises 33–34 use the inverse method to solve each system for the given values of $a$, $b$, and $c$.

**33.** $\begin{aligned} x \qquad - z &= a \\ x + y \qquad &= b \\ y + 2z &= c \end{aligned}$
**(a)** $a = 5$, $b = 2$, $c = -6$ **(b)** $a = 0$, $b = 5$, $c = 6$ **(c)** $a = -3$, $b = -4$, $c = 0$

**34.** $\begin{aligned} x + y - z &= a \\ 2x \qquad + z &= b \\ x - y \qquad &= c \end{aligned}$
**(a)** $a = 4$, $b = 10$, $c = 6$ **(b)** $a = 5$, $b = 7$, $c = 0$ **(c)** $a = 0$, $b = -5$, $c = 1$

In Exercises 35–36 find the inverse of the given $4 \times 4$ matrix.

**35.** $\begin{bmatrix} 1 & 0 & 0 & -1 \\ 0 & 2 & 0 & 1 \\ 1 & 0 & 1 & -1 \\ 0 & 1 & -1 & 0 \end{bmatrix}$

**36.** $\begin{bmatrix} 1 & -1 & 0 & -2 \\ 0 & 1 & 0 & 0 \\ -1 & 0 & 0 & 3 \\ 2 & -1 & 1 & 3 \end{bmatrix}$

**37.** If $A$ is a nonsingular $n \times n$ matrix, and if $B$ and $C$ are $n \times n$ matrices such that $AB = AC$, prove that $B = C$.

**38.** If $A = \begin{bmatrix} a & b \\ c & d \end{bmatrix}$ is a matrix with the property that $ad - bc \neq 0$, prove that $A^{-1} = \dfrac{1}{ad - bc}\begin{bmatrix} d & -b \\ -c & a \end{bmatrix}$.

In Exercises 39–42 use the formula for $A^{-1}$ given in Exercise 38 to find each inverse.

**39.** $\begin{bmatrix} 2 & 3 \\ -1 & 5 \end{bmatrix}$ **40.** $\begin{bmatrix} 7 & -2 \\ 3 & 4 \end{bmatrix}$ **41.** $\begin{bmatrix} 2 & -2 \\ 1 & -1 \end{bmatrix}$ **42.** $\begin{bmatrix} 3 & 6 \\ 1 & 2 \end{bmatrix}$

**43.** Suppose that $A = \begin{bmatrix} 1 & 3 \\ 2 & 7 \end{bmatrix}$ and $B = \begin{bmatrix} -3 & 2 \\ 2 & -1 \end{bmatrix}$.

**(a)** Find $AB$. **(b)** Find $(AB)^{-1}$. **(c)** Find $A^{-1}$.
**(d)** Find $B^{-1}$. **(e)** Find $A^{-1}B^{-1}$. **(f)** Find $B^{-1}A^{-1}$.
**(g)** What might be concluded in view of **(a)**–**(f)**?

In Exercises 44–45 solve by the inverse method.

**44. BUSINESS** Ponderosa Stoves sells two models of wood-burning stoves, the Sierra selling for \$300 and the Aspen selling for \$500. During the four weeks in January, the owner sets a goal of selling the total number of stoves, earning the weekly revenue, shown in the following table. How many of each model must be sold during each week in order to reach that goal?

| | Week 1 | Week 2 | Week 3 | Week 4 |
|---|---|---|---|---|
| Total stoves sold | 12 | 15 | 10 | 20 |
| Total revenue earned | \$5400 | \$7500 | \$3400 | \$9000 |

**45. BIOLOGY** A biologist has three groups of animals in an experiment. He wishes to prepare a different diet for each group by specifying the amounts of protein and fat. He purchases two food mixes: Supergrow, which contains 30% protein and 6% fat by weight, and Healthy Mix, which contains 20% protein and 2% fat by weight. How many ounces of each food should be combined in order to obtain the mixtures shown in the following table? Give answers correct to the nearest tenth.

| | Mix 1 | Mix 2 | Mix 3 |
|---|---|---|---|
| Protein | 13.5 ounces | 14 ounces | 16 ounces |
| Fat | 2.4 ounces | 2.6 ounces | 3.1 ounces |

## For Review

In Exercises 46–48 solve each system using the Gaussian method.

**46.** $\begin{aligned} 4x - 5y &= -38 \\ 2x + 3y &= 14 \end{aligned}$

**47.** $\begin{aligned} 6x - 2y &= 1 \\ -3x + y &= 1 \end{aligned}$

**48.** $\begin{aligned} x - 2y + z &= -4 \\ 2x \qquad - z &= 7 \\ -3x + 4y + z &= -8 \end{aligned}$

## 11.5 Determinants and Cramer's Rule

### The Determinant of a 2 × 2 Matrix

Associated with every square matrix is a real number called its *determinant*. There are numerous applications of determinants, the most famous of which is a method of solving systems of $n$ linear equations in $n$ variables called *Cramer's rule,* named after Gabriel Cramer (1704–52). Throughout this section and the next, all matrices under consideration will be square matrices. We begin by defining the determinant of a $2 \times 2$ matrix.

Let $A = \begin{bmatrix} a & b \\ c & d \end{bmatrix}$ be a $2 \times 2$ matrix. The **determinant** of $A$ is given by

$$|A| = \begin{vmatrix} a & b \\ c & d \end{vmatrix} = ad - bc.$$

NOTE The notation $|A|$ should not be confused with the absolute value notation used for real numbers. Also, the square brackets used to designate a matrix are replaced with the vertical bars when we wish to denote the determinant of the matrix. The formula for a $2 \times 2$ determinant can be remembered by considering the following diagram.

$$\begin{vmatrix} a & b \\ c & d \end{vmatrix} \qquad \begin{vmatrix} a & b \\ c & d \end{vmatrix}$$

$$ad - bc$$ ■

### EXAMPLE 1

Evaluate the following determinants.

**(a)** $\begin{vmatrix} 1 & 2 \\ 3 & 4 \end{vmatrix} = (1)(4) - (2)(3) = 4 - 6 = -2$

**(b)** $\begin{vmatrix} 3 & 9 \\ 2 & 6 \end{vmatrix} = (3)(6) - (9)(2) = 18 - 18 = 0$

**(c)** $\begin{vmatrix} a_1 & b_1 \\ a_2 & b_2 \end{vmatrix} = a_1b_2 - a_2b_1$ ■

### Cramer's Rule For Systems of Two Equations

In order to understand the motivation for the definition of a $2 \times 2$ determinant, suppose we solve a general system of two linear equations in two variables $x$ and $y$,

$$a_1x + b_1y = c_1$$
$$a_2x + b_2y = c_2,$$

where $a_1$, $b_1$, $c_1$, $a_2$, $b_2$, and $c_2$ are real numbers. Multiply the first equation by $b_2$ and the second by $b_1$.

$$a_1\mathbf{b_2}x + b_1\mathbf{b_2}y = c_1\mathbf{b_2} \qquad b_2 \text{ times the first equation}$$
$$a_2\mathbf{b_1}x + b_2\mathbf{b_1}y = c_2\mathbf{b_1} \qquad b_1 \text{ times the second equation}$$

Subtract the second from the first to eliminate $y$.

$$a_1b_2\mathbf{x} - a_2b_1\mathbf{x} = c_1b_2 - c_2b_1 \qquad b_1b_2y - b_2b_1y = 0 \cdot y = 0$$
$$(a_1b_2 - a_2b_1)\mathbf{x} = c_1b_2 - c_2b_1 \qquad \text{Factor out } x$$
$$x = \frac{c_1b_2 - c_2b_1}{a_1b_2 - a_2b_1} \qquad \text{Divide by } a_1b_2 - a_2b_1, \text{ if } a_1b_2 - a_2b_1 \neq 0$$

Similarly, we may first eliminate $x$ and then solve for $y$.

$$y = \frac{a_1c_2 - a_2c_1}{a_1b_2 - a_2b_1}$$

Consider the coefficient matrix for this system,

$$A = \begin{bmatrix} a_1 & b_1 \\ a_2 & b_2 \end{bmatrix}.$$

The denominator in each fractional expression for $x$ and $y$ is the determinant of the coefficient matrix $A$,

$$|A| = \begin{vmatrix} a_1 & b_1 \\ a_2 & b_2 \end{vmatrix} = a_1b_2 - a_2b_1.$$

Also, since the numerator of each expression has the appearance of a determinant, we define the matrices $A_x$ and $A_y$ with determinants

$$|A_x| = \begin{vmatrix} \mathbf{c_1} & b_1 \\ \mathbf{c_2} & b_2 \end{vmatrix} = c_1b_2 - c_2b_1, \qquad \text{The numerator for } x$$

and

$$|A_y| = \begin{vmatrix} a_1 & \mathbf{c_1} \\ a_2 & \mathbf{c_2} \end{vmatrix} = a_1c_2 - a_2c_1. \qquad \text{The numerator for } y$$

Notice that $A_x$ can be obtained from $A$ by replacing the column of coefficients of $x$ with the constants, and that $A_y$ is obtained in a similar way replacing the coefficients of $y$ with the constants. Then

$$x = \frac{c_1b_2 - c_2b_1}{a_1b_2 - a_2b_1} = \frac{\begin{vmatrix} \mathbf{c_1} & b_1 \\ \mathbf{c_2} & b_2 \end{vmatrix}}{\begin{vmatrix} a_1 & b_1 \\ a_2 & b_2 \end{vmatrix}} = \frac{|A_x|}{|A|},$$

and

$$y = \frac{a_1c_2 - a_2c_1}{a_1b_2 - a_2b_1} = \frac{\begin{vmatrix} a_1 & \mathbf{c_1} \\ a_2 & \mathbf{c_2} \end{vmatrix}}{\begin{vmatrix} a_1 & b_1 \\ a_2 & b_2 \end{vmatrix}} = \frac{|A_y|}{|A|}.$$

We have just proved the following theorem.

### Cramer's Rule

The system

$$a_1x + b_1y = c_1$$
$$a_2x + b_2y = c_2$$

has solutions

$$x = \frac{|A_x|}{|A|} \quad \text{and} \quad y = \frac{|A_y|}{|A|}, \quad \text{if } |A| \neq 0,$$

where $|A| = \begin{vmatrix} a_1 & b_1 \\ a_2 & b_2 \end{vmatrix}$, $|A_x| = \begin{vmatrix} c_1 & b_1 \\ c_2 & b_2 \end{vmatrix}$, and $|A_y| = \begin{vmatrix} a_1 & c_1 \\ a_2 & c_2 \end{vmatrix}$.

**EXAMPLE 2** Solve the system using Cramer's rule.

$$3x - 4y = -5$$
$$5x + y = 7$$

$$|A| = \begin{vmatrix} 3 & -4 \\ 5 & 1 \end{vmatrix} = (3)(1) - (-4)(5) = 23$$ Determinant of the coefficient matrix

$$|A_x| = \begin{vmatrix} -5 & -4 \\ 7 & 1 \end{vmatrix} = (-5)(1) - (-4)(7) = 23$$ Replace $x$-coefficients in $A$ with the constants

$$|A_y| = \begin{vmatrix} 3 & -5 \\ 5 & 7 \end{vmatrix} = (3)(7) - (-5)(5) = 46$$ Replace $y$-coefficients in $A$ with the constants

Then $$x = \frac{|A_x|}{|A|} = \frac{23}{23} = 1 \text{ and } y = \frac{|A_y|}{|A|} = \frac{46}{23} = 2.$$

The solution to the system is (1, 2). Check this in both equations. ■

If the determinant of the coefficient matrix, $|A|$, is zero, Cramer's rule does not apply. In such cases, the system has either no solution or infinitely many solutions, and the solution techniques given in the previous chapter should be used.

## Determinants of Matrices of Higher Order

To extend Cramer's rule to systems of three equations in three variables, the determinant of a $3 \times 3$ matrix must first be defined. Rather than concentrating on the $3 \times 3$ case, we will introduce the notions of *minors* and *cofactors* of the elements of a square matrix, and use them to obtain the determinant of an arbitrary $n \times n$ matrix for $n \geq 3$. In fact, as we shall see, a $3 \times 3$ determinant is given in terms of three $2 \times 2$ determinants, a $4 \times 4$ determinant in terms of four $3 \times 3$ determinants, and so forth.

### The Minor of an Element

Let $A$ be a square matrix of order $n \geq 3$. The **minor** of an element $a_{ij}$, denoted by $M_{ij}$, is the determinant of the matrix obtained by deleting the $i$th row and the $j$th column of $A$.

**EXAMPLE 3** Consider the matrix

$$A = \begin{bmatrix} 1 & 2 & 3 \\ 4 & 5 & 6 \\ 7 & 8 & 9 \end{bmatrix}.$$

**(a)** Find $M_{12}$.

$M_{12}$ is the minor of the element $a_{12} = 2$. Eliminate the 1st row and the second column of $A$ and evaluate the determinant of the result.

$$\begin{bmatrix} 1 & 2 & 3 \\ 4 & 5 & 6 \\ 7 & 8 & 9 \end{bmatrix} \qquad M_{12} = \begin{vmatrix} 4 & 6 \\ 7 & 9 \end{vmatrix} = 36 - 42 = -6$$

**(b)** Find $M_{31}$.

$M_{31}$ is the minor of $a_{31} = 7$. Eliminate the 3rd row and the 1st column of $A$.

$$\begin{bmatrix} 1 & 2 & 3 \\ 4 & 5 & 6 \\ 7 & 8 & 9 \end{bmatrix} \qquad M_{31} = \begin{vmatrix} 2 & 3 \\ 5 & 6 \end{vmatrix} = 12 - 15 = -3 \quad ■$$

A number closely related to the minor of an element is its cofactor.

### The Cofactor of an Element

Let $A$ be a square matrix of order $n \geq 3$. The **cofactor** of an element $a_{ij}$, denoted by $A_{ij}$, is given by

$$A_{ij} = (-1)^{i+j} M_{ij},$$

where $M_{ij}$ is the minor of $a_{ij}$.

**NOTE** The minor and cofactor of an element differ at most in the sign of the number. For the element $a_{ij}$,

$A_{ij} = M_{ij}$ if $i + j$ is even. Then $(-1)^{i+j} = +1$

$A_{ij} = -M_{ij}$ if $i + j$ is odd. Then $(-1)^{i+j} = -1$

The following "checkerboard" array gives the signs applied to corresponding minors to obtain the cofactors for a $3 \times 3$ matrix.

$$\begin{bmatrix} + & - & + \\ - & + & - \\ + & - & + \end{bmatrix} \quad \text{Signs for cofactors}$$

**EXAMPLE 4** Consider the matrix

$$A = \begin{bmatrix} \overset{+}{1} & \overset{-}{2} & \overset{+}{3} \\ \overset{-}{4} & \overset{+}{5} & \overset{-}{6} \\ \overset{+}{7} & \overset{-}{8} & \overset{+}{9} \end{bmatrix}.$$

**(a)** Find $A_{21}$.

$A_{21}$ is the cofactor of the element $a_{21} = 4$. Eliminate the 2nd row and 1st column to determine $M_{21}$.

$$A_{21} = (-1)^{2+1}M_{21} = (-1)^{2+1}\begin{vmatrix} 2 & 3 \\ 8 & 9 \end{vmatrix} = (-1)^3[18 - 24] = (-1)(-6) = 6$$

Since the element $a_{21} = 4$ is in a "$-$" position in the checkerboard of signs, we could have shortened the work as follows.

$$A_{21} = -M_{21} = -\begin{vmatrix} 2 & 3 \\ 8 & 9 \end{vmatrix} = -(18 - 24) = 6$$

**(b)** Find $A_{13}$.

$A_{13}$ is the cofactor of the element $a_{13} = 3$. Since $a_{13}$ is in a "$+$" position,

$$A_{13} = +M_{13} = +\begin{vmatrix} 4 & 5 \\ 7 & 8 \end{vmatrix} = +[32 - 35] = +[-3] = -3.$$

The determinant of a square matrix of order $n \geq 3$ can be found using a method called **expansion by cofactors.**

### Expansion by Cofactors

Let $A$ be a square matrix of order $n \geq 3$. The determinant of $A$, $|A|$, is found by adding the products of the elements in *any* row or column and their respective cofactors.

**EXAMPLE 5** Use expansion by cofactors to find $|A|$ if

$$A = \begin{bmatrix} 1 & 2 & -3 \\ -1 & 0 & 4 \\ 5 & 2 & 3 \end{bmatrix}.$$

To illustrate the method and show that any row or column yields the same result, we find $|A|$ using **(a)** elements in the second column and **(b)** elements in the first row.

**(a)** $A = a_{12}A_{12} + a_{22}A_{22} + a_{32}A_{32}$

$$= (2)(-1)^{1+2}\begin{vmatrix} -1 & 4 \\ 5 & 3 \end{vmatrix} + (0)(-1)^{2+2}\begin{vmatrix} 1 & -3 \\ 5 & 3 \end{vmatrix} + (2)(-1)^{3+2}\begin{vmatrix} 1 & -3 \\ -1 & 4 \end{vmatrix}$$

$$= (-2)(-3 - 20) + (0)(3 - (-15)) + (-2)(4 - 3)$$

$$= (-2)(-23) + 0 + (-2)(1) = 46 - 2 = 44$$

**(b)** $A = a_{11}A_{11} + a_{12}A_{12} + a_{13}A_{13}$

$$= (1)(-1)^{1+1}\begin{vmatrix} 0 & 4 \\ 2 & 3 \end{vmatrix} + (2)(-1)^{1+2}\begin{vmatrix} -1 & 4 \\ 5 & 3 \end{vmatrix} + (-3)(-1)^{1+3}\begin{vmatrix} -1 & 0 \\ 5 & 2 \end{vmatrix}$$

$$= (1)(0 - 8) + (-2)(-3 - 20) + (-3)(-2 - 0)$$

$$= (1)(-8) + (-2)(-23) + (-3)(-2) = -8 + 46 + 6 = 44$$

Thus, we obtain the same result in both expansions. ■

By selecting a row or column with one or more zero entries, the work can be shortened since a zero makes the corresponding term in the expansion equal to zero.

**EXAMPLE 6** Find

$$|A| = \begin{vmatrix} 2 & 1 & 0 \\ -1 & 4 & 0 \\ 3 & 2 & -1 \end{vmatrix}.$$

In this case it would be wise to choose the third column and take advantage of the zero entries.

$$|A| = a_{13}A_{13} + a_{23}A_{23} + a_{33}A_{33}$$

$$= 0A_{13} + 0A_{23} + (-1)A_{33}$$

$$= (-1)A_{33} = (-1)(-1)^{3+3}\begin{vmatrix} 2 & 1 \\ -1 & 4 \end{vmatrix}$$

$$= (-1)(+1)[8 - (-1)]$$

$$= (-1)(9) = -9$$ ■

**EXAMPLE 7** Find

$$|A| = \begin{bmatrix} \overset{+}{-1} & \overset{-}{2} & \overset{+}{0} & \overset{-}{3} \\ \overset{-}{2} & \overset{+}{4} & \overset{-}{3} & \overset{+}{-1} \\ \overset{+}{0} & \overset{-}{1} & \overset{+}{0} & \overset{-}{-2} \\ \overset{-}{-3} & \overset{+}{-1} & \overset{-}{1} & \overset{+}{2} \end{bmatrix}.$$

Suppose we expand along the third row taking advantage of the two zeros. The signs for the cofactors have been placed in color above each element.

$$|A| = 0 + 1(-1)\begin{vmatrix} -1 & 0 & 3 \\ 2 & 3 & -1 \\ -3 & 1 & 2 \end{vmatrix} + 0 + (-2)(-1)\begin{vmatrix} -1 & 2 & 0 \\ 2 & 4 & 3 \\ -3 & -1 & 1 \end{vmatrix}$$

$$= 0 + (-1)\left[(-1)\begin{vmatrix} 3 & -1 \\ 1 & 2 \end{vmatrix} + (3)\begin{vmatrix} 2 & 3 \\ 3 & 1 \end{vmatrix}\right] + 0 + (2)\left[(-1)\begin{vmatrix} 4 & 3 \\ -1 & 1 \end{vmatrix} + (-2)\begin{vmatrix} 2 & 3 \\ -3 & 1 \end{vmatrix}\right]$$

Expand along first row — Expand along first row

$$= 0 + (-1)[(-1)(6 + 1) + (3)(2 + 9)] + 0 + (2)[(-1)(4 + 3) + (-2)(2 + 9)]$$
$$= 0 + (-1)[26] + 0 + (2)[-29] = -26 - 58 = -84 \quad ■$$

## Cramer's Rule for Systems of Three Equations

The definition of the value of a third-order determinant would seem reasonable if we were to solve the following general system of three equations in the three variables $x$, $y$, and $z$.

$$\begin{aligned} a_1x + b_1y + c_1z &= d_1 \\ a_2x + b_2y + c_2z &= d_2 \\ a_3x + b_3y + c_3z &= d_3 \end{aligned}$$

The solutions to this system have the following form.

$$x = \frac{\begin{vmatrix} d_1 & b_1 & c_1 \\ d_2 & b_2 & c_2 \\ d_3 & b_3 & c_3 \end{vmatrix}}{\begin{vmatrix} a_1 & b_1 & c_1 \\ a_2 & b_2 & c_2 \\ a_3 & b_3 & c_3 \end{vmatrix}}, \quad y = \frac{\begin{vmatrix} a_1 & d_1 & c_1 \\ a_2 & d_2 & c_2 \\ a_3 & d_3 & c_3 \end{vmatrix}}{\begin{vmatrix} a_1 & b_1 & c_1 \\ a_2 & b_2 & c_2 \\ a_3 & b_3 & c_3 \end{vmatrix}}, \quad z = \frac{\begin{vmatrix} a_1 & b_1 & d_1 \\ a_2 & b_2 & d_2 \\ a_3 & b_3 & d_3 \end{vmatrix}}{\begin{vmatrix} a_1 & b_1 & c_1 \\ a_2 & b_2 & c_2 \\ a_3 & b_3 & c_3 \end{vmatrix}}$$

This result is also known as Cramer's rule, this time for a system of three linear equations in three variables. The determinant of the coefficient matrix appears in the denominator of each fraction, and the determinant in each numerator is formed by replacing the coefficients of the respective variable with the constants. As with systems of two equations in two variables, we use the following notation.

$$|A| = \begin{vmatrix} a_1 & b_1 & c_1 \\ a_2 & b_2 & c_2 \\ a_3 & b_3 & c_3 \end{vmatrix} \quad |A_x| = \begin{vmatrix} \mathbf{d_1} & b_1 & c_1 \\ \mathbf{d_2} & b_2 & c_2 \\ \mathbf{d_3} & b_3 & c_3 \end{vmatrix} \quad |A_y| = \begin{vmatrix} a_1 & \mathbf{d_1} & c_1 \\ a_2 & \mathbf{d_2} & c_2 \\ a_3 & \mathbf{d_3} & c_3 \end{vmatrix} \quad |A_z| = \begin{vmatrix} a_1 & b_1 & \mathbf{d_1} \\ a_2 & b_2 & \mathbf{d_2} \\ a_3 & b_3 & \mathbf{d_3} \end{vmatrix}$$

The solutions to the system, using Cramer's rule, are

$$x = \frac{|A_x|}{|A|}, \quad y = \frac{|A_y|}{|A|}, \quad z = \frac{|A_z|}{|A|}.$$

**EXAMPLE 8** Solve the system using Cramer's rule.

$$\begin{aligned} x + \phantom{2}y - z &= 4 \\ 2x + \phantom{2}y + z &= 1 \\ 3x - 2y - z &= 3 \end{aligned}$$

It is easy to verify that the four determinants are

$$|A| = \begin{vmatrix} 1 & 1 & -1 \\ 2 & 1 & 1 \\ 3 & -2 & -1 \end{vmatrix} = 13 \qquad |A_x| = \begin{vmatrix} \mathbf{4} & 1 & -1 \\ \mathbf{1} & 1 & 1 \\ \mathbf{3} & -2 & -1 \end{vmatrix} = 13$$

$$|A_y| = \begin{vmatrix} 1 & \mathbf{4} & -1 \\ 2 & \mathbf{1} & 1 \\ 3 & \mathbf{3} & -1 \end{vmatrix} = 13 \qquad |A_z| = \begin{vmatrix} 1 & 1 & \mathbf{4} \\ 2 & 1 & \mathbf{1} \\ 3 & -2 & \mathbf{3} \end{vmatrix} = -26.$$

Thus,

$$x = \frac{|A_x|}{|A|} = \frac{13}{13} = 1,\ y = \frac{|A_y|}{|A|} = \frac{13}{13} = 1,\ z = \frac{|A_z|}{|A|} = \frac{-26}{13} = -2. \quad ■$$

We were able to solve the system in Example 8 in less time by using the elimination method from the preceding chapter. However, the real value of any computational numerical technique, such as Cramer's rule, becomes apparent when a computer or programmable calculator is used. Once the method of evaluating a determinant has been programmed into the device, the various coefficients and constants can be entered and solutions to systems are available at the push of a button.

## 11.5 EXERCISES

Evaluate the determinants in Exercises 1–8.

**1.** $\begin{vmatrix} 3 & 5 \\ 2 & 4 \end{vmatrix}$ **2.** $\begin{vmatrix} 5 & 7 \\ 6 & 1 \end{vmatrix}$ **3.** $\begin{vmatrix} -1 & 0 \\ 3 & 4 \end{vmatrix}$ **4.** $\begin{vmatrix} 0 & -2 \\ 4 & 8 \end{vmatrix}$

**5.** $\begin{vmatrix} 2 & 5 \\ -3 & -1 \end{vmatrix}$ **6.** $\begin{vmatrix} -2 & -7 \\ -3 & -4 \end{vmatrix}$ **7.** $\begin{vmatrix} -2 & -3 \\ b & a \end{vmatrix}$ **8.** $\begin{vmatrix} a & b \\ b & a \end{vmatrix}$

Use the matrix $A = \begin{bmatrix} 2 & -1 & 3 \\ 1 & -2 & 4 \\ -3 & 5 & -4 \end{bmatrix}$ to answer Exercises 9–16.

**9.** Find the minor of the element 3.

**10.** Find the minor of the element 4.

**11.** Find the cofactor of the element 3.

**12.** Find the cofactor of the element 4.

**13.** Find $|A|$ by expanding along the first row.

**14.** Find $|A|$ by expanding along the second row.

**15.** Find $|A|$ by expanding along the first column.

**16.** Find $|A|$ by expanding along the second column.

Evaluate the determinants in Exercises 17–22.

**17.** $\begin{vmatrix} 0 & 2 & -1 \\ 1 & -3 & 0 \\ 0 & 1 & -2 \end{vmatrix}$ **18.** $\begin{vmatrix} 2 & 0 & -1 \\ 0 & 1 & 0 \\ -1 & 3 & -2 \end{vmatrix}$ **19.** $\begin{vmatrix} 3 & 1 & 2 \\ 4 & 0 & -2 \\ -2 & 1 & 3 \end{vmatrix}$

**20.** $\begin{vmatrix} -1 & 0 & 2 \\ 3 & -1 & 1 \\ 1 & -2 & -3 \end{vmatrix}$ **21.** $\begin{vmatrix} 1 & -2 & 1 \\ 2 & 3 & 2 \\ 3 & 1 & 3 \end{vmatrix}$ **22.** $\begin{vmatrix} 1 & 1 & 1 \\ -1 & 2 & -2 \\ 3 & 4 & -3 \end{vmatrix}$

In Exercises 23–32 solve each system using Cramer's rule.

**23.** $\begin{aligned} x + 2y &= -1 \\ 2x - 3y &= 12 \end{aligned}$ **24.** $\begin{aligned} 2x - y &= 8 \\ 3x + 5y &= -14 \end{aligned}$ **25.** $\begin{aligned} 3x + 2y &= -3 \\ 6x - 3y &= 8 \end{aligned}$ **26.** $\begin{aligned} 3x + 4y &= 1 \\ 5x - 2y &= 6 \end{aligned}$

**27.** $\begin{aligned} 4x - y &= -3 \\ 2x - 3y &= 1 \end{aligned}$ **28.** $\begin{aligned} 7x - 3y &= -6 \\ 4x + 2y &= 4 \end{aligned}$ **29.** $\begin{aligned} 3x - y \phantom{+ 4z} &= -11 \\ 2x \phantom{- 3y} + 4z &= -2 \\ -3y + 5z &= -1 \end{aligned}$ **30.** $\begin{aligned} 2x \phantom{- 3y} + 3z &= -8 \\ y - 4z &= 9 \\ x - 3y \phantom{- 4z} &= -4 \end{aligned}$

**31.** $\begin{aligned} 2x + y - z &= -3 \\ x - y + z &= 0 \\ -3x + 2y - 4z &= -3 \end{aligned}$ **32.** $\begin{aligned} x - 2y - z &= -1 \\ 2x + y + z &= 2 \\ -x - y + 2z &= 7 \end{aligned}$

Solve for $x$ in Exercises 33–38.

**33.** $\begin{vmatrix} x & 2 \\ 3 & 1 \end{vmatrix} = 3$ **34.** $\begin{vmatrix} 2 & x \\ 1 & -3 \end{vmatrix} = 5$ **35.** $\begin{vmatrix} x & 1 \\ 1 & x \end{vmatrix} = 3$

**36.** $\begin{vmatrix} 3 & x \\ x & 1 \end{vmatrix} = -6$ **37.** $\begin{vmatrix} x & 0 & 0 \\ 2 & x & -1 \\ -3 & 2 & 1 \end{vmatrix} = 3$ **38.** $\begin{vmatrix} 0 & 0 & x \\ x & 1 & -3 \\ 3 & -1 & 7 \end{vmatrix} = -10$

Evaluate the determinants in Exercises 39–40.

**39.** $\begin{vmatrix} 2 & 3 & 0 & -2 \\ 1 & -2 & -1 & 0 \\ 0 & 1 & 0 & 0 \\ -1 & 4 & -2 & 3 \end{vmatrix}$ **40.** $\begin{vmatrix} 2 & -3 & 0 & -1 \\ 0 & -1 & 0 & 1 \\ 5 & 4 & 1 & -2 \\ -2 & 1 & 0 & 2 \end{vmatrix}$

**41.** **(a)** Evaluate. $\begin{vmatrix} 0 & 0 \\ 0 & 0 \end{vmatrix}$ **(b)** Evaluate. $\begin{vmatrix} 0 & 0 & 0 \\ 0 & 0 & 0 \\ 0 & 0 & 0 \end{vmatrix}$

**(c)** Evaluate. $\begin{vmatrix} 0 & 0 & 0 & 0 \\ 0 & 0 & 0 & 0 \\ 0 & 0 & 0 & 0 \\ 0 & 0 & 0 & 0 \end{vmatrix}$ **(d)** What is the determinant of any $n \times n$ zero matrix?

**42.** **(a)** Evaluate. $\begin{vmatrix} 1 & 0 \\ 0 & 1 \end{vmatrix}$ **(b)** Evaluate. $\begin{vmatrix} 1 & 0 & 0 \\ 0 & 1 & 0 \\ 0 & 0 & 1 \end{vmatrix}$

**(c)** Evaluate. $\begin{vmatrix} 1 & 0 & 0 & 0 \\ 0 & 1 & 0 & 0 \\ 0 & 0 & 1 & 0 \\ 0 & 0 & 0 & 1 \end{vmatrix}$

**(d)** What is the determinant of any identity matrix?

**43. (a)** Evaluate. $\begin{vmatrix} a & 0 \\ 0 & b \end{vmatrix}$

**(b)** Evaluate. $\begin{vmatrix} a & 0 & 0 \\ 0 & b & 0 \\ 0 & 0 & c \end{vmatrix}$

**(c)** Evaluate. $\begin{vmatrix} a & 0 & 0 & 0 \\ 0 & b & 0 & 0 \\ 0 & 0 & c & 0 \\ 0 & 0 & 0 & d \end{vmatrix}$

**(d)** Matrices with determinants shown in **(a), (b),** and **(c)** are called **diagonal matrices.** What is the determinant of any diagonal matrix?

**44. (a)** Evaluate. $\begin{vmatrix} a & b \\ 0 & c \end{vmatrix}$

**(b)** Evaluate. $\begin{vmatrix} a & b & c \\ 0 & d & e \\ 0 & 0 & f \end{vmatrix}$

**(c)** Evaluate. $\begin{vmatrix} a & b & c & d \\ 0 & e & f & g \\ 0 & 0 & h & i \\ 0 & 0 & 0 & j \end{vmatrix}$

**(d)** Matrices with determinants shown in **(a), (b),** and **(c)** are called **triangular matrices.** What is the determinant of any triangular matrix?

**45. (a)** Evaluate. $\begin{vmatrix} a & 0 \\ b & 0 \end{vmatrix}$

**(b)** Evaluate. $\begin{vmatrix} a & b & 0 \\ c & d & 0 \\ e & f & 0 \end{vmatrix}$

**(c)** Evaluate. $\begin{vmatrix} a & b & c & 0 \\ d & e & f & 0 \\ g & h & i & 0 \\ j & k & l & 0 \end{vmatrix}$

**(d)** What is the determinant of any matrix that has a column of zeros?

**46. (a)** Evaluate. $\begin{vmatrix} a & b \\ 0 & 0 \end{vmatrix}$

**(b)** Evaluate. $\begin{vmatrix} a & b & c \\ d & e & f \\ 0 & 0 & 0 \end{vmatrix}$

**(c)** Evaluate. $\begin{vmatrix} a & b & c & d \\ e & f & g & h \\ i & j & k & l \\ 0 & 0 & 0 & 0 \end{vmatrix}$

**(d)** What is the determinant of any matrix that has a row of zeros?

**47.** If $A = \begin{vmatrix} a & b \\ c & d \end{vmatrix}$, find $|A|$ and $|A^T|$ and compare the results.

**48.** If $A = \begin{vmatrix} a & b & c \\ d & e & f \\ g & 0 & 0 \end{vmatrix}$, find $|A|$ and $|A^T|$ and compare the results.

**49.** Show that the equation of the line passing through the two points $(x_1, y_1)$ and $(x_2, y_2)$ is given by the equation

$$\begin{vmatrix} x & y & 1 \\ x_1 & y_1 & 1 \\ x_2 & y_2 & 1 \end{vmatrix} = 0.$$

**50.** Use the equation in Exercise 49 to find the equation of the line passing through the two points (2, 0) and (−1, 3).

**51.** Show that the area of a triangle with vertices $(x_1, y_1)$, $(x_2, y_2)$, and $(x_3, y_3)$ is the absolute value of

$$\frac{1}{2}\begin{vmatrix} x_1 & y_1 & 1 \\ x_2 & y_2 & 1 \\ x_3 & y_3 & 1 \end{vmatrix}.$$

**52.** Use the equation in Exercise 51 to find the area of the triangle with vertices (0, 1), (3, 5), and (−1, 2).

**53.** If $A = \begin{bmatrix} a & b \\ c & d \end{bmatrix}$, show that $|A^{-1}| = \dfrac{1}{|A|}$, provided $|A| \neq 0$.

**54.** If $A = \begin{bmatrix} a & b \\ c & d \end{bmatrix}$ and $B = \begin{bmatrix} e & f \\ g & h \end{bmatrix}$, show that $|AB| = |A||B|$.

**55.** If $A$ and $B$ are $2 \times 2$ matrices such that $|AB| = 0$, must $|A| = 0$ or $|B| = 0$?

**56.** If $A$ and $B$ are $2 \times 2$ matrices such that $|AB| = 0$, must $A = 0$ or $B = 0$?

**57.** Show that $\begin{vmatrix} a & b \\ na & nb \end{vmatrix} - 0.$

**58.** Show that $\begin{vmatrix} a & b \\ c & d \end{vmatrix} = -\begin{vmatrix} c & d \\ a & b \end{vmatrix}.$

**59.** Show that $\begin{vmatrix} a & b \\ c & d \end{vmatrix} = -\begin{vmatrix} b & a \\ d & c \end{vmatrix}.$

**60.** Show that $\begin{vmatrix} na & nb \\ nc & nd \end{vmatrix} = n^2\begin{vmatrix} a & b \\ c & d \end{vmatrix}.$

## For Review

In Exercises 61–62 find the inverse of each matrix.

**61.** $\begin{bmatrix} 2 & -4 \\ -1 & 3 \end{bmatrix}$

**62.** $\begin{bmatrix} 1 & -1 & 1 \\ 0 & 2 & -1 \\ 2 & 3 & 0 \end{bmatrix}$

Use the inverse method to solve each system in Exercises 63–64. The inverse of the coefficient matrix was found in Exercises 61–62.

**63.** $\begin{aligned} 2x - 4y &= -20 \\ -x + 3y &= 15 \end{aligned}$

**64.** $\begin{aligned} x - y + z &= 0 \\ 2y - z &= -2 \\ 2x + 3y &= -4 \end{aligned}$

**65.** **BUSINESS** Recreation Enterprises manufactures two types of picnic tables, the standard model and the deluxe model. Costs are \$20 for labor and \$25 for materials to make one standard model, and \$40 for labor and \$80 for materials to make one deluxe model. During each week, the total resources available for labor and material costs fluctuate due to employee vacations and variations in delivery schedules. The owner has determined that during the next 3 weeks, his available resources are limited according to the schedule in the table below. Use the inverse method to determine the number of tables of each type he should schedule for construction during each week.

| | Week 1 | Week 2 | Week 3 |
|---|---|---|---|
| Labor | \$1800 | \$1400 | \$2200 |
| Materials | \$3300 | \$2200 | \$3950 |

## 11.6 Properties of Determinants

The theorems presented in this section are helpful when finding the determinants of square matrices of order 3 or greater. They also have numerous applications in more advanced courses in matrix theory. Due to difficulties with notation, proofs of these theorems for general $n \times n$ matrices are somewhat involved. As a result, only informal proofs are given using $3 \times 3$ matrices. For easy reference, the theorems in this section are assigned numbers.

**THEOREM 1** If $A$ is a square matrix of order $n$, then $|A| = |A^{\mathrm{T}}|$.

In Exercise 47 of the preceding section, we showed that $|A| = |A^{\mathrm{T}}|$ if $A$ is of order 2. Suppose that $A$ is of order 3 given by

$$A = \begin{bmatrix} a_1 & a_2 & a_3 \\ a_4 & a_5 & a_6 \\ a_7 & a_8 & a_9 \end{bmatrix}.$$

Then

$$A^{\mathrm{T}} = \begin{bmatrix} a_1 & a_4 & a_7 \\ a_2 & a_5 & a_8 \\ a_3 & a_6 & a_9 \end{bmatrix}.$$

Expand $|A|$ along the first row and $|A^{\mathrm{T}}|$ along the first column.

$$|A| = a_1\begin{vmatrix} a_5 & a_6 \\ a_8 & a_9 \end{vmatrix} - a_2\begin{vmatrix} a_4 & a_6 \\ a_7 & a_9 \end{vmatrix} + a_3\begin{vmatrix} a_4 & a_5 \\ a_7 & a_8 \end{vmatrix}$$

$$\updownarrow \qquad\qquad \updownarrow \qquad\qquad \updownarrow$$

$$|A^{\mathrm{T}}| = a_1\begin{vmatrix} a_5 & a_8 \\ a_6 & a_9 \end{vmatrix} - a_2\begin{vmatrix} a_4 & a_7 \\ a_6 & a_9 \end{vmatrix} + a_3\begin{vmatrix} a_4 & a_7 \\ a_5 & a_8 \end{vmatrix}$$

Since the corresponding $2 \times 2$ determinants are equal, $|A| = |A^T|$.

Theorem 1 allows us to conclude that any property of determinants involving rows is also true for the columns.

**THEOREM 2** If $A$ is a square matrix of order $n$, and $B$ is a matrix formed by multiplying all elements in a row (or column) of $A$ by a constant $k$, then $|B| = k|A|$.

Suppose $A$ is a general matrix of order 3 given by

$$A = \begin{bmatrix} a_1 & a_2 & a_3 \\ a_4 & a_5 & a_6 \\ a_7 & a_8 & a_9 \end{bmatrix},$$

and $B$ is formed from $A$ by multiplying all elements in row 1 by $k$.

$$B = \begin{bmatrix} ka_1 & ka_2 & ka_3 \\ a_4 & a_5 & a_6 \\ a_7 & a_8 & a_9 \end{bmatrix}.$$

Find $|B|$ by expanding along the first row.

$$\begin{aligned} |B| &= ka_1 \begin{vmatrix} a_5 & a_6 \\ a_8 & a_9 \end{vmatrix} - ka_2 \begin{vmatrix} a_4 & a_6 \\ a_7 & a_9 \end{vmatrix} + ka_3 \begin{vmatrix} a_4 & a_5 \\ a_7 & a_8 \end{vmatrix} \\ &= k\left( a_1 \begin{vmatrix} a_5 & a_6 \\ a_8 & a_9 \end{vmatrix} - a_2 \begin{vmatrix} a_4 & a_6 \\ a_7 & a_9 \end{vmatrix} + a_3 \begin{vmatrix} a_4 & a_5 \\ a_7 & a_8 \end{vmatrix} \right) \\ &= k(|A|) = k|A| \end{aligned}$$

A similar argument will prove the theorem for $k$ multiplied by any other row or column.

**EXAMPLE 1** Given that

$$A = \begin{bmatrix} 1 & 2 & -3 \\ -1 & 0 & 4 \\ 5 & 2 & 3 \end{bmatrix}$$

with $|A| = 44$, find $|B|$ if

$$B = \begin{bmatrix} 3 & 6 & -9 \\ -1 & 0 & 4 \\ 5 & 2 & 3 \end{bmatrix}.$$

Since the elements in row 1 of $B$ are 3 times the elements in row 1 of $A$,

$$|B| = 3|A| = 3(44) = 132. \quad \blacksquare$$

Theorem 2 can also be used in reverse order to factor out a number common to all the elements in any row or column. For instance, consider the matrix $B$ given in Example 1. If we knew that $|B| = 132$, then

$$|B| = 3|A| = 132$$

would imply that $|A| = \frac{132}{3} = 44$.

**THEOREM 3** If $A$ is a square matrix of order $n$, and every element in a row (or column) of $A$ is 0, then $|A| = 0$.

This theorem follows immediately from Theorem 2 by factoring out the common factor 0 from the elements in the row (or column) of $|A|$ that are all 0.

**EXAMPLE 2** Given that

$$A = \begin{bmatrix} -3 & 0 & 5 \\ 2 & 0 & -7 \\ 4 & 0 & 8 \end{bmatrix},$$

find $|A|$.

Since the elements in the second column of $A$ are all 0, by Theorem 3, $|A| = 0$. This could also be seen directly by expanding $|A|$ along the second column. ■

**THEOREM 4** If $A$ is a square matrix of order $n$, and $B$ is a matrix formed by interchanging two rows (or two columns) of $A$, then

$$|B| = -|A|.$$

This theorem was verified for $2 \times 2$ determinants in Exercises 58 and 59 in the previous section. The proof for $3 \times 3$ determinants is essentially the same, but the notation is somewhat involved.

**EXAMPLE 3** Given that

$$A = \begin{bmatrix} 1 & 2 & -3 \\ -1 & 0 & 4 \\ 5 & 2 & 3 \end{bmatrix}$$

with $|A| = 44$, show that $|B| = -44$ if

$$B = \begin{bmatrix} 1 & 2 & -3 \\ 5 & 2 & 3 \\ -1 & 0 & 4 \end{bmatrix}.$$

Since $B$ is formed from $A$ by interchanging the second and third rows, $|B| = -|A| = -44$. ■

**THEOREM 5** Let $A$ be a square matrix of order $n$. If the corresponding elements in two rows (or columns) of $A$ are equal, then $|A| = 0$.

Let $A$ be an $n \times n$ matrix with two equal rows (or columns). If these two rows (or columns) are interchanged, the resulting matrix is the same as $A$. But by Theorem 4, the determinant of the new matrix, which is $|A|$, must be equal to the negative of the determinant of the original matrix, $-|A|$. Thus,

$$\begin{aligned} |A| &= -|A| \\ 2|A| &= 0 \\ |A| &= 0. \end{aligned}$$

**EXAMPLE 4** Find

$$\begin{vmatrix} 1 & 2 & 2 \\ 3 & 2 & 2 \\ -7 & 2 & 2 \end{vmatrix}.$$

Since the second and third columns are the same, by Theorem 5 the determinant is 0. ■

**THEOREM 6** Let $A$ be a square matrix of order $n$. If all the elements in one row (or column) of $A$ are a multiple of the elements in another row (or column) of $A$, then $|A| = 0$.

Suppose

$$A = \begin{bmatrix} ka_1 & ka_2 & ka_3 \\ a_1 & a_2 & a_3 \\ a_4 & a_5 & a_6 \end{bmatrix},$$

in which the elements of row 1 are all $k$-multiples of the elements in row 2. Then by Theorem 2,

$$|A| = k \begin{vmatrix} a_1 & a_2 & a_3 \\ a_1 & a_2 & a_3 \\ a_4 & a_5 & a_6 \end{vmatrix}.$$

Since two rows of this determinant are equal, by Theorem 5 the determinant is 0. Thus,

$$|A| = k(0) = 0.$$

**EXAMPLE 5** Find

$$\begin{vmatrix} 1 & -5 & 3 \\ -1 & 5 & -2 \\ 3 & -15 & 7 \end{vmatrix}.$$

Since the elements in the second column are $-5$ times the corresponding elements in the first column, by Theorem 6 the determinant is 0. ■

**THEOREM 7** If $A$ is a square matrix of order $n$, and $B$ is a matrix formed by replacing a row (or column) of $A$ by the sum of that row (or column) and a constant multiple of another row (or column) then $|B| = |A|$.

Suppose that $A$ is of order 3 given by

$$A = \begin{bmatrix} a_1 & a_2 & a_3 \\ a_4 & a_5 & a_6 \\ a_7 & a_8 & a_9 \end{bmatrix},$$

and

$$B = \begin{bmatrix} a_1 + ka_4 & a_2 + ka_5 & a_3 + ka_6 \\ a_4 & a_5 & a_6 \\ a_7 & a_8 & a_9 \end{bmatrix}.$$

Expand $|B|$ along the first row.

$$|B| = (a_1 + ka_4)\begin{vmatrix} a_5 & a_6 \\ a_8 & a_9 \end{vmatrix} - (a_2 + ka_5)\begin{vmatrix} a_4 & a_6 \\ a_7 & a_9 \end{vmatrix} + (a_3 + ka_6)\begin{vmatrix} a_4 & a_5 \\ a_7 & a_8 \end{vmatrix}$$

$$= a_1\begin{vmatrix} a_5 & a_6 \\ a_8 & a_9 \end{vmatrix} - a_2\begin{vmatrix} a_4 & a_6 \\ a_7 & a_9 \end{vmatrix} + a_3\begin{vmatrix} a_4 & a_5 \\ a_7 & a_8 \end{vmatrix}$$

$$+ k\left(a_4\begin{vmatrix} a_5 & a_6 \\ a_8 & a_9 \end{vmatrix} - a_5\begin{vmatrix} a_4 & a_6 \\ a_7 & a_9 \end{vmatrix} + a_6\begin{vmatrix} a_4 & a_5 \\ a_7 & a_8 \end{vmatrix}\right)$$

$$= \begin{vmatrix} a_1 & a_2 & a_3 \\ a_4 & a_5 & a_6 \\ a_7 & a_8 & a_9 \end{vmatrix} + k\begin{vmatrix} a_4 & a_5 & a_6 \\ a_4 & a_5 & a_6 \\ a_7 & a_8 & a_9 \end{vmatrix}$$

$$= \quad |A| \quad + k \quad (0)$$ Two rows are equal so the determinant is 0

$$= |A|$$

Notice the similarity between the Gaussian method for solving linear systems and the method indicated in Theorem 7. We use Theorem 7 to transform a determinant into equal determinants with 0 entries in all but one position in a row or column. By expanding along this row or column, an $n \times n$ determinant can be found by computing a single $(n - 1) \times (n - 1)$ determinant. The next example will clarify the process.

**EXAMPLE 6** Use Theorem 7 to find

$$\begin{vmatrix} 2 & 3 & -1 \\ 4 & -2 & 1 \\ -5 & -1 & 3 \end{vmatrix}.$$

When using Theorem 7 it is helpful to key on an element equal to 1, if possible. Suppose we use 1 in the second row and third column and force the remaining elements in the third column (−1 and 3) to be 0.

$$\begin{vmatrix} 2 & 3 & -1 \\ 4 & -2 & 1 \\ -5 & -1 & 3 \end{vmatrix} = \begin{vmatrix} 6 & 1 & 0 \\ 4 & -2 & 1 \\ -17 & 5 & 0 \end{vmatrix} \begin{matrix} \leftarrow \text{Replaced with (1)(Row 2) + (Row 1)} \\ \\ \leftarrow \text{Replaced with } (-3)\text{(Row 2) + (Row 3)} \end{matrix}$$

$$= (-1)\begin{vmatrix} 6 & 1 \\ -17 & 5 \end{vmatrix} \quad \text{Expanding down the third column}$$

$$= (-1)(30 + 17) = -47$$ ■

## 11.6 EXERCISES

In Exercises 1–18 determine the theorem in this section that justifies each statement. Do not evaluate the determinants.

1. $\begin{vmatrix} 1 & 3 \\ -1 & 5 \end{vmatrix} = \begin{vmatrix} 1 & -1 \\ 3 & 5 \end{vmatrix}$

2. $\begin{vmatrix} 0 & 2 \\ -5 & 4 \end{vmatrix} = \begin{vmatrix} 0 & -5 \\ 2 & 4 \end{vmatrix}$

3. $\begin{vmatrix} 0 & 7 \\ 0 & -8 \end{vmatrix} = 0$

4. $\begin{vmatrix} 6 & -2 \\ 0 & 0 \end{vmatrix} = 0$

5. $\begin{vmatrix} 4 & -4 \\ -1 & 3 \end{vmatrix} = 4\begin{vmatrix} 1 & -1 \\ -1 & 3 \end{vmatrix}$

6. $\begin{vmatrix} -2 & 1 \\ 0 & 5 \end{vmatrix} = \frac{1}{2}\begin{vmatrix} -4 & 1 \\ 0 & 5 \end{vmatrix}$

7. $\begin{vmatrix} -5 & 0 \\ 1 & 6 \end{vmatrix} = -\begin{vmatrix} 0 & -5 \\ 6 & 1 \end{vmatrix}$

8. $\begin{vmatrix} -3 & 2 \\ 6 & 7 \end{vmatrix} = -\begin{vmatrix} 6 & 7 \\ -3 & 2 \end{vmatrix}$

9. $\begin{vmatrix} 1 & 1 & -6 \\ 2 & 2 & 7 \\ 3 & 3 & -8 \end{vmatrix} = 0$

10. $\begin{vmatrix} -1 & 0 & 5 \\ 7 & 4 & -3 \\ -1 & 0 & 5 \end{vmatrix} = 0$

11. $\begin{vmatrix} 1 & 3 \\ 4 & -5 \end{vmatrix} = \begin{vmatrix} 1 & 3 \\ 4-4 & -5-12 \end{vmatrix}$

12. $\begin{vmatrix} 7 & 5 \\ 1 & -2 \end{vmatrix} = \begin{vmatrix} 7 & 5+14 \\ 1 & -2+2 \end{vmatrix}$

13. $\begin{vmatrix} -2 & -6 \\ 1 & 3 \end{vmatrix} = 0$

14. $\begin{vmatrix} -4 & 12 \\ 1 & -3 \end{vmatrix} = 0$

15. $3\begin{vmatrix} -1 & 2 & 0 \\ 2 & -1 & 5 \\ -1 & 2 & 0 \end{vmatrix} = \begin{vmatrix} -1 & 2 & 0 \\ 6 & -3 & 15 \\ -1 & 2 & 0 \end{vmatrix}$

16. $\begin{vmatrix} 5 & 0 & 1 \\ -3 & 2 & 6 \\ 1 & 4 & -2 \end{vmatrix} = -\begin{vmatrix} 1 & 0 & 5 \\ 6 & 2 & -3 \\ -2 & 4 & 1 \end{vmatrix}$

17. $\begin{vmatrix} 5 & 0 & 7 \\ -6 & 0 & 5 \\ 1 & 0 & -4 \end{vmatrix} = 0$

18. $\begin{vmatrix} 1 & -3+3 & 0 \\ 2 & -4+6 & 1 \\ -3 & 2-9 & -1 \end{vmatrix} = \begin{vmatrix} 1 & -3 & 0 \\ 2 & -4 & 1 \\ -3 & 2 & -1 \end{vmatrix}$

In Exercises 19–22 the determinant on the left was transformed to the determinant on the right using Theorem 7. Find the value of $x$.

**19.** $\begin{vmatrix} 1 & 4 \\ -3 & 5 \end{vmatrix} = \begin{vmatrix} 1 & 0 \\ -3 & x \end{vmatrix}$

**20.** $\begin{vmatrix} 2 & 6 \\ -1 & 3 \end{vmatrix} = \begin{vmatrix} 0 & x \\ -1 & 3 \end{vmatrix}$

**21.** $\begin{vmatrix} 1 & -1 & 3 \\ 0 & 2 & -2 \\ 4 & -3 & 5 \end{vmatrix} = \begin{vmatrix} 1 & -1 & 3 \\ 0 & 2 & -2 \\ x & 0 & -4 \end{vmatrix}$

**22.** $\begin{vmatrix} 0 & 2 & 4 \\ 1 & -3 & -1 \\ 2 & 5 & -2 \end{vmatrix} = \begin{vmatrix} 0 & 2 & 4 \\ 1 & -3 & -1 \\ x & 11 & 0 \end{vmatrix}$

In Exercises 23–24 use Theorem 7 to introduce two zeros in row 1 of each determinant.

**23.** $\begin{vmatrix} 2 & 1 & -5 \\ -3 & 4 & -2 \\ 1 & 3 & -1 \end{vmatrix}$

**24.** $\begin{vmatrix} 2 & -3 & 1 \\ -1 & -2 & 3 \\ 4 & 1 & -2 \end{vmatrix}$

In Exercises 25–26 use Theorem 7 to introduce two zeros in column 2 of each determinant.

**25.** $\begin{vmatrix} 2 & -3 & 1 \\ -1 & -2 & 3 \\ 4 & 1 & -2 \end{vmatrix}$

**26.** $\begin{vmatrix} 2 & 1 & -5 \\ -3 & 4 & -2 \\ 1 & 3 & -1 \end{vmatrix}$

In Exercises 27–34 use Theorem 7 to introduce two zeros in a row or column of each determinant and evaluate the determinant.

**27.** $\begin{vmatrix} 0 & 1 & 2 \\ -1 & 2 & 5 \\ 2 & -3 & 1 \end{vmatrix}$

**28.** $\begin{vmatrix} 1 & 0 & 3 \\ -2 & 4 & -1 \\ 2 & 1 & 5 \end{vmatrix}$

**29.** $\begin{vmatrix} 2 & -1 & 5 \\ 3 & 1 & -2 \\ 4 & -3 & 2 \end{vmatrix}$

**30.** $\begin{vmatrix} 2 & -3 & 5 \\ -1 & -2 & 4 \\ 6 & 3 & 1 \end{vmatrix}$

**31.** $\begin{vmatrix} 4 & -5 & 7 \\ 2 & 3 & -3 \\ -6 & -2 & -4 \end{vmatrix}$

**32.** $\begin{vmatrix} 2 & 5 & -2 \\ -3 & 6 & 3 \\ -5 & 2 & 4 \end{vmatrix}$

**33.** $\begin{vmatrix} 2 & 3 & 0 & -2 \\ 1 & 4 & -1 & 0 \\ 3 & 2 & 2 & 5 \\ -2 & 0 & 1 & -3 \end{vmatrix}$

**34.** $\begin{vmatrix} -3 & 1 & 0 & 5 \\ 2 & -4 & -1 & 3 \\ -1 & 0 & -2 & 2 \\ 3 & -1 & 4 & -2 \end{vmatrix}$

**35.** Show without evaluating that $(-1, 3)$ and $(2, 7)$ are solutions to the equation

$$\begin{vmatrix} x & -2 & y \\ -1 & -2 & 3 \\ 2 & -2 & 7 \end{vmatrix} = 0.$$

**36.** Show without evaluating that $(4, 1)$ and $(-2, 3)$ are solutions to the equation

$$\begin{vmatrix} -1 & -1 & -1 \\ x & 4 & -2 \\ y & 1 & 3 \end{vmatrix} = 0.$$

**37.** In Exercise 51 of the previous section we showed that the area of a triangle with vertices $(x_1, y_1)$, $(x_2, y_2)$, and $(x_3, y_3)$ is the absolute value of

$$\frac{1}{2}\begin{vmatrix} x_1 & y_1 & 1 \\ x_2 & y_2 & 1 \\ x_3 & y_3 & 1 \end{vmatrix}.$$

If this determinant is zero, what can be said about the three points?

**38.** Suppose that $(x_1, y_1)$, $(x_2, y_2)$, and $(x_3, y_3)$ are the coordinates of three points on the same straight line. What is the value of the following determinant?

$$\begin{vmatrix} x_1 & y_1 & 1 \\ x_2 & y_2 & 1 \\ x_3 & y_3 & 1 \end{vmatrix}$$

## For Review

In Exercises 39–40 solve each system using Cramer's rule. You may wish to use Theorem 7 when evaluating the determinants.

**39.** $$\begin{aligned} 2x \qquad - z &= -6 \\ 3y + 5z &= 29 \\ x - y + z &= 0 \end{aligned}$$

**40.** $$\begin{aligned} 3x - 5y + 7z &= 4 \\ 2x + 3y - 4z &= -6 \\ 4x - 7y - 8z &= -12 \end{aligned}$$

# CHAPTER 11 REVIEW EXERCISES

Exercises 1–14 refer to the following matrices.

$$A = \begin{bmatrix} -1 & 2 \\ 3 & 4 \end{bmatrix} \qquad B = \begin{bmatrix} 0 & -3 \\ 1 & 5 \end{bmatrix} \qquad C = \begin{bmatrix} -2 & 4 & 5 \\ 0 & -1 & 2 \end{bmatrix}$$

$$D = [1 \quad -3 \quad 4] \qquad E = \begin{bmatrix} 2 \\ -1 \\ 0 \end{bmatrix} \qquad F = \begin{bmatrix} 2 & 0 & -1 \\ 4 & 1 & 3 \\ -2 & -1 & 5 \end{bmatrix}$$

**1.** What is the order of matrix $C$?

**2.** Give the zero matrix with the same order as $A$.

**3.** Identify the element $a_{21}$ in matrix $A$.

**4.** Give the matrix $-B$.
**5.** Give the matrix $C^T$.
**6.** Find $A + B$.
**7.** Find $A - B$.

**8.** Find $-3C$.
**9.** Find $2A - 3B$.
**10.** Find $D \cdot E$.
**11.** Find $AC$.

**12.** Find $CA$.
**13.** Find $CF$.
**14.** Find $|C|$.

**15. BUSINESS** Island Rentals has two car rental agencies on the island of Jamaica. The total number of compact, intermediate, and luxury cars rented during the months of July and August are given in matrices $J$ and $A$.

$$J = \begin{array}{c} \text{Compact} \quad \text{Intermediate} \quad \text{Luxury} \\ \begin{bmatrix} 200 & 180 & 20 \\ 300 & 120 & 40 \end{bmatrix} \end{array} \begin{array}{l} \text{Agency 1} \\ \text{Agency 2} \end{array}$$

$$A = \begin{bmatrix} 150 & 120 & 30 \\ 220 & 100 & 10 \end{bmatrix} \begin{array}{l} \text{Agency 1} \\ \text{Agency 2} \end{array}$$

Suppose that the average revenue per rental is \$80 for a compact car, \$120 for an intermediate car, and \$150 for a luxury car. The revenue matrix for such rentals is given by

$$C = \begin{bmatrix} 80 \\ 120 \\ 150 \end{bmatrix}.$$

**(a)** Find the matrix that gives the two-month totals in each category.
**(b)** Find the matrix $\frac{1}{2}(J + A)$ and discuss what it represents.
**(c)** Find $[1 \quad 1]J$ and discuss what it represents.
**(d)** Find $[1 \quad 1](J + A)$ and discuss what it represents.
**(e)** Find $([1 \quad 1](J + A)) \cdot \begin{bmatrix} 1 \\ 1 \\ 1 \end{bmatrix}$ and discuss what it represents.
**(f)** Find $JC$ and interpret the result.
**(g)** Find $(J + A)C$ and interpret the result.
**(h)** Find $[1 \quad 1] \cdot ((J + A)C)$ and interpret the result.

In Exercises 16–17 solve each system using the Gaussian method.

**16.** $\begin{aligned} x - 3y &= 4 \\ 4x + 5y &= -1 \end{aligned}$

**17.** $\begin{aligned} x - 2y + 3z &= 7 \\ -x + 3y + 2z &= 8 \\ 3x - 4y - z &= -9 \end{aligned}$

In Exercises 18–19 find the inverse of each matrix.

**18.** $\begin{bmatrix} -5 & -2 \\ 3 & 1 \end{bmatrix}$

**19.** $\begin{bmatrix} 2 & -1 & 1 \\ 1 & 0 & 2 \\ 0 & 1 & -5 \end{bmatrix}$

In Exercises 20–21 use the inverse method to solve each system. The inverse of the coefficient matrix was found in Exercises 18–19.

**20.** $\begin{aligned} -5x - 2y &= 7 \\ 3x + y &= -5 \end{aligned}$

**21.** $\begin{aligned} 2x - y + z &= 9 \\ x + 2z &= 8 \\ y - 5z &= 17 \end{aligned}$

**22. BUSINESS** The Fun and Fitness Center is conducting a three-month membership drive. The annual individual membership fee is \$300 and the annual family membership fee is \$400. During the three months, the owners have set a goal of selling the total number of memberships and earning the annual revenue shown in the following table. How many of each type of membership must be sold during each month in order to reach their goal?

| | January | February | March |
|---|---|---|---|
| Total memberships | 30 | 50 | 65 |
| Annual revenue produced | \$10,000 | \$17,000 | \$23,000 |

Evaluate the determinants in Exercises 23–24.

**23.** $\begin{vmatrix} -3 & 5 \\ 1 & 7 \end{vmatrix}$

**24.** $\begin{vmatrix} 2 & -1 & 3 \\ -2 & 2 & 4 \\ 1 & -3 & 5 \end{vmatrix}$

**25.** Find the cofactor of the element 4 in the determinant of Exercise 24.

In Exercises 26–27 solve each system using Cramer's rule.

**26.** 
$$\begin{aligned} -2x + y &= 9 \\ 3x + 4y &= 14 \end{aligned}$$

**27.** 
$$\begin{aligned} 3x - y + z &= -12 \\ -2x + y + 3z &= 9 \\ x - y - 2z &= -6 \end{aligned}$$

Solve for $x$ in Exercises 28–29.

**28.** $\begin{vmatrix} 2 & x \\ x & 5 \end{vmatrix} = 6$

**29.** $\begin{vmatrix} x & 1 & 3 \\ 0 & 2 & x \\ 0 & 1 & -1 \end{vmatrix} = -3$

In Exercises 30–33 state a theorem that justifies each statement. Do not evaluate the determinants.

**30.** $\begin{vmatrix} 0 & 0 \\ 5 & -7 \end{vmatrix} = 0$

**31.** $\begin{vmatrix} 3 & -2 & 3 \\ 5 & 0 & 5 \\ 1 & 6 & 1 \end{vmatrix} = 0$

**32.** $\begin{vmatrix} 2 & -3 \\ 1 & 7 \end{vmatrix} = -\begin{vmatrix} -3 & 2 \\ 7 & 1 \end{vmatrix}$

**33.** $\begin{vmatrix} 3 & -3 \\ 0 & 4 \end{vmatrix} = 3\begin{vmatrix} 1 & -1 \\ 0 & 4 \end{vmatrix}$

**34.** If the determinant $\begin{vmatrix} 0 & 1 \\ x & 5 \end{vmatrix}$ was obtained from $\begin{vmatrix} 2 & 1 \\ -3 & 5 \end{vmatrix}$ using Theorem 7 of the section on properties of determinants, find the value of $x$.

**35.** Evaluate the following determinant by introducing two zeros in column 2.

$$\begin{vmatrix} -2 & 5 & 2 \\ 4 & 1 & 7 \\ 1 & -3 & -1 \end{vmatrix}$$

**36.** Suppose that

$$A = \begin{bmatrix} a_1 & a_2 & 0 & 0 \\ a_3 & a_4 & 0 & 0 \\ 0 & 0 & b_1 & b_2 \\ 0 & 0 & b_3 & b_4 \end{bmatrix},$$

show that

$$|A| = \begin{vmatrix} a_1 & a_2 \\ a_3 & a_4 \end{vmatrix} \begin{vmatrix} b_1 & b_2 \\ b_3 & b_4 \end{vmatrix}.$$

CHAPTER

# 12 SEQUENCES, SERIES, AND PROBABILITY

The topics in this chapter not only are of interest in college algebra but give a foundation for study in more advanced mathematics. First we give examples of the numerous applications.

### BANKING ►

Gloria Bell borrows \$1000.00 at 12% interest compounded annually. If she repays the loan in full at the end of three years, how much does she pay?

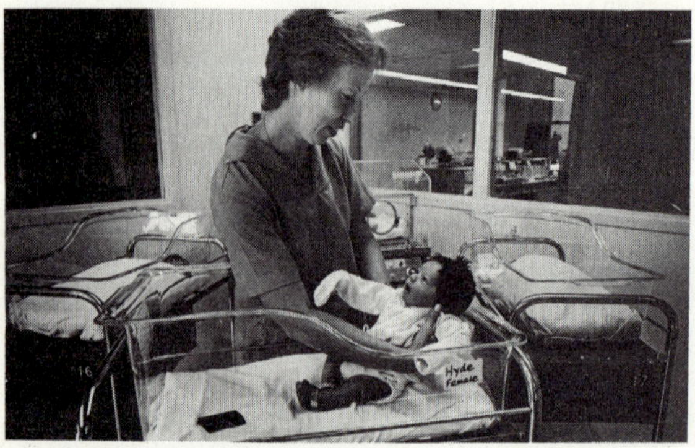

### ◄ BIOLOGY

A couple plans to have two children. Assuming that it is equally likely that a boy or a girl will be born, what is the probability that the couple will have two girls? What is the probability of two boys? What is the probability of one boy and one girl?

To solve the first of these applications we use a geometric sequence (see Example 5 in Section 12.3). The biological application gives us an idea of the likelihood of certain events (see Example 5 in Section 12.7).

In this chapter we first cover sequences and series and introduce mathematical induction. We then discuss permutations and combinations before concluding with the binomial theorem and probability.

## 12.1 Sequences and Series

### Sequences

Informally we think of a sequence as a collection of numbers arranged in a particular order. This means there is a first number, a second number, a third number, and so forth. Thus, each number in the sequence corresponds to a natural number. This suggests the following formal definition.

**Infinite Sequence**

An **infinite sequence** is a function with domain the set of natural numbers.

As an example consider the function defined by

$$f(n) = 2n. \qquad n = 1, 2, 3, 4, \ldots$$

The three dots mean that the pattern continues. Instead of using the usual function notation we usually write

$$x_n = 2n. \qquad n = 1, 2, 3, 4, \ldots$$

That is, a letter with a subscript, such as $x_n$, is used instead of $f(n)$. For the sequence defined by $x_n = 2n$,

$$\begin{aligned} x_1 &= 2(\mathbf{1}) = 2 \\ x_2 &= 2(\mathbf{2}) = 4 \\ x_3 &= 2(\mathbf{3}) = 6 \\ x_4 &= 2(\mathbf{4}) = 8 \\ &\vdots \\ x_n &= 2(\boldsymbol{n}) = 2n \\ &\vdots \end{aligned}$$

A sequence is frequently defined by indicating its range. Thus, the sequence above can be written

$$2, 4, 6, 8, \ldots, 2n, \ldots.$$

Each number is called a **term** of the sequence, and the terms are written in order of increasing $n$. The term $2n$ is called the ***n*th term** or the **general term.** Note the use of the three dots after the general term to indicate an infinite sequence. We will also study sequences which are finite.

### Finite Sequence

A **finite sequence** with $m$ terms is a function with domain the set of natural numbers from 1 to $m$.

For example,

$$2, 4, 6, 8$$

indicates a sequence with four terms, and

$$3, 6, 9, \ldots, 3n$$

is a sequence with $n$ terms. The three dots used in this example indicate the presence of the terms between 9 and $3n$.

The following examples illustrate the use of sequence notation.

**EXAMPLE 1** Find the first four terms and the seventh term of each sequence. Assume the domain in each case is the set of natural numbers.

**(a)** $x_n = \frac{1}{n+1}$. Use $n = 1, 2, 3, 4$, and 7 in order.

$$x_1 = \frac{1}{1+1} = \frac{1}{2}$$

$$x_2 = \frac{1}{2+1} = \frac{1}{3}$$

$$x_3 = \frac{1}{3+1} = \frac{1}{4}$$

$$x_4 = \frac{1}{4+1} = \frac{1}{5}$$

$$x_7 = \frac{1}{7+1} = \frac{1}{8}$$

The complete sequence can be written

$$\frac{1}{2}, \frac{1}{3}, \frac{1}{4}, \frac{1}{5}, \ldots, \frac{1}{n+1}, \ldots.$$

**(b)** $a_n = 2^n - 1$. Note that $a_n$ is used in this example.

$a_1 = 2^1 - 1 = 2 - 1 = 1$

$a_2 = 2^2 - 1 = 4 - 1 = 3$

$a_3 = 2^3 - 1 = 8 - 1 = 7$

$a_4 = 2^4 - 1 = 16 - 1 = 15$

$a_7 = 2^7 - 1 = 128 - 1 = 127$

The infinite sequence can be written

$$1, 3, 7, 15, \ldots, 2^n - 1, \ldots.$$

**(c)** $b_n = (-1)^n(n^2 + 1)$

$$b_1 = (-1)^1(1^2 + 1) = (-1)(2) = -2$$
$$b_2 = (-1)^2(2^2 + 1) = (1)(5) = 5$$
$$b_3 = (-1)^3(3^2 + 1) = (-1)(10) = -10$$
$$b_4 = (-1)^4(4^2 + 1) = (1)(17) = 17$$
$$b_7 = (-1)^7(7^2 + 1) = (-1)(50) = -50$$

Notice that the factor $(-1)^n$ changes the sign on each term. When $n$ is odd, $(-1)^n = -1$, and when $n$ is even, $(-1)^n = 1$. ■

Consider the following way of defining a sequence.
Let $x_1 = 5$ and $x_{n+1} = 3x_n$. Then

$$x_1 = 5$$
$$x_2 = x_{1+1} = 3x_1 = 15 \qquad n = 1 \text{ and } n + 1 = 2$$
$$x_3 = x_{2+1} = 3x_2 = 45 \qquad n = 2 \text{ and } n + 1 = 3$$
$$x_4 = x_{3+1} = 3x_3 = 135 \qquad n = 3 \text{ and } n + 1 = 4$$
$$\vdots$$
$$x_{n+1} = 3x_n$$
$$\vdots$$

With $x_1$ given, and $x_{n+1}$ defined in terms of $x_n$, we can determine the whole sequence. This method of defining a sequence is called a **recursive definition.**

**EXAMPLE 2** Let $a_1 = -2$ and $a_n = 2a_{n-1} + 3$ for $n \geq 2$. Determine $a_2$, $a_3$, $a_4$, and $a_5$. Notice that in this recursive formula $a_n$ is defined in terms of $a_{n-1}$ for $n \geq 2$.

$$a_2 = 2a_{2-1} + 3 = 2a_1 + 3 = 2(-2) + 3 = -1$$
$$a_3 = 2a_{3-1} + 3 = 2a_2 + 3 = 2(-1) + 3 = 1$$
$$a_4 = 2a_{4-1} + 3 = 2a_3 + 3 = 2(1) + 3 = 5$$
$$a_5 = 2a_{5-1} + 3 = 2a_4 + 3 = 2(5) + 3 = 13 \quad ■$$

## Series

Associated with each sequence is a **series,** the indicated sum of the terms of the sequence. For example, associated with the sequence

$$1, 3, 5, 7, 9, 11, 13$$

is the series

$$1 + 3 + 5 + 7 + 9 + 11 + 13.$$

Notice that terms of a sequence are separated by commas, but the word *series* means that terms are added.

The Greek letter $\Sigma$ (sigma), called the **summation symbol,** is used to abbreviate a series.

**Summation Notation**

The sum of the sequence $a_1, a_2, a_3, \ldots, a_n$ can be written

$$\sum_{k=1}^{n} a_k = a_1 + a_2 + a_3 + \cdots + a_n,$$

where $k$ is the **index** on the summation, 1 is the **lower limit,** and $n$ is the **upper limit of summation.**

For example, if $a_k = 2k$ and $n = 7$, we write

$$\begin{aligned}\sum_{k=1}^{n} a_k &= \sum_{k=1}^{7} 2k \\ &= 2(1) + 2(2) + 2(3) + 2(4) + 2(5) + 2(6) + 2(7) \\ &= 2 + 4 + 6 + 8 + 10 + 12 + 14 = 56.\end{aligned}$$

**EXAMPLE 3** Determine the value of each series.

**(a)** $$\begin{aligned}\sum_{k=1}^{5} (k^2 - k) &= (1^2 - 1) + (2^2 - 2) + (3^2 - 3) + (4^2 - 4) + (5^2 - 5) \\ &= 0 + 2 + 6 + 12 + 20 = 40\end{aligned}$$

**(b)** $$\begin{aligned}\sum_{k=2}^{6} (-1)^k\sqrt{k} &= (-1)^2\sqrt{2} + (-1)^3\sqrt{3} + (-1)^4\sqrt{4} + (-1)^5\sqrt{5} + (-1)^6\sqrt{6} \\ &= (1)\sqrt{2} + (-1)\sqrt{3} + (1)(2) + (-1)\sqrt{5} + (1)\sqrt{6} \\ &= \sqrt{2} - \sqrt{3} + 2 - \sqrt{5} + \sqrt{6}\end{aligned}$$ ■

Notice in Example 3(**b**) that the lower limit of the summation is 2. In general the lower limit can be any whole number less than or equal to the upper limit. Thus,

$$\sum_{k=3}^{8} a_k = a_3 + a_4 + a_5 + a_6 + a_7 + a_8.$$

If $a_k$ is a constant, that is $a_k = c$ for all $k$, then

$$\begin{aligned}a_1 + a_2 + a_3 + \cdots + a_n &= c + c + c + \cdots + c \\ &= nc.\end{aligned}$$

This gives the following theorem.

**Sum of Constants**

If $a_k = c$ for all $k$, then

$$\sum_{k=1}^{n} a_k = nc.$$

Thus, if $a_k = 5$ for $k = 1, 2, \ldots, 8$, then

$$\sum_{k=1}^{8} a_k = \sum_{k=1}^{8} 5 = 8(5) = 40.$$

Other properties can be established which involve sums of terms, differences of terms, and a constant times each term of a series.

**Properties of Summation Notation**

If $a_1, a_2, a_3, \ldots, a_n$ and $b_1, b_2, b_3, \ldots, b_n$ are sequences and $c$ is a constant, then

$$\sum_{k=1}^{n} (a_k + b_k) = \sum_{k=1}^{n} a_k + \sum_{k=1}^{n} b_k,$$

$$\sum_{k=1}^{n} (a_k - b_k) = \sum_{k=1}^{n} a_k - \sum_{k=1}^{n} b_k,$$

$$\sum_{k=1}^{n} c\, a_k = c \sum_{k=1}^{n} a_k.$$

The proof of this theorem involves the repeated use of the commutative and associative laws as well as the distributive law. For example,

$$\begin{aligned}\sum_{k=1}^{n} (a_k + b_k) &= (a_1 + b_1) + (a_2 + b_2) + (a_3 + b_3) + \cdots + (a_n + b_n)\\ &= (a_1 + a_2 + a_3 + \cdots + a_n) + (b_1 + b_2 + b_3 + \cdots + b_n)\\ &= \sum_{k=1}^{n} a_k + \sum_{k=1}^{n} b_k.\end{aligned}$$

## 12.1 EXERCISES

In Exercises 1–8 give the first five terms, the eighth term, and the twelfth term of each sequence.

**1.** $a_n = 4n$

**2.** $x_n = 3n + 1$

**3.** $x_n = \left(-\frac{1}{2}\right)^n$

**4.** $x_n = (n - 1)(n + 2)(n - 3)$

**5.** $b_n = (-1)^n(n^2 + 5)$

**6.** $a_n = (-1)^{n+1}2^n$

**7.** $x_n = (-1)^n + (-1)^{n+1}$

**8.** $b_n = [1 + (-1)^n]\frac{n + 1}{n}$

In Exercises 9–14 determine the second, third, fourth, and fifth terms of each sequence.

**9.** $a_1 = 3;\ a_{n+1} = 4a_n$

**10.** $x_1 = -2;\ x_{n+1} = x_n + 3$

**11.** $b_1 = 8;\ b_n = -3b_{n-1}$

**12.** $a_1 = \frac{1}{2};\ a_n = 2a_{n-1} - 3$

**13.** $x_1 = \frac{2}{3};\ x_{n+1} = 9x_n + 5$

**14.** $b_1 = 2;\ b_{n+1} = b_n^2$

Write out each series in Exercises 15–20.

**15.** $\sum_{k=1}^{5} x_k$

**16.** $\sum_{k=0}^{4} a_k$

**17.** $\sum_{m=1}^{4} \frac{1}{2m + 1}$

**18.** $\sum_{m=0}^{5} \pi m$

**19.** $\sum_{k=0}^{6} \frac{(-1)^{k+1}}{k + 1}$

**20.** $\sum_{k=1}^{4} \frac{(-1)^k}{3k}$

In Exercises 21–26 write each series using sigma summation notation.

**21.** $1 + 2 + 3 + 4 + \cdots + n$

**22.** $\frac{1}{2} + \frac{2}{3} + \frac{3}{4} + \cdots + \frac{n}{n + 1}$

**23.** $1 + \frac{1}{\sqrt{2}} + \frac{1}{\sqrt{3}} + \cdots + \frac{1}{\sqrt{n}}$

**24.** $1 + \frac{1}{4} + \frac{1}{9} + \cdots + \frac{1}{n^2}$

**25.** $2 + 4 + 6 + 8$

**26.** $-1 - \frac{1}{2} - \frac{1}{3} - \frac{1}{4} - \frac{1}{5}$

Evaluate each series in Exercises 27–38.

**27.** $\sum_{k=1}^{6} (2k - 1)$

**28.** $\sum_{k=1}^{4} (k - 1)^2$

**29.** $\sum_{i=0}^{3} (i^2 + 1)$

**30.** $\sum_{k=3}^{4} (-1)^{2k}2^{-k}$

**31.** $\sum_{m=1}^{5} [1 + (-1)^m]m^2$

**32.** $\sum_{m=1}^{3} [1 - (-1)^m]\frac{2m}{m + 1}$

**33.** $\sum_{k=1}^{50} 3$

**34.** $\sum_{j=4}^{24} 100$

**35.** $\sum_{k=1}^{4} (k + 1)(k - 2)$

**36.** $\sum_{k=1}^{3} \frac{k + 1}{k + 2}$

**37.** $\sum_{m=1}^{5} \left(\frac{1}{m} - \frac{1}{m + 1}\right)$

**38.** $\sum_{m=1}^{4} \left[\frac{1}{m^2} - \frac{1}{(m + 1)^2}\right]$

Compare the series in each of Exercises 39–40.

**39.** $\sum_{k=2}^{5} (k + 1)(k + 2)$ and $\sum_{k=1}^{4} (k + 2)(k + 3)$

**40.** $\sum_{m=3}^{5} \frac{m - 2}{m + 1}$ and $\sum_{k=1}^{3} \frac{k}{k + 3}$

In Exercises 41–42 determine the approximate value of $a_{30}$ by repeated use of the square root key on a calculator.

**41.** $a_1 = 6;\ a_{n+1} = \sqrt{a_n}$

**42.** $a_1 = 8;\ a_n = \sqrt{a_{n-1}}$

If $\sum_{k=1}^{n} k = \frac{n(n+1)}{2}$ and $\sum_{k=1}^{n} k^2 = \frac{n(n+1)(2n+1)}{6}$, determine each sum in Exercises 43–48.

**43.** $\sum_{k=1}^{6} k$

**44.** $\sum_{k=1}^{6} k^2$

**45.** $\sum_{k=1}^{n} (k^2 + k)$

**46.** $\sum_{k=1}^{n} (k - k^2)$

**47.** $\sum_{k=1}^{n} (2k^2 + 3k)$

**48.** $\sum_{k=1}^{n} \left(\frac{1}{2} - k\right)$

**49.** Prove that $\sum_{k=1}^{n} (a_k - b_k) = \sum_{k=1}^{n} a_k - \sum_{k=1}^{n} b_k$.

**50.** Prove that $\sum_{k=1}^{n} ca_k = c \sum_{k=1}^{n} a_k$.

**STATISTICS** The arithmetic mean $\bar{x}$ of a collection of data is given by $\bar{x} = \frac{1}{n} \sum_{k=1}^{n} x_k$. Use this series and the series used in Exercises 43–48 to determine $\bar{x}$ for the data in Exercises 51–52.

**51.** $1, 2, 3, \ldots, n$

**52.** $1, 4, 9, \ldots, n^2$

The slope $m$ and $y$-intercept $b$ in the least squares regression line $y = mx + b$ which best approximates a collection of $n$ points $(x_k, y_k)$ can be determined by solving the following system of equations.

$$nb + \left(\sum_{k=1}^{n} x_k\right)m = \sum_{k=1}^{n} y_k$$

$$\left(\sum_{k=1}^{n} x_k\right)b + \left(\sum_{k=1}^{n} x_k^2\right)m = \sum_{k=1}^{n} x_k y_k$$

Use this system to find the regression line for the data given in Exercises 53–54.

**53.** $(-1, 2), (1, -1), (5, -4)$

**54.** $(-4, -1), (-2, 2), (0, 1), (3, 5)$

In Exercises 55–56 use the sequence defined by

$$a_1 = a_2 = 1 \quad \text{and} \quad a_{k+1} = a_k + a_{k-1}.$$

This is a well-known sequence in mathematics called the Fibonacci sequence.

**55.** Determine the first eight terms of the sequence.

**56.** If $r_k = \frac{a_{k+1}}{a_k}$, determine $r_8$ to three decimal places. As $k$ increases the sequence $r_k$ gives better and better approximations for the *golden ratio,* the ratio of the sides of a rectangle thought to be most pleasing to the eye.

## For Review

Evaluate the determinants in Exercises 57–58.

**57.** $\begin{vmatrix} \sum_{k=1}^{3} k & \sum_{k=1}^{4} k^2 \\ \sum_{k=3}^{6} k & \sum_{k=4}^{5} k^2 \end{vmatrix}$

**58.** $\begin{vmatrix} \sum_{k=1}^{2} k^2 & -5 & 0 \\ \sum_{k=1}^{5} k & -9 & \sum_{k=2}^{3} 5k \\ \sum_{k=1}^{4} k(k+1) & 7 & 0 \end{vmatrix}$

## 12.2 Arithmetic Sequences

There are several special types of sequences which have a wide range of applications. The first of the two types that we will consider is illustrated by the following:

$$1, \quad 4, \quad 7, \quad 10, \quad 13, \quad 16, \ldots .$$

Notice that $4 - 1 = 3$, $7 - 4 = 3$, $10 - 7 = 3$, $13 - 10 = 3$, and $16 - 13 = 3$. That is, the successive terms of the sequence differ by a constant.

### Arithmetic Sequence

An **arithmetic sequence (progression)** is a sequence in which the successive terms differ by some constant $d$, called the **common difference.**

### Formula for $a_n$

Since $a_n - a_{n-1} = d$ for all $n$, we have the recursive formula

$$a_n = a_{n-1} + d$$

For arithmetic sequences we can develop a formula for $a_n$ in terms of $a_1$, $n$, and $d$. Consider the following:

$$\begin{aligned} \mathbf{1}\text{st term} &= a_1 = a_1 + \mathbf{0}d \\ \mathbf{2}\text{nd term} &= a_2 = a_1 + d = a_1 + \phantom{2}d \\ \mathbf{3}\text{rd term} &= a_3 = a_2 + d = (a_1 + d) + d = a_1 + \mathbf{2}d \\ \mathbf{4}\text{th term} &= a_4 = a_3 + d = (a_1 + 2d) + d = a_1 + \mathbf{3}d \\ \mathbf{5}\text{th term} &= a_5 = a_4 + d = (a_1 + 3d) + d = a_1 + \mathbf{4}d \end{aligned}$$

Notice that the $n$th term $a_n$ is the first term $a_1$ plus $n - 1$ times the common difference $d$. Thus,

$$\boldsymbol{a_n = a_1 + (n - 1)d.}$$

**EXAMPLE 1** Find the eighth and the twelfth terms of the arithmetic sequence 2, 7, 12, 17, 22. . . . .

We have $a_1 = 2$ and $d = 5$. Thus,

$$\begin{aligned} a_8 &= a_1(8 - 1)d = 2 + (8 - 1)5 \\ &= 2 + 35 = 37 \end{aligned}$$

and

$$\begin{aligned} a_{12} &= a_1 + (12 - 1)d = 2 + (11)5 \\ &= 2 + 55 = 57. \end{aligned}$$

Notice that the $n$ in $a_n$ is the same as the $n$ in $(n - 1)$. ■

**EXAMPLE 2** Find $x$ so that $x + 3$, $2x + 8$, and $4x + 15$ form a three-term arithmetic sequence in the given order. Also, give the sequence.

We use the fact that the difference between successive terms is equal to the common difference $d$.

$$(2x + 8) - (x + 3) = d \quad \text{and} \quad (4x + 15) - (2x + 8) = d$$

Set both expressions for $d$ equal to each other.

$$\begin{aligned} (2x + 8) - (x + 3) &= (4x + 15) - (2x + 8) \\ x + 5 &= 2x + 7 \\ -2 &= x \end{aligned}$$

Then $x + 3 = -2 + 3 = 1$, $2x + 8 = 2(-2) + 8 = 4$, $4x + 15 = 4(-2) + 15 = 7$, so that the desired arithmetic sequence is 1, 4, 7. ■

## Formulas for $S_n$

The sum of the first $n$ terms in an arithmetic sequence can also be determined by a formula. Let $S_n$ denote the sum of the first $n$ terms of an arithmetic sequence. Then we have the series

$$\begin{aligned} S_n &= a_1 + a_2 + a_3 + a_4 + \cdots + a_n \\ &= a_1 + a_1 + d + a_1 + 2d + a_1 + 3d + \cdots + a_1 + (n - 1)d. \end{aligned}$$

Reversing the order of addition, we obtain

$$\begin{aligned} S_n &= a_n + a_{n-1} + a_{n-2} + \cdots + a_1 \\ &= a_1 + (n - 1)d + a_1 + (n - 2)d + a_1 + (n - 3)d + \cdots + a_1. \end{aligned}$$

Add corresponding terms in both representations of $S_n$.

$$\begin{aligned} S_n &= a_1 + a_1 + d + a_1 + 2d + \cdots + a_1 + (n - 1)d \\ S_n &= a_1 + (n - 1)d + a_1 + (n - 2)d + a_1 + (n - 3)d + \cdots + a_1 \\ \hline 2S_n &= 2a_1 + (n - 1)d + 2a_1 + (n - 1)d + 2a_1 + (n - 1)d + \cdots + 2a_1 + (n - 1)d \\ 2S_n &= n[2a_1 + (n - 1)d] \quad \text{There are } n \text{ terms of the form } 2a_1 + (n - 1)d \end{aligned}$$

$$S_n = \frac{n}{2}[2a_1 + (n - 1)d]$$

**EXAMPLE 3** Find the fifteenth term and the sum of the first fifteen terms of the arithmetic sequence $-2, 1, 4, 7, 10, \ldots$.

We have $a_1 = -2$ and $d = 3$, and we must calculate $a_{15}$ ($n = 15$) and $S_{15}$.

$$a_{15} = a_1 + (15 - 1)d = -2 + (14)(3) = -2 + 42 = 40$$

$$\begin{aligned} S_{15} = \frac{n}{2}[2a_1 + (n-1)d] &= \frac{15}{2}[2(-2) + (15 - 1)3] \\ &= \frac{15}{2}[-4 + (14)3] = \frac{15}{2}[-4 + 42] \\ &= \frac{15}{2}[38] = 285 \end{aligned}$$ ■

An alternate form for $S_n$ is easily derived from

$$S_n = \frac{n}{2}[2a_1 + (n-1)d].$$

We write $2a_1$ as $a_1 + a_1$ and observe that $a_1 + (n-1)d = a_n$.

$$S_n = \frac{n}{2}[a_1 + \underbrace{a_1 + (n-1)d}_{a_n}]$$

$$\mathbf{S_n = \frac{n}{2}[a_1 + a_n]}$$

In the preceding example, since we had already calculated $a_{15} = 40$, it would have been easier to substitute into the new formula for $S_n$.

$$S_{15} = \frac{15}{2}[a_1 + a_{15}] = \frac{15}{2}[-2 + 40] = \frac{15}{2}[38] = 285$$

The following theorem summarizes the formulas that have been derived for arithmetic sequences.

**Formulas for Arithmetic Sequences**

For an arithmetic sequence the general or $n$th term is given by

$$a_n = a_1 + (n-1)d.$$

The sum of the first $n$ terms is given by

$$S_n = \frac{n}{2}[a_1 + a_n] \quad \text{or} \quad S_n = \frac{n}{2}[2a_1 + (n-1)d].$$

**EXAMPLE 4** Suppose that the fifteenth term of an arithmetic sequence is 71, and that the twenty-first term is 101. Find $a_1$, $d$, the first five terms, and the sum of the first five terms.

We have $\quad a_{15} = 71 = a_1 + (15 - 1)d = a_1 + 14d$

and $\quad a_{21} = 101 = a_1 + (21 - 1)d = a_1 + 20d.$

To find $a_1$ and $d$, we must solve the following system.

$$a_1 + 14d = 71$$
$$a_1 + 20d = 101$$

Subtract the first equation from the second.

$$6d = 30$$
$$d = 5$$

Then substitute this value for $d$ in the first equation.

$$a_1 + 14 \cdot 5 = 71$$
$$a_1 = 71 - 70 = 1$$

The first five terms are 1, 6, 11, 16, 21, and

$$S_5 = \frac{5}{2}[a_1 + a_5] = \frac{5}{2}[1 + 21] = \frac{5}{2}[22] = 55. \quad \blacksquare$$

**Arithmetic Means**

We now consider a property of arithmetic sequences which is a generalization of the average or arithmetic mean of two numbers. Notice that since the average $\frac{a+b}{2}$ is midway between the numbers $a$ and $b$ on a number line,

$$a, \frac{a+b}{2}, b$$

is an arithmetic sequence. We define **arithmetic means** to be all the terms of an arithmetic sequence between $a_1$ and $a_n$.

**EXAMPLE 5**

Insert four arithmetic means between the numbers 8 and $-7$.

Since four means must be inserted between 8 and $-7$, there will be six terms in the sequence, with $a_1 = 8$ and $a_6 = -7$. Use $a_n = a_1 + (n - 1)d$ with $n = 6$, $a_1 = 8$, and $a_6 = -7$ to find $d$.

$$a_6 = a_1 + (6 - 1)d$$
$$-7 = 8 + 5d$$
$$-3 = d$$

Since $8 + (-3) = 5$, $5 + (-3) = 2$, $2 + (-3) = -1$, and $-1 + (-3) = -4$, the four arithmetic means between 8 and $-7$ are 5, 2, $-1$, and $-4$. $\blacksquare$

There are many applications of arithmetic sequences. Consider the following depreciation problem.

**EXAMPLE 6**
**CONSTRUCTION**

An earth mover is purchased for $260,000. Assume that it depreciates 7.0% the first year, 6.5% the second year, 6.0% the third year, and continues in the same manner for ten years. If all depreciations apply to the original cost, what is the value of the earth mover in ten years?

We calculate the sum of the depreciations over the ten years with $a_1 = 7.0$, $a_2 = 6.5$, $a_3 = 6.0$, . . . and $d = -0.5$.

$$S_{10} = \frac{10}{2}[2(7.0) + (10 - 1)(-0.5)]$$
$$= 5[14.0 - 4.5] = 47.5$$

Thus the total percentage depreciation in ten years is 47.5%. This means that the equipment is worth 52.5% of its original value.

$$0.525(260{,}000) = 136{,}500$$

The earth mover is worth \$136,500 in ten years. ■

## 12.2 EXERCISES

In Exercises 1–6 determine if each sequence is arithmetic. If it is, give the common difference $d$.

**1.** 4, 8, 12, 16, . . .
**2.** 4, 8, 10, 12, . . .
**3.** 9, −1, −11, −21, . . .
**4.** $\frac{2}{3}, 2, \frac{10}{3}, \frac{14}{3}, \ldots$
**5.** 1, −1, 1, −1, . . .
**6.** log 2, log 4, log 8, log 16, . . .

Find the first six terms of each arithmetic sequence in Exercises 7–12.

**7.** $a_1 = 2$ and $d = 7$
**8.** $a_1 = 8$ and $d = -5$
**9.** $a_1 = -2$ and $a_2 = 5$
**10.** $a_1 = 2$ and $a_6 = -13$
**11.** $a_1 = \sqrt{3}$ and $d = 4\sqrt{3}$
**12.** $a_1 = \ln 10$ and $a_3 = \ln 1000$

**13.** Find $x$ so that $x$, $x + 4$, and $2x$ form a three-term arithmetic sequence in the given order. Determine the sequence.

**14.** Find $y$ so that $2y$, $3y + 7$, and $5y + 1$ form a three-term arithmetic sequence in the given order. Determine the sequence.

In Exercises 15–20 find the indicated sum of the arithmetic sequence.

**15.** −8, −1, 6, 13, . . . ; $S_{10}$
**16.** $a_1 = 3$ and $a_{10} = 57$; $S_{10}$
**17.** $a_4 = 6$ and $a_8 = 26$, $S_{12}$
**18.** $a_3 = 9$ and $a_5 = -1$; $S_{20}$
**19.** $\sum_{k=1}^{6} (k + 2)$
**20.** $\sum_{k=1}^{30} (1 - 2k)$

In Exercises 21–32, some of the numbers $n$, $a_1$, $a_n$, $d$, and $S_n$ are given. Find the missing ones.

**21.** $a_1 = 2$, $n = 17$, $d = 3$
**22.** $a_1 = 24$, $a_n = 3$, $n = 8$
**23.** $a_n = 27$, $S_n = 63$, $a_1 = -9$
**24.** $a_1 = -40$, $S_{21} = 210$
**25.** $a_1 = \frac{5}{3}$, $d = \frac{1}{6}$, $n = 12$
**26.** $a_1 = -7$, $d = 8$, $S_n = 225$
**27.** $a_{15} = 4$, $S_{15} = 30$
**28.** $a_1 = 10$, $a_n = -8$, $S_n = 7$
**29.** $a_1 = -9$, $a_7 = 21$
**30.** $a_1 = \frac{3}{4}$, $a_5 = -\frac{1}{4}$
**31.** $a_1 = \log 7$, $d = \log 49$, $n = 5$
**32.** $a_1 = \ln 3$, $a_4 = \ln 81$

**33.** Insert six arithmetic means between 11 and 32.

**34.** Insert eight arithmetic means between 47 and 11.

**35.** How many integers between 39 and 146 are divisible by 5?

**36.** How many integers between $-92$ and 261 are divisible by 7?

**37.** Find the sum of all the even integers between 1 and 201.

**38.** Find the sum of all the integers divisible by 5 from 25 through 350.

**39.** Prove that the sum of the sequence 2, 4, 6, . . . , $2n$ is $n^2 + n$.

**40.** Prove that the sum of the first $n$ odd natural numbers is $n^2$.

Solve.

**41.** CONSUMER A new car costs \$8400. Assume that it depreciates 2.1% the first year, 1.8% the second year, 1.5% the third year, and continues in the same manner for twelve years. If all depreciations apply to the original cost, what is the value of the car (rounded to the nearest dollar) in twelve years?

**42.** BUSINESS A man earned \$3500 the first year he worked. If he received a raise of \$750 at the end of each year for 20 years, what was his salary during his twenty-first year of work? How much income did he have during the first 21 years of work?

**43.** RECREATION A theater has 40 rows with 20 seats in the first row, 23 in the second row, 26 in the third row, and so forth. How many seats are in the theater?

**44.** A collection of nickels is arranged in a triangular array with 15 coins in the base row, 14 in the next, 13 in the next, and so forth. Find the value of the collection.

**45.** BANKING If a woman puts \$1 in a bank on the first day of September, \$2 on the second day of September, \$3 on the third day, and so forth, how much money will be in the bank at the end of the month?

**46.** FORESTRY There are 190 logs to be stacked in such a way that one log will be on top, 2 logs on the second row, 3 logs on the third, and so on. How many logs should be put on the first row on the bottom of the stack?

**47.** PHYSICS A rock is dropped from the top of a tall cliff and falls 16 ft during the first second, 48 ft during the second, 80 ft during the third, and so on. How many feet does the rock fall during the eighth second? (Do not take friction into account.)

**48.** DEMOGRAPHY The population of a town is decreasing by 500 inhabitants each year. If its population at the beginning of 1960 was 20,135, what was its population at the beginning of 1970?

## For Review

In Exercises 49–50 determine the fourth and fifth terms of each general sequence.

**49.** $x_n = (-1)^{n+1}3^{-n}$

**50.** $x_1 = 6$ and $x_{k+1} = 1 - x_k$

**51.** Write $\frac{1}{2} + \frac{2}{5} + \frac{3}{10} + \cdots + \frac{n}{n^2 + 1}$ in sigma summation notation.

**52.** Evaluate. $\sum_{k=2}^{4} \frac{1}{3k - 2}$

## 12.3 Geometric Sequences

The second type of special sequence that we consider is illustrated by

$$1, \quad \frac{1}{2}, \quad \frac{1}{4}, \quad \frac{1}{8}, \quad \frac{1}{16}, \quad \frac{1}{32}, \ldots .$$

For this sequence

$$\frac{a_2}{a_1} = \frac{1}{2}, \quad \frac{a_3}{a_2} = \frac{1}{2}, \quad \frac{a_4}{a_3} = \frac{1}{2}, \quad \frac{a_5}{a_4} = \frac{1}{2}, \quad \frac{a_6}{a_5} = \frac{1}{2}, \quad \text{and so on.}$$

### Geometric Sequence

A **geometric sequence (progression)** is a sequence with the property that

$$\frac{a_n}{a_{n-1}} = r$$

for all $n$. The number $r$ is called the **common ratio.**

**Formula for $a_n$**

Since $\dfrac{a_n}{a_{n-1}} = r$, then $a_n = a_{n-1}r$. We can use this relation to derive a formula for the $n$th or general term of a geometric sequence.

$$\begin{aligned}
1\text{st term} &= a_1 = a_1r^0 \\
2\text{nd term} &= a_2 = a_1 \cdot r = a_1r^1 \\
3\text{rd term} &= a_3 = a_2 \cdot r = (a_1r) \cdot r = a_1r^2 \\
4\text{th term} &= a_4 = a_3 \cdot r = (a_1r^2) \cdot r = a_1r^3 \\
5\text{th term} &= a_5 = a_4 \cdot r = (a_1r^3) \cdot r = a_1r^4
\end{aligned}$$

Notice that the $n$th term $a_n$ is equal to the first term $a_1$ times $r^{n-1}$. Thus,

$$a_n = a_1r^{n-1}.$$

**EXAMPLE 1**

Find the seventh and tenth terms of the geometric sequence

$$-2, \quad 1, \quad -\frac{1}{2}, \quad \frac{1}{4}, \quad -\frac{1}{8}, \ldots .$$

We have $a_1 = -2$ and $r = -1/2$.

$$a_7 = a_1r^{7-1} = (-2)\left(-\frac{1}{2}\right)^6 = -\frac{1}{32}$$

$$a_{10} = a_1r^{10-1} = (-2)\left(-\frac{1}{2}\right)^9 = \frac{1}{256}$$

Notice that the $n$ in $a_n$ is the same as the $n$ in the exponent $n - 1$. ■

**EXAMPLE 2** Find $x$ so that $x - 3$, $x - 1$, and $2x + 1$ form a three-term geometric sequence in the given order. Also, give the sequence.

Use the fact that the ratio of successive terms is equal to the common ratio $r$.

$$\frac{x-1}{x-3} = \frac{2x+1}{x-1}$$
$$(x-1)(x-1) = (x-3)(2x+1)$$
$$x^2 - 2x + 1 = 2x^2 - 5x - 3$$
$$0 = x^2 - 3x - 4$$
$$0 = (x+1)(x-4)$$
$$x + 1 = 0 \quad \text{or} \quad x - 4 = 0$$
$$x = -1 \qquad\qquad x = 4$$

If $x = -1$:

$$x - 3 = -1 - 3 = -4$$
$$x - 1 = -1 - 1 = -2$$
$$2x + 1 = 2(-1) + 1 = -1;$$

the sequence is

$$-4, \quad -2, \quad -1 \quad \left(r = \frac{1}{2}\right).$$

If $x = 4$:

$$x - 3 = 4 - 3 = 1$$
$$x - 1 = 4 - 1 = 3$$
$$2x + 1 = 2(4) + 1 = 9;$$

the sequence is

$$1, \quad 3, \quad 9 \quad (r = 3).$$ ■

**Formulas for $S_n$** We now derive a formula for calculating the sum of the first $n$ terms of a geometric sequence. Let $S_n$ denote the sum of the first $n$ terms. The series is

$$S_n = a_1 + a_1r + a_1r^2 + a_1r^3 + \cdots + a_1r^{n-2} + a_1r^{n-1}.$$

Multiply by $r$.

$$rS_n = a_1r + a_1r^2 + a_1r^3 + \cdots + a_1r^{n-2} + a_1r^{n-1} + a_1r^n$$

Subtract.

$$\begin{aligned} S_n &= a_1 + a_1r + a_1r^2 + a_1r^3 + \cdots + a_1r^{n-2} + a_1r^{n-1} \\ -rS_n &= \phantom{a_1} - a_1r - a_1r^2 - a_1r^3 - \cdots - a_1r^{n-2} - a_1r^{n-1} - a_1r^n \\ \hline S_n - rS_n &= a_1 \phantom{+ a_1r + a_1r^2 + a_1r^3 + \cdots + a_1r^{n-2} + a_1r^{n-1}} - a_1r^n \\ S_n(1-r) &= a_1 - a_1r^n \\ S_n &= \frac{a_1 - a_1r^n}{1-r} \end{aligned}$$

**EXAMPLE 3** Find the ninth term and the sum of the first nine terms of the following geometric sequence.

$$\frac{2}{5}, \frac{1}{5}, \frac{1}{10}, \frac{1}{20}, \frac{1}{40}, \ldots$$

We have $a_1 = 2/5$, $r = 1/2$, and we must calculate $a_9$ ($n = 9$) and $S_9$.

$$a_9 = a_1 r^{9-1} = \left(\frac{2}{5}\right)\left(\frac{1}{2}\right)^8 = \left(\frac{2}{5}\right)\left(\frac{1}{256}\right) = \frac{1}{640}$$

$$S_9 = \frac{a_1 - a_1 r^9}{1 - r} = \frac{\frac{2}{5} - \frac{2}{5}\left(\frac{1}{2}\right)^9}{1 - \frac{1}{2}} = \frac{\left(\frac{2}{5}\right)\left[1 - \left(\frac{1}{2}\right)^9\right]}{\frac{1}{2}} = \frac{\frac{2}{5}\left(1 - \frac{1}{512}\right)}{\frac{1}{2}} = \frac{2}{5}\left(\frac{511}{512}\right)\left(\frac{2}{1}\right) = \frac{511}{640}$$ ■

An alternate form for $S_n$ is easily derived from

$$S_n = \frac{a_1 - a_1 r^n}{1 - r}$$

by writing $a_1 r^n = r(a_1 r^{n-1}) = r a_n$.

$$\mathbf{S_n = \frac{a_1 - r a_n}{1 - r}}$$

In the preceding example, since we had already calculated $a_9 = 1/640$, it would have been easier to substitute into the new formula for $S_n$.

$$S_9 = \frac{a_1 - r a_9}{1 - r} = \frac{\frac{2}{5} - \left(\frac{1}{2}\right)\left(\frac{1}{640}\right)}{1 - \frac{1}{2}} = \frac{\frac{2}{5} - \frac{1}{1280}}{\frac{1}{2}} = \frac{\frac{512}{1280} - \frac{1}{1280}}{\frac{1}{2}} = \frac{511}{640}$$

A summary of the formulas derived for geometric sequences is given in the following theorem.

**Formulas for Geometric Sequences**

For a geometric sequence the general or $n$th term is given by

$$a_n = a_1 r^{n-1}.$$

The sum of the first $n$ terms is given by

$$S_n = \frac{a_1 - r a_n}{1 - r} \quad \text{or} \quad S_n = \frac{a_1 - a_1 r^n}{1 - r}.$$

**EXAMPLE 4** Suppose that the third term of a geometric sequence is 27 and the fifth term is 243. Find $a_1$, $r$, and $S_5$.

$$a_3 = a_1 r^{3-1} = a_1 r^2 = 27 \quad \text{and} \quad a_5 = a_1 r^{5-1} = a_1 r^4 = 243$$

Solve the first equation for $a_1$, $a_1 = 27/r^2$, and substitute into the second.

$$a_1 r^4 = \frac{27}{r^2} \cdot r^4 = 27r^2 = 243$$

$$r^2 = 9$$

$$r = \pm\sqrt{9} = \pm 3$$

We obtain two different solutions since there are two values for $r$.

**If $r = 3$:**

$$a_1r^2 = 27$$
$$a_1(3)^2 = 27$$
$$a_1 = 3$$

The first sequence is 3, 9, 27, 81, 243, . . .

$$S_5 = \frac{a_1 - ra_5}{1 - r} = \frac{3 - 3 \cdot 243}{1 - 3} = \frac{3(1 - 243)}{-2} = 363$$

**If $r = -3$:**

$$a_1r^2 = 27$$
$$a_1(-3)^2 = 27$$
$$a_1 = 3$$

The second sequence is 3, −9, 27, −81, 243, . . .

$$S_5 = \frac{a_1 - ra_5}{1 - r} = \frac{3 - (-3)(243)}{1 - (-3)} = \frac{3(1 + 243)}{4} = 183$$

The sum in the first case, where all signs are positive, is greater than the second sum, which includes negative terms. ■

We now solve the first applied problem stated in the introduction to this chapter.

**EXAMPLE 5**

**BANKING**

Gloria Bell borrows \$1000.00 at 12% interest compounded annually. If she repays the loan in full at the end of three years, how much does she pay?

At beginning of first year

$$a_1 = 1000$$

At beginning of second year (or end of first year)

$$a_2 = 1000 + 0.12(1000) = [1 + 0.12](1000) = (1.12)(1000)$$

At beginning of third year (or end of second year)

$$a_3 = (1.12)(1000) + (0.12)(1.12)(1000) = [1 + 0.12](1.12)(1000) = (1.12)^2(1000)$$

To obtain the next term of the sequence, multiply the preceding term by 1.12. Thus

$$a_n = a_1(1.12)^{n-1}.$$

To find $n$ we must count the amount borrowed as the first term in the sequence

$$1000,\ 1000(1.12),\ 1000(1.12)^2,\ 1000(1.12)^3.$$

There are four terms ($n = 4$) in the sequence for the three-year period. Thus,

$$a_4 = 1000(1.12)^{4-1} = 1000(1.12)^3 \approx 1404.93.$$

She must pay \$1404.93 at the end of the three years. ■

## Infinite Sequences

If the definition of the sum of a sequence is generalized, we can find the sum of certain infinite geometric sequences. Write the sum formula as follows:

$$S_n = \frac{a_1 - a_1r^n}{1 - r} = \frac{a_1}{1 - r} - \frac{a_1}{1 - r}r^n.$$

Consider the case $r = \frac{1}{2}$.

$$S_n = \frac{a_1}{1 - r} - \frac{a_1}{1 - r}\left(\frac{1}{2}\right)^n$$

As $n$ becomes large, $\left(\frac{1}{2}\right)^n$ becomes small and $\frac{a_1}{1 - r}\left(\frac{1}{2}\right)^n$ approaches zero. Thus, the sum is approximately $\frac{a_1}{1 - r}$. In fact it can be shown that for $|r| < 1$ the sum of all the terms of an infinite geometric sequence is exactly $\frac{a_1}{1 - r}$. There is no sum for an infinite geometric sequence with common ratio $r$ satisfying $|r| \geq 1$.

### Sum of Infinite Geometric Sequence

An infinite geometric sequence, with first term $a_1$ and common ratio $r$ satisfying $|r| < 1$, has a sum given by

$$S = \frac{a_1}{1 - r}.$$

Suppose we consider the infinite sequence

$$1, \frac{1}{2}, \frac{1}{4}, \frac{1}{8}, \frac{1}{16}, \ldots.$$

The corresponding infinite series

$$1 + \frac{1}{2} + \frac{1}{4} + \frac{1}{8} + \frac{1}{16} + \cdots$$

can be evaluated by noting that $a_1 = 1$, $r = 1/2$, and $|r| = 1/2 < 1$ and then using the formula for $S$.

$$S = \frac{a_1}{1 - r} = \frac{1}{1 - \frac{1}{2}} = \frac{1}{\frac{1}{2}} = 2$$

Thus the sum of all the terms of the infinite sequence is exactly 2.

**EXAMPLE 6** Find the first five terms of an infinite geometric sequence with $a_1 = 5/2$ and $S = 5$.

Substitute into the formula.

$$S = \frac{a_1}{1 - r}$$

$$\mathbf{5} = \frac{\frac{\mathbf{5}}{\mathbf{2}}}{1 - r}$$

$$5(1 - r) = \frac{5}{2}$$

$$5 - 5r = \frac{5}{2}$$

$$r = \frac{1}{2}$$

Since $|r| < 1$, there is an infinite sequence with the conditions stated and the first five terms are

$$\frac{5}{2}, \frac{5}{4}, \frac{5}{8}, \frac{5}{16}, \frac{5}{32}. \quad ■$$

CAUTION In Example 6 if $r$ had been such that $|r| \geq 1$, there would have been no sequence satisfying the given properties. ■

**EXAMPLE 7**

Convert $2.3\overline{4}$ to a fraction.

First write $2.3\overline{4}$ as follows:

$$\begin{aligned} 2.3\overline{4} &= 2.3 + 0.04 + 0.004 + 0.0004 + \cdots \\ &= \frac{23}{10} + (0.04 + 0.004 + 0.0004 + \cdots). \end{aligned}$$

The series $(0.04 + 0.004 + 0.0004 + \cdots)$ is an infinite geometric series with $a_1 = 0.04$ and $r = 0.1$.

$$S = \frac{a_1}{1 - r} = \frac{0.04}{1 - 0.1} = \frac{0.04}{0.9} = \frac{4}{90}$$

$$2.3\overline{4} = \frac{23}{10} + \frac{4}{90} = \frac{207}{90} + \frac{4}{90} = \frac{211}{90} \quad ■$$

**EXAMPLE 8**

PHYSICS

A ball rebounds 4/5 as far as it falls. If the ball is dropped from a height of 40 ft, how far does it travel (up and down) before coming to rest?

Use an infinite geometric series to approximate the total distance traveled. Make a sketch like the one in Figure 1 describing the situation.

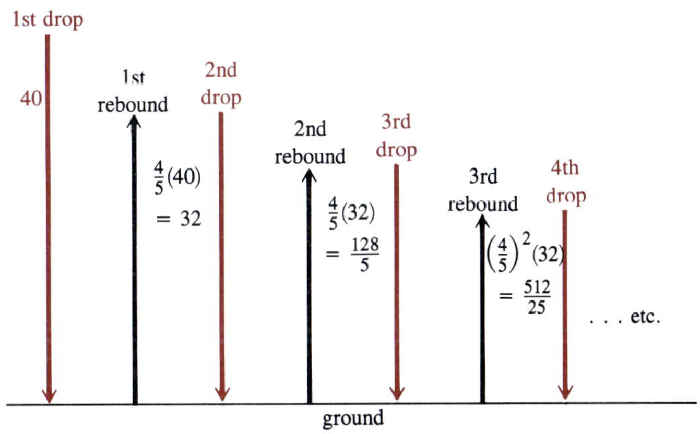


**Figure 1 Rebounding Ball**

We must find the following sum.

$$\begin{aligned} &40 + 32 + 32 + \frac{128}{5} + \frac{128}{5} + \frac{512}{25} + \frac{512}{25} + \cdots \\ &= 40 + 2\left[32 + \frac{128}{5} + \frac{512}{25} + \cdots\right] \end{aligned}$$

Since there are two of each number after the first drop, we simply double the sum of the "ups" from that point on to obtain the sum of the "ups" and "downs." Then the formula is applied only to the series in the brackets,

$$32 + \frac{128}{5} + \frac{512}{25} + \cdots,$$

with $a_1 = 32$ and $r = 4/5$.

$$S = \frac{a_1}{1 - r} = \frac{32}{1 - \frac{4}{5}} = \frac{32}{\frac{1}{5}} = 5 \cdot 32 = 160$$

Then $40 + 2[32 + 128/5 + 512/25 + \cdots] = 40 + 2[160] = 40 + 320 = 360$. The total distance traveled by the ball is 360 ft. ■

## 12.3 EXERCISES

Find the first six terms of each geometric sequence in Exercises 1–6.

**1.** $a_1 = 4$ and $r = 2$

**2.** $a_1 = -\frac{1}{8}$ and $r = -2$

**3.** $a_1 = -16$ and $r = -\frac{1}{2}$

**4.** $a_1 = 64$ and $a_6 = \frac{1}{16}$

**5.** $a_1 = 25$ and $S = 125$

**6.** $a_1 = 4$ and $S = -7$

**7.** Find $x$ so that $x + 7$, $x - 3$, and $x - 8$ form a three-term geometric sequence in the given order. Determine the sequence.

**8.** Find $y$ so that $2y + 5$, $y + 7$, and $3y - 7$ form a three-term geometric sequence in the given order. Determine the sequence.

In Exercises 9–14 find the indicated sum of each geometric sequence. The symbol $\sum_{k=1}^{\infty}$ means an infinite series.

**9.** $\frac{1}{6}, \frac{1}{12}, \frac{1}{24}, \ldots; S_9$

**10.** $a_1 = -\frac{1}{9}$ and $a_6 = -27$; $S_6$

**11.** $a_3 = \frac{1}{6}$ and $a_5 = \frac{1}{24}$; $S_5$

**12.** $a_2 = 2$ and $a_3 = 1$; $S$

**13.** $\sum_{k=1}^{\infty} \left(\frac{3}{4}\right)^k$

**14.** $\sum_{k=1}^{\infty} 3^{-k}$

In Exercises 15–20 some of the numbers $n$, $a_1$, $a_n$, $r$, and $S_n$ are given. Find the missing ones.

**15.** $a_1 = 2$, $n = 6$, $r = 2$

**16.** $a_1 = 1$, $r = -2$, $a_n = 64$

**17.** $r = \frac{1}{2}$, $a_9 = 1$

**18.** $a_1 = 2$, $a_n = 32$, $S_n = 62$

**19.** $a_7 = \frac{1}{5}$, $r = \frac{1}{5}$

**20.** $r = -2$, $S_n = -63$, $a_n = -96$

The terms between $a_1$ and $a_n$ of a geometric sequence are called the **geometric means** of $a_1$ and $a_n$. Use this definition in Exercises 21–22.

**21.** Insert four geometric means between 5 and $-160$.

**22.** Insert four geometric means between $-8$ and $\frac{1}{4}$.

In Exercises 23–30 find the sum of each infinite geometric sequence.

**23.** $\frac{1}{2}, \frac{1}{4}, \frac{1}{8}, \frac{1}{16}, \ldots$

**24.** $3, 1, \frac{1}{3}, \frac{1}{9}, \ldots$

**25.** $\frac{1}{16}, \frac{1}{8}, \frac{1}{4}, \frac{1}{2}, \ldots$

**26.** $\frac{1}{81}, \frac{1}{27}, \frac{1}{9}, \frac{1}{3}, \ldots$

**27.** $-14, 8, -\frac{32}{7}, \frac{128}{49}, \ldots$

**28.** $2, \frac{2}{\sqrt{2}}, 1, \frac{1}{\sqrt{2}}, \frac{1}{2}, \ldots$

**29.** 15, 1.5, 0.15, 0.015, . . .

**30.** 40, 28, 19.6, 13.72, . . .

In Exercises 31–38 convert each decimal to a fraction.

**31.** $0.\overline{3}$

**32.** $0.\overline{8}$

**33.** $0.\overline{21}$

**34.** $0.\overline{36}$

**35.** $0.\overline{123}$

**36.** $2.\overline{6}$

**37.** $2.1\overline{5}$

**38.** $4.2\overline{3}$

**39.** Prove that 0.999. . . = 1.

**40.** Prove that $4.\overline{9} = 5$.

Solve.

**41.** **CONSUMER** A new car costing \$6400 depreciates 20% of its value each year. How much is the car worth at the end of six years?

**42.** **CONSUMER** A new car costing \$5800 depreciates 25% of its value each year. How much is the car worth at the end of five years?

**43.** **BANKING** Lori Wade borrows \$2000.00 at 11% interest compounded annually. If she pays off the loan in full at the end of four years, how much does she pay?

**44.** **BANKING** Rafael Mendez borrows \$1000.00 at 14% interest compounded annually. If Rafael pays off the loan in full at the end of five years, how much does he pay?

**45.** **ECONOMICS** Louis was offered a job for the month of June (thirty days), and was told he would be paid 1¢ at the end of the first day, 2¢ at the end of the second day, 4¢ at the end of the third day, and so forth, doubling each previous day's salary. However, Louis refused the job thinking that the pay was inferior. Would you take the job? Why?

**46.** **ECONOMICS** Eddie Petrowski receives \$5000 on the day of his birth, and 3/5 as much on each birthday as on the previous one. Approximately how much will Eddie receive in his lifetime (assuming he lives a long life)?

**47.** **PHYSICS** The tip of a pendulum sweeps out an arc of 20 cm on its first pass. If each succeeding pass the length traveled is 4/5 of the length of the preceding pass, how far has it traveled by the end of the fourth arc?

**48.** **PHYSICS** The tip of a pendulum moves back and forth so that it sweeps out an arc 12 inches in length, and on each succeeding pass, the length of the arc traveled is 7/8 of the length of the preceding pass. What is the total distance traveled by the tip of the pendulum?

**49.** **PHYSICS** A ball is dropped from a height of 12.0 feet. If on each rebound it rises to a height of 3/4 the distance from which it fell, how far (up and down) will the ball have traveled when it hits the ground for the eighth time?

**50.** **PHYSICS** A ball is dropped from a height of 18.0 ft. If on each rebound it rises to a height of 2/3 the distance from which it fell, how far (up and down) will the ball have traveled when it hits the ground for the sixth time?

51. **PHYSICS** A ball dropped from a height of 15 m always rebounds 3/5 of the height of the previous drop. How far does it travel (up and down) before coming to rest?

52. **PHYSICS** A ping pong ball dropped from a height of 32 ft always rebounds 1/4 of the distance of the previous fall. What distance does it rebound the seventh time?

53. **PHYSICS** A ball dropped from a height of 40 ft always rebounds 1/2 the length of the preceding fall. **(a)** What is the length of the fifth rebound? **(b)** What is the total distance traveled (up and down) when the ball hits the ground for the sixth time? **(c)** What is the total distance traveled (up and down) if the ball is allowed to continue bouncing until it comes to rest? **(d)** How might you approximate the total distance traveled when the ball hits the ground for the hundredth time?

54. **RECREATION** A child on a swing traverses an arc of 22 ft. Each pass thereafter, she traverses an arc that is 5/7 the length of the previous arc. How far does she travel before coming to rest?

55. **DEMOGRAPHY** The population of a town is increasing by 20% each year. If its present population is 1250, what will be its population seven years from now?

56. **GEOMETRY** A square has area 64 in$^2$ (each side is 8 inches). A second square is constructed by connecting in order the midpoints of the sides of the first square, a third by connecting in order the midpoints of the sides of the second square, and so forth. Calculate the sum of the areas of all these squares.

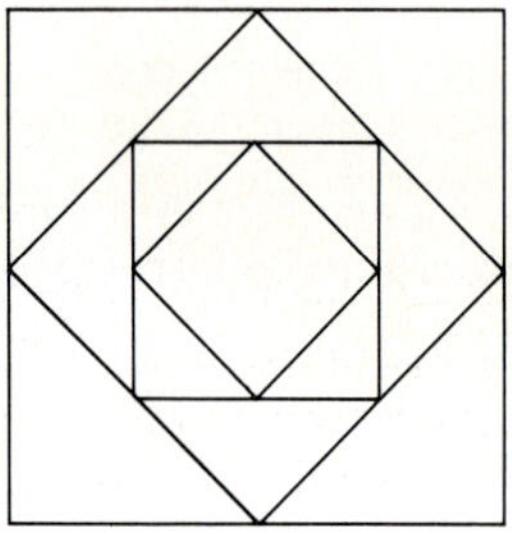

## For Review

In Exercises 57–58 find the first five terms of the arithmetic sequence.

57. $a_1 = 7$ and $d = -4$

58. $a_1 = -12$ and $a_6 = 23$

In Exercises 59–60 find the sum of the first nine terms of the arithmetic sequence.

59. $-\frac{1}{2}, 0, \frac{1}{2}, 1, \frac{3}{2}, \ldots$

60. $a_1 = -18$ and $a_9 = 38$

61. Insert four arithmetic means between 12 and $-13$.

62. Find the sum of the integers divisible by 12 between $-111$ and 25.

63. **RECREATION** A theater has 50 rows with 15 seats in the first row, 17 in the second row, 19 in the third row, and so forth. How many seats are in the theater?

64. **DEMOGRAPHY** The population of a town is increasing by 700 inhabitants each year. If its present population is 1250, what will be its population seven years from now?

## 12.4 Mathematical Induction

Certain statements or formulas which involve positive integers, such as the sum formulas for arithmetic and geometric sequences, can be proved by a method called **mathematical induction.** The process is used when each positive integer gives rise to a particular statement which must be established. For example, we might wish to show that the sum of the first $n$ positive integers is equal to $\frac{1}{2}n(n + 1)$, that is,

$$1 + 2 + 3 + \cdots + n = \frac{1}{2}n(n + 1).$$

We can verify by substitution that the formula holds for the first few positive integers.

| *Substitution* | *Left side* | *Right side* |
|---|---|---|
| $n = 1$ | $1 = 1$ | $\frac{1}{2}(1)(1 + 1) = \frac{1}{2} \cdot 2 = 1$ |
| $n = 2$ | $1 + 2 = 3$ | $\frac{1}{2}(2)(2 + 1) = \frac{1}{2} \cdot 6 = 3$ |
| $n = 3$ | $1 + 2 + 3 = 6$ | $\frac{1}{2}(3)(3 + 1) = \frac{1}{2} \cdot 12 = 6$ |
| $n = 4$ | $1 + 2 + 3 + 4 = 10$ | $\frac{1}{2}(4)(4 + 1) = \frac{1}{2} \cdot 20 = 10$ |
| $n = 5$ | $1 + 2 + 3 + 4 + 5 = 15$ | $\frac{1}{2}(5)(5 + 1) = \frac{1}{2} \cdot 30 = 15$ |

Although we know that the formula is true for the first five positive integers, we have not established it for every possible positive integer. In many situations a statement may be true for the first few positive integers, but not true for all integers. For example, 1, 2, 3, 4, 5, and 6 are all divisors of 60. However, to conclude that all integers are divisors of 60 would be incorrect since, for example, 7 is not such a divisor.

In general, the method of proof by mathematical induction involves three parts.

1. *Verification* of truth for one particular value. Usually we show that the statement is true when $n = 1$.
2. *Induction,* or extension of truth from one particular value to the next. This step involves showing that the statement is true for the next positive integer $(k + 1)$ under the **induction assumption** that it is true for a given positive integer $(k)$. That is, if we assume the statement is true when $n = k$, then we must show it is also true when $n = k + 1$ (the next integer after $k$).
3. *Conclusion* of the truth of the statement for all positive integers. This part essentially combined the results of the first two parts. Having shown that the result is true for a particular value of $n$ (say $n = 1$), by the induction step we know that it is true for the *next* value of $n$ $(n = 2)$. Then again by induction, we know it is now true for the next value of $n$ $(n = 3)$. Since this process can be continued indefinitely, we know that the result is true for any choice of positive integer $n$.

A proof by mathematical induction could be compared to the child's game of tipping over dominoes. The dominoes can be placed so that if one falls over ($n = k$), the next one will also fall over ($n = k + 1$). Thus, if the first one is tipped ($n = 1$), the entire collection of dominoes will fall.

**Principle of Mathematical Induction**

Let $S(n)$ be a statement involving the positive integer $n$. Then $S(n)$ is true for every positive integer provided:

1. $S(n)$ is true when $n = 1$. ($S(1)$ is true.)
2. Under the assumption that $S(n)$ is true when $n = k$, we can show that $S(n)$ is true when $n = k + 1$. (If $S(k)$ is true, then $S(k + 1)$ is true.)

**EXAMPLE 1** Show that $$S(n):\quad 1 + 2 + 3 + \cdots + n = \frac{1}{2}n(n + 1)$$

is true for every positive integer $n$.

*Verification:* Clearly $S(n)$ is true when $n = 1$, since in this case it becomes

$$S(1):\quad 1 = \frac{1}{2} \cdot 1 \cdot (1 + 1) = 1.$$

*Induction:* Assume that the statement is true when $n = k$.

$$S(k):\quad 1 + 2 + 3 + \cdots + k = \frac{1}{2} \cdot k \cdot (k + 1) \qquad \text{Induction assumption}$$

We must show that the statement is true when $n = k + 1$. That is, we must show that

$$S(k + 1):\quad 1 + 2 + 3 + \cdots + k + (k + 1) = \frac{1}{2} \cdot (k + 1)[(k + 1) + 1]$$

is true. The left side of this equation can be rewritten as

$$[1 + 2 + 3 + \cdots + k] + (k + 1).$$

Substitute $(1/2) \cdot k \cdot (k + 1)$ for $[1 + 2 + 3 + \cdots + k]$.

$$\begin{aligned}
[1 + 2 + 3 + \cdots + k] + (k + 1) &= \left[\frac{1}{2} \cdot k \cdot (k + 1)\right] + (k + 1) \\
&= \left(\frac{1}{2} \cdot k + 1\right)(k + 1) && \text{Factor out } k + 1 \\
&= \left(\frac{1}{2}k + \frac{1}{2} \cdot 2\right)(k + 1) && 1 = \frac{1}{2} \cdot 2 \\
&= \frac{1}{2}(k + 2)(k + 1) && \text{Factor out } \frac{1}{2} \\
&= \frac{1}{2}(k + 1)(k + 2) && \text{Commute}
\end{aligned}$$

$$= \frac{1}{2}(k + 1)[(k + 1) + 1]$$

Thus, using the induction assumption, we have shown that

$$S(k + 1): \quad 1 + 2 + 3 + \cdots + k + (k + 1) = \frac{1}{2}(k + 1)[(k + 1) + 1]$$

is true; that is, that $S(n)$ is true when $n = k + 1$.

*Conclusion:* Since $S(n)$ is true when $n = 1$ (by verification), it is true when $n = 2$ (by induction). Then, since $S(n)$ is true when $n = 2$, it is true when $n = 3$ (by induction). Since this process can be extended indefinitely, we know $S(n)$ is true for every positive integer $n$. ■

Usually, in a proof by mathematical induction, the conclusion is shortened somewhat as illustrated in the next example.

**EXAMPLE 2** Show that $\quad S(n): \quad 2 + 4 + 6 + \cdots + 2n = n(n + 1)$

is true for every positive integer $n$.

*Verification:* When $n = 1$, the left side is 2 (there is only one term on the left) and the right side is $1(1 + 1) = 2$. Thus, $S(n)$ is true when $n = 1$.

*Induction:* Assume that $S(n)$ is true when $n = k$. That is, assume that

$$S(k): \quad 2 + 4 + 6 + \cdots + 2k = k(k + 1) \qquad \text{Induction assumption}$$

is true. Then we must show that $S(n)$ is true when $n = k + 1$. That is, we must show that

$$S(k + 1): \quad 2 + 4 + 6 + \cdots + 2(k + 1) = (k + 1)[(k + 1) + 1]$$

is true. But the left side of this equation is really

$$\underbrace{[2 + 4 + 6 + \cdots + 2k]} + 2(k + 1)$$

which becomes $\quad [k(k + 1)] + 2(k + 1)$

when we substitute $k(k + 1)$ for $(2 + 4 + 6 + \cdots + 2k)$. Factoring out $(k + 1)$, we have

$$(k + 2)(k + 1)$$

which is equal to $\quad (k + 1)[(k + 1) + 1]$.

Thus, using the induction assumption, we have shown that

$$2 + 4 + 6 + \cdots + 2(k + 1) = (k + 1)[(k + 1) + 1].$$

*Conclusion:* By the principle of mathematical induction, $S(n)$ is true for every positive integer $n$. ■

**EXAMPLE 3** Show that $S(n)$: $\left(\frac{a}{b}\right)^n = \frac{a^n}{b^n}$

is true for every positive integer $n$.

*Verification:* When $n = 1$,

$$\left(\frac{a}{b}\right)^1 = \frac{a}{b} = \frac{a^1}{b^1}.$$

Thus, $S(n)$ is true when $n = 1$.

*Induction:* Assume that $S(n)$ is true when $n = k$. Then,

$$S(k): \quad \left(\frac{a}{b}\right)^k = \frac{a^k}{b^k}. \quad \text{Induction assumption}$$

We must now show that $S(n)$ is true when $n = k + 1$.

$$\begin{aligned}\left(\frac{a}{b}\right)^{k+1} &= \left(\frac{a}{b}\right)^k\left(\frac{a}{b}\right)\\ &= \frac{a^k}{b^k} \cdot \frac{a}{b} \quad \text{Since } \left(\frac{a}{b}\right)^k = \frac{a^k}{b^k}\\ &= \frac{a^k \cdot a}{b^k \cdot b} = \frac{a^{k+1}}{b^{k+1}}\end{aligned}$$

*Conclusion:* By the principle of mathematical induction, $S(n)$ is true for every positive integer $n$. ■

**EXAMPLE 4** Show that $S(n)$: $n < 3^n$

is true for every positive integer $n$.

*Verification:* Clearly, when $n = 1$, $1 < 3^1 = 3$ so that $S(1)$ is true.

*Induction:* Assume that $S(n)$ is true when $n = k$. That is, assume that

$$S(k): \quad k < 3^k \quad \text{Induction assumption}$$

is true. Then we must show that $S(n)$ is true when $n = k + 1$. That is, we must show that

$$S(k + 1): \quad k + 1 < 3^{k+1}$$

is true. Since $k < 3^k$ we know that

$$\begin{aligned}3 \cdot k &< 3 \cdot 3^k \quad \text{Multiply both sides by 3}\\ 3k &< 3^{k+1}.\end{aligned}$$

But since $k$ is a positive integer,

$$1 \le k$$

so that

$$\begin{aligned}k + 1 &\le k + k \quad \text{Add } k \text{ to both sides}\\ k + 1 &\le 2k.\end{aligned}$$

Also, $$\begin{aligned} 2 &< 3 \\ 2k &< 3k. \quad \text{Multiply by } k \end{aligned}$$

Combining these facts, we have

$$k + 1 \leq 2k \leq 3k < 3^{k+1}.$$

Hence, $S(n)$ is true when $n = k + 1$.

*Conclusion:* By the principle of mathematical induction, $S(n)$ is true for every positive integer $n$. ∎

NOTE In all of our examples the statements have been true for all $n \geq 1$. Some statements are not true for $n = 1$ but are true for $n \geq m$ for some $m > 1$. For example $2^n > 10$ is true for all $n \geq 4$. In cases like this we show $S(m)$ to be true and consider the induction assumption for $k \geq m$. ∎

## 12.4 EXERCISES

In Exercises 1–6 the statements are not true for all $n$. Show that $S(1)$, $S(2)$, and $S(3)$ are true, then find the first positive integer $n$ for which $S(n)$ is false.

**1.** $n < 5$

**2.** $n > n^2 - 100$

**3.** $n$ divides 420

**4.** $n^2 - n + 5$ is prime

**5.** $\dfrac{|n - 10|}{n - 10} = 1$

**6.** $n^3 - 6n^2 + 11n - 6 = 0$

In Exercises 7–12 write out the statements $S(1)$, $S(k)$, and $S(k + 1)$.

**7.** $S(n)$: $1 + 3 + 5 + \cdots + (2n - 1) = n^2$

**8.** $S(n)$: $3 + 6 + 9 + \cdots + 3n = \frac{3}{2}n(n + 1)$

**9.** $S(n)$: $n < 2^n$

**10.** $S(n)$: $n < n + 1$

**11.** $S(n)$: $(ab)^n = a^n b^n$

**12.** $S(n)$: 5 divides $6^n - 1$

In Exercises 13–18 prove that each statement in Exercises 7–12 is true for all positive integers $n$.

**13.** $1 + 3 + 5 + \cdots + (2n - 1) = n^2$

**14.** $3 + 6 + 9 + \cdots + 3n = \frac{3}{2}n(n + 1)$

**15.** $n < 2^n$

**16.** $n < n + 1$

**17.** $(ab)^n = a^n b^n$

**18.** 5 divides $6^n - 1$

In Exercises 19–34 prove that each statement is true for all positive integers $n$.

**19.** $5 + 10 + 15 + \cdots + 5n = \frac{5}{2}n(n + 1)$

**20.** $1 + 5 + 9 + \cdots + (4n - 3) = n(2n - 1)$

**21.** $1^2 + 2^2 + 3^2 + \cdots + n^2 = \frac{n}{6}(n + 1)(2n + 1)$

**22.** $1^3 + 2^3 + 3^3 + \cdots + n^3 = \frac{1}{4}n^2(n + 1)^2$

**23.** $2 \leq 2^n$

**24.** $2n \leq 2^n$

**25.** $2 + 2^2 + 2^3 + \cdots + 2^n = 2^{n+1} - 2$

**26.** $1 + \frac{1}{2} + \frac{1}{2^2} + \cdots + \frac{1}{2^{n-1}} = \frac{2^n - 1}{2^{n-1}}$

**27.** $\left(1 + \frac{1}{1}\right)\left(1 + \frac{1}{2}\right) \ldots \left(1 + \frac{1}{n}\right) = n + 1$

**28.** $\frac{1}{1 \cdot 2} + \frac{1}{2 \cdot 3} + \frac{1}{3 \cdot 4} + \cdots + \frac{1}{n(n+1)} = \frac{n}{n+1}$

**29.** $1 + 2n < 3^n$, $n \geq 2$

**30.** $n + 7 < n^2$, $n \geq 4$

**31.** 2 divides $n^2 + n$

**32.** $x - 1$ divides $x^{2n} - 1$ if $x \neq 1$

**33.** $a_1 + (a_1 + d) + (a_1 + 2d) + \cdots + [a_1 + (n - 1)d] = \frac{n}{2}[2a_1 + (n - 1)d]$

**34.** $a_1 + a_1r + a_1r^2 + \cdots + a_1r^{n-1} = \frac{a_1 - a_1r^n}{1 - r}$

## For Review

In Exercises 35–36 find the first four terms of the geometric sequence.

**35.** $a_1 = -21$ and $r = -\frac{1}{7}$

**36.** $a_1 = 32$ and $a_6 = 1$

In Exercises 37–38 find the sum of the first seven terms of the geometric sequence.

**37.** $-8, 2, -\frac{1}{2}, \frac{1}{8}, \ldots$

**38.** 9, 36, 144, 576, . . .

**39.** Convert $3.6\overline{25}$ to a fraction.

**40.** **PHYSICS** The tip of a pendulum moves back and forth in such a way that it sweeps out an arc 18 inches in length, and on each succeeding pass, the length of the arc traveled is 7/9 of the length of the preceding pass. What is the total distance traveled by the tip of the pendulum?

# 12.5 Permutations and Combinations

## Counting Techniques

The techniques for counting the number of ways that a collection of objects or acts can be arranged, ordered, combined, chosen, or occur in succession come under the heading of **combinatorial algebra.** As an example consider a student who wishes to take three courses. She must choose one math course from algebra (A) or trigonometry (T), one English course from technical writing (W), literature (L), or rhetoric (R), and one science course from chemistry (C), geology (G), or physics (P). The **tree diagram** in Figure 2 will help count the number of ways the student can select the three courses.

Each branch of the tree represents one possible selection of courses. For example A, L, G means algebra, literature, and geology were chosen. Notice that 18 choices can be made. But $18 = 2 \cdot 3 \cdot 3$ is the product of the number of math courses, the number of English courses, and the number of science courses. This example leads to the **fundamental counting principle.**

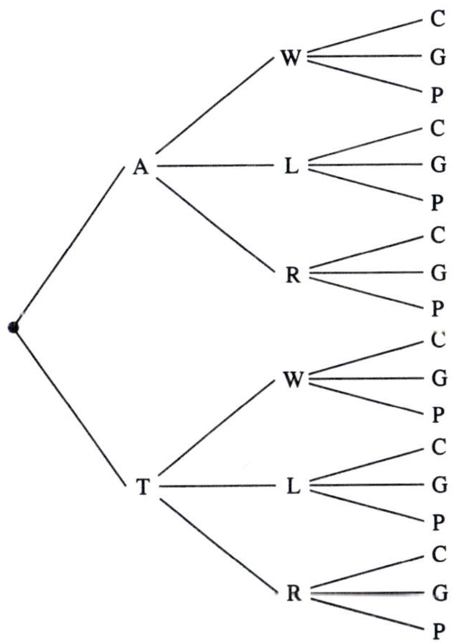


**Figure 2 Counting**

### Fundamental Counting Principle

If event $E_1$ can occur in $m_1$ ways, $E_2$ in $m_2$ ways, $E_3$ in $m_3$ ways, etc., then the total number of ways $E_1, E_2, \ldots, E_k$ can occur is

$$m_1 m_2 \ldots m_k.$$

**EXAMPLE 1** Consider making three-digit numbers from the digits 1, 2, 3, 4, 5, 6, and 7.

**(a)** How many ways can this be done if repetition of digits is allowed?

Since there are three digits to select, fill three blanks with the number of ways each digit can be selected. Since digits may be repeated, each of the three blanks will contain a 7.

$$\underline{\quad 7 \quad} \quad \underline{\quad 7 \quad} \quad \underline{\quad 7 \quad}$$

Thus, there are $7 \cdot 7 \cdot 7 = 343$ different three-digit numbers possible when repetition is allowed.

**(b)** How many ways are there when repetition is not allowed?

There are seven choices for the first slot, only six choices for the second (without repetition, the digit used first is not available now), and only five choices remain for the third slot.

$$\underline{\quad 7 \quad} \quad \underline{\quad 6 \quad} \quad \underline{\quad 5 \quad}$$

Thus, there are $7 \cdot 6 \cdot 5 = 210$ possible different three-digit numbers when repetition is not allowed. ■

**EXAMPLE 2** License plates are to be made by using three letters followed by a three-digit number. How many such license plates are possible?

Since there are twenty-six letters and ten numerals (0, 1, 2, . . . , 9) and repetitions are allowed, the number of such plates can be derived by filling slots

$$\underline{\ 26\ }\quad \underline{\ 26\ }\quad \underline{\ 26\ }\quad \underline{\ 10\ }\quad \underline{\ 10\ }\quad \underline{\ 10\ }$$

to obtain $26^3 \cdot 10^3 = 17{,}576{,}000$ license plates. ■

We now concentrate on the problem of choosing a collection of objects from a given set of objects.

**Permutation**

A **permutation** is any arrangement of a collection of objects in a particular order.

Consider the permutations (arrangements) of the letters A, B, and C.

ABC, ACB, BAC, BCA, CAB, CBA

There are six permutations of three objects. This number can be determined by filling three slots. There are 3 choices for the first position, 2 for the second, and 1 for the third.

$$\underline{\ 3\ } \cdot \underline{\ 2\ } \cdot \underline{\ 1\ } = 6$$

The number of permutations of 5 distinct objects is

$$\underline{\ 5\ } \cdot \underline{\ 4\ } \cdot \underline{\ 3\ } \cdot \underline{\ 2\ } \cdot \underline{\ 1\ } = 120.$$

The number of permutations of 8 distinct objects is

$$\underline{\ 8\ } \cdot \underline{\ 7\ } \cdot \underline{\ 6\ } \cdot \underline{\ 5\ } \cdot \underline{\ 4\ } \cdot \underline{\ 3\ } \cdot \underline{\ 2\ } \cdot \underline{\ 1\ } = 40{,}320.$$

In general, the number of permutations of $n$ distinct objects is

$$n(n-1)(n-2)(n-3)\cdots 2 \cdot 1.$$

**Factorial Notation**

Since products of successive integers from a given integer down to 1 occur so often in our work with permutations, we adopt the **factorial notation**

$$\boldsymbol{n! = n(n-1)(n-2)(n-3)\cdots 2 \cdot 1} \quad \text{(read ``} n \text{ factorial'')}.$$

Thus, $3! = 3 \cdot 2 \cdot 1 = 6$, $5! = 5 \cdot 4 \cdot 3 \cdot 2 \cdot 1 = 120$, and $8! = 8 \cdot 7 \cdot 6 \cdot 5 \cdot 4 \cdot 3 \cdot 2 \cdot 1 = 40{,}320$. Notice that $1! = 1$.

The phrase "the number of permutations of $n$ objects using all $n$ of the objects" or briefly "the number of permutations of $n$ objects" occurs repeatedly, so we use $_nP_n$ to symbolize this phrase.

### Permutations of $n$ Objects

The number of permutations of $n$ objects, ${}_nP_n$, is given by

$${}_nP_n = n! = n(n-1)(n-2)(n-3)\cdots 2\cdot 1.$$

**Formula for ${}_nP_r$**

Given a collection of objects, we may wish to consider all arrangements of subcollections of a particular size. For example, consider the permutations of the letters A, B, C, D, and E, taken two at a time.

| | | | | |
|---|---|---|---|---|
| AB | BA | CA | DA | EA |
| AC | BC | CB | DB | EB |
| AD | BD | CD | DC | EC |
| AE | BE | CE | DE | ED |

In the array, there are twenty permutations of five distinct objects using two of them at a time. This number could have been obtained using the slot-filling technique since there are 5 choices for the first position and 4 for the second.

$$\underline{\quad 5 \quad}\cdot\underline{\quad 4 \quad} = 20$$

In general, we are concerned with finding "the number of permutations of $n$ objects using $r$ of the objects at a time." This phrase is symbolized by ${}_nP_r$. Consider filling $r$ slots. There are $n$ choices for the first slot, $n-1$ for the second, $n-2$ for the third, and $n-(r-1)$ for the $r$th slot

$$\underset{1}{\underline{\quad n \quad}}\ \ \underset{2}{\underline{\quad n-1 \quad}}\ \ \underset{3}{\underline{\quad n-2 \quad}}\ \ \underset{4}{\underline{\quad n-3 \quad}}\ \cdots\ \underset{r}{\underline{\quad n-(r-1) \quad}}.$$

Thus $${}_nP_r = n(n-1)(n-2)(n-3)\cdots[n-(r-1)].$$

Suppose we multiply on the right by $(n-r)!/(n-r)!$, which is 1.

$$\begin{aligned}{}_nP_r &= n(n-1)(n-2)(n-3)\cdots[n-(r-1)]\frac{(n-r)!}{(n-r)!}\\ &= \frac{n(n-1)(n-2)(n-3)\cdots[n-(r-1)](n-r)!}{(n-r)!}\\ &= \frac{n!}{(n-r)!}\end{aligned}$$

This proves the following theorem.

### Permutations of $n$ Objects Using $r$

The number of permutations of $n$ objects using $r$ of the objects at a time ($r < n$) is given by

$${}_nP_r = \frac{n!}{(n-r)!}.$$

In the previous example, $n = 5$ and $r = 2$ so, using the formula,

$$ {}_5P_2 = \frac{5!}{(5-2)!} = \frac{5!}{3!} = \frac{5 \cdot 4 \cdot \cancel{3} \cdot \cancel{2} \cdot \cancel{1}}{\cancel{3} \cdot \cancel{2} \cdot \cancel{1}} = 5 \cdot 4 = 20. $$

We do not evaluate the factorials in the numerator and denominator, we just indicate the products and cancel common factors. Since $5! = 5 \cdot 4 \cdot 3! = 5 \cdot 4 \cdot 3 \cdot 2 \cdot 1$, we could have abbreviated further.

$$ {}_5P_2 = \frac{5!}{(5-2)!} = \frac{5!}{3!} = \frac{5 \cdot 4 \cdot \cancel{3!}}{\cancel{3!}} = 5 \cdot 4 = 20 $$

In addition, if we define $0! = 1$, we can use the formula

$$ {}_nP_r = \frac{n!}{(n-r)!} $$

in the case when $n = r$ [$n - r = 0$ so $(n - r)! = 0! = 1$], and the result is the formula for ${}_nP_n$.

**EXAMPLE 3**

**(a)** $ {}_5P_5 = \frac{5!}{(5-5)!} = \frac{5!}{0!} = \frac{5!}{1} = 5! = 5 \cdot 4 \cdot 3 \cdot 2 \cdot 1 = 120 $

**(b)** $ {}_{10}P_4 = \frac{10!}{(10-4)!} = \frac{10!}{6!} = \frac{10 \cdot 9 \cdot 8 \cdot 7 \cdot \cancel{6!}}{\cancel{6!}} = 10 \cdot 9 \cdot 8 \cdot 7 = 5040 $

**(c)** $ {}_5P_0 = \frac{5!}{(5-0)!} = \frac{5!}{5!} = 1 $ ■

**EXAMPLE 4**

How many ways can a president, vice president, and secretary be chosen for a club with ten members?

Since order is important in the selection, this is a permutation problem.

$$ {}_{10}P_3 = \frac{10!}{(10-3)!} = \frac{10 \cdot 9 \cdot 8 \cdot \cancel{7!}}{\cancel{7!}} = 720 $$ ■

## Distinguishable Permutations

Consider the number of arrangements of the seven letters in the word ARIZONA. Since two letters are the same, there are not 7! different arrangements. Each possible arrangement has an identical twin formed by interchanging the two A's. Since the two A's can be arranged 2! ways, divide 7! by 2! to obtain the number of distinguishable arrangements. This illustrates the following theorem.

### Distinguishable Permutations

The number of distinguishable permutations of $n$ objects of which $n_1$ are the same, $n_2$ are the same, $n_3$ are the same, and so forth, is

$$ \frac{n!}{n_1!n_2!n_3!\cdots}. $$

**EXAMPLE 5** Determine the number of distinguishable permutations of the letters in MISSISSIPPI.

In this case there are eleven letters: one M, four I's, four S's, and two P's. That is, $n_1 = 1$, $n_2 = 4$, $n_3 = 4$, and $n_4 = 2$. The number of distinguishable permutations is

$$\frac{n!}{n_1!n_2!n_3!n_4!} = \frac{11!}{1!4!4!2!} = \frac{11 \cdot 10 \cdot 9 \cdot 8 \cdot 7 \cdot 6 \cdot 5 \cdot \cancel{4!}}{1 \cdot 4 \cdot 3 \cdot 2 \cdot 1 \cdot \cancel{4!} \cdot 1 \cdot 2}$$

$$= \frac{11 \cdot 10 \cdot 9 \cdot \cancel{8} \cdot 7 \cdot \cancel{6} \cdot 5}{\cancel{4} \cdot \cancel{3} \cdot \cancel{2} \cdot \cancel{2}} = 34{,}650.$$ ■

## Circular Permutations

We have been arranging objects in a straight line, and have seen that four objects can be arranged in a straight line in ${}_4P_4 = 4! = 24$ different ways. Suppose we arranged these same four objects in a symmetric circular pattern. For example, let us arrange A, B, C, and D around a circle. One such arrangement is shown in Figure 3, and three others in Figure 4.

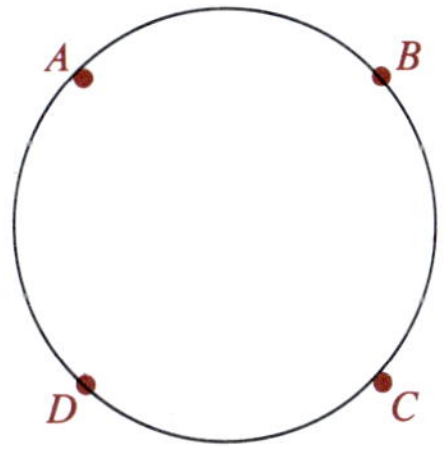


**Figure 3 Circular Permutations**

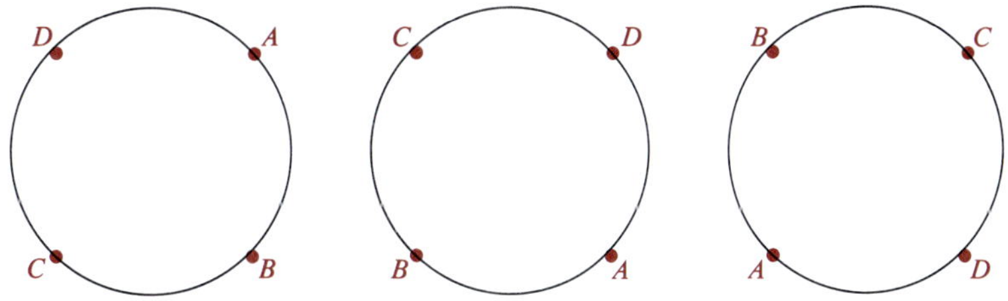


**Figure 4 Circular Permutations**

Are the arrangements in Figure 4 different or distinguishable from the one in Figure 3? If we ignore the positions of the letters and consider only their relationship to each other, we see that these four arrangements are the same. That is, we can begin at A and move clockwise around any of the circles to get the same arrangement, A B C D and then back to A again. Thus, the four different arrangements ABCD, BCDA, CDAB, and DABC are not distinguishable in a circular arrangement. In general, if there are $x$ distinct circular arrangements of four objects, there would be $4 \cdot x$ arrangements of these objects along a straight line. But since the number of arrangements along a straight line is 4!, we have $4 \cdot x = 4!$ or

$$x = \frac{4!}{4} = \frac{\cancel{4} \cdot 3!}{\cancel{4}} = 3!$$

This is a special case of the following theorem.

### Circular Permutations

The number of distinguishble **circular permutations** (arrangements evenly around a circle) of $n$ objects is $(n-1)!$.

**EXAMPLE 6**

In how many ways can six people be seated around a circular table?

In this case $n = 6$, so that

$$(n-1)! = (6-1)! = 5! = 120$$

is the number of possible circular permutations of the six people. ■

### Combination

A **combination** is any collection of objects. Order of the selection is not important.

**Formula for $_nC_r$**

Consider the $_4P_3 = 24$ permutations of the letters A, B, C, and D, using three of them at a time.

| | | | |
|---|---|---|---|
| ABC | ABD | ACD | BCD |
| ACB | ADB | ADC | BDC |
| BAC | BAD | CAD | CBD |
| BCA | BDA | CDA | CDB |
| CAB | DAB | DAC | DBC |
| CBA | DBA | DCA | DCB |

Notice that the six permutations in the first column all involve the same three letters A, B, and C, so they are counted as a single combination (order is not counted for a combination). Similarly, the six permutations listed in the second, third, and fourth columns give rise to the single combinations

A, B, D, A, C, D B, C, D.

As a result, there are only four combinations of the four letters A, B, C, and D using three of them at a time. In general, there are fewer combinations of a group of objects than there are permutations of the group.

We use $_nC_r$ to represent the number of combinations of $n$ objects using $r$ of them at a time. In the preceding example, we found that $_4C_3 = 4$. Since each combination of three objects can be resolved into $6 = 3!$ permutations of the three objects, we have that

$$_4P_3 = (3!) \cdot {_4C_3} = 3! \cdot 4 = 6 \cdot 4 = 24$$

or
$$_4C_3 = \frac{_4P_3}{3!} = \frac{\frac{4!}{(4-3)!}}{3!} = \frac{4!}{(4-3)!3!}$$

which is a special case of the next theorem.

**Combinations of *n* Objects Using *r* of Them**

The number of combinations of $n$ objects using $r$ of the objects at a time ($r \leq n$) is given by

$$_nC_r = \frac{n!}{(n-r)!r!}.$$

**EXAMPLE 7**

**(a)** $_5C_3 = \frac{5!}{(5-3)!3!} = \frac{5!}{2!3!} = \frac{5 \cdot \overset{2}{\cancel{4}} \cdot \cancel{3!}}{1 \cdot \cancel{2} \cdot \cancel{3!}} = 10$

**(b)** $_5C_5 = \frac{5!}{(5-5)!5!} = \frac{\cancel{5!}}{0!\cancel{5!}} = \frac{1}{1} = 1$

**(c)** $_5C_0 = \frac{5!}{(5-0)!0!} = \frac{\cancel{5!}}{\cancel{5!}0!} = \frac{1}{1} = 1$ ■

Parts **(b)** and **(c)** of Example 7 do make sense. If we have five objects, there is one way to choose all five of them (simply take them all), and one way to choose none of them (simply do not take them). In fact, there is nothing special about five in these examples;

$$_nC_n = 1 \qquad \text{and} \qquad _nC_0 = 1$$

for any natural number $n$.

**EXAMPLE 8**

How many five-card poker hands consisting of three kings and two cards that are not kings are possible in a fifty-two card deck?

There are

$$_4C_3 = \frac{4!}{1!3!} = 4$$

ways to get three of the four kings and

$$_{48}C_2 = \frac{48!}{46!2!} = \frac{\overset{24}{\cancel{48}} \cdot 47 \cdot \cancel{46!}}{\cancel{46!} \cdot \cancel{2}} = 1128$$

ways to have two of the forty-eight cards that are not kings. Thus, in total, there are

$$_4C_3 \cdot {}_{48}C_2 = 4 \cdot 1128 = 4512$$

ways to have the prescribed hand. ■

## 12.5 EXERCISES

Evaluate ${}_nP_r$ and ${}_nC_r$ in Exercises 1–8.

**1.** ${}_7P_3$ **2.** ${}_6P_6$ **3.** ${}_{10}P_2$ **4.** ${}_8P_0$

**5.** ${}_9C_0$ **6.** ${}_{10}C_3$ **7.** ${}_{10}C_7$ **8.** ${}_{20}C_{19}$

Solve.

**9.** Suppose there are five roads connecting town A to town B and three roads connecting town B to town C. In how many ways can a person travel from A to C via B?

**10.** How many three-digit symbols can be formed from the digits 0, 1, 2, 3, 4, 5, 6, 7, 8, 9, if repetition of digits is allowed? If repetition of digits is not allowed?

**11.** **MANUFACTURING** License plates are to be made using two letters followed by four digits. How many plates can be made if repetition of both letters and digits is allowed? How many plates can be made if repetition of letters is possible, but repetition of digits is not allowed? How many plates can be made if neither letters nor digits can be repeated?

**12.** **BUSINESS** The Spear Shirt Company has four basic designs for shirts, in either short or long sleeves, with eight different color patterns. How many different shirts must be made to insure that one of each possible kind is made?

**13.** **MILITARY** A signal is made by placing three flags, one above the other, on a flagpole. If there are eight different flags available, how many possible signals can be flown?

**14.** **EDUCATION** In how many ways can three freshmen and four sophomores line up at the registrar's office if no other restrictions are imposed? If the four sophomores insist on standing first in line? If the sophomores and freshmen must alternate in line? If a sophomore must stand first and no other restrictions are imposed? If the sophomores insist on standing together and the freshmen insist on standing together?

**15.** In how many ways can the letters in the word SHELF be arranged using all five letters?

**16.** In how many ways can the letters in the word SHELF be arranged using four of the letters?

**17.** **LAW ENFORCEMENT** In how many ways can a police department arrange eight suspects in a lineup?

**18.** **MUSIC** A musician plans to perform nine selections. In how many ways can she arrange the program?

**19.** **SPORTS** In how many ways can the manager of a baseball team arrange his batting order under the following conditions? **(a)** Any player can bat in any position. [Hint: There are nine players on a baseball team.] **(b)** The pitcher must bat last. **(c)** The pitcher must bat last, and the centerfielder must bat in the clean-up position (fourth). **(d)** The pitcher must bat last, and the three outfielders must bat in the first three positions.

**20.** A penny, a nickel, a dime, and a quarter are to be arranged in a straight line. **(a)** How many ways can this be done? **(b)** If we distinguish between heads and tails, in how many ways can this be done?

**21.** How many distinguishable permutations of the letters in the given word are possible? **(a)** SHEEP **(b)** LETTER **(c)** TENNESSEE **(d)** AARDVARK

**22.** In how many distinct ways can $x^4y^3z^5$ be expressed without exponents?

**23.** MILITARY How many distinct signals can be made on a vertical flagpole with fourteen flags if five of them are white, four are red, three are green, and two are black?

**24.** RECREATION If Judith Crist, one of the leading movie critics, has viewed 500 movies this year, in how many ways could she list, in order of preference, the ten best movies she has seen? [Hint: You need not evaluate your answer.]

**25.** In how many ways can eight people be arranged around a circular table?

**26.** In how many ways could the twelve knights of the round table be seated at King Arthur's round table?

**27.** MILITARY In how many ways can a detail of 5 soldiers be chosen from a group of 14?

**28.** EDUCATION On a test, a student must answer 8 out of 10 questions. In how many ways can she choose the 8 questions?

**29.** MILITARY In how many ways can a detail of 5 soldiers be chosen from a group of 14 and be presented with 5 distinct awards?

**30.** How many ways can 8 people be chosen and arranged in a straight line if there are 10 people to choose from?

**31.** EDUCATION In how many ways can a committee of 5 be chosen from 8 seniors and 10 juniors under the following conditions? **(a)** The committee must consist of exactly 4 seniors. **(b)** The committee must consist of at least 4 seniors. [Hint: Find the number of ways of having 4 seniors and 1 junior and add to this the number of ways of having 5 seniors.] **(c)** The committee must consist of exactly 3 juniors. **(d)** The committee must consist of at least 3 juniors.

**32.** SPORTS A group of 16 students are to play a tennis tournament with 4 four-man teams. **(a)** In how many ways can the first team be formed? **(b)** In how many ways can the second team be formed once the first is formed? **(c)** In how many ways can the third team be formed once the first two are formed? **(d)** In how many ways can the final team be made once the first three are formed?

**33.** GEOMETRY Points on the same straight line are **collinear.** How many lines are determined by seven points, no three of which are collinear? [Hint: How many points determine a line?]

**34.** GEOMETRY Three noncollinear points determine a circle. How many circles are determined by 5 such points?

**35.** EDUCATION A student must answer 7 of the first 10 questions on a test. On the next 5 questions, he must answer 4. In how many ways can this be done?

**36.** How many different sums of money can be obtained using any 3 of 5 coins consisting of a penny, a nickel, a dime, a quarter, and a half-dollar?

**37.** RECREATION How many 5-card poker hands dealt from a 52-card deck are possible if they must consist of the following cards? **(a)** 5 hearts **(b)** 4 aces and a seven **(c)** 3 aces and two cards which are not aces **(d)** 3 hearts and 2 clubs **(e)** 3 face cards and 2 sevens.

**38.** RECREATION How many 5-card poker hands dealt from a 52-card deck are possible if they must consist of the following cards? **(a)** 5 black cards **(b)** 4 aces and a joker **(c)** An ace and 4 cards which are not aces **(d)** An ace, a queen, a king, and two cards which are not any of these three **(e)** 2 face cards, 2 aces, and an eight.

**39.** CONSUMER Fred wishes to purchase a new car. He has narrowed the choices to one of 3 models, one of 4 colors, and one of 5 upholstery colors. How many possible combinations are still in the running?

**40.** EDUCATION A professor plans to give three A's, six B's, ten C's, five D's, and two F's to his class of twenty-six students. In how many ways can he do this?

**41.** Give an example to show that $(n!)(m!) \neq (n \cdot m)!$

**42.** Give an example to show that $n! + m! \neq (n + m)!$

**43.** Prove that ${}_nC_r = {}_nC_{n-r}$.

**44.** Prove that ${}_nP_0 = 1$, ${}_nC_0 = 1$, and ${}_nC_n = 1$.

## For Review

In Exercises 45–46 use mathematical induction to prove the statement.

**45.** $3 + 3^2 + 3^3 + \cdots + 3^n = \frac{3}{2}(3^n - 1)$

**46.** 6 divides $7^n - 1$

# 12.6 The Binomial Theorem

Another application of combinations is in the expansion of $(a + b)^n$ where $a + b$ is any binomial and $n$ is a nonnegative integer. First let us expand several such expressions, and search for a pattern.

$$\begin{aligned}
(a + b)^0 &= 1 \\
(a + b)^1 &= a + b \\
(a + b)^2 &= a^2 + 2ab + b^2 \\
(a + b)^3 &= a^3 + 3a^2b + 3ab^2 + b^3 \\
(a + b)^4 &= a^4 + 4a^3b + 6a^2b^2 + 4ab^3 + b^4 \\
(a + b)^5 &= a^5 + 5a^4b + 10a^3b^2 + 10a^2b^3 + 5ab^4 + b^5 \\
(a + b)^6 &= a^6 + 6a^5b + 15a^4b^2 + 20a^3b^3 + 15a^2b^4 + 6ab^5 + b^6
\end{aligned}$$

In each case we observe the following:

1. There are always $n + 1$ terms in the expansion.
2. The exponents on $a$ start with $n$ and decrease to 0.
3. The exponents on $b$ start with 0 and increase to $n$.
4. The sum of the exponents in each term is always $n$.

## Pascal's Triangle

To discover the pattern of the numerical coefficients of each term, write the coefficients in the same arrangement as in the preceding expansions.

| | |
|---|---|
| Row 0 | 1 |
| Row 1 | 1 1 |
| Row 2 | 1 2 1 |
| Row 3 | 1 3 3 1 |
| Row 4 | 1 4 6 4 1 |
| Row 5 | 1 5 10 10 5 1 |
| Row 6 | 1 6 15 20 15 6 1 |

This triangular array is known as **Pascal's triangle.** The row number corresponds to the exponent $n$ in the expansion of $(a + b)^n$. The numbers in any row, other than the first and the last which are always 1, can be determined by adding the two numbers immediately above and to the left and right of it. For example, as indicated, 15 is $5 + 10$. Thus, Pascal's triangle gives us one way to determine the coefficients in the expansion of a given binomial.

## EXAMPLE 1

Expand $(x + 2y)^5$.

In this example, $n = 5$, $a = x$, and $b = 2y$. Row 5 of Pascal's triangle has the following numbers for the coefficients.

**1 5 10 10 5 1**

We thus substitute $a = x$ and $b = 2y$ into

$$a^5 + 5a^4b + 10a^3b^2 + 10a^2b^3 + 5ab^4 + b^5$$

to obtain

$$\begin{aligned} & x^5 + 5x^4(2y) + 10x^3(2y)^2 + 10x^2(2y)^3 + 5x(2y)^4 + (2y)^5 \\ &= x^5 + 5x^4(2y) + 10x^3(4y^2) + 10x^2(8y^3) + 5x(16y^4) + 32y^5 \\ &= x^5 + 10x^4y + 40x^3y^2 + 80x^2y^3 + 80xy^4 + 32y^5. \quad \blacksquare \end{aligned}$$

CAUTION Make sure that you evaluate $b^3 = (2y)^3$ as $2^3y^3 = 8y^3$ and *not* $2y^3$. This is perhaps the most common mistake made with this expansion technique.

## EXAMPLE 2

Expand $(2y - 3)^4$.

Express $(2y - 3)^4$ as $(2y + (-3))^4$; then $n = 4$, $a = 2y$, and $b = -3$. In Row 4 of Pascal's triangle, we find the following coefficients.

**1 4 6 4 1**

Substituting $a = 2y$ and $b = -3$ into

$$a^4 + 4a^3b + 6a^2b^2 + 4ab^3 + b^4,$$

we have

$$\begin{aligned} & (2y)^4 + 4(2y)^3(-3) + 6(2y)^2(-3)^2 + 4(2y)(-3)^3 + (-3)^4 \\ &= 16y^4 + 4(8y^3)(-3) + 6(4y^2)(9) + 4(2y)(-27) + 81 \\ &= 16y^4 - 96y^3 + 216y^2 - 216y + 81. \quad \blacksquare \end{aligned}$$

When $n$ is large, or when we are interested in determining only one particular term in a binomial expansion, it is helpful if we make one further observation relative to the binomial coefficients in Pascal's triangle. If is easy to verify that Pascal's triangle is in fact an array of combinations.

| | |
|---|---|
| Row 0 | $1$ |
| Row 1 | ${}_1C_0 \quad {}_1C_1$ |
| Row 2 | ${}_2C_0 \quad {}_2C_1 \quad {}_2C_2$ |
| Row 3 | ${}_3C_0 \quad {}_3C_1 \quad {}_3C_2 \quad {}_3C_3$ |
| Row 4 | ${}_4C_0 \quad {}_4C_1 \quad {}_4C_2 \quad {}_4C_3 \quad {}_4C_4$ |
| Row 5 | ${}_5C_0 \quad {}_5C_1 \quad {}_5C_2 \quad {}_5C_3 \quad {}_5C_4 \quad {}_5C_5$ |
| Row 6 | ${}_6C_0 \quad {}_6C_1 \quad {}_6C_2 \quad {}_6C_3 \quad {}_6C_4 \quad {}_6C_5 \quad {}_6C_6$ |

For example,

$$ {}_6C_1 = \frac{6!}{(6-1)!1!} = \frac{6!}{5!1!} = 6, \qquad {}_5C_3 = \frac{5!}{2!3!} = 10, \qquad {}_4C_2 = \frac{4!}{2!2!} = 6. $$

As a result, the entries in any row $n$ ($n \geq 1$) can be found by evaluating ${}_nC_r$ where $r$ has values $0, 1, 2, \ldots, n$.

## Binomial Theorem

We now state the binomial theorem using a new notation for ${}_nC_r$ which is commonly used in this context.

$$ \binom{n}{r} = {}_nC_r $$

Thus $\binom{5}{3} = {}_5C_3 = \frac{5!}{2!3!} = 10$ and $\binom{7}{0} = {}_7C_0 = \frac{7!}{0!7!} = 1$.

### Binomial Theorem

For any binomial $a + b$ and any positive integer $n$,

$$ (a+b)^n = \binom{n}{0}a^n + \binom{n}{1}a^{n-1}b^1 + \binom{n}{2}a^{n-2}b^2 + \binom{n}{3}a^{n-3}b^3 + \cdots + \binom{n}{n}b^n. $$

Using the sigma summation notation,

$$ (a+b)^n = \sum_{r=0}^{n} \binom{n}{r} a^{n-r}b^r. $$

We call $\binom{n}{r}$ the **binomial coefficient** of the $(r+1)$th term of the expansion.

$$ (r+1)\text{th term} = \binom{n}{r} a^{n-r}b^r $$

CAUTION Notice that since $r$ varies from 0 to $n$, a particular term is called the $(r + 1)$th term. For example, when $r = 0$ we have the $(0 + 1)$th or 1st term. The 3rd term is the $(2 + 1)$th term which is the $(r + 1)$th term for $r = 2$. Thus, if you are asked for a particular term, identify $n$, $a$, and $b$, and simply remember that *$r$ is 1 less than the number of the requested term.*

**Proof of Binomial Theorem**

The binomial theorem is proved using mathematical induction. Remember that we must show that $S(1)$ is true, and then prove that if $S(k)$ is true then $S(k + 1)$ is true.

First show that $S(1)$ is true, that is, the binomial theorem holds for $n = 1$.

$$\sum_{r=0}^{1} \binom{1}{n} a^{1-r}b^r = \binom{1}{0} a^{1-0}b^0 + \binom{1}{1} a^{1-1}b^1 = a + b = (a + b)^1$$

Thus, the theorem is true for $n = 1$. That is,

$$(a + b)^1 = \sum_{r=0}^{1} \binom{1}{r} a^{1-r}b^r.$$

Now assume that $(a + b)^k = \sum_{r=0}^{k} \binom{k}{r} a^{k-r}b^r$, and prove that $(a + b)^{k+1} = \sum_{r=0}^{k+1} \binom{k+1}{r} a^{k+1-r}b^r$. To do this start with the assumed equality and multiply both sides by $a + b$.

$$(a + b)^k = \sum_{r=0}^{k} \binom{k}{r} a^{k-r}b^r$$

$$\begin{aligned}(a + b)^k(a + b) &= \left[\sum_{r=0}^{k} \binom{k}{r} a^{k-r}b^r\right](a + b)\\ &= \sum_{r=0}^{k} \binom{k}{r} a^{k-r}b^r a + \sum_{r=0}^{k} \binom{k}{r} a^{k-r}b^r b\\ &= \sum_{r=0}^{k} \binom{k}{r} a^{k+1-r}b^r + \sum_{r=0}^{k} \binom{k}{r} a^{k-r}b^{r+1}.\end{aligned}$$

When these sums are expanded and like terms are collected the following series results.

$$\begin{aligned}(a + b)^{k+1} = \binom{k}{0} a^{k+1} &+ \left[\binom{k}{0} + \binom{k}{1}\right] a^k b + \left[\binom{k}{1} + \binom{k}{2}\right] a^{k-1}b^2 + \cdots\\ &+ \left[\binom{k}{k-1} + \binom{k}{k}\right] ab^k + \binom{k}{k} b^{k+1}.\end{aligned}$$

In the exercises you are asked to prove that

$$\binom{k}{0} = \binom{k+1}{0}, \quad \binom{k}{k} = \binom{k+1}{k+1}, \quad \text{and} \quad \binom{k}{r-1} + \binom{k}{r} = \binom{k+1}{r}.$$

Using these equations we have

$$(a+b)^{k+1} = \binom{k+1}{0}a^{k+1} + \binom{k+1}{1}a^k b + \binom{k+1}{2}a^{k-1}b^2 + \cdots + \binom{k+1}{k}ab^k + \binom{k+1}{k+1}b^k$$

$$= \sum_{r=0}^{k+1}\binom{k+1}{r}a^{k+1-r}b^r.$$

Thus, by mathematical induction the binomial theorem is established for all positive integers $n$.

## EXAMPLE 3

**(a)** Find the 5th term in the expansion of $(x+2y)^7$.

In this example $n = 7$, $a = x$, $b = 2y$, and since we want the 5th term, $r$ will be 4.

$$\binom{n}{r}a^{n-r}b^r = \binom{7}{4}(x)^{7-4}(2y)^4$$

$$= \frac{7!}{3!4!}x^3(16y^4)$$

$$= 35x^3(16y^4) = 560x^3y^4$$

**(b)** Find the 7th term in the expansion of $(a^2 - 3y)^{10}$.

We have $n = 10$, $a = a^2$, $b = (-3y)$ [be careful to write $(a^2 - 3y)^{10}$ as $(a^2 + (-3y))^{10}$], and $r = 6$ (1 less than 7).

$$\binom{n}{r}a^{n-r}b^r = \binom{10}{6}(a^2)^{10-6}(-3y)^6$$

$$= \frac{10!}{4!6!}(a^2)^4(729y^6)$$

$$= 210a^8(729y^6) = 153{,}090a^8y^6$$ ■

## 12.6 EXERCISES

In Exercises 1–6 evaluate each binomial coefficient.

**1.** $\binom{4}{2}$ **2.** $\binom{6}{4}$ **3.** $\binom{5}{0}$

**4.** $\binom{5}{5}$ **5.** $\binom{10}{8}$ **6.** $\binom{12}{7}$

Expand each binomial in Exercises 7–18.

**7.** $(x+y)^5$ **8.** $(x-y)^5$ **9.** $(3a-1)^5$ **10.** $(x^2-2)^4$

**11.** $(3x-y)^4$ **12.** $(x+3y)^4$ **13.** $(u^2+v^2)^6$ **14.** $(u^2-v^2)^6$

**15.** $(a+a^{-1})^7$ **16.** $(a^{-2}-a^2)^7$ **17.** $(x^{1/2}-y^{1/2})^4$ **18.** $(x-1)^9$

In Exercises 19–30 find the indicated term in each binomial expansion.

**19.** 3rd; $(x+2)^5$ **20.** 5th; $(x-2)^5$ **21.** 4th; $(x+2y)^5$ **22.** 4th; $(x-2y)^5$

**23.** 4th; $(x + y^2)^5$ **24.** 4th; $(x - y^2)^5$ **25.** 6th; $(3a - b)^7$ **26.** 3rd; $(4a - 2)^6$

**27.** middle; $(4a - 2)^6$ **28.** middle; $(2a - 3b)^8$ **29.** 6th; $(a - a^{-1})^8$ **30.** 4th; $(1 - \sqrt{x})^7$

In Exercises 31–34 find the value of the real or complex number to a power.

**31.** $(1 + 1)^7$ **32.** $(1 + \sqrt{2})^3$ **33.** $(1 + i)^5$ **34.** $(1 - 3i)^4$

**35.** Find the term containing $x^5$ in $(3x - 5y)^6$.

**36.** Find the term containing $y^5$ in $(5x - 2y)^6$.

**37.** Use the first four terms of $(1 - 0.1)^6$ to approximate $(0.9)^6$. Compare your results with a calculator value.

**38.** Use the first four terms of $(1 + 0.1)^6$ to approximate $(1.1)^6$. Compare your results with a calculator value.

**39.** Prove that $\binom{n}{0} = \binom{n+1}{0}$ and $\binom{n}{n} = \binom{n+1}{n+1}$.

**40.** Prove that $\binom{n}{r-1} + \binom{n}{r} = \binom{n+1}{r}$.

## For Review

**41.** SPORTS How many ways can the five members of a basketball team be introduced to the crowd if the center must be introduced first?

**42.** How many distinguishable permutations of the letters in the word ALABAMA are possible?

**43.** MANUFACTURING How many ways can 4 light bulbs be selected from 24 light bulbs for testing?

**44.** EDUCATION How many ways can 4 essays be selected from 24 to be given first, second, third, and fourth place awards?

**45.** How many ways can 6 people be seated for dinner around a round table?

# 12.7 Probability

Probability originated with problems related to games of chance, but today it has grown to include applications in such areas as genetics, insurance, physics, social science, and medicine.

Determining the likelihood of a particular event is the substance of probability. To make this precise, we need a variety of terms. We use the term **experiment** to mean the performing of some activity such as flipping a coin, drawing a card, or rolling dice. If an experiment is performed, the collection of all possible **outcomes** or results is called the **sample space** of the experiment. It is assumed that any particular outcome is just as likely to occur as any other, that is, the outcomes are **equally likely.** Any subset of the sample space is called an **event.** Elements in a particular event are called **favorable outcomes** or **successes,** while outcomes not in the event are called **unfavorable outcomes** or **failures.**

The experiment of flipping two fair coins has sample space

$$S = \{hh, ht, th, tt\},$$

where *hh* corresponds to the outcome of obtaining a head on each coin, and *ht* corresponds to the outcome of obtaining a head on the first coin and a tail on the second. Note that *ht* and *th* are different. This is clear if one coin is thought to be a dime, and the other a quarter, for then certainly there are two ways to obtain one head and one tail. The event

$$E = \{tt\}$$

is the event of obtaining two tails, while

$$F = \{ht, th\}$$

is the event of obtaining one head and one tail.

The experiment of rolling two unloaded dice has a sample space consisting of thirty-six pairs of numbers, each number varying from 1 through 6. This is because we have insisted that each outcome be as likely as any other. We cannot use {2, 3, 4, 5, 6, 7, 8, 9, 10, 11, 12} to represent the sample space since the total of 7 can be obtained in six ways

$$(1, 6), \quad (2, 5), \quad (3, 4), \quad (4, 3), \quad (5, 2), \quad (6, 1)$$

whereas the total of 2 is less likely since it can only be obtained in one way, (1, 1).

Observe that (2, 5) and (5, 2) are different outcomes. This is perhaps easier to understand if we think of the dice as being of different colors, one red and the other white. A 2 with the red die and a 5 with the white is clearly a different outcome than a 5 with the red and a 2 with the white.

We show the sample space for this experiment as follows.

| | | | | | |
|---|---|---|---|---|---|
| (1, 1) | (1, 2) | **(1, 3)** | (1, 4) | (1, 5) | (1, 6) |
| (2, 1) | **(2, 2)** | (2, 3) | (2, 4) | (2, 5) | **(2, 6)** |
| **(3, 1)** | (3, 2) | (3, 3) | (3, 4) | **(3, 5)** | (3, 6) |
| (4, 1) | (4, 2) | (4, 3) | **(4, 4)** | (4, 5) | (4, 6) |
| (5, 1) | (5, 2) | **(5, 3)** | (5, 4) | (5, 5) | (5, 6) |
| (6, 1) | **(6, 2)** | (6, 3) | (6, 4) | (6, 5) | (6, 6) |

*F*   *E*

The event $E$ of obtaining a total of 8 is

$$E = \{(6, 2), (5, 3), (4, 4), (3, 5), (2, 6)\}$$

while the event $F$ of obtaining a total of 4 is

$$F = \{(3, 1), (2, 2), (1, 3)\}.$$

The pairs that give a particular total are all found along a diagonal from lower left to upper right.

We are now able to give a precise definition of probability along with several important properties. We use $n(S)$, read "$n$ of $S$," to represent the number of elements in set $S$, and $n(E)$ for the number of elements in $E$.

### Probability

Let $S$ be the sample space of an experiment, and $E$ an event ($E$ is a subset of $S$). The **probability that $E$ will occur,** or simply the **probability of $E$,** denoted by $P(E)$, is given by

$$P(E) = \frac{n(E)}{n(S)} = \frac{\text{the number of favorable outcomes}}{\text{the total number of outcomes}}.$$

Since $E$ is a subset of $S$, $0 \leq n(E) \leq n(S)$, so that dividing through by $n(S)$, we obtain

$$\frac{0}{n(S)} \leq \frac{n(E)}{n(S)} \leq \frac{n(S)}{n(S)}$$

or

$$\mathbf{0 \leq P(E) \leq 1.}$$

Hence the probability of an event is always a number between 0 and 1, inclusive. If $P(E) = 0$, we say that $E$ is **impossible** ($E$ cannot occur). If $P(E) = 1$, we say that $E$ is **certain** ($E$ must occur). If $E$ and $F$ are two events such that $P(E) < P(F)$, we say that $E$ is **less likely to occur** than $F$, or that $F$ is **more likely to occur** than $E$. If $P(E) = P(F)$, the events are called **equally likely.**

**EXAMPLE 1** The sample space for the experiment of flipping two fair coins is $S = \{hh, ht, th, tt\}$.

**(a)** If $E$ is the event of obtaining two heads, what is $P(E)$?
Since $E = \{hh\}$,

$$P(E) = \frac{n(E)}{n(S)} = \frac{1}{4}.$$ Remember that $hh$ is one outcome

**(b)** If $F$ is the event of obtaining at least one head, find $P(F)$.

$$F = \{hh, ht, th\}.$$

Thus,

$$P(F) = \frac{n(F)}{n(S)} = \frac{3}{4}. \quad \blacksquare$$

**EXAMPLE 2** Consider the sample space for the experiment of rolling two dice.

**(a)** Determine the probability of rolling a total of 8.
The sample space was given earlier in this section.

$$E = \{(6, 2), (5, 3), (4, 4), (3, 5), (2, 6)\}$$

Thus $P(E) = \dfrac{n(E)}{n(S)} = \dfrac{5}{36}$.

**(b)** Determine the probability of rolling a total of 3.
From the array of outcomes given in this section, the number of ways of obtaining a three is 2. Thus,

$$P(3) = \frac{2}{36} = \frac{1}{18}. \quad \blacksquare$$

**NOTE** The probabilities in Example 2 do not mean that if the dice are thrown 36 times exactly 5 sums of 8 and 2 sums of 3 will be obtained. It does mean that if the dice are thrown hundreds of times it is likely that about 5/36 of the tosses will result in a sum of 8, and about 1/18 of the tosses will give a sum of 3. ■

**EXAMPLE 3** Find the probability of drawing a five-card hand consisting of 3 aces and two cards that are not aces from a deck of 52 cards.

Let $E$ represent the event. There are ${}_4C_3$ ways of drawing 3 aces, ${}_{48}C_2$ ways of drawing two non-aces, and ${}_{52}C_5$ total ways of drawing 5 cards. Thus,

$$P(E) = \frac{{}_4C_3 \cdot {}_{48}C_2}{{}_{52}C_5} = \frac{\dfrac{4!}{3!1!}\,\dfrac{48!}{46!2!}}{\dfrac{52}{5!47!}} = \frac{94}{54145} \approx 0.0017. \quad ■$$

## Probability of A or B

When two or more events are joined by the word *or*, we say that the result is the **disjunction** of the events. If $A$ is the event of rolling a 5 with one die, and $B$ is the event of rolling a 3 with one die, then $A$ *or* $B$ is the event of rolling a 3 or a 5 with one die. In this case, it is clear that

$$\begin{aligned} P(A \text{ or } B) &= P(A) + P(B) \\ \frac{2}{6} &= \frac{1}{6} + \frac{1}{6}. \end{aligned}$$

The **conjunction** of two or more events is formed by joining them with the word *and*. In the above, it is impossible for both $A$ and $B$ to occur on one trial. When this happens, we say that $A$ and $B$ are **mutually exclusive,** and observe that the probability that both $A$ and $B$ occur is 0, that is,

$$P(A \text{ and } B) = 0.$$

Thus, the event ($A$ *and* $B$) is impossible.

Suppose that $A$ is the event of drawing an ace and $B$ is the event of drawing a spade in the experiment of drawing one card from a well-shuffled deck of 52 cards. Then ($A$ *or* $B$) is the event of drawing an ace or a spade on one draw of a card. Since there are 13 spades and 4 aces, we are tempted to add to obtain 17 as the number of outcomes favorable to ($A$ *or* $B$). However, the ace of spaces has then been counted twice (both as a spade and as an ace) so that 17 is 1 too many for the number of favorable outcomes. Thus, $P(A \text{ or } B) = 16/52$. This same result can be obtained by the following formula.

$$\begin{aligned} P(A \text{ or } B) &= P(A) + P(B) - P(A \text{ and } B) \\ &= \frac{4}{52} + \frac{13}{52} - \frac{1}{52} \qquad P(A \text{ and } B) \text{ is } \frac{1}{52} \\ &= \frac{16}{52} \end{aligned}$$

When $A$ and $B$ are mutually exclusive $P(A \text{ and } B) = 0$ so that the formula still applies. The above examples are specific cases of the next theorem.

**Probability of $A$ or $B$**

Let $A$ and $B$ be events. Then

$$P(A \text{ or } B) = P(A) + P(B) - P(A \text{ and } B).$$

If $A$ and $B$ are mutually exclusive, $P(A \text{ and } B) = 0$ and the formula reduces to

$$P(A \text{ or } B) = P(A) + P(B).$$

**EXAMPLE 4**

If a card is drawn from a well-shuffled deck of 52 cards, find the probability of drawing the following cards.

**(a)** An ace or a king

$$P(\text{ace } or \text{ king}) = P(\text{ace}) + P(\text{king}) - P(\text{ace } and \text{ king})$$
$$= \frac{4}{52} + \frac{4}{52} - \frac{0}{52} = \frac{8}{52} = \frac{2}{13}$$

**(b)** A face card or a heart

$$P(\text{face } or \text{ heart}) = P(\text{face}) + P(\text{heart}) - P(\text{face } and \text{ heart})$$
$$= \frac{12}{52} + \frac{13}{52} - \frac{3}{52} \qquad \text{3 heart face cards}$$
$$= \frac{22}{52} = \frac{11}{26}$$

**(c)** An ace and a jack

$P(\text{ace } and \text{ jack}) = 0$ since the two events are mutually exclusive. ■

We now solve the second applied problem in the introduction to this chapter.

**EXAMPLE 5**

BIOLOGY

A couple plans to have two children. Assuming that it is equally likely that a boy or a girl will be born, what is the probability that the couple will have two girls? What is the probability of two boys? What is the probability of one boy and one girl?

The sample space for this experiment is

$$S = \{gg, gb, bg, bb\},$$

where $gb$ means girl-then-boy and $bg$ means boy-then-girl. The probability of two girls is

$$P(gg) = \frac{n(gg)}{n(S)} = \frac{1}{4}.$$

The probability of two boys is the same.

$$P(bb) = \frac{n(bb)}{n(S)} = \frac{1}{4}$$

However, the probability of one boy and one girl is

$$P(1 \text{ boy and } 1 \text{ girl}) = P(gb) + P(bg) = \frac{1}{4} + \frac{1}{4} = \frac{1}{2}. \quad ■$$

**Complementary Events**

If $E$ is an event and $F$ is the event described as not $E$, then $E$ and $F$ are **complementary events.** Since either $E$ or $F$ must be true, $P(E \text{ or } F) = 1$. Also, $P(E \text{ and } F) = 0$. Thus,

$$P(E \text{ or } F) = P(E) + P(F) = 1$$
$$\text{or} \quad P(F) = 1 - P(E).$$

### Complementary Events

If $E$ is an event and $F$ is the event that $E$ will not occur, then $E$ and $F$ are complementary and

$$P(F) = 1 - P(E).$$

## EXAMPLE 6

MEDICINE

During an outbreak of both measles and mumps, health officials warned parents to have children vaccinated. If they did not, the probability of a child contracting measles was 0.012, the probability of contracting the mumps was 0.031, and the probability of contracting both was 0.005.

**(a)** Find the probability of a child who was not vaccinated contracting the measles or the mumps.

$$\begin{aligned} P(\text{Me or Mu}) &= P(\text{Me}) + P(\text{Mu}) - P(\text{Me and Mu}) \\ &= 0.012 + 0.031 - 0.005 \\ &= 0.038 \end{aligned}$$

**(b)** Find the probability of a child contracting neither of the diseases.

Let $F$ be the event of contracting neither of the diseases.

$$\begin{aligned} P(F) &= 1 - P(\text{Me or Mu}) \\ &= 1 - 0.038 = 0.962 \end{aligned}$$ ■

## 12.7 EXERCISES

In Exercises 1–6 answer *yes* if the number could be a probability and *no* if it could not.

**1.** 0

**2.** $\frac{2}{3}$

**3.** $-\frac{1}{2}$

**4.** 1

**5.** 0.00001

**6.** $\frac{5}{4}$

RECREATION Let $S = \{1, 2, 3, 4, 5, 6\}$ be the sample space of rolling one unloaded die. Determine the probability of each event in Exercises 7–12.

**7.** Rolling a 5

**8.** Rolling a number less than 3

**9.** Rolling a number greater than 0

**10.** Rolling a multiple of 3

**11.** Rolling a number $n \geq 4$

**12.** Rolling a number $n < 1$

Let $S$ be the sample space for the experiment of rolling two unloaded dice. Find the probability of each event in Exercises 13–18.

**13.** Rolling a total of 7

**14.** Rolling a total of 11

**15.** Rolling a total of 7 or 11

**16.** Rolling a total of 5 or 2

**17.** Rolling a total $>3$

**18.** Rolling a total $\leq 11$

**RECREATION** In the game of craps, on the first roll a player wins with a total of 7 or 11 and loses with a total of 2, 3, or 12 Any other total becomes his point, and he must continue to roll until he rolls his point and wins, or rolls a 7 and loses. Find the probability of the event in Exercises 19–21, and answer the question in Exercise 22.

**19.** Winning on the first roll

**20.** Losing on the first roll

**21.** Rolling a total other than a winning or losing total on the first roll

**22.** Of all the points 4, 5, 6, 8, 9, 10, which two are the most likely to be rolled? Least likely to be rolled?

**RECREATION** A roulette wheel contains 38 slots numbered 00, 0, 1, 2, 3, 4, . . . , 35, 36. Eighteen of the slots numbered 1 through 36 are colored black and the rest are colored red. The 00 and 0 slots are colored green and are called house numbers. The wheel is spun and a ball is rolled around the rim in the opposite direction. Eventually the ball falls into a slot. Find the probability of each event in Exercises 23–30.

**23.** Ball falls into the number 3 slot

**24.** Ball falls into a black slot

**25.** Ball falls into a red slot

**26.** Ball falls into a green slot

**27.** Ball falls into a blue slot

**28.** Ball falls into a black or red slot

**29.** Ball falls into an odd-numbered slot

**30.** Ball falls into the number 0 slot

**RECREATION** A card is drawn from a well-shuffled deck of 52 cards. In Exercises 31–44 find the probability of each event.

**31.** Drawing an ace

**32.** Drawing a heart

**33.** Drawing a black card

**34.** Drawing a face card

**35.** Drawing the queen of hearts

**36.** Drawing a joker

**37.** Drawing a 7 or a queen

**38.** Drawing a 7 or a black card

**39.** Drawing a 7 or a diamond

**40.** Drawing a face card or a black card

**41.** Drawing a face card or a heart

**42.** Drawing a face card or a 3

**43.** Drawing a king and a heart

**44.** Drawing a face card and a heart

If 5 cards are dealt from a well-shuffled deck of 52 cards, find the probability of the hands in Exercises 45–50.

**45.** 5 hearts

**46.** 5 black cards

**47.** 4 aces and a card that is not an ace

**48.** 4 aces and a joker

**49.** 3 aces and 2 kings

**50.** 2 aces, 2 kings, and a card that is not an ace or a king

**51. RECREATION** If the probability of winning a game is $\frac{4}{7}$, what is the probability of losing?

52. SPORTS If the probability of a horse losing a race is $\frac{9}{11}$, what is the probability that the horse will win?

53. A child picks up a telephone and dials seven digits at random; what is the probability that he dials his own number?

54. Claude has flipped a fair coin fifty times and it has fallen heads each time. He conjectures that on the fifty-first flip the likelihood of a tail will certainly increase. Do you agree? What is the probability of a tail on the fifty-first flip?

55. GENETICS It has been determined that the number of possible genetic combinations that a child of one couple can have is $2^{48}$. Assuming that Mr. and Mrs. Levin have one child, what is the probability that a second child will have the same genetic makeup?

56. GENETICS A couple plans to have three children. **(a)** Give the sample space representing these children according to sex. **(b)** What is the probability of having three boys? **(c)** What is the probability of having three children of the same sex? **(d)** What is the probability of two girls and one boy? **(e)** What is the probability of at least two boys?

## For Review

57. Use the binomial theorem to expand $(2x - 5y)^4$.

58. Find the value of the complex number $(2 - i)^5$.

59. Find the 5th term of $(x^2 - 2y)^7$.

60. Find the term with no $x$ in $(x^{-1} + x)^6$.

# CHAPTER 12 REVIEW EXERCISES

In Exercises 1–2 the $n$th term of a sequence is given. Find the first five terms and the eighth term of each.

1. $a_n = \dfrac{n^2 - 1}{n}$

2. $x_n = \dfrac{(-1)^n}{3n + 1}$

3. Write $\sum_{k=0}^{3} \sqrt{k^2 + 1}$ without using sigma summation notation.

4. Write $\frac{1}{2} + \frac{4}{3} + \frac{9}{4} + \frac{16}{5} + \cdots + \frac{n^2}{n + 1}$ using sigma summation notation.

In Exercises 5–6 determine the second and third terms of each sequence.

5. $a_1 = -3$; $a_{n+1} = 1 - 2a_n$

6. $x_1 = \frac{1}{3}$; $x_n = 9x_{n-1} - 5$

Evaluate each series in Exercises 7–8.

7. $\sum_{k=1}^{5} (-1)^{k+1}2^k$

8. $\sum_{m=3}^{6} \dfrac{m + 5}{m - 2}$

9. Write the first six terms of an arithmetic sequence with $a_1 = -9$ and $d = 4$, and find $S_{12}$.

10. For an arithmetic sequence, $a_1 = 5$, $a_n = 19$, and $S_n = 96$. Find $n$ and $d$.

11. Insert three arithmetic means between 17 and 5.

12. Find $x$ so that $x + 1$, $2x + 3$, $4x + 1$ forms a three-term arithmetic sequence. Also, give the sequence.

13. Find the sum of the integers divisible by 4 between $-25$ and 125.

14. **DEMOGRAPHY** The population of a town is decreasing by 400 inhabitants each year. If its population is presently 8500, what will its population be in nine years?

15. **CONSUMER** A new car costing \$9000.00 depreciates 24% the first year, 20% the second year, 16% the third year and continues in that manner for five years. If all depreciations apply to the original cost, what is the value of the car in five years?

16. A collection of dimes is arranged in a triangular array with 20 coins in the base row, 19 in the next, 18 in the next, and so forth. Find the value of the collection.

17. Write the first six terms of a geometric sequence with $a_1 = \frac{1}{6}$ and $r = 2$, and find $S_6$.

18. Find the sum of the first eight terms of the geometric sequence with $a_1 = \frac{1}{27}$ and $a_8 = 81$.

19. Insert four geometric means between $-12$ and $\frac{3}{8}$.

20. Find a positive number $x$ so that $3x - 1$, $x + 3$, $x - 2$ forms a three-term geometric sequence in the given order. Also, give the sequence.

21. Find the sum of the infinite geometric sequence $8, -2, 1/2, -1/8, \ldots$.

22. Find the first five terms of an infinite geometric sequence with $a_1 = 30$ and $S = 36$.

In Exercises 23–24 convert each decimal to a fraction.

23. $1.\overline{2}$

24. $5.\overline{15}$

25. **BUSINESS** Michael Healy borrows \$1500.00 at 9% interest compounded annually. If he pays off the loan in full at the end of five years, how much does Michael pay?

26. **PHYSICS** A ball is dropped from a height of 30.0 ft. If, on each rebound, it rises to a height of $\frac{4}{5}$ the distance from which it fell, how far does it rise on the fifth rebound, and how far up and down has it traveled when it hits the ground for the sixth time?

27. **RECREATION** A child on a swing traverses an arc of 8 m. Each pass thereafter, he traverses an arc that is $\frac{8}{9}$ the length of the previous arc. How far does he travel before coming to rest?

28. **PHYSICS** A ball dropped from a height of 27 ft always rebounds $\frac{2}{3}$ of the height of the previous drop. How far does it travel (up and down) before coming to rest?

In Exercises 29–30 use the principle of mathematical induction to prove that each statement is true for every positive integer $n$.

29. $S(n)$: $4 + 8 + 12 + \cdots + 4n = 2n(n + 1)$

30. $S(n)$: $1 \cdot 2 + 2 \cdot 3 + 3 \cdot 4 + \cdots + n(n + 1) = \frac{1}{3}n(n + 1)(n + 2)$

Evaluate each permutation or combination in Exercises 31–36.

31. ${}_3P_2$

32. ${}_3C_2$

33. ${}_5P_5$

34. ${}_5C_5$

35. ${}_8P_0$

36. ${}_8C_0$

**37.** How many serial numbers formed by a letter followed by a three-digit numeral are possible under the following conditions? **(a)** No other restrictions are imposed. **(b)** The letter must not be an A and repetition of digits is not allowed. **(c)** The letter must be an A, B, or C and the first digit cannot be a zero.

**38.** In how many ways can four officers be chosen from an organization consisting of thirteen members?

**39.** In how many ways can nine people be placed in a line?

**40.** In how many ways can nine people be seated around a circular table?

**41.** How many distinguishable permutations of the letters in the word FLAGSTAFF are there?

**42.** GEOMETRY How many lines are determined by six points, no three of which are collinear?

**43.** RECREATION In how many ways can a picnic committee of three men and two women be selected from eight men and eleven women?

**44.** In how many ways can a committee of four be chosen from an organization consisting of thirteen members?

**45.** How many 5-card poker hands consisting of 3 aces and 2 eights are possible with an ordinary 52-card deck? What is the probability of being dealt such a hand?

**46.** EDUCATION In how many ways can a student choose 10 out of 15 problems on a quiz?

**47.** Expand $(3a - z)^5$.

**48.** Expand $(y^{-1} + y)^4$.

**49.** Find the 4th term in the binomial expansion of $(2x - y^2)^6$.

**50.** Find the 5th term in the binomial expansion of $(a - 3b)^7$.

One marble is selected, sight unseen, from a bag containing 8 red and 7 black marbles. In Exercises 51–54 determine the probability of the given event.

**51.** Selecting a red marble

**52.** Selecting a black marble

**53.** Selecting a blue marble

**54.** Selecting a marble that is not blue

RECREATION A single card is drawn from a well-shuffled deck of 52 cards. In Exercises 55–66 determine the probability of the event.

**55.** Drawing a spade

**56.** Drawing a face card

**57.** Drawing a black card

**58.** Drawing a card that is not a face card

**59.** Drawing the jack of hearts

**60.** Drawing a red ace

**61.** Drawing a joker

**62.** Drawing a heart or a club

**63.** Drawing a ten or a queen

**64.** Drawing a ten and a queen

**65.** Drawing a red card or an ace

**66.** Drawing a face card or a club

**67.** RECREATION If the probability of winning first prize in a contest is $10^{-4}$, what is the probability of not winning?

**68.** EDUCATION Given that the probability of passing English is 0.92, the probability of passing math is 0.85, and the probability of passing them both is 0.78, find the probability of passing English or math. What is the probability of failing both?

# APPENDIX A Logarithmic Tables and Interpolation

In the past few years, scientific calculators have all but replaced logarithm tables. However, when a calculator is unavailable, a table such as Table 1 giving common logarithms can be used. Such a table is limited in accuracy to three significant digits (four if interpolation is used).

If $n$ is any positive number, $n$ can be written in scientific notation as

$$n = m \times 10^c,$$

where $1 \leq m < 10$ and $c$ is an integer. Using the product rule,

$$\log n = \log (m \times 10^c) = \log m + \log 10^c = \log m + c.$$

Log $m$, the **mantissa** of log $n$, is a decimal greater than or equal to 0 and less than 1. The **characteristic** $c$ of log $n$ is an integer. Since any positive number can be written in scientific notation, Table 1, with logarithms of numbers between 1.00 and 9.99 in increments of 0.01, gives us logarithms of any number with three significant digits.

**EXAMPLE 1**

**(a)** Find log 1.23.

Since $1.23 = 1.23 \times 10^0$,

$$\begin{aligned} \log 1.23 &= \log (1.23 \times 10^0) \\ &= \log 1.23 + 0 \quad \text{The characteristic is 0} \end{aligned}$$

The left-hand column of Table 1 shows the numbers 1.0 through 9.9, while the numbers 0 through 9 lead each of the other columns. To find log 1.23, we look down the left column to 1.2, read across to the column headed 3, and find the decimal, .0899, the mantissa of log 1.23. Thus,

$$\log 1.23 = .0899 + 0 = 0.0899.$$

| $n$ | 0 | 1 | 2 | 3 | 4 | 5 | 6 | 7 | 8 | 9 |
|---|---|---|---|---|---|---|---|---|---|---|
| 1.0 | .0000 | .0043 | .0086 | .0128 | .0170 | .0212 | .0253 | .0294 | .0334 | .0374 |
| 1.1 | .0414 | .0453 | .0492 | .0531 | .0569 | .0607 | .0645 | .0682 | .0719 | .0755 |
| 1.2 | .0792 | .0828 | .0864 | .0899 | .0934 | .0969 | .1004 | .1038 | .1072 | .1106 |
| 1.3 | .1139 | .1173 | .1206 | .1239 | .1271 | .1303 | .1335 | .1367 | .1399 | .1430 |
| 1.4 | .1461 | .1492 | .1523 | .1553 | .1584 | .1614 | .1644 | .1673 | .1703 | .1732 |

**(b)** Find log 123.
Since $123 = 1.23 \times 10^2$,

$$\begin{aligned}\log 123 &= \log (1.23 \times 10^2)\\ &= \log 1.23 + \log 10^2\\ &= \log 1.23 + 2\\ &= .0899 + 2 = 2.0899\end{aligned}$$

**(c)** Find log 0.000123.
Since $0.000123 = 1.23 \times 10^{-4}$,

$$\begin{aligned}\log (0.000123) &= \log (1.23 \times 10^{-4})\\ &= \log 1.23 + \log 10^{-4}\\ &= .0899 + (-4)\\ &= -3.9101. \quad ■\end{aligned}$$

We could have left log 0.000123 in the form $0.0899 - 4$ rather than simplifying it to $-3.9101$. Actually, for use of a table this form is preferred since it shows the positive mantissa (0.0899) and the characteristic ($-4$).

Up to now we have illustrated how to find a logarithm of a given number. Now we show how to find a number when its logarithm is given, that is, how to find an **antilogarithm** (**antilog**). For use of Table 1, the given logarithm must be in standard form showing the positive mantissa and the integer characteristic.

## EXAMPLE 2

**(a)** Find the antilog of 2.7679.
In effect we are given that

$$\log n = 2.7679,$$

and we are asked to find $n$. Since $\log n = 2.7679 = 0.7679 + 2$, the mantissa is .7679 and the characteristic is 2. In the body of Table 1 find .7679 in the row headed by 5.8 under the column headed by 6. Thus,

$$\begin{aligned}n &= 5.86 \times 10^2\\ &= 586.\end{aligned}$$

**(b)** Find the antilog of $-2.2321$.
We must first express

$$\log n = -2.2321$$

in standard form by adding and subtracting 3.

$$\begin{array}{r} 3.0000 - 3 \\ -2.2321 \quad\; \\ \hline 0.7679 - 3 \end{array}$$

Thus, the mantissa is .7679 and the characteristic is $-3$. From Table 1,

$$\begin{aligned}n &= 5.86 \times 10^{-3}\\ &= 0.00586. \quad ■\end{aligned}$$

To find the common logarithm of a number with four significant digits, such as 5138, we can use a process called **linear interpolation.** The word *linear* appears because the process uses a straight line to approximate the portion of the graph of $y = \log x$ between two values in the table. Consider Figure 1.

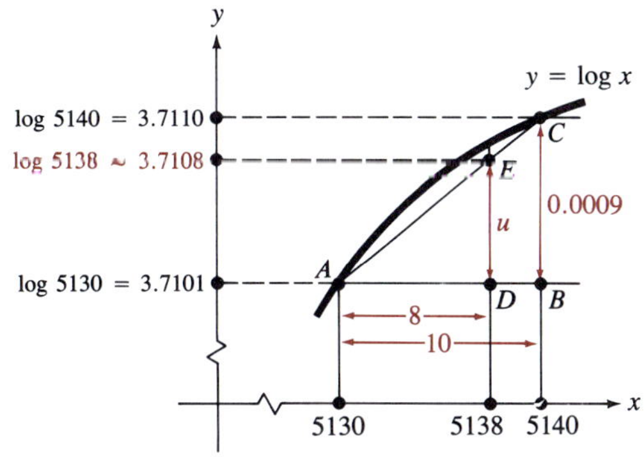


**Figure 1 Interpolation**

Noting that triangles $ABC$ and $ADE$ are similar, we conclude that

$$\frac{8}{10} = \frac{u}{0.0009}.$$

Thus,
$$u = \frac{8}{10}(0.0009) \approx 0.0007.$$

To obtain the approximation for log 5138, add 0.0007 to log 5130.

$$\log 5138 \approx \log 5130 + 0.0007 = 3.7101 + 0.0007 = 3.7108.$$

**EXAMPLE 3** Use linear interpolation to find log 0.006412. The characteristic is $-3$ so we concentrate on finding the mantissa for log 6.412.

| | | $x$ | $\log x$ | | |
|---|---|---|---|---|---|
| 0.010 | 0.002 | 6.410 | 0.8069 | $u$ | 0.0006 |
| | | 6.412 | ? | | |
| | | 6.420 | 0.8075 | | |

$$\frac{0.002}{0.010} = \frac{2}{10} = \frac{u}{0.0006}$$

$$u = \frac{2}{10}(0.0006) \approx 0.0001$$

$$\begin{aligned}\log 6.412 &\approx \log 6.41 + 0.0001\\ &= 0.8069 + 0.0001\\ &= 0.8070\end{aligned}$$

$$\log 0.006412 = 0.8070 - 3 = -2.1930$$ ■

If a calculation requires finding the antilog of a logarithm whose mantissa is not in Table 1, linear interpolation can be used to approximate the four-digit answer.

**EXAMPLE 4** Find the antilog of 2.4705.

The characteristic 2 tells us where to place the decimal point in the final answer. Thus we can concentrate on 0.4705. In Table 1, 0.4705 is between log 2.95 = 0.4698 and log 2.96 = 0.4713.

| | | $x$ | $\log x$ | | |
|---|---|---|---|---|---|
| 0.010 | $u$ | 2.950 | 0.4698 | 0.0007 | 0.0015 |
| | | ? | 0.4705 | | |
| | | 2.960 | 0.4713 | | |

$$\frac{u}{0.010} = \frac{0.0007}{0.0015} = \frac{7}{15}$$

$$u = \frac{7}{15}(0.010) \approx 0.005$$

$$\log 2.955 \approx 0.4705$$

Thus, the antilog of 2.4705 is approximately

$$2.955 \times 10^2 = 295.5. \quad \blacksquare$$

## APPENDIX A EXERCISES

In Exercises 1–6 use Table 1 to find the common logarithm of each number.

**1.** 4.68 **2.** 46.8 **3.** 0.0468

**4.** 0.000279 **5.** 279 **6.** 2.79

In Exercises 7–12 use Table 1 to find the antilogarithm of each number.

**7.** 0.7364 **8.** 0.9827 **9.** 3.5855

**10.** 4.7896 **11.** $-2.3799$ **12.** $-3.2692$

In Exercises 13–18 use Table 1 and linear interpolation to approximate the logarithm of each number.

**13.** 3.278 **14.** 6.157 **15.** 437.6

**16.** 249.7 **17.** 0.003972 **18.** 0.0004256

In Exercises 19–24 use Table 1 and linear interpolation to approximate the antilogarithm of each number.

**19.** 1.6974 **20.** 1.2390 **21.** $0.5409 - 2$

**22.** 2.6754 **23.** $-3.1155$ **24.** $-2.8118$

APPENDIX B

# Trigonometric Function Tables and Interpolation

As with logarithm tables, trigonometric function tables have nearly become obsolete with the advent of the handheld calculator. However, if your calculator does not have the trigonometric functions, Table 2, giving trigonometric function values, can be used to calculate angles correct to the nearest ten minutes. Angles from 0°00′ to 45°00′ are found *down* the left-hand column while each trigonometric function of these angles is listed across the top row. Angles from 45°00′ to 90°00′ are found *up* the right-hand column while each trigonometric function of these angles is listed across the bottom row. The table can also be used for angles in radian measure by referring to the columns labeled "Radians."

**EXAMPLE 1** Use Table 2 to find each trigonometric function value.

**(a)** sin 52°50′

Looking up the right-hand column, proceed to 52°50′ as shown below. Move across to the column labeled sin $\theta$ in the bottom row to find 0.7969. Thus,

$$\sin 52°50' = 0.7969.$$

| | | | | | | | | | |
|---|---|---|---|---|---|---|---|---|---|
| **37°00′** | .6458 | .6018 | 1.662 | .7536 | 1.327 | 1.252 | .7986 | .9250 | **53°00′** |
| 10 | 487 | 041 | 655 | 581 | 319 | 255 | 969 | 221 | 50 |
| 20 | 516 | 065 | 649 | 627 | 311 | 258 | 951 | 192 | 40 |
| 30 | .6545 | .6088 | 1.643 | .7673 | 1.303 | 1.260 | .7934 | .9163 | 30 |
| 40 | 574 | 111 | 636 | 720 | 295 | 263 | 916 | 134 | 20 |
| 50 | 603 | 134 | 630 | 766 | 288 | 266 | 898 | 105 | 10 |
| **38°00′** | .6632 | .6157 | 1.624 | .7813 | 1.280 | 1.269 | .7880 | .9076 | **52°00′** |
| | | **cos $\theta$** | **sec $\theta$** | **cot $\theta$** | **tan $\theta$** | **csc $\theta$** | **sin $\theta$** | **Radians** | **Degrees** |
| | | | | | | | | **Angle $\theta$** | |

**(b)** cot 11°40′

Look down the degrees column to find 11°40′, then across to 4.843 in the column headed "cot $\theta$." Thus,

$$\cot 11°40' = 4.843. \quad ■$$

Table 2 can also be used to determine an angle with a given trigonometric function value.

**EXAMPLE 2** Find $A$ in degrees to the nearest 10′ using Table 2.

**(a)** $\tan A = 0.7907$

We look for 0.7907 in the tangent columns (there are two such columns, one labeled "tan $\theta$" at the top and the other labeled "tan $\theta$" at the bottom). The value 0.7907 is found in the column labeled "tan $\theta$" at the top so we look horizontally to the left to find 38°20′. Thus $A = 38°20'$.

**(b)** $\sec A = 4.560$

Locate 4.560 in the secant column (in the bottom row) and move horizontally to the right to find 77°20′. Make sure you read *up* the column from 77°, *not down* from 78°. Thus, $A = 77°20'$. ■

Although Table 2 can be used to find trigonometric functions of angles given to the nearest ten minutes, it can also be used to approximate angles to the nearest minute with **linear interpolation.** Suppose we illustrate the process by approximating sin 28°43′. Consider Figure 1.

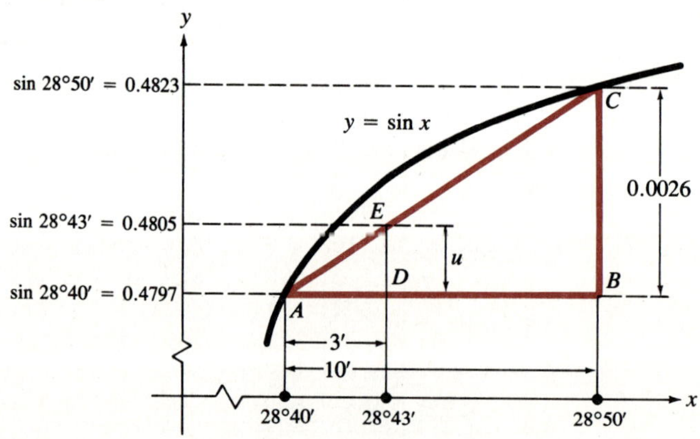


**Figure 1 Interpolation**

Noting that triangles $ABC$ and $ADE$ are similar, we may conclude that

$$\frac{3}{10} = \frac{u}{0.0026}$$

Thus,

$$u = \frac{3}{10}(0.0026) \approx 0.0008.$$

Since $\sin 28°40' < \sin 28°50'$, we add 0.0008 to $\sin 28°40'$ to obtain the approximation for $\sin 28°43'$.

$$\begin{aligned}\sin 28°43' &\approx \sin 28°40' + 0.0008\\ &= 0.4797 + 0.0008\\ &= 0.4805 \quad ■\end{aligned}$$

Linear interpolation can be applied to any trigonometric function. If we have a decreasing function such as $\cos x$, $\cot x$, or $\csc x$, we must remember to *subtract* the value of $u$. This process is illustrated in the following example.

**EXAMPLE 3** Use linear interpolation to find $\cot 36°26'$.

| | | $x$ | $\cot x$ | | |
|---|---|---|---|---|---|
| 10 | 6 | 36°20′ | 1.360 | $u$ | 0.009 |
| | | 36°26′ | ? | | |
| | | 36°30′ | 1.351 | | |

$$\frac{6}{10} = \frac{u}{0.009}$$

$$u = \frac{6}{10}(0.009) \approx 0.005$$

Since $\cot 36°20' > \cot 36°30'$, we subtract 0.005 from $\cot 36°20'$, which is 1.360.

$$\begin{aligned}\cot 36°26' &\approx \cot 36°20' - \mathbf{0.005}\\ &= 1.360 - \mathbf{0.005}\\ &= 1.355 \quad ■\end{aligned}$$

If in the process of a calculation we obtain a value of a trigonometric function which is not in the table, linear interpolation can be used to find the angle to the nearest minute.

**EXAMPLE 4** Find $x$ to the nearest minute given that $\cos x = 0.3001$.

Using Table 2, we find that 0.3001 is between $\cos 72°30' = 0.3007$ and $\cos 72°40' = 0.2979$.

| | | $x$ | $\cos x$ | | |
|---|---|---|---|---|---|
| 10 | $u$ | 72°30′ | 0.3007 | 0.0006 | 0.0028 |
| | | ? | 0.3001 | | |
| | | 72°40′ | 0.2979 | | |

$$\frac{u}{10} = \frac{0.0006}{0.0028} = \frac{6}{28}$$

$$u = \frac{6}{28}(10) \approx 2$$

$$x \approx 72°32' \quad ■$$

## APPENDIX B EXERCISES

In Exercises 1–6, use Table 2 to find each trigonometric function value.

**1.** sin 79°20′ **2.** cos 48°50′ **3.** tan 26°30′

**4.** cot 31°10′ **5.** csc 51°40′ **6.** sec 63°20′

In Exercises 7–12, use Table 2 to find each angle with the given trigonometric function value to the nearest 10′.

**7.** $\cos A = 0.8936$ **8.** $\sin A = 0.5568$ **9.** $\cot A = 0.8342$

**10.** $\tan A = 1.411$ **11.** $\sec A = 1.431$ **12.** $\csc A = 2.166$

Use linear interpolation in Exercises 13–18 to find the value of each trigonometric function.

**13.** sin 41°17′ **14.** cos 18°42′ **15.** tan 81°26′

**16.** cot 58°14′ **17.** sec 40°49′ **18.** csc 23°25′

Use linear interpolation in Exercises 19–24 to find $x$ to the nearest minute.

**19.** $\sin x = 0.9215$ **20.** $\cos x = 0.1122$ **21.** $\tan x = 2.131$

**22.** $\cot x = 1.739$ **23.** $\sec x = 1.404$ **24.** $\csc x = 3.159$

# Table 1 Common Logarithms

| $n$ | 0 | 1 | 2 | 3 | 4 | 5 | 6 | 7 | 8 | 9 |
|---|---|---|---|---|---|---|---|---|---|---|
| 1.0 | .0000 | .0043 | .0086 | .0128 | .0170 | .0212 | .0253 | .0294 | .0334 | .0374 |
| 1.1 | .0414 | .0453 | .0492 | .0531 | .0569 | .0607 | .0645 | .0682 | .0719 | .0755 |
| 1.2 | .0792 | .0828 | .0864 | .0899 | .0934 | .0969 | .1004 | .1038 | .1072 | .1106 |
| 1.3 | .1139 | .1173 | .1206 | .1239 | .1271 | .1303 | .1335 | .1367 | .1399 | .1430 |
| 1.4 | .1461 | .1492 | .1523 | .1553 | .1584 | .1614 | .1644 | .1673 | .1703 | .1732 |
| 1.5 | .1761 | .1790 | .1818 | .1847 | .1875 | .1903 | .1931 | .1959 | .1987 | .2014 |
| 1.6 | .2041 | .2068 | .2095 | .2122 | .2148 | .2175 | .2201 | .2227 | .2253 | .2279 |
| 1.7 | .2304 | .2330 | .2355 | .2380 | .2405 | .2430 | .2455 | .2480 | .2504 | .2529 |
| 1.8 | .2553 | .2577 | .2601 | .2625 | .2648 | .2672 | .2695 | .2718 | .2742 | .2765 |
| 1.9 | .2788 | .2810 | .2833 | .2856 | .2878 | .2900 | .2923 | .2945 | .2967 | .2989 |
| 2.0 | .3010 | .3032 | .3054 | .3075 | .3096 | .3118 | .3139 | .3160 | .3181 | .3201 |
| 2.1 | .3222 | .3243 | .3263 | .3284 | .3304 | .3324 | .3345 | .3365 | .3385 | .3404 |
| 2.2 | .3424 | .3444 | .3464 | .3483 | .3502 | .3522 | .3541 | .3560 | .3579 | .3598 |
| 2.3 | .3617 | .3636 | .3655 | .3674 | .3692 | .3711 | .3729 | .3747 | .3766 | .3784 |
| 2.4 | .3802 | .3820 | .3838 | .3856 | .3874 | .3892 | .3909 | .3927 | .3945 | .3962 |
| 2.5 | .3979 | .3997 | .4014 | .4031 | .4048 | .4065 | .4082 | .4099 | .4116 | .4133 |
| 2.6 | .4150 | .4166 | .4183 | .4200 | .4216 | .4232 | .4249 | .4265 | .4281 | .4298 |
| 2.7 | .4314 | .4330 | .4346 | .4362 | .4378 | .4393 | .4409 | .4425 | .4440 | .4456 |
| 2.8 | .4472 | .4487 | .4502 | .4518 | .4533 | .4548 | .4564 | .4579 | .4594 | .4609 |
| 2.9 | .4624 | .4639 | .4654 | .4669 | .4683 | .4698 | .4713 | .4728 | .4742 | .4757 |
| 3.0 | .4771 | .4786 | .4800 | .4814 | .4829 | .4843 | .4857 | .4871 | .4886 | .4900 |
| 3.1 | .4914 | .4928 | .4942 | .4955 | .4969 | .4983 | .4997 | .5011 | .5024 | .5038 |
| 3.2 | .5051 | .5065 | .5079 | .5092 | .5105 | .5119 | .5132 | .5145 | .5159 | .5172 |
| 3.3 | .5185 | .5198 | .5211 | .5224 | .5237 | .5250 | .5263 | .5276 | .5289 | .5302 |
| 3.4 | .5315 | .5328 | .5340 | .5353 | .5366 | .5378 | .5391 | .5403 | .5416 | .5428 |
| 3.5 | .5441 | .5453 | .5465 | .5478 | .5490 | .5502 | .5514 | .5527 | .5539 | .5551 |
| 3.6 | .5563 | .5575 | .5587 | .5599 | .5611 | .5623 | .5635 | .5647 | .5658 | .5670 |
| 3.7 | .5682 | .5694 | .5705 | .5717 | .5729 | .5740 | .5752 | .5763 | .5775 | .5786 |
| 3.8 | .5798 | .5809 | .5821 | .5832 | .5843 | .5855 | .5866 | .5877 | .5888 | .5899 |
| 3.9 | .5911 | .5922 | .5933 | .5944 | .5955 | .5966 | .5977 | .5988 | .5999 | .6010 |
| 4.0 | .6021 | .6031 | .6042 | .6053 | .6064 | .6075 | .6085 | .6096 | .6107 | .6117 |
| 4.1 | .6128 | .6138 | .6149 | .6160 | .6170 | .6180 | .6191 | .6201 | .6212 | .6222 |
| 4.2 | .6232 | .6243 | .6253 | .6263 | .6274 | .6284 | .6294 | .6304 | .6314 | .6325 |
| 4.3 | .6335 | .6345 | .6355 | .6365 | .6375 | .6385 | .6395 | .6405 | .6415 | .6425 |
| 4.4 | .6435 | .6444 | .6454 | .6464 | .6474 | .6484 | .6493 | .6503 | .6513 | .6522 |
| 4.5 | .6532 | .6542 | .6551 | .6561 | .6571 | .6580 | .6590 | .6599 | .6609 | .6618 |
| 4.6 | .6628 | .6637 | .6646 | .6656 | .6665 | .6675 | .6684 | .6693 | .6702 | .6712 |
| 4.7 | .6721 | .6730 | .6739 | .6749 | .6758 | .6767 | .6776 | .6785 | .6794 | .6803 |
| 4.8 | .6812 | .6821 | .6830 | .6839 | .6848 | .6857 | .6866 | .6875 | .6884 | .6893 |
| 4.9 | .6902 | .6911 | .6920 | .6928 | .6937 | .6946 | .6955 | .6964 | .6972 | .6981 |
| 5.0 | .6990 | .6998 | .7007 | .7016 | .7024 | .7033 | .7042 | .7050 | .7059 | .7067 |
| 5.1 | .7076 | .7084 | .7093 | .7101 | .7110 | .7118 | .7126 | .7135 | .7143 | .7152 |
| 5.2 | .7160 | .7168 | .7177 | .7185 | .7193 | .7202 | .7210 | .7218 | .7226 | .7235 |
| 5.3 | .7243 | .7251 | .7259 | .7267 | .7275 | .7284 | .7292 | .7300 | .7308 | .7316 |
| 5.4 | .7324 | .7332 | .7340 | .7348 | .7356 | .7364 | .7372 | .7380 | .7388 | .7396 |
| $n$ | 0 | 1 | 2 | 3 | 4 | 5 | 6 | 7 | 8 | 9 |

# Table 1 Common Logarithms (cont.)

| *n* | 0 | 1 | 2 | 3 | 4 | 5 | 6 | 7 | 8 | 9 |
|---|---|---|---|---|---|---|---|---|---|---|
| 5.5 | .7404 | .7412 | .7419 | .7427 | .7435 | .7443 | .7451 | .7459 | .7466 | .7474 |
| 5.6 | .7482 | .7490 | .7497 | .7505 | .7513 | .7520 | .7528 | .7536 | .7543 | .7551 |
| 5.7 | .7559 | .7566 | .7574 | .7582 | .7589 | .7597 | .7604 | .7612 | .7619 | .7627 |
| 5.8 | .7634 | .7642 | .7649 | .7657 | .7664 | .7672 | .7679 | .7686 | .7694 | .7701 |
| 5.9 | .7709 | .7716 | .7723 | .7731 | .7738 | .7745 | .7752 | .7760 | .7767 | .7774 |
| 6.0 | .7782 | .7789 | .7796 | .7803 | .7810 | .7818 | .7825 | .7832 | .7839 | .7846 |
| 6.1 | .7853 | .7860 | .7868 | .7875 | .7882 | .7889 | .7896 | .7903 | .7910 | .7917 |
| 6.2 | .7924 | .7931 | .7938 | .7945 | .7952 | .7959 | .7966 | .7973 | .7980 | .7987 |
| 6.3 | .7993 | .8000 | .8007 | .8014 | .8021 | .8028 | .8035 | .8041 | .8048 | .8055 |
| 6.4 | .8062 | .8069 | .8075 | .8082 | .8089 | .8096 | .8102 | .8109 | .8116 | .8122 |
| 6.5 | .8129 | .8136 | .8142 | .8149 | .8156 | .8162 | .8169 | .8176 | .8182 | .8189 |
| 6.6 | .8195 | .8202 | .8209 | .8215 | .8222 | .8228 | .8235 | .8241 | .8248 | .8254 |
| 6.7 | .8261 | .8267 | .8274 | .8280 | .8287 | .8293 | .8299 | .8306 | .8312 | .8319 |
| 6.8 | .8325 | .8331 | .8338 | .8344 | .8351 | .8357 | .8363 | .8370 | .8376 | .8382 |
| 6.9 | .8388 | .8395 | .8401 | .8407 | .8414 | .8420 | .8426 | .8432 | .8439 | .8445 |
| 7.0 | .8451 | .8457 | .8463 | .8470 | .8476 | .8482 | .8488 | .8494 | .8500 | .8506 |
| 7.1 | .8513 | .8519 | .8525 | .8531 | .8537 | .8543 | .8549 | .8555 | .8561 | .8567 |
| 7.2 | .8573 | .8579 | .8585 | .8591 | .8597 | .8603 | .8609 | .8615 | .8621 | .8627 |
| 7.3 | .8633 | .8639 | .8645 | .8651 | .8657 | .8663 | .8669 | .8675 | .8681 | .8686 |
| 7.4 | .8692 | .8698 | .8704 | .8710 | .8716 | .8722 | .8727 | .8733 | .8739 | .8745 |
| 7.5 | .8751 | .8756 | .8762 | .8768 | .8774 | .8779 | .8785 | .8791 | .8797 | .8802 |
| 7.6 | .8808 | .8814 | .8820 | .8825 | .8831 | .8837 | .8842 | .8848 | .8854 | .8859 |
| 7.7 | .8865 | .8871 | .8876 | .8882 | .8887 | .8893 | .8899 | .8904 | .8910 | .8915 |
| 7.8 | .8921 | .8927 | .8932 | .8938 | .8943 | .8949 | .8954 | .8960 | .8965 | .8971 |
| 7.9 | .8976 | .8982 | .8987 | .8993 | .8998 | .9004 | .9009 | .9015 | .9020 | .9025 |
| 8.0 | .9031 | .9036 | .9042 | .9047 | .9053 | .9058 | .9063 | .9069 | .9074 | .9079 |
| 8.1 | .9085 | .9090 | .9096 | .9101 | .9106 | .9112 | .9117 | .9122 | .9128 | .9133 |
| 8.2 | .9138 | .9143 | .9149 | .9154 | .9159 | .9165 | .9170 | .9175 | .9180 | .9186 |
| 8.3 | .9191 | .9196 | .9201 | .9206 | .9212 | .9217 | .9222 | .9227 | .9232 | .9238 |
| 8.4 | .9243 | .9248 | .9253 | .9258 | .9263 | .9269 | .9274 | .9279 | .9284 | .9289 |
| 8.5 | .9294 | .9299 | .9304 | .9309 | .9315 | .9320 | .9325 | .9330 | .9335 | .9340 |
| 8.6 | .9345 | .9350 | .9355 | .9360 | .9365 | .9370 | .9375 | .9380 | .9385 | .9390 |
| 8.7 | .9395 | .9400 | .9405 | .9410 | .9415 | .9420 | .9425 | .9430 | .9435 | .9440 |
| 8.8 | .9445 | .9450 | .9455 | .9460 | .9465 | .9469 | .9474 | .9479 | .9484 | .9489 |
| 8.9 | .9494 | .9499 | .9504 | .9509 | .9513 | .9518 | .9523 | .9528 | .9533 | .9538 |
| 9.0 | .9542 | .9547 | .9552 | .9557 | .9562 | .9566 | .9571 | .9576 | .9581 | .9586 |
| 9.1 | .9590 | .9595 | .9600 | .9605 | .9609 | .9614 | .9619 | .9624 | .9628 | .9633 |
| 9.2 | .9638 | .9643 | .9647 | .9652 | .9657 | .9661 | .9666 | .9671 | .9675 | .9680 |
| 9.3 | .9685 | .9689 | .9694 | .9699 | .9703 | .9708 | .9713 | .9717 | .9722 | .9727 |
| 9.4 | .9731 | .9736 | .9741 | .9745 | .9750 | .9754 | .9759 | .9763 | .9768 | .9773 |
| 9.5 | .9777 | .9782 | .9786 | .9791 | .9795 | .9800 | .9805 | .9809 | .9814 | .9818 |
| 9.6 | .9823 | .9827 | .9832 | .9836 | .9841 | .9845 | .9850 | .9854 | .9859 | .9863 |
| 9.7 | .9868 | .9872 | .9877 | .9881 | .9886 | .9890 | .9894 | .9899 | .9903 | .9908 |
| 9.8 | .9912 | .9917 | .9921 | .9926 | .9930 | .9934 | .9939 | .9943 | .9948 | .9952 |
| 9.9 | .9956 | .9961 | .9965 | .9969 | .9974 | .9978 | .9983 | .9987 | .9991 | .9996 |
| *n* | 0 | 1 | 2 | 3 | 4 | 5 | 6 | 7 | 8 | 9 |

# Table 2 Values of Trigonometric Functions

| Angle $\theta$ | | | | | | | | | |
|---|---|---|---|---|---|---|---|---|---|
| **Degrees** | **Radians** | **sin $\theta$** | **csc $\theta$** | **tan $\theta$** | **cot $\theta$** | **sec $\theta$** | **cos $\theta$** | | |
| **0°00′** | .0000 | .0000 | No value | .0000 | No value | 1.000 | 1.0000 | 1.5708 | **90°00′** |
| 10 | 029 | 029 | 343.8 | 029 | 343.8 | 000 | 000 | 679 | 50 |
| 20 | 058 | 058 | 171.9 | 058 | 171.9 | 000 | 000 | 650 | 40 |
| 30 | .0087 | .0087 | 114.6 | .0087 | 114.6 | 1.000 | 1.0000 | 1.5621 | 30 |
| 40 | 116 | 116 | 85.95 | 116 | 85.94 | 000 | .9999 | 592 | 20 |
| 50 | 145 | 145 | 68.76 | 145 | 68.75 | 000 | 999 | 563 | 10 |
| **1°00′** | .0175 | .0175 | 57.30 | .0175 | 57.29 | 1.000 | .9998 | 1.5533 | **89°00′** |
| 10 | 204 | 204 | 49.11 | 204 | 49.10 | 000 | 998 | 504 | 50 |
| 20 | 233 | 233 | 42.98 | 233 | 42.96 | 000 | 997 | 475 | 40 |
| 30 | .0262 | .0262 | 38.20 | .0262 | 38.19 | 1.000 | .9997 | 1.5446 | 30 |
| 40 | 291 | 291 | 34.38 | 291 | 34.37 | 000 | 996 | 417 | 20 |
| 50 | 320 | 320 | 31.26 | 320 | 31.24 | 001 | 995 | 388 | 10 |
| **2°00′** | .0349 | .0349 | 28.65 | .0349 | 28.64 | 1.001 | .9994 | 1.5359 | **88°00′** |
| 10 | 378 | 378 | 26.45 | 378 | 26.43 | 001 | 993 | 330 | 50 |
| 20 | 407 | 407 | 24.56 | 407 | 24.54 | 001 | 992 | 301 | 40 |
| 30 | .0436 | .0436 | 22.93 | .0437 | 22.90 | 1.001 | .9990 | 1.5272 | 30 |
| 40 | 465 | 465 | 21.49 | 466 | 21.47 | 001 | 989 | 243 | 20 |
| 50 | 495 | 494 | 20.23 | 495 | 20.21 | 001 | 988 | 213 | 10 |
| **3°00′** | .0524 | .0523 | 19.11 | .0524 | 19.08 | 1.001 | .9986 | 1.5184 | **87°00′** |
| 10 | 553 | 552 | 18.10 | 553 | 18.07 | 002 | 985 | 155 | 50 |
| 20 | 582 | 581 | 17.20 | 582 | 17.17 | 002 | 983 | 126 | 40 |
| 30 | .0611 | .0610 | 16.38 | .0612 | 16.35 | 1.002 | .9981 | 1.5097 | 30 |
| 40 | 640 | 640 | 15.64 | 641 | 15.60 | 002 | 980 | 068 | 20 |
| 50 | 669 | 669 | 14.96 | 670 | 14.92 | 002 | 978 | 039 | 10 |
| **4°00′** | .0698 | .0698 | 14.34 | .0699 | 14.30 | 1.002 | .9976 | 1.5010 | **86°00′** |
| 10 | 727 | 727 | 13.76 | 729 | 13.73 | 003 | 974 | 981 | 50 |
| 20 | 756 | 756 | 13.23 | 758 | 13.20 | 003 | 971 | 952 | 40 |
| 30 | .0785 | .0785 | 12.75 | .0787 | 12.71 | 1.003 | .9969 | 1.4923 | 30 |
| 40 | 814 | 814 | 12.29 | 816 | 12.25 | 003 | 967 | 893 | 20 |
| 50 | 844 | 843 | 11.87 | 846 | 11.83 | 004 | 964 | 864 | 10 |
| **5°00′** | .0873 | .0872 | 11.47 | .0875 | 11.43 | 1.004 | .9962 | 1.4835 | **85°00′** |
| 10 | 902 | 901 | 11.10 | 904 | 11.06 | 004 | 959 | 806 | 50 |
| 20 | 931 | 929 | 10.76 | 934 | 10.71 | 004 | 957 | 777 | 40 |
| 30 | .0960 | .0958 | 10.43 | .0963 | 10.39 | 1.005 | .9954 | 1.4748 | 30 |
| 40 | 989 | 987 | 10.13 | 992 | 10.08 | 005 | 951 | 719 | 20 |
| 50 | 1018 | .1016 | 9.839 | .1022 | 9.788 | 005 | 948 | 690 | 10 |
| **6°00′** | .1047 | .1045 | 9.567 | .1051 | 9.514 | 1.006 | .9945 | 1.4661 | **84°00′** |
| | | **cos $\theta$** | **sec $\theta$** | **cot $\theta$** | **tan $\theta$** | **csc $\theta$** | **sin $\theta$** | **Radians** | **Degrees** |
| | | | | | | | | **Angle $\theta$** | |

## Table 2 Values of Trigonometric Functions (cont.)

| Angle $\theta$ | | | | | | | | | |
|---|---|---|---|---|---|---|---|---|---|
| **Degrees** | **Radians** | **sin $\theta$** | **csc $\theta$** | **tan $\theta$** | **cot $\theta$** | **sec $\theta$** | **cos $\theta$** | | |
| **6°00′** | .1047 | .1045 | 9.567 | .1051 | 9.514 | 1.006 | .9945 | 1.4661 | **84°00′** |
| 10 | 076 | 074 | 9.309 | 080 | 9.255 | 006 | 942 | 632 | 50 |
| 20 | 105 | 103 | 9.065 | 110 | 9.010 | 006 | 939 | 603 | 40 |
| 30 | .1134 | .1132 | 8.834 | .1139 | 8.777 | 1.006 | .9936 | 1.4573 | 30 |
| 40 | 164 | 161 | 8.614 | 169 | 8.556 | 007 | 932 | 544 | 20 |
| 50 | 193 | 190 | 8.405 | 198 | 8.345 | 007 | 929 | 515 | 10 |
| **7°00′** | .1222 | .1219 | 8.206 | .1228 | 8.144 | 1.008 | .9925 | 1.4486 | **83°00′** |
| 10 | 251 | 248 | 8.016 | 257 | 7.953 | 008 | 922 | 457 | 50 |
| 20 | 280 | 276 | 7.834 | 287 | 7.770 | 008 | 918 | 428 | 40 |
| 30 | .1309 | .1305 | 7.661 | .1317 | 7.596 | 1.009 | .9914 | 1.4399 | 30 |
| 40 | 338 | 334 | 7.496 | 346 | 7.429 | 009 | 911 | 370 | 20 |
| 50 | 367 | 363 | 7.337 | 376 | 7.269 | 009 | 907 | 341 | 10 |
| **8°00′** | .1396 | .1392 | 7.185 | .1405 | 7.115 | 1.010 | .9903 | 1.4312 | **82°00′** |
| 10 | 425 | 421 | 7.040 | 435 | 6.968 | 010 | 899 | 283 | 50 |
| 20 | 454 | 449 | 6.900 | 465 | 6.827 | 011 | 894 | 254 | 40 |
| 30 | .1484 | .1478 | 6.765 | .1495 | 6.691 | 1.011 | .9890 | 1.4224 | 30 |
| 40 | 513 | 507 | 6.636 | 524 | 6.561 | 012 | 886 | 195 | 20 |
| 50 | 542 | 536 | 6.512 | 554 | 6.435 | 012 | 881 | 166 | 10 |
| **9°00′** | .1571 | .1564 | 6.392 | .1584 | 6.314 | 1.012 | .9877 | 1.4137 | **81°00′** |
| 10 | 600 | 593 | 277 | 614 | 197 | 013 | 872 | 108 | 50 |
| 20 | 629 | 622 | 166 | 644 | 084 | 013 | 868 | 079 | 40 |
| 30 | .1658 | .1650 | 6.059 | .1673 | 5.976 | 1.014 | .9863 | 1.4050 | 30 |
| 40 | 687 | 679 | 5.955 | 703 | 871 | 014 | 858 | 1.4021 | 20 |
| 50 | 716 | 708 | 855 | 733 | 769 | 015 | 853 | 992 | 10 |
| **10°00′** | .1745 | .1736 | 5.759 | .1763 | 5.671 | 1.015 | .9848 | 1.3963 | **80°00′** |
| 10 | 774 | 765 | 665 | 793 | 576 | 016 | 843 | 934 | 50 |
| 20 | 804 | 794 | 575 | 823 | 485 | 016 | 838 | 904 | 40 |
| 30 | .1833 | .1822 | 5.487 | .1853 | 5.396 | 1.017 | .9833 | 1.3875 | 30 |
| 40 | 862 | 851 | 403 | 883 | 309 | 018 | 827 | 846 | 20 |
| 50 | 891 | 880 | 320 | 914 | 226 | 018 | 822 | 817 | 10 |
| **11°00′** | .1920 | .1908 | 5.241 | .1944 | 5.145 | 1.019 | .9816 | 1.3788 | **79°00′** |
| 10 | 949 | 937 | 164 | 974 | 066 | 019 | 811 | 759 | 50 |
| 20 | 978 | 965 | 089 | .2004 | 4.989 | 020 | 805 | 730 | 40 |
| 30 | .2007 | .1994 | 5.016 | .2035 | 4.915 | 1.020 | .9799 | 1.3701 | 30 |
| 40 | 036 | .2022 | 4.945 | 065 | 843 | 021 | 793 | 672 | 20 |
| 50 | 065 | 051 | 876 | 095 | 773 | 022 | 787 | 643 | 10 |
| **12°00′** | .2094 | .2079 | 4.810 | .2126 | 4.705 | 1.022 | .9781 | 1.3614 | **78°00′** |
| 10 | 123 | 108 | 745 | 156 | 638 | 023 | 775 | 584 | 50 |
| 20 | 153 | 136 | 682 | 186 | 574 | 024 | 769 | 555 | 40 |
| 30 | .2182 | .2164 | 4.620 | .2217 | 4.511 | 1.024 | .9763 | 1.3526 | 30 |
| 40 | 211 | 193 | 560 | 247 | 449 | 025 | 757 | 497 | 20 |
| 50 | 240 | 221 | 502 | 278 | 390 | 026 | 750 | 468 | 10 |
| **13°00′** | .2269 | .2250 | 4.445 | .2309 | 4.331 | 1.026 | .9744 | 1.3439 | **77°00′** |
| | | **cos $\theta$** | **sec $\theta$** | **cot $\theta$** | **tan $\theta$** | **csc $\theta$** | **sin $\theta$** | **Radians** | **Degrees** |
| | | | | | | | | **Angle $\theta$** | |

# Table 2 Values of Trigonometric Functions (cont.)

| Angle θ | | | | | | | | | |
|---|---|---|---|---|---|---|---|---|---|
| **Degrees** | **Radians** | **sin θ** | **csc θ** | **tan θ** | **cot θ** | **sec θ** | **cos θ** | | |
| **13°00′** | .2269 | .2250 | 4.445 | .2309 | 4.331 | 1.026 | .9744 | 1.3439 | **77°00′** |
| 10 | 298 | 278 | 390 | 339 | 275 | 027 | 737 | 410 | 50 |
| 20 | 327 | 306 | 336 | 370 | 219 | 028 | 730 | 381 | 40 |
| 30 | .2356 | .2334 | 4.284 | .2401 | 4.165 | 1.028 | .9724 | 1.3352 | 30 |
| 40 | 385 | 363 | 232 | 432 | 113 | 029 | 717 | 323 | 20 |
| 50 | 414 | 391 | 182 | 462 | 061 | 030 | 710 | 294 | 10 |
| **14°00′** | .2443 | .2419 | 4.134 | .2493 | 4.011 | 1.031 | .9703 | 1.3265 | **76°00′** |
| 10 | 473 | 447 | 086 | 524 | 3.962 | 031 | 696 | 235 | 50 |
| 20 | 502 | 476 | 039 | 555 | 914 | 032 | 689 | 206 | 40 |
| 30 | .2531 | .2504 | 3.994 | .2586 | 3.867 | 1.033 | .9681 | 1.3177 | 30 |
| 40 | 560 | 532 | 950 | 617 | 821 | 034 | 674 | 148 | 20 |
| 50 | 589 | 560 | 906 | 648 | 776 | 034 | 667 | 119 | 10 |
| **15°00′** | .2618 | .2588 | 3.864 | .2679 | 3.732 | 1.035 | .9659 | 1.3090 | **75°00′** |
| 10 | 647 | 616 | 822 | 711 | 689 | 036 | 652 | 061 | 50 |
| 20 | 676 | 644 | 782 | 742 | 647 | 037 | 644 | 032 | 40 |
| 30 | .2705 | .2672 | 3.742 | .2773 | 3.606 | 1.038 | .9636 | 1.3003 | 30 |
| 40 | 734 | 700 | 703 | 805 | 566 | 039 | 628 | 974 | 20 |
| 50 | 763 | 728 | 665 | 836 | 526 | 039 | 621 | 945 | 10 |
| **16°00′** | .2793 | .2756 | 3.628 | .2867 | 3.487 | 1.040 | .9613 | 1.2915 | **74°00′** |
| 10 | 822 | 784 | 592 | 899 | 450 | 041 | 605 | 886 | 50 |
| 20 | 851 | 812 | 556 | 931 | 412 | 042 | 596 | 857 | 40 |
| 30 | .2880 | .2840 | 3.521 | .2962 | 3.376 | 1.043 | .9588 | 1.2828 | 30 |
| 40 | 909 | 868 | 487 | 994 | 340 | 044 | 580 | 799 | 20 |
| 50 | 938 | 896 | 453 | .3026 | 305 | 045 | 572 | 770 | 10 |
| **17°00′** | .2967 | .2924 | 3.420 | .3057 | 3.271 | 1.046 | .9563 | 1.2741 | **73°00′** |
| 10 | 996 | 952 | 388 | 089 | 237 | 047 | 555 | 712 | 50 |
| 20 | .3025 | 979 | 357 | 121 | 204 | 048 | 546 | 683 | 40 |
| 30 | .3054 | .3007 | 3.326 | .3153 | 3.172 | 1.048 | .9537 | 1.2654 | 30 |
| 40 | 083 | 035 | 295 | 185 | 140 | 049 | 528 | 625 | 20 |
| 50 | 113 | 062 | 265 | 217 | 108 | 050 | 520 | 595 | 10 |
| **18°00′** | .3142 | .3090 | 3.236 | .3249 | 3.078 | 1.051 | .9511 | 1.2566 | **72°00′** |
| 10 | 171 | 118 | 207 | 281 | 047 | 052 | 502 | 537 | 50 |
| 20 | 200 | 145 | 179 | 314 | 018 | 053 | 492 | 508 | 40 |
| 30 | .3229 | .3173 | 3.152 | .3346 | 2.989 | 1.054 | .9483 | 1.2479 | 30 |
| 40 | 258 | 201 | 124 | 378 | 960 | 056 | 474 | 450 | 20 |
| 50 | 287 | 228 | 098 | 411 | 932 | 057 | 465 | 421 | 10 |
| **19°00′** | .3316 | .3256 | 3.072 | .3443 | 2.904 | 1.058 | .9455 | 1.2392 | **71°00′** |
| 10 | 345 | 283 | 046 | 476 | 877 | 059 | 446 | 363 | 50 |
| 20 | 374 | 311 | 021 | 508 | 850 | 060 | 436 | 334 | 40 |
| 30 | .3403 | .3338 | 2.996 | .3541 | 2.824 | 1.061 | .9426 | 1.2305 | 30 |
| 40 | 432 | 365 | 971 | 574 | 798 | 062 | 417 | 275 | 20 |
| 50 | 462 | 393 | 947 | 607 | 773 | 063 | 407 | 246 | 10 |
| **20°00′** | .3491 | .3420 | 2.924 | .3640 | 2.747 | 1.064 | .9397 | 1.2217 | **70°00′** |
| | | **cos θ** | **sec θ** | **cot θ** | **tan θ** | **csc θ** | **sin θ** | **Radians** | **Degrees** |
| | | | | | | | | **Angle θ** | |

## Table 2 Values of Trigonometric Functions (cont.)

| Angle $\theta$ | | | | | | | | | |
|---|---|---|---|---|---|---|---|---|---|
| **Degrees** | **Radians** | **sin $\theta$** | **csc $\theta$** | **tan $\theta$** | **cot $\theta$** | **sec $\theta$** | **cos $\theta$** | | |
| **20°00′** | .3491 | .3420 | 2.924 | .3640 | 2.747 | 1.064 | .9397 | 1.2217 | **70°00′** |
| 10 | 520 | 448 | 901 | 673 | 723 | 065 | 387 | 188 | 50 |
| 20 | 549 | 475 | 878 | 706 | 699 | 066 | 377 | 159 | 40 |
| 30 | .3578 | .3502 | 2.855 | .3739 | 2.675 | 1.068 | .9367 | 1.2130 | 30 |
| 40 | 607 | 529 | 833 | 772 | 651 | 069 | 356 | 101 | 20 |
| 50 | 636 | 557 | 812 | 805 | 628 | 070 | 346 | 072 | 10 |
| **21°00′** | .3665 | .3584 | 2.790 | .3839 | 2.605 | 1.071 | .9336 | 1.2043 | **69°00′** |
| 10 | 694 | 611 | 769 | 872 | 583 | 072 | 325 | 1.2014 | 50 |
| 20 | 723 | 638 | 749 | 906 | 560 | 074 | 315 | 985 | 40 |
| 30 | .3752 | .3665 | 2.729 | .3939 | 2.539 | 1.075 | .9304 | 1.1956 | 30 |
| 40 | 782 | 692 | 709 | 973 | 517 | 076 | 293 | 926 | 20 |
| 50 | 811 | 719 | 689 | .4006 | 496 | 077 | 283 | 897 | 10 |
| **22°00′** | .3840 | .3746 | 2.669 | .4040 | 2.475 | 1.079 | .9272 | 1.1868 | **68°00′** |
| 10 | 869 | 773 | 650 | 074 | 455 | 080 | 261 | 839 | 50 |
| 20 | 898 | 800 | 632 | 108 | 434 | 081 | 250 | 810 | 40 |
| 30 | .3927 | .3827 | 2.613 | .4142 | 2.414 | 1.082 | .9239 | 1.1781 | 30 |
| 40 | 956 | 854 | 595 | 176 | 394 | 084 | 228 | 752 | 20 |
| 50 | 985 | 881 | 577 | 210 | 375 | 085 | 216 | 723 | 10 |
| **23°00′** | .4014 | .3907 | 2.559 | .4245 | 2.356 | 1.086 | .9205 | 1.1694 | **67°00′** |
| 10 | 043 | 934 | 542 | 279 | 337 | 088 | 194 | 665 | 50 |
| 20 | 072 | 961 | 525 | 314 | 318 | 089 | 182 | 636 | 40 |
| 30 | .4102 | .3987 | 2.508 | .4348 | 2.300 | 1.090 | .9171 | 1.1606 | 30 |
| 40 | 131 | .4014 | 491 | 383 | 282 | 092 | 159 | 577 | 20 |
| 50 | 160 | 041 | 475 | 417 | 264 | 093 | 147 | 548 | 10 |
| **24°00′** | .4189 | .4067 | 2.459 | .4452 | 2.246 | 1.095 | .9135 | 1.1519 | **66°00′** |
| 10 | 218 | 094 | 443 | 487 | 229 | 096 | 124 | 490 | 50 |
| 20 | 247 | 120 | 427 | 522 | 211 | 097 | 112 | 461 | 40 |
| 30 | .4276 | .4147 | 2.411 | .4557 | 2.194 | 1.099 | .9100 | 1.1432 | 30 |
| 40 | 305 | 173 | 396 | 592 | 177 | 100 | 088 | 403 | 20 |
| 50 | 334 | 200 | 381 | 628 | 161 | 102 | 075 | 374 | 10 |
| **25°00′** | .4363 | .4226 | 2.366 | .4663 | 2.145 | 1.103 | .9063 | 1.1345 | **65°00′** |
| 10 | 392 | 253 | 352 | 699 | 128 | 105 | 051 | 316 | 50 |
| 20 | 422 | 279 | 337 | 734 | 112 | 106 | 038 | 286 | 40 |
| 30 | .4451 | .4305 | 2.323 | .4770 | 2.097 | 1.108 | .9026 | 1.1257 | 30 |
| 40 | 480 | 331 | 309 | 806 | 081 | 109 | 013 | 228 | 20 |
| 50 | 509 | 358 | 295 | 841 | 066 | 111 | 001 | 199 | 10 |
| **26°00′** | .4538 | .4384 | 2.281 | .4877 | 2.050 | 1.113 | .8988 | 1.1170 | **64°00′** |
| 10 | 567 | 410 | 268 | 913 | 035 | 114 | 975 | 141 | 50 |
| 20 | 596 | 436 | 254 | 950 | 020 | 116 | 962 | 112 | 40 |
| 30 | .4625 | .4462 | 2.241 | .4986 | 2.006 | 1.117 | .8949 | 1.1083 | 30 |
| 40 | 654 | 488 | 228 | .5022 | 1.991 | 119 | 936 | 054 | 20 |
| 50 | 683 | 514 | 215 | 059 | 977 | 121 | 923 | 1.1025 | 10 |
| **27°00′** | .4712 | .4540 | 2.203 | .5095 | 1.963 | 1.122 | .8910 | 1.0996 | **63°00′** |
| | | **cos $\theta$** | **sec $\theta$** | **cot $\theta$** | **tan $\theta$** | **csc $\theta$** | **sin $\theta$** | **Radians** | **Degrees** |
| | | | | | | | | **Angle $\theta$** | |

# Table 2 Values of Trigonometric Functions (cont.)

| Angle $\theta$ | | | | | | | | | |
|---|---|---|---|---|---|---|---|---|---|
| **Degrees** | **Radians** | **sin $\theta$** | **csc $\theta$** | **tan $\theta$** | **cot $\theta$** | **sec $\theta$** | **cos $\theta$** | | |
| **27°00′** | .4712 | .4540 | 2.203 | .5095 | 1.963 | 1.122 | .8910 | 1.0996 | **63°00′** |
| 10 | 741 | 566 | 190 | 132 | 949 | 124 | 897 | 966 | 50 |
| 20 | 771 | 592 | 178 | 169 | 935 | 126 | 884 | 937 | 40 |
| 30 | 4800 | .4617 | 2.166 | .5206 | 1.921 | 1.127 | .8870 | 1.0908 | 30 |
| 40 | 829 | 643 | 154 | 243 | 907 | 129 | 857 | 879 | 20 |
| 50 | 858 | 669 | 142 | 280 | 894 | 131 | 843 | 850 | 10 |
| **28°00′** | .4887 | .4695 | 2.130 | .5317 | 1.881 | 1.133 | .8829 | 1.0821 | **62°00′** |
| 10 | 916 | 720 | 118 | 354 | 868 | 134 | 816 | 792 | 50 |
| 20 | 945 | 746 | 107 | 392 | 855 | 136 | 802 | 763 | 40 |
| 30 | .4974 | .4772 | 2.096 | .5430 | 1.842 | 1.138 | .8788 | 1.0734 | 30 |
| 40 | .5003 | 797 | 085 | 467 | 829 | 140 | 774 | 705 | 20 |
| 50 | 032 | 823 | 074 | 505 | 816 | 142 | 760 | 676 | 10 |
| **29°00′** | .5061 | .4848 | 2.063 | .5543 | 1.804 | 1.143 | .8746 | 1.0647 | **61°00′** |
| 10 | 091 | 874 | 052 | 581 | 792 | 145 | 732 | 617 | 50 |
| 20 | 120 | 899 | 041 | 619 | 780 | 147 | 718 | 588 | 40 |
| 30 | .5149 | .4924 | 2.031 | .5658 | 1.767 | 1.149 | .8704 | 1.0559 | 30 |
| 40 | 178 | 950 | 020 | 696 | 756 | 151 | 689 | 530 | 20 |
| 50 | 207 | 975 | 010 | 735 | 744 | 153 | 675 | 501 | 10 |
| **30°00′** | .5236 | .5000 | 2.000 | .5774 | 1.732 | 1.155 | .8660 | 1.0472 | **60°00′** |
| 10 | 265 | 025 | 1.990 | 812 | 720 | 157 | 646 | 443 | 50 |
| 20 | 294 | 050 | 980 | 851 | 709 | 159 | 631 | 414 | 40 |
| 30 | .5323 | .5075 | 1.970 | .5890 | 1.698 | 1.161 | .8616 | 1.0385 | 30 |
| 40 | 352 | 100 | 961 | 930 | 686 | 163 | 601 | 356 | 20 |
| 50 | 381 | 125 | 951 | 969 | 675 | 165 | 587 | 327 | 10 |
| **31°00′** | .5411 | .5150 | 1.942 | .6009 | 1.664 | 1.167 | .8572 | 1.0297 | **59°00′** |
| 10 | 440 | 175 | 932 | 048 | 653 | 169 | 557 | 268 | 50 |
| 20 | 469 | 200 | 923 | 088 | 643 | 171 | 542 | 239 | 40 |
| 30 | .5498 | .5225 | 1.914 | .6128 | 1.632 | 1.173 | .8526 | 1.0210 | 30 |
| 40 | 527 | 250 | 905 | 168 | 621 | 175 | 511 | 181 | 20 |
| 50 | 556 | 275 | 896 | 208 | 611 | 177 | 496 | 152 | 10 |
| **32°00′** | .5585 | .5299 | 1.887 | .6249 | 1.600 | 1.179 | .8480 | 1.0123 | **58°00′** |
| 10 | 614 | 324 | 878 | 289 | 590 | 181 | 465 | 094 | 50 |
| 20 | 643 | 348 | 870 | 330 | 580 | 184 | 450 | 065 | 40 |
| 30 | .5672 | .5373 | 1.861 | .6371 | 1.570 | 1.186 | .8434 | 1.0036 | 30 |
| 40 | 701 | 398 | 853 | 412 | 560 | 188 | 418 | 1.0007 | 20 |
| 50 | 730 | 422 | 844 | 453 | 550 | 190 | 403 | 977 | 10 |
| **33°00′** | .5760 | .5446 | 1.836 | .6494 | 1.540 | 1.192 | .8387 | .9948 | **57°00′** |
| 10 | 789 | 471 | 828 | 536 | 530 | 195 | 371 | 919 | 50 |
| 20 | 818 | 495 | 820 | 577 | 520 | 197 | 355 | 890 | 40 |
| 30 | .5847 | .5519 | 1.812 | .6619 | 1.511 | 1.199 | .8339 | .9861 | 30 |
| 40 | 876 | 544 | 804 | 661 | 501 | 202 | 323 | 832 | 20 |
| 50 | 905 | 568 | 796 | 703 | 1.492 | 204 | 307 | 803 | 10 |
| **34°00′** | .5934 | .5592 | 1.788 | .6745 | 1.483 | 1.206 | .8290 | .9774 | **56°00′** |
| | | **cos $\theta$** | **sec $\theta$** | **cot $\theta$** | **tan $\theta$** | **csc $\theta$** | **sin $\theta$** | **Radians** | **Degrees** |
| | | | | | | | | **Angle $\theta$** | |

# Table 2 Values of Trigonometric Functions (cont.)

| Angle θ | | | | | | | | | |
|---|---|---|---|---|---|---|---|---|---|
| Degrees | Radians | sin θ | csc θ | tan θ | cot θ | sec θ | cos θ | | |
| **34°00′** | .5934 | .5592 | 1.788 | .6745 | 1.483 | 1.206 | .8290 | .9774 | **56°00′** |
| 10 | 963 | 616 | 781 | 787 | 473 | 209 | 274 | 745 | 50 |
| 20 | 992 | 640 | 773 | 830 | 464 | 211 | 258 | 716 | 40 |
| 30 | .6021 | .5664 | 1.766 | .6873 | 1.455 | 1.213 | .8241 | .9687 | 30 |
| 40 | 050 | 688 | 758 | 916 | 446 | 216 | 225 | 657 | 20 |
| 50 | 080 | 712 | 751 | 959 | 437 | 218 | 208 | 628 | 10 |
| **35°00′** | .6109 | .5736 | 1.743 | .7002 | 1.428 | 1.221 | .8192 | .9599 | **55°00′** |
| 10 | 138 | 760 | 736 | 046 | 419 | 223 | 175 | 570 | 50 |
| 20 | 167 | 783 | 729 | 089 | 411 | 226 | 158 | 541 | 40 |
| 30 | .6196 | .5807 | 1.722 | .7133 | 1.402 | 1.228 | .8141 | .9512 | 30 |
| 40 | 225 | 831 | 715 | 177 | 393 | 231 | 124 | 483 | 20 |
| 50 | 254 | 854 | 708 | 221 | 385 | 233 | 107 | 454 | 10 |
| **36°00′** | .6283 | .5878 | 1.701 | .7265 | .1376 | 1.236 | .8090 | .9425 | **54°00′** |
| 10 | 312 | 901 | 695 | 310 | 368 | 239 | 073 | 396 | 50 |
| 20 | 341 | 925 | 688 | 355 | 360 | 241 | 056 | 367 | 40 |
| 30 | .6370 | .5948 | 1.681 | .7400 | 1.351 | 1.244 | .8039 | .9338 | 30 |
| 40 | 400 | 972 | 675 | 445 | 343 | 247 | 021 | 308 | 20 |
| 50 | 429 | 995 | 668 | 490 | 335 | 249 | 004 | 279 | 10 |
| **37°00′** | .6458 | .6018 | 1.662 | .7536 | 1.327 | 1.252 | .7986 | .9250 | **53°00′** |
| 10 | 487 | 041 | 655 | 581 | 319 | 255 | 969 | 221 | 50 |
| 20 | 516 | 065 | 649 | 627 | 311 | 258 | 951 | 192 | 40 |
| 30 | .6545 | .6088 | 1.643 | .7673 | 1.303 | 1.260 | .7934 | .9163 | 30 |
| 40 | 574 | 111 | 636 | 720 | 295 | 263 | 916 | 134 | 20 |
| 50 | 603 | 134 | 630 | 766 | 288 | 266 | 898 | 105 | 10 |
| **38°00′** | .6632 | .6157 | 1.624 | .7813 | 1.280 | 1.269 | .7880 | .9076 | **52°00′** |
| 10 | 661 | 180 | 618 | 860 | 272 | 272 | 862 | 047 | 50 |
| 20 | 690 | 202 | 612 | 907 | 265 | 275 | 844 | .9018 | 40 |
| 30 | .6720 | .6225 | 1.606 | .7954 | 1.257 | 1.278 | .7826 | .8988 | 30 |
| 40 | 749 | 248 | 601 | .8002 | 250 | 281 | 808 | 959 | 20 |
| 50 | 778 | 271 | 595 | 050 | 242 | 284 | 790 | 930 | 10 |
| **39°00′** | .6807 | .6293 | 1.589 | .8098 | 1.235 | 1.287 | .7771 | .8901 | **51°00′** |
| 10 | 836 | 316 | 583 | 146 | 228 | 290 | 753 | 872 | 50 |
| 20 | 865 | 338 | 578 | 195 | 220 | 293 | 735 | 843 | 40 |
| 30 | .6894 | .6361 | 1.572 | .8243 | 1.213 | 1.296 | .7716 | .8814 | 30 |
| 40 | 923 | 383 | 567 | 292 | 206 | 299 | 698 | 785 | 20 |
| 50 | 952 | 406 | 561 | 342 | 199 | 302 | 679 | 756 | 10 |
| **40°00′** | .6981 | .6428 | 1.556 | .8391 | 1.192 | 1.305 | .7660 | .8727 | **50°00′** |
| 10 | .7010 | 450 | 550 | 441 | 185 | 309 | 642 | 698 | 50 |
| 20 | 039 | 472 | 545 | 491 | 178 | 312 | 623 | 668 | 40 |
| 30 | .7069 | .6494 | 1.540 | .8541 | 1.171 | 1.315 | .7604 | .8639 | 30 |
| 40 | 098 | 517 | 535 | 591 | 164 | 318 | 585 | 610 | 20 |
| 50 | 127 | 539 | 529 | 642 | 157 | 322 | 566 | 581 | 10 |
| **41°00′** | .7156 | .6561 | 1.524 | .8693 | 1.150 | 1.325 | .7547 | .8552 | **49°00′** |
| | | cos θ | sec θ | cot θ | tan θ | csc θ | sin θ | Radians | Degrees |
| | | | | | | | | Angle θ | |

# Table 2 Values of Trigonometric Functions (cont.)

| Angle $\theta$ | | | | | | | | | |
|---|---|---|---|---|---|---|---|---|---|
| **Degrees** | **Radians** | **sin $\theta$** | **csc $\theta$** | **tan $\theta$** | **cot $\theta$** | **sec $\theta$** | **cos $\theta$** | | |
| **41°00′** | .7156 | .6561 | 1.524 | .8693 | 1.150 | 1.325 | .7547 | .8552 | **49°00′** |
| 10 | 185 | 583 | 519 | 744 | 144 | 328 | 528 | 523 | 50 |
| 20 | 214 | 604 | 514 | 796 | 137 | 332 | 509 | 494 | 40 |
| 30 | .7243 | .6626 | 1.509 | .8847 | 1.130 | 1.335 | .7490 | .8465 | 30 |
| 40 | 272 | 648 | 504 | 899 | 124 | 339 | 470 | 436 | 20 |
| 50 | 301 | 670 | 499 | 952 | 117 | 342 | 451 | 407 | 10 |
| **42°00′** | .7330 | .6691 | 1.494 | .9004 | 1.111 | 1.346 | .7431 | .8378 | **48°00′** |
| 10 | 359 | 713 | 490 | 057 | 104 | 349 | 412 | 348 | 50 |
| 20 | 389 | 734 | 485 | 110 | 098 | 353 | 392 | 319 | 40 |
| 30 | .7418 | .6756 | 1.480 | .9163 | 1.091 | 1.356 | .7373 | .8290 | 30 |
| 40 | 447 | 777 | 476 | 217 | 085 | 360 | 353 | 261 | 20 |
| 50 | 476 | 799 | 471 | 271 | 079 | 364 | 333 | 232 | 10 |
| **43°00′** | .7505 | .6820 | 1.466 | .9325 | 1.072 | 1.367 | .7314 | .8203 | **47°00′** |
| 10 | 534 | 841 | 462 | 380 | 066 | 371 | 294 | 174 | 50 |
| 20 | 563 | 862 | 457 | 435 | 060 | 375 | 274 | 145 | 40 |
| 30 | .7592 | .6884 | 1.453 | .9490 | 1.054 | 1.379 | .7254 | .8116 | 30 |
| 40 | 621 | 905 | 448 | 545 | 048 | 382 | 234 | 087 | 20 |
| 50 | 650 | 926 | 444 | 601 | 042 | 386 | 214 | 058 | 10 |
| **44°00′** | .7679 | .6947 | 1.440 | .9657 | 1.036 | 1.390 | .7193 | .8029 | **46°00′** |
| 10 | 709 | 967 | 435 | 713 | 030 | 394 | 173 | .7999 | 50 |
| 20 | 738 | 988 | 431 | 770 | 024 | 398 | 153 | 970 | 40 |
| 30 | .7767 | .7009 | 1.427 | .9827 | 1.018 | 1.402 | .7133 | .7941 | 30 |
| 40 | 796 | 030 | 423 | 884 | 012 | 406 | 112 | 912 | 20 |
| 50 | 825 | 050 | 418 | 942 | 006 | 410 | 092 | 883 | 10 |
| **45°00′** | .7854 | .7071 | 1.414 | 1.000 | 1.000 | 1.414 | .7071 | .7854 | **45°00′** |
| | | **cos $\theta$** | **sec $\theta$** | **cot $\theta$** | **tan $\theta$** | **csc $\theta$** | **sin $\theta$** | **Radians** | **Degrees** |
| | | | | | | | | Angle $\theta$ | |

# Answers to Selected Exercises

## For the Student

*To help you study and understand the course material, a* Solutions and Study Guide, *by Joseph Mutter, is available from your college bookstore. This book provides complete, step-by-step solutions to more than half of the odd-numbered exercises in the text, detailed chapter summaries, and practice chapter tests with complete solutions.*

## Chapter 1

### [1.1]

**1.** true **3.** false **5.** false **7.** true **9.** true **11.** false **13.** true **15.** false **17.** true **19.** true **21.** true **23.** false **25.** commutative property of addition **27.** symmetric law of equality **29.** additive identity **31.** $(x + 3) + 7$ **33.** $a < 9$ **35.** $-15$ **37.** $-3$ **39.** $10 - x$ **41.** $-\frac{4}{9}$ **43.** $-\frac{1}{2}$ **45.** $\frac{3}{4}$ **47.** $y - x$ **49.** 3 **51.** $-3$ **53.** 4 **55.** 10 **57.** 5 **59.** $y - 7$ **61.** $y - x$ **63.** One set of numbers is $a = 1$, $b = 2$, and $c = 3$. **65.** Start with $0 - a$, use the definition of subtraction and the additive identity property. **67.** Start with $a + c = a + c$ and use the substitution axiom. **69.** Start with $-a + (-(-a)) = 0$ and $0 = (-a) + a$. Use the transitive law, then use the addition property of equality. **71.** Start by using the fact that $-(a + b) = (-1)(a + b)$.

### [1.2]

**1.** $x^7$ **3.** $6x$ **5.** $b^6$ **7.** $\frac{a^3}{b^2}$ (cannot be simplified) **9.** 1 **11.** $\frac{1}{8x}$ **13.** $2x^4$ **15.** 1 **17.** $\frac{xy}{y + x}$ **19.** $\frac{1}{4x^4y^6}$ **21.** $\frac{x^7}{4y^4}$ **23.** $\frac{x^{24}y^6}{27}$ **25.** $\frac{x^{12}y^4}{4}$ **27.** $\frac{4a^4}{b^{10}}$ **29.** $\frac{a^8}{4b^4}$ **31.** $4.56 \times 10^8$ **33.** $1 \times 10^{-2}$ **35.** 235,000,000 **37.** 0.000 000 041 7 **39.** $1.17 \times 10^2$ **41.** $1.18 \times 10^9$ **43.** $\$2.95 \times 10^3 = \$2950$ **45.** $1.50 \times 10^{11}$ m **47.** true **48.** false **49.** true **50.** true

### [1.3]

**1.** binomial; 3 **3.** trinomial; 7 **5.** $x^2 + 2x$ **7.** $10a^2b^2 + 5ab - 3$ **9.** $8x^2 - 7x + 5$ **11.** $9a^2b^2 - 12ab$ **13.** $a^3 + 2a$ **15.** $2y^5 + 2y^2 - 8$ **17.** $-a^4b^2 - 13a^2b^4 + 10$ **19.** $7a^2b^2 + 2$ **21.** $18x^6y^4$ **23.** $-12a^3b^5 + 20a^2b^4$ **25.** $a^2 + ab - 2b^2$ **27.** $10x^2y^2 - 29xy - 21$ **29.** $x^2 - 4y^2$ **31.** $x^2 + 4xy + 4y^2$ **33.** $x^2 - 4xy + 4y^2$ **35.** $9a^4 - b^4$ **37.** $12x^2 - 32xy + 21y^2 + 4x - 6y$ **39.** $12a^3 - 34a^2b + 35ab^2 - 49b^3$ **41.** $x^2 + 4xy + 4y^2 - 10x - 20y + 25$ **43.** $x^2 - 4xy + 4y^2 - 25$ **45.** $x^{3n}y^n - x^{2n}y^{2n}$ **47.** $x^{2n} - y^{2n}$ **49.** $x^{2n+2} + 2x^{n+1}y^{n+1} + y^{2n+2}$ **51.** $2x^4 - 2x^3y - 11x^2y^2 + 17xy^3 - 6y^4$ **53.** $y^3 - 6y^2 + 5y - \frac{1}{y^2}$ **55.** $-3x^2 + xy - 2x^3y^2$ **57.** $a^2 + a + 2 - \frac{1}{a - 1}$ **59.** $x + 3y$ **61.** $x^3 - x^2 + 3x - 1$ **63.** $3xy(2xy - 1)$ **65.** $(5ab + 3)(a - 2)$ **67.** $(2a^2b^2 - 3c^2)(4ab - 1)$ **69.** $(x + 1)(x + 7)$ **71.** $(u + 4)(u + 5)$ **73.** $(3x + 1)(x - 2)$ **75.** $(5u - v)(u + 5v)$ **77.** $(2x + 3y)(2x - 3y)$ **79.** $(2x + 3y)^2$ **81.** $(3u - v)(9u^2 + 3uv + v^2)$ **83.** $6(x + y)(x - y)$ **85.** $2(2x - 5y)(7x + 3y)$ **87.** $(uv - 2)(5uv - 1)$

**89.** $3(u - 10v)^2$ **91.** $4(2x + y^3)(4x^2 - 2xy^3 + y^6)$ **93.** prime **95.** $-5(u - 6v)(u - 8v)$ **97.** $(u + 2v)(3u - 8v)$ **99.** $(u - v + 4)(u - v - 4)$ **101.** $(3u^3 - 5v^2)^2$ **103.** $(3x + 5y)(5x + 3y)$ **105.** $(x + y + u + v)(x + y - u - v)$ **107.** $(x^2 + 5y^2)(x^2 - 3y^2)$ **109.** $(2u + v)(u - v)(4u^2 - 2uv + v^2)(u^2 + uv + v^2)$ **111.** $(2x^n + y^n)(2x^n - y^n)$ **113.** $(3x^n - 2y^n)(9x^{2n} + 6x^n y^n + 4y^{2n})$ **115.** $(x - 2y)(x + 2y + 6)$ **117.** $(u + 5v + w)(u + 5v - w)$ **119.** $(x + y)(x^2 + xy + y^2)$ **121.** **(a)** $6u^2 + 11$ **(b)** \$1361 **123.** **(a)** $\pi(R + r)(R - r)$ **(b)** 255.47 $m^2$ **(c)** approximately 341 **125.** $\dfrac{1}{9x^6y^4}$ **126.** $\dfrac{4}{a^8b^{12}}$ **127.** $\dfrac{2x^7y^2}{5}$ **128.** $\dfrac{a^6}{2b^{14}}$ **129.** $\dfrac{9y^6}{x^{20}}$ **130.** $\dfrac{9}{a^{10}b^5}$

## [1.4]

**1.** $x = -7$; $y = 2, 3$ **3.** $a = 0$; $a = -b$ **5.** yes **7.** $\dfrac{7y^2}{3x}$ **9.** $\dfrac{x(2x + 3)}{3 - x}$ **11.** $\dfrac{xy(x + y)}{2}$ **13.** $\dfrac{5xy}{12(x^2 - y^2)}$ **15.** $\dfrac{5(x - 1)}{(x - 3)(x + 4)}$ **17.** 1 **19.** $\dfrac{(y - 1)(y + 2)}{2(y - 2)}$ **21.** $\dfrac{x(u + x)}{v - w}$ **23.** $x - 2y$ **25.** $x^2y^2$ **27.** $5x^2(x - 3)(x^2 + 3x + 9)$ **29.** $\dfrac{3y - 1}{y}$ **31.** $\dfrac{-2}{a - 2}$ **33.** $\dfrac{3(7x - 6)}{5(x - 2)(x + 2)}$ **35.** $\dfrac{1}{y + 1}$ **37.** $\dfrac{21x - 16}{(x - 6)(x - 1)}$ **39.** $\dfrac{11x^2 + 46x + 80}{12(x - 2)(x + 2)(x + 1)}$ **41.** $\dfrac{y^2 + yx + x^2}{y^2}$ **43.** $\dfrac{x - 2}{x}$ **45.** $\dfrac{x + 4}{x + 2}$ **47.** $-a^2$ **49.** $\dfrac{-2x - h}{x^2(x + h)^2}$ **51.** $\dfrac{6(2x + 1)(x + 1)}{(4x + 3)^2}$ **53.** $\dfrac{ab}{a + b}$ **55.** $\dfrac{v + u}{uv}$ **57.** $\dfrac{ab}{a + 3b}$ **59.** $\dfrac{u(1 + uv)}{1 + u^2 + uv}$ **60.** $10(x + 10y)(x - 10y)$ **61.** $(3u - 7v)(2u + 5v)$ **62.** $(3u - 4v)(9u^2 + 12uv + 16v^2)$ **63.** $4a^3b^2 - 3a^2b^3 + 9$ **64.** $27x^3 + 125y^3$ **65.** $4x^3 - 3x^2 + 2x - 5$ **66.** true **67.** true **68.** true **69.** false **70.** **(a)** $y^2$ **(b)** $y^3$ **71.** **(a)** $4x^2$ **(b)** $8x^3$ **72.** **(a)** $9z^4$ **(b)** $27z^6$

## [1.5]

**1.** 9 **3.** $-5$ **5.** $2a^3$ **7.** $-3ab^2$ **9.** $x + 5$ **11.** 100 **13.** $28\sqrt[3]{2}$ **15.** $2xy\sqrt[3]{x^2y}$ **17.** $\dfrac{5x}{y}$ **19.** $5xy$ **21.** $x^2y^3$ **23.** $2xy^2\sqrt[4]{x^2y}$ **25.** $5x\sqrt{5}$ **27.** $2a^3b^4$ **29.** $5x^2y^3\sqrt{3xy}$ **31.** $\dfrac{5a^3b}{2}$ **33.** $2(x + y)\sqrt[4]{2(x + y)}$ **35.** $\dfrac{y^6}{8x^9}$ **37.** $27\sqrt{3}$ **39.** $21\sqrt[4]{3}$ **41.** $16x\sqrt{2xy}$ **43.** $2 + \sqrt{3}$ **45.** $\dfrac{5\sqrt{30}}{9}$ **47.** $\dfrac{2\sqrt{3xy}}{3y}$ **49.** $\dfrac{\sqrt[3]{12xy^2}}{2y}$ **51.** $\dfrac{2a\sqrt[4]{ab}}{b}$ **53.** $-(\sqrt{3} + \sqrt{5})$ **55.** $\dfrac{x + 2\sqrt{x} + 1}{x - 1}$ **57.** $\dfrac{28\sqrt{5a}}{5}$ **59.** $\dfrac{x - y}{x\sqrt{y} + y\sqrt{x}}$ **61.** 0.819 **63.** 2.2 **65.** $\dfrac{1}{x - 8y}$ **66.** $\dfrac{-2a - 13}{(a + 3)(a - 3)(a - 4)}$ **67.** $\dfrac{y^2}{x^2}$ **68.** $\dfrac{u + v}{uv - 1}$ **69.** $9u^2 - 12uv + 4v^2 + 6u - 4v + 1$ **70.** $-x^2y^2 - 3xy^2 + 4xy$

## Chapter 1 Review Exercises

**1.** false **2.** true **3.** false **4.** true **5.** associative law of addition **6.** transitive law of $<$ **7.** b **8.** $-7$ **9.** $-x$ **10.** $\dfrac{1}{2}$ **11.** $4\sqrt{7}$ **12.** $\dfrac{5}{x^7}$ **13.** $\dfrac{3}{y}$ **14.** $x^2y^6$ **15.** $\dfrac{y^3}{125x^9}$ **16.** $3.29 \times 10^{-13}$ **17.** $1.57 \times 10^{10}$ **18.** $11x^2y^2 - 4xy + 9$ **19.** $-x^2 - 7y^2 - x - 2y$ **20.** $8u^3v^2 - 4u^2v^3 - u^2v^2 + 2$ **21.** $-10x^4y^3 + 35x^5y^3$ **22.** $49x^2 - 28xy + 4y^2$ **23.** $9a^2 - 64b^2$ **24.** $10a^4 + a^2b - 2b^2$ **25.** $10u^3 + u^2v - 19uv^2 + 8v^3$ **26.** $9x^2 - 12xy + 4y^2 - 30x + 20y + 25$ **27.** $-2ab + 4 - \dfrac{7b^2}{a}$ **28.** $x^2 - xy - y^2 - \dfrac{6y^3}{x - 3y}$ **29.** $5xy^2(2xy - 3)$ **30.** $(2x - 3)(x - 2y)$ **31.** $(3a - 4b)(3a + 4b)$ **32.** $(3a + 7b)(a - 2b)$ **33.** $2(3x + y^2)(9x^2 - 3xy^2 + y^4)$ **34.** $(5x + 3y)^2$ **35.** $(2u - 3v)(4u + 5v)$ **36.** $(5x - 4y)(25x^2 + 20xy + 16y^2)$ **37.** $(4x - 7y)^2$ **38.** $(u + v)(u - v)(2u^2 + v^2)$ **39.** $(x + y - 5)(x - y + 5)$ **40.** $(3x^n + y^n)(3x^n - y^n)$ **41.** 1, 2 **42.** $a = 2, -2$; $b = -7$ **43.** yes **44.** no **45.** $\dfrac{12a}{25b^2}$ **46.** $-1$ **47.** $\dfrac{x + 8}{x - 7}$ **48.** $\dfrac{a - 2}{a - 5}$ **49.** $\dfrac{(a - b)(a + 5)}{a - 1}$ **50.** $(a - b)^2(a^2 + ab + b^2)$ **51.** $(a - 1)(a - 6)^2$ **52.** $\dfrac{3 - 2y}{y(y + 1)}$ **53.** $\dfrac{3x + 4y}{x + y}$ **54.** $-\dfrac{a}{b}$ **55.** $\dfrac{a - 2}{a - 5}$ **56.** $-3$ **57.** not a real number **58.** $3a^2$ **59.** $3a^2b^5$ **60.** $12xy\sqrt{3}$ **61.** $2ab^3$ **62.** $\dfrac{2ab\sqrt[4]{b}}{3}$ **63.** $40\sqrt{5}$ **64.** $-14y\sqrt[3]{3x}$ **65.** $3\sqrt{6} - 7$ **66.** $\dfrac{2x + 3\sqrt{xy} + y}{4x - y}$ **67.** $\dfrac{58\sqrt{10a}}{5}$ **68.** $-\dfrac{3x^2}{y^2}$ **69.** $2a^3b^3\sqrt[3]{b}$ **70.** $\dfrac{1}{729x^5y^{12}}$ **71.** $9.46 \times 10^{12}$ km **72.** $\dfrac{-2x^2 + 3}{4x^4}$ **73.** $1.12s$; \$27,440 **74.** 611 ft

## Chapter 2

### [2.1]

**1.** $\frac{1}{2}$ **3.** 6.625 **5.** $\frac{7}{2}$ **7.** $\frac{13}{11}$ **9.** no solution **11.** every real number **13.** 3 **15.** $-7$ **17.** 3 **19.** 2 **21.** 4 **23.** $-1$ **25.** no solution **27.** 3 **29.** 1 **31.** $-11$ **33.** 20 **35.** no solution **37.** $\frac{2}{3}$ **39.** 2, $-4$ **41.** $-2$ **43.** 4, $-\frac{2}{3}$ **45.** $-\frac{1}{2}$ **47.** All numbers $y \geq 1$ **49.** $a = \frac{5c - 2b}{3}$ **51.** $y = \frac{x}{z^2}$ **53.** $a = c - 2\sqrt{cb}$ **55.** $h = \frac{2A}{b_1 + b_2}$ **57.** $r = \frac{A - P}{Pt}$ **59.** $r_2 = \frac{Rr_1}{r_1 - R}$ **61.** $g = \frac{2(h - vt)}{t^2}$ **63.** $-9.531$ **65.** 2.218 **67.** 2 **69.** $-2$ **71.** no **73.** $b^{12}$ **74.** $\frac{1}{3y}$ **75.** $\frac{3}{y}$ **76.** $24u^2 + 2uv - v^2$ **77.** $-24u^3v^3 + 2u^2v^2 + 47uv - 15$ **78.** $-x$ **79.** $\frac{7}{(x - 7)^2}$ **80.** $\frac{5a^3b^2\sqrt{5a}}{2}$

### [2.2]

**1.** $\frac{17}{3}$ **3.** 35, 37, 39 **5.** 84 **7.** Sam is 28, Harry is 14 **9.** $\frac{7}{12}$ **11.** \$2.10 **13.** \$23,000 **15.** \$1450 **17.** 116°, 37°, 27° **19.** 420 m by 460 m **21.** 4.2 days **23.** 31.5 min **25.** 2.9 days **27.** 2.5 hr **29.** 480 mph; 530 mph **31.** 25 mph **33.** 90 fliers per minute, 60 fliers per minute **35.** 81 lb **37.** 105 ft **39.** $\pm 15$ **41.** 3 **43.** 4 **45.** 330 hertz, 396 hertz **47.** \$6000 **49.** 1.5 hr **51.** $-\frac{1}{2}$ **52.** $\frac{22}{3}$ **53.** 1, $\frac{7}{5}$ **54.** $r = \frac{S - a_n}{S}$ **55.** $(x - 6)^2$ **56.** $\left(x + \frac{1}{3}\right)^2$ **57.** 3 **58.** $\sqrt{17}$

### [2.3]

**1.** 5, $-2$ **3.** 6, $-1$ **5.** $-\frac{1}{2}$ **7.** 0, $-\frac{3}{2}$ **9.** 0, $-\frac{3}{5}$ **11.** $\pm 6$ **13.** $\pm 5$ **15.** 2, $-4$ **17.** 16 **19.** $\frac{1}{4}$ **21.** 8, $-3$ **23.** $\frac{5 \pm \sqrt{13}}{6}$ **25.** $-8$, 3 **27.** $\frac{5 \pm \sqrt{13}}{6}$ **29.** $-1 \pm \sqrt{2}$ **31.** $\frac{1}{2}$, $-5$ **33.** $\pm 4$ **35.** 0, 3 **37.** $-\frac{1}{3}$, $-4$ **39.** $\frac{3 \pm \sqrt{21}}{2}$ **41.** $2 \pm \sqrt{3}$ **43.** $\pm 6a$ **45.** $\pm(a + b)$ **47.** $2a$, $-a$ **49.** 3.37, $-1.27$ **51.** 0.89, $-2.42$ **53.** 67 **54.** \$1200 **55.** 62 mph, 50 mph **56.** 320 mi **57.** 22 km/hr **58.** 2 **59.** 16 **60.** 6

### [2.4]

**1.** $\pm 1$, $\pm 3$ **3.** $-2$, $-1$ **5.** $-125$, 343 **7.** $-1$, 1, 3 **9.** $\frac{1}{3}$, $-\frac{1}{5}$ **11.** $\frac{2}{3}$, $\frac{1}{6}$ **13.** $\frac{16}{9}$, 9 **15.** 64 **17.** 3, $-1$ **19.** $-2$, 1 **21.** $\frac{1 \pm \sqrt{3}}{2}$ **23.** $-2$, 4 **25.** 12 **27.** no solution **29.** 4, 7 **31.** 5, 8 **33.** $\pm 3$ **35.** $-1$ **37.** 2 **39.** 3 **41.** 1 **43.** 2.41, $-2.41$, 0.41, $-0.41$ **45.** 4.46, 0.34 **47.** 7, $-\frac{3}{2}$ **48.** $\pm 10$ **49.** 0, $-3$ **50.** $\frac{1 \pm \sqrt{21}}{10}$

### [2.5]

**1.** 13, 15 **3.** 12 **5.** 2 or $\frac{1}{2}$ **7.** 5 in by 16 in **9.** 4 ft **11.** 18 in **13.** 36 in, 64 in **15.** 11% **17.** 9 **19.** 40 **21.** 11, 15 **23.** Jim: 45 hr; Father: 36 hr **25.** 22.8 hr **27.** 30 mph, 40 mph **29.** 40 mph **31.** 9.6 sec **33.** 2.97 sec **35.** 1.26 sec **37.** 61.2 mph **39.** approximately 36 mph **41.** Solve the equation using the quadratic formula to obtain the two solutions $\frac{-b + \sqrt{b^2 - 4c}}{2}$ and $\frac{-b - \sqrt{b^2 - 4c}}{2}$. Adding these we obtain $-b$. **43.** $-5$, $-\frac{5}{4}$ **44.** 5, $-\frac{5}{2}$ **45.** no solution **46.** 4, $-1$, $\frac{5 \pm \sqrt{41}}{2}$ **47.** Every real number is a solution. **48.** $\frac{3 \pm \sqrt{41}}{2}$ **49.** $\frac{4\sqrt{2} - 5}{7}$ **50.** $\frac{-7 - \sqrt{5}}{4}$

**[2.6]**

**1.** $2i\sqrt{2}$ **3.** $-\sqrt{35}$ **5.** $\sqrt{5}$ **7.** $-i$ **9.** $i\sqrt{29}$ **11.** $-3i$ **13.** $x = 1; y = 0$ **15.** $x = 2; y = 2$ **17.** $2 + 7i$ **19.** $2 + 2i$ **21.** $-2 + 11i$ **23.** $10 - 15i$ **25.** $117 - 44i$ **27.** $8i$ **29.** $-3 - i\sqrt{5}$ **31.** $\frac{9}{13} + \frac{19}{13}i$ **33.** $\frac{7}{26} - \frac{17}{26}i$ **35.** $\frac{1}{3}i$ **37.** $1 - i$ **39.** $\frac{-1 \pm i\sqrt{2}}{3}$ **41.** $\pm i\sqrt{3}, \pm 2$ **43.** $\frac{3 \pm i\sqrt{7}}{2}$ **45.** $1, \frac{-1 \pm i\sqrt{3}}{2}$ **47.** $0, 2i, -2i$ **49.** $\pm 1, \pm i$ **51.** one real solution **53.** $x_1 + x_2 = 10; x_1 \cdot x_2 = 25$ **55.** $x_1 + x_2 = 2; x_1 \cdot x_2 = -\frac{2}{3}$ **57.** Yes, 2 and 1 are solutions. **59.** No, 4 and $-\frac{1}{4}$ are not solutions. **61.** **(a)** $-1$ **(b)** $-i$ **(c)** 1 **(d)** $i$ **63.** **(a)** 5 **(b)** $\sqrt{2}$ **(c)** 9 **65.** Let $x = a + bi$ and $y = c + di$. Then $\overline{x - y} = \overline{(a + bi) - (c + di)} = \overline{(a - c) + (b - d)i} = (a - c) - (b - d)i$. Also, $\bar{x} - \bar{y} = \overline{(a + bi)} - \overline{(c + di)} = (a - bi) - (c - di) = (a - c) + (d - b)i = (a - c) - (b - d)i$. Thus, $\overline{x - y} = \bar{x} - \bar{y}$. **67.** $\bar{x} \cdot \bar{y} = \overline{(a + bi)} \cdot \overline{(c + di)} = (a - bi)(c - di) = (ac - bd) + (-ad - bc)i$. Also, $\overline{x \cdot y} = \overline{(a + bi)(c + di)} = \overline{(ac - bd) + (ad + bc)i} = (ac - bd) - (ad + bc)i = (ac - bd) + (-ad - bc)i$. Thus, $\bar{x} \cdot \bar{y} = \overline{x \cdot y}$. **69.** If $b^2 - 4ac = 0$, then the solutions to the equation are $x = \frac{-b \pm \sqrt{0}}{2a} = -\frac{b}{2a}$. Since $-\frac{b}{2a}$ is a rational number, we have the desired result. **71.** **(a)** Since $b^2 - 4ac = 1 - 4(1)(-56) = 225 = 15^2$, by Exercise 70, $x^2 + x - 56$ can be factored. In fact, $x^2 + x - 56 = (x - 7)(x + 8)$. **(b)** Since $b^2 - 4ac = 9 - 4(7)(9) = -243$, and $-243$ is not a perfect square, $7x^2 - 3x + 9$ cannot be factored. **(c)** Since $b^2 - 4ac = 3600 - 4(36)(25) = 0$, by Exercise 69, $36x^2 - 60x + 25$ can be factored. In fact, $36x^2 - 60x + 25 = (6x - 5)^2$. **73.** 4 cm, 11 cm **74.** 20 shares **75.** 5.5 days **76.** **(a)** 8 sec **(b)** at 3 seconds on the way up and at 5 seconds on the way down **(c)** 4 sec; this is the maximum height of the rocket **(d)** The rocket will not reach a height of 300 ft (the maximum height is 256 ft). When solving the equation with $h = 300$, the solutions for $t$ are not real numbers. **77.** $\pm 1, \pm 3$ **78.** $\pm 3, \pm i\sqrt{7}$

**[2.7]**

**1.** $x \leq 2; (-\infty, 2]$

**3.** $y > 7; (7, \infty)$

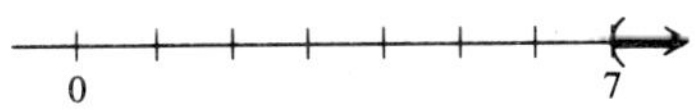


**5.** $z > \frac{1}{4}; \left(\frac{1}{4}, \infty\right)$

**7.** $y \leq -\frac{2}{11}; \left(-\infty, -\frac{2}{11}\right]$

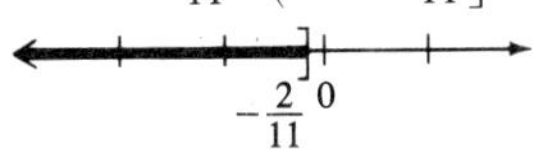


**9.** $y > -2; (-2, \infty)$

**11.** $z > -4; (-4, \infty)$

**13.** $y \geq -1; [-1, \infty)$

**15.** $5 \leq x < 8; [5, 8)$

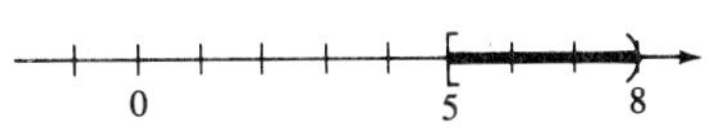


**17.** $-1 \leq z < 1; [-1, 1)$

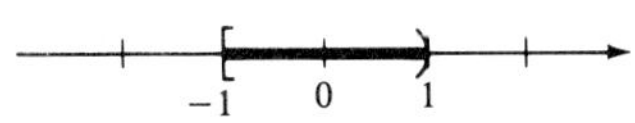


**19.** $y < 5$ or $y \geq 8; (-\infty, 5)$ or $[8, \infty)$

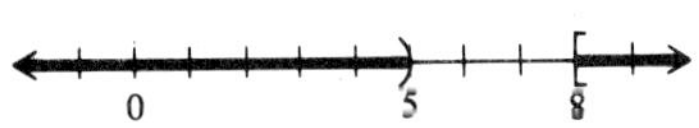


**21.** $z < -3$ or $z > -1; (-\infty, -3)$ or $(-1, \infty)$

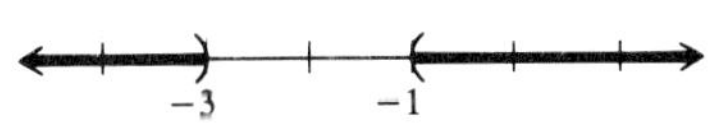


**23.** $-1 < y < 0; (-1, 0)$

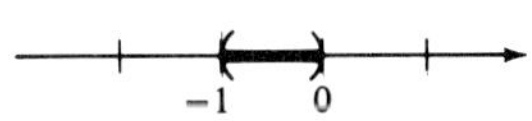


**25.** $x > -5$ **27.** all values of $x$ except $x = -5$, that is, $x < -5$ or $x > -5$ **29.** $x > -5$ **31.** $5° \leq F \leq 86°$ **33.** $145 \leq x \leq 185$; which is impossible if the highest score possible is 100% **35.** $2 \leq I \leq 21$ **37.** $-7 < x < 7; (-7, 7)$ **39.** $z = 0$ **41.** every real number except 0, that is, $y < 0$ or $y > 0$; $(-\infty, 0)$ or $(0, \infty)$ **43.** $x > 1$ or $x < -5$; $(1, \infty)$ or $(-\infty, -5)$ **45.** $-1 < z < \frac{7}{5}; \left(-1, \frac{7}{5}\right)$ **47.** $y > 2$ or $y < -16$; $(2, \infty)$ or $(-\infty, -16)$ **49.** no solution **51.** $z = \frac{7}{2}$ **53.** $m > 1$ **55.** $m < \frac{1}{8}$ **57.** Since $a < b$, $b - a$ is positive. Also, since $a > 0$ and $b > a$, both $a$ and $b$ are positive making $b + a$ positive. Thus, $(b + a)(b - a) = b^2 - a^2$ is positive so that $a^2 < b^2$. If $0 < a$ is deleted, the result is not true. For example, $-2 < 1$ but $(-2)^2 > 1^2$. **58.** $-6 - 2i$ **59.** $14 - 2i$ **60.** $\frac{3}{5} + \frac{16}{5}i$ **61.** $\frac{1}{6} + \frac{1}{6}i$ **62.** $\frac{1 \pm i\sqrt{11}}{6}$ **63.** $0, \frac{5 \pm i\sqrt{7}}{4}$ **64.** $x_1 + x_2 = 2; x_1 \cdot x_2 = \frac{1}{6}$ **65.** Since $b^2 - 4ac = -19 < 0$, the equation has two complex solutions.

## [2.8]

**1.** $x < -5$ or $x > 3$; $(-\infty, -5)$ or $(3, \infty)$ **3.** $x \le -4$ or $x \ge \frac{5}{2}$; $(-\infty, -4]$ or $\left[\frac{5}{2}, \infty\right)$ **5.** $x < 1 - \sqrt{3}$ or $x > 1 + \sqrt{3}$; $(-\infty, 1 - \sqrt{3})$ or $(1 + \sqrt{3}, \infty)$ **7.** every real number; $(-\infty, \infty)$ **9.** $x = \frac{5}{2}$ **11.** $\frac{1 - \sqrt{6}}{5} \le x \le \frac{1 + \sqrt{6}}{5}$; $\left[\frac{1 - \sqrt{6}}{5}, \frac{1 + \sqrt{6}}{5}\right]$ **13.** $x < -1$ or $x \ge 3$; $(-\infty, -1)$ or $[3, \infty)$ **15.** $-3 < x < 4$; $(-3, 4)$ **17.** $x \le -8$ or $x > 3$; $(-\infty, -8]$ or $(3, \infty)$ **19.** $-2 \le x \le 2$ or $x \ge 5$; $[-2, 2]$ or $[5, \infty)$ **21.** $x \le -1$ or $x \ge 4$; $(-\infty, -1]$ or $[4, \infty)$ **23.** $x < -6$ or $1 < x < 3$; $(-\infty, -6)$ or $(1, 3)$ **25.** $x \le -5$ or $-3 \le x \le 0$ or $x \ge 2$; $(-\infty, -5]$ or $[-3, 0]$ or $[2, \infty)$ **27.** $-1 \le x \le 0$ or $x \ge 1$; $[-1, 0]$ or $[1, \infty)$ **29.** $x \le -5$ or $x \ge 0$; $(-\infty, -5]$ or $[0, \infty)$ **31.** $-8 < x \le 3$; $(-8, 3]$ **33.** $-4 < m < 4$ or $(-4, 4)$ **35.** $t > 5$ **37.** $0 \le t \le 5$ sec **39.** 6 sec $< t <$ 9 sec **41.** 4 sec $< t <$ 7 sec **43.** $x < -2.05$ or $3.55 < x \le 6.25$ **44.** $y \le -1$; $(-\infty, -1]$ **45.** $z < -3$; $(-\infty, -3)$ **46.** $x < -1$ or $x > \frac{3}{2}$; $(-\infty, -1)$ or $\left(\frac{3}{2}, \infty\right)$ **47.** $-\frac{14}{3} \le z \le 2$; $\left[-\frac{14}{3}, 2\right]$ **48.** $x \ge 175$; that is, he must sell at least 175 mice. **49.** 15 days **50.** 5.93, −7.85

## Chapter 2 Review Exercises

**1.** $\frac{3}{2}$ **2.** 3 **3.** 10 **4.** 3 **5.** $\frac{5}{7}$ **6.** no solution **7.** 2, −9 **8.** $\frac{1 \pm \sqrt{13}}{6}$ **9.** $\frac{-1 \pm 2i}{5}$ **10.** −1 **11.** $-5, -\frac{7}{3}$ **12.** $\frac{1}{2}, -8$ **13.** $1 \pm \sqrt{2}$ **14.** 4, −1 **15.** 5 **16.** −3, 0 **17.** $\frac{3}{4}$ **18.** 0 **19.** $y = \frac{xz}{z - x}$ **20.** every real number **21.** $x < -5$ or $3 < x < 7$ **22.** $-2 \le x \le \frac{5}{3}$ **23.** $x \le \frac{3 - \sqrt{5}}{2}$ or $x \ge \frac{3 + \sqrt{5}}{2}$ **24.** no solution **25.** $x \le -3$; $(-\infty, -3]$

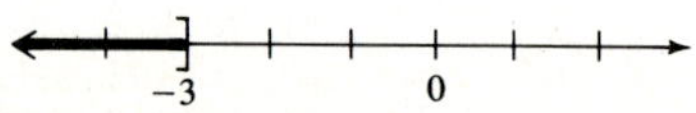

**26.** $-2 \le x < \frac{3}{2}$; $\left[-2, \frac{3}{2}\right)$

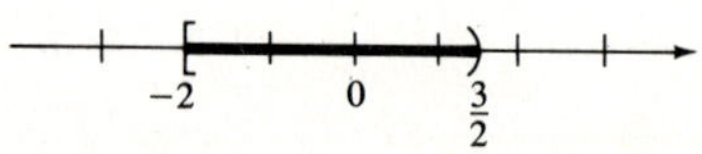

**27.** $1 < x < 3$; $(1, 3)$

**28.** $x < -2$ or $x > \frac{5}{3}$; $(-\infty, -2)$ or $\left(\frac{5}{3}, \infty\right)$

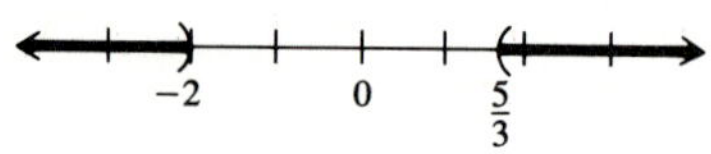

**29.** $11 - 7i$ **30.** $-\frac{11}{5} + \frac{7}{5}i$ **31.** two complex solutions **32.** $x_1 + x_2 = 2$; $x_1 \cdot x_2 = 5$ **33.** 42, 44, 46 **34.** 75 kg **35.** \$28,000 **36.** 23 m by 12 m **37.** 25°, 145°, 10° **38.** 10.3 min **39.** 9:30 A.M. **40.** 120 km/hr **41.** 44 ft, 20 ft **42.** 50 ft **43.** 45 mph, 60 mph **44.** 6 **45.** Jan: 12 min; Mother: 24 min **46.** 38.6 mi **47.** 1 sec $< t <$ 3 sec **48.** 1000 antelope **49.** $x > 3$, that is, a profit will be made when more than 3 stoves are sold per week. **50.** 1.31, −0.88

## Chapter 3

## [3.1]

**1.**

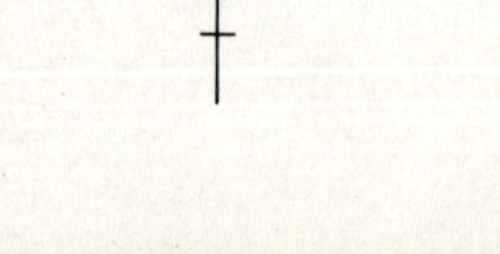

**3.** II **5.** III

**7.**

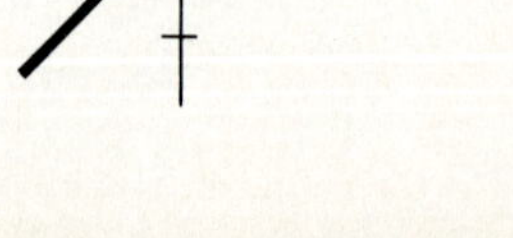

**9.**

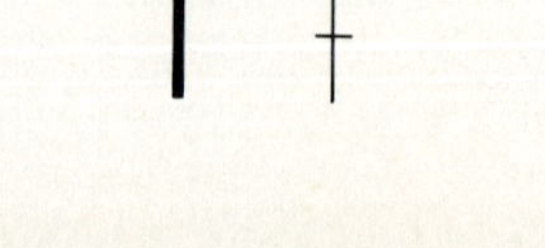

**11.**

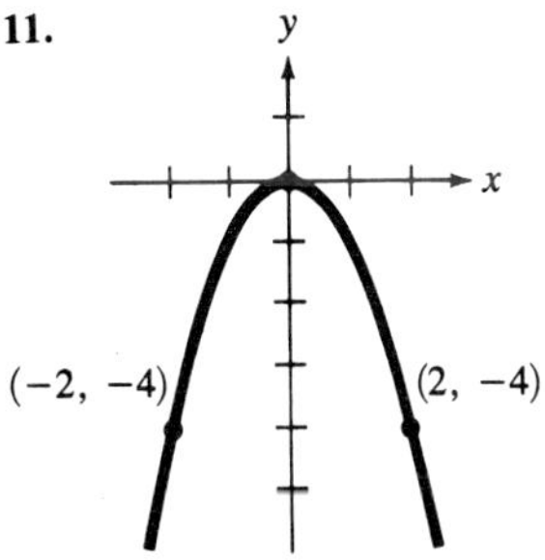


**13.**

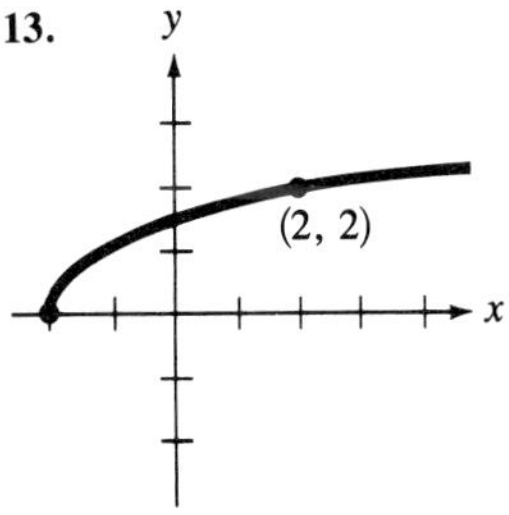


**15.** 5; $\left(\frac{5}{2}, 5\right)$ **17.** $\sqrt{85}$; $\left(6, \frac{7}{2}\right)$ **19.** 7; $\left(-6, -\frac{3}{2}\right)$ **21.** $a = -1$, $b = -4$

**23.** 4, 12 **25.** sides: $7\sqrt{2}$, $2\sqrt{5}$, $3\sqrt{10}$; not a right triangle **27.** 6.4 mi **29.** (J, 2.5), (F, 3.7), (M, 6.3) **31.** $x < -3$ or $x > -2$; $(-\infty, -3)$ or $(-2, \infty)$ **32.** Every real number is a solution; $(-\infty, \infty)$. **33.** $-\frac{3}{2} < x \leq 5$; $\left(-\frac{3}{2}, 5\right]$ **34.** $x \leq \frac{1}{2}$ or $x \geq 1$; $\left(-\infty, \frac{1}{2}\right)$ or $(1, \infty)$ **35. (a)** $y = 3$ **(b)** $x = 6$ **(c)** $y = -\frac{1}{2}x + 3$

[3.2]

**1.**

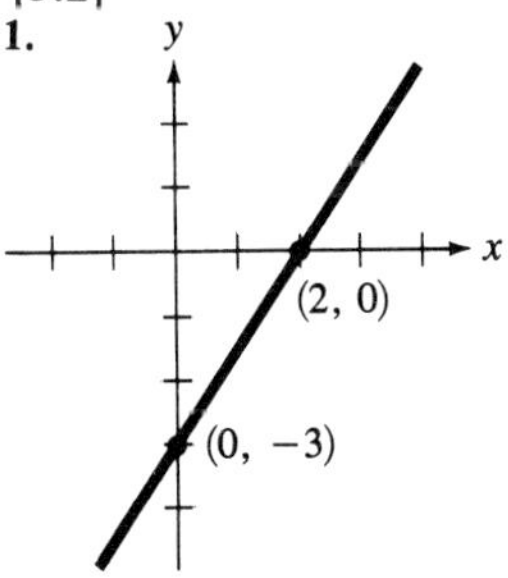


**3.**

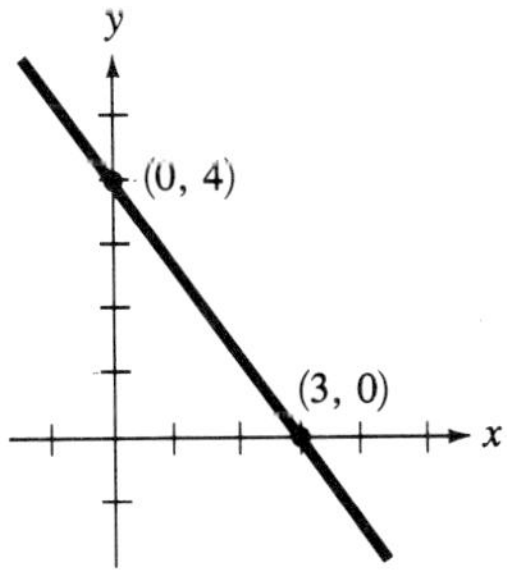


**5.**

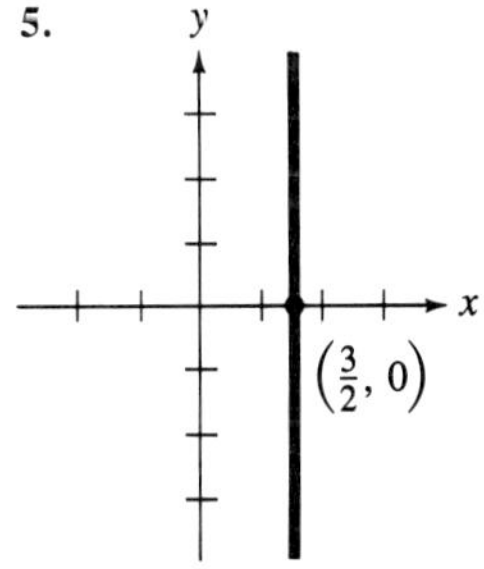


**7.**

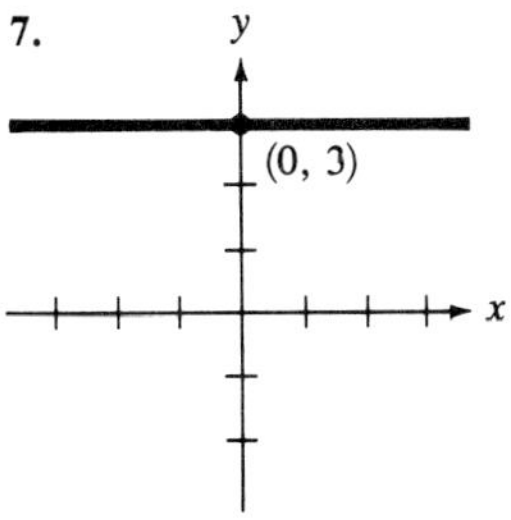


**9.**

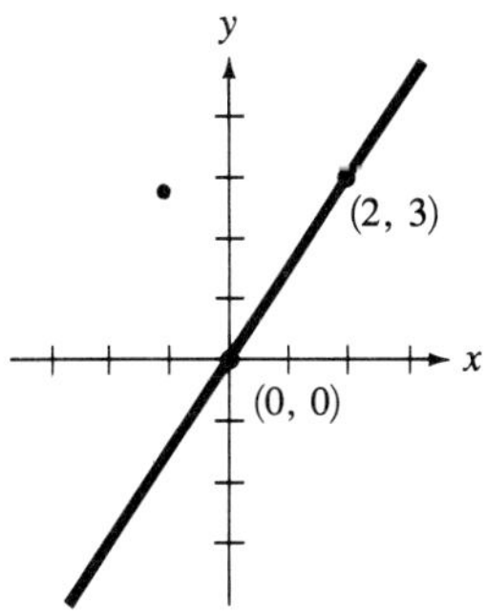


**11.** $-\frac{3}{2}$ **13.** undefined **15.** 0 **17.** $2x + y + 1 = 0$ **19.** $4x - 3y - 12 = 0$ **21.** $y + 5 = 0$ **23.** $2x - 5y + 26 = 0$ **25.** $2x + y - 1 = 0$ **27.** $y - 3 = 0$ **29.** $x + 5 = 0$ **31.** $\frac{2}{7}$, $\left(0, \frac{5}{7}\right)$ **33.** $\frac{8}{3}$; $\left(0, \frac{7}{3}\right)$ **35.** 0; (0, 2) **37.** perpendicular **39.** parallel **41.** perpendicular **43.** Yes. The slope of the line through $P$ and $Q$ is the same as the slope of the line through $Q$ and $R$. **45.** let $m = \frac{y_2 - y_1}{x_2 - x_1}$; $2x + y = 0$ **47.** $x + 3y - 2 = 0$ **49.** $y = 4000x + 85{,}000$; \$113,000 **51.** 10 yr **53. (a)** \$1524 **(b)** 5 yr **55. (a)** $g = \frac{1}{4}t - 1$ ($g$ is used for $y$ and $t$ for $x$) **(b)** 4 gm **(c)** The grams produced at $t = 0$ should not be $-1$. **57.** $\sqrt{41}$; $\left(\frac{3}{2}, 4\right)$ **58.** $\sqrt{113}$; $\left(0, -\frac{3}{2}\right)$ **59.** $x = -1, 5$ **60.** $x \leq -3$

**[3.3]**

**1.** function **3.** function **5.** function **7.** $\{2, 7, 12\}$ **9.** all nonnegative real numbers **11.** all real numbers except 0 **13.** all real numbers **15.** $x \le 2$ **17.** $x \ge -8$ and $x \ne -2$ and $x \ne 3$ **19.** identity **21.** linear **23.** constant **25.** $-3, 2, -18, 5a - 3, 5a - 8$ **27.** 7, 7, 7, 7, 7 **29.** $5, 4, 8, |a - 5|, |a - 6|$ **31.** $3a^2 - 7, (3a - 7)^2, \frac{3}{a} - 7, \frac{1}{3a - 7}$ **33.** $\frac{1}{a^2} + 6, \left(\frac{1}{a} + 6\right)^2, a + 6, \frac{a}{1 + 6a}$ **35.** $h + 6, 1, x + h + 5, 1$ **37.** $h^2 + 2h + 3, h + 2, x^2 + 2xh + h^2 + 2, 2x + h$ **39.** **(a)** $S(c) = 1.285c$, **(b)** $71.96 **41.** **(a)** $S(x) = 50(1 - x)$, **(b)** $T(x) = 52.5(1 - x)$, **(c)** 20% **43.** **(a)** $T(r) = 5\pi r^2$ **(b)** $1447.65 **45.** **(a)** Range is given by $\$112.77 \le c(x) \le \$203.17$ **(b)** For $x = 0.30$ the cost would be $-\$12.87$ which is clearly not reasonable. **47.** $x - 3y + 10 = 0$ **48.** $x - 2y - 6 = 0$ **49.** $x - 8 = 0$ **50.** $4x + 10y + 21 = 0$

**51.**

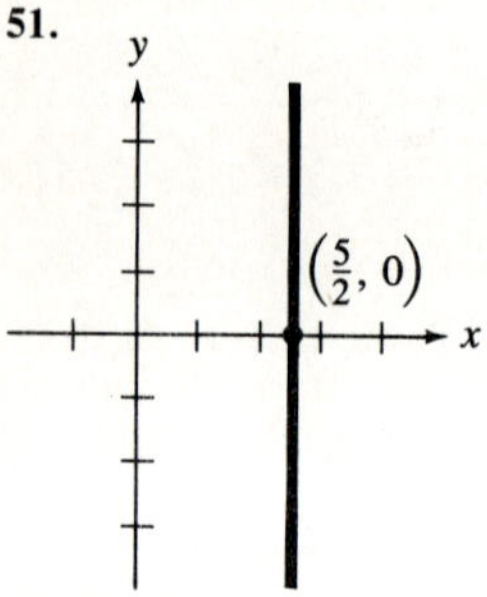


**52.** $\frac{7}{6}$

**[3.4]**

**1.** function **3.** not a function **5.** function

**7.** increases for all $x$

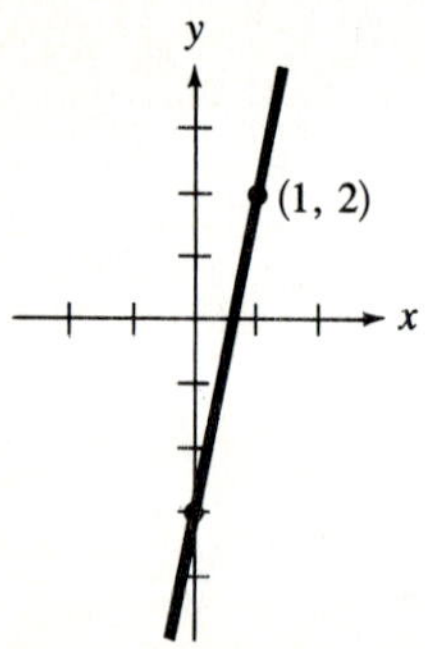


**9.** constant for all $x$

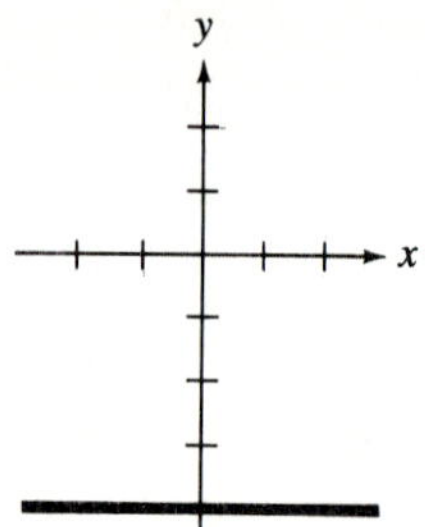


**11.** decreases for $x \le 0$
increases for $x \ge 0$

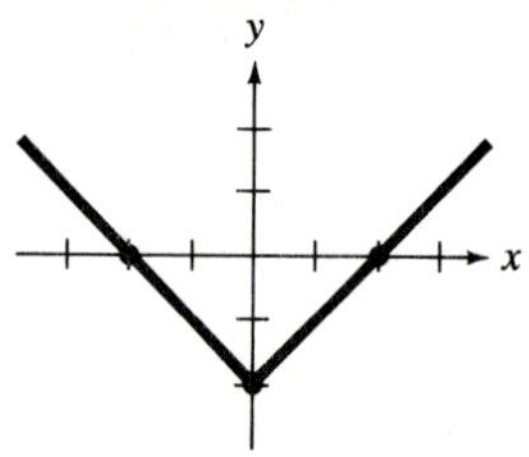


**13.** increases for $x \le 3$
decreases for $x \ge 3$

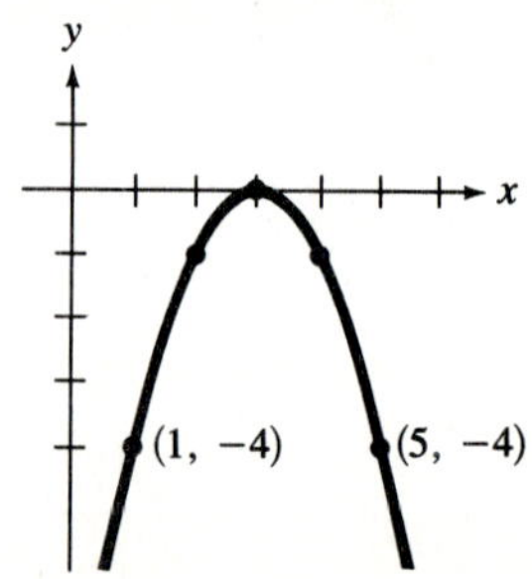


**15.**

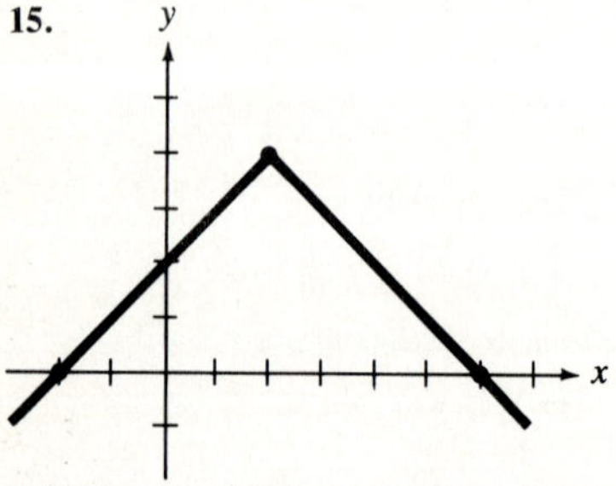


**17.**

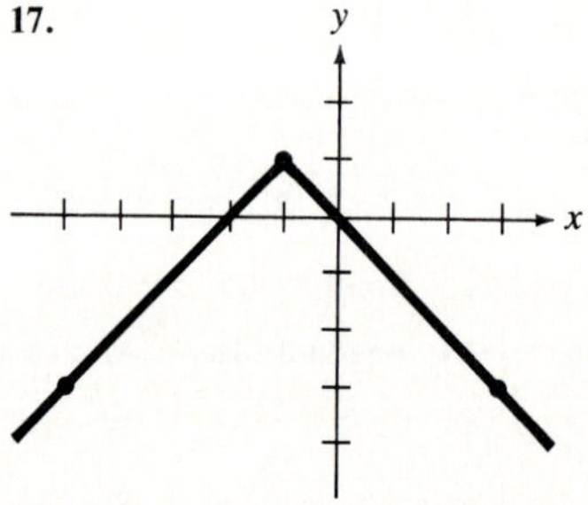


**19.**

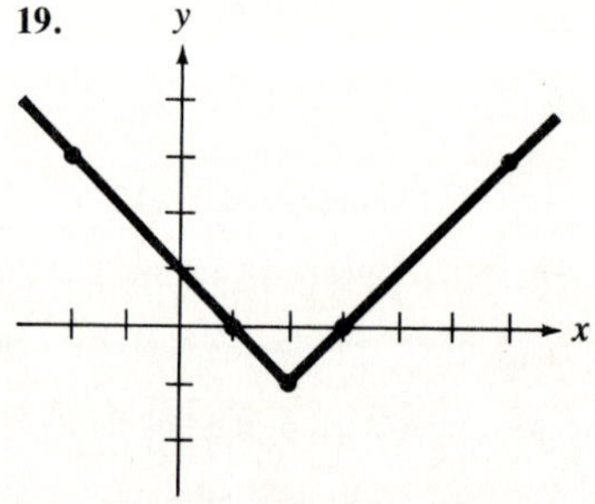

21.
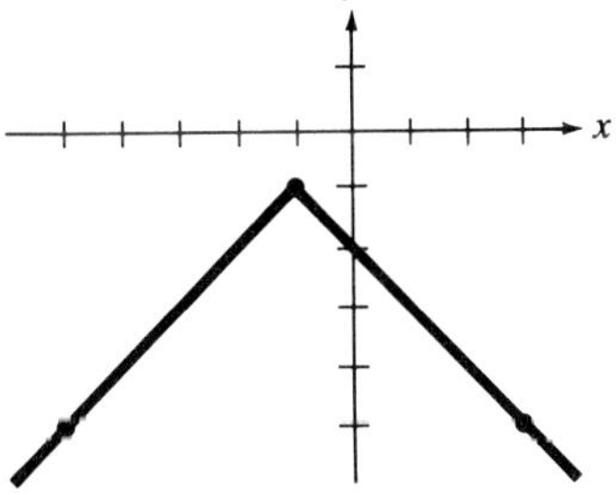


23.
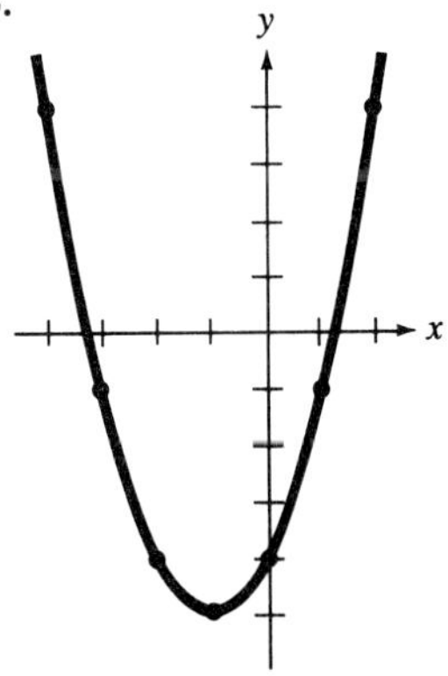


25.
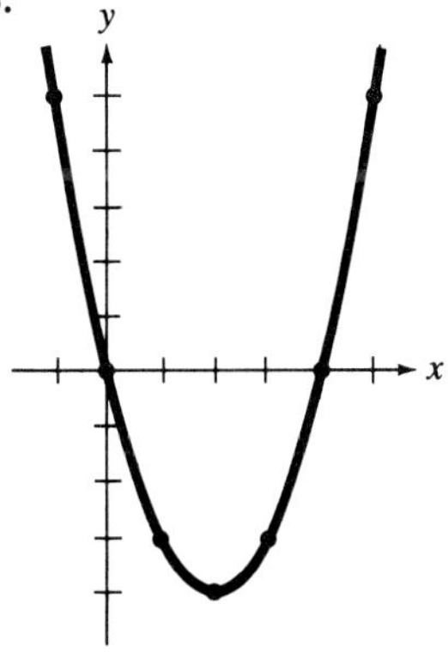


27.
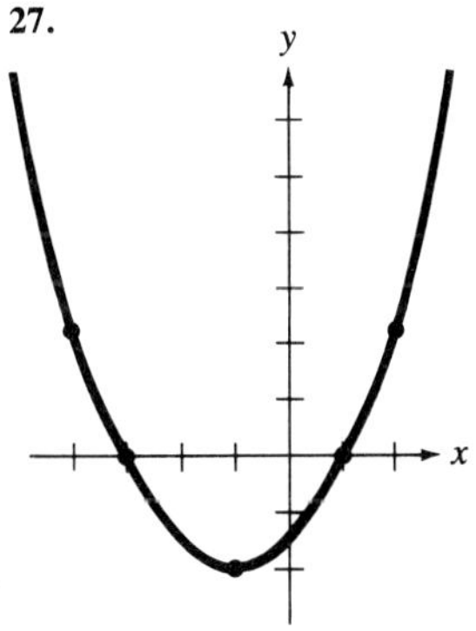


29.
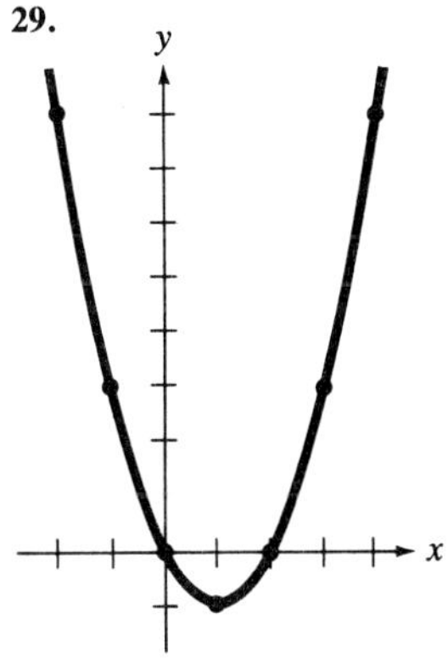


**31.** $g(x) = f(x) - 3$ **33.** $h(x) = g(x - 4) + 3$ **35.** odd **37.** both even and odd **39.** even **41.** $x$-axis, $y$-axis, origin
**43.** $y$-axis **45.** none **47.** origin

49.
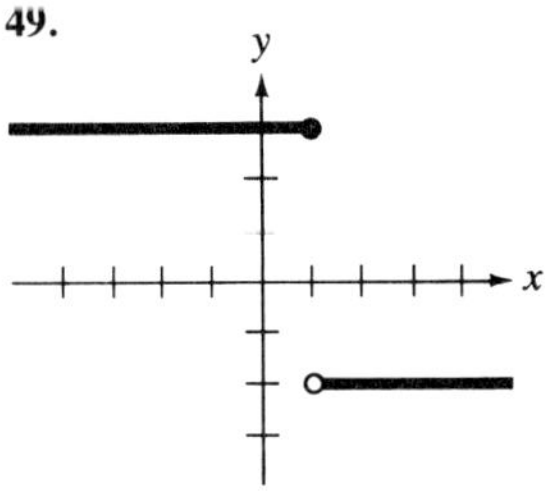


51.
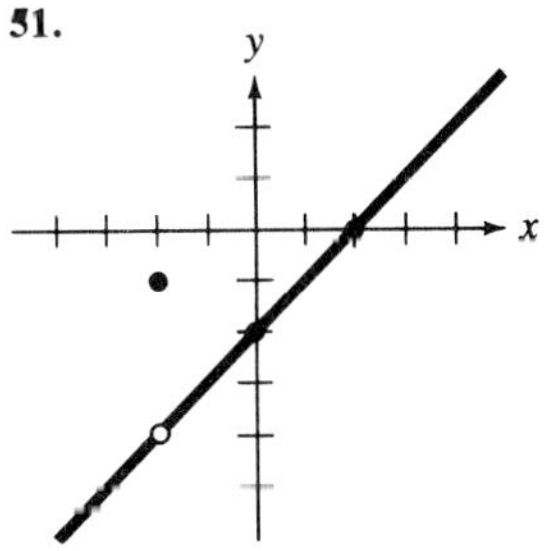


53.
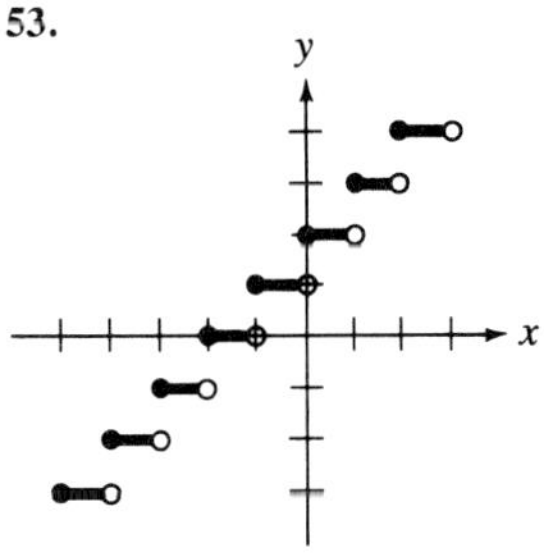


**55.** 1, $x^2 - 2x - 2$, $x^4 - 3$ **56.** 25, $(x - 4)^2$, $(x^2 - 3)^2$ **57.** **(a)** $S(x) = 1.08x + 1200$ **(b)** \$21,180
**58.** **(a)** $V(x) = 4x(8 - x)(6 - x)$ **(b)** 192 in$^3$ **59.** $\sqrt{58}$ **60.** $(3, -5)$

[3.5]
**1.** 20, $-17$, $15x - 10$, $15x - 2$ **3.** 8, $-2$, $2x^2$, $-2x^2$ **5.** $-11$, 11, $-2x^2 - 3$, $4x^2 - 4x + 3$ **7.** 2, $-1$, $x$, $x$
**9.** one-to-one **11.** not one-to-one **13.** one-to-one

**15.** yes

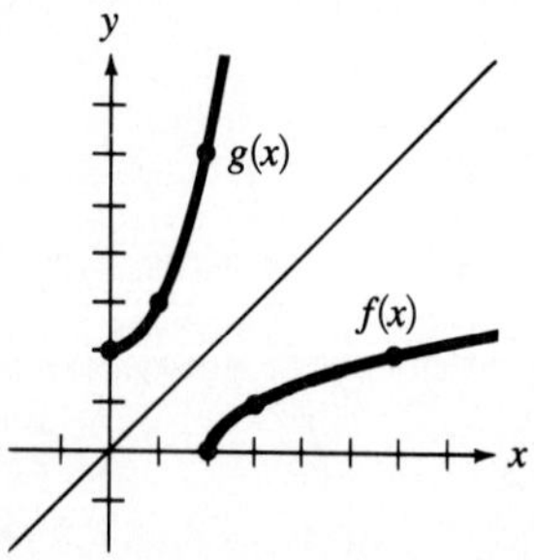


**17.** yes

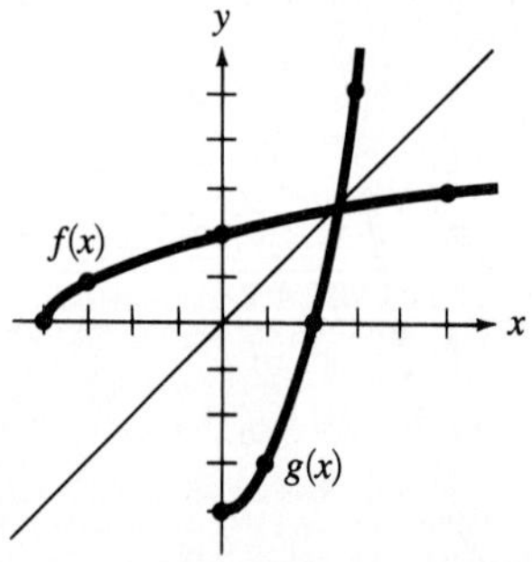


**19.** $f^{-1}(x) = \frac{x-3}{2}$ **21.** $h^{-1}(x) = x^2 + 3,\ x \geq 0$ **23.** $k^{-1}(x) = \sqrt[3]{\frac{x}{2}}$ **25.** $g^{-1}(x) = \sqrt{1 - x^2},\ 0 \leq x \leq 1$
**27.** $f^{-1}(x) = \frac{x-b}{m},\ m \neq 0$ **29.** **(a)** $C(t) = 2\pi(2t^2 - 1)$ **(b)** $A(t) = \pi(2t^2 - 1)^2$ **31.** **(a)** $22{,}000(3t^2 + 2t + 1)$
**(b)** 374,000 $m^2$ **33.** right 4 units and upward 2 units **34.** left 4 units and downward 2 units **35.** odd **36.** even
**37.** $3x - 2y + 22 = 0$ **38.** $3x + 2y + 2 = 0$ **39.** **(a)** 16 **(b)** $\frac{9}{4}$ **40.** $\frac{1}{2}, -5$

**[3.6]**

**1.** $\left(\frac{5}{2}, -\frac{1}{4}\right)$; (2, 0) and (3, 0) **3.** $(-1, 1)$; no $x$-intercepts **5.** $(-4, 16)$; (0, 0) and $(-8, 0)$ **7.** $(0, -4)$; $(\sqrt{2}, 0)$ and $(-\sqrt{2}, 0)$ **9.** $(-2, -13)$; $\left(\frac{-4 + \sqrt{26}}{2}, 0\right)$ and $\left(\frac{-4 - \sqrt{26}}{2}, 0\right)$ **11.** (3, 2); no $x$-intercepts

**13.**

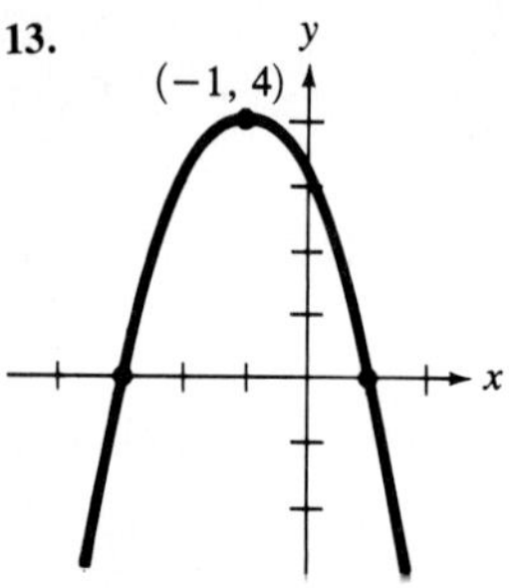


**15.**

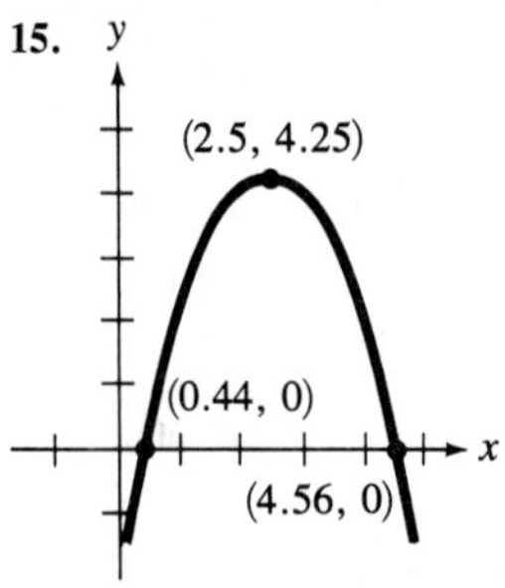


**17.**

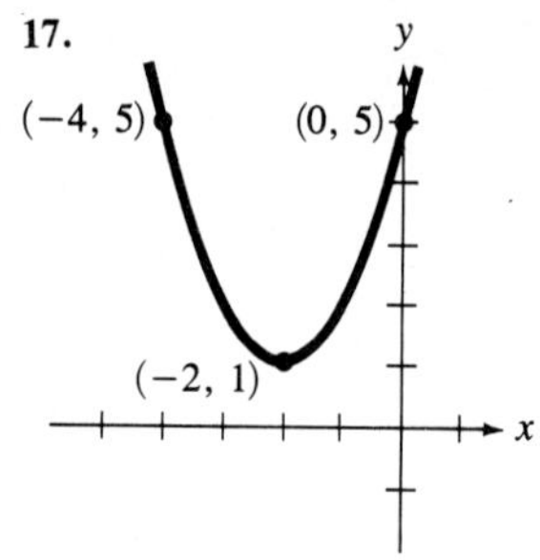


**19.**

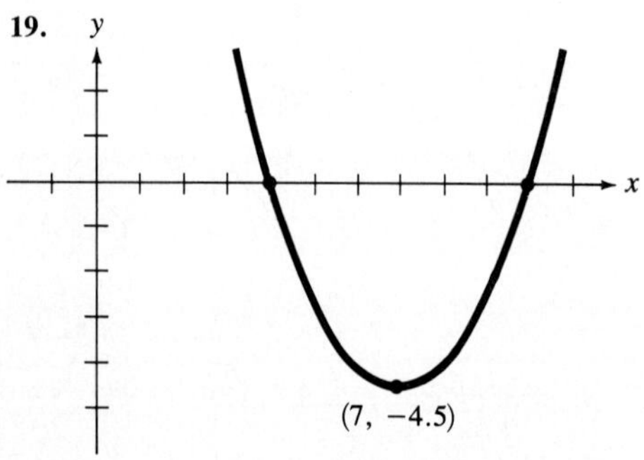


**21.**

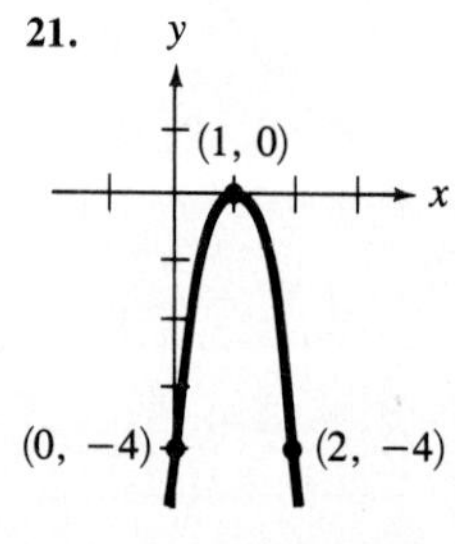


**23.**

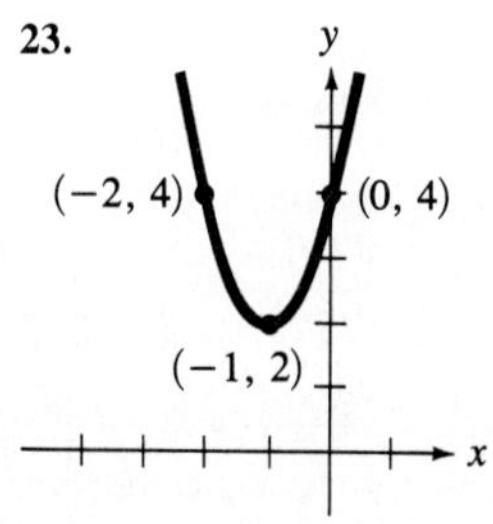


**25.** 250 yd by 250 yd; 62,500 $yd^2$ **27.** 45 ft **29.** **(a)** 8 sec **(b)** 1024 ft **(c)** 16 sec **31.** 60; \$2600
**33.** $C(x) = 2x^2 - 120x + 2000$ **35.** 300 units **36.** 19, 15, $-6x + 19$, $-6x + 3$ **37.** 5, 9, $x^2 + 5$, $(x + 5)^2$

**38.** $f^{-1}(x) = \frac{x-8}{7}$ **39.** $g^{-1}(x) = x^2 - 4,\ x \geq 0$ **40.** $h^{-1}(x) = x^3 - 1$

**41.** decreases for all $x$

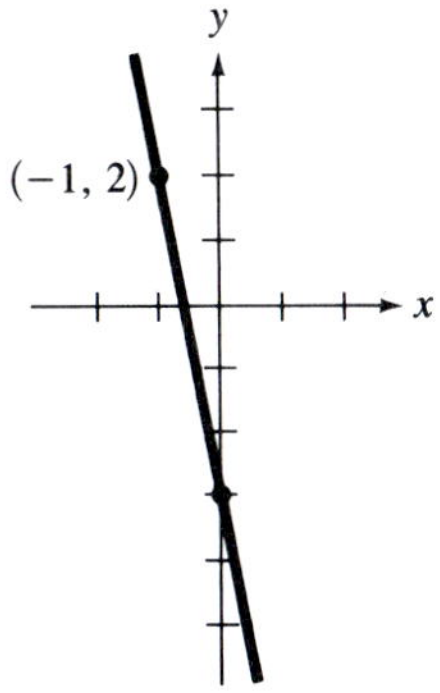


**42.** decreases for $x \leq 0$
increases for $x \geq 0$

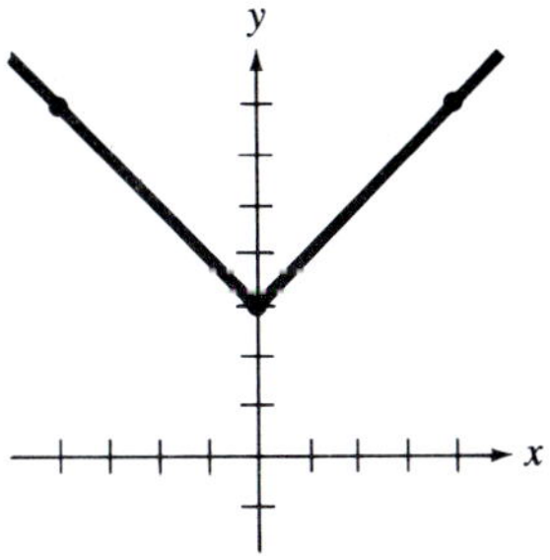


## [3.7]

**1.** $y = \frac{1}{2}x$ **3.** $z = \frac{60}{w}$ **5.** $z = 3xy$ **7.** $a = \frac{225b^2}{\sqrt{d}}$ **9.** $w = \frac{0.825x^2\sqrt[3]{y}}{\sqrt{z}}$ **11.** \$1156.25 **13.** \$39,337

**15.** 13.2 **17.** 120 in$^3$ **19.** weight is divided by 4 **21.** 2.10 sec **23.** distance must be multiplied by $\frac{1}{\sqrt{2}}$

**25.** 110 sec **27.** 61 mph **29.** $M = \frac{5}{12}m$ **31.** $\left(\frac{7}{2}, -\frac{25}{4}\right)$; (1, 0), (6, 0) **32.** (−6, 25); (−1, 0), (−11, 0)

**33.** $\left(\frac{3}{4}, \frac{169}{8}\right)$; (4, 0), $\left(-\frac{5}{2}, 0\right)$ **34.** $\left(-\frac{2}{3}, -\frac{16}{3}\right)$; (−2, 0), $\left(\frac{2}{3}, 0\right)$ **35.** $\left(\frac{1}{2}, 0\right)$; $\left(\frac{1}{2}, 0\right)$ **36.** $\left(\frac{1}{4}, -\frac{7}{4}\right)$; none

**37.** $8x + y + 32 = 0$ **38.** $2x - 3y - 14 = 0$ **39.** $5x + 2y - 15 = 0$ **40.** $x - 3y - 24 = 0$

## Chapter 3 Review Exercises

**1.**

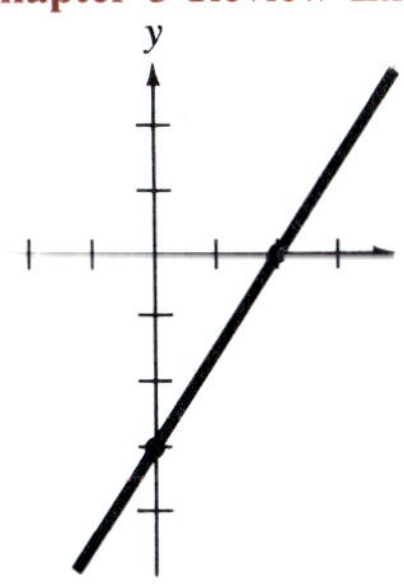


**2.** $3\sqrt{10}$ **3.** (6, −1) **4.** (6, 2), (−2, 2) **5.** 9 **6.** $9x - 2y + 13 = 0$

**7.** $4x - 3y + 8 = 0$ **8.** $\frac{7}{2}$; (0, −2) **9.** parallel **10.** perpendicular **11.** yes

**12.** $x + 2y - 5 = 0$ **13.** function **14.** not a function **15.** $x > -1$ **16.** $x \neq \pm 3$

**17.** linear **18.** constant **19.** 0, 2, 2, $2a^2 + 3a$, $2a^2 + 7a + 5$ **20.** $3a^3 - 5$, $\frac{3-5a}{a}$, $3\sqrt{a} - 5$ **21.** not a function **22.** function **23.** decreases for all $x$ **24.** decreases for all $x$ where it is defined

**25.**

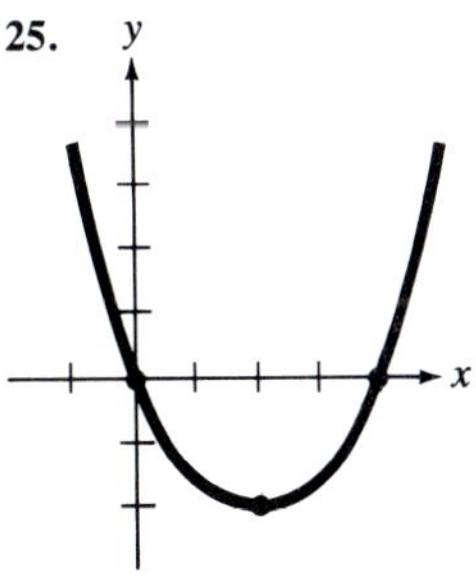


**26.**

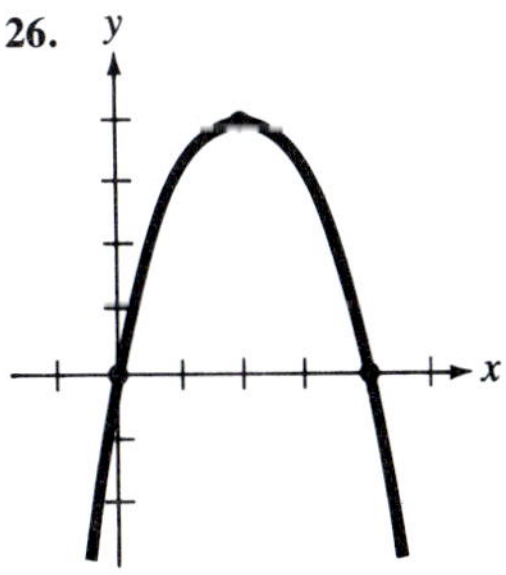


**27.**

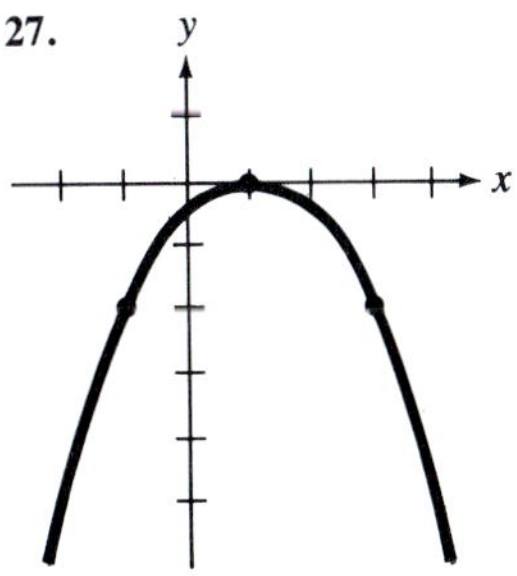


**28.**

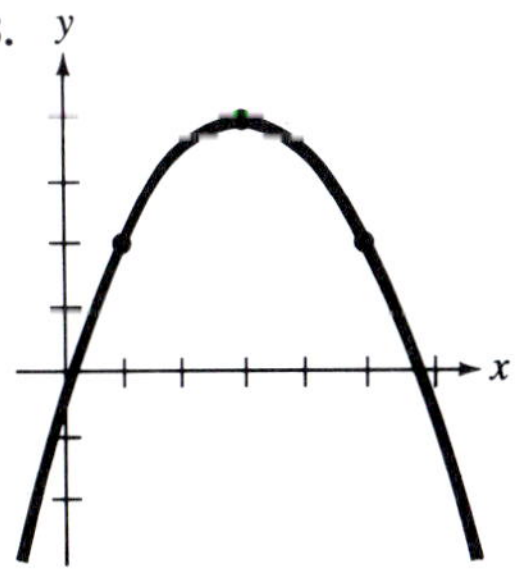

**29.** even **30.** odd **31.** neither **32.** $x$-axis, $y$-axis, origin **33.** origin **34.** $x$-axis

**35.**

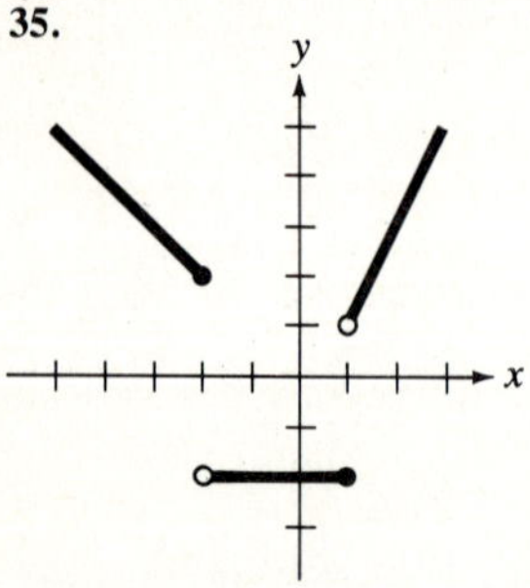


**36.**

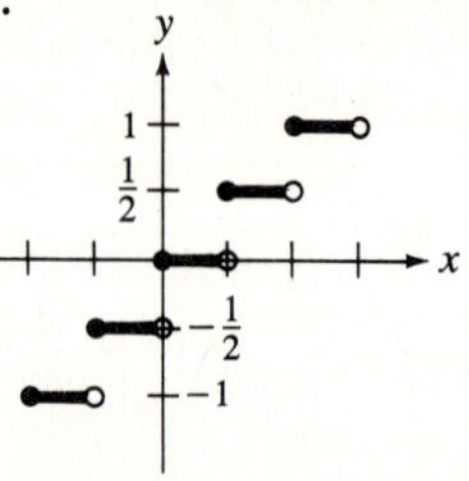


**37.** 14, 24, $3x^2 - 30x + 77$, $3x^2 - 3$ **38.** 10, 0, $x^2 + 1$, $-x^2 + 4x - 3$ **39.** one-to-one **40.** not one-to-one **41.** $f^{-1}(x) = 2x - 6$ **42.** $g^{-1}(x) = x^2 - 5$, $x \geq 0$ **43.** $h^{-1}(x) = \sqrt{2x + 4}$ **44.** $k^{-1}(x) = \sqrt[3]{2x}$ **45.** $\left(-\frac{9}{4}, -\frac{121}{8}\right)$; $\left(\frac{1}{2}, 0\right)$, $(-5, 0)$ **46.** $\left(-\frac{1}{3}, \frac{25}{3}\right)$; $\left(\frac{4}{3}, 0\right)$, $(-2, 0)$

**47.**

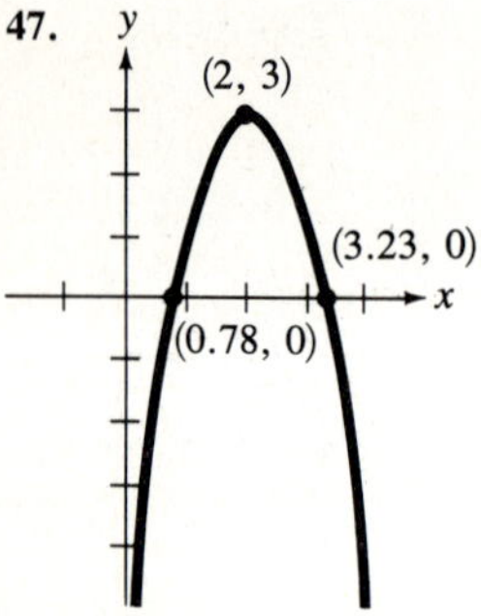


**48.**

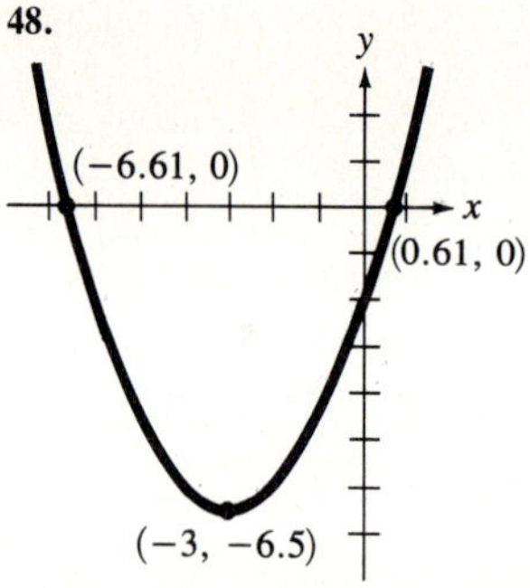


**49.** $S = 1.33x$; \$2926 **50.** $163.8(1 - x)$; \$98.28 **51.** $x = 20$ ft, $y = 60$ ft **52.** 20; \$1400 **53.** $f(x) = -2x^2 + 600x - 44{,}150$ **54.** \$7187.50 **55.** $C(t) = 120t^2 + 120t + 8030$; \$9470 **56.** 95.1% **57.** $C(r) = 12.8\pi r^2$, \$411.77 **58.** 144 ft, 6 sec **59.** 4.8 horsepower **60.** $l$ would have to be 9 times as long

## Chapter 4

### [4.1]

**1.** **(a)** 3 **(b)** 1 **(c)** −7 **(d)** −10 **3.** **(a)** 1 **(b)** −2 **(c)** $-4 + \sqrt{2}$ **(d)** $2 + \sqrt{2}$ **5.** **(a)** 4 **(b)** −1 **(c)** −9 **(d)** 0 **7.** **(a)** 0 **(b)** 8 **(c)** 8 **(d)** 8 **9.** **(a)** 12 **(b)** 3 **(c)** 12,255 **(d)** 0 **11.** $f(1) = 0$, $f(-1) = 0$, $f(2) = 39$; 1 and −1 are zeros **13.** 0 and 1 are solutions **15.** $\frac{3 + i\sqrt{3}}{2}$ and $\frac{3 - i\sqrt{3}}{2}$ are solutions **17.** Since $h(20) = 0$, the height of the object after 20 seconds is zero which means that it takes 20 seconds for the object to hit the ground. **19.** Q: $3y^2 + 4y + 9$; $R$: 13 **21.** Q: $3z^3 + 9z^2 + 25z + 76$; R: 225 **23.** Q: $2x^4 + 3x^3 + 3x^2 + 9x + 28$; R: 92 **25.** Q: $y^3 - 1$; R: −2 **27.** Q: $6x^2 + 3x + \frac{1}{2}$; R: $\frac{17}{4}$ **29.** Q: $t^2 + (3 - i)t - 3i$; R: 0 **31.** Q: $7x^4 - 39x^3 + 158x^2 - 486x + 1277$; R: 46 **33.** Q: $6.5x^2 - 26.17x + 105.625$; R: −836.579 **35.** Since all odd powers of $x$ have a negative coefficient, each of these terms would be positive for a negative value of $x$. Since all the even-powered terms would also be positive in this case, the polynomial will always be positive, never zero, for a negative value of $x$. **37.** $m = -1$ **39.** **(a)** 3 **(b)** 3 **(c)** The remainder is equal to $P(1)$. **41.** $\left(\frac{9}{2}, 2\right)$ **42.** 0 **43.** $2x - y - 11 = 0$ **44.** Q: $3x^2 + 2x + 5$; R: 3 **45.** **(a)** $3 + 2i$ **(b)** $5 - i\sqrt{7}$

### [4.2]

**1.** 0 **3.** 56 **5.** −2 **7.** −320 **9.** 0 **11.** 37 **13.** 17 **15.** 87 **17.** $P(x) = (x^2 + x - 2)(x - 7) + 0$ **19.** $P(x) = (x^2 - 7x - 2)(x + 1) + 16$ **21.** yes **23.** no **25.** $P(r)$ **27.** $x - r$ **29.** 1 **31.** $(x - 5)^2$ **33.** 5 **35.** $3 - 2i$ **37.** $1 + 3\sqrt{2}$ **39.** Substitution shows that 3 is a solution. The other solutions, 1 and −1, are solutions to $x^2 - 1 = 0$, the polynomial obtained when the original is divided by $x - 3$. **41.** 5; 1; 5 **43.** $P(x) = x^4 - 8x^3 + 26x^2 - 48x + 45$ **45.** $P(x) = x^4 - 2x^3 - 2x^2$ **47.** $P(-2.6) = -1518.713$ **49.** 3; 1, $\frac{-1 \pm i\sqrt{3}}{2}$ **51.** Consider the polynomial $P(x) - Q(x) = 0$, and use Exercise 50. **53.** Direct substitution shows that $P(1 + \sqrt{2}) = 0$. No; the theorem in this section is true for rational coefficients not for real (irrational) ones. **54.** **(a)** 9 **(b)** −1 **(c)** 7 **55.** The only break-even point is 15, $P(15) = 0$. **56.** $x^7 + 8x^6 - 7x^5 - 3x^3 + 2x^2 + 5$ **57.** **(a)** $P(-x) = 15x^4 + 3x^3 + x^2 - 7x - 6$ **(b)** ±1, ±3, ±5, ±15 **(c)** ±1, ±2, ±3, ±6

## [4.3]

In Exercises 1–7, the various possibilities for solutions are listed in the order: negative, positive, nonreal. **1.** either 1, 2, 0 or 1, 0, 2 **3.** either 0, 0, 4; 0, 2, 2; or 0, 4, 0 **5.** either 1, 1, 4; 1, 3, 2; 3, 1, 2; or 3, 3, 0 **7.** upper bound: 1; lower bound: −1 **9.** upper bound: 3; lower bound: −4 **11.** upper bound: 3; lower bound: −2 **13.** 1, −1, −2 **15.** 4, $\frac{1 \pm \sqrt{13}}{2}$ **17.** $-\frac{1}{3}$, $\pm 2i$ **19.** 0, $\frac{1}{3}$, $-\frac{2}{3}$, $\frac{-1 \pm i\sqrt{3}}{2}$ **21.** 2 would have to divide the leading coefficient 1 for $\frac{1}{2}$ to be a solution. **23.** Since the number of sign variations is 0 for both $f(x) = x^4 + a$ and $f(-x)$, there are no positive and no negative real solutions. Also, clearly 0 is not a solution; so $x^4 + a = 0$ cannot have a real root. **25.** 30 in by 10 in by 8 in **27.** 9 ft **29.** 5 **30.** $P(-3) = 314$ **31.** Since $P(5) = 0$, $x - 5$ is a factor of $P(x)$. **32.** 6; 1; 6 **33.** $5 + 2i$ **34.** $1 - \sqrt{7}$ **35.** $P(x) - x^6 - 8x^5 + 25x^4 - 38x^3 + 26x^2 - 8$

## [4.4]

**1.** **(d)** **3.** **(c)** **5.** **(a)** 3 **(b)** 2 **(c)** (0, 7) **(d)** upward **(e)** downward **7.** **(a)** 3 **(b)** 2 **(c)** (0, 12) **(d)** downward **(e)** upward **9.** **(a)** 4 **(b)** 3 **(c)** (0, −1) **(d)** upward **(e)** upward **11.** **(a)** 6 **(b)** 5 **(c)** (0, −5) **(d)** downward **(e)** downward

**13.**

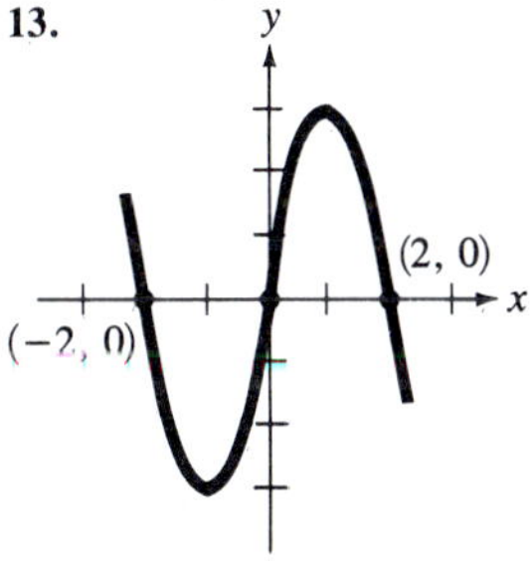


**15.**

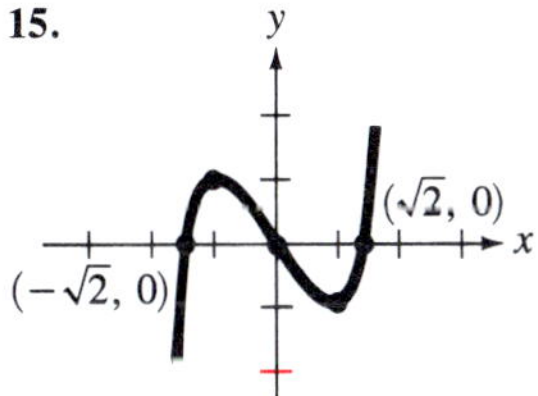


**17.**

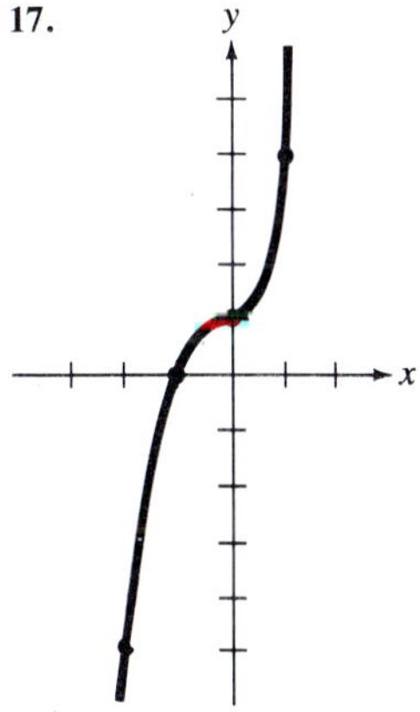


**19.**

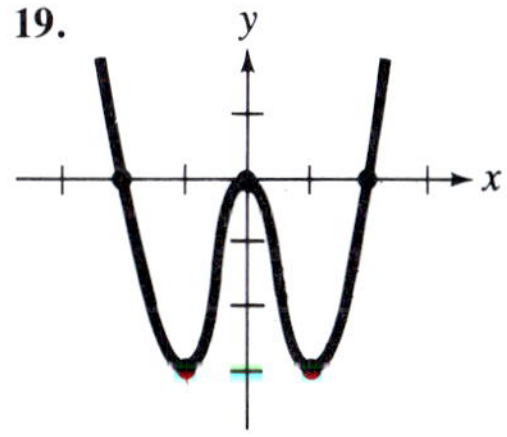


**21.** $P(1) = -6$ and $P(2) = 14$; so by the intermediate value theorem there is a zero between 1 and 2. **23.** $P(-3) = 8$ and $P(-2) = -12$; so by the intermediate value theorem there is a zero between −3 and −2. **25.** 1.73 **27.** 2.24

**29.**

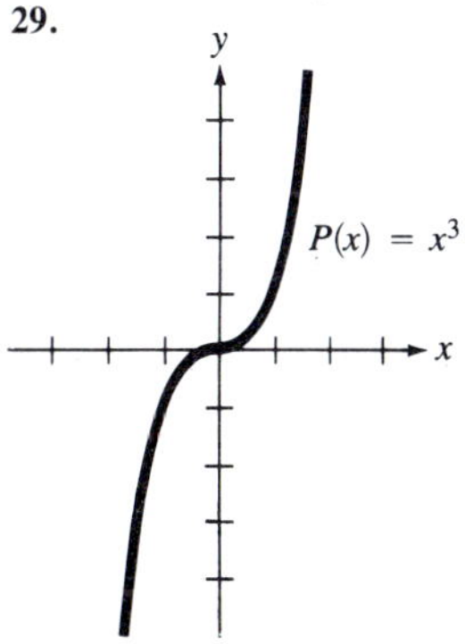


**(a)** The graph of $P(x) = x^3 + 2$ is the graph of $P(x) = x^3$ moved up 2 units.

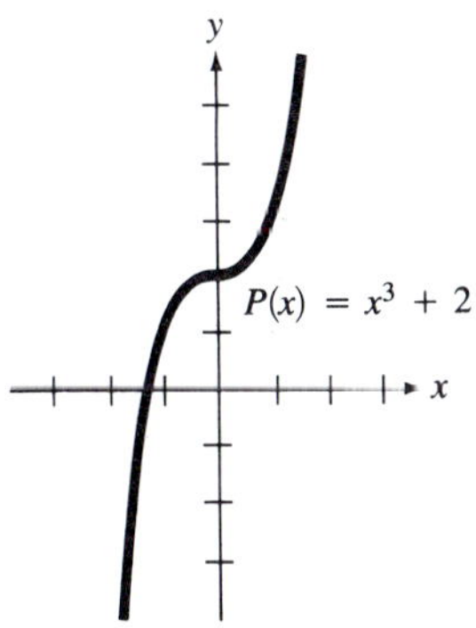


**(b)** The graph of $P(x) = x^3 - 3$ is the graph of $P(x) = x^3$ moved down 3 units.

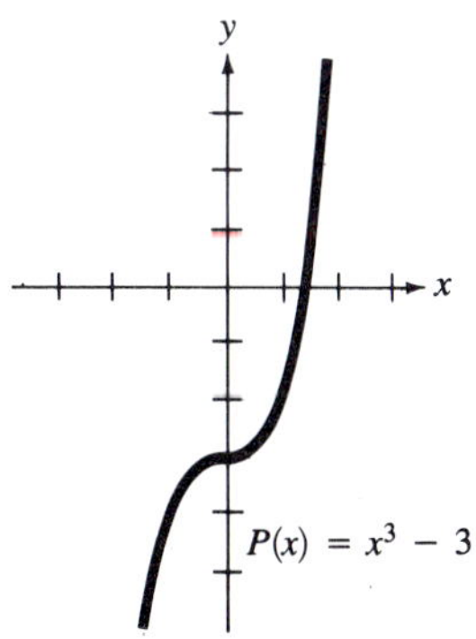


**(c)** The graph of $P(x) = (x + 1)^3$ is the graph of $P(x) = x^3$ moved left 1 unit.

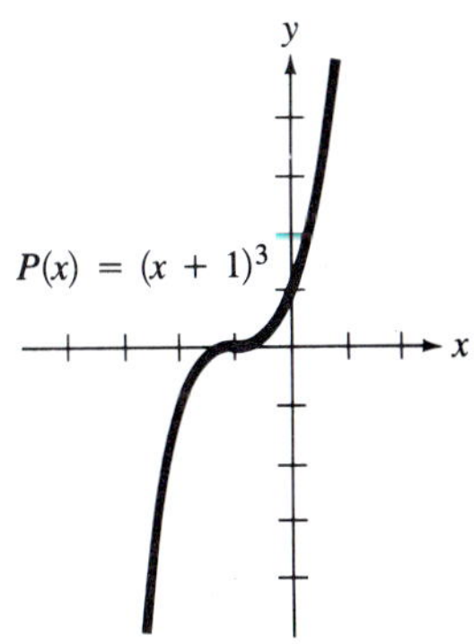

**(d)** The graph of $P(x) = (x - 3)^3$ is the graph of $P(x) = x^3$ moved right 3 units.

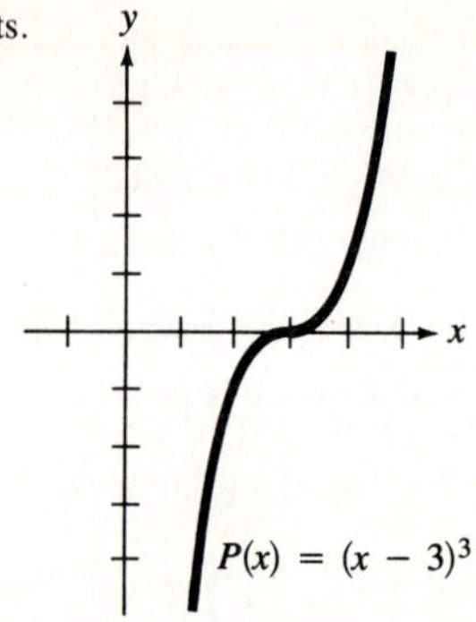


**31.** $m = -15$ **33.** 5, $i$, $-i$ **34.** 3, $-\frac{1}{2}$, $\sqrt{5}$, $-\sqrt{5}$ **35.** Q: $y^4 - 4y^3 + 5y^2 - 12y + 13$; R: $-15$ **36.** $P(4.1) = 0$
**37.** 3; $-2$, $1 \pm i\sqrt{3}$ **38.** all real numbers except $x = -3$ **39.** all real numbers

**[4.5]**

**1. (a)** $x = -2$ **(b)** $y = 0$ **(c)** none **(d)** none **(e)** $\left(0, \frac{5}{2}\right)$ **(f)** none **3. (a)** $x = 5$ **(b)** $y = 4$ **(c)** none **(d)** $(0, 0)$ **(e)** $(0, 0)$ **(f)** none **5. (a)** $x = 1$ and $x = -2$ **(b)** $y = 0$ **(c)** none **(d)** $(-1, 0)$ **(e)** $\left(0, -\frac{1}{2}\right)$ **(f)** none **7. (a)** $x = 5$ **(b)** none **(c)** $y = x + 6$ **(d)** $(2, 0)$ and $(-3, 0)$ **(e)** $\left(0, \frac{6}{5}\right)$ **(f)** none **9. (a)** none **(b)** $y = 5$ **(c)** none **(d)** $(0, 0)$ **(e)** $(0, 0)$ **(f)** $y$-axis

**11.**

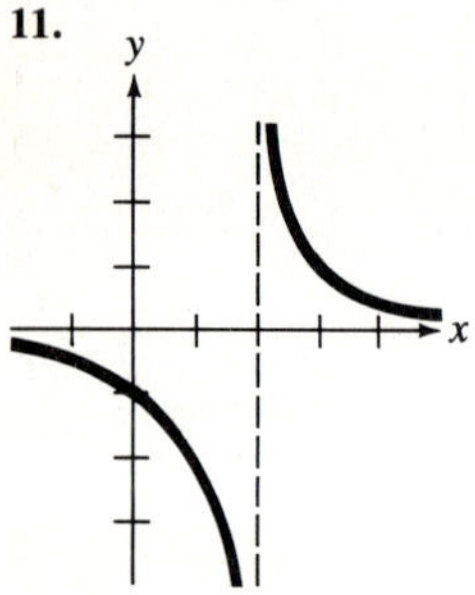


**13.**

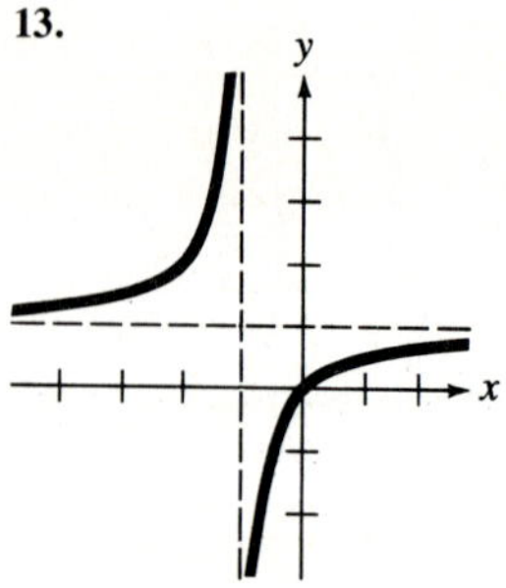


**15.**

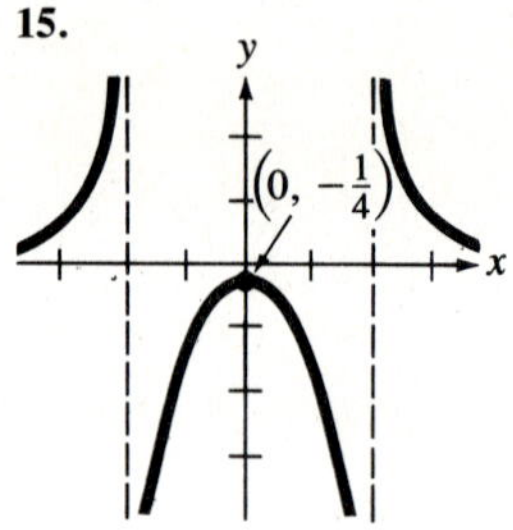


**17.**

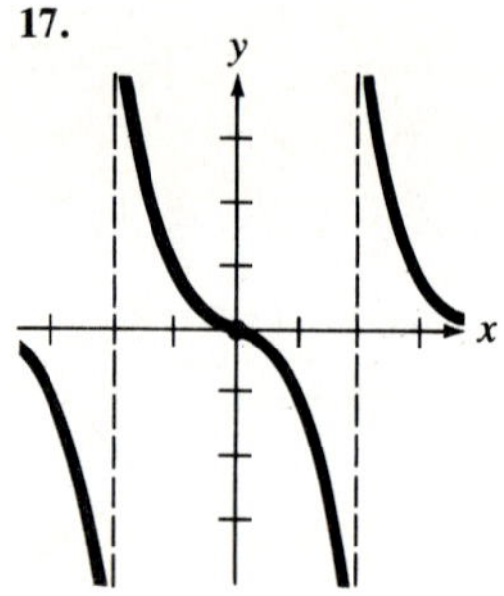


**19.**

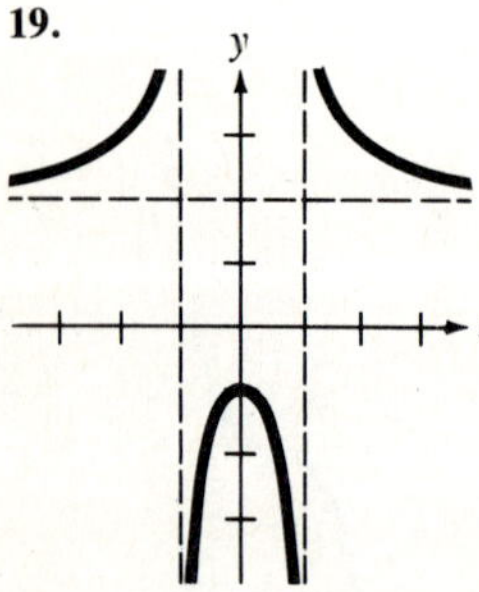


**21.**

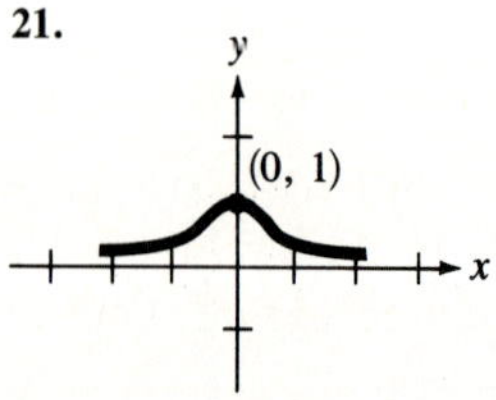


**23.** The average cost of production decreases to a minimum of \$8 per item after 12 hours.

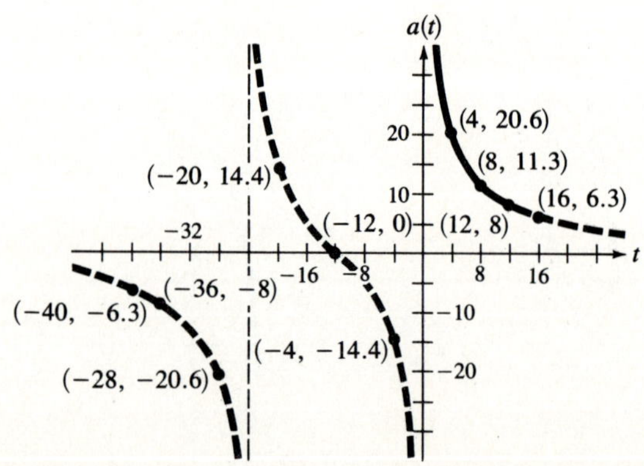

**25.** $f(x) = \frac{3}{x-1}$ **27.** Yes, see Example 7.

**29.**

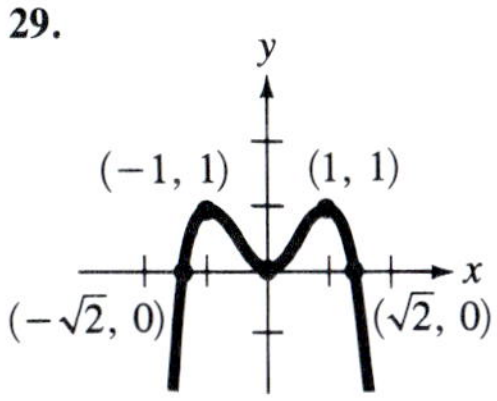


**30.**

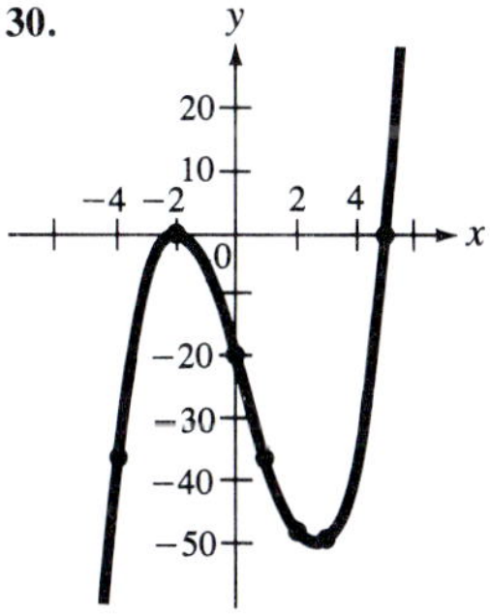


**31.** either 0, 0, 4; 2, 0, 2; or 4, 0, 0 (in order negative, positive, nonreal) **32.** upper bound: 2; lower bound: −2

## Chapter 4 Review Exercises

**1.** **(a)** 4 **(b)** 3 **(c)** 0 **(d)** −14 **2.** All three are zeros. **3.** $x^3 - 6x^2 + 18x - 53$, 157 **4.** $P(5)$ **5.** remainder theorem **6.** $x + 3$ **7.** factor theorem **8.** Since $P(100) = 0$, 100 is a break-even point. **9.** $m = 25$; 0 **10.** **(a)** $P(1) = -8$ **(b)** $P(-1) = -10$ **(c)** $P(2) = 14$ **11.** **(a)** $x - 4$ is a factor **(b)** $x + 1$ is a factor **(c)** $x + 2$ is not a factor **12.** $(x + 4)^3$ **13.** $2 + 7i$ **14.** $3 - \sqrt{11}$ **15.** 7; 1; 7 **16.** $P(x) = x^6 - 6x^4 + 4x^3 + 5x^2 - 28x - 12$ **17.** either 1, 3, 2 or 1, 1, 4 **18.** 1, 0, 4 **19.** upper bound: 3; lower bound: −1 **20.** upper bound: 1; lower bound: −5 **21.** **(a)** 5 **(b)** 4 **22.** If $\frac{p}{q}$ is a rational solution, then $q = 1$ since $q$ must be a factor of 1. **23.** $\frac{1}{3}, \frac{3 \pm i\sqrt{7}}{4}$ **24.** −1, −1, 5 **25.** $r = 3$ **26.** **(a)** 3 **(b)** 2 **(c)** (0, −5) **(d)** upward **(e)** downward **27.** **(a)** 4 **(b)** 3 **(c)** (0, 8) **(d)** downward **(e)** downward

**28.**

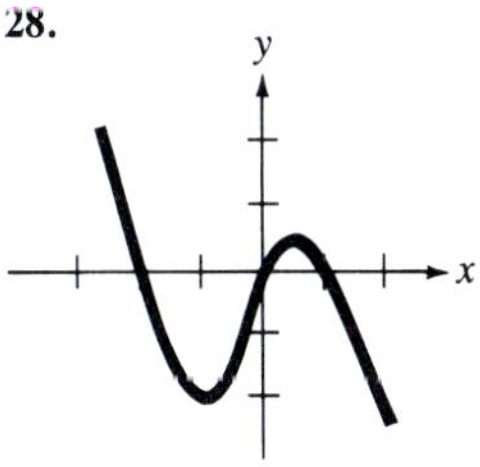


**29.**

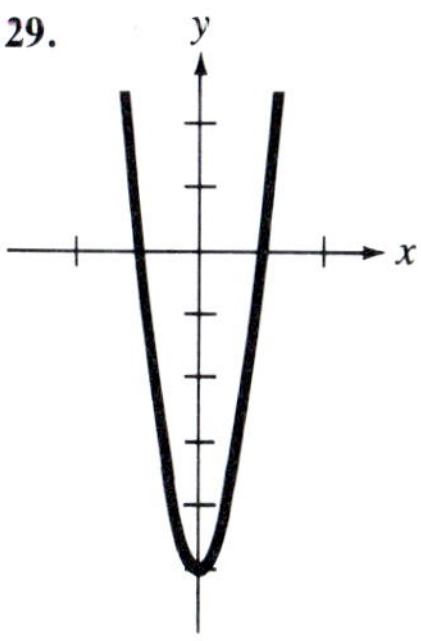


**30.** 3.32 **31.** **(a)** $x = 9$ **(b)** $y = 0$ **(c)** none **(d)** none **(e)** $\left(0, \frac{-10}{9}\right)$ **(f)** none **32.** **(a)** $x = \frac{1}{3}$ and $x = -\frac{1}{3}$ **(b)** $y = \frac{2}{3}$ **(c)** none **(d)** (0, 0) **(e)** (0, 0) **(f)** none

**33.**

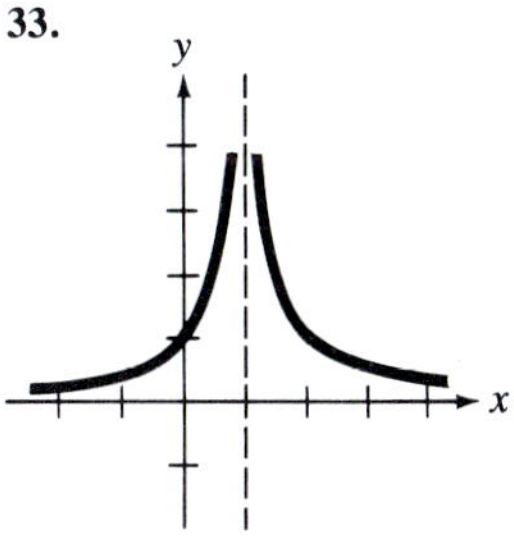


**34.**

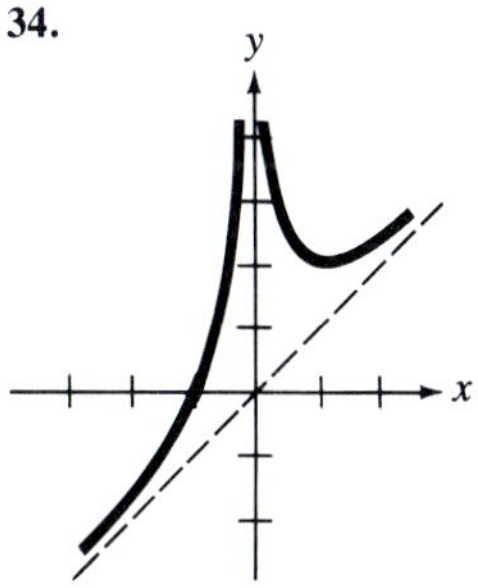


**35.** 6.5 in

## Chapter 5

### [5.1]

**1.** $\log_2 8 = 3$ **3.** $\log_5 9 = v$ **5.** $\log_a c = -3$ **7.** $\log_u w = -v$ **9.** $3^2 = 9$ **11.** $a^7 = b$ **13.** $3^b = \frac{1}{27}$

**15.** $a^c = 6$ **17.** 5 **19.** 5 **21.** $\frac{1}{2}$ **23.** 49 **25.** $\frac{1}{36}$ **27.** 3 **29.** 81 **31.** 3 **33.** $-1$ **35.** $\pm 2$

**37. (a)** 10 **(b)** 20 **39. (a)** 5 **(b)** 4 **(c)** $(0, -9)$ **(d)** downward **(e)** upward

**40.** $C(95) = \$19{,}000$

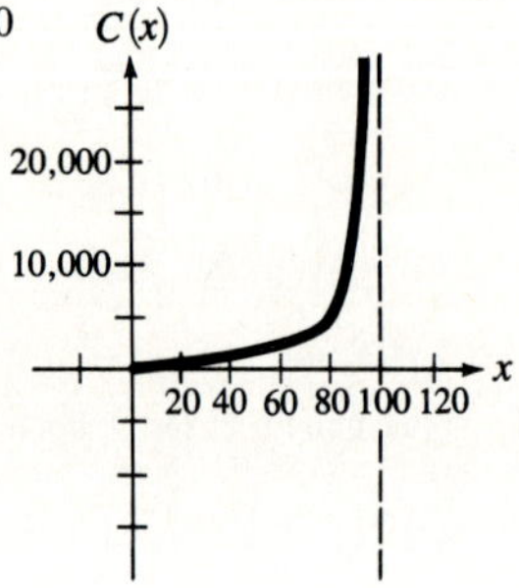


### [5.2]

**1.**

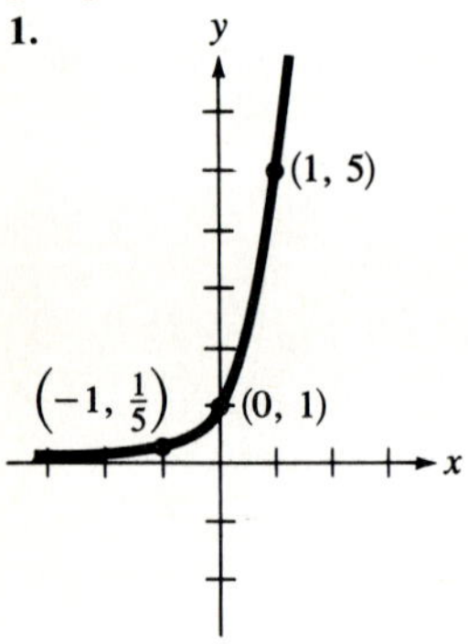


**3.**

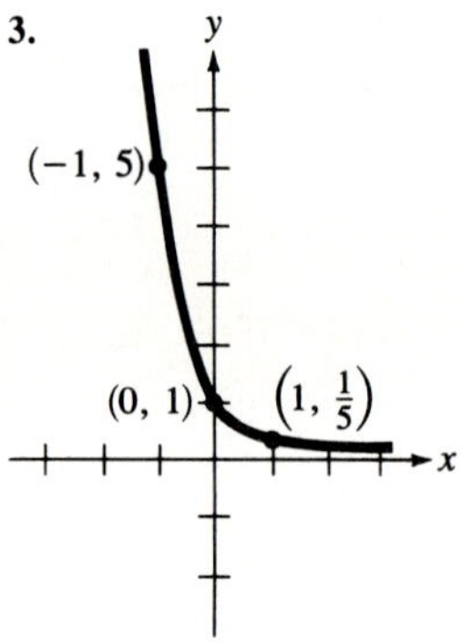


**5.**

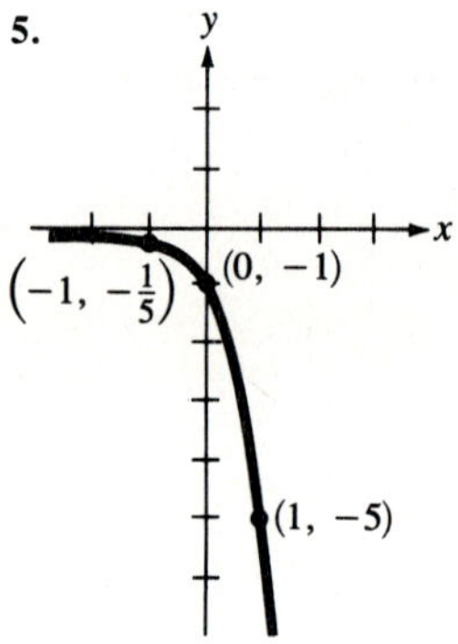


**7.**

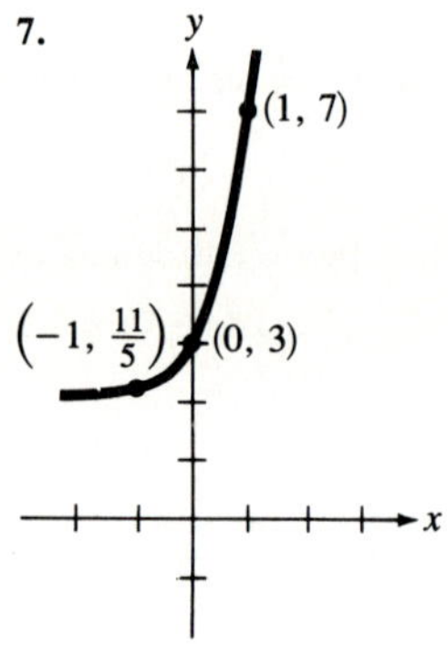


**9.**

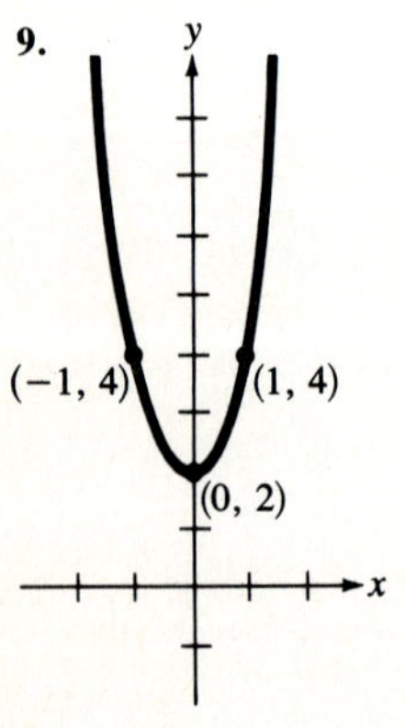


**11.**

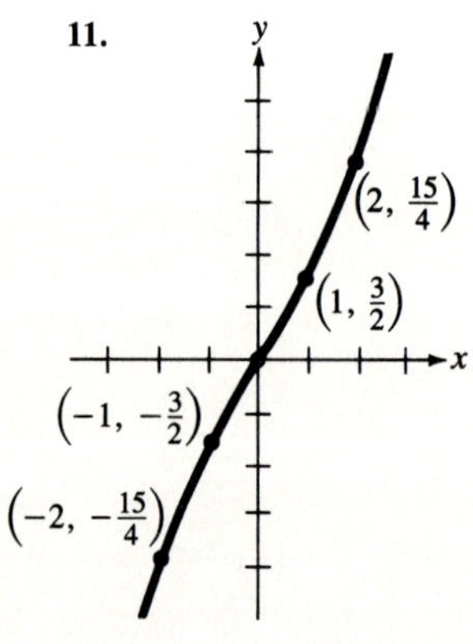


**13.**

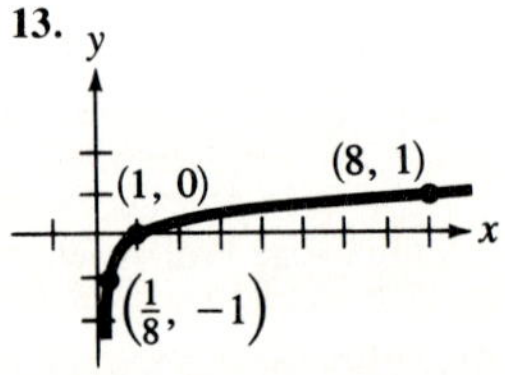


**15.**

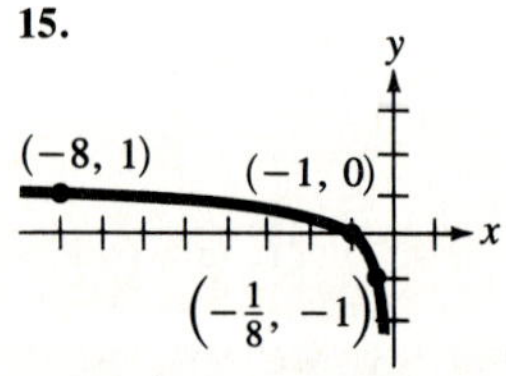


**17.**

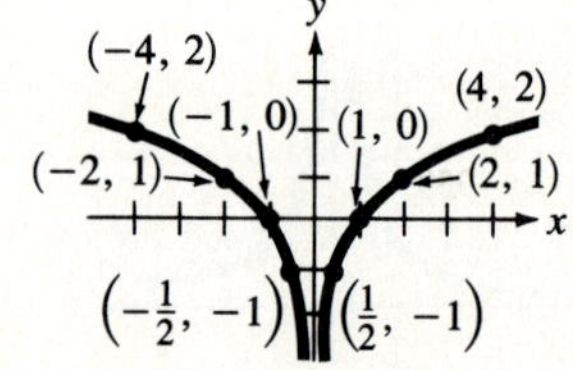

**19.** 3 **21.** −3 **23.** $x + 7$ **25.** 16.2425 **27.** −7.2625 **29.** 0.0793 **31.** 3.2810 **33.** (0, 1) **35.** all real numbers **37.** all positive real numbers **39.** $a = 3$ **41.** If $a = 1$, $\log_1 x = y$ is equivalent to $1^y = x$, or $x = 1$ for all values of $y$, which would not be a function. **43.** $f^{-1}(x) = -1 + \log_2 (x - 1)$; for all $x > 1$ **45.** $f^{-1}(x) = 3^{x/2} + 4$; for all real numbers $x$ **47.** **(a)** 5000 **(b)** 160,000 **(c)** 47,568 **49.** **(a)** 35.36 g **(b)** 0.000 001 5 g **(c)** 19.28 g

**51.**

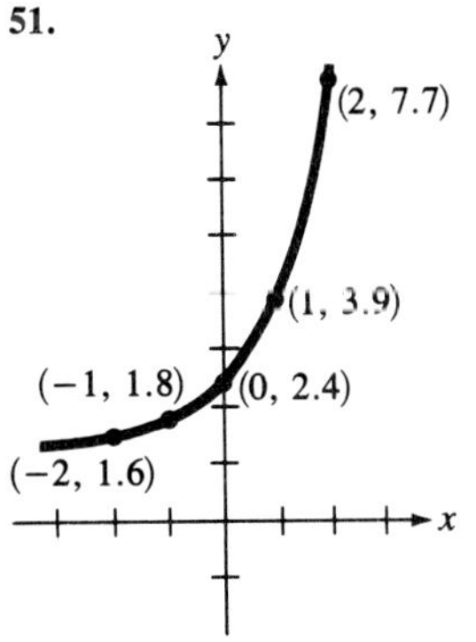


**53.** 2 **54.** 3 **55.** 0.01 **56.** 4 **57.** 1 **58.** 9, −9 **59.** true **60.** false **61.** true **62.** true

[5.3]

**1.** 4 **3.** 4, −1 **5.** 2 **7.** 2.6990 **9.** 1.4313 **11.** −0.3980 **13.** −0.9542 **15.** 0.3891 **17.** 0.0682 **19.** 0.6825 **21.** 3.0143 **23.** 0.2552 **25.** $\log_a x + \log_a y + \log_a z$ **27.** $\log_a x + 2 \log_a z - \log_a y$ **29.** $\log_a y + \frac{1}{2} \log_a x - 3 \log_a z$ **31.** $\log_a z + 3 \log_a (x + 1)$ **33.** $2 \log_a x + \frac{1}{2} \log_a y - \frac{1}{2} \log_a z$ **35.** $\frac{1}{2} \log_a x + \frac{1}{4} \log_a y$ **37.** $\log_a x\sqrt{y}$ **39.** $\log_a \frac{x}{y^2}$ **41.** $\log_a \frac{z^3\sqrt{x}}{y^3}$ **43.** $\log_a \frac{y^x}{z^3}$ **45.** $\log_a(x + 1)$ **47.** $\log_a \sqrt{\frac{x}{y\sqrt{z}}}$ **49.** true **51.** false **53.** true **55.** true **57.** false **59.** $\frac{3}{5}$ **61.** $\frac{1}{3}$ **63.** $\frac{3}{2}$ **65.** $\log_2 (8 + 8) = \log_2 16 = 4$; but $\log_2 8 + \log_2 8 = 3 + 3 = 6$ **67.** $\log_3 (1 \cdot 9) = \log_3 9 = 2$, but $(\log_3 1)(\log_3 9) = (0)(2) = 0$ **69.** $a^z = \frac{x}{y}$

**71.** 5.3A71

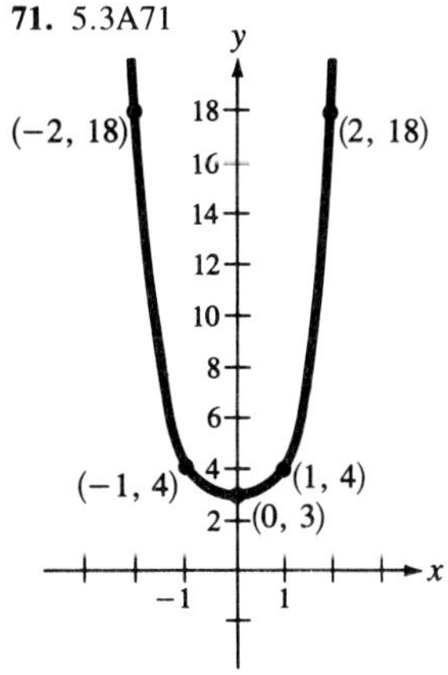


**72.** 5.3A72

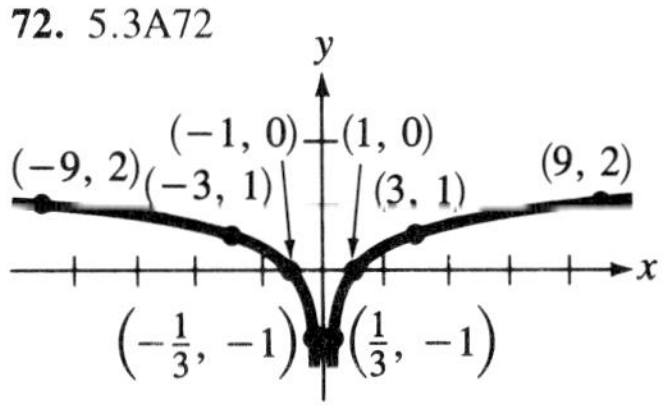


**73.** $3x + 2$ **74.** $5x^2$ **75.** 35

[5.4]

**1.** 2.7959 **3.** 0.3011 **5.** −4.1746 **7.** 11.7090 **9.** −8.1302 **11.** 1.9817 **13.** 4.9079 **15.** 16.8875 **17.** 22.7 **19.** 0.000 614 **21.** $4.50 \times 10^{14}$ **23.** 0.992 **25.** $5.94 \times 10^{-36}$ **27.** 6.7405 **29.** −7.5521 **31.** −10.1597 **33.** 35.517 **35.** 1.1447 **37.** 11.4 **39.** 0.0157 **41.** $2.78 \times 10^9$ **43.** 0.993 **45.** $1.02 \times 10^{-18}$

**47.**
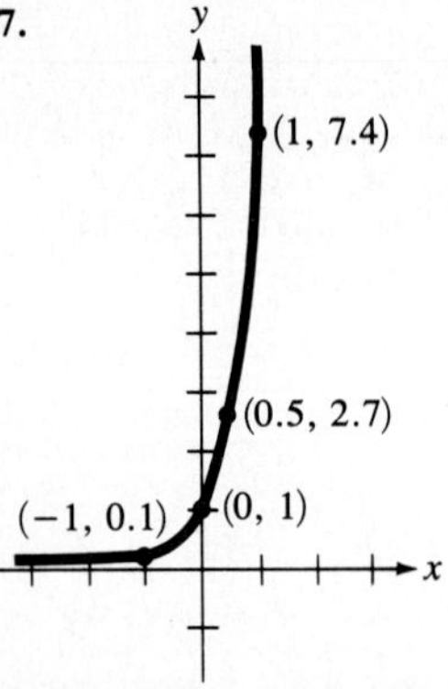


**49.**
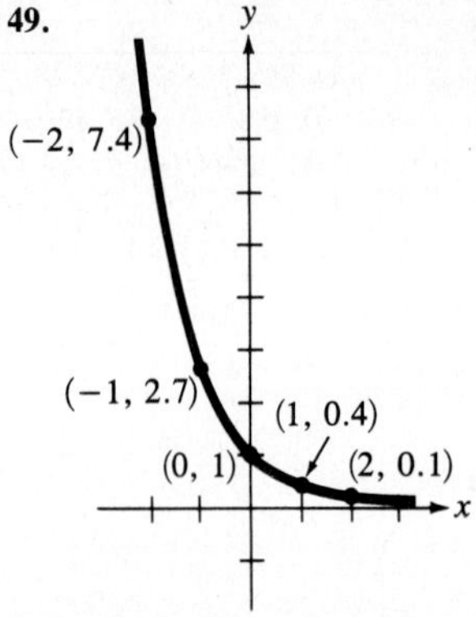


**51.**
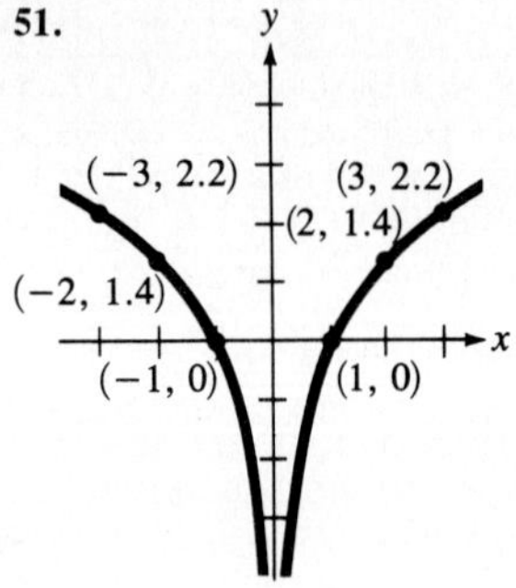


**53.** **(a)** \$99.67 **(b)** \$70.32 **(c)** 921 **(d)** 1060 **55.** **(a)** 14.7 lb/in$^2$ **(b)** 4.2 lb/in$^2$ **(c)** 3.47 mi **(d)** 6.93 mi
**57.** **(a)** 6.10 mg **(b)** 0.01 mg **(c)** 3.66 hr **(d)** 2.31 hr **59.** $1 + \frac{1}{2}\ln z - \frac{1}{3}\ln x$ **60.** $\frac{1}{2}\ln x + \frac{1}{4}\ln z - 2\ln y$
**61.** $\frac{1}{2} + \frac{1}{4}\ln y$ **62.** $\ln x^3y^3\sqrt{z}$ **63.** $\ln\sqrt{x-y}$ **64.** $\ln\frac{x^3}{z^5\sqrt{y}}$ **65.** $x^2y = e^5$ **66.** 1, 2

**67.**
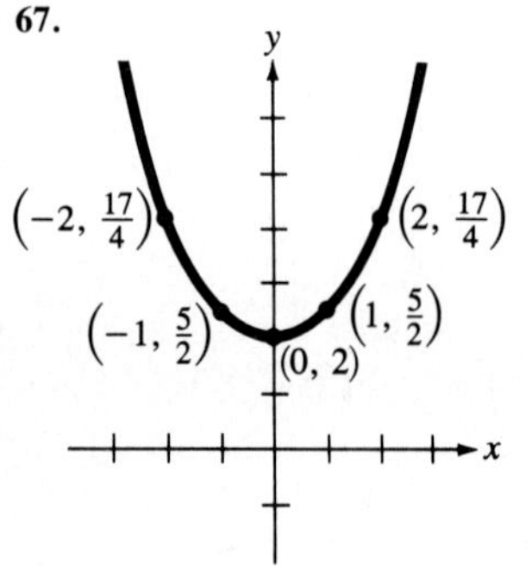


**68.**
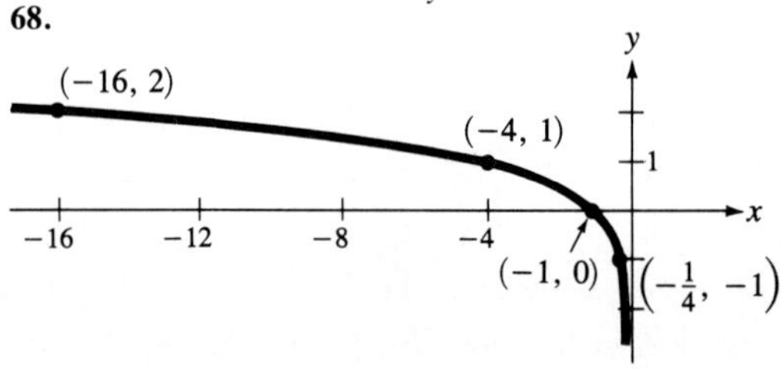


**69.** −3 **70.** 2

**[5.5]**

**1.** $\frac{3}{2}$ **3.** $\frac{15}{2}$ **5.** 0, 4 **7.** 0.936 **9.** 2.74 **11.** 59.8 **13.** 5 **15.** 2, 3 **17.** 0.461 **19.** 6 **21.** 27
**23.** 5 **25.** 8 **27.** 5.82 **29.** 1 **31.** 1, 25 **33.** 1, 5.65, 0.177 **35.** 7.39, 0.368 **37.** $10^{10}$ **39.** 0
**41.** 1, 10,000 **43.** $e$, $-e$ **45.** 100, 0.01 **47.** 0.549 **49.** 210°F **51.** $k = 0.0654$; 125°F
**53.** $t = -\frac{1}{k}\ln\left[\frac{T - T_c}{T_0 - T_c}\right]$ **55.** 14 ft

**57.**
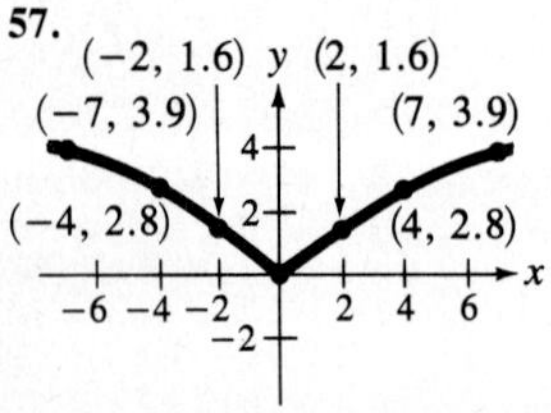


**58.**
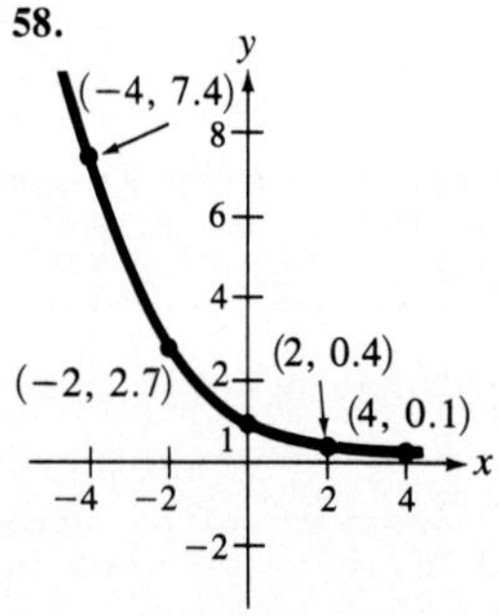


**59.** 6.5 **60.** \$248.78

## [5.6]

Many answers are given using an approximate value. **1. (a)** \$1411.58 **(b)** \$1422.10 **(c)** \$1427.62 **(d)** \$1431.41 **(e)** \$1432.88 **(f)** \$1433.26 **(g)** \$1433.33 **3.** 6.12 years **5.** 5.95 years **7.** 5.78 years **9.** \$400.40; \$6024.00 **11.** 33 years **13.** 271 million **15.** 22.4 years **17.** 1644 **19.** 23.1 years **21.** 685 years **23.** 4783 years **25.** 4.27 mg **27.** 6.4 **29.** $10^{7.2}$ **31.** 66 decibels **33.** $10^{-6}$ watt/m$^2$ **35.** No; it will increase by about 20 decibels. **37.** 1.53 **38.** 5 **39.** 100,000; 1 **40.** $e^5$; $e^{-5}$ **41.** 6.6830 **42.** cannot find the natural logarithm of a negative number **43.** $3.17 \times 10^{-10}$ **44.** 558,000 **45.** $\ln x + \frac{1}{2} \ln y - 4 \ln z$

## Chapter 5 Review Exercises

**1. (a)** 1 **(b)** 0 **2.** $x = \log_2 5$ **3.** $3^w = a$ **4.** 4 **5.** $\frac{1}{3}$ **6.** $x$ **7.** $x$

**8.**

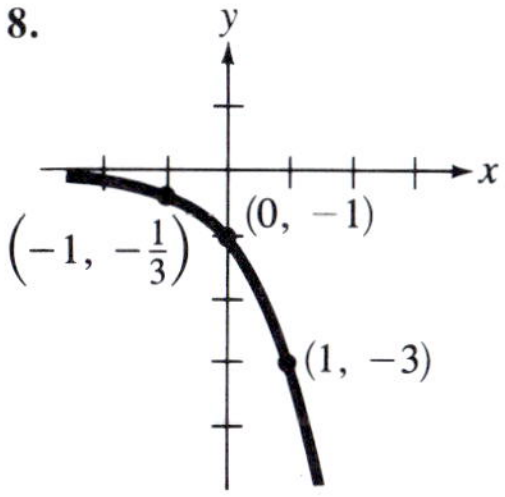


**9.**

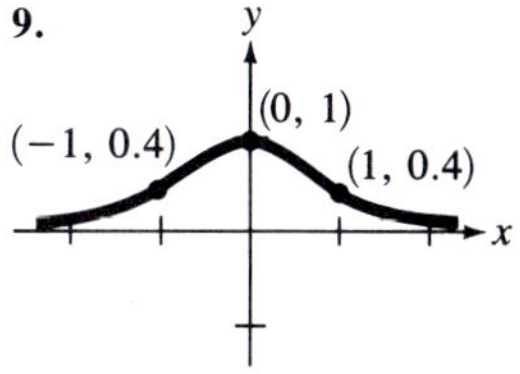


**10.**

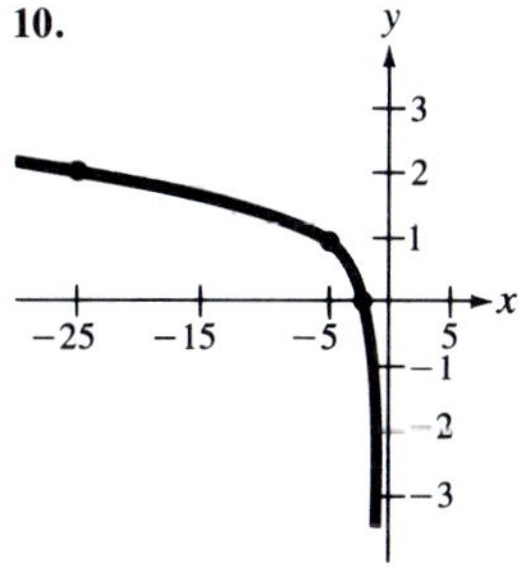


**11.** 12.5297 **12.** 0.3715 **13.** 1.4641 **14.** 0.4075 **15.** 0.6113 **16.** 1.7713 **17.** $\frac{2}{3} \log_a x + \frac{2}{3} \log_a y - \frac{2}{3} \log_a z$ **18.** $5 \log_a y + \frac{2}{3} \log_a x - \frac{2}{3} \log_a z$ **19.** $\log_a x^3y^3\sqrt{z}$ **20.** $\log_a (x - y)^2(x + y)$ **21.** false **22.** true **23.** −15.17 **24.** −4.5564 **25.** $2.39 \times 10^{-9}$ **26.** $1.19 \times 10^{18}$ **27.** 9 **28.** 5 **29.** 10 **30.** 20.6 **31.** 78.1 **32.** $\frac{1}{2}$ **33.** 1; 1.62 **34.** $e$ **35. (a)** 192,000 **(b)** 10 hr **36. (a)** 4.43 lb/in$^2$ **(b)** 5.5 mi **37. (a)** \$997.76 **(b)** 1901 **38. (a)** 6.6 **(b)** $7.94 \times 10^{-6}$ **39. (a)** 11.6 **(b)** 6.7 in **40. (a)** 5.5 mg **(b)** 1.62 hr **41. (a)** 45.8 **(b)** 19 months **42.** $k = 0.056$; 82.8°F **43.** $k = 0.0347$ **44. (a)** \$2621.59 **(b)** \$2641.97 **(c)** \$2646.26 **45.** 13.9 yr **46.** \$360.05; \$34,809.03 **47.** 34,774 **48.** 150.7 years **49.** 5.4 **50.** Yes; the decibel level is approximately 107.4, which exceeds 90 decibels by 17.4.

# Chapter 6

## [6.1]

Answers may vary somewhat. Calculations were made using the calculator value of $\pi$ with no rounding until the final step. **1.** no **3.** yes **5.** yes **7.** yes **9.** 38.69° **11.** 52.59° **13.** 85°25′12″ **15.** −48°10′48″ **17.** 540° **19.** 401.07° **21.** −81.57° **23.** 286°28′44″ **25.** −48°50′41″ **27.** $\frac{\pi}{3}$ **29.** $-\frac{2\pi}{3}$ **31.** 0.6720 **33.** 0.9657 **35.** −1.9345

## [6.2]

Answers may vary somewhat. Calculations were made using the calculator value of $\pi$ with no rounding until the final step. **1.** 4.2 cm **3.** 11.82 ft **5.** 101°59′ **7.** 4328 mi **9.** 49°25′N **11.** 672 rev **13.** 44 ft **15.** 130,000 mi **17.** 3.6 rev **19.** The area of the circle is $\pi r^2$. The area of the sector is $\frac{\theta}{2\pi}$ times the area of the circle. Thus, $A = \frac{\theta}{2\pi}\pi r^2 = \frac{1}{2}r^2\theta$. **21.** 19.6 ft/sec **23.** 9.4 ft/sec **25.** 1047 mph **27.** $N_1$: 1.0 yd/min; $N_2$: 1.6 yd/min **29.** 257°49′52″ **30.** −1.4958

## [6.3]

**1.** $\sin A = \frac{5}{13} = \cos B$; $\cos A = \frac{12}{13} = \sin B$; $\tan A = \frac{5}{12} = \cot B$; $\csc A = \frac{13}{5} = \sec B$; $\sec A = \frac{13}{12} = \csc B$; $\cot A = \frac{12}{5} = \tan B$

**3.** $\sin A = \frac{7}{15} = \cos B$; $\cos A = \frac{4\sqrt{11}}{15} = \sin B$; $\tan A = \frac{7\sqrt{11}}{44} = \cot B$; $\csc A = \frac{15}{7} = \sec B$; $\sec A = \frac{15\sqrt{11}}{44} = \csc B$; $\cot A = \frac{4\sqrt{11}}{7} = \tan B$ **5.** $\sin A = 0.6 = \cos B$; $\cos A = 0.8 = \sin B$; $\tan A = 0.75 = \cot B$; $\csc A = 1.\bar{6} = \sec B$; $\sec A = 1.25 = \csc B$; $\cot A = 1.\bar{3} = \tan B$ **7.** $\boxed{\sin \theta = \frac{5}{7}} = \cos (90° - \theta)$; $\cos \theta = \frac{2\sqrt{6}}{7} = \sin (90° - \theta)$; $\tan \theta = \frac{5\sqrt{6}}{12} = \cot (90° - \theta)$; $\csc \theta = \frac{7}{5} = \sec (90° - \theta)$; $\sec \theta = \frac{7\sqrt{6}}{12} = \csc (90° - \theta)$; $\cot \theta = \frac{2\sqrt{6}}{5} = \tan (90° - \theta)$
**9.** $\sin \theta = \frac{4\sqrt{3}}{7} = \cos (90° - \theta)$; $\cos \theta = \frac{1}{7} = \sin (90° - \theta)$; $\tan \theta = 4\sqrt{3} = \cot (90° - \theta)$; $\csc \theta = \frac{7\sqrt{3}}{12} = \sec (90° - \theta)$; $\boxed{\sec \theta = 7} = \csc (90° - \theta)$; $\cot \theta = \frac{\sqrt{3}}{12} = \tan (90° - \theta)$ **11.** $\sin \theta = \frac{\sqrt{7}}{3} = \cos (90° - \theta)$; $\boxed{\cos \theta = \frac{\sqrt{2}}{3}} = \sin (90° - \theta)$; $\tan \theta = \frac{\sqrt{14}}{2} = \cot (90° - \theta)$; $\csc \theta = \frac{3\sqrt{7}}{7} = \sec (90° - \theta)$; $\sec \theta = \frac{3\sqrt{2}}{2} = \csc (90° - \theta)$; $\cot \theta = \frac{\sqrt{14}}{7} = \tan (90° - \theta)$

**13.**

| Function | $30°\left(\frac{\pi}{6}\right)$ | $45°\left(\frac{\pi}{4}\right)$ | $60°\left(\frac{\pi}{3}\right)$ |
|---|---|---|---|
| sine | $\frac{1}{2}$ | $\frac{\sqrt{2}}{2}$ | $\frac{\sqrt{3}}{2}$ |
| cosine | $\frac{\sqrt{3}}{2}$ | $\frac{\sqrt{2}}{2}$ | $\frac{1}{2}$ |
| tangent | $\frac{\sqrt{3}}{3}$ | 1 | $\sqrt{3}$ |
| cosecant | 2 | $\sqrt{2}$ | $\frac{2\sqrt{3}}{3}$ |
| secant | $\frac{2\sqrt{3}}{3}$ | $\sqrt{2}$ | 2 |
| cotangent | $\sqrt{3}$ | 1 | $\frac{\sqrt{3}}{3}$ |

**15.** $\frac{1 + \sqrt{3}}{2}$ **17.** 1 **19.** $\frac{1}{2}$ **21.** false **23.** true **25.** false **27.** true **29.** false **31.** true **33.** 0.9063 **35.** 0.3249 **37.** 0.7357 **39.** 1.3254 **41.** 7.1952 **43.** 1.7013 **45.** 56.29° **47.** 75.97° **49.** 72.20° **51.** 56.56° **53.** 0.6087 **55.** 1.2639 **57.** 1.3444 **59.** 0.5707 **61.** $\sin \theta = 0.7341$; $\cos \theta = 0.6790$; $\tan \theta = 1.0811$; $\csc \theta = 1.3622$; $\sec \theta = 1.4727$; $\cot \theta = 0.9250$ **63.** $\sin \theta = 0.9915$; $\cos \theta = 0.1300$; $\tan \theta = 7.6260$; $\csc \theta = 1.0086$; $\sec \theta = 7.6913$; $\cot \theta = 0.1311$ **65. (a)** 0.5707 **(b)** 0.4293 **(c)** 1.3292 **(d)** 1.7523 **(e)** 2.3292 **(f)** 0.7523 **67.** 802 ft-lb **69.** 2.9 ft **70.** 7.16° **71.** 157.5 ft/sec

## [6.4]

**1.** $c = 52$; $A = 22°$; $B = 68°$ **3.** $b = 4.2$; $A = 65°40'$; $B = 24°20'$ **5.** $B = 63°20'$; $a = 5.4$; $b = 10.7$ **7.** $A = 79°10'$; $a = 16.7$; $c = 17.0$ **9.** Any similar triangle with the given angles is a possible solution; at least one side must be given for a triangle to be solvable. **11.** $A = 8°18'$; $a = 3.15$; $b = 21.62$ **13.** 45.7 ft **15.** 133.7 ft **17.** 92 m **19.** 1008 ft **21.** 9.4 mi **23.** 129 km east; 153 km south **25.** 9° **27.** 20.9 mi **29.** 31.6 ft **31.** 22.68 km/hr **33.** 1294 mph **35.** 26°50′ **37.** 86°20′ **39.** 1383 ft **41.** 26°30′; 2 hr 5 min **43. (a)** 0.9501 **(b)** 0.3120 **(c)** 3.0451 **(d)** 1.0525 **(e)** 3.2051 **(f)** 0.3284 **(g)** 0.9027 **(h)** 10.2729 **44. (a)** 0.8885 **(b)** 0.4588 **(c)** 1.9367 **(d)** 1.1254 **(e)** 2.1796 **(f)** 0.5164 **(g)** 0.7895 **(h)** 4.7506 **45. (a)** 14.89° **(b)** 0.2598 **(c)** 14°53′10″ **46. (a)** 13.53° **(b)** 0.2361 **(c)** 13°31′47″ **47.** 21.0 rev **48.** A function is a relation with the property that to each real number there corresponds one and only one element in the range.

## [6.5]

**1.** $P\left(\frac{3}{5}, \frac{4}{5}\right)$, $Q\left(-\frac{3}{5}, \frac{4}{5}\right)$, $R\left(-\frac{3}{5}, -\frac{4}{5}\right)$, $S\left(\frac{3}{5}, -\frac{4}{5}\right)$ **3.** $A(1, 0)$, $B\left(\frac{\sqrt{2}}{2}, \frac{\sqrt{2}}{2}\right)$, $C(0, 1)$, $D(0, -1)$, $E\left(-\frac{\sqrt{2}}{2}, -\frac{\sqrt{2}}{2}\right)$
**5.** $2\pi$ **7.** counterclockwise **9. (a)** $(a, -b)$ **(b)** $(-a, b)$ **(c)** $(-a, -b)$ **(d)** $(-a, -b)$ **(e)** $(-b, a)$ **(f)** $(b, -a)$

**11.** (0, 1) **13.** (0, −1) **15.** (0, 1) **17.** (1, 0) **19.** $\left(\frac{\sqrt{2}}{2}, \frac{\sqrt{2}}{2}\right)$ **21.** $\left(-\frac{\sqrt{2}}{2}, -\frac{\sqrt{2}}{2}\right)$ **23.** $\left(\frac{1}{2}, \frac{\sqrt{3}}{2}\right)$ **25.** $\left(-\frac{1}{2}, -\frac{\sqrt{3}}{2}\right)$ **27.** $\left(\frac{\sqrt{3}}{2}, \frac{1}{2}\right)$ **29.** $\left(-\frac{\sqrt{3}}{2}, -\frac{1}{2}\right)$ **31. (a)** 0 **(b)** $\frac{\pi}{2}$ **(c)** $\pi$ **(d)** $\frac{3\pi}{2}$ **(e)** $\frac{\pi}{4}$ **(f)** $\frac{3\pi}{4}$ **(g)** $\frac{5\pi}{4}$ **(h)** $\frac{7\pi}{4}$ **33. (a)** 0 **(b)** $-\frac{3\pi}{2}$ **(c)** $-\pi$ **(d)** $-\frac{\pi}{2}$ **(e)** $-\frac{7\pi}{4}$ **(f)** $-\frac{5\pi}{4}$ **(g)** $-\frac{3\pi}{4}$ **(h)** $-\frac{\pi}{4}$ **35. (a)** $2\pi$, 0, or $-2\pi$ **(b)** $\frac{\pi}{2}, -\frac{3\pi}{2}$ **(c)** $\pi, -\pi$ **(d)** $\frac{3\pi}{2}, -\frac{\pi}{2}$ **(e)** $\frac{\pi}{4}, -\frac{7\pi}{4}$ **(f)** $\frac{\pi}{6}, -\frac{11\pi}{6}$ **(g)** $\frac{\pi}{3}, -\frac{5\pi}{3}$ **(h)** $\frac{5\pi}{4}, -\frac{3\pi}{4}$ **(i)** $\frac{5\pi}{6}, -\frac{7\pi}{6}$
**37. (a)** −1 **(b)** 0 **(c)** undefined **(d)** −1 **(e)** undefined **(f)** 0 **39. (a)** 1 **(b)** 0 **(c)** undefined **(d)** 1 **(e)** undefined **(f)** 0 **41. (a)** $\frac{\sqrt{2}}{2}$ **(b)** $\frac{\sqrt{2}}{2}$ **(c)** 1 **(d)** $\sqrt{2}$ **(e)** $\sqrt{2}$ **(f)** 1 **43. (a)** $-\frac{\sqrt{2}}{2}$ **(b)** $-\frac{\sqrt{2}}{2}$ **(c)** 1 **(d)** $-\sqrt{2}$ **(e)** $-\sqrt{2}$ **(f)** 1 **45. (a)** $\frac{\sqrt{3}}{2}$ **(b)** $\frac{1}{2}$ **(c)** $\sqrt{3}$ **(d)** $\frac{2\sqrt{3}}{3}$ **(e)** 2 **(f)** $\frac{\sqrt{3}}{3}$ **47. (a)** $\frac{\sqrt{3}}{2}$ **(b)** $-\frac{1}{2}$ **(c)** $-\sqrt{3}$ **(d)** $\frac{2\sqrt{3}}{2}$ **(e)** −2 **(f)** $-\frac{\sqrt{3}}{3}$ **49. (a)** $\frac{1}{2}$ **(b)** $\frac{\sqrt{3}}{2}$ **(c)** $\frac{\sqrt{3}}{3}$ **(d)** 2 **(e)** $\frac{2\sqrt{3}}{3}$ **(f)** $\sqrt{3}$ **51. (a)** $-\frac{1}{2}$ **(b)** $\frac{\sqrt{3}}{2}$ **(c)** $-\frac{\sqrt{3}}{3}$ **(d)** −2 **(e)** $\frac{2\sqrt{3}}{3}$ **(f)** $-\sqrt{3}$ **53.** All trigonometric functions are positive in quadrant I. **55.** tangent, cotangent **57. (a)** 0.6570 **(b)** 0.7539 **(c)** 0.8714 **(d)** 1.5221 **(e)** 1.3264 **(f)** 1.1475 **59. (a)** 0.0663 **(b)** 0.9978 **(c)** 0.0665 **(d)** 15.0780 **(e)** 1.0022 **(f)** 15.0448 **61. (a)** 0 **(b)** −1 **(c)** 0 **(d)** undefined **(e)** −1 **(f)** undefined **63. (a)** −0.7071 **(b)** 0.7071 **(c)** −1 **(d)** −1.4142 **(e)** 1.4142 **(f)** −1 **65.** $A = 35°33'$; $b = 13.3$; $c = 16.3$ **66.** approximately 530 mph **67.** 47.8 mi **68.** 17.47 mph **69.** decreasing: $x \leq 1$; increasing: $x \geq 1$ **70.** decreasing: $x \geq -2$; increasing: $x \leq -2$

**[6.6]**

**1.**

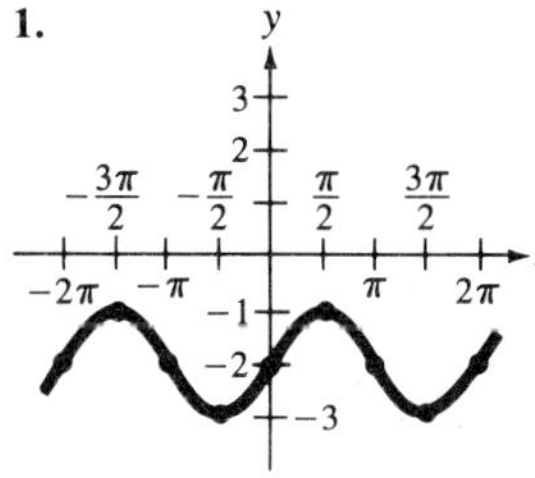

**3.**

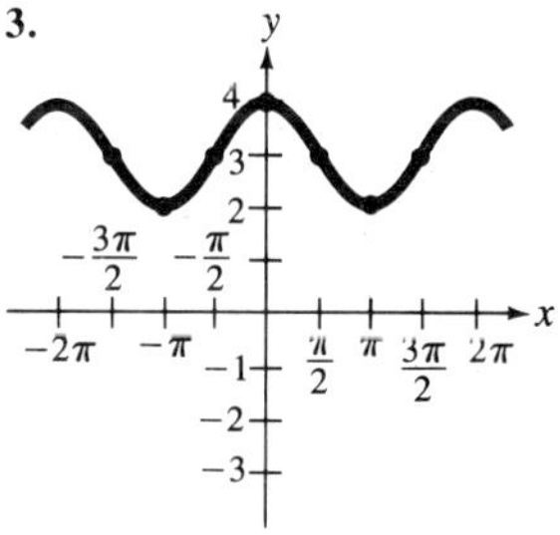

**5.**

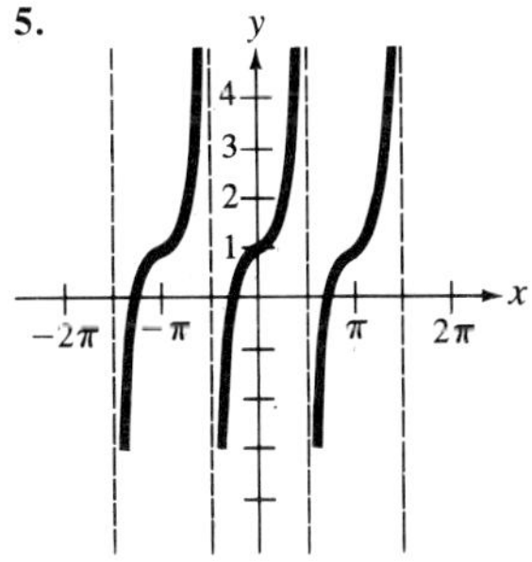

**7. (a)** 3 **(b)** $2\pi$

**(c)**

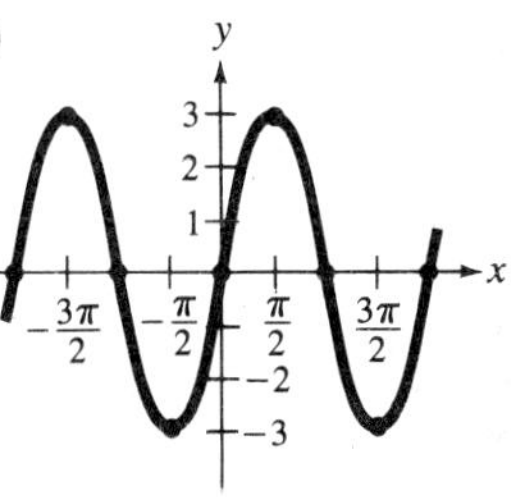

**9. (a)** 2 **(b)** $2\pi$

**(c)**

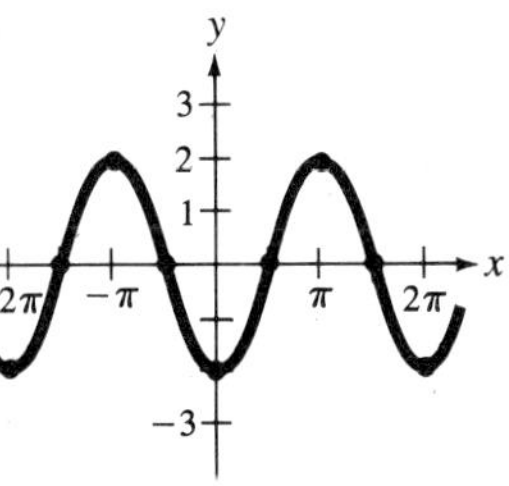

**11. (a)** $\frac{1}{2}$ **(b)** $2\pi$

**(c)**

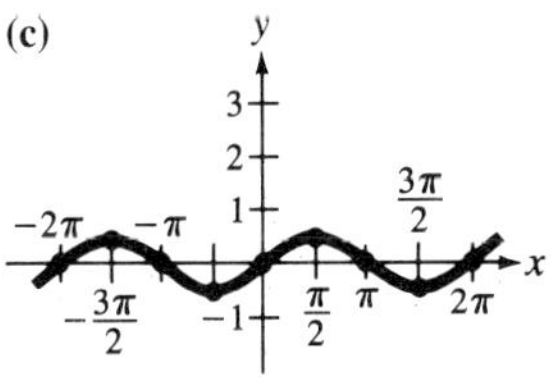

**13. (a)** $\frac{1}{2}$ **(b)** $2\pi$

**(c)**

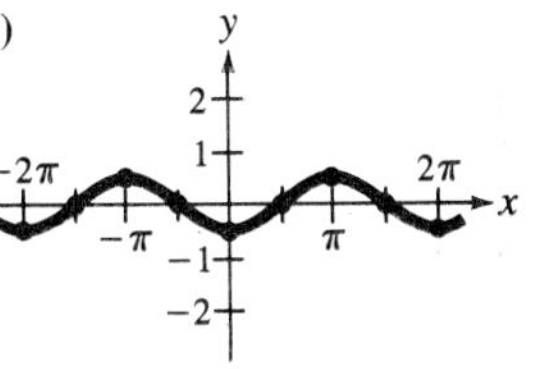

**15. (a)** 1 **(b)** $\frac{\pi}{2}$

**(c)**

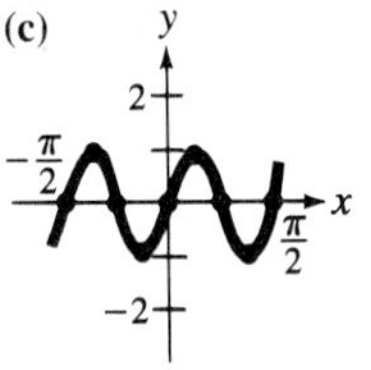

**17. (a)** 1 **(b)** $6\pi$

**(c)**

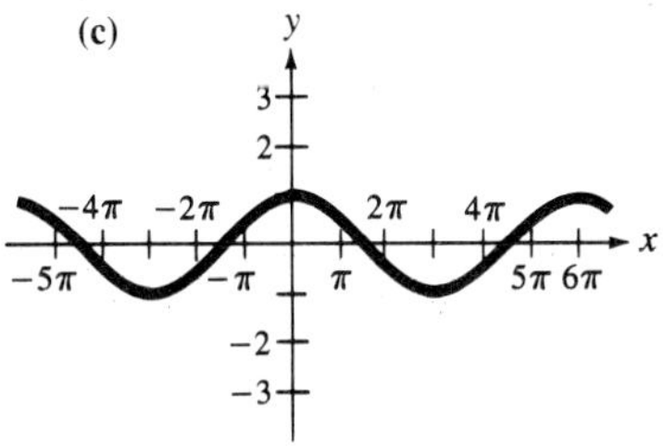

**19.** **(a)** no amplitude **(b)** $\pi$
**(c)**

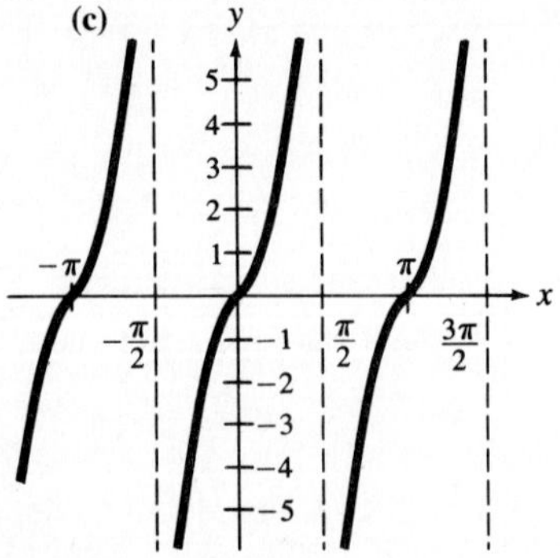


**21.** **(a)** no amplitude **(b)** $\pi$
**(c)**

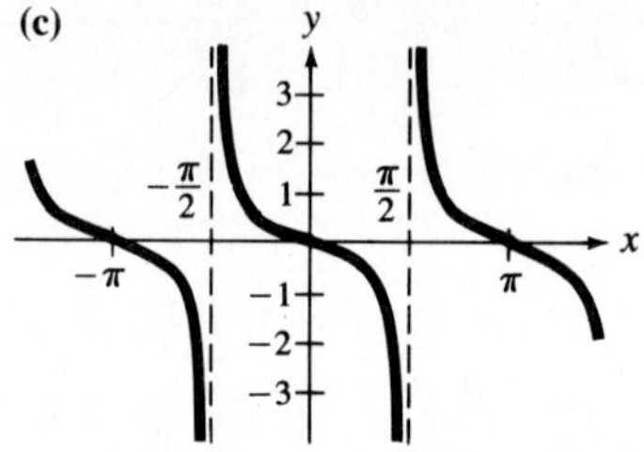


**23.** **(a)** no amplitude **(b)** $2\pi$
**(c)**

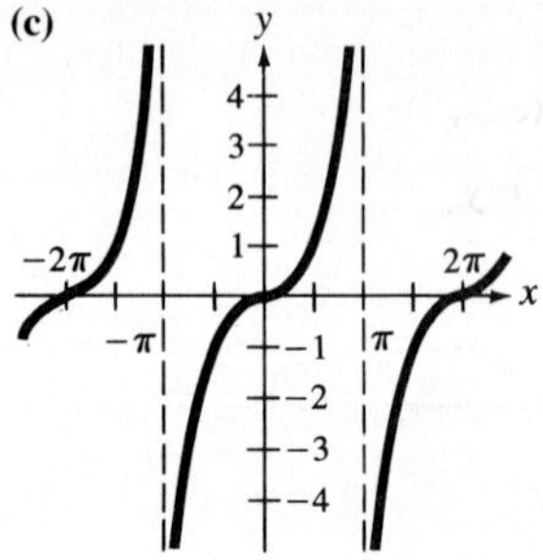


**25.** **(a)** 3 **(b)** $4\pi$
**(c)**

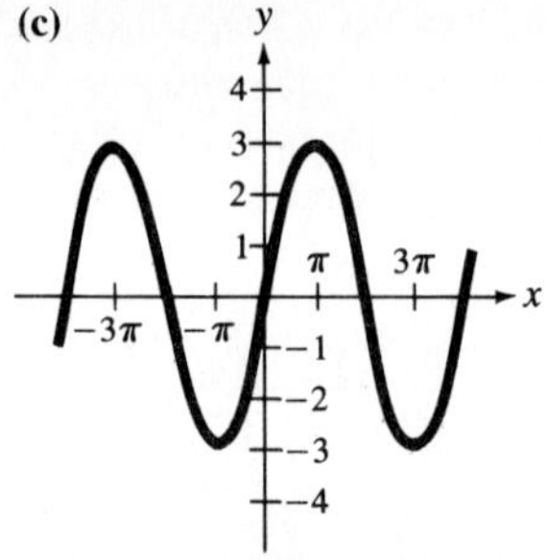


**27.** **(a)** 4 **(b)** $\frac{2\pi}{3}$
**(c)**

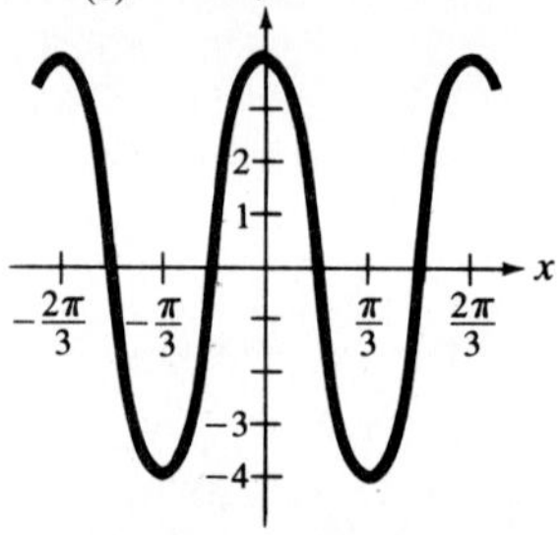


**29.** **(a)** 2 **(b)** $6\pi$
**(c)**

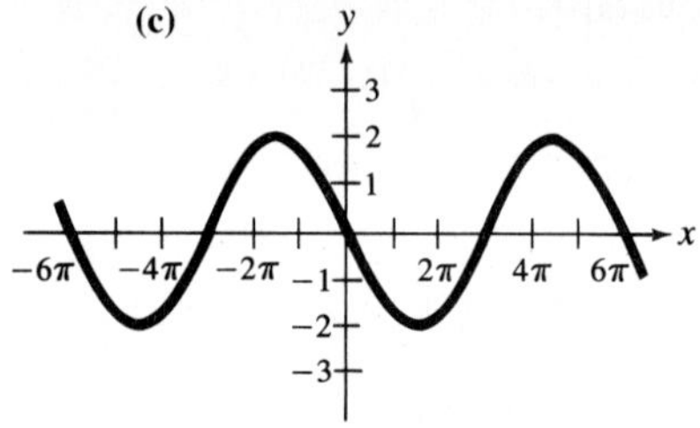


**31.** **(a)** all real numbers $x$ **(b)** $-1 \le y \le 1$ **33.** **(a)** all real numbers $x$ except $\pm\frac{\pi}{2}, \pm\frac{3\pi}{2}, \pm\frac{5\pi}{2}, \pm\frac{7\pi}{2}, \ldots$ **(b)** all real numbers $y$ **35.** **(a)** all real numbers $x$ except $\pm\frac{\pi}{2}, \pm\frac{3\pi}{2}, \pm\frac{5\pi}{2}, \pm\frac{7\pi}{2}, \ldots$ **(b)** $y \le -1$ or $y \ge 1$ **37.** Show that $\cos(-x) = \cos x$, for every real number $x$. Suppose $T(a, b)$ is the point on the unit circle determined by $x$. Then $S(a, -b)$ is the point determined by $-x$. Thus, $\cos(-x) = a$ and $\cos x = a$ so that $\cos(-x) = \cos x$. **39.** Show that $\sec(-x) = \sec x$, for every real number $x$ for which $\sec x$ is defined. Using the reciprocal relationship between the cosine and secant functions and the result of Exercise 37, we have $\sec(-x) = \frac{1}{\cos(-x)} = \frac{1}{\cos x} = \sec x$. **41.** Show that $\tan(-x) = -\tan x$, for all real numbers $x$ for which $\tan x$ is defined. Suppose $T(a, b)$ is the point on the unit circle determined by $x$. Then $S(a, -b)$ is the point determined by $-x$. Thus, $\tan(-x) = \frac{-b}{a} = -\frac{b}{a} = -\tan x$. **43.** Assume that the period of the sine function is not $2\pi$, but some number $a > 0$. Since $\sin(x + 2\pi) = \sin x$, for every real number $x$, if $a$ is the period of the sine function, then $0 < a < 2\pi$, and $\sin(x + a) = \sin x$ for every real number $x$. Then in particular, if $x = 0$, $\sin a = \sin(0 + a) = \sin 0 = 0$. But if $\sin a = 0$ and $0 < a < 2\pi$, then $a = \pi$, and $\sin(x + \pi) = \sin x$, for every real number $x$. But again in particular, for $x = \frac{\pi}{2}$, $\sin\left(\frac{\pi}{2} + \pi\right) = \sin\frac{3\pi}{2} = \sin\frac{\pi}{2}$, a contradiction since $\sin\frac{3\pi}{2} = -1$ and $\sin\frac{\pi}{2} = 1$. Thus, our assumption that $a$ was the period of the sine function is incorrect and the period must be $2\pi$. **45.** **(a)** $\left[0, \frac{\pi}{2}\right]$ and $\left[\frac{3\pi}{2}, 2\pi\right]$ **(b)** $\left[\frac{\pi}{2}, \frac{3\pi}{2}\right]$ **47.** **(a)** The tangent function is increasing everywhere on $[0, 2\pi]$; $x \ne \frac{\pi}{2}, \frac{3\pi}{2}$. **(b)** The tangent function is never decreasing on $[0, 2\pi]$; $x \ne \frac{\pi}{2}, \frac{3\pi}{2}$. **49.** **(a)** $\left[0, \frac{\pi}{2}\right)$ and $\left(\frac{\pi}{2}, \pi\right]$ **(b)** $\left[\pi, \frac{3\pi}{2}\right)$ and $\left(\frac{3\pi}{2}, 2\pi\right]$ **51.** **(a)** $\left(-\frac{\sqrt{2}}{2}, \frac{\sqrt{2}}{2}\right)$ **(b)** $\frac{\sqrt{2}}{2}$ **(c)** $-\frac{\sqrt{2}}{2}$ **(d)** $-1$ **(e)** $\sqrt{2}$ **(f)** $-\sqrt{2}$ **(g)** $-1$ **52.** **(a)** $\left(-\frac{1}{2}, \frac{\sqrt{3}}{2}\right)$ **(b)** $\frac{\sqrt{3}}{2}$ **(c)** $-\frac{1}{2}$ **(d)** $-\sqrt{3}$ **(e)** $\frac{2\sqrt{3}}{3}$ **(f)** $-2$ **(g)** $-\frac{\sqrt{3}}{3}$ **53.** **(a)** $\left(-\frac{\sqrt{3}}{2}, -\frac{1}{2}\right)$ **(b)** $-\frac{1}{2}$ **(c)** $-\frac{\sqrt{3}}{2}$ **(d)** $\frac{\sqrt{3}}{3}$ **(e)** $-2$ **(f)** $-\frac{2\sqrt{3}}{3}$ **(g)** $\sqrt{3}$ **54.** **(a)** 0.3739 **(b)** 0.9275 **(c)** 0.4031 **(d)** 2.6747 **(e)** 1.0782 **(f)** 2.4807 **55.** **(a)** 0.8660 **(b)** 0.5 **(c)** 1.7321 **(d)** 1.1547 **(e)** 2 **(f)** 0.5774
**56.** 57°30′

**[6.7]**

**1.** **(a)** 1 **(b)** $2\pi$ **(c)** $\frac{\pi}{2}$ (left) **(d)**

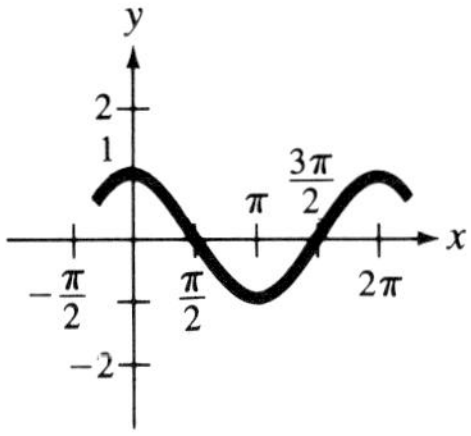

**3.** **(a)** 1 **(b)** $2\pi$ **(c)** $\frac{\pi}{4}$ (right) **(d)**

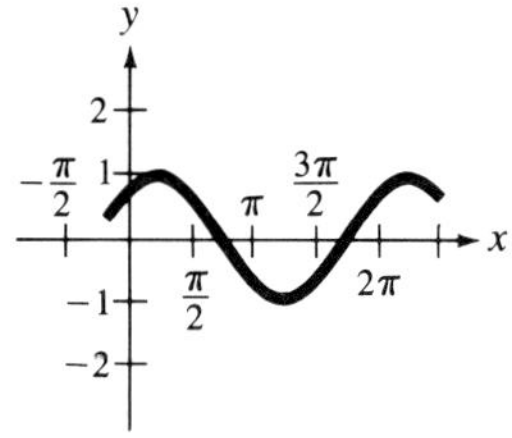

**5.** **(a)** 2 **(b)** $2\pi$ **(c)** $\frac{\pi}{2}$ (left) **(d)**

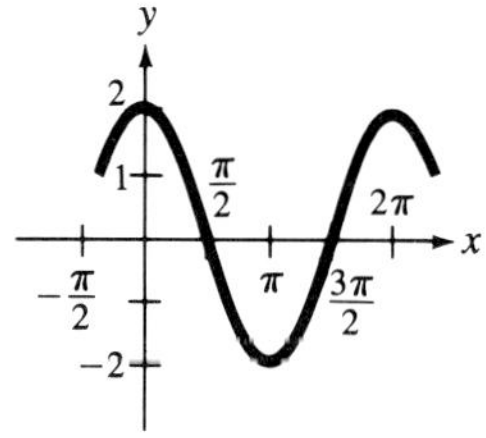

**7.** **(a)** 3 **(b)** $2\pi$ **(c)** $\frac{\pi}{2}$ (right) **(d)**

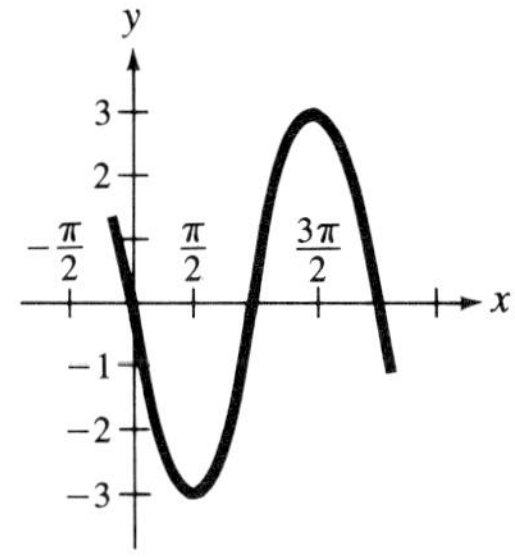

**9.** **(a)** 3 **(b)** $\pi$ **(c)** $\frac{\pi}{2}$ (right) **(d)**

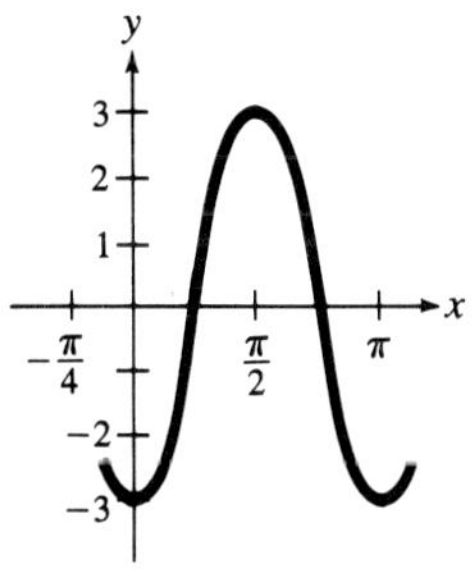

**11.** **(a)** 4 **(b)** $\pi$ **(c)** $\frac{\pi}{2}$ (left) **(d)**

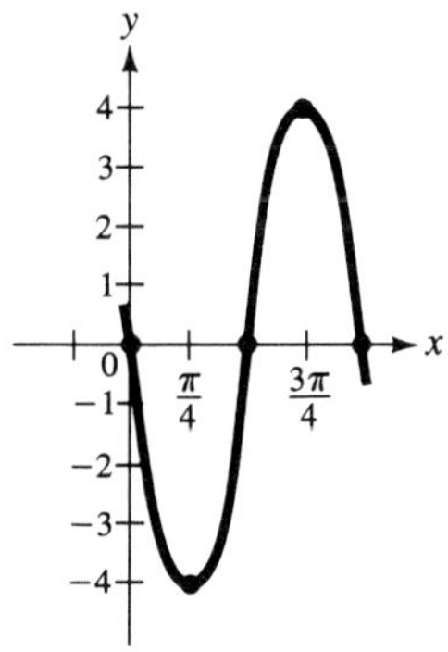

**13.** **(a)** 3 **(b)** $4\pi$ **(c)** $\pi$ (right) **(d)**

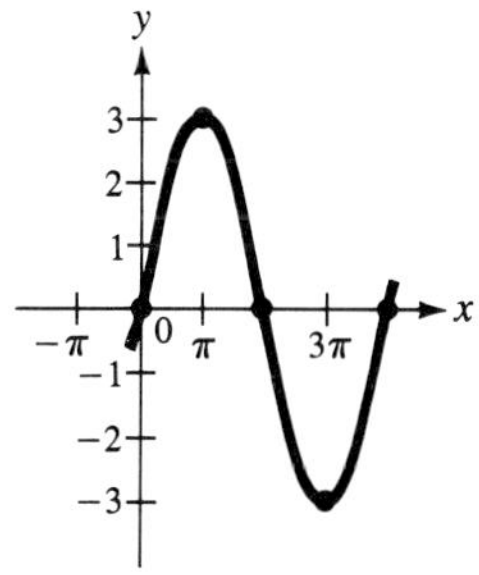

**15.**

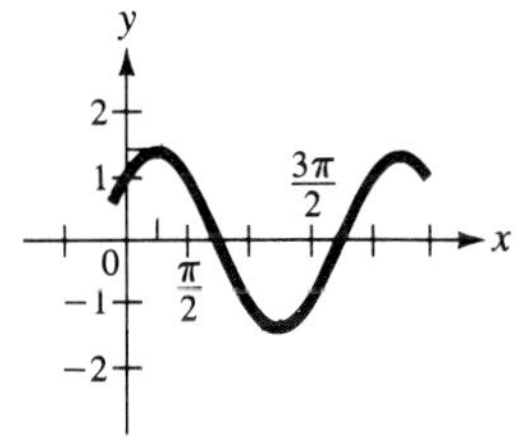

**17.**

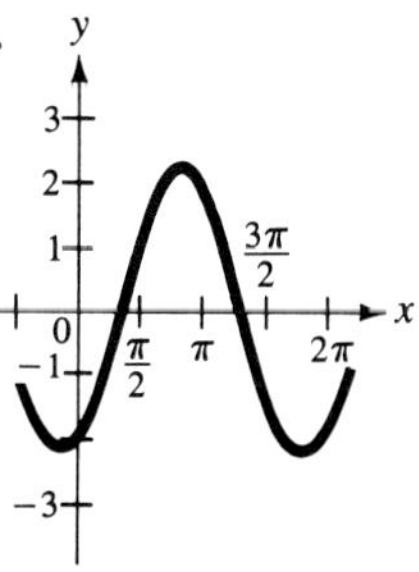

**19.**

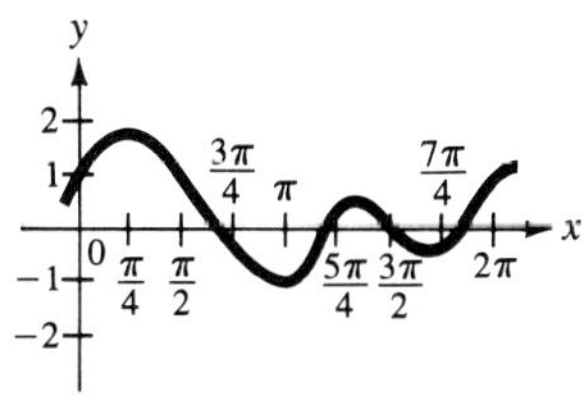

**21.**

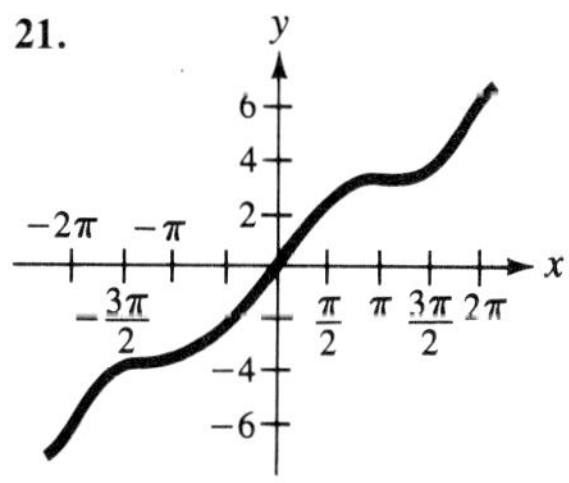

**23.**

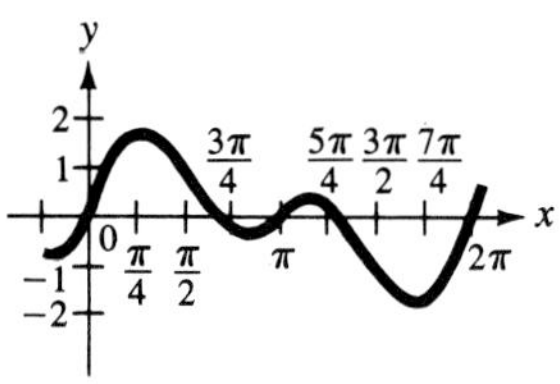

**25.**

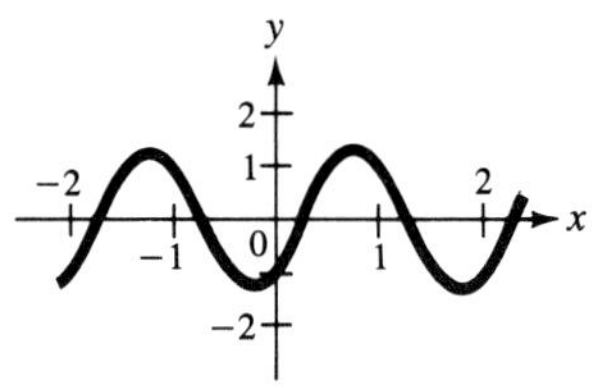

**27.**

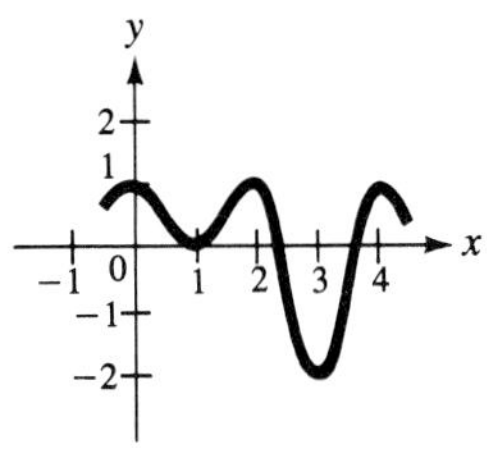

**29.**

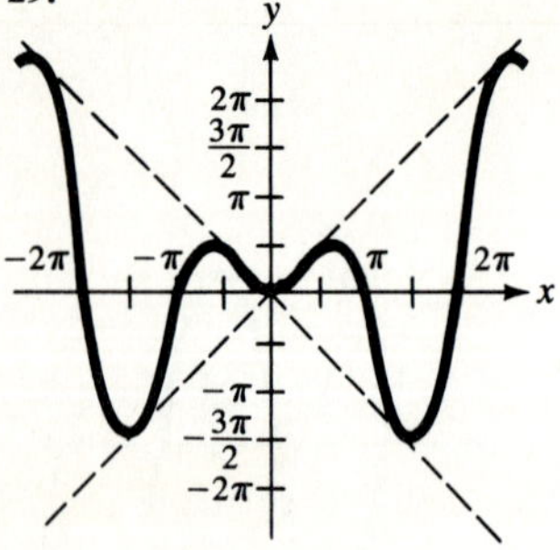


**31.**

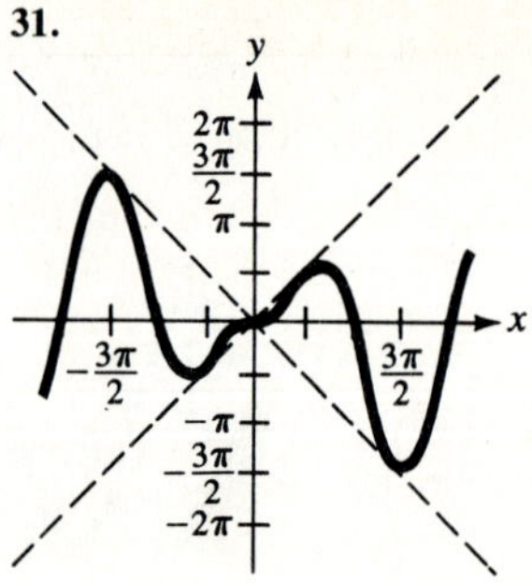


**33.**

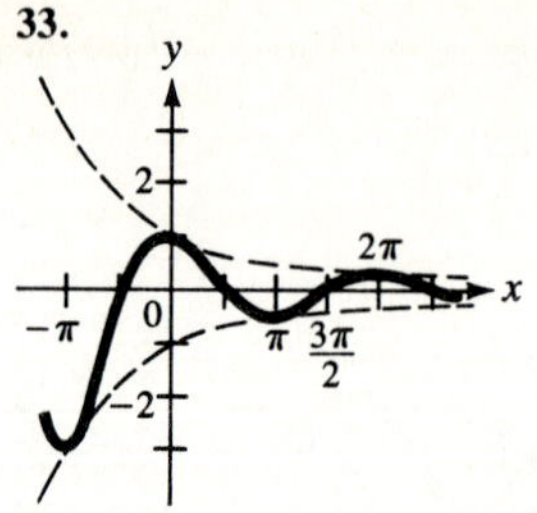


**35.**

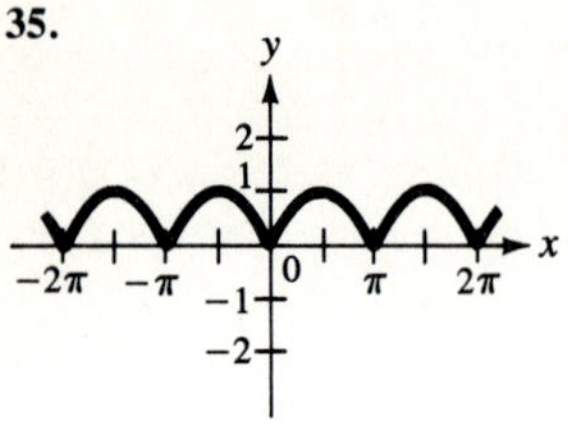


**37.**

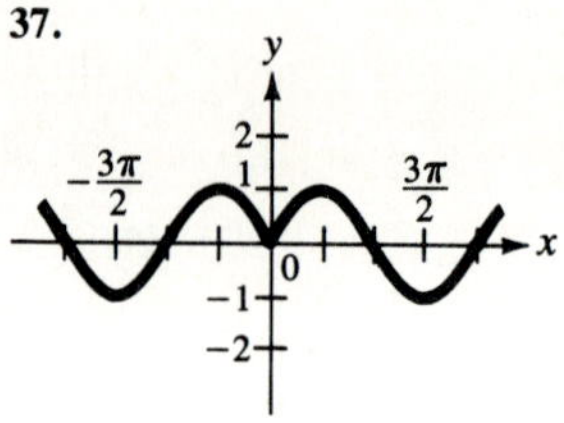


**39.** The month of least sales is September.

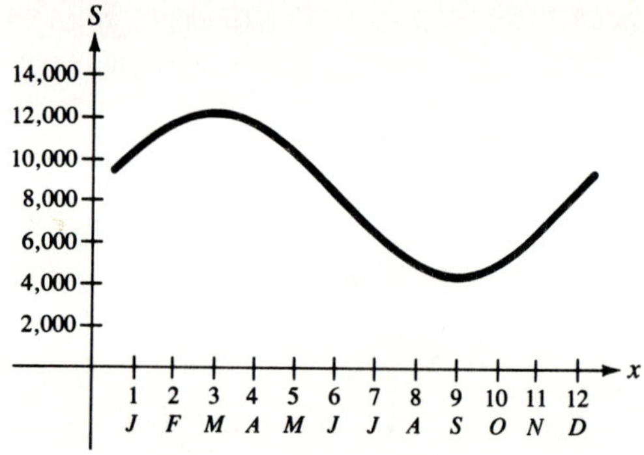


**41.** approximately 7.5 minutes **42.** 39°00′

**[6.8]**

**1.** $\sec t = \frac{1}{x} = \frac{1}{\cos t}$ **3.** $\tan t = \frac{y}{x}$, $1 + \tan^2 t = 1 + \frac{y^2}{x^2} = \frac{x^2 + y^2}{x^2} = \frac{1}{x^2} = \frac{1}{\cos^2 t} = \sec^2 t$ **5.** $\tan t = \frac{y}{x} = \frac{\sin t}{\cos t}$

**7.** Suppose $(x, y)$ is the point corresponding to $t$, then $(-y, x)$ is the point corresponding to $t + \frac{\pi}{2}$. Then $\sin\left(t + \frac{\pi}{2}\right) = x = \cos t$. Thus, $\sin\left(t + \frac{\pi}{2}\right) \equiv \cos t$. **9.** If $(x, y)$ corresponds to $t$, then $(y, -x)$ corresponds to $t - \frac{\pi}{2}$. Then $\cos\left(t - \frac{\pi}{2}\right) = y = \sin t$. Thus, $\cos\left(t - \frac{\pi}{2}\right) \equiv \sin t$. **11.** $\cot\left(t - \frac{\pi}{2}\right) \equiv \frac{\cos\left(t - \frac{\pi}{2}\right)}{\sin\left(t - \frac{\pi}{2}\right)} \equiv \frac{\sin t}{-\cos t} \equiv -\frac{\sin t}{\cos t} \equiv -\tan t$

**13.** $\sec\left(\frac{\pi}{2} - t\right) \equiv \frac{1}{\cos\left(\frac{\pi}{2} - t\right)} \equiv \frac{1}{\sin t} \equiv \csc t$ **15.** If $(x, y)$ is the point corresponding to $t$, then $(-x, -y)$ is the point corresponding to $t + \pi$. Then $\cos(t + \pi) = -x = -\cos t$. Thus, $\cos(t + \pi) \equiv -\cos t$. **17.** If $(x, y)$ corresponds to $t$, then $(-x, y)$ corresponds to $\pi - t$. Then $\cos(\pi - t) = -x = -\cos t$. Thus, $\cos(\pi - t) \equiv -\cos t$. **19.** If $(x, y)$ corresponds to $t$, then $(x, -y)$ corresponds to $-t$. Then $\sin(-t) = -y = -\sin t$. Thus, $\sin(-t) \equiv -\sin t$. **21.** $\cos^2 t$ **23.** 1 **25.** 1 **27.** 1 **29.** $\cot^2 y$ **31.** $\csc t$ **33.** $\cot t$ **35.** $\cot u$ **37.** $\sin t$ **39.** $\sin t$ **41.** $\sec^2 x + \cos^2 x + 2$ **43.** $\csc^2 t + 2 \cot t$ **45.** $\tan^2 u$ **47.** 1 **49.** $-\tan^2 y - \sec^2 y$ **51.** $(1 - \sin t)(1 + \sin t + \sin^2 t)$ **53.** 1 **55.** $\sin z\,(1 - \tan z)$ **57.** $\sec t = \frac{-1}{\sqrt{1 - \sin^2 t}}$ **59.** $\cot t = -\frac{\cos t}{\sqrt{1 - \cos^2 t}}$ **61. (a)** $-\frac{3}{5}$ **(b)** $-\frac{4}{5}$ **(c)** $-\frac{5}{4}$

**63.** Since $\tan x = \frac{\sin x}{\cos x}$, taking the common logarithm of both sides, $\log \tan x = \log \frac{\sin x}{\cos x} = \log \sin x - \log \cos x$. **65.** Since

$\sec x = \dfrac{1}{\cos x}$, taking the natural logarithm of both sides, $\ln \sec x = \ln \dfrac{1}{\cos x} = \ln 1 - \ln \cos x = 0 - \ln \cos x = -\ln \cos x$.

**66.** **(a)** 3 **(b)** $\frac{\pi}{2}$ **(c)** $3\pi$ (right) **67.** **(a)** $\frac{1}{2}$ **(b)** $4\pi$ **(c)** $6\pi$ (left)

**68.**

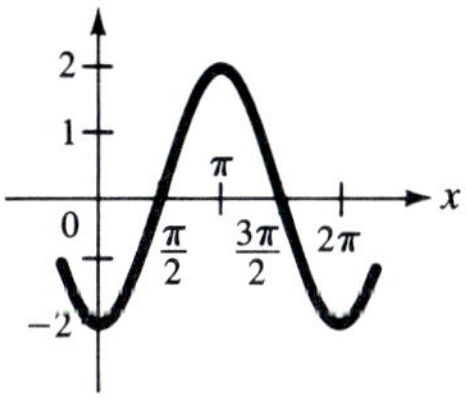


**69.**

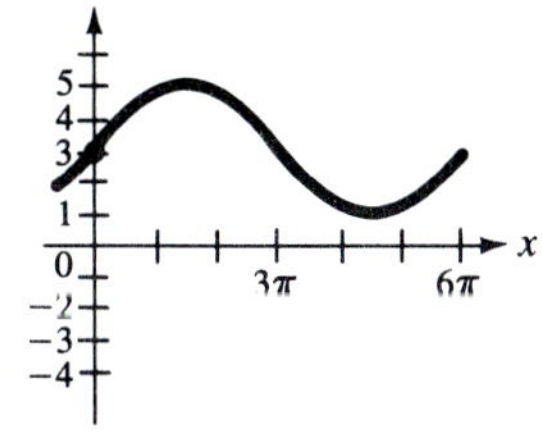


**70.**

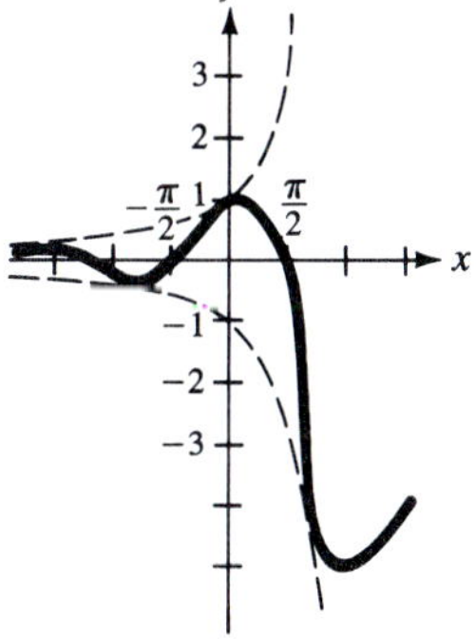


## Chapter 6 Review Exercises

**1.** no **2.** 148.66° **3.** 0.5166 **4.** 408°51′25″ **5.** 8.46 m **6.** 4933 mi **7.** 2 ft **8.** 358 rpm **9.** $\sin\theta = \frac{\sqrt{5}}{3} = \cos(90° - \theta)$; $\cos\theta = \frac{2}{3} = \sin(90° - \theta)$; $\tan\theta = \frac{\sqrt{5}}{2} = \cot(90° - \theta)$; $\csc\theta = \frac{3\sqrt{5}}{5} = \sec(90° - \theta)$; $\sec\theta = \frac{3}{2} = \csc(90° - \theta)$; $\cot\theta = \frac{2\sqrt{5}}{5} = \tan(90° - \theta)$ **10.** $\sin 30° = \frac{1}{2}$; $\cos 30° = \frac{\sqrt{3}}{2}$; $\tan 30° = \frac{\sqrt{3}}{3}$; $\csc 30° = 2$; $\sec 30° = \frac{2\sqrt{3}}{3}$; $\cot 30° = \sqrt{3}$ **11.** $\sin 45° = \frac{\sqrt{2}}{2}$; $\cos 45° = \frac{\sqrt{2}}{2}$; $\tan 45° = 1$; $\csc 45° = \sqrt{2}$; $\sec 45° = \sqrt{2}$; $\cot 45° = 1$ **12.** $\sin 60° = \frac{\sqrt{3}}{2}$; $\cos 60° = \frac{1}{2}$; $\tan 60° = \sqrt{3}$; $\csc 60° = \frac{2\sqrt{3}}{3}$; $\sec 60° = 2$; $\cot 60° = \frac{\sqrt{3}}{3}$; **13.** 0.5085 **14.** 4.0151 **15.** 3.2361 **16.** **(a)** 52.33° **(b)** 0.9133 **17.** **(a)** 17.08° **(b)** 0.2981 **18.** **(a)** 62.35° **(b)** 1.0882 **19.** $c = 15.8$; $A = 18°30'$; $B = 71°30'$ **20.** 58.3 m **21.** 8.4 mi from $B$; 6.6 mi from $A$ **22.** 17 mi west; 47 mi south **23.** 543 km/hr **24.** 31 mph **25.** 32°40′ **26.** **(a)** 0 **(b)** $\frac{\pi}{2}$ **(c)** $\frac{3\pi}{2}$ **(d)** $\frac{\pi}{4}$ **(e)** $\frac{\pi}{6}$ **(f)** $\pi$ **(g)** $\frac{7\pi}{4}$ **(h)** $\frac{5\pi}{3}$ **(i)** $\frac{3\pi}{4}$ **(j)** $\frac{7\pi}{6}$ **27.** **(a)** 0 **(b)** $-\frac{3\pi}{2}$ **(c)** $-\frac{\pi}{2}$ **(d)** $-\frac{7\pi}{4}$ **(e)** $-\frac{11\pi}{6}$ **(f)** $-\pi$ **(g)** $-\frac{\pi}{4}$ **(h)** $-\frac{\pi}{3}$ **(i)** $-\frac{5\pi}{4}$ **(j)** $-\frac{5\pi}{6}$ **28.** **(a)** $-1$ **(b)** 0 **(c)** undefined **(d)** $-1$ **(e)** undefined **(f)** 0 **29.** **(a)** $-\frac{\sqrt{2}}{2}$ **(b)** $\frac{\sqrt{2}}{2}$ **(c)** $-1$ **(d)** $-\sqrt{2}$ **(e)** $\sqrt{2}$ **(f)** $-1$ **30.** **(a)** $-\frac{1}{2}$ **(b)** $-\frac{\sqrt{3}}{2}$ **(c)** $\frac{\sqrt{3}}{3}$ **(d)** $-2$ **(e)** $-\frac{2\sqrt{3}}{3}$ **(f)** $\sqrt{3}$ **31.** III **32.** IV

**33.**

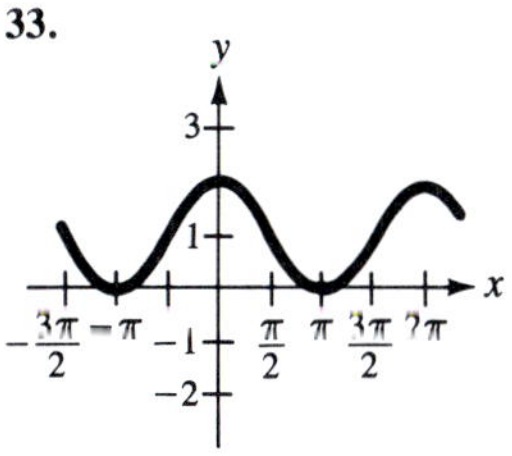


**34.**

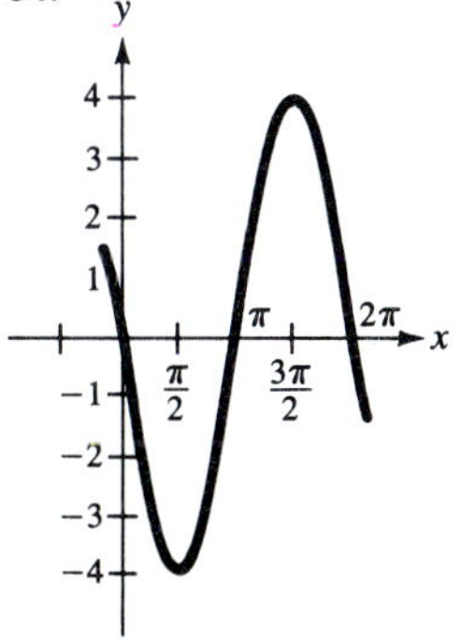


**35.**

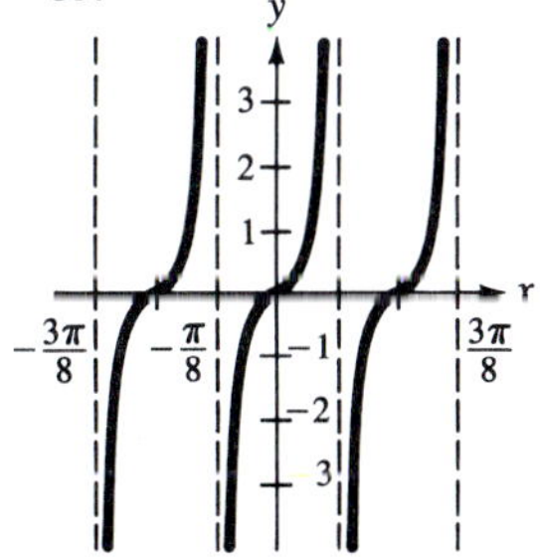


**36.**

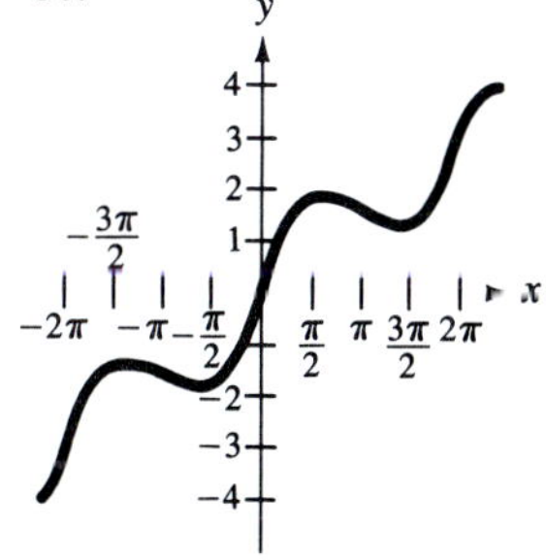

**37.**

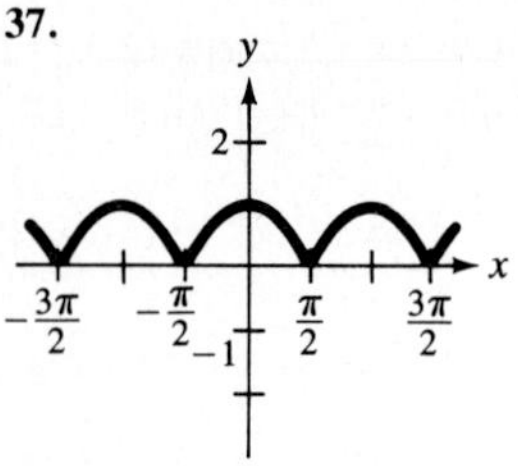

**38.**

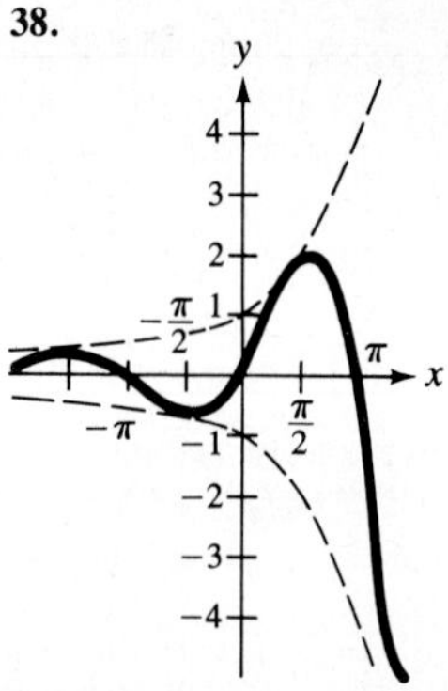

**39.** **(a)** 3 **(b)** $\pi$ **(c)** $\frac{\pi}{2}$ (right) **(d)**

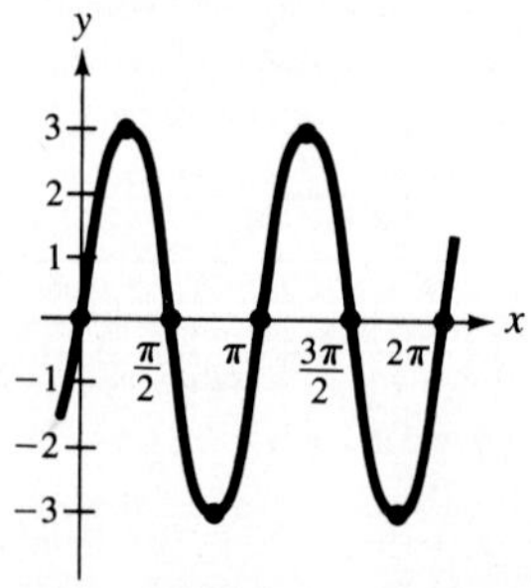

**40.** **(a)** 3 **(b)** $4\pi$ **(c)** $\pi$ (left) **(d)**

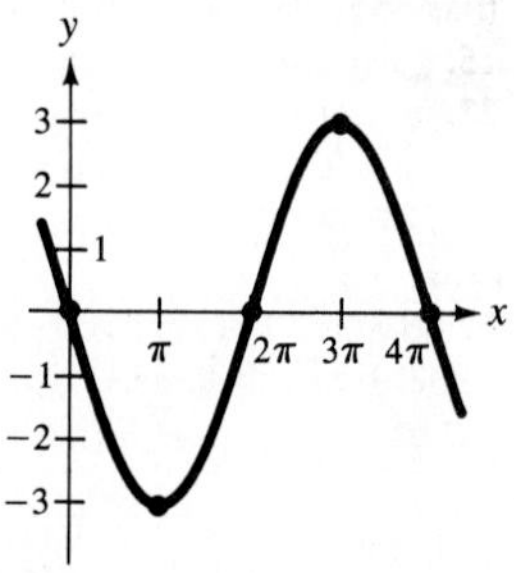

**41.** false **42.** true **43.** true **44.** true **45.** false **46.** true **47.** true **48.** true **49.** false **50.** true **51.** true **52.** false **53.** $\cos^2 x$ **54.** $\sin y - \tan y$ **55.** $\tan^3 y + \cot^3 y$ **56.** $1 + 2 \sin x \cos x$ **57.** $\sin x(1 - \tan x)$ **58.** $\sec^2 y + 2 \tan^2 y$ or $1 + 3 \tan^2 y$ **59.** $\sin y - 1$ **60.** $\sin x + 1$

## Chapter 7

### [7.1]

**1.** $\sin A = \frac{4}{5}$; $\cos A = \frac{3}{5}$; $\tan A = \frac{4}{3}$; $\csc A = \frac{5}{4}$; $\sec A = \frac{5}{3}$; $\cot A = \frac{3}{4}$ **3.** $\sin A = -\frac{4}{5}$; $\cos A = -\frac{3}{5}$; $\tan A = \frac{4}{3}$; $\csc A = -\frac{5}{4}$; $\sec A = -\frac{5}{3}$; $\cot A = \frac{3}{4}$ **5.** $\sin A = \frac{5\sqrt{26}}{26}$; $\cos A = \frac{\sqrt{26}}{26}$; $\tan A = 5$; $\csc A = \frac{\sqrt{26}}{5}$; $\sec A = \sqrt{26}$; $\cot A = \frac{1}{5}$ **7.** $\sin A = \frac{2\sqrt{5}}{5}$; $\cos A = -\frac{\sqrt{5}}{5}$; $\tan A = -2$; $\csc A = \frac{\sqrt{5}}{2}$; $\sec A = -\sqrt{5}$; $\cot A = -\frac{1}{2}$ **9.** $\sin A = \frac{\sqrt{5}}{5}$; $\cos A = \frac{2\sqrt{5}}{5}$; $\tan A = \frac{1}{2}$; $\csc A = \sqrt{5}$; $\sec A = \frac{\sqrt{5}}{2}$; $\cot A = 2$ **11.** $\sin A = \frac{\sqrt{5}}{5}$; $\cos A = -\frac{2\sqrt{5}}{5}$; $\tan A = -\frac{1}{2}$; $\csc A = \sqrt{5}$; $\sec A = -\frac{\sqrt{5}}{2}$; $\cot A = -2$ **13.** $\sin A = \frac{\sqrt{2}}{2}$; $\cos A = \frac{\sqrt{2}}{2}$; $\tan A = 1$; $\csc A = \sqrt{2}$; $\sec A = \sqrt{2}$; $\cot A = 1$
**15.** $\sin A = 0$; $\cos A = 1$; $\tan A = 0$; $\csc A$ is undefined; $\sec A = 1$; $\cot A$ is undefined **17.** $\sin A = 0$; $\cos A = -1$; $\tan A = 0$; $\csc A$ is undefined; $\sec A = -1$; $\cot A$ is undefined **19.** same answers as Exercise 15 **21.** same answers as Exercise 17
**23.** I **25.** II **27.** II **29.** 43° **31.** 40.2° **33.** 34°45′ **35.** 1°15′ **37.** $\frac{\pi}{4}$ **39.** $\frac{\pi}{6}$ **41.** $\frac{\pi}{3}$ **43.** $\frac{\pi}{3}$
**45.** $\sin(-50°) = -\sin 50° = -0.7660$; $\cos(-50°) = \cos 50° = 0.6428$; $\tan(-50°) = -\tan 50° = -1.1918$; $\csc(-50°) = -\csc 50° = -1.3054$; $\sec(-50°) = \sec 50° = 1.5557$; $\cot(-50°) = -\cot 50° = -0.8391$
**47.** $\sin 920° = -\sin 20° = -0.3420$; $\cos 920° = -\cos 20° = -0.9397$; $\tan 920° = \tan 20° = 0.3640$; $\csc 920° = -\csc 20° = -2.9238$; $\sec 920° = -\sec 20° = -1.0642$; $\cot 920° = \cot 20° = 2.7475$ **49.** $A = 204°56'$ and $A = 335°04'$ **51.** $A = 65°44'$ and $A = 245°44'$ **53.** $A = 193°22'$; and $A = 346°38'$ **55.** $\sin A = -\frac{3}{5}$; $\tan A = -\frac{3}{4}$; $\csc A = -\frac{5}{3}$; $\sec A = \frac{5}{4}$; $\cot A = -\frac{4}{3}$ **57.** $\cos A = -\frac{\sqrt{5}}{3}$; $\tan A = \frac{2\sqrt{5}}{5}$; $\csc A = -\frac{3}{2}$; $\sec A = -\frac{3\sqrt{5}}{5}$; $\cot A = \frac{\sqrt{5}}{2}$
**59.** impossible since $\sin A$ cannot be greater than 1 **61.** 1495 ft

### [7.2]

**1.** $C = 87°10'$; $b = 12.0$; $c = 15.5$ **3.** $A = 66.00°$; $b = 10.85$; $c = 11.95$ **5.** $B = 90.0°$; $C = 60.0°$; $c = 10.4$
**7.** no solution **9.** first solution: $a = 16.5$; $C = 54°30'$; $A = 96°40'$; second solution: $a = 7.2$; $C = 125°30'$; $A = 25°40'$
**11.** $B = 18°30'$; $A = 110°20'$; $a = 36.4$ **13.** first solution: $a = 16.43$, $B = 34°38'$, $A = 124°40'$; second solution: $a = 4.81$, $B = 145°22'$, $A = 13°56'$ **15.** 14.3 km **17.** 39.4 mi from lighthouse, 35.9 mi from island **19.** 0.75 mi **21.** 3106 ft
**23.** The Wilson is 91.2 nautical miles from the pilot, and the Lincoln is 104.4 nautical miles away. The Wilson should arrive first in approximately 2.4 hours. **25.** approximately 593 ft **27.** 71°40′ **28.** $\sin A = \frac{4\sqrt{17}}{17}$; $\cos A = -\frac{\sqrt{17}}{17}$; $\tan A = -4$; $\csc A = \frac{\sqrt{17}}{4}$; $\sec A = -\sqrt{17}$; $\cot A = -\frac{1}{4}$ **29.** $A = 103°35'$ and $A = 283°35'$ **30.** $\sin A = -\frac{5\sqrt{41}}{41}$; $\cos A = \frac{4\sqrt{41}}{41}$; $\tan A = -\frac{5}{4}$; $\csc A = -\frac{\sqrt{41}}{5}$; $\sec A = \frac{\sqrt{41}}{4}$

**[7.3]**

**1.** $a = 6.4$; $B = 56.0°$; $C = 81.5°$ **3.** $A = 15°20'$; $B = 20°40'$; $C = 144°00'$ **5.** $A = 64°20'$; $B = 64°20'$; $C = 51°20'$ [Did you make use of the fact that $A = B$ since the triangle is isosceles?] **7.** no solution [Do you see that $a + b < c$ so that no triangle is possible?] **9.** $B = 37°10'$; $a = 3.6$; $c = 12.6$ **11.** $b = 17.0$; $A = 35.0°$; $C = 24.6°$ **13.** 5.6 and 12.8 **15.** 7.07 and 12.73 **17.** 107 mi **19.** 82 mi **21.** 52 m **23.** 857 ft **25.** approximately 350 ft **27.** approximately 81.8 mi **29.** 102°20′ **31. (a)** 5.5 mi **(b)** 22°30′ **33.** $C = 82°00'$; $a = 1.6$; $c = 3.7$ **34.** $B = 25°10'$; $C = 106°00'$; $c = 18.8$ **35.** 579 m **36.** 12,700 ft

**[7.4]**

**1.** 46.3 $in^2$ **3.** 48.90 $yd^2$ **5.** 48.3 $ft^2$ **7.** 178.93 $in^2$ **9.** 3.73 $yd^2$ **11.** 7.3 $m^2$ **13.** 31.7 $ft^2$ **15.** Three angles do not determine a unique triangle so no single solution is possible. **17.** \$1715.17 **19.** 55.2 $yd^2$ **21.** 93.5 $in^2$ **23.** 42.6 $in^3$ **25.** 14.0 $m^2$ **27.** The area of triangle $ABD$ is $\frac{1}{2}bd \sin A$. Since the parallelogram is made up of two triangles with this area, Area $= 2\left(\frac{1}{2}bd \sin A\right) =$ bd $\sin A$. **29.** Find the area of triangle $ABC$ using Heron's formula and set the result equal to the sum of the areas of triangles $ABP$, $BCP$, and $ACP$. Solving for $r$ gives the desired formula. **31.** 55.1 m **32.** 119 nautical miles **33.** (a) 3 **(b)** $\pi$ **(c)** $\frac{\pi}{2}$ (right) **34. (a)** 5 **(b)** $\frac{\pi}{2}$ **(c)** $\frac{\pi}{8}$ (left)

**[7.5]**

**1. (a)** $h = -30 \cos \frac{\pi}{5}t + 35$ **(b)** 65 ft **3. (a)** 15 ft from ceiling **(b)** $\frac{1}{2}$ second **(c)** 5 ft **(d)** 15 ft **(e)** 10 ft (the equilibrium position) **(f)** 5 **(g)** 2 **(h)** $\frac{1}{2}$ **(i)** $\frac{1}{2}$ (right) **5.** When $t = 0$, then $y = -5$. If the weight is pulled down 5 ft and then released it should be 5 ft above the equilibrium position 1 second later, an impossibility since $k = 4$. Thus, released in this way the problem has no solution. **7. (a)** 12 ft from the ceiling (the weight is pushed upward to start the motion) **(b)** 2 seconds **(c)** 12 ft **(d)** 28 ft **(e)** approximately 23 ft **(f)** 8 **(g)** 8 **(h)** $\frac{1}{8}$ **(i)** 2 (left) **9. (a)** 0 **(b)** 0 **(c)** $5\pi$ **(d)** 12.7 **11.** $y = -2 \cos\left(\frac{\pi}{6}t\right)$ **13. (a)** 180 **(b)** $\frac{1}{60}$ **(c)** 60 **15.** $y = 4 + 1.5 \sin (4\pi t)$ **17.** 3 ft **19. (a)** 72 **(b)** 140 **(c)** January 1, 1977 $\left(t = \frac{3}{2}\right)$ **(d)** 60 **(e)** January 1, 1979 **(f)** 140 coyotes ($t = 9.5$) **(g)** The highs and lows in the coyote population occur 6 months after the highs and lows in the rabbit population. **20.** \$5550 **21.** 43.7 cm

**[7.6]**

**1. (a)** 5 **(b)** 53°10′ **(c)** $\langle 4, 7\rangle$ **(d)** $\langle -2, -1\rangle$ **(e)** $\langle 9, 12\rangle$ **(f)** $\langle 11, 18\rangle$ **(g)** $\langle -1, 7\rangle$ **(h)** $\langle 1, 3\rangle$ **(i)** $\langle -3, -4\rangle$ **3. (a)** 4 **(b)** 180°00′ **(c)** $\langle -4, 2\rangle$ **(d)** $\langle 4, 2\rangle$ **(e)** $\langle -12, 0\rangle$ **(f)** $\langle -12, 4\rangle$ **(g)** $\langle 8, 10\rangle$ **(h)** $\langle 0, 2\rangle$ **(i)** $\langle 4, 0\rangle$ **5. (a)** $\sqrt{26}$ **(b)** 101°20′ **(c)** $\mathbf{i} + 8\mathbf{j}$ **(d)** $3\mathbf{i} - 2\mathbf{j}$ **(e)** $-3\mathbf{i} + 15\mathbf{j}$ **(f)** $\mathbf{i} + 21\mathbf{j}$ **(g)** $12\mathbf{i} + 5\mathbf{j}$ **(h)** $2\mathbf{i} + 3\mathbf{j}$ **(i)** $\mathbf{i} - 5\mathbf{j}$ **7. (a)** 5 **(b)** 0°00′ **(c)** $5\mathbf{i} - 2\mathbf{j}$ **(d)** $-5\mathbf{i} - 2\mathbf{j}$ **(e)** $15\mathbf{i}$ **(f)** $15\mathbf{i} - 4\mathbf{j}$ **(g)** $-10\mathbf{i} - 10\mathbf{j}$ **(h)** $-2\mathbf{j}$ **(i)** $-5\mathbf{i}$

**9.** $\mathbf{u} + \mathbf{v} = \langle a, b\rangle + \langle c, d\rangle = \langle a + c, b + d\rangle = \langle c + a, d + b\rangle = \langle c, d\rangle + \langle a, b\rangle = \mathbf{v} + \mathbf{u}$

**11.** $\mathbf{u} + \mathbf{0} = \langle a, b\rangle + \langle 0, 0\rangle = \langle a + 0, b + 0\rangle = \langle a, b\rangle = \mathbf{u}$ **13.** $0\mathbf{u} = 0\langle a, b\rangle = \langle 0 \cdot a, 0 \cdot b\rangle = \langle 0, 0\rangle = \mathbf{0}$

**15.** $(kn)\mathbf{u} = (kn)\langle a, b\rangle = \langle (kn)a, (kn)b\rangle = \langle k(na), k(nb)\rangle = k\langle na, nb\rangle = k(n\langle a, b\rangle) = k(n\mathbf{u})$ **17.** $k(\mathbf{u} + \mathbf{v}) = k(\langle a, b\rangle + \langle c, d\rangle) = k\langle a + c, b + d\rangle = \langle k(a + c), k(b + d)\rangle = \langle ka + kc, kb + kd\rangle = \langle ka, kb\rangle + \langle kc, kd\rangle = k\langle a, b\rangle + k\langle c, d\rangle = k\mathbf{u} + k\mathbf{v}$

**19.** $|k\mathbf{v}| = |k\langle c, d\rangle| = |\langle kc, kd\rangle| = \sqrt{(kc)^2 + (kd)^2} = \sqrt{k^2c^2 + k^2d^2} = \sqrt{k^2(c^2 + d^2)} = \sqrt{k^2}\sqrt{c^2 + d^2} = |k||\mathbf{v}|$

**21.** Consider the sketch of $\mathbf{v}$ given in the figure with $\theta$ as indicated (we show $\mathbf{v}$ in quadrant I but the result is true for any position of $\mathbf{v}$). Then $\cos \theta = \frac{a}{|\mathbf{v}|}$ and $\sin \theta = \frac{b}{|\mathbf{v}|}$ so that $a = |\mathbf{v}| \cos \theta$ and $b = |\mathbf{v}| \sin \theta$.

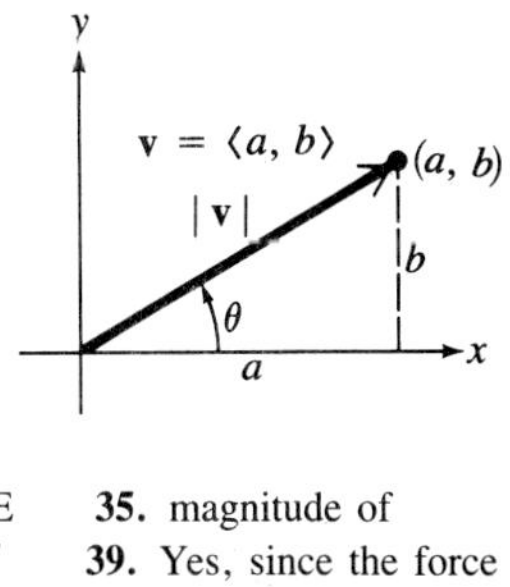


**27.** $\mathbf{v} = \langle 12.9, 27.1\rangle$ **29.** $\mathbf{v} = \langle -3.3, -11.7\rangle$ **31.** 48.9 lb; S80°50′E **33.** 18.9 kg; N26°10′E **35.** magnitude of horizontal component: 15.3 ft/sec; magnitude of vertical component: 12.9 ft/sec **37.** 74.3 lb; 19°40′ **39.** Yes, since the force of the barrel down the ramp is approximately 100 lb, the man can overcome this and roll the barrel up the ramp.

**41.** 96°10′; 276.6 mph **43.** 147°; 247 mph **45.** 8 mph; 29°40′ **47.** 386 mi **49.** 22.7 lb
**50.** **(a)** 7.5 ft from ceiling **(b)** $\frac{1}{2}$ second **(c)** 2.5 ft **(d)** 7.5 ft **(e)** 5 ft **(f)** 2.5 **(g)** 2 **(h)** $\frac{1}{2}$ **(i)** $\frac{1}{2}$ (right)

## Chapter 7 Review Exercises

**1.** $\sin A = \frac{3\sqrt{10}}{10}$; $\cos A = -\frac{\sqrt{10}}{10}$; $\tan A = -3$; $\csc A = \frac{\sqrt{10}}{3}$; $\sec A = -\sqrt{10}$; $\cot A = -\frac{1}{3}$ **2.** $\sin A = -\frac{2\sqrt{5}}{5}$; $\cos A = -\frac{\sqrt{5}}{5}$; $\tan A = 2$; $\csc A = -\frac{\sqrt{5}}{2}$; $\sec A = -\sqrt{5}$; $\cot A = \frac{1}{2}$ **3.** III **4.** 25° **5.** 55°40′ **6.** $\frac{\pi}{3}$ **7.** $\frac{\pi}{4}$
**8.** $A = 72°26'$ and $A = 252°26'$ **9.** $A = 76°29'$ and $A = 283°31'$ **10.** $\sin A = -\frac{4}{5}$; $\cos A = -\frac{3}{5}$; $\csc A = -\frac{5}{4}$; $\sec A = -\frac{5}{3}$; $\cot A = \frac{3}{4}$ **11.** $C = 80°20'$; $b = 19.0$; $c = 21.3$ **12.** no solution **13.** $a = 5.53$; $B = 46°08'$; $C = 62°07'$
**14.** $A = 40°40'$; $B = 28°50'$; $C = 110°30'$ **15.** 60 yd **16.** 18.8 km **17.** 74 mi **18.** 93 mi **19.** 5.6 yd$^2$
**20.** 49.75 in$^2$ **21.** approximately 0.9 acres **22.** approximately 1.5 bags will be needed **23.** **(a)** 120 **(b)** $\frac{1}{50}$ **(c)** 50
**24.** $y = 10 \sin\left(\frac{\pi}{2}t\right)$ **25.** **(a)** 13 **(b)** 112°40′ **(c)** $\langle -4, 10\rangle$ **(d)** $\langle -6, 14\rangle$ **(e)** $\langle 10, -24\rangle$ **(f)** $\langle -23, 54\rangle$ **(g)** $\langle 1, -2\rangle$ **(h)** $\langle 5, -12\rangle$ **26.** **(a)** 7 **(b)** 0°00′ **(c)** $6\mathbf{i} + \mathbf{j}$ **(d)** $8\mathbf{i} - \mathbf{j}$ **(e)** $-14\mathbf{i}$ **(f)** $31\mathbf{i} - 3\mathbf{j}$ **(g)** $-\mathbf{i} + \mathbf{j}$ **(h)** $-7\mathbf{i}$ **27.** magnitude of horizontal component: 50.3 ft/sec; magnitude of vertical component: 32.7 ft/sec **28.** 182°30′; 418 mph **29.** 35 ft/sec; 31°
**30.** 70 lb to hold; 187 lb against ramp

## Chapter 8

### [8.1]

**55.** $0 \le x \le \pi$ **57.** $0 < x < \pi$ **59.** $0 < x < \frac{\pi}{2}$ **61.** $a^5 \sin^5 \theta$ **63.** $\frac{1}{a} \sec\theta \tan\theta$ **65.** $\frac{1}{a}\cos\theta$
**67.** $\dfrac{1}{\cos\theta\sqrt{1-\cos^2\theta}}$ **69.** $\dfrac{1-\cos^2\theta}{\cos\theta}$ **71.** Let $A = 0$ and $B = 0$, then $\cos(A - B) = \cos 0 = 1$, but $\cos A - \cos B = 1 - 1 = 0$. **73.** Let $A = \frac{\pi}{2}$ and $B = \frac{\pi}{2}$, then $\sin(A + B) = \sin\pi = 0$, but $\sin A + \sin B = 1 + 1 = 2$.
**75.** $8\mathbf{i} - 23\mathbf{j}$ **76.** $7\mathbf{i} - 16\mathbf{j}$

### [8.2]

**1.** **(a)** $\dfrac{\sqrt{2}(\sqrt{3}+1)}{4}$ **(b)** $\dfrac{\sqrt{2}(1-\sqrt{3})}{4}$ **(c)** $-2-\sqrt{3}$ **3.** **(a)** $\dfrac{-\sqrt{2}(\sqrt{3}+1)}{4}$ **(b)** $\dfrac{\sqrt{2}(1-\sqrt{3})}{4}$ **(c)** $2+\sqrt{3}$
**5.** **(a)** $\dfrac{\sqrt{2}(\sqrt{3}-1)}{4}$ **(b)** $\dfrac{\sqrt{2}(1+\sqrt{3})}{4}$ **(c)** $2-\sqrt{3}$ **7.** $\cos(S - T)$ **9.** $\sin(T + S)$ **11.** $\tan(S - T)$ **13.** cos 170°
**15.** sin 210° **17.** $\cos 3x$ **19.** sin 4 **21.** tan 3 **23.** **(a)** $\frac{7}{25}$ **(b)** $-1$ **(c)** $-\frac{24}{25}$ **(d)** 0 **(e)** $-\frac{24}{7}$ **(f)** 0
**55.** $\dfrac{f(x+h)-f(x)}{h} = \dfrac{\sin(x+h)-\sin x}{h} = \dfrac{\sin x\cos h + \cos x \sin h - \sin x}{h} = \dfrac{\sin x(\cos h - 1) + \cos x \sin h}{h} = \sin x\left(\dfrac{\cos h - 1}{h}\right) + \cos x\left(\dfrac{\sin h}{h}\right)$ **57.** We know $l_1$ is perpendicular to $l_2$ if and only if $\theta_2 - \theta_1 = 90°$ if and only if $\tan(\theta_2 - \theta_1)$ is undefined if and only if $1 + m_1m_2 = 0$ if and only if $m_1 = -\frac{1}{m_2}$.

### [8.3]

**1.** **(a)** $\frac{24}{25}$ **(b)** $\frac{7}{25}$ **(c)** $\frac{24}{7}$ **(d)** I **3.** **(a)** $-\frac{120}{169}$ **(b)** $\frac{119}{169}$ **(c)** $-\frac{120}{119}$ **(d)** IV **5.** **(a)** $\frac{24}{25}$ **(b)** $\frac{7}{25}$ **(c)** $\frac{24}{7}$ **(d)** I
**7.** **(a)** $-\frac{120}{169}$ **(b)** $-\frac{119}{169}$ **(c)** $\frac{120}{119}$ **(d)** III **9.** **(a)** $\dfrac{\sqrt{2-\sqrt{3}}}{2}$ **(b)** $\dfrac{\sqrt{2+\sqrt{3}}}{2}$ **(c)** $2-\sqrt{3}$ **11.** **(a)** $\dfrac{\sqrt{2-\sqrt{2}}}{2}$ **(b)** $\dfrac{\sqrt{2+\sqrt{2}}}{2}$ **(c)** $\sqrt{2}-1$ **13.** **(a)** $\dfrac{\sqrt{2+\sqrt{3}}}{2}$ **(b)** $-\dfrac{\sqrt{2-\sqrt{3}}}{2}$ **(c)** $-2-\sqrt{3}$ **15.** 2 **17.** $\cos 2x$ **19.** $\tan x$
**21.** $\cos^4 A - 6\sin^2 A\cos^2 A + \sin^4 A$ **23.** $\cos^3 A - 3\sin^2 A\cos A$ **25.** $\frac{1}{8}\cos 4A - \frac{1}{2}\cos 2A + \frac{3}{8}$

**27.** $\frac{1}{16} + \frac{1}{16}\cos 2A - \frac{1}{16}\cos 4A - \frac{1}{16}\cos 2A\cos 4A$ **49.** 459 ft **51. (a)** $-\frac{33}{65}$ **(b)** $-\frac{63}{65}$ **(c)** $\frac{56}{65}$ **(d)** $\frac{16}{65}$ **(e)** $-\frac{56}{33}$ **(f)** $-\frac{16}{63}$ **54.** $y = A\cos Bx\cos BC + A\sin Bx\sin BC$ **55.** $y = A\sin Bx\cos BC - A\cos Bx\sin BC$

**[8.4]**

**1.** $\frac{1}{2}[\sin 3x + \sin x]$ **3.** $\cos 5u + \cos 3u$ **5.** $2[\sin 7x + \sin x]$ **7.** $\frac{2+\sqrt{3}}{4}$ **9.** $\frac{1}{4}$ **11.** $\frac{1}{2}$

**13.**

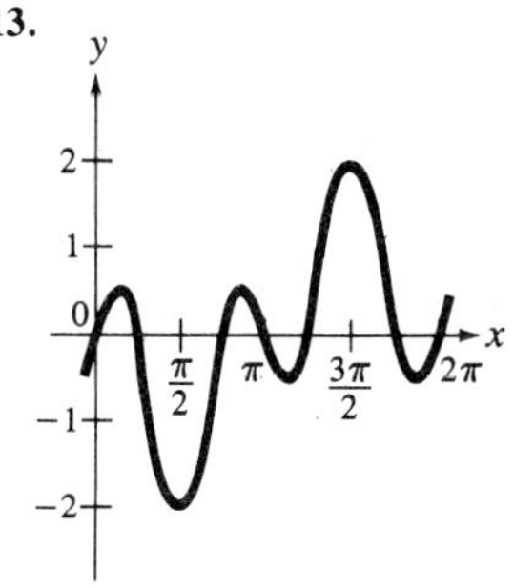


**15.**

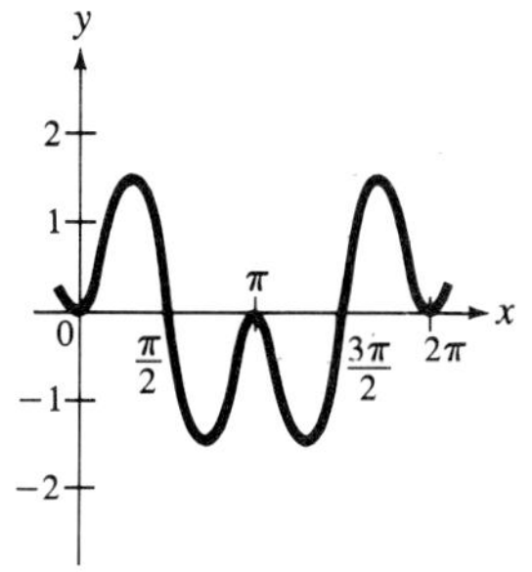


**17.** $2\cos 3x\sin 2x$ **19.** $2\cos\left(\frac{3u}{2}\right)\cos\left(\frac{u}{2}\right)$ **21.** $-2\sin\left(\frac{7x}{2}\right)\sin\left(\frac{x}{2}\right)$ **23.** $\frac{\sqrt{6}}{2}$ **25.** $\frac{\sqrt{2}}{2}$ **27.** $\frac{\sqrt{2}}{4}$

**29.** $y = \sqrt{2}\cos\left(x + \frac{\pi}{4}\right)$; amplitude $= \sqrt{2}$; period $= 2\pi$; phase shift $= \frac{\pi}{4}$ to the left **31.** $y = 2\cos 2\left(x + \frac{\pi}{6}\right)$; amplitude $= 2$; period $= \pi$; phase shift $= \frac{\pi}{6}$ to the left **33.** $y = 5\cos 2\pi(x - 0.352)$; amplitude $= 5$; period $= 1$; phase shift $= 0.352$ to the right

**35.** $y = 2\cos\left(x - \frac{\pi}{3}\right)$
amplitude $= 2$
period $= 2\pi$
phase shift $= \frac{\pi}{3}$ (right)

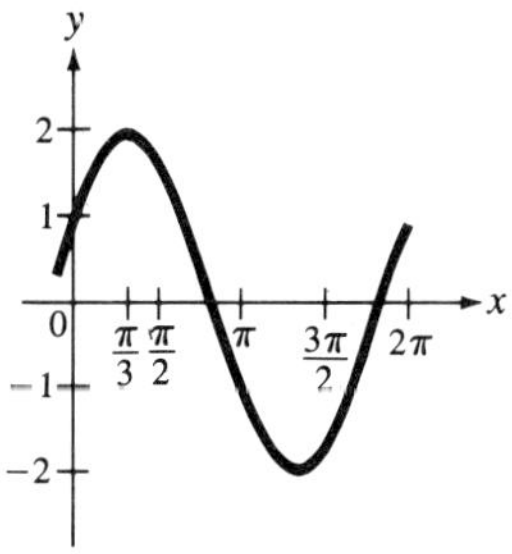


**37.** $y = \sqrt{2}\cos\left(x - \frac{\pi}{4}\right)$
amplitude $= \sqrt{2}$
period $= 2\pi$
phase shift $= \frac{\pi}{4}$ (right)

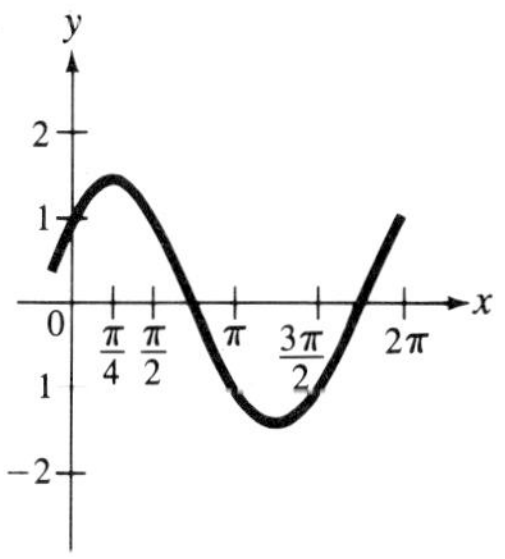


**51.** Both sides are approximately equal to $-0.4598$. **53.** Both sides are approximately equal to 1.5020. **55. (a)** $\frac{24}{25}$ **(b)** $-\frac{7}{25}$ **(c)** $-\frac{24}{7}$ **(d)** II **56. (a)** $\pm\frac{3\sqrt{13}}{13}$ **(b)** $\pm\frac{2\sqrt{13}}{13}$ **(c)** $\frac{3}{2}$ **(d)** I or III

**[8.5]**

**1.** $\frac{\pi}{6}, \frac{5\pi}{6}$ **3.** $\frac{\pi}{3}, \frac{4\pi}{3}$ **5.** $\frac{\pi}{6}, \frac{\pi}{2}, \frac{3\pi}{2}, \frac{11\pi}{6}$ **7.** $\frac{\pi}{2}, \frac{2\pi}{3}, \frac{4\pi}{3}, \frac{3\pi}{2}$ **9.** $0, \pi$ **11.** $\frac{\pi}{4}, \frac{3\pi}{4}, \frac{5\pi}{4}, \frac{7\pi}{4}$ **13.** $\frac{\pi}{6}, \frac{5\pi}{6}, \frac{3\pi}{2}$ **15.** 1.3181, $\pi$, 4.9651 **17.** $0, \pi$ **19.** $0, \pi$ **21.** 30°, 150° **23.** 0°, 90°, 180°, 270° **25.** 90° **27.** 15°, 75°, 195°, 255° **29.** 45°, 105°, 165°, 225°, 285°, 345° **31.** 0°, 120°, 240° **33.** 34°20′, 145°40′ **35.** 15°, 75°, 195°, 255° **37.** 0°, 45°, 90°, 135°, 180°, 225°, 270°, 315° **39.** 0°, 90°, 180°, 270° **41.** 0°, 60°, 90°, 120°, 180°, 240°, 270°, 300° **43.** 0°, 120°, 180°, 240° **45.** no solution **47.** 109°28′, 138°35′, 221°25′, 250°32′ **49.** 78°41′, 153°26′, 258°41′, 333°26′ **51.** 60° **53.** $\frac{1}{720}$ second **55.** $\sin 8x - \sin 4x$ **56.** $2\sin 4x\cos 3x$ **57.** $y = 2\cos 2\left(x - \frac{\pi}{3}\right)$, amplitude 2, period $\pi$, phase shift $\frac{\pi}{3}$ to right **59.** $f^{-1}(x) = \frac{x}{5} + \frac{3}{5}$ **60.** $y = x^2 - 3$, $x \geq 0$

## [8.6]

**1.** $\frac{\pi}{6}$ **3.** $\frac{3\pi}{4}$ **5.** $-\frac{\pi}{3}$ **7.** undefined **9.** $\frac{\pi}{3}$ **11.** 0.5511 **13.** 0.7592 **15.** $\frac{\sqrt{3}}{2}$ **17.** $\sqrt{3}$ **19.** $\frac{5\pi}{6}$ **21.** 0.9051 **23.** 0.7137 **25.** no value exists **27.** 1.9539 **29.** 0.1140 **31.** $\frac{3}{5}$ **33.** $\frac{3}{4}$ **35.** $\frac{4}{5}$ **37.** $\frac{2\sqrt{2}}{3}$ **39.** $\frac{\sqrt{3}}{3}$ **41.** $\frac{7}{25}$ **43.** $\frac{\sqrt{2+\sqrt{3}}}{2}$ **45.** $\frac{\sqrt{3}}{3}$ **47.** 0 **49.** $\frac{24}{25}$ **51.** $\frac{x}{\sqrt{1+x^2}}$ **53.** $\sqrt{1-x^2}$ **55.** $\frac{\pm\sqrt{1-x^2}}{x}$ **61.** Let $x = \frac{1}{2}$, for example, then the left side is $\frac{\pi}{6}$, but the right side is approximately 2.0858. **63.** Let $x = \frac{1}{2}$, for example, then the left side is 0.5 but the right side is approximately 0.4636. **65.** If $y = \text{arcsec } x$ for $x$ in $(-\infty, -1]$ or $[1, \infty)$, then $x = \sec y$ and $y$ is in one of the intervals $\left[-\pi, -\frac{\pi}{2}\right)$ or $\left[0, \frac{\pi}{2}\right)$. **67.** 1.1416; since 2 is not in the range of the inverse sine function, arcsin (sin 2) will not equal 2. **69.** The value must be $-1$ since $-1$ is in the range of the inverse sine function. **71.** 30°, 150°, 210°, 330° **72.** 0°, 180° **73.** 68°30′, 248°30′ **74.** 14°29′, 165°31′, 191°32′, 348°28′ **75.** 2°9′

## Chapter 8 Review Exercises

**15.** $\frac{\sqrt{1-\sin^2\theta}}{\sin\theta(1-\sin^2\theta)}$ **16.** $\frac{63}{65}$ **17.** $\frac{16}{65}$ **18.** $\frac{63}{16}$ **19.** $\frac{33}{65}$ **20.** $-\frac{56}{65}$ **21.** $-\frac{33}{56}$ **22.** $-\frac{120}{169}$ **23.** $-\frac{119}{169}$ **24.** $\frac{120}{119}$ **25.** $\pm\frac{3\sqrt{13}}{13}$ **26.** $\pm\frac{2\sqrt{13}}{13}$ **27.** $\frac{3}{2}$ **28.** $2[\sin 8x + \sin 2x]$ **29.** $\frac{1+\sqrt{3}}{4}$ **30.** $-2\sin 4x \sin x$ **31.** 1 **32.** $y = 2\cos\left(t + \frac{\pi}{6}\right)$; amplitude = 2; period = $2\pi$; phase shift = $\frac{\pi}{6}$ to the left **33.** $\frac{\pi}{6}, \frac{7\pi}{6}$, **34.** $\frac{\pi}{4}, \frac{\pi}{2}, \frac{3\pi}{4}, \frac{3\pi}{2}$ **35.** $\frac{\pi}{6}, \frac{5\pi}{6}, \frac{3\pi}{2}$ **36.** 15°, 75°, 105°, 165°, 195°, 255°, 285°, 345° **37.** 65°30′, 245°30′ **38.** 60°, 180°, 300° **39.** 0°, 45°, 90°, 135°, 180°, 225°, 270°, 315° **40.** 90° **41.** $\frac{\pi}{4}$ **42.** $-\frac{\pi}{4}$ **43.** $\frac{2\pi}{3}$ **44.** $\frac{\pi}{3}$ **45.** $-\frac{\pi}{3}$ **46.** $\frac{3\pi}{4}$ **47.** $\frac{\sqrt{3}}{2}$ **48.** 0 **49.** $\frac{\sqrt{10}}{10}$ **50.** $\frac{4\sqrt{2}}{9}$ **51.** $\frac{\sqrt{3}}{2}$ **52.** $-\frac{7}{25}$ **53.** $\frac{2x}{\sqrt{1+4x^2}}$ **54.** $\sqrt{1-4x^2}$ **55.** 29°40′

## Chapter 9

### [9.1]

**1.** $(x+3)^2 + (y-2)^2 = 1^2$; $x^2 + y^2 + 6x - 4y + 12 = 0$ **3.** $\left(x - \frac{1}{4}\right)^2 + (y+1)^2 = 3^2$; $16x^2 + 16y^2 - 8x + 32y - 127 = 0$ **5.** $(x-0)^2 + (y-0)^2 = 1^2$ or $x^2 + y^2 = 1$; $x^2 + y^2 - 1 = 0$ **7.** $x^2 + y^2 = 25$ **9.** $(x-1)^2 + (y+5)^2 = 100$ **11.** $(x-4)^2 + (y+1)^2 = 3$ **13.** $(x-6)^2 + (y-8)^2 = 64$ **15.** $(x-3)^2 + (y+1)^2 = 13$ **17.** $\left(x + \frac{3}{2}\right)^2 + \left(y - \frac{15}{2}\right)^2 = \frac{61}{2}$

**19.**
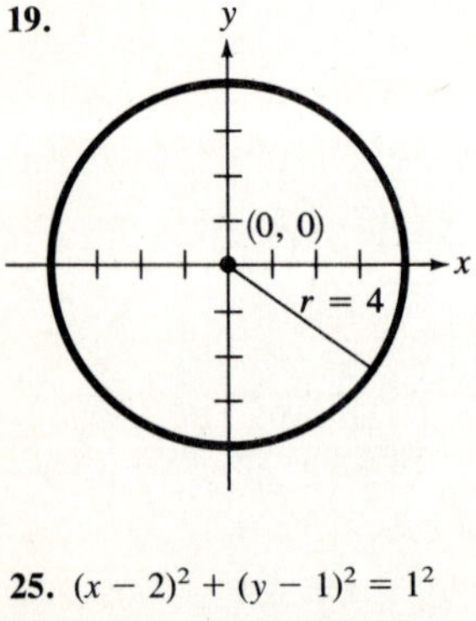


**21.**
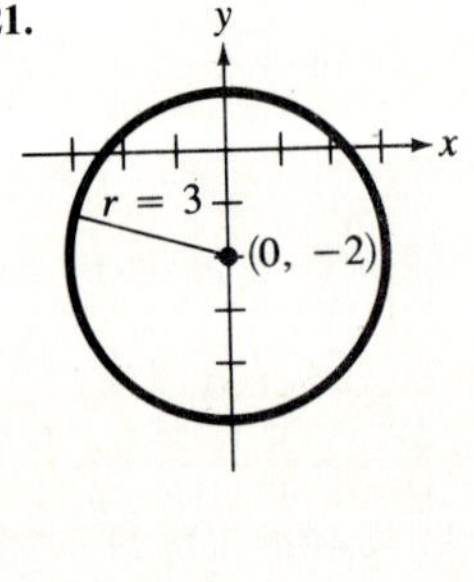


**23.**
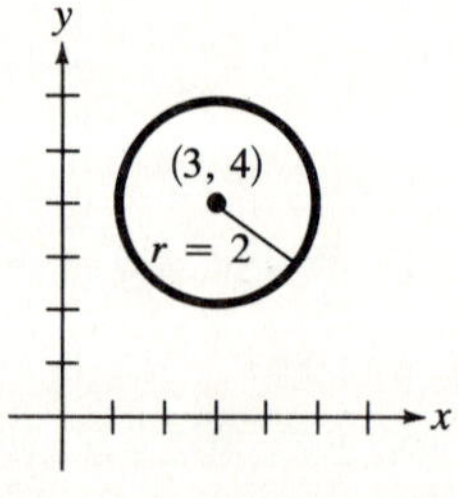


**25.** $(x-2)^2 + (y-1)^2 = 1^2$ **27.** $\left(x - \frac{1}{2}\right)^2 + (y+3)^2 = 5^2$ **29.** $\left(x + \frac{1}{3}\right)^2 + \left(y - \frac{1}{3}\right)^2 = 4^2$ **31.** $y = -\sqrt{100 - x^2}$, 8 ft

**[9.2]**

**1.** $\frac{x^2}{49} + \frac{y^2}{36} = 1$ **3.** $\frac{x^2}{25} + \frac{y^2}{16} = 1$ **5.** $\frac{x^2}{16} + \frac{y^2}{25} = 1$ **7.** $\frac{x^2}{36} + \frac{y^2}{25} = 1$ **9.** $\frac{x^2}{100} + \frac{y^2}{4} = 1$

**11.** $\frac{(x-5)^2}{16} + \frac{(y+2)^2}{12} = 1$ **13.** $\frac{x^2}{25} + \frac{(y-3)^2}{9} = 1$ **15.** $\frac{(x-7)^2}{55} + \frac{(y+4)^2}{64} = 1$

**17.**

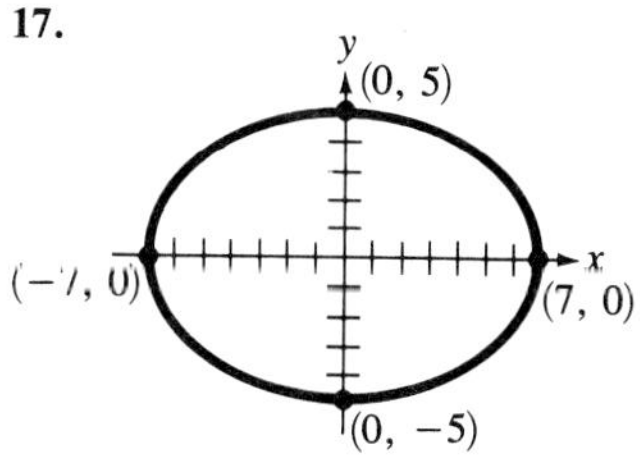


**19.**

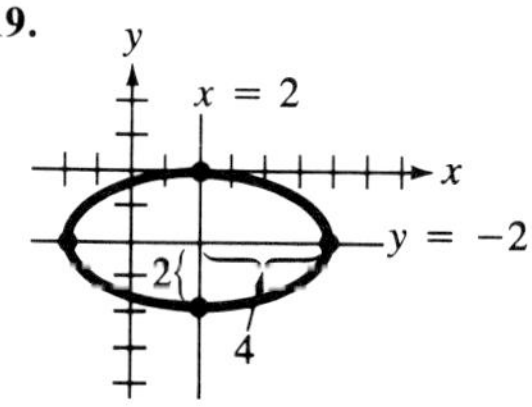


**21.**

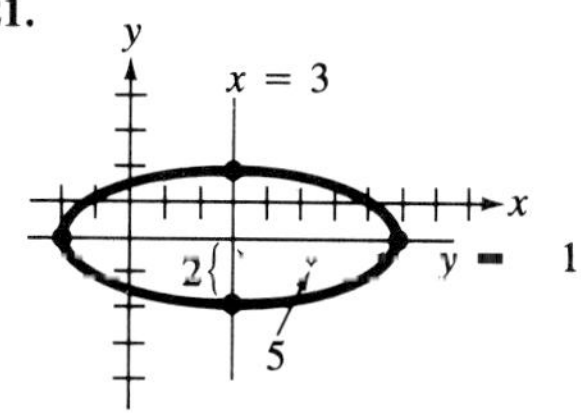


**23.**

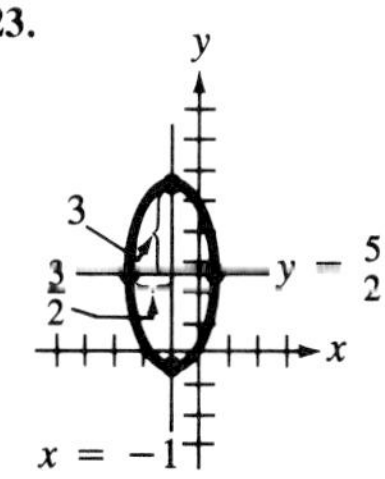


**25.** $\frac{x^2}{7} + \frac{y^2}{8} = 1$ **27.** $\frac{(x+3)^2}{25} + \frac{(y-3)^2}{\frac{9}{2}} = 1$ **29.** not possible, there is no graph **31.** $\frac{x^2}{9} + \frac{4y^2}{9} = 1$

**33.** $\frac{x^2}{1309} + \frac{y^2}{900} = 1$ **35.** $\frac{1}{2}(m - \sqrt{m^2 - n^2})$ **37.** $(x-2)^2 + (y+5)^2 = 20$ **38.** $\left(x - \frac{3}{2}\right)^2 + \left(y - \frac{3}{2}\right)^2 = \frac{41}{2}$

**39.** $(x+2)^2 + (y-8)^2 = 3^2$ **40.** $\left(x - \frac{1}{2}\right)^2 + \left(y + \frac{1}{3}\right)^2 = 2^2$

**[9.3]**

**1.** $\frac{x^2}{9} - \frac{y^2}{16} = 1$ **3.** $\frac{y^2}{9} - \frac{x^2}{9} = 1$ **5.** $\frac{x^2}{25} - \frac{y^2}{24} = 1$ **7.** $\frac{y^2}{36} - \frac{x^2}{64} = 1$ **9.** $\frac{(y-2)^2}{16} - \frac{(x+3)^2}{20} = 1$

**11.** $\frac{x^2}{16} - \frac{y^2}{12} = 1$

**13.**

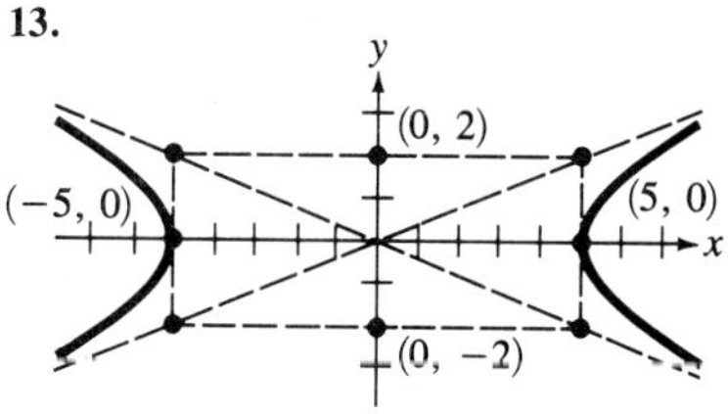


**15.**

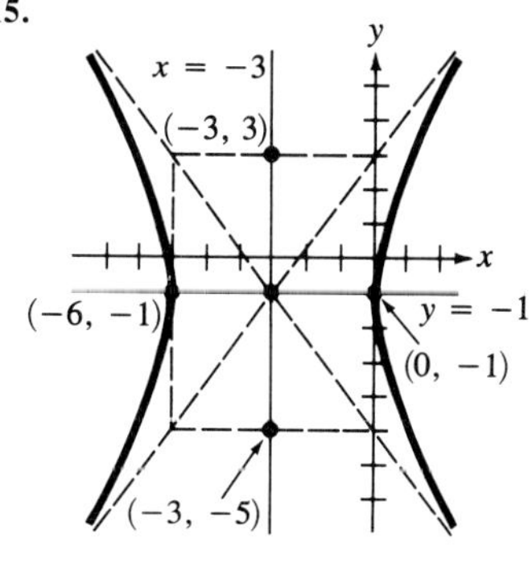


**17.**

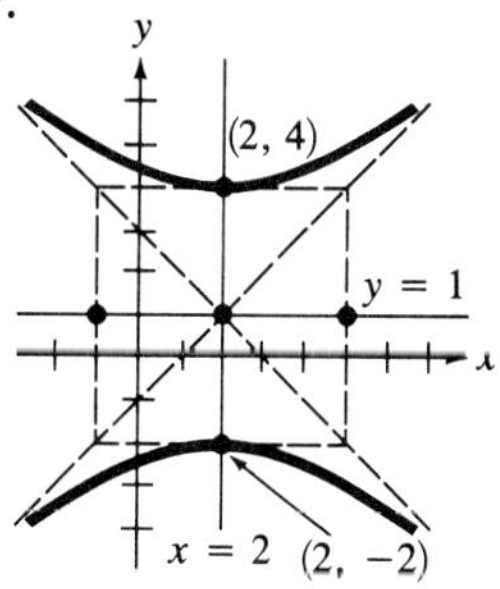


**19.**

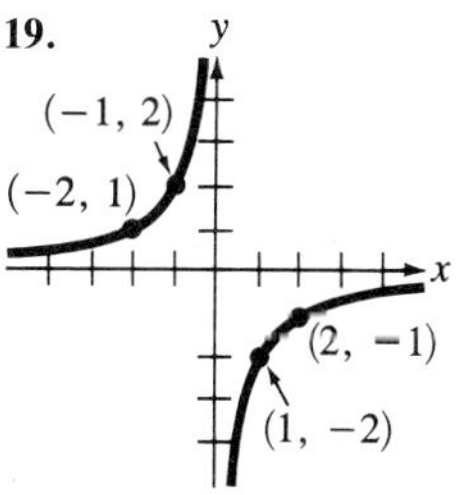


**21.** $\frac{y^2}{25} - \frac{x^2}{25} = 1$ **23.** $\frac{(y-6)^2}{9} - \frac{(x+3)^2}{16} = 1$ **25.** degenerate hyperbola $(x+4)^2 - \frac{(y-1)^2}{4} = 0$, graph is two intersecting lines $y = 2x + 9$ and $y = -2x - 7$ **27.** 7.8 m **29.** $A = b^2$, $B = 0$, $C = -a^2$, $D = -2b^2h$, $E = 2a^2k$, $F = b^2h^2 - a^2k^2 - a^2b^2$ **31.** $\frac{x^2}{4} + \frac{y^2}{15} = 1$ **32.** $\frac{x^2}{45} + \frac{(y-2)^2}{9} = 1$ **33.** $\frac{x^2}{4} + \frac{(y-3)^2}{8} = 1$

**34.** $\frac{(x-2)^2}{7} + \frac{(y+4)^2}{15} = 1$

**[9.4]**

**1.** $x^2 = 6y$ **3.** $y^2 = 14x$ **5.** $y^2 = 12x$ **7.** $(x - 4)^2 = 12(y - 2)$ **9.** $(x - 4)^2 = -8(y + 6)$

**11.** $(y + 7)^2 = -\frac{1}{2}(x - 8)$

**13.**

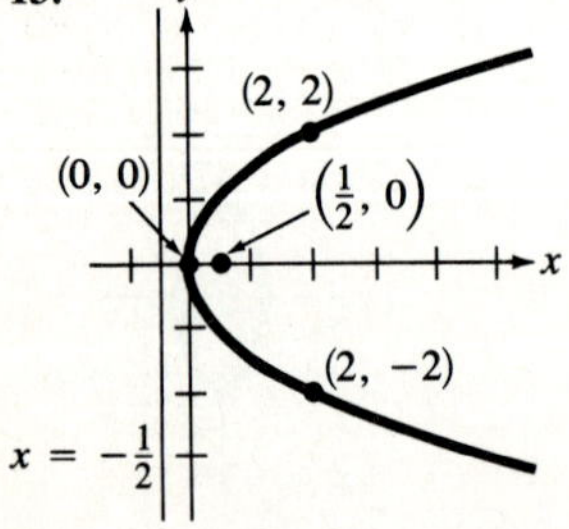


**15.**

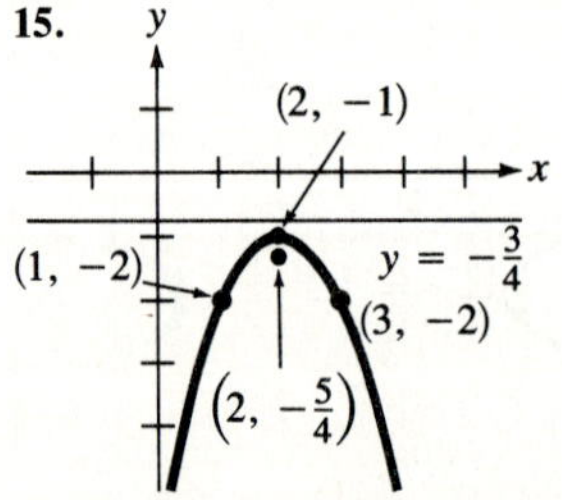


**17.**

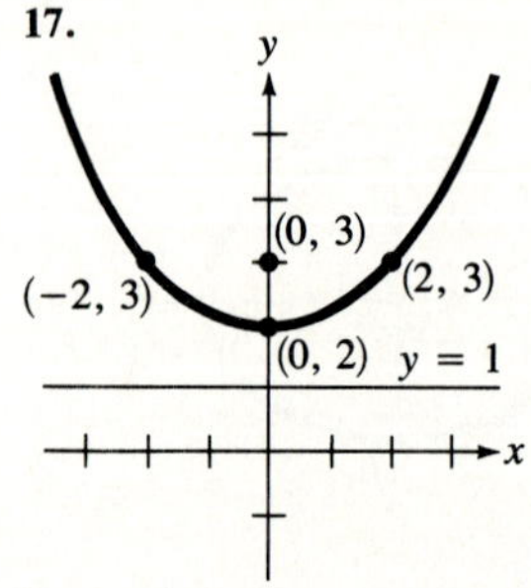


**19.**

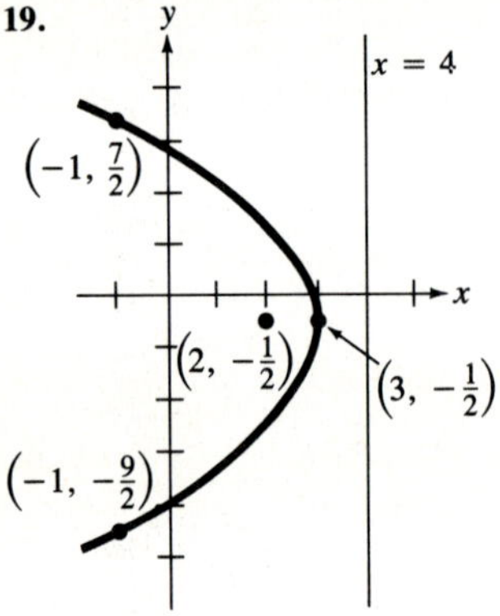


**21.** $y^2 = 12(x - 2)$; $(2, 0)$; $(5, 0)$; $x = -1$

**23.** $(x + 4)^2 = -9(y + 1)$; $(-4, -1)$; $\left(-4, -\frac{13}{4}\right)$; $y = \frac{5}{4}$

**25.** $\left(x - \frac{3}{2}\right)^2 = \frac{1}{2}(y + 6)$; $\left(\frac{3}{2}, -6\right)$; $\left(\frac{3}{2}, -\frac{47}{8}\right)$; $y = -\frac{49}{8}$

**27.** hyperbola

**29.** circle

**31.** degenerate conic (graph is two intersecting lines)

**33.** ellipse

**35.** $x^2 = 4.5y$

**37.** 6.4 m

**39.** $\frac{(y + 2)^2}{9} - \frac{(x - 4)^2}{16} = 1$ **40.** $\frac{x^2}{25} - \frac{y^2}{12} = 1$ **41.** $\frac{(x - 2)^2}{5} - \frac{(y + 3)^2}{6} = 1$ **42.** $\frac{(y + 5)^2}{8} - \frac{(x + 9)^2}{10} = 1$

**[9.5]**

**1.** $\left(\frac{4\sqrt{3} - 3}{2}, \frac{4 + 3\sqrt{3}}{2}\right)$ **3.** $\left(\frac{5\sqrt{2}}{2}, \frac{\sqrt{2}}{2}\right)$ **5.** $(3\sqrt{2}, \sqrt{2})$ **7.** $\left(\frac{\sqrt{3} - 3}{2}, \frac{3\sqrt{3} + 1}{2}\right)$ **9.** $y'^2 - x'^2 = 1$

**11.** $3x'^2 - y'^2 = 16$

**13.** $\frac{y'^2}{2} - \frac{x'^2}{2} = 1$ **15.** $\frac{x'^2}{36} + \frac{y'^2}{6} = 1$ **17.** $x' = \pm 2$ (two parallel lines)

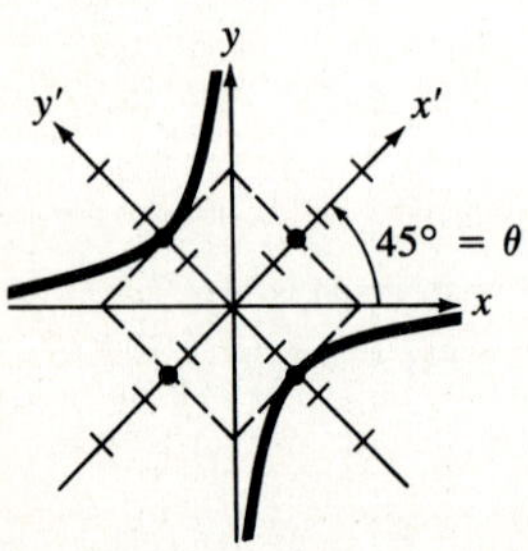


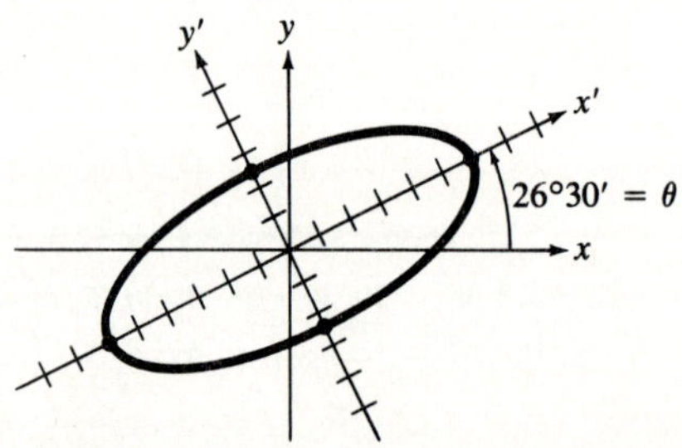


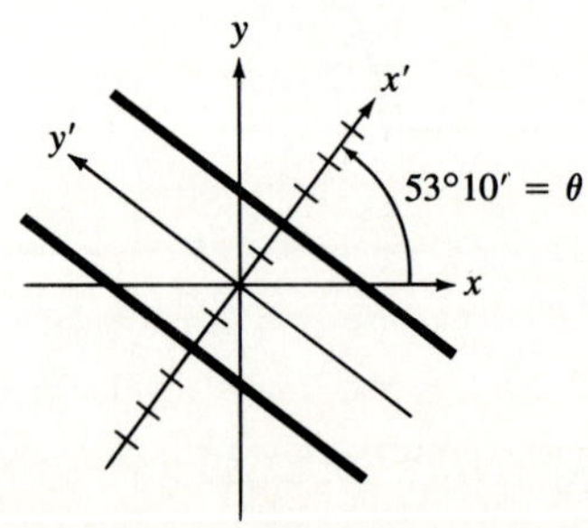

**19.** $\frac{x'^2}{24} - \frac{y'^2}{8} = 1$

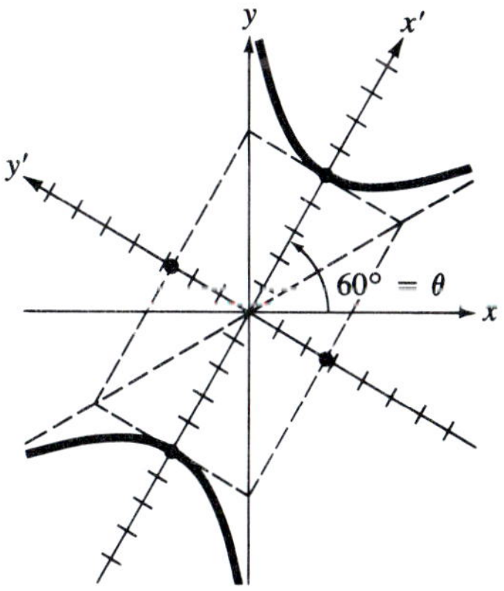


**21.** $y'^2 = 4(x' - \sqrt{2})$

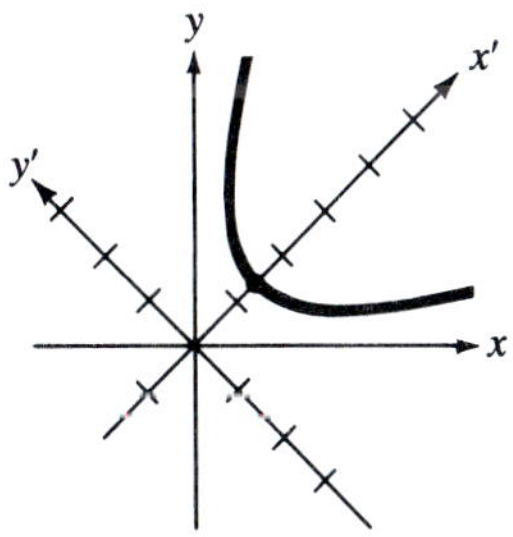


**25.** ellipse **27.** hyperbola **29.** $(x - 3)^2 = -20(y + 1)$ **30.** $(y + 2)^2 = 8(x - 4)$ **31.** $(x - 5)^2 = 20(y - 2)$
**32.** $(y + 8)^2 = -2(x + 3)$ **33.** hyperbola **34.** circle **35.** ellipse **36.** parabola

## [9.6]

**1–11.**

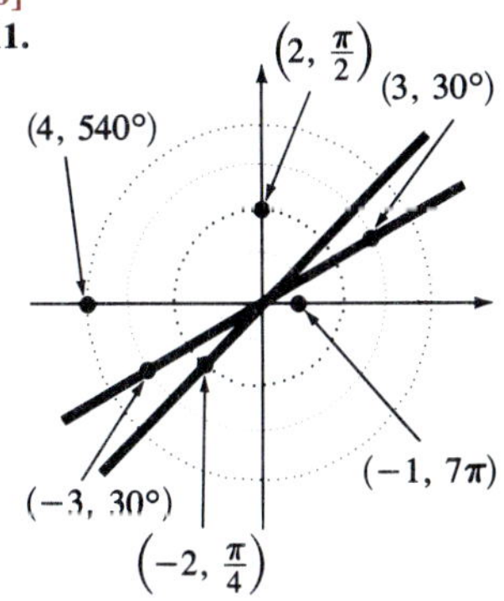


**13.** $(3\sqrt{3}, 3)$ **15.** $(4\sqrt{2}, 4\sqrt{2})$ **17.** $(-1, -\sqrt{3})$ **19.** $(0, 5)$ **21.** $(5, 126.9°)$ **23.** $\left(6, -\frac{\pi}{2}\right)$ **25.** $\left(7\sqrt{2}, \frac{5\pi}{4}\right)$
**27.** $(13, -22.6°)$ **29.** $r = 5$ **31.** $r \sin^2 \theta = 3 \cos \theta$ **33.** $r = -8 \sin \theta$ **35.** $y = 3$ **37.** $(x - 2)^2 + y^2 = 4$
**39.** $y = \sqrt{3}x$

**41.**

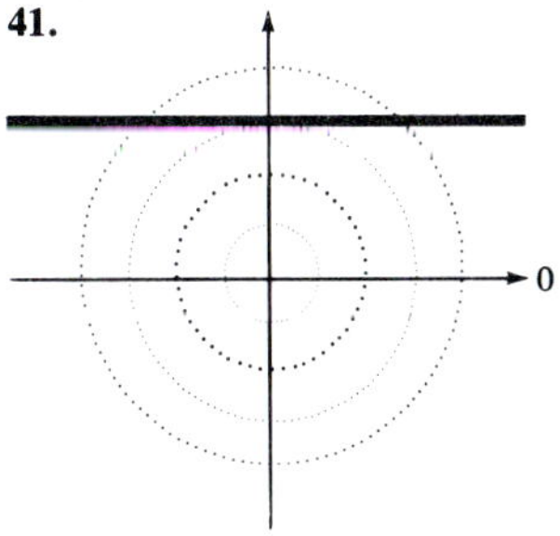


**43.**

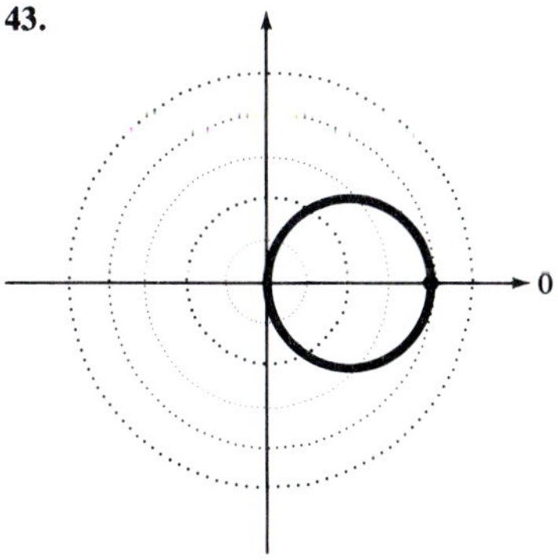


**45.**

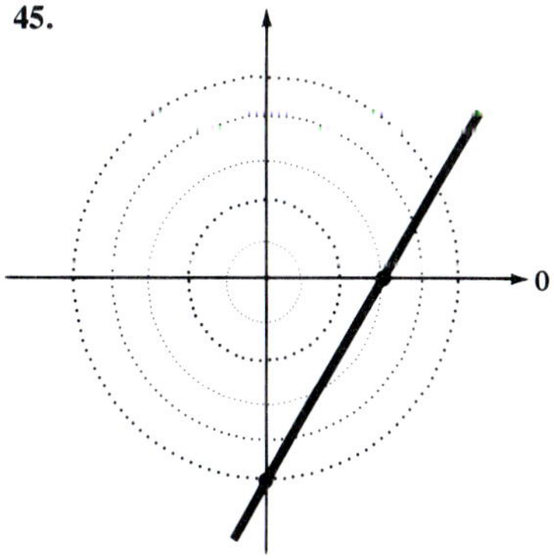

**47.**

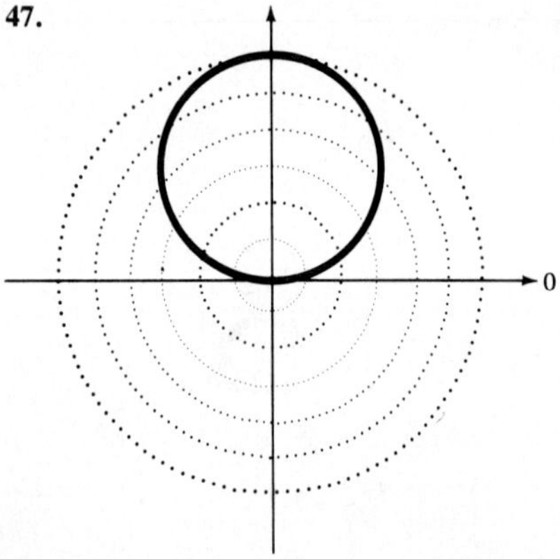


**49.**

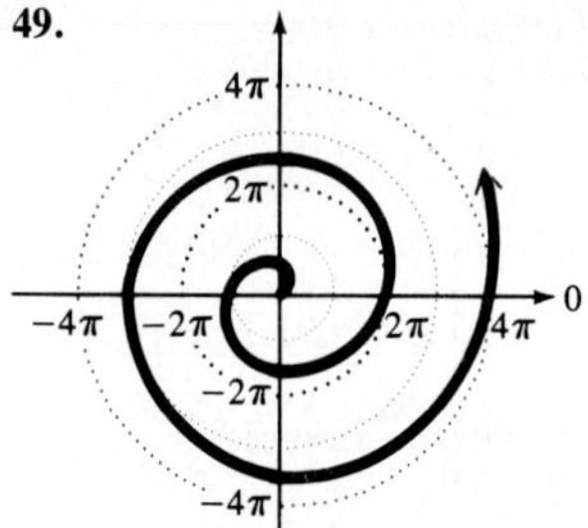


**51.**

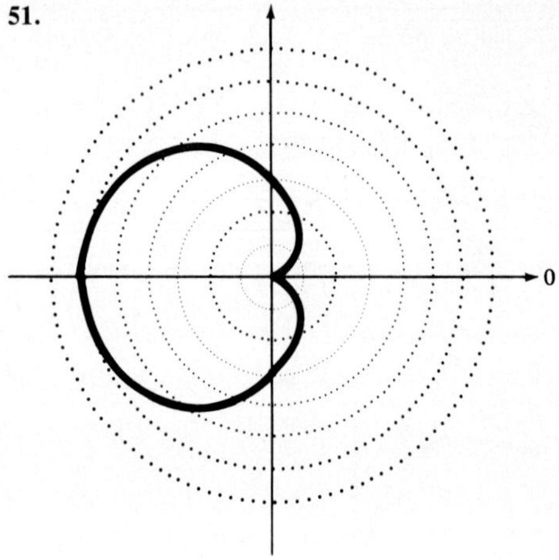


**53.** $4y'^2 - 2x'^2 = 5$ **54.** $x'^2 = -y'$

**[9.7]**

Answers to Exercises 1–11 are the indicated point in the complex plane. **1.** (5, 2) **3.** (−4, 6) **5.** (−2, 1) **7.** (−4, 0) **9.** (0, 1) **11.** (3, 4) **13.** $z = (1)(\cos 0 + i \sin 0)$ **15.** $z = (1)\left(\cos \frac{\pi}{2} + i \sin \frac{\pi}{2}\right)$ **17.** $z = \sqrt{2}\left(\cos \frac{\pi}{4} + i \sin \frac{\pi}{4}\right)$ **19.** $z = 4\left(\cos \frac{7\pi}{4} + i \sin \frac{7\pi}{4}\right)$ **21.** $z = 4\left(\cos \frac{\pi}{6} + i \sin \frac{\pi}{6}\right)$ **23.** $z = \frac{1}{2}\left(\cos \frac{2\pi}{3} + i \sin \frac{2\pi}{3}\right)$ **25.** $z = \frac{3}{2} + \frac{3\sqrt{3}}{2}i$ **27.** $z = -2\sqrt{3} - 2i$ **29.** $z = \frac{\sqrt{6}}{2} - \frac{\sqrt{6}}{2}i$ **31.** $6i, \frac{\sqrt{3}}{3} - \frac{1}{3}i$ **33.** $-3, -\frac{1}{2} - \frac{\sqrt{3}}{2}i$ **35.** $6 + 6i, -\frac{3}{2} - \frac{3}{2}i$ **37.** $5, \frac{3}{5} - \frac{4}{5}i$ **39.** $-1$ **41.** $-8i$ **43.** $-128 + 128\sqrt{3}i$ **45.** $-1024$ **47.** $1, i, -1, -i$ **49.** $\frac{\sqrt{3}}{2} + \frac{1}{2}i, -\frac{\sqrt{3}}{2} + \frac{1}{2}i, -i$ **51.** $2\left(\cos \frac{\pi}{16} + i \sin \frac{\pi}{16}\right), 2\left(\cos \frac{9\pi}{16} + i \sin \frac{9\pi}{16}\right), 2\left(\cos \frac{17\pi}{16} + i \sin \frac{17\pi}{16}\right),$ $2\left(\cos \frac{25\pi}{16} + i \sin \frac{25\pi}{16}\right)$ **53.** $-\frac{\sqrt{2}}{2} + \frac{i\sqrt{6}}{2}, \frac{\sqrt{2}}{2} - \frac{i\sqrt{6}}{2}$ **55.** $i, -i$ **57.** $-\frac{\sqrt{2}}{2} + \frac{\sqrt{2}}{2}i, \frac{\sqrt{2}}{2} - \frac{\sqrt{2}}{2}i$ **59.** $4r^2 = 1$ **60.** $\cos \theta = 2 \sin \theta$ **61.** $r = 2(\cos \theta + \sin \theta)$ **62.** $x + 5y = 2$ **63.** $x^2 + (y - 3)^2 = 9$ **64.** $3x^2 - 3y^2 = 1$

## Chapter 9 Review Exercises

**1.** $(x + 3)^2 + \left(y - \frac{2}{3}\right)^2 = 16$ **2.** $\frac{x^2}{9} + \frac{y^2}{49} = 1$ **3.** $\frac{x^2}{9} - \frac{y^2}{7} = 1$ **4.** $(y - 2)^2 = 16(x - 1)$ **5.** $(x - 2)^2 + (y - 2)^2 = 41$ **6.** $(x - 2)^2 + \frac{(y + 3)^2}{5} = 1$ **7.** $\frac{y^2}{25} - \frac{x^2}{3} = 1$ **8.** $(x - 1)^2 = 4(y + 2)$

**9.**

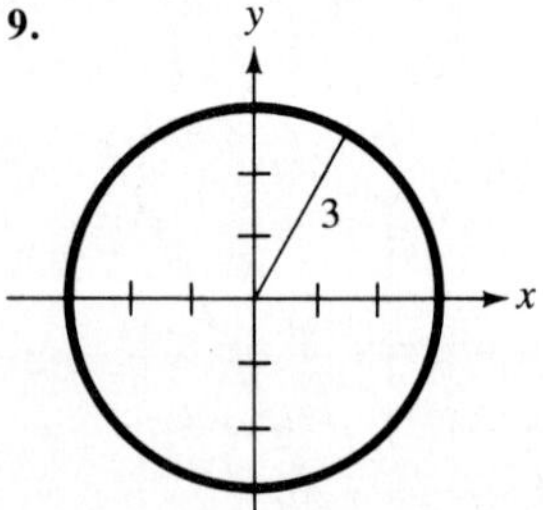


**10.**

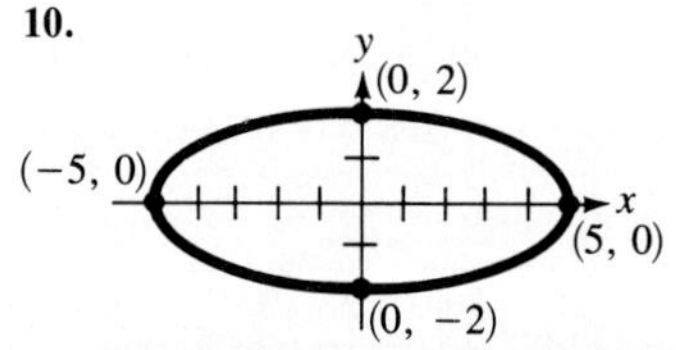


**11.**

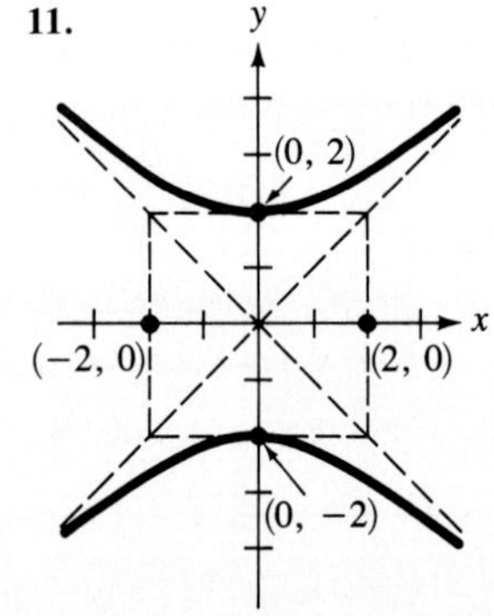

**12.**

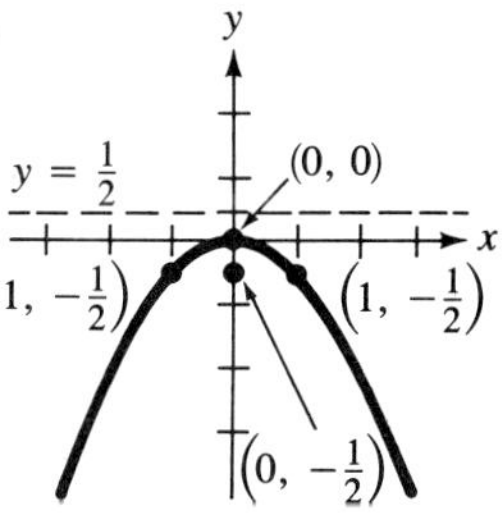


**13.**

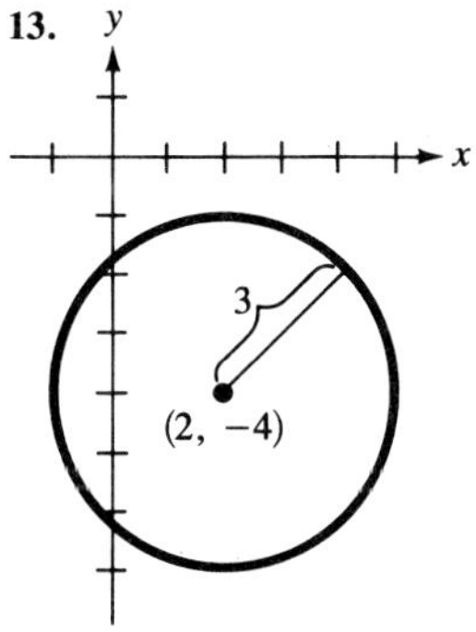


**14.**

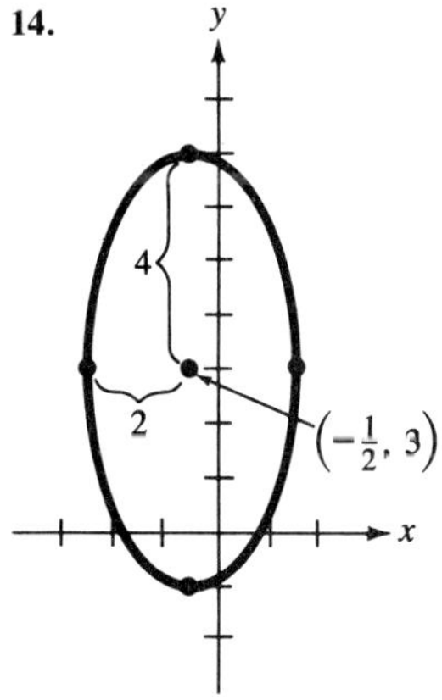


**15.**

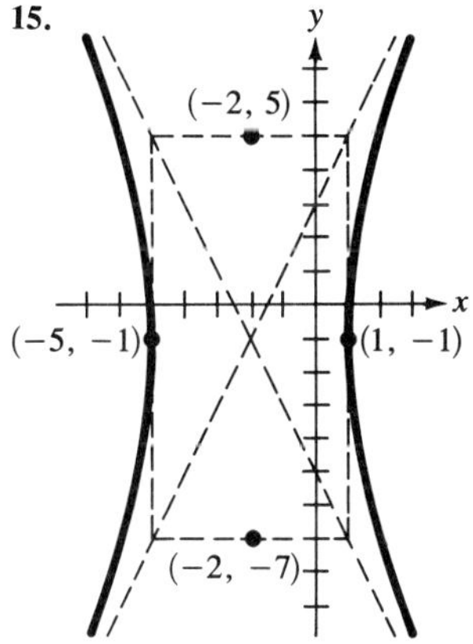


**16.**

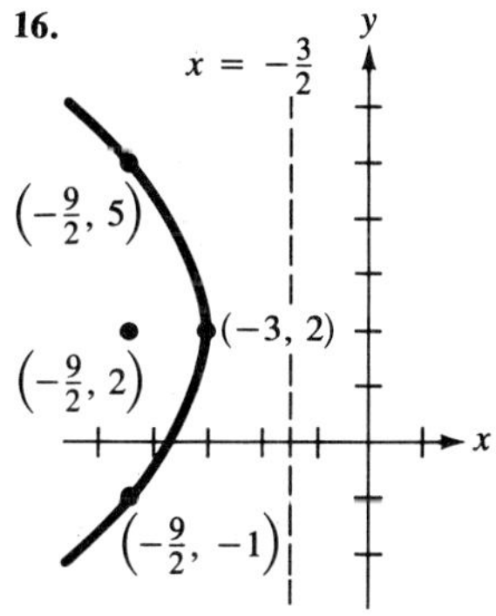


**17.** circle **18.** hyperbola **19.** degenerate conic (a point) **20.** degenerate conic (parallel lines) **21.** parabola

**22.** ellipse **23.** $\left(\frac{-\sqrt{3}-5}{2}, \frac{5\sqrt{3}-1}{2}\right)$ **24.** $\left(-\frac{\sqrt{2}}{2}, -\frac{9\sqrt{2}}{2}\right)$ **25.** $\frac{x'^2}{4} + \frac{y'^2}{9} = 1$

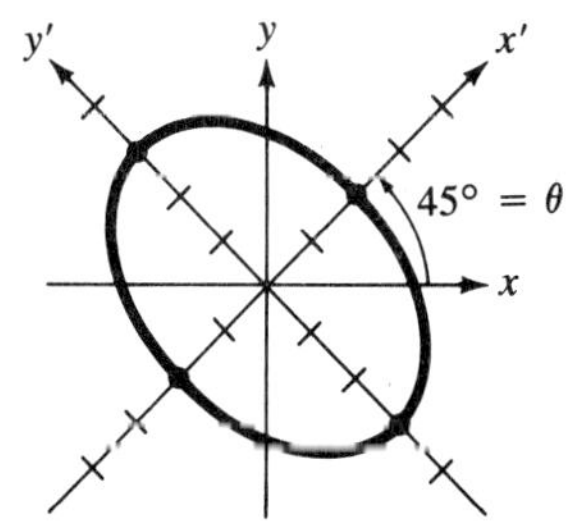


**26.** $\frac{x'^2}{5} + \frac{y'^2}{1} = 0$; graph is the point $(0, 0)$ **27.** $(2\sqrt{3}, 2)$ **28.** $(1, -\sqrt{3})$ **29.** $(5, 143.1^\circ)$ **30.** $\left(5\sqrt{2}, \frac{5\pi}{4}\right)$

**31.** $r(2 \cos \theta - 5 \sin \theta) = -1$ **32.** $2xy = 7$ **33.** The graph is the line with equation $x = 2$ in rectangular coordinates.

**34.** The graph is the circle with radius 4 centered at $(0, 4)$ in rectangular coordinates. **35.** $z = \sqrt{2}\left(\cos \frac{7\pi}{4} + i \sin \frac{7\pi}{4}\right)$

**36.** $z = 4\left(\cos \frac{5\pi}{6} + i \sin \frac{5\pi}{6}\right)$ **37.** $-3\sqrt{3} - 3i$ **38.** $10i, \frac{\sqrt{3}}{5} + \frac{1}{5}i$ **39.** 16 **40.** $2i, -\sqrt{3} - i, \sqrt{3} - i$

**41.** 11.4 m **42.** $x^2 = -\frac{32}{3}(y - 20)$

## Chapter 10

### [10.1]

**1.** **(a)** yes **(b)** yes **3.** **(a)** exactly one **(b)** intersect **(c)** consistent and independent **5.** **(a)** infinitely many **(b)** coincide **(c)** dependent **7.** $(-1, 4)$ **9.** $\left(x, \frac{5}{2}x - \frac{3}{2}\right)$ $x$ any real number **11.** $(3, 0)$ **13.** $(-2, -1)$ **15.** $(1, 1)$ **17.** no solution **19.** $\left(\frac{4}{5}, \frac{9}{4}\right)$ **21.** $(-1, 3)$ **23.** $\left(\frac{1}{2}, -\frac{1}{3}\right)$ **25.** $\left(-\frac{1}{a}, \frac{2}{b}\right)$ **27.** $a = 1$; $b = -4$ **29.** 12, 32 **31.** 3 hr at 40 km/hr; 5 hr at 50 km/hr **33.** 22°, 158° **35.** shirts cost \$15; socks cost \$3 **37.** 40 lb of 90¢ candy; 20 lb of \$1.50 candy **39.** 22 dimes; 8 quarters **41.** 10 L of 25% solution: 15 L of 50% solution **43.** 53.6 g of Control K; 35.7 g of Special Diet **45.** jog 10 times; play tennis 5 times **47.** Since the slope of the line corresponding to the first equation is $-2$, and the slope of the line corresponding to the second equation is $\frac{1}{4}$, the lines intersect. Thus, the system has exactly one solution no matter what value $m$ assumes. **49.** $a = 4$; $b = -3$ **51.** ellipse **52.** parabola **53.** hyperbola **54.** circle

### [10.2]

**1.** $(-1, 1, 3)$ **3.** $(1, 2, 3)$ **5.** no solution **7.** $(x, 2x, -5x)$ $x$ any real number **9.** $(0, 0, 0)$ **11.** $\left(x, -\frac{1}{3}x + 1, 1 - x\right)$ $x$ any real number **13.** no solution **15.** $(8, 6, 4)$ **17.** $(1, 0, -1, 2)$ **19.** $\left(x, -\frac{1}{2}x - \frac{7}{2}, -x - 5, -\frac{1}{2}x - \frac{19}{2}\right)$ $x$ any real number **21.** no solution **23.** 2, $-3$, 5 **25.** Milt is 18; Lew is 20; Jenny is 15 **27.** 3000 **29.** 12 nickels; 20 dimes; 8 quarters **31.** \$1500 in bonds; \$1500 in certificates; \$2000 in mutual fund **33.** Sierra: 15; San Juan: 20; Blue Ridge: 10 **35.** $a = 1$; $b = -1$; $c = 5$ **37.** $y = -x^2 + 3x - 2$ **39.** $x^2 + y^2 - 2x + 2y - 7 = 0$ **41.** $\left(\frac{2}{a}, 0\right)$ **42.** $m = 9$ **43.** If $m$ is any real number except $-\frac{5}{4}$, the system will be inconsistent. **44.** 32 **45.** 17 ft by 8 ft **46.** $a = 2$; $b = -1$

**47.**

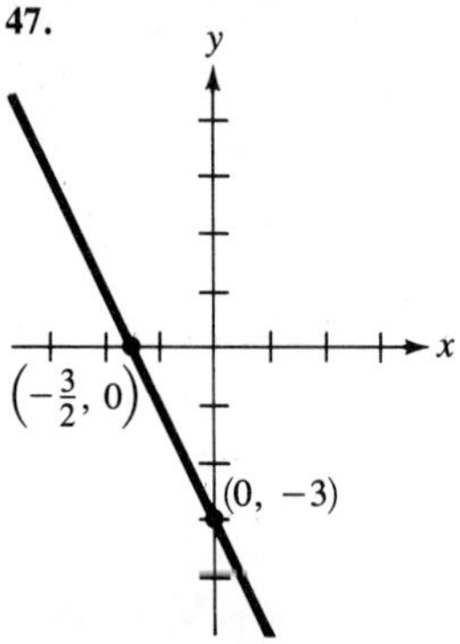


**48.**

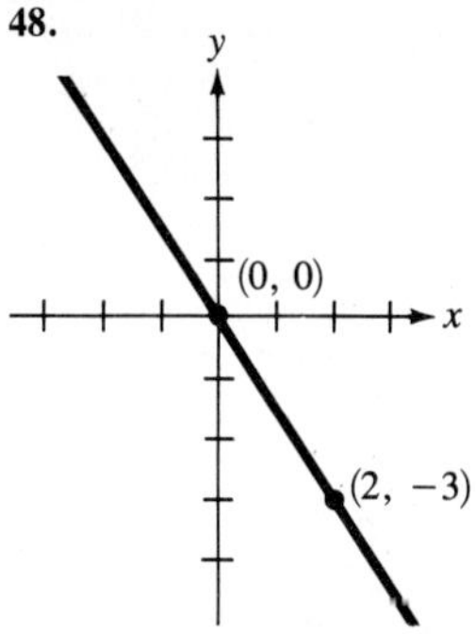


**49.**

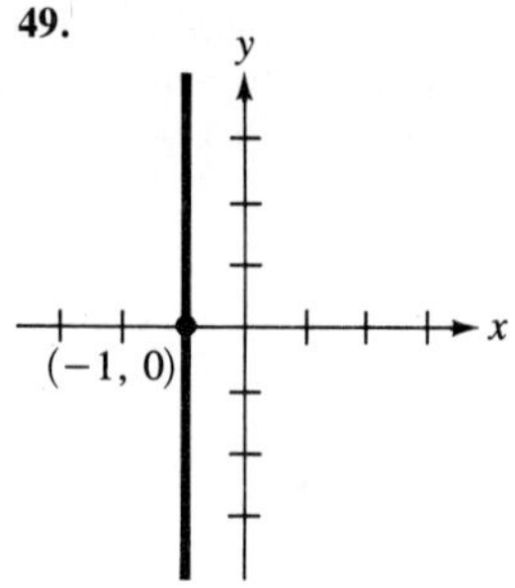


**50.**

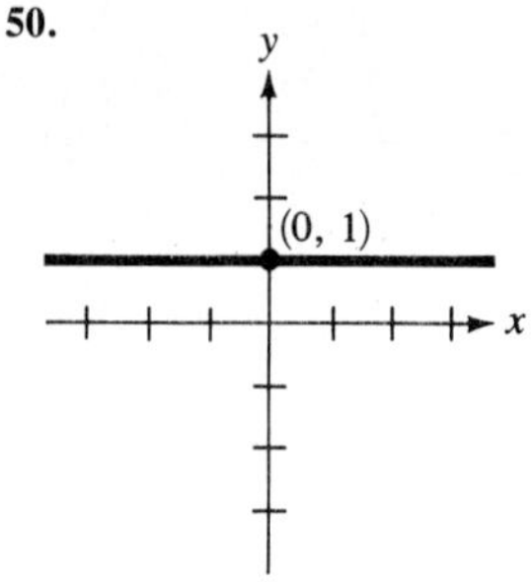


### [10.3]

**1.**

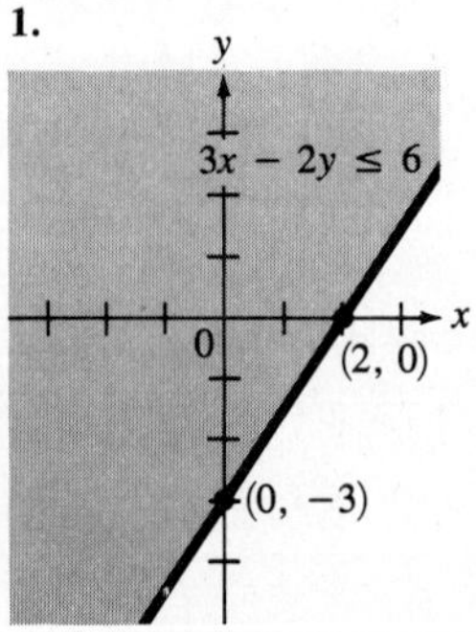


**3.**

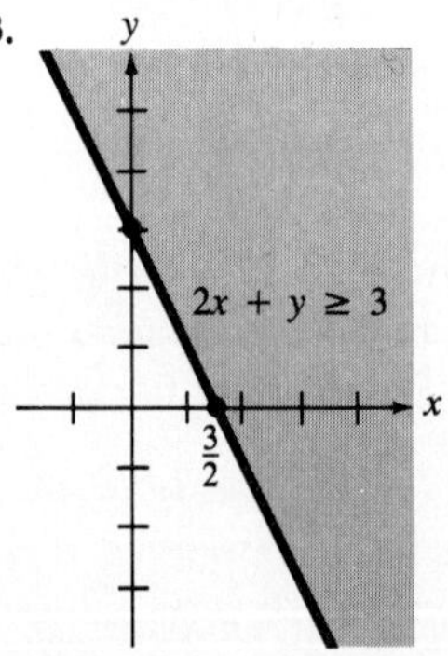


**5.**

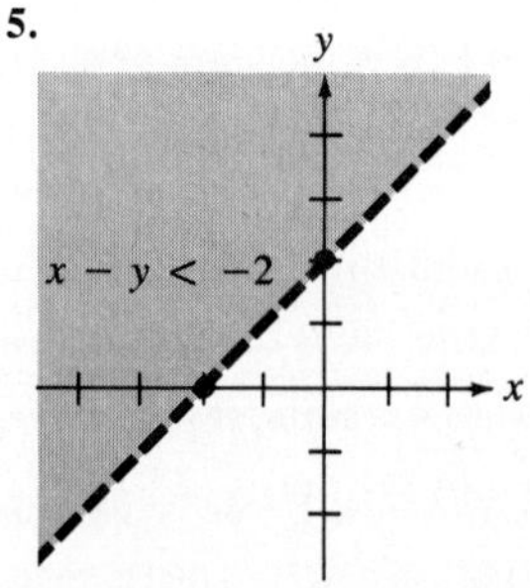


**7.**

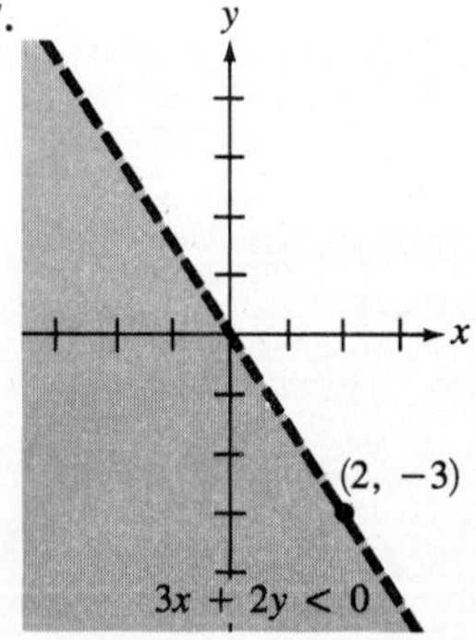

**9.**

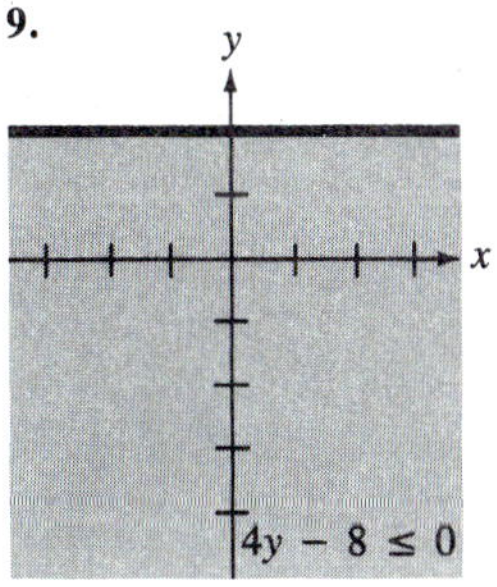


**11.**

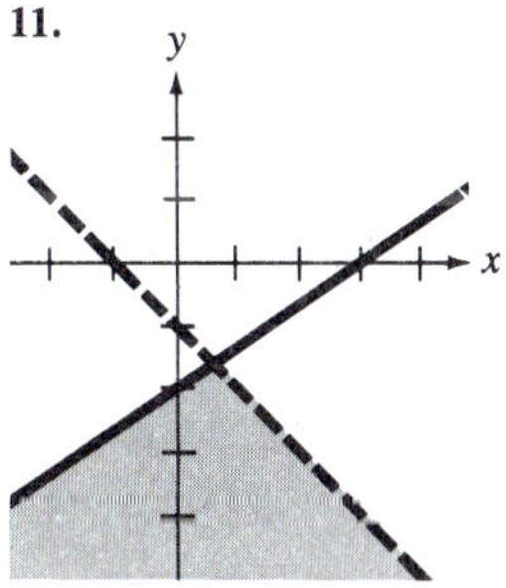


**13.**

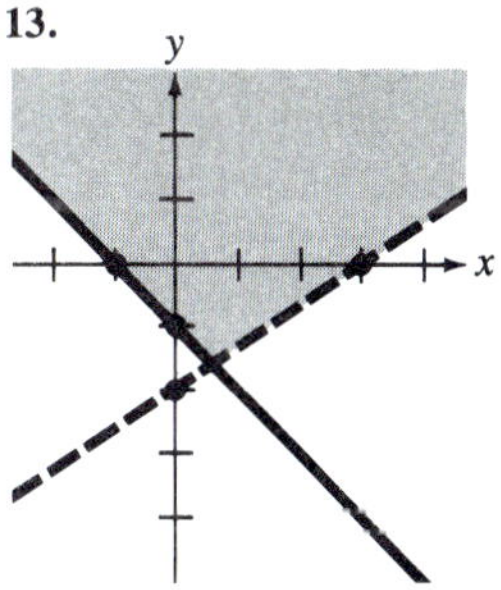


**15.**

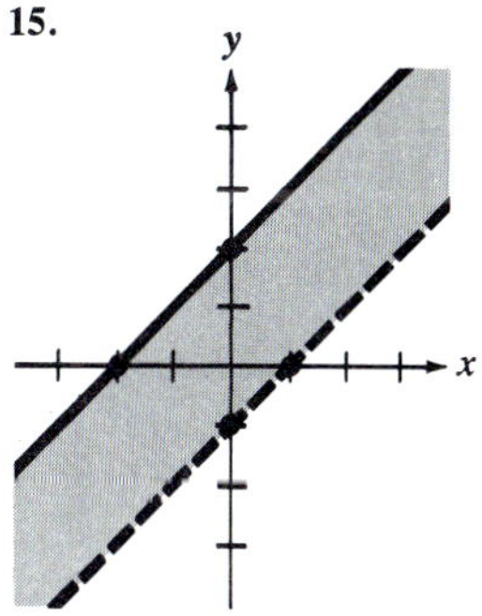


**17.**

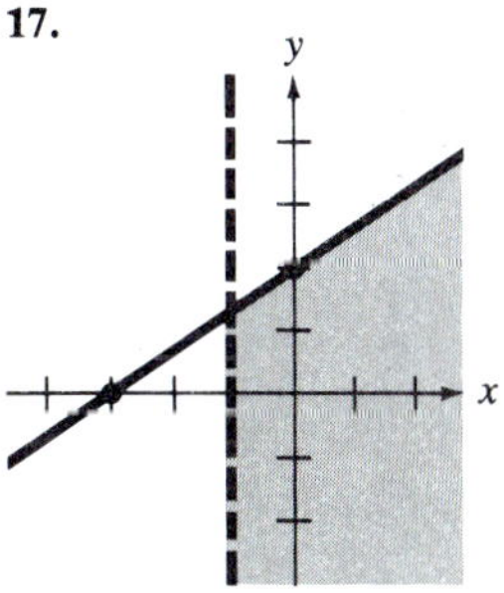


**19.**

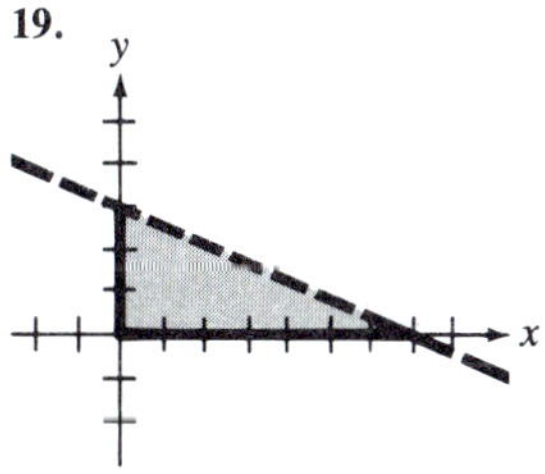


**21.**

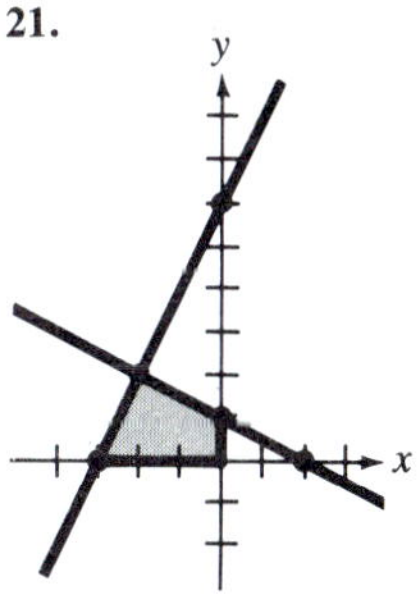


**23.** Let $x$ = the number of Princess homes,
$y$ = the number of Knight homes.

$$x \geq 6$$
$$y \geq 2$$
$$x - 3y \geq 0$$
$$30{,}000x + 20{,}000y \leq 600{,}000$$

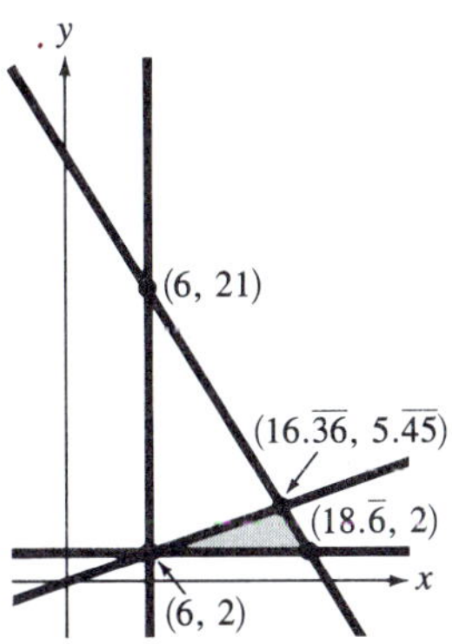


**25.** candy costs 80¢/lb; nuts cost \$1.10/lb **26.** 16 L of 5% solution: 24 L of 10% solution **27.** 42°, 68°, 70° **28.** \$4000 at 11%: \$12,000 at 10%; \$4000 at 8% **29.** $(x, -7x, -4x)$ $x$ any real number **30.** $(x, 8x - 1, 21x - 4)$ $x$ any real number

## [10.4]

**1.** maximum value: 46; minimum value: 6 **3.** maximum value: none; minimum value: 350 **5.** maximum value: 150; minimum value: 0 **7.** maximum value: none; minimum value: 75 **9.** 100 True-Shot models and 50 True-Bounce models **11.** 3500 barrels of type G and 1500 barrels of type H **13.** 32 acres of corn and 68 acres of wheat **15.** 7 units of A and 4 units of B **17.** all 1000 books shipped from P to CTU; all 750 books shipped from C to MC

**19.**

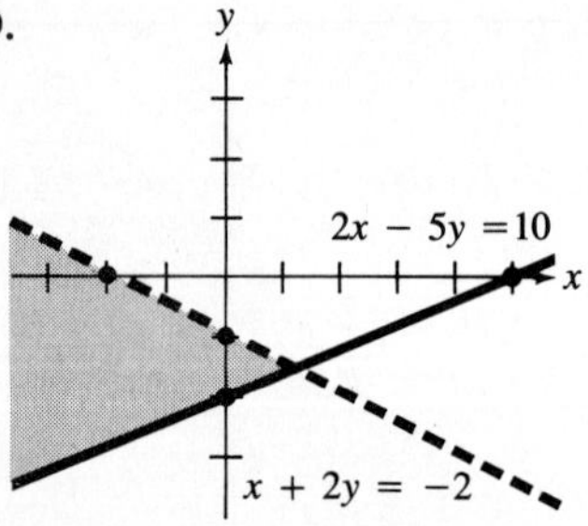


**20.**

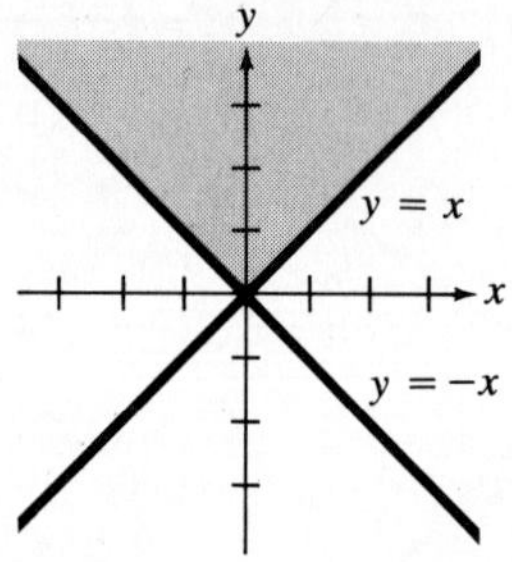


**21.** Let $x$ = the number of \$20 seats,
$y$ = the number of \$10 seats.

$$x + y \leq 1000$$
$$20x + 10y \geq 16{,}000$$
$$x \geq 0$$
$$y \geq 300$$

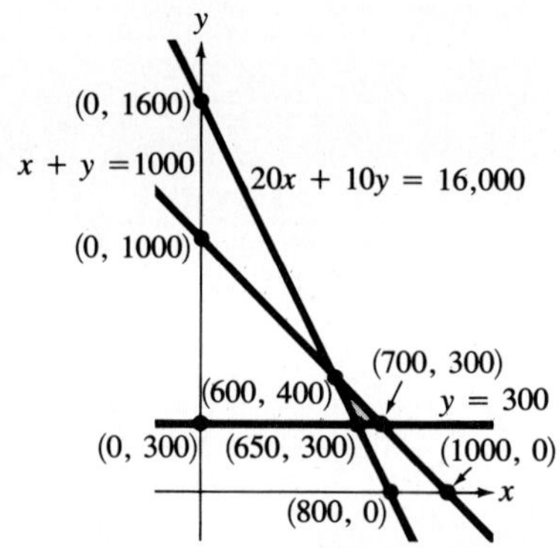


**22.**

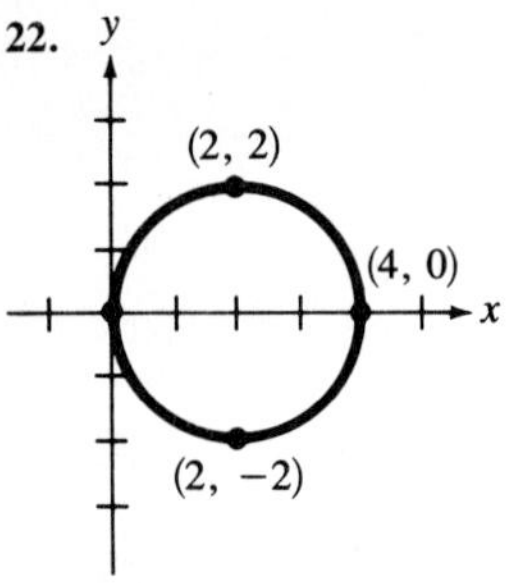


**23.**

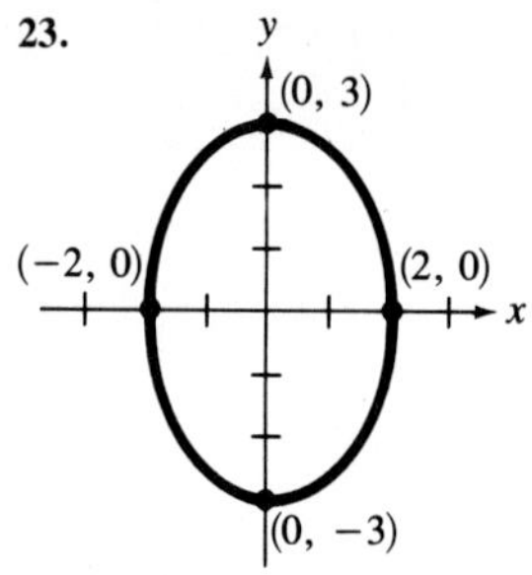


**24.**

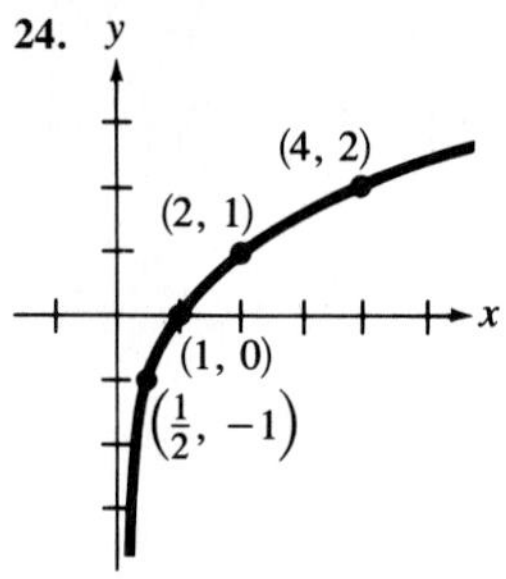


**25.**

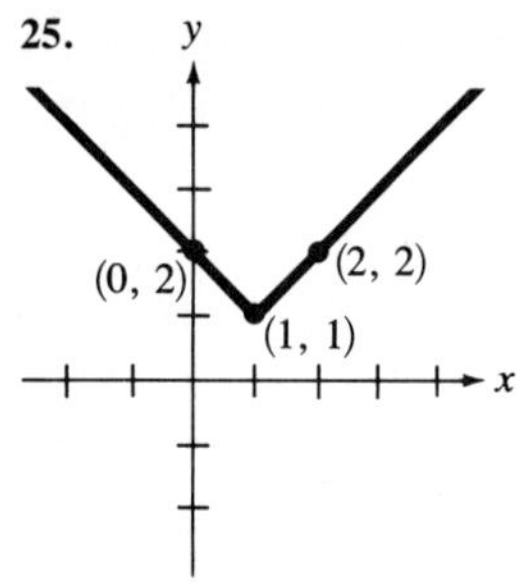


**[10.5]**

**1.** (3, 1), (−1, −3) **3.** (2, 1) **5.** (0, 1), (1, 3) **7.** (5, 2), (5, −2), (−5, 2), (−5, −2) **9.** (2, 1), (−2, −1), (1, 2), (−1, −2) **11.** (3, 0), (3, −6), (−3, 0), (−3, 6) **13.** (4, −1), (−4, 1), $(\sqrt{2}, \sqrt{2})$, $(-\sqrt{2}, -\sqrt{2})$ **15.** (1, 1) **17.** (1, 5) **19.** (3, 4), (3, −4)

**21.**

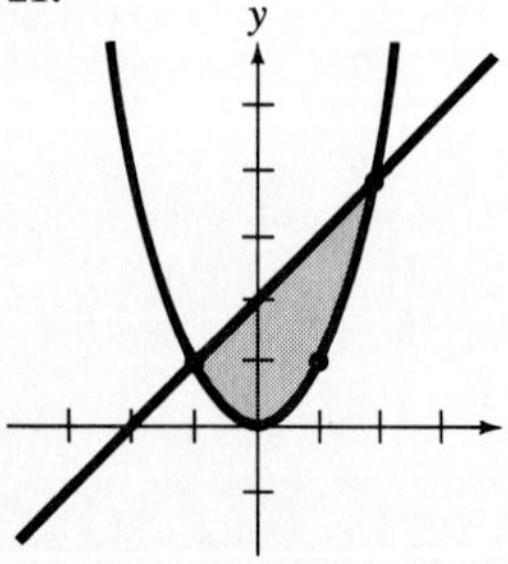


**23.**

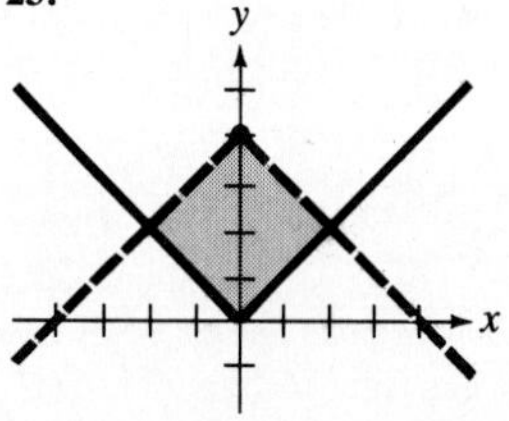


**25.**

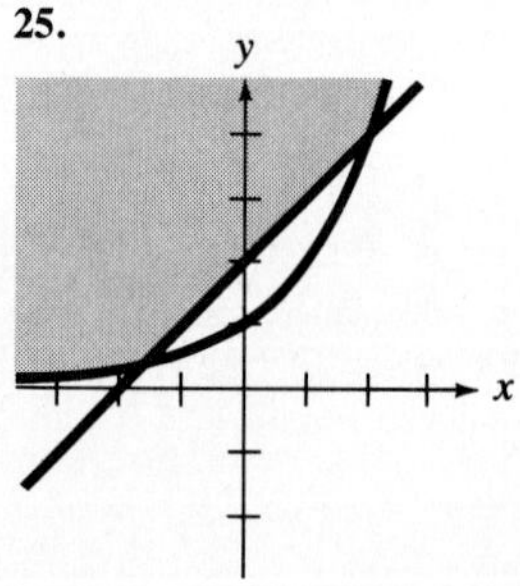


**27.**

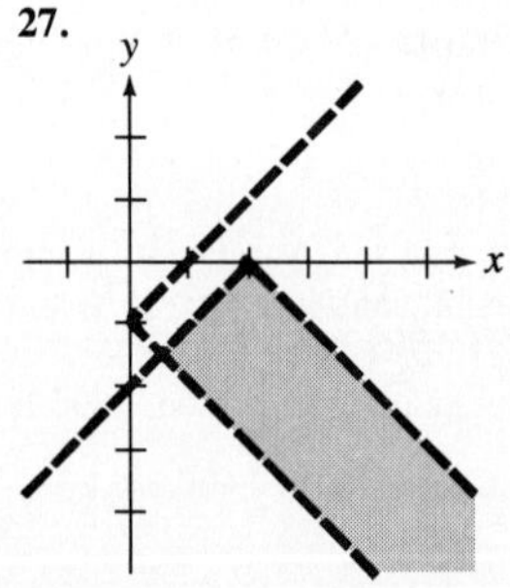

**29.**

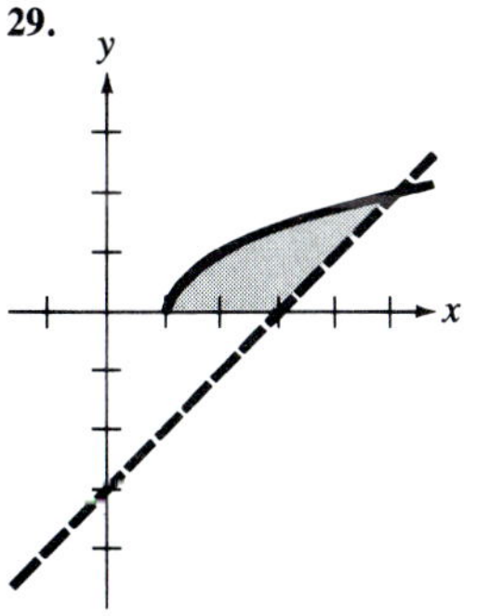

**31.**

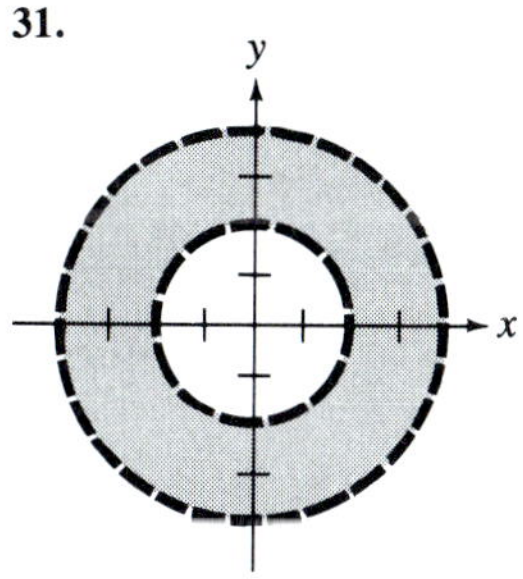

**33.** 6 yd by 4 yd **35.** 12 people; approximately \$8.33 **37.** maximum value: 6; minimum value: $-17$ **38.** maximum value: 60; minimum value: $-20$ **39.** 50 Pro-Model racquets and 30 Hacker-Model racquets **40.** 5 **41.** $2 - 3i$ **42.** $x^2 + x + 1$ **43.** $f(x) = \frac{3x - 2}{x^2 - 4}$ **44.** (1, 2) **45.** (1, 2, 3)

## [10.6]

**1.** $\frac{3x + 2}{(x - 1)(x + 5)} = \frac{A}{x - 1} + \frac{B}{x + 5}$ **3.** $\frac{6x^2 - 7}{(x + 3)^2(x + 5)} = \frac{A}{x + 3} + \frac{B}{(x + 3)^2} + \frac{C}{x + 5}$ **5.** $\frac{x^3 - 5}{(x^2 + 2x + 10)^2} = \frac{Ax + B}{x^2 + 2x + 10} + \frac{Cx + D}{(x^2 + 2x + 10)^2}$ **7.** $\frac{x + 1}{x^3 - 2x^2 - 3x} = \frac{A}{x} + \frac{B}{x - 3}$ **9.** $\frac{x - 3}{x^2(x + 2) - 2x(x + 2) - 3(x + 2)} - \frac{A}{x + 2} + \frac{B}{x + 1}$ **11.** $\frac{x^2 + 5x + 5}{x^4 + 5x^2 + 5} - \frac{Ax + B}{x^2 + 1} + \frac{Cx + D}{x^2 + 4}$ **13.** $A - 2$; $B - 3$ **15.** $A = -1$; $B = 2$ **17.** $A = 1$; $B = 2$; $C = 5$ **19.** $A = 1$; $B = 1$; $C = 1$; $D = -1$ **21.** $A = 0$; $B = 7$; $C = -2$ **23.** $A = -1$; $B = -1$; $C = -2$; $D = 2$; $E = 1$ **25.** $f(x) = \frac{1}{x - 3} + \frac{2}{x + 3}$ **27.** $f(x) = \frac{1}{x + 2} + \frac{-3}{(x + 2)^2} + \frac{5}{x + 1}$ **29.** $f(x) = \frac{2x + 1}{x^2 + 1} + \frac{-2}{x - 5}$ **31.** $f(x) = \frac{3}{x + 2} + \frac{-3}{(x + 2)^2} + \frac{4}{x - 2}$ **33.** $f(x) = \frac{2x - 1}{x^2 + 1} + \frac{x}{(x^2 + 1)^2}$ **35.** $f(x) = x + \frac{1}{x - 1} + \frac{5x - 1}{(x - 1)^2}$ **37.** (2, 1), (−2, −1), (1, 2), (−1, −2) **38.** $\left(\sqrt{6}, \frac{\sqrt{6}}{3}\right)$, $\left(-\sqrt{6}, -\frac{\sqrt{6}}{3}\right)$, (2, 1), (−2, −1)

**39.**

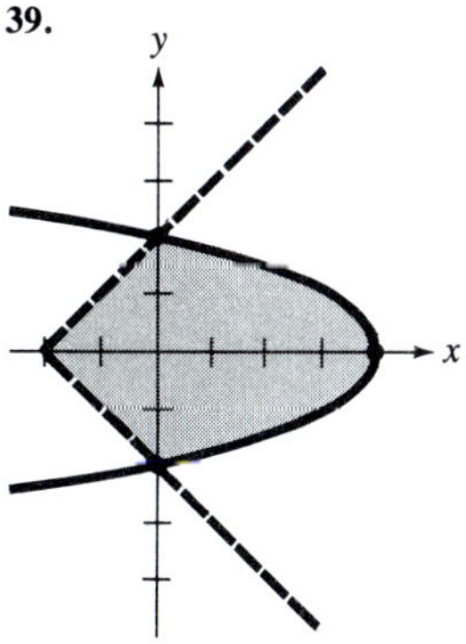

**40.**

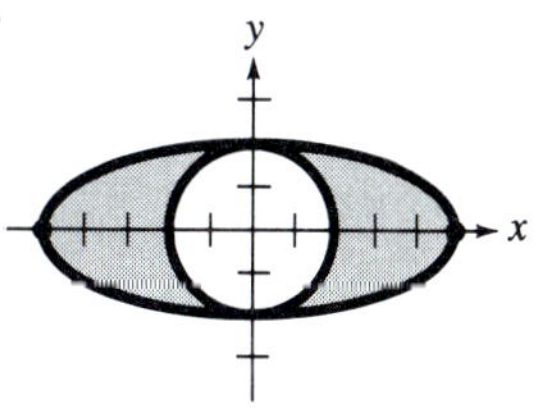

## Chapter 10 Review Exercises

**1.** yes **2.** (−2, 5), **3.** (1, −2) **4.** no solution **5.** (1, −1) **6.** $(x, 3 - x)$ $x$ any real number **7.** (0, −4) **8.** $m = -\frac{5}{2}$ **9.** (2, 0, −1) **10.** $\left(x, \frac{9}{2}x, 2x\right)$ $x$ any real number **11.** $(2 - 2z, 3z + 3, z)$ $z$ any real number **12.** $a = 2$, $b = -3$, $c = 5$ **13.** 3 hr at 40 mph; 5 hr at 50 mph **14.** 12°, 78° **15.** 32 lb of \$1.60/lb candy; 48 lb of \$1.20/lb candy **16.** 40 gal of 15% solution; 60 gal of 20% solution **17.** 25 nickels; 30 dimes; 15 quarters **18.** \$2000 in bonds; \$5000 in certificates; \$2000 in stocks **19.** 25°, 75°, 80° **20.** $a = -5$, $b = 8$

**21.**

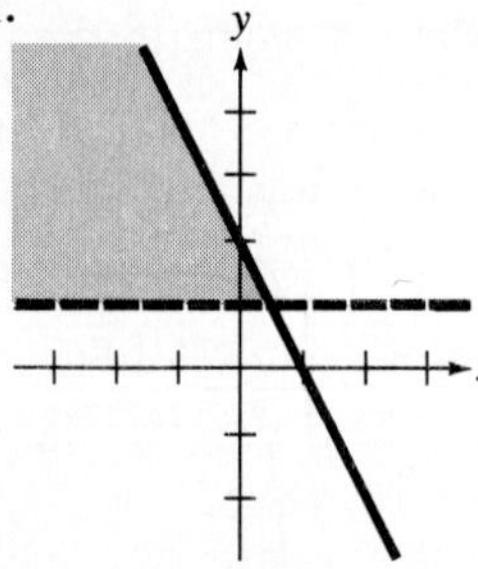


**22.**

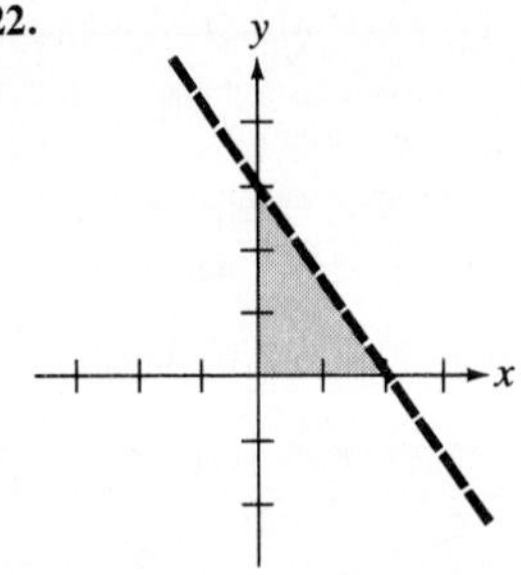


**23.** maximum value: 159; minimum value: 33
**24.** maximum value: 10; minimum value: −48
**25.** 60 hamburgers and 420 tacos; $109.20
**26.** $\left(\frac{1}{3}, -\frac{5}{3}\right)$, $(1, -1)$
**27.** $(2, 4), (-2, -4), (4, 2), (-4, -2)$
**28.** $(2, 1), (-2, -1), (2i, 2i), (-2i, -2i)$
**29.** $(2, 1)$

**30.**

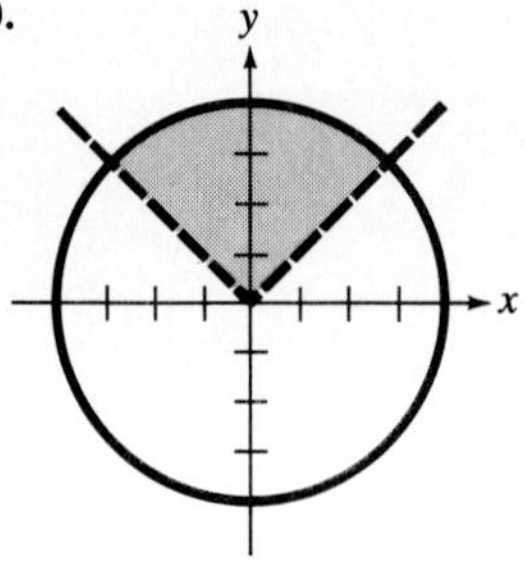


**31.**

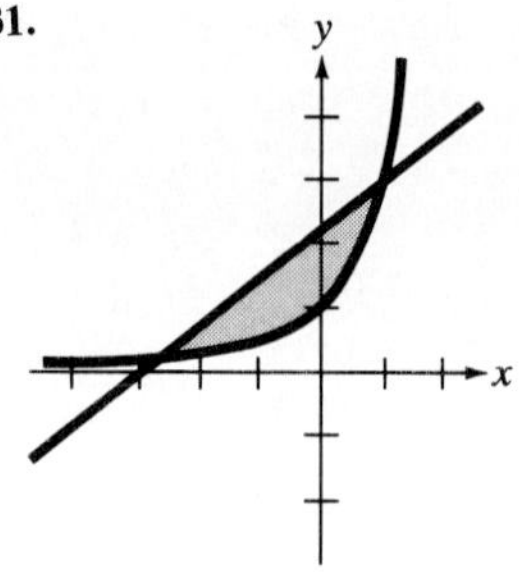


**32.** $A = -2; B = 3$
**33.** $A = 1; B = 1; C = 2; D = 0$
**34.** $f(x) = \dfrac{3x - 1}{x^2 + 3} + \dfrac{5}{x - 1}$
**35.** $f(x) = 2x + \dfrac{1}{x - 2} + \dfrac{3}{x - 6}$

## Chapter 11

### [11.1]

**1.** $2 \times 2$ **3.** $1 \times 4$ **5.** $A$, $B$, and $C$ **7.** $G$ **9.** yes **11.** −1 **13.** $\begin{bmatrix} 0 & 0 \\ 0 & 0 \end{bmatrix}$ **15.** 3 **17.** $\begin{bmatrix} -1 & -2 \\ -3 & -4 \end{bmatrix}$

**19.** $\begin{bmatrix} 6 & 0 \\ 12 & 0 \end{bmatrix}$ **21.** undefined **23.** $\begin{bmatrix} 5 & -4 & 4 \\ -3 & -4 & -2 \end{bmatrix}$ **25.** undefined **27.** $\begin{bmatrix} -2 & -4 \\ -6 & -8 \end{bmatrix}$ **29.** $\begin{bmatrix} 10 & -4 \\ 18 & -8 \end{bmatrix}$

**31.** $\begin{bmatrix} -13 & 10 \\ -21 & 20 \end{bmatrix}$

**33.** $a = 7; b = -5; c = 5; d = -1$

**35.** $a = -3; b = 2$

**37.** **(a)** $F + S = \begin{bmatrix} 40 & 9 & 68 \\ 44 & 17 & 62 \\ 18 & 17 & 54 \\ 16 & 8 & 62 \\ 32 & 24 & 60 \end{bmatrix}$ **(b)** $\frac{1}{2}(F + S) = \begin{bmatrix} 20 & 4.5 & 34 \\ 22 & 8.5 & 31 \\ 9 & 8.5 & 27 \\ 8 & 4 & 31 \\ 16 & 12 & 30 \end{bmatrix}$

The matrix $\frac{1}{2}(F + S)$ represents the average points, rebounds, and minutes played for each player in the tournament.

**39.** $(-2, 3)$ **40.** no solution **41.** $(0, 3, -1)$ **42.** $(1, -6, -4)$
**43.** $(5, 12), (12, 5)$ **44.** $(5, 0), (-5, 0)$

**45.**

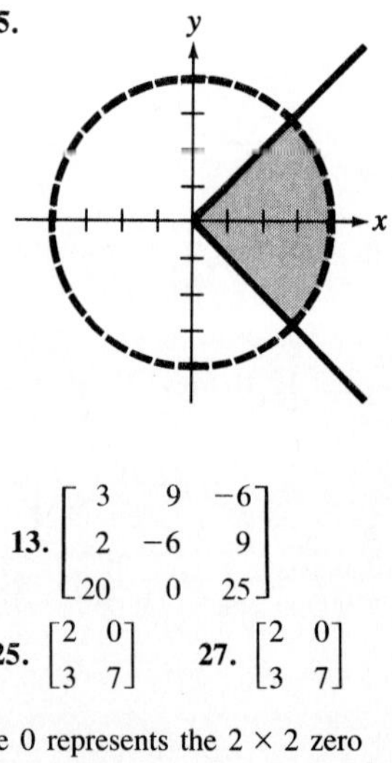


### [11.2]

**1.** −7 **3.** 8 **5.** $m = 4; n = 3$ **7.** $m = 2; n = 3$ **9.** $\begin{bmatrix} 5 & 16 \\ 3 & 14 \end{bmatrix}$ **11.** $\begin{bmatrix} 5 & 3 & 3 \\ 5 & 3 & 3 \end{bmatrix}$ **13.** $\begin{bmatrix} 3 & 9 & -6 \\ 2 & -6 & 9 \\ 20 & 0 & 25 \end{bmatrix}$

**15.** $\begin{bmatrix} 8 \\ -3 \end{bmatrix}$ **17.** $\begin{bmatrix} 0 & 1 \\ -2 & 4 \end{bmatrix}$ **19.** $\begin{bmatrix} -2 & -8 \\ 4 & 14 \end{bmatrix}$ **21.** $\begin{bmatrix} 3 & -2 & 0 \\ 0 & 1 & 5 \end{bmatrix}$ **23.** $\begin{bmatrix} 20 & 8 \\ 12 & -4 \end{bmatrix}$ **25.** $\begin{bmatrix} 2 & 0 \\ 3 & 7 \end{bmatrix}$ **27.** $\begin{bmatrix} 2 & 0 \\ 3 & 7 \end{bmatrix}$

**29.** $AB = \begin{bmatrix} 7 & 0 \\ 5 & -10 \end{bmatrix} \neq \begin{bmatrix} 6 & 8 \\ 2 & -9 \end{bmatrix} = BA$ **31.** $I = \begin{bmatrix} 1 & 0 \\ 0 & 1 \end{bmatrix}$ and $AI = A$ **33.** Since in this case 0 represents the $2 \times 2$ zero

matrix, $A + 0 = A$. **35.** $(B + C)A = \begin{bmatrix} 6 & 28 \\ -2 & -6 \end{bmatrix} = BA + CA$ **37. (a)** $[1 \quad 1 \quad 1 \quad 1 \quad 1]F = [72 \quad 36 \quad 157]$ represents the total points, rebounds, and minutes played by the starters in the first game. **(b)** $[1 \quad 1 \quad 1 \quad 1 \quad 1]\,(F + S) = [150 \quad 75 \quad 306]$ represents the total points, rebounds, and minutes played by the starters in the tournament. **(c)** $\frac{1}{2}[1 \quad 1 \quad 1 \quad 1 \quad 1]\,(F + S) = [75 \quad 37.5 \quad 153]$ represents the average number of points, rebounds, and minutes played by the starters in the tournament.
**39. (a)** $JC = \begin{bmatrix} 8450 \\ 13{,}900 \end{bmatrix}$ represents the total revenue in rentals at each store in the month of January. **(b)** $FC = \begin{bmatrix} 13{,}440 \\ 17{,}550 \end{bmatrix}$ represents the total revenue in rentals at each store in the month of February. **(c)** $(J + F)C = \begin{bmatrix} 21{,}890 \\ 31{,}450 \end{bmatrix}$ represents the total revenue in rentals at each store during the two-month period. **(d)** $[1, \quad 1] \cdot ((J + F)C) = \$53{,}340$ represents the total rental revenue at both stores during the two-month period. **41.** $u = -5$; $v = 0$; $w = 2$; $z = 7$ **42.** $a_{12} = 0$ **43.** $A$
**44.** $\begin{bmatrix} -5 & 3 \\ 0 & 8 \end{bmatrix}$ **45.** $\begin{bmatrix} -5 & -3 \\ 4 & 6 \end{bmatrix}$ **46.** $\begin{bmatrix} -15 & -21 \\ 20 & 14 \end{bmatrix}$ **47.** $(2, -3)$ **48.** $(-1, 0, 3)$ **49.** no solution
**50.** $(z - 3, 4 - 4z, z)$ $z$ any real number

## [11.3]

**1. (a)** $\begin{matrix} (B) \\ (A) \end{matrix} \begin{bmatrix} 1 & -2 & 5 \\ 2 & 3 & -4 \end{bmatrix}$ **(b)** $\begin{matrix} (B) \\ (C) \end{matrix} \begin{bmatrix} 1 & -2 & 5 \\ 0 & 7 & -14 \end{bmatrix}$ **(c)** $\begin{matrix} (B) \\ (D) \end{matrix} \begin{bmatrix} 1 & -2 & 5 \\ 0 & 1 & -2 \end{bmatrix}$ **(d)** $\begin{matrix} (E) \\ (D) \end{matrix} \begin{bmatrix} 1 & 0 & 1 \\ 0 & 1 & -2 \end{bmatrix}$

**3. (a)** $\begin{matrix} (D) \\ (B) \\ (C) \end{matrix} \begin{bmatrix} 1 & -2 & 1 & -1 \\ 3 & -1 & 0 & 3 \\ -4 & 2 & 3 & -10 \end{bmatrix}$ **(b)** $\begin{matrix} (D) \\ (E) \\ (F) \end{matrix} \begin{bmatrix} 1 & -2 & 1 & -1 \\ 0 & 5 & -3 & 6 \\ 0 & -6 & 7 & -14 \end{bmatrix}$ **(c)** $\begin{matrix} (D) \\ (G) \\ (F) \end{matrix} \begin{bmatrix} 1 & -2 & 1 & -1 \\ 0 & 1 & -4 & 8 \\ 0 & -6 & 7 & -14 \end{bmatrix}$
**(d)** $\begin{matrix} (H) \\ (G) \\ (I) \end{matrix} \begin{bmatrix} 1 & 0 & -7 & 15 \\ 0 & 1 & -4 & 8 \\ 0 & 0 & -17 & 34 \end{bmatrix}$ **(e)** $\begin{matrix} (H) \\ (G) \\ (J) \end{matrix} \begin{bmatrix} 1 & 0 & -7 & 15 \\ 0 & 1 & -4 & 8 \\ 0 & 0 & 1 & -2 \end{bmatrix}$ **(f)** $\begin{matrix} (L) \\ (K) \\ (J) \end{matrix} \begin{bmatrix} 1 & 0 & 0 & 1 \\ 0 & 1 & 0 & 0 \\ 0 & 0 & 1 & -2 \end{bmatrix}$ **5.** $(4, -1)$
**7.** $\left(\frac{1}{2}, \frac{5}{2}\right)$ **9.** $(-1, -1)$ **11.** $\left(\frac{1}{3}, 0\right)$ **13.** $(1 - 2y, y)$ for $y$ any real number **15.** $(1, -1, 2)$ **17.** $(-2, 1, 3)$
**19.** $(2, 1, -1)$ **21.** $(x, 5x - 4, x - 1)$ for $x$ any real number **23.** $(1, 0, -1, 2)$ **25.** $(4, 1, -1)$ or $(4, -1, -1)$
**27.** $2 \times 3$ **28.** 2
**29.** $\begin{bmatrix} 2 & -5 \\ -1 & -4 \end{bmatrix}$ **30.** $\begin{bmatrix} 3 & 0 \\ -1 & -2 \\ 2 & 4 \end{bmatrix}$ **31.** $\begin{bmatrix} -1 & 5 \\ -1 & 9 \end{bmatrix}$ **32.** not defined **33.** $\begin{bmatrix} 10 & -20 \\ -8 & -6 \end{bmatrix}$ **34.** 1 **35.** $\begin{bmatrix} -2 & 9 \\ 5 & 10 \end{bmatrix}$
**36.** $[10 \quad -15]$ **37.** $A$ **38.** $C$ **39.** $\begin{bmatrix} 6 & 3 & -3 \\ 0 & 0 & 0 \\ 10 & 5 & -5 \end{bmatrix}$ **40.** $C \cdot N = 4600$. Thus, the gross sales of the four models on Saturday was \$4600.

## [11.4]

**1.** Since $AB = BA = I$, $B = A^{-1}$. **3.** Since $AB = BA = I$, $B = A^{-1}$. **5.** $\begin{bmatrix} -5 & 4 \\ -4 & 3 \end{bmatrix}$ **7.** $-\frac{1}{2}\begin{bmatrix} -3 & 1 \\ -4 & 2 \end{bmatrix}$ **9.** the inverse does not exist **11.** $\begin{bmatrix} 1 & 3 & 2 \\ 2 & 6 & 5 \\ 1 & 2 & 2 \end{bmatrix}$ **13.** $-\frac{1}{2}\begin{bmatrix} 0 & -2 & 2 \\ -1 & 1 & -1 \\ 1 & -3 & 1 \end{bmatrix}$ **15.** $-\frac{1}{10}\begin{bmatrix} 1 & -3 & 1 \\ -3 & -1 & -3 \\ 2 & -6 & -8 \end{bmatrix}$
**17.** $x = -8$; $y = 14$ **19.** $x = 12$; $y = 7$; $z = 4$ **21.** $(2, 1)$ **23.** $(4, -3)$ **25.** $(2, 0, -3)$ **27.** $(4, 0, 6)$
**29.** $(3, 3, -1)$ **31. (a)** $(2, -3)$ **(b)** $(0, 6)$ **(c)** $(-3, 5)$ **33. (a)** $(2, 0, -3)$ **(b)** $(1, 4, 1)$ **(c)** $(-2, -2, 1)$
**35.** $\begin{bmatrix} 3 & 1 & -2 & -2 \\ -1 & 0 & 1 & 1 \\ -1 & 0 & 1 & 0 \\ 2 & 1 & -2 & -2 \end{bmatrix}$ **37.** If $A$ is nonsingular, then $A^{-1}$ exists. Then $A^{-1}(AB) = A^{-1}(AC)$, $(A^{-1}A)B = (A^{-1}A)C$, $IB = IC$, $B = C$. **39.** $\frac{1}{13}\begin{bmatrix} 5 & -3 \\ 1 & 2 \end{bmatrix}$ **41.** The inverse does not exist since $ad - bc = 0$. **43. (a)** $\begin{bmatrix} 3 & -1 \\ 8 & -3 \end{bmatrix}$ **(b)** $\begin{bmatrix} 3 & -1 \\ 8 & -3 \end{bmatrix}$
**(c)** $\begin{bmatrix} 7 & -3 \\ -2 & 1 \end{bmatrix}$ **(d)** $\begin{bmatrix} 1 & 2 \\ 2 & 3 \end{bmatrix}$ **(e)** $\begin{bmatrix} 1 & 5 \\ 0 & -1 \end{bmatrix}$ **(f)** $\begin{bmatrix} 3 & -1 \\ 8 & -3 \end{bmatrix}$ **(g)** It would appear that $(AB)^{-1} = B^{-1}A^{-1}$ and $(AB)^{-1} \neq A^{-1}B^{-1}$.

**45.** Mix 1: 35 oz of Super Grow and 15 oz of Healthy Mix; Mix 2: 40 oz of Super Grow and 10 oz of Healthy Mix; Mix 3: 50 oz of Super Grow and 5 oz of Healthy Mix **46.** $(-2, 6)$ **47.** no solution **48.** $(1, 0, -5)$

## [11.5]

**1.** 2 **3.** $-4$ **5.** 13 **7.** $-2a + 3b$ **9.** $-1$ **11.** $-1$ **13.** $-19$ **15.** $-19$ **17.** 3 **19.** 6 **21.** 0 **23.** $(3, -2)$ **25.** $\left(\frac{1}{3}, -2\right)$ **27.** $(-1, -1)$ **29.** $(-3, 2, 1)$ **31.** $(-1, 1, 2)$ **33.** 9 **35.** $\pm 2$ **37.** $1, -3$ **39.** 0 **41.** **(a)** 0 **(b)** 0 **(c)** 0 **(d)** 0 **43.** **(a)** $ab$ **(b)** $abc$ **(c)** $abcd$ **(d)** The determinant is the product of the elements along the main diagonal, $a_{11}\, a_{22}\, a_{33} \ldots a_{nn}$. **45.** **(a)** 0 **(b)** 0 **(c)** 0 **(d)** 0 **47.** $|A| = ad - bc = |A^T|$ **49.** Show that the determinant equation is equivalent to $y - y_1 = \dfrac{y_2 - y_1}{x_2 - x_1}(x - x_1)$, the equation of the line passing through $(x_1, y_1)$ and $(x_2, y_2)$. **51.** Make a sketch using three points in quadrant I and show that the area of the triangle can be found by considering the areas of three trapezoids. **53.** $A^{-1} = \begin{bmatrix} \frac{d}{ad - bc} & \frac{-b}{ad - bc} \\ \frac{-c}{ad - bc} & \frac{a}{ad - bc} \end{bmatrix}$, and $|A^{-1}| = \dfrac{1}{ad - bc} = \dfrac{1}{|A|}$. **55.** By Exercise 54, $|AB| = |A||B|$ so if $|AB| = 0$, then $|A||B| = 0$ which means that $|A| = 0$ or $|B| = 0$ by the zero-product rule for real numbers.

**57.** $\begin{bmatrix} a & b \\ na & nb \end{bmatrix} = anb - bna = 0$ **59.** $\begin{bmatrix} a & b \\ c & d \end{bmatrix} = ad - bc$ and $-\begin{bmatrix} b & a \\ d & c \end{bmatrix} = -[bc - ad] = ad - bc.$ **61.** $\begin{bmatrix} \frac{3}{2} & 2 \\ \frac{1}{2} & 1 \end{bmatrix}$

**62.** $\begin{bmatrix} 3 & 3 & -1 \\ -2 & -2 & 1 \\ -4 & -5 & 2 \end{bmatrix}$ **63.** $(0, 5)$ **64.** $(-2, 0, 2)$ **65.** week 1: 20 standard models and 35 deluxe models; week 2: 40 standard models and 15 deluxe models; week 3: 30 standard models and 40 deluxe models

## [11.6]

**1.** Theorem 1 **3.** Theorem 3 **5.** Theorem 2 **7.** Theorem 4 **9.** Theorem 5 **11.** Theorem 7 **13.** Theorem 6 **15.** Theorem 2 **17.** Theorem 3 **19.** $x = 17$ **21.** $x = 1$ **23.** $\begin{vmatrix} 0 & 1 & 0 \\ -11 & 4 & 18 \\ -5 & 3 & 14 \end{vmatrix}$ **25.** $\begin{vmatrix} 14 & 0 & -5 \\ 7 & 0 & -1 \\ 4 & 1 & -2 \end{vmatrix}$ **27.** 9

**29.** $-59$ **31.** $-104$ **33.** $-98$ **35.** When $-1$ is substituted for $x$ and 3 is substituted for $y$, the first and second rows are equal making the determinant 0. Similarly when $(2, 7)$ is substituted, the first and third rows are equal making the determinant 0. **37.** Since the area of the triangle would then be 0, there is no triangle meaning the three points are on the same straight line. **39.** $(-1, 3, 4)$ **40.** $(-1, 0, 1)$

## Chapter 11 Review Exercises

**1.** $2 \times 3$ **2.** $\begin{bmatrix} 0 & 0 \\ 0 & 0 \end{bmatrix}$ **3.** 3 **4.** $\begin{bmatrix} 0 & 3 \\ -1 & -5 \end{bmatrix}$ **5.** $\begin{bmatrix} -2 & 0 \\ 4 & -1 \\ 5 & 2 \end{bmatrix}$ **6.** $\begin{bmatrix} -1 & -1 \\ 4 & 9 \end{bmatrix}$ **7.** $\begin{bmatrix} -1 & 5 \\ 2 & -1 \end{bmatrix}$

**8.** $\begin{bmatrix} 6 & -12 & -15 \\ 0 & 3 & -6 \end{bmatrix}$ **9.** $\begin{bmatrix} -2 & 13 \\ 3 & -7 \end{bmatrix}$ **10.** 5 **11.** $\begin{bmatrix} 2 & -6 & -1 \\ -6 & 8 & 23 \end{bmatrix}$ **12.** undefined **13.** $\begin{bmatrix} 2 & -1 & 39 \\ -8 & -3 & 7 \end{bmatrix}$

**14.** undefined **15.** **(a)** $J + A = \begin{bmatrix} 350 & 300 & 50 \\ 520 & 220 & 50 \end{bmatrix}$ **(b)** $\frac{1}{2}(J + A) = \begin{bmatrix} 175 & 150 & 25 \\ 260 & 110 & 25 \end{bmatrix}$ represents the average number of rentals per agency in each category over the two-month period. **(c)** $[1 \quad 1]\, J = [500 \quad 300 \quad 60]$ represents the total number of rentals in each category at both agencies during July. **(d)** $[1 \quad 1]\, (J + A) = [870 \quad 520 \quad 100]$ represents the total number of rentals in each category at both agencies during the two-month period. **(e)** $([1 \quad 1]\, (J + A)) \cdot \begin{bmatrix} 1 \\ 1 \\ 1 \end{bmatrix} = 1490$ represents the total number of rentals in every category at both agencies during the two-month period. **(f)** $JC = \begin{bmatrix} 40{,}600 \\ 44{,}400 \end{bmatrix}$ represents the total revenue produced by each agency during July. **(g)** $(J + A)C = \begin{bmatrix} 71{,}500 \\ 75{,}500 \end{bmatrix}$ represents the total revenue produced by each agency during the two-month period. **(h)** $[1 \quad 1] \cdot ((J + A)C) = \$147{,}000$ represents the total revenue produced by both agencies during the two-month period.

**16.** $(1, -1)$ **17.** $(-2, 0, 3)$ **18.** $\begin{bmatrix} 1 & 2 \\ -3 & -5 \end{bmatrix}$ **19.** $-\frac{1}{8}\begin{bmatrix} -2 & -4 & -2 \\ 5 & -10 & -3 \\ 1 & -2 & 1 \end{bmatrix}$ **20.** $(-3, 4)$ **21.** $(2, -2, 3)$

**22.** January: 20 individual and 10 family memberships; February: 30 individual and 20 family memberships; March: 30 individual and 35 family memberships **23.** $-26$ **24.** 42 **25.** 5 **26.** $(-2, 5)$ **27.** $(-3, 3, 0)$ **28.** $\pm 2$ **29.** $-3, 1$ **30.** Since a row is all zeros, the determinant is 0. **31.** Since two columns are equal, the determinant is 0. **32.** Since two columns were interchanged, the determinant is negated. **33.** When a row is multiplied by 3, the determinant is multiplied by 3.

**34.** $x = -13$ **35.** $-11$ **36.** $|A| = \begin{vmatrix} a_1 & a_2 & 0 & 0 \\ a_3 & a_4 & 0 & 0 \\ 0 & 0 & b_1 & b_2 \\ 0 & 0 & b_3 & b_4 \end{vmatrix} = a_1 \begin{vmatrix} a_4 & 0 & 0 \\ 0 & b_1 & b_2 \\ 0 & b_3 & b_4 \end{vmatrix} - a_3 \begin{vmatrix} a_2 & 0 & 0 \\ 0 & b_1 & b_2 \\ 0 & b_3 & b_4 \end{vmatrix} = a_1a_4\begin{vmatrix} b_1 & b_2 \\ b_3 & b_4 \end{vmatrix} - a_3a_2\begin{vmatrix} b_1 & b_2 \\ b_3 & b_4 \end{vmatrix} =$
$(a_1a_4 - a_3a_2)\begin{vmatrix} b_1 & b_2 \\ b_3 & b_4 \end{vmatrix} = \begin{vmatrix} a_1 & a_2 \\ a_3 & a_4 \end{vmatrix}\begin{vmatrix} b_1 & b_2 \\ b_3 & b_4 \end{vmatrix}$

## Chapter 12

### [12.1]

**1.** 4, 8, 12, 16, 20; 32; 48 **3.** $-\frac{1}{2}, \frac{1}{4}, -\frac{1}{8}, \frac{1}{16}, -\frac{1}{32}; \frac{1}{256}; \frac{1}{4096}$ **5.** $-6, 9, -14, 21, -30; 69; 149$ **7.** 0, 0, 0, 0, 0; 0; 0 **9.** 12, 48, 192, 768 **11.** $-24, 72, -216, 648$ **13.** 11, 104, 941, 8474 **15.** $x_1 + x_2 + x_3 + x_4 + x_5$ **17.** $\frac{1}{3} + \frac{1}{5} + \frac{1}{7} + \frac{1}{9}$ **19.** $-1 + \frac{1}{2} - \frac{1}{3} + \frac{1}{4} - \frac{1}{5} + \frac{1}{6} - \frac{1}{7}$ **21.** $\sum_{k=1}^{n} k$ **23.** $\sum_{k=1}^{n} \frac{1}{\sqrt{k}}$ **25.** $\sum_{k=1}^{4} 2k$ **27.** 36 **29.** 18 **31.** 40 **33.** 150 **35.** 12 **37.** $\frac{5}{6}$ **39.** each has value 104 **41.** 1 **43.** 21 **45.** $\frac{n(n+1)(n+2)}{3}$ **47.** $\frac{n(n+1)(4n+11)}{6}$ **51.** $\frac{n+1}{2}$ **53.** $m = -\frac{27}{28}; b = \frac{17}{28}$ **55.** 1, 1, 2, 3, 5, 8, 13, 21 **57.** $-294$ **58.** $-5875$

### [12.2]

**1.** yes; 4 **3.** yes; $-10$ **5.** no **7.** 2, 9, 16, 23, 30, 37 **9.** $-2, 5, 12, 19, 26, 33$ **11.** $\sqrt{3}, 5\sqrt{3}, 9\sqrt{3}, 13\sqrt{3}, 17\sqrt{3}, 21\sqrt{3}$ **13.** $x = 8$; 8, 12, 16 **15.** 235 **17.** 222 **19.** 33 **21.** $a_{17} = 50$; $S_{17} = 442$ **23.** $n = 7$; $d = 6$ **25.** $a_{12} = \frac{7}{2}$; $S_{12} = 31$ **27.** $n = 15$; $d = \frac{2}{7}$; $a_1 = 0$ **29.** $n = 7$; $d = 5$; $S_7 = 42$ **31.** $a_5 = 9 \log 7$; $S_5 = 25 \log 7$ **33.** 14, 17, 20, 23, 26, 29 **35.** 22 **37.** 10,100 **41.** \$7946 **43.** 3140 **45.** \$465 **47.** 240 ft **49.** $x_4 = -\frac{1}{81}$; $x_5 = \frac{1}{243}$ **50.** $x_4 = -5$; $x_5 = 6$ **51.** $\sum_{k=1}^{n} \frac{k}{k^2+1}$ **52.** $\frac{69}{140}$

### [12.3]

**1.** 4, 8, 16, 32, 64, 128 **3.** $-16, 8, -4, 2, -1, \frac{1}{2}$ **5.** $25, 20, 16, \frac{64}{5}, \frac{256}{25}, \frac{1024}{125}$ **7.** $x = 13$; 20, 10, 5 **9.** $\frac{511}{1536}$ **11.** $\frac{11}{24}$ or $\frac{31}{24}$ **13.** 3 **15.** $a_6 = 64$; $S_6 = 126$ **17.** $n = 9$; $a_1 = 256$; $S_9 = 511$ **19.** $n = 7$; $a_1 = 3125$; $S_7 = \frac{19531}{5}$ **21.** $-10, 20, -40, 80$ **23.** 1 **25.** no sum, $|r| = 2 > 1$ **27.** $-\frac{98}{11}$ **29.** $\frac{50}{3}$ **31.** $\frac{1}{3}$ **33.** $\frac{7}{33}$ **35.** $\frac{41}{333}$ **37.** $\frac{97}{45}$ **41.** \$1678 **43.** \$3036.14 **45.** yes, if you would work for approximately \$10,700,000 for the month **47.** $\frac{1476}{25}$ cm **49.** 74.4 ft **51.** 60 m **53.** **(a)** $\frac{5}{4}$ ft **(b)** 117.5 ft **(c)** 120 ft **(d)** Use $S = 120$ ft for an approximation **55.** 4479 **57.** $7, 3, -1, -5, -9$ **58.** $-12, -5, 2, 9, 16$ **59.** $\frac{27}{2}$ **60.** 90 **61.** $7, 2, -3, -8$ **62.** $-504$ **63.** 3200 **64.** 6150

**[12.4]**

**1.** $n = 5$ **3.** $n = 8$ **5.** $n = 10$ **7.** $S(1)$: $1 = 1^2$; $S(k)$: $1 + 3 + 5 + \cdots + (2k - 1) = k^2$; $S(k + 1)$: $1 + 3 + 5 + \cdots + (2k - 1) + [2(k + 1) - 1] = (k + 1)^2$ **9.** $S(1)$: $1 < 2^1$; $S(k)$: $k < 2^k$; $S(k + 1)$: $k + 1 < 2^{k+1}$ **11.** $S(1)$: $(ab)^1 = a^1b^1$; $S(k)$: $(ab)^k = a^kb^k$; $S(k + 1)$: $(ab)^{k+1} = a^{k+1}b^{k+1}$ **35.** $-21, 3, -\frac{3}{7}, \frac{3}{49}$ **36.** 32, 16, 8, 4 **37.** $-\frac{3277}{512}$ **38.** 49,149 **39.** $\frac{3589}{990}$ **40.** 81 in

**[12.5]**

**1.** 210 **3.** 90 **5.** 1 **7.** 120 **9.** 15 **11.** 6,760,000; 3,407,040 **13.** 336 **15.** 120 **17.** 40,320 **19.** **(a)** 362,880 **(b)** 40,320 **(c)** 5040 **(d)** 720 **21.** **(a)** 60 **(b)** 180 **(c)** 3780 **(d)** 3360 **23.** 2,522,520 **25.** 5040 **27.** 2002 **29.** 240,240 **31.** **(a)** 700 **(b)** 756 **(c)** 3360 **(d)** 5292 **33.** 21 **35.** 600 **37.** **(a)** 1287 **(b)** 4 **(c)** 4512 **(d)** 22,308 **(e)** 1320 **39.** 60

**[12.6]**

**1.** 6 **3.** 1 **5.** 45 **7.** $x^5 + 5x^4y + 10x^3y^2 + 10x^2y^3 + 5xy^4 + y^5$ **9.** $243a^5 - 405a^4 + 270a^3 - 90a^2 + 15a - 1$ **11.** $81x^4 - 108x^3y + 54x^2y^2 - 12xy^3 + y^4$ **13.** $u^{12} + 6u^{10}v^2 + 15u^8v^4 + 20u^6v^6 + 15u^4v^8 + 6u^2v^{10} + v^{12}$ **15.** $a^7 + 7a^5 + 21a^3 + 35a + 35a^{-1} + 21a^{-3} + 7a^{-5} + a^{-7}$ **17.** $x^2 - 4x^{3/2}y^{1/2} + 6xy - 4x^{1/2}y^{3/2} + y^2$ **19.** $40x^3$ **21.** $80x^2y^3$ **23.** $10x^2y^6$ **25.** $-189a^2b^5$ **27.** $-10{,}240a^3$ **29.** $-56a^{-2}$ **31.** 128 **33.** $-4 - 4i$ **35.** $-7290x^5y$ **37.** 0.53 **41.** 24 **42.** 210 **43.** 10,626 **44.** 255,024 **45.** 120

**[12.7]**

**1.** yes **3.** no **5.** yes **7.** $\frac{1}{6}$ **9.** 1 **11.** $\frac{1}{2}$ **13.** $\frac{1}{6}$ **15.** $\frac{2}{9}$ **17.** $\frac{11}{12}$ **19.** $\frac{2}{9}$ **21.** $\frac{2}{3}$ **23.** $\frac{1}{38}$ **25.** $\frac{9}{19}$ **27.** 0 **29.** $\frac{9}{19}$ **31.** $\frac{1}{13}$ **33.** $\frac{1}{2}$ **35.** $\frac{1}{52}$ **37.** $\frac{2}{13}$ **39.** $\frac{4}{13}$ **41.** $\frac{11}{26}$ **43.** $\frac{1}{52}$ **45.** $\frac{33}{66{,}640}$ **47.** $\frac{1}{54{,}145}$ **49.** $\frac{1}{108{,}290}$ **51.** $\frac{3}{7}$ **53.** $10^{-7}$ **55.** $2^{-48}$ **57.** $16x^4 - 160x^3y + 600x^2y^2 - 1000xy^3 + 625y^4$ **58.** $-38 - 41i$ **59.** $560x^6y^4$ **60.** 20

## Chapter 12 Review Exercises

**1.** $0, \frac{3}{2}, \frac{8}{3}, \frac{15}{4}, \frac{24}{5}; \frac{63}{8}$ **2.** $-\frac{1}{4}, \frac{1}{7}, -\frac{1}{10}, \frac{1}{13}, -\frac{1}{16}; \frac{1}{25}$ **3.** $1 + \sqrt{2} + \sqrt{5} + \sqrt{10}$ **4.** $\sum_{k=1}^{n} \frac{k^2}{k+1}$ **5.** 7, −13 **6.** −2, −23 **7.** 22 **8.** $\frac{223}{12}$ **9.** −9, −5, −1, 3, 7, 11; $S_{12} = 156$ **10.** $n = 8$, $d = 2$ **11.** 14, 11, 8 **12.** $x = 4$; 5, 11, 17 **13.** 1900 **14.** 4900 **15.** $1800 **16.** $21 **17.** $\frac{1}{6}; \frac{1}{3}; \frac{2}{3}; \frac{4}{3}; \frac{8}{3}, \frac{16}{3}; S_6 = \frac{21}{2}$ **18.** $\frac{3280}{27}$ **19.** $6, -3, \frac{3}{2}, -\frac{3}{4}$ **20.** $x = 7$; 20, 10, 5 **21.** $\frac{32}{5}$ **22.** $30, 5, \frac{5}{6}, \frac{5}{36}, \frac{5}{216}$ **23.** $\frac{11}{9}$ **24.** $\frac{170}{33}$ **25.** $2307.94 **26.** 9.8 ft; 191.4 ft **27.** 72 m **28.** 135 ft **31.** 6 **32.** 3 **33.** 120 **34.** 1 **35.** 1 **36.** 1 **37.** **(a)** 26,000 **(b)** 18,000 **(c)** 2700 **38.** 17,160 **39.** 362,880 **40.** 40,320 **41.** 30,240 **42.** 15 **43.** 3080 **44.** 715 **45.** 24; $\frac{1}{108{,}290}$ **46.** 3003 **47.** $243a^5 - 405a^4z + 270a^3z^2 - 90a^2z^3 + 15az^4 - z^5$ **48.** $y^{-4} + 4y^{-2} + 6 + 4y^2 + y^4$ **49.** $-160x^3y^6$ **50.** $2835a^3b^5$ **51.** $\frac{8}{15}$ **52.** $\frac{7}{15}$ **53.** 0 **54.** 1 **55.** $\frac{1}{4}$ **56.** $\frac{3}{13}$ **57.** $\frac{1}{2}$ **58.** $\frac{10}{13}$ **59.** $\frac{1}{52}$ **60.** $\frac{1}{26}$ **61.** 0 **62.** $\frac{1}{2}$ **63.** $\frac{2}{13}$ **64.** 0 **65.** $\frac{7}{13}$ **66.** $\frac{11}{26}$ **67.** $\frac{9999}{10{,}000}$ **68.** 0.99; 0.01

## Appendix A

**1.** 0.6702 **3.** 0.6702 − 2 = −1.3298 **5.** 2.4456 **7.** 5.45 **9.** 3850 **11.** 0.00417 **13.** 0.5156 **15.** 2.6411 **17.** 0.5990 − 3 = −2.4010 **19.** 49.82 **21.** 0.03475 **23.** 0.0007665

## Appendix B

**1.** 0.9827 **3.** 0.4986 **5.** 1.275 **7.** 26°40′ **9.** 50°10′ **11.** 45°40′ **13.** 0.6598 **15.** 6.639 **17.** 1.322 **19.** 67°09′ **21.** 64°52′ **23.** 44°35′

# Index of Applications

Numbers give the pages where applications are found. If a number is in parentheses, it is an exercise number on the indicated page.

# Index